World – Physical

GREAT BASIN	Land features
Caribbean Sea	Water bodies
Aleutian Trench	Underwater features

ARCTIC OCEAN
Queen Elizabeth Islands
Ellesmere Island
GREENLA
Beaufort Sea
Victoria Island
Baffin Island
Baffin Bay
Davis Strait
80°N
Bering Strait
Yukon R.
Great Bear Lake
MACKENZIE MTS.
Mackenzie R.
Hudson Bay
LABRADOR
Labrador Sea
60°N
Bering Sea
Mt. McKinley 20,320 ft (6,194 m)
Gulf of Alaska
Saskatchewan R.
NORTH AMERICA
CANADIAN SHIELD
Newfoundland
Aleutian Islands
Aleutian Trench
Vancouver I.
ROCKY MOUNTAINS
CASCADE RANGE
Columbia R.
Great Slave Lake
Lake Winnipeg
Great Lakes
Missouri R.
Mississippi R.
Ohio R.
APPALACHIAN MTS.
Sohm Plain
40°N
Northeast
Mendocino Fracture Zone
GREAT BASIN
GREAT PLAINS
ATLANTIC OCEAN
SIERRA NEVADA
Murray Fracture Zone
Colorado R.
Rio Grande
Hatteras Plain
Bermuda Rise
Mid-Atlantic Ridge
Hawaiian Ridge
Tropic of Cancer
Molokai Fracture Zone
BAJA CALIFORNIA
SIERRA MADRE
MEXICAN PLATEAU
Gulf of Mexico
Bahama Is.
Pacific
20°N
Hawaiian Is.
Cuba
Clarion Fracture Zone
CENTRAL AMERICA
Puerto Rico Trench
Johnston Atoll
Greater Antilles
West Indies
Basin
Middle America Trench
Caribbean Sea
Central Pacific Basin
PACIFIC OCEAN
Line Islands
Clipperton Fracture Zone
Orinoco R.
Demerara Plain
ANDES
GUIANA HIGHLANDS
0°
Equator
Galápagos Is.
AMAZON
Amazon R.
Phoenix Is.
BASIN
SOUTH AMERICA
POLYNESIA
Marquesas Is.
East Pacific Rise
MATO GROSSO PLATEAU
BRAZILIAN HIGHLAND
Samoa Is.
Tuamotu Archipelago
Peru-Chile
20°S
Tonga Is.
Cook Is.
Society Is.
Tahiti
Tropic of Capricorn
ANDES
ATACAMA DESERT
GRAN CHACO
Paraná R.
Tonga Trench
Austral Islands
Pitcairn I.
Sala y
Gómez Ridge
Nazca Ridge
Mt. Aconcagua 22,834 ft (6,960 m)
Rio Gra
Rise
Kermadec Tr.
Louisville Ridge
Southwest Pacific Basin
Easter I.
Challenger Fracture Zone
Juan Fernández Is.
Trench
PAMPAS
Rio de la Plata
Argentine Plain
40°S
PATAGONIA
0 1000 2000 Miles
0 1000 2000 Kilometers
Southeast Pacific Basin
Humboldt Plain
Falkland Is.
So
Geor
Eltanin Fracture Zone
Strait of Magellan
CAPE HORN
South Georgia Ridge
Udintsev Fracture Zone
60°S
Drake Passage
Pacific-Antarctic Ridge
Antarctic Circle
160°W 120°W 80°W

World Regional Geography

A Development Approach

World Regional Geography

A Development Approach

11th Edition

Edited by

Douglas L. Johnson
Clark University

Viola Haarmann
Clark University

Merrill L. Johnson
University of New Orleans

Contributors

Christopher A. Airriess
Ball State University

Robert Argenbright
University of Utah

Samuel Aryeetey-Attoh
Loyola University

Simon Batterbury
University of Melbourne, Australia

Corey Johnson
University of North Carolina at Greensboro

Brad D. Jokisch
Ohio University

William C. Rowe
Independent Scholar and Development Consultant

PEARSON

Boston   Columbus   Indianapolis   New York   San Francisco   Upper Saddle River
Amsterdam   Cape Town   Dubai   London   Madrid   Milan   Munich   Paris   Montréal   Toronto
Delhi   Mexico City   Sao Paulo   Sydney   Hong Kong   Seoul   Singapore   Taipei   Tokyo

Senior Geography Editor: Christian Botting
Senior Marketing Manager: Maureen McLaughlin
Program Manager: Anton Yakovlev
Director of Development: Jennifer Hart
Development Editor: Veronica Jurgena
Assistant Editor: Bethany Sexton
Senior Marketing Assistant: Nicola Houston
Media Producers: Laura Tommasi / Ziki Dekel
Project Manager, Instructor Media: Eddie Lee
Team Lead, Geosciences and Chemistry: Gina M. Cheselka
Production Project Manager: Connie M. Long
Full Service/Composition: PreMediaGlobal
Full-Service Project Manager: Jared Sterzer
Illustrations: International Mapping
Image Lead: Maya Melenchuk
Project Manager (Supplements): Kristen Sanchez
Photo Researchers: Lauren McFalls/Abdul Khader, PreMediaGlobal
Text Permissions Manager: Timothy Nicholls
Design Manager: Mark Ong
Interior and Cover Designer: Gary Hespenheide
Operations Specialist: Christy Hall
Cover Photo Credit: Chinch Gryniewicz/ecoscene

Credits and acknowledgments borrowed from other sources and reproduced, with permission, in this textbook appear on the appropriate page within the text or on the credits page beginning on page C-1.

Library of Congress Cataloging-in-Publication Data
World regional geography : a development approach / edited by Douglas L. Johnson,
  Clark University, Viola Haarmann, Clark University, Merrill L. Johnson, University of New Orleans. —Eleventh edition.
    pages cm.
  ISBN-13: 978-0-321-93965-4
  ISBN-10: 0-321-93965-4
  1. Economic development.   2. Economic history.   3. Economic geography.
  4. Developing countries—Economic conditions.   I. Johnson, Douglas L.
  HD82.G39.W67 2015
  330.9—dc23
                                                              2014044388

24 2021

www.pearsonhighered.com

ISBN-10: 0-321-93965-4; ISBN-13: 978-0-321-93965-4 (Student Edition)
ISBN-10: 0-321-96892-1; ISBN-13: 978-0-321-96892-0 (Instructor's Review Copy)

KD 01.22.2021 1401

Brief Contents

About Our Sustainability Initiatives

Pearson recognizes the environmental challenges facing this planet, as well as acknowledges our responsibility in making a difference. This book has been carefully crafted to minimize environmental impact. The binding, cover, and paper come from facilities that minimize waste, energy consumption, and the use of harmful chemicals. Pearson closes the loop by recycling every out-of-date text returned to our warehouse.

Along with developing and exploring digital solutions to our market's needs, Pearson has a strong commitment to achieving carbon neutrality. As of 2009, Pearson became the first carbon- and climate-neutral publishing company. Since then, Pearson remains strongly committed to measuring, reducing, and offsetting our carbon footprint.

The future holds great promise for reducing our impact on Earth's environment, and Pearson is proud to be leading the way. We strive to publish the best books with the most up-to-date and accurate content, and to do so in ways that minimize our impact on Earth. To learn more about our initiatives, please visit **www.pearson.com/responsibility**.

PEARSON

Contents

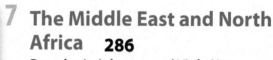

8　Africa South of the Sahara　342
Samuel Aryeetey Attoh

The Paradox of Sub-Saharan Africa's Mineral Wealth　342

9　South Asia　406
Christopher A. Airriess

Energy and the Development Challenge　406

South Asia's Environmental and Historical Contexts　409

Preface

We live in a remarkable age when satellites circle the globe in a matter of minutes and messages are sent halfway around the world in seconds. For many, these are the best of times, with progress in health care extending human life expectancies to record levels and with material comforts unheard of even a generation ago. It is easy in such conditions to overlook the fact that most of the world's population still lives in regions characterized by hunger, malnutrition, poor health, and limited educational and economic opportunities. Many also live under restricted personal freedom owing to political, racial, religious, or gender-based prejudice. Even more troubling is the fact that the income gap between technologically advanced countries and those struggling to catch up is increasing.

The purpose of *World Regional Geography* is to introduce students to the geographical foundations of development and to help them recognize the contributions that the study of geography can make to environmentally and culturally sustainable development. As we study the lives of others through this text, we will not only learn about them but also come to better understand ourselves. We will come to realize that development, or the improvement of the human condition, consists of far more than increased economic output and that each person and society contributes to the cultural diversity and richness of the global community.

College students are in a unique position to increase their understanding of the world and to use that knowledge to benefit themselves and others. Through the study of world regional geography, we can begin to comprehend the issues involved in the pursuit of world peace, preservation of the environment, improved health, and higher levels of living. The *Association of American Geographers*, the *National Geographic Society*, the *American Geographical Society*, and the *National Council for Geographic Education* have devoted significant resources and effort to improving geographic awareness. The U.S. Congress has cited geographic education as critical to understanding our increasingly interdependent world.

World Regional Geography is dedicated to college students who are seeking a better understanding of our complex world. It is written for both majors and nonmajors and does not require a background in geography. Its foundation is a basic regional structure. A multiple author approach permits each region to be discussed by an authority in that area. Although our regional specializations vary, we are united in our dedication to expanding geographical awareness and to contributing to the knowledge of the peoples of the world's diverse regions. The ultimate purpose of this book is to help students develop an increased understanding of our world's geographic diversity, both cultural and physical, and a knowledge of how each of us can contribute to the betterment of humankind.

New to the Eleventh Edition

- New **Exploring Environmental Impacts** case studies in each chapter discuss specific environmental challenges to the region, such as tropical deforestation or the Fukushima nuclear disaster.

- New **Focus on Energy** features in each chapter key in on critical energy issues of the regions.

- New **Visualizing Development** features cover a defining development in each world region, supported by engaging visualizations of economic and spatial data.

- New **Environmental Challenges** sections cover major environmental issues in the regions.

- **Chapter Opening Vignettes** with panoramic photographs highlight the human face of each region and draw students into the region's geography and development discussion.

- New **Read & Learn** sections at the beginning of each chapter feature region-specific learning outcomes.

- New **Stop & Think** review questions integrated at the end of thematic sections help students check their comprehension as they read.

- End of chapter review includes a concise bullet point summary, and **Understanding Development** and **Geographers @ Work** questions.

- **Current data** are incorporated from the latest Censuses in each region and from the latest available Population Reference Bureau's *World Population Data Sheet*.

- **New cartography** with modern styles and the latest data, including various recent economic and geopolitical events.

- **MasteringGeography™** is an online homework, tutorial, media, and assessment platform that helps students master concepts. Visual and media-rich tutorials feature immediate wrong-answer feedback and hints that emulate the office-hour experience.

- **Learning Catalytics,** a bring-your-own-device student engagement, assessment, and classroom intelligence system.

The 11th edition also reflects some changes in contributors. We welcome Brad Jokisch as the new author of the Latin America chapter and Corey Johnson as the new author of the Europe chapter, adding fresh perspectives and regional expertise to the discussion of these regions. We want to take this opportunity to express our heartfelt appreciation to our colleague and friend David L. Clawson, who devoted many years of dedicated work as author and editor to *World Regional Geography*.

Organization and Features

After a brief introduction that surveys the basic concepts of geography, *World Regional Geography* opens with a chapter that examines issues related to the nature of development, the impact of globalization, and the importance of taking care of our resources in a sustainable manner. The following eleven chapters are structured within a regional framework.

Although the use of this text as a discrete and unified entity is encouraged, its organization allows a variety of teaching strategies. As time requires, sections treating historical or environmental processes, or specific regions, can be selectively emphasized or omitted. In a two-term sequence, the book facilitates consideration along continental or regional divisions. Table and figure references can encourage students to pursue external data sources and analysis.

World Regional Geography contains numerous features that are designed to assist and stimulate students:

- More than 225 full-color maps and diagrams have been rendered by a professional cartographic studio and thoroughly updated for this edition, incorporating the latest boundary and name changes, as well as physical, cultural, and economic data. The maps are carefully integrated into the text to convey spatial relationships and strengthen our understanding of geographic patterns and concepts.

- More than 450 specially chosen color photographs, over three-quarters of which are new to this edition, help communicate the physical, cultural, and economic nature of individual regions.

- Informative tables and graphs supplement textual material.

- Boxed features highlight subjects of special regional and topical significance, such as cultural diversity, gender and development, environmental issues, migration, natural resource utilization, and distinctive regional events and characteristics.

- Key terms are presented in boldface type within the text and at the end of each chapter; a convenient glossary is provided at the end of the book.

- An end of chapter review includes a concise summary and critical thinking questions that can be used for homework, projects, group work, or in discussion sections.

- MasteringGeography and Learning Catalytics help extend the book with rich media, assessment, and lecture enrichment tools.

Acknowledgments

Every new edition of a textbook is like running a marathon, with many unsung heroes working hard behind the scenes, many supporters lending a hand from the sidelines, and a select few on constant call for advice, help, and crisis management. We want to express our gratitude to all who assisted with the preparation of the 11th edition of *World Regional Geography*. The comments of reviewers and previous readers have been indispensable in helping to give fresh focus to revisions. We owe a debt of gratitude to the numerous reviewers who have generously given their time to offer input over the course of many editions, and specifically want to thank the reviewers who took the time to critique the 10th edition of the book and contributed many constructive suggestions: Gary Brown, *Lonestar Community College*; Craig Campbell, *Youngstown State University*; Patricia Dennis, *Mississippi University for Women*; Robert Dennison, *Heartland Community College*; Dawn Drake, *University of Tennessee*; Alison Feeney, *Shippensburg University*; Chad Garick, *Jones County Junior College*; Anthony Ijomah, *HACC—Harrisburg*; Heidi Lannon, *Santa Fe College*; James Leonard, *Marshall University* (also 11th edition accuracy reviewer); Terry Nelson, Mount Marty College (also 11th edition accuracy reviewer); Kirk White, *Harrisburg Area Community College—Lancaster*.

We also thank the reviewers of the 8th and 9th editions: Brian W. Blouet, *the College of William and Mary*; Patricia Boudinot, *George Mason University*; Stanley D. Brunn, *University of Kentucky*; L. Scott Deaner, *Kansas State University*; Greg Gaston, *University of North Alabama*; Tarek A. Joseph, *Henry Ford Community College*; Robert G. Kremer, *Metropolitan State College of Denver*; Paul R. Larson, *Southern Utah University*; Elena Lioubimtseva, *Grand Valley State University*; Franklin Long, *Coastal Carolina Community College*; James Penn, *Grand Valley State University*; Stephen E. Podewell, *Western Michigan University*; Robert Rohli, *Louisiana State University*; Susan C. Slowey, *Bline College*; Jacob R. Sowers, *Kansas State University*.

We are especially grateful to the Pearson publishing team for their patience, friendship, and professionalism. Christian Botting and Anton Yakovlev kept us grounded and on the marathon course with their expert guidance and e-mail and telephone lifelines. Veronica Jurgena's meticulous close-reading of the chapters and many perceptive comments have also made this book a better product. We would also like to thank Jennifer Hart, Bethany Sexton, Connie Long, Gina Cheselka, Kristen Sanchez, and Tim Nicholls at Pearson, as well as Jared Sterzer, Lauren McFalls, and Abdul Khader at PreMedia Global for their work on the project.

World Regional Geography lives by its regional authors! We want to convey our deep appreciation for their hard work, responsiveness in matters large and small, and endurance in working through the revision process.

Help and advice from colleagues and professionals is essential to a book's successful progress from conception to print. But no project like this can be accomplished without family and friends providing a sustaining network of support and strength. We have been blessed with an abundance of empathy and understanding and are deeply grateful for their help in pulling this book across the finish line.

Douglas L. Johnson,
Viola Haarmann,
Merrill L. Johnson

About the Authors

Douglas L. Johnson *Editor*

Chapter One, Geography and Development in an Era of Globalization, *and Chapter Seven,* The Middle East and North Africa

Douglas Johnson is a Professor of Geography Emeritus at Clark University. His geographical career has focused on the North African and Middle Eastern culture realm. Both his master's thesis and doctoral dissertation at the University of Chicago dealt with the history and spatial implications of nomadism. Studying North Africa and the Middle East and teaching students about the complexities of this region have been a central focus of his work, including lengthy field periods in Libya, Sudan, and Morocco and visiting appointments in the Middle Eastern Center at the University of California, Berkeley and Al-Akhawayn University in Morocco. His research has addressed issues of land degradation and desertification, arid land management, pastoral nomadism, and the cultural ecology of animal keeping. He is the coauthor of *Land Degradation: Creation and Destruction*, 2nd ed. (Rowman & Littlefield, 2007) and from 2003 to 2006 he served as co-editor of the *Geographical Review*. In light of events in the Middle East, contributing to a world regional geography textbook that helps promote better understanding of one of the most conflicted areas in the world seems more important than ever.

Viola Haarmann *Editor*

Chapter One, Geography and Development in an Era of Globalization, *and Chapter Seven,* The Middle East and North Africa

Viola Haarmann is a Research Fellow at the George Perkins Marsh Institute at Clark University. From an early age she was at home both in Europe and North America as she grew up in Germany and Canada. She earned a dual master's degree in English and Geography, and received her D.Sc. from Hamburg University, Germany, after carrying out fieldwork on land-use potential and change in the southern Sahel of Sudan's Darfur Province. For many years she held a Sahel project coordinator position at the Hamburg geography department. Since then she has operated primarily as an independent editor of academic research. She served on the editorial board of *The Columbia Gazetteer of the World* (Columbia University Press, 1998), and from 2003 to 2006 as co-editor of the *Geographical Review*. She currently provides English language editorial services to the *International Journal of Disaster Risk Science*. Her special interest is the geography of food and agriculture.

Merrill L. Johnson *Editor*

Chapter One, Geography and Development in an Era of Globalization, *and Chapter Two,* United States and Canada

In addition to serving as a Professor of Geography, Merrill Johnson is the founding Executive Director for Global UNO at the University of New Orleans. He has a B.A. in International Relations from West Texas A&M University, an M.A. in Geography from Arizona State University, and a Ph.D. in Geography from the University of Georgia. He joined the faculty of the University of New Orleans in 1981, was Chair of the Department of Geography from 1989 to 2000, Associate Dean of the College of Liberal Arts from 2001 to 2008, and Associate Provost/Associate Vice President of Academic Affairs from 2008 to 2013. His long-term academic interests have focused on economic and political geography, particularly in the U.S. South, Canada, and Latin America, with technical interests in cartography and geographic information systems. More recently, he has begun to explore the role of geographers in the development and use of 3-D Internet worlds. While Dr. Johnson has taught a variety of undergraduate and graduate courses in geography, he looks back with special affection at all of the students that he has been privileged to work with in the many, many introductory world regional geography courses that he has taught.

Brad D. Jokisch

Chapter Three, Latin America and the Caribbean

Brad Jokisch is Associate Professor at Ohio University, where he has been teaching since

1997. He received his Ph.D from Clark University (1998), and did his dissertation work with Billie Lee Turner II before Turner moved to Arizona State. Jokisch is a specialist in Andean geography, and has carried out most of his fieldwork in Ecuador. He has written on environment and development issues in highland South America and in the Upper Amazon, although his main research interest is in migrants who seek employment outside their native region, particularly in the United States. How these people stay connected with their home community and how their remittances affect development and the environment is the focus of much of his research.

At Ohio University Brad teaches a regional course on Latin America and the Caribbean and other courses (Population, Agriculture) that focus on the region at both the undergraduate and graduate levels. He founded Ohio University's Ecuador study abroad language program, and is widely traveled in Central and South America.

Corey Johnson

Chapter Four, Europe

Corey Johnson is an Assistant Professor of Geography at the University of North Carolina at Greensboro, where he also held a Candace Bernard and Robert Glickman Dean's Professorship in 2012-13. His research and teaching areas include the political geography of Europe and Eurasia, borders and border security, natural resources and energy geopolitics, and Germany. In 2011-12 he was the Joachim Herz Fellow at the Transatlantic Academy in the German Marshall Fund of the United States in Washington, D.C. Originally from Emporia, Kansas, Corey holds a Ph.D. in geography from the University of Oregon and a BA in Geography and German from the University of Kansas.

Corey Johnson wishes to thank Shane Canup for research assistance and Professor Ron Wixman for many invaluable insights on how to teach and explain the European region.

Robert Argenbright

Chapter Five, Northern Eurasia

Robert Argenbright is an Assistant Professor (Lecturer) of Geography at the University of Utah. He earned a B.A., M.A., and Ph.D., all in Geography, from the University of California, Berkeley. His research focuses on the historical geography of the Soviet Union and the current transformation of Moscow. A brief tour of Moscow and Leningrad in 1975 first piqued his interest in Russia. Since then, he has returned to the region 23 times to conduct research. He has published articles on the U.S.S.R. and post-Soviet Russia in such journals as *Eurasian Geography and Economics, The Geographical Review, Political Geography, Revolutionary Russia, The Russian Review,* and *Urban Geography.* He is writing a book with the working title *Moscow under Construction.* Included among the many courses he has taught are Geography of Post-Soviet Eurasia and History of the Soviet Union.

William C. Rowe

Chapter Six, Central Asia and Afghanistan

William Rowe, a native of upper East Tennessee and sixth-generation Appalachian tobacco farmer, has traveled, worked, and studied in the Muslim world for over 25 years. He received his Bachelor of Science in Languages (BSLA) from Georgetown University, with concentrations in Arabic and French. He worked at the American University in Cairo before returning to the United States for his master's on water and population in Southern and Eastern Morocco at the University of Texas at Austin. For his dissertation his focus shifted to the newly independent Muslim nations of Central Asia with an emphasis on Tajikistan and its language, Tajiki. He spent 2 years on research in Tajikistan as it was emerging from its devastating civil war. Since receiving his Ph.D., he has continued his work in Central Asia, most notably in Uzbekistan, Tajikistan, and the Tajik regions of Afghanistan and has been able to witness the changes in the region since American involvement in the wake of 9/11. His research concentrates on the economic and environmental transformations that have occurred since independence in Central Asia and Afghanistan as well as Muslim identity in post-Soviet Central Asia.

Samuel Aryeetey Attoh

Chapter Eight, Africa South of the Sahara

Samuel Aryeetey Attoh is Dean of the Graduate School and Associate Provost for Research at Loyola University Chicago. He received his Ph.D. from Boston University and his M.A. from Carleton University, Ottawa. He also earned a B.A. with honors from the University of Ghana, Legon. His research and teaching interests are in Urban and Regional Planning, Housing and Community Development, and the Geography of Development in Africa. He is the author of *Geography of Sub-Saharan Africa,* 3rd ed. (Prentice Hall, 2009), has published many articles in geographical journals, and received numerous research grants. He is past chair of the African Specialty Group of the Association of American Geographers, past treasurer of the AAG, and past president of the Illinois Association of Graduate Schools. He has also served on the editorial boards of the *Professional Geographer* and the *African Geographical Review,* advisory boards of World Education Services and the Council of Graduate Schools, and review panels for the National Science Foundation. He is a member of the American Council on Education Fellows, the Council of Graduate Schools, and the American Planning Association. He believes a course in World Regional Geography presents teachers with an opportunity to showcase the integrative and holistic human and physical dimensions of geography and demonstrate how they relate to real world situations—whether social, political, environmental, or economic in nature.

Christopher A. Airriess

Chapter Nine, South Asia, Chapter Ten, East Asia, and Chapter Eleven, Southeast Asia.

Christopher Airriess is a Professor of Geography at Ball State University, Muncie, Indiana. He earned his B.A. and M.A. in Geography at Louisiana State University, and his Ph.D. in Geography at the University of Kentucky in 1989. Born on Long Island, New York, he spent his childhood in Singapore and Malaysia. While he has traveled throughout much of the western Pacific Rim, his favorite places are southern China, Indonesia, and Malaysia. His research interests include development, the geography of ports and maritime transport, and the human dimensions of ethnic-Southeast Asians in North America. He is the recipient of two Fulbright Awards that allowed extended visits to Indonesia in 1987, and Hong Kong in 2000. In addition to regularly teaching World Regional Geography, he also teaches Human Geography, and Geography of Asia. He believes that a World Regional Geography course provides an essential piece to the undergraduate educational experience by imparting an understanding of the connections between the environment, culture, society, and economic and political systems within the context of real places where the process of globalization directly and indirectly impacts people's lives.

Simon Batterbury

Chapter Twelve, Australia, New Zealand, and the Pacific Islands.

Simon Batterbury teaches environmental studies and geography at the University of Melbourne, Australia, where he is Associate Professor and former Director of the Master's Program in Environment. His main interest is in how rural households and their livelihood systems alter, and rely upon, local landscapes and environments in countries marginal to the world economy. Born in southeast London, he studied human and physical geography at Reading University and earned an M.A. and Ph.D. at the Graduate School of Geography, Clark University, on the political ecology of natural resource management in Burkina Faso, West Africa. In Burkina he worked for two years with a German aid program helping farmers manage drought and soil erosion. He has taught at the University of Arizona, London School of Economics, Brunel University, and Roskilde University. In 2007–2008 he was a James Martin Fellow, University of Oxford. He has conducted long-term collaborative research on rural development and environmental management in West Africa (Burkina Faso and Niger), Southeast Asia (East Timor), and the Pacific (New Caledonia).

Digital and Print Resources

For Teachers and Students

MasteringGeography™ with Pearson eText. The **Mastering** platform is the most widely used and effective online homework, tutorial, and assessment system for the sciences. It delivers self-paced tutorials that provide individualized coaching, focus on course objectives, and are responsive to each student's progress. The Mastering system helps teachers maximize class time with customizable, easy-to-assign, and automatically graded assessments that motivate students to learn outside of class and arrive prepared for lecture. MasteringGeography offers:

- **Assignable activities** that include MapMaster™ interactive map activities, *Encounter* Google Earth Explorations, video activities, Geoscience Animation activities, Map Projections activities, GeoTutor coaching activities on the toughest topics in geography, Dynamic Study Modules that provide each student with a customized learning experience, end-of-chapter questions and exercises, reading quizzes, Test Bank questions, and more.

- **A student Study Area** with MapMaster™ interactive maps, videos, Geoscience Animations, web links, glossary flashcards, "In the News" RSS feeds, chapter quizzes, PDF downloads of outline maps, an optional Pearson eText including versions for iPad and Android, and more.

Pearson eText gives students access to the text whenever and wherever they can access the Internet. The eText pages look exactly like the printed text and include powerful interactive and customization functions, including links to the multimedia.

Television for the Environment *Earth Report* Geography Videos on DVD (0321662989). This three-DVD set helps students visualize how human decisions and behavior have affected the environment and how individuals are taking steps toward recovery. With topics ranging from the poor land management promoting the devastation of river systems in Central America to the struggles for electricity in China and Africa, these 13 videos from Television for the Environment's global *Earth Report* series recognize the efforts of individuals around the world to unite and protect the planet.

Television for the Environment *Life* World Regional Geography Videos on DVD (013159348X). From the Television for the Environment's global *Life* series, this two-DVD set brings globalization and the developing world to the attention of any world regional geography course. These 10 full-length video programs highlight matters such as the growing number of homeless children in Russia, the lives of immigrants living in the United States and trying to aid family still living in their native countries, and the European conflict between commercial interests and environmental concerns.

Television for the Environment *Life* Human Geography Videos on DVD (0132416565). This three-DVD set is designed to enhance any human geography course. These DVDs include 14 full-length video programs from Television for the Environment's global *Life* series, covering a wide array of issues affecting people and places in the contemporary world, including the serious health risks of pregnant women in Bangladesh, the social inequalities of the "untouchables" in the Hindu caste system, and Ghana's struggle to compete in a global market.

Geoscience Animation Library 5th edition DVD-ROM (0321716841) Created through a unique collaboration among Pearson's leading geoscience authors, this resource offers over 100 animations covering the most difficult-to-visualize topics in physical geology, physical geography, oceanography, meteorology, and earth science. The animations are provided as Flash files and pre-loaded into PowerPoint(R) slides for both Windows and Mac.

***Practicing Geography: Careers for Enhancing Society and the Environment* by Association of American Geographers (0321811151).** This book examines career opportunities for geographers and geospatial professionals in the business, government, nonprofit, and education sectors. A diverse group of academic and industry professionals shares insights on career planning, networking, transitioning between employment sectors, and balancing work and home life. The book illustrates the value of geographic expertise and technologies through engaging profiles and case studies of geographers at work.

***Teaching College Geography: A Practical Guide for Graduate Students and Early Career Faculty* by Association of American Geographers (0136054471).** This two-part resource provides a starting point for becoming an effective geography teacher from the very first day of class. Part One addresses "nuts-and-bolts" teaching issues. Part Two explores being an effective teacher in the field, supporting critical thinking with GIS and mapping technologies, engaging learners in large geography classes, and promoting awareness of international perspectives and geographic issues.

***Aspiring Academics: A Resource Book for Graduate Students and Early Career Faculty* by Geographers Association of American Geographers (0136048919).** Drawing on several years of research, this set of essays is designed to help graduate students and early career faculty start their careers in geography and related social and environmental sciences. *Aspiring Academics* stresses the interdependence of teaching, research, and service—and the importance of achieving a healthy balance of professional and personal life—while doing faculty work. Each chapter provides accessible, forward-looking advice on topics that often cause the most stress in the first years of a college or university appointment.

For Teachers

Learning Catalytics

Learning Catalytics™ is a "bring your own device" student engagement, assessment, and classroom intelligence system. With Learning Catalytics you can:

- Assess students in real time, using open-ended tasks to probe student understanding.
- Understand immediately where students are and adjust your lecture accordingly.
- Improve your students' critical-thinking skills.
- Access rich analytics to understand student performance.
- Add your own questions to make Learning Catalytics fit your course exactly.
- Manage student interactions with intelligent grouping and timing.

Learning Catalytics is a technology that has grown out of twenty years of cutting edge research, innovation, and implementation of interactive teaching and peer instruction. Available integrated with MasteringGeography.

Instructor Resource Manual (Download) (0321968794). The *Instructor Resource Manual*, by William Bailey (Auburn University), follows the new organization of the main text. It includes a sample syllabus, chapter learning objectives, lecture outlines, a list of key terms, and answers to the textbook's Review and end of chapter questions. Discussion questions, classroom activities, and advice about how to integrate visual supplements (including MasteringGeography and Learning Catalytics resources) are integrated throughout the chapter lecture outlines.

TestGen/Test Bank (Download) (0321968816). TestGen is a computerized test generator that lets teachers view and edit *Test Bank* questions, transfer questions to tests, and print the test in a variety of customized formats. Authored by James Leonard (Marshall University), this *Test Bank* includes approximately 1,500 multiple-choice, true/false, and short answer/essay questions. Questions are correlated against the book's Learning Objectives, the revised U.S. National Geography Standards, chapter-specific learning outcomes, and Bloom's Taxonomy. Available for download from www.pearsonhighered.com/irc, and in the Instructor Resources area of MasteringGeography. The *Test Bank* is also available in Microsoft Word® and is importable into Blackboard.

Instructor Resources (Download) (0321968808). The *Instructor Resources (Download)* provide a collection of resources to help teachers make efficient and effective use of their time. All digital resources can be found in one well organized, easy-to-access place. The IRC download includes:

- All textbook images as JPEGs, PDFs, and PowerPoint™ Presentations
- Pre-authored Lecture Outline PowerPoint™ Presentations (by Richard Smith, Harford Community College), which outline the concepts of each chapter with embedded art and can be customized to fit teachers' lecture requirements

- CRS "Clicker" Questions (by Bob Dennison, Heartland Community College) in PowerPoint™ format, which correlate to the book's Learning Objectives, the U.S. National Geography Standards, chapter-specific learning outcomes, and Bloom's Taxonomy
- The TestGen software, *Test Bank* questions, and answers
- Electronic files of the *IRM* and *Test Bank*

This Instructor Resource content is available online via the Instructor Resources section of MasteringGeography and www.pearsonhighered.com/irc.

For Students

***Goode's World Atlas, 22nd Edition* (0321652002).** *Goode's World Atlas* has been the world's premiere educational atlas since 1923—and for good reason. It features over 250 pages of maps, from definitive physical and political maps to important thematic maps that illustrate the spatial aspects of many important topics. The 22nd Edition includes 160 pages of digitally produced reference maps, as well as thematic maps on global climate change, sea-level rise, CO_2 emissions, polar ice fluctuations, deforestation, extreme weather events, infectious diseases, water resources, and energy production.

Pearson's Encounter Series provides rich, interactive explorations of geoscience concepts through GoogleEarth™ activities, covering a range of topics in regional, human, and physical geography. For those who do not use MasteringGeography, all chapter explorations are available in print workbooks, as well as in online quizzes at www.mygeoscienceplace.com. Each exploration consists of a worksheet, online quizzes whose results can be emailed to teachers, and a corresponding Google Earth™ KMZ file.

- *Encounter World Regional Geography* by Jess C. Porter (0321681754)
- *Encounter Human Geography* by Jess C. Porter (0321682203)
- *Encounter Physical Geography* by Jess C. Porter and Stephen O'Connell (0321672526)
- *Encounter Geosystems* by Charlie Thomsen (0321636996)
- *Encounter Earth* by Steve Kluge (0321581296)

***Dire Predictions: Understanding Global Warming* by Michael Mann, Lee R. Kump (0136044352).** *Dire Predictions* is appropriate for any science or social science course in need of a basic understanding of the reports from the Intergovernmental Panel on Climate Change (IPCC). These periodic reports evaluate the risk of climate change brought on by humans. But the sheer volume of scientific data remains inscrutable to the general public, particularly to those who may still question the validity of climate change. In just over 200 pages, this practical text presents and expands upon the essential findings in a visually stunning and undeniably powerful way to the lay reader. Scientific findings that provide validity to the implications of climate change are presented in clear-cut graphic elements, striking images, and understandable analogies.

Development of the World's Regions

The book introduces students to the geographic foundations of development across the world's diverse regions.

▲ Girls reading by candlelight in Assam state during the massive 2012 power outages in India.

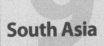

South Asia

Christopher A. Airriess

Energy and the Development Challenge

In July 2012 India experienced the largest power blackout in human history, affecting some 600 million people in 20 northern states that comprise the traditional economic core of the country. Trains ground to a halt, traffic signals went haywire, and hospitals lost essential power for a day or more. While the national power company restored 90 percent of the power grid within half a day in some areas, this event was a catastrophic culmination of years of daily power cuts and outages. India, the world's second-fastest growing economy and the sixth greatest consumer of electricity, cannot produce enough power. The blackout was the result of one or more states drawing excess electricity from the national power grid, leading to the collapse of the entire system.

India's blackout tells us much about larger development issues directly or indirectly associated with globalization. In addition to energy needs for industrialization, growing affluence of Indians benefiting from the country's engagement with the global economy means increased use of energy-intensive household appliances, especially air conditioners, which have become the new status symbol. But the main challenge is state-owned power firms that sell electricity at low, subsidized prices. Without assurances of acceptable profits by power providers, there is no incentive to invest in needed infrastructure to secure a dependable transmission network. The process of globalization is generally accompanied by the neoliberal philosophy of privatizing state-owned assets to create adequate financial incentives for private capital to thrive. The moral quandary in India, and other poor countries, is what happens to the hundreds of millions of poor people who cannot afford higher electricity costs resulting from privatization. It is difficult to envision India keeping pace with the other BRIC countries (Brazil, Russia, and China), defined as newly advanced economies, as long as a basic economic good such as electricity is in short supply.

Read & Learn

▶ Describe the climate characteristics that explain the influence of wet and dry monsoons.
▶ Identify the region's primary environmental and energy challenges.
▶ Explain how historical movements of peoples and cultures resulted in South Asia's current cultural conflicts.
▶ Link British colonial rule to the current economic contours of South Asia.
▶ Identify social and economic forces that reinforce gender inequality in India.

▶ Describe how the process of globalization affects the lives of India's poor rural population.
▶ Compare India's urban-industrial regions and link their development to globalization and the government's economic policies since 1990.
▶ Contrast the political and economic constraints to greater levels of economic development in Pakistan, Nepal, Bangladesh, and Sri Lanka.

407

GEOGRAPHY IN ACTION

Achieving Gender Parity in Sub-Saharan Africa

The 2012 World Development Report on Gender Equality and Development focuses on three key dimensions of gender equality: the accumulation of endowments (education, health, and physical assets), the use of those endowments to take advantage of economic opportunities and generate incomes, and the application of those endowments to take action and exercise control—the ability to make effective choices in the household and in society at large. This is consistent with the 2010 Millennium Development Goal (MDG) Summit's call for action to ensure that women have equal access to education, health care, and economic opportunities, and can be involved in development policy decision-making.

The need for improved gender equality in sub-Saharan Africa is demonstrated by the relatively poor performance of these countries on the United Nations Development Program's (UNDP) Gender Inequality Index and the Gender Empowerment Measure. Most of the progress made on gender parity has been with primary and secondary education, but with few gains in tertiary or post-secondary education. In 2009, the majority of sub-Saharan African countries had a Gender Parity Index (GPI) of more than 0.90 in primary and secondary education, putting them on track to achieve parity by 2015. Malawi (100 girls in school for every 100 boys), Rwanda (100.3), Namibia (103), and Lesotho (107) have already achieved parity. In tertiary education, GPIs over 0.90 have been achieved only by Botswana, Cape Verde, Lesotho, Mauritius, and South Africa. Not much progress has been made toward women's economic empowerment as measured by the share of women employed in the nonagricultural sector. The highest shares are in Ethiopia (47%), South Africa (44%), and Namibia (42%) (Figure 8-3-1). Further gender gaps remain in terms of access to assets (land rights and credit) and in wage earnings. Limited progress has occurred with women's representation in national parliaments. The highest performing countries in 2011 were Rwanda (51%), South Africa (43%), Mozambique (39%), Uganda (37%), and Burundi (36%) (Figure 8-3-2). In two-thirds of sub-Saharan African countries less than 20 percent of the parliamentarians are women.

The Convention on the Elimination of all Forms of Discrimination against Women (CEDAW) and the MDGs show that women's empowerment and gender equality are both global and African priorities. Various programs aimed at empowering African women have been instituted. Microfinancing and cell phone-based mobile banking pr...now provide women with access...business start-ups, and money tra...services through such organizatio...World Women's Banking (WWB), \...Opportunity Network (WON), Acc...Bank in Gambia and Rwanda, and...Lease and Finance in Tanzania. Eth...which previously did not have an...rights for women, now issues join...titles for wives and husbands thro...its land certification program. Wo...in Burundi, South Sudan, and Uga...have increasingly engaged in pea...reconstruction efforts following c...The Adolescent Girls Initiative (AG...public-private partnership, has de...a set of programs in Liberia, Rwan...South Sudan to assist adolescent...their transition from school to pro...employment through job and vo...training, mentoring, and basic bu...skills training.

Sources: World Bank, World Development...(2012); Gender Equality and Development (\...DC: International Bank for Reconstruction...opment/World Bank, 2011); United Nation...ment Program, Assessing Progress in Africa...Millennium Development Goals: MDG Repo...York: United Nations Development Progra...

▲ FIGURE 8-3-2 **Gender parity in Rwanda.** At 52 percent Rwanda has the highest percentage of women parliamentarians in the world.

▲ FIGURE 8-3-1 **Gender reforms in Ethiopia.** Ethiopia is engaging in a number of policy reforms to improve gender equality. This is a call center in Addis Ababa with a largely female workforce.

▲ **NEW! Chapter Opening Vignettes** with panoramic photographs highlight the human face of each region and draw students into the region's geography and development discussion.

◀ Updated **Geography in Action** case studies zero in on specific applications of development, populations, and subregions.

GEOGRAPHY IN ACTION

The Outsiders: Historical Minorities in Japan

Japan's human resources and culture are not as homogeneous as popular perception would suggest. Ethnic and social minorities make up about 4 percent of the population, or roughly 5 million people. The ethnic minorities consist of Koreans, Chinese, Okinawans, Ainu, and foreign residents. The social minorities, however, are composed of burakumin, persons with disabilities, and children of interracial ancestry.

Historically, Japan's cultural homogeneity stems from practices that limit opportunities for people outside the cultural and social mainstream. The Japanese traditionally regard themselves as a unique people, sometimes referring to themselves as the Yamato people, in reference to the Yamato Plain around Kyoto where the Japanese culture developed in centuries past and from which the ancestry of the imperial family is derived. There is still a strong current in Japanese society to preserve the purity of the Yamato majority; anyone else is an outsider and can never hope to be fully accepted into the mainstream. The 1947 constitution expressly prohibits discrimination based on race, creed, sex, social status, or family origin. But that U.S.-imposed provision has not fundamentally altered...

guarantee citizenship, and the government makes it very difficult for Koreans to obtain citizenship although most have Japanese names, speak fluent Japanese, and have attempted to integrate into Japanese society. The Koreans remain mired at the lower end of the economic ladder, victims of social and economic discrimination, and tend to live in ghettos in the larger cities.

The Ainu were among Japan's earliest inhabitants (Figure 10-6-1). Racially different and almost exclusively a hunting and fishing people, they also were treated as aliens by the Yamato Japanese. Only 25,000 pureblood are left, mostly in a few locations in Hokkaido. The Ainu have been gradually assimilated into Japanese culture since the early 1800s. In the early 1900s, "native schools" were established in Hokkaido in the hope of making Ainu children more Japanese by destroying their cultural identity. Like Native Americans struggling to maintain some of their identity, there has been an upsurge of cultural pride over the past several decades.

The Okinawans, on the Ryukyu Islands south of Kyushu, were not politically incorporated into Japan until early in the seventeenth century, even though they are...

▲ FIGURE 10-6-1 **Elderly Ainu males in traditional dress at the Marimo Festival at Lake Akan, Hokkaido.** The festival's function is to celebrate nature, and Ainu culture is just one aspect of the larger festival's events to promote tourism.

xx

Sustainability, Energy, and the Environment

Critical issues related to resources and the environment are explored in depth.

▼ **NEW! Exploring Environmental Impacts** discuss specific environmental challenges to the region.

▼ **NEW! Focus on Energy** examine critical energy issues of the regions.

 EXPLORING ENVIRONMENTAL IMPACTS — The Battle for New Orleans

Since its founding, New Orleans has been waging a war against nature to counter the related threats of soil subsidence, flooding, loss of coastal wetlands, and increasingly potent hurricanes. In a sense, New Orleans should never have been located where it was. The first French settlers were attracted to the relatively high and fertile natural riverbanks, or levees, along the Mississippi River. Settlers initially avoided the swamps and marshes that lay beyond the levees, but as the city expanded, the land in the backswamps and marshes was settled. Drainage of these muck soils—roughly 90 percent water by volume—triggered soil subsidence. The remaining 10 percent of these soils comprises mostly organic matter that oxidizes when exposed to the atmosphere, compounding the subsidence problem. Subsidence poses especially serious problems for construction, as unstable soils cause foundations to tilt and roads to buckle. The solution is to drive numerous pilings into the ground at the building site, and then pour the structure's cement slab on top of the pilings. This process results in a building that "floats" on its pilings while the ground beneath and around it continues to sink. Thereafter, fill is added to the lawns around the building in a never-ending cycle to keep the "ground" at grade. Subsidence has left much of the bowl-shaped city 5–15 feet (2–5 meters) below sea level. Man-made levees keep the Mississippi River and Lake Pontchartrain from pouring into the city, and an elaborate pumping system removes the average 60 inches (1,500 millimeters) of rainfall received annually. For New Orleans residents, the threat of heavy rain presents all sorts of frightening possibilities.

As if living in a bowl-shaped depression below sea level were not threatening enough, New Orleans is losing its natural hurricane protection through erosion of nearby coastal marshes and barrier islands.

River control projects, wetland drainage, channel dredging, construction of petroleum exploration canals, and other factors have substantially reduced land recharge through deposition of river sediment. Coastal Louisiana is losing land to the ocean at the rate of 25–30 square miles (65–78 square kilometers) a year, an area the size of Manhattan. The loss of river sediment also has caused barrier islands to shrink as longshore currents no longer carry sediment loads sufficient to replace land as it erodes away. These wetlands and islands historically have helped to protect New Orleans from deadly hurricane-produced surges of water. Scientists have long argued that a major hurricane is a human catastrophe waiting to happen.

With Hurricane Katrina, theory suddenly became reality. Early in the morning of 29 August 2005, the eye of Hurricane Katrina passed over St. Bernard Parish, brushed against eastern New Orleans and St. Tammany Parish, and slammed with full force into Waveland, Bay St. Louis, Pass Christian, Biloxi, and other coastal Mississippi towns. Katrina had weakened to a strong Category 3 storm before hitting land, but still packed winds greater than 100 miles per hour (160 kilometers per hour) and produced storm surges in excess of 25 feet (8 meters). Government leaders in New Orleans breathed a sigh of relief when Katrina did not directly hit the

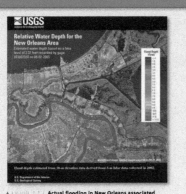

▲ Figure 2-2-1 **Actual flooding in New Orleans associated with Hurricane Katrina.** This image shows where the water was deepest in New Orleans following Hurricane Katrina and the breach of the levees. Areas along the Mississippi River, where the natural levees are highest, remained relatively dry. Included here are the French Quarter, the West Bank of the river, and an area called Uptown. Areas adjacent to Lake Pontchartrain also remained relatively dry but badly windblown. Those parts in the interior "bowl" of the city, and in eastern New Orleans, received in excess of 11 feet (3.5 meters) of water. Only floodwaters within New Orleans are shown; adjacent parishes and outlying areas were also badly flooded.

Source: United States Geological Service (October 5, 2005), http://eros.usgs.gov/katrina/products.html; background image from Landsat 7 (2000).

city. But relief quickly turned into horror when hurricane-protection levees were overtopped and broke under the pressure of storm surges, releasing millions of gallons of water into the city.

Katrina caused extreme damage throughout New Orleans and the Mississippi Gulf Coast. Perhaps 80 percent of New Orleans lay under water (Figure 2-2-1). Storm surge ripped apart the Interstate

---

 FOCUS ON ENERGY — Biofuels in Brazil

Brazil's energy consumption is remarkable because the country produces about 45 percent of its energy from renewable sources. Some observers consider Brazil to be the most sustainable large economy in the world. Hydroelectric power contributes part, but biofuels account for the largest portion (29%). This is in stark contrast to the United States where only 7 percent of energy consumption comes from renewable sources, and less than 1 percent from biofuels. A biofuel is any fuel that comes from an organic or biological source. The vast majority of Brazil's biofuel mix is composed of ethanol (ethyl alcohol), which is then blended with gasoline to run much of the country's automobile fleet. Ethanol can be made from a wide variety of crops, but in Brazil nearly all comes from sugarcane (Figure 3-5-1). Brazil produces one-third of the world's sugarcane, occupying approximately 9 million hectares (22 million acres), mostly in South-Central Brazil (São Paulo state). Just over half of the sugar produced is converted to ethanol, totaling over 27 billion liters. Brazil is the second largest producer of ethanol, and the world's largest exporter. The sugar/ethanol

industry is now a substantial part of Brazil's economy ($48 billion) and employs over 1 million people. Ethanol accounts for half of the gasoline market in Brazil (as opposed to 10% in the United States). Practically all cars sold in Brazil today are flex-fuel vehicles, which means that they can run on either low-ethanol or high-ethanol gasoline, up to 100 percent. One of the most important contributions of biofuels in Brazil is that it reduces greenhouse gas emissions substantially compared to gasoline consumption, perhaps by 80–90 percent.

How has Brazil been successful in substituting a renewable energy source for gasoline? First, the savanna climate of South-Central Brazil and fertile soils of São Paulo state are ideal for growing sugarcane. Sugar is also efficiently processed into ethanol; it has a very positive energy balance, meaning that it produces much more energy than it takes to produce the sugar. Converting sugarcane to ethanol is seven times more efficient than converting corn to ethanol, as is done in the United States. The Brazilian government has a long history of supporting ethanol production

and consumption, in large part to reduce petroleum imports and support domestic farmers and processers. Starting in 1975 Brazil began the *Programa Nacional do Álcool*, or the National Alcohol Program, which encouraged converting sugar into ethanol by mandating that gasoline be sold as a gasoline/ethanol blend. By 1993 gasoline/ethanol blends were required to be at least 22 percent ethanol; the figure was increased to 25 percent in 2003. The government also provided subsidies to the ethanol industry, including a guaranteed market, and has taxed gasoline so that it is more expensive compared to a gasoline/ethanol blend. A downturn in the cost of gasoline and an increase in the cost of ethanol in the late 2000s prompted the Brazilian government to announce a $38 billion plan to increase ethanol production and support the ethanol industry. Brazil hopes to export more ethanol and to provide a steady supply of ethanol to keep it competitive with the cost of gasoline.

The expansion of biofuels in Brazil has provoked controversy. In fact, ethanol has become a central issue in the debate about biofuels, food security, and land use change. Oxfam and other groups have argued that the expansion of biofuels has contributed to the increased cost of food globally. That criticism may apply elsewhere in the world (using corn in the United States), but there is little evidence that expanding sugarcane production came at the expense of raising other crops or increased food prices. Sugarcane occupies less than 5 percent of Brazil's arable land, and most of the sugarcane expansion in São Paulo state replaced pasture for cattle, not food crops. The larger concern is that expansion of sugarcane has an indirect effect on land use, displacing cattle ranching and other land uses northward into the Amazon rain forest.

Sources: José Goldemberg, "The Brazilian biofuels industry," *Biotechnology for Biofuels* 1 (May 2008): 1–6. Ethan Goffman "Biofuels: What Place in our Energy Future?" 2009. www.csa.com/discoveryguides/discoveryguides-main.php.

▼ Figure 3-5-1 **Sugarcane harvest, São Paulo state.** More sugarcane is planted than any other crop in the world, and Brazil grows and produces the most.

▼ **NEW! Environmental Challenges** sections cover major environmental issues in the regions.

Environmental Challenges

It would be difficult to find a region where the human modification of the landscape has been more extensive and dramatic than in Europe. The region's economic success over centuries, even millennia, is often a result of natural resource exploitation and agricultural and manufacturing innovations that profoundly altered the natural environment. Each historical epoch in Europe can be linked to major environmental modifications, and the contemporary landscape of Europe still reflects those changes over time. The Vikings needed wood for their ships and open fields for agriculture and grazing, and they cut down trees wherever they conquered, from the British Isles to Greenland. The Romans did much the same around the Mediterranean; places such as Sicily to this day are largely devoid of forests. Some scholars have attributed the fall of the Roman Empire to self-inflicted economic decline caused by ecological destruction. Five thousand years ago, Scotland's hilly terrain was covered mostly by mixed pine, aspen, birch, and oak forest, but that gave way to the stark, treeless landscapes familiar from films such as *Braveheart* and *Skyfall*. That landscape was largely a result of successive waves of deforestation that continued into the early twentieth century. The legendary forests covering most of the southern two-thirds of Germany, the stuff of Grimms' fairy tales, met a similar fate, and the large tracts of forested land along the *Autobahn* are typically heavily managed, second- or third-growth tree stands.

Not all environmental change is caused by humans. Natural climate cycles in past eras caused both human expansions into previously uninhabited areas as well as retrenchment when the climate became less favorable to established livelihoods. Changes in precipitation and temperature can also be linked to periods of political and social unrest as well as disease outbreaks (Figure 4-7). Archeologists have identified a period from around 300 to 800 C.E., when changing precipitation regimes and below-normal temperatures

and erosion. The scale and, above all, speed of that modification reached its height as a result of the **Industrial Revolution,** which began in northern Britain in the late eighteenth century and spread through much of Europe by the late nineteenth century. Massive industrialization of production required raw materials, transportation networks, and housing for the factory workers. The fuel of choice was coal, which was abundant in the areas most closely tied to industrialization such as northwest England, the Ruhr River valley and Saxony in Germany, and Silesia in Poland. Bituminous, or black coal, typically comes from underground mines and so the surface disturbance is relatively minor, but the low-quality lignite or brown coal is typically strip-mined (Figure 4-8). Large open pits, many now recreational lakes, in parts of eastern Germany

▼ Figure 4-6 **Persenbeug castle, Austria.** Cycling along the Danube in Austria is a very popular vacation activity.

▼ An updated **Introduction** refocuses on geography fundamentals, including information on the history and contemporary relevance of geography, map interpretation, and modern geospatial tools.

Introduction

There is a good chance you will use Google Maps today. The most popular smartphone app in the world—ahead of Facebook, YouTube, and Twitter—Google Maps has become the essential tool of the decade. Never before have so many people been interested in knowing where things are, how to find them, and making connections between what they see in real life with what is mapped on their devices. Geography is seemingly at our fingertips. Yet geography is more than just features on a map. Geography is fundamentally the study of location—the location of physical features, economic activities, human settlement patterns and cultural attributes, and anything else that a person can find on a map—and the connections between those things and places. In other words, geographers have a keen interest in understanding what defines terrestrial space (as opposed to "outer" space), and the interactions between people and their environments within that space. For that reason, geography is known as a "spatial" science that, in a fundamental way, studies the *why* of *where*.

At the most basic level, students of geography are called on to address three questions:

What is located where? This is the map or the "location list" question that most of us want to answer when we use Google Maps or Google Earth. For example, we may ask where the mountains are found in the eastern United States (answer: the Appalachian system running from Maine to Alabama), or where in Europe ethnic tensions have created a fractured political geography (answer: the Balkan Peninsula). Asking where people, places, and activities are located provides context and creates a basic knowledge of locations that is a starting point for more detailed examination of places later. It also teaches students what is geographically correct, so they know that Brazil is a Portuguese-speaking Latin American country, or that the Mississippi River is the major river system of the U.S. Midwest. Knowing where things are is important, but geography is much more than a rote memory exercise more appropriate in long car trips than serious study.

Why are things located where they are? Geographers primarily want to understand *why* things are located where they are. This is the "explanation" question that looks for the processes that produce a particular geographic pattern. To continue with the examples used above, we know that the Appalachian mountain

require examining human-environment interactions. Geographers might study the relationship between cropping patterns and the expansion of deserts, or the role of city structures in modifying climate (for example, creating urban heat islands).

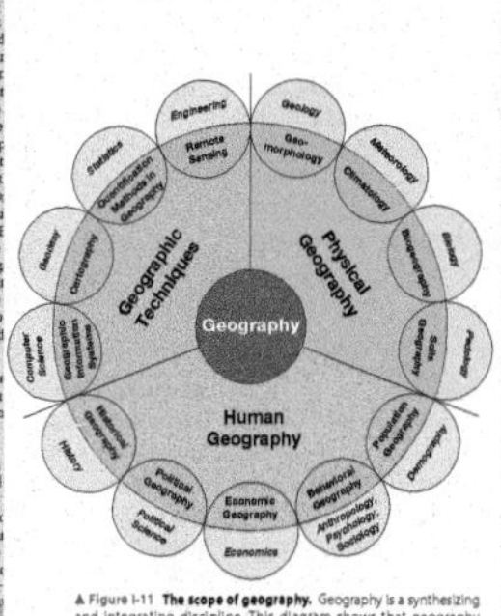

[text continues]

▲ Figure I-11 **The scope of geography.** Geography is a synthesizing and integrating discipline. This diagram shows that geography intersects with many fields of study, including the physical sciences, engineering, social sciences, and humanities.

central city as opposed to suburban locations, examine the purchasing power of the citizens of a particular community or neighborhood, assess traffic patterns to ensure a new outlet is accessible to a high volume of potential customers, evaluate the likely competition that rival firms might represent in the area, and predict the direction

▲ Figure I-13 **The National Geographic Bee in 2013.** For twenty-five years and counting, National Geographic and state Geographic Alliances have sponsored a competition involving "bees" at ascending geographical scales. The 50 state winners meet in Washington, D.C. for the ultimate showdown, won in 2013 for the first time by a middle school student from Massachusetts.

geographic information science analysis, travel agent, and consumer behavior services to firms and individuals wishing to conduct business outside the North American continent.

Government

Second to teaching, more geographers probably apply their skills to government agencies than to any other area. At a local level, many are municipal or regional planners, charged with facilitating orderly residential, business, and industrial growth and redevelopment. On a national level, knowledge of distant places and cultures, often in combination with remote sensing, map interpretation, and GIS skills, provide geographers with analytical abilities that are much in demand in government agencies. The Office of the Geographer in the Department of State is a focus of geographic activity, as is service in the diplomatic corps. The United States Agency for International Development (USAID) also includes many geographers whose expertise in resource analysis, re-

Guided Learning

An integrated learning path guides and engages students through active learning and applied activities.

▶ **NEW!** **Read & Learn** sections at the beginning of each chapter feature region-specific learning outcomes, helping students prioritize key concepts and information..

▼ **NEW!** **Stop & Think** review questions are integrated throughout the chapter text to help students check their comprehension and reflect on the concepts as they read.

Read & Learn

▶ Describe the climate characteristics that explain the influence of wet and dry monsoons.

▶ Identify the region's primary environmental and energy challenges.

▶ Explain how historical movements of peoples and cultures resulted in South Asia's current cultural conflicts.

▶ Link British colonial rule to the current economic contours of South Asia.

▶ Identify social and economic forces that reinforce gender inequality in India.

▶ Describe how the process of globalization affects the lives of India's poor rural population.

▶ Compare India's urban-industrial regions and link their development to globalization and the government's economic policies since 1990.

▶ Contrast the political and economic constraints to greater levels of economic development in Pakistan, Nepal, Bangladesh, and Sri Lanka.

Stop & Think

▶ Why is tourism such a major component of the economy in Central American and Caribbean countries?

Stop & Think

▶ Can Mexico achieve both economic growth and significant poverty reduction?

▼ A **Consistent Thematic Structure** includes general overviews of the region, followed by reviews of the sub-regions. **Environmental and Historical Contexts** of the region begin each chapter.

▼ Expanded **end of chapter review** includes a concise bullet point **Summary** and **Key Terms** followed by **Understanding Development** and **Geographers @ Work** critical thinking questions.

Europe's Environmental and Historical Contexts

So where is Europe, and maybe more importantly, *what* is it? These questions have been pondered for a very long time, and the answers rarely satisfy students of Europe. Eurasia is the largest landmass on Earth, and looking at imagery from space (Figure 4-1) we can see at its western edge a series of peninsulas and large islands. The word peninsula means "almost an island" and refers to land surrounded on three sides by water. A distinguishing feature of most of what we consider Europe is proximity to oceans or seas and the resulting moderating influences on climate.

This lack of clear borders and identity makes the story of Europe as a world region compelling. We all "know" of Europe, its centrality to world affairs over centuries, its high culture, marvelous cities, and economic prosperity. It is by and large the most developed world region, with the highest literacy rates, longest lifespans, and smallest disparity between rich and poor. But its varied human and physical geography makes the region full of surprises even to the seasoned traveler and student.

Environmental Setting: Physical Geography Enables Development

Europe's physical geography helps to explain the remarkable trajectory of this region as a major population center and global economic and political powerhouse. Navigable waterways and relatively easy access to the seas promoted trade, while a mild climate and productive agricultural land allowed the region to sustain a large population.

▼ Figure 4-1 **Western Eurasia.** Geologically speaking, Europe is not a continent but rather a subcontinent of Eurasia consisting of several peninsulas. The main ones are the Scandinavian, Iberian, Italian, and Balkan peninsulas.

Landforms

Europe is characterized by mountainous zones, plains, and river valleys (see chapter opener map). Most of Europe receives ample precipitation for agriculture, so that very little agricultural land in Europe requires irrigation outside of some areas of the far southern Mediterranean climate region. The major river systems of the region historically have been transportation and communication arteries and sources of drinking water or water for industrial uses rather than the source of vital irrigation waters, as in the case of the Nile, Colorado, or Mekong rivers. Europe's most extensive river system is the Danube, which begins in Germany and passes through eight other countries before emptying into the Black Sea in Romania. The Danube drains a basin encompassing 315,000 square miles (816,000 square kilometers), an area slightly larger than Turkey. Europe's best-known rivers flow mostly toward the Atlantic (Rhine, Seine, Loire, Elbe) or Mediterranean/Adriatic Seas (Ebro, Po). Managing river flow as well as traffic on the navigable river systems has always required some degree of cross-border cooperation, and a recent development that illustrates the political and economic integration of Europe through the European Union is the challenge of cross-border river management, as two-thirds of EU land lies within river drainage basins that cross international borders (see *Exploring Environmental Impacts:* Transboundary Water).

Summary

▶ Latin America and the Caribbean were created out of Europe's accidental discovery of two large continents connected by an isthmus. These continents held millions of people, speaking hundreds of languages in numerous environmental and social settings. The Spanish and Portuguese were the first European powers to reshape the human geography of the Americas, but the British and Dutch would soon follow.

▶ Dramatic change continued as the European powers imported between 10 and 12 million Africans for slave labor, and indigenous peoples were concentrated into towns with plazas and Roman Catholic cathedrals. By the early 1820s the colonial era ended and geopolitical wrangling ensued. In general the region's economy and population languished until the end of the nineteenth century, when both grew tremendously, reshaping the region's economies, population structure, and cultural makeup. By the start of the twentieth century the influence of Great Britain waned and the influence of the United States grew, especially in Central America and the Caribbean. Panama was created out of an American desire to control a canal that still connects the Atlantic and Pacific realms.

▶ Latin America's economy underwent two "shocks" and eras of restructuring during the twentieth century. The Great Depression ushered in an era of protected economies and the debt crisis and economic restructuring of the 1980s led to the "lost decade." Most countries reoriented their economies to export primary products, assemble manufactured goods (EPZs), and attract tourists. Thirty years of neoliberal economic policies have brought mixed results, and dissatisfaction has led to the repeal of some neoliberal policies by "Pink Tide" presidents. Brazil has emerged as the largest economy and population, followed distantly by Mexico, which continues to rely heavily on the United States and has yet to become a leader in the region the way Brazil has.

▶ Latin America has never been isolated, but through trade, tourism, and migration it likely has never been more integrated into the global economy. Free trade agreements are common and China's role has increased. Although progress has been made since the 1980s, income inequality remains high and more than one-quarter of the region's people live in poverty. Millions of Latin Americans left out of the region's economic progress decided to migrate to the United States and Europe, where as a group they remit billions of dollars.

▶ Latin America and the Caribbean can be divided into coherent, if problematic, subregions. Vast economic and environmental differences exist between and even within the subregions, but there are also strong similarities brought about by a shared history of Iberian conquest and cultural traits. The region faces many economic and environmental challenges. Globalization has created wealth and helped many people, but has left others behind and threatened ecosystems and vital natural resources.

▶ Like all places, Latin America and the Caribbean are still in the process of becoming. This world region's landscapes, economies, and environments will reflect global processes such as climate change and economic globalization, but also reflect the creativity and determination of its residents as they respond to these processes and participate in the reshaping of their own landscapes and livelihoods.

Key Terms

altitudinal life zones 118	encomiendas 126	Isthmus of Panama 147	quilombos 159
Aztec 121	export processing zones	land concentration 125	rain shadow 120
banana republic 146	(EPZ) 133	lost decade 129	remittances 140
Bolivian gas wars 157	free trade zones (FTZ) 133	maquiladora industry 133	syncretism 126
BRIC (Brazil, Russia, India, and	haciendas 126	Maya 121	thermal inversion 121
China) countries 158	home town associations	Media Luna 157	trade blocs 130
business process outsourcing	(HTA) 140	neoliberalism 129	transnational migration 138
(BPO) 133	import substitution	payment for environmental	Treaty of Tordesillas 125
chain migration 138	industrialization (ISI) 129	services (PES) 148	voluntourism 136
chinampas 123	Inca 121	Pentecostalism 139	Zapatista movement 142
ecotourism 136	informal economy 130	primate city 144	
El Niño 118	internally displaced people 158	pristine myth 121	

Understanding Development in Latin America and the Caribbean

1. What are the three great landform divisions of Latin America?
2. How does altitude influence temperature, precipitation, and agricultural options in the Andes?
3. What major indigenous civilizations ruled in Latin America before the arrival of Europeans and where were those civilizations located?
4. Why did Latin America experience a major population decline after the European conquest?
5. Where did the slave labor used to produce sugar and other agricultural products come from and how have society and economy adjusted to the end of slavery in the nineteenth century?
6. Why is Panama such an important part of the global transportation system, and will that role continue in the future?
7. What are the gains and the losses produced by NAFTA (North American Free Trade Agreement) in Mexico's economy?
8. What are the advantages and disadvantages of the varying types of tourism in Central America and the Caribbean?
9. Why has Brazil emerged as an important global economy?
10. Why is Evo Morales such an important voice in South America?

Geographers @ Work

1. Explain the role traditionally played by coca in highland Andean culture in contrast to the impact that drugs derived from this plant have had on both local economies and the global community.
2. Latin America is characterized by many different physical environments. Demonstrate how this diversity influences the planning of agricultural development strategies in the region.
3. Remittances are a vital part of Latin American economies. Evaluate how that money is invested in recipient countries and what strategies might be developed to increase the benefits of those investments.
4. Explain why Paraguayans are such a prominent part of Argentina's immigrant community.
5. Investigate the benefits and costs of mining lithium on the Bolivian Altiplano.

Geographic Visualization

The 11th edition features a strong focus on compelling visualization, with new features and illustrations throughout, and dramatic updates to the visual program.

▶ **NEW! Visualizing Development** features cover a defining development in each world region, supported by engaging visualizations of economic and spatial data.

▼ **A completely overhauled cartographic program** incorporates the latest data and GIS techniques and modern styles in scores of engaging and dynamic maps.

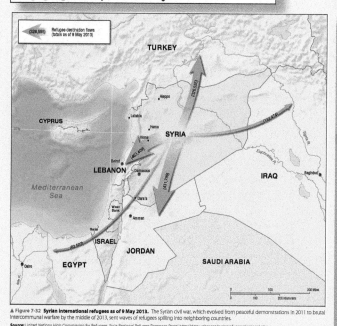

▲ Figure 7-32 **Syrian international refugees as of 9 May 2013.** The Syrian civil war, which evolved from peaceful demonstrations in 2011 to brutal intercommunal warfare by the middle of 2013, sent waves of refugees spilling into neighboring countries.
Source: United Nations High Commission for Refugees, Syria Regional Refugee Response Portal http://data.unhcr.org/syrianrefugees/regional.php.

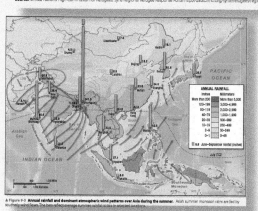

▲ Figure 9-3 **Annual rainfall and dominant atmospheric wind patterns over Asia during the summer.** Asia's summer monsoon rains are fed by southerly wind flows. The bars reflect average summer rainfall totals in selected locations.

Community, Gender, and Regional Variations in Population Growth

Population growth rates within India vary considerably by ethnic, religious, and caste groups. Although culture certainly influences fertility rates, economic status better explains these differences—Hindus, for example, have fewer children than the lowest castes and Muslims but regional population patterns can also be explained by economic factors. The states with the most rapid growth during the 2001–2011 period were a handful of poorer tribal northeast states. Above-average growth rates also characterize the highly populated and poorer northern "Hindi belt" states of Bihar, Uttar Pradesh, Madhya Pradesh, and Rajasthan. This is in sharp contrast to the richer southern states of Andhra Pradesh and Kerala, where growth rates are significantly below the national average. As an indicator of relative levels of poverty, urban fertility rates in 2010 stood at 1.9 per woman, while in rural regions, 2.8 was the norm.

These statistics express regional variation in population growth but mask the regional **gender bias** in India's population structure (Figure 9-1-1). Gender bias exists when one or the other sex represents an abnormally larger percentage of the population. In 2010, there were only 940 Indian females per 1,000 males. Whether for reasons of infanticide, abortion, or nutritional and medical neglect, India has, like China and a number of other developing countries, a deficit in females. In India the problem is extreme. A 2011 United Nations report claims that India exhibits the highest female child mortality rates of any country in the world, and that girls between the ages of one and five are 75 percent more likely to die when compared to boys. At the regional level, the degree of bias favoring males is substantial. Northwest and northern India exhibit stronger gender bias than do southern states. In 2010, the northwestern states of Punjab, Haryana, and Uttar Pradesh were together characterized by a ratio of 892 females per 1,000 males, whereas the far southern states of Karnataka, Kerala, and Tamil Nadu averaged a 1,015 females per 1,000 males. While male in- and out-migration rates impact the female ratio of a given state, female gender bias is strongly correlated with high rates of female literacy and labor force participation. In southern states, women tend to have greater social and economic freedoms, such as owning land and engaging in a wide variety of empowering economic activities. In the northwest, a patrilocal social structure exists whereby a bride moves to her husband's parents' village, is secluded within the household, and denied access to land and political participation.

Sources: Census of India, 2011; Ravinder Kaur, "Across-Region Marriages: Poverty, Female Migration and the Sex Ratio," Economic and Political Weekly 39, no. 25 (2004): 19–25; Mahendra Premi, "Religion in India: A Demographic Perspective," Economic and Political Weekly 39, no. 39 (2004): 4297–4302.

▼ Figure 9-1-1 **Gender bias by Indian political units, 2011.** Much of India is experiencing a severe deficit in females. This map shows the number of females per 1,000 males.

Source: Census of India, 2011.

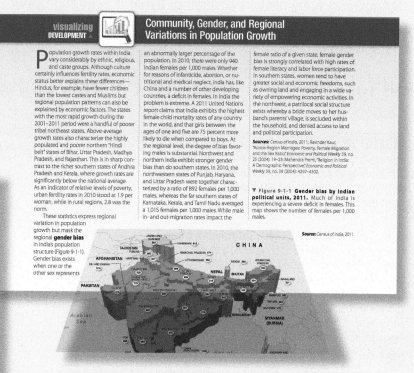

▼ **Over 75% new photos** ensure currency and provide a clear connection with callouts in the text, with geographic IDs provided in all captions.

MasteringGeography™

MasteringGeography delivers engaging, dynamic learning opportunities—focusing on course objectives and responsive to each student's progress—that are proven to help students absorb world regional course material and understand difficult geographic concepts.

Tools for improving geographic literacy and exploring Earth's dynamic landscape

MapMaster is a powerful interactive map tool that presents assignable layered thematic and place name interactive maps at world and regional scales for students to test their geographic literacy and spatial reasoning skills, and explore the modern geographer's tools.

▶ **MapMaster Layered Thematic Interactive Map Activities** act as a mini-GIS tool, allowing students to layer various thematic maps to analyze spatial patterns and data at regional and global scales and answer multiple-choice and short-answer questions organized by region and theme. Includes zoom and annotation functionality, and hundreds of map layers with current data from sources such as the U.S. Census, United Nations, CIA, World Bank, and Population Reference Bureau.

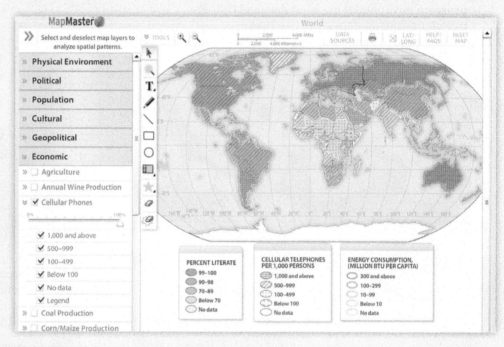

▶ **MapMaster Place Name Interactive Map Activities** have students identify place names of political and physical features at regional and global scales, explore select recent country data from the CIA World Factbook, and answer associated assessment questions.

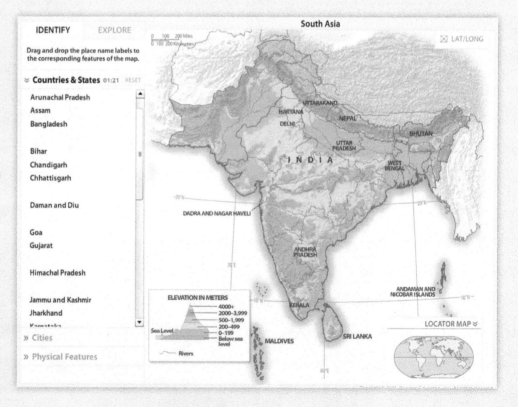

Help students develop spatial reasoning and a sense of place

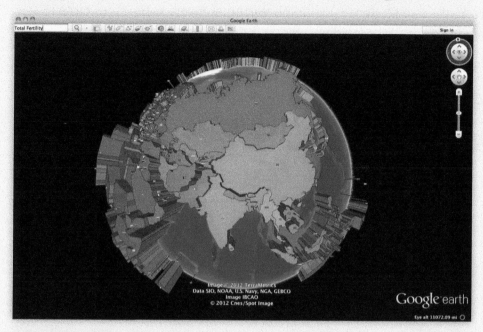

◀ **Encounter Activities** provide rich, interactive explorations of geoscience concepts through Google Earth™ activities, exploring a range of topics in world regional geography. Dynamic assessment includes questions related to core world regional geography concepts. All explorations include corresponding Google Earth KMZ media files, and questions include hints and specific wrong-answer feedback to coach students towards mastery of the concepts.

▶ **Geography videos** provide students a sense of place and allow them to explore a range of locations and topics related to world regional and physical geography. Covering issues of economy, development, globalization, climate and climate change, culture, etc., there are 10 multiple choice questions for each video. These video activities allow teachers to test students' understanding and application of concepts, and offer hints and wrong-answer feedback.

▼ **Thinking Spatially and Data Analysis** and NEW **GeoTutor** Activities help students master the toughest concepts and develop spatial reasoning and critical thinking skills by identifying and labeling features from maps, illustrations, graphs, and charts. Students then examine related data sets, answering multiple-choice and increasingly higher-order conceptual questions, which include hints and wrong-answer feedback.

Drag the appropriate labels to their respective targets. Place the blue age-cohort description labels in the blue boxes and the pink population growth descriptions in the pink boxes. You must use all labels and fill all targets.

High population growth pyramid | Relatively large reproductive population | Relatively small productive population | Large young population | Small aging population | Large aging population | Negative population growth pyramid

Small young population

Laredo, TX   Naples, FL

85+
80-84
75-79
70-74
65-69
60-64
55-59
50-54
45-49
40-44
35-39
30-34
25-29
20-24
15-19
10-14
5-9
0-4
Age

6 4 2 0 2 4 6   6 4 2 0 2 4 6

Submit   Hints My Answers Give Up Review Part

reset ? help

Dynamic Study Modules

Personalize each student's learning experience with Dynamic Study Modules. Created to allow students to study on their own and be better prepared to achieve higher scores on their tests. Mobile app available for iOS and Android devices for study on the go.

Student Study Resources in MasteringGeography include:

- MapMaster™ interactive maps
- Practice quizzes
- Geography videos
- Select Geoscience Animations
- "In the News" RSS feeds
- Glossary flashcards
- Optional Pearson eText and more

Callouts to MasteringGeography appear at the end of each chapter to direct students to extend their learning beyond the textbook.

MasteringGeography™

With the Mastering gradebook and diagnostics, you'll be better informed about your students' progress than ever before. Mastering captures the step-by-step work of every student—including wrong answers submitted, hints requested, and time taken at every step of every problem—all providing unique insight into the most common misconceptions of your class.

Quickly monitor and display student results

The **Gradebook** records all scores for automatically graded assignments. Shades of red highlight struggling students and challenging assignments.

Diagnostics provide unique insight into class and student performance. With a single click, charts summarize the most difficult questions, vulnerable students, grade distribution, and score improvement over the duration of the course.

With a single click, **Individual Student Performance Data** provides **at-a-glance statistics** into each individual student's performance, including time spent on the question, number of hints opened, and number of wrong and correct answers submitted.

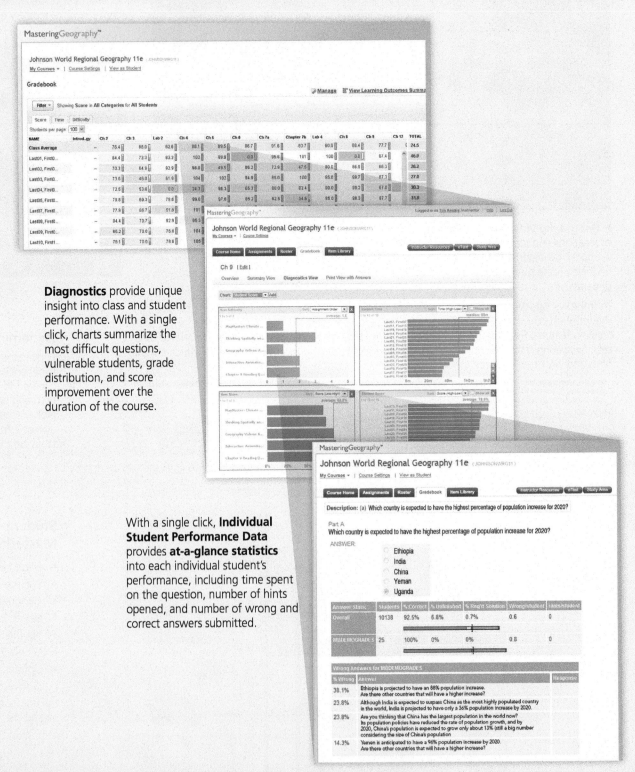

Easily measure student performance against your Learning Outcomes

Learning Outcomes

MasteringGeography provides quick and easy access to information on student performance against your learning outcomes and makes it easy to share those results.

- Quickly add your own learning outcomes, or use publisher-provided ones, to track student performance and report it to your administration.
- View class and individual student performance against specific learning outcomes.
- Effortlessly export results to a spreadsheet that you can further customize and/or share with your chair, dean, administrator, and/or accreditation board.

Easy to customize

Customize publisher-provided items or quickly add your own. MasteringGeography makes it easy to edit any questions or answers, import your own questions, and quickly add images, links, and files to further enhance the student experience.

Upload your own video and audio files from your hard drive to share with students, as well as record video from your computer's webcam directly into MasteringGeography—no plug-ins required. Students can download video and audio files to their local computer or launch them in Mastering to view the content.

learning | catalytics

Learning Catalytics is a "bring-your-own-device" student engagement, assessment, and classroom intelligence system. With Learning Catalytics you can:

- Assess students in real time, using open-ended tasks to probe student understanding.
- Understand immediately where students are and adjust your lecture accordingly.
- Improve your students' critical-thinking skills.
- Access rich analytics to understand student performance.
- Add your own questions to make Learning Catalytics fit your course exactly.
- Manage student interactions with intelligent grouping and timing.

Learning Catalytics is a technology that has grown out of twenty years of cutting edge research, innovation, and implementation of interactive teaching and peer instruction. Available integrated with MasteringGeography or standalone.

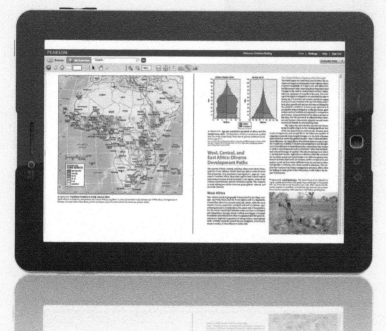

Pearson eText gives students access to *World Regional Geography: A Development Approach,* **11th Edition** whenever and wherever they can access the Internet. The eText pages look exactly like the printed text, and include powerful interactive and customization functions. Users can create notes, highlight text in different colors, create bookmarks, zoom, click hyperlinked words and phrases to view definitions, and view as a single page or as two pages. Pearson eText also links students to associated media files, enabling them to view an animation as they read the text, and offers a full-text search and the ability to save and export notes. The Pearson eText also includes embedded URLs in the chapter text with active links to the Internet. **The Pearson eText app** is a great companion to Pearson's eText browser-based book reader. It allows existing subscribers who view their Pearson eText titles on a Mac or PC to additionally access their titles in a bookshelf on the iPad or an Android tablet either online or via download.

Pearson Choices
for texts with Mastering

Pearson offers content in a variety of formats designed to meet the needs of today's cost-conscious students.

BOUND TEXTBOOK

Pearson texts are written by authors you trust, and undergo a rigorous editorial process to ensure quality and accuracy.

A LA CARTE

A binder-ready, loose-leaf, 3-hole-punched version of the text is available as a budget-friendly option for students.

eTEXT IN MASTERING

Our new eTexts in Mastering have dynamic interactive features such as note-taking, searching, bookmarking, and more.

········· CUSTOM OPTIONS ·········

CUSTOM EDITION

School-specific editions help reduce costs while aligning closely with your course goals.

PEARSON CUSTOM LIBRARY®

Pearson Custom Library is an option for many science titles, allowing you to build a custom text for as few as 25 students.
www.pearsoncustomlibrary.com

ALWAYS LEARNING

PEARSON

World Regional Geography

A Development Approach

▲ This Blue Marble image by NASA is a composite of six separate orbits taken on January 23, 2012 by the Suomi National Polar-orbiting Partnership satellite using the Visible Infrared Imaging Radiometer Suite (VIIRS).
Source: NASA/NOAA.

Introduction

There is a good chance you will use Google Maps today. The most popular smartphone app in the world—well ahead of Facebook, YouTube, and Twitter—Google Maps has become the essential tool of the decade. Never before have so many people been interested in knowing where things are, how to find them, and making connections between what they see in real life with what is mapped on their devices. Geography is seemingly at our fingertips. Yet geography is more than just features on a map. **Geography** is fundamentally the study of *location*—the location of physical features, economic activities, human settlement patterns and cultural attributes, and anything else that a person can find on a map—and the connections between those things and places. In other words, geographers have a keen interest in understanding what defines terrestrial space (as opposed to "outer" space), and the interactions between people and their environments within that space. For that reason, geography is known as a "spatial" science that, in a fundamental way, studies the *why* of *where*.

At the most basic level, students of geography are called on to address three questions:

What is located where? This is the map or the "location list" question that most of us want to answer when we use Google Maps or Google Earth. For example, we may ask where the mountains are found in the eastern United States (answer: the Appalachian system running from Maine to Alabama), or where in Europe ethnic tensions have created a fractured political geography (answer: the Balkan Peninsula). Asking where people, places, and activities are located provides context and creates a basic knowledge of locations that is a starting point for more detailed examination of places later. It also teaches students what is geographically correct, so they know that Brazil is a Portuguese-speaking Latin American country, or that the Mississippi River is the major river system of the U.S. Midwest. Knowing where things are is important, but geography is much more than a rote memory exercise more appropriate to long car trips than serious study.

Why are things located where they are? Geographers primarily want to understand *why* things are located where they are. This is the "explanation" question that looks for the processes that produce a particular geographic pattern. To continue with the examples used above, we know that the Appalachian mountain system runs the length of the eastern United States. Why are these mountains where they are? The answer lies in the history of Earth's colliding continents, a story of plate tectonics that has contributed to the world's numerous mountain chains. Along the same lines, the Balkan Peninsula has a long tradition of ethnic strife leading to political fracturing, which stems from the highly conflicted territorial histories of the south Slavs and the influences of outside empires. Other "why" questions of interest to geographers may require examining human-environment interactions. Geographers might study the relationship between cropping patterns and the expansion of deserts, or the role of city structures in modifying climate (for example, creating urban heat islands).

So what? Now that we know where things are and why, we might ask "Who cares?" This is the "significance" question. Geographers find meaning in the geographical phenomena they have identified—we may know where the Appalachians are located and how they were formed, but then we examine their significance to settlement patterns in the early United States (that is, colonial clustering along the eastern seaboard), to weather and climate patterns (orographic effects producing massive rainfall quantities in the southern Appalachians), or to cultural patterns (think of the very different cultures that developed in the remote valleys of the Appalachians). By the same token, we should be reminded that the fractured political geography of the Balkans contributed to the outbreak of World War I, and that political distress following the collapse of communism in Eastern Europe led to the stationing of U.S. troops in parts of the Balkan Peninsula as peacekeepers to separate ethnic groups engaged in bloody conflict. For many Americans, the significance is quite personal. Geographers can also apply their knowledge about the Balkans to understand other parts of the world where upheaval allows ethnic groups to resurrect age-old conflicts, as in the recent Buddhist attacks on Muslims in Myanmar (Burma).

The goal of this book is to help students know and understand where places are, why they are there, and their significance. This goal is addressed using an economic development perspective. Above all, we want students to know that geography matters.

Geography's Roots

Geography has a rich, ancient, and varied heritage. Its solid foundation rests on the works of ancient scholars, who recorded the physical and cultural characteristics of lands near and far. The study of geography evolved in many civilizations, with the first surviving maps appearing on clay tablets in ancient Iraq six millennia before the present. Folk cultures also developed their own pragmatic body of geographic knowledge—for example, early Polynesians produced maps of wave patterns around South Pacific islands that, combined with their knowledge of the stars and constellations, allowed them to navigate across great stretches of otherwise featureless ocean. But it was the Greeks who made the most enduring contributions to geography's early formal development. The term "geography" comes to us from the Greek words *geo* ("the earth") and *graphos* ("to write about" or "describe").

Contributions of the Greeks

The early Greeks studied the same kinds of geographic problems that confront us today, but without the benefit of modern knowledge and technologies. **Herodotus** (ca. 484-425 B.C.), called by some the father of geography as well as the father of history, placed historic events in their geographic settings in his famous *Historia* (ca. 450 B.C.). He described and explained the physical and human geography of his day, with particular emphasis on the seasonal flows of the Nile and Ister (Danube) rivers. Herodotus was also one of the earliest Greek geographers to map and name the continents of Europe, Asia, and Africa, which he called Libya (Figure I-1). **Aristotle** (384-322 B.C.) discussed the physical characteristics of the earth, including temperature, wind, alluvial or stream deposition, and vulcanism, in his *Meteorologica*. Aristotle was also the first to divide the world into three broad climatic zones, which he called the "torrid" (tropical), the "ekumene" (literally, "the home of man," which corresponded to the mid-latitudes), and the "frigid," or polar realms. Although Aristotle erroneously believed that neither the torrid nor frigid zones permitted the full development of human potential, his discussion of the influences of physical environments on humans reflected a principal concern of geographic inquiry in all ages.

Other Greek scholars examined the size and shape of the earth and its relationship to the rest of the cosmos. Which methods, they wondered, could be used to show where places are in relation to one another and what people do in the various parts of the world? The Greeks did not answer all of their questions, but furthered knowledge of the world. **Eratosthenes** (ca. 276-195/194 B.C.) a Greek living in Alexandria, Egypt measured the earth's circumference. He had learned that on only one day each year (the Northern Hemisphere's summer solstice) did the noon sun shine directly down a well near what is now the city of Aswan. On that special day Eratosthenes measured the noon sun's angle at Alexandria, some 500 miles (805 kilometers) north of Aswan, and found that there the sun's rays were not vertical but cast a shadow of 7.2° from a pole projecting straight up from the earth. Using geometry, he concluded that the distance between Aswan and Alexandria of 500 miles (805 kilometers) must be equal to a 7.2° arc of the earth's surface. He then computed the value of a full circle's 360°, estimating the earth's circumference to be 25,200 miles (40,554 kilometers). Today we know that the circumference at the equator is quite close to that estimate: It is actually 24,901.5 miles (40,073.9 kilometers).

Eratosthenes and other Greeks recognized the need for maps to show the relationships between one place or region and another—a way to locate themselves on Earth and describe their location to other people. To understand the challenge of their task, think of a mark on a smooth, uniform ball and then try to describe the position of the mark. Fortunately, the earth rotates around an axis that intersects the surface at two known points—the North Pole and the South Pole. With those points, the Greek geographers established two reference lines: the equator halfway between the poles and another line extending from pole to pole. They then drew a grid of latitude and

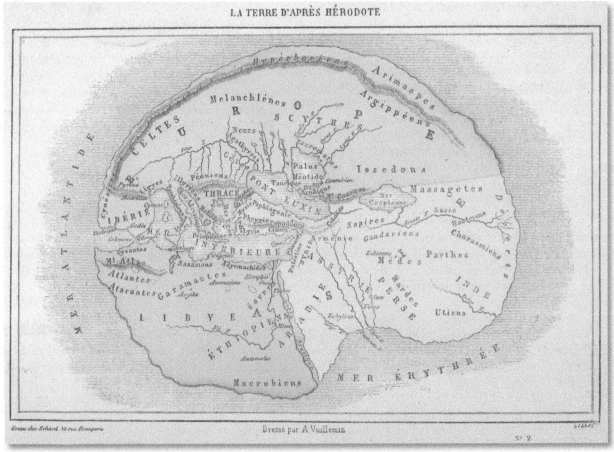

▲ **Figure I-1 The World through the eyes of Herodotus.** We will never know exactly how Herodotus depicted the world in cartographic form; no copy of his famous map survives. This is a French effort to imagine that map based on the places described in the geographical and historical writings of Herodotus.

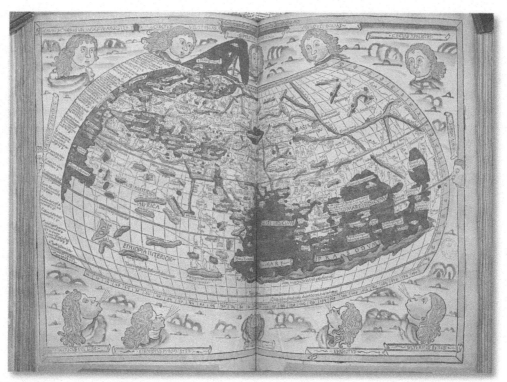

▲ **Figure I-2 Ptolemy's World Map.** Ptolemy brought the classical period of Greek and Roman geography to a close. His world map, while containing numerous errors, was nevertheless a reflection of the remarkable geographical achievements of the Greek geographers. It stood as the standard reference map of the Western world for more than 1,300 years, until the Age of Exploration brought increased geographical knowledge of the world.

longitude lines from those geographic reference points, thereby locating any point on Earth using just two numbers. For example, the location of Washington, D.C., is 38°50 N, 77°00 W.

The Greeks' next step was to use the **geographic grid** to construct a map whereby all or part of the earth could be drawn on a flat surface and the relative positions of places and regions could be marked. Maps became the tools used by geographers to depict **spatial relationships**. About A.D. 150, a Greek astronomer from Alexandria, **Ptolemy**, (A.D. 90–168), designed a map of the world and then, using a coordinate system, compiled the location of 8,000 known places (Figure I-2). This early work of the Greeks was not flawless; Christopher Columbus, for example, using Ptolemy's estimate of the earth's circumference, thought the world to be much smaller than it actually is and believed he had reached the coast of Asia rather than the Americas.

Geography thrived during Greek and Roman times. New lands were discovered, and inventories of their resources and characteristics had great practical importance. One compiler was **Strabo** (64 B.C.–ca. A.D. 23), whose *Geographia* described the then known world and revealed a fascination with the distinctive physical and human characteristics that make each location unique. Part gazetteer, part travel guide, part handbook for government officials, Strabo's work was an effort to view places in a holistic, multipurpose fashion. The accomplishments of Strabo and other Greeks established geographers as leading figures of their times. Their most enduring contribution was the development of a scholarly approach that emphasized the importance of describing the world from a **spatial perspective** and recognizing the interdependence of the physical and cultural elements of the world. Geography's position was one of acclaim until the Roman Empire began to decline.

Mainstream Western geographical thought can be traced to the ancient Greeks, but other centers of geographical thought existed as well. Islamic explorers and scholars, such as the famed geographer **Muhammad al-Idrisi** (A.D. 1099–1166), served as a bridge from ancient to modern thought. Ancient China was a major center of geographical scholarship and exploration as well, with travel books dating back to A.D. 1000. Marco Polo (c.1254 – 1324), a Venetian merchant, undertook a 24-year journey of discovery to China and East Asia that inspired later European explorers in the same way as did Moroccan explorer Ibn Battuta (1304–1368), who traveled as far as China and Southeast Asia and expanded the horizons of the Muslim world. Although Chinese geography did not eclipse the work of Greek scholars during the classical period, it thrived from the fifth to the fifteenth centuries. During that millennium, Chinese geographers traveled through southern Asia, the Mediterranean, and western Europe. They established human geography, completed regional studies inside and outside China, studied geomorphological processes, and wrote geographical encyclopedias.

The Renaissance and the Age of Discovery

After a decline in the Middle Ages, the Renaissance marked a resurgence of geography and other sciences. New routes to Asia and the Americas were opened, first under the sponsorship of **Prince Henry the Navigator** (A.D. 1394–1460) of Portugal and later by seamen sailing for Spain, Holland, England, and France. Geographer-explorers, including Christopher Columbus, Ferdinand Magellan,

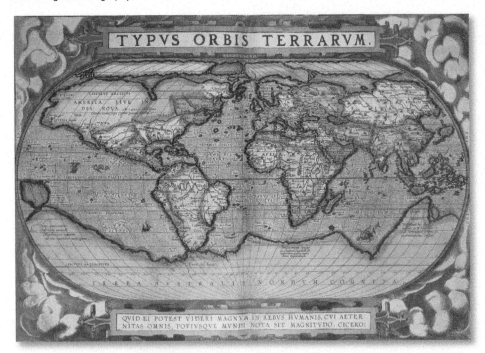

▲ **Figure I-3 A world map (1571) by Flemish cartographer (mapmaker) Abraham Ortelius (1527–1598).** By the late sixteenth century, increased exploration was leading to more accurate renderings of world maps. By comparing this view of the east and west coasts of the Americas or the west coast of Africa with the South Pacific, we can see the effect of exploration.

Salem, Massachusetts, Nathaniel Bowditch (1773–1838), to compute, correct, and publish absolutely accurate navigation tables for mariners (Figure I-4). Bowditch's *The New American Practical Navigator*, first published in 1802, remains the bible of navigators to this day. An age of scientific reasoning began, with experimentation and the testing of hypotheses. New explanations in the natural sciences challenged old ideas about the origin of continents and oceans, the formation of landforms, and the evolution of plants and animals. Scientific travelers—students of natural history—looked for explanations for the varied world around them. Among those was the great German geographer **Alexander von Humboldt** (1769–1859), who traveled widely in Europe and Latin America. His curiosity, careful observation, and extensive knowledge of botany, physics, chemistry, Greek, archaeology, and geology enabled him to synthesize information from a variety of fields into a coherent geographic composite. In his most celebrated work, *Kosmos* (1845–1862), he attempted a comprehensive description of the earth.

Karl Ritter (1779–1859), a contemporary of Humboldt's, first studied geography as a basis for understanding history, but eventually found that geography itself could provide an understanding of the human dimension of the world. His great uncompleted work, *Die Erdkunde*, included nineteen volumes on Africa and Asia. Ritter is generally recognized as having held the first chair of geography at the University of Berlin in 1820. By the middle of the nineteenth century, geography was a respected discipline in

and James Cook, ushered in an Age of Discovery. Renaissance geographers produced increasingly accurate maps (Figure I-3). The physical and cultural characteristics of exotic foreign lands were described, and the processes that created differences and similarities between one place and another were analyzed in greater detail.

European scholars began to question age-old concepts in light of discoveries in other parts of the world, although it took a self-taught mathematician, astronomer, and geographer-navigator from

(a)

(b)

▲ **Figure I-4 Nineteenth Century Geographers @ Work.** Both Nathaniel Bowditch and Alexander von Humboldt engaged in profoundly important research that continues to have great impact and enhanced the reputation of geography. Bowditch (a) was a theoretical and mathematical scholar who applied his skills to the solution of practical navigational problems. His likeness graces the landscape of Mt. Auburn Cemetery in Cambridge, Massachusetts. Humboldt (b) linked extensive field collection of data to a complex understanding of ecosystems and place. Arguably the most famous scholar of his day, he is widely commemorated by statues in Germany and around the world, with one of the more famous located in front of the Berlin university that bears his name.

European universities, and geographical societies served as important meeting places for scholars of all disciplines who were interested in the world around them. From Europe, a new age of geography spread around the world, fostered in part by geographical societies founded in France (1821), Germany (1828), and Great Britain (1830). Geography found a particularly receptive audience in the western hemisphere, where the earliest geographical societies were established in Mexico (1839) and the United States (1851). Americans were eager for knowledge about their country, especially the frontier regions, and geographical literature was particularly popular. Geography as an academic field of study began to flourish in the latter part of the nineteenth century, and in subsequent years spread from a few centers to almost every major college and university.

The Modern Practice of Geography

Modern geography has grown beyond description of the earth. Today's geographers not only describe through words, maps, and statistics, but also analyze interrelationships between physical and cultural phenomena to explain why things are distributed over the earth as they are.

Modern Geography

Modern geography is best understood as the study of how the physical and cultural attributes of the earth interact to form spatial or regional patterns. Modern geography has improved our ability to explain the world by utilizing four traditional areas of study:

1. the location of physical and cultural features and activities (spatial distributions);
2. the relationships between people and the lands that support them;
3. the existence of distinctive areas or regions, including analysis and explanation of how they came to be formed; and
4. the physical characteristics of the earth, perhaps the oldest of all geographic traditions.

The focus of each of these traditions as described by William Pattison[1] is evident in the work of the early Greek scholars. The **spatial tradition**, with its concern for distance, geometry, and movement, can be seen in the work of Ptolemy. The writings of Hippocrates (ca. 460–370 B.C.) were concerned with the relationship of human health to the surrounding environment, a theme common to the **man-land tradition.** The **area studies tradition**, with its concern for the nature of places and for understanding the "where" of places, is evident in Strabo's *Geographia*. The **earth science tradition**, as a study of the earth and its environments, is identifiable in the work of Aristotle and his students. These classical traditions have been joined by other **perspectives**, such as the **behavioral** or **feminist** or **postmodern**, which enrich traditional approaches to geographic scholarship by incorporating new knowledge and insights into the geographer's understanding of how the world works.

[1]William D. Pattison, "The Four Traditions of Geography," *Journal of Geography* 63 (May 1964): 211–216; reprinted in the *Journal of Geography* 89 (September–October 1990): 202–206.

The Subdivisions of Geography

Geography has many subdivisions that encompass one or more of the traditional areas of study described. The principal ones are physical geography, human geography, systematic geography, and regional geography.

Physical geography is the study of the environment from the viewpoint of distribution and process. For example, landform geographers, or geomorphologists, are concerned with the location of terrain features and with the ways in which those features have acquired their shapes and forms. Geomorphologists might study the impact of stream deposition in a floodplain, the effect of wind erosion in a dryland, or the formation of coral reefs around a tropical island. Biogeographers are interested in the distribution of plants and animals, the ways organisms live together, the processes (both natural and people-induced) that affect the biological Earth, and the effect of environmental changes on human life. Climatologists study the long-term characteristics of the atmosphere and any climatic differences created by temperature or energy and moisture conditions in various parts of the Earth. Physical geography emphasizes the interdependence of people and the physical earth. Such contemporary problems as ozone depletion, acid precipitation, desertification, rain-forest removal, global warming, and sea level rise are of particular interest (Figure I-5).

Human geography consists of the study of various aspects of our occupancy of the Earth. Urban geographers, for example, examine the location and spatial structure of cities to explain why urban areas are distributed as they are and to account for patterns of settlement and economic and cultural activity within cities. Urban geographers are interested in the process of urban growth and decline, in the types of activities carried on in cities, and in the movement of goods and people within and between cities and metropolitan areas (Figure I-6). Cultural geographers examine the ways in which groups of people organize themselves; they study the distribution and diffusion of

▼ Figure I-5 **Deforestation in the Brazilian Amazon.** Roads open the rain forest to lumbering operations and agricultural settlement. Clearcutting reduces the ability of the forest to grow back, and farmers often find that maintaining cultivation in the face of declining soil fertility is difficult. Invasion by grasses and the expansion of cattle herding are often the result.

▲ **Figure I-6 New light rail tram in the square of the United Nations in Casablanca, Morocco.** The urban transportation system in Moroccan cities is undergoing a massive upgrade as efficient light rail systems are constructed to move large numbers of people in and out of city centers.

such cultural institutions as language and religion, as well as social and political structures. Economic geography involves the study of systems of livelihood, especially the distribution of related activities and explanations for such distribution. Economic geographers are concerned with the analysis of natural and cultural resources, with

their utilization, and with the structures of power and control over and patterns of access to resources at various scales that determine the equity with which resources are shared.

Systematic, or topical, **geography** consists of the study of specific subjects. Historical geographers, for example, study past landscapes and the changes that have taken place. How did the people of the Great Plains organize themselves in 1870, as compared with their organization in 1935 in the midst of the Great Depression? What past characteristics have persisted, and what effect do they have on present-day distributional patterns? Historical geography thus adds depth perception to time, facilitating an explanation of present patterns and their reasons for being. Systematic geographers normally study one aspect of the field—landforms, economic activities, urban places, or natural hazards, for example (Figure I-7) .

Regional geography involves the analysis of environmental and human patterns within a single area. A wide variety of facts are placed into a coherent form in order to explain how a region is organized and how it functions. A regional geographer is an expert on a particular area of the world, applying systematic approaches to an understanding of that area (Figure I-8). This textbook applies a regional approach, dividing the globe into 11 broad regions, each with its own distinctive set of shared characteristics, in order to provide the context needed to understand and compare how globalization and development processes, among many others, operate around the world.

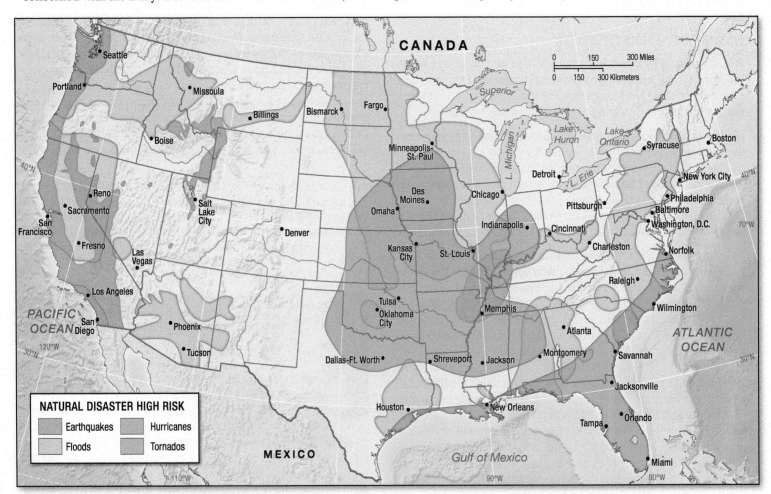

NATURAL DISASTER HIGH RISK

| ■ Earthquakes | ■ Hurricanes |
| ■ Floods | ■ Tornados |

▲ **Figure I-7 The regional hazard threat in the United States.** Natural hazards are profoundly regional, representing different problems in different parts of the country. How much risk the people of each area will accept determines the type and effectiveness of the management strategies they adopt.

Source: Data from American Red Cross and NOAA.

▲ Figure I-8 **Alpine villagers move cows to high elevation pastures.** Moving cows to high Alpine pastures in spring is a festive occasion throughout the region as decorated animals, festooned with flowers and massive bells, parade through the streets of a village in the canton of Fribourg, Switzerland.

All fields of geography, despite focusing on different sets of phenomena, share the geographic viewpoint; that is, all geographers analyze spatial arrangements (distributions) and search for explanations of the patterns and interrelationships among those and other phenomena. All geographers rely on maps as analytical tools and have added computers and remote-sensing techniques to aid in recording and analyzing data.

The Geographic Information Science Explosion

Twenty-first-century maps are more than just maps—they are analytical tools referred to as **geographic information systems (GIS)** and are part of a larger field of study called geographic information science (GIScience). Broadly defined, GIS is a digital representation of the earth's surface (a site, region, or country) that can be used to describe landscape features (roads, boundaries, mountains, rivers) and can support analysis of these features. In a sense, a GIS is like having a whole atlas in a single computer presentation with the ability to relate different pages of the atlas to each other. The emergence of GIS in the United States is a recent phenomenon. It began when Jack and Laura Dangermond established ESRI, the Environmental Systems Research Institute, in 1982. ESRI's first software program, ARC/INFO, has since developed into a suite of different GIS programs that now dominate the GIS marketplace. A research-oriented GIS software tool called IDRISI was first created by Ron Eastman in 1987, and is a specialized system that is particularly popular with research scientists and conservationist NGOs.

Key to our understanding of GIS is the information layer—a map page showing a specific type of information, such as political boundaries, physical features, economic activities, cultural attributes, or any of a large number of other possibilities (Figure I-9). A GIS project will have multiple layers of information that can be called on to answer a specific question or to address a specific issue. For example, layers may be combined to show electoral districts and income levels in a city (Figure I-10);

another set of layers may be used to show fire-hazard potential in a national forest, given certain vegetation types and climatic conditions. Distances, areas, and volumes can be computed; searches can be conducted; optimal routes can be selected; and facilities can be located at the most suitable sites. In a fully integrated GIS implementation, some layers may contain satellite imagery or other remotely sensed information, and other layers may contain line data (road networks or county boundaries, for instance) that can be superimposed on the imagery.

The use of GIS is exploding. Governments use GIS to track everything from power lines to demographic profiles. Businesses find geographic information helpful to locate facilities and to develop markets. Law enforcement agencies employ electronic maps to identify crime "hot spots" and to build "geographic profiles" of criminals. Militaries of the twenty-first century rely on GIS for terrain analysis and battlefield information. Political campaigns have discovered the value of mapped profiles of potential voters and the issues that interest them. The list is almost endless. Geographers find that their GIS skills are in high demand by employers in both the public and private sectors, not to mention the everyday life uses like mapping and Google Earth.

Geography and Its Disciplinary Neighbors

Geography is a spatial science that focuses on the location of human and physical phenomena, and their interactions within terrestrial space, which explains geographers' fascination with maps. Geographers are as interested in understanding earth space itself as they are in examining the institutional or physical characteristics of objects that occupy that space. This spatial emphasis has two implications for understanding geography's relationship with neighboring disciplines. First, geography is a bridge between the social and physical sciences, and possesses many characteristics of both areas. Take a look at a typical road map, for example, and note the references to mountains, rivers, cities, boundaries, and roads—all on the same map. While most maps are not so encompassing, the road map demonstrates how geography requires an understanding of both human and physical phenomena. In a different example, a geographer who studies the growing of wheat on the Great Plains needs information on the climate, soils, and landforms of the region (the physical sciences), as well as knowledge of the farmers' cultural characteristics, the transport network, the costs of wheat farming in relation to other economic opportunities, and a host of other socioeconomic factors (the social sciences). For that reason, geography is referred to as a **holistic discipline** that synthesizes knowledge from many fields.

Consequently, geography also touches many related disciplines in both the social and physical sciences (Figure I-11). For example, a political geographer may have interests shared with political scientists, a historical geographer with historians, a biogeographer with biologists, and a GIS person with computer scientists. The list goes on. Remember, however, that the geographer is distinguished from these neighboring disciplines by his or her spatial perspective; other disciplines do not have a spatial starting point, although they may deal with some of the same objects or phenomena.

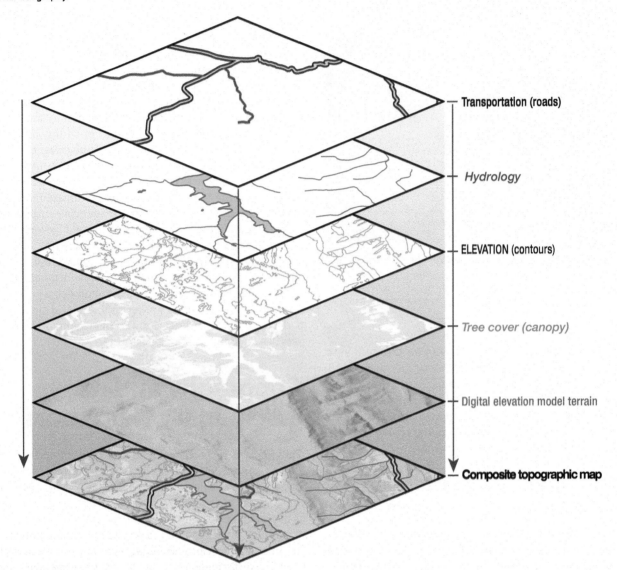

- Transportation (roads)
- *Hydrology*
- ELEVATION (contours)
- *Tree cover (canopy)*
- Digital elevation model terrain
- **Composite topographic map**

▲ **Figure I-9 A hypothetical GIS design.** Within a GIS, environmental data attached to a common terrestrial reference system, such as latitude/longitude, can be stacked in layers for spatial comparison and analysis.

Here is an example of how geographers interact with related disciplines, using the case of the economic geographer. Economists are interested in the production, distribution, and consumption of goods and services. They study how people use resources to earn a livelihood, investigating such topics as the costs and benefits of resource allocation, the causes of changes in the economy, the impact of monetary policies, the workings of different economic systems, the problem of supply and demand, and the dynamics of business cycles and forecasting. Economic geographers are also concerned with how people earn their livelihoods, but economic geographers look at where the economic activity takes place and what factors—such as the availability of labor, raw materials, and markets—along with certain physical attributes, influence that location. Economic geographers are less interested in learning about the economics of the auto industry than in understanding why auto manufacturing is located where it is, keeping in mind that economic considerations cannot be ignored in understanding location. They want to know about the historical, physical, social, as well as economic contexts of auto plant location.

Geography is not the only integrating discipline; so is the study of history. But history uses a **chronological** (time) **framework**, whereas

geography's perspective involves a **chorological** (place) **framework**. Neither can be studied effectively without a knowledge of the other. Isaiah Bowman (1878–1950), former president of Johns Hopkins University, emphasized that "a man [or woman] is not educated who lacks a sense of time [history] and place [geography]."[2] By integrating information in a regional context, as we do in this textbook, the geographer pulls together knowledge shared with a variety of disciplines into a single, all-encompassing, whole. In so doing, geography can provide insights and understanding that would not be available through separate study of the individual elements.

Applied Geography

Modern geographers differ from their late nineteenth and early twentieth century predecessors in emphasizing explanation rather than description. They ask not only "where" questions but "why" questions. This shift in emphasis has increased geography's utility

---

[2]Quoted in Alfred H. Meyer and John H. Strietelmeyer, *Geography in World Society* (Philadelphia: Lippincott, 1963), 31.

▲ **Figure I-10 A hypothetical GIS application.** This illustration shows how various types of map information can be electronically combined into a single presentation.

in solving many problems of our contemporary world. The result is increased employment in roles such as market and location analysts, urban or regional planners, cartographers, environmental analysts, and teachers.

Education

Traditionally many geographers have been employed as teachers, although geography as a specific subject is taught with varying emphasis in different parts of the country and at different levels in the curriculum. Many Americans have little contact with geography as a formal subject after middle school, and the general public's geographic knowledge and skills are often inadequate. Reading a map is a basic skill, yet many people have difficulties performing this fundamental task. Many citizens find it difficult to locate other countries on a map and frequently know little about the cultures, values, and beliefs of other communities. The *Goals 2000: The Educate America Act,* passed by Congress in 1994, designated geography as one of several fundamental subjects that deserved more attention in school curricula. National Geography Awareness Week has been promoted since the late 1980s as a way to encourage a wider positive perception of geography's contemporary importance. Since 1989, the National Geographic Society has sponsored the National Geographic Bee; this contest tests the geographic knowledge of middle school students in a series of local and state competitions that culminates in a "final exam" between the 50 state winners (Figure I-12). Beginning in 2001, the College Board has included geography in its advanced placement course system and has offered advance placement geography exams as part of its SAT (scholastic aptitude test) array. Where there are increasing numbers of students taking geography courses, there are also opportunities for larger numbers of geographically trained teachers! This development is good for both American education and for newly minted teachers, since the ways in which the United States is connected to the larger world—from the need for products and materials not readily available in the United States, to migrants fleeing adverse economic conditions in their homelands, to foreign policy entanglements—will only grow greater as we move deeper into the twenty-first century.

Business

Many of the skills that geographers possess, particularly those related to location, are very useful in the business world. Many geographers have found employment as locational analysts or environmental consultants. Banks would be foolish to make loans to new businesses or to established companies wishing to start new outlets or franchises without first determining whether the proposed location is an economically advantageous one. It is as important to the financial institution lending start-up money as it is to the business owner to know that the new restaurant or comic book store or gift shop will make money. How else can the bank be sure that the principle and interest on its loan will be repaid? Many large companies retain their own internal group of locational analysts who travel from site to site, study the relative benefits of

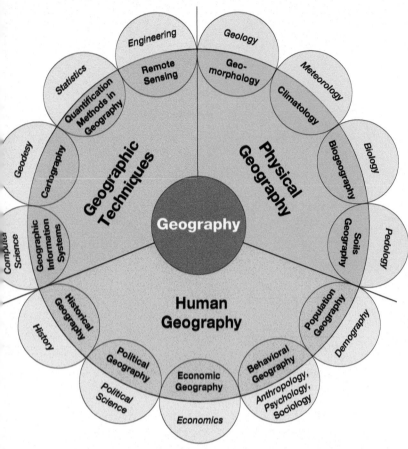

▲ Figure I-11 **The scope of geography.** Geography is a synthesizing and integrating discipline. This diagram shows that geography interrelates with many fields of study, including the physical sciences, engineering, social sciences, and humanities.

central city as opposed to suburban locations, examine the purchasing power of the citizens of a particular community or neighborhood, assess traffic patterns to ensure a new outlet is accessible to a high volume of potential customers, evaluate the likely competition that rival firms might represent in the area, and predict the direction that patterns of growth and decline in the regional economy are likely to experience. Once all the data are assembled, it is possible to determine whether a proposed location is a good one or not.

The skills of physical geographers and students of environmental hazards also have application to real world problems. Concerns about the environmental impacts of economic development require the preparation of environmental impact statements before a project can proceed. Federal law requires that changes with impacts on the environment, such as new housing developments, industrial plant expansion plans, road construction, wetland drainage, and many other activities, must be identified, their scope assessed, and potential damages mitigated before a project can begin. Geographers have often played a major role in environmental impact assessment as well as in the emergency management of hazards, both natural and human-caused (Figure I-13).[3] Geographers have also employed their general knowledge of foreign areas by helping companies understand the challenges and opportunities that businesses in those places are likely to encounter, as well as providing cartographic, climatological,

---

[3]David Alexander, *Principles of Emergency Planning and Management* (New York: Oxford University Press, 2002).

▲ Figure I-12 **The National Geographic Bee in 2013.** For twenty-five years and counting, National Geographic and state Geographic Alliances have sponsored a competition involving "bees" at ascending geographical scales. The 50 state winners meet in Washington, D.C. for the ultimate showdown, won in 2013 for the first time by a middle school student from Massachusetts.

geographic information science analysis, travel agent, and consumer behavior services to firms and individuals wishing to conduct business outside the North American continent.

Government

Second to teaching, more geographers probably apply their skills to government agencies than to any other area. At a local level, many are municipal or regional planners, charged with facilitating orderly residential, business, and industrial growth and redevelopment. On a national level, knowledge of distant places and cultures, often in combination with remote sensing, map interpretation, and GIS skills, provide geographers with analytical abilities that are much in demand in government agencies. The Office of the Geographer in the Department of State is a focus of geographic activity, as is service in the diplomatic corps. The United States Agency for International Development (USAID) also includes many geographers whose expertise in resource analysis, regional development, planning, and sustainability science provides

▼ Figure I-13 **Tornado devastation in Moore, Oklahoma.** A tornado with winds over 200 miles per hour hit Moore on 20 May 2013, killing 23 people and causing over $2,000,000 in damages. Despite a history of severe tornadoes in this area, few shelters exist in private or public buildings to reduce loss of life.

valued services. Internationally, geographers often work closely with agencies of the United Nations such as FAO (Food and Agriculture Organization), UNEP (United Nations Environment Program), and the United Nations University, as well as with international financial institutions such as the World Bank and the Inter-American Development Bank. For geographers who view governmental agencies as excessively bureaucratic and too "top down" in their approach to promoting economic development, NGOs (non-governmental organizations) engaged in grassroots development activities have provided a productive outlet for energy and initiative.

Humans suffer from two types of economic poverty: sudden disaster, such as war, hurricane, or famine and lack of economic development caused by complex historical and environmental factors. MapAction creates maps in disaster zones that present information about disaster situations on critical subjects, such as concentrations of people, emergency health care, logistics, or emergency shelter. Disaster victims need food, shelter, and medical care, and these maps help practical relief organizations such as the Red Cross to accurately deliver aid to the people in greatest need.

Global MapAid (GMA) compliments disaster relief by mapping longer-term solutions to poverty, often starting immediately after the disaster relief phase ends. These maps show victims of poverty or disaster returnees where to get help from medical facilities, government offices, and anything else of practical value. Vitally, GMA also makes poverty solution maps to promote sustainable employment in poverty zones, with maps combining overlays that match up vocational educational schools and microcredit organizations with their student catchment areas. These maps are used by both victims of poverty and NGOs interested in getting afflicted people into gainful employment, which in turn empowers them to take care of themselves rather than relying on others. In this way, all stakeholders—media, relief agencies, donors, and those in need—can coordinate the daunting task of rebuilding homes, reconstructing lives, and promoting development.

The work of GMA demonstrates one way in which geography can be applied to real-world situations and can make a difference in human lives. The GMA map in Figure I–14 shows the distribution of

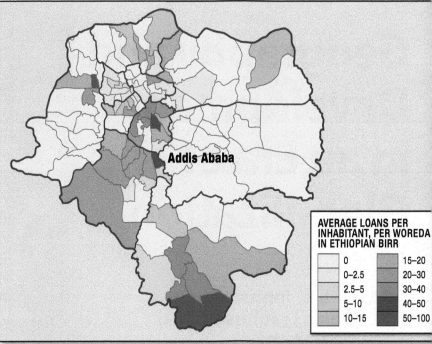

▲ **Figure I-14 Geography @ Work in disaster aid.** Lending money to marginalized people with good ideas for a new business helps promote initiative and development. Identifying areas where loans have been made helps clarify the parts of Addis Ababa most likely to benefit from future microfinance programs in this example of GlobalMapAid's work.

Source: copyright © Global MapAid. Reprinted by permission.

microcredit across the capital city of Ethiopia, Addis Ababa in 2013. The lighter areas illustrate areas of greater need where the ratio of loan money to the underlying population is lower. These areas would probably be worthy of consideration for more microcredit by donors and microfinance institutions.

Increasingly large numbers of geographers are engaged in employing geographical skills to the pragmatic world of improving people's daily lives. And increasingly large numbers of people, geographers or not, are engaged with geography through GPS devices, storm tracking, Google Maps, maps on Facebook, and many other applications.

Summary

▶ This overview explains what geography is about. Geography is a spatial science that focuses on the location of human and physical phenomena, and their interactions. These interactions include relationships between people and their environments.

▶ Central issues for geographers include the "where," "why," and "so what" questions. Geography is a holistic discipline that is integrative in its spatial approach and touches on a variety of sister disciplines. Above all, geography matters!

Key Terms

Alexander von Humboldt I-6
area studies tradition I-7
Aristotle I-4
behavioral perspective I-7
chorological framework I-10
chronological framework I-10
earth science tradition I-7
Eratosthenes I-4

feminist perspective I-7
geographic grid I-5
geographic information systems (GIS) I-9
geography I-3
Herodotus I-4
holistic discipline I-9
human geography I-7

Karl Ritter I-6
man-land tradition I-7
Muhammad al-Idrisi I-5
physical geography I-7
postmodern perspective I-7
Prince Henry the Navigator I-5
Ptolemy I-5
regional geography I-8

spatial perspective I-5
spatial relationships I-5
spatial tradition I-7
Strabo I-5
systematic geography I-8

Geography and Development in an Era of Globalization

**Merrill L. Johnson,
Douglas L. Johnson, and
Viola Haarmann**

Give a Man an Animal...

Small-scale development initiatives have an astonishing ability to multiply and spread rapidly. Their transformative power is directly related to how closely they address immediate local needs. Securing adequate nutrition, particularly protein, is a universal human requirement. But poor people find this need difficult to meet, particularly when social and environmental constraints are great and are deeply embedded.

Fortunately, a potential solution is readily available—animals. Many nongovernmental organizations (NGOs) feature animals among their development programs because each female animal has the ability to reproduce many times. A model and leader in this style of development is Heifer International, which began operation in 1944, and has since expanded from large stock (cows, camels, water buffalo, llama) to small stock (goats, sheep, pigs, rabbits, guinea pigs) to still smaller animals (chickens, ducks, honeybees). The operating strategy is simple: once a potential development community is identified, animals suitable to local needs and culture are identified, eligible recipients are trained, and each recipient is given an animal. Beginning initially with heifers (young, full-grown cows, often pregnant with their first calf) intended for World War II relief in postwar Europe, the program has grown to include over 120 countries. The recipient of a heifer is required to give the first female offspring to a properly vetted neighbor. Decisions on who receives a gift heifer are informed by advice from a local village council, as well as local Heifer staff. Each person who receives an animal is obligated to give an animal to someone else until everyone in the village who wants livestock or poultry has access to their own animals.

Each animal keeper's herd or flock can grow until it reaches a limit imposed by available fodder or family needs. Because the offspring of the original gift animal and their offspring can be sold to meet domestic expenses, the animal population never increases exponentially. Male animals are sold for their meat or muscle power, female animals generate surplus milk sold fresh or converted into dairy products, and hair and wool from both are valuable for domestic crafts that also have income-generating prospects. Animal owners find themselves owning a capital resource that grows faster than the compound interest earned by deposits in a bank. This development strategy has become an important grassroots approach used by many NGOs as part of their development project portfolio.

Read & Learn

- ▶ Define "development" and list the various ways in which it can be interpreted.
- ▶ Describe the major economic revolutions and show how they inform our understanding of development.
- ▶ Explain population growth through human history and link this to development.
- ▶ Identify and characterize the major "players" in the globalization process.
- ▶ Discuss the environmental challenges facing the world today and connect these challenges to human activities and the development process.

- ▶ Define "environmental stewardship" and outline this concept in the context of development.
- ▶ Describe cultural attributes and processes and how these influence development.
- ▶ Evaluate the role of per capita income in the development process and suggest other measures.
- ▶ Characterize the actual and potential role women play in development.
- ▶ Examine the role of different energy sources in sustainable energy futures, particularly for populations currently without access to electricity.

Contents

3

Geography and Development in an Era of Globalization

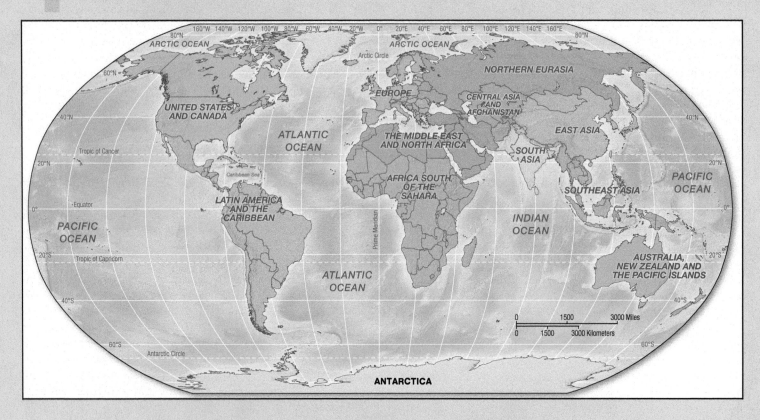

Figure 1-0 **The major regions of the world.**

World regional geography explains where places and activities are located, and why. We identify 11 major regions in which broad natural, cultural, historical, and economic patterns show many similarities. Within each large region, smaller regions of different sizes exist, each possessing features similar to, as well as distinct from, those of their larger region. Australia is a very different place from Fiji! People living in each place have different origins, grow different crops, have different cuisines, can access different resources, are exposed to different hazards, have created different cultures and technologies, understand their world and find meaning for their existence in different ways, and communicate through different languages and media. Why this complex mosaic of place-based regions exists, how these regions are developing as globalization brings distant places into increasingly close contact, and what prospects people have in each region to improve the material and nonmaterial quality of their lives is the focus of this book.

The word **development** is used in a variety of ways. Basically, it denotes a progressive improvement of the human condition, in both material and nonmaterial ways. Economic development signifies a process of long-term advancement in the physical and material quality of life. Development also has nonmaterial dimensions, many of which relate to the achievement of personal fulfillment through the exercise of individual freedoms. Because most nonmaterial aspects of development cannot be easily measured in a numerical or statistical sense, this volume is organized primarily on the basis of economically developed and less-developed regions. Economic development can be measured in several ways—by income, energy use, employment, and various other indicators. Such measures must be kept in proper perspective, for each may tell us little about other aspects of an area, a country, or a people. For example, differences in income may simply reflect more basic differences in cultural goals and values. Therefore we also address other aspects of development, including social class structure, health, educational achievement, gender relations, and political and religious freedom.

After an overview of the meaning of development in an age of globalization, we consider the natural environment, primarily from the standpoint of resources and how people use those resources in more or less sustainable ways. We also pay attention to elements of culture, especially those that influence development. We use measures of economic well-being to define the more-developed and less-developed regions of the world, discuss their characteristics, and present some theories of development. This discussion sets the stage for further exploration of the more-developed and less-developed regions of the world.

Development and Globalization

It is trite but true to remark that the world is getting smaller. Not only are we bombarded daily with news from around the world, but events in other countries also influence our daily lives. The attack by radical Islamic terrorists on the World Trade Center in New York in 2001 claimed nearly 3,000 lives, and brought home to Americans the intensity of conflicts in the Middle East. In less violent and catastrophic ways, but with serious consequences in terms of job loss, today most clothes are made in Asia rather than in the United States, and call center jobs are frequently outsourced to India and other countries. A coffee crop failure, the development of a new high-yielding variety of wheat, the discovery of a new chemical process, an outbreak of a new flu strain, and scores of other distant events originating in distant places may all materially affect the way we live.

The Meaning of Development

This book takes a developmental approach to the understanding of world regional geography. One of the most serious ethical issues of our time is the great disparity in material and nonmaterial well-being that exists among the world's societies. With a more intimate world brought about by better communication and transportation, knowledge of how others live is at our fingertips. Economically less fortunate individuals and communities often aim to follow more rapidly the development path of their materially richer neighbors, but may be frustrated in their attempts by cultural, social, and political constraints (Figure 1-1). The disparity in economic achievement is widening, and the social and political ramifications affect us all. The causes and consequences of that development disparity provide the focus and theme of this volume. We begin by examining the various meanings of development, how the term is used in this book, and the manner in which development can be examined in an era of globalization.

Development—What's in a Name?

The term "development" is complex in its meanings and implications. Development embraces more than economic improvement, although the economic component figures prominently and normally comes to mind when hearing the word "development." Scholars also study issues related to social and political development, and

▼ **Figure 1-1 Favela Morumbi in Sao Paulo.** Urban areas often display sharp landscape contrasts, but few are more stark than the close location of wealth and poverty in many cities.

on occasion consider a psychological component. Development is only one of many terms applied to the processes of change, or lack of change, that have come to distinguish the economic and political circumstances of different countries. Indeed, there is a tyranny of labels—"undeveloped," "stagnant," "emerging," "less developed," to list but a few—with each producing meanings that may or may not be helpful in understanding the change process. Although an industrializing country's economic performance may be deficient when measured by these standards in comparison to postindustrial states, the same country may be very highly developed in other dimensions of human activity, such as art, homeopathic medicine, environmental values, or care of the elderly. In this book, we generally use expressions such as "developed," "developing," or "less developed" to describe material and nonmaterial aspects of growth and change in countries and regions. In addition to material indicators, such as gross national income, industrial output, or foreign trade balance, commonly employed nonmaterial parameters might include nutritional status, adult literacy rate, demographic trends, and life expectancy at birth.

Development—Toward a Definition

Development means different things to different people; but for our purposes development refers to a progressive improvement of the human condition in both material and nonmaterial ways. We should be aware of certain assumptions that guide our understanding of what this path to improvement looks like.

First, development is a process of change that may or may not embrace the developmental ideals associated with North America and Europe. The paths to modern economies followed by Americans and Europeans may not be workable and/or desirable in other countries, given their specific cultural and environmental contexts. For example, American farmers use tractors and combines to achieve a high level of agricultural productivity per farm. Are tractors and combines really needed to increase productivity and improve well-being in many peasant-farm settings? If so, what will be the consequences for densely populated rural areas?

Second, development is a process of change that is not necessarily synonymous with economic growth. In market economies, a high priority is placed on increasing output—more production is good! But improvement of the human condition may or may not be immediately associated with growth in the production of goods and services. Improvements in welfare may come from such simple expedients as providing clean water to an isolated rural community, which may make village families happy and healthy but is unlikely to cause a spike in the local stock market. Some cultures may object to the materialistic impulses that create economic growth.

Third, development also is a process of change that is sustainable in the dual sense that improvements in human welfare today should not be achieved at the expense of future generations and should be made in a manner sensitive to environmental impacts. For example, exploitation of natural resources to extract as much value in as short a time as possible without regard to long-term implications is not a sustainable development strategy, even though short-term profits and economic growth might be impressive. Overusing an aquifer in an arid environment to promote agricultural development would only be acceptable if the resulting development and profits are part of a conscious strategy to produce a set of activities that could endure for a long time in the future.

We can define development as a process of change that leads to improved well-being in people's lives, takes into account the needs of future generations, and is compatible with local cultural and environmental contexts. Consistent with this definition and the assumptions that support it, we can identify four components—people, natural environment, culture, and history—that figure prominently in the development process, particularly economic development. An important component of the economic development process is people—their numbers, distribution, consumption, production, and technology base. Improved sanitation and medical science have lowered death rates sharply over the past two centuries and caused unprecedented world population growth. If the additional workers prove, through the adoption of improved technologies, to be more productive than their ancestors, material levels of living will increase despite population growth. If, however, production levels remain constant or decrease, living conditions will decline.

The natural environment of a country or region provides both the stage for development and the materials necessary to achieve developmental goals. Some environments have abundant supplies of raw materials that can be used for economic gain. For example, the well-watered alluvial floodplain of a river, coupled with a long growing season, provides many opportunities for productive agriculture (Figure 1-2).

Culture—the way in which a society organizes itself in terms of beliefs, values, customs, and lifestyles—greatly influences both the direction and the degree of economic development. Many cultures embrace materialistic values, wherein the acquisition of material wealth is viewed as an index of individual worth and success. Other cultures do not place so high a priority on material achievement (Figure 1-3). The social and political structure of a society also has a direct influence on development. Some countries have achieved a relatively high level of social and political equality, assuring the fullest possible development of their human resources. Others are characterized by rigid social stratification, gender inequalities, and political control by elites that prevent large portions of their populations from achieving their true potential. Similarly, the collective values of some cultures

▼ **Figure 1-2 China's Yangtze River delta and alluvial plain.** Rich farmland has long dominated land use in coastal and riverine China. This tradition continues as the new bullet train crosses the Yangtze delta on an elevated track that conserves precious farmland.

▲ **Figure 1-3 Children explore a laptop in Australia.** Traditional values and modern technology are integrated seamlessly in the lives of these children as they examine their world and its possibilities.

encourage the adoption of new ideas and technologies, while those of others discourage experimentation and change.

Concurrently, never forget history. That the past is a key to the present and a guidepost to the future is well demonstrated in the formation of the world's various cultures and their economic activities. Economic development is not a short-term process. In most places that are now undergoing rapid change or have attained a high level of economic well-being, the necessary foundations or prerequisites for economic development were laid decades, even centuries, ago. For example, the cornerstones of Europe's Industrial Revolution, which began in the middle of the eighteenth century, were formed during the Renaissance, with beginnings in Roman and Greek times and even earlier.

Stop & Think

→ The "development" process is not strictly economic and does not necessarily mean "growth." Explain.

Human Transitions and Development Processes

Current interest in economic development obscures the reality of the long-term nature of development and change. For every contemporary development, there is a deep history of growth and change in values, technologies, and understandings of the world we live in. Transitions that appear abrupt are only observable in retrospect, and the economic revolutions that we tend to emphasize are invariably preceded by long, slow developments in social, political, and cultural spheres.

The Significance of Economic Revolutions

The development process involves a change or a transition of some sort—preferably one leading to economic, political, and/or social improvement. Typically such change is slow and plodding, and hardly perceptible over time. At times, however, change occurs rapidly and massively, leading to a fundamental transformation of society. Such

a transformation is commonly referred to as a revolution. It may be a political, economic, or social revolution; but in every instance it involves a transition that is fundamental, a change that transforms society to its core. The American Revolution, for example, led to a fundamental change in the structure of U.S. politics and government.

Perhaps 10,000 years ago, people learned how to domesticate plants and animals, leading to one of the world's most transforming economic developments—the **Agricultural Revolution**. Instead of wandering around eating whatever Mother Nature's bounty provided, people gradually settled down and started raising crops. Rather than relying on hunting game and fish, they began herding livestock and breeding animals. New technologies were introduced, from the simple hoe to more complex irrigation systems. The ancient Egyptians, for example, developed lift devices to divert water from the Nile River to irrigate some of their fields during low-flow periods of the year, an innovation that thoroughly transformed their capacity to produce crops and create surpluses.

A key to understanding the meaning of the Agricultural Revolution is to understand the significance of **agricultural surplus**. Before the domestication of plants and animals, people lived literally hand-to-mouth, and everybody was involved in the process of finding food. With domestication, agricultural productivity soared to the point that some farmers produced more than they consumed, creating surpluses. Consequently, not everyone had to live on a farm and devote all their time to providing food and fiber. Towns and cities appeared, supported by the farmers' surpluses.

The first real towns and cities arose perhaps 8,000 years ago in parts of the world, often the floodplains of rivers, where agricultural productivity was greatest. Life in these settlements was more congested but was somewhat easier and more secure than life in the countryside. Permanent homes, even substantial houses, replaced the crude huts or caves that may have served as temporary lodgings for more mobile hunting and gathering peoples. Many tools and other large and small luxuries were acquired, such as chairs, tables, and beds, which previously were impractical because of the migratory way of life. Materialism may have had its true beginning with the development of agriculture. Possessions could be accumulated and passed on to new generations.

Thanks to surpluses and the growth of cities, economies and societies became more complex. Expanding populations, production, and interpersonal contact required increased group action and led to the growth of secular leadership organizations. Political organizations were established to settle disputes, govern, and provide leadership for collective action in warfare and in such public works as irrigation, drainage, and road building. The formation of a priestly class helped formalize religion. Religious leaders were frequently the holders of both philosophical and practical knowledge, often serving as medical practitioners and weather forecasters. In the Mayan civilization of southern Mexico and Guatemala, for example, priests developed an agricultural calendar based on the progression of the sun, the planets, and the stars. It predicted the beginning of wet and dry seasons and told farmers when to prepare the land for planting to take full advantage of the seasonal rains. As increased production per worker yielded more than a family unit needed, a portion of the labor force was freed not only for government and religious activities but also for activities such as pottery making, metallurgy, and weaving (Figure 1-4).

The connection between the production of surpluses and the appearance of complex, urban-oriented societies occurred only in the most environmentally favored locations—the Nile River valley, the valley of the Tigris and Euphrates rivers, and the Indus River valley to name only a few locations. Elsewhere, the benefits of the Agricultural Revolution were limited to a more sedentary version of a subsistence existence, which was still an improvement over the migratory hunter–gatherer societies that existed previously.

In a sense, the agricultural revolutions have never completely ended. A second revolution began in Europe in the mid-seventeenth century in which new technologies and procedures were applied to farming, partly to increase farming's commercial potential. By the twentieth century, still another revolution occurred that replaced most farm labor with machinery and eventually led to a corporate takeover of many family farms. A single farm could now support dozens of non-farmers. In our lifetimes, a biosciences revolution has altered the nature of agriculture by introducing fundamentally new crop varieties capable of generating dramatically higher yields. The **Green Revolution** combined selected high-yielding crop varieties with technological packages of fertilizer, pesticides, water management, and capital to boost yields. Detailed knowledge of the genetic makeup of plants and animals has permitted microscopic cutting and splicing of DNA chains and the transfer of genes from one species to another. The result is the creation of genetically modified crops with potentially much improved yield characteristics but unknown potential impact on the biosphere.

Remember what a revolution is: a fundamental change in the way that something is done. By the mid-eighteenth century, it was time for a fundamental change in the way that goods were made—a manufacturing or **Industrial Revolution**. Manufacturing is a human activity that has existed since before the beginning of civilization. The prehistoric cave dweller who chipped a piece of flint into a spear point was involved in manufacturing. So were the ancient Greeks and Romans when they shaped pottery out of clay and transformed ore into metal ornaments. But this manufacturing process relied on manual labor (hence the term "manu" facturing) and relatively unsophisticated forms of technology. In the medieval period, for example, the shoemaker depended on his hammer, shaping tools, and a strong arm to make shoes; the blacksmith had his fire, bellows, anvil, hammer, and an even stronger arm to bend iron; and the tailor made use of his needle, thread, and the dexterity of his hands to stitch cloth. This type of handicraft manufacturing was slow and inefficient, although quality could be, and often still is, very high.

Everything changed beginning in the mid-1700s. Europe and North America entered an age of mass production of manufactured goods, using much more sophisticated technology and large industrial workforces. This fundamental change in manufacturing represented a response to at least two major forces: First, an age of innovation had dawned thanks to the thinking and discoveries of the Renaissance, which led to the introduction of new technologies to speed up the production of goods. In 1769, for example, James Watt developed a practical steam engine. His innovation paved the way for a wholesale mechanization of the manufacturing process and for the steam engine to become the main source of power for the Industrial Revolution in its first century and a half. Coal became the industrial fuel of choice, replacing biofuels such as wood or charcoal. New furnace-based technologies emerged to replace the time-tested, but highly inefficient, charcoal method of iron processing. New spinning and weaving technologies made the volume manufacturing of cloth much more efficient. Machinery replaced muscle power and inanimate energy replaced animate energy. Second, the way in which production was organized changed drastically. Manufacturing no longer was a single activity done by a craftsman in his shop, but a volume enterprise carried out in a factory—a stand-alone building containing a full-time industrial workforce and an array of machines. Demand for processed goods exceeded what the cottage industries and craft shops of the day could produce, and the more efficient factory was the only way to supply this demand. Over time, craftsmen and small guilds gradually disappeared or remained in isolated pockets as quaint reminders of a pre-Revolutionary past. The factory came to epitomize all that the Industrial Revolution was about, as well as to symbolize the social and environmental price paid in the name of production efficiency.

The Industrial Revolution spawned companion revolutions. Development of the steam engine led to a transportation revolution. Steam power was first applied to maritime shipping in 1807; by 1829, steam railway locomotives were being built. Steam power significantly diminished the **friction of distance**, or the difficulty of moving from place to place. Before railroads, a traveler walking the 948 miles (1,525 kilometers) from New York to St. Louis took more than six weeks to make the journey; by 1870, with railroads, the trip became a matter of about three days; today a commercial airliner can cover the distance in under three hours! Transportation efficiency contributed to production efficiency in manufacturing centers. The Industrial Revolution promoted the agricultural revolutions by creating a new way of doing things down on the farm. Tractors, combines, and other equipment quickly replaced sweat and muscle in the cultivation of crops, which created a large farm population looking for another line of work. Most of that work took the form of factory jobs in cities, which contributed to a massive population transfer as people

▼ Figure 1-4 **Mayan ruins at Palenque, Chiapas, Mexico.** Surpluses supported the specialization in production and crafts that sustained increasingly elaborate material forms and urban structures.

swarmed into the cities in search of work. The impact often was profound. The village of Essen, Germany, for example, had 4,000 people in 1800; by 1920, there were 439,000 people working in a highly industrialized city. Similar stories can be told for the United States. In 1800, the vast majority of people worldwide lived in the countryside. By 2000, an overwhelming majority of people in developed countries lived in towns and cities.

Currently an **Information Revolution** is reshaping societies, economies, and personal lives. Who in the twenty-first century can imagine living without a cell phone or laptop computer? We are part of what increasingly is called the Information Age and/or the New Economy. We are really looking at the second great information revolution. The first such revolution we can trace back to Johann Gutenberg, a German, who in 1450 built the first printing press. It would be hard to overstate the fundamental changes that occurred because of the printing press.

Communicating information is a basic human activity—even prehistoric cave dwellers conveyed information when they painted pictures of animals on cave walls. It is a matter of how the information is produced, stored, accessed, and applied that creates a revolution; in the Information Age, it all boiled down to a single transforming technology: the microprocessor. Mechanical computers have existed for some time—even the ancient abacus can be considered a type of computer—but it took the miniaturization and integration of electronic circuits to create the fundamental changes that led to the current Information Revolution. The first true electronic computers appeared after World War II and required a whole room to do what a good smartphone does today. More space-efficient mainframe computers followed, and by the 1990s desktops and laptops became everyday appliances. In the twenty-first century, thanks to constant innovation, hardly anything is done without computer involvement (Figure 1-5).

In partnership with the microchip, we must also credit innovations in communications technology for the revolutionary changes of the late twentieth century. The Internet permits nearly

instantaneous communication worldwide between connected computers. Satellites and cellular phone technology have made wireless voice and data communication the norm rather than the exception. Time and distance constraints have not disappeared. They have been modified by the ability to contact a significant proportion of the rest of the world by simply punching in a number.

> ### Stop & Think
> ➢ What is the power of the "surplus" to effect change, not just in economic terms but in social conditions?

Where Does Population Change Fit In?

Revolutions create many fundamental changes, but some of the most pronounced and enduring transformations occur in the dynamic behavior of human populations. We briefly review how global population numbers have changed over time, and then examine selected population concepts and growth models that help us understand the pivotal role of population in economic change and development.

At the dawn of the Agricultural Revolution, when animals and crops were first domesticated, the world's population was extremely low (Figure 1-6). Population clusters, usually associated with areas of agricultural surplus, later spread throughout the world, along with the diffusion of crops and animals. In a few places, such as Australia, the diffusion process was delayed until the coming of European colonists. Today, only in polar zones, remote drylands, and other harsh physical environments do small and dwindling numbers of people still live by the age-old occupations of hunting, fishing, and gathering.

At the beginning of the Christian era, world population totaled little more than 260 million people, most of whom were located in the Old World. The majority of those people lived within three great empires: the Roman Empire, around the fringe of the Mediterranean Sea and northward into Europe; the lands of the Han dynasty of China, which extended into Southeast Asia; and the Mauryan Empire of northern India. In those empires, the simpler political, economic, and social organizations of agricultural villages vied with the more complex, integrating structures of the empires and the newly created cities. Urbanism became a way of life. By 1650, the world's population had grown to more than 500 million, despite interruptions of famine, plague, and warfare. Most of the growth was in and around the preexisting centers, with gradual expansion of the populace into areas once sparsely settled. Productive capacity expanded with improved technology and new resources. Urbanization, the process by which towns and cities are formed, became more pronounced, though agriculture remained the primary livelihood for most people.

Since 1650, the world's population has increased rapidly. It took an estimated 1,650 years for the population to double from 260 million to 500 million. By 1850, just 200 years later, the population had doubled again to approximately 1.175 billion. Within the next 100 years, the population nearly doubled a third time, reaching 2 billion. By 1975, it had doubled yet again, to 4 billion. As of mid-2013, the world had more than 7.1 billion people; by 2025 it may exceed 7.9 billion.

▼ **Figure 1-5 Auto assembly plant in Wuhan, China.** Highly automated and computerized assembly lines typify Chinese automobile manufacturing. The growth of industry in Hubei province, focused on Wuhan, reflects the steady development of inland centers of industrial development in China.

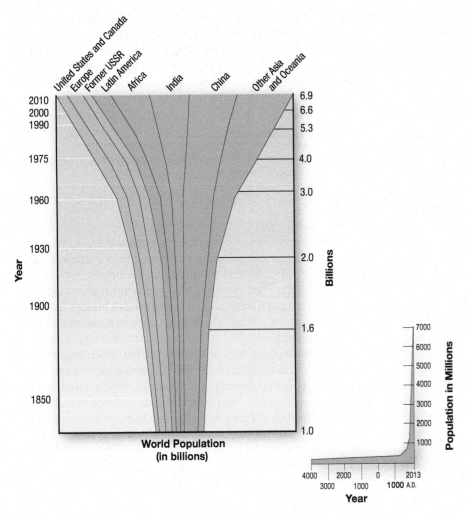

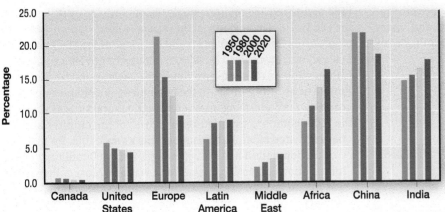

▲ **Figure 1-6 World population growth.** For most of human history, the world held relatively few people. In the last 400 years, however, the total number of people has expanded greatly and at an increasingly rapid rate. The term "population explosion" is often used to describe such rapid growth. The top two graphics illustrate the manner in which population has increased since 1850, whereas the bottom graphic indicates how population growth is shifting.

Note: Percentages do not sum to 100 since not all countries were included.

Source: U.S. Bureau of the Census, International Data Base, *http://www.census.gov/cgi-bin/ipc/idbagg.*

people per unit area) show strong ties with the past (Figure 1-7). Three principal centers of dense population are apparent: the Indian subcontinent, eastern China and adjacent areas, and Europe. China and India represent old areas of large populations, stemming both from an early start in the Agricultural Revolution and from empire building. Today, about half of the world's population lives in southern and eastern Asia, where agriculture and village life remain important. Yet modern cities, with their service and manufacturing functions, are also present. Population density in India and China varies considerably, usually in association with the relative productivity of the land. **Physiologic density**, the number of people per square mile of arable (farmable) land, is a useful expression of the density relationship in agricultural societies. On the coastal and river plains, where alluvial soils are rich and water is abundant, rural densities of 2,000 people per square mile (772 per square kilometer) are not uncommon. Away from well-watered lowlands, such as the Huang He (Yellow River) or Chang Jiang (Yangtze River) valleys, densities diminish but may still be in the range of 250–750 people per square mile (97–290 per square kilometer).

Europe's high population density can be traced back to technological developments originating in the Middle East, which were adopted by the Greeks and Romans and much later expanded by the Industrial Revolution. Further increase in the European population is readily associated with developments in technology. Many of the high-density areas in western Europe are urban regions associated with coalfields or advantageous water transportation, indicating the importance of those assets in the Industrial Revolution (Figure 1-8). Even though Europe's population density is high, it is significantly less than that of comparable Indian and Chinese areas. Agricultural villages and agriculture itself are overshadowed in Europe by modern metropolises and manufacturing.

Secondary centers of high population density are more numerous worldwide. The northeast quadrant of the United States and the adjacent parts of Canada are a principal cluster, though total population numbers are smaller than those of Europe and far less than those of East Asia. High densities also occur in Africa along the Guinea Coast and the Nile River and in the eastern highlands, but the total number of people involved in each cluster is relatively small. Around major urban centers of Latin America and in the old Aztec, Mayan, and Inca realms, locally dense population

An increasingly large proportion of the population is living in cities, some of which have the populations of medium-sized countries (Table 1-1). The world's **population distribution** (the spatial arrangement of people) and **population density** (the number of

Table 1-1	The World's 10 Largest Urban Agglomerations, 2013				
		Population (in thousands)			
Agglomeration	**Country**	**1700**	**1800**	**1900**	**2013**
Tokyo	Japan	500	492	1,497	37,239
Jakarta (Batavia)	Indonesia	52	92	115	26,746
Seoul	South Korea	170	190	195	22,868
Delhi (Dilli)	India	500	125	207	22,826
Shanghai	China	45	100	837	21,766
Manila	Philippines		77	190	21,241
Karachi	Pakistan				20,877
New York	United States		63	4,242	20,673
Sao Paulo	Brazil			205	20,568
Mexico City	Mexico	100	128	344	20,032

Sources: The 2013 figures are for urban areas and are from *Demographia World Urban Areas: 9th Annual Edition.* See the comments in *http://www.demographia.com/db-worldua.pdf.* See also *http://www.worldatlas.com/citypops.htm#.UgqK3flo6Uk.* The historical figures are from T. Chandler and G. Fox, *3000 Years of Urban Growth* (New York: Academic Press, 1974).

Note: The figures for 1900 are for urban areas; the remaining historical figures are for cities (except 2013). All historical figures are estimates. Blank spaces indicate no data provided. The world's largest cities are found outside North America and Europe, with the exception of New York.

centers are common. Other pockets of high density are found in Indonesia, the Malay Peninsula, Japan, the Philippines, and parts of the Middle East.

Overall, the **population growth rate** for the world is 1.1 percent a year, but that growth is by no means uniform. One explanation for the varied growth pattern is the theory of **demographic transition**, which is based on four population stages (Figure 1-9). Stage I postulates an agrarian society in which **birthrates** and **death rates** are high, creating a stable or very slowly growing population. Productivity per person is limited. Large families (or many births) are an economic asset, particularly since life expectancy is low and security depends on family members, including young children. Employment opportunities outside of agriculture are few in this stage, and technology develops very slowly.

In Stage II, the cultural custom of large families persists, and the

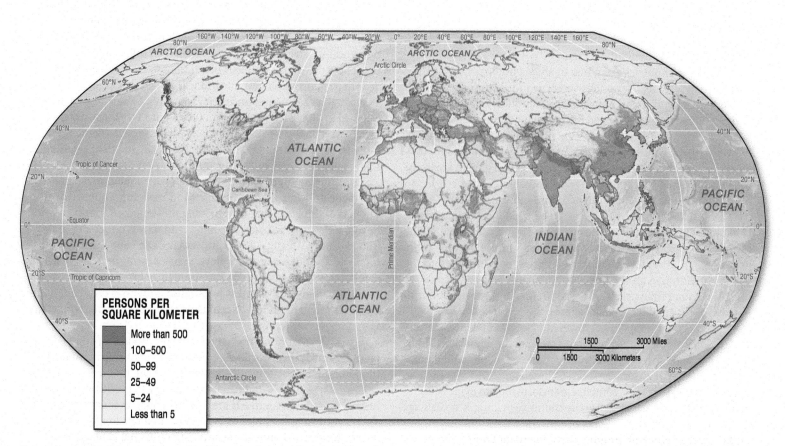

▲ **Figure 1-7 World population distribution.** The world's population is unevenly distributed. Three large areas of dense population are China, the Indian subcontinent, and Europe. Most of the sparsely populated areas have environmental impediments such as aridity, extreme cold, or mountainous terrain.

Source: Population Reference Bureau, *2013 World Population Data Sheet, http://www.prb.org/pdf13/2013-population-data-sheet_eng.pdf.*

▲ **Figure 1-8 Duisburg, Germany, and the heavy industry of the Ruhr.** Duisburg and iron and steel manufacturers like ThyssenKrupp rose to prominence at the junction of the Ruhr and Rhine River valleys because massive coal fields and river transportation were in close association. Despite economic difficulties and changing technologies, these assets continue to support industrial and transport activities and the workers who staff them.

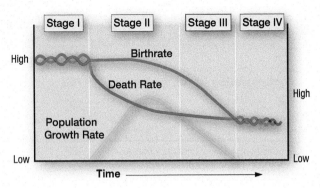

▲ **Figure 1-9 Model of demographic transition.** The demographic transition model is based on the European experience. It may not represent what will happen elsewhere, especially in areas of non-European cultures. The model assumes an initial period of traditional rural life with high birthrates and death rates and no population growth. With increased production and improved sanitation, the death rate declines, creating a condition of rapid population growth. As the population becomes more urbanized, the birthrate also declines until low birthrates and death rates finally prevail and the population no longer grows.

birthrate remains high. However the death rate drops dramatically due to better sanitation and medical treatment and because of greater productivity. Productivity may increase in the agricultural sector, but more important is the appearance of alternative economic activity resulting from industrialization. With industry come urbanization and labor specialization. The principal feature of Stage II is rapid population increase.

Numerous countries remained in Stage I until World War II. In the postwar years, death rates all over the world were greatly reduced through the introduction of improved sanitation and health care. Several African countries moved into Stage II only in recent decades. Lower death rates, which have contributed to faster population growth in these countries, are likely to decline further before they bottom out, so that even higher rates of population growth are likely. Uganda, Mali, and Niger illustrate such a situation (Figure 1-10).

Stage III is characterized by continued urbanization, industrialization, and other economic trends begun in previous stages. Demographic conditions, however, show a significant change. The birthrate begins to drop rapidly as smaller families become more prevalent. The shift toward smaller families may be related to the fact that children in an urban environment are generally economic liabilities rather than assets. In Stage III, the population continues to grow, but at an ever-slowing pace.

In Stage IV, the population growth rate is stable or increasing very slowly. Both birthrates and death rates are low. The population is now urbanized, and birth control is widely practiced. Population density may be quite high. Countries such as Brazil and Mexico remain in Stage III; the time of their complete transformation to Stage IV still is an open question. Virtually all of the countries in Europe are in Stage IV, as are the United States, Canada, and Japan, countries that exhibit European-style industrialization and urbanization.

The difficulty of predicting exactly when a country will enter a particular stage is illustrated by the United States. Birthrates reached a very low level during the 1930s (comparable to current rates) but then accelerated sharply during the postwar years, creating growth conditions equivalent to those in Stage III. Birthrates are sociological phenomena that are strongly influenced by cultural value systems, unlike death rates, which can be reduced by control of disease, proper diet, and sanitation efforts. The model of demographic transition is based on analysis of Europe's experience with urbanization and industrialization; other areas with different cultures may not follow the European example. If the model is valid, population growth in many parts of the world will slow down as Stages III and IV are reached. Whether and when world population growth will be reduced or stabilized is uncertain (see Figure 1-10). The differences experienced in the rate of population growth, though seemingly minute, have enormous impacts on the absolute growth of population from region to region and on the utilization of resources (Table 1-2).

Malthusian theory is another model of population change that has received widespread attention and has numerous advocates. Thomas Malthus, an Englishman, first presented his theory in 1798, basing it on two premises: Humans tend to reproduce prolifically, that is, geometrically—2, 4, 8, 16, 32; and the capacity to produce food and fiber expands more slowly, that is, arithmetically—2, 3, 4, 5, 6. Thus Malthus believed that population growth eventually exceeds the food supply unless population growth is checked by society. If growth continues, surplus populations will be reduced by war, disease, and famine.

If we plot Malthus's idea on a graph, we can identify three different stages and outcomes flowing from the relationship between population and production (Figure 1-11). In Stage I, human needs are not as great as production capacity. By Stage II, production capacity and increased human needs are roughly equal. In Stage III, population has grown to the point where its needs can no longer be met.

When Stage III occurs, the population dies off, and we cannot be sure what follows. One idea is that Stage III is simply a repetition of Stage II, with alternating periods of growth and die-off, represented

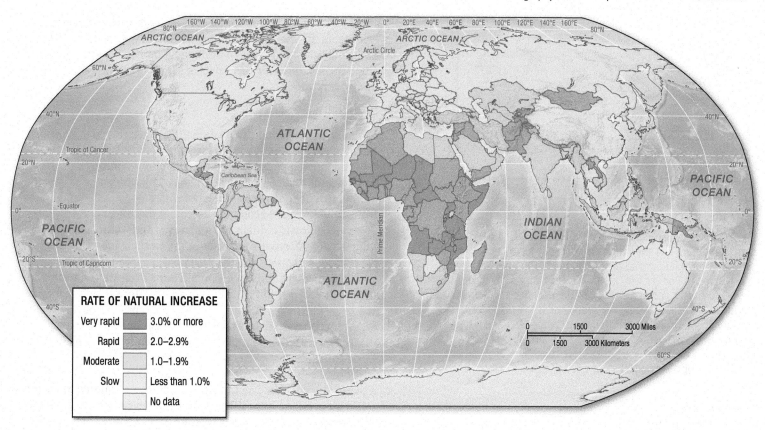

▲ **Figure 1-10 World population growth rates.** There are great differences in population growth from one part of the world to another. Areas of the world that are highly urbanized and have more-industrialized economies tend to have low birthrates and death rates and low rates of population growth. Areas that are less urbanized and less industrialized have high birthrates, declining death rates, and moderately to rapidly growing populations.

Source: Population Reference Bureau, *2013 World Population Data Sheet, http://www.prb.org/pdf13/2013-population-data-sheet_eng.pdf.*

Table 1-2	Population Doubling Time by Region		
Region	**Population (in millions)**	**Natural Growth Rate**	**Approximate Years to Double**
Africa	1,072	2.5	28
United States and Canada	349	1.0	70
Asia	4.260	1.1	63
Europe	740	0.0	N/A
Latin America and Caribbean	599	1.3	54
Oceania	37	1.1	63
World	7.058	1.2	58

Source: Population Reference Bureau, "Population, Health, and Environmental Data and Estimates for Countries and Regions of the World," *2012 World Population Data Sheet, http://www.prb.org/pdf12/2012-population-data-sheet_eng.pdf* (accessed August 13, 2013). Doubling times gathered by author.

Malthus assumed that people would reject birth control on moral grounds, and he could not foresee the impact of the Industrial Revolution. Since the 1950s and 1960s, **neo-Malthusians** (present-day advocates of Malthus) have argued that the population/production crisis has merely been delayed, and disaster may yet strike.

All theories aside, the fact remains that the population of the world is large and is growing. This growth is not evenly distributed across the global landscape. Increasing numbers of people live in the poor world or the developing nations (Figure 1-12). Since 1900 the higher rates of population growth have shifted dramatically to less-developed regions as the rate of natural increases has slowed significantly in more-developed areas. These changing population proportions are likely to continue until at least the end of the twenty-first century.

by line (a) in Figure 1-11. Another idea is that the die-off is so great that Stage I is reproduced, as represented by line (b). Malthus's theoretical die-off has not occurred, owing in part to the enormous increase in production associated with the Industrial Revolution.

Stop & Think

▶ How can one argue that increasing productive capacity in a country is a bad thing?

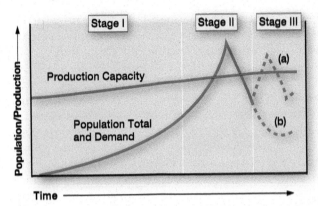

▲ **Figure 1-11 Malthusian theory.** The Malthusian theory can be illustrated by a three-part diagram. In Stage I, the needs of the population are less than production capacity. In Stage II the increase in population is so great that needs soon exceed production capacity. For a while, the population continues to grow by using surpluses accumulated from the past and by overexploiting resources. Eventually, the pressure on the resource base is too great, and the population begins to die off. Stage III may be a continual repetition of Stage II with similar cycles of population growth and decay (a) or a return close to Stage I (b) until habitat recovery supports another major growth spurt.

Globalization as a New Economic Revolution?

Up to this point, we have focused on several of the great revolutions in human history—the Agricultural, Industrial, and Information Revolutions—and how these transitions contributed to the types of improvement that we associate with the development process. We now look at the latest transition, globalization, which has emerged as the defining social and economic force of the late twentieth and twenty-first centuries. While it is easy to overstate the significance

▼ **Figure 1-12 Housing availability in Jakarta, Indonesia.** This shelter on a railroad right-of-way in Jakarta is symptomatic of the plight of the poor and homeless in all countries, but is particularly widespread in poorer places with a limited social safety net.

of globalization—far too many people invoke the term to explain everything that ails and benefits the world as well as ignore the role that globalization has played throughout human history—this transition process is having an impact of some sort on just about everybody, everywhere.

What Is Globalization?

Defining globalization can be difficult, given its many dimensions, but in its broadest sense, **globalization** refers to a growing integration and interdependence of world communities through a vast network of trade and communication links. This process of integration and interdependence is associated with a wide range of technological, cultural, and economic outcomes that affect our daily lives: communication is virtually instantaneous and worldwide; a visit to the mall is a visit to the factories of the world; a job gained or lost may reflect a decision made by a global corporation; a car assembled in the United States will contain parts from around the world; the music and fashion decisions of an American teenager may be shared by a teenager in South Asia; and perhaps most obvious of all, a trip abroad does not mean giving up a fast-food hamburger.

The increased integration and interdependence of a globalizing world constitute a response to two major forces: technology change and global capitalism. The technological fruits of the Information Age are everywhere. The Internet, satellite communications, and other innovations have shrunk time and space to such an extent that virtually no point in the world is farther than a mouse click away. Such virtual proximity has made it much easier to conduct global commerce and to spread ideas. Computing technologies make it possible to create, access, store, and process extraordinarily large quantities of data with breathtaking speed, thereby making all aspects of society and business more productive.

The embrace of capitalist, or free-market **neoliberal policies** in their various forms is equally important to understanding the dynamics of a globalizing world. Since the early 1980s, government leaders have joined their business counterparts in expressing a renewed faith in the power of freely operating markets and unimpeded trade to create prosperity. Their argument rests on the proposition that less government leads to greater economic efficiency. Proponents advocate decreased government regulation and ownership of economic activities, reduced government spending, and elimination of trade barriers. A rising tide of prosperity will lift all boats, they argue, and will do what no government policy has yet done in overcoming the scourge of global poverty. The neoliberal faith has become worldwide in scope. Even such bastions of socialist conviction as the former Soviet Union have become converts (ignoring for the moment the somewhat rickety and occasionally corrupt market mechanisms that characterize the Russian marketplace). Neoliberal thinking had become so pervasive that local adoption of neoliberal policies was mandated by the International Monetary Fund (IMF) and other organizations as a condition for lifting many less-developed countries from suffocating debt burdens in the 1980s.

Although the free-market mantra has helped generate extraordinary pockets of wealth, critics contend that the costs of global capitalism outweigh the benefits. These costs include destruction of small farms and manufacturers unable to compete in the global economy; the instability of businesses and worker livelihoods; the breakdown of social fabrics and consequent outbreaks of violence; the sudden and disruptive movements of production and capital; and a "race to the bottom" as communities and countries compete for investment based on low wages, low taxes, and minimal labor and environmental regulation. Neoliberal policies came under even more intense scrutiny with the global recession of 2008 and the collapse of many corporate institutions associated with the unrestrained pursuit of the neoliberal ideal.

Globalization is not new. As Europe emerged from its Dark-Ages slumber, it embarked on a journey of world discovery that led to increased global interdependence. These discoveries led to huge Spanish, French, and British colonial empires that created global trade connections and a worldwide diffusion of European culture in its various forms. The explosive growth of mass production during the Industrial Revolution led to a global search for industrial raw materials (for instance, cotton for British textile factories) and global markets for manufactured goods (often in overseas colonies). But it is possible to argue that today's globalization is different in a qualitative sense from past globalizing events—it is more complex and deeply rooted. Global economic interaction in the twenty-first century involves huge transnational corporations and elaborate production networks made possible by Information Age technologies. Much global trade occurs within the worldwide internal network of a single industry or corporation, as opposed to trade among independent firms around the world.

Globalization's Players

So who is leading the charge in the globalization of societies and economies? We can identify several major players. First and foremost is the **transnational corporation (TNC)** (Figure 1-13). These corporations have offices, production facilities, and other activities in multiple countries. They are geographically mobile and can take advantage of lower labor costs in one country or a more lenient regulatory environment in another country to minimize production costs and/or to maximize revenues. Transnational corporations vary in size

and shape, and in corporate and national culture. Japanese and U.S. corporations, for example, typically have very different organizational and cultural styles, but both are players in a globalizing world.

The semiconductor industry is an early example of a manufacturing activity with distinct global geographical separation of functions. The more routine assembly operations were located in developing countries where low-wage female labor was abundant; higher-level design, research, and development operations were located in the United States and other developed countries. A computer purchased in the United States probably had components manufactured in Asia. The first U.S. overseas assembly plant was located in Hong Kong in 1962; by the late 1960s, plants had been opened in Taiwan, South Korea, Singapore, Malaysia, and northern Mexico. In each case, firms were looking for low wages to keep production costs down. Over time, however, Asian producers became more sophisticated. A production network appeared that was geographically specific and less oriented toward simple component assembly: Bangalore, India, focused on software design, Singapore on process engineering, South Korea on semiconductor memory, and Taiwan on digital design. Assembly remained important everywhere, but most particularly in Malaysia and coastal China. The global network of semiconductor and related firms had become far more complex, but it was still very much a global operation (Figure 1-14).

Highly mobile global corporations have the means to circumvent the power of national governments, and they can to some degree. Some observers have commented that this mobility and power have led to the "hollowing out" of nation-states, making countries and their governments less relevant to world affairs than in the past. But in reality, countries remain major players because they can still limit access to their spaces, impose regulations on corporations operating in their spaces, and come together (as in the European Union) to create multistate authorities. Countries still have power in a globalizing world.

Labor is another important variable, but is more or less fixed in space and has limited bargaining ability in the global economy. If unskilled labor becomes too expensive in one location, the corporation

▼ **Figure 1-13 Athletic shoe factory near Hanoi, Vietnam.** Shoe factories such as this village plant outside Hanoi have made Vietnamese produced footwear a major contributor to the profits of transnational corporations.

▼ **Figure 1-14 Microchip production and electronics assembly in Asia.** Operating in a scrupulously clean environment, this Shanghai microchip facility helps make Semiconductor Manufacturing International Corporation of Taiwan, and Asia, one of the world's largest producers.

will simply move production to a less-expensive place. This has been the fate of many U.S. assembly plants, which have closed as parent corporations seek lower wage levels in Latin America and Asia. In theory, successful global organization of labor would counter this trend, but as a practical matter, organizing workers on an international scale is difficult to achieve and unlikely in the near future.

In a real sense, consumers call the shots. If they stop buying, corporations will start dying. We now live in a world in which products can be marketed to global consumers. Although this may appear to be an advantage to corporations, cultural differences can make marketing a single product across cultural boundaries frustratingly difficult.

Regulatory organizations and civil movements influence the way in which global society and economy function. They include economic institutions such as the International Monetary Fund (IMF), the World Bank, and the World Trade Organization (WTO), which were created to provide at least limited multinational oversight of and support for global economic activities. Also important are regional institutions, such as the Association of Southeast Asian Nations (ASEAN), that promote economic growth of and free trade between their member states. Other organizations and groups, such as grassroots, nongovernmental organizations representing women workers or environmental causes, that pressure corporations and governments to act in certain ways are included as well. In some instances, transnational social movements have appeared to challenge the perceived excesses of globalization, usually countered by equally vocal and partisan defenders of the globalization faith. Since the late 1990s, demonstrations, occasionally degenerating into violence, have occurred at almost every international economic conference site.

Stop & Think

➤ What is a "revolution"? Can the globalization process be considered a revolution?

Globalization's Geographic and Developmental Footprints

Now that we have defined globalization and have identified the major players, it is important to understand the geographic and developmental footprints of the processes behind globalization. From geographical and developmental standpoints, what does globalization really mean?

A popular assumption is that globalization is leading to a worldwide unity of tastes and values, and a destruction of local cultures. It is true that a McDonald's hamburger and a Coca-Cola can be bought almost anywhere in the world; Walmart, beginning in 1991, began an aggressive campaign of locating stores throughout the world; and jeans have become a global style statement. North American standards related to music, television programming, journalism, and some aspects of political campaigning have spread to the rest of the world. Then there is the popularity of the English language, its global spread considered by many observers as the ultimate expression of increasing global homogeneity—probably 400 million people speak English in countries in which English is the native language, but another billion or more speak English as some form of second, third, or fourth language. English is closely associated with the growth of global business, although it should not come as

a surprise if Mandarin Chinese eventually gives English a run for its money. English is also the primary language of social media and the Internet—many foreign firms and institutions have English-language pages (see *Visualizing Development:* We Do "Like" Our Social Media).

A certain amount of homogenization has occurred—some call it a transnational, postmodern, or postnationalist culture—but in reality this assumption is easily exaggerated. What appears to be homogenization often is little more than a veneer—a penetration of local cultures by outside influences that is only skin deep; the essential cultural characteristics that distinguish people have not been obliterated. Local cultures tend to "domesticate," "indigenize," or "tame" imported consumer culture by giving it a local flavor. South Asians eat pizza but it may be a "Punjabi Pizza" that includes local toppings and spices. They may add garlic and chili to McDonald's hamburgers. In the Netherlands mayonnaise is the condiment of choice for french fries; in Peru, fries come with a local sauce called "*aji*"; in Taiwan, rice patties can be ordered in place of hamburger buns; and in Canada, some stores offer "*poutine*"—a mix of fries, cheese curds, and gravy. Fries in Britain are called chips and may be free of salt but doused in vinegar; what Americans call chips, the British call crisps—and they may still be soaked in vinegar. A fast-food outlet in Spain may include beer on the menu, along with its own version of a hamburger and french fries. And in European countries where planning standards are often strict, rigid rules that control signage to preserve traditional streetscapes often force a global corporation such as McDonald's to minimize its visual impact, diminish its golden arches logo to very modest scale, and adapt its operations to fit within existing buildings rather than construct stand-alone facilities (Figure 1-15).

Even the use of English is not as homogenous as one might think. We all know about the different accents—British, American, Australian, Canadian—and the different forms of expression (British cars have "bonnets," whereas American cars have "hoods"). The structure of the language has taken on changes reflective of local cultures as well. In India, the word "may" is used to express obligation in a polite way—whereas Americans would say, "you will do it

▼ **Figure 1-15 McDonald's in Paris' Latin Quarter.** In order to gain access to this corner location along the Boulevard Saint-Germain, in a pedestrian-scale district with abundant foot and pedal traffic, McDonald's has complied with local signage restrictions and adapted to the spaces and window/door openings of existing structures.

Visualizing DEVELOPMENT

We Do "Like" Our Social Media

By June 2013, Facebook boasted 1.15 billion monthly active users and 699 million daily active users. Eighty percent of its activity was outside the United States. The daily number of "likes" numbered in the billions. This massive online presence, which has even been the subject of a movie, only began in 2004 (Figure 1.1.1). Facebook is not alone in the world of social media. We use other platforms like LinkedIn, YouTube, Twitter, and Google+. Social media allow us to use the Internet to network with people worldwide, to establish communities, and to post to friends and associates everything from major news items to what we ate for lunch.

Social media touch almost all of us. The Pew Research Center reported a study in August 2013 showing that 72 percent of online U.S. adults 18 years of age or older used social networking sites. Yet in February 2005 when social media were still new, only 8 percent of online adults used social media. While nearly 90 percent of Internet users under 30 use social media, 43 percent of online adults 65 or older also use social media. The Pew study found little difference in percentages based on gender, education, income, or urban-rural status. Race/ethnicity was another matter. Hispanics showed the highest percentages, followed by African Americans and whites. Rapid and widespread adoption of cell phones and social media is also characteristic of many Middle Eastern, Latin American, and East Asian countries, and is beginning to accelerate in Sub-Saharan Africa, but remains largely restricted to the economic and social elite in South Asia.

The range of social media applications is immense. Not only are these outlets good for keeping in touch, but increasingly they are used by businesses to advertise products or check out prospective employees, by researchers to conduct surveys, by medical professionals to screen students for depression, and by political movements to mobilize supporters. In Sub-Saharan Africa generally, and particularly in Eastern Africa, formerly isolated rural populations are able to access financial, health, and educational information and advice unobtainable before the dawn of the Internet Age. Banking is particularly benefiting from the ability of rural residents to access and manage their accounts remotely. Successful opposition in Egypt to the entrenched dictatorship of Hosni Mubarak and to the regime of his successor Mohamed Morsi was organized and coordinated largely through social media.

But there are downsides. Social media have taken the place of the family room, business hallway, or water cooler for gossipy conversation. The difference is that online conversation has the potential not only to last forever, but to be spread to all eyes and ears. In the United States, nearly three-quarters of corporate recruiters acknowledge that they have reviewed social media sites before hiring an employee, making the cyber water cooler a potentially toxic location for indiscrete job hunters. Governments are trying to assess where the line should be drawn regarding privacy.

An equally large concern is whether the hyperconnected world where it is possible to "friend" people from everywhere and at all hours may be making us less social. Does Facebook, for example, actually increase loneliness? Are social media part of a larger "Internet paradox" in which broader connectivity corresponds with a heightened sense of isolation? Loneliness is unquestionably growing, but social media probably do not make us less social or more lonely unless we are that way already. For people who already struggle socially, Facebook "friends" may seem to compensate for a lack of companionship and a fear of social interaction. But these friends may produce only an "ersatz intimacy" that provides a poor substitute for true friendship, even though some of the unpleasantness of face-to-face encounters, such as awkward pauses in conversations or spilled drinks at parties, may be avoided. For people less challenged socially, Facebook

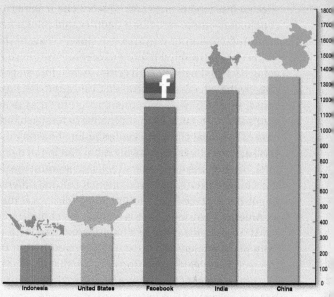

▲ FIGURE 1-1-1 **If Facebook Were a Country.** If Facebook were a country, it would have more monthly active users than the populations of every country in the world except China and India.

may be used to strengthen face-to-face contacts and to build deeper relationships.

The rise of social media reflects the role of technology in the globalization process. Thanks to social media, we now have on our computer screens the cyber equivalent of family rooms, lounges and water-cooler spaces. They redefine our personal geographies and create electronic landscapes impervious to the friction of distance. At the same time, we are witnessing the evolution of new social dynamics that present both challenges and opportunities.

Sources: See http://newsroom.fb.com/Key-Facts for official 2013 Facebook data. An interesting secondary source of information is the blog of Craig Smith, http://expandedramblings.com/index.php/by-the-numbers-17-amazing-facebook-stats/; J. Brenner and A. Smith, "72% of Online Adults Are Social Networking Site Users," Pew Research Center, August 5, 2013, http://www.pewinternet.org/~/media//Files/Reports/2013/PIP_Social_networking_sites_update.pdf; S. Marche, "Is Facebook Making Us Lonely?" The Atlantic, May 2012, 60–69,http://news.cnet.com/8301-1023_3-57558851-93/developing-nations-adopting-social-media-quickly/; Donna Tam, "Developing Nations Adopting Social Media Quickly," CNET News, December 12, 2012.

tomorrow," Indians are more likely to state firmly but politely, "you may do it tomorrow." Both expressions have the same meaning.

Many countries promote a consumer nationalism that encourages local over "foreign" goods. For instance, while the Chinese appear to have welcomed globalization wholeheartedly—and there are enough fast-food outlets and Western chain stores to support this conclusion—there remains a deep-seated reluctance to embrace imported consumer cultures too closely. Part of this reluctance stems from a troubled history of contact with European cultures, going back to the nineteenth century when foreign governments indirectly but effectively controlled Chinese affairs. Regardless of the cause, some Chinese entrepreneurs go so far as to advertise their goods as the genuine local alternative to foreign offerings, as in the case of the local fried chicken company boasting that its chicken was so much better than the big American brand named after a colonel, because the local eatery took into account the peculiarities of the Chinese palate. The fundamental cultural differences that distinguish one place from another do not disappear the moment that an American fast-food store is built on the corner.

If homogenization has been overstated, and if differences from place to place still exist, then geography still matters. While the technological foundations of globalization may have helped to annihilate space and time in a functional sense, technology has not destroyed the importance of place in a geographic sense. For example, corporate strategies may be global in reach, but rely on close associations with specific localities to succeed. A corporation setting up assembly plants in a country with low wages is clearly establishing a close association with a certain place, and is acknowledging the importance of place in its corporate decision making. The importance of place is also revealed by the way a community competes with other communities to attract an automobile assembly plant or some other large employer, touting local benefits or differences in order to win the corporation's approval. Competition among communities can be intense, almost ruthless, in pursuit of the prize (Figure 1-16). The geographic questions—where? why? and so what?—still count in an era of globalization.

▼ **Figure 1-16 Castroville, California, artichoke center of the world.** Castroville's self-proclaimed preeminence in artichoke production is more than local boosterism and a cry for attention. It reaffirms that place is important in identity and states why anyone interested in their agricultural product should do business with them.

Although place still matters, technology has changed the meaning of distance between places. Thanks to the ease of contacting different places across the world through elaborate communications and transportation networks, a person may be "closer" to a chatroom partner or business associate living a continent away than to a next-door neighbor. Functional proximity may be more important to people than physical proximity.

Contemporary geographic footprints of globalization include those of both winners and losers. The forces of globalization have led to an unevenness in the geography of development. Winners include **world cities**, or those centers of global finance, corporate decision making, and creativity that have worldwide reach, such as New York City, Hong Kong, and London. Winners also include communities that have secured a piece of global commerce (for instance, an automobile assembly plant that provides jobs for the local community). Winners additionally include consumers who pay less for goods coming from low-cost production abroad, as well as workers who owe their positions and success to the global expansion of corporate activities. These workers may be part of what is increasingly referred to as a **transnational capitalist class** (see *Geography in Action:* The Transnational Capitalist Class). Finally, winners include those countries that have been able to transform their low-wage economies into destinations for industries that pay higher wages and are more technologically sophisticated, a process that is at least partly due to trade ties developed as a part of the globalization process. Good examples include some of the Asian countries that have emerged from low-wage assembly economies to become powerhouse players on the world economic scene. South Korea clearly fits into this category; China and India may be following closely.

Losers include people who have lost jobs due to wage competition elsewhere, including in some cases skilled professionals; people too poor to take advantage of the consumer benefits of globalization; people affected by the pollution and other harmful environmental outcomes of industrial expansion; and people who emigrate because of poverty at home but become economically marginalized in their destination countries. Also included are many cities that are just "ordinary"—not world class—and countries that suffer resource depletion to support the needs of global corporations. An especially destructive outcome is an expectations gap that is created when impoverished people in developing countries are exposed to advertising and other messages that broadcast lifestyle ideals that are both alien to the local cultural context and hopelessly out of reach from an income perspective. The participation of these people in a globalizing world takes the form of a vicarious consumption of all that globalization is supposed to offer, a situation that can breed frustration and resentment.

Winners and losers are not always easy to identify, and one person's winner may be another person's loser. For example, the teenage girl working in an assembly plant in a developing country for very low wages and in an unsafe environment may be viewed by U.S. counterparts as clearly a loser in the globalization process. The teenage girl, in contrast, may see her situation as a marked improvement over what she faces at home in her village or elsewhere in the city; she may

The Transnational Capitalist Class

When we think of "citizens," we normally think of people who "belong" to a specific community or country. Marx and Lenin wrote about citizenship in an economic class—the proletariat versus the bourgeoisie. Some observers now believe that the globalization process has created a new opportunity for citizenship: the transnational capitalist class.

Who are these people? They constitute a "super-elite" of extraordinarily wealthy entrepreneurs who bring together the labor, financing, and other components necessary for innovation, whether it be the development of revolutionizing computer technologies, new communications devices, or highly profitable financial instruments. They are corporate executives, financiers, and, to some extent, media people, globalizing bureaucrats, and politicians who find reasons to associate with the group. This is the globe-trotting, jet-setting crowd who one day may rub shoulders with their counterparts at a global conference in Davos, Switzerland (Figure 1-2-1), and the next day attend to business in London or visit an assembly plant in Thailand. They are the success stories of the globalization process. Rather than inheriting their wealth, members of this group tend to earn it through hard work, converting the promise of the globalization process to marketable goods and services. They believe firmly in global capitalism and rewarding meritorious work. They have taken full advantage of the distance-collapsing effects of communication technologies.

Members of the transnational capitalist class tend to be from everywhere and from nowhere. They hold national passports, but view themselves as global citizens who identify more with their peers than their countries. The passport may say France, but their true affinity is for people from all over the world who share their success and values, rather than their cultural traditions.

The good news is that these people have changed our lives in overwhelmingly positive ways. The technologies that we use daily, from cell phones to social media to search engines, have roots in the creativity of this class. They have helped to make former economic backwaters into productive members of the new global system, particularly in the developing world. Members of this class are just as eager to direct their entrepreneurial impulses to improving the world as to making money. Creating a philanthropic foundation has become a virtual rite of passage for true aspirants to transnational capitalist class status, using the Bill and Melinda Gates Foundation as a source of inspiration. These foundations are extraordinarily focused on new and innovative solutions to old problems, particularly in health and education, and rarely regard simply throwing money at a problem as wise or productive.

The bad news is that many elements of society have been left out in the process of innovation and wealth creation. Some observers argue that the transnational capitalist class has become a standard-bearer for growing income inequality, and for the decline of middle classes in Europe and North America. Perhaps 65 percent of all income growth in the United States between 2002 and 2007 benefited the top 1 percent of the population. Members are unlikely to be sympathetic, believing that incomes are declining because of middle class listlessness, a failure to understand the twenty-first century economy, and lack of entrepreneurial zeal. Transnational capitalists have become increasingly defensive since the financial meltdowns of the 2008 Great Recession where they, among others, were seen as villains who too often had to be bailed out by governments.

Regardless of how one feels about transnational capitalists, they personify the best and the worst of the globalization process. It remains to be seen whether they will become increasingly isolated in their corporate towers and villas, or become the engines for more widespread wealth creation.

Sources: C. Freeland, "The Rise of the New Global Elite," *The Atlantic,* January/February 2011, 44–55; P. Dicken, *Global Shift: Mapping the Changing Contours of the World Economy,* 6th ed. (New York: Guilford Press, 2011), Chapter 16.

▼ FIGURE 1-2-1 **The World Economic Forum in Davos, Switzerland.** Many international conferences meet at the conference center in the ski resort town of Davos. But the World Economic Forum is the best known gathering. Every January for more than forty years, business, political, academic, and journalistic leaders have gathered to build greater international understanding and make the world a better place by finding common ground on solutions for common problems. In 2013 about 2,500 people participated.

see herself as a winner. In addition, winners and losers may appear in the same country or locality. Southeastern China, for example, with its explosive economic growth and vast network of connections, epitomizes all that globalization promises, whereas western China largely remains a globalization backwater.

Most globalization debates focus on two scales: the global and the local. In reality, middle levels should be taken into account in identifying the geographic footprints of globalization. For example, the country or the state constitutes a sort of middle level involved in the processes defining globalization, and sometimes joins in partnership with a global corporation to market the country's products or secure some other economic advantage. Scholars have also identified geographic super-regions, such as global cores and peripheries. The United States, Canada, the European Union, Australia/New Zealand, Japan, and Israel currently compose the global core. The periphery is just about everyone else. Cores and peripheries exist at different scales. The twin cities of Khartoum-Omdurman and their immediate hinterland constitute the core region of Sudan; the rest of the country is the periphery to this capital district. Cores and peripheries are not permanent. Spain, the core to a peripheral global empire in the fifteenth and sixteenth centuries, was a geopolitical and economic backwater by the end of the nineteenth century. Israel's recent development has propelled it into global core membership. Other countries, such as Brazil, China, India, and South Korea, may soon approach or attain core status.

In terms of developmental footprints, it is important to recognize that globalization does not necessarily lead to development, and whether it ever will is still subject to debate. Proponents of the neoliberal policies driving the globalization process argue that eventually the benefits will, speaking metaphorically, lift all boats and benefit everyone; critics counter that globalization will lift all yachts, swamping the rest of the boats. There is no question that the developmental landscape shows peaks of great prosperity that owe their existence to globalization. At the same time, this landscape reveals regions of deprivation and poverty that may or may not find their ultimate salvation in globalizing societies and economies.

Thus far our story about globalization has focused on cultural, technological, and economic issues, but the interdependence that defines globalization contains other dimensions. There is increased concern about the globalization of environmental problems. Atmospheric warming offers a classic illustration. The greenhouse gases that are the primary culprits behind **global warming** are produced everywhere fossil fuels are used, making countries dependent on each other's motivations and fossil fuel policies in order to attack the global warming problem (Figure 1-17). China primarily uses coal to produce large amounts of iron and steel and meet many energy needs. The blast furnaces and manufacturing equipment that create industrial products are often antiquated technologies no longer used in the United States and Europe, where new technology makes cleaner air possible. Beijing's poor air quality and high levels of particulate matter are not only problems for the Chinese. Emissions that reach higher levels in the atmosphere over China are frequently carried across the Pacific, where they have an impact on atmospheric conditions in the United States. The solution to global warming will have to be a global solution.

▲ **Figure 1-17 Air Pollution in metropolitan Manila.** The black jeepney in the foreground, an extremely popular mode of public transportation in the Philippines, spews black smoke into the Quezon City, Manila, sky in defiance of environmental regulations, adding its substantial bit to a dangerously polluted atmosphere.

Another problem is the disposal of wastes, especially hazardous materials. In some cases, developing countries have found it desirable to accept and dispose of other countries' wastes as a source of revenue. In the past, for example, India received large quantities of toxic wastes from Europe and North America for purposes of "recycling." Unfortunately, many developing countries do not have the resources to recycle or dispose of hazardous materials without creating environmental damage, thereby creating a new set of environmental hazards with potentially global ramifications.

A still darker side of globalization takes the form of globalized terror. As the world has discovered in the last decade, certainly since 2001, religious and other fanatics are able to use Information Age technologies to launch terror campaigns in very different parts of the world. Herein lies a paradox: religious fundamentalists who have sworn to destroy modern societies end up using the technologies and networks of the modern world in their effort to destroy it. The world has become globalized in ways unforeseen and unwanted.

Environment, Society, and Development

In some parts of the world, people live under prosperous conditions. In others, many struggle to secure the necessities of life. Two principal factors affect the standard of living in an area: the physical environment and its natural resources; and the political, educational, economic, and social systems found in a place. This section focuses on the role that physical environment and human cultural systems play in development. The complex interplay of environment and society influences how people use and abuse the

▲ **Figure 1-18 Deforestation creates agricultural space in Vang Vieng, Laos.** Cutting and burning opens up plots for subsistence agriculture; the cut vegetation, when burned, provides a nutrient-rich fertilizer for crops for several years. But only through long fallow cycles is fertility restored and soil erosion prevented.

resources available to them as well as the degree to which over time they are able to successfully achieve their development goals and establish sustainable systems of resource management.

The first concern of all peoples is providing food, shelter, and other economic necessities through some form of production that necessarily involves a relationship with the natural environment. The nature of that relationship depends on the skills and resources that a society accumulates and on the value system that motivates it. For example, the way people in the United States use a desert environment differs substantially from the way the Bushmen of southern Africa do. Likewise, development takes different paths and results in different outcomes in different countries and regions, depending on their physical environments and cultural attributes.

Using the environment means modifying it, and modification can upset natural balances and relationships (Figure 1-18). Recent experience with water and air pollution and solid waste disposal encourages the idea that landscape modification and environmental degradation are new phenomena. But that notion is inaccurate. Soil salinization (salt accumulation in irrigated soil) in Iraq and deforestation in China have been serious problems for thousands of years. **Landscape modification** is as old as humankind. Today few landscapes are truly natural. Most are largely cultural, and have been shaped and formed by the societies that have occupied and used them. Culture is a part of the environment, just as the physical world is, and recognizing this relationship is essential to understanding the condition of humankind in various regions of the world.

Environmental Setting: Complex Landscapes and Evolving Challenges

The elements of the physical environment—rocks, soils, landforms, climate, vegetation, animal life, minerals, and water—are interrelated. For instance, climate is partially responsible for variations in vegetative patterns, soil formations, and landforms; and organic matter from decaying vegetation is essential for soil development. Environmental elements combine to form intricately interconnected **ecosystems** in which changes in one or more parts produce corresponding variations in other elements. Recognizing the relationships among natural processes is vital, as is recognizing that human activities frequently impact those processes. Although natural forces continue to play a major role in shaping the world's ecosystems, the single most active agent of environmental change today is people. Understanding the structure and dynamic forces operating in the natural world is extremely important because nature is the foundation upon which human livelihood and survival depend. Human activity has steadily intensified, and humankind's pressure on the natural world is becoming a serious threat to the health of many ecosystems. Whenever economic development interacts with nature, it reinforces the impact of natural processes, often in unexpected ways.

Landforms

The surface of Earth is usually divided into four categories of landforms: (1) plains, with little slope or local relief (that is, few variations in elevation); (2) plateaus, or level land at high elevations; (3) hills, with moderate to steep slopes and moderate local relief; and (4) mountains, with steep slopes and great local relief (see chapter opener map).

Plains are the landform most widely used for settlement and production when other environmental characteristics permit. Large areas of land with little slope or relief are well suited to agriculture. Ease of movement over plains, particularly those with grassland vegetation, has facilitated interaction and exchange between societies. Not all exchange is peaceful, and the features of plains that contribute to their use in peacetime can be handicaps in wartime because these regions possess few natural barriers to provide protection. Great population densities can occur on plains where intensive agriculture is practiced or where industrial and commercial activity is concentrated. Many early high-density agricultural populations were associated with plains. Sparsely settled plains are generally less desirable for climatic reasons.

Hills and mountains are a very different habitat. In mountainous regions, small basins and valleys become the focus of settlement and, because they are difficult to penetrate, may lead to the formation of distinct cultures and can also define country or regional borders. Although such areas provide security from attack, they may also be less prosperous than more favored regions.

Major differences and often conflicts between highland and lowland inhabitants are common and are a part of regional history in many areas of the world. Separatist movements are an expression of those differences, and isolated highland areas provide excellent bases for guerrilla organizations. The mountain-dwelling Kurds of Iran, Iraq, and Turkey have been politically at odds with the governments of all three states; and in Myanmar, the Karen of the Shan Plateau have raised communist and indigenous insurrections against a government that is controlled by the lowland river plain majority. Unifying a country that incorporates such contrasting environments and cultures remains difficult.

Climate

Climate directly affects human efforts to produce food and industrial crops. All plants have specific requirements for optimal growth; rice needs substantial amounts of moisture, whereas wheat is more drought tolerant. Coffee requires a year-round growing season and grows best in the tropics at elevations that provide cooler temperatures. Even though humans have modified the character of many plants, climatic differences influence which crops are grown in specific geographic regions (Figure 1-19).

The most important climatic elements are precipitation and temperature. Precipitation is water vapor that condenses in the atmosphere and then drops to the surface. Rainfall is the most common precipitation type, but the solid forms—snow, sleet, and hail—are also significant. Melting snow is an important source of water for streams or as soil moisture that can be used in later seasons.

The average annual precipitation map shows that areas with the highest rainfall are concentrated in parts of the tropics and the middle latitudes (Figure 1-20). In the middle latitudes of the western hemisphere, where prevailing air movement is from west to east, precipitation is particularly heavy on the western (windward) sides of mountain ranges and continents between 40° and 60° latitude. The eastern (leeward) sides of midlatitude mountains and continents often receive less rain. The southern, midlatitude Pacific coast of Latin America in southern Chile is extremely rainy, but in Argentina's Patagonia, desert conditions dominate vast areas east of the Andes Mountains. As one moves eastward away from the barrier effect of the Andes and Rocky mountain ranges toward the Atlantic coast, drier conditions moderate. Storms drawing moisture from tropical waters bring greater rainfall to the eastern half of North America. A similar phenomenon occurs in East Asia, but is less prominent in South America.

Continental interiors, notably central Asia, northern Africa, central Australia, and inland North America, and areas on the leeward, or downwind, side of mountains are dry locations so far removed from sources of moisture that rainfall is a rare event. Some subtropical areas on the western side of continents—northwestern and southwestern coastal Africa (Figure 1-21, p. 26), northern coastal Chile and Peru, northwestern Mexico, and coastal southwestern United States—are equally dry. Subsiding air associated with high pressure systems and cold offshore currents flowing toward the equator drastically reduce the likelihood of precipitation. But upwelling of cold water along these coasts brings nutrients to the surface and produces an environment teeming with fish and the birds that feed on them.

Seasonal rainfall patterns are as important to land use as yearly rainfall totals. Some equatorial areas, including the tropical rain-forest zones, receive a significant amount of rainfall in every season. Much of the tropics, however, alternate wet summers with dry winters. Although temperatures in the wet and dry tropics are high enough for the growing season to be yearlong, that advantage is partly offset by the

(a)

(b)

◀ Figure 1-19 **Climatic adaptations of food crops.** All plants, including those used as food by humans, are adapted to specific environmental conditions. Shown here are two crops that thrive under very different climatic and soil conditions. Sugarcane (a) is a perennial grass adapted to warm tropical to temperate climates. The sugar beet (b) is a root crop that grows best in cool, temperate conditions in the midlatitudes.

seasonality of rainfall. Some subtropical climates, particularly areas around the Mediterranean and analogs elsewhere (southern California or the Cape of Good Hope in South Africa), also experience distinct seasonal variations, with dry summers and autumns and wet winters and springs.

Variability of precipitation is the percentage of departure from the annual average, which is derived from a 30- or 50-year record. The lower the rainfall, the greater the annual variability, and the less dense is settlement unless a special local source of water is available. Transitional areas (steppe or savanna grasslands) between humid and dry regions are frequently important settlement zones with significant grain production. Under normal conditions, the Great Plains of the United States, the North China Plain, the Sahel (on the southern edge of the Sahara), and the Black Earth region of Russia, Ukraine, and Kazakhstan typify such regions. Unfortunately, these areas are prone to periodic droughts and can be risky habitats for agriculture.

Temperature has a big influence on plant growth. Low temperature and frost limit plant growth to part of the year or prevent it completely. High temperature, especially in areas with limited rainfall, increases evaporation and plant transpiration and produces a high **evapotranspiration rate**. Low precipitation and high evapotranspiration means that plant growth will be limited. Deserts represent the extreme cases of this moisture-temperature relationship.

Perhaps the most important aspect of temperature is the length of the **frost-free period**, which is related to latitude, altitude, and large bodies of water. The modifying influence of water on temperature can be seen in coastal locations, where growing seasons are longer than would normally be expected at those latitudes. The mild winters of northwestern Europe, moderated by the warm Gulf Stream, compared to similar latitudes in Russia or eastern North America is an example. Winter temperatures are important as well. Many midlatitude fruit trees require a specific dormancy period with temperatures below a certain level. Cold temperatures are necessary to ensure the activation of a new flowering cycle.

In addition to moisture requirements, each plant variety has a specific range of daily low and high temperatures and a photoperiod (the length of the day, or the active period of photosynthesis) within which growth can occur. Barley requires a long daily photoperiod to flower or set seed, while soybeans or rice benefit from shorter photoperiods. Although less-than-optimum conditions do not prevent a particular crop from being cultivated, these conditions reduce crop production efficiency and lower yields. Efficiency, viewed as cost per unit of production, is a major concern in commercial agriculture.

Climatic classification of places is based on both temperature and precipitation conditions (Table 1-3, p. 27). Although conditions may vary within a climatic region, a generalized classification system is useful for comparing the characteristics of different areas. Figure 1-22 (p. 28) shows the global distribution of climates and reveals a close relationship between latitude, continental position, and climate. The similar continental location for humid subtropical climates in the United States and China is just one example. The distinction between dry and humid climates is fundamental to classification, but no specific precipitation limit separates the dry from the humid. For example, areas are classified as desert or steppe if the potential for evaporation exceeds actual precipitation.

In the midlatitudes, deserts normally receive less than 10 inches (250 millimeters) of precipitation; steppes (semiarid grassland regions) receive between 10 and 20 inches (250 and 500 millimeters). In the higher midlatitudes, 25 inches (650 millimeters) of precipitation may provide a humid climate and forest growth, but that same amount in tropical areas may result in a semiarid, treeless environment. This is because the higher temperatures of the tropics place greater transpiration stress on plants, and the fact that most rainfall in tropical areas generally occurs during the high sun period of the year when evapotranspiration is greatest.

Along with the other controls of climate—such as latitude, ocean exposure, prevailing winds, and atmospheric pressure systems—elevation is another controlling factor. Highland climates (see Figure 1-22) feature variable temperature and rainfall conditions according to elevation and position within mountains. Even though many mountainous areas are sparsely settled and provide a meager resource base for farmers, highland settlement is important in Latin America, East Africa, and South and Southeast Asia.

Vegetation

Natural vegetation has been an important component of human settlement patterns. Our ancestors moved easily across grass-covered plains, whereas dense forests functioned as barriers, isolating culture groups and providing refuge for those who wanted to remain apart. The grasslands south of the Sahara represent one such cultural transition zone, with features of both African and Arab cultures. In contrast, the equatorial rain forests of Central Africa have served to keep peoples apart, primarily because travel through them is difficult and the soils are generally poor.

Vegetation patterns are closely associated with climate, as can be seen by comparing world regions with similar locational attributes (see Figures 1-22 and 1-23, p. 30). For example, the southeastern United States and southeastern China exhibit similarities in both climate and vegetation patterns. Less obvious on maps of this scale is the locational distinction between grasslands (herbaceous plants) and forests (woody plants). Grasslands exist as part of midlatitude prairies, mediterranean woodland shrub vegetation, tropical savannas, and steppes and are directly attributable to lower moisture totals as well as seasonal and interannual variation in precipitation. A very cold climate (for example, in the tundra) has little woody vegetation because the growing season is short and the subsoil is permanently frozen. The more humid climates are capable of supporting forest vegetation, which usually requires a minimum annual rainfall of 15–20 inches (375–500 millimeters), depending on evapotranspiration rates.

Natural vegetation is what would be expected in an area if **vegetation succession** (replacement of one plant community by others over time) was allowed to proceed over a long period without human interference. But over the centuries, humankind has greatly altered the world's natural vegetation and few areas of truly natural vegetation remain. Most vegetative cover today reflects the results of human activity as succession has been diverted toward plants that humans find useful. In midlatitude Asia, North America, and Europe, for example, population growth and changes in agricultural technology have caused deforestation (Figure 1-24, p. 32) of vast areas so agriculture could expand. High-latitude coniferous forests and tropical rain forests still exist more widely because their habitats limit possibilities for permanent settlement or large-scale modern agriculture. But in many countries situated within the Latin American, African, and Asian tropics, rapid population growth, the need for space, and government economic policies have contributed to accelerated rain-forest clearing. Some people view the expansion of Brazilian settlements in the Amazon rain forest as an expression of national pride and unity for Brazilians, a means of easing population pressure, and a source of new economic wealth. Others fear that short-term gains may bring long-term damage to the forest, the soils, the streams, and the atmosphere. History is full of examples in which short-term or immediate need has clouded our vision in matters of **environmental stewardship**—the imperative to manage resources wisely.

We now recognize that natural vegetation is significant to many aspects of life and is related to other components of our environment, such as soil and air. Forest vegetation is more critical as our numbers grow and as we consume ever-greater amounts of lumber and paper. Sustained-yield forestry—harvesting no more than the annual growth rate of trees can replace—is a more common practice as we realize that wasteful use results in greater resource problems. Forests are valuable for more than their wood. Not only do they represent an important botanical gene pool, but they also help reduce erosion, which can destroy land and lead to the silting of rivers, streams, and reservoirs. Forests contribute to flood prevention by reducing water runoff and potentially slow down global warming by drawing carbon out of the atmosphere (carbon sinking). In wealthier countries, forests are prized as recreation areas.

Soils

Plants need nutrients as well as moisture and energy from sunlight. Nutrients come from minerals in the earth and from humus—organic materials added to the soil by decomposed vegetation. The nutrients that a soil contains depend on the kind of rock lying beneath it, the slope of the land, the vegetative cover, the microorganisms within the soil, and the soil's age.

Three processes are particularly important in the formation of soil and greatly affect its supply of nutrients and therefore its fertility: laterization, podzolization, and calcification. These processes are strongly associated with climate type. **Laterization** is a process by

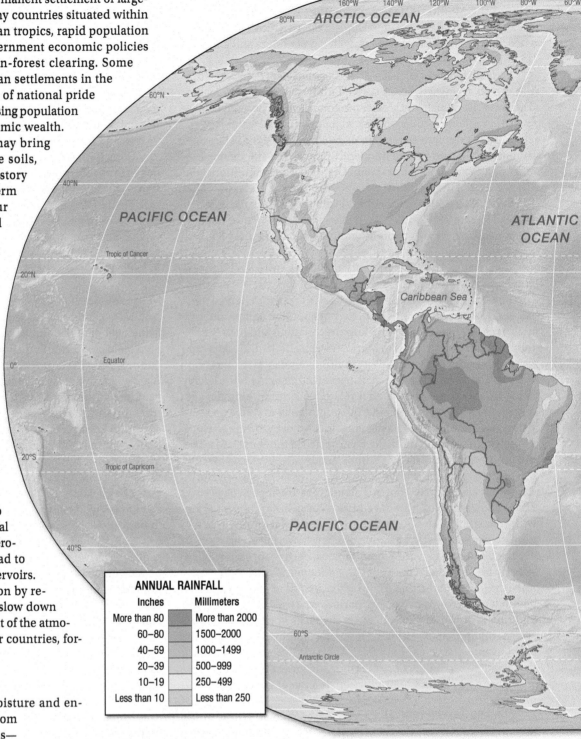

ANNUAL RAINFALL	
Inches	Millimeters
More than 80	More than 2000
60–80	1500–2000
40–59	1000–1499
20–39	500–999
10–19	250–499
Less than 10	Less than 250

▲ **Figure 1-20 World mean annual precipitation.** Precipitation varies greatly from one part of the world to another. Moreover, there is considerable variability in precipitation from one year to the next. Variability is usually greatest in areas of limited precipitation.

which infertile soils are formed in the humid tropics. Plentiful rainfall leaches (dissolves) the important soluble minerals—nutrients such as calcium, phosphorus, and nitrogen—in the soil and carries them away. The iron and aluminium compounds that remain give the soil a characteristic red color; these soils need humus to remain fertile. Once rain-forest vegetation—the

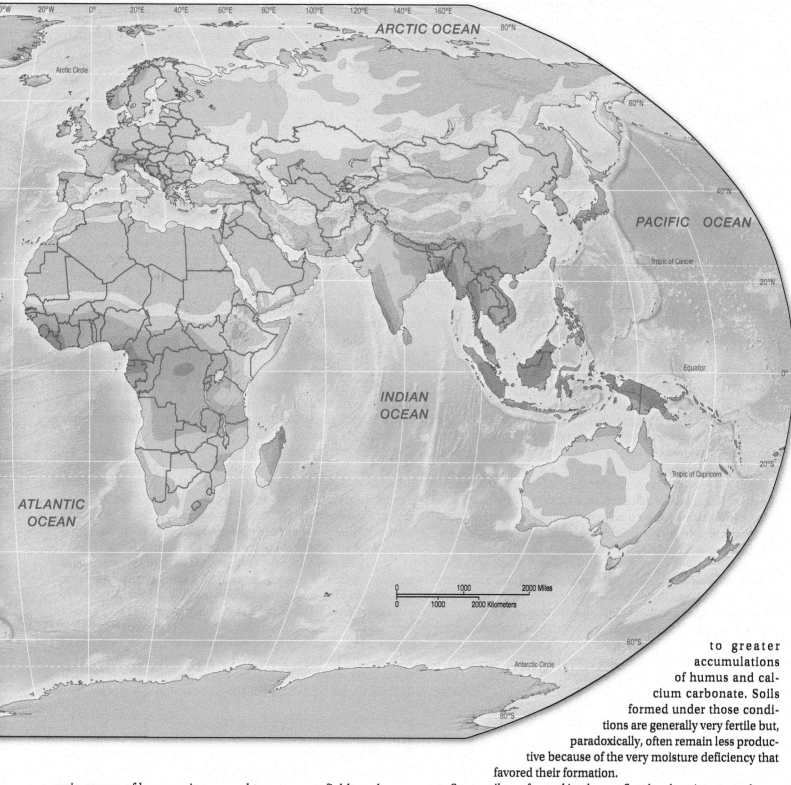

to greater accumulations of humus and calcium carbonate. Soils formed under those conditions are generally very fertile but, paradoxically, often remain less productive because of the very moisture deficiency that favored their formation.

main source of humus—is removed to open new fields and pastures, soils in areas of high rainfall and high temperature rapidly decrease in fertility. **Podzolization** occurs in high latitudes or at high altitudes, where normal leaching is limited because rainfall is low or the ground is frozen for part of the year; poor drainage and partly decomposed vegetation produces an acidic soil with low fertility. **Calcification** occurs on the drier margins of humid regions and in arid zones where limited precipitation leads

Some soils are formed in place, reflect local environmental conditions, and are strongly influenced by laterization, podzolization, or calcification. Other soils are formed from transported materials and have characteristics unrelated to local environmental conditions. **Alluvium** is soil that is transported and deposited by water, whereas **loess** is transported by wind. Many alluvial soils, loess, and soils formed from some volcanic materials are exceptionally fertile. In humid areas, they are often the most preferred soils for

▲ Figure 1-21 **The Northwest coast of Africa and the Canary Current.** The cold Canary Current and stable, descending air create an extremely arid landscape along the southern Moroccan and northern Mauretanian coast. But offshore, nutrient-rich, upwelling cold water supports a very productive fishing resource exploited as much by foreign as by local fishermen.

cultivation; but in dry areas, they may require great amounts of labor or capital to cultivate using irrigation.

Humans have modified soils in many ways, sometimes improving poor soils and at other times harming fertile ones. Soils damaged by depletion, erosion, or other forms of misuse may be restored by appropriate management practices. Adding inorganic chemical fertilizers can overcome the declining fertility that results from prolonged use, but may harm soil structure. The addition of lime reduces acidity that can be detrimental to many plants, but may also kill beneficial soil organisms such as earthworms. Terraces can prevent erosion or help distribute irrigation water, and in many parts of the world they are used to transform steep slopes into productive, small fields. Crops that quickly use up soil nutrients can be rotated with less demanding plants or with leguminous plants, which add nitrogen. Or fields can lie fallow (rest) for a season or more between periods of cultivation so that fertility builds up again through natural processes. Farmers in many areas of the United States installed tile drains to remove excess moisture from poorly drained areas, and other drainage techniques have been used in Asia, Europe, and the humid tropics to transform unproductive soils.

Efforts to reduce soil destruction or improve fertility involve labor and capital inputs, but the long-term needs of society require that we accept the costs of maintaining the environment. Unfortunately, people are not always willing or able to bear such costs, particularly when short-term profits or immediate survival is the main concern. Exploiting soils may lead to maximum return over a short period, but it often results in the rapid destruction of one of our most basic resources. Soil erosion is largely preventable through the adoption of agricultural practices that reduce soil exposure to wind and running water, better management of irrigation projects, and the more selective and reduced use of agricultural chemicals. Individual farmers and gardeners can often improve their soils through increased use of organic soil additives, by plowing along contours rather than in straight lines, as well as by constructing simple, low

stone barriers built along the contour that capture soil particles and water. It is difficult to overstate the importance of protecting and enhancing soil quality, because only with good quality soils can food be produced abundantly.

The experience of the United States illustrates the environmental damage and human hardship that can result from less-than-cautious use of resources. High prices and above-average rainfall throughout World War I and into the 1920s encouraged farmers on the western margin of the Great Plains to convert their grasslands into wheat fields. When several years of severe drought occurred, coinciding with the Depression of the 1930s, unprotected cultivated lands were heavily damaged by wind erosion, and the farmers suffered terribly (Figure 1-25, p. 32). Many farming families moved away from the region that came to be known as the "Dust Bowl," abandoning their land. Banks failed, businesses declared bankruptcy, and small towns declined. But the Dust Bowl did not become a permanent landscape feature because farmers and government institutions accepted the permanent reality of the region's drought hazard and made adjustments: farm sizes became larger; marginal land was shifted into less intensive uses; better land-management practices reduced erosion; and government programs paid farmers to take erosion-vulnerable land out of production. These improvements in agricultural practice and increased institutional support stabilized the situation, although little changed in basic attitudes toward land management. When central pivot irrigation emerged after 1953 as a technology that could exploit the deep groundwater resources of the Ogallala aquifer (Figure 1-26, p. 33), a new round of intensive development began. Heavy exploitation of the Ogallala's groundwater reserves threatens to increase land degradation and diminish intensification, and many people may have to shift from irrigation to herding or to cultivation of rain-fed crops.

The experience of trying (and often failing) to match land use and available technology to an unpredictable dryland ecosystem and achieve long-term sustainable results is a generic, worldwide problem. The U.S. Dust Bowl has parallels in the overextension of farming characteristic of the former Soviet Union's Virgin Lands program in southwest Siberia and northern Kazakhstan, the land degradation that is accompanying intensified development of China's Inner Mongolia region, and the Sahelian drought that afflicted the semiarid fringe of Sub-Saharan Africa in the 1970s.

Minerals and Energy Sources

Today's politically and economically powerful countries have built up their industrial structures by using huge amounts of **fossil fuels** (coal, petroleum, and natural gas) to process large quantities of minerals. Any country that wants to be considered technologically advanced and industrialized must either possess such resources or acquire them. No other alternative energy source so far has surpassed fossil fuels in importance, although nuclear energy contributes a large share of power in a few countries and hydroelectricity is important in some places but contributes little in many others. Wind, tidal, geothermal, and solar power sources have great potential to tap clean, renewable, and inexhaustible energy, and advances in their development are

Table 1-3   Characteristics of World Climate Types

Climate	Location, by Latitude (continental position, if distinctive)	Temperature	Precipitation, in Inches Per Year (millimeters per year)
Tropical rainy	Equatorial	Warm, range[a] less than 5°F (3°C); no cold season	60+(1500); no distinct dry season
Tropical wet and dry (Savanna)	5°–20°	Warm, range[a] 5°–15°F (3°–8°C); no cold season	25–60 (650–1500); summer rainy, low-sun period dry
Semiarid (Steppe)	Subtropics and middle latitudes (sheltered and interior continental positions)	Hot and cold seasons; dependent on latitude	Normally 10–20 (250–500)
Arid (Desert)	Subtropics and middle latitudes (sheltered and interior continental and some cold current coastal positions)	Hot and cold seasons in interior areas; constantly cool in subtropical coastal zones	Normally less than 10 (250)
Dry summer subtropical (mediterranean)	30°–40° (western and subtropical coastal portions of continents)	Warm to hot summers; mild but distinct winters	20–30 (500–750); dry summer; maximum precipitation in winter
Humid subtropical	20°–35° (eastern and southeastern subtropical portions of continents)	Hot summers; mild but distinct winters	30–65 (750–1,650); rainy throughout the year; occasional dry winter (Asia)
Marine west coast	40°–60° (west coasts of midlatitude continents)	Mild summers and mild winters	Highly variable, 20–100 (500–2,500); rainfall throughout the year; tendency to winter maximum
Humid continental (warm summer)	35°–45° (continental interiors and east coasts, Northern Hemisphere only)	Warm to hot summers; cold winters	20–45 (500–1,150); summer concentration; no distinct dry season
Humid continental (cool summer)	45°–60° (continental interiors and east coasts, Northern Hemisphere only)	Short, mild summers; severe winters	20–45 (500–1,150); summer concentration; no distinct dry season
Subarctic	50°–70° (Northern Hemisphere only)	Short, mild summers; long, severe winters	20–45 (500–1,150); summer concentration; no distinct dry season
Arctic (Tundra)	60° and poleward	Frost anytime; short growing season, vegetation limited	Limited moisture, 5–10 (125–250), except at exposed locations
Ice cap	Polar areas	Constant winter	Limited precipitation, but surface accumulation
Highland (undifferentiated) (see Figure 1-22)			

[a]The difference in average daily temperature between the warmest month and the coldest month.

promising. But while their importance is steadily rising, their share in the overall power mix is still relatively small in most countries.

The use of petroleum and natural gas has increased rapidly in the world's leading industrialized regions since World War II. That trend has given less-industrialized countries with oil wealth a special importance in international politics, an importance far greater than their size or military strength could have provided. Although

huge quantities of coal are available—especially in the United States, Russia, and China—in the second half of the twentieth century petroleum increasingly replaced coal use because petroleum and natural gas are relatively cleaner sources of energy, cost less to transport, and contribute less to habitat destruction and air pollution. The increasing cost of petroleum has begun to generate a move back to coal. The impact of this trend is uncertain, but it is unlikely to represent good news for the environment, particularly in terms of atmospheric pollution and an acceleration of the global warming that promotes **climate change** (see *Geography in Action:* The Vulnerable Arctic, p. 34).

Iron, aluminum, and copper are the most important metallic minerals used in industry. Others include chromium, zinc, lead, gold, and silver. Nonmetallic minerals such as nitrogen, calcium, potash, and phosphate are used for chemical fertilizers. Salt, building stones, lime, sulfur, and sand are other commonly used nonmetallic minerals.

Future discoveries of significant mineral deposits in parts of the world that are currently not large-scale producing regions may greatly affect our evaluation of the industrial potential of such areas. At present the United States and Russia appear to possess a broader array of mineral and fossil fuel resources than most countries. Their influential position in the global economy is, in part, a product of the richness of their industrial and energy resources. Europe and Japan must import many essential resources, but export high-value industrial products that are much in demand, and are making major efforts to develop **renewable energy** sources (for example, energy derived from water, wind, and solar sources) and to confront global warming issues.

Resource Management Opportunities and Constraints

One glance at a map of where the world's population is located (see Figure 1-7) reveals the unevenness of population distribution. Only about one-fifth of Earth's land surface is densely occupied by humankind. There are good reasons why people live where they do, reasons that at least in part are rooted in the geography and the resources of specific areas. Many of these densely populated sites have great advantages from an agricultural standpoint. It is no accident that river valleys where water is abundant, soils are rich, and transport by boat of bulky goods is easy are sites where people concentrate; the danger of floods is something they learn to cope with or avoid. Rich, fertile volcanic soils attract people despite the risk of eruption or earthquake. Dense populations also are found in many places where water and land meet. The people who live there are able to exploit the fish of the sea as well as the crops of the land. These populations have other important economic opportunities that are a result of their location. They can engage in trade with distant places, and historically have harnessed the wind to carry cargo across the low-resistance surface of the water. Yet coastal dwellers are also

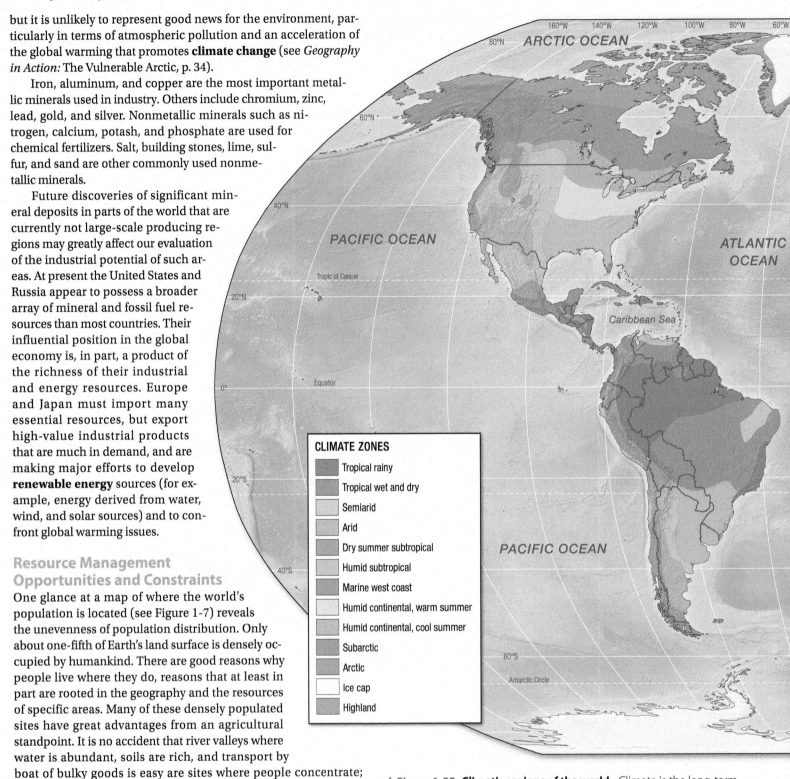

▲ Figure 1-22 **Climatic regions of the world.** Climate is the long-term condition of the atmosphere. Although many elements of climate exist, most classifications use only the two most important: temperature (level and seasonality) and precipitation (amount and seasonality).

exposed to the unbridled fury of cyclonic storms, high tides and storm surges, as well as tsunamis (Figure 1-27, p. 34). The coastal wetlands near their settlements are spawning ground for the fish in adjacent seas, but also are habitats for disease-bearing insects and microbes that undermine health and well-being.

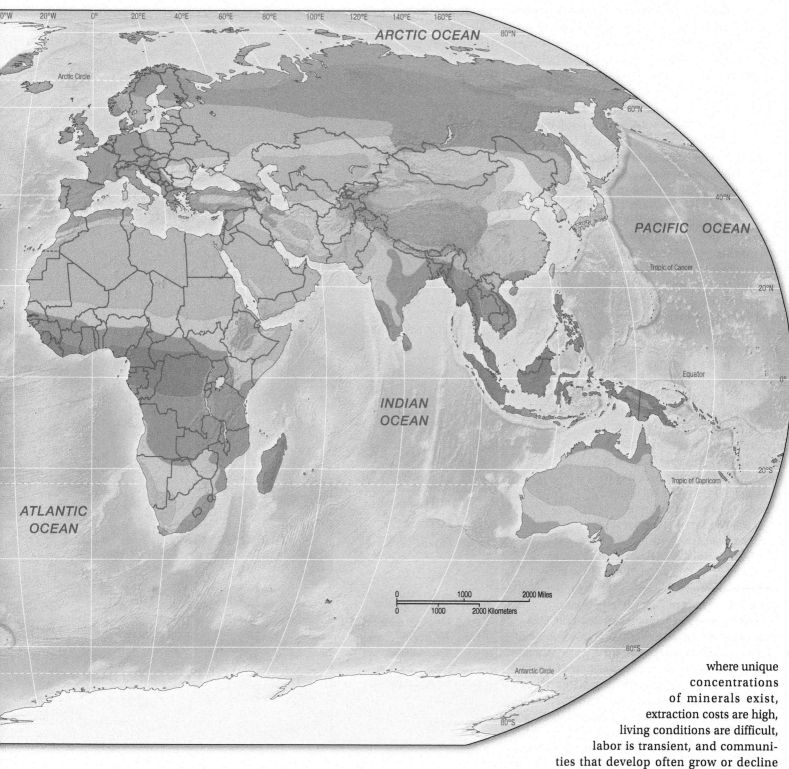

where unique concentrations of minerals exist, extraction costs are high, living conditions are difficult, labor is transient, and communities that develop often grow or decline depending on economic and technological factors outside their control. It is tempting to relate population density to broad physical patterns, but a correlation between population and physical environment is an oversimplification. Technology and political organization are additional factors to consider, as are other aspects of culture such as desired family size and economic organization. As world population growth increases and levels of technology rise, it is possible that many less densely settled regions will become more populous.

Just as people concentrate into favorable sites for a reason, so too are Earth's relatively empty places lacking in inhabitants for a reason. Nearly 80 percent of the world's land surface falls into this category. These voids are usually too steep, wet, dry, or cold to support many people. Trying to use these places imposes tremendous burdens on the people who decide to live there. The sparse population of other areas, such as some of the humid tropics of South America and Africa, can be attributed to fragile soils of low natural fertility. Even

Resource-abundant as well as resource-limited places are faced with similar environmental management issues. Each place is a complex mix of opportunities and liabilities that has to be managed. Failure to do so means that one set of uses begins to compete with others, to the long-term impoverishment of all. If cities sprawl ever outward without limits to growth, farmland, forests, recreational open space, surface and ground water, and wildlife habitat are sacrificed. The result is a more homogeneous, less resilient system that is less able to cope with extreme natural events.

Environmental setting, and how people use it, plays a significant role in the development process. Avoiding environmental deterioration at a time when increasing population and expanding expectations place ever-greater demands on environmental systems is difficult. An important, often overlooked, resource is the knowledge base (ethnoscience) amassed over generations by local communities. The longer a group of people has lived in a particular place, however inhospitable, the more likely they are to possess useful knowledge about how to use local resources successfully. The ethnoscience of local farmers and herders is essentially a long-term plan, and it can become a departure point for careful management of local resources in the future.

Those cultures that have regarded themselves as protectors of their environments, charged with a responsibility to pass on an improved habitat to future generations, have managed their available resources more successfully than others. The Bedouin of Egypt's Red Sea hills, like many other nomadic groups, are an example of this stewardship. Living in a very dry environment, perennial plants, particularly acacia trees, are crucial to survival. The leaves of woody plants not only supplement seasonal grasses for the herders' basic subsistence animal, goats, but also become the primary fodder source in times of drought. Each valley segment has a guardian, a person who protects trees against individual use to the disadvantage of the group. For these Bedouin, every place is also more than a resource site; it is a homeland full of memories of people and events, imbued with spirit and meaning. Both a focus for individual and group identity, each site supports a way of life that, if managed responsibly, can continue deep into a multigenerational future without exhausting the resources it is built on.

When the Egyptian government decided in 1988 to establish a national park in the southern Sinai Peninsular near St. Katherine's

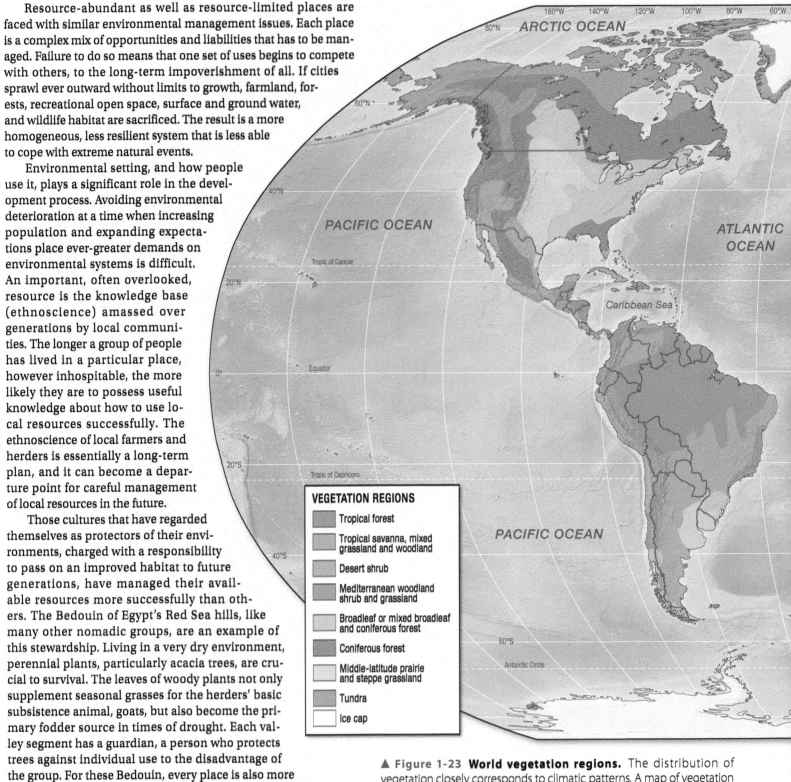

▲ Figure 1-23 **World vegetation regions.** The distribution of vegetation closely corresponds to climatic patterns. A map of vegetation can be used to determine an area's agricultural potential, since crops and natural vegetation use the same environmental elements for growth.

Monastery, it began by incorporating local Bedouin groups into the planning process. The Bedouin were willing supporters of the St. Katherine Protectorate, and put their local knowledge at the

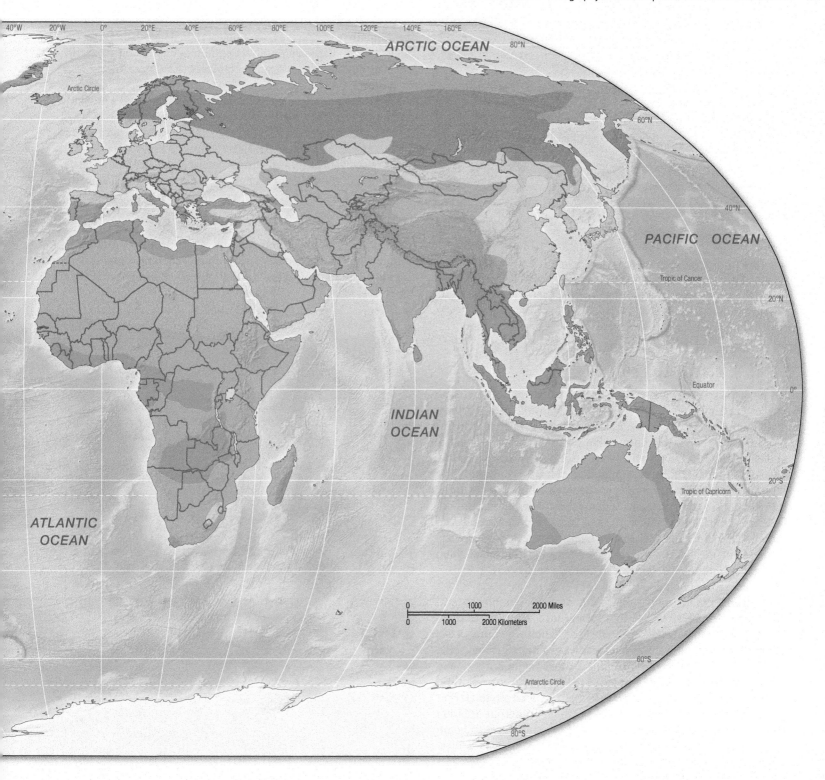

disposal of park managers, because jobs were offered to them. By using local knowledge to both defend park resources and protect wild animal populations that would attract tourists, the Bedouin were able to support their livelihood and contribute productively to local economy and environment. This goal has been realized as local Bedouin tribes have translated their conservation ethic into employment as park guards and ecotourism guides to groups and individuals interested in exploring the unique biodiversity and spirituality of a region sacred to Judaism, Christianity, and Islam.

Stop & Think

▶ Where are the densest population concentrations in the world? What desirable environmental attributes attract people to these places?

▲ Figure 1-24 **Farming in central Luxembourg.** For centuries, farmers have cleared woodland from the best agricultural soils and restricted forests to steeper slopes and poorer soils. These forest areas are intensely managed, without clear-cutting, for a sustained yield of forest products.

Historical Background: Culture and Technology

Humans believe they are unique among living creatures because they can accumulate learned behavior and transmit it to successive generations. This claim may not be as distinctive as once thought. But what is certain is that humankind's learned behavior patterns and cultural heritages have helped to ensure survival, sustenance, and social order. With that accumulation of learned behavior (culture) people make decisions and create ways of life. Styles of dress, forms of houses, food preferences, settlement patterns, religious beliefs, languages, and much more vary. Some behavior is based on inherited culture, a society's own earlier experiences; other actions are based on ideas that originated elsewhere and diffused to blend with a local culture. All elements that form a society's way of life—from values to technology—constitute its culture.

Culture is made up of a hierarchy of traits, complexes, and realms. A **culture trait** is the way a society deals with a single activity—for example, how people plant seeds. A **culture complex** is a group of traits that are employed together in a more general activity, such as agricultural production. A **culture realm** is a region in which most of the population shares a similar culture. Not all world regions have easily definable culture realms. Some straddle transition zones between realms, zones in which numerous different cultures have met and clashed. Even within a single political unit, cultural pluralism is frequently evident and may complicate achieving national unity. A **culture hearth** is a source area in which a culture complex has become so well established and advanced that

its attributes are passed to future generations within and outside the immediate hearth area. No single, original hearth exists, as was once thought. Rather, a number of different hearths have contributed various culture traits and complexes to cultures distant from each source area.

Societies advance technologically at uneven rates and along different paths. The culture complexes of food production in south China and northwestern Europe historically have involved very different traits. The Chinese intensively cultivated rice, vegetables, and pond-raised fish; European farmers focused on dairy cattle and cereal grains. Cultural distinctiveness is not necessarily a difference in level of achievement; it is most frequently a difference in kind.

Centers of Origin

The world's present cultural patterns reflect the major human accomplishments of the past. Early cultural accomplishments common to all human societies included the use of fire, the making of

(a)

(b)

▶ Figure 1-25 **Dust and drought hazard in the American Southwest.** Dust storms and drought that ravaged farmland in the "Dust Bowl" of the southern American Great Plains during the 1930s continue today. The historical image (a) shows the approach of a severe dust storm in the Texas Panhandle in 1935. Contemporary suburban settings throughout the Southwest experience similar storms (b) to the one that afflicted Phoenix, Arizona in August, 2013. Soil picked up by dust storms can be carried for hundreds of miles by the wind.

(a)

(b)

▲ **Figure 1-26 Central-pivot irrigation in the western Great Plains.** Rotating around a central well, this overhead sprinkler boom (a) delivers groundwater to a soybean field in the Texas Panhandle. Viewed from above, the result of this kind of irrigation is perfect circular fields of green in eastern Colorado (b), but high rates of extraction from the Ogallala Aquifer threaten to make seasonally brown fallow circles a permanent reality in much of the aquifer area.

tools, and the construction of shelters. They also included the domestication of food plants and animals that today are the basis for humankind's food supply.

The Middle East is one of the world's earliest and most influential culture hearths (Figure 1-29, p. 36). Known as the Fertile Crescent, the region consisted of several hearths: the Nile River valley; the hills of southern Turkey and western Iran; and the alluvial plain of the Tigris and Euphrates rivers. The earliest domesticated plants and first agricultural villages appeared in the hills of eastern Turkey and western Iran 10,000 years ago. By 5,000 years ago city-states and empires flourished in the Tigris and Euphrates lowlands of Iraq. Among other major achievements, those civilizations codified laws, used metals, put the wheel to work, established mathematics, and contributed three of the world's great religions—Judaism, Christianity, and Islam.

Another of the world's early culture hearths developed in the Indus River valley (Pakistan) by 2500 B.C. Exchanges of ideas and materials with Mesopotamia (ancient Iraq) took place from 3000 B.C. to 1000 B.C. The Indus valley experienced invasions and migrations of people from northwestern and central Asia, who brought an infusion of new culture traits with each invasion. The Indus and Ganges valleys became the source areas for culture traits that eventually spread

throughout India. This particular culture hearth made significant contributions in literature, architecture, metalworking, and city planning. Philosophy also evolved and contributed to Hinduism and Buddhism.

The Huang He (Yellow River) valley was the site of China's most important culture hearth. Wheat and oxen may have had Middle Eastern origins; rice, pigs, poultry, and water buffalo may have originated in Southeast Asia. Yet the assemblage and development of culture traits are distinctively Chinese achievements. Crop domestication, including soybeans, bamboo, peaches, and tea; village settlement; distinctive architecture; early manufacturing; and metalworking were in evidence at the time of the Shang dynasty (1700 B.C.). These culture traits eventually spread into South China, Manchuria, Korea, and, later, Japan.

Two areas in the Americas that began to develop around 3000 B.C. became hearths for major civilizations. One appeared in northern Central America and southern Mexico (later becoming the Mayan civilization) and extended to central Mexico (later becoming the Aztec civilization). This large Mesoamerican region supported a sizable urban population, large cities, monumental architecture, political-religious hierarchies, the use of numeric systems, and the domestication of maize, beans, squash, and cotton.

The Middle Andean area of Peru and Bolivia gave rise to the Inca civilization. By the sixteenth century A.D., the Incas through their conquests had developed irrigation, worked metals, domesticated the white potato, established complex political and social systems, set up a transportation network, and built an empire (Figure 1-30, p. 36).

Several other areas also functioned as culture hearths with more limited regional influence. The Bantu language family had its source area in West Africa. The Ethiopian highlands were a domestication center for wheat, millets, and sorghums, and Great Zimbabwe in southern Africa achieved high levels of mining and metallurgy and the manufacturing of clay and wood products and clothing. Central Asia was a domestication center for grains and animals, and the Silk Road was a trade route along which ideas and commodities were exchanged among major civilizations, such as the Middle East and China.

The initial developments in culture hearths may have begun in relative isolation, but contact between hearths also increasingly took place. Good ideas and inventions are readily recognized by people, adopted, and improved on. Thus the history of innovations developed in one culture hearth is their spread to other places. This process of diffusion was slow and often interrupted, but it gradually brought adjacent parts of the world into increasingly closer contact.

When Islam spread rapidly in the first two centuries following its emergence after A.D. 610, the exchange of ideas and technology was facilitated. Sugarcane, a tall tropical grass originating in Southeast Asia, was brought westward by Muslim traders and agricultural innovators. Spaniards became familiar with sugarcane through centuries of conflict and culture contact between Muslims and Christians in Spain and North Africa. Where Columbus successfully sailed in 1492, sugarcane followed quickly. The Spaniards also carried many other items to the New World! Cattle, horses, pigs, wheat, chickens, and much more were imported to fill unoccupied agricultural niches and to provide the conquerors with familiar food. Tomatoes, corn, beans, cacao, cacti, tobacco, potatoes, and other New World domesticates moved eastward and enriched the culinary traditions of Europe.

GEOGRAPHY IN ACTION · The Vulnerable Arctic

The icy deserts of the polar regions are much more significant for the planet's well-being than their inhospitable environment might suggest. Bordered by permafrost land areas of Russia, Canada, the United States, and five Scandinavian countries, the ice-covered Arctic Ocean is at the forefront of the accelerating consequences of man-made climate change caused by global industrialization and fossil fuel consumption. The Antarctic constitutes a separate, remote, much colder, ice-covered continent twice the size of Australia surrounded by the southern Pacific, Atlantic, and Indian oceans. Used only by scientists based at research stations and visiting fishermen, Antarctica could contribute mightily to rising sea levels if global warming continues unabated.

Since the middle of the twentieth century, average global temperatures have gone up 1.3°F (0.7°C). But they have increased by 2.6°F (1.5°C) in the Arctic. The massive accumulation of carbon dioxide (CO_2) and other so-called **greenhouse gases** released by the use of fossil fuels traps heat in the atmosphere, causing a greenhouse-like effect. Receding glaciers and melting sea ice uncover increasing areas of darker ocean and land surfaces in the summer season. These darker surfaces absorb more sunlight and accelerate the warming and melting process (Figure 1-3-1). This loss of ice and snow cover is further enhanced by air pollution from industrialized regions. Fine, dark particles like soot settle on ice surfaces and increase the absorption of warming sunlight.

Dangerous consequences and potential benefits result from Arctic warming. Meltwaters from glaciers and ice caps, especially the Greenland ice cap, are raising sea levels. Thawing permafrost leads to the breakdown of previously frozen organic matter and results in the release of methane, adding more greenhouse gases to the atmosphere. Many polar animal species are threatened with extinction, while the opening seas are not expected to become major fishing grounds. Because warm and fresh waters are lighter than cold and salt waters, increasing stratification of the seawater is likely. Greater stratification hinders vertical nutrient exchange, the movement of deeper water toward the surface and sunlight where it can support marine food chains. Equally serious is the increasing acidification of Arctic ocean waters, because the absorption of CO_2 may hinder primary nutrient production necessary for viable fish habitats.

Opportunities will come from easier access to mineral resources, particularly oil and natural gas, in locations no longer covered year-round by ice, and from seasonal opening of shipping routes that have been all but impassable. Access to the Northeast and Northwest Passages for more months each year could shave off major time and distance requirements for shipping to destinations currently routed through the Suez Canal. Of the world's undiscovered oil and gas reserves, more than 10 percent and 30 percent, respectively, are estimated to be found in the Arctic. There is diabolic irony to the fact that fossil fuel–driven global warming may lead to a new Arctic oil and gas boom and more burning of fossil fuels, undermining the road to cleaner, renewable energy sources.

None of the opportunities will come easy. The Arctic will remain a harsh environment, hard to conquer, expensive to harness, and full of potential for disaster. An experiment with controlled oil spills, for example, revealed how ice and oil freeze together in layers; the floating ice spreads the oil in the ocean, and melting ice disperses it further. Shipping will depend on expensive back-up from ice breakers, reinforced ships able to withstand ice pressure, and specially trained operators. A short summer, long dark winter, severe storms, and ice-strewn shipping lanes will remain challenging. The Arctic is not going to become the next major world economic boom region any time soon. But how climate change affects the health of the planet through environmental responses in the Arctic should not be underestimated.

Source: James Astill, "The Melting North," *The Economist*, Special Report, The Arctic, June 16, 2012.

In this last great human integration of global ecosystems and cultures, Europe played a central role. Europe had long been a peripheral outlier of the Middle Eastern culture realm. The increasingly global maritime dominance of European countries after A.D. 1500, and their initiating role in the development of industrial technology after A.D. 1700, particularly in England, touched off a competition over access to and control of resources in virtually all parts of the world. Between the sixteenth and early twentieth centuries, Europeans extended their influence and culture around the world, spurred by internal competition and a new interest in science, exploration, and trade. They explored, traded, conquered, and claimed

◀ **Figure 1-27 Vulnerability of coastal sites to storm damage.** The impact of Hurricane Sandy in October, 2012 affected more than just a few seasonal structures built just above the high tide line. Entire communities along the New Jersey shore were inundated when Sandy's storm surge flooded residential neighborhoods. Both recovery from damage and protection from future storms promise to be expensive tasks.

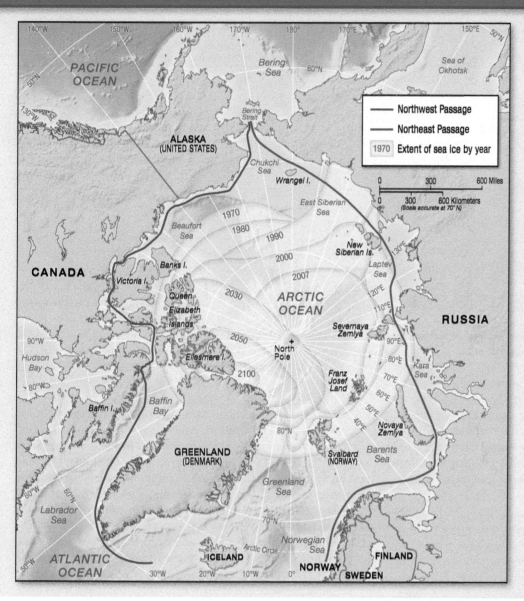

◄ Figure 1-3-1 **Estimated sea ice extension in the Arctic Ocean.** For several decades now Arctic sea ice melt during the summer season has opened up greater expanses of ocean.
Sources: Data from http://mapoftheweek. blogspot.com/2011/11/melting-sea-ice-in-arctic-ocean.html; http://www. wunderground.com/climate/SeaIce.asp

new territories in the name of their homelands. Modern European states emerged with the capability of extending their power over other areas, and many European people resettled in newly discovered lands in the Americas, southern Africa, Australia, and New Zealand. Many of our contemporary global conflicts have origins in the confrontation between European culture—represented not only by individual nation-states but also their colonial offspring—and the traditional systems found elsewhere.

► Figure 1-28 **Chuño production near Cuzco, Peru.** One little-known component of highland Andean Indian potato use is the production of chuño, or sun-dried slices of potatoes. This ingenious form of naturally dehydrated food enables the farmers to extend the "shelf life" of one of their staple crops. The potato is native to western South America and the smallholder farmers there continue to cultivate literally thousands of varieties, each differing in physical and environmental attributes.

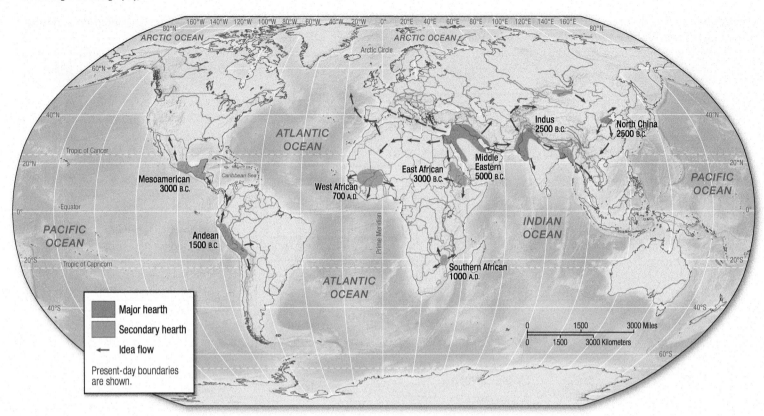

▲ **Figure 1-29 Early culture hearths of the world.** Three major culture hearths are recognized as the principal contributors to modern societies throughout the world. From those hearths, plants, animals, ideas, religions, and other cultural characteristics have spread. Minor hearths, although locally important, have not had as much impact outside their source areas.

Acculturation, the process by which a group takes on some of the cultural attributes of another society, does not occur without conflict. European cultural and political expansion encouraged a form of cultural convergence as elements of Western culture (food preferences, clothing styles, architecture, and political

▼ **Figure 1-30 Incan agriculture in the Andes.** Spectacular terraced field systems were created by ancient civilizations in the Andes. More the achievement of subjects of the Incas than the Incas themselves, these walled fields near Cusco, Peru, made very steep slopes productive and continue to operate today.

systems) appeared in local landscapes. In certain respects political organization, food production, industrialization, energy use, and consumption patterns in the independent former colonies have retained similarities to those of their former colonial rulers. Simultaneously, in the former centers of colonial power, descendants of migrants from former colonies play a prominent, if at times uneasy, role in the economic and social institutions of their acquired homeland.

This blending of peoples and cultures will continue as migrant flows increase around the world. The technological aspects of Western culture are the most widely accepted. Japan provides one example of the way in which European traits have been integrated in modified form as Western industrial technology has fused with uniquely Japanese work values and modes of social and industrial organization. Other countries have and likely will continue to experience acculturation in a manner that reflects their own cultural heritage. While cultural convergence occurs everywhere, regional differences in economic, social, and environmental conditions will produce new and unique systems that are place specific.

Acculturation is not a smooth process. Laotian Hmong refugees from the Vietnam War arrived in the United States in a series of waves after 1976. The goal of immigration agencies was to spread the refugees broadly in the country to assist acculturation and assimilation. But for cultural and clan reasons, many quickly relocated to places with existing, larger Hmong populations. Today most of the 260,000 Hmong in the United States are concentrated in California, where one-third of the total is found in the major towns of the Central Valley, and one-quarter is located in Minnesota, primarily in the Minneapolis–St. Paul metropolitan area. This concentration helps to preserve Hmong identity and culture, but drastically slows the pace of acculturation and creates problems in access to health care, education, and job opportunities. Long experience with Turkish and North African migration streams to Europe does not prepare the EU for the current flood of migrants to Western Europe from surrounding regions (Albanians to Italy; Lebanese and Syrians to France; Kurds to Scandinavia; sub-Saharan West Africans to Spain and Italy). Often caught on or shortly after arrival, these migrants are held in temporary encampments while decisions about whether to deport them and, if so, where to send them are completed. Fleeing from extremely difficult economic conditions in their homeland, these economic refugees are viewed more as a burden than a labor benefit for their host country.

The patterns of acculturation and convergence that emerge when culture realms come into contact are affected by a number of factors. The most important as symbols of group identity and least susceptible to change are language, religion, and value systems. These cultural elements provide a sense of identity to their adherents and express primary differences in values and thought processes that are difficult to bridge.

Language

Language gives a set of meanings to various sounds used in common by a number of people. It is the basic means by which culture is transferred from one generation to the next. Because of the relative isolation in which societies evolved, a great number of languages formed, frequently with common origins but without being mutually understandable. Linguistic differences can function as barriers to the exchange of ideas, the acceptance of common goals, and the achievement of national unity and allegiance. Most members of most societies are not bilingual and do not speak the language of a neighboring society if it is different from their own.

Sometimes linguistic differences are overcome through the use of a **lingua franca**, a language that is used throughout a wide area for commercial or political purposes by people with different native tongues. Swahili is the lingua franca of Eastern Africa; English, of India (Figure 1-31); and Urdu, of Pakistan. In many countries that acquired their independence after World War II, the political leadership has found national unity and stability difficult to achieve. Internal problems and conflict frequently stem, in part, from cultural differences, one of which is often linguistic variation. The situation in East Pakistan (now Bangladesh) and West Pakistan (now Pakistan) is a case in point. From 1947 to 1972, those two regions functioned as one political unit. Political leaders, seeking to promote a common culture and common goals, deemed a single national language necessary, even though several mother tongues were in use. Bengali, the language of East Pakistan, was derived centuries ago from Sanskrit; it provided East Pakistanis with a unifying cultural element. But West Pakistan encompassed a number of languages: Baluchi, Pashto, Punjabi, Sindhi, and Urdu, the lingua franca. That linguistic diversity was just one of the many challenges that eventually led to the establishment of Pakistan and Bangladesh as separate and independent states.

Pluralistic societies often lack a common language within a country, and that pluralism frequently produces political instability or hinders development. Switzerland is noteworthy because it has largely overcome the problem of linguistic pluralism by maximizing local autonomy. Others, such as, Belgium, Canada, and the former Yugoslavia, have been less successful at overcoming regional differences, although only Yugoslavia separated into independent states over linguistic and religious issues.

Religion

Belief systems originate in the desire to understand life's purposes and share support from and with others of similar views. Animism, the worship of natural objects believed to have souls or spirits, is an early form of religion. It includes rituals and sacrifices to appease or pacify spirits, but it usually lacks complex organization. Most animistic religions were probably localized; certainly, the few that have survived are found within small, isolated culture groups.

Modern religions are either ethnic (limited to a group with a common heritage) or universalizing (wanting to be accepted by everyone). Ethnic religions originate in a particular area and involve people with common customs, language, and social views. Examples of ethnic religions include Shintoism (in Japan), Judaism, and Hinduism (in India). Even when adherents of the faith move to distant places, no special effort is made to convert local populations. The ethnic religion remains exclusively a fundamental part of the transplanted population's identity system. Members of universalizing religions believe their religion is essential for the salvation of everyone. Buddhism, Christianity, and Islam all had ethnic origins, but became universal as they lost their association with a single ethnic group. So compelling is their faith to adherents of these religions that spreading "the word" is an absolute obligation. Universalizing religions spread from one person or group to another by active conversion efforts, not only by official missionaries but

▼ **Figure 1-31 Newspaper vendor in India.** English language newspapers are a distinct part of the offerings of this street stand and bridge the barriers of communication between many different languages and dialects in this vast country.

also by the efforts and example of ordinary people such as traders, migrants, and military personnel (Figure 1-32).

Religious belief systems exert a great impact on culture. Sacred structures help shape rural and urban landscapes, and religions have influenced many of the routine aspects of daily life. The spread of citrus growing throughout the Mediterranean basin was directly related to Jewish observances in which citrus was required for one of the ceremonial activities of Sukkot (the Feast of Booths or Tabernacles). Religious food restrictions account for the absence of pigs in the agricultural systems of Jewish and Muslim peoples in the Middle East. The Hindu taboo on eating beef is a response to their respect for the cow. This practice has resulted in a very large cattle population that requires space and feed but also returns material benefits (manure, milk, and draft power) vital to rural Indians (Figure 1-33). In the United States, the impact of religion is seen in prayers offered before public ceremonies and events, work taboos on specific holidays, attitudes toward materialism and work, taxation policies (church buildings used for worship are not taxed), and so on.

One implied function of religion is to promote cultural norms, which may be positive. Societies can benefit from the stability that cohesiveness and unity of purpose bring. But conflict also can arise when religion-based differences evoke intolerance, suppression of minorities, and conflict. Examples include the Crusades of the Middle Ages when Christians attempted to recapture the Holy Land from Muslims; the partition of the Indian subcontinent in 1947 in response to fighting between Hindus and Muslims; the conflict between Catholics and Protestants in Northern Ireland; and the tensions between Christian and Muslim communities in the Balkans that caused the former Yugoslavia to break up.

Values and Political Systems

Value systems and political ideologies have major implications for social unity, stability, and the use of land and resources. Such ideologies need not be common to the majority of a population; many people may be apathetic, or they may be unable to resist an imposed system

▼ **Figure 1-32 Ubadiah mosque in Kuala Kangsar, Malaysia.** The dominance of Islam in the southeast Asian nations of Brunei, Malaysia, and Indonesia is an expression of the diffusion of a religion far from its point of origin. It is also evidence of the power of religion to shape and unify, to a degree, the lives of peoples whose cultures are otherwise distinct.

▲ **Figure 1-33 Sacred cow in Rajasthan, India.** Cattle are allowed to roam freely in Hindu-dominated India. This scene is from Chittaugarh, Rajasthan, and shows a decorated cow outside the Pratap Palace.

of rules and decision making. In some instances, oligarchies, where control is exercised by a small group, or dictators have made their particular philosophies basic to the functioning of their societies. Most modern governments assume some responsibility for the well-being of the people in their states. Some governments assume complete control over the allocation of resources, the investment of capital, and even the use of labor, thereby severely limiting individual decision making. In other societies, governments provide an environment in which individuals or corporations may own and determine the use of resources. National and economic development programs reflect these differences in political systems and heavily influence the way particular countries respond to the issue of sustainable economic development.

Stop & Think

➡ Give examples of how cultural values can impact the direction of development.

Environmental Health and Development

Without a healthy environment, development is impossible. Rapid change is an enemy of environmental health because most adjustments to change take time. Only animals and plants with closely spaced generations are quick to adjust biologically to change. Humans can only cope effectively with change by forethought, improved knowledge, and careful planning. In the past, most people lived in one place for much of their lives, could see the changes that were taking place locally, and were able to make the adjustments needed to keep their habitat healthy and productive.

The world has changed with increasing speed over the last several centuries. Improvements in transportation, communication systems, and technology have tied places closer and closer together. The earthquake and tsunami that destroy a nuclear power plant in Japan also distribute pollutants to locations in the Hawaiian Islands and the west coast of North America. We live in an interconnected, globalized world in which human impacts on the environment have caught up to and interact with the natural hazards and disasters that have always been part of the human experience. A strong sense of stewardship for the health and well-being of Earth and all of its inhabitants is needed to navigate successfully into the uncharted future.

Stewardship and the Future

All parts of the natural world experience change. Some changes are very slow or take place infrequently. Although pieces of the earth's crust "drift" over the surface, their movement is very slow, measured in thousands and millions of years of geological time. Plants, animals, and people do not notice this movement at all. It has been 20,000 years since the last ice age ended; humans view a largely ice free globe as "normal." Earthquakes, tornadoes, volcanic eruptions, tsunamis, and hurricanes are all violent natural events that cause great damage in relatively small areas. The destruction they cause may take 100 years for nature to repair, but slowly the scars in the landscape fade and few readily visible signs of the original disturbance remain.

Humans are a part of nature, but they also have the ability to change the natural world drastically. This capacity is not new. Once fire was controlled 500,000 years ago, for example, by accident as much as by design, people used fire to create environments that were more useful to them than the habitats nature originally provided. Early communities did not have an elaborate theory about practicing environmental stewardship (careful management of resources), but they did know that it was unwise to take more from their environment than they needed for immediate and future maintenance. As long as populations were relatively small and lived in close association with and dependence on their immediate surroundings, relatively little massive, abrupt change occurred. But over thousands of years, people gradually developed technologies and management strategies that had greater and greater ability to change the environment. Until quite recently, changes introduced by people took place relatively slowly. Nature and humankind had an opportunity to adjust to each other and develop along parallel

paths as part of coevolved ecosystems (Figure 1-34). But accelerating technological change is increasing the spatial extent and intensity of human influence on the natural world. The speed with which signs of such impacts appear in the landscape also is increasing. Many of the changes introduced by humans now take place at a pace too rapid for the environment to adapt to quickly. One result of this excessive pressure is land degradation.

When an area becomes less biologically productive or its usefulness to humans is reduced, **land degradation** has occurred. In the United States, interest in efficient, rapid regrowth of trees after harvesting was motivated in the early 1800s by the realization that New England faced a serious fuel crisis in an economy that was almost completely dependent on wood to meet domestic fuel and construction needs. Only when railroads were developed after 1830, and coal could be imported to New England from Pennsylvania as a wood substitute, did this energy crisis diminish.

By the late nineteenth century, scholars and environmentalists began to compare the problems of deforested lands, erosion, siltation, animal extinction, and disease (such as malaria) that affected the Greeks and Romans of the Mediterranean 2,000 years before with the forest clear-cutting, dust storms, gully erosion, and infertile, "worn out" farmland emerging in North America. Demands that nature should be preserved untransformed in national parks and managed sustainably on farms grew. Despite the achievement of some of these goals, the list of land degradation issues has grown longer rather than shorter.

Coastlines where land and sea come together illustrate the basic problem. Offshore sand bars, barrier islands, coastal marshes and tidal lagoons, and sand dunes are nature's defensive structures, flexible shock absorbers of the power of waves and storm surges. Yet these same sites are attractive places for second homes and vacation cottages despite the high risk of damage from storms that accompany the location. In the United States, building in vulnerable coastal locations is an expectation for anyone with land near the shore or the resources to acquire it (Figure 1-35). Dunes and the grasses that trap sand are leveled for building sites as close to the beach as possible. When storm damage occurs, the political pressure to support rebuilding with subsidies is enormous even though property owners often cannot afford government or private insurance. An alternative is publicly funded infrastructure that hardens the coast with concrete barrier walls, breakwaters, groins, and jetties. These structures never provide complete security, and are subject to undermining or bypassing and landward erosion on any part of the coast not similarly hardened. Overtopping occurs in extreme storm events made more common by global warming. A better approach is to withdraw settlement and defensive infrastructure from the high-risk beach environment, and restore as much of nature's flexible buffer as possible. Politically this approach is unpalatable. The solution is to find a better way to mix technological expedients with natural buffers in order to balance human needs and expectations with the opportunities and risks that nature presents.

Finding a Balance

To live and prosper in a place for a long time, people need to find a balance between meeting their material and spiritual needs and the rhythms, hazards, opportunities, and extremes of the

▼ **Figure 1-34 Goats in an argan tree in southern Morocco.** Goats, trees, and people have developed together. Goats eat leaves and fruit; time in the goat's intestine prepares seeds for germination; people protect seedlings from goat browsing, use goat milk, meat, and hair, and process argan fruits for edible oil. Slowly over time there are more goats, argan trees, and people; everyone benefits and is mutually interdependent.

(a)

(b)

▲ **Figure 1-35 Hurricane Sandy's impact on Seaside Heights, New Jersey.** Before Hurricane Sandy struck the New Jersey coast in Fall, 2012, the Amusement Pier at Seaside Heights (a), with ferris wheel, roller coaster, and go-cart track, was a popular summer recreation destination. After the hurricane's furious assault ended (b), only devastation remained.

theories of development professionals cannot substitute completely for this local knowledge. Techniques and ideas developed in other places can be introduced selectively to support local knowledge, but the practices derived from other places seldom provide easily transferable solutions for local problems.

The transition to agriculture is a good example of development based on local knowledge. When hunting and gathering cultures began to shift from collecting what their local habitat produced seasonally, and concentrated on specific plants and animals that prospered in close association with people, a dramatic process of creative transformation began. This positive type of **creative destruction** transforms nature and produces systems that can potentially support human communities for hundreds of years. Humans do this every time they remove one plant community in an area in favor of another. They cut down trees to plant a cornfield, or remove a grassland to grow wheat. The plants people want to eat require space and sunlight to grow successfully. We destroy the competition to create the conditions that favor crop growth. We work hard to remove weeds and other invaders that threaten to "steal" nutrients from the plants we want.

Humans destroy in order to create! The Chinese have destroyed river delta and wetland vegetation to create elaborate pond-dike farmlands that combine fish farming and crop production. The Dutch have created farmland from freshwater wetlands and the former Zuider Zee, once an arm of the North Sea (Figure 1-36). Nabateans who lived in the northern Negev Desert 2,000 years ago encouraged soil erosion on hill slopes to transfer soil and water to terraced fields created in nearby valleys. Quechua and Aymara farmers, beginning some 3,500 years ago in the Lake Titicaca basin of Peru, gradually converted wetlands into raised fields of exceptional agricultural productivity and stability. The critical feature of all of these examples of creative destruction is that the transformed agroecosystem must be sustainable and endure for a long time.

All too often the change that people introduce does not last. The effort and intention is to create a positive result, but destruction is the major outcome. Things go wrong because people miscalculate: they may want a quick return on their investment and engage in practices that erode soil or damage fertility. They might set production estimates too high and when an unexpectedly severe drought occurs, the development efforts of a generation may be destroyed. Frequently planners calculate the profits and losses that are created within the boundaries of a development project, but fail to include gains and losses that occur away from the project site. When a large dam is constructed, it is easy to put a monetary value on the electrical power generated by turbines in the dam, the profits made from crops irrigated with water stored behind the dam, and

environment. This requires that human settlement and land use in a particular place must be based on a clear understanding of local ecology and the limits to which technology is able to modify the habitat safely.

Cultures that have grown and developed for centuries in a place have created novel ways of using their habitat successfully. They often used a combination of technology and natural forces to change the environment drastically, but what they created also lasted for a long time. They could do this because they became increasingly knowledgeable about the limits of their environment and how to operate within those limits. Think of a rubber band: you can stretch it for many purposes and it snaps back unharmed; stretch the band beyond its tensile strength and it is irreversibly destroyed. Deep experiential knowledge is built up over generations as the experiences, trial and error experiments, and lessons of living in a place are passed on from generation to generation. The technologies and

▲ Figure 1-36 **The Afsluitdijk barrier and the transformation of nature.** Completed in 1933, the Afsluitdijk barrier cut off the Zuiderzee, an arm of the North Sea, and turned it into a freshwater lake, the IJsselmeer. Large parts of this lake have since been drained. A statue of Cornelius Lely, the engineer who conceived and supervised construction of the barrier dike, commemorates his role in creating dry land from former seabed.

the savings that result from preventing floods. It is much more difficult to assess the damage inflicted on people whose villages are buried under the new reservoir, the access lost by herders whose fodder and grazing lands are converted to fields, or the increased occurrence of disease caused by mosquito and snail habitats that have expanded along with irrigation systems. The environmental impacts and financial costs that harm people and places distant from a change are **sacrifice zones** that need to be linked to their source—but seldom are!

Minimizing environmental disruption and maximizing **sustainability** (long-term survival and use) requires employing the **precautionary principle**. This rule requires that any land use involving significant change or unclear long-term consequences must be brought into existence in stages, has to proceed slowly, and needs to be examined carefully for likely impacts and available remedial measures. Open public discussion of the issues involved also must take place. Potentially damaging practices (such as excessive carbon emissions) have to be regulated and controlled to avoid damage that is too costly to repair or reverse. Using the precautionary principle acts to slow down the pace of change until consequences are better understood and practices for limiting damage exist. The precautionary principle is not a favorite policy and management tool of developers interested in a quick profit, politicians demanding quick results to present to their communities before the next election, and vested interests who expect to benefit from a project. But many projects would not be undertaken if this principle was employed. For example, nuclear power plants would have a difficult time getting built if energy companies had to assume responsibility for the cost of safe, long-term disposal of the waste material produced by every nuclear reactor.

Almost all human activity and every change that affects people involves some risk. Absolutely complete and unquestionable knowledge about the future is impossible to assemble. We are unlikely to convince everyone, for example, that climate change is taking place and that human activities are part of the cause. But waiting until everyone agrees that a problem exists would be a false use of the precautionary principle. If measures to reduce global warming are not introduced on a meaningful, world-wide scale soon, the chance to address the problem may be lost or may impose an extreme economic and social burden on future generations.

Stop & Think

▶ Why is it important to find a balance between economic development and maintaining the health of environmental resources?

Tracking Change and Managing Wastes

Keeping track of change in the global environment—whether from farms, industry, urban economies, and many other livelihood pursuits—and supporting positive economic development is a large-scale, complex activity. But in principle, it is similar in many ways to managing a household garden, a familiar small-scale enterprise. The process begins with deciding where to locate the garden and assessing the essential conditions of sun, soil (texture and nutrients), and drainage. At every stage in plant growth, the gardener must monitor changes in the state of the garden that indicate when to water, fertilize, weed, and combat pests. All of these steps are similar to the ones any farmer or manager takes to ensure healthy growth environments on a larger scale.

For example, if the fertilizer that a gardener buys from a local supplier turns out to be defective, the plants will die rather than thrive. But because the supplier is local, the gardener can identify the point source of the pollution and fix the problem. Locating the point at which polluting wastes originate is difficult at regional, continental, and global scales, but it is important to do so. Garbage, junk, and toxic substances are much easier to see, collect, and process locally (Figure 1-37). Wastes that are out of sight or a long distance away are particularly dangerous because these are easy to ignore until it is too late to stop the damage that they cause. When global population was small, it was possible to rely on natural processes to eliminate or disperse the more manageable volume of wastes. But today too much waste is produced and builds up in the air, water, and soil that sustains us. We must reduce the amount of waste that we produce, capture wastes that poison or otherwise harm the environment, recycle used materials to reduce pressure on basic resources, and design information-gathering systems that monitor environmental health at scales from the local to the global.

Reducing waste is a major issue. Our household gardener can compost organic material from food residues and garden debris for fertilizer and catch rainwater to reduce runoff and water the plants. On a larger scale we can use renovated waste water from sewage treatment plants to irrigate public green spaces, saving valuable

▲ **Figure 1-37 Seymore Green recycling van in Mayfair, central London.** Seymore Green positions a fleet of mobile electric vans curbside at various places in inner city areas. Stationary, they provide collection points in facility-deficit districts. Mobile, the vans advertise the advantages of being green.

clean water for domestic and other high value uses. Eliminating the use of materials that only break down slowly in nature is a way of keeping human waste aligned with natural processes and rhythms.

Capturing and removing wastes before they harm the environment is important. Scrubbing carbon and heavy metals from the smokestack emissions of industrial plants prevents potentially harmful substances from entering the air and from producing damaging acid rain far from the factory site. Removing unwanted carbon from the atmosphere by capturing it (carbon sequestration) by natural processes is evolving as an important environmental management tool. Planting trees or any other long-term vegetation that captures carbon as part of normal growth creates an improved habitat and helps keep the global environment in balance. Encouraging recycling reduces demand for basic primary minerals and resources as well as reduces waste. These resurrected raw materials can then be refabricated into new products that reduce the need to extract primary materials from the environment. Steel production based in part on recycled iron and steel from cars and trucks is an increasingly common technology. When scrap metal becomes a basic input, it is no longer a waste product but becomes a fundamental resource. Salt applied to roads to counter slippery conditions enters runoff and damages groundwater and the surface environment. Use of sand and the grit from mine tailings on roads instead of salt increases

▶ **Figure 1-38 Deepwater Horizon oil spill viewed from a MODIS satellite.** Sunglint reveals the partial extent of the oil spill from the offshore oil platform that was the source of the spill. Bands of oil are widely dispersed northward and eastward by surface currents.

traction; coupled with changed driver behavior in the form of reduced speed, the result is safer driving, an improved environment, and the reuse of a sterile mining by-product. Natural gas, which is often "flared off" (burned) as a waste product at petroleum wells, also can be pumped back into underground reservoirs to maintain pressure and extract larger total amounts of petroleum. The "waste" gas can also be liquefied and shipped to distant markets rather than burned. In both cases fewer gases and chemicals are released to pollute the atmosphere.

Knowing what is happening in the environment, being able to identify the direction of change, having time and opportunity to respond to bad developments and reinforce the direction of positive changes is essential to resource management that minimizes harmful side effects. This is important to undertake at all scales: local, regional, and global. The most fundamental monitoring occurs at a very local scale where changes in environmental status are observable to ordinary citizens. More formal monitoring of basic environmental parameters can be readily undertaken by local public institutions, schools, and universities. At regional and national scales observations from platforms in the air and space are needed. The spatial extent of oil spills and similarly sweeping accidents that impact environmental quality can be assessed (Figure 1-38). The extent of damaged property and ecosystem impact caused by tsunamis, tornados, and hurricanes are revealed and mitigation plans can be developed. Global scale monitoring is essential to observe changes in the status of larger ecosystems. Monitoring the extent of Arctic and Antarctic ice caps and ocean ice sheets is an important component of tracking changing coastal environments globally (see *Exploring Environmental Impacts:* Sea Level on the Rise). Based on these data, intervention potentially can take place before changes that threaten the health and well-being of environments large and small reach critical and, in the worst case, irreversible levels.

Stop & Think

➤ Explain the significance of monitoring environmental change and the need to control wastes from economic activities.

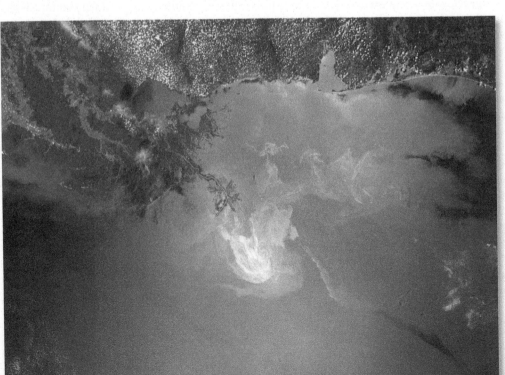

Geographic Dimensions of Development

We have defined development as an improvement of the human condition in both material and nonmaterial ways that involves a variety of sustainable cultural and economic changes. In this section we examine the specific criteria used to identify levels of development in different parts of the world. We then use these criteria here and throughout this book to describe the characteristics of the developed and less-developed regions of the world, and briefly review some of the theories employed to explain differences in development.

Measures of Wealth and Their Unequal Geographies

Several measures are traditionally used to quantify differences between the developed and less-developed countries of the world. While useful in its own way, each measure has weaknesses that prevent it from serving as an exclusive indicator of well-being.

Per Capita Income

A popular measure of development is the amount of income earned per person. In general, high levels of economic development are associated with high per capita incomes. Since the end of World War II, many less industrialized countries have shown a significant increase in per capita income. In combination with indicators such as death rates, infant mortality rates, and dietary consumption, these increases would initially appear to suggest considerable progress for residents of those countries.

A major weakness of the income measure is that it is an average and does not account for income distribution. For example, a few extraordinarily wealthy people or industries can skew upward a per capita income amount, making an area appear wealthier and more developed than it really is. A second weakness relates to the true value of income in a given country. In other words, how much can a person's income really buy? Some countries are prohibitively expensive to live in and other countries are extraordinarily cheap. Inflation can decrease effective buying power. Consequently, we increasingly use a measure called **Gross National Income in Purchasing Power Parity (GNI PPP)**, divided by population to provide a per capita figure. This measure indicates national income on the basis of "international" dollars, which take into account a country's purchasing power relative to other countries. Although countries and regions are positioned all along the income continuum, most are clustered at the lower end of the scale. A few regions have achieved the middle range; even fewer are found at the upper end of the scale. Many countries in Africa average less than $2,700; Latin America and the Caribbean about $10,760; and Western Europe more than $37,000. The United States has a figure of $47,310. Unfortunately, even when comparing purchasing power, the world still has rich countries and poor countries. More than 3 billion people remain in extreme poverty, surviving by traditional economic systems in countries where the per capita GNI PPP may be only $4,000 to $5,000 per year.

The data presented in Figure 1-39 show the various world regions and their different development experiences. These

▼ Figure 1-39 **World per capita GNI PPP, 2013.** Per capita Gross National Income is considered the best single measure of economic well-being. High per capita GNIs are closely associated with areas of high levels of technological achievement and with oil-exporting countries. Low per capita GNIs are associated with southern, southeastern, and eastern Asia and much of Africa.

Source: Population Reference Bureau, "Population, Health, and Environmental Data and Estimates for Countries and Regions of the World," *2013 World Population Data Sheet,* http://www.prb.org/pdf13/2013-population-data-sheet_eng.pdf.

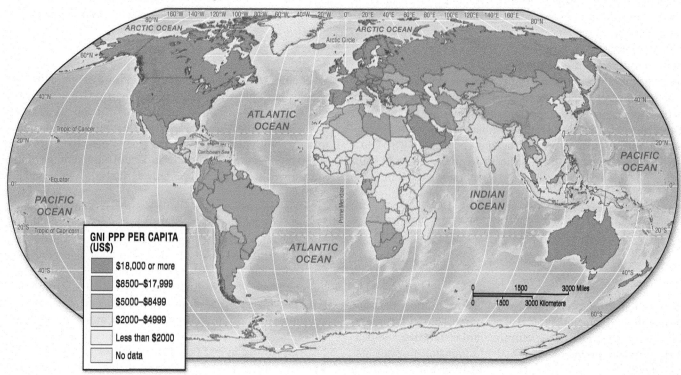

EXPLORING ENVIRONMENTAL IMPACTS

Sea Level on the Rise

At the Pacific Island Forum's Majuro Summit in the Marshall Islands in September 2013, 13 Pacific island countries as well as Australia and New Zealand signed the Majuro Declaration for Climate Leadership and presented it later in the same month to the annual United Nations assembly in New York. The Pacific islands are at the forefront of climate change threats related to sea level rise. The Marshall Islands capital, Majuro, had been submerged by high tides just months earlier. The document states each country's concrete commitment to carbon emission reductions and urges the global community to finally come together and move decisively on the issue of lowering global carbon emissions. The warming atmosphere and the resulting rises in sea level can no longer be ignored. The global community postpones dealing with the threat, both in terms of slowing the warming trend and devising adaptive solutions for increasingly violent storms and inexorably creeping sea level rise, at its own peril.

Sea level rise occurs when ice on land melts (glaciers, ice caps, and ice sheets, the Greenland ice sheet, for example) and when warming ocean waters expand. Overall global sea level rise has been eight inches since 1900, currently adding another eighth of an inch every year and accelerating with every year of additional warming. In fact, even if the whole world would stop adding warming greenhouse gases to the atmosphere tomorrow, the world would still be locked into centuries of additional sea level rise because of the length of time that these gases persist in the atmosphere. Think of a fast moving car—when you put on the brakes it takes a substantial distance for the brakes to counter the momentum and the car to come to a full stop. Both aggressive emissions reductions and bold adaptive planning are needed to cope with the situation.

Coastal areas around the world are affected, particularly major metropolitan areas. The devastation caused by Hurricane Sandy in the New York area in October 2012 foreshadows the increasing vulnerability of densely populated coastal cities from the combination of more intense storms and increasing sea level rise. A 1.6 foot (0.5 meter) sea level rise in New York City by 2070 would endanger 1.5 million people and more than $2 trillion in property values. Averaging 6.6 feet (2 meters) above sea level, New York, one of the 10 most populous cities in the world, is also the world's third most vulnerable city to sea level rise (Figure 1-4-1).

In addition to dealing with the issue of global warming, awareness, proactive disaster planning, and engineered protections will be necessary. Various protection strategies are being debated for New York City. Unlike many other major port cities, Rotterdam in the Netherlands, for example, New York has no storm surge barriers and levees. That will have to change, together with many other necessary measures. New buildings, particularly hospitals and other vital infrastructure, will have to be constructed to much more stringent building codes, for example, to withstand more frequent and intense storms and floods. While the cost of any preparedness package at $20 billion and more would be high, it is nothing against what Hurricane Sandy already cost the city—$19 billion in damages—and what future calamities will cost.

Sources: Tim Folger, "Rising Seas," *National Geographic* 224, no. 3 (September 2013): 30–59; http://www.majurodeclaration.org/.

---

experiences include the long-industrialized states of western Europe, the United States, Canada, and Japan; transitional countries such as former centrally planned economies (for instance, Hungary, Poland, and Russia) and members of what are still occasionally referred to as newly industrializing countries (NICs); and a relatively large group of countries, mostly in Africa, Asia, and Latin America, that appear hopelessly mired in underdevelopment.

Stop & Think

 Why may simply increasing per-capita income not lead to improved development in a region?

Agricultural Production

Employment in primary economic activities is a common yardstick of development. The **primary level of economic activity** focuses on extractive activities such as agriculture, mining, forestry, and fishing; the **secondary level** includes activities transforming raw materials into usable goods, such as manufacturing; and the **tertiary level** embraces the provision of goods and services. Countries in which a large part of the labor force is engaged in primary activities do not produce much income and use relatively small amounts of energy per capita. Conversely, countries with strong secondary and tertiary components usually have greater per capita GNIs and consume more energy. Use of this indicator—percentage of the labor force in primary activities (mainly agriculture)—is based on those relationships.

Countries in which primary production dominates offer limited opportunities for labor specialization, especially if the economy depends on subsistence agriculture and most efforts go toward meeting local food needs. In theory at least, labor specialization and production diversity are basic to economic growth. Prospects for high levels of individual production are diminished if workers must not only grow the crops but also process, transport, and market them, in addition to providing their own housing, tools, and clothing.

Agriculture is by far the most important primary activity. Roughly 98 percent of the world's primary-sector labor force is

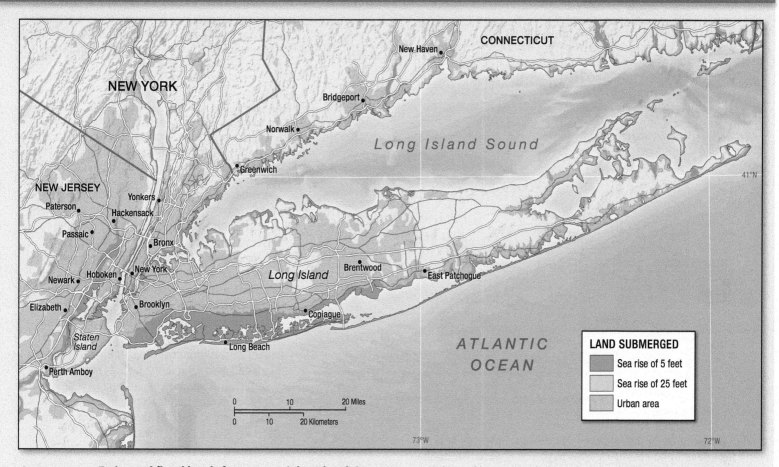

▲ Figure 1-4-1 **Estimated flood levels from potential sea level rise.** Major coastal areas like metropolitan New York and Long Island are particularly vulnerable to rising ocean waters.
Source: USGS, the National Hydrography Dataset (NHD).

involved in agriculture; about 1 percent is engaged in hunting and fishing, and the other 1 percent in mining. Labor force in agriculture (Figure 1-40) gives a good representation of the place of primary occupations at different levels of economic development. The higher the proportion of the labor force in agriculture, the lower the per capita GNI is likely to be.

Part of the explanation for the gap between more-developed and less-developed countries lies in different levels of agricultural productivity. If we use coarse grains (corn, barley, oats) as an example, the *FAO Statistical Yearbook 2012, Part 3* shows that developed countries produce twice as much per hectare as developing countries do. More-developed nations utilize technological innovations, plant high-yielding hybrid seed, and extensively apply chemical fertilizers, pesticides, and irrigation. Those technologies have not been applied as uniformly in the less-developed nations, many of which for economic and cultural reasons cannot afford and/or are reluctant to embrace these technologies.

Figure 1-41 charts changes in per capita food production. For most regions, as well as for the world as a whole, per capita food production has increased, which is no small feat when we consider continuing population growth. The gains achieved vary greatly from place to place, with the greatest increases occurring in some Asian and South American regions. Much of Africa continues to lag in its struggle to attain higher food productivity per capita. Europe's continuing decline may be attributable to its shrinking agricultural sector rather than to decreasing yields per unit of farmland.

Industrial Production

Traditionally, as employment in primary production declined, the percentage of the workforce in manufacturing increased. Consequently, the industrial workforce of a country could be used as an indicator of development. But by the twenty-first century, the industrial sector had lost ground relative to the services sector. In addition, technology had eliminated the need for many industrial workers. Employment in manufacturing is less valuable as an indicator of development than in the past but the strength of the industrial sector as a whole is still relevant.

Most less-developed countries have a limited industrial sector, the growth of which has often been driven in recent times by

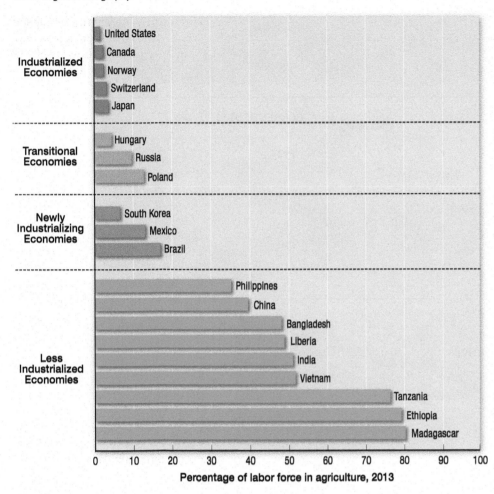

Percentage of labor force in agriculture, 2013

◄ **Figure 1-40 Percentage of labor force in agriculture for selected more- and less-developed countries as of 2013.** The percentage of the labor force employed in the primary sector, almost all of which is in farming, shows the degree of economic diversity of a nation. If a large percentage of the population is engaged in agricultural pursuits, manufacturing and services are limited in their development. Conversely, if only a small percentage of the labor force is in agriculture, manufacturing and services are well staffed.

Source: Data produced by the United Nations Food and Agriculture Organization, *FAO Statistical Yearbook 2013 (Rome, 2013) See Table 6 in http://www.fao.org/docrep/018/i3107e/i3107e01.pdf.*

use measures production in terms of power expended.

Thus, energy consumption is closely related to economic activity. Low per capita energy consumption is associated with subsistence and other nonmechanized agricultural economies; high per capita energy use is associated with industrialized societies. Intermediate levels of energy use characterize regions that have both industrialized urban centers and more traditional nonmechanized rural areas (see *Focus on Energy*: Living Off the Grid, p. 51). The distribution of per capita inanimate energy consumption is shown in Figure 1-44; its similarity to the distribution of per capita GNI (see Figure 1-39) is readily apparent.

foreign investment. In the last several decades, however, at least some less-developed countries have begun to distinguish themselves industrially, producing higher growth percentages than in already industrialized countries (Figure 1-42). Much of this industrial progress is concentrated in the small group of transitioning countries that form the upper tier of developing nations. No specific definition exists for inclusion in this group; the usual list includes Hong Kong (special status part of China), Singapore, the Republic of Korea (South Korea), and Taiwan. These have been referred to as the "four tigers" of Asia because of their dramatic industrial and income growth, and no longer should still be considered "developing." Other countries that may be included are Mexico, Brazil, and Chile in Latin America and other Asian countries such as Malaysia and Thailand. The absolute increase in production is often higher in the more-developed countries because they have a much larger industrial base to begin with.

Per Capita Inanimate Energy Consumption

One of the characteristics of the Industrial Revolution was a shift from animate power (human or beast) to inanimate energy (Figure 1-43)—initially mineral fuels and hydroelectricity and, more recently, nuclear, solar, and wind power. The degree to which a country is able to supply inanimate energy from internal sources or to import it is an important indicator of applied modern technology and of productivity. Just as per capita GNI measures productivity in terms of value, per capita inanimate energy

Other Measures of Development

Although GNI, energy use, and agricultural labor force are most often used to identify more-developed and less-developed countries, other measures are also employed. Two that indicate quality of life are life expectancy and food supply.

Life expectancy would seem to be the ultimate indicator of development. Because some cultures are less materialistic than others, standard measures of wealth may mask some important cultural attitudes of a society. All cultures, however, value the preservation of life and, at least in part, life expectancy is a measure of the end result of economic activity. It tells how well a system functions to support life by providing medical care and improved sanitation. A great range of average life expectancies characterizes the nations of the world. Due to the ravages of the AIDS epidemic and continuing poverty, average life expectancies in some southern African nations have fallen back to less than 40 years—a shocking and extremely troubling development that none would have predicted just a generation ago (Figure 1-45). At the same time, life expectancy for the entire world has climbed to 67 years, and in a number of countries has now reached the upper 70s and low 80s. Perhaps no single measure better expresses the growing gap between the more- and less-developed nations.

Two measures of food supply are also fundamental: the number of calories available is an indicator of dietary quantity,

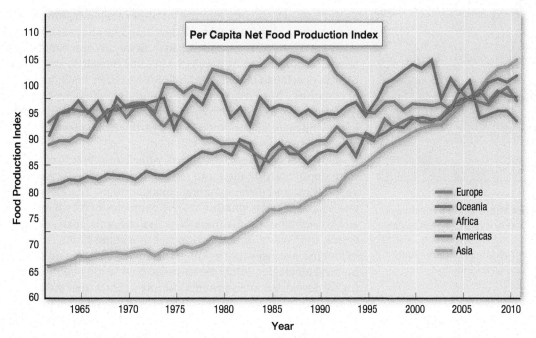

▲ **Figure 1-41 Per capita food production in major regions, 1960 to 2011.** Per capita food production has increased meaningfully throughout the period in the Americas and Asia, due in part to the Green Revolution. Despite difficulties increasing per capita productivity in some regions, there has been an increase in overall world food supplies despite continuing population growth.

Source: Data produced by the United Nations Food and Agriculture Organization (FAO), *FAOSTAT Online Statistical Service* (Rome: FAO, 2006). See also *FAO Statistical Yearbook, 2012* (Rome: United Nations Food and Agriculture Organization, 2012), Part 3, Chart 65, p. 175, *http://www.fao.org/docrep/015/i2490e/i2490e03a.pdf*; see also *http://www.fao.org/docrep/015/i2490e/i2490e03d.pdf*.

and protein supply is an indicator of dietary quality. Adequate quantity is at least 2,400 available calories per person per day. Adequate protein supply is attained if at least 60 grams of protein are available per person per day. Figure 1-46 depicts the prevalence of undernourishment by country as of 2012, where caloric intake is not sufficient.

Combined Development Measures

No single measure tells the full story of development. Many scholars combine measures to provide aggregate indices of development that, it is hoped, are more informative. One quantitative index widely used to measure and compare overall development levels of the countries and regions of the world is the **Human Development Index (HDI)**. The index is derived from three measurable variables: life expectancy at birth, educational attainment, and income, with the score for the highest possible level of development being 1.000 and the lowest 0.000. Figure 1-47 presents HDI scores for the less- and more-developed regions of the world. Among the less-developed regions of the world, Africa south of the Sahara is ranked

▶ **Figure 1-42 Percentage of value added accounted by manufacturing for selected more- and less-developed countries as of 2010.** Note that truly less-developed countries have low percentages (for instance Ethiopia), but other countries still considered less developed have unexpectedly high percentages (such as Bangladesh). See the *United Nations Statistical Yearbook 2010* (New York: Statistics Division, Department of Economic and Social Affairs, 2011), Table 20. See also *http://unstats.un.org/unsd/syb/syb55/SYB_55.pdf*.

lowest and Latin America the highest. While these figures change slightly from year to year, they are useful indicators of comparative levels of development.

Patterns of Development

Using the measures and indices just described, we can construct a highly generalized understanding of the more-developed and less-developed countries in the world. These patterns show the United States and Canada, Europe, Russia, and the Eurasian states of the former Soviet Union, Australia and New Zealand, and Japan are regions that most fully exhibit the attributes of technological development. Those world regions that collectively are characterized by lower levels of the attributes of technological development compose the less-developed world. Included in this realm are parts of East Asia, Southeast Asia, South Asia,

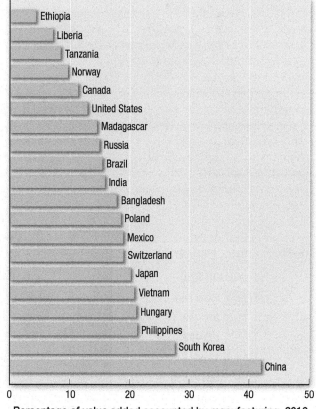

Percentage of value added accounted by manufacturing, 2010

▲ Figure 1-43 **Many modes of transport share India's roadways.** Elephants, bicycles, scooters, cars, buses, and trucks all share the same travel space. The mixture of animate and inanimate energy users reflects both pragmatism and tolerance that accommodates multiple needs in many developing nations. This scene is in New Delhi, India.

the Middle East and North Africa, Africa south of the Sahara, and Latin America and the Caribbean.

These patterns are highly generalized, so some more-developed nations—Israel, South Korea, and Taiwan, for example—are in some analyses lumped with less-developed countries; and some less-developed nations—such as Albania, Uzbekistan, and

Kyrgyzstan—are included in the more-developed regions. Some countries are classified as more developed by all measures, and many are classified as less developed in all ways. A number of countries fall among the more developed in some categories and among the less developed in others, with China and India leading a group of countries whose recent rate of economic growth places them at the forefront of a cluster of transitional countries generating rapidly improving material conditions for large, but by no means all, segments of their populations.

One final comment is required. After World War II countries were unofficially divided into separate "worlds," reflecting both geopolitical and developmental considerations. The First World was a Cold War designation for the countries of North America and Western Europe, Australia, New Zealand, and Japan—in other words, the "West," or those countries that stood as the front line of defense against exportation of the Soviet Union's and China's political ideologies. While the so-called First World countries were highly developed, their geopolitical role defined their status. The Second World referred to the old Cold War "enemies" of the West, particularly the Soviet Union, its client states of Eastern Europe, and the People's Republic of China. The former Soviet Union was a significant industrial power in its own right, particularly in terms of heavy industry, and could only readily be distinguished from the First World if one focused on differences in economic philosophy—capitalist versus communist. The term Third World was constructed

▼ Figure 1-44 **Total Primary Energy Consumption per Capita (Million Btu per Person) as of 2010.** Per capita energy consumption is a measure of development that indicates the use of technology. Because most forms of modern technology use large amounts of inanimate energy, countries that use small amounts of that type of energy must rely principally on human or animal power.

Source: Retrieved from U.S. Energy Information Administration, Independent Statistics and Analysis, 2010.

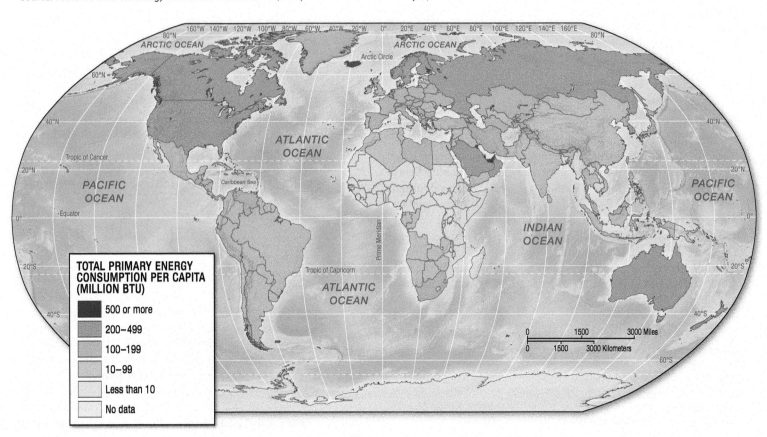

TOTAL PRIMARY ENERGY CONSUMPTION PER CAPITA (MILLION BTU)

- 500 or more
- 200–499
- 100–199
- 10–99
- Less than 10
- No data

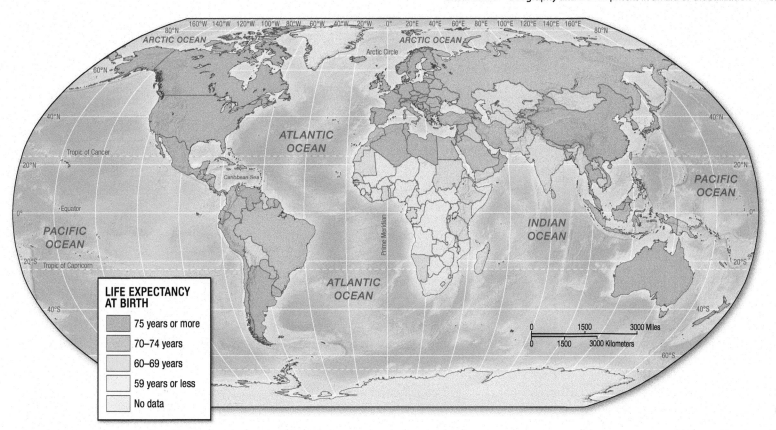

▲ **Figure 1-45** **World life expectancy at birth, ca. 2013.** Life expectancy is a measure of how well a nation is able to care for its population. Long life expectancy is closely associated with the other indicators of developed nations.

Source: Population Reference Bureau, *2013 World Population Data Sheet*, "Population, Health, and Environmental Data and Estimates for Countries and Regions of the World," *http:// www.prb.org/pdf13/2013-population-data-sheet_eng.pdf.*

almost solely in response to economic considerations to refer to those countries in Latin America, Africa, Asia, and the Middle East that were viewed from a Western perspective as less developed or economically distressed. Some scholars added a Fourth World to include countries in which economic conditions were especially desperate.

With the collapse of the Soviet Union and the end of the Cold War, a division of the globe into "worlds" makes less sense. Many so-called Second World countries of what used to be called Eastern Europe—countries that were tied to the Soviet bloc—now have political and economic relationships with the West that make them very much a part of the First World; many have become members of the European Union, including the Baltic states of Estonia, Latvia, and Lithuania once part of the Soviet Union. These terms have outlived their usefulness, but are still in use; the term Third World, in particular, remains a popular synonym for countries with lower levels of economic development.

To avoid the Cold War connotations of the various "worlds" described above, scholars have found new ways to group countries based more strictly on developmental conditions. For example, terms such as industrialized (United States), transitional (Vietnam), newly industrializing (Brazil), and less industrialized (Ethiopia) are used. In some cases, we see expressions such as old industrial countries (OICs) juxtaposed with newly industrializing countries (NICs), with the clear connotation that the former have grown old and stodgy and that the latter are young and dynamic.

Summary Characteristics of More-Developed Regions

There are several characteristics of more-developed countries: high per capita GNI and energy use, a small part of the labor force in primary activities, an emphasis on secondary and tertiary occupations, longer life expectancy, and a better and more abundant food supply. More-developed countries also tend to have a lower rate of population growth, which is primarily the result of having a highly urbanized population for many generations. To a large degree, those countries are found within Stage IV of the demographic transition. In addition, more-developed countries share other economic and cultural characteristics.

The most basic economic characteristic of the more-developed world is a widespread use of technology. In those regions, the fruits of the Agricultural, Industrial, and Information Revolutions are widely applied, and new techniques are quickly adopted and diffused. New technologies create new resources and expand existing ones, thus increasing opportunity for still greater generation of wealth. These technologies also lead to increased labor productivity and improved infrastructure: roads, communications, energy and water supply, sewage disposal, as well as credit institutions, schools, housing, and medical services. Those support facilities are necessary for accelerated economic activity and for specialization of production, which further enhances productivity.

The heavy dependence of the more-developed countries on minerals further differentiates them from the less-developed countries. Not only are industrialized countries in an iron-and-steel

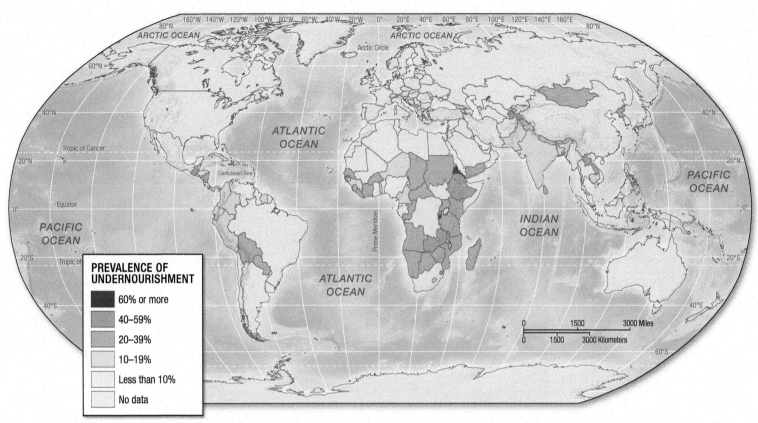

▲ **Figure 1-46 Prevalence of undernourishment by country as of 2013.** Food supply is a measure of well-being in the most basic sense. Daily caloric intake varies greatly according to level of national economic development.

Source: Data obtained United Nations Food and Agriculture Organization, *FAO Statistical Yearbook 2013*, Part 2: Hunger Dimensions, Table 12. See *http://www.fao.org/docrep/018/ i3107e/i3107e02.pdf*.

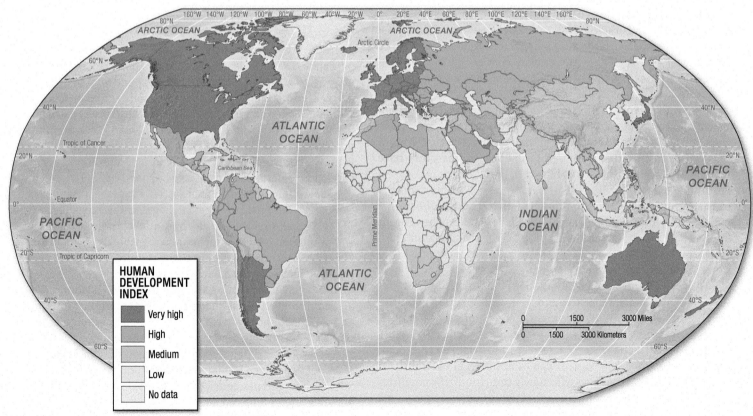

▲ **Figure 1-47 Human Development Index values by country as of 2013.** At a global scale, the HDI conveys a useful impression of national development status. But general national progress masks populations and regions in each country that exceed or fall short of the national norm.

Source: UNDP Human Development Report, 2013.

FOCUS ON ENERGY

Living Off the Grid

Imagine living in a place without electricity. For most of the urbanized world, and for almost all rural areas in rich countries, this is just a bad dream. But for one fifth of the world's population no electricity is a fundamental reality of life. In much of the world, vast districts are dark at night because these areas are totally unconnected to the grid or are very unreliably linked. Without reliable light after sunset, businesses must close, household production of craft goods stops, and students are unable to read and study. Productivity is limited, and the improvement of a difficult material situation is blocked.

For many people supposedly connected to the grid, two problems reduce the significance of that breakthrough: the reality of connectivity and its reliability. Many have attained only a theoretical connection to modern power and illumination. In India, bureaucratic definitions of connectivity consider a village to be linked if 10 percent of its households have electrical service. This virtual connectivity may be great for statistics of bureaucratic achievement, but is a disaster for those with insufficient resources or political clout to realize the latent reality of near connection. Left behind as connectivity funds are shifted elsewhere, grid linkage becomes even more difficult to attain. For those who are connected, overload breakdowns and rolling blackouts reduce the reliability of linkage to the grid. Power failures and cutoffs occur sporadically, eliminate power for hours at a time, and introduce into electricity use a factor of randomness that is exceedingly disruptive and frustrating.

The solution is to develop disaggregated systems of energy production and storage. These systems improve the quality and material prosperity of life precisely because they can be placed at sites where disadvantaged and disconnected people are located. The energy these systems generate is distributed to consumers in their immediate vicinity. Solar power provides one of the best options for success. Solar panels and cells have become available in all sizes and can operate efficiently in most environments, but work best in sites with maximum sun exposure. With locally generated electricity, people can replace dangerous and health-threatening kerosene lanterns with rechargeable lamps, swap wood-burning stoves or open dung fires for solar cookers (see front cover), keep stores and workshops open longer hours, acquire cell phones whose towers are powered by larger but local solar collector arrays, recharge their phones with energy produced by household solar panels, and provide an after-dark reading environment for students (Figure 1-5-1). Better, brighter work and home environments mean better health, increased output, greater earnings, and more opportunities for investment in productive local enterprises.

The search for renewable energy sources that are sustainable and cost-effective will characterize the twenty-first century. Fossil fuels rule the global energy mix today, as they have since the beginning of the Industrial Revolution. But their dominance will increasingly diminish throughout the century because of their finite nature and damaging environmental impact. The International Energy Agency (IEA) estimates that renewable energy sources will provide a third of total energy production in two decades. Specifically directing a substantial proportion of that shift in energy sources toward the 1.3 billion people currently without access to electricity would go a long way toward making a significant improvement in people's lives.

Sources: International Energy Agency (IEA), *World Energy Outlook 2012.* Paris: OECD/IEA, 2012; "Out of the Gloom," *The Economist,* July 20, 2013, *http://www.economist.com/news/asia/21582043-villagers-enjoy-sunlight-after-dark-out-gloom*; "Waiting for the Sun," *The Economist,* April 28, 2012, *http://www.economist.com/node/21553480.*

▼ Figure 1-5-1 **Solar power in isolated areas.** In this Tibetan village, completely without electricity from the grid, solar power is the only modern power source. At the Ge-Sang Solar store, bringing light, and delight, to fellow villagers is big business.

age, but they are also in a fossil-fuel age, a cement age, a copper-and-aluminum age, a silicon age—the list is almost endless. To a greater and greater degree, more-developed countries are importing minerals from less-developed countries and then exporting manufactured goods and, in some cases, food. That practice has led to trade surpluses for many of the more-developed countries and trade deficits for most of the less-developed countries because the price of raw materials (with the exception of petroleum) has increased more slowly than the price of manufactured goods. More-developed countries gain even more revenues by investing capital in less-developed countries. The resulting flow of wealth out of less-developed countries has led some of these countries to nationalize or restrict foreign investments.

New technologies, high productivity, and a favorable trade balance result in higher personal and corporate incomes. Consequently, many individuals in more-developed countries spend only a part of their income for food and shelter; other income can be used for services, products, and savings—where wealth stimulates further economic growth. For entrepreneurs, more-developed countries offer numerous infrastructural advantages.

Cultural characteristics play a major role in development. One cultural attribute of more-developed nations is a high level of educational achievement (Figure 1-48), enhanced by advanced communications technologies that support the rapid diffusion of new ideas. Education is geared partly toward economic advancement; disciplines such as engineering, geography, economics, agronomy, and chemistry have obvious and direct applicability to resource development. An educated populace is an extremely important resource, for it more readily accepts change. Acceptance of change

means that technology is more easily adopted, new products and services are welcomed, and the mobile population can more readily take advantage of opportunities in other areas of the country. Other cultural attributes common to the more-developed countries include a large and economically strong middle class and an expectation that government officials will be honest and accountable to the public for their actions.

Summary Characteristics of Less-Developed Regions

We have already learned that less-developed nations are characterized by low per capita GNI and energy use, a high proportion of the labor force in primary pursuits, a comparatively short life span, and a diet often deficient in quantity and quality. In addition, less-developed countries commonly display a variety of other characteristics.

Most less-developed countries have a relatively high rate of population growth, the result of a continuing high birthrate and a declining death rate. As a consequence, the age structure of the population is different from that of a more-developed country. In less-developed countries, a substantial portion of the population is youthful and only a small segment is elderly. In this sense per capita comparisons are unfair because all people are counted equally, yet in the less-developed nations, a smaller proportion of the population is made up of mature adult workers. A large youthful population also foretells continued population growth.

The status of women in many of the diverse cultures of less-developed countries often excludes them from many occupations (see *Geography in Action*: Gender Inequality as a Barrier to

▼ **Figure 1-48 Literacy by country as of 2010.** Literacy is an increasingly widespread skill, with two-thirds of the population of most countries being able to read and write

Source: Data collected from CIA World Factbook and http://en.wikipedia.org/wiki/List_of_countries_by_literacy_rate.

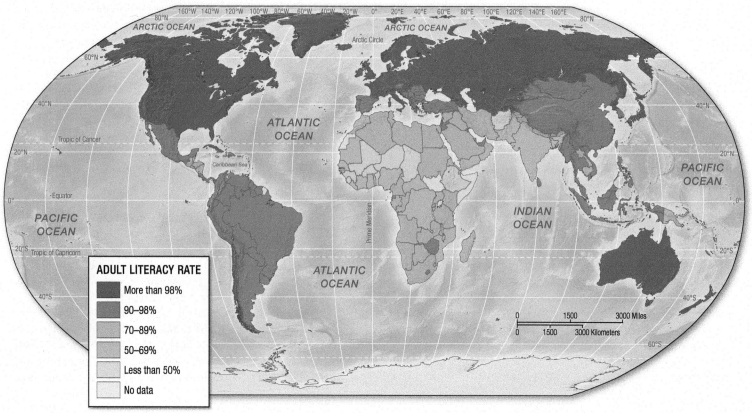

Development). In the rural sector, women are often expected to work with men in the fields during periods of peak labor requirement while still maintaining the home and caring for the children. While at home, women may engage in craft industry such as weaving for household use and for sale in the local market. In addition, many rural women in the less-developed world play a pivotal role in marketing the family's surplus on a daily or weekly basis (Figure 1-49). In the cities, women find employment as domestics, secretaries, teachers, nurses, and, more recently, industrial workers. For many illiterate women, however, nondomestic employment opportunities are not available.

The literacy rate in less-developed countries is generally lower than in more-developed nations. Many older rural inhabitants of developing nations cannot read or write or can do so only at minimal levels. Although literacy campaigns have made great strides in cities and among the younger generations in recent decades, large numbers of urban poor remain functionally illiterate. The inability to read limits great numbers of people from learning new technologies and finding better-paying jobs.

The cultures of many less-developed nations are also often more conservative and resistant to change. This presents a fundamental paradox. Those cultures, by and large, wish to preserve their customs and mores, yet want the material benefits affluent societies enjoy. Clearly, economic development leads to cultural change, to the possible destruction of at least some traditions, and to the acquisition of new behavior patterns. Sometimes the family loses part of its cohesiveness, villages become subservient to larger urban centers, labor specialization and regional specialization lead to commercialization of the economy, and loyalties to the local community give way to national allegiances. These changes are often inevitable when economic development occurs. Consequently, social and cultural disruption is a common characteristic of less-developed countries, as old and new ways of life come into conflict.

▼ **Figure 1-49** **Street vendor in Speightstown, Barbados.** Selling food commodities in developing countries is often a female occupation. The weekend market in Speightstown, the second largest town on the Caribbean island of Barbados, usually has the most vendors and buyers as well as best vegetables..

Lower levels of development often are considered to be associated with specific attitudes and personality characteristics. One attribute is a distrust of change and the people who promote change. Subsistence agriculturalists, for example, must avoid miscalculations in their crop production for fear of starving. They are understandably reluctant to take a risk on a new idea or procedure that could lead to crop failure with devastating consequences. A second characteristic is fatalism about the world. A common belief is that lives are controlled by outside forces, such as a vengeful god or nature's wrath. The ability to empathize is sometimes absent—how can the subsistence farmer ever dream of doing something else? Yet the steady flow of migrant populations from rural countryside to urban areas would seem to challenge this presumed inability to imagine a different life. Third, thinking tends to be ethnoscientific rather than technoscientific in character. Understanding the connection between a physical cause and a physical effect is based on the empirical observations of generations of resource users, which are passed on as part of a slowly evolving oral tradition. Often the understanding of the cause of a problem is partial or wrong from the standpoint of modern technoscience, but makes perfect sense from a local perspective. Indigenous farmers may recognize several different soil types in a field and plant their crops accordingly using hand technologies; the modern soil scientist may see only one soil type fit to grow one commercial crop using mechanized equipment. Putting the two worldviews together in productive ways is extremely difficult.

Finally, the stereotypic conservative, slow-to-adapt, change-resistant person from a less-developed district is assumed to have a restricted worldview and limited aspirations, and lacks innovativeness and the ability to defer gratification. In fact, few people living in the developing world embrace all, or even some, of these traits. In the local context there is often a perfectly rational explanation for what seems to the outsider to be unreasonable behavior. Because local situations may often make risk-taking and innovation-adopting behavior difficult or impossible, reluctance to change can create even more challenges in the effort to promote development.

Although we generalize about the more- and less-developed worlds, individual countries are not so easily classified. The progression from less developed to developed occurs on a rough continuum, and the real world is more complex than suggested by a two-world model. The classification problem is especially difficult in the twenty-first century as more countries acquire characteristics of both worlds. China and India are hailed as models of capitalistic progress, and boast booming economic regions with growth rates that can only be imagined elsewhere in the developed world. Yet both countries are still plagued by widespread poverty, unemployment, and many of the other conditions that characterize the developing world. Nicaragua has a highly literate population that shops in beautifully appointed and abundantly stocked malls and displays all the trappings of the developed world. Yet these same people compete with donkey carts for space on the highways. While economic and social inequities have always been present, their

GEOGRAPHY IN ACTION

Gender Inequality as a Barrier to Development

People are the greatest resource of any country or region, and it is important that the status and treatment of women, more than half the world's population, be considered in assessing quality of development. Women continue to be the targets of discrimination and prejudice in many societies. In the past, women were treated as little more than the personal property of men, to be acquired, used, and discarded at the whim of the husband, father, or other male family members.

Attempts to limit the development potential of women remain widespread even today. In many traditional societies, women are prohibited from holding offices of authority, and females worldwide occupy only 20 percent of the seats in national parliaments or congresses. Although much progress has been made in recent decades, education has often been denied to girls by male authorities who fear that knowledge removes women from their traditional home-centered lives. Literacy rates among adult women are far lower than those of their male counterparts in many less-developed countries, and men continue to be given preference in admissions to trade schools and universities. Because women in many countries are treated legally as minors in the care of their husbands, they are also subject to laws that limit or even prohibit their ability to travel, file for divorce, inherit land, or secure loans and other financial services.

Most women in less-industrialized countries have traditionally worked in the informal economic sector—jobs such as street and market vending, small business operation, and domestic work not covered by national labor and employment laws such as cooking, cleaning, and child care. Limited in their earning potential, many women have no hope for economic advancement. Even those fortunate enough to receive an education and secure good career positions tend to be paid less for the same work than their male counterparts. The majority of the world's adults living in poverty are women, a condition referred to as the feminization of poverty. Because women are the primary caregivers to children, this situation has grave implications worldwide for child health and nutrition. Another consequence of these trends is the so-called female double-day workload in which women are expected to do their normal levels of cooking, laundry, cleaning, and child rearing after putting in a full day of work outside the home.

The United Nations Development Program's Gender Inequality Index (GII) assesses the disparity between men and women in 148 countries, taking into account women's reproductive health, female empowerment (as reflected in share of parliamentary seats and higher educational attainment), and participation in the workforce. National GII rankings range from 0 (full gender equality) to 1 (total absence of gender equality) (Figure 1-6-1). Although there are differences between GII rankings and those of other measures of development, the broad regional patterns are similar. Many technologically less advanced countries of Africa, Asia, and Latin America also rank lower in gender equality. This suggests a correlation between economic development and development in other areas. Less-developed countries will never achieve their full development potential until they remove the barriers to female opportunity and advancement.

Source: United Nations Development Program (UNDP), "Gender Inequality Index (GII)," *Human Development Reports*, 2013, *http://hdr.undp.org/en/statistics/gii/*.

---

prominence and visibility has risen along with the new global economy. Change in the global economy continues to be an uneven spatial and social process, both among and within regions.

Selected Theories of Development

To understand and explain the characteristics and global dimensions of development, we also need to examine some of the major theories scholars advance to help us understand why development occurs in some cases and is especially difficult to achieve in others. A vast literature attempts to explain development; this is a brief summary of some of the more prominent theories. No theory is accepted by everyone, and some are strongly opposed, but their diversity offers insight into the nuances of development. The theories and policies associated with what is loosely called "neoliberalism," which until the Great Recession of 2008 was ascendant nearly everywhere, are discussed earlier in the context of globalization and are omitted here.

Some of the earlier and more colorful theories we can summarize as "control" or "deterministic" theories. Environmental determinism is based on the premise that the physical environment, especially climate, controls or predetermines human behavior and levels of development. If a country is too cold, for example, it can never expect to be developed. Cultural determinism uses the same logic, with cultural characteristics claimed as the control. Adherents of cultural determinism might assume that a country was beyond help because of an antidevelopment religious tradition. While the environment and religion may play a role in the development process, it is almost never a determining role. Other theories focus more on colonialism and trade. Mercantilism was a policy employed by colonial powers from the sixteenth century. Colonies were permitted to trade only with the mother country; this ensured the prosperity of the mother country at the expense of the colonies. Spain was especially artful in the use of this policy. Colonialism and neocolonialism refer to modern policies that appear to privilege past colonial powers by reinforcing the economic dependency of former colonies on the major industrial powers for capital, skills, and processed goods. In return, the former colonies remain suppliers of raw materials. This dependency model, derived from the world-systems theory

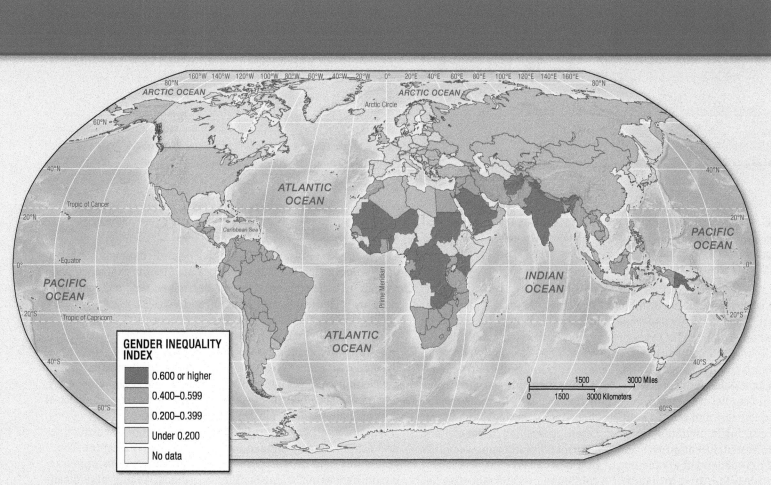

▲ Figure 1-6-1 **The Gender Inequality Index (GII).** Gender inequality remains a major issue in many parts of the world.
Source: UNDP Human Development Report, 2013.

of Immanuel Wallerstein,[1] argues that the nations of the world can be divided into two groupings: a "core," which consisted originally of the countries of Western Europe, and a "periphery," comprised initially of Africa, Asia, and Latin America. As with dependency theory, the core-periphery model holds that the economic development of the industrialized core has been, and continues to be, achieved through trade relationships that work to the disadvantage of the less-industrialized regions of the periphery (Figure 1-50).

Still other theories focus more on the specific circumstances involved in the development process. Circular causation theory holds that development succeeds or fails depending on the inputs available over time. An example is the downward spiral in a farm family that produces barely enough to feed itself. With little or no savings, this family has nothing to fall back on if a minor crop failure reduces the harvest. Its members simply have less to eat, so they work less, produce less, and then have even less to eat. Perhaps providing fertilizer or a new tool would be the inputs necessary to reverse the downward

spiral and create an upward spiral. The geographer Lakshman Yapa[2] goes beyond specific economic circumstances in the development process and argues that lack of material well-being among the poor is "socially constructed," brought on by people and their governments who are influenced by factors such as values and language. These decision makers focus too much on improving the "poverty sector" through economic means: investment, job creation, and general income growth, with little attention paid to technical, social, cultural, political, ecological, and academic factors. Yapa suggests that "development" is a far more complex process, with a different starting point than most development experts are willing to concede.

Economist Walt Rostow[3] looked at historical economic data and proposed an economic development model consisting of five **stages of economic growth**. The first stage is traditional society in which most workers are engaged in agriculture, have limited savings, and use age-old production methods, all characteristics of a

[1]Immanuel Wallerstein, *The Modern World-System: Capitalist Agriculture and the origins of the European World-Economy in the Sixteenth Century* (New York: Academic press, 1974); *The Capitalist World-Economy* (Cambridge: Cambidge University Press. 1979).

[2]Lacshman Yapa, "What Causes Poverty? A Postmodern View," *Annals of the Association of American Geographers* 86 (1996): 706–728.
[3]Walt W. Rostow, *The Stages of Economic Growth: A Non-Communist Manifesto.* 2nd ed. (Cambridge: Cambridge University Press, 1971).

▲ **Figure 1-50 Multinational corporate influence in the global economy.** Even in the heart of Buenos Aires, where consumption of prime beef is a national obsession, the international fast food industry is able to penetrate and establish a foothold.

truly poor society are exhibited. "Preconditions for takeoff" are established in the second stage, initiated internally by the desire for a higher standard of living or externally by forces that intrude into the region. In either case, production increases enough to cause fundamental changes in attitudes and alter individual and national goals. Takeoff occurs in the third stage, when new technologies and capital are applied and production is greatly increased. Manufacturing and service activities become increasingly important, resulting in migration from rural areas to bustling urban centers. Infrastructure improves and expands, and political power shifts from the landed aristocracy to an urban-based elite. The fourth stage is the "drive to maturity," a continuation of the processes begun in the previous stage. Urbanization progresses, and manufacturing and service activities become increasingly important. The rural sector loses much of its population, but those who remain use

mechanized equipment and modern technology to produce large yields. The final stage is one of high mass consumption: personal incomes are high, and abundant goods and services are readily available. Individuals no longer worry about securing the basic necessities of life and can devote more of their energies to noneconomic pursuits.

Stop & Think

➤ Name five countries that you think belong to the global "core" and five that are found in the global "periphery."

World Regional Geography: A Development Approach

The concepts, perspectives, and background about environment, development, and globalization presented in this chapter provide a gateway to the regional chapters that follow. Each regional chapter is organized around an initial section that systematically develops the environmental and historical characteristics of that region. And each chapter has a unique story to tell in which regional combinations of common factors have produced highly individual development trajectories. Country studies and subregional groupings within chapters add diversity to the larger regional tapestry, and provide a composite image of the variable patterns and pace of regional paths to development in a globalized world.

In our analysis of world regions, keep in mind the principles and explanations provided in this introductory chapter. These guidelines help us recognize the highly variable developmental circumstances of the globe's different regions. We may periodically revisit the introductory chapter for reminders about concepts and processes that enable us better to understand real-world geographical situations.

Summary

▶ In this chapter, our concern is to apply the geographer's approach to understanding what development involves within a world region framework; to understand the revolutionary kinds of change that reshape societies and economies, with special emphasis on globalization; and to identify the growing need to guide change so it is sustainable and beneficial.

▶ Globalization plays an increasingly significant role in the development process, although the blending, culturally transforming effects of globalization should not be overstated. We also examined how long-term development requires a commitment by people to understand the environments of which they are a part, and to act creatively in directing the impulses that determine whether human habitats will be carelessly degraded in the interest of short-term gain or planned in favor of long-term benefit. The combination of landforms, climate, vegetation, soils, and resource endowments create positive places that favor development. Humans tend to concentrate in these places, and over time these

places reflect millennia of cultural diffusion and inheritance from initial culture hearths. They also reflect a continuing accumulation of technological innovations and cultural practices that shape the development settings of countries and regions.

▶ Sustainable development promotes change that leads to improved well-being in people's lives, takes into account the needs of future generations, is based on principles of stewardship, and is compatible with local cultural and environmental contexts. Sustainable use of the planet in support of peoples' lives—environmental stewardship—requires knowledge of the physical and cultural setting of development.

▶ Per capita income, agricultural employment, energy use, and other measures are potential indicators of development, but each measure has drawbacks. Although a number of theories exist to explain the development process, each is limited in its explanatory power and none is a complete explanation of development.

Key Terms

acculturation 36
Agricultural Revolution 7
agricultural surplus 7
alluvium 25
birthrate 11
calcification 25
climate change 28
creative destruction 40
culture complex 32
culture hearth 32
culture realm 32
culture trait 32
death rate 11
demographic transition 11
development 5
ecosystems 21

environmental stewardship 24
evapotranspiration rate 23
fossil fuels 26
friction of distance 8
frost-free period 23
globalization 14
global warming 20
greenhouse gases 34
Green Revolution 8
Gross National Income in
   Purchasing Power Parity
   (GNI PPP) 43
Human Development Index
   (HDI) 47
Industrial Revolution 8
Information Revolution 9

land degradation 39
landscape modification 21
laterization 24
lingua franca 37
loess 25
Malthusian theory 12
neoliberal policies 14
neo-Malthusians 13
physiologic density 10
podzolization 25
population density 10
population distribution 10
population growth rate 11
precautionary principle 41
primary level of economic
   activity 44

renewable energy 28
sacrifice zones 41
sea level rise 44
secondary level of economic
   activity 44
stages of economic growth 55
sustainability 41
tertiary level of economic
   activity 44
transnational capitalist class 18
transnational corporation
   (TNC) 15
vegetation succession 23
world cities 18

Understanding Development in an Era of Globalization

1. Why is the term "development" often difficult to define and explain?
2. How has the development process been facilitated by several major human transitions? In what ways have these transitions contributed to the improvement of the human condition?
3. How is population change related to the development process?
4. What is the globalization process and who are its major "players"?
5. Why are river valleys and coastlines desirable locations for many economic activities?
6. Can technology expand the limits of land use in a given environment?
7. How are global warming, melting ice sheets, and sea level rise linked?
8. What measures have been used to define development? What are their strengths and weaknesses?
9. What theories do geographers draw on to explain how the development process works?
10. In what ways does gender inequality affect development?

Geographers @ Work

1. You have been retained by a government to design a national development plan. Use your knowledge of geography to give your plan a character that is different from one created by an economist or engineer.
2. You want to become a member of the "Transnational Capitalist Class." Design a strategy that will take you to this goal.
3. Plan a tourist resort that will draw visitors but not overwhelm and destroy the natural and cultural assets that are the region's primary attractions.
4. Devise strategies that city planners might use to cope with storm and flood hazards.
5. An international agency wants you to think "outside the box" to find ways to lift a country out of poverty. List your recommendations.

MasteringGeography™

Looking for additional review and test prep materials? Visit the Study Area in MasteringGeography™ to enhance your geographic literacy, spatial reasoning skills, and understanding of this chapter's content by accessing a variety of resources, including MapMaster® interactive maps, videos, RSS feeds, flashcards, web links, self-study quizzes, and an eText version of *World Regional Geography*.

United States and Canada

Merrill L. Johnson

Canadians Are Really Americans ... Eh?

The United States and Canada share a continent and the longest unguarded border in the world. Both countries have a similar political history as British colonies, though Canadians are quick to point out that their separate status evolved peacefully rather than through a bloody revolution. And yes, the Canadian and American peoples see themselves as great friends.

But Canadians do not view themselves as simply Americans who wear heavier coats in winter, have stricter gun control laws and universal health care, or who call their dollars by strange names (dollar coins in Canada are called "loonies" and two-dollar coins "toonies"). Canadians have a different culture. This difference can be seen in a variety of ways, most notably in the distinctive Canadian accent (even French speakers have a North American accent). Americans pronounce "schedule" as "skedule" but for Canadians it is "shedule." Americans say they have "been" somewhere, pronouncing it "bin," whereas Canadians say "bean." And there is the ever-present "eh" at the end of an expression—as in, "He's a real charmer, eh?"—which has an affirming quality that is roughly similar to the American "ya know?," or "right?" And it is not just a matter of accent. Words have different meanings. The American "sofa" in Canada is a "chesterfield." Canadians go on holiday; Americans take a vacation. Canadians light their homes thanks to "hydro"; Americans use electricity or "power."

The differences extend beyond accent and word usage. Canadians have a higher median family worth than Americans; they spend proportionally less on housing and cars, but more on alcoholic drinks; they work fewer hours per week and take longer vacations; they are less inclined to get married and wait longer when they do; they are far less likely to say that religion is important to their lives. Canadians pay much higher taxes (there is a steep national sales tax) but they spend half as much per person on health care; life expectancies are higher and obesity rates much lower; gun murders per 100,000 people annually are only a fraction of those in the United States due to highly restrictive gun laws, but arson, break-and-enter, and auto theft rates are higher. There is also a perception (probably unfair) that Canadians are less entrepreneurial and exciting than their southern cousins.

It is important to remember that Canada and the United States are different countries. While many generalizations pertain to both countries (particularly in terms of physical geography), people in one country should be careful about stumbling into convenient judgments and stereotypes about people in the other country.

Read & Learn

- Describe the general differences and similarities between the United States and Canada in terms of culture and history.
- Organize correctly the location of the general landform regions in Canada and the United States.
- Locate climatic regions in the United States and Canada, and explain the general processes involved in their formation.
- Explain major environmental challenges facing this world region, from water resources to the impacts of resource development.
- Outline the territorial evolution of what is now the United States and Canada.

- Compare the benefits and costs of the various forms of energy production in Canada and the United States.
- Relate the geographies of agriculture and manufacturing in the United States and Canada to the larger processes of globalization and economic restructuring.
- Explain the meaning of diversity to the population geographies of Canada and the United States, and how this diversity serves as both an advantage and disadvantage to the development process.

▲ Waterton Lakes Park in Alberta and Glacier National Park in Montana form a binational park that is a UNESCO World Heritage Site and Biosphere Reserve. This view across Middle Waterton Lake shows the Prince of Wales Hotel against a mountain backdrop dusted by an early snowfall.

Contents

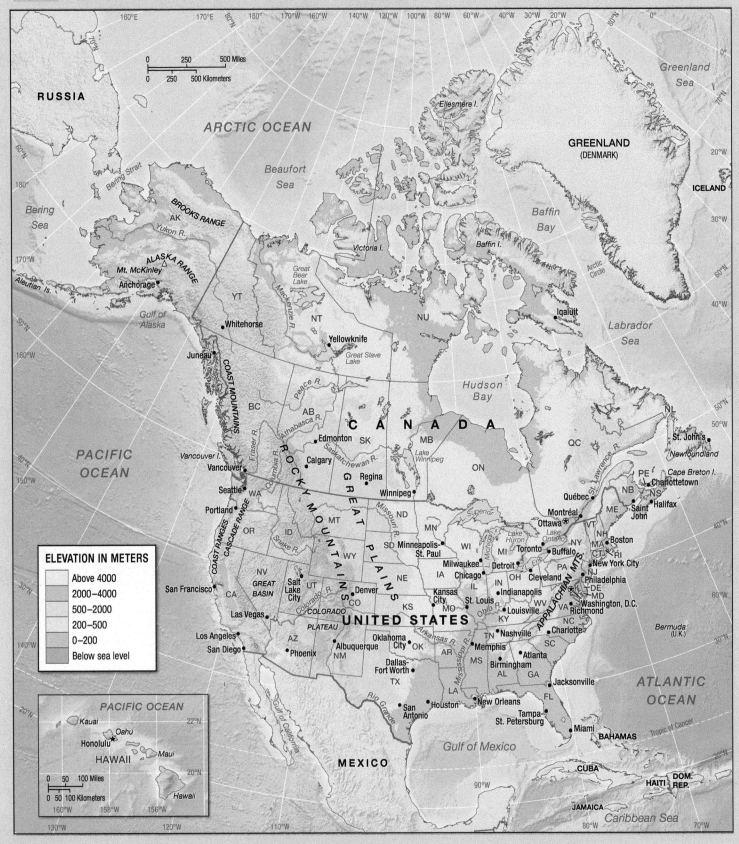

Figure 2-0 **The major physical features of the United States and Canada.**

In the process of settlement and development, the United States and Canada have marshaled the rich resources of an enormous domain. As a result, the two countries are among the most prosperous and most highly developed in the world, and we examine the physical, territorial, and demographic contexts in which these two countries evolved. An analysis of their respective territorial evolutions features the importance of immigration, since both countries are populated by immigrants who have shaped the economy and ethnic composition of their nations. Throughout, we focus on issues related to economic development and globalization, including the spatial dimensions of agricultural change, industrial development, and urban evolution.

The highly productive agricultural systems of both countries emerged from interaction with a rich resource base, but are experiencing major adjustments in the twenty-first century. Industrialization in this world region occurred first in a concentrated core, but now is increasingly dispersed. Today outsourcing of production has given North American industrial output and consumption a highly international character. Industrialization also promoted first widespread urbanization and then suburbanization, processes that pose new challenges in an era of economic globalization.

Neither economic development nor globalization occurs without social and cultural costs. Income disparities, economic stagnation in regions bypassed by growth, the difficulties experienced by minorities, neighborhood ethnic change, and the often bumpy relationships between economically interdependent neighboring countries create serious disagreements about policies to generate social and economic justice for all citizens. Both countries face complex challenges, but there is reason to hope they can be overcome.

The United States' and Canada's Environmental and Historical Contexts

Together, the United States and Canada cover more than 7.5 million square miles (19.4 million square kilometers), an area that exceeds 14 percent of the world's land surface. Canada is the larger of the two by more than 130,000 square miles (336,700 square kilometers), but much of this territory lies in physically harsh, high-latitude environments poorly suited for settlement. Yet Canada's population is scarcely more than one-tenth that of the United States. Both countries have highly diverse physical endowments and human legacies that have influenced North American political, settlement, and economic patterns. We examine the significance of these endowments in creating the physical and human contexts for development.

Environmental Setting: The Lay of the Land and the Geography of the Climate

Taken together, Canada and the United States embrace virtually every type of landform and every climate classification that exists (with the exception of truly tropical climates). The environmental setting provides a physical framework in which the economic development process could take place. In some cases, this framework has offered great advantages, as in the Corn Belt of the U.S. Midwest; in other cases, it has presented nearly insurmountable obstacles, as in the high latitudes of Canada and Alaska.

Landforms

The United States and Canada lie on the North American continent, a crustal mass that includes Mexico and Central America. For cultural rather than physical reasons, the United States and Canada are treated as a separate world region—they derive mainly from British political origins, whereas Mexico and Central America have primarily Spanish roots. But many generalizations made about the physical structure of the United States and Canada also pertain to Mexico and Central America.

Our starting point is the great **Canadian Shield**. A shield is a piece of the earth's crust that is very old and geologically very stable—a continental nucleus—around which mountains form. The Canadian Shield extends outward from Hudson Bay to include much of Quebec and Labrador, most of Ontario and Manitoba, and a substantial part of Arctic Canada (Figures 2–1 and 2–2). Exposed pieces of the shield extend into the northern United States. Farming the shield has always been difficult, due to thin and rocky soils, but its forests once supported a profitable fur trade. Today the shield is an important source of metallic raw materials and timber.

To the east and south of the Canadian Shield lies the eastern mountain backbone of the United States and Canada, the **Appalachian Highlands**. Since the adjacent lowlands owe their existence, at least partially, to the degradation of nearby mountains, we will examine the Appalachian Highlands first. This relatively low-lying, yet imposing, system of mountains extends from Newfoundland to northern Alabama. By global standards, the 6,000-foot (1,800-meter) peaks that mark the highest points are relatively low in elevation (Figure 2–3). Nonetheless, these mountains created a significant barrier to westward movement in the early years of the United States.

These Highlands contain distinct landform regions. The Piedmont is an old, highly eroded plateau forming the eastern margin of the Appalachians from Pennsylvania to central Georgia. A series of river and stream rapids marks the edge of the Piedmont where it descends onto the coastal plain. These rapids, known as the Fall Line, forced early colonists to transfer their loads to smaller boats

▲ Figure 2-1 **Landform regions of the United States and Canada.** Physiographic diversity—together with climate, vegetation, and soils—provides highly varied environmental regions.

upstream or to portage around difficult spots. Many towns were founded along the Fall Line to accommodate this transition, and some towns subsequently grew into important southern cities such as Richmond, Virginia. The Piedmont itself emerged as an important agricultural and industrial region in the South.

The Blue Ridge Mountains, or Great Smokies as they are known in North Carolina, Tennessee, and Georgia, constitute a favored tourist destination today. The Ridge and Valley province, a strikingly folded landscape of long, parallel ridges and valleys, extends from New York to northern Alabama and includes such familiar locations as the Hudson and Shenandoah Valleys. In terms of geologic age, the Blue Ridge formation is part of the relatively "old" eastern Appalachians and the Ridge and Valley formation is part of the relatively "new" western Appalachians.

▲ **Figure 2-2 The Canadian Shield.** Because of repeated continental glaciation, the Canadian Shield has little arable land but provides a wealth of timber and minerals. The Canadian Shield covers much of Canada and parts of the northern United States and constitutes a physical core area for North America.

Despite the name, the western Appalachian Plateau has been severely eroded into hills and mountains, especially in West Virginia. Other areas, such as the Cumberland Mountains in Tennessee and the Allegheny Mountains in Pennsylvania, retain features characteristic of plateaus. This region contains many of the vast coal reserves of North America.

The New England section of the Appalachian Highlands differs from its southern counterparts in having been significantly modified by glaciation. This section includes the White Mountains of New Hampshire and Maine and the Green Mountains of Vermont. The Appalachians continue northward into Canada to form a Maritimes–Newfoundland extension.

As the name suggests, the **Gulf-Atlantic Coastal Plain** lies along the Gulf of Mexico and Atlantic coasts of the United States. In the east, the plain extends from Cape Cod to Florida, with the Fall Line serving as the western boundary. In the south, the plain embraces Florida, coastal Texas, and much of the lower Mississippi River valley. Most of this Coastal Plain consists of geologically young rock

▼ **Figure 2-3 Appalachian Mountains.** The Appalachians are not high by global standards with only a few peaks exceeding 6,000 feet (1,800 meters), but they figure prominently in the development of the United States. Today, the Appalachians provide a variety of recreational opportunities.

that inclines gently toward the coast. Sometimes the slopes of the coastal plains are so gentle it is difficult to discern where the land ends and the sea begins, due to the proliferation of lagoons, sand bars, and barrier islands. Poorly drained wetlands are abundant (and often spectacularly beautiful), as exemplified by Georgia's Okefenokee Swamp, Louisiana's Atchafalaya Swamp, and North Carolina's Dismal Swamp (Figure 2–4).

To the west and southwest of the Canadian Shield lie the mountain ranges and plateaus of North America's western backbone. These features extend southward from Alaska to Central America and westward to the Pacific Ocean. Unlike their Appalachian counterparts, the western mountains are geologically younger, higher, and more rugged.

When we think of the West, the **Rocky Mountains** and their postcard vistas inevitably come to mind, since they contain much of the region's scenic grandeur (Figure 2–5). But the Rockies constitute only the eastern third of the western ranges. The major ranges begin in northern New Mexico and continue northward through Colorado and Wyoming. A break in the trend appears in central Wyoming, which provided a pass through which nineteenth-century pioneers traveled on their way to Oregon and California. In northern Wyoming the full extent of the Rocky Mountains reappears and continues northward into Canada and, ultimately, as the Brooks Range of northern Alaska. The present location of the Rockies was once a vast sedimentary plain. Upward pressure on this plain led to folding and fracturing of the rock cover that, along with volcanic activity, produced the mountains we see today.

The majestic buttes and mesas that we associate with the "Old West" are not in the Rockies, but in the **Interior Plateaus** region to the west. The Interior Plateaus contain several components. The Colorado Plateau lies more than a mile high in southwestern Colorado, eastern Utah, northern Arizona, and New Mexico (Figure 2–6). The plateau is an uplifted piece of Earth's crust comprising sedimentary layers that were downcut by the Colorado River and its tributaries, thereby producing some of America's great canyonlands, like the Grand Canyon.

▼ **Figure 2-4 Okefenokee Swamp, Georgia.** The Okefenokee is one of many large wetlands along the Gulf-Atlantic Coastal Plain. One of the most common tree species in the southern swamps is bald cypress, a deciduous needleleaf tree that is known for its excellent quality timber that resists rotting and termite infestation.

▲ **Figure 2-5 Pikes Peak, Colorado.** One of the many peaks in Colorado exceeding 14,000 feet (4,200 meters), Pikes Peak rises abruptly on the edge of the Great Plains, signaling the beginning of the Rocky Mountains. Note the barren summit, where harsh, tundra-like conditions prevent the growth of trees and all but the hardiest of plants. On the way to the summit, the traveler will pass through various vegetation zones reflecting the changing microclimates produced by increasing elevation. Rock outcrops are characteristic of the broken terrain that gives the Rocky Mountain chain its name.

The Basin and Range region lies to the west and south of the Colorado Plateau, occupying much of Nevada and western Utah as the Great Basin, and parts of southern California and Arizona. Bordered by the higher Colorado Plateau to the east and the lofty Sierra Nevada Mountains to the west, the Basin and Range region consists of often parallel mountain ridges separated by valleys or basins, as exemplified by the rows of north–south-trending mountain ranges in Nevada. The broken terrain and relative isolation of the Basin and Range region have created drainage conditions characterized by rivers without exits to the sea and isolated saline lakes, such as the Great Salt Lake. Death Valley, the lowest point in the United States

▼ **Figure 2-6 The Colorado Plateau and the Grand Canyon.** The Colorado Plateau has a rich geologic history and a dramatic landscape characterized by buttes and mesas. The Grand Canyon is one of America's most popular tourist destinations, in large measure because of the sensational vistas created by the Colorado River as it cuts into the plateau.

at 282 feet (86 meters) below sea level, is situated in the California portion of the Basin and Range region.

To the north of the Basin and Range region lie the Columbia Plateau of eastern Oregon and Washington and the Snake River region of southern Idaho. The plateau consists of weathered lava fields associated with some of the richest agricultural soils in the United States: eastern Washington is a major wheat- and apple-producing region, and Idaho is known for its potatoes and sugar beets. Farther north, the Interior Plateaus region narrows to a series of plateaus and isolated mountain ranges in British Columbia and the Yukon, finally ending as the broad Yukon River Valley of central Alaska.

The **Pacific Coastlands** constitute a system of mountains and valleys that extends along the western edge of North America. On the eastern side lie two great interior ranges that contain peaks that rival the Rockies in grandeur and elevation. The Sierra Nevada Mountains run from north to south in eastern California and have more than 11 peaks in excess of 14,000 feet (4,200 meters). To the north of the Sierra Nevada lie the volcanic Cascade Mountains of central Oregon and Washington. These interior ranges continue northward as the Coast Mountains of British Columbia and as the Alaska Range. It is within the Alaska Range that we find Mt. McKinley, the highest peak in the United States and Canada at 20,320 feet (6,194 meters).

The smaller Coast Ranges run the length of the Pacific coast. In California, these mountains rise gently from the Pacific but by the time they reach Washington's Olympic Peninsula, they exceed 7,500 feet (2,300 meters) in elevation. This mountain trend continues into Canada as an island series—the floors of the Coast Ranges now lie below sea level—dominated by Vancouver Island.

Several lowland areas separate the Coast Ranges from the Sierra Nevada and the Cascade Mountains, including the Great Valley of California, a fertile alluvial trough that is home to one of the most productive agricultural regions in the United States. Farther north, the Willamette Valley of Oregon and the Puget Sound Lowland of Washington have emerged as important agricultural regions and contain major urban centers, such as the Seattle metropolitan area.

North America's western rim is one of the most geologically active regions on the continent. The infamous San Andreas Fault separates two crustal plates, making much of California susceptible to earthquakes. In Oregon and Washington and northward into Alaska, the threat includes volcanoes as well as earthquakes, thanks to subduction of the Juan de Fuca Plate beneath the North American Plate. No better evidence of this risk can be found than the eruption of Mount St. Helens in 1980, an eruption whose force literally decapitated the mountain (Figure 2–7).

Between North America's mountain backbones of the east and west lies a middle zone of sedimentary accumulation called the **Interior Lowlands**. These lowlands reach as far north as the Arctic plains of Alaska, as far south as central Texas, and as far east as the Great Lakes. With several notable exceptions, the terrain is flat to rolling and consists of sediments that have washed out of the eroding Rocky Mountains and Appalachian Highlands. The Great Plains east of the Rockies feature one of the world's flattest surfaces, interrupted only by the occasional eastward-flowing river (Figure 2–8).

Much of the northern interior lowlands were glaciated. Minnesota is called the "Land of 10,000 Lakes" because of the countless tiny ponds and lakes that formed following the retreat of the ice

▲ **Figure 2-7 Mount St. Helens, Washington.** The beauty of the Cascade Mountains is the product of massive crustal disturbances caused by earthquakes and volcanic eruptions. Mount St. Helens (now 8,364 feet or 2,549 meters) lost 1,314 feet (400 meters) in height when a violent eruption decapitated the mountaintop on May 18, 1980. The 0.67 cubic mile (2.8 cubic kilometer) of material ejected into the atmosphere showered debris over a 230-square-mile (596-square-kilometer) area and killed 57 people.

▲ **Figure 2-9 Mt. Rushmore, South Dakota.** The presidential images of Mt. Rushmore are part of the Black Hills of western South Dakota. From left to right are carvings of George Washington, Thomas Jefferson, Theodore Roosevelt, and Abraham Lincoln.

sheets. The most visible remnants of glaciation are the **Great Lakes** once scoured by glaciers and subsequently filled with meltwater.

The generally flat terrain of the lowlands is interrupted dramatically by the Ozark Plateau of Missouri and the Ouachita Mountains of Arkansas. Another interruption is the Black Hills region of western South Dakota, with peaks more than 7,000 feet (2,100 meters) high. These mountains are part of a very old, highly resistant dome of crystalline rock protruding through the land's surface. The sculptors of the presidential faces on Mt. Rushmore were, no doubt, grateful for this resistance to erosion (Figure 2–9).

Stop & Think

▶ Given your knowledge of North American landforms, how could you argue that it was not "natural" for early settlers to move from east to west?

▼ **Figure 2-8 The Great Plains.** The region of the interior United States known as the Great Plains is one of the world's foremost agricultural settings. Wheat farming and cattle grazing are the principal agricultural activities.

Climate

Several general factors influence North American climates (Figure 2–10). First, most of the United States and Canada lie in the middle and high latitudes, with clearly defined seasonal changes in temperature. Second, for much of the United States and Canada the prevailing wind direction is from the west, thanks to the westerly wind system and its embedded jet streams that push weather systems from west to east. Third, the north–south mountain ranges in the West modify air masses as they move eastward, often drying them out. Fourth, the configuration of mountain systems and the size of the continental landmass create a condition called **continentality** in the interior of the continent, in which the atmosphere takes on the more extreme heating and cooling characteristics of land rather than water; winters tend to be cold and summers hot. Finally, the **Gulf of Mexico** provides an important source of moisture for the Gulf Coast and the Interior Lowlands.

Anyone traveling along the U.S. Gulf Coast in the summer notices how hot and sticky it can be. This is the **humid subtropical climate** at its best (or its worst!). This climate covers much of the southeastern United States. Tropical air from the Gulf of Mexico settles over the area in summer, bringing heat and abundant moisture that is released in the afternoon as brief but intense convectional thunderstorms. In winter a combination of cold, dry continental air masses from the north, and cool, moist air masses from the Pacific, compete for space with warm and humid air masses from the south. Temperatures swing from hot and muggy one day to chilly and dry the next. Winter rainfall is mainly frontal in origin. Snowfalls are relatively infrequent or, in the case of the Gulf Coast and Florida, virtually nonexistent. The dominant vegetation types in this climate include rapidly growing mixed hardwood and softwood forests that yield great quantities of cut timber and pine pulp for paper.

On the Pacific Coast we encounter the dry subtropical or **mediterranean climate**, a relatively narrow belt along the California

▲ Figure 2-10 **Climate regions in the United States and Canada.** Note the diversity of climates, from the humid subtropical climate of the Gulf Coast to the Arctic tundra.

coast from San Diego to north of San Francisco. Summer temperatures are relatively cool and winter temperatures are relatively warm due to the moderating effects of the Pacific Ocean. Precipitation levels are low, with clearly defined winter wet and summer dry seasons. Winter storm fronts can be intense enough to erode coastlines, undermine homes, and cause mudslides. Snow is virtually unheard of in the low elevations because, thanks to the ocean, temperatures rarely dip below freezing. Summer is a time of drought as the fronts shift northward, and the desert climate to the east invades and takes control.

North of the dry subtropical climate is the **marine west coast climate**, which extends along the coast to Alaska. Winter temperatures are warmer than expected at such a high latitude because of the moderating effects of the ocean. For the same reason, summer temperatures are relatively cool. Overcast and drippy conditions prevail throughout the year and support a lush array of vegetation types. The Cascade Mountains exert a blocking effect on the marine west coast climate (see *Geography in Action:* Landforms and Climate: Orographic Effects in the United States).

Landforms and Climate: Orographic Effects in the United States

Mountains can profoundly influence climate through what is called the **orographic effect**. Mountains create a barrier to a moving air mass that must be overcome, modifying temperature and moisture characteristics of the air mass in the process. The result is excess precipitation on the windward side of the mountain and drought on the leeward side.

The orographic process is quite straightforward. An air mass is forced against a mountain's windward side (Figure 2-1-1). As the air mass rises, it cools (the stable air lapse rate is 3.5°F per 1,000 feet, or roughly 6°C per 1,000 meters). Because cool air holds less water vapor than warm air, the ascending air loses its capacity to hold water vapor, which is condensed out. This leads to the formation of clouds and, often, rain or snow. Once the air mass passes the summit and starts to descend the mountain's leeward side, the reverse process begins. As the air descends, its rising temperature exceeds the condensation point. The drying air thus produces what is called a rainshadow—that is, an area that is shielded from rainfall.

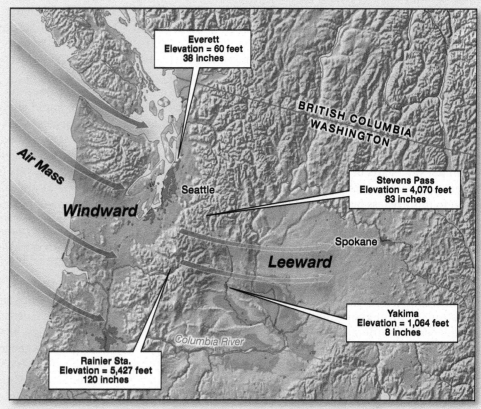

▲ FIGURE 2-1-2 **The Cascade Mountains effect (precipitation in inches per year).** Notice how much drier it is on the eastern (leeward) side of the Cascades than on the western (windward) side.

Source: Elevation and precipitation information from National Climatic Data Center, *Climatography of the United States*, no. 81 (February 2002).

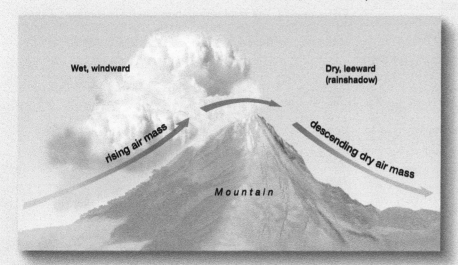

▲ **FIGURE 2-1-1 The orographic effect.** As air ascends against the windward side of the mountain, the air cools and condenses, producing precipitation. On the leeward side of the mountain, descending air warms and dries, creating a relatively rainless rainshadow.

The United States and Canada contain countless examples of the orographic effect. Many of Hawaii's major islands have mountains high enough to trap water from the trade winds on the windward sides, producing rainfall totals that exceed 100 inches (2,500 millimeters) annually, but leaving near-drought conditions on the leeward sides of the same mountains only a few miles away. Everett, Washington, on the windward side of the Cascade Mountains, receives 38 inches (965 millimeters) of rainfall a year (Figure 2-1-2), whereas Yakima, on the rainshadow side, may expect only 8 inches (203 millimeters).

East of the Cascades lies one of the two large areas of mid-latitude **steppe climate**. This area embraces eastern Oregon and Washington and extends as far south as the Colorado Plateau. The second area lies to the east of the Rocky Mountains (in the rainshadow of the Rockies) and corresponds roughly to the Great Plains. A steppe climate is semiarid, but not as dry as a desert—moist enough to sustain grasses and shrubs. Temperatures tend toward extremes thanks to continentality, with frigid winters and scorching summers.

The true **deserts** of North America are concentrated in the U.S. Southwest, the result of lower latitudes producing hotter summers, along with mountain barriers and isolation from moisture sources. Here the gap between evapotranspiration—water evaporation plus water released by vegetation to the atmosphere—and the amount of precipitation received is extreme. Phoenix, Arizona, for example, receives an average of only 8 inches (203 millimeters) of precipitation per year. Parts of southern California are fortunate to get 3 inches (76 millimeters). Natural vegetation such as the cactus adapt to this dryness by storing moisture internally and slowing down moisture loss—for example, by growing needles instead of leaves (Figure 2–11).

Eastward of the steppes and deserts, we enter the **humid continental climate**, which dominates the northeastern quarter of the United States and southern Canada. Humidity is a hallmark of this environment, especially in the east. Precipitation is moderate to abundant throughout the year, falling as snow in winter. Continentality contributes to major seasonal temperature swings, particularly on the western edges of the region—for example, the subzero January temperatures of Winnipeg may be matched by more than 90°F (30°C) temperatures in July.

Poleward of the humid continental climate lie the **subarctic** and **polar climates**. The subarctic climate occupies a wide swath of central Canada and Alaska. It is characterized by long, intensely cold winters and short, warm summers. Precipitation is relatively low, but more than high enough to meet the needs of a cooler environment with less evaporation. Most precipitation falls as summer rain rather than winter snow. The snow that falls does not thaw until spring and can be whipped around by the wind to create the false impression of constant snowfall. The subarctic is home to large stands of softwood boreal forest and many fur-bearing animals, but few people.

Conditions are even more extreme in the polar, or tundra, climate found along the northern edges of Canada and Alaska. Freezing conditions last for most of the year and, in the depths of winter, are accompanied by days without sunlight. Conditions are so harsh that a thin cover of grass, moss, and lichens replaces trees (Figure 2–12). Very few people other than aboriginal groups live in the far north, and this region's economic potential is limited.

Stop & Think

▶ Define what a "favorable" climate is in your thinking and describe where in Canada or the United States this favorable climate would be found.

Environmental Challenges

The United States and Canada have always experienced environmental extremes, particularly in the continental climates. Global climate change, however, is intensifying extreme weather and rearranging the geography of one of humankind's most important assets, water. It is a story of growing shortage on some parts of the continent and worrisome inundations elsewhere. There is little doubt today that human activities promote climate change.

It often seems that some part of the continent is in the grip of "record-breaking" dryness, usually combined with "record-setting" summer high temperatures. Temperatures in general have been rising. Since 1895, the U.S. average temperature has increased by

▼ **Figure 2-11 Desert vegetation in Arizona.** Desert vegetation must be able to withstand long periods of drought. The tall saguaro cactus embodies the adaptations required to survive desert conditions, as for example, relying on spines instead of leaves to reduce moisture loss and a barrel-like trunk to store water. No more plant area than necessary is exposed to the intense evaporative power of the desert sun. Shade is at a premium. The saguaro cactus is a slow-growth plant that Arizonans increasingly attempt to protect as they build out into the desert.

▼ **Figure 2-12 The treeless summer tundra in Nunavut, Canada.** In the short tundra summer daylight hours are long, but in the long dark winter days are short and temperatures can reach −50°F(−46°C).

about 1.5°F, with 80 percent of that increase occurring since 1980 (these and other figures, and most of the predictions, are taken from the January 2013 *Draft Climate Assessment Report* of the National Climate Assessment and Development Advisory Committee). Precipitation has also decreased and/or become more sporadic, particularly in the Southwest and parts of the Southeast United States. Higher temperatures have led to increased levels of transpiration and evaporation—that is, a higher atmospheric demand for water—that worsens the dryness problem.

The consequences of prolonged drought are considerable, particularly for the U.S. Southwest, which has experienced rapid population growth in recent decades:

- Mountain snowpack is reduced and, with increasing temperatures, what snowpack there is melts more quickly (along with centuries-old glaciers). Less water flows into the streams and rivers that support southwestern cities and irrigated agriculture. Water usage can be so intense that there are years when the Colorado River is dry by the time it reaches the ocean.
- As surface water becomes less available, the demand for underground water from **aquifers**, water-bearing rock strata, increases at a time when the aquifers cannot recharge because of precipitation shortages.
- The wildfire hazard increases, with larger, more frequent fires occurring due to the abundance of dried vegetation as fuel.
- Competition for water is intensified among different users—farmers engaged in irrigated agriculture versus city dwellers who enjoy not only hot showers but manicured golf courses and swimming pools. Competition also has a political dimension, as governments compete for water rights; this is most evident in the United States and Mexico's tense relationship over use of the lower reaches of the Colorado River (Figure 2–13).

Whereas the Southwest and parts of the Southeast anticipate more frequent drought conditions, the U.S. Midwest and New England anticipate more precipitation. Average precipitation over the continental United States as a whole increased by two inches between 1895 and 2011. In some cases, this is good news. Agriculture in parts of the upper Midwest may benefit from the additional rainfall (though a devastating "flash drought" also struck the Midwest in 2012), along with the longer growing seasons that may accompany global warming. But increasingly, precipitation is falling as part of violent weather events, leading to flooding and sometimes catastrophic wind damage. Extreme precipitation events are projected to increase everywhere, to the point that storms predicted to occur once every 20 years may occur as often as once every 5 years by the end of this century. Heavy downpours wash out roads, destroy bridges, and flood fields and homes. The ultimate extreme storm is the Gulf-Atlantic late-summer hurricane that can produce a wall of water—a storm surge—that can be as high as 20 or 30 feet (6 or 9 meters) in

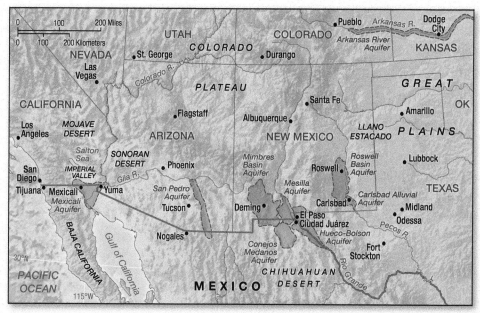

▲ **Figure 2-13 Water Competition in the Southwest.** An aquifer is a water-bearing rock layer that lies beneath the surface of the earth. Many parts of the United States and Canada rely on these layers to provide groundwater. Note how in the southwestern United States these aquifers overlap political boundaries, creating the potential for conflict and litigation in an increasingly dry part of the country. (See Deborah Hathaway, "Transboundary Groundwater Policy: Developing Approaches in the Western and Southwestern United States," *Journal of the American Water Resources Association* 47 (February 2011): 103–113.)

the case of a major hurricane that flows inland and literally destroys everything in its path. It is not unusual for 90 percent of the fatalities in a hurricane to come from storm surge.

The effects of increased storm intensity are made worse by rising sea levels in coastal areas, a process associated with global warming and the melting of polar ice packs. The estimated 5 million people in the United States living between sea level and 4 feet above sea level are especially at risk. Whereas sea level rose only about 8 inches during the last century, estimates of sea level increases during the twenty-first century range from 1–6 feet (0.3–1.8 meters), depending on global warming trends. Major metropolitan areas such as Boston, New York City, the San Francisco Bay area, New Orleans (much of which actually lies below sea level) and the adjacent Gulf Coast, and the cities of south Florida are particularly vulnerable. Hurricanes Katrina (2005) and Sandy (2012) have already tested the preparedness of the New Orleans and the New York / New Jersey areas (see *Exploring Environmental Impacts:* The Battle for New Orleans).

The twenty-first century will inevitably witness changes to existing physical and human geographies in response to where water is increasingly abundant and where it is increasingly absent. In addition, coastal population centers will continue to armor themselves against more intense storms and rising sea levels, including sea walls and elevated structures.

Stop & Think

▶ Explain why New Orleans is fighting an ongoing battle against the elements.

The Battle for New Orleans

Since its founding, New Orleans has been waging a war against nature to counter the related threats of soil subsidence, flooding, loss of coastal wetlands, and increasingly potent hurricanes. In a sense, New Orleans should never have been located where it was. The first French settlers were attracted to the relatively high and fertile natural riverbanks, or levees, along the Mississippi River. Settlers initially avoided the swamps and marshes that lay beyond the levees, but as the city expanded, the land in the backswamps and marshes was settled. Drainage of these muck soils—roughly 90 percent water by volume—triggered soil subsidence. The remaining 10 percent of these soils comprises mostly organic matter that oxidizes when exposed to the atmosphere, compounding the subsidence problem. Subsidence poses especially serious problems for construction, as unstable soils cause foundations to tilt and roads to buckle. The solution is to drive numerous pilings into the ground at the building site, and then pour the structure's cement slab on top of the pilings. This process results in a building that "floats" on its pilings while the ground beneath and around it continues to sink. Thereafter, fill is added to the lawns around the building in a never-ending cycle to keep the "ground" at grade. Subsidence has left much of the bowl-shaped city 5–15 feet (2–5 meters) below sea level. Man-made levees keep the Mississippi River and Lake Pontchartrain from pouring into the city, and an elaborate pumping system removes the average 60 inches (1,500 millimeters) of rainfall received annually. For New Orleans residents, the threat of heavy rain presents all sorts of frightening possibilities.

As if living in a bowl-shaped depression below sea level were not threatening enough, New Orleans is losing its natural hurricane protection through erosion of nearby coastal marshes and barrier islands.

River control projects, wetland drainage, channel dredging, construction of petroleum exploration canals, and other factors have substantially reduced land recharge through deposition of river sediment. Coastal Louisiana is losing land to the ocean at the rate of 25–30 square miles (65–78 square kilometers) a year, an area the size of Manhattan. The loss of river sediment also has caused barrier islands to shrink as longshore currents no longer carry sediment loads sufficient to replace land as it erodes away. These wetlands and islands historically have helped to protect New Orleans from deadly hurricane-produced surges of water. Scientists have long argued that a major hurricane is a human catastrophe waiting to happen.

With Hurricane Katrina, theory suddenly became reality. Early in the morning of 29 August 2005, the eye of Hurricane Katrina passed over St. Bernard Parish, brushed against eastern New Orleans and St. Tammany Parish, and slammed with full force into Waveland, Bay St. Louis, Pass Christian, Biloxi, and other coastal Mississippi towns. Katrina had weakened to a strong Category 3 storm before hitting land, but still packed winds greater than 100 miles per hour (160 kilometerrs per hour) and produced storm surges in excess of 25 feet (8 meters). Government leaders in New Orleans breathed a sigh of relief when Katrina did not directly hit the

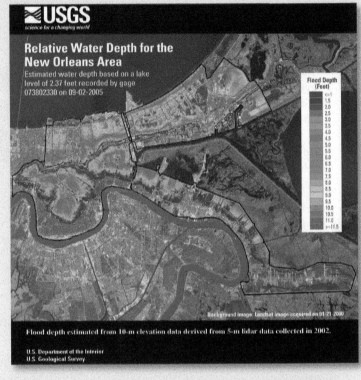

▲ FIGURE 2-2-1 **Actual flooding in New Orleans associated with Hurricane Katrina.** This image shows where the water was deepest in New Orleans following Hurricane Katrina and the breach of the levees. Areas along the Mississippi River, where the natural levees are highest, remained relatively dry. Included here are the French Quarter, the West Bank of the river, and an area called Uptown. Areas adjacent to Lake Pontchartrain also remained relatively dry but badly windblown. Those parts in the interior "bowl" of the city, and in eastern New Orleans, received in excess of 11 feet (3.5 meters) of water. Only floodwaters within New Orleans are shown; adjacent parishes and outlying areas were also badly flooded.

Source: United States Geological Service (October 5, 2005), *http://eros.usgs.gov/katrina/products.html*; background image from Landsat 7 (2000).

city. But relief quickly turned into horror when hurricane-protection levees were overtopped and broke under the pressure of storm surges, releasing millions of gallons of water into the city.

Katrina caused extreme damage throughout New Orleans and the Mississippi Gulf Coast. Perhaps 80 percent of New Orleans lay under water (Figure 2-2-1). Storm surge ripped apart the Interstate

10 bridge, a 5-mile (8-kilometer) concrete twin span connecting New Orleans with the north shore of Lake Pontchartrain. Some Mississippi towns largely disappeared from the map, swept away by record-breaking surges. Trees were left lying with roots pointing toward the sky. Most tragically of all, more than 1,800 people were killed in Mississippi and Louisiana.

Perhaps 100,000 New Orleans residents had not heeded the mandatory evacuation orders issued before the storm. In some cases, people had no transportation; in other cases, they decided to do the "New Orleans thing" and hunker down. When the levees collapsed, people were trapped and many drowned in their own homes. It took heroic efforts to evacuate survivors huddled in public buildings, on highway overpasses, on roofs, anywhere they could escape the rising flood waters. By mid-September, New Orleans was largely abandoned. The city, known for its raucous party life, gastronomic delights, and "what, me care?" culture was plunged into a surreal silence. There was no electricity. Most TV and radio stations were forced off the air. The city's main newspaper, the *Times-Picayune,* had to publish from other cities. The airport closed and Orleans Parish public schools suspended operations for the year. The New Orleans Saints football team played "home" games in San Antonio and Baton Rouge.

By late September 2005, residents of New Orleans and the surrounding parishes were allowed to inspect the damage to their homes and businesses—many discovered that they could not live in their dwellings for some time to come. By December 2007, perhaps two-thirds of the city's pre-storm population had returned, leaving tens of thousands still in a New Orleans "diaspora" that extended to virtually every state in the country.

▲ Figure 2-2-2 **Armoring of the levees in New Orleans.** Massive amounts of money have been spent to fortify the levees protecting New Orleans. The levees are the walls that protect the city from outside flooding, especially storm surge, and it was a failed levee after Hurricane Katrina that was responsible for the immense devastation of the city—almost 80 percent of the city was left under water. The "de-watering" process, as it was called, took weeks. Rebuilding of flooded areas took months and years.

By 2010, about 71 percent of the city's year-2000 population was present.

While the population has yet to recover fully, the city's economy is surprisingly healthy. The "Great Recession" was less "great" than in many other parts of the country, with unemployment rates lower than national levels and wages approaching the national average. Entrepreneurship has surged, with local business startups of 427 per 100,000 people exceeding the national average of 333. One clear post-Katrina geographic trend: the "suburban" parishes of Jefferson and St. Tammany now account for more jobs than New Orleans proper. In keeping with the city's great cultural traditions, the number of relatively large arts and culture nonprofit agencies is about double the national average. But problems remain, not the least of which are poverty, crime, lack of education, abandoned lots and dwellings, and the gradually disappearing hurricane buffer zone created by nearby marshes.

It will be years before anybody knows how the "new" New Orleans will look. To limit flood damage, new homes and structures are increasingly raised and rest on "basements" (which are the same as basements everywhere, except in New Orleans they lie aboveground!). Recently constructed and fortified levee systems will protect the city from future hurricanes (at least that is the hope) (Figure 2-2-2). The city is somewhat less African American and more Hispanic in makeup. Tourism and shipping remain prominent income sources, despite the appearance of more "knowledge-based" industries. People will always be at least a little traumatized by Katrina, uneasy about future weather threats. Hurricane Katrina remains a loud wake-up call for environmentally endangered cities everywhere.

Sources: See the annual "New Orleans Index" published by the Brookings Institute. For more on Katrina, see Merrill Johnson, "Geographical Reflections on the 'New' New Orleans in the Post-Hurricane Katrina Era," *The Geographical Review* 96 (2006): 139–156.

Historical Background: The Spatial Evolution of Settlement in Canada and the United States

Canada and the United States occupy a continent inhabited for thousands of years by indigenous groups, but only relatively recently by people of European and African ancestry, and even more recently by Asian and other ancestries. In this section, we examine the spatial evolution of settlement, focusing on the spread of European populations and the gradual elimination of frontiers, as well as establish the context that led to the formation of two countries with increasingly diverse populations.

Early Settlement: European Culture Cores

Europeans in North America were preceded by numerous aboriginal or indigenous peoples, whose numbers exceeded 2 million when the first Europeans arrived. Warfare and exposure to new diseases led initially to drastic depopulation, although the twentieth century saw a bit of a rebound. Today's Native American population accounts for less than 1 percent of that of the United States and Canada. While the historical significance of indigenous peoples should not be understated, these groups ultimately were not sufficiently populous in core areas to develop and spread their cultural characteristics over large areas (see *Geography in Action: North America's First Inhabitants*).

Of particular interest to geographers is how groups form cores or **culture hearths**, or centers of cultural innovation, and then diffuse these innovations across the landscape. Often this **diffusion process** takes the form of settlers moving into new territories, or **settlement frontiers**, and establishing their cultures in these territories. They gradually create trade and other ties with the core area and surrounding communities, and eliminate the frontier through a process of **spatial integration**. European core areas on the Atlantic Seaboard provided important early sources of settlers for the spatial integration process in the United States and Canada.

While the Spanish attempted early settlements in eastern North America (for example, St. Augustine, Florida, in 1565), their impact was limited. The first permanent non-Spanish settlements founded along the Atlantic Seaboard included the English Jamestown colony in 1607, the French settlement in Quebec in 1608, the English Plymouth colony in 1620, and the Dutch settlement along the Hudson River in 1625, which soon came under English control. These settlements provided the starting points for core areas that, through the diffusion process, created lasting imprints on the human geography of the United States and Canada (Figure 2-14).

The **New England Core** area, which consisted of parts of modern Massachusetts, Rhode Island, and Connecticut, began as a patchwork of small, subsistence-oriented farms surrounding small villages. While farming was necessary for survival, glaciated soils and hilly terrain made cultivation difficult. There was no single crop that provided great wealth or formed a basis for trade, as was the case with tobacco in the southern colonies. This core was initially a destination for religious and political dissidents who preached thrift, hard work, and piety—traits that helped settlers achieve agricultural self-reliance despite harsh environmental conditions. These traits also contributed to New England's later success at developing its considerable nonagricultural resource base. White pine provided lumber for shipbuilding; codfish from offshore banks was abundant and easily harvested. By the late 1700s, the combination of capital and labor from nonagricultural activities, an abundance of streams to turn waterwheels, and a growing maritime trading tradition contributed to New England's early success in the Industrial Revolution, and its development of innovations such as new firearm technologies that were later diffused by westward expansion. New England became increasingly urbanized and developed a large commercial middle class, with Boston figuring prominently in this process.

From the start, the colonies in Virginia and southward, the **Southern Core**, differed from those in New England. Like the Spanish in Middle America, the first Virginia settlers believed that they would become wealthy from gold and silver. Metallic wealth was never found in abundance; instead, settlers became wealthy by

▼ Figure 2-14 **Early European settlement areas in Anglo America.** Culture traits evolved in these four areas and were later diffused to other parts of Anglo America. For a more detailed analysis, see W. Zelinsky, *The Cultural Geography of the United States: A Revised Edition* (Englewood Cliffs, NJ: Prentice Hall, 1992), 80–84.

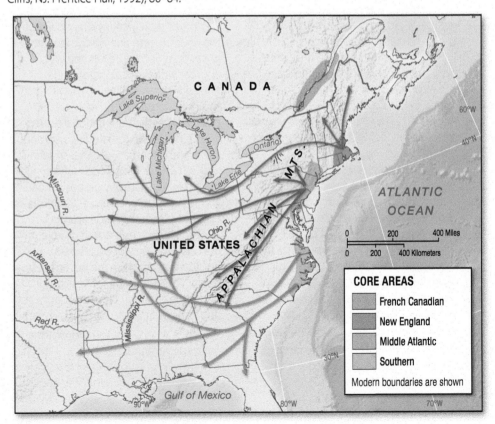

CORE AREAS
- French Canadian
- New England
- Middle Atlantic
- Southern

Modern boundaries are shown

North America's First Inhabitants

Scholars believe that 10,000–25,000 years ago a land bridge connected Alaska with Asia, over which the first Americans traveled to populate what is now North and South America. The land bridge was a result of sea-level drops caused by the last of the great Ice Ages. As the continental glaciers melted the land bridge was flooded, but corridors between ice sheets to the interior of the Americas were opened through which people could pass. By the time the Europeans arrived, more than 2 million people (and maybe up to ten times that number; estimates vary wildly) were already in what is now the United States and Canada, many of whom lived in highly organized societies. The early Spaniards referred to them as "indios" or "Indians," in the mistaken belief that the Europeans had reached the "Indies," or Asia.

Terminology is important. In the United States, "Native American" is perhaps the most widely accepted term. "American Indian," while commonly used, is suggestive of a disagreeable European past. In Canada, the terms "First Nations" and "aboriginal" groups tend to be most common and acceptable. Within Canada's official aboriginal groups are North American Indians, Metís (a blend of mainly French and Indian), and Inuit (native groups in the far north). Americans tend to think of "tribes"; Canadians of "bands." The word Canada derives from the Huron term for village, "Kanata," which says something about the importance of North America's first citizens to Canada's very being.

How to relate to its first citizens was a dilemma faced by Europeans in both countries. Europeans first arrived with communicable diseases such as measles and smallpox (to which they had some degree of immunity but Native Americans had none) and witnessed the decimation of the native populations. Where disease did not make space, Europeans often pushed violently into territories occupied by Native Americans, leading to Indian wars. In other cases, some form of accommodation was attempted. In the early 1800s, the U.S. government determined that "removal" was the answer, and whole eastern tribes were relocated to the Great Plains that other tribes already called home. In 1838–1839, for example,

thousands of Cherokee were forcibly uprooted from Georgia and the Carolinas, and settled in Indian Territory (now Oklahoma) after great suffering along the "Trail of Tears." Later in the nineteenth century, Native Americans were relocated to reservations in the United States and reserves in Canada. By the twentieth century, a widespread effort was made to "civilize" (Europeanize) Native Americans and assimilate them into mainstream North American society, including encouraging Native Americans to move to urban areas. In most cases, these policies did more harm than good; in the worst cases, they reflected an incredible cultural arrogance.

In both countries, tribes and bands were viewed as some variation of "nations" with which treaties could be negotiated by the respective federal governments. These negotiations led to the creation of reservations and reserves that today have special sovereign status within the American and Canadian federal systems. Although subject to federal authorities, native groups in both countries have wide latitude to formulate their own laws and in some cases are exempt from statutes and tax requirements created by provinces, states, and other local jurisdictions.

The reservation system, its special status notwithstanding, has been associated with some of the worst conditions in the United States and Canada. The corruption of nineteenth and early twentieth-century U.S. Indian agents was legendary as they took advantage of the second-class status of their "conquered" subjects. To this day, reservations are linked to high levels of poverty, substance abuse, unemployment, and other problems. As people leave the reservation and assimilation occurs, there is concern that the cultural and linguistic heritage of Native Americans may be lost. New attempts are being made in schools and elsewhere to preserve indigenous languages.

▲ Figure 2-3-1   **Native American casino in the Pacific Northwest.** Casinos have become major income earners for Native Americans such as the Coquille Tribe of Oregon.

Life on some reservations is improving thanks to an unanticipated development: gambling. In the late 1980s, the U.S. Supreme Court ruled that states could not restrict gambling activities on Native American lands. This led to a land rush to fill reservations with casinos, creating what some refer to as "Native American, Inc." (Figure 2-3-1). Native groups in both countries also have sought redress for past injustices from the courts and legislative bodies, resulting in significant cash payouts by governments and, in some cases, territorial adjustments. Most of these claims were settled during the second half of the twentieth century, but adjudication continues.

It is easy to treat the story of first citizens as just a footnote in American and Canadian history. In reality, not only did they begin the settlement process in North America, they enriched the cultural fabric that emerged. In the twenty-first century we may also witness the rise of some tribal groups as business powerhouses—a supreme twist of irony emerging from a tortured past.

Sources: Adapted from Elyse Ashburn, "A Race to Rescue Native Tongues," *Chronicle of Higher Education* 54 (September 28, 2007): B15–B16; "Census Reveals Aboriginals Fastest Growing Population," *Windspeaker*, (February 2008): 11; Walter C. Fleming, "Getting Past Our Myths and Stereotypes about Native Americans," *The Education Digest*, March 2007, 51–57.

cultivating subtropical crops that could be exported to Europe and other parts of the Americas. Tobacco became a commercial success almost immediately; indigo, rice, and cotton were added later.

Most of the South's commercial agricultural production relied on the plantation system. Spatially, socially, and operationally, this system could not have been more different from the New England farming system. Indentured and slave labor led to a clear distinction between farm workers and farm managers or owners, with attendant distinctions in wealth, social status, and the built environment of the plantation. The economic emphasis was on the production of one or two crops for export. While subsistence-oriented activities were present, they were not the plantation's reason for being. Plantations generally were located on navigable rivers or along the coast to make shipment of commodities easy. In the plantation South, few cities other than ports appeared; a commercial middle class was largely missing. Inland from the coastal agricultural settlements, and away from convenient water transport routes, were smaller, free-labor farms run by yeoman farmers; these were quite different in economic mission from the plantations.

New York and Pennsylvania, together with portions of New Jersey and Maryland, made up the **Middle Atlantic Core**. A variety of peoples settled the middle colonies: English, Dutch, German, Scots-Irish, and Swedish. Neither the cash crops of the South nor the lumbering and fishing of New England were significant sources of income; instead, a mixed agricultural system focused on the fattening of hogs and cattle and on Indian maize, now called corn. Settlers also made tools, guns, and wagons, and worked the deposits of iron ore found in eastern Pennsylvania. The diffusion of their culture, especially its mixed farming system, had a significant impact on the American Middle West and parts of the Appalachians.

Of special note is the eighteenth-century migration of people from the middle colonies into southern Appalachia, including Virginia and West Virginia, and into the Blue Ridge Mountains and the Ridge and Valley region. Descendants of the Scots-Irish, Germans, and English who peopled the middle colonies, these settlers carved out small subsistence farms as slaveless yeoman farmers, and contributed traits (an overwhelmingly white, Protestant population of small farmers) that still distinguish southern Appalachia from the rest of the lower South today. Eventually, population pressures forced these people to migrate to the highlands of Arkansas and Missouri and to the hill country of central Texas.

The earliest European settlements in Canada were French in origin and formed the **French Canada Core**. Only part of this core survived. Acadians, for example, settled the valleys of western Nova Scotia under the French flag, only to come under British rule in 1713. Resistance to British rule led the British to transport much of the Acadian population to other parts of North America in 1755. Today's "Cajuns" of southern Louisiana trace their roots to the Acadians of Nova Scotia—Cajun is an Anglicized corruption of the French word from which Acadian is derived. Modern Nova Scotia is an English-speaking province with only traces of its Acadian past in evidence.

By far the most enduring French presence in Canada appeared in New France, focusing on Montreal and Quebec City. Immense wealth in the form of furs was extracted from New France's seemingly endless forests. Later, farmers settled on the shores of the St. Lawrence River and created an agricultural system that was sufficient, though not extraordinary, in its bounty (Figure 2–15). Farms

▲ Figure 2-15 **French Canadian agricultural landscape.** French Canadian agriculture along the St. Lawrence River has always focused on the use of long lots that provide every farmer with a narrow frontage to this important transportation corridor—narrow to maximize the number of farms with river access. The lots extend lengthwise back from the river, in some cases a considerable distance, to give each farmer access to a range of soil types and forest products.

were arranged in long strips to provide each farmer with access to the river, the principal transportation corridor. French Canadian life was hierarchical in structure, with the Catholic Church, government officials, and landowners at the top and the mass of tenant farmers at the bottom. A large commercial middle class was absent.

By 1763, there were only about 65,000 people in New France, compared with more than 2 million in the neighboring English colonies. New France was unable to defend itself against the British, who defeated the French colonials on the battlefield in 1759 and acquired New France by treaty in 1763. In spite of British rule, however, the French Canadian population increased to several million by the twentieth century, and today French speakers account for roughly a fifth of Canada's population. But little spatial diffusion of the culture occurred out of this core, apart from small pockets in western Canada.

While the eastern core areas were of considerable importance to the spatial evolution of North America, it is important not to overlook what amounts to a Hispanic core in the southwestern United States (Figure 2–16). After the conquest of the Aztecs by Hernando Cortés in the 1520s, the Spanish turned their attention to other parts of Mexico, particularly the north. In 1598, the Oñate expedition, with hundreds of people, cattle, and wagons, made a daring penetration of the northern frontier to establish a Spanish presence in the Pueblo Indian regions of modern New Mexico. By the early 1600s, a capital was established in Santa Fe and connected settlements appeared in such locations as Socorro, New Mexico, and El Paso, Texas. Spanish missions were established later in southern Arizona and along the California coast, leading to a second region of Spanish settlement. These two thrusts into what is now the United States were so remote as never to escape "frontier" status relative to the rest of what became Mexico, and occasionally these frontier communities were besieged by hostile indigenous groups. In 1680, for example, an uprising led to considerable loss of life and the abandonment of Santa Fe; it took 16 years for Spanish authority to be reasserted. But the Spanish and later Mexican presence in the American Southwest were sufficiently large and persistent to create a mostly Hispanic borderland stretching from Texas to California.

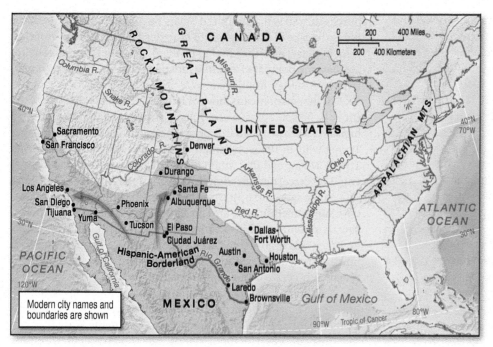

▲ **Figure 2-16 The Hispanic American borderland.** Centuries of settlement by Spanish and later Mexicans left an indelible cultural imprint on the southwestern United States, one that is being reinforced by modern migrations. There were originally two migration destinations: the Santa Fe area, and Arizona and California.

Westward Expansion, Immigration, and Receding Settlement Frontiers

After the American Revolution, the new United States consisted of the original 13 colonies and territory extending westward to the Mississippi River (ceded by Britain to the United States in 1783). Barring struggles with indigenous groups, these western territories were available for settlement (Figure 2-17). To the north lay British North America, or Canada, which constituted the bulk of Britain's remaining North American territories. Territories to the west and south were held by a variety of European powers.

It was not long before Americans were looking with interest at foreign-held territories that lay beyond the initial U.S. borders. Many believed that America was entitled to these lands, partly because of their proximity to the United States and partly because of America's growing sense of **manifest destiny**—the conviction that God had willed the continent to the United States to be civilized by Americans and their ennobling institutions.

This sense of American destiny was not shared by those countries claiming territories lying in the path of American aspirations. Consequently, westward settlement—its spatial integration process—is largely a history of U.S. negotiation and conflict with external powers over the control of western and southern territories.

An early negotiation for land involved the French-controlled Louisiana Territory, just west of the Mississippi River and extending to the Rocky Mountains. This region included the valuable port of New Orleans near the mouth of the river, the strategic significance of which was apparent to British, French, Spanish, and Americans alike. After a disastrous French military campaign against slave rebellions in Hispaniola, however, and fearful of British intentions in the Americas, Napoleon, Emperor of France, negotiated a deal that gave the Louisiana Territory to the Americans. While President Thomas Jefferson was interested mainly in securing New Orleans, his agents consented to an arrangement in 1803 that sold all of the Louisiana Territory to the United States for $15 million—a bargain deal at 3 cents an acre that nearly doubled the size of U.S. territory.

The Louisiana Purchase was only the first in a series of territorial acquisitions that gave the U.S. map its familiar shape. Major acquisitions included:

▼ **Figure 2-17 Selected major acquisitions that defined the territorial evolution of the contiguous United States.** By 1783, the United States occupied territory westward to the Mississippi River. Over time, however, the United States acquired lands that extended to the Pacific Ocean and, eventually, to Alaska and Hawaii (not shown).

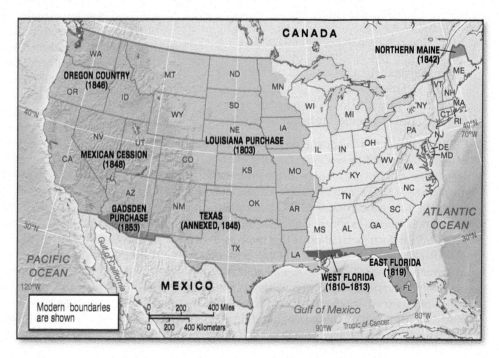

▲ **Figure 2-18 The Alamo, San Antonio.** Texas was originally conquered and settled by the Spanish, who lost the territory in the 1830s. Although the battles of the Alamo and elsewhere contributed to the failure of Mexico to retain the region, distance from the Mexican heartland and the ever-increasing number of Anglo settlers were long-term geographic factors that favored American political and economic interests. Today the historic monument is a popular tourist destination.

- East and West Florida from the Spanish, partly in response to military action and offers of money (East Florida cost the United States $5 million).
- Texas from Mexico, mainly through military action that initially involved only an American insurgency within Texas but ultimately led to war between the United States and Mexico in 1846 (Figure 2-18).
- California and the U.S. Southwest, from Mexico. The war with Mexico resulted in a treaty in 1848 that not only formalized U.S. acquisition of Texas, but ultimately led to a transfer of what is now California, Arizona, Nevada, Utah, and parts of neighboring states for $15 million. The final piece was a purchase from Mexico of a strip of territory in southern Arizona in 1853 (Gadsden Purchase) for a railroad route.
- Oregon and the Pacific Northwest, from Britain. While the United States originally claimed territory well into the modern Canadian province of British Columbia, negotiation with Britain eventually gave the United States only the territory south of the 49th parallel of latitude (the modern boundary).

Other boundary adjustments were made to create the modern United States, including settlement in 1842 of a final boundary between Maine and Quebec. The last major territorial acquisitions were the purchase of Alaska from Russia in 1867 and annexation of Hawaii in 1898. In the case of Hawaii, the manifest destiny that propelled America to become a continental power in the mid-1800s was transformed into an international cause that motivated overseas expansion in the late 1800s. Even Cuba and Canada were targeted by some American politicians. The international political culture of the late nineteenth century valued such aggressiveness, and U.S. territorial expansion paled in comparison with that of many European countries.

By 1870 the spatial evolution of the United States was largely complete (excluding Alaska and Hawaii). Through negotiation, purchase, and war, the United States had acquired territory many times the size of the original 13 colonies. The settlement frontiers were in place; now they had to be filled. And once western territories became populated and tied functionally to the rest of the country, they lost their frontier status.

For many Americans, however, the frontier was more than a territory to be settled—it was a state of mind to be understood. Perhaps no one articulated this position better than Frederick Jackson Turner who, in 1893, presented what became known as the Turner Thesis. He argued that much of the meaning of American history could be discerned through the prism of the frontier experience. The challenges and opportunities of the frontier molded character, shaped institutions, and provided Americans with a constant source of rebirth. While scholars have generally been reluctant to embrace this argument, Turner's perception of the frontier took on almost mythical qualities and is important as a reflection of the nineteenth-century American mindset.

Settlement frontiers were populated by various groups, including people moving from, or otherwise influenced by, the core areas on the Atlantic Seaboard. As settlers migrated to the West, they carried their cultures with them, setting into motion a spatial diffusion process that shaped frontier landscapes.

The spatial diffusion and integration processes initially proceeded slowly. Westward migration meant traveling by land, and before the mid-nineteenth century, overland travel was done mainly by foot or beast of burden and was painfully slow. In 1800, for example, traveling overland from New York to what is now St. Louis could take a month. Overcoming this **friction of distance**—that is, the time and effort required to travel across the continent—required the technical innovations of America's Industrial Revolution.

The nineteenth century was a period of transforming innovations in transport technology, all of which helped integrate frontiers. Canal building in the early 1800s took advantage of the cost advantages of water transport and led to a series of canals in the east. Particularly prominent was the 1825 opening of the Erie Canal that connected the Great Lakes and the Hudson River. With the Erie Canal, the cost and time required for shipping goods from Buffalo to New York City plummeted from $100 a ton and 20 days to $5 a ton and 5 days. The Erie Canal facilitated the movement of grains, forest products, and minerals from west of the Appalachians to the East. New York became the center of a new commercial universe, largely at the expense of Boston.

The railroad era began even before the short canal boom ended. Although occasionally competing with canals, railroads generally complemented water transportation and greatly increased the significance of the Great Lakes as an interior waterway. By the mid-nineteenth century, new transportation focal points had been established. Rail networks focused on selected coastal ports, such as New York, and converged on interior ports, such as Chicago and St. Louis. By the end of the nineteenth century, both the United States and Canada had transcontinental networks. The integrating effects of the railroad can hardly be overstated; its relative speed easily overcame the friction of distance for both passengers and freight, and it was only a matter of time before the railroad, more than anything else, would eliminate the frontier. By the 1860s, the trip from New York to St. Louis took a mere 2–3 days.

The American Midwest—defined roughly as the Great Lakes states, the Ohio River valley, and the upper Mississippi River valley—became a popular frontier destination after the American Revolution. The gently rolling, forest-covered landscape possessed some of the best agricultural soils found anywhere. Settlers from the Middle Atlantic Core region brought their knowledge of mixed farming, which was easily adapted to middle western environmental conditions. The result of this diffusion process was an agricultural heartland that quickly overshadowed the Middle Atlantic region and produced a surplus of wheat, corn, pork, and beef.

The frontiers of the American South—defined generally as the territory from Alabama to east Texas—were settled at about the same time as the Middle West. Settlers brought with them, or were influenced by, the culture of the Southern Core area, including reliance on the plantation system and its focus on cultivating commercial export crops. This diffusion process, however, almost did not happen. Prior to 1800, southern commercial agriculture was based on tobacco, a type of cultivation that depleted soils of nutrients and was beginning to erode the economic foundations of the plantation system. Had it not been for Eli Whitney's development of the cotton gin in 1793, the plantation system and slavery might have slipped into economic obscurity. The cotton gin efficiently separated seeds from fibers in short-staple cotton, and encouraged cotton cultivation in both upland and lowland regions in the South. Driven by a growing demand for cotton fiber in British textile mills, cotton production increased tenfold in the South within a decade. Cotton revived the plantation system and spread as far west as the 24-inch (600-millimeter) precipitation line in Texas, the line beyond which it was too dry to grow cotton profitably without irrigation.

Continued westward expansion beyond these two initial settlement frontiers was more difficult. The "Great American Desert," what we now call the Great Plains, lay in the path. While not actually a desert, its extreme temperatures, relative dryness, hard soils, and largely treeless landscape nonetheless made it unappealing to settlers, especially when more attractive lands lay to the West. By the late 1840s, the Great American Desert was being "leapfrogged" by settlers on the way to the farmlands of Oregon's Willamette Valley and the gold fields and fertile soils of northern California.

The migration, of the Mormons stands out as an exceptional cultural event in the spatial integration of America's settlement frontiers. In effect, this migration produced a new American culture core in Utah from which a separate round of spatial diffusion later occurred. Originating in western New York in the 1820s, the Mormon culture emphasized hard work and thrift, combined with a strong sense of community life and civic obligation. This culture contrasted markedly with the buccaneering individualism that characterized many of the Mormons' gentile counterparts. Public opposition to Mormon beliefs (and Mormon economic successes) led the Mormon community to move progressively farther to the West, eventually settling in Illinois. The murder of their leader, Joseph Smith, and growing harassment by the secular community convinced the Mormons that they needed to leave the more densely populated part of the United States if they were going to practice their religion in peace. Beginning in the late 1840s, tens of thousands of Mormons made the difficult trek to the deserts of the Great Salt Lake Basin in Utah, where they developed a thriving irrigation-based agricultural economy largely isolated from the eastern culture cores. To this day, Salt Lake City remains the heart of this culture group (Figure 2-19).

▲ Figure 2-19 **Salt Lake City, Utah.** The Mormon Tabernacle, the spiritual center of the faith's adherents, is overtopped by the skyscrapers of Salt Lake City's central business district as the Wasatch Mountains loom in the background.

Eventually, even the Great American Desert yielded to the settler's plow. While hostile environments and persistent conflicts with indigenous groups delayed the early spatial integration of this settlement frontier, by the 1870s and 1880s the Great Plains were succumbing to the land appetites of both European immigrants and Americans, and to the technologies of industrial America. These technologies included railroads to transport settlers to this frontier, barbed wire to fence in land where wood was unavailable, windmills to pump water out of the ground, and steel plows to break the heavy prairie sod. By the end of the 1890s, the frontier (excluding Alaska) was officially closed.

Westward expansion was also important to Canada's nation-building experience, although there are important differences in the way that Canada became a country and opened its western regions. To begin with, the people of Quebec in the 1770s were largely indifferent to the colonial uprising against the British Crown in the south. In spite of (and perhaps because of) colonial attacks on Montreal and Quebec City, Quebec and the maritime colonies remained loyal to Britain. After the Revolution, what was left of British North America became an important destination for British sympathizers, known as **Loyalists**, who no longer felt welcome in the lower 13 colonies. Several tens of thousands of Loyalists made the journey northward, many in Royal Navy sea evacuations.

The Loyalists established communities in Nova Scotia, New Brunswick, and the Great Lakes region, creating an English-speaking counterweight to French-speaking Quebec and contributing to an Anglicization process that eventually put French speakers in the minority. In 1791, this linguistic divide was formally recognized with the establishment of French-speaking Lower Canada on the lower reaches of the St. Lawrence River and the Gulf of St. Lawrence (modern Quebec), and English-speaking Upper Canada on the upper reaches of the St. Lawrence and in the Great Lakes region (modern Ontario). The promise of free or low-cost land in the Great Lakes region led to additional substantial immigration by Irish, Scots, English, and Welsh to Upper Canada. Cultural conflicts and less attractive land inhibited similar immigration into Quebec.

Canada officially became a self-governing country with the implementation of the British North America Act on July 1, 1867. The new country embraced parts of modern Ontario, Quebec, New Brunswick, and Nova Scotia. The new government combined an American type of federalism—a confederation of provinces—with a British parliamentary form of government—a House of Commons and a Senate.

Unlike in the United States, the western lands into which Canada expanded were already under Canadian or British control. While a sense of manifest destiny may have been part of the Canadian expansion process, its expression was far more subdued than in the United States—there were no foreign powers controlling territory toward which Canadians might have felt a sense of destiny. By 1870, Manitoba (the Red River area) was brought into the Confederation, as were British Columbia in 1871, Prince Edward Island in 1873, and Alberta and Saskatchewan in 1905. Newfoundland did not join the Confederation until 1949 (Figure 2-20).

As in the United States, Canadian frontiers contained indigenous people that often stood in the way of European settlement. Some argue that the presence of the scarlet-coated North West Mounted Police, who preceded the settlers, created a civilizing influence that limited the number and severity of conflicts. Armies had to be raised to subdue rebellions only twice, in 1870 and 1885. Others argue that, the relatively peaceful settlement process notwithstanding, the long-term plight of Canadian Indians was only marginally better than that of their U.S. counterparts.

The United States and Canada are thus lands largely peopled by immigrants and their descendants. Much of the spatial integration process would not have occurred had it not been for immigrants, and the texture of North American culture would have evolved much differently.

U.S. geographer Wilbur Zelinsky observed that immigration into the United States can be divided into two major eras.[1] The first, or "colonial" era occurred between 1607 and 1775. Most European migrants were from the British Isles, followed by Germans, Dutch, and Swedes. These migrants generally came with the expectation that they would remain English, or Scottish, or German once they arrived in the New World and that there would be no acculturation process. Large numbers of Africans were brought involuntarily to the United States as slaves, until the slave trade was prohibited in 1808—prohibited as a matter of law, but not always as a matter of fact. The African acculturation process was painful. The second, or "national," era lasted from roughly 1820 to the present. These migrants entered a dynamic culture that required major adjustments in terms of language and lifestyle—they experienced "culture shock." The national era can be divided into three smaller segments, each with its own characteristics: (1) the northwestern European wave (1820–1870) was still heavily British, Irish, German, and Dutch, but included some of the first migrants from Asia and Latin America; (2) the "Great Deluge" (1870–1920) witnessed the migration of more than 26 million people to the United States, many from traditional northwestern European sources, but many more from Scandinavia, eastern and southern Europe, China, Japan, and Latin America; and (3) the miscellaneous influx from 1920 to the present has been made up of a wide variety of origins, especially Asian and Latin American.

Immigrants settling in the United States tended to cluster in regions, many of which are still identifiable (Figure 2-21). This clustering reflected multiple processes, including "environmental affinity" in the words of Zelinksy, or the selection of a familiar physical habitat. Such affinity helps explain the presence of Scandinavians in the upper Midwest, Dutch farmers in southwestern Michigan, and Italians and Armenians in parts of California's Central Valley (Figure 2-22). Relative proximity to home led Asian immigrants to focus on the West, Hispanic immigrants the Southwest and south

▼ Figure 2-20 **The territorial evolution of Canada.** By 1912, after final adjustments to the territories of Manitoba, Ontario, and Quebec, the current political geography of Canada was largely in place. A new territory, Nunavut, was carved out of the Northwest Territories in 1999.

[1] W. Zelinsky, *The Cultural Geography of the United States: A Revised Edition* (Englewood Cliffs, NJ: Prentice Hall, 1992), Chapter 1.

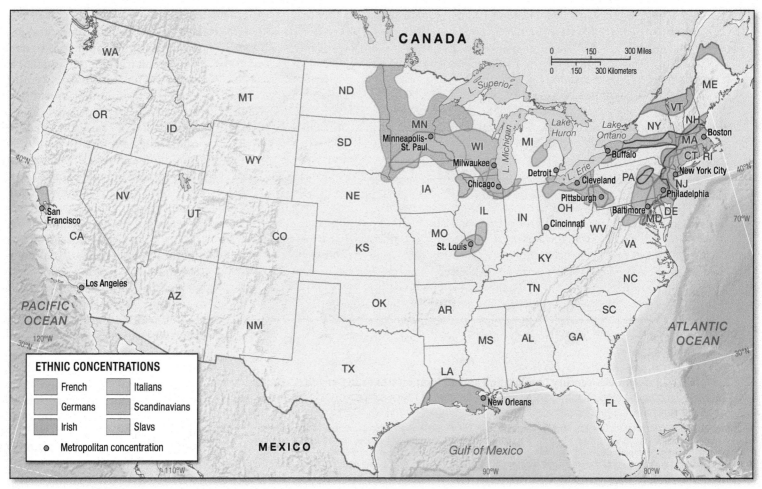

▲ **Figure 2-21 Generalized areas of settlement for selected European groups.** While immigrants ultimately settled throughout the United States, specific regions became associated with specific groups of people. For more detailed information, including non-European groups, see W. Zelinsky, *The Cultural Geography of the United States: A Revised Edition* (Englewood Cliffs, NJ: Prentice Hall, 1992), 30–31.

Florida, and European immigrants on Atlantic port cities. Economic circumstances also influenced clustering patterns, as in the case of Japanese horticulturists in southern California, Portuguese fishermen in Massachusetts, and Basque shepherds in the arid West. Most European immigrants during the great waves of the nineteenth and early twentieth centuries avoided the U.S. South.

Immigration into Canada was just as intense and nation-altering as in the United States, but began somewhat later (Figure 2–23). Prior to Confederation, Canada was populated mainly by people from the British Isles, by Americans (Loyalists and others who came looking for cheap land), and by the French in Quebec. Most of these people lived in the east. Few Europeans inhabited the great expanse of land that lay between the Great Lakes and the Rocky Mountains, that is, the northern extension of the Great Plains and the Interior Lowlands that Canadians call the Prairies. As with America's Great American Desert, the Prairies were initially perceived as hostile and remote. With the completion of the Canadian Pacific Railway main line in 1885, the Prairies became easily accessible.

Settlers soon discovered that the Prairies contained the last large expanse of good farmland in North America. Migrants came from

▼ **Figure 2-22 Tulip festival of Holland, Michigan.** Large numbers of Dutch immigrants settled southwestern Michigan in the mid-nineteenth century. The region continues to bear the Dutch cultural imprint.

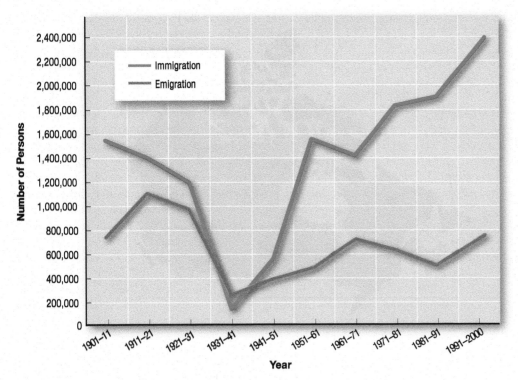

▲ **Figure 2-23 Twentieth-century immigration and emigration trends in Canada.** Canada, like the United States, is a land of immigrants, especially during the twentieth century. Unlike the United States, Canada has experienced considerable emigration, or out-migration, often to the United States.

Source: Adapted from Statistics Canada, http://www.statcan.ca/english/Pgdb/People/Population/demo03.htm. Statistics Canada information is used with the permission of Statistics Canada. Users are forbidden to copy the data and redisseminate them, in an original or modified form, for commercial purposes, without the expressed permission of Statistics Canada. Information on the availability of the wide range of data from Statistics Canada can be obtained from Statistics Canada's Regional Offices, its World Wide Web site at http://www.statcan.ca, and its toll-free access number, 800-263-1136.

Population Contours

In 1776, the United States had perhaps 3 million people, and Canada only a small fraction of that number. During the nineteenth and twentieth centuries, the populations of both countries increased substantially (Figure 2–24). By early 2013, the U.S. population had risen to more than 315 million and the Canadian population to an estimated 35 million. By 2020, both countries may see substantial growth, due partly to natural increases but mainly to immigration.

The rapid population growth experienced by the United States after 1800 was due to not only immigration but also high birthrates and declining death rates. Although precise figures are not available, annual birthrates and death rates in the early nineteenth century may have exceeded 5 percent and 2 percent, respectively. A century later (during the Great Depression of the 1930s) the birthrate had decreased to 1.8 percent per year, but rose sharply after World War II, possibly to compensate for wartime postponements of childbearing and in response

everywhere to make claims. Between 1896 and 1911, about 2.5 million immigrants—from the Ukraine, Hungary, Iceland, Sweden, Germany, Italy, Finland, and China, as well as Great Britain and the United States—arrived in Canada, many on the way to the Prairies. Canada's population increased by a third in the first decade of the twentieth century. After World War I immigration continued, particularly to urban areas, and included large numbers of people not only from traditional European and U.S. sources but also from Commonwealth Asia and Africa. Today, the people of Canada embrace a cultural spectrum of such breadth that Ukrainian, Sri Lankan, and Portuguese surnames (among many others) are almost as common as British and French. Canada, like its neighbor to the south, has become a cultural mosaic.

Stop & Think

▶ What do you think the cultural geography of North America would look like today if there had been a major spatial diffusion of people out of the French Canada Core?

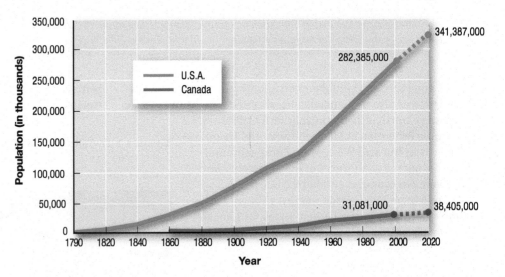

▲ **Figure 2-24 Population growth trends in the United States and Canada, with projections to 2020.** Note the upward trend in both cases, especially in the United States. The U.S. trend line begins in 1790. Due to data limitations, the Canadian line begins in 1861. The actual year of the Canadian census is 1 year after the U.S. census—the 1980 Canadian data shown in the table, for example, were actually for 1981.

Sources: U.S. Bureau of the Census, *Statistical Abstract of the United States: 2001*, no. 1, 3 (Washington, D.C.: Government Printing Office, 2001); *Statistical Abstract of the United States: 2008*, Population, Table 3; Statistics Canada, Population and Growth Components, and *http://www.statcan.ca/english/Pgdb/People/Population/demo02.htm*. See also the Statistical Abstract for 2012, Population section, Table 3, for the most recent 2020 projections. For Canada, see Statistics Canada, *http://www.statcan.gc.ca/tables-tableaux/sum-som/l01/cst01/demo08b-eng.htm* (extracted March, 2013) for the latest "medium growth" 2020 projection (actually, 2011).

to postwar economic prosperity. During the baby boom of 1946–1965, the birthrate reached a new modern-era high; by 1955, it was 2.5 percent. Subsequently it dipped to a new low of 1.46 percent in 1975. When the baby boomers themselves reached childbearing age, a very modest upturn took place, and by the early 1990s the birthrate was slightly above 1.6 percent. Because of the increased population, that rate represents nearly 4.1 million births per year, not so different from the 4.3 million at the peak of the baby boom. In 2008, there were 4.2 million births and a birth rate of 1.4 percent.

The Canadian demographic experience has been generally similar to that of the United States. Canada grew mainly by natural increase between 1867 and 1900, but its natural increase was limited by a low fertility rate. Net migration was negative during most of that early period, but after 1900 a large influx of immigrants arrived from Europe, slowing the decline of birthrates. Like the United States, Canada experienced a low birthrate during the Depression and a sharp rise after World War II. The Canadian baby boom was also followed by a decline in the birthrate during the late 1960s and early 1970s. The birthrate in 2009–2010 was slightly more than 1.1 percent. Unlike in the United States, emigration, or

out-migration, from Canada in the postwar years has been substantial. Roughly 3 percent of Canadians live abroad.

Most of the U.S. population presently lives east of the Mississippi River (Figure 2–25). The greatest concentrations are in the northeastern quadrant of the country, the area bounded by the Mississippi and Ohio rivers, the Atlantic Ocean, and the Great Lakes. Population densities are somewhat lower in the South, except in such growth areas as the Piedmont and southern Florida. Most of the West Coast population is clustered in lowland areas, such as the Los Angeles Basin, the Great Valley of California, the valleys of the Coastal Ranges near San Francisco, and the Willamette Valley and Puget Sound Lowland. The remainder of the western United States is sparsely populated, particularly west of the 100th meridian. Exceptions include metropolitan areas such as Phoenix, Arizona; Denver, Colorado; and Salt Lake City, Utah.

Given the harsh environments of the Canadian north, most Canadians live within 200 miles (320 kilometers) of the U.S. border (Figure 2–26). Were it not for the interrupting effect of the Canadian Shield (which supports few people), Canada's population distribution would appear as a ribbon of settlements stretched across the

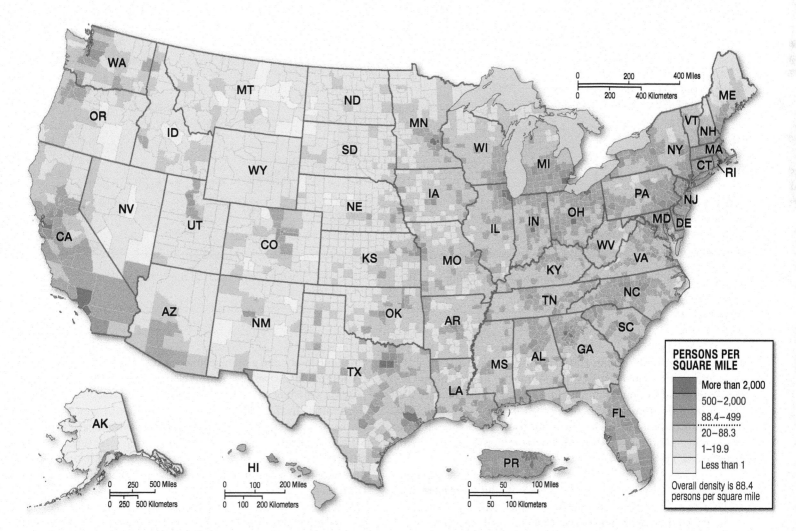

▲ **Figure 2-25 Population distribution in the United States, 2010.** The unevenness of population in the United States reflects numerous influences, including natural environments, early settlement areas, level of urban and industrial growth, and ongoing redistribution (mobility).

Source: Bureau of the Census, *2010 Census of Population*. See *http://www.census.gov/geo/maps-data/maps/pdfs/thematic/us_popdensity_2010map.pdf* (extracted March 2013).

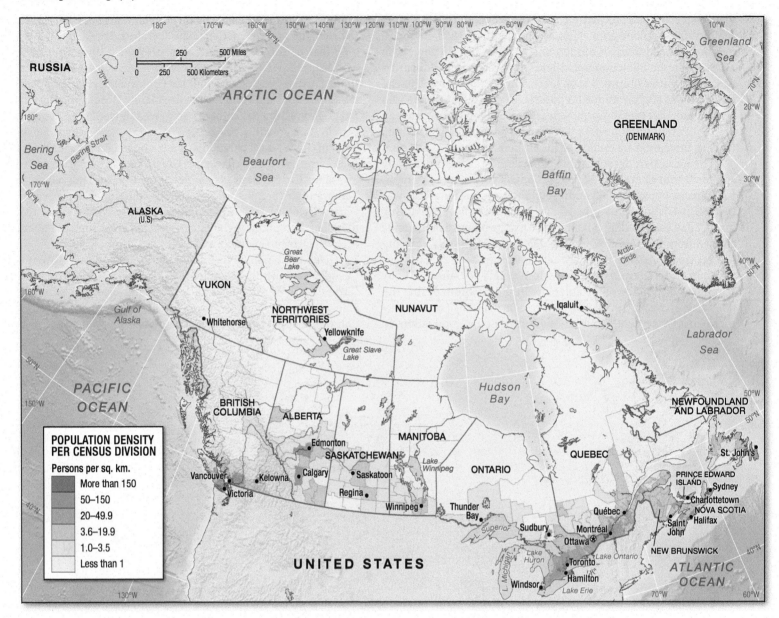

▲ **Figure 2-26 Population distribution in Canada, 2006.** Note the proximity of the Canadian population to the border with the United States. Environmental conditions farther north are too harsh to support large population concentrations.

Source: Government of Canada. Natural Resources Canada, Earth Sciences Sector. *The Atlas of Canada.* See *http://atlas.nrcan.gc.ca/data/english/maps/population/population_density_2006 _cd_map.jpg* (extracted March 2013).

southern edge of the country. The majority (around 60 percent) of Canadians live between Windsor, Ontario, and Quebec City, Quebec. The other major population concentration is on the West Coast around the cities of Vancouver and Victoria, British Columbia. The only exceptions to this southern population ribbon are selected small concentrations in the prairie provinces, most notably Alberta's capital city, Edmonton.

For a long time, population movement in the United States and Canada was focused on westward migration. But there were other migrations. After the American Revolution, British sympathizers from the United States migrated to Canada. Later in the nineteenth century and into the twentieth century, African Americans from the U.S. South fled difficult social and economic circumstances for the promise of jobs and improved living conditions in northern

cities. Job losses and general economic distresses played a role. The "Okies" made the trek to the promised land of California during the 1930s in search of opportunities denied them in Depression-weary Oklahoma. In the 1970s and 1980s, the South was a draw for workers displaced by struggling industrial economies in the upper Midwest. Throughout the latter half of the twentieth century, climatic amenities lured retirees to Florida and Arizona, among other southern destinations. This rediscovery of the South and West continued into the 1990s and 2000s, generally at the expense of the Midwest and Northeast (see Figure 2-27 for recent trends). This process is part of a larger Sun Belt migration stream. Canada has no "Sun Belt," unless the relatively snow-free (but overcast) conditions of Vancouver qualify as such; and British Columbia has been one of Canada's major growth provinces (Table 2-1). Much of Canada's

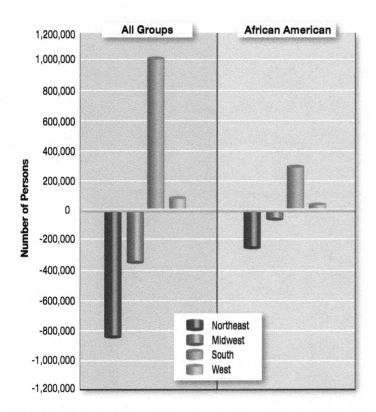

◀ **Figure 2-27 Net migration by United States region: 2005–2010.** Between 2005 and 2010, more people left the Northeast and Midwest than moved in. The South and the West experienced the opposite migration, with the South especially strong in terms of population gain. The South is no longer experiencing a net loss of African Americans and has recently emerged as an important destination.

Source: Adapted from U.S. Bureau of the Census, *Current Population Survey, 2010 Annual Social and Economic Supplement,* Table 11 in *http://www.census.gov/hhes/migration/data /cps/cps2010-5yr.html* (extracted March 2013).

recent interprovincial migration, however, has focused on Alberta and Ontario, due in no small measure to favorable economic conditions. Massive numbers of Canadians do flock to Florida's beaches every winter, a holiday tradition among many residents of Ontario and Quebec. In effect, Florida is also Canada's "Sun Belt."

An old but continuing process of population redistribution involves the migration of people from farms to cities and then outward to suburbs. The shift from an agrarian to an industrial society started early in the nineteenth century and continued into the twentieth century. Factories attracted people from the countryside, swelling the sizes of cities. Urban growth continued during the twentieth century but was spurred more by expansion of tertiary activities than by industrial growth. The 1970s saw movement to smaller metropolitan and nonmetropolitan areas, suggesting a counterflow away from large cities that continues today.

Table 2–1   Canadian Population Data

	Population, 2011	Population, 2006	Percent Estimated Population Change, 2006–2011	Net Interprovincial Migration, 2011–2012
Canada	33,476,688	31,612,897	5.9	N/A
Newfoundland	514,536	505,469	1.8	−1,556
Prince Edward Island	140,204	135,851	3.2	−1,252
Nova Scotia	921,727	913,462	0.9	−3,008
New Brunswick	751,171	729,997	2.9	−2,182
Quebec	7,903,001	7,546,131	4.7	−3,886
Ontario	12,851,821	12,160,282	5.7	−8,091
Manitoba	1,208,268	1,148,401	5.2	−4,675
Saskatchewan	1,033,381	968,157	6.7	2,846
Alberta	3,645,257	3,290,350	10.8	28,170
British Columbia	4,400,057	4,113,487	7.0	−4,648
Yukon	33,897	30,372	11.6	265
Northwest Territories	41,462	41,464	0.0	−1,491
Nunavut	31,906	29,474	8.3	−492

Sources: Adapted from Statistics Canada population sources, including *Focus on Geography Series, 2011 Census, http://www12.statcan.gc.ca/census-recensement/2011 /as-sa/fogs-spg/Facts-pr-eng.cfm?Lang=Eng&GK=PR&GC=35;* for migration data, *http://www.statcan.gc.ca/tables-tableaux/sum-som/l01/cst01/demo33a-eng.htm,* which makes reference to CANSIM, Table 051-0004. See also *Canada Yearbook 2012* (Ottawa: Ministry of Industry, 2012).

Note: Canadian census figures reveal meaningful growth in population between 2006 and 2011. Stunning growth has occurred in the West, particularly in Alberta. The Maritime provinces tend to have the lowest growth rates. The interprovincial migration data indicate the extent to which Canadians moved to or out of a particular province, with negative numbers indicating more people leaving than coming. Even though the data are estimates for only 1 year and hardly establish a trend, they do show that, for the most part, people are leaving the slow-growth provinces. Alberta is far and away the most popular destination, followed by Saskatchewan. Ontario and British Columbia have been more of a long-term destination than the 1-year figures suggest.

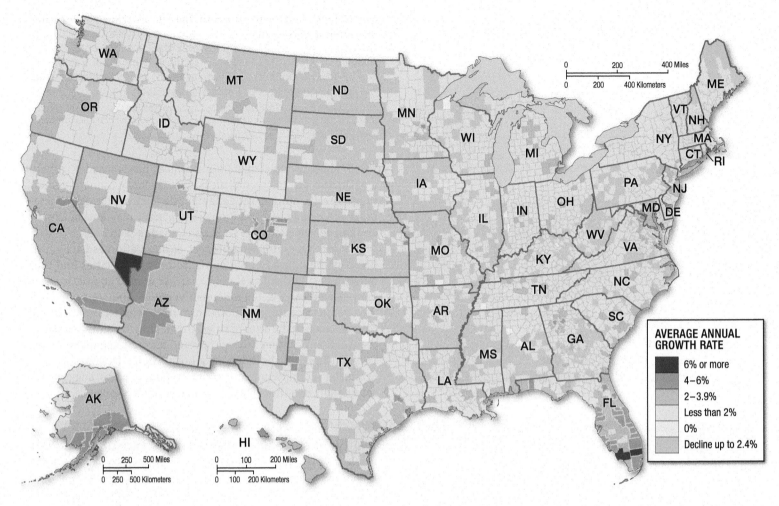

▲ Figure 2-28 **Population growth by county, 1930–2000.** During the twentieth century, population growth in the United States focused on the urban areas of the South (particularly Florida and Texas) and the West. Regions of population decline included portions of the Great Plains and the lower Mississippi Delta.

Sources: *U.S. data from 2010 Census of Population: Geography.* Map adapted from *http://www.census.gov/geo/maps-data/maps/thematic.html* (extracted March 2013).

County-level data for the 1930–2000 period provide a clearer picture of U.S. population change (Figure 2–28). First, central cities, while not necessarily losing population, generally did not show meaningful population growth, but surrounding counties often recorded exceptional growth (the Chicago and Atlanta metropolitan areas provide good illustrations). Second, the metropolitan South and a variety of counties in the West were areas of growth. Note, for example, the growth in most of Florida, much of the Piedmont from North Carolina to central Alabama, the Dallas-Houston-San Antonio triangle in Texas, most of Arizona, the Albuquerque-to-Denver corridor, Nevada, and the Pacific Northwest. Also visible are clearly defined pockets of population decline. For example, the Great Plains experienced substantial declines that reflect dramatically changing agricultural circumstances. A second pocket of decline shows the other face of the Sun Belt, one of continuing economic distress in the lower Mississippi Delta and much of the coastal plain. The pattern of decline extends northward along the western side of the Appalachians from Kentucky to Pennsylvania.

Some of the same trends can be identified for Canada (see Table 2–1). Recent population change in Canada (2006–2011) has focused on its central and western regions. Alberta experienced significant growth during the 1970s and early 1980s in response to the development of petroleum resources, and remained an attractive destination for migrants between 2006 and 2011. British Columbia is experiencing relatively high population growth as a result of both international and interprovincial migration, and its growing global prominence. The only provinces that showed very low growth rates were Newfoundland and Nova Scotia, although in the case of Newfoundland it is no longer losing population.

Stop & Think

▶ Migration of people from one region to another can occur for a variety of reasons. Describe some of the reasons for interregional migration in Canada and the United States.

The United States and Canada: Profiles of a Developed Realm and Its Challenges

Commercial activity began in the United States and Canada soon after the first European settlers arrived. Wherever transportation was suitable—which usually meant water transportation—commercial activity soon appeared. Agricultural commodities, lumber, furs, and fish were produced or gathered for exchange. Primary production and tertiary activities (that is, trade) were important long before the settlement of North America was complete and before manufacturing activities became significant. Agriculture quickly became the economic mainstay, and remained so for more than two centuries. Most people today, however, are not farmers (Figure 2–29). Eventually, a growing population, the development of thriving markets, and revolutions in manufacturing and transportation technologies led to a largely urban-industrial economy in both countries, followed by what we now see: a largely urbanized continent with people increasingly working in service-related occupations. In this section we examine the geographic patterns of agricultural, manufacturing, and urban activity, and show how they have shaped a developed realm.

Agricultural Issues and Regions

The story of agriculture in the United States and Canada is one of extraordinary success. This success can be seen in the billions of dollars of farm output each year—nearly $300 billion in the United States by 2009—and in the extent to which both countries supply the food needs of other countries. In 2010, the United States supplied over 53 percent of the world's corn. Agriculture was an early engine of economic development and remains critically important to both countries today. Several factors have contributed to this success:

- The United States and Canada have an abundance of good land. Approximately one-fifth of all U.S. land is classified as cropland, though not all of it is cultivated in a given year. The United States has about 5 acres (2 hectares) per person for agricultural production, either foodstuffs or industrial raw materials, which is a highly favorable ratio. Canada's northern latitudes give it far less usable land, from 4–5 percent of the total land area; but Canada's small population relative to the United States means that the ratio of farmland to people is about the same as in the United States at slightly fewer than 5 acres per person.

- Agriculture in Canada and the United States is highly mechanized. The days of the farmer and his mule are over; now it is the farmer and his tractor with the air-conditioned cab and the global positioning system unit. In 1910, approximately 1,000 tractors were in use on American farms, along with roughly 24 million horses and mules; by the late 1990s, the figure approached 4 million tractors, and not nearly as many horses and mules. In addition to machinery, farmers use hybrid seeds, pesticides, herbicides, biotechnology, and scientific farming techniques to increase output per acre. Output per farmer is immeasurably greater than a century ago, thereby requiring fewer people to supply the food and fiber needed to maintain a healthy economy. Today's farmer can produce enough food to support many tens of families.

- Both countries possess large-scale, diverse environments that promote regional specialization in agriculture. Over time, farmers discovered that it was more productive to concentrate efforts on what they grow best relative to other farmers in other regions. The Corn Belt can produce more bushels per acre of corn than wheat and produces more per acre of both crops than the Wheat Belt; thus Corn Belt farmers focus on corn production and let Wheat Belt farmers grow wheat. In the end, there is more corn and wheat for everyone, and trade makes up for any regional deficiencies. The principle at work is called comparative advantage (at least for the wheat farmers—it would be an "absolute" advantage for the corn farmers) and points out that specialization and trade in the agricultural economy lead to greater productivity overall.

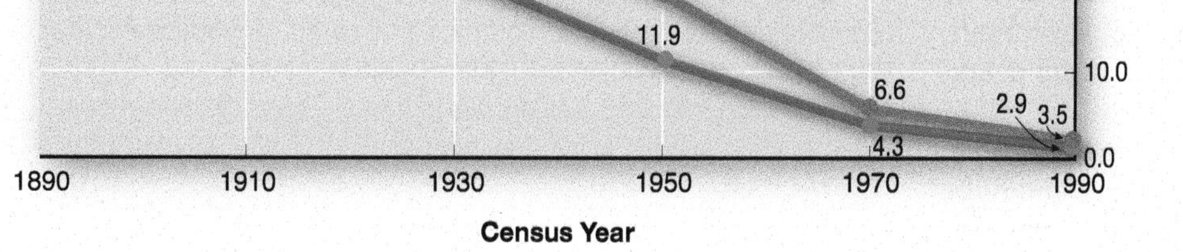

Agricultural Regions

The largest expanse of highly productive land in the United States and Canada is the **Corn Belt**, which extends from central Ohio to eastern Nebraska, and from Minnesota and South Dakota south to eastern Kansas (Figure 2–30). Here lie some of the most fertile soils in the world—rich in organic content and neither too wet nor too dry. The relatively flat terrain is easy to work, rainfall occurs at the right times of the year (late spring and midsummer for corn cultivation), and farmers

▲ Figure 2-29 **Percentage of economically active population employed in agriculture during the 20th Century.** Both the United States and Canada have witnessed major declines in the proportions of their workforces engaged in agriculture since 1890. It is estimated that by 2020 in the U.S., the percentage will be about 1.2 percent.

Source: Data from B. R. Mitchell, *International Historical Statistics: The Americas 1750–1993*, 4th ed. (London: Macmillan, 1998), Table B1.

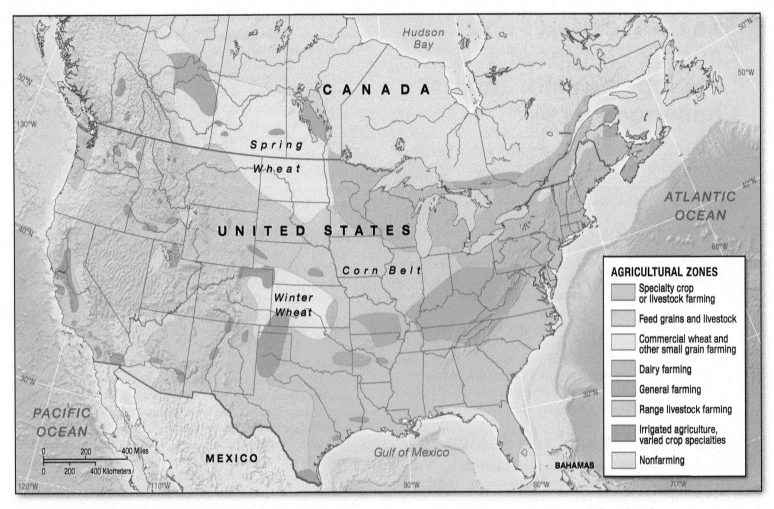

▲ **Figure 2-30 Agricultural regions of the United States and Canada.** A diversity of natural environments has contributed to the great variety in types of agricultural specialties possible in the United States and Canada.

can expect a frost-free period of at least 150 days. The natural advantages of the Corn Belt are enhanced by its proximity to some of the densest railroad and waterway networks anywhere, which facilitate shipment of farm commodities to domestic and foreign markets.

Historically, corn or maize has been favored by farmers because of its multiple uses—food for people and feed for animals—and its high yield per acre (Figure 2-31). But winter wheat, oats, hay crops, and soybeans are grown by Corn Belt farmers for additional income, to support livestock operations, and to keep soils fertile. The versatile soybean is now a major source of income as it is used for animal feed, as protein-rich human food (tofu, soymilk, soyburgers, cooking oil), and as a component in the manufacturing of everything from paint to plastics.

Given the name "Corn Belt," it may come as a surprise that much of the region's income actually comes from selling cattle and hogs. The Corn Belt began as a mixed farming region in which equal emphasis was given to crop cultivation and livestock production, with each activity supporting the other. Over time, farmers discovered that they could make more money from corn by feeding it to livestock and then selling the livestock for meat. Increasingly, the Corn Belt became a "Pork Belt" and a "Beef Belt." Today cattle and hogs are prepared for market in large feedlots, especially in the western part of the Corn Belt. Livestock production accounts for about three-fourths of Corn Belt income, even though most of the farmland is used for cultivation.

To the north of the Corn Belt is the **Dairy Belt** stretching from Nova Scotia and New England to Wisconsin and Minnesota. While physical conditions are less favorable for agriculture—soils are thinner and less fertile, and growing seasons are shorter—the cool and moist environments promote the growth of hay crops and the productivity of dairy cattle. Depending on proximity to markets, dairy farmers concentrate on producing fluid milk or milk products. Farmers in the east focus on fluid milk production to meet the

▼ **Figure 2-31 Corn cultivation.** Corn is widely cultivated in the Corn Belt; but so are wheat, oats, hay crops, and particularly soybeans. In addition, livestock production is a major source of income.

demand for milk in nearby cities. Fluid milk is heavy and perishable and generally is produced close to the market to avoid excessive transportation costs. The term **milkshed** refers to a city's adjacent milk-producing region; cities everywhere, not just in the Dairy Belt, have milksheds. Farther west, away from major urban markets, dairy farmers focus on manufactured milk products such as cheese in Wisconsin and butter in Minnesota. Cheese and butter are less perishable and easier to transport than fluid milk.

The **Specialty Crop and Livestock region** extends from southern New England to eastern Texas and embraces a wide array of activities. In the Northeast, with its hilly terrain and thin, sometimes infertile soils, agriculture was always difficult except in such favored locations as the Connecticut and Hudson river valleys. As much better western lands became available to farmers, the Northeast lost whatever earlier advantages it had, except for one: proximity to large cities. Northeastern farmers discovered that city dwellers demanded fruits and vegetables, commodities that often are perishable and difficult to transport and are best grown near consumers. Farmers focused on cultivating apples, cherries, asparagus, beans, tomatoes, and related crops—commodities that could be trucked to nearby cities. This type of high-value, land-intensive, market-oriented agriculture is called truck farming. A similar market-gardening region extends from southern Virginia to the Georgia coast. While perhaps not truck farming in the same sense of the word, the Annapolis Valley of Nova Scotia is famous for apples while parts of Maine, New Brunswick, and especially Prince Edward Island are known for potatoes.

The story in the U.S. South is one of both disappointment and good fortune. Large parts of southern Appalachia are too hilly to farm profitably; the relatively acidic soils of the Piedmont have lost much of its already limited fertility because of repeated row-crop production; and the swamps and sandy pine barrens of coastal areas are not well suited to intensive agriculture. But farmers in the more environmentally favored parts of the South, such as the black-soil belts of Texas and Alabama, the alluvium-rich Mississippi River valley, and the Nashville and Bluegrass basins are fortunate: specialized cultivation includes tobacco in North Carolina and Kentucky; peanuts in southern Georgia; rice in Louisiana, Arkansas, and coastal Texas; cane sugar in Florida and Louisiana; and citrus in central Florida and south Texas. Cotton, which came to define the antebellum South, no longer dominates except in a few locations, particularly the lower Mississippi Valley, and central and west Texas. Disease, insects (the boll weevil), depleted soils, and mechanization were the culprits. Despite attempts to revive production in selected areas, cotton has given way to soybeans in much of the South.

The South also has become a livestock region. While cattle are found throughout the South, the rise of poultry broiler production is most noteworthy. Much of the chicken meat consumed in the United States comes from a series of broiler production areas that begin on the Delmarva Peninsula east of Baltimore, continue through north Georgia, and end in Arkansas. Most poultry farms are huge, producing tens of thousands of chickens at a time in a factory-like setting. These farms often are located where traditional farming is no longer profitable and have reinvigorated declining agricultural areas. The lay of the land and problems with soil fertility—historic challenges to southern farmers—are largely irrelevant to poultry producers (Figure 2–32).

To the west of the Corn Belt lie the **Wheat Belts**, characterized by large farms that are highly mechanized. In the winter wheat belt of Kansas, Oklahoma, Colorado, and north Texas, wheat is planted in the fall, allowed to lie dormant in the winter, and harvested in the late spring before the onset of summer's scorching heat. Winter wheat is low-gluten soft wheat that is used for spaghetti, crackers, and pastries. In the spring wheat belt of the Dakotas, Montana, and Saskatchewan, wheat is planted in the spring and harvested in the late summer. Winters in the spring wheat belt are too cold for wheat seedlings to survive, but the summers are not so hot as to stress the wheat. The high-gluten hard wheat of this region is excellent for bread. A third wheat-producing region is the Columbia River basin of eastern Washington, which takes advantage of fertile volcanic soils and the relatively dry climates found on the rainshadow side of the Cascades. Wheat is related to natural grassland vegetation and can withstand relatively dry conditions, although irrigation may be required on the especially drought-sensitive western margins of the Great Plains.

Water is the big issue facing western farmers. Where water is pumped from underground aquifers, or where river water is available, a thriving oasis type of agriculture is possible. Arid soils generally are not infertile soils, just dry soils. Numerous examples of

▼ **Figure 2-32 Chicken farming in Maryland.** Poultry is big business in Maryland, but at very different scales. (a) Large-scale, high-density, confinement systems such as this farm in Snow Hill, Worcester County, dominate Maryland's Eastern Shore. (b) Smaller-scale, more flexible systems such as this low-density, organic operation in Adamstown, Frederick County, characterize other parts of the state.

(a)

(b)

▲ **Figure 2-33 Center-pivot irrigation, Washington State.** In much of the western United States, center-pivot irrigation systems draw water from underground aquifers and distribute it through giant, rotating sprinklers, creating great circles of irrigated crops.

thriving agriculture can be identified: the irrigated cotton lands of west Texas and central Arizona (a new "Cotton Belt"?); the potato lands of the Snake River plains in southern Idaho; the highly diversified vegetable, fruit, and wheat regions of California's interior; the Salt Lake basin of Utah; and the corn and sugar beet areas of eastern Colorado. In the case of Colorado and many other locations, the use of irrigation is apparent from the large green circles of cultivated land produced by center-pivot watering systems (Figure 2-33).

Where irrigation is not possible, little will grow beyond certain types of grasses and xerophytic (drought-resistant) plants like cacti. This land is best used for livestock ranching, a land-extensive grazing activity that normally focuses on cattle. Western ranches are very large, in many cases exceeding 100,000 acres (40,000 hectares), because it takes so much land to support a single animal. In the driest parts of the West, a steer may require more than 100 acres (40 hectares) to find sufficient forage to survive. Increasingly, ranchers send their cattle to nearby feedlots for finishing. As in the Corn Belt, feedlots concentrate large numbers of animals in small areas for intensive feeding and fattening in preparation for slaughter and transport to market. Because of weight, it is more cost-effective to ship processed beef to market rather than live animals. Western feedlots often purchase feed from nearby farms that grow irrigated corn and sugar beets, creating a symbiotic relationship between rancher and farmer. Some of the largest feedlots are found in northern Colorado and in the Texas Panhandle.

An exception to this western pattern is the well-watered marine west coast climate of the Pacific Northwest, where irrigation is not necessary. The Puget Sound Lowland and the Willamette Valley, for example, focus on dairy production, grains, orchard crops, and berries. In the somewhat drier mediterranean climate of coastal California, a variety of crops are grown, including wine grapes in the famous Napa Valley region.

Stop & Think

Which parts of the United States and Canada have environments naturally favorable for agriculture, which parts are successful at agriculture with technological interventions, and which parts offer little promise of successful agriculture?

Continuing Adjustments in Agriculture

The U.S. and Canadian agricultural systems, despite their relatively short histories, have been forced to make many adjustments to accommodate changing economic and technological circumstances.

Farm employment as a percentage of total employment in both countries declined precipitously in the twentieth century. In 1890, between 40 and 50 percent of the U.S. and Canadian workforces were engaged in agriculture; by 1990, the figures hovered between 3 and 4 percent. A trend that has accompanied mechanization is the growth of corporate farming or agribusiness. The corporate farm is capital- and energy-intensive, occupies about twice the acreage of the individual farm, and is often part of a food conglomerate. A cereal corporation may own the farms producing the grain crops that are converted into breakfast food sold to consumers—a "seedling-to-supermarket" concept that is referred to as vertical integration. The rise of the corporate farm, however, does not necessarily mean the death of the traditional family farm. Most farms are still individually owned and operated, and most corporate farms have family connections—90 percent in 2007. In an era of declining, farm populations and growing agribusiness, the number of farms has declined, and their average size has increased. In 1910, the United States had more than 6 million farms; in 2007, the number was down to 2.2 million. Average farm size in 1910 was 138 acres (56 hectares); in 2007, it was 418 acres (169 hectares), with the average corporate farm at 1,469 acres (594 hectares). Rural geography is changing from an intricate mosaic of small farms to a coarse quilt of large corporate holdings.

Success in producing food and fiber has not always led to financial prosperity for the North American farmer. Farm output has increased dramatically, but demand has not necessarily kept up. At the same time, the modern farmer spends increasingly larger quantities of money on equipment, fertilizer, insecticides, herbicides, transportation, and other production costs. The ratio of the prices farmers receive for their products to the costs of inputs has declined precipitously since the 1970s. Only the most efficient farms can survive, creating another reason for the farmer to leave the farm.

Some farmers' fortunes may be changing. In response to growing demand for biofuels to supplement fossil fuels, the price of corn has skyrocketed and the corn grower's cost-price squeeze has been eliminated. Food planners, development economists, and everyday consumers, however, worry that converting corn to fuel is occurring at the expense of already stressed food and feed markets, especially as demand for foodstuffs grows in Asia. Growing corn for fuel could lead to higher food prices at home and starvation abroad.

Natural Resources and Manufacturing in a Postindustrial Era

Canada is especially well endowed with the types of natural resources traditionally important to economic development. The United States has many important resources but is deficient in many others and must rely on global markets for supplies. A relative lack of certain key resources is one of multiple factors that have made the United States a major player in the global economy. Both countries emerged as major industrial powers, thanks initially to large endowments of coal, and remain global manufacturing cores despite a dispersal of production worldwide in response to technology and labor costs.

Energy and Power

Energy consumption often is used as a measure of economic development. Because highly industrialized economies have large energy appetites, the United States and Canada are large consumers of energy. Both countries have large energy supplies, though even these supplies are sometimes inadequate to meet demand.

Coal was the initial source of power for the industrial expansion of the United States, but now provides less than one-fourth of the country's energy supply (Table 2–2). The major coal-producing states are Wyoming, Kentucky, West Virginia, and Pennsylvania (Figure 2-34). Coal is abundant throughout the western states, but production is handicapped by the small local need and the great distances to major eastern markets. Coal from states such as Wyoming offers the advantages of lower sulfur content and easy extraction through relatively low-cost strip-mining techniques. Large

Table 2–2	Energy Consumption (percent) in the United States, by Source		
	1980	**2000**	**2010**
Petroleum	34.2	38.3	36.0
Natural Gas	20.2	23.8	24.6
Coal	15.4	22.6	20.8
Nuclear Power	2.7	7.9	8.4
Renewable Energy	5.4	6.1	8.1

Source: U.S. Census Bureau, *Statistical Abstract of the United States: 2012*, 131st ed. (Washington, DC, 2011); http://www.census.gov/compendia/statab/. Data retrieved from Table 925.

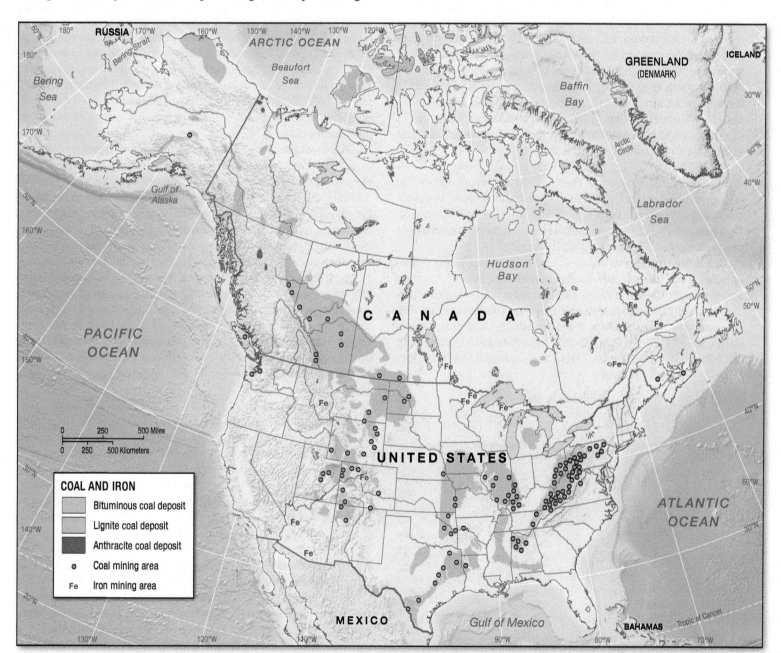

▲ **Figure 2-34 Coal and iron ore in the United States and Canada.** Major coalfields are found in many regions. The Appalachian fields were major contributors of power for U.S. industrial expansion during the late nineteenth century.

quantities of bituminous coal remain, enough to meet the energy needs of the United States for several hundred years. Most coal is used to generate electricity; the rest serves mainly as industrial fuel.

Canadian coal reserves are also large, but coal is less important as a source of power compared to the United States. Most of Canada's reserves are in two maritime provinces (New Brunswick and Nova Scotia) and two prairie provinces (Alberta and Saskatchewan), which are far from the major areas of need (the urban-industrial regions of Ontario and Quebec). Canadians find it more practical to import coal from the Appalachian region of the United States. Another result of the unfavorable location of Canadian coal is that Canada places a greater emphasis on petroleum (which costs less to transport) and hydroelectric power (which is both clean and renewable) as energy sources.

Coal is a good example of how environmental concerns can affect the economic feasibility of a specific resource. Any effort to expand coal use for power will be affected by the restrictions and regulations placed on industries and individuals to control air quality, and by concern for the scarred landscapes that result from surface mining of bituminous coal (Figure 2–35). The negative health impacts, especially of black lung disease, associated with the tunnel or shaft mining of coal are especially troubling. Large-scale use of low-sulfur coal likely will depend on an economical means of removing sulfurous pollutants from high-sulfur coal or finding an acceptable way to use western coal.

Equally troubling are the long-term consequences of continued coal use for global climate change. Coal is fundamentally some combination of carbon, water, and other combustibles. As people are called on to reduce carbon footprints, coal use will be targeted. Coal users will be under considerable pressure to reduce emissions in a manner that helps to slow global warming.

Both the United States and Canada are major producers of petroleum and natural gas. Canada accounted for about 4.3 percent and the United States for about 12.5 percent of world crude oil production in 2012, somewhat higher percentages than in the past. U.S. production is concentrated in Texas, Louisiana, Oklahoma, Kansas, California, and Alaska (Figure 2–36). Some of America's oldest fields are located in an arc from western Pennsylvania to Illinois, but

▼ **Figure 2-35** **Surface coal mining in West Virginia.** The landscape devastation resulting from the surface mining of coal has raised environmental issues, which affect the feasibility of continued use of these lands.

these fields have been eclipsed by much more productive areas to the south and west. Historically, Canadian production has focused on the province of Alberta, with secondary fields in neighboring Saskatchewan. More recently, Canada has drilled into potentially large reserves off the Newfoundland coast in the Jeanne d'Arc Basin—a tough challenge given the winter cold and the high winds. The first fields came into production in 1997.

Both countries are also major consumers of petroleum and natural gas, relying on oil and gas for over three-fifths of their power (see Table 2–2 for U.S. trends). Canadian demand for oil, which accounts for about 2.6 percent of world consumption, can easily be met by domestic production (with a substantial amount left for export). In the United States demand (about 22 percent of world consumption) cannot be satisfied by domestic production, and the shortfall must be filled by imports. In 2006, the United States imported almost 60 percent of its petroleum, mainly from Canada, Saudi Arabia, Venezuela, Nigeria and Mexico (in order of importance); by 2011, the United States imported only about 45 percent of its oil supplies, suggesting a trend toward less dependence on foreign producers that analysts believe will continue into the future. This decline can be explained by the economic downturn that began in 2008, energy efficiency improvements throughout the economy, changes in consumer behavior, and the growth of domestic production.

The most familiar sources of petroleum are the reservoirs that lie in sedimentary basins beneath the land or the sea, and these areas are extensive in both countries. But petroleum is also found in oil shale and in tar sands. The solid organic materials in the shale formations of Utah, Colorado, and Wyoming (the Green River formation) represent one of the world's largest deposits of hydrocarbons, with an immense energy potential, as do the tar sands along the Athabasca River of Alberta. Technological and economic factors will limit their contribution to the energy supply in the near future, but advances in technology or changes in the price of petroleum could increase their role dramatically during the twenty-first century. Any production, however, will have to take into account environmental issues related to waste generation and the use of large quantities of water in relatively dry climates (see *Focus on Energy: North America's New Oil Boom and the Quest for a Sustainable Energy Future*).

Other sources of energy are available, particularly for electricity generation. Hydroelectric sources currently provide 60 percent of Canada's electricity but only 8 percent of U.S. power. Canada has a large quantity of water compared with coal to use for electricity generation, whereas in the United States there is relatively more coal than water available. Areas of waterpower generation in North America include the Columbia River Basin, the Tennessee River Valley and southern Piedmont, and the St. Lawrence Valley (Figure 2–37). The use of nuclear power in the United States increased rapidly during the 1970s and 1980s, and by 2010 was estimated to account for over 19 percent of net electricity generation; in Canada, the figure was only 14 percent.

Every energy source has disadvantages and threats, some more worrisome than others. The greatest concerns are associated with nuclear power. While there have been few major disasters, the prospect of radioactive clouds wafting over cities induces fear in the

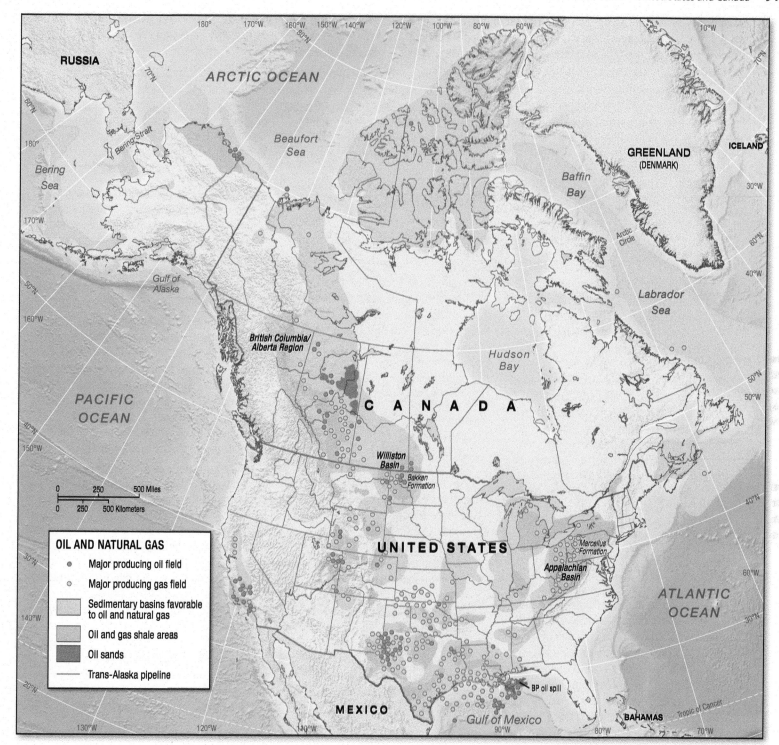

▲ **Figure 2-36 Petroleum and natural gas in the United States and Canada.** The United States and Canada have had the advantage of large oil and gas supplies. But the level of development, substitution of oil and gas for coal, and high per capita consumption have made the cost and availability of oil a significant issue, particularly for the United States. Technology advances in "fracking" may be providing Canada and the United States with a new oil boom.

most hardened observer. A major disaster at Three-Mile-Island, Pennsylvania, was averted in 1979, but it demonstrated the vulnerability that comes with nuclear-power generation. Even more disastrous was the Chernobyl (Russia) meltdown in 1986 and the Fukushima (Japan) disaster in 2011. Interest in nuclear power was gaining momentum until the 2011 earthquake and tsunami in Japan exposed the much greater than expected susceptibility to disaster of nuclear facilities. Many countries are thinking about nuclear-plant phase-outs, although phase-outs have not figured prominently in the U.S. energy discussion.

North America's New Oil Boom and the Quest for a Sustainable Energy Future

Given North America's long-running campaign to reduce fossil fuel dependence, it is ironic that the continent is witnessing a new oil and gas boom. It is all about hydraulic fracturing, or "fracking" of difficult-to-reach oil and gas formations. In the old days, an oil company drilled a well into a pool of oil and/or a bubble of gas, and the resource came flowing to the surface. This form of extraction is increasingly unproductive as the easily reached oil and gas reserves become exhausted. Abundant oil and gas remain, but tend to be trapped in rock formations, particularly shale and some coal beds, that need to be shaken up to release their treasure—hence, fracking.

The basics of fracking are not hard to grasp. Oil and gas-bearing formations, several thousand feet below the surface, are blasted under high pressure with large quantities of water, sand, and chemicals, cracking the rock to release trapped oil and gas. Increasingly the drilling is horizontal as well as vertical, travelling lengthwise through the formation, fracturing it further (Figure 2-4-1).

Fracking has led to new energy booms in locations never considered as oil regions or considered depleted of oil and gas. The Marcellus Shale area of Pennsylvania, for example, which had fewer than 20 gas wells prior to 2008, acquired 800 wells in 2009, and 5,500 in 2012. Permits were issued for another 5,500. The value of recoverable gas in Pennsylvania alone may be $500 billion. The number of wells drilled in New York State doubled to 13,687 between 2000 and 2008, with a potential for 80,000 wells total (although the state has been reluctant to issue new drilling permits pending additional study of environmental impacts). In Canada, the western provinces have embraced fracking to extract what may be 200 trillion cubic feet of natural gas in northeastern British Columbia and northwestern Alberta.

In the last several decades the fossil fuel potential of western North Dakota and eastern Montana has also become evident. The initial focus was on lignite coal, but shifted to oil production in the Bakken shale formation that underlies northwestern North Dakota (particularly around Williston), and parts of neighboring Montana, Saskatchewan, and Manitoba.

Fracking technology is used to dislodge nearly 660,000 barrels a day, making North Dakota second only to Texas as a domestic oil producer. Estimates suggest that the Bakken field may eventually produce up to 2 million barrels a day, and that the number of wells may increase from roughly 8,000 in 2012 to perhaps 40,000 when production becomes fully developed. Cities such as Williston are exploding in population, with the attendant demand for housing and public services. During the height of the recession of 2008, Williston's unemployment rate was less than 1 percent.

The problems associated with fracking may well affect the physical and settlement landscapes of production areas like the Bakken field. The problems, some known and some only worried about at this point, include:

- Water consumption: A single well may require 2–4 million gallons of water and several thousand gallons of chemicals to provide the fracturing force. Where does the water come from in dry climates, and where does the used fluid go? Used fluid that flows back to the surface often is stored in open lined pits that dot the landscape.
- Leaks: Improperly lined pits may leak dangerous contaminants into drinking water supplies. Boreholes created while drilling are lined with cement to keep water and chemicals from escaping as the well is being drilled, but cracks in the cement may allow contaminated water and gas to escape into the surroundings, including drinking water. Turn on the tap and the gas in the water catches fire!
- Fracking control: The fracking process is violent and is meant to crack rock. If the cracks expand beyond the designated rock formation and link up with other fissures and old wells, there could be "fracture communication" causing gas to migrate in unwanted and dangerous ways.

The growth of oil and gas production using new technologies like fracking should not distract decision makers from exploring renewable power sources (Figure 2-4-2), although these are geographically limited. For example, hydropower relies on a dam to back up water which then travels through turbines in the dam to generate electricity. But hydropower is terrain dependent—it needs a flowing river that can be backed up—and creating the reservoir can damage ecosystems and disrupt settlements. Wind power is cited as tomorrow's technology, but requires a windy region—for example, the western plains—and the propeller blades are a flight hazard for birds.

▼ FIGURE 2-4-1 **How fracking is done.** Fracking is an inherently violent process that involves coaxing oil and gas out of rock by cracking the formation that contains these resources.

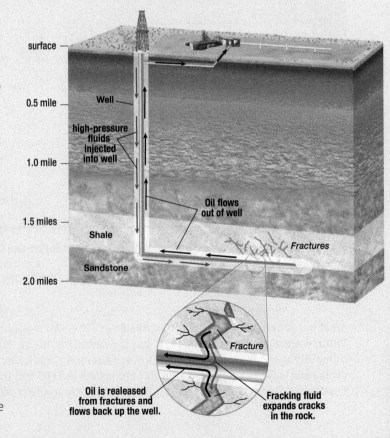

The main problem with solar power, apart from the need for a sunny climate, is the expense of purchasing and installing equipment; electricity from fossil fuels often is still less expensive. Geothermal power makes use of hot spots close to the earth's surface that can be used as a source of steam to drive electricity-producing turbines at the surface. It is a clean energy source but limited in use to hot spots, which tend to be in tectonically active regions.

Biomass power is becoming more important. Plants, most commonly crops (corn, sugar cane), residues from forestry, and the organic materials in municipal and industrial wastes are converted into alcohol that can

be used as a fuel (ethanol in North America) or methane for power generation. Biopower is largely carbon-neutral and may help slow global warming, but diverts food crops to fuel production. Nuclear power is a zero-carbon option, but disadvantages include huge start-up costs and the potentially catastrophic outcome of a plant failure.

As long as fossil fuels are available and affordable, they will continue to account for a substantial percentage of U.S. and Canadian energy consumption. As long as technologies such as those associated with fracking can be deployed in a cost-effective manner, creating new oil booms, the incentives to develop different power sources

will be reduced. But as the connection between carbon-based pollutants and global warming is clarified, and the environmental impacts of hydraulic fracturing become apparent, the allure of a "fracked" energy landscape may be diminished.

Sources: U.S. Energy Information Administration, *Energy in Brief, http://www.eia.gov/energy_in_brief /article/foreign_oil_dependence.cfm.*; U.S, Department of the Interior, "Renewable Energy Sources in the United States," *National Atlas of the United States, http://www .nationalatlas.gov/articles/people/a_energy.html*. For additional reading, see also H. Jacoby and S. Paltsev, "Nuclear Exit, the US Energy Mix, and Carbon Dioxide Emissions," *Bulletin of the Atomic Scientists* 69 (2013): 34–43; and D. LePoire, "Exploring New Energy Alternatives," *The Futurist*, September/October 2011, 34–38.

▼ Figure 2-4-2 **Alternative energy possibilities in the United States.** Oil and gas hardly represent the only options available to the American energy-consuming public. These maps show where alternative energy options can best be deployed.
Source: Adapted from United States Department of the Interior, "Renewable Energy Sources in the United States," *National Atlas of the United States.* http://www.nationalatlas.gov/articles/people/a_energy.html

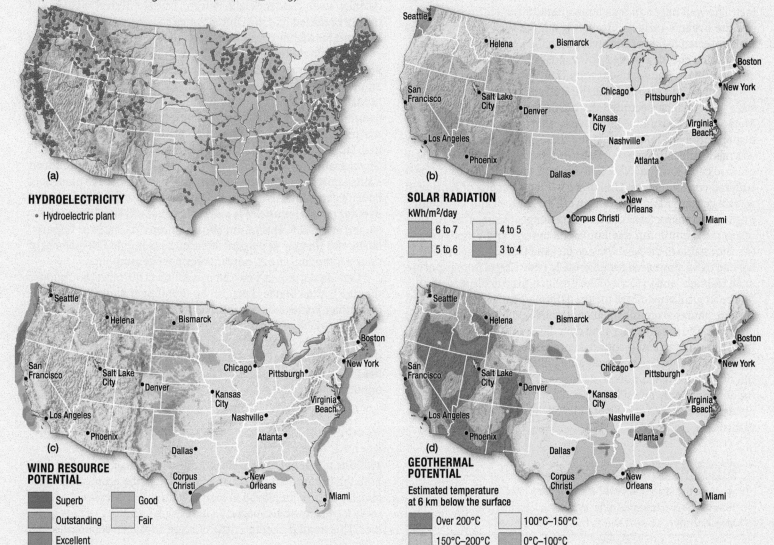

(a)
HYDROELECTRICITY
• Hydroelectric plant

(b)
SOLAR RADIATION
kWh/m²/day
■ 6 to 7 □ 4 to 5
■ 5 to 6 ■ 3 to 4

(c)
WIND RESOURCE POTENTIAL
■ Superb   □ Good
□ Outstanding □ Fair
■ Excellent

(d)
GEOTHERMAL POTENTIAL
Estimated temperature at 6 km below the surface
■ Over 200°C   □ 100°C–150°C
□ 150°C–200°C □ 0°C–100°C

▲ **Figure 2-37 Hydroelectric power generation in Quebec.**
Canada has an abundance of rivers, especially in the east, that can be used for hydroelectric power. Canada relies much more heavily than the United States on this type of electricity source.

Then there are the spills and leaks associated with oil production, which often leave devastation that lasts for years. One of the largest spills in history occurred in the Gulf of Mexico in April 2010. An oil platform off the coast of Louisiana, British Petroleum's Deep Water Horizon, exploded into flames and 11 lives were lost. The well began to leak large quantities of oil from deep below the ocean's surface, producing a large slick that eventually contaminated fisheries and swept onto the Louisiana, Mississippi, and Florida coasts. Shutting down the well, which had to be done at great depths, took 87 days. By that time, perhaps 5 million barrels of oil had escaped into the Gulf and a slick larger than the state of Louisiana occupied the northeastern part of the Gulf of Mexico (see Figure 2-36). Immense effort was made to contain, disperse, remove, and destroy the slick, and while great progress was made, the damage was still considerable. In an eerie reminder of Hurricane Katrina, New Orleans and much of the Gulf Coast suffered consequences. Unlike the wind and floods of Katrina, it was oil contamination affecting work and health, and threatening fragile ecosystems. Fishermen were no longer fishing. New Orleans restaurants could not use Louisiana oysters and other seafood (a major staple of the New Orleans diet and tourist industry). Workers had to be monitored for potentially toxic effects from exposure to oil residues, and to some of the products used to disperse the oil. The ecological impact was major: deformed and mutated fish and shrimp, potential groundwater contamination from oil seepage on the coast, and the prospect of oil globules resurfacing after a storm. British Petroleum and associated corporations remain in clean-up mode and eventually will have spent billions of dollars addressing the damage. The Gulf oil crisis reminds consumers that energy production is only one side of the coin. With production, vigilance needs to be directed at the potential environmental consequences of mistakes.

Stop & Think

➤ What are the major sources of energy in the United States and Canada? How do the two countries differ in the resources that they use to generate electricity?

Metals

Without convenient access to iron ore and coal, the Industrial Revolution in the United States and Canada would have been substantially diminished in its size and scope. Today, both countries remain major producers and consumers of iron ore. Canada is in a position to export ore; the United States is a net importer.

In 2007, an estimated 95 percent of the usable iron consumed in the United States came from the Lake Superior area, especially the great Mesabi Range in northern Minnesota (see Figure 2-34), which has been yielding its iron wealth since the late nineteenth century. The Adirondack Mountains of upstate New York and the area around Birmingham, Alabama, are other locations of historic importance. Numerous additional iron ore deposits are scattered throughout western states. Canadian ore is available in the Lake Superior district at Steep Rock Lake, Ontario; other major deposits have been developed in Labrador.

During the 1940s, high-grade iron ores became less readily available in the United States, and the next two decades saw U.S. dependence on foreign ores grow. Almost half of the iron ore used in the United States during the 1970s came from foreign sources but by the late 1980s, the United States imported less than one-fourth of its iron and by 2010, only 7 percent. One reason for that decrease is that technological advancements now allow the use of ores such as taconite, a very hard rock with low iron content. Today, about half of the ore imported into the United States comes from Canada; Brazil accounts for slightly over a third of imports.

The United States has only limited domestic production of a number of metals critical to manufacturing. Aluminum, for example, is used extensively in the transportation and construction industries and for consumer products. Although aluminum is a common earth element, its occurrence in the form of bauxite, the raw material from which aluminum is extracted, is relatively limited. Most bauxite comes from countries such as Jamaica, Suriname, Guyana, and Australia. For some time, the United States produced a small quantity of bauxite in Arkansas; now almost 100 percent of bauxite consumed in the United States is imported. The situation in other metals industries is less critical, but still significant. About 56 percent of U.S. chromium consumption originates as imports, along with roughly 43 percent of nickel consumption, 69 percent of tin consumption, and 30 percent of copper consumption.

Canada is a metals powerhouse. Much of the nickel and copper that the United States lacks can be found near Sudbury, Ontario. Iron mines are located nearby in Ontario and farther to the east in Labrador. Much of this production occurs on the Canadian Shield, which confirms the importance of shield formations worldwide to the location of metallic resources. Like the United States, Canada has little bauxite and produces large quantities of aluminum using imports.

The Spatial Evolution of Manufacturing in the United States and Canada

Millions of people are employed in the manufacturing sector in both the United States and Canada, although the proportion of the total labor force employed in manufacturing has decreased as relatively faster growth has occurred in services and other tertiary activities. This trend is characteristic of more advanced economies, but manufacturing remains basic to the stability of the American and Canadian economies.

By the mid-nineteenth century, the great Industrial Revolution was spreading across much of Europe and North America. The U.S. version took root beginning in the late eighteenth century in southeastern New England and the Middle Atlantic area from New York City to Philadelphia and Baltimore. New England in particular had an abundance of rivers and streams for the waterwheels that powered the machinery in the new factories, as well as the investment capital needed to build factories, due to earlier commercial successes in lumber, ocean shipping, and fishing. The agricultural sector, which had always struggled with poor soils and sloped land, and existed more out of necessity than choice, had a labor surplus that could be easily converted into an industrial workforce. These workers were critical for the labor-oriented textile industry that came to dominate New England. Other important industries to appear by 1900 included leather and shoes, machine tools, shipbuilding, furniture, ironworking, papermaking, and printing.

Widespread adoption of coal furnaces and steam power in the early 1800s quickly turned the United States into an industrial economy that reached maturity by the end of the century. Mass production of steel beginning in the 1850s hastened the process as did the explosive growth of the railway network that began a decade earlier. The nineteenth century was a time of coal-based industrial development—iron and steel producers, factories with steam engines, and an assortment of related activities relied on coal as a fuel. The geographical consequence was a clustering of factories as close to coal as possible, which favored locations near the coalfields of the western Appalachians and later, the interior lowlands. In some cases, intermediate break-of-bulk locations—those locations where goods changed from one mode of transportation to another—proved ideal, as in the case of the iron and steel industry. Iron ore, transported from the Mesabi Range of Minnesota very inexpensively by Great Lakes vessels, and coal by rail from the Appalachians, contributed to the growth of industrial cities such as Chicago, Erie, and Cleveland. Perhaps the most famous steel city of all, Pittsburgh, initially took advantage of the nearby presence of both coal and iron ore. In eastern Pennsylvania, an iron industry appeared near local anthracite coal deposits.

The South, remote from the new national transportation routes, continued to produce agricultural goods for external markets. Regional differences had led the southern colonies to develop a commercial system that was financially and socially rewarding for a select few, who tended to invest their economic surplus in more land, but this system failed to develop fully the talents and skills of African Americans and poor whites. To some extent, the South was like a colonial appendage. Manufacturing did exist, but never achieved the rate of growth or the dominance that it did in

the North. Following the Civil War (1861–1865), during which the weakness of southern industry was made evident, postwar conditions made a reversal of the traditional economy even more difficult. By the 1930s, the northeastern quarter of the United States had evolved as a major urban-industrial region, with numerous specialized urban-industrial districts interspersed among prosperous agricultural regions. This industrial coreland was characterized by relatively high urbanization, high industrial and agricultural output, high income, and immense internal exchange. The South was a region of low urbanization, limited industrial and agricultural growth, and poverty for great numbers of whites and African Americans in underdeveloped rural areas.

Manufacturing in the Coreland

The primary manufacturing region in Anglo America is a large area in the northeastern United States and southern Canada that consists of numerous urban-industrial districts, within which distinct industrial specialties can be identified (Figure 2-38). That coreland contains a large proportion of the manufacturing capacity of the United States and Canada and the majority of the Anglo American market. The complementary transportation system of waterways, roads, and railroads provides great advantages for both assembling materials and distributing finished products.

Southern New England remains an important U.S. industrial district. New England was early to embrace the Industrial Revolution with its focus on machinery, firearms, and the labor-intensive textile industry. As times changed, the original advantages of New England deteriorated, and much of its Industrial Revolution manufacturing base moved elsewhere. New England was increasingly peripheral to the westward-moving core of the United States and isolated from important markets. It also had become a high-cost region in terms of taxes, power, and labor.

▼ **Figure 2-38 Manufacturing regions of the United States and Canada.** Industrial regions and districts show as much variety in specialties as agriculture does. Industrial specialties reflect the influences of markets, materials, labor, power, and historic forces.

New England reemerged as an altogether different type of industrial region emphasizing high-value products—for example, electronic equipment, firearms, and tools—that can withstand high transportation, power, and labor costs. A "high-tech" region developed to the west of Boston, partly due to the presence of highly regarded centers of learning such as the Massachusetts Institute of Technology and Harvard University. The industrial history of New England illustrates how the industrial structure of a region can change, not always by choice.

Metropolitan New York contains the largest manufacturing complex in the United States, thanks to the city's location at the mouth of the Hudson River, and its function as the major port for the rich interior. This district is characterized by diversified manufacturing, including apparel (particularly high fashion), printing, publishing, machinery making, food processing, metal fabricating, and petroleum refining. There are few "smokestack" industries such as iron and steel.

Inertia, immense capital investments, and linkages with other industries once ensured considerable locational stability for steel industries. But now, all three manufacturing districts with prominent steel industries are undergoing major industrial change. A massive steel-producing capacity exists near Baltimore, Maryland, and the area around Philadelphia, Bethlehem, and Harrisburg, Pennsylvania, supporting shipbuilding (along the Delaware River and the Chesapeake Bay) and many other machinery industries. The steel industry expanded there because of proximity to large eastern markets (that is, other manufacturers of fabricated metals and machinery) and accessibility to external waterways. The importance of waterway accessibility grew as U.S. dependence on foreign sources of iron ore increased.

A second major steel district is a large triangle with points at Pittsburgh and Erie, Pennsylvania, and Toledo, Ohio. The **initial advantage**—that is, early factors that propelled development—of America's oldest steel-producing center stems from its location between the Appalachian coalfields and the Great Lakes. But the locational advantages have changed, particularly for Pittsburgh. Now South American iron ore is shipped to East Coast works, and Canadian ore comes by way of the St. Lawrence Seaway and the Great Lakes. The Baltimore–Philadelphia steel district is closer to both eastern markets and foreign ores, and Detroit and Chicago have better access to Canadian ore and midwestern markets.

The third steel area lies around the southwestern shore of Lake Michigan. It includes Gary, Indiana; Chicago, Illinois; and Milwaukee, Wisconsin, which have a vast array of machinery-manufacturing plants that are supplied by the massive steelworks nearby. Chicago and Gary steel benefited from a superb location; ore moving southeast across the Great Lakes meets coal from Illinois, Indiana, Ohio, Kentucky, and West Virginia. By the late nineteenth century, Chicago had also become a major transportation center; railroads met and complemented freighters, making the southern Lake Michigan area an excellent site for assembling materials and distributing manufactured products.

Southern Michigan and neighboring parts of Indiana, Ohio, and Ontario are distinguished by their emphasis on automotive production, both parts and assembly. Those industries are linked not only to the Detroit steel industry but also to steel manufacturers in the Chicago area and along the shores of Lake Erie (cities such as Toledo, Lorain, and Cleveland, Ohio). The automotive industry serves as a huge market for major steel-producing districts on either side of the international border.

Since the 1970s, the coreland has faced unprecedented challenges to its industrial preeminence, particularly in the steel and automobile sectors (Figure 2-39). Deregulation and competition from lower-cost areas in North America and abroad have led to a restructuring of industry—a shift in what is produced and how it is produced—often with job loss not only to core industries, but to the suppliers and service providers linked to these industries. Workers who survive the cuts constantly worry about the erosion

(a)

(b)

▲ Figure 2-39 **(a) A modern American automobile assembly plant utilizing robotics in Chattanooga, Tennessee; (b) abandoned steel plant in Bethlehem, Pennsylvania.** Manufacturing enterprises that survive do so by emphasizing capital-intensive technology and reduced labor requirements. The result is that, while many industries survive, the residual labor force suffers from unemployment.

of wages and benefits as management responds to competition by outsourcing activities, moving production overseas, and substituting technology for labor.

The same restructuring processes benefited other industries and regions. For example, by the mid-1980s new "silicon" regions—such as the Silicon Valley in California, the Research Triangle of North Carolina, and the outskirts of Boston—introduced whole new ways and types of production and created a "high-tech" North America. Massive assembly districts appeared in labor-abundant and low-wage regions of the world—particularly in Latin America and Asia—to supply North American consumers with low-cost goods.

Restructuring led to regional changes in the distribution of U.S. manufacturing employment. Many states in the industrial core have lower employment percentages than only a few decades ago. By comparison, southern and western states have realized percentage increases, with Texas and California standing out. These shifts have occurred while the overall percentage of the workforce employed in manufacturing has been declining.

Other Manufacturing Regions in the United States

The beginning of the economic revolution in the South is difficult to pinpoint. It probably started in the 1880s with the **New South** advocates, who believed that industrialization was necessary for the region's revitalization. Recent increases in the southern (and western) share of U.S. manufacturing is really the continuation of an industrial dispersion that has been under way for many decades.

The South's first major manufacturing activity was the textile industry, which had evolved as a dominant force in the nineteenth-century industrial growth of New England. By the early twentieth century, however, the industrial maturity of New England was reflected in high wages, unionization, costly fringe-benefit programs, high power costs (imported coal), and obsolete equipment and buildings. As old New England textile plants closed, new ones were established in the South.

Because the textile industry did not continue to grow on a nationwide basis after 1920, the regional shift benefited one region at the expense of another. Firms in the South found a major advantage in the availability of relatively low-cost labor. The agrarian South had a surplus of landless rural people willing to switch from farming to manufacturing. Other advantages included better access to raw materials (cotton), lower cost of power, and lower taxes. By 1930, more than half of the nation's textile industry was located in the South. At one time more than 90 percent of cotton, 75 percent of synthetic fiber, and 40 percent of woolen textiles were of southern manufacture. Globalization, however, is shrinking the southern textile industry, along with its sister apparel industry, as U.S. consumers increasingly buy clothing assembled in other countries.

In addition to the labor-oriented textile and apparel industries, material-oriented pulp-and-paper, food-processing, and forest industries have been historically prominent in the South. In Texas and Louisiana, the petroleum-refining and petrochemical industries have contributed much to Gulf Coast industrial expansion. Furthermore, the South itself provides a significant regional market

for its goods, generating further industrial growth in response to population growth in a variety of industrial sectors, including auto assembly. The transformation from agrarian to urban work has brought higher incomes and new consumption patterns that have greatly increased the market importance of a formerly rural populace (Figure 2-40).

The southeastern manufacturing region coincides with much of the southern Piedmont and the neighboring parts of Alabama (see Figure 2-38), and is dominated by light industry: textiles, apparel, food processing, and furniture. Automobile assembly and electronic products manufacturing are also expanding. The chief attraction has been the availability of suitable labor at costs below industry wage scales elsewhere. An exception to the general pattern is the Birmingham, Alabama, steel area, which started because of an unusual nearby supply of coal, iron ore, and limestone. The Atlanta region is a second exception, focusing on an array of manufacturing types from auto assembly to aircraft construction. The Gulf Coast region emphasizes processing of oil, natural gas, wood pulp, and agricultural commodities that include sugar and rice. Finally, there has been significant growth in manufacturing in central Florida. Especially important are electronic products and electrical machinery.

The Pacific Coast has a significant manufacturing presence, particularly around such metropolitan areas as Los Angeles and Seattle. California's productive agriculture and commercial fishing stimulated the food-processing activities that became the state's first major industry and dominated until the 1940s. World War II generated the defense industries, aircraft manufacturing, and shipbuilding; defense has continued to be a major employer in the Los Angeles area. Automobiles, electronic parts, petrochemicals, and apparel are important to California's industrial structure. Especially in the Los Angeles area, the growing Hispanic population is a major source of labor for the large apparel industry. Much of California's early industrial growth was based on local material resources and rapidly growing local markets. Since the 1950s, however, industries that serve national markets have grown enormously. The

▼ Figure 2-40 **Flood light assembly at the Lumitec production facility in Delray Beach, Florida.** Lumitec develops and manufactures high-quality extreme environment LED lighting.

▲ **Figure 2-41 Jet aircraft manufacturing at a Boeing Aircraft Company plant in Renton, Washington.** Aircraft manufacturing is one of several manufacturing industries that have experienced highly cyclical employment.

high-technology clusters of Silicon Valley in the San Francisco Bay region illustrate this point.

Manufacturing successes in the Pacific Northwest include food processing, forest-products industries, primary metals processing (aluminum), and aircraft factories (Figure 2-41). The emphasis is on processing local primary resources and the use of hydropower from the Columbia River system. Distance from eastern markets and the smaller size of local markets have historically inhibited the region's growth, but the economic expansion of many Pacific Rim countries is proving to be a major stimulus to the entire West Coast, including the Portland and Seattle areas.

> ### Stop & Think
>
> ▶ How did the presence of coal influence the geography of the Industrial Revolution in the United States?

The Spatial Evolution and Distribution of Manufacturing in Canada

Canada is an industrial power as well. Canada's historical and geographical circumstances parallel to some degree those of the United States, except that Canada's industrial coming of age was somewhat later, and Canada relied more heavily on its immense resource base. Proximity to the United States has been both a benefit and source of concern. Early twentieth-century growth in Canada was a response to markets and capital in the United States, and Canadians feared that their location would make them little more than a supplier of raw materials for the United States—a kind of economic colony. Tariffs were adopted to encourage manufacturing and export, thereby ensuring that primary production and secondary processing would take place in Canada. Some Canadians have argued that the tariff policy has resulted in higher prices for the commodities they consume, and therefore a lower level of living. But tariffs have forced the use of Canadian resources, both human and material, by stimulating manufacturing at home and attracting an immense investment of capital in the Canadian economy by U.S. and other foreign companies. In 1987, the United States and Canada agreed to work toward reducing trade restraints. That beginning culminated in the **North American Free Trade Agreement (NAFTA)** implemented in 1994.

Canada, Mexico, and the United States are now members of a single trade arrangement intended to lead eventually to totally free trade and greater interaction between the member countries. Canada and the United States are each other's most important trading partners.

The St. Lawrence Valley and the Ontario Peninsula form the manufacturing heartland of Canada, producing perhaps 75 percent of the nation's industrial output. The Canadian area adjoins the industrial core of the United States and may be thought of as its northern edge, specializing in the production and processing of materials from Canadian mines, forests, and farms. Montreal is the site of a large proportion of Canadian industry; in some ways, it parallels New York City. Both produce a variety of consumer items (chiefly foodstuffs, apparel, and books and magazines) intended for local and national markets. Both cities function as significant ports for international trade.

Outside Montreal, the Canadian industrial structure is more specialized. The immense hydroelectric potential of Quebec is a major source of power for industries along the St. Lawrence River and has provided the basis for Canada's important aluminum production. With bauxite shipped in from Jamaica and Guyana, the Saguenay and St. Maurice rivers, tributaries of the St. Lawrence, have powered aluminum refining and smelting in multiple nearby locations. Because Canada's production of aluminum far exceeds its consumption, it is a leading exporter. The aluminum industry is an example of Canada's role as a processor and supplier for other countries, using both national and international resources in that capacity. Other important industries include processing of nickel, copper, and magnesium—Canada is a leading global producer of each. Not to be left out is the iron production associated with the Canadian Shield. Finally, the valleys of the St. Lawrence and its tributaries also represent Canada's major area of pulp-and-paper manufacturing, with the boreal forests providing the resource base. Canada is the world's leading supplier of newsprint, most of which is sent to the United States and Europe.

Beyond Montreal and its vicinity, the most intense concentration of industry is found in southern Ontario from Windsor to Ottawa, but most notably in the **golden horseshoe**, a district that extends from Toronto and Hamilton around the western end of Lake Ontario to St. Catharines (Figure 2–42). That district

▼ **Figure 2-42 Steel mill in Hamilton, Ontario.** The golden horseshoe district on the western end of Lake Ontario is the source of much of Canada's output of steel, automobiles, and agricultural machinery.

produces most of Canada's steel and a great variety of other industrial goods, such as automobile parts, assembled automobiles, electrical machinery, and agricultural implements. It is one of the most rapidly growing industrial districts in Canada, partly because a significant majority of Canada's market is found along the southern edges of Ontario and Quebec provinces. Canadian tariffs helped to solidify the importance of the golden horseshoe as tariff-avoiding firms from abroad sought established industrial areas to locate factories.

In Canada as in the United States, manufacturing employment has drifted away from established core areas. Increasingly, manufacturers are locating in western Canada, particularly Alberta, which has an abundance of natural resources.

A Postindustrial and Globalized North America

The Industrial Revolution that reshaped the economic geography of North America is being replaced by what is often referred to as a postindustrial society, with an emphasis on providing services and knowledge rather than manufactured goods, a process that began in the 1950s but accelerated in the 1970s and beyond, aided by the explosion of new computer technologies. Workers increasingly are employed in retailing, information and business services, banking, and related activities. In 1950 only 11.9 percent of the U.S. workforce was employed in the service sector, whereas 33.7 percent was engaged in manufacturing; by 2010 this relationship had been reversed, with an estimated 53.7 percent in services and only 10.1 percent in manufacturing (Table 2–3). Within urban areas, service providers increasingly are attracted to suburban locations, often to suburban "downtowns." As with manufacturing, the service industry has become global in its reach; large legal firms, for example, often have offices abroad. Having stressed the importance of services, however, it is important to emphasize that manufacturing has not disappeared. Instead, the manufacturing process has been restructured to take on a more technology intensive and globally involved form. With all the changes to the workforce and improvements in technology, some observers have noted the appearance of "smart cities" in the United States and Canada (see *Visualizing Development:* Does America Have "Smart" Cities?).

This globally involved form of economic activity is not without its detractors. Perhaps the most public complaint is that globalization means transfer of jobs to low-wage labor markets abroad, and pressure to lower wages on jobs left behind. Relatively low-skill assembly jobs moved away first, but now even skilled occupations are at risk as evidenced by the growth of software design in India. Especially troubling are jobs lost to locations where not only wages are lower but working conditions are inferior. As a consequence, critics argue, globalization has promoted a loss of community control over economic activities and has led to a fierce competition for jobs, setting into motion "place wars" in which communities compete in the often expensive search for factories and other job providers. These same critics contend that the large corporations providing these jobs have become increasingly stateless—able to mobilize resources and manage operations irrespective of national boundaries and laws—in ways that hurt labor conditions, incomes, and the environment. Globalization has contributed to almost unimaginable concentrations of wealth among the globally connected at the expense of the middle class and poor, and has worsened environmental conditions worldwide.

Proponents of North America's role in the globalization process hope that there will not be a rush to judgment regarding the efficacy of free-market policies. These proponents believe that painful short-term adjustments are part of the transformation process leading to a worldwide economic system that is sustainable and wealth creating. While the emphasis on free trade and open markets may result in job losses in the United States, the final outcome will be the rationalization of production and workforce capacity worldwide, and the creation of value for consumers and workers everywhere. Advocates of free-market policies prompt us to remember how little we pay for clothing assembled in Haiti, computers made in Malaysia, or the huge array of goods manufactured in China. They remind us that new types of jobs will be created in North America that will replace those lost. Over the long term, they argue, North America will be a better place economically because of globalization.

Table 2–3 Employment Percentages for Selected United States Sectors, 1930–2010									
	1930	**1940**	**1950**	**1960**	**1970**	**1980**	**1990**	**2000**	**2010**
Agriculture	22.0	20.0	11.9	8.3	4.3	3.4	2.9	1.8	1.6
Mining	3.4	2.9	2.0	1.3	.7	1.0	.6	.4	.5
Manufacturing	32.5	33.9	33.7	31.0	26.4	22.1	18.0	14.4	10.1
Transportation/ Utilities	12.5	9.4	8.9	7.4	6.7	6.6	6.9	5.4	5.1
Services	—	11.3	11.9	13.6	25.9	29.0	33.1	48.8	53.7

Sources: Percentages from 1970 to 1990 were computed from the U.S. Bureau of the Census, *Statistical Abstract of the United States: 1999* (Washington, D.C.: Government Printing Office, 1999), Table 578; earlier data were derived from the same source, Table 1432, using slightly different counts; the 2000–2010 data were computed from the *Statistical Abstract of the United States: 2012*, Table 620. The services sector data are not strictly comparable over time, but the trends are correct. The 2000 and 2010 services category includes financial, professional and business, education and health, leisure, and other services.

Note: Not all activities are included, so column sums do not equal 100 percent.

visualizing DEVELOPMENT ▲

Does America Have "Smart" Cities?

Richard Florida is known for his innovative take on development. He argues that economic developers have placed too much emphasis on technical skills at the expense of "softer" abilities that bring people together to work on new ideas, cultivate teamwork, and find value in social perceptiveness and empathy.

In a recent *The Atlantic* article, Florida described how America's most prosperous cities have evolved beyond the industrial age (where technical and managerial skills were of highest importance) to centers of innovation that rely on people with "analytic" skills—college graduates focusing on problem-solving and "social" skills—persuasion, social perceptiveness, empathy, and a capacity to function in teams. His calculations revealed that the monetary benefits are significant: occupations requiring greater analytic and/or social skills pay between $25,000 and $35,000 more than occupations that require other skills. People with analytic and social skills tend to congregate in the largest cities, whereas people with more traditional "physical" skills, as he calls them, cluster in smaller metropolitan areas.

Where are these cities? As the map (Figure 2-5-1) indicates, America's "smart" cities tend to cluster in the northeast (Boston to Washington, D.C.) and the Pacific Coast (California and the Seattle region). Secondary clusters can be found in the

Dallas to San Antonio corridor in Texas, in Minneapolis, and in the Atlanta and Raleigh/Durham areas of the South. The cities in the traditionally industrial Midwest tend not to have similar concentrations of analytic and social occupations, reflecting their past and current manufacturing prominence.

In a similar vein, *PCWorld* recently identified the "tech-friendliest" cities in the United States. Tech-friendly was measured using variables that included the number of information-technology jobs, the speed and cost of Wi-fi, availability of public wireless services and government apps, and even the number of "tweets" (as in Twitter) originating in a given city. Tech-friendly leaders included metropolitan areas like San Jose and San Francisco, Atlanta, Boston, Minneapolis, and the New York region. Lagging cities included El Paso, Texas; Fresno, California; Memphis, Tennessee, and Oklahoma City, Oklahoma. The tech-friendliest cities tend to correlate with Florida's "smart" cities.

What do smart cities say about the economic development process? First, they illustrate how innovation and prosperity increasingly lie in the hands of people who can conceptualize and implement new ideas. According to Florida, today's students need to integrate liberal arts with technological literacy, understand the importance of persuasion and teamwork, and build the

social intelligence that "makes for creative collaboration and leadership." Second, size matters—smart cities tend to be larger cities. Florida refers to "urban metabolism," which is the basic functioning of the city system. Unlike in the biological world, where size tends to slow metabolism, the larger a city grows the faster its metabolism becomes, and the greater the innovation and wealth created per person. Highly skilled people are attracted to the dynamism of this accelerated urban metabolic state, taking advantage of the kinetic energy that proximity produces and sharing ideas. Innovation is born, new businesses are created, and economic development is enhanced.

What lies ahead for North American cities remains a subject of debate. Energy costs will play a role. Expensive fuel will lead to higher commuting costs, which will almost certainly reshape cities. At this point, we can only speculate: Will people move to city centers closer to jobs? Will jobs increasingly flow out to the suburbs? Will public transportation become more important? Possible innovations in transport technology (for example, electric commuting vehicles) could provide cost-effective substitutes for gasoline, permitting a continuation of current urban trends.

Sources: Richard Florida, "Where the Skills Are," *The Atlantic*, October 2011, 75–78; and "America's Most Tech-Friendly Cities," *PCWorld*, April 2013, 77–84.

Urban Dynamics in a Twenty-First Century Setting

Rapid industrial growth in North America was associated with the explosive growth of cities. From an economic perspective, this association made good sense. Cities offered manufacturers the agglomeration economies that come from locating near other, linked activities. Firms discovered that they could reduce costs by sharing services, suppliers, labor forces, and other common needs. Many manufacturers were attracted to cities because of the markets created by the large numbers of people living in close proximity. Industrial jobs created multiplier effects that generated employment opportunities in wholesaling, retailing, education, government, the professions, and a host of other urban-oriented activities. The United States and Canada have always had cities and towns,

growing in response to a variety of factors. Industrialization created new reasons for cities to exist and to expand, which they did in an unprecedented manner beginning in the early nineteenth century.

By 2000, both countries were highly urbanized. Fully four-fifths of the U.S. and Canadian populations were classified as urban in the first decade of the twenty-first century. Added to the already high percentage is the vast majority of rural residents who do not farm and lead a largely urban existence.

As North American cities grew in response to the Industrial Revolution, their shapes became more complex. Changing transportation technologies and evolving urban cultures led to a sequence of events in the spatial expansion of the North American city that are still visible today (Figure 2-43). In the 1970s geographer John Adams argued that the form of the modern North American city was the result of four transportation-based stages: (1) a post-Civil

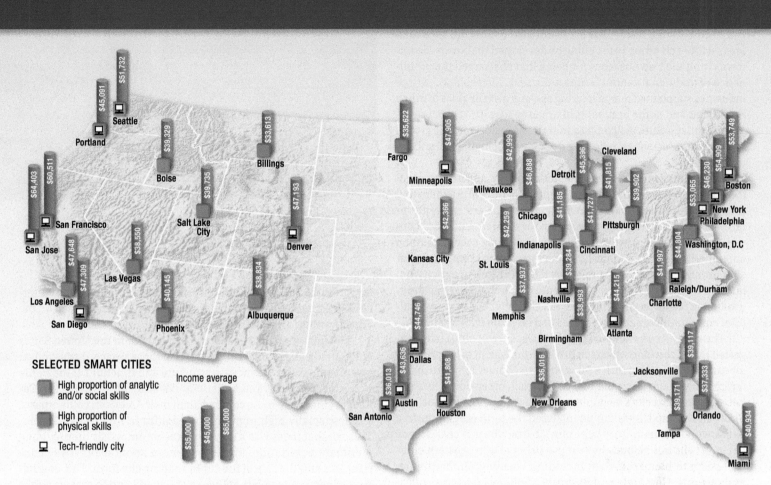

▲ **Figure 2-5-1 Does America Have "Smart" Cities?** According to Richard Florida, multiple cities stand out as centers of innovation and high-tech attraction.

Sources: Based loosely on analyses presented in: Richard Florida, "Where the Skills Are," *The Atlantic*, October, 2011, 75–78; and "America's Most Tech-Friendly Cities," *PCWorld*, April 2013, 77–84.

War period in which cities were relatively small in area but densely packed, with people living literally on top of each other because the foot and animal transportation of the day did not permit long commutes to work; (2) an electric streetcar stage beginning in the late 1880s in which relatively fast streetcars (moving at 15–20 mph) enabled an outbound middle-class migration along streetcar lines, producing spoke-like "streetcar suburbs" radiating outward from the city center; (3) a road-building stage for the increasingly popular new

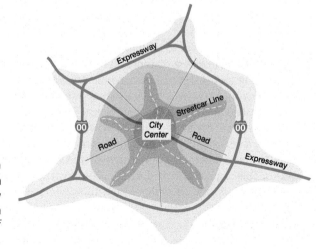

▶ **Figure 2-43 Generalized spatial expansion of the American city, late nineteenth and twentieth centuries.** As transportation technologies improved, cities took on distinctive shapes from a compact core in the early years when transportation was slow and difficult, to a sprawling city by the late twentieth century with the construction of expressways. For additional information, see J. Adams, "Residential Structure of Midwestern Cities," *Annals of the Association of American Geographers* 60 (1970): 56.

form of transportation, cars, by the 1920s, and the resulting filling in of suburban areas between the streetcar lines, further stimulating the middle-class exodus to the edges of the city; and (4) a post-World War II boom in expressway construction that not only tied the city center to outlying residential areas but also connected suburban areas with each other. Later, suburbs developed their own "downtowns" and their own reasons for being that challenged the prominence of traditional central business districts. Increasingly, service industries supplanted manufacturing as the growth engines of urban economies and as the occupants of urban spaces. The spatial structures of today's cities are complex mosaics of socioeconomic groups and economic activities, reflecting decades of technological and social change.

Regardless of the complexity of today's cities, the one constant found throughout urban North America is outward expansion, often described as urban sprawl. New suburbs are created ever farther from city cores as people search for the tranquility and convenience of low-density living, while maintaining access to urban amenities. Three geographical consequences stem from this expansion process. First, urban sprawl has led to the loss of agricultural land; cornfields are replaced by suburbs, thereby removing food-producing resources and changing the character of nearby agricultural communities that may have existed for more than a century.

The second consequence is the coalescence of cities into congested urban corridors. A person driving from Miami to Jacksonville, for example, may never drive through open country for more than a few miles. As cities grow, they bump into each other—one city's end becomes the next city's beginning, even to the point of sharing city-limits signs. Jean Gottmann popularized the term **megalopolis**, a Greek word meaning "very large city," to describe this coalescence of cities.[2] His initial focus was on the string of northeastern cities that seem to blend together, including Washington, Baltimore, Philadelphia, New York, and Boston. Together with a large collection of nearby cities, they comprise a giant urban region with tens of millions of people. Similar megalopolitan areas, or "spread" cities, can be identified elsewhere: "San-San" (San Francisco to San Diego), "Chi-Pitts" (Chicago to Pittsburgh), and "Ja-Mi" (Jacksonville to Miami), to name only three of potentially many such U.S. agglomerations.

An unfortunate corollary of urban expansion has been the abandonment and decay of city centers, not only due to middle-class migration to the suburbs but also many businesses frustrated by downtown congestion. Left behind are empty buildings, densely populated low-income residential areas (the "ghettos"), and too often crime and other social problems—often within the shadow of a bustling central business district that depends on workers from the suburbs (Figure 2-44). A major economic development goal of many cities is downtown revival, using a variety of investment incentives, tax breaks, and other policies. While central cities are still distressed, signs of renewal are appearing. Many warehouse districts, for example, are experiencing a "gentrification" process as dilapidated buildings are transformed into high-priced condominiums. For some people, accessibility to the urban culture of downtown is more

▲ Figure 2-44 **Central city decay, Detroit.** Virtually all cities have at least some central districts that have fallen into economic decline. One of our greatest challenges is to devise ways to revitalize these areas.

important than life in the suburbs and the stresses of commuter-choked expressways. Recent news reports have even focused on the attractiveness to retirees of gentrified downtown areas.

The focus here has been on urban growth in the United States and Canada as a whole. Do the cities of these two countries reflect any meaningful differences in structure and function? If the frame of reference is the 30 years after World War II, the answer is yes. Canadian cities were more compact than their American counterparts, with especially high population densities in inner-city suburbs. This compactness was a result of greater emphasis on providing public transportation, and less concern about expressway construction and the role of the car in shaping city form. This emphasis was partly due to less affluence in Canada and to greater public acceptance of government planning. Beginning in the 1970s, however, the differences receded, as the influence of the car in shaping the Canadian city became more pronounced, bringing with it the growth of expressways and the explosion of outer suburbs.

Stop & Think

➤ What is meant by "megalopolis" and where in the United States do we find good examples of megalopolis?

Poverty on a Continent of Plenty

Even industrialized and prosperous countries such as the United States and Canada have significant and sometimes pressing challenges. We have already discussed the complexities of supplying and consuming enormous quantities of resources. In addition, the United States and Canada must deal with unbalanced economic growth, ineffective integration of various regions into a national economy, environmental degradation, and the social, economic, and political circumstances of minority groups.

Large numbers of people in both countries cannot acquire the material benefits necessary for an acceptable level of living. Exactly how many poor people live in the United States and Canada

[2] J. Gottmann, *Megalopolis: The Urbanized Northeastern Seaboard of the United States* (Cambridge, MA: MIT Press, 1961).

depends, of course, on exactly how poverty is defined. In general, poverty is a material deprivation that affects biologic and social well-being. Low income levels typically point to poverty, although other factors can be associated with deprivation.

Throughout the United States and Canada, significant income disparities—that is, differences in the amount of money people earn—exist among groups and regions. In 2011, about 15 percent of the overall U.S. population could be considered poor (Table 2–4). Among African Americans, the figure increased to 27.6 percent nationally, and to a stunning 32.3 percent in the Midwest—about one in four African Americans. Poverty levels among Hispanics were only slightly lower than African Americans, with the white and Asian populations showing the lowest figures. It is interesting to observe, however, that the absolute number of whites in poverty was nearly double that of African Americans—19 million whites (non-Hispanic) compared with almost 11 million African Americans—confirming that poverty among the white population is still a serious matter, the lower percentages notwithstanding.

At one time, the poor in the United States were almost equally divided between metropolitan and nonmetropolitan areas. Now 83 percent of all poor people live in metropolitan areas, with most concentrated in central cities. Often the poor represent a smaller proportion of a metropolitan population than they do of a nonmetropolitan population, but the metropolitan poor are more geographically concentrated and frequently live in inner-city ethnic communities that social scientists refer to as "ghettos" (Figure 2–45). Poor people in nonmetropolitan areas are generally scattered and are therefore less visible, adding to their problems of securing employment and services.

The map in Figure 2–46 shows clearly defined pockets of poverty. Numerous poor are found throughout the southern Coastal Plain, especially in the lower Mississippi basin and in the upland South, particularly the Appalachian Plateau and the Ozarks. Rural poverty is not limited to the South, although it is most widespread there. The rural poor are also found in New England, the upper Great Lakes, the northern Great Plains, and the Southwest. Poverty is particularly chronic in the Native American lands of the West and along the border with Mexico.

▲ **Figure 2-45 Urban poverty.** In sharp contrast to most of the periods of American history, today the majority of poor people live in large urban areas—most often in inner-city communities. This scene is from Los Angeles, where neighborhood residents are collecting holiday food boxes from a local food pantry.

Canada also has regional differences in income and poverty (Figure 2–47). In 2002, approximately 9.5 percent of all persons fell into the "low-income after tax" category, a figure somewhat lower than the U.S. poverty rate. Other figures suggest slightly higher rates, such as the 2010 estimate of 15.9 percent of all persons in Canada earning less than $10,000 (Can.) per year. (Different countries use slightly different measures of poverty, making direct comparisons difficult.) The regional variations are clear. Recently, Alberta and Saskatchewan have emerged as the provinces with the lowest levels of extremely low incomes, whereas traditionally high levels remain in Quebec and the Maritime provinces. Surprisingly high levels in Ontario represent a recent development.

Why does poverty exist in developed countries? Of considerable importance is the effect of economic and technological change. As new technologies are introduced and economies are restructured, new types of production are created, demanding different resources and workers. The result often is a transformed geography of wealth and poverty, with certain groups and regions benefiting from these

Table 2–4 Poverty in the United States, by region and group, 2011					
Percentage of All People Below 100 Percent of Poverty					
Region	**Total Population**	**White**	**African American**	**Hispanic[a]**	**Asian**
United States	**15.0**	**9.8**	**27.6**	**25.3**	**12.3**
Northeast	13.1	8.4	25.9	26.1	15.2
Midwest	14.0	10.0	32.3	27.3	11.8
South	16.0	10.4	26.7	24.7	9.9
West	15.8	10.0	27.7	25.2	12.2

Source: Bureau of the Census, *Current Population Survey, Annual Social and Economic (ASEC) Supplement: POV41*, Poverty Status for All People, Family Members and Unrelated Individuals. Extracted from *http://www.census.gov/hhes/www/cpstables/032012/pov/POV41_100.htm*, various tables, March 2013.

[a]The white population does not include Hispanics. Hispanics are defined as persons of Puerto Rican, Cuban, Central American, South American, or some other Spanish culture of origin, regardless of race.

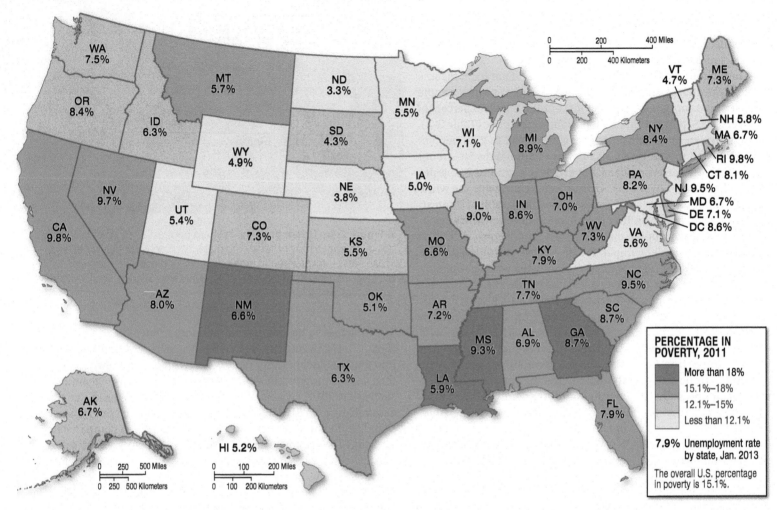

▲ **Figure 2-46 Poverty and unemployment in the United States, percentage of people in poverty by state, using 2010–2011 averages.** The percentage of population in poverty varies significantly by region and reflects differences in the development process over many decades. Poverty is closely associated with high unemployment rates. With some exceptions, particularly in the Midwest, the map shows strikingly similar patterns of occurrence of high poverty and high unemployment rates.

Sources: United States Bureau of the Census, *2010 Census of Population, Percentage of People in Poverty by State Using 2- and 3-year Averages: 2008–2009 and 2010–2011*. Table selected from *http://www.census.gov/hhes/www/poverty/data/incpovhlth/2011/tables.html* (extracted March 2013). United States Department of Labor, Bureau of Labor Statistics, Civilian labor force and unemployment by state and selected area, seasonally adjusted, Table 3 in the *Economic News Release*, as of March 2013. See *http://www.bls.gov/news.release/laus.t03.htm*

changes, and others falling by the wayside. For example, in portions of Appalachia, small-scale farming on poor land has been the basic activity since the area was first settled, but new technologies and the rise of corporate farming have left these areas impoverished agricultural backwaters. Miners from Pennsylvania to Kentucky lost jobs as mines were automated and the demand for coal fell. High unemployment levels are consistently associated with high poverty levels.

The historical response to high poverty and unemployment has been to "go where the jobs are." Often this meant leaving the rural areas and moving to the city, or even leaving one region for another. While such migration has improved the material well-being of many people, additional rounds of technological change and economic restructuring have reduced many of these migrants and/ or their descendants to lives of poverty in crowded urban ghettos.

Lack of skills and opportunities create cultures of poverty that are persistent and, for too many people, inherited.

The Power and Challenges of Diversity in the United States and Canada

It is not unusual for countries and regions to contain subgroups distinguishable by race, ethnic and linguistic differences, and levels of economic achievement. While enriching to society as a whole, such divisions may be obstacles to a unified political organization, and social and economic satisfaction. The difficulty of integrating diverse groups into a larger society stems not only from outward cultural differences but also from basic human nature.

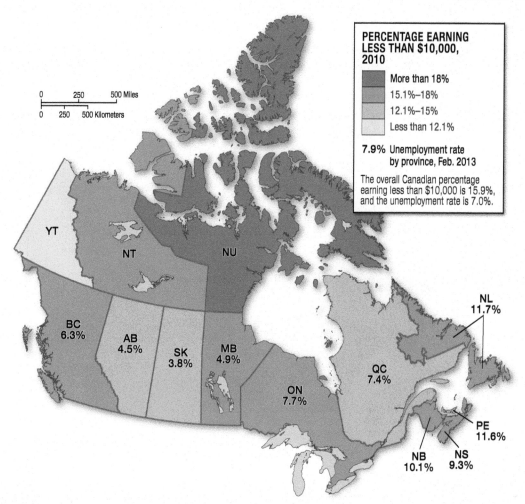

▲ **Figure 2-47 Poverty and unemployment in Canada.** As in the United States, poverty in Canada has a regional expression. The lowest levels are in the West. Unemployment rates have a similar expression.

Sources: Statistics Canada, *Individuals by Total Income Level, by Province and Territory.* See the tables in *http://www.statcan.gc.ca/tables-tableaux/sum-som/l01/cst01/famil105a-eng.htm* (extracted March 2013). The unemployment data were taken from Statistics Canada, *Labour Force Characteristics, Seasonally Adjusted by Province* (February 2013). See the tables in *http://www .statcan.gc.ca/tables-tableaux/sum-som/l01/cst01/lfss01a-eng.htm* (extracted March 2013).
Note: While Canada acknowledges "low income," Statistics Canada does not have an official poverty level. The figures used here show the proportion of persons earning below $10,000 in 2010.

African Americans

In the United States, initial patterns of African American residence and African American–white relationships grew out of the diffusion of the plantation system across the lower South. A small proportion of African Americans, about one in seven, were freedmen living outside the South or in southern urban areas, where slightly less rigid social pressures allowed them to be artisans. The rest of the African American population formed the backbone of the plantation system.

The original center for the plantation system was Tidewater Virginia and parts of Maryland and North Carolina. As settlement proceeded southward, slavery spread to coastal areas. In the more isolated and subsistence-oriented hill lands of the South, as well as northward into the Middle Atlantic and New England colonies, slavery never became important and was eventually declared illegal, but the diffusion of plantation agriculture in the lower South established the initial pattern of African American residence across the South.

The distribution of African Americans did not change immediately after the Civil War. Although no longer slaves, but freedmen, they were still agricultural laborers with no land, money or capital, and little training beyond their farming skills. Consequently,

the newly freed slaves stayed where they were and entered into a system of tenancy or sharecropping with white landowners who continued to disdain manual labor. With few exceptions, the rural parts of today's South with large numbers of African Americans are the same areas in which the plantation system once flourished.

By 1900, the South had a large, landless tenant labor force of rural African Americans. As opportunities for employment outside the South grew after World War I and mechanization of southern agriculture began, African Americans moved in large numbers from rural to urban areas of both the North and South. Although this **African American migration** slowed somewhat during the 1930s, when economic conditions were as bad in urban areas as they were in rural areas, the northward movement continued into the 1970s. Today, African Americans live in urban areas of the North as well as urban and rural areas of the South (Figure 2-48). The proportion of African Americans living in the South seems to have stabilized and hovers around 53–57 percent (Table 2-5); more African Americans are now moving to the South than are leaving it.

Patterns of migration have implications that go far beyond mere redistribution of population. Many African Americans have clearly

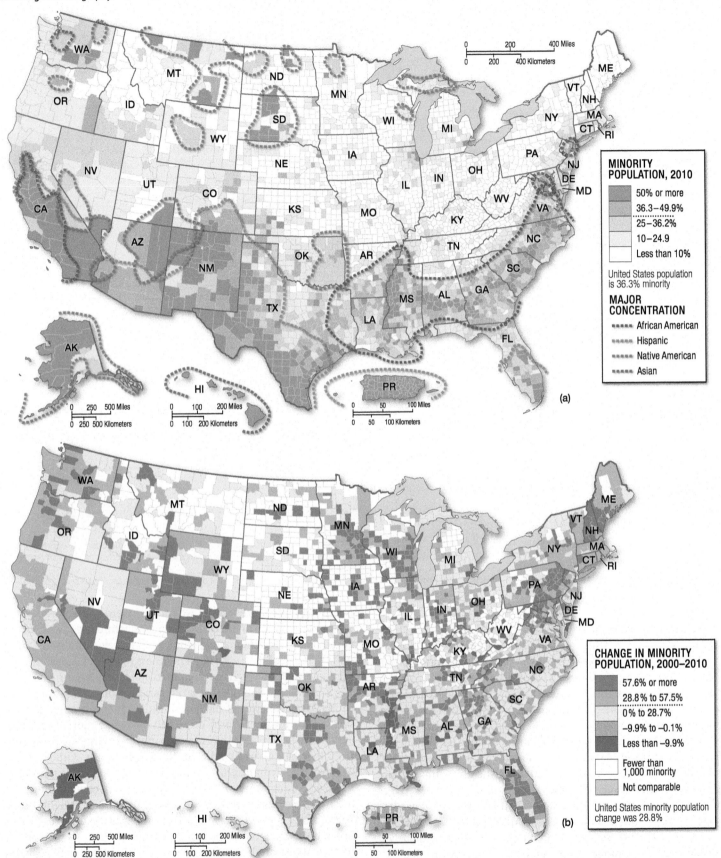

▲ Figure 2-48   (a) Minority population as a percentage of county population in the United States in 2010 and (b) change in minority population, 2000–2010. Minority populations in the United States tend to have a regional component with, for example, the African American population historically located in the South and the Hispanic population in the Southwest and in Florida. This map obscures much of the diversity found in metropolitan areas. Many cities outside the South, for example, have large African American populations. Note the growth of minority populations in states without traditionally large minority populations. The base maps were prepared by the U.S. Census and are found at *http://www.census.gov /prod/cen2010/briefs/c2010br-02.pdf*, extracted March 2013.

Table 2–5   African American Population of the Conterminous United States

		Regional Distribution (percent)			
	Total (thousands)	Northeast	Midwest	South	West
1900	8,834	4.4	5.6	89.7	0.3
1910	9,828	4.9	5.5	89.0	0.5
1920	10,463	6.5	7.6	85.2	0.8
1930	11,891	9.6	10.6	78.7	1.0
1940	12,886	10.6	11.0	77.0	1.3
1950	15,042	13.4	14.8	68.0	3.8
1960	18,860	16.1	18.3	60.0	5.7
1970	22,580	19.2	20.2	53.0	7.5
1980	26,505	18.3	20.1	53.0	8.5
1990	29,986	18.7	19.1	52.8	9.4
2000	34,658	17.6	18.8	54.8	8.9
2010	38,929	16.8	17.9	56.5	8,8

Sources: U.S. Bureau of the Census, *Census of Population* (Washington, D.C.: Government Printing Office, 1910–1980); U.S. Bureau of the Census, Current Population Reports, Population Estimates and Projections, *Projections of the Population of States by Age, Sex, and Race: 1988 to 2010*, Series P-25, No. 1017; Bureau of the Census, *Statistical Abstract of the United States, 1996* (Washington, D.C.: Government Printing Office, 1996). Data for 2010 extracted from *http://www.census.gov/prod/cen2010/briefs/c2010br-06.pdf*, Table 5 (March 2013).

Note: Because these figures have been rounded, the totals by year may not equal 100 percent.

improved their economic and social positions in urban areas (Figure 2–49), but the improvement has been a group achievement not easily won and not without failure for many individuals, as evidenced by the poverty-ridden, central-city neighborhoods of many large U.S. cities. Many migrants to the cities have been the better educated and more motivated representatives of rural areas, but have also lacked the skills needed to do well in urban areas. Racial and social biases have further added to the difficulty of obtaining suitable housing, proper education, and access to economic opportunity. Many African Americans have ended up in ghettos, with an unemployment rate consistently higher than it is for other segments of society.

The residential pattern for urban African Americans stems from their economic and social position. Highly concentrated African American neighborhoods appear in older residential areas, often vacated as economically progressive African Americans and whites flee to the city's periphery or to suburban communities. This process has progressed to the point that many cities have become more African American than white, as evidenced by the increasing number of African Americans elected mayors of large cities.

Despite political successes and tangible improvements to the lives of many African Americans, challenges remain as the African American population pursues its aspirations within American society. The progress that has been achieved could only have come, however, with a break from the old system of doing things, especially in the South; and migration, whether to northern or southern cities, was indicative of that break.

Hispanic Americans

Hispanic Americans trace their backgrounds to a Spanish American culture, regardless of whether it is in the United States or Latin America. The term Hispanic does not refer to a racial group.

▼ **Figure 2-49 African American socioeconomic mobility.** Many African Americans have improved their economic circumstances in recent decades. This middle-class African American family is enjoying an outing in a park on the perimeter of the downtown area of Charlotte, North Carolina.

There are established African components to the populations of many Latin American countries, especially in the Caribbean, and these people easily fit into a Hispanic setting. The same can be said for Latin America's Asian populations.

The Hispanic American population in the United States is large and rapidly growing. The U.S. Census set the 2000 Hispanic population at 35.3 million people; by 2010, the figure was 50.5 million, constituting over 16 percent of the total U.S. population (see Figure 2-48). Most Hispanic Americans claim Mexican origins, perhaps 30 million people in 2010; but most Latin American countries are represented. Some countries have witnessed especially large shares of their populations migrate to the United States. For example, 932,000 Salvadoreans lived in the United States in 2000, about 12 percent of the population of the small country of El Salvador itself; by 2011, that number increased to well over a million. By the middle of 2003, the Hispanic American population had grown to the point that it had overtaken the African American population as the country's "majority" minority (Table 2-6).

A large proportion of the illegal immigration to the United States is comprised of migrants from Latin America. According to the Pew Hispanic Center, of the estimated 11.2 million illegal aliens in the United States in 2010, approximately 58 percent were from Mexico; other countries in Latin America accounted for about 23 percent. In the last several years, considerably fewer unauthorized immigrants from Mexico have made their way to the United States, down from about 500,000 in the early 2000s to perhaps 150,000 annually by 2010. Most of these migrants are young, with about one in six under 18 years of age.

How the United States should respond to this illegal migration has produced intense debate. Advocates of tighter controls argue that illegal migrants have a depressing effect on wage levels and impair the ability of states to provide for the welfare and safety of their citizens; illegal migrants should not be allowed the benefits of citizenship without legal status. On the other side of the issue are those who argue that illegal aliens do work that legal residents often refuse to do; that these migrants contribute to the U.S. economy through their work; and that sanctions against these people, such as denial of licenses, educational opportunities, and welfare benefits, will only drive undocumented citizens farther underground and deeper into illegal activities. This debate will not go away soon, but in all probability will result in a changed legal landscape for undocumented Hispanic migrants.

Regardless of the legal status of the Hispanic population, their economic circumstances tend to be worse than for the U.S. population as a whole. Hispanic incomes are only about two-thirds that of white non-Hispanics. A disproportionate percentage of Hispanics work in the lower-paying construction, manufacturing, and service industries (particularly food and lodging). As of 2011, home ownership among the Hispanic population was only about 47 percent, compared with 65 percent for the population as a whole. The percentage of Hispanics without health insurance in 2011 was twice the national average, the highest gap of any of the major racial/ethnic groups. Recent arrivals often send parts of their paychecks back home, creating a business of remittances that has become a big industry, with a quarter of the population of countries like El Salvador receiving funds from friends and family in the United States.

The majority of the Hispanic population resides in the southwestern United States, a region with a long history of involvement in the Spanish-speaking world (see the earlier discussion of the Hispanic core). As of 2011, according to the Pew Hispanic Center, Texas was 38 percent Hispanic; New Mexico, 47 percent; Arizona, 30 percent; and California, 38 percent. People with Spanish surnames dominate many smaller communities and even some sizable cities, such as San Antonio, Texas (63 percent in 2010). The initial Hispanic infusion into what is now the southwestern United States resulted from the region's inclusion in the expanding Spanish Empire in the late sixteenth and seventeenth centuries. Thinly scattered Spanish settlements eventually extended from Texas to California. The Spanish were unable, however, to prevent a flood of Anglo-American settlement in the nineteenth century. Initial growth of those settlements diminished the proportion of the population that was Hispanic, but did not erase the long-established cultural imprints. These imprints were reinforced during the twentieth century by the tide of both legal and illegal immigration from neighboring Mexico. This migration process, combined with a higher-than-the-national-average fertility rate among Hispanics, is contributing to a renewed Hispanicization of the southwestern United States, and is creating a Hispanic American (largely Mexican) borderland that is at least a partial reprise of an earlier time in history.

Several large metropolitan areas outside the Southwest also have significant Hispanic populations (Figure 2-50). In New York and Chicago (approximately 29 percent for both cities in 2010), those populations are largely of Puerto Rican and Mexican descent. In southern Florida the largest Hispanic group is Cuban, a result of the widespread migration of Cuba's middle class after the Cuban Revolution; Miami, for example, is more than half Hispanic. In New

Table 2–6	**Population Percentages by Race and Hispanic Origin**			
	Percentages			
	1980	**1990**	**2000**	**2010**
White	85.9	83.9	75.1	72.4
African American	11.8	12.3	12.3	12.6
Asian or Pacific Islander	1.6	3.0	3.6	4.8
Hispanic Origin	6.4	9.0	12.5	16.3

Sources: Adapted from U.S. Bureau of the Census, *Statistical Abstract of the United States: 2001—Population* (Washington, D.C.: Government Printing Office, 2001), Tables 14 and 15. The 2000 and 2010 data are taken from the *2010 Census of Population*, extracted from *http://www.census.gov/prod/cen2010/briefs/c2010br-02.pdf* (March 2013). These data show those persons declaring only one race as a percentage of total population, which increasingly has a depressing effect on the percentages shown as people identify themselves as multi-racial. The number of people identifying as multi-racial increased by nearly 2.2 million between 2000 and 2010; the number of people in racial groups other than the major Census categories increased by 3.7 million.

Note: Persons of Hispanic origin may be of any race.

▲ Figure 2-50 **Hispanic cultural imprint in New York City.** Hispanics now constitute a large and rapidly growing ethnic group in many cities of the United States and Canada.

Orleans, the Hispanic population was largely of Central American origin before Hurricane Katrina, with approximately half of Honduran ancestry; but massive rebuilding has attracted a large number of Mexican workers.

There was also considerable movement of minority populations, mainly Hispanics, within the United States between 2000 and 2010 (see Figure 2–48). While growth occurred in expected locations like Florida, Arizona, Nevada, and California, meaningful percentage increases can be found in states not normally considered traditional Hispanic settlement areas. New Hampshire, for example, saw its Hispanic population increase from 1.7 to 2.8 percent of the total population—still small, but part of a noteworthy trend. North Carolina grew from 4.7 to 8.4 percent. Significant concentrations are also found in Washington and Oregon, along with numerous counties throughout the Midwest. The promise of economic opportunity accounts for much of this dispersal.

Other Expressions of Diversity

The Hispanic American culture region is also home to a third of the 2.9 million Native Americans in the United States as of 2010 (see Figure 2–48). Although the Native Americans of the Southwest have maintained their tribal structure better than other Native Americans have, they endure greater socioeconomic disparities than any other minority. The 2006 Canadian Census estimate is 1.2 million people in aboriginal groups, spread across much of Canada.

Another rapidly growing segment of the United States includes people of Asian descent, whose numbers have increased considerably since the 1970s in response to immigration (see Figure 2–48). The estimated 7.5 million Asians in the United States in 1990 increased to almost 14.4 million by 2010 and will exceed 16 million in 2015. Growth has been particularly significant in the urban centers of the West, focusing on California, although a substantial presence is found in a number of metropolitan regions outside of the West. Clustering of ethnic groups in cities has a long history in the United States and Canada, involving a wide range of groups (see *Geography in Action:* The "Dearborn Effect": Ethnic Clustering in Contemporary American Cities).

Canada's Search for Identity and Unity

Canada is a pluralistic society as well. While people normally think first of the large French minority, Canada actually embraces many other groups, creating a complex cultural mosaic. In addition, Canadians of all backgrounds struggle to create an identity that is uniquely Canadian, which can be a daunting task in the shadow of the American colossus to the south.

Canada was organized as a federation over 140 years ago because the descendants of French settlers insisted that any system of union preserve French identity and influence. French Canadian distinctiveness is not only linguistic but religious, as French Canadians are overwhelmingly Roman Catholic, in contrast to the largely Protestant English Canadians. French Canadians continue to struggle with stereotypes that characterize their community as quaint, rural, agrarian, unchanging, and with high birthrates. While some of these traits have a basis in history, the province of Quebec is highly urbanized and part of Canada's economic core, with birthrates at or below the national average. A significant rural population remains, with many people in the countryside cherishing traditional values, but the reality of twenty-first century Quebec is far more complex than the stereotypes suggest.

Creating a single Canadian identity acceptable to both French- and English-speaking Canadians has been difficult, to the point that French-Canadian nationalists have periodically called for a separate, independent Quebec. In 1970 the debate over an independent Quebec took a violent turn, leading to the occupation of parts of Quebec by the Canadian military and the first-ever peacetime imposition of Canada's War Measure Act. With the exception of these events, French-Canadian separatists have relied on persuasion and the ballot box to advance their cause. A 1980 referendum in Quebec was conducted by the ruling Parti Quebecois, the separatist party, to determine whether the provincial government should move toward separation from the rest of Canada. The referendum failed, but 42 percent of Quebec voters voted in the affirmative, as did 52 percent of the French-speaking population in Quebec. In 1982, Canada attempted to strengthen its union by revising its constitution to include a bill of rights. The government of Quebec, still under the control of the Parti Quebecois, refused to sign the new document, ostensibly because it would not permit Quebec to secure its cultural identity as a distinct society within the larger framework of Canada. A conference in 1987 produced a document, the Meech Lake Accord, designed to address Quebec's concerns. The accord was rejected by the governments of Newfoundland and Manitoba, which argued that the accord gave Quebec cultural rights that the other provinces did not have. The accord did not survive its ratification deadline of June 23, 1990. Partly as a consequence, the separatist cause found new strength. A second referendum was held in 1995 and the federalist position won by a very slim margin. The twenty-first century has been a largely quiescent time for Quebec separatism, though the spark is never extinguished completely. French and English are both official languages throughout Canada, several recent prime ministers have been from Quebec (including but not limited to Pierre Trudeau, Brian Mulroney, and Jean Chrétien), and the separatist party in Quebec has controlled the provincial government on more than one occasion.

GEOGRAPHY IN ACTION

The "Dearborn Effect": Ethnic Clustering in Contemporary American Cities

Dearborn, Michigan, located on the southwestern edge of Detroit, is a small city known for its historic role in the U.S. automobile industry. Henry Ford lived and worked in Dearborn, and it continues to house the world headquarters of the Ford Motor Company. One of the great auto plants of all time—the Rouge facility—is located in Dearborn and, at its peak, employed several tens of thousands of workers.

Today, Dearborn is known not just for its role in the auto industry, but as the poster child for ethnic clustering in modern America. Of the estimated 98,153 residents of Dearborn in 2010, approximately 40 percent claim a Middle Eastern connection, mainly Lebanon and Iraq. Southeastern Michigan is home to about 250,000 people from this part of the world. The presence of the auto industry and a Middle Eastern population is not entirely coincidental. Henry Ford initially was reluctant to hire African Americans to work in his car factories (he had equally unenlightened views about other groups), but he was willing to hire immigrants from the Middle East. A Middle Eastern core area was thus created—initially populated by Lebanese—that became a magnet for other immigrants. Later immigrants had little or no connection to the auto industry; civil war in Lebanon and other regional conflicts swelled the number of migrants.

The changes to Dearborn have been profound. Local schools teach in both English and Arabic. The local branch of the University of Michigan offers a curriculum in Arabic and has a Center for Arab-American Studies. The Wal-Mart store sells halal meats, falafel, mango juice from Egypt, and vine leaves from Turkey for stuffed grape leaves.

▲ Figure 2-6-1 **Dearborn, Michigan.** This city is not only a heart of the auto industry but a core region for Middle Eastern immigrants to the United States.

The sights and sounds of the city reflect an immigrant culture (Figure 2-6-1).

Ethnic clustering in Dearborn—what we conveniently call here the "Dearborn Effect"—is only the latest variation on a recurring theme in American history. The United States is a country of immigrants and typically these immigrants concentrated in specific locations, like the Irish in Boston and the Poles in Chicago. What we see today, however, is a large number of clusters of migrants with origins other than Europe. The Dearborn Effect can be found in Fremont, California, where a quarter of the city's population of 200,000 was born in Asia, mainly China and India; especially perplexing for Fremont is the large number of different linguistic groups within the Asian community. We find a similar story of clustering in Miami with its large Cuban influx. New Orleans has a large pocket of Vietnamese. New York City has 250,000 people born in China and 115,000 born in Ecuador.

Ethnic clustering is reshaping the cultural landscapes of American cities, presenting new challenges and offering new opportunities for city leaders. The extent to which these clusters lose their identities as their people become acculturated, and the degree to which these clusters imprint their home cities, remains to be seen. Given present trends, the Dearborn Effect will become even more pronounced in future years.

Sources: K. Naughton, "Arab-America's Store: Wal-Mart Stocks Falafel, Olives, and Islamic Greeting Cards to Attract Dearborn's Ethnic Shoppers," *Newsweek*, March 10, 2008, 42.

While the conflict between English and French Canada monopolizes the debate for most Canadians, this conflict often obscures how ethnically diverse English Canada is and how few French speakers live outside of Quebec. The population of the maritime provinces, for example, consists of persons of German, English, Scottish, and Irish descent. New Brunswick is an exception since about one-third of its population are French speakers, most living in the northern half of the province close to Quebec. Farther west there are very few pockets of native French speech (Figure 2–51), but these provinces are likely to have significant minorities whose native language is neither French nor English. Toronto, Ontario, is one of the most multicultural cities in North America. Of its 4.6 million people,

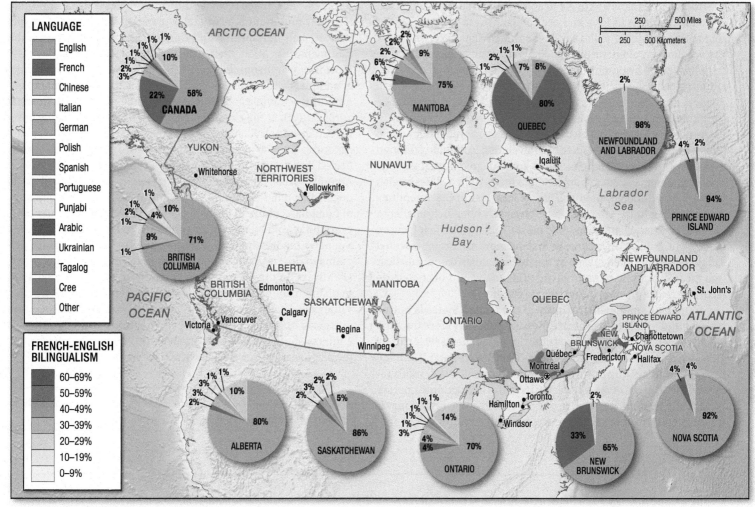

▲ **Figure 2-51 The several Canadas.** The extraordinary attention given Quebec and its relationship to the remainder of Canada sometimes overshadows the numerous other important differences that exist among the several regions of the vast Canadian territory. Note especially the diversity of languages spoken in Canada in addition to the official languages of English and French as of 2006. The area of predominant French speech is indicated on the map.

Source: Adapted from Statistics Canada, *Population by Mother Tongue, by Province and Territory: 2006. Provincial data for 2011 can be found at http://www.statcan.gc.ca/tables-tableaux/sum-som/l01/cst01/demo11d-eng.htm* See also Atlas of Canada.

only 58 percent speak English as a mother tongue and a mere 1.2 percent speak French; the remainder speak Chinese (348,000), Italian (195,000), Portuguese (108,000), Punjabi (96,000), Tagalog (Filipino) (77,000), and a wide variety of other languages. In British Columbia, 9 percent of the population speak Chinese as a native language, but only a scant 1 percent speak French; there are several times as many Punjabi speakers from South Asia as there are French speakers. These language distributions reflect shifting immigration patterns. In 2003, the largest source of immigrants to Canada, 16 percent, came from the People's Republic of China, compared with a meager 2.3 percent from Great Britain. There were more immigrants from the Philippines than from the United States. Finally, more than 1 million Canadians are at least partially of aboriginal origin—largely North American Indian, Metis, or Inuit. These "First Nation" people are the majority population in the Northwest Territories and Nunavut, and are significantly represented in the prairie provinces. It was the issue of special privileges granted to Quebec in the Meech Lake Accord, but not to the aboriginal population of Manitoba, that led the Manitoba government to oppose ratification.

Even within French Canada, there is considerable cultural diversity. While 80 percent of the population speak French as a mother tongue, visitors to the province will hear English, Italian, Spanish, and even Arabic. Montreal is especially multicultural, with only 67 percent of its population speaking French natively—more than 400,000 people speak English, 120,000 speak Italian, and 70,000 speak Arabic.

The considerable economic diversity of Canada also contributes to a sense of fragmentation. The maritime provinces traditionally have been relatively poor and depend on fishing, farming, and mining (all primary activities), as well as subsidies from the central government. Quebec is an industrial province, but slow-growth industries and political uncertainty have clouded the province's economic future. Ontario represents the Canadian heartland, with its strong commercial and industrial base. The prairie provinces are producers of wheat and cattle. Alberta has an abundance of oil and the wealth that goes with it. British Columbia is a western growth center known for its lumbering and trade, but it is a province somewhat removed from the remainder of Canada both in distance and spirit.

Adding to Canada's struggle to achieve internal cohesion and a national identity is its long and occasionally exasperating relationship with its neighbor to the south, the United States. Not only does Canada embody often misunderstood cultural differences like those described at the beginning of this chapter, but deeper and occasionally threatening differences persist. Canadians have a distinct political history, not only in terms of how they achieved independence ("evolution" instead of "revolution") and structured its government (a parliamentary system) but also in the way they viewed international issues. Beginning with the American Revolution, continuing through the War of 1812, and reappearing to some degree during the U.S. Civil War, Canada stood in opposition to many United States foreign policies. In addition, Canadians have a long fear of U.S. intentions toward their country, which is larger physically but much smaller in terms of population and economic power. Initially, this fear focused on outright annexation, a concern that the more expansion-oriented American leaders often fueled; by the mid-twentieth century the fear was more economic in nature—that Canada would become a de facto colony of a growing U.S. corporate empire.

To many Canadians, proximity to the United States is both a blessing and a curse. It has been a blessing politically since the early twentieth century when the two countries found themselves closely allied in major global struggles such as World Wars I and II and the Korean conflict. Only during the Vietnam conflict and, more recently, the U.S.-led invasion of Iraq, have the two countries parted ways, but never in a way that might jeopardize the world's longest unguarded border. It has been a blessing economically in that resource-rich Canada provides the United States with a secure origin for many important industrial raw materials (for example, copper, iron, nickel, newsprint), and Canada has a wealthy and stable trading partner for all variety of goods. Each country is the other's most important source of trade. But proximity has been a curse economically to the extent that Canadians have felt vulnerable to American economic policies and the controlling power of American corporations, at the expense of developing a more broadly based Canadian-owned economy. Canadians want a status beyond that of American "branch plant," and Canadian tariff policies often have reflected that desire, although Canada became a partner in the North American Free Trade Agreement. It has been a curse culturally to the degree that it has been difficult to cultivate and sustain expressions of Canadian culture in the face of overwhelming U.S. influences in everything from television programming to literature to movies to the news media; the Canadian government has set in place programs to promote the arts and humanities in Canada. Canadians are always on the lookout for ways to "Canadianize" Canada and to distinguish their identity from that of their neighbors, but they value their American friendship. As some Canadian politicians have expressed it, sharing a continent with the United States is a little like a mouse sharing a bed with an elephant: no matter how friendly the elephant, any move will be felt by the mouse.

Stop & Think

 Canada's population is quite diverse culturally. Describe this diversity in terms of the groups involved and their geography.

Summary

▶ The United States and Canada occupy one of the largest land masses in the world, and one with a wide array of environments. Climate types range from subtropical in the southern United States (tropical in Hawaii and Puerto Rico) to arctic in northern Canada and Alaska, and from humid in the east and coastal northwest, to arid in the mountain west. Virtually every type of landform is present, and these features help to define regional character, for example, the mountains of the American West.

▶ Both countries evolved out of a spatial integration process in which frontiers were pushed back by advancing populations. People of European origin accounted for much of this advance. But today's population geography includes many immigrants from other parts of the world as well as, in the U.S. case, a large, involuntary infusion of people from Africa. The result is a racial and cultural mosaic that enriches their contemporary geographies, while presenting challenges to governments that struggle to accommodate competing groups. Not to be forgotten is the agonizing displacement and loss experienced by indigenous populations as the modern settlement geographies emerged. Population growth rates and densities in both countries remain low compared with other parts of the world, with Canada having some of the lowest overall densities anywhere.

▶ Each country is an economic powerhouse that occupies a prominent position in the world economy. An abundance of fertile land has contributed to astonishing agricultural success. Immense and varied natural resources, available labor and capital, and an eager entrepreneurial spirit stimulated the development of sophisticated industrial economies and sprawling urban landscapes by the mid-twentieth century. The late twentieth century witnessed an information revolution spawned by advances in microelectronics, which intensified an ongoing globalization of economic activities. Americans and Canadians, as a consequence, are among the wealthiest citizens of the world, although noteworthy pockets of poverty seem stubbornly resistant to remedy in both countries.

▶ Progress has come with costs in terms of environmental degradation, global warming, employment insecurity among North American workers, increased reliance on foreign sources for important natural resources, and a growing fear that economic progress may not be sustainable. Petroleum costs can rise to over 100 U.S. dollars per barrel, and in the United States close to half of the needed petroleum is imported. Vital to transportation as a fuel and a wide range of industries as a raw material, oil is not only expensive but figures prominently in the global-warming debate. Will there ultimately be a post-petroleum or post–fossil fuel future? If so, how will development be defined? Are there potential global crises related to food and water that will impact the geography of North America? Nobody knows for sure, but the future geography of development in the United States and Canada will reflect accommodation to economic and social challenges arising from new technologies, globalization, and resource exhaustion—possibly creating the fundamental changes associated with the word "revolution."

Key Terms

African American migration 105
Appalachian Highlands 61
aquifers 69
Canadian Shield 61
continentality 65
Corn Belt 85
culture hearths 72
Dairy Belt 87
deserts 68
diffusion process 72
French Canada Core 74

friction of distance 76
Great Lakes 65
golden horseshoe 98
Gulf-Atlantic Coastal Plain 63
Gulf of Mexico 65
humid continental climate 68
humid subtropical climate 65
initial advantage 96
Interior Lowlands 64
Interior Plateaus 63
Loyalists 77

manifest destiny 75
marine west coast climate 66
mediterranean climate 65
megalopolis 102
Middle Atlantic Core 74
milkshed 87
New England Core 72
New South 97
North American Free Trade
  Agreement (NAFTA) 98
orographic effect 67

Pacific Coastlands 64
polar climate 68
Rocky Mountains 63
settlement frontiers 72
Southern Core 72
spatial integration 72
Specialty Crop and Livestock
  region 87
steppe climate 68
subarctic climate 68
Wheat Belts 87

Understanding Development in the United States and Canada

1. The physical structure of the United States and Canada is characterized by several mountain backbones separated by lowlands. Where are these mountain and lowland features located?
2. Why does most of Canada's population live within 200 miles (320 kilometers) of the United States border?
3. How do mountains influence the climate in the western United States?
4. Who are the major European immigrant groups that came to the United States during the nineteenth and early twentieth centuries, and where did they become concentrated?
5. Where in the United States and Canada has recent population loss been most pronounced?
6. How has the rural landscape of Canada and the United States changed dramatically over the last 100 years and why?
7. Where is the historic manufacturing core of the United States and Canada, and why did this core form?
8. How do the natural resource endowments of Canada and the United States differ?
9. Where is the Hispanic-American borderland located? In which other parts of the country do we find large Hispanic concentrations?
10. The United States and Canada may be experiencing a new oil and gas boom. Why is this occurring? Where do we find evidence of this boom?

Geographers @ Work

1. Geographers often describe the process of pushing back the frontiers of the United States and Canada as one of spatial evolution. What do you think is meant by spatial evolution? Describe how this evolutionary process occurred in Canada and the United States.
2. You are a consulting geographer for a travel agency. What would you say to an American tourist planning to visit Canada about the differences between the two countries?
3. You are still a consulting geographer for a travel agency. You are asked to identify where the best ocean beaches and the sunniest climates are located. How would you respond to this request?
4. As a geographer, you are asked to explain why Canada faces challenges with ethnic separatism. How would you respond?
5. You are in charge of a committee that has been asked to find a new name for the Corn Belt. What criteria would you use to select the new name and what name would you choose?

MasteringGeography™

Looking for additional review and test prep materials? Visit the Study Area in MasteringGeography™ to enhance your geographic literacy, spatial reasoning skills, and understanding of this chapter's content by accessing a variety of resources, including MapMaster interactive maps, videos, RSS feeds, flashcards, web links, self-study quizzes, and an eText version of *World Regional Geography*.

Latin America and the Caribbean

Brad D. Jokisch

The Heart of the Latin American City

Arequipa, Peru, like most large cities in Latin America, has a historic district where the Spanish conquerors and early colonists transformed indigenous settlements into Spanish and Roman Catholic spaces.

These plazas were the center of cultural and economic life and held the most important iconic monuments, such as statues to national heroes, and significant buildings, including the cathedral and colonial government buildings. Arequipa's plaza commemorates its founding in 1540 and both the plaza and the entire historic district were added to the United Nation's Educational, Scientific, and Cultural Organization's (UNESCO) list of world heritage sites. Quito, Ecuador's Metropolitan Cathedral, on the Plaza de Independencia (Independence Plaza) holds the remains of national heroes, including Antonio José de Sucre, the leader of the country's independence fight against Spain.

Other plazas, such as that of Cusco, Peru, display icons or statues of indigenous leaders, and Mexico City's main plaza (zocolo) sits atop the Aztec's holiest temples. Some of these colonial plazas remain the center of the community, but most are now historic districts and tourist attractions. The economic and cultural center of the city has moved into high-rise office buildings, where millions of Latin Americans work in service jobs. Quito and dozens of other cities grew tremendously during the second half of the twentieth century as migrants from the countryside streamed to cities, where jobs and amenities were more plentiful. Urbanization made Latin America the most urban part of the developing world. Millions of people in the region are still engaged in subsistence agriculture, but even more work in cities in a wide array of jobs, from low-paying informal sector jobs to business executives overseeing exports of copper or soybeans to China. Latin America's colonial past can still be seen and felt, but the economy and culture continue to change as Latin America and the Caribbean have become integrated into the global economy through trade, migration, and numerous other exchanges made possible by globalization. Latin America's change and development have been uneven. Some have prospered from the opportunities presented by globalization, but many have not.

- Describe the role major environmental factors play in the location and success of development in Latin America.

- Explain why Latin America is predominantly Roman Catholic in religious affiliation, but is experiencing a rapid growth in Pentecostalism today.

- Characterize the increasingly important role played by China in Latin America's economy and trade.

- Explain the significance of indigenous social movements in economic development throughout Latin America.

- Indicate the importance of mineral exploitation in the economies of Latin America.

- Identify the strongest and the weakest economies in Latin America and the Caribbean and explain the reasons for these national differences.

- Show how migration is reshaping Latin American economies and population distribution.

- List the reasons for the high level of urbanization characteristic of most Latin American countries.

▲ This cathedral at Plaza de Armas in Arequipa, Peru, was constructed with sillar, a white volcanic stone taken from the nearby extinct volcano. Earthquakes, which are common in southern Peru, have damaged this cathedral and other colonial buildings numerous times. Arequipa is Peru's second largest city and has a strong regional identity.

Contents

115

Latin America and the Caribbean

Figure 3–0 The major physical features and countries of Latin America and the Caribbean.

The cultural, economic, and environmental diversity of Latin America and the Caribbean is not well understood by the American public, despite the region's proximity to the United States. South America, for example, contains both the driest desert in the world—the Atacama in Chile and Peru—and the largest tropical rain forest in the world—the Amazon. It also contains the world's seventh largest economy—Brazil—and various struggling economies still dependent on exporting primary products, where more than 30 percent of the population lives in poverty. Culturally, the contrast between a Portuguese-speaking wealthy businessman in São Paulo, Brazil, and an indigenous Shuar woman in eastern Ecuador could not be much greater. Or compare the cultural reality of a cab driver in Salvador, Brazil, whose ancestors were African slaves, and a hotel operator in southern Chile, who continues to speak German, three generations after his ancestors immigrated. These contrasts also apply to Central America and the Caribbean where a similar diversity exists. Yet, much of Latin America and the Caribbean share a similar history of European conquest and economic exploitation, the imposition and diffusion of European cultural traits such as the Spanish and Portuguese languages, Roman Catholicism, and even the iconic urban grid design. Most of the region also shares a strong, if troubled, economic relationship with the United States, even as the presence of China is increasing.

Latin America and the Caribbean's Environmental and Historical Contexts

Is Latin America really *Latin*? And, why is the Caribbean always treated as an appendage to or separate from Latin America? Are there historic or biophysical traits that both distinguish it from other regions and create coherence over such a large area? Latin America and the Caribbean have complex histories that bring together European expansion, African slavery, and the indigenous civilizations of two continents. The physical environment did not make history and does not define the region, but the mountains, oceans, plains, and biodiversity of the region are integral parts of contemporary Latin American cultures and economies.

Environmental Setting: Tropical Hazards and Opportunities

Most of Latin America is tropical or subtropical; only southern Argentina and Chile are located in the midlatitudes. This location greatly affects the weather and climates in the region. Many places experience a wet and a dry season rather than a warm and a cold season. Latin America and the Caribbean also have several tectonic plates colliding into each other, producing impressive mountain ranges and several volcanic "hotspots," and earthquakes which occasionally cause devastation and considerable loss of life. The following section describes the most prominent landforms and climates of the region and briefly discusses how they influence the region's development.

Landforms

Latin America and the Caribbean contain numerous mountain ranges, most of which result from the collision of tectonic plates along the western edge of the region (see chapter opener map). The sudden movement of these plates causes earthquakes and occasionally tsunamis, a large ocean wave that causes rapid flooding along coasts. Where one plate is pushed beneath another, the result is a mountain range, usually with active volcanoes.

Mexico's most prominent mountain ranges are the Sierra Madre Oriental (east) and Occidental (west), the Trans-Mexican Volcanic Belt, and the Sierra Madre del Sur (south). The Sierra Madre Oriental and Occidental Ranges run from Central Mexico northward; between the ranges is the Central Mexican Plateau, a large elevated plateau that covers much of northern Mexico and extends to the U.S. border. Much of Mexico's mineral wealth, especially silver and bismuth, and part of its oil reserves are found in these ranges. The Trans-Mexican Volcanic Belt stretches east-west from the Caribbean Sea to the Pacific Ocean, south of Mexico City. This belt contains at least five active volcanoes, including Popocatépetl, a volcano that reaches 17,802 feet (5,426 meters) and can be seen steaming from Mexico City. The Sierra Madre del Sur is an older and lower range that extends through Michoacán and Guerrero states in southern Mexico.

The Pacific or western side of Central America is dominated by the Central American Volcanic Arc. This chain of volcanoes stretches from the Guatemala/Mexico border to Panama. It contains numerous active volcanoes, including Pacaya (8,273 feet / 2,521 meters), a popular tourist attraction in southeast Guatemala.

The Lesser Antilles, a chain of small islands that run north-south in the southeastern Caribbean, owe their existence to volcanic activity. This chain of more than 30 islands contains numerous active volcanoes, including the Soufriere Hills Volcano on Montserrat that erupted in 1999, destroying the capital city and forcing the evacuation of most of the island's residents.

The Andes Mountains run north-south for more than 4,300 miles (6,900 kilometers) through seven countries in western South America. This range contains several volcanic hotspots found in southern Colombia / northern Ecuador, southern Peru / western Bolivia, and much of southern Chile. It also contains remarkable diversity and very high peaks; more than 50 mountains exceed 19,000 feet (5,800 meters). Aconcagua, in western Argentina, reaches 22,841 feet (6,962 meters) and is the highest peak in all of the Americas. The highest peak in the tropics anywhere in the world is Huascarán (22,205 feet or 6,768 meters), in the Cordillera Blanca (white range) in Peru (Figure 3–1).

Volcanoes can be very deadly and costly when they erupt. The 1985 eruption of the Nevado Del Ruiz volcano in Colombia, for example, killed more than 23,000 and damages cost tens of millions of dollars. Volcanoes do not erupt very often, sometimes not for tens of thousands of years, but even in the tropics and subtropics they can receive a lot of snowfall and some host large glaciers. The meltwater from these glaciers and snowpack is an important source of water for irrigated agriculture and drinking water for rural and urban populations. Furthermore, volcanoes emit ash and other materials that help form fertile soils in the valleys between the volcanoes. The flower industry in southern Colombia and Ecuador, for example,

▲ **Figure 3–1 Huascarán in the Peruvian Andes.** The highest mountain in Peru and all of the tropics is named after the Inca ruler Huascar. The glaciers and snowpack provide valuable meltwater for irrigation and human consumption.

benefits from having fertile volcanic soils and a good source of irrigation water. Another benefit of volcanoes is the ecotourism it creates. Throughout Latin America, tourists from the United States, Europe, and elsewhere in Latin America flock to volcanic regions to appreciate their beauty, engage in mountain climbing and scientific education, and explore the biodiversity created by their stature.

In all of Latin America and the Caribbean there are only three major river systems, all of which are found in South America. By far the largest and most impressive is the Amazon River system, which encompasses much of northern South America (Figure 3–2). The river is the second longest in the world and drains more than 2.7 million square miles (5.1 million square kilometers) in parts of seven countries. The second most important river system is the

▼ **Figure 3–2 The Amazon in Brazil.** The Amazon River is responsible for 20 percent of the freshwater that flows into the world's oceans. Millions of people in Brazil and elsewhere depend on the river and its tributaries for their livelihood.

Paraná River, which drains from southern Brazil through Paraguay and Uruguay before emptying into the Río de la Plata in Argentina. The Orinoco River system drains from Colombia through Venezuela before emptying into the Atlantic Ocean. All three rivers are very important transportation corridors.

Stop & Think

➡ Why are all of Latin America's major rivers east of the Andes?

Climate

Four major climates encompass much of Latin America and the Caribbean (Figure 3–3). First, the tropical wet climate has warm, wet conditions in all months and provides the conditions for biodiverse tropical rain forests in South and Central America. This climate regime is found close to the equator in South America and next to the warm waters of the Caribbean Sea in Central America. Immediately adjacent (north and south) of this climatic zone is the tropical savanna climate, marked by a hot and rainy summer season and a cool, mostly dry winter season. This climate zone is found in parts of Central America and Mexico, most of the Caribbean, and much of central Brazil. Third, the humid subtropical climate has hot, humid summers and cool and mild winters; this climate zone is found in a swath from southern Brazil southward through Uruguay, Paraguay and northeastern Argentina. The fourth major climate type, arid/semiarid, is marked by low precipitation. Much of interior Argentina (and neighboring countries) and northern Mexico are arid or semiarid.

The Atacama Desert, the driest desert on earth, runs along the coast of Chile through Peru (called Sechura in Peru) to the Ecuador border. This amazing strip of aridity is caused by the cold Humboldt ocean current that flows northward from Chile to Ecuador (Figure 3–4). The cold current creates stable atmospheric conditions that prevent rainfall. When this current weakens, warm water from the western Pacific region (Southeast Asia) pushes eastward toward South America. This phenomenon, called **El Niño**, has tremendous climatic and economic consequences for South America. Parts of Ecuador and Peru receive much more rain than normal, causing flooding, damage to infrastructure and crops, and increased disease. The large and lucrative fishing industries in Chile and Peru are damaged as well because cold-water fish (anchovies in particular) flee the area. The 1972–1973 and 1982 El Niño events were particularly severe; the most recent significant event in 1997–1998 crippled Ecuador's banana exports and caused more than $2 billion of economic damage.

Understanding the climates of Latin America is complicated by mountains, because elevation has a strong influence on temperature and precipitation (Figure 3–5). As elevation increases, the temperature decreases 3.5°F per 1,000 feet (in motionless air; about 6°C per 1,000 meters), creating **altitudinal life zones**. In Ecuador, for example, the coast is hot and wet and part of the *tierra caliente* (hot land) zone. Sugarcane, bananas, and other tropical fruit are grown in this zone. Climbing higher in the Andes leads one to the next life zone—*tierra templada* (temperate land), where temperatures are cooler and it is less humid; coffee and maize (corn), fruit trees, and vegetables are grown at these elevations. Continuing up the Andes would lead one to the *tierra fría* (cold land), where nights are chilly or cold and the days start chilly, but the sun's strong rays warm the air until late afternoon, when it begins to cool again. Potatoes,

▲ **Figure 3–3 Climate regions of Latin America and the Caribbean.** Although most of Latin America is located in the tropics, mountain ranges and both warm and cold ocean currents have resulted in the presence of nearly every climate on Earth.

▲ Figure 3–4 **The Atacama or Sechura Desert north of Lima, Peru.**
This barren landscape extends for hundreds of miles, but numerous communities line the beach where the desert meets the Pacific Ocean.

flooding, destroying crops, tens of thousands of homes, and killing approximately 20,000 people in these countries and Nicaragua. In some places it rained more than five feet (1.5 meters) in just a few days. Although this disaster had a devastating and long-lasting effect on Central America, preparedness and emergency planning can minimize loss of life. For example, Cuba gets hit by hurricanes quite frequently, but has few fatalities because the country has an effective and mandatory preparedness system.

barley, wheat, beans, and some maize are grown here. Above approximately 12,500 feet (3,800 meters) exists the *tierra helada* (frozen land), where few crops besides potatoes and quinoa will grow, but alpaca and cattle can graze on the native grasses and pasture. Because of this zonation almost every environment on earth can be found in a single tropical region, making Ecuador and its Andean neighbors extremely rich in biodiversity.

Elevation influences the climate of mountainous areas in another important way. The side of the mountain that faces the prevailing winds—the windward side—is likely to receive much more precipitation than the side facing away from the winds—the leeward side. Where mountains block rain from falling in a significant area, a **rainshadow** results. This effect is what helps create the Argentine Desert and numerous small pockets of aridity (microclimates) in the Andes and other mountainous areas in the region.

Another weather phenomenon that affects Central America (except Panama), Mexico, and most of the Caribbean are hurricanes, which are well-organized tropical storms with winds 74 mph (119 kmph) or greater. Once these storms develop over the warm ocean waters in late summer or early fall, they can be destructive and deadly when they hit land. One of the worst hurricane disasters ever occurred in October 1998, when Hurricane Mitch slowly made its way through Honduras, Guatemala, and Mexico. The storm caused tremendous

Stop & Think

How do climate and livelihood options change with elevation in the Andes?

Environmental Challenges

Latin America and the Caribbean face numerous environmental challenges, including deforestation, environmental degradation, water and air quality problems, natural hazards, and the uncertainties caused by global climate change. Some of these challenges are caused by forces beyond people's control, such as hurricanes, volcanic eruptions, and earthquakes, but there are many environmental issues that people either create or can influence. These challenges have multiple and complex causes and to understand the causes (and hopefully solutions) we must consider the physical environment and the numerous pressures placed on resources. It is paramount to understand that these environmental challenges are health challenges as well.

Deforestation is not new in Latin America, but it is continuing, especially in tropical areas in Central America and the Amazon basin (see *Exploring Environmental Impacts:* The Threatened Amazon Rain Forest). In the 1970s and 1980s much of the tropical deforestation was caused by colonization projects and the roads that accompanied the settlers. Poor farmers commonly followed the roads into the Amazon or forested areas in Guatemala or Costa Rica and replaced tropical forests with agriculture. Much of the recent deforestation has been caused by expansion of cattle ranching, and in the Amazon, expansion of soybean cultivation. As more people live in cities and their income increases, China and other Asian countries demand more soy—much of it fed to pigs as Chinese meat consumption rises. The result has been deforestation, which leads to degraded environments and loss of biodiversity.

Water quality problems can be found in many places in Latin America; the causes vary by place. Massive mining operations in Brazil, Peru, and Bolivia have threatened water supplies, mostly because these activities increase erosion and use toxic chemicals to extract the desired minerals. Elsewhere water problems are caused by cities' inability to procure and deliver clean water. Sometimes agriculture and mining interests compete with urban residents for water, and many cities simply do not have the necessary infrastructure to filter and deliver clean water on demand. At the same time people in these cities suffer from air pollution. Excessive automobile traffic (high-emission vehicles) and industrial pollution, combined with poor regulations and policy

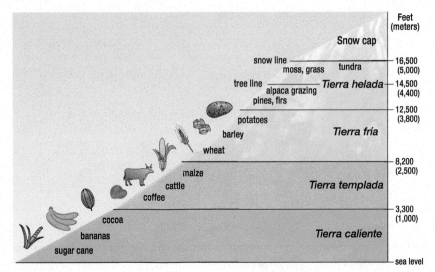

▲ Figure 3–5 **Altitudinal life zones of the tropical Andes.** Elevation influences what can be grown and is a critical part of the region's agricultural ecology.

enforcement, mean that an estimated 100 million people in the region routinely breathe polluted air, leading to health complications and, in some cases, early death. Air pollution problems are particularly severe in Mexico City and Santiago, Chile; their location in basins traps air pollutants by a process called **thermal inversion** (Figure 3–6). As air descends from nearby mountains, the air warms and provides a "lid" of warmer air over cooler air below. Warm air over cool air (the inversion) prevents pollutants from rising and dispersing away from the city, and escaping over the surrounding uplands.

Global climate change also poses tremendous challenges for Latin America and the Caribbean. Given the immense size and diversity of Latin America, the consequences of climate change and people's ability to respond to the change will vary by location. It is projected that average temperatures may rise 4°C (more than 7°F) during the twenty-first century. If that occurs, it will place greater pressure on already taxed ecosystems and agricultural systems. There may be more numerous and stronger storms, including hurricanes; more disease and droughts; and rising oceans, which will increase flooding and damage coastal ecosystems. Small islands in the Caribbean would face the greatest threats from rising sea levels.

Historical Background: The Origins of a Region

When Christopher Columbus arrived in what is now the Bahamas on October 12, 1492, he had no idea that he was on the edge of a vast territory that would soon be labeled "America," and later Latin America and the Caribbean. Contrary to what is commonly taught in high schools, Columbus did not discover a lightly populated wilderness, where human modification of the environment was minimal. This misconception has been labeled the "**pristine myth**," and has persisted despite evidence to the contrary. This territory, in fact, contained thousands of civilizations speaking hundreds of languages, with complicated cosmologies, agricultural systems, and communication systems. Indigenous peoples reached the Americas approximately 15,000 B.P. from northeast Asia in various waves,

and by 1492 the population of Latin America and the Caribbean was likely between 45 and 55 million people, including 16 million people in what is now Mexico and 10 million in the Andes. The term "Indian" was given to these peoples by Europeans even though they had their own identity as Arawak, Mexica, or Cañari because Columbus was convinced he had landed in the region of his intended destination, the Indies of Southeast Asia.

The Americas before Europeans

Some places in Latin America had not been modified much by people, if at all, but much of the region showed signs of human modification, and serious environmental transformation had occurred in some locations. Numerous groups had carved agricultural terraces into hillsides in Mexico and Peru, created cities, deforested large areas, and deliberately burned other areas in an effort to modify the vegetation. In the Amazon basin we know that people were able to create soils in an otherwise difficult environment by diverting and impounding waterways and concentrating organic matter, including waste. Recent archeological evidence shows clearly that the Amazon basin was home to numerous civilizations, mostly in river valleys, but also on the plains above the river valleys.

The three largest civilizations thriving in Latin America when the Spanish and later Portuguese arrived were the **Aztec** of Central Mexico, the **Maya** of the Yucatan Peninsula and Guatemala, and the **Inca**, whose empire stretched more than 2,000 miles (3,200 kilometers) from the present-day border of Colombia and Ecuador through the Andes to central Chile (Figure 3–7). These

▼ Figure 3–6 **Mexico City's pollution.** The city is plagued by severe air pollution because thermal inversions trap the abundant contaminants emitted in the densely populated valley.

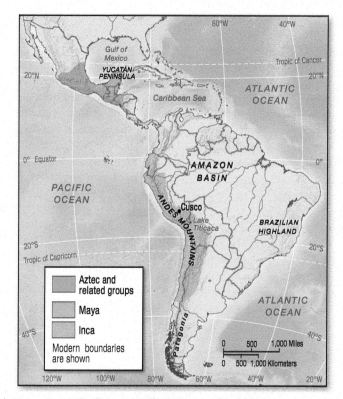

▲ Figure 3–7 **The largest civilizations in the Americas at European arrival.** When Europeans arrived there were three large empires, but hundreds of smaller cultural groups and languages. Archeological evidence indicates that the Amazon Basin was populated, mostly in the plains near waterways.

EXPLORING ENVIRONMENTAL IMPACTS

The Threatened Amazon Rain Forest

The Amazon rain forest is the largest forest in the world, occupying about 2.1 million square miles (5.4 million square kilometers) of the Amazon basin in South America. Most of the forest is located in Brazil, but sizable portions are found in Peru and Colombia and smaller sections in Ecuador, Bolivia, Venezuela, Guyana, Suriname, and French Guiana. The Amazon rain forest accounts for about half of the world's rain forests and is the most biologically diverse, with thousands of species of mammals and many times more species of plants, insects, and other forms of life. In addition to biological diversity, the rain forest performs other environmental services. It stores a vast amount of the greenhouse gas, carbon. One researcher estimated the forest holds 100 gigatons (a gigaton equals 100 billion tons) of carbon that if released would substantially increase carbon levels in the atmosphere and translate into warmer global temperatures. The Amazon rain forest also feeds the Amazon River, the largest river in the world.

The forest likely originated at least 50 million years before present (B.P.), with the first people entering the forest about 11,000 B.P. Francisco de Orellana was the first European to travel the length of the Amazon River in 1542. He reported encountering large civilizations along the banks and up some of its tributaries. Archeological evidence now confirms that there was a substantial human presence in the Amazon at the time of European arrival. Disease and European mistreatment is blamed for drastically reducing the indigenous population.

People have placed pressure on the Amazon to extract resources numerous times since Europeans arrived. The infamous rubber boom of 1879–1912 led to widespread loss of life, but did not lead to significant deforestation. It was not until the 1960s that large-scale deforestation began. The Andean republics (Colombia, Ecuador, Peru, and Bolivia) each had colonization projects in their portion of the Amazon. Road building, oil extraction, and colonization all led to deforestation and, in

▲ FIGURE 3-1-1 **Deforestation in Ecuador's Amazon region.** Road construction into rain forests commonly facilitates colonization of the forest from outsiders and can allow gold mining, timber extraction, and other land uses that reduce forests, increase pollution, and displace people who already lived there.

some cases, drastic reduction in the numbers of indigenous groups (Figure 3-1-1). Most of the deforestation, however, resulted from Brazil's efforts to incorporate its vast Amazonian territory into the national economy by building highways and sending thousands of people into the forest as colonists/settlers. The government encouraged large cattle ranches and required colonists to clear part of their land to claim ownership. The development of roads, including the Trans-Amazonian highway (1970s) and the BR-364 highway (1980s), facilitated the movement of people and cattle into the Amazon and timber out of the forest. Most of the deforestation of the 1960s–1980s is attributed to policies that encouraged cattle ranchers and small farmers/settlers to cut their forests. The rate of deforestation during the 1980s and 1990s fluctuated with the performance of the Brazilian economy: when the economy suffered, deforestation eased, but picked up again when the economy improved.

Since the 1990s the causes of deforestation have multiplied, and include loggers, gold miners, squatters, clandestine drug operations, and others. The greatest culprits are the expansion of soybean cultivation in Mato Grosso state, and the expansion of cattle ranching. Soybeans have replaced only a small forested area, but they have displaced cattle ranching and attracted a soybean infrastructure of highways and railways. This land use change to soybeans has pushed more cattle ranches north into the Amazon forest, and the soybean infrastructure has facilitated more ranching and logging of the forest. As of 2007 it was estimated that about 18 percent of the Amazonian forest that existed in 1500 has been deforested. Most of that deforestation has occurred in an arc on the eastern and southern edges of the forest.

Sources: Philip Fearnside, et al. "Amazonian Forest Loss and the Long Reach of China's Influence," *Environment, Development, and Sustainability* 15 (2013): 325–338; Philip Fearnside, "The Roles and Movements of Actors in the Deforestation of Brazilian Amazonia," *Ecology and Society* 13, no. 1 (2008): 23.

▲ Figure 3-8 **Teotihuacan, north of Mexico City.** The Sun Temple and the Avenue of the Dead were important sacred landscapes of the Toltec civilization. Although little is known about this civilization, it is believed that the Toltecs influenced the architecture and cultures of groups in central and southern Mexico, including the Aztec and Mayan.

groups were the largest and best organized; they dominated or incorporated numerous other ethnic groups, and hundreds of groups existed independent of their empires. Many cultural and technological achievements commonly attributed to these peoples were borrowed from other ethnic groups, or had their origins elsewhere, prior to the rise of these empires. Teotihuacan, for example, was a large empire north of present-day Mexico City (Figure 3-8). It thrived from approximately 100 B.C. to the seventh or eighth century, and its architecture and art are known to have influenced both the Maya and Aztec civilizations.

The Aztec were a Nahuatl-speaking ethnic group that dominated much of Central Mexico from the mid-1300s until the Spanish conquest in 1521. They had a complex cosmology, built enormous and impressive temples, and were skilled farmers. Their crowning achievement was their capital city, Tenochtitlan. The largest in the Americas, this city was actually on an island in the middle of present-day Mexico City. The *templo mayor*, the most important temple complex, was located there, as were the residences of the nobility and the ruler. Rather than incorporating other ethnic groups into their empire, the Aztec intimidated other peoples with their military capabilities and demanded tribute in the form of food and other valuables. The Aztec and neighboring groups were able to grow much of their food by converting marshy areas adjacent to lakes into highly productive raised agricultural fields called **chinampas**.

The Mayan civilization was actually a series of city-states rather than a unified empire (Figure 3-9). These cities were dotted throughout Mexico's Yucatan Peninsula, present-day Guatemala, Belize, and small parts of El Salvador and Honduras. The Maya are known especially for having the only sophisticated writing system in the Americas, their knowledge of astronomy, and their complex calendar system. The Mayan city-states were most prominent from approximately A.D. 250–900, known as the Classic Period, but a smaller number persisted until the arrival of the Spanish. Although the city-states had some cultural traits in common, they were independent enough to speak separate languages and they occasionally went to war against each other. One of the largest and best reconstructed sites is Tikal, in northern Guatemala. This city-state likely

had a population greater than 50,000 and was the dominant political and military force of the region during much of the Classic Period; it appears to have collapsed nearly 500 years before the Spanish arrived.

The Incan Empire, based in Cusco, Peru, was controlled by a relatively small ethnic group that expanded from approximately the mid-1400s to 1533 to create the largest empire in the Americas prior to European arrival. The Quechua speakers controlled numerous other peoples in most of the Andean and coastal regions of modern-day Ecuador, Peru, and Bolivia, and smaller parts of Argentina and Chile. As they expanded, they conquered numerous groups, and assimilated others into the empire. The Inca Empire was predominantly an agrarian society, but they were masterful administrators and architects; they oversaw nearly 25,000 miles (40,000 kilometers) of roads (trails) that covered the Andes and allowed for efficient transportation and the redistribution of people and food. The Inca moved rebellious or recently conquered peoples to distant corners of the empire and redistributed food from areas of abundance to areas of need. Their best-known architectural achievement is Machu Picchu, an impressive

▼ Figure 3-9 **Tikal, Guatemala.** This Mayan city-state was one of the largest and most impressive. It reached its peak from approximately A.D. 200–900 and is one of Guatemala's chief tourist attractions.

estate and religious site perched atop a ridge overlooking their sacred valley (Figure 3–10). As with the Maya and Aztec, many of the Inca cultural and technological achievements had their origins in conquered or ancestral ethnic groups.

European Arrival and Conquest: Making Latin America Latin

The Aztec, Maya, and Inca civilizations, as well as most others, were profoundly and negatively impacted by the arrival of Europeans. Spanish soldiers, African slaves, and others carried with them diseases unknown to indigenous peoples of the Americas. They had no immunity to, and no experience in treating smallpox, influenza, measles, mumps, typhus, whooping cough, and other diseases. Smallpox, which also killed a smaller number of Europeans, was particularly deadly and important in shaping European-indigenous interactions and the population decline of indigenous peoples. The disease was first reported on the island of Hispaniola in 1507 where Spanish settlers forced African slaves and indigenous peoples to work in mines and cultivate sugarcane. The disease spread throughout the island, killing many thousands of the Taino people, contributing to the extinction of this and other Caribbean cultures.

Smallpox helped Spanish military forces conquer the Aztec and Inca Empires despite having a much smaller number of soldiers. By 1519 the Spanish, led by Hernán Cortés, and smallpox had reached present-day Mexico. The disease spread through the Aztec Empire and killed many soldiers and advisors to Montezuma, the Aztec leader. Cortés and his soldiers had other advantages: technology, strategic alliances, and ocean-going ships that brought supplies and more soldiers. They had horses, which were unknown in the Americas; steel armor and swords; and primitive canons, which meant they had superior fighting capabilities. Most importantly, the Spanish created alliances with the enemies of the Aztec; the Totonacs and Tlaxcaltecas fought with the Spanish against the Aztec Empire, and after an eight-month siege of Tenochtitlan, the Aztec Empire fell in 1521 and the holy city was razed. The *templo mayor* was destroyed and the Spanish erected a Roman Catholic

cathedral nearby. The Aztec Empire's organization dissolved, and a Spanish elite soon appeared in present-day Mexico.

Spanish forces, led by Francisco Pizarro, similarly were aided by disease, technological advantages, and the enemies of the Inca to topple the Inca Empire. In 1525 the ruler of that empire, Huayna Capac, who chose to establish a second imperial residence in Quito (Ecuador), died suddenly of symptoms many scholars believe was caused by smallpox. The leader had never seen a European, but the disease may have made its way to the Andes through trade routes. By 1527 the Inca Empire had fallen into civil war, with two of Huayna Capac's sons fighting over who would rule the empire. When Pizarro arrived in 1531 Atahualpa had recently defeated his brother's forces and had him imprisoned. Pizarro tricked the unsuspecting Atahualpa and kidnapped him, demanding that he produce rooms full of gold and silver. Even after satisfying the ransom, Atahualpa was executed. The empire was severely weakened, and the Spanish took control of Cusco. The Inca resisted, but in 1572 the last Inca ruler was executed and by that time the Spanish had dismantled most of the empire's organization.

Disease continued to devastate indigenous populations throughout Latin America and the Caribbean. Outbreaks of smallpox, measles, and later yellow fever and malaria led to drastic reductions in populations and to the wholesale loss of cultural groups. By approximately 1650 the indigenous population of Latin America and the Caribbean had declined by 70–90 percent. Even in more remote areas, such as the Amazon basin, disease took its toll. Jesuit missionaries set up Catholic missions throughout the Amazon basin in the late 1500s. By concentrating scattered indigenous peoples into settlements in order to readily access their labor and unintentionally introducing disease, the consolidated central sites led to epidemics that killed thousands of indigenous Amazonians as well as Andean and Central American populations. These epidemics were accompanied by massive social dislocation. In many cases, mistreatment at the hands of wealthy Spaniards and others who exploited the labor of indigenous peoples contributed to high mortality rates.

Stop & Think

With the arrival of the European conquerors, what became the deadliest battle that confronted indigenous populations?

▼ **Figure 3–10 Machu Picchu, Peru.** This spectacular site is the most common icon of the Inca Empire. Built in the fifteenth century, it is located high in the Andes near Cusco and was never plundered by the Spanish.

Colonial Latin America

The colonial era in Latin America stretched from the Spanish conquests until the early 1820s, when Spanish and Portuguese territories gained independence and republics replaced colonial viceroyalties. The cultural and economic institutions established by the Spanish and Portuguese have had a lasting effect on Latin America and the Caribbean. Both Spain and Portugal established mercantile economies, which laid the foundation for an export economy based on primary products. Another colonial legacy was land concentration, first through *encomiendas* (labor/land grants) and later through *haciendas* (rural estates). The importation of millions of African slaves to work the land and the displacement of indigenous peoples has significantly impacted the cultural geography of Latin America, as did a racial hierarchy that put African and indigenous peoples on the lowest rungs.

Diffusion of the Spanish and Portuguese languages and Roman Catholicism is one of the most important legacies of the colonial era. A final legacy was the increased involvement of other European powers in the Caribbean and in Central America; Great Britain, the Netherlands, and France all challenged Spanish and Portuguese control of the region, and the cultural impacts of that time are still felt today. Soon after Columbus's first trip, his report of the lands he encountered reignited a dispute with Portugal over which kingdom would have the right to occupy and control newly encountered territories. In 1494 the kingdoms of Spain and Portugal, in a deal brokered by Pope Alexander VI, agreed on a meridian (about 46° longitude) that would divide the non-European world into Spanish and Portuguese spheres of influence (Figure 3–11). Portugal, with a greater interest in India and lands further east, received the eastern sphere; Spain received everything to the west of the meridian. Knowledge of geography was sufficiently weak that no one realized the meridian cut through the eastern portion of present-day Brazil, which allowed the Portuguese to claim a foothold in the New World. Although the **Treaty of Tordesillas** had little significance for Latin America at the time, it explains why Brazil became a Portuguese colony and Portuguese speaking, as opposed to most of the rest of Latin America, where Spain was the colonial ruler and Spanish became the predominant language.

The Colonial Economy and European Interlopers

The Spanish attempted to administer their colonies through viceroyalties; New Spain encompassed North and Central America and Peru, most of northern South America. Through these administrative units

▲ **Figure 3–11 Treaty of Tordesillas.** In 1494, the Pope divided the Americas between Spain and Portugal.

the Spanish Crown sought to enrich the Crown by encouraging mineral and agricultural exports, and taxing the profits. Gold and silver were the most valuable mineral exports, although mercury and other minerals were mined as well. Most of the silver exploited in Mexico came from the northern plateau and the Sierra Madre Oriental Mountains; Zacatecas and San Luis Potosí states became the largest producers by the end of the 1500s. South America contained silver deposits in places such as Cerro Potosí, Bolivia, but the Andes held more gold than silver. Spaniards opened mines throughout the Andes, especially in Peru and Chile. The mine owners profited from the mineral exports; the Crown also profited by charging the *quinto real,* or one-fifth the value of the minerals mined. Mine owners and the Crown crafted numerous ways to exploit indigenous labor. In Peru and Bolivia they used the Inca system of drafting labor; in Mexico other labor obligations were enforced. Outright slavery of indigenous peoples also occurred, even though the Spanish Crown did not approve.

Agricultural exports, especially sugar, eventually eclipsed the value of mineral exports from colonial Latin America. Plantation economies based primarily on sugar developed first in Brazil during the late 1500s. Growing and processing sugarcane required a lot of labor, and when labor demands could not be met by enslaving indigenous peoples from the Amazon and elsewhere, slaves were imported from Africa. By the late 1600s sugar had diffused to the Caribbean and came to dominate most of the islands' economy. Similarly, African slaves were imported, mostly from West and Southwest Africa, to meet the labor demand, created in part by the drastic decline in the indigenous population. In Jamaica, African slaves outnumbered whites by 5 to 1 during the 1700s, the height of the sugar-growing era. The vast majority of the 10–12 million Africans who were imported as slaves to the Americas ended up in the Portuguese, Spanish, or English colonies of Latin America and the Caribbean; Brazil alone received 30–40 percent. Slave rebellions occurred, but the institution persisted until the early- to mid-1800s. The country of Haiti is the product of a slave rebellion in 1803.

The value of sugar and the decline in the strength of the Spanish Empire invited other European countries to challenge Spain's grip on the Caribbean and Central America and Portugal's hold on Brazil during the 1600s. The English took control of Jamaica, Belize, and several smaller Caribbean Islands; the Dutch and French also seized numerous islands in the Caribbean. The Dutch even controlled a good portion of coastal Brazil from 1630 to 1654 before they were evicted. The European powers imported slaves and made considerable profit from sugar, just as the Spanish and Portuguese Empires had done. The Dutch, English, and French were able to hold onto most of their island territories and this "scramble for the Caribbean" is what made the Caribbean a distinctive cultural and economic region.

Stop & Think

▶ Why was slave labor, as was also the case in the United States, such a big part of the Latin American and Caribbean colonial economies?

Colonial Legacies

Another important economic legacy of the colonial era was **land concentration**; a small elite came to control much of the productive land whereas the majority of the population owned little. This first occurred through *encomiendas*, grants of indigenous people's labor and tribute to Spaniards as a reward for their service to the Crown. During the 1700s large *haciendas* developed. These were large, semiautonomous rural estates that usurped indigenous people's lands and became status symbols for their owners. On many *haciendas* semifeudal labor relations developed, whereby indigenous peoples worked the owner's land in exchange for rights to cultivate subsistence plots of land.

The Roman Catholic Church was an important institution during the colonial era, mostly because of the missions it established that spread Roman Catholicism and the Spanish language. The Franciscan and Jesuit Orders aggressively sought to convert indigenous peoples to Christianity and to eradicate indigenous belief systems. The missions spread Christianity and Spanish, but in areas with large indigenous populations, syncretic (new belief systems resulting from the combination of introduced and local belief systems) mixes developed, such as in southern Mexico and Guatemala among Mayan peoples and in Peru among Quechua speakers. The most dramatic examples of this **syncretism** can be found in the Caribbean and Brazil, where the greatest number of African slaves were found. Candomblé and Umbanda are belief systems found mostly in Brazil that have their origins in Yoruban (West African) traditions. Similarly, Santeria is a mixture of Roman Catholicism and Yoruban beliefs found throughout the Caribbean, but most prominently in Cuba (Figure 3–12).

Spanish colonial practices also left an enduring imprint on the form of most cities and towns in Latin America. The Spanish built their cities, commonly in the same location as indigenous towns, in a grid pattern. At the center of this grid was the plaza, surrounded by the most important buildings of the time—the Roman Catholic cathedral and colonial government buildings (Figure 3–13). By design the cathedral was usually the largest and most conspicuous building in town. The wealthiest families lived closest to the plaza, and one's status declined with distance from the plaza. In larger cities multiple plazas, usually with a church, developed over time. Today in many Latin American cities this original part of town is the historic district and no longer the business or economic center. In San Juan, Puerto Rico, the original Spanish plaza (Old San Juan) is both a historic and tourist district, catering to tourists who disembark from cruise ships docked in the nearby port.

▲ Figure 3–13 **The Cathedral of Morelia, Michoacán, Mexico.** This baroque cathedral is considered one of Mexico's most beautiful cathedrals. It was finished in 1744 and is the dominant colonial building of the Plaza de Armas.

The End of Colonialism: New Development and Geopolitical Challenges

The colonial era for most of Latin America and the Caribbean ended between 1821 and 1822 when colonial armies were defeated by various rebel armies of the Americas. Independence movements had been under way for more than a decade, and by the early 1820s armies led by such heroes as Simón Bolívar and José de San Martin were able to force Spain to abdicate control in South America. Mexico, Central America, and Brazil, which also gained independence in 1822, followed suit.

Independence initiated geopolitical struggles that created republics out of the newly independent territories. Central America was governed as one republic until 1829 and the present-day countries of Venezuela, Colombia, and Ecuador were governed as Gran Colombia until 1830, when the union broke apart; the similarities in their contemporary flags date to this 1820s unity. It took many years for some border disputes to be reconciled, and some borders remain contested.

During the 1800s Latin America and the Caribbean continued to export primary products, commonly in the form of a series of export booms. Peru exported massive amounts of guano (bird droppings) from their "guano islands" immediately off the coast. The high concentration of nitrogen and phosphorous made the droppings an excellent fertilizer and an important source of nitrates for explosives. Other countries experienced similar export booms with important national consequences. Chile had a nitrate export boom, which instigated the War of the Pacific with Peru and Bolivia in 1879, when Chile resisted efforts to tax or nationalize Chilean companies exploiting mineral resources in the Atacama Desert. Both states lost territory to Chile, and Bolivia became Latin America's second land-locked country, joining Paraguay. Brazil experienced a rubber boom in the Amazon forest starting in 1879, which led to the enslavement and death of many indigenous peoples. Brazil's other major export boom during this time was coffee; by the end of the 1800s it was

▼ Figure 3–12 **Santeria ceremony and shrine in a Cuban household.** Santeria is practiced in elaborate ceremonies and as part of everyday life in Cuba, Puerto Rico, and elsewhere in the Caribbean.

the world's largest coffee exporter. This export was critical for Brazil because it moved much of the economic power from the sugar-growing regions in the north to São Paulo state in the south. It also prompted the immigration of nearly 2 million European immigrants, mostly Italians. These workers were needed to replace slave labor when slavery was finally abolished in 1888. This wave of European immigration during the late 1800s and early 1900s extended to the "Southern Cone" countries of Uruguay, Argentina, and Chile. Argentina's growing economy at the end of the 1800s also prompted massive immigration of Spaniards and Italians; in 1914 half of the population of Buenos Aires was composed of immigrants or their descendants, as was nearly 30 percent of the country's total population. The current Pope of the Roman Catholic Church, Francis, descends from this migratory wave from Italy to Argentina. Jorge Mario Bergoglio, the first priest in all of the Americas to ascend to the Papacy, was born in 1936 near Buenos Aires. His father was an Italian immigrant and his mother was the daughter of an Italian immigrant. Pope Francis is fluent in Italian and generally speaks Italian in public.

By the beginning of the 1900s the United States was becoming the most important military and economic power in the hemisphere and began to throw its weight around in Central America and the Caribbean. Wealthy U.S. businessmen took control of the most valuable land in Costa Rica and Guatemala to develop large banana plantations. The United States went to war with Spain in 1898 and took possession of Puerto Rico and, briefly, Cuba. Following a policy of "gunboat diplomacy" the United States intervened in eight Central American and Caribbean countries and occupied Haiti, the Dominican Republic, Nicaragua, and Cuba. The United States also helped a coup that resulted in the separation of present-day Panama from Colombia. This led to the United States digging the Panama Canal in 1914 and establishing the Panama Canal Zone, a swath of land in Panama that the United States controlled outright until 1977 and partially until 1999. The crude interventionist gunboat diplomacy of the United States ended by the 1930s in favor of a less brutal policy, but the economic, military, and cultural power of the United States continued to loom large throughout the twentieth century and continues to influence Latin America and the Caribbean today.

Population Contours

Latin America's population reached its low point somewhere in the mid-1600s, and grew slowly until the end of the 1800s, when the arrival of European immigrants and a falling death rate increased population growth. By the late 1800s, about 400 years after the European arrival, the population of the region finally reached the pre-European population of approximately 50 million people (Figure 3-14). Latin America experienced rapid population growth during the 1900s, doubling more than three times and increasing tenfold to over 500 million people by 2000. Since the 1980s, and earlier in the Caribbean and the Southern Cone, birth rates have declined considerably. Latin American women now typically have two to three children instead of five to six common in the 1960s. This trend toward a lower total fertility rate applies everywhere in the region, but women in the poorest countries (Haiti, Honduras, Bolivia) still have three to four children on average. A declining growth rate will mean that by 2050 Latin America and the Caribbean is expected to have about 740 million people (compared to 600 million in 2012) and will have virtually no population growth.

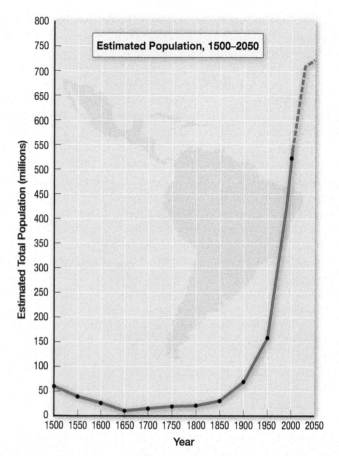

▲ Figure 3–14 **Total estimated Latin American and Caribbean population, 1500–2050.** The region's population did not return to preconquest levels until the late nineteenth century. Public health improvements lowered mortality rates and the population grew rapidly. By 2050 the population is expected to level off near 750 million.

This trend also means that numerous countries no longer face the consequences of rapid population growth; rather, they face the beginnings of an aging population. Brazil and Chile are projected to have 20 percent of its population 60 years of age or older by 2025; by 2050 nearly one-fourth of the population of Latin America and the Caribbean will be 60 years of age or older.

Pre-Columbian settlement patterns and access to oceans or other major waterways are important factors influencing the distribution of Latin America's population (Figure 3-15). As it was during the Aztec Empire, the population of Mexico is highly concentrated in the Valley of Mexico City and surrounding states in Central Mexico, including Guadalajara. With the exception of Monterrey and border cities (Matamoros, Mexicali, Ciudad Juarez), northern Mexico is lightly populated with low population densities. Central America's population is highly concentrated in primate cities such as Guatemala City, Guatemala; San Jose, Costa Rica; and Panama City, Panama. In all six countries at least 25 percent of the population is concentrated in the capital city's metropolitan area. Only Panama City is located on the Pacific Ocean. Northern Honduras, eastern Nicaragua and the Petén region of Guatemala are lightly populated. Population is even more concentrated in South America. The interior valleys of Colombia (Bogotá, Medellín, and Cali) have a high population density as do northern portions of

▲ **Figure 3–15 Population distribution of Latin America and the Caribbean.** The highest population densities in the region are found in the Caribbean and around the megacities of Latin America that have more than 10 million inhabitants, including metro Buenos Aires, Argentina, Rio de Janeiro and Sao Paulo, Brazil, and Mexico City, Mexico.

the Andes surrounding Quito, Ecuador. More than half of Ecuador's population is located on the southern coast close to the port city of Guayaquil. The people of Venezuela (Caracas), Peru (Lima), Chile (Santiago), and Argentina (Buenos Aires) are clustered in the primate city of each country. Brazilians are concentrated in São Paulo and Rio de Janeiro states where the two largest cities are located. But Brazil has several other population clusters, such as the northern coast between Salvador and Fortaleza, which was the colonial center of Portuguese Brazil, and Brasilia, the capital. The desert regions of Argentina and Chile and the Amazon Basin are in general lightly populated, with important exceptions; Lima is located both on the Pacific Coast and in the Sechura (Atacama) Desert, and Manaus, Brazil in the heart of the Amazon has a population of more than 2 million. Callao, the port city adjoining Lima, was the most important port during most of the Spanish colonial era. Manaus, located at the union of the Rio Negro and Solimões, which is what Brazilians call the Amazon River upstream from Manaus, has been important for river travel and transportation for centuries and remains an important ocean-going port. The density of the Amazon Basin is quite low despite more than 40 years of colonization efforts by the governments of several countries.

Contemporary Latin American and Caribbean Development

The economies of Latin America and the Caribbean underwent two crises in the twentieth century, the Great Depression of the 1930s and the "debt crisis" of the 1980s. Each crisis was the impetus for significant and long-lasting economic restructuring of the region's economies. The depression of the 1930s was the motivation for widespread adoption of an economic model called **import substitution industrialization (ISI)**, which started in the 1950s. When the depression occurred most countries' economies were based on exporting primary products, either an agricultural commodity or a mineral, including petroleum, and the United States was the most important trading partner. During the 1930s coffee accounted for approximately 92 percent of El Salvador's exports; bananas accounted for 64 percent of Honduras' exports; copper accounted for 52 percent of Chile's exports; and petroleum accounted for 92 percent of Venezuela's exports. In return, Latin America imported processed and manufactured goods from the United States and Europe. When the depression occurred, demand for the region's exports fell and economies suffered badly.

The problem was diagnosed as an overreliance on one or two exports and a weak manufacturing sector. As long as Latin America remained an exporter of primary products and an importer of processed goods, it would not develop economically. This interpretation of Latin America's position in the global economy was known as dependency theory, because the region was dependent on the United States and other wealthy countries and stuck in a pattern of underdevelopment. Import substitution industrialization was intended to remedy this situation. It was a complex economic model designed to substitute a country's imports with its own industrial production. Instead of cutting trees in Mexico for export to the United States, where they would be processed into furniture, and exported back to Mexico, the idea was that Mexican industry could process the lumber in Mexico. The hope was that Mexico's industrial sector would grow, its economy would advance, and more people would work in better-paid manufacturing jobs rather than living as poor farmers.

Under ISI the state became very important in designing the national economic plan. In most countries the state owned many services deemed necessary for economic development, including electrical facilities, communication services (phone companies), energy production (oil companies), and transportation (railroads and airlines). Domestic companies were protected from international competition, and many services were subsidized by the government. To maintain this strong governmental role many countries borrowed tens of millions of dollars from banks in the United States and Europe. In many countries the decision to borrow so much money was not done through democratic processes; rather, many countries in the 1960s and 1970s were ruled by military dictators (Argentina, Brazil, Chile, Peru, Ecuador, and others) who were largely unaccountable to their people.

For several decades ISI produced positive results. From the 1950s to late 1970s Latin America saw considerable economic growth; poverty was reduced and infrastructure such as road, water, and sewer projects were developed. Mexico and Brazil's manufacturing sectors in particular grew significantly. But by the late 1970s, excessive borrowing and the global energy crisis and economic slowdown doomed ISI. When energy prices increased and the global economy slowed, the economies of Latin America once again suffered. Their exports decreased and as countries could no longer pay the money they owed on their loans a serious and long-lasting "debt crisis" developed. Economic and social conditions worsened and much of the progress made in reducing poverty during the previous three decades was lost. Many countries experienced hyperinflation; Argentina suffered an average inflation rate of over 300 percent per year from 1975 to 1991, and Brazil and Peru's rates exceeded 1,000 percent during the late 1980s. Because the economic crisis lasted for most of the 1980s, the decade has been dubbed the "**lost decade**."

The debt crisis was the impetus for the second major economic restructuring of the twentieth century. The World Bank and the International Monetary Fund offered economic assistance during the crisis, but required countries to radically restructure their economies, which in the short run further worsened the economic suffering. ISI was dismantled, replaced with a new economic model touting free markets, small government, and economic globalization. Many of the enterprises owned by the state were sold to private, usually international, investors. The logic of this model, commonly called **neoliberalism**, is that Latin American economies must compete in the global economy and that an open economy would be more efficient and lead to economic growth.

Stop & Think

> Explain the rationale for import substitution industrialization as an improvement over relying on the export of primary products for revenues.

Thirty Years of Globalization and Export-Led Development

The economic restructuring brought on by the end of ISI and the rise of neoliberal economic models is critical to understanding the economies and development of Latin America and the Caribbean.

Most countries in the region reoriented their economies to export products they could sell better than other countries (comparative advantage), raising the question of whether Latin American economies are in a position much like their pre-1930s economies. Indeed, there are similarities. The United States remains the most important trade partner for most countries outside of the Southern Cone and Brazil, and many economies are based on exporting primary products. Many oil and mineral exporting countries increased their exports during the "commodity boom" of the 2000s. However, many countries now have a wider array of export products, and trade within Latin America and with China is notable and growing. There have been other development initiatives, such as export processing zones (EPZ) in Central America and the Caribbean, and more recently business process outsourcing (BPO), both designed to take advantage of cheap labor and proximity to the United States and Canada. Multiple forms of tourism have been promoted as sustainable alternatives to agriculture and mining. The most dangerous and one of the most lucrative economic activities has been the illegal drug industry. Billions of dollars have flowed into Latin America and the Caribbean as cocaine, heroin, marijuana, and other drugs travel northward to the United States and Europe. The resulting corruption and violence negatively impacts economic and social development, as well as transforms ecosystems. International migration to the United States and Europe has surged in importance economically, and remittances transform households, regional landscapes, and even national economies.

The strongest economies in Latin America are Brazil, Mexico, and Argentina, in that order (Table 3-1). Brazil has the world's sixth largest economy, approximately twice as large as Mexico's, and about five times as large as Argentina's. Based on health data, life expectancy, Gross Domestic Product (GDP) per capita, and poverty rate, Chile, Argentina, and Uruguay have higher standard of living indicators than either Brazil or Mexico, as measured by the Human Development Index (HDI), a composite statistic measuring life expectancy, education, and income. The weakest economy (excluding small Caribbean Islands) and the poorest population in the region is Haiti, followed by the Central American republics excluding Costa Rica and Panama, and Bolivia in South America.

The economy for this world region as a whole is dominated by the service sector, accounting for both 62 percent of the region's GDP and employment. This sector includes a wide range of economic activities, from highly paid financial, technical, and legal jobs to poorly paid cleaning services and the **informal economy**. The informal economy is loosely defined as economic activities that are not taxed or regulated by the government. One study estimated that all Latin American countries have a minimum of 15 percent of their economic activity in the informal sector; 13 countries in the region have 40-60 percent in this sector. The informal economy offers plenty of flexibility, but work conditions are not regulated and the government fails to collect taxes it needs for services it provides.

Manufacturing accounts for about 32 percent of the region's GDP and about 22 percent of employment. Manufacturing ranges from EPZs in the Dominican Republic and elsewhere in the Caribbean and Central America, to steel production in Brazil. The vast majority of manufacturing in the region—over 70 percent—occurs in the three largest economies, Brazil, Mexico, and Argentina. High-tech manufacturing, including pharmaceuticals, aircraft, automobiles, computers, and communication devices, has grown substantially

but is concentrated in Brazil and Mexico; Argentina, Colombia, and Costa Rica have smaller high-tech sectors.

Agriculture remains an important economic activity in Latin America and the Caribbean and, given the diversity of the region, there exists a wide range of agricultural systems. Agriculture accounts for 6 percent of the region's GDP and about 15 percent of the workforce, but in much of Central America and the Andes, where subsistence agriculture is still common, between 20 and 40 percent of the population is engaged in agriculture. Argentina has the most mechanized agriculture in the region and this sector employs less than 2 percent of Argentines. Agriculture is a larger part of the economy in both major exporting countries (Argentina) and poor, rural countries (Honduras, Paraguay).

Four **trade blocs** have become powerful institutions in the region since the start of the neoliberal era. Trade blocs are "free" trade agreements among several countries (Figure 3-16). These agreements reduce the obstacles to and expenses of trading, such as quotas (restricted imports) and tariffs (taxes on imports). The North American Free Trade Agreement (NAFTA), which began in 1994 and includes Canada, the United States, and Mexico, is the most powerful bloc in the region because it gives companies in Mexico access to two very strong economies. The *Mercado Común del Sur* or Southern Common Market (MERCOSUR), established in 1991, is the second largest trade bloc in region, with five members—Brazil, Argentina, Venezuela, Paraguay, and Uruguay. The two strongest economies, Brazil and Argentina, account for nearly half of the trade within the bloc. The Andean Community of Nations (CAN), composed of Colombia, Ecuador, Peru, and Bolivia, has been important at facilitating trade among the member states and, contrary to the other blocs, has no single dominant economy. The Dominican Republic-Central America Free Trade Agreement (CAFTA-DR), encompassing the United States, the Dominican Republic, Costa Rica, El Salvador, Nicaragua, Honduras, and Guatemala, is the newest (2004) and most controversial trade bloc. Central American countries are wary of the United States dumping subsidized agricultural products onto their markets, thereby hurting farmers and the farm economy, and U.S. manufacturers are concerned that they cannot compete with lower environmental standards and cheap labor found in Central America. The trade bloc has increased trade among the countries, but it is too soon to determine the consequences to particular economic sectors in each country.

Petroleum, Natural Gas, and Mining Exports

Petroleum, natural gas, and mineral exports are vital to the economy of a number of Latin American and Caribbean countries. Venezuela, Ecuador, and Mexico rely heavily on oil exports; Trinidad and Tobago and Bolivia rely on natural gas exports. Chile, Peru, and to a lesser extent the other Andean republics depend on mineral exports. The commodity boom that started in 2000 and diminished with the global recession in 2008 increased these countries' export revenues, contributed substantially to their economic growth during that time, but also increased their reliance on these exports.

Venezuela and Ecuador stand out because of the exaggerated importance of oil exports to their economies. Venezuela is the sixth largest oil exporter in the world and the value of oil exports increased from $22 billion in 2001 to $88 billion in 2012. Oil exports account for approximately one-fifth of the country's GDP and over half of its exports. Ecuador's earnings and reliance on oil exports

Table 3-1 Economic and Social Indicators of Latin America and the Caribbean

	GDP (2010)	GDP per capita	Percent Poverty (2010)	Human Development Index (2012)	UN Classification
Antigua and Barbuda	$1,245	$15,635		0.760	
Argentina	$369,992	$9,131	8.6	0.811	Upper middle
Bahamas	$7,700	$22,350	9.0	0.794	
Barbados	$4,110	$14,858		0.825	High
Belize	$1,401	$4,226		0.702	
Bolivia	$19,810	$1,900	42.4	0.675	Lower middle
Brazil	$2,090,314	$10,816	24.9	0.730	Upper middle
Chile	$203,299	$11,827	11.5	0.819	Upper middle
Colombia	$289,433	$6,360	37.3	0.719	Upper middle
Costa Rica	$35,789	$7,701	18.5	0.773	Upper middle
Dominica	$476	$6,632	29.0	0.745	
Dominican Republic	$51,626	$5,227	41.4	0.702	Upper middle
Ecuador	$57,978	$3,921	37.1	0.724	Upper middle
El Salvador	$21,215	$3,618	46.6	0.680	Lower middle
Grenada	$789	$7,571	38.0	0.770	
Guatemala	$41,178	$2,867	54.8	0.581	Lower middle
Guyana	$2,258	$2,923		0.636	Lower middle
Haiti	$6,575	$667	80.0	0.456	Low
Honduras	$15,347	$1,908	67.4	0.632	Lower middle
Jamaica	$13,356	$4,915	17	0.730	Upper middle
Mexico	$1,034,308	$9,522	36.3	0.775	Upper middle
Nicaragua	$6,551	$1,127	58.3	0.559	Lower middle
Panama	$26,808	$7,601	25.8	0.780	Upper middle
Paraguay	$18,427	$2,878	54.8	0.669	Lower middle
Peru	$153,802	$5,205	31.3	0.741	Upper middle
Saint Lucia	$1,198	$7,233		0.725	
St. Kitts and Nevis	$626	$12,263		0.745	
St. Vincent and Grenadines	$684	$7,233		0.733	
Suriname	$3,682	$6,975		0.684	
Trinidad and Tobago	$20,375	$15,463	17.0	0.760	High
Uruguay	$40,272	$11,998	8.6	0.792	Upper middle
Venezuela	$293,268	$10,049	27.8	0.748	Upper middle

Sources: GDP data from the International Monetary Fund

Poverty data from the Economic Commision for Latin Amcerica and the Caribbean (ECLAC)

Poverty data for Bahamas (2004), Dominica (2009), Grenada (2008), Haiti (2003), Jamaica (2009), Trinidad and Tobago (2007) from the CIA World Factbook.

HDI data from the United Nations.

REGIONAL TRADE BLOCS

- NAFTA (North American Free Trade Association)
- CAFTA-DR (Central America Free Trade Agreement, Dominican Republic)
- Caricom (Caribbean Community)
- SICA (Central American Integration System)
- Andean Community
- Mercosur (Southern Cone Common Market)
- No Latin American trade bloc

Note: Bolivia is an accessing member and Chile, Bolivia, Peru, Ecuador, Guyana, Suriname, and Colombia are associate members of Mercosur. Chile and Panama are not members of a Latin American trade bloc, but have free trade agreements with numerous countries in the region and elsewhere. Paraguay was suspended from Mercosur in 2012, but is seeking to rejoin.

▲ **Figure 3–16 Regional Trade Blocs.** Economic and trade agreements among countries have increased trade and economic efficiencies, but when large economies and smaller economies are part of the same bloc it causes many controversies in both the wealthier and poorer countries.

increased dramatically from 2000 to 2012. Oil revenues rose from about $2 billion in 2000 to nearly $18 billion in 2012. Oil accounts for between 40 and 60 percent of its exports.

Chile and Peru also experienced a large increase in the value of their mineral exports during the commodity boom. Peru primarily exports six metals, with copper and gold the most lucrative. Export of these minerals account for about 60 percent of exports (15% of GDP) and their value increased eightfold to $27 billion from 2000 to 2011. Peru has relied on foreign investors, mostly Canadian, Chinese, and American, to operate many of its mines. Chile's reliance on copper exports increased dramatically during the boom, going from 36 percent ($7.8 billion) of total exports in 2003 to 57 percent ($40.3 billion) in 2010. China's demand for copper accounts for much of the increase in demand. Chile now accounts for about 40 percent of world copper exports (see *Visualizing Development: The Rise of China in Latin America*).

Mining has sparked numerous controversies, especially in Peru, Bolivia, and Ecuador. Much of the protest centers on the massive environmental alterations that occur when enormous mines are opened and operated, and on who benefits from the mining activities. In Peru, mountain tops have been removed to extract gold, copper, tin, and other minerals (Figure 3-17). Toxic chemicals are used to extract the minerals and massive amounts of water are used in the processes. There are also equity concerns about local populations having to live with the consequences of the mines while the national government collects revenues and foreign companies and investors make sizable profits.

Stop & Think

> What are the cornerstones of Latin America's economies today and what is their relationship with China?

Export Processing Zones and Business Process Outsourcing

Since the 1960s a development effort in Latin America has focused on export processing in **free trade zones (FTZ)**. These zones, sometimes called **export processing zones (EPZ)**, are enclaves in countries where materials are imported, processed, and reexported as finished products, usually to Europe and the United States. A large number of products are processed in these zones including shoes, clothes, and electronics. The logic behind EPZs was that industries would locate in a country like Mexico, Haiti, Nicaragua, or Honduras with an inexpensive and generally poorly trained workforce. Companies would benefit from the cheap labor and proximity to two large markets, the United States and Canada, and would then hire more workers, jump-starting Latin American and Caribbean economies through linkages to other sectors. Export processing zones were a significant part of the economic restructuring of the 1980s.

The **maquiladora industry** in Mexico began in 1965, when it designated a small area of land next to the U.S. border as a FTZ; the industry gained new life in 1994 when NAFTA went into effect. Hundreds of American and Asian companies were located in Nuevo Laredo, Ciudad Juarez, Tijuana, Nogales, and other borderland cities. Employment peaked in the late 1990s to early 2000s at approximately 1.3 million employees and has decreased slightly since then.

In total 22 of the region's countries adopted EPZs, mostly in the Caribbean and Central America. Outside of Mexico, the largest impact has been in the Dominican Republic, Honduras, El Salvador, and Costa Rica. In the Dominican Republic employment in EPZs has exceeded 150,000 people, and accounts for more than half of Dominican exports.

With a few exceptions EPZs have not been the economic "jump-start" planners hoped they would be. They have provided fewer linkages in the economy and have not employed as many people as anticipated. Many analysts have been critical of EPZs because the host country commonly invested considerable money in attracting companies, but the benefits have not been as high as expected. Many companies in EPZs employ predominantly women, who commonly earn less than men, and the workplace conditions are substandard. Export processing zones continue to be an important part of the region's economy, but the economic downturn since 2008 and increased global competition, especially from China, has diminished their importance.

An emerging trend in the service sector in free trade zones in Latin America is "near-shore" outsourcing, where many business practices are outsourced—the so-called **business process outsourcing (BPO)**. These business practices include call centers where business customer services are directed, and a variety of needs such as data processing and telemarketing. India and the Philippines, both countries with a large pool of English speakers, lead in this type of economic activity, but Latin America is considered an emerging center for U.S. and European companies looking for cheap labor to satisfy their customer service needs. Latin America's niche appears to be proximity to the United States, similar time zones, cultural affinity, and the ability to serve the estimated 45 million U.S. Spanish speakers (Hispanophones). Costa Rica, Chile, and Argentina appear to be the most likely places to attract these businesses, although many transnational corporations are organizing to attract businesses to various other countries. Several transnational consulting companies have located business centers in the Honduran free trade zone, in San Pedro Sula.

Agriculture

Many patterns of agricultural production have held constant for several decades. The tropical and subtropical countries export plantation crops, such as bananas, oranges, coffee, and sugarcane; subsistence agriculture is prevalent in Mexico, Central America, the Andes, and the Amazon; the Southern Cone / Brazil have complex agricultural systems and both import and export agricultural commodities; and raising cattle is common in most non-Caribbean countries.

▼ **Figure 3–17 Gold mining and mountain top removal in the Cordillera Negra (Black Range), Peru.** Mining has become an important part of the economy of every Andean country, but it can have negative environmental and health consequences.

Visualizing DEVELOPMENT

The Rise of China in Latin America

China's economic presence in Latin America has grown considerably since the early 2000s, especially in terms of trade. As China's economy has expanded rapidly, the Chinese government and Chinese investors have looked to Latin America for raw materials, investment opportunities, and most importantly, as a market for its products.

Trade with China has grown dramatically from 2000 to 2012 (Figure 3-2-1). China is now the second most important trade partner for the region, behind the United States, and recently has surpassed the European Union. The pattern of China's trade with Latin America reveals much about the region's integration into the global economy. China imports raw materials from some of the region's most important exporters of raw materials—Brazil, Chile, and Peru. China is now the most important destination of exports from these countries. Argentina and Brazil ship large quantities of soy and Peru, Chile, and Brazil ship minerals (copper, tin, iron ore) to China. A good part of the commodity boom that Latin America experienced from 2003 to 2008 is attributed to increased Chinese demand for raw materials. China's exports to the region, which account for less than 10 percent of the region's imports, are nearly all manufactured goods, mostly garments, shoes, electronics, and other inexpensive consumer items, although high-tech exports have increased. These exports pose a threat to Central American countries, Mexico, and the Dominican Republic, whose exports compete directly with Chinese exports. Many of the region's EPZs have suffered due in part to Chinese competition.

Chinese foreign direct investment (FDI) in the region has not kept pace with its trade growth. When a foreign company or state-owned enterprise purchases, develops,

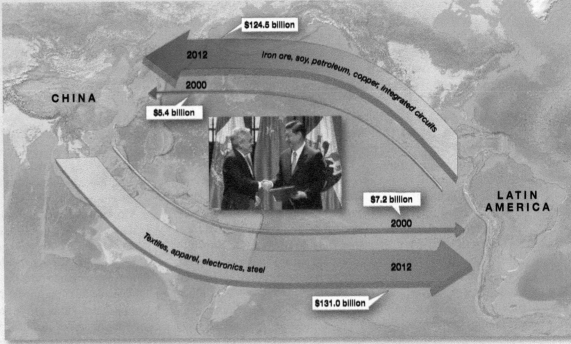

▲ FIGURE 3-2-1 **The spectacular growth in trade between Latin America and China.** Trade increased nearly 2,000% in just 12 years. (Inset) Chilean President Sebastian Pinera shakes hands with Chinese Vice President Xi Jinping in 2011. Formal events where South American presidents announce agreements with Chinese officials have become common.

or expands a company, foreign direct investment occurs. In Latin America and the Caribbean, FDI has surged to $153 billion, with well over half coming from the European Union and the United States. China's annual investment in the region appears to be over $10 billion, but most of that money goes to the Cayman Islands and the British Virgin Islands for banking purposes. Actual investments in productive ventures amount to less than $1 billion annually, and those investments are strategically targeted. Most of the investments are in economic sectors that provide Chinese industry with minerals and other resources they need. Chinese companies have invested in railways, mines, Bolivian natural gas, Venezuelan and Ecuadoran oil, and Peruvian and Chilean copper mining. The Chinese state company Minmetals signed a deal with the largest copper producer in the world, the Chilean state-owned mining company CODELCO (National Copper Corporation

of Chile). In exchange for US$500 million, CODELCO agreed to sell 55,750 metric tons of copper per year at a fixed price over a 15-year period.

China has become an important purchaser of primary products, mostly from South American exporters, and a source of competition for countries that produce manufactured goods, mostly in Central America, Mexico, and parts of the Caribbean. Like in many other places of the world, the region's economic future will increasingly depend on the Chinese economy.

Sources: Rhys Jenkins, "China's Global Expansion and Latin America," *Journal of Latin American Studies* 42 (2010): 809–837. Economic Commission for Latin America and the Caribbean (ECLAC), "Direct Investment by China in Latin America and the Caribbean," in *Foreign Direct Investment in Latin America and the Caribbean* (2010), 99–130 . http://www.eclac .org/publicaciones/xml/0/43290/Chapter_III._Direct _Investment_by_China_in_Latin_America and the Caribbean.pdf

▲ Figure 3–18 **Guinea pigs (cuy) in a Peruvian house.** Guinea pigs are an important food source in the Andean highlands, including in many urban households. They commonly run freely and are fed food scraps and grass.

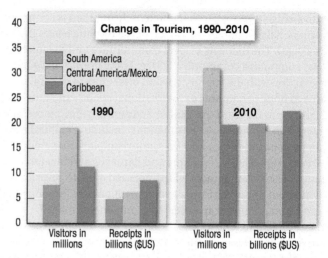

▲ Figure 3–19 **Tourism in Latin America and the Caribbean, 1990 and 2010.** Both the number of tourists and revenue generated from tourism have grown rapidly throughout the region. South America has shown the most growth. **Source:** United Nations World Tourism Organization (UNWTO) 2012.

Globalization and the economic restructuring of the 1980s have had significant impact on agriculture. Many countries chose to pursue alternative and/or niche agricultural exports. Guatemala, Mexico, Chile, and Peru export off-season fruits and vegetables to the United States. Ecuador is still the world's largest exporter of bananas, but also exports shrimp, cocoa beans (chocolate), and palm oil. Guatemala continues to export sugar and bananas, but now exports broccoli, cauliflower, and melons, mostly to the United States. Costa Rica's most valued agricultural export is no longer coffee or bananas, but pineapple. Soy has had a great impact in Brazil, Uruguay, Argentina, and Paraguay. Chinese demand for soy products has made soybeans (or soy products) the number one agricultural export of all four countries; these countries are now four of the six largest agricultural exporters in the world.

Urban agriculture has increased in many of the region's cities. Urban agriculture ranges from households growing a small part of their food in backyards/patios (or raising small animals) to organized and extensive gardening projects (Figure 3–18). Both forms are an important way for urban residents to get access to fresh produce. This phenomenon became particularly important in Cuba during the 1990s when the country lost financial support from the collapsed Soviet Union and entered very difficult economic times.

Tourism

The economic and cultural impact of tourism in Latin America and the Caribbean has grown remarkably in the past 25 years. In 2011 more than 78 million international visitors (not all were tourists) traveled to or within Latin America and the Caribbean, generating nearly $66 billion in revenues. This is a tremendous increase; in 1990 the region witnessed only 38 million visitors and revenue of less than $20 billion. The greatest growth has been in Central America, from less than 2 million visitors to more than 8.3 million, and in South America where visitors more than tripled from 7.7 million to nearly 26 million (Figure 3–19). The economic and environmental impact is greatest in the Caribbean, where numerous island countries rely heavily on tourism. More than 25 percent of St. Lucia's, Antigua's, and Barbuda's GDP is derived from tourism. Many stronger and diverse economies receive many times more tourists and revenue from tourism, but this accounts for a smaller percentage of their GDP; Argentina and Brazil each receive more than 5 million visitors and tourism revenue exceeds $6 billion each year, but this makes up less than 1 percent of the GDP.

The Dominican Republic, Cuba, Costa Rica, and Jamaica are the most common tourist destinations in Central America and the Caribbean.

These numbers include many types of tourism, including backpackers, commonly young people on tight budgets who go to tourist attractions but also get off the beaten path and visit nonglamorous sites, and students on study abroad programs, whose trips include tourist excursions. The most important form of tourism in the region, however, is mass tourism, or the travel of large numbers of people to a given place. Mass tourism includes cruise ship tours, where (relatively) wealthy tourists from the United States, Europe, and a smaller number of Latin Americans board enormous floating cities in any number of ports (Miami; San Juan, Puerto Rico; New York) for a trip to several destinations in the region (Figure 3–20). A four-day tour from Miami may take thousands of travelers to Cozumel, Mexico where they will spend two days on the beach, snorkeling among reefs, drinking rum, and traveling to Chichen-itza, a Mayan archeological site. Cruises travel to Rio de Janeiro, Brazil; Buenos Aires, Argentina; and the Galapagos Islands of Ecuador, but Central America and the Caribbean are the most common destinations, where the attractions are "sun, rum, and fun."

Mass tourism has been criticized for several reasons. The largest concern, especially for Caribbean and Central American resorts

▼ Figure 3–20 **Cruise ship docked in San Juan, Puerto Rico.** Mass tourism is a common way for many tourists from Europe and the United States to visit the Caribbean and Central America, but many tourists see little of the everyday life of the country.

and cruise-ship stopovers, is that much of the tourism is controlled by transnational companies and much of the revenue leaves the region (leakage). The interests of local communities and the interests of tourist companies commonly diverge. Many tourists have minimal interaction with the country beyond brief tours to cultural sites; they dine at hotel restaurants or international chains, and purchase materials imported from the United States and other places outside the region. Many tourist destinations, including cultural sites, exclude local populations, and mass tourism places tremendous pressure on tropical environments and damages endangered ecosystems. Wealth is generated, and the national or state government may benefit but not much local development occurs; local populations can be marginalized and their environments damaged.

As a result of these criticisms a variety of alternative forms of tourism have developed, many of which have labeled themselves as a form of **ecotourism**. The idea behind ecotourism is a win-win proposition; countries receive economic benefits from tourism and promote conservation at the same time. Instead of cutting tropical forests for revenue, tourists travel to the forests to appreciate their beauty and to learn about their ecology. But the term has become overused and nearly every form of tourism now claims to have an ecotourism component. As a result, some ecotourism operations may not be much different than mass tourism. Other alternatives include "volunteer tourism," or **voluntourism**. This entails visitors traveling to tourist destinations, usually at a discounted rate, but also spending some time working for a touted good cause such as building schools, engaging in conservation activities, or teaching. This form of tourism occurs in numerous countries, but mostly in areas where cultural and natural attractions are combined with an impoverished population such as Peru, Ecuador, the Amazon, or Costa Rica. Included in these alternatives is "cultural tourism," where the focus is on the heritage and customs of a contemporary or historic people. Most cultural tourism is focused on the places with a long history of indigenous civilizations—Guatemala, Mexico, and the Andes.

Even alternatives such as cultural tourism can be problematic if not done well. Tourists commonly want to see the most "authentic" people and places, especially in indigenous areas of Central America, the Andes, and the Amazon. Indigenous culture is seen as an interesting historical remnant and commonly distorted. Indigenous people are expected to act in particular ways, even if the behavior that tourists expect is no longer part of their culture or livelihood. Indigenous culture is presented in a romantic and paternalistic way and their local political and economic realities are ignored because they do not have tourist appeal. In this way indigenous culture becomes a prop for tourist operators to generate revenue, not for promoting equitable cultural and economic development.

Illegal Drugs: The Most Dangerous Export

Most of the cocaine and much of the marijuana, methamphetamine, and heroin sold in the United States originates in, or travels through, Latin America (see *Geography in Action*: The Geography of Drugs). The United Nations estimated that over 600 tons of illegal drugs with a street value of $38 billion are grown or produced in Latin America. Much of that money stays within the United States and Europe, but billions go to traffickers and a lesser amount to producers in Latin America. Colombia is the most important source of cocaine and Mexico has recently emerged as the most important transshipment country of illegal drugs and producer of methamphetamines. But either through the production or shipment of illegal drugs, every country in the region participates in this multibillion dollar business. The U.S. Department of State lists 17 countries in Latin America and the Caribbean as major money-laundering countries. Most of the money likely comes from illegal drugs, but some also comes from other illegal activities. Costa Rica is used primarily as a "bridge" for criminal organizations to move profits from cocaine sales from one country to another. The State Department also lists 17 countries in the region as illegal drug-producing or major transit countries and six countries as major sources of chemicals used to produce illegal drugs. Argentina and Chile, for example, produce very few drugs, but are major transit countries for cocaine going to Europe.

Demographic Change and Migration in Latin America and the Caribbean

The second half of the twentieth century witnessed Latin America's population grow from approximately 165 million people in 1950 to 518 million in 2000. Population increase has slowed considerably, growing at only about 1.0 percent per year. Two of the most striking features of demographic change in the region have been rapid and widespread urbanization and the growth of international migration. Latin America and the Caribbean was only 33 percent urban in 1940; by 1990 it had increased to an impressive 72 percent. Currently only a few non-Caribbean countries are under 50 percent urban and the region as a whole is nearly 80 percent urban. Urbanization took off in the late 1940s and continued until the debt crisis of the 1980s due to a combination of factors. As populations grew in the countryside, where there was a long-standing land concentration, more people had less land. Import substitution industrialization programs promoted cheap food programs, which further encouraged people to leave the countryside to go to cities, where there was greater availability of basic needs (clean water, doctors, electricity, schools) and amenities. Millions of young Latin Americans moved to cities, where they had on average 4–5 children. Numerous cities in the region grew very rapidly. Mexico City grew more than 5 percent annually from 1950 to 1970; Bogotá, Colombia and São Paulo, Brazil grew at over 6 percent annually during those two decades. This rapid population growth produced enormous and sprawling urban centers in nearly every South American country and Mexico. It also gave rise to makeshift shantytowns, known as *pueblos jóvenes* in Hispanic Latin America and *favelas* in Brazil (Figure 3–21).

▼ **Figure 3–21 Shantytown in Guayaquil, Ecuador.** As Latin America's cities grew many informal communities with high population density and lack of services developed, commonly on the outskirts of towns or on hillsides.

The Geography of Drugs

The geography of drug production and transportation has fluctuated in recent decades depending on military and police pressure from the United States and other countries. The Andean republics, especially Colombia, have produced nearly all of the cocaine and a lesser amount of the marijuana entering the United States and Europe, but production and shipment has changed in recent years. Cocaine is derived from the coca bush, which grows in several environments in western South America, mostly the Andes. Coca, however, is very different from cocaine. Coca has been used as a medicine, tea, for religious purposes, and to stave off hunger and thirst for at least 3,000 years in the Andes. Coca is still chewed by many indigenous people in Bolivia and Peru; used in this way it is not addictive and does not have adverse health effects. The stimulant has been used in many products, including Coca-Cola until 1903. To this day Coca-Cola uses extracts from the coca leaf but without the cocaine.

It was only in the mid-1800s that an alkaloid in the plant's leaves—cocaine—was identified and separated. By the 1970s most coca cultivated in the Andes was for cocaine production, grown by small farmers or peasants in Peru and Bolivia and sold to intermediaries who would extract the cocaine from the leaves and make a paste. The paste was shipped clandestinely to laboratories in Colombia where it was made into cocaine. The drug traveled onward by boat and plane to Florida, commonly with a stopover in Panama or a Caribbean island. Drug "cartels" with incredibly wealthy, powerful and violent leaders emerged in Medellin and Cali, Colombia. The most famous and wealthy leader was Pablo Escobar, head of the Medellin cartel. At the height of its power the cartel controlled 80 percent of the cocaine that went to the United States, and Escobar had a net worth of perhaps $3 billion. He was finally killed by Colombian police forces in 1993. Colombian cartel leaders were either killed or extradited to the United States, where they are serving long prison sentences.

By the early 1990s the Colombian cartels were severely weakened, but demand for the drug in the United States and Europe remained strong. The power vacuum created by the demise of the Colombian cartels was filled by equally violent and shrewd cartels and other criminal organizations in Mexico. At the same time drug routes from Colombia to Florida became increasingly dangerous because the U.S. Coast Guard and military successfully put enough pressure on the routes to force the traffickers to look for alternatives. The result of these changes has been a shift in the geography of drug production and transshipment.

Currently half of the world's cocaine originates in Colombia, one-third from Peru, and the rest from Bolivia. Ecuador has been a transshipment point but has played a minor role in coca cultivation and cocaine production. Mexico has emerged as the most important transshipment point for cocaine and marijuana entering the United States; much of the cocaine first travels to a Central American country, usually Honduras, El Salvador, or Guatemala (Figure 3-3-1).

One of the most important criminal organizations in Mexico, the Gulf Cartel, became powerful and wealthy when it filled the power vacuum and linked Colombian producers with U.S. buyers. Its chief rivals, the Sinaloa Cartel and the Zetas, are equally powerful. All three groups use violence, extortion, bribery, and other forms of coercion to terrorize Mexicans and control the police and judicial system in much of northern Mexico. The Zetas started as armed security for the Gulf Cartel, but broke away in 2010 to become a sadistic criminal organization, specializing in kidnappings and extortion in addition to drug production and trafficking. They are blamed for thousands of killings in Mexico including the San Fernando massacre, when 72 migrants traveling from Central and South America were killed after they were unable to pay ransom. Although cartel leaders (and others) are routinely imprisoned or killed, the underlying problem remains the same. Drug money flowing in from the United States and Europe has fueled criminal organizations that intimidate Mexican citizens, undermine the judicial system, and destroy trust in law enforcement and the military.

Sources: United Nations World Drug Report, 2013. http://www.unodc.org/unodc/secured/wdr /wdr2013/World_Drug_Report_2013.pdf; Wide Angle: An Honest Citizen. 2004 PBS documentary: http://www.pbs.org/wnet/wideangle/episodes /an-honest-citizen/map-colombia-cocaine-and-cash /colombia/536/; Nik Steinberg, "The Monster and Monterrey," *The Nation*, June 11, 2011, 27–33.

▼ FIGURE 3-3-1 **Cocaine Submarine in Colombia.** Colombian soldiers guard a submarine that could have carried 8 tons of cocaine to Mexico. Smugglers respond to increased surveillance with their own technology to avoid detection.

International migration has had a remarkable effect on Latin America and the Caribbean in the past three decades (see *Geography in Action*: Migration and Development). The vast majority of international migration has been emigration from the region to the United States and Europe, but there also have been notable movements of people within and to Latin America. Brazil is the largest destination for immigrants, mostly from Europe and the United States. Migration patterns within Latin America are generally explained by movements of people from poorer countries to neighboring or nearby countries with higher wages and a demand for cheap labor. Argentina and Chile are the main destination countries in South America. Argentina attracts migrants from Peru, Bolivia, and Paraguay. Chile receives migrants from neighboring Peru and Bolivia and a smaller number of Colombians and Ecuadorans. Because of the prolonged civil war and, more recently, coca fumigation in Colombia, both Venezuela and Ecuador have received a sizable number of Colombian immigrants. Mexico, Costa Rica, and Panama receive a fair number of migrants, both from other Latin American countries and from North Americans retiring to (sub)tropical locations where retirement savings go further. The favored destination of retirees from the United States and Canada has changed from Guadalajara, Mexico to Costa Rica, and more recently to Panama and Ecuador. Although the number of immigrants is not large, they have a significant local impact where they cluster and tend to spend considerably more money than local residents. Many of the immigrants have North American expectations on a variety of cultural matters including speaking English, noise and animal control, and other issues that sometimes clash with local customs.

The United States is the most important extraregional migrant destination for Latin Americans. In 2011 there were more than 50 million people in the United States (17% of the U.S. population) who considered themselves Latino or Hispanic. Of those, 18 million were born in Latin America. Somewhere between 9 and 11 million immigrants live and work in the United States without legal permission—the so-called illegal immigrant population. About two-thirds of the Latino population in the United States is Mexican or Mexican American; this figure also applies to the "illegal immigrant" population. Mexican emigration is unique in at least two ways, in addition to the fact that it has more emigrants (12 million) than any other country. Mexico is the only developing country with a long border with a wealthy country; in Europe the Mediterranean Sea separates it from Africa. The fact that Mexico has a large impoverished population, and part of the U.S. economy depends on cheap labor, has fueled one of the largest migration flows in the world. Second, Mexicans are overwhelmingly concentrated in just the one country; approximately 97 percent of emigrants are in the United States.

Latin American migration to the United States increased significantly during the 1980s when several Central American countries (Guatemala, El Salvador, and Nicaragua) were embroiled in civil war and most countries went through the pain of structural adjustment and hyperinflation. Immigration has continued in large part because U.S. migration policy encourages family unification. When Latinos in the United States gain residency or citizenship they are able to sponsor the legal immigration of immediate family members, developing what is called **chain migration**. Nearly two-thirds of all legal Latin American immigrants join immediate family members.

Latinos are overwhelmingly concentrated in metropolitan New York, California, Florida, Arizona, and Illinois (Chicago). Many of the migrants go to the United States planning to earn enough money to make their lives better back in Latin America and do not plan to stay in the United States. Many of those migrants participate in **transnational migration**, meaning that they use communication and transportation technologies (cell phones, videoconferencing, Skype, e-mail) to stay in touch with family and friends. They also send home money and other goods, commonly through multipurpose agencies that cater to migrants (Figure 3–22). Many thousands of migrants who planned to stay for a short time ended up staying much longer or are still in the United States. When family members join them in the United States and/or when they have children in the United States their departure is commonly delayed, sometimes permanently.

Mexican migration changed dramatically starting in about 2008. It appears likely that for the first time in 60 years more Mexicans left the United States than entered it. The recession of 2008 and growing hostility toward immigrants combined to discourage Mexican migration to the United States and encouraged some of those living in the United States to return to Mexico. Between 2005 and 2010 only about 1.4 million Mexicans entered the United States and about the same number left. Mexican migration will likely increase as the U.S. economy grows, but it will be much smaller than the large influx of 1980–2005.

Stop & Think

➤ Why does migration play such an important part in many of Latin America's communities?

▼ **Figure 3–22 A multipurpose migrant agency in Queens, New York.** Agencies like Delgado Travel facilitate transnational migration and the movement of millions of dollars from the United States to migrants' families in Latin America and the Caribbean. These two long-term immigrants were campaigning for Rafael Correa (Dale Correa) in 2006. It was the first time Ecuadorans living overseas could vote in the Ecuadoran presidential election.

Contemporary Social and Cultural Trends

Latin America and the Caribbean have experienced considerable social and cultural changes in recent decades. One of the most important cultural changes has been religious upheaval, caused mostly by the growth of Protestantism. The role of women in society has also undergone considerable change. Latin America remains heavily patriarchal, but women's roles and legal rights have expanded along with educational and employment opportunities. A third major change has been the rise of indigenous movements in several countries. Indigenous peoples have demanded and received titles to ancestral land and greater autonomy. They have also made important gains in formal political processes.

Is Latin America Becoming Protestant?

Since the conquest at the hands of the Spanish and Portuguese, Roman Catholicism has been the predominant religion in Latin America. It is important to understand that there are many forms of Roman Catholicism and many people are culturally Catholic rather than strong believers of the faith. In areas with a strong indigenous or African presence, Roman Catholicism has mixed with indigenous and African faiths to create religious hybrids. Where the British and Dutch settled in the Caribbean and Central America, Protestant faiths, rather than Catholicism, are more prominent. Jamaica is well over 90 percent Protestant. Nevertheless, Latin America and the Caribbean have more followers of Roman Catholicism than any other world region; nearly 40 percent of the world's Catholics are found here, compared to 24 percent in Europe and 16 percent in Africa. Brazil and Mexico alone account for one-fifth of the world's Catholics (230 million). The region also has the greatest percentage of its population Catholic, at approximately 76 percent. Yet the percentage of the population (not total numbers) identifying as Catholic was much higher in 1950. Brazil was 94 percent Catholic in 1950, and while Brazil has the second largest Catholic population in the world, behind the United States, it is only 74 percent Catholic now. This decline in the relative dominance of Catholicism is explained by the rise of Protestantism in the region during the second half of the twentieth century. About 19 percent or nearly 111 million people in the region identify as Protestant, including mainline Protestant denominations, numerous evangelical and Pentecostal groups, and adherents to other non-Catholic Christian faiths such as the Church of Jesus Christ of Latter-day Saints (Mormons) and Jehovah's Witnesses.

Numerous Latin American countries such as Chile and Brazil have had Protestant populations since the mid-nineteenth century, due to Protestant immigration and missionary activities among mainline churches such as Lutherans, Methodists, and others. By1950, however, the region's Protestant population was likely no more than 2 percent. Starting in the 1960s growth of Protestantism was widespread and rapid. Much of this rapid growth has been among Evangelical or Pentecostal groups. **Pentecostalism** is distinct from mainline Christian denominations because it focuses on the literal interpretation of the Bible, a strong "personal relationship with Jesus Christ," and a strong desire to convert others to Pentecostalism. As much as 85 percent of the Protestant growth in the region has been Pentecostal. Brazil has by far the largest Protestant population, between 39 and 41 million adherents, followed distantly by Mexico at 9–10 million followers. Central America has by far the greatest percentage of Protestants in Hispanic Latin America. Guatemala, El Salvador, and Honduras are between 35 and 38 percent Protestant and Nicaragua's Protestants number between 24 and 27 percent (Figure 3–23). There is a great deal of diversity among the Pentecostal faiths and many are led by local, lay clergy. Some churches have strong ties to Pentecostal organizations in the United States, but many more have loose ties or none at all. Critics of the Pentecostal movement have labeled these organizations as "sects," or splinter groups, led by charismatic and authoritarian leaders.

What explains this dramatic rise of Pentecostal Protestantism in Latin America in the past 50 years? Many explanations exist, ranging from the highly political to the spiritual. Some observers, including many priests and leftists, have blamed outside intervention as the chief cause. Some even interpret the rise of Pentecostalism as another form of American imperialism, this time in the form of missionary activities. These observers see the growth of Pentecostalism as destroying or dividing communities, betraying long-held cultural traditions, and promoting conservative politics. Others interpret the rise as an understandable result of the massive social changes that occurred after World War II, such as land reform and the demise of the hacienda system in the countryside, population growth, massive rural-to-urban migration, and the development of large *pueblos jóvenes* and *favelas*. Still others see the rise of Pentecostalism as a result of the failure of the Catholic Church to meet the spiritual and social needs of millions of poor people. Pentecostalism offers the poor a spiritual alternative and a strong sense of community.

The Role of Women in Society

The idea that women are passed from their father to their husband at marriage and their most important tasks were to be wife and mother has been the enduring stereotype of the Latin American woman. The man of the house was authoritarian, women were subordinate, and women's space and role was in the house tending to children and the couple's parents. She worked outside the home only when necessary. Women who defied these expectations by working independently and/or shunning marriage faced the wrath of a society with rigid gendered roles. This depiction, of course, is

▼ Figure 3–23 **A Pentecostal church in rural Guatemala.** Many adherents of Pentecostal faiths in Central America attend services in informal buildings and the preachers are self-trained, and commonly charismatic.

Migration and Development

Migration is an important aspect of development in Latin America and the Caribbean, mostly because of remittances, but also because of return migration. **Remittances** are money or goods sent from the migrant destination to the home country. Latin Americans living in the United States and Europe remit an estimated $61 billion every year (Figure 3-4-1). Most of the money and goods are sent by individuals or families to their loved ones—children, spouse, parents—at home. These remittances have a significant impact at the individual or household level and on the national economy.

A number of households in Central America, Mexico, the Caribbean, and to a lesser extent the Andes rely on remittances as a significant source of income. It is estimated that between 10 and 20 percent of households in Paraguay, Bolivia, Colombia, Nicaragua, and Honduras received remittances in 2010. Between 5 and 10 percent received remittances in Mexico, Ecuador, and Peru. More than 20 percent received remittances in Nicaragua, Haiti, and the Dominican Republic. The amount sent home depends on whether the migrant has employment, how much debt he/she has, how long the migrant has been gone, and whether or not the migrant has children. Some households receive only $100–$200 per year while others commonly receive $1,000–$2,000 or more.

For households receiving remittances, the money made up between 30 and 50 percent of income in most countries. Remittances commonly pay for basic needs such as food, medicine, clothing, and education. Many migrants send home thousands of dollars to pay off a loan taken to pay for illegal transportation to the United States or elsewhere. Beyond basic needs and to cancel debts, remittances may pay for a loved one to migrate, commonly wife or child, or to build a modern home. Some money pays for conspicuous items, such as a large shiny truck and cowboy boots in Mexico, or a house much larger than needed in Ecuador, but most money pays for basic needs and lifting families out of poverty.

Remittances are also sent home for community purposes. Migrants in the United States, for example, commonly organize and form **home town associations (HTA)**. These HTAs may be based on the migrants' home community, but they may also be soccer clubs. HTAs have raised money for fiestas, schools, roads, soccer fields, and other events or items. Many small, rural communities in southern Ecuador have new or renovated churches, mostly paid for by money raised by HTAs in New York or Chicago.

Mexico receives by far the most remittances. Of the estimated $61 billion in remittances to the region, Mexico received nearly $23 billion. Seven other countries received between $2 and $4.4 billion. Yet, because Mexico's economy is strong, that figure accounts for only about 2 percent of its GDP. This situation is quite different in smaller economies in the region. Remittances account for nearly 30 percent of Haiti's economy and more than 15 percent of the economies of Jamaica, Honduras, El Salvador, Nicaragua, and Guyana. Remittances to the region reached their peak in 2008 at $65 billion, declined for two years during the global recession, but have since regained momentum.

A number of countries have sought to use their overseas migrant population for development purposes at home. Mexico, for example, has reached out to migrants through a *tres por uno* (three for one) program. The program allows HTAs to send money for approved projects and the state, federal, and municipal government will match the contribution. Since its inception in the early 1990s, the program has funneled tens of millions of dollars into over 2,000 development projects in Zacatecas, Michoacán, and other states. Most of the projects have been directed at health and education or infrastructure initiatives. In 2004 Ecuador's National Secretary for Migration began a program, *Todos Somos Migrantes* (we are all migrants) to reach out to migrants and encourage them to return. Ecuador offered to allow migrants to bring their belongings home duty free and to help them with business start-ups. Like other countries, Ecuador and Mexico see migrants not as abandoning their country, but citizens with skills and resources to help the home country.

Sources: "Remittances to Latin America and the Caribbean in 2012: Differing Behavior across subregions," Inter-American Development Bank, *http://www.iadb.org/en/topics/remittances/remittances,1545.html*; Jonathan Fox and Xochitl Bada, "Migrant Organization and Hometown Impacts in Rural Mexico," *Journal of Agrarian Change* 8, no. 2 (2008): 435–461.

too generalized as there have always been other types of households and gender relations in Latin America have been far more complex. But the stereotype certainly contains very real and truthful elements. Women in most of Latin American society have had fewer rights and opportunities and faced greater domestic violence from men, who had considerable control over women, than women experienced in North America or Western Europe. In recent decades this patriarchal structure has been challenged by numerous women's (feminist) movements and the role of women has greatly expanded. Many, but not all, women have benefited from these changes, although many obstacles remain.

Since the 1970s the number of women's movements has grown throughout Latin America. These movements have been diverse, both in scale and the problems they identify in society. Some movements were very local and addressed education, health care, job security, and domestic violence. Others, commonly based in the middle and upper

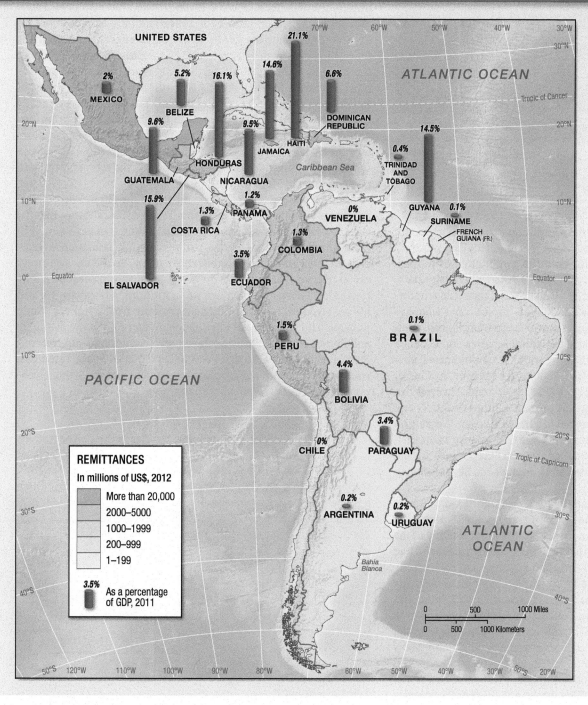

◀ FIGURE 3-4-1 **The economic importance of remittances.** Most of the remittances to Latin America and the Caribbean originate in the United States, but since the late 1990s Europe has become an important source as well.

Source: IADB (Inter-American Development Bank).

classes, became national movements to change legislation and give women more rights in society. These movements demanded changes to patriarchal systems of power that disadvantage women. They commonly sought greater educational and employment opportunities; changes to laws on marriage, divorce, and contraception; and greater political representation. One of the most famous movements led by women was the Mothers of the Plaza de Mayo in Argentina. Between 1976 and 1983 the military dictators of Argentina conducted a "dirty war" against people they perceived to be enemies of the state. Tens of thousands of people were persecuted, detained, and tortured. Between 12,000 and 20,000 people, mostly young leftists, disappeared and were killed. The Mothers of the Plaza de Mayo was a group of women who demanded the government reveal the whereabouts of their sons and daughters. They marched weekly in Buenos Aires's Plaza de Mayo in front of the presidential palace. The movement gained international attention and placed greater pressure on the military leaders. After the

▲ **Figure 3-24 Bananeras in Colombia.** Women make up much of the labor force in banana plantations, and efforts at unionization have been led by women in many countries.

military stepped down in 1983 the organization played an important role in bringing to justice many of the perpetrators of the dirty war.

Central America has witnessed the emergence of numerous working-class women's organizations. The 1979 revolution in Nicaragua and the subsequent 11 years of Sandinista rule empowered women in many respects and helped inspire other women's organizations, such as the *bananera* movement, a transnational feminist movement of banana workers in Central America and Ecuador. Since the 1980s women have organized within male-dominated labor unions to demand better working conditions, education, and better pay for both men and women (Figure 3-24). The women have reached across borders to join the efforts of women in other countries and consider gender equity to be integral to labor organizing. Working on banana plantations in Latin America and the Caribbean remains a difficult and precarious job, with few benefits and little job security, but the bananera movement has worked to improve the lives of women and empower banana workers.

Women now account for about 40 percent of the workforce of Latin America and the Caribbean, compared to only 18 percent in 1950. Women's involvement in the workforce reflects many of the social and economic changes that have occurred in the second half of the twentieth century. Rural-to-urban migration led the majority of women to cities, and by the 1970s women postponed marriage by a few years and fertility rates declined as women's education improved. The economic crisis and restructuring of the 1980s, as well as the increased importance of service sector jobs and the informal economy, all influenced what jobs were available to women. Many women have found professional jobs in business, medicine, and law, but many others are stuck in low-paid, insecure, and sometimes dangerous jobs. Many of the export and business processing jobs are dominated by women, especially in the maquiladora jobs in Mexico. Greater participation in the workforce has led to greater empowerment for many women, but has led to more work for many women who still do most of the work at home.

Indigenous Social Movements

One of the most important social movements in Latin America has been the mobilization of indigenous peoples in numerous countries. This mobilization has taken many forms: protest against economic and social injustice, greater participation in the formal political process, and efforts to secure title to land and indigenous territories.

The most dramatic uprising has been the **Zapatista movement** in Mexico. Comprised mostly of indigenous peoples (Mayan and others) the Zapatista (*Ejército Zapatista de Liberación Nacional*) movement began on January 1, 1994, when the North American Free Trade Agreement (NAFTA) went into effect. The Zapatistas timed their emergence to protest NAFTA because they believed that the trade agreement would further impoverish rural and indigenous peoples in Mexico, especially southern Mexico where most of the country's indigenous peoples reside. The movement has mostly been nonviolent and is avowedly antiglobalization. It promotes participatory democracy and the right of indigenous peoples to determine their own politics and economy. Globalization is seen as a top-down, totalitarian system. The Zapatista movement remains a political force in Mexico.

Indigenous peoples have also made progress in formal political processes as well. In 2005 Evo Morales, an Aymaran from Bolivia, was the first indigenous person elected president of any country in Latin America. Morales was born into a family of subsistence farmers and later grew coca and became a leader in the coca-growers union (Figure 3-25). He became active politically, protesting efforts to eradicate coca and privatize water for mining interests. He and his government have led Bolivia away from neoliberal economic policies. The government has a much greater role in the economy and Bolivia even nationalized the natural gas, mining, and communications industries. Morales was reelected in 2009 and remains a strong and controversial advocate for the poor and indigenous peoples and an enemy of globalization.

▼ **Figure 3-25 Bolivian President Evo Morales at the United Nations in 2006.** The president points to a coca leaf to emphasize his point that coca is a valuable crop and because there are many positive uses of the plant it should not be considered illegal by the United Nations.

Indigenous peoples of Latin America have also been successful at securing land titles and regaining territories, especially in tropical rain forests. Numerous countries in the Amazon basin have ceded traditional lands to indigenous groups. Nearly two-thirds of Venezuela's, Colombia's, and Ecuador's Amazonian region has been legally granted to indigenous peoples and more than 20 percent has been granted in Brazil and Bolivia. In Mexico, Panama, and Nicaragua large areas of historically indigenous territories have been mapped and ceded to indigenous peoples.

Mexico, Central America, and the Caribbean

Mexico and Central America are commonly considered "Middle" or Meso-America (Figure 3–26). For some purposes, such as NAFTA, Mexico is considered part of North America while for other purposes Mexico is considered part of Middle America. Economically Mexico is strongly linked to the United States, but culturally it (especially southern Mexico) has more in common with Central America. By most classifications Guatemala, Honduras, Nicaragua, El Salvador, Costa Rica, and Panama make up Central America,

leaving Belize as part of the Caribbean. Belize and the Mosquito Coast (Caribbean coast of Nicaragua and eastern Honduras) are a hybrid between Central America and the Caribbean because they have Hispanic cultural features, but their history of being part of the British Empire gives them cultural and political features distinct from the rest of Central America. Panama, part of Colombia until 1904, also has some Caribbean cultural characteristics, and because of its geopolitical significance, stands out from the rest of Central America.

Mexico

Mexico is the third largest country in Latin America and has four major mountain ranges. The Sierra Madre Oriental (east) and Occidental (west) ranges run southeast to northwest from approximately Mexico City to the U.S.-Mexico border. The Trans-Mexican Volcanic Belt extends east-west from the Gulf of Mexico to the Pacific Ocean, south of Mexico City. The Sierra Madre del Sur runs through Michoacán and Guerrero states in southern Mexico. Because Mexico stretches nearly 2,000 miles (3,200 kilometers) from Tijuana on the border with California, to the border with Guatemala in the south, the country experiences numerous climates. Much of the Yucatan Peninsula and southern Mexico have a tropical savanna climate, receiving heavy rains from May to October; and a portion of this

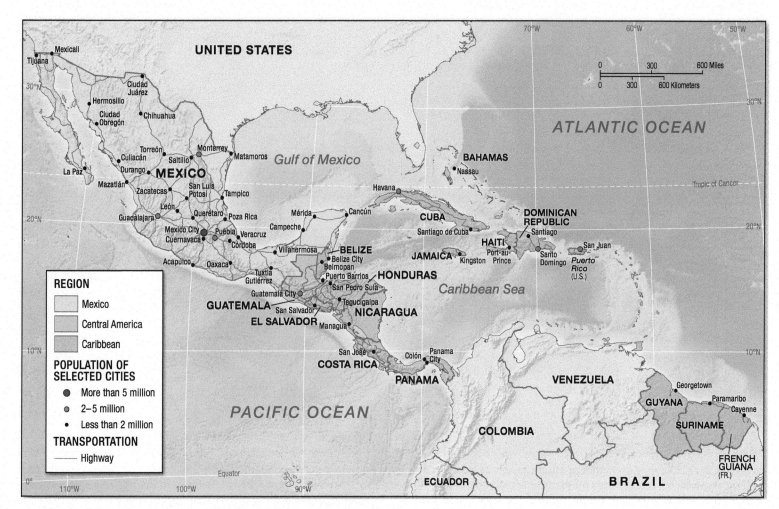

▲ Figure 3–26 **Countries and major cities and highways in Mexico, Central America, and the Caribbean.**

region has a tropical wet climate. As a rule of thumb, the climate of Mexico becomes drier going north. The Sonoran and Chihuahuan Deserts of Northern Mexico (and the southwestern United States) are very dry and hot. Exceptions are some elevated areas in the Sierra Madres and the Trans-Mexican ranges, which receive enough rainfall to have a subtropical climate.

Ranked immediately behind Brazil, Mexico has the second largest economy and second largest population in Latin America. Many of the estimated 117 million residents are concentrated in Central Mexico. Greater Metropolitan Mexico City, a classic **primate city** because it has a disproportionate share of the country's population and economic power, has approximately 20 million people, the largest urban concentration anywhere in North or South America (New York is second). Guadalajara, located in a valley in the western portion of the Trans-Mexican Volcanic Belt, is a distant second, with 4.3 million in its metropolitan region. Most Mexicans are mestizo, but about 15 percent of the population is indigenous (Figure 3–27). Most of the indigenous population is concentrated in the southern states of Oaxaca and Chiapas, with a sizable population in the Yucatan Peninsula, Michoacán, and Puebla. Nahua, Mayan, Zapotec, and Mixtec peoples are the most numerous. More than 60 indigenous

languages are spoken in Mexico. Widespread poverty and lack of education and opportunities have encouraged some indigenous peoples to migrate to Mexico City, Guadalajara, tourist centers such as Cancun, and the United States; as many as half a million indigenous people live in metropolitan Mexico City. When indigenous migration to the United States increased during the 1990s, it created communities of Mexicans whose first language is not Spanish, complicating their ability to interact with not only the non-Hispanic population of the United States but other Mexicans and Latinos as well.

There are many notable regional disparities in Mexico's large and diverse economy. Mexico's economy has transformed substantially since the 1980s when neoliberalism began in the region. The economy opened to international investment, state-owned enterprises were sold, and tariffs were reduced. Asian countries, especially China, are important to Mexico's economy. But through trade, including NAFTA, and migration, Mexico's economy is still strongly linked to the U.S. economy and to a lesser extent, Canada's. One of the most important economic changes in the last two decades of the twentieth century was the massive expansion of manufacturing exports, facilitated by NAFTA. The United States and Canada now receive over 85 percent of Mexico's exports, predominantly manufactured goods. Much of the

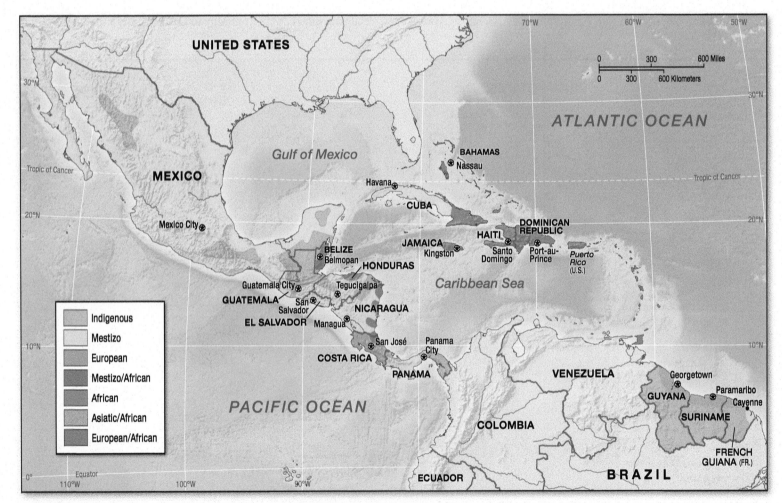

▲ Figure 3–27 **Ancestry of dominant ethnic groups of Mexico, Central America, and the Caribbean.** Middle America has considerable ethnic and cultural diversity. Mexico is mostly mestizo, but indigenous peoples dominate southern Mexico and Guatemala. Populations with African ancestry are found in numerous locations in Central America, but most commonly in the Caribbean. Ethnic minorities live in many places where their ethnicity is not dominant, which is important to keep in mind when interpreting maps of ethnicity and ancestry.

manufacturing and assemblage of these products occurs in the north, along the U.S. border, and in Monterrey, Mexico's third largest metropolitan area. Monterrey is perhaps Mexico's wealthiest city and hosts numerous Mexican and international businesses.

Mexico's economic and political power remains concentrated in Mexico City, one of Latin America's most important economic centers. Mexico City is an important manufacturing center for export products and domestic consumption. Services, including high-tech services, are the most important part of the region's economy. Many exporters, transnational companies, insurers, and banks are located in Mexico City, many of which both employ and cater to Mexico's middle and upper classes. This concentration of economic activity in the north and Mexico City has heightened regional differences. The economic power of the north has been rooted since the Spanish Conquest in the mineral resources found in the mountains and deserts of the region (Figure 3-28). Northern dominance was further enhanced in the twentieth century by the discovery of petroleum and natural gas deposits in and along the coast of the Mexican Gulf. Northern Mexico on average is wealthier, more urban, and more industrialized than southern Mexico, which is much poorer, rural, and agricultural.

Despite the strength of the economy, roughly one-third of Mexico's population and nearly 90 percent of the indigenous population lives in poverty. As Mexico has become integrated into the global economy, its middle and upper classes have grown, but its economy has failed to provide jobs to lift nearly 40 million people out of poverty. If more than 33 million Mexicans (or Mexican Americans) did not remit nearly $20 billion annually, Mexican poverty would be much higher than it is.

Tourism is also important to the Mexican economy, and takes many forms. One extreme is mass tourism to beach resorts such as Cancun and Acapulco. Both cities are popular cruise ship destinations and tourist enclaves in the sense that the infrastructure is built for tourists, many of whom have little interaction with Mexicans or Mexico beyond the resort and a few nearby sites. Tourists also go to historical and cultural sites in the Yucatan to see Mayan ruins or to Mexico City to see Aztec sites and the pre-Hispanic city of Teotihuacan, a UNESCO World Heritage Site. Mexico has a sufficiently large middle and upper class that at many tourist destinations Mexicans mix with tourists from the United States and Europe.

In 2000, the Institutional Revolutionary Party (Partido Revolucionario Institucional or PRI) lost its first presidential

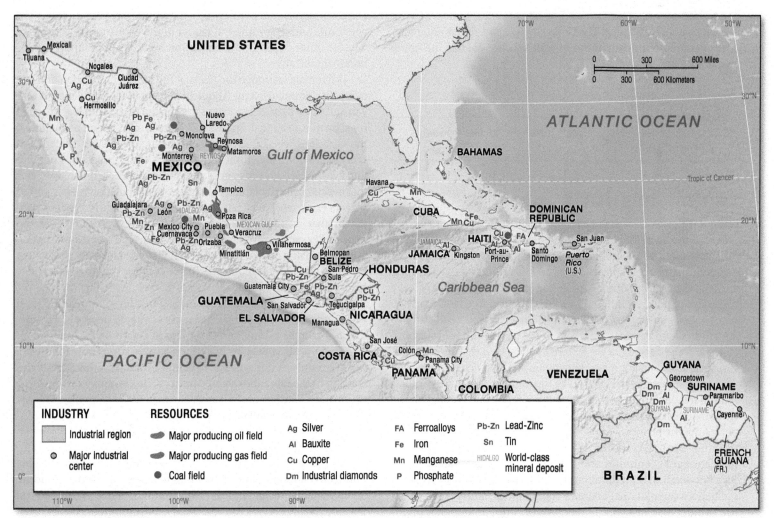

▲ Figure 3-28 **Mineral resources in Mexico, Central America, and the Caribbean.** The major minerals in this region include petroleum and silver in Mexico and bauxite in Jamaica, Guyana, and Suriname. Mexico also contains sufficient coal and iron ore to sustain a steel industry.

election in 71 years. Since the end of the Mexican Revolution in 1920 the PRI held Mexico in its political grip, but became known for corruption, mismanagement, and making deals with drug cartels.

In late 2006, Felipe Calderon of the National Action Party (Partido Acción Nacional or PAN) was elected president for a six-year term. He immediately declared war on the drug cartels and dedicated federal troops and resources to capture cartel leaders and disrupt their operations. Mexican troops and police captured or killed numerous cartel leaders and members, but the war unleashed a massive increase in drug-related killings, kidnappings, and struggles with and among the cartels. More than 50,000 people died during Calderon's "war on drugs," and the violence and chaos has negatively affected millions of Mexicans, most of whom have no involvement with the cartels or drug trafficking.

Why Mexico has not emerged as a stronger economy with a higher development status has been the subject of much debate among economists and development experts. Many observers have argued that its rich endowment of resources, location next to the United States, and rising education levels should have propelled Mexico into a leadership role in Latin America and the developing world. The young (47) and conservative president of Mexico, Enrique Peña Nieto, came into office in 2012 promising to reform the economy to encourage more investment and economic growth and to start a new era for Mexico. The title of his 2011 book, *Mexico, la gran esperanza* (Mexico, The Great Hope), captures his and many other people's optimism about Mexico's future despite the many obstacles.

Stop & Think

▶ Can Mexico achieve both economic growth and significant poverty reduction?

Central America

Central America consists of seven small republics with a population of about 45 million. Despite occupying a small territory with a relatively small population, Central America exhibits considerable economic and cultural diversity. Guatemala, Honduras, El Salvador, and Nicaragua have a number of shared characteristics: they all have high rates of poverty, ranging from roughly 40 percent to nearly 70 percent, a sizable portion of the population is engaged in subsistence agriculture, and at least 25 percent of the population is Protestant. Honduras and Nicaragua are extremely poor, with weak economies. Guatemala, El Salvador, and Nicaragua suffered civil wars, which were most intense in the 1980s. The repercussions of these conflicts are still felt, including the large number of refugees who fled to the United States. All three countries rely on remittances for 10 to 16 percent of the economy. Costa Rica and Panama are distinct from the other Central American republics in many respects. They both have stronger economies and a different ethnic geography. Costa Rica's population is predominantly white/mestizo, with a very small indigenous population. Panama's population is mostly mestizo, but the country also has an indigenous and Afro-Panamanian population. Costa Rica has had a stable democracy for many decades, a high education rate, and has diversified its economy to attract

tourists and high-tech companies. Panama is unique in all of Latin America because of the presence of the Panama Canal, which provides considerable employment and is critical to its economy. Panama is also the only Central American country whose population is closer to 15 rather than 20 percent Protestant.

Guatemala

Guatemala can be divided into three major physical regions: the coastal plain bordering the Pacific Ocean, the highlands, and the Petén. The coastal plain, which is *tierra caliente,* is where most of Guatemala's sugarcane is grown and cattle are raised. Going north/northeast, and up in elevation, leads to a series of climatic transitions (see Figure 3-5); most of Guatemala's bananas and coffee are cultivated in *tierra templada.* The southern portion of Guatemala's highlands is part of the Central American Volcanic Arc. Four of Guatemala's 37 volcanoes are active and many, such as Agua and Atitlan, are important tourist attractions. The northern highlands range from 5,000 to 8,000 feet (1,500–2,400 meters) in elevation, although a few peaks exceed 11,000 feet (3,300 meters). The Petén region is mostly flat with a tropical wet climate. The geology of the Petén encourages rapid drainage (and sinkholes), so despite plentiful rainfall there are no large rivers, and only one notable lake.

Most of Guatemala's population and nearly all of the cities are concentrated in the highlands. Guatemala City (3 million) is the economic and governmental center of the country and its primate city. No other city in Guatemala approaches its population or economic importance.

Guatemala was once the heart of the Mayan civilization and has complex cultural diversity. About 40 percent of the population is Mayan and more than 22 languages are spoken in the country, although most people also speak Spanish. Despite proximity and belonging to the same language family, many Mayan languages such as Mam and Q'eqchi are mutually unintelligible. Much of the indigenous population lives in the highlands and is poverty-stricken; one report estimates that over 80 percent of the rural indigenous population is poor and 30 percent extremely poor. The 30 percent of the national indigenous population that lives in cities fares somewhat better, with half considered poor and 7 percent extremely poor. This same report classified extreme poverty as living on less than 67 cents (United States) per day, which was the amount required to purchase enough nutritionally adequate food. Poverty is based on living on less than $1.52 per day.

Sugar, coffee, and bananas are the most important agricultural export crops upon which Guatemala's economy depends. The history of bananas is long and troubling in Guatemala. United Fruit Company, a U.S.-based transnational corporation (now Chiquita), had tremendous influence in several Central American countries starting in the late nineteenth century. The company owned large tracts of land and controlled the railways and the shipping industry. The company also had a disproportionate amount of political influence. In 1904 the term "**banana republic**" was coined by American writer O. Henry to describe countries that were dependent on a single crop for export, had a highly stratified society, and an unstable political system dominated by foreign companies and a domestic elite. Guatemala's land concentration and influence of United Fruit became deadly serious in 1954 when U.S.-supported troops implemented a coup and removed the democratically elected President Jacobo Árbenz. Two years earlier, Árbenz had conducted a land reform that threatened the interests of United Fruit, which owned over 40 percent of the arable land in Guatemala. The coup ended the land reform and helped create the conditions that started a civil war in 1960. The war pitted leftist insurgents against the government, lasted until 1996, and killed more than 250,000 people. In 1982 military general Efrain Ríos Montt came to power through another coup. He conducted a massive and brutal campaign to stop the leftist insurgents. The military killed thousands of indigenous people, many of whom were not combatants. President Reagan supported Ríos Montt, arguing that he was a staunch fighter of communism. In 2013, Ríos Montt was convicted of genocide and crimes against humanity, although the decision is under appeal.

Widespread poverty and the civil war encouraged large-scale emigration from Guatemala to the United States and Mexico. Most migrants to Mexico returned after the war, but far fewer have returned from the United States. In 2011 it was estimated that more than 1.2 million people living in the United States were either born in Guatemala or had Guatemalan ancestry. Over one-third of these Guatemalans live in California, but there are clusters in New York, Miami, Houston, Cincinnati, and numerous other cities. Remittances exceed $4 billion annually and account for about 10 percent of the economy.

Since the peace accords were signed in 1996, Guatemala's economy has improved, including tourism. Tourists commonly visit indigenous communities in the highlands, such as Quetzaltenango or Chichicastenango, which has the most celebrated "Indian market," or Antigua, which is a well-preserved colonial city. Tourists also go to Mayan archeological sites in the Petén, especially Tikal or El Mirador.

Panama

Panama is a tropical country located on the **Isthmus of Panama**, a narrow strip of land that connects two larger land masses (Central and South America) and is surrounded by water (Caribbean Sea and Pacific Ocean). At its narrowest, only about 40 miles (65 kilometers) of land separate the Pacific Ocean from the Caribbean Sea. The Central American Volcanic Arc ends in western Panama, and a relatively small part of the country is elevated. The Cordillera de Talamanca stretches across western Panama into southern Costa Rica. This range is mostly low-lying, with two peaks exceeding 11,000 feet (3,300 meters); Barú Volcano (Chiriqui Volcano locally) reaches 11,400 feet (3,500 meters) and is both the highest point in Panama and the country's only active volcano. The far eastern portion of the country, both on the Caribbean and Pacific coasts, has low-lying but rugged hills. Nearly all of Panama has a tropical wet climate. The rainy season occurs from May through November, but no month is dry or cold. Panama is the only Central American or Caribbean country that does not get hit by hurricanes. Its latitude, mostly between 7 and 9 degrees north, is too low for hurricanes to develop. Much of Darien Province, located on the border with Colombia, contains biodiverse tropical rain forests.

Panama did not exist as a country until 1903. After independence from Spain in 1821 the Isthmus was part of Gran Colombia, and then was a Colombian province after Gran Colombia dissolved. Panama was born from the desire of the United States for a canal to link the Pacific Ocean and Caribbean Sea / Atlantic Ocean. The United States had plans to develop a canal in Nicaragua and even conducted detailed geological studies in the late 1800s. Nicaragua was in favor of developing the canal, which would have connected Lake Nicaragua

and the Pacific Ocean with a canal and then used the San Juan River to connect to the Caribbean Sea. The plan was abandoned when the U.S. Congress became concerned about volcanic activity in Nicaragua and Panama looked like a better option. The French attempted to construct a canal during 1881–1889, but due to disease (malaria, yellow fever) and engineering problems the project failed. In 1903, the United States proposed to Colombia that the United States build the canal and create a canal zone that the United States would control. When the Colombian Senate refused the offer the United States quickly supported a plan for the province of Panama to secede from Colombia, and sent warships to prevent Colombia from stopping the secession. The United States recognized Panamanian independence and within a few days the Panamanian Ambassador signed a treaty granting the United States the right to build the canal and control more than a five-mile swath on both sides of the canal—the Panama Canal zone. The ambassador was actually a French citizen; no Panamanians signed the treaty. The canal was built from 1905 to 1914, a technological wonder that revolutionized ocean travel. An important part of building the canal was controlling mosquito-borne diseases. It had only recently been discovered that mosquitos carried malaria, yellow fever, and other diseases. A massive effort to kill mosquitos, eliminate their habitat, and protect workers ensued. As disease declined, the canal proceeded. Many techniques learned in Panama were applied elsewhere in the tropics.

The terms by which the United States controlled the canal were resented in Panama and periodically protests ensued. In 1977, the United States agreed to turn over operations of the canal in 1999. The canal generates over $1.5 billion in revenue annually. Ships pay a toll depending on the type of ship and its weight; a container ship will pay over $50,000 for passage through the canal (Figure 3–29). One of the major problems with the canal is that the locks are too small. Panamax, or the maximum size of a ship that can pass through the canal, is too small for many contemporary ships. Therefore, Panama is undertaking a $5–$6 billion expansion of the canal system, creating new locks for the largest ships, deepening the canals, and

▼ **Figure 3–29 Entering the Panama Canal.** A German cargo ship enters the Panama Canal. The trip from the Pacific Ocean to the Atlantic Ocean will take less than one day.

implementing water-saving techniques. The "post-Panamax" canal is scheduled to open by 2015.

A Nicaraguan route connecting the two oceans was revived in 2013 when the Nicaraguan government came to an agreement with a Chinese company to build a canal starting in late 2014. The Hong Kong Nicaraguan Canal Development Investment Company received a 50-year concession to build the canal (and another 50 once it is built). This canal would be large enough to allow passage for the world's largest ships, some so large that they will not be able to pass through the "post-Panamax" canal. The exact route has yet to be determined, but it would surely run through Lake Nicaragua. Many Nicaraguans are concerned about the environmental consequences of routing thousands of ships through Central America's largest lake because it is the source of fresh water for many Nicaraguans.

Prior to the construction of the Panama Canal, transit from one ocean to the other was possible on a railway built in the 1850s, immediately after the discovery of gold in California. Both the railway and canal relied on servile labor from numerous places including China, Colombia, Ireland, and especially Jamaica, Barbados, and other Caribbean Islands. These immigrants contributed to the ethnic diversity seen in Panama today.

The canal has allowed Panama to create a number of complementary services that makes it distinct from other Latin American economies. Panama's economy depends overwhelmingly on international services, and unlike many countries in Latin America, does not rely on agricultural or mineral exports. Panama has always used the U.S. dollar as its currency, maintains an open economy, and because of the business the canal has created, encourages international banking. Numerous international banks have offices in Panama City (another primate city), and billions of dollars flow through these banks with little regulation, making Panama the largest international banking center in Latin America (Figure 3–30). Panama also has the world's second largest free trade zone, the Colón Free Trade Zone at the mouth of the Atlantic side of the canal. Another economic idiosyncrasy of Panama is that it registers more ships than any other country in the world. Japanese companies in particular register their ships in Panama as their "flag of convenience," rather than fly their national flag. When companies register their ships with Panama they generally pay lower taxes and have fewer worker or safety regulations than in their home country.

Costa Rica

Bordering Panama to the south and Nicaragua to the north, Costa Rica is a small country with three major physical regions. The Cordillera de Talamanca and the Central Cordillera separate the Caribbean and Pacific coastal plains. The Central Cordillera is part of the Central American Volcanic Arc and has about 10 volcanoes, several of which are active. Irazú Volcano reaches 11,200 feet (3,400 meters) and is immediately east of the capital, San José. Arenal, in northern Costa Rica, has been active since its eruption in 1968. These two volcanoes and the parks that surround them are important tourist attractions. The coastal plains are a combination of pasture, forest, and export agriculture, mostly bananas and pineapples. Most of Costa Rica experiences a tropical wet climate, but the mountains create variability and the northern portion of the Pacific plain is considerably drier than the rest of the country.

Costa Rica stands out as the most prosperous Central American country, although nearly one-fifth of its population lives in poverty. It relies on traditional agricultural exports such as coffee and bananas, but has also diversified its agricultural exports to include pineapples and vegetables. More than 70 percent of its economy comes from services and only about 15 percent from agriculture. Costa Rica is a destination for international migrants, especially Nicaraguans who work in the banana plantations or in San José. Only a small percentage of Costa Ricans live abroad and remittances are a small part of its economy. Costa Rica abolished its army in 1948 after a civil war, and has been a stable democracy since then—earning the nickname of the "Switzerland of Central America." The country also stands out for implementing a large-scale conservation plan that has placed about 25 percent of its territory in protected conservation areas. It also pioneered a **payment for environmental services (PES)** program. A PES program charges people who benefit from environmental "services" such as clean water, biodiversity, reducing carbon emissions, and the like. Costa Rica's efforts to "go green" are part of its plan to attract more ecotourism. Its efforts have been rewarded with a large and growing tourist industry. More than 2 million tourists now visit Costa Rica each year, more than any other Central American country.

With a stable democracy, good education system, nonunion workforce, and business-friendly government, Costa Rica has been able to attract high-tech companies such as Intel to its free trade zones. Intel's large $300 million plant just outside of San José assembles and tests microprocessors, and employs more than 3,000 people. Its arrival in 1997 led other companies, such as Abbot Laboratories, Proctor and Gamble, and Microsoft, to open offices. Most of the companies investing in Costa Rica come from the United States. These businesses were a welcome alternative to apparel firms, which can no longer compete with China. Exports from these high-tech companies make up a considerable portion of Costa Rica's GDP, and unlike the apparel industry, seem to be creating a number of good jobs. Corporations such as Intel like to brag about their contributions to education (over $1 million per year) and job training. The International Labor Organization reports that EPZs

▼ **Figure 3–30 The Panama City skyline.** Tourists peruse indigenous artisan weavings in the historic district of Panama City. The buildings in the background are transnational companies and high-rise apartments.

tend to pay higher wages than comparable jobs elsewhere in society, but the high-tech sector does not create many linkages to other parts of the economy and most corporate profits are repatriated and not invested in Costa Rica.

The Caribbean

Most of the countries of the Caribbean depend on tourism, exporting bananas or sugar, subsistence agriculture, and exporting apparel and electronic items manufactured or assembled in EPZs. This is to some extent also true for both the Dominican Republic and Cuba, but their economies are distinct from the rest of the Caribbean. The Dominican Republic has the largest economy and exports a disproportionately large amount of manufactured goods in addition to agricultural products. Cuba has been a socialist state run by the Communist Party since the early 1960s. The economy is controlled by the state and has endured a U.S. trade embargo for more than 50 years (Table 3-2).

Table 3-2   Caribbean Political History and Status

Political Entity	Original European Colonizer	Subsequent Colonizer/Cultural Influence	Political Status	Date of Independence or Autonomy
Anguilla	Spanish	British	BOT[a]	
Antigua and Barbuda	Spanish	British	Independent	1981
Aruba	Dutch		Autonomous[b]	1986
Bahamas	British		Independent	1973
Barbados	British		Independent	1966
Belize	British		Independent	1901
Cuba	Spanish		Independent	1981
Dominica	Spanish	British	Independent	1978
Dominican Republic	Spanish		Independent	1844
French Guiana	French		FAR[c]	1974
Grenada	French	British	Independent	1974
Guadeloupe	French		Independent	1974
Guyana	Dutch	British	Independent	1970
Haiti	Spanish	French	Independent	1804
Jamaica	Spanish	British	Independent	1962
Martinique	French		FAR[c]	1974
Montserrat	French	British	BOT[a]	
Netherlands Antilles (Bonaire, Curacao, St. Eustatius, St. Maarten, Saba)	Various	Dutch	Part of the Kingdom of the Netherlands	
Puerto Rico	Spanish	United States	American Commonwealth	1952
St. Christopher (St. Kitts) and Nevis	British		Independent	1983
St. Lucia	Spanish	British	Independent	1979
St. Vincent and the Grenadines	Various	British	Independent	1979
Suriname	Dutch		Independent	1975
Trinidad and Tobago (unified)	Spanish	British		1962
Virgin Islands	Danish	U.S./British	Various	

Sources: *The Statesman's Yearbook 2005* (New York: St. Martin's Press, 2004); John MacPherson, *Caribbean Lands,* 4th ed. (London: Longman Caribbean, 1980); *CIA Worldfact Book,* www.cia/gov/cia/publications/factbook/geos/av.html.

[a]British Overseas Territory, considered part of the United Kingdom.

[b]Aruba is an autonomous part of the Kingdom of the Netherlands.

[c]French Administrative Region, considered part of France.

Cuba

Cuba is the largest island in the Caribbean, and has a savanna climate, receiving most of its rainfall during the summer months. It is located immediately west of the island of Hispaniola and less than 100 miles (160 kilometers) south of Florida. The Sierra Maestra is Cuba's only notable mountain range. The rugged but small range is located in the far eastern part of the country, and its highest peak (Turquino) reaches only 6,476 feet (1,974 meters). This range contains some minerals including copper and manganese.

With more than 11 million people, Cuba is also the most populous island in the Caribbean. Cuba's ethnic makeup was shaped largely by European immigration and African slavery. The indigenous people of Cuba were devastated by disease and enslavement during the early colonial years. The Taino lived on numerous islands in the Caribbean when the Spanish arrived. They spoke an Arawakan language, indicating that their ancestors migrated from northern South America. The decline of the Taino population and the value of sugarcane cultivation prompted the importation of thousands of West African slaves, especially in the nineteenth century. Slavery was not fully abolished until the 1880s. Most Cubans descend from European people, or a combination of European and African. Unlike the rest of Hispanic Latin America, Cuba (and Puerto Rico) remained a colony of Spain; it was not until the Spanish-American War of 1898 that Cuba separated from Spain and eventually gained its independence. This long-standing relationship with Spain facilitated an immigration of over 1 million Spaniards to Cuba during the late nineteenth and early twentieth centuries. Cuba is the only country in Central America and the Caribbean to receive a large number of European immigrants.

Cuba was ruled by a dictator during the 1950s. A popular uprising became a revolution and in January 1959 the Batista dictatorship was toppled and replaced with Fidel Castro's revolutionary government. The Castro government initiated radical changes, including literacy and health campaigns and giving Black Cubans equal rights. The government also nationalized numerous businesses and properties owned by Cubans and foreigners. The government's land reforms resulted in the confiscation of most arable land, including the large tracts owned by U.S. citizens. President Eisenhower responded with a trade embargo and the seizure of Cuban assets in the United States. The embargo tightened over the next few years and remains in effect (2013). The Castro government grew close to the Soviet Union and U.S. efforts to oust or kill Castro failed. Castro remained in power until 2008 when he transferred authority to his brother, Raul.

Until 1991, Cuba relied heavily on its relationship with the Soviet Union. Cuba had a guaranteed market for its agricultural products and the Soviet Union provided food, medicine, and especially petroleum. When the Soviet Union collapsed in 1991, Cuba entered an economic depression because the special arrangement with the Soviet Union collapsed. This led to a series of economic and social changes called the "Special Period," which has continued with varying degrees of severity for over 20 years. The Special Period has been a time when Cuba has had to learn how to live on less petroleum. Organic agriculture (pesticides and some fertilizers are derived from petroleum) with little or no mechanization has been required to grow enough food (Figure 3–31). Mass

▲ Figure 3–31 **Cuban farmer and a plow team of oxen.** Without access to inexpensive petroleum Cuban farmers have had to rely on animals and human labor to plant tobacco, sugarcane, and food crops.

transit systems are a necessity. The country has promoted tourism with European and South American countries. It also sought out a special relationship with Venezuela to get cheap petroleum when President Hugo Chávez assumed power in 1999. Venezuela continues to provide subsidized oil in exchange for technical and medical help. Thousands of Cuban doctors and technicians work in Venezuela.

Literacy is nearly universal in Cuba and its health and education systems are free and controlled by the government. Life expectancy is high (78 years) and Cuba's infant mortality rate is the lowest anywhere in North or South America, including the United States. Many Latin Americans travel to Cuba for health care.

Cuba has been strongly criticized for human rights violations. The Human Rights Watch, a nongovernmental organization, has criticized Cuba for its record on arbitrary imprisonments and repressing nearly all political freedoms and freedom of expression. This oppression and persistent economic problems have led to several waves of refugees leaving the country, mostly for the United States. For decades international observers have expected Cuba's government to either be overthrown or change due to an apparently unsustainable economy and oppressive political conditions. Yet Cuba remains the anomaly in Latin America despite changes in the late 2000s. Tourism has increased, and under President Raul Castro reforms have been made. Limited capitalism and a real estate market are permitted and it is easier for Cubans to travel abroad. But the economy is still highly planned and the politics are controlled by the Communist Party. Cuba exports sugar and nickel to China, Canada, and a few European countries. Cuba's government insists that it would negotiate a better relationship with the United States, but all efforts to remove the U.S. embargo are met with stiff resistance from various constituencies in the United States, especially Cuban-Americans in Florida. Cuba will undoubtedly face more political change as Raul Castro is over eighty and cannot be expected to serve very many more years. Many older Cubans still believe in the ideals of the revolution, but the future of Cuba will soon be decided by a generation born after the revolution.

Dominican Republic

The Dominican Republic occupies the eastern two-thirds of the island of Hispaniola. The country is divided by the Cordillera Central (Central Range), which is rugged and has three peaks just over 10,000 feet (3,050 meters). The Cibao Valley to the north of the range is one of two important agricultural regions of the country. Elevation moderates the tropical climate and this region has been the center of coffee, tobacco, and cocoa production, mostly cultivated by smallholders. The southeast plain of the country is hotter and wetter and is the center of sugar production. During the colonial era African slaves produced sugar on plantations that were commonly foreign-owned. The plantation economy was altered when General Rafael Trujillo, who ruled as a dictator from 1930 to 1961, nationalized the country's largest plantations. Sugar profits went to fuel import substitution industrialization and make the capital Santo Domingo an industrial and commercial primate city.

With 10 million people, the Dominican Republic has the second largest population in the Caribbean. Most of the population has mixed African and European heritage, although few self-identify as black or Afro-Dominican. As in Cuba the Taino population that existed prior to 1492 was nearly exterminated, but a number of Dominicans have Taino ancestry. Many Dominicans are considered to be black when they travel to the United States, showing how race is a social category created by history, place, and power, not a biological fact.

The Dominican Republic is the Caribbean's largest economy, and while it exhibits many of the elements found elsewhere in the region, its large EPZ sector distinguishes it from other countries (Figure 3–32). The country relies heavily on the United States for trade and remittances, and exports manufactured products from more than 50 EPZs, agricultural products, and a few minerals. But tourism and services are an increasingly important part of the economy.

The first EPZs developed in the early 1970s, and by 1980 the Dominican Republic was the most important EPZ center in Latin America. The growth of EPZs increased dramatically during the 1980s and 1990s. By 2000, companies in EPZs employed nearly 200,000 workers and exported more than $4.7 billion worth of goods, mostly apparel (clothes and shoes). Over 80 percent of the country's exports came from EPZs, and 70 percent of the companies in EPZs were foreign-owned. These EPZs face greater competition from Chinese producers, but with the Dominican Republic's entry into the CAFTA-DR, these businesses have more access to markets in the United States.

Although EPZ exports dominate the Dominican export economy, sugar remains the most important agricultural export. Many families rely on more than $3 billion in remittances sent from the estimated 1.5 million people with Dominican heritage living in the United States. Tourism, however, now exceeds remittances; more than 4.3 million visitors spend more than $4.3 billion annually. Tourism ranges from cruise ship stopovers and Club Med luxury enclaves to sex tourism, which commonly exploits young women.

Baseball is a very important cultural and economic activity in the Dominican Republic. At the start of the 2013 Major League Baseball (MLB) season in the United States, nearly 25 percent of the players were born in Latin America or the Caribbean; the Dominican Republic led this category with 89 players, followed by Venezuela with 63. Hundreds more Dominican young men were on minor league teams around the country (Figure 3–33). The game was introduced from Cuba at the end of the nineteenth century, and soon baseball leagues appeared, including at sugar plantations and refineries. In 1956 the first Dominican-born player made his debut in MLB; since then more than 500 Dominicans have played for MLB. The names of standout players like Felipe Alou, Sammy Sosa, David Ortiz, and Albert Pujols are well-known in the country, and inspire thousands of young Dominicans to emulate them. Every MLB team has a baseball "academy" in the Dominican Republic. These academies hire scouts (*buscones*) to search for young players, who are then recruited, trained, and groomed for the MLB minor leagues. Few recruited players advance to the minor leagues and even fewer make it to the major leagues. Many of the players sacrifice their education or other job training in the

▼ **Figure 3–32 Clothing factory in Santo Domingo, Dominican Republic.** The lingerie factory in the Hainamosa free trade zone employs mostly women, which is common in many EPZ businesses.

▼ **Figure 3–33 Baseball in the Caribbean.** These boys play a game of baseball in the streets of Havana, Cuba. The success of players like the Los Angeles Dodgers' Yasiel Puig and others encourages young men to dream of playing baseball in the major leagues in the United States.

process, leaving them with broken dreams, but no career. Some successful Dominican baseball players, such as David Ortiz and Albert Pujols, have foundations that donate money in the country, but the overall development impact is more about what young men sacrifice in pursuit of their dream rather than the money or development that results from the relatively few successful players.

As the Dominican economy turned to EPZs and tourism, it diminished its reliance on sugar and the country now has a more diverse economy. Its infrastructure and telecommunication networks have improved markedly, and access to education has increased. The Dominican Republic should remain the largest economy in the Caribbean and will likely increase efforts to attract tourists and may look to diversify its exports.

Stop & Think

▷ Why is tourism such a major component of the economy in Central American and Caribbean countries?

South America

South America is commonly divided between the Andean republics, the Southern Cone, and Brazil (Figure 3-34). The Andean countries include Venezuela, Colombia, Ecuador, and Peru, and although there is considerable variation among the countries they are clearly distinct from the Southern Cone or Brazil. The Southern Cone includes Chile, Argentina, Paraguay, and Uruguay. Brazil is treated separately because its history of Portuguese colonialism is one of numerous economic and cultural characteristics that distinguish it from every other country in Latin America.

The Andean Republics

The Andean countries of Ecuador, Peru, and Bolivia have their own identity and complex history, but they also have much in common. They are multiethnic with notable indigenous populations and they have strong regional differences and identities (Figure 3-35). The population is primarily Roman Catholic, with a small number of Pentecostal converts, and syncretic belief systems in indigenous regions. While the Andes Mountains do not determine cultural identities or economic outcomes, they nonetheless have a strong influence over the economic opportunities, constraints, and livelihoods of each country. The economies of all three countries, like so many South American states, are based on exporting primary products to wealthier countries (primarily Brazil, United States, and China). Minerals are widely distributed throughout the region, but are primarily concentrated in the mountains or adjacent lowlands (Figure 3-36). From both primary mineral and agricultural exports great wealth is created, and the Andean republics face the challenge of how to make export-led development help their poorest, usually indigenous, citizens. Colombia's economy likewise is based on exporting primary products, mostly to the United States. Colombia, however, is distinct because it has a larger Afro-Colombian population but a very small indigenous population. It also endured more than 30 years of civil war and has been the center of illegal drug

production and exports and a central focus of U.S. efforts to stop cocaine production.

Ecuador

Ecuador is a small Andean republic of more than 15 million people, with remarkable cultural and biophysical diversity. The country is predominantly mestizo (mixed European and indigenous ancestry), although somewhere between 15 and 30 percent of its population is indigenous, and a considerably smaller portion of the population is Afro-Ecuadoran. Almost everyone speaks Spanish, including those who also speak one or more of the 11 indigenous languages found in the country. Quichua, a dialect of the Peruvian/Incan Quechua, is the most commonly spoken indigenous language in the country. The economy depends heavily on exporting primary products; oil, bananas, shrimp, gold, and cut flowers are the most important. Very few illegal drugs are grown or processed in Ecuador, but it is a transshipment country.

Ecuador is situated on the equator (that is what Ecuador means) and is divided by two North–South ranges of the Andes Mountains. Ecuadorans commonly perceive their country as four regions—the Galapagos Islands, the Coastal Lowlands (la costa), the Sierra (highlands), and El Oriente (the east). The Oriente is the eastern part of Ecuador, but the western extreme of the Amazon drainage basin. The Oriente is home to numerous indigenous groups, including Shuar, Kichwa, Huaoroni, and many smaller groups. It is also home to incredible biodiversity. Yasuni National Park is one of the most biologically diverse places on earth; it is estimated that in one hectare (2.5 acres) there are roughly 100,000 insect species, comparable to the number found in all of North America. Ecuador's interest in the Oriente increased in the 1940s and again in the 1960s with discoveries of oil. The border between Ecuador and Peru was disputed, resulting in numerous skirmishes over the years. The two countries reached an agreement in 1999, and Peru controls much of the Amazonian region Ecuador once claimed. Like the other Andean republics, Ecuador sought to colonize and exploit the Amazon region, beginning in earnest in the 1960s. The roads that were built for highland colonists and oil companies allowed settlers to penetrate the region and claim and deforest tropical rain forests.

One of the most damaging and long-lasting effects of this colonization effort was the environmental and health mess made by the Texaco Petroleum Company, now owned by Chevron, and Petroecuador, Ecuador's national petroleum company. Texaco and Petroecuador extracted large quantities of petroleum from 1972 to 1992, and left much of the area surrounding the oil fields as an environmental and health disaster, mostly because toxic materials were dumped into the environment. After a lengthy legal battle, an Ecuadoran court ruled in 2011 that Chevron owed $18 billion in damages. Chevron denies responsibility and the battle continues. The result of this struggle has been far greater safeguards for oil extraction, environmental monitoring, fewer roads to deter spontaneous colonization, and indigenous wariness of oil extraction. Nearly all of Ecuador's petroleum continues to come from the Oriente. In 2007, President Rafael Correa pledged not to allow oil development in Yasuni National Park if Ecuador was paid $3.6 billion, roughly half the value of the oil reserves beneath the park. The funds, to be raised from international donors and foreign governments, are to go to

▲ **Figure 3–34 Countries and major cities and highways in South America.** Most of the highways of South America are concentrated on the coast and connect large cities. More recently constructed roads, however, connect less populated places such as in the Amazon basin and the interior of Argentina.

renewable energy sources and sustainable development. The plan, however, has faltered and not been realized.

The Sierra is home to roughly 42 percent of the country's population, including Quito, the capital and second largest city (2.3 million), and Cuenca (500,000), the third largest city. Most of the population lives in the valleys between two cordilleras of the Andes.

Quito is situated at approximately 9,350 feet above sea level (2,850 meters). It is the administrative capital of the country and receives thousands of tourists annually who visit Quito's historic district, nearby volcanoes, and indigenous markets. The Sierra also is home to the country's largest indigenous population, many of whom engage in mixed livelihoods, combining subsistence agriculture with

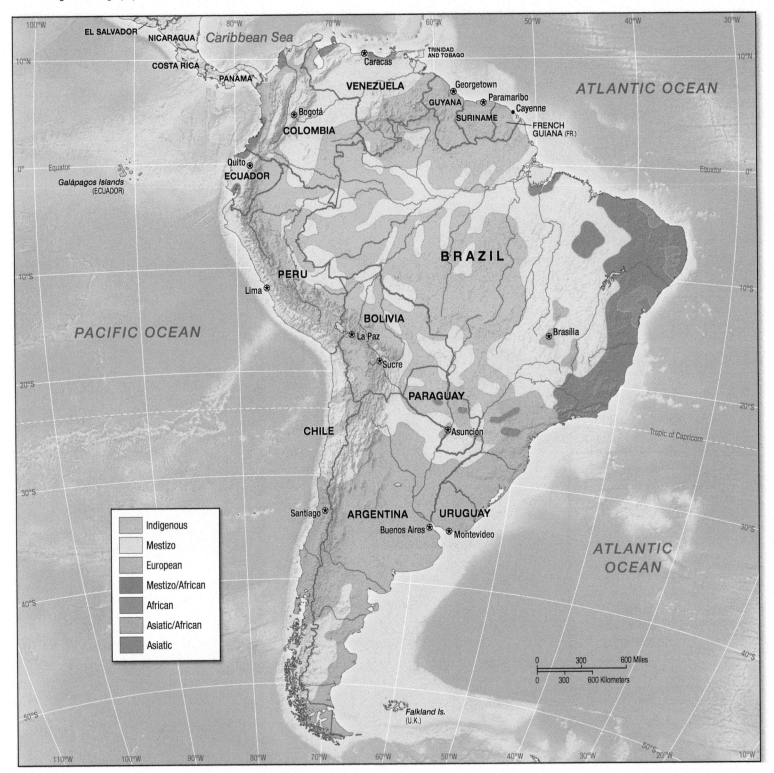

▲ **Figure 3–35 Ancestry of dominant ethnic groups of South America.** Most countries in South America are predominantly mestizo, but indigenous groups are most numerous in many parts of the Ecuadoran, Peruvian, and Bolivian Andes and in rural Amazonia. Due to immigration, much of the population of southern Brazil and the Southern Cone have European ancestry. Brazil is highly diverse and in the Northeast, much of the population has African ancestry.

off-farm employment. The most famous and well-traveled indigenous group are the Otavaleños. They are known as excellent weavers and travel around the world (and likely to your campus!) to sell their sweaters and other textiles. The northern half of Ecuador's Sierra contains numerous active volcanoes. The snow-covered volcanoes provide a tourist attraction, water for irrigation, and the ash from previous eruptions contributes to fertile soil for agriculture, including vast tracts of greenhouses full of flowers. The flowers are grown under precise industrial conditions (Figure 3–37), with the application of substantial amounts of fungicides and other chemicals. The

▲ Figure 3–36 **Mineral resources in South America.** Petroleum production in Venezuela, Brazil, and Ecuador is of world importance, as is iron ore production in Brazil and Venezuela. Equally significant is coal in Colombia; tin in Peru, Brazil, and Bolivia; and silver in Peru. Copper in Chile and Peru, as well as bauxite in Brazil and Venezuela, also play a major role in global markets.

flowers are patented and exporters must pay a royalty to the owner of the patent. The flowers are shipped out of Quito's international airport to Miami and a few European cities.

The southern Sierra is dominated by Cuenca, a UNESCO World Heritage Site. Cuenca is known for its artisans, Panama hats, historic district, and as one of the largest sources of international migrants to the United States. Tens of thousands of people from this region migrated to the United States, starting as a slow trickle in the 1950s, and increasingly rapidly during the lost decade (1980s). The migration, much of it illegal, overwhelmingly targeted

▲ **Figure 3-37 Ecuadoran flower exports.** Workers in a flower processing plant cut and prepare specialty flowers for export. Ecuador is the third largest exporter of flowers following the Netherlands and neighboring Colombia.

metropolitan New York, where over 200,000 people born in Ecuador live. The mental map of many people from the region is so focused on New York that it is not uncommon for a North American visitor to be asked what part of New York he or she comes from! Remittances have transformed the area's economy and landscape. Large homes, some of which sit empty, can be found in many rural communities surrounding Cuenca (Figure 3-38). Ecuador experienced a mass emigration in the late 1990s / early 2000s when it entered a political and economic crisis. Most of the migrants went to Spain, where nearly 500,000 Ecuadorans live today. Relatively few people from the southern Sierra went to Spain; the United States remains the most important destination for households in the region.

The Coastal Lowlands are dominated by export agriculture and Guayaquil, Ecuador's largest city (3.75 million) and most important port and industrial center. Ecuador is the world's largest exporter of bananas, and most of the agricultural land of the southern coastal lowlands (Los Rios, El Oro Provinces) are dedicated to banana plantations. Many of the plantation owners are Ecuadoran entrepreneurs who grow bananas on contract to large transnational fruit companies such as Del Monte, Dole, and Bonita. The bananas must be grown without blemishes or the growers will lose their contract with the company and be forced to sell them domestically or as feed for cattle. One of the wealthiest men in Ecuador, Alvaro Noboa, owns large tracts of banana plantations, and controls the Bonita Company. The second major agricultural export of the Coastal Lowlands is shrimp. Shrimp ponds have replaced tens of thousands of hectares of ecologically valuable mangrove forests (trees that grow in saltwater on the coast). The ponds are shallow and receive tidal water daily. When the shrimp are harvested they are frozen and exported to the United States and Europe.

Guayaquil is a large, bustling port. Many agricultural exporters are based in Guayaquil and much of Ecuador's industrial base is located there. Culturally and economically the Coastal Lowlands are quite distinct from the Sierra. The coast is considered socially liberal, but politically conservative, mostly because the region relies on agricultural exports. The Sierra is the opposite, especially Cuenca, which tends to vote for leftist politicians, but is considered socially very conservative.

The Galapagos Islands are located approximately 700 miles (1,100 kilometers) west of Ecuador in the Pacific Ocean and are of volcanic origin. Their isolation has produced impressive biological diversity and biological oddities such as the blue-footed booby, which is a flightless bird with no fear of people. The islands attract thousands of tourists annually. The area is under pressure from local fishermen who want to exploit the marine resources surrounding the islands and tourists whose presence threatens to "love the place to death."

Bolivia

Bolivia has just over 10 million people, divided between its two major regions: the highlands in the west and south and the lowlands to the east and north. Bolivia is distinctive for several reasons. It is the only country in Latin America with a majority (55%) indigenous population, divided between Ayamara and Quechua and numerous smaller ethnic groups. It also became one of two landlocked countries (Paraguay is the other) when it lost the War of the Pacific (1879–1883), and had to cede to Chile its mineral-rich corridor to the sea in the northern Atacama Desert. Another distinctive feature of Bolivia is that it contains one of the largest elevated plateaus in the world—the altiplano. With nearly half the population living in poverty, Bolivia is the poorest country in South America and one of the poorest in all of Latin America. It is deeply divided, as it has had a long history of being ruled by a small white, European upper class. The election of Evo Morales in 2005 and the reversal of neoliberal economic reforms have led many in the lowlands to demand autonomy from the indigenous-dominated highlands. The Bolivian economy, long dependent on exporting minerals such as tin and silver, relies now on exporting natural gas, petroleum, and minerals such as zinc and iron ore. The location of the natural gas reserves in the eastern portion of the country has contributed to regional and cultural divisions between the highlands and lowlands.

The highlands of Bolivia are marked by two high ranges of the Andes Mountains, the Cordillera Occidental and the Cordillera Central. Both ranges have peaks that exceed 19,600 feet (6,000 meters). The Cordillera Occidental has numerous active volcanoes and forms the border with Chile to the west. The Cordillera Central is a large, complex range where most of Bolivia's mineral deposits have been mined, including tin, zinc, and silver. Sucre,

▼ **Figure 3-38 Migrant house near Cuenca, Ecuador.** This large home was built with remittances from metropolitan New York. Despite the migrant's plan to return, he sponsored the legal migration of his entire immediate family to the United States, leaving this house to the care of his parents.

the constitutional capital of Bolivia, and Cochabamba, an agricultural center, are located in valleys on the eastern side of the Cordillera Central. Between these two ranges is the altiplano, or high plateau. This plateau stretches nearly 680 miles (1,100 kilometers) from north to south and averages over 12,000 feet (3,700 meters) above sea level. Water drains into the altiplano from the surrounding mountains, creating Lakes Poopó and Titicaca. Lake Titicaca is the highest navigable lake in the world and has been a sacred place and a central resource for indigenous populations for at least 3,000 years. This internal drainage has created, over thousands of years, the *Salar de Uyuni,* the world's largest salt flat, thought to contain 50–70 percent of the world's lithium supply. Lithium, an alkali metal, is critical to electric and hybrid batteries and it has been suggested that Bolivia may become the "Saudi Arabia of the green world." Whether or not Bolivia should mine this unique and fascinating landscape has generated considerable controversy.

Culturally, the highlands are predominantly indigenous (Aymara and Quechua). La Paz, Bolivia's second largest city (2.3 million), is situated on the altiplano at the base of the Cordillera Central. The people have endured racism, poverty, and political exclusion for most of Bolivia's history. Many families engage in subsistence agriculture, cultivating numerous varieties of potatoes and other tubers, and raising cattle, sheep, and alpaca in the very difficult altiplano environment (Figure 3-39). Many indigenous folks now live in La Paz, El Alto, and Cochabamba and no longer engage in subsistence agriculture.

The eastern lowlands are distinct physically and culturally from the highlands. The two regions are separated by the Yungas, a transitional region of valleys on the northeastern side of the Cordillera Central. Much of Bolivia's coca cultivation occurs in this ecoregion. The southern portion of the lowlands is part of the dry Gran Chaco. Further north, the climate becomes more humid. The central portion of the lowlands, near Santa Cruz, has a tropical savanna climate and has considerable agriculture, including soy and cotton. The northern lowlands, bordering Peru and Brazil, are a tropical rain forest.

▲ **Figure 3-40 The skyline of Santa Cruz, Bolivia.** The center of Bolivia's largest city is home to residential and government buildings in addition to new office complexes. Santa Cruz is the center of eastern Bolivia's economy and has experienced considerable population and economic growth.

With the exception of Santa Cruz (2.7 million) the population of the lowlands is sparse. The phrase **Media Luna** (half-moon) refers to the shape of the departments in the eastern portion of the country, including all of the lowlands and a small part of the southern highlands (Figure 3-40). In this region most people see themselves as mestizo and do not identify with Bolivia's indigenous history or its current political agenda.

The cultural and economic divide between the highlands and lowlands has intensified since Bolivia has reversed and modified its economy away from neoliberalism. In 2003, President Sánchez de Lozada announced a plan to sell natural gas to the United States by shipping it through Chile. Popular protest in the highlands was so intense that the plan was abandoned and the president resigned. Evo Morales, an Aymaran, was elected president in 2005 and immediately instituted social programs, land redistribution, and welfare that aided Bolivia's large impoverished population, and nationalized the natural gas industry to pay for those programs. These events have been dubbed the "**Bolivian gas wars.**" Successfully resisting a recall referendum in 2008, and achieving reelection to a second term in 2009, Morales and his political party continue to change many neoliberal economic policies. Bolivia still exports natural gas, mostly to Brazil through a 2,000 mile (3,200 kilometer) pipeline. But the government earns much more than it used to and has spent much of the revenue on anti-poverty programs.

These changes have heightened the regional divide between highlands and lowlands, especially in Santa Cruz, where a separatist movement has formed. The Movimiento Nación Camba de Liberación (Movement for the Liberation of the Camba Nation) has sought to gain greater autonomy for the Media Luna lowland departments, a development that would shift resources away from the highland departments. Many people in the Media Luna feel that this region, with its agricultural exports and natural gas reserves, provides much of the country's wealth but is dominated politically by the highlands.

▼ **Figure 3-39 Potato processing on the Bolivian Altiplano.** This Aymara woman is trampling potatoes to make a freeze-dried form of potatoes called chuño. The trampling and intense sun at more than 12,500 feet (3,800 meters) removes the moisture and the night time freezing allows potatoes to be preserved for many months.

Colombia

With nearly 50 million citizens, Colombia contains the third largest population in Latin America (after Brazil and Mexico), and possesses an economy much larger than the other Andean republics. Colombia has only a small indigenous population, located primarily in the Amazonian region of the country, and not in the Andean highlands. Spanish is spoken by nearly everyone and more than 80 percent of the population considers itself either mestizo or white. About 10 percent of the population is Afro-Colombian. Colombia's economy is more diverse than the other Andean countries, and much like Ecuador, has very close economic ties to the United States and a sizable migrant population in the United States, especially in New York and Florida, and Spain.

Colombia's physical geography is similar to the other Andean republics, in that it is divided by Andean ranges that run roughly north-south and has a large area of tropical rain forest east of the Andes that drains into the Amazon River. Colombia, however, does not have a dry climatic region, and is the only country in South America to have both a Pacific and Caribbean coast. Southern Colombia shares many physical attributes with northern Ecuador, as they both have active volcanoes and fertile valleys. The northern part of the vast territory east of the Andes transitions into a savanna climate and forms part of a large inland plain (Llanos) that extends into Venezuela. This region drains northeastward into the Orinoco River.

Colombia's population is overwhelmingly concentrated in the Andean highlands and on the Caribbean coast. All three of Colombia's largest cities, Bogotá (10.7 million), Medellín (3.6 million), and Cali (3.2 million) are located in intermontane valleys of the Andes. Barranquilla (2.1 million), the fourth largest city, is the country's main Caribbean port. Most of Colombia's petroleum is extracted from the Magdalena basin in central/ northern Colombia and from fields in the Llanos base northeast of Bogotá. Coffee is grown on the Andean hillsides in numerous valleys in the *tierra templada* (Figure 3–41).

Like the other Andean republics Colombia relies on exporting primary products, especially petroleum products. High oil prices

▼ **Figure 3–41 Coffee harvest in Colombia.** A man picks "cherries" from coffee bushes on a large field. Coffee remains an important agricultural export for numerous Latin American countries, including Brazil, the world's largest exporter.

have earned Colombia considerable revenue from oil exports. The country also exports traditional (coffee, bananas) and nontraditional agricultural products (cut flowers). But manufacturing and exporting manufactured goods are more important to its economy.

Colombia has been engaged in a battle to control the cocaine trade that has gripped the country since the 1980s. Most of Colombia's coca is grown in the tropical lowlands, both near the Ecuadoran border and in more remote tropical areas. While the cocaine trade is controlled by transnational criminal networks, much of the coca is grown by small farmers. With more than $8 billion in aid from the United States, Colombia sought to eradicate coca fields by spraying herbicide over vast swaths of agricultural lands. This effort, part of Plan Colombia, did not immediately lessen coca production because it caused farmers to look for new lands and to find ways to avoid detection. A leftist guerrilla group, the Revolutionary Armed Forces of Colombia (*Fuerzas Armadas Revolucionarias de Colombia*—FARC), has earned hundreds of millions of dollars by protecting growers and taxing various aspects of the industry, although some FARC members have also been involved in production and trafficking. Because of drug eradication programs and FARC activities, as well as struggles against right-wing paramilitary groups, there are many Colombian refugees in Venezuela and Ecuador and Colombia has a large number of **internally displaced people**, estimated at between 3 and 5 million. Internally displaced people are those forced to flee their homes, but remain within their native country. This huge internal refugee population gave Colombia the dubious distinction of having more internally displaced people than any other country in the world—until the Syrian civil war that began in 2011 produced an internal refugee population of comparable scale.

Brazil

Brazil stands out in Latin America for many reasons. It is the only country in the Americas colonized by the Portuguese, and where that language is spoken. At about the size of the continental United States, it occupies nearly half of South America and is by far the largest country in the region. It extends from the tropical wet climate of the equator to the humid subtropics, bordering on the midlatitudes. Not only does Brazil have the largest economy and population in Latin America, it also has more cultural diversity than any other country in the region. Brazil is the "B" in the group of **BRIC (Brazil, Russia, India, and China) countries**, identified as emerging economies with increasing global economic power. This recognition is understandable because the country's economy has grown so much that it is now the world's sixth largest and accounts for half of all of South America's GDP. Despite these promising indicators, the country also has a large poor population, with over 20 percent living in poverty, and considerable income inequalities.

Most of Brazil's territory is within one of two major physiographic regions, the Amazon Basin and the Brazilian Highlands. The eastern side of the Andes provides much of the runoff for the Amazonian river system, but several large rivers (Tapajós and Xingu) drain from central Brazil northward into the Amazon River. The Madeira River, the Amazon's largest tributary, drains both from the Peruvian and Bolivian Andes, and the Brazilian Highlands. The highlands, commonly referred to as plateaus, occupy nearly half of Brazil's territory in the southern and eastern parts of the country. They are low-lying, in some places rugged, but with no volcanic or seismic (earthquake) activity. Most of the country's minerals (iron ore, bauxite, gold, diamonds, and others) are located throughout the highlands.

Brazil's climate becomes drier, cooler, and more seasonal south of the tropical wet climate found in much of the Amazon Basin. The Northeast is a semiarid region with frequent droughts. The southern portion of the country has warm, wet summers (December to February) and cool, dry winters (May to August).

Brazil was a colony of Portugal from 1500 to 1815, and gained full independence in 1822. The Portuguese Crown granted a few wealthy men rights to exploit a rectangular swath of land, extending westward from the coast, and the colony was originally governed by 15 private captaincy colonies. The captaincies were short-lived because most were poorly managed and failed financially. The Portuguese reorganized their management of Brazil and in 1621 created states. The original economic basis for the colonial economy was harvesting the brazilwood tree (where Brazil gets its name), which produced a valuable red dye. By 1530 sugarcane was grown and its importance increased over 150 years. The labor demands for sugarcane cultivation led to the importation of millions of Africans, commonly from Angola and Mozambique, which were Portugal's African colonies. The value of sugar attracted the Dutch as they became a colonial power in the Americas. In 1630 they seized a large part of the Brazilian coast, made Recife their capital, and took over much of the sugar exporting business. By 1654 they were evicted and the coast returned to Portuguese control. Sugar's profitability declined during the seventeenth and eighteenth centuries, when competition from Caribbean Islands (Jamaica and others) increased. By the eighteenth century gold extracted mostly from Minas Gerais (General Mines) state became the most valuable Brazilian commodity. The gold rush created the city of Ouro Preto (Black Gold), one of Brazil's best preserved colonial cities, now a university town, World Heritage Site, and tourist attraction (Figure 3–42).

▼ **Figure 3–42 A view of Ouro Preto, Brazil.** This small city in Minas Gerais Province is a unique combination of baroque architecture, museums, a university, and a rich history of gold mining.

By 1600, runaway slaves created settlements in Brazil's hinterlands. Called **quilombos**, these communities became home to thousands of former African slaves and in some cases indigenous and other marginalized peoples. The largest community, Palmares, had a population of over 25,000 before the Portuguese conquered it. Zumbi, one of Palmares's leaders, was killed in 1695, and the date of his execution (November 20) is now recognized as "Black Awareness Day" in Brazil.

Portugal granted Brazil kingdom status in 1815, but it was short-lived and Brazil declared independence from Portugal in 1822 and became a constitutional monarchy. During the nineteenth century Brazil struggled economically, but was able to attract immigrants from the American South after the U.S. Civil War (1861–1865). Brazil's monarch, Dom Pedro II, recruited American southerners by offering various incentives to entice them because he wanted them to help develop cotton plantations. Perhaps 10,000 southerners went to Brazil, mostly to São Paulo and Rio de Janeiro states. Known as *Confederados,* many became successful farmers, although about half eventually returned to the United States. The descendants of those who remained integrated into Brazilian society, but some cultural traits of the American South, such as Baptist churches, southern food, and even the place name "Americana," can still be found.

Brazil's economy and economic geography changed at the end of the nineteenth century, when coffee plantations expanded in São Paulo state. Coffee had been grown since the beginning of the century, but Brazil's economy, especially its exports, grew in the second half of the century. Coffee accounted for more than half of those exports. Although slavery was not completely banned until 1888, acquiring servile labor was increasingly difficult by 1850. To meet the demand for labor, European immigration increased dramatically during the last few decades of the nineteenth century when Brazil received more than 2 million immigrants, mostly from Europe; Italians and Portuguese were most numerous. Brazil actively recruited Italian migrants. Not unexpectedly, many migrants eventually returned to Europe; in Portugal some successful migrants built notable homes with their earnings and painted them green and yellow, the colors of Brazil's flag. That experience is similar to Latin American migrant experiences today. Metropolitan São Paulo also has a large Japanese–Brazilian population, descendants of early- to mid-twentieth century Japanese immigrants.

Brazil's economy continued to expand during the first half of the twentieth century, but it was after World War II that its manufacturing sector increased substantially. The country successfully used the import substitution industrialization (ISI) model to develop its industrial sector, especially in the 1960s and 1970s. Brazil borrowed tens of billions of dollars to finance its development, especially for state-owned enterprises. Like most Latin American countries, Brazil faced numerous economic problems during the 1980s. The country endured stabilization programs to control hyperinflation, and austerity measures to control its debt. By the mid-1990s most ISI elements of the economy had been eliminated and many national companies had been sold. Two companies stand out as successfully making the transition from publicly owned to mostly privately owned. Embraer (Empresa Brasileira de Aeronáutica) is an airline manufacturer started in 1969 in an effort to jump-start the airline manufacturing sector. The company was privatized in 1994, although the government retains some control. By 2012 it employed more than 18,000 people and had become one of the most important jet manufacturers in the world. The next time you board a commercial flight on a relatively small plane, you will likely board an Embraer jet or a jet made by its chief rival, Bombardier, a Canadian airline manufacturer. The Companhia Vale do Rio Doce (Freshwater Valley Company) was a mining and mineral processing enterprise founded by the government in 1942. It expanded to include many other industrial processes until it was privatized in 1997. Vale, as the company is now called, is one of the world's largest mining companies. Although most of its operations are in Brazil, it has operations in more than 30 countries and annual revenues of more than $40 billion.

Brazil is functionally divided into five regions by the Brazilian Institute of Geography and Statistics. These regions are recognized and used by the government for statistical purposes, and many people refer to the regions colloquially, but they have no administrative power and only loosely reflect cultural or physical characteristics. The regions include the North, the Northeast, the South, the Southeast, and the Central-West region.

Brazil's population and economic power are concentrated in the Southeast region, in a triangle that connects the cities of São Paulo, Rio de Janeiro, and Belo Horizonte, the capital of the mining state Minas Gerais. The three states of São Paulo, Rio de Janeiro, and Minas Gerais generate about half of the country's GDP and almost one-quarter of South America's GDP. This part of Brazil is also active in developing "green" technology (see *Focus on Energy*: Biofuels in Brazil). São Paulo is an industrial and agricultural powerhouse (Figure 3–43). Rio de Janeiro also has industry and agriculture, but is best known for its thriving tourist industry. These three states have a population of nearly 80 million, about 40 percent of the national total. São Paulo state has 42 million people and the city of São Paulo has more than 11 million, making it one of the world's largest cities.

▼ **Figure 3–43 Cathedral Square, Sao Paulo, Brazil.** The Metropolitan Cathedral (right) and the Palace of Justice (left) feature prominently on Sao Paulo's best known plaza. The plaza, located at the center of this large city, has been the site for many historic events.

▲ **Figure 3–44 Sao Joaquim market, Salvador, Brazil.** Salvador was the first colonial capital of Brazil, and is Brazil's third largest metropolitan area. It is known for its colonial Portuguese architecture and as the center of Afro-Brazilian culture.

The Northeast region of Brazil was the most important region economically until the twentieth century. The region still produces sugar, cotton, cocoa, has an industrial base, and receives millions of tourists annually, many from Brazil. The three largest cities are Recife, Fortaleza, and Salvador (Figure 3–44). This region today has the lowest per capita income and the highest rate of poverty in the country. The population of the region is nearly one-third Afro-Brazilian, a remnant of being the center of colonial sugar production.

The North region contains over 40 percent of Brazil's territory, most of Brazil's portion of the Amazon Basin, and is lightly populated. Manaus (2.3 million), situated where the Negro River flows into the Amazon River, is the only large metropolitan area. It is a free trade zone and an important port for ocean-going vessels trading with the interior of Brazil. The North also contains most of the country's indigenous population. Although small (about 1 million, or 0.5% of the national population), the population exhibits tremendous diversity, with more than 300 ethnic groups and 200 languages. Most indigenous people live in one of the country's 600 Indian reserves, which are also concentrated in the North.

As the world's sixth largest economy, Brazil's economy is complex. It is a major industrial and agricultural producer, and its relationship with China continues to grow. Brazil has made progress lifting people out of poverty (and extreme poverty) and lessening inequality to some degree, but millions of people have not benefited much from Brazil's improved economic status. Many of those people are ethnic minorities living in the slums of São Paulo, Rio de Janeiro, and the Northeast, and the indigenous population.

The Southern Cone

The Southern Cone refers to the southernmost countries in South America: Chile, Argentina, Uruguay, and Paraguay. The term comes from geographic proximity and convenience; together these countries form the southern portion of South America, which can

FOCUS ON ENERGY

Biofuels in Brazil

Brazil's energy consumption is remarkable because the country produces about 45 percent of its energy from renewable sources. Some observers consider Brazil to be the most sustainable large economy in the world. Hydroelectric power contributes part, but biofuels account for the largest portion (29%). This is in stark contrast to the United States where only 7 percent of energy consumption comes from renewable sources, and less than 1 percent from biofuels. A biofuel is any fuel that comes from an organic or biological source. The vast majority of Brazil's biofuel mix is composed of ethanol (ethyl alcohol), which is then blended with gasoline to run much of the country's automobile fleet. Ethanol can be made from a wide variety of crops, but in Brazil nearly all comes from sugarcane (Figure 3-5-1). Brazil produces one-third of the world's sugarcane, occupying approximately 9 million hectares (22 million acres), mostly in South-Central Brazil (São Paulo state). Just over half of the sugar produced is converted to ethanol, totaling over 27 billion liters. Brazil is the second largest producer of ethanol, and the world's largest exporter. The sugar/ethanol industry is now a substantial part of Brazil's economy ($48 billion) and employs over 1 million people. Ethanol accounts for half of the gasoline market in Brazil (as opposed to 10% in the United States). Practically all cars sold in Brazil today are flex-fuel vehicles, which means that they can run on either low-ethanol or high-ethanol gasoline, up to 100 percent. One of the most important contributions of biofuels in Brazil is that it reduces greenhouse gas emissions substantially compared to gasoline consumption, perhaps by 80–90 percent.

How has Brazil been successful in substituting a renewable energy source for gasoline? First, the savanna climate of South-Central Brazil and fertile soils of São Paulo state are ideal for growing sugarcane. Sugar is also efficiently processed into ethanol; it has a very positive energy balance, meaning that it produces much more energy than it takes to produce the sugar. Converting sugarcane to ethanol is seven times more efficient than converting corn to ethanol, as is done in the United States. The Brazilian government has a long history of supporting ethanol production and consumption, in large part to reduce petroleum imports and support domestic farmers and processers. Starting in 1975 Brazil began the *Programa Nacional do Álcool*, or the National Alcohol Program, which encouraged converting sugar into ethanol by mandating that gasoline be sold as a gasoline/ethanol blend. By 1993 gasoline/ethanol blends were required to be at least 22 percent ethanol; the figure was increased to 25 percent in 2003. The government also provided subsidies to the ethanol industry, including a guaranteed market, and has taxed gasoline so that it is more expensive compared to a gasoline/ethanol blend. A downturn in the cost of gasoline and an increase in the cost of ethanol in the late 2000s prompted the Brazilian government to announce a $38 billion plan to increase ethanol production and support the ethanol industry. Brazil hopes to export more ethanol and to provide a steady supply of ethanol to keep it competitive with the cost of gasoline.

The expansion of biofuels in Brazil has provoked controversy. In fact, ethanol has become a central issue in the debate about biofuels, food security, and land use change. Oxfam and other groups have argued that the expansion of biofuels has contributed to the increased cost of food globally. That criticism may apply elsewhere in the world (using corn in the United States), but there is little evidence that expanding sugarcane production came at the expense of raising other crops or increased food prices. Sugarcane occupies less than 5 percent of Brazil's arable land, and most of the sugarcane expansion in São Paulo state replaced pasture for cattle, not food crops. The larger concern is that expansion of sugarcane has an indirect effect on land use, displacing cattle ranching and other land uses northward into the Amazon rain forest.

▼ FIGURE 3-5-1 **Sugarcane harvest, Sao Paulo state.** More sugarcane is planted than any other crop in the world, and Brazil grows and produces the most.

Sources: José Goldemberg, "The Brazilian biofuels Industry," *Biotechnology for Biofuels* 1 (May 2008): 1–6. Ethan Goffman "Biofuels: What Place in our Energy Future?" 2009, www.csa.com/discoveryguides /discoveryguides-main.php.

be interpreted to be shaped as a cone. The term distinguishes these countries from Brazil to the northeast and the Andean countries to the northwest. People commonly view the Southern Cone as a Hispanic region with strong economies and populations with European ancestry/ethnicity, making it culturally distinct from Brazil and culturally and economically distinct from the Andes. That perception is mostly accurate, but as with any region there are significant differences within the region and important similarities with neighboring countries not included in the region. It is true that most of the people of Chile, Uruguay, and Argentina have European ancestry and identify with that ancestry. But most Paraguayans are mestizo, in the sense that they have both European and Guaraní ancestry. Their identity is closely linked to their bilingualism; most speak both Spanish and Guaraní. Chile also has notable indigenous and mestizo populations. The Mapuche form about 9 percent of Chile's population, and about one-quarter consider themselves to be mestizo. Economically, Uruguay, Chile, and Argentina are well-off compared to their Andean neighbors and the rest of Latin America, but Paraguay's economy is considerably smaller, with a much higher poverty rate. Chile's economy is stronger than Peru and Bolivia (based on GDP per capita), but all three rely on mining for a considerable part of their economy and both Chile and Peru also depend on fishing.

Argentina

Argentina has the largest economy of the Southern Cone and culturally many Argentines see their country as quite distinct from the rest of Latin America. At the beginning of the twentieth century Argentina entered the global economy as a major agricultural exporter and emerging industrial power. This economic progress did not last, and despite recent economic growth, the country has not become the world leader many thought it would. The vast majority of Argentina's people descend from Europeans who arrived between 1850 and 1950. The indigenous population, which is mostly Mapuche, accounts for less than 2 percent of the country's population. From official data Argentina has about 1.8 million immigrants, yet there are no large ethnic minority enclaves in the country. The most significant colloquial regions in the country are Patagonia, Mesopotamia, the Gran Chaco, and the Pampas. Buenos Aires is Argentina's dominant economic and cultural influence.

Argentina extends from the Andes Mountains in the west to the Atlantic Ocean in the east. The country is very narrow at its southernmost point on the island of Tierra del Fuego and grows in width northward. Nearly all of the country is within the subtropics or midlatitudes, and only a small part of the northwest is in the tropics. The Argentine Andes are the highest in the range, with four peaks over 22,000 feet (6,700 meters), including Aconcagua at 22,829 feet (6,958 meters), the highest peak in the western hemisphere. Less than 10 of the approximately 35 volcanoes in Argentina are active.

Argentina's climate in the northeast is warm and wet during the summer months (December–February) and mild during the winter months (May–August). This area between the Paraná and Uruguay Rivers is known colloquially as Mesopotamia ("land between rivers") because Spanish colonists compared this fertile swath of land to the land between the Euphrates and Tigris rivers in what is now contemporary Iraq. The Gran Chaco is a large, featureless, often poorly drained region stretching westward toward the Andean foothills and northward into Paraguay and eastern Bolivia. Going south from the Argentine Mesopotamia, the climate becomes cooler and drier, with greater seasonal differences.

In the northern half of the country rainfall is higher closest to the Atlantic Ocean and diminishes going westward. This temperate climate and year-round rainfall helps create the Pampas, a large fertile plain stretching in a crescent from extreme southern Brazil, through Uruguay, and into east-central Argentina. West and south of the Pampas the climate becomes drier, and much of southern Argentina is arid or semiarid. The aridity in western Argentina is caused by its increased distance from the Atlantic Ocean, which is the source of most of Argentina's precipitation. This distance decay effect is similar to the transition from the humid Midwest to the semiarid Great Plains in the United States. The aridity in the south is caused mostly by the Andes Mountains blocking precipitation from the west and producing an enormous dry area (rainshadow) called the Patagonian Desert. This desert occupies about one-quarter of the country's territory.

Buenos Aires, the capital and primate city, was also the capital of the Viceroyalty of Río de la Plata, which was carved out of Spain's southern territory in 1776. The region declared independence from Spain in 1816 after several years of warfare. Argentina soon fell into a series of civil wars and a war with the Empire of Brazil. They fought to a stalemate and in 1828 their peace treaty called for Brazil's southern province "Cisplatine" to become a separate country, Uruguay.

Argentina was consolidated into a nation-state in the latter half of the nineteenth century. Together with Brazil and Uruguay, Argentina defeated Paraguay in the War of the Triple Alliance in 1870. More than half of Paraguay's population was killed and the country remained landlocked. In the late 1870s and early 1880s General Julio Argentino Roca conducted the Conquest of the Desert, which extended Argentine sovereignty through the Pampas and Patagonia by conquering, killing, and displacing thousands of indigenous inhabitants. *Gauchos*—cattle herders or cowboys—filled the Pampas and Patagonia during the nineteenth century, creating Argentina's enduring iconic image of a rugged independent herder.

From 1880 to 1916 Argentina was ruled by the Generation 80, a group of conservative elites who sought to make Argentina an international power. During the late 1800s and early 1900s Argentina's economy transformed and expanded quickly. It became a major exporter of livestock and grains (wheat) to Europe and attracted British and other European investment. Industry also grew and Argentina appeared to have entered the ranks of more developed countries. As the economy expanded, the demand for laborers increased. As a result European immigration grew exponentially, radically changing the country's ethnic makeup. Most of the immigrants were either Italian or Spanish, but others arrived from Great Britain, Poland, Russia, Germany, and elsewhere in Europe. A number of eastern European immigrants were Jewish, and gave Argentina Latin America's largest Jewish population. Many immigrants settled in ethnic agricultural communities or colonies. The best-known, but not largest, group were the Welsh communities of Patagonia. Between one-quarter and half of Italian and Spanish immigrants eventually returned to Europe.

Argentina's economic success did not continue and by the 1930s economic growth stagnated. In 1946 Juan Perón was elected president, and he and his wife Eva remain important if controversial figures in Argentine history. Perón embarked on an ambitious economic and social program to create economic independence for Argentina and to increase living standards for the working classes. His government implemented an import substitution industrialization (ISI) model that lasted until the late 1970s. His nationalistic economic and social policies were controversial, and he was loved by many and despised by many others. Perón was ousted in a coup in 1955, but returned briefly in 1973–1974.

Since the 1970s, Argentina has undergone significant economic and political upheavals. A military coup in 1976 led to the dirty war, during which thousands of young people considered to be leftists were kidnapped, tortured, and killed. Military dictatorships in Chile and elsewhere in South America paralleled the Argentine experience, with mixed results economically, socially, and politically (see *Geography in Action*: Chile and Venezuela: Controversial Development Agendas). The Argentine military regime ended in 1982 after the armed forces performed badly against Great Britain during their brief war over the Falklands/Malvinas Islands. The 1980s saw serious economic problems with debt, inflation, and the implementation of austerity measures. The ISI model was mostly eliminated in the early to mid-1990s and the economy improved until 1998 when it entered another recession.

Beginning in the late 1990s, Argentina's economy improved, helped substantially by the commodity price boom that started in 2003. Argentina is one of the largest exporters of soybeans and soy products, sunflower seeds, and beef, but also exports fruit and wine. Argentina's wine district is centered near Mendoza in a semiarid region between 2,000 and 3,500 feet (600–1,000 meters) (Figure 3–45). Argentina also has a diverse manufacturing sector, which accounts for over 20 percent of its GDP. Argentine industry processes agricultural products, but also produces automobiles, biodiesel fuel, and many other products. Most of Argentina's industry is concentrated in

▲ Figure 3–46 **Plaza of the Republic, Buenos Aires, Argentina.** This obelisk was built at the intersection of two major roads in 1936 to commemorate the 400th anniversary of the founding of the city. Buenos Aires is sometimes referred to as the Paris of South America.

metropolitan Buenos Aires, but Córdoba and Rosario are also important manufacturing centers. Argentina's exports to China, especially soy products, have increased, but Brazil remains its most important trade partner.

Argentina's population and economic activity are overwhelmingly concentrated in metropolitan Buenos Aires (Figure 3–46). With nearly 13 million people, about one-third of the country's population, Buenos Aires has one of the largest metropolitan areas in the world, second only to São Paulo in Latin America. Buenos Aires was an original core because its location on the Río de la Plata (a large estuary, not a river) made it an important port. Its port functions continue to grow as agricultural and manufacturing exports have grown. Residents of Buenos Aires are called "Porteños," and are known for their pride in their city.

During the economic crisis of the late 1990s as many as 300,000 Argentines left the country, many going to Spain. That migration diminished after the crisis, and Argentina is a net receiver of immigrants. About 60 percent of its 1.8 million immigrants come from its bordering countries; Paraguayans are most numerous (over 550,000), followed by Bolivians and Chileans.

▼ Figure 3–45 **Vineyards near Mendoza, Argentina.** Argentina is one of the largest producers of wine in the world, but only began exporting red wine in the 1990s. Meltwater from the Andes Mountains is used to irrigate many vineyards in this high-altitude, semiarid region.

Stop & Think

▶ Is large-scale settlement in and development of the empty areas of South American countries a likely prospect?

Chile and Venezuela: Controversial Development Agendas

Chile and Venezuela's controversial political and economic systems have generated fierce debate about human rights, democracy, and economic equity. In Chile a conservative dictator, General Augusto Pinochet, imposed a free-market, export-led development program starting in the mid-1970s, resulting in economic growth during the latter half of the dictatorship (Figure 3-6-1). Chile returned to democracy in 1990 and was governed by center-left or democratic socialist presidents until 2010. Venezuela was led by democratically elected leftist Hugo Chávez from 1999 until his death in 2013. Some observers have linked Hugo Chávez's populist administration with Chile's socialist president Michelle Bachelet as part of Latin America's "new left" or "Pink Tide." Since the end of the 1990s other Latin American countries (Brazil, Uruguay,

Argentina, Ecuador, and Nicaragua) have elected leftist presidents, most of whom were critical of neoliberalism. Including Chile and Venezuela in this category may appear commonsense, but masks important differences in politics and economic development between the two countries.

In 1973 socialist president Salvador Allende was killed in a military coup. General Pinochet's subsequent dictatorship killed about 3,000 political opponents, tortured many others, and forced tens of thousands of people to flee the country. Chile's economic policies were based on the neoliberal model and focused on trade liberalization, privatization, and returning properties and businesses that had been expropriated by previous presidents. The economic progress made during the dictatorship is controversial, but many businesses, transnational companies,

and some of the middle class prospered while the working classes saw their wages decrease and income inequality increase. In 1990 Patricio Aylwin became the first elected president of Chile in 20 years. He and his successors kept many of the economic policies in place, but made concerted efforts to reduce poverty by increasing spending on social programs, health, and education, and raising the minimum wage. The poverty rate fell from about 40 percent in 1989 to less than 20 percent in 2000 and about 14 percent in 2013. Chile's economy is strong and though it relies heavily on copper exports to China, it has diversified its exports (fruit, fish, wine) and its economy as a whole. Although income inequality remains high, Chile is recognized as a strongly democratic country that has used its economic success, especially during the commodity

▲ FIGURE 3-6-1 **The harbor and port of Valparaiso, Chile.** Valparaiso is Chile's most important port and one of its most important tourist destinations. Numerous container ships export fruit and other products and cruise ships bring visitors from around the world.

boom of 2003–2008, to increase the well-being of its less fortunate citizens.

Venezuelans elected former military officer Hugo Chávez three times since 1998 and wrote a new constitution while he was president. Few people in contemporary Latin America are more controversial than Chávez. Called "Bolivarianism" after Simón Bolivar, the region's most famous liberator during the independence wars from Spain, Chávez's policies gave the government a much greater role in the economy and sought to eradicate poverty with massive social programs, called Bolivarian Missions. He also pursued political cooperation throughout South America and was staunchly anti-imperialistic, often directing criticism at the U.S. government. These missions and other programs have reduced poverty substantially and improved health and other social indicators. Ironically, these programs have been paid for by oil revenues (Figure 3-6-2), generated mostly from sales to the United States. Chávez was incredibly popular among the poor and working class who benefited most from his programs and who mistrust traditional Venezuelan politicians and the upper classes. But he was despised by the upper class and much of the middle class, who considered him to be authoritarian and hostile to business interests. They argue that his social programs are not sustainable and were about ensuring Chávez's power. Chávez was allied with President Correa of Ecuador, President Morales of Bolivia, and made special arrangements with Cuba to sell the country subsidized oil. Chávez even sold subsidized heating oil in the Boston area in 2005 through Citgo, the Venezuelan state-owned oil company that operates in the United States. This act followed through on a Chávez pledge to help the poor throughout the Americas, which is consistent with his populist rhetoric. Chávez was last elected in October 2012, even though he was ill and Venezuela had a struggling economy, crumbling infrastructure, and a high crime rate. His failures and imminent death did not deter the popular masses from supporting him. Hugo Chávez died of cancer in March 2013 and questions remain whether "Chavismo" will persist or if the institutions and policies he helped promote will fade away.

Sources: Michael Shifter and Cameron Combs, "Chávez Stays, Again," *The International Spectator* 47, no. 4 (December 2012): 69–75; Kurt Weyland, "The Rise of Latin America's Two Lefts," *Comparative Politics*, January 2009, 145–164.

▲ FIGURE 3-6-2 **Venezuela's state-owned oil company, PDVSA (Petroleum of Venezuela, South America).** Rafael Ramirez, the president of PDVSA, meets with Igor Sechin, the executive chairman of Rosneft, Russia's state-owned oil company in 2013. Venezuela relies on oil exports more than any other country in Latin America.

Summary

▶ Latin America and the Caribbean were created out of Europe's accidental discovery of two large continents connected by an isthmus. These continents held millions of people, speaking hundreds of languages in numerous environmental and social settings. The Spanish and Portuguese were the first European powers to reshape the human geography of the Americas, but the British and Dutch would soon follow.

▶ Dramatic change continued as the European powers imported between 10 and 12 million Africans for slave labor, and indigenous peoples were concentrated into towns with plazas and Roman Catholic cathedrals. By the early 1820s the colonial era ended and geopolitical wrangling ensued. In general the region's economy and population languished until the end of the nineteenth century, when both grew tremendously, reshaping the region's economies, population structure, and cultural makeup. By the start of the twentieth century the influence of Great Britain waned and the influence of the United States grew, especially in Central America and the Caribbean. Panama was created out of an American desire to control a canal that still connects the Atlantic and Pacific realms.

▶ Latin America's economy underwent two "shocks" and eras of restructuring during the twentieth century. The Great Depression ushered in an era of protected economies and the debt crisis and economic restructuring of the 1980s led to the "lost decade." Most countries reoriented their economies to export primary products, assemble manufactured goods (EPZs), and attract tourists. Thirty years of neoliberal economic policies have brought mixed results, and dissatisfaction has led to the repeal of some neoliberal

policies by "Pink Tide" presidents. Brazil has emerged as the largest economy and population, followed distantly by Mexico, which continues to rely heavily on the United States and has yet to become a leader in the region the way Brazil has.

▶ Latin America has never been isolated, but through trade, tourism, and migration it likely has never been more integrated into the global economy. Free trade agreements are common and China's role has increased. Although progress has been made since the 1980s, income inequality remains high and more than one-quarter of the region's people live in poverty. Millions of Latin Americans left out of the region's economic progress decided to migrate to the United States and Europe, where as a group they remit billions of dollars.

▶ Latin America and the Caribbean can be divided into coherent, if problematic, subregions. Vast economic and environmental differences exist between and even within the subregions, but there are also strong similarities brought about by a shared history of Iberian conquest and cultural traits. The region faces many economic and environmental challenges. Globalization has created wealth and helped many people, but has left others behind and threatened ecosystems and vital natural resources.

▶ Like all places, Latin America and the Caribbean are still in the process of becoming. This world region's landscapes, economies, and environments will reflect global processes such as climate change and economic globalization, but also reflect the creativity and determination of its residents as they respond to these processes and participate in the reshaping of their own landscapes and livelihoods.

Key Terms

altitudinal life zones 118
Aztec 121
banana republic 146
Bolivian gas wars 157
BRIC (Brazil, Russia, India, and China) countries 158
business process outsourcing (BPO) 133
chain migration 138
chinampas 123
ecotourism 136
El Niño 118

encomiendas 126
export processing zones (EPZ) 133
free trade zones (FTZ) 133
haciendas 126
home town associations (HTA) 140
import substitution industrialization (ISI) 129
Inca 121
informal economy 130
internally displaced people 158

Isthmus of Panama 147
land concentration 125
lost decade 129
maquiladora industry 133
Maya 121
Media Luna 157
neoliberalism 129
payment for environmental services (PES) 148
Pentecostalism 139
primate city 144
pristine myth 121

quilombos 159
rain shadow 120
remittances 140
syncretism 126
thermal inversion 121
trade blocs 130
transnational migration 138
Treaty of Tordesillas 125
voluntourism 136
Zapatista movement 142

Understanding Development in Latin America and the Caribbean

1. What are the three great landform divisions of Latin America?
2. How does altitude influence temperature, precipitation, and agricultural options in the Andes?
3. What major Indigenous civilizations ruled in Latin America before the arrival of Europeans and where were those civilizations located?
4. Why did Latin America experience a major population decline after the European conquest?
5. Where did the slave labor used to produce sugar and other agricultural products come from and how have society and economy adjusted to the end of slavery in the nineteenth century?
6. Why is Panama such an important part of the global transportation system, and will that role continue in the future?
7. What are the gains and the losses produced by NAFTA (North American Free Trade Agreement) in Mexico's economy?
8. What are the advantages and disadvantages of the varying types of tourism in Central America and the Caribbean?
9. Why has Brazil emerged as an important global economy?
10. Why is Evo Morales such an important voice in South America?

Geographers @ Work

1. Explain the role traditionally played by coca in highland Andean culture in contrast to the impact that drugs derived from this plant have had on both local economies and the global community.
2. Latin America is characterized by many different physical environments. Demonstrate how this diversity influences the planning of agricultural development strategies in the region.
3. Remittances are a vital part of Latin American economies. Evaluate how that money is invested in recipient countries and what strategies might be developed to increase the benefits of those investments.
4. Explain why Paraguayans are such a prominent part of Argentina's immigrant community.
5. Investigate the benefits and costs of mining lithium on the Bolivian Altiplano.

MasteringGeography™

Looking for additional review and test prep materials? Visit the Study Area in MasteringGeography™ to enhance your geographic literacy, spatial reasoning skills, and understanding of this chapter's content by accessing a variety of resources, including MapMaster™ interactive maps, videos, RSS feeds, flashcards, web links, self-study quizzes, and an eText version of *World Regional Geography*.

▶ **Figure 3–47 Latin America and the Caribbean from space.**

Europe

4

Corey Johnson

City Walls in the Twenty-First Century

No trip to Europe is complete without a visit to a medieval walled city. Nearly every European country has one, and visitors flock to walled cities such as York, England, and Sibiu, Romania, to snap photos of their narrow alleyways, cobblestone streets, and encircling walls. Their large fortifications, symbols of how leaders many centuries ago tackled the problems of their times, enchant tourists. City walls clarified the divide between urban and rural, and often between enviable prosperity and dismal poverty. They divided citizens from noncitizens. Who would be allowed to enter the city gates depended on whether their goods or their services were needed or desired. When the gates closed at night, all visitors were expected to be gone, and the bustle of the deliveries, street markets, and construction was replaced by a relative calm.

Cities in Europe no longer need walls for protection. By the eighteenth century towns and cities were integral parts of the political territories that surrounded them, and nation-states took on the roles of defending borders. But the symbol of city walls as marking inside and outside is still a potent one. Like the great, guarded medieval cities, Europe today is itself a market for goods and ideas, a productive economic engine that requires energy, supplies, and sources of labor from beyond the region. Europeans also seek in various ways to define the borders of Europe and control who and what can enter. Europe is thoroughly integrated in many ways in its neighborhood and with the world at large, but like a medieval city, there is also a dynamic tension between openness and access to the outside world and a desire by some to strictly regulate how, where, and with whom those interactions take place.

Read & Learn

► Describe the physical geography of the European region, and explain its effects on the climate of the region.

► List factors defining Europe as a cultural and physical region.

► Discuss the region's major environmental problems and identify strategies to address these challenges.

► Chart Europe's transition from an agricultural to an industrial and postindustrial region.

► Explain the political-territorial order of Europe, including the origins of the nation-state and sovereignty.

► Discuss the role of language and religion in shaping European geography.

► Locate the key nodes in the European integration project.

► Understand the role of cities in the development of Europe and their distinctive geographic features.

► Describe the factors shaping Europe's role in the global economy and its geopolitical neighborhood.

▲ Dubrovnik, Croatia, is a popular tourist destination on the Dalmatian coast. On July 1, 2013, Croatia became a member of the European Union.

Contents

Europe

Figure 4-0 The major physical features and countries of Europe.

In January 2013, the European Central Bank began releasing its "Europa" series, the second generation of Euro paper currency. The new banknotes differ from their first-generation predecessors because of the likeness of Europa printed on them, part of an elaborate anticounterfeiting design. According to Greek mythology, Europa was a Phoenician princess who was abducted by the god Zeus, bore his children, and became queen of Crete. How she became the namesake for Europe—a geographical term adopted by nearly all languages spoken in the region—is heavily disputed by scholars. What is clear is that the term Europe originated in Greece, a country today considered the geographical, economic, and political periphery of the European Union. The redesigned currency, celebrating the linkage of European identity to ancient Greece, coincides with daily news reports on modern Greece as the outcast of Europe for its fiscal woes.

Europe's Environmental and Historical Contexts

So where is Europe, and maybe more importantly, *what* is it? These questions have been pondered for a very long time, and the answers rarely satisfy students of Europe. Eurasia is the largest landmass on Earth, and looking at imagery from space (Figure 4-1) we can see at its western edge a series of peninsulas and large islands. The word peninsula means "almost an island" and refers to land surrounded on three sides by water. A distinguishing feature of most of what we consider Europe is proximity to oceans or seas and the resulting moderating influences on climate.

▼ **Figure 4-1 Western Eurasia.** Geologically speaking, Europe is not a continent but rather a subcontinent of Eurasia consisting of several peninsulas. The main ones are the Scandinavian, Iberian, Italian, and Balkan peninsulas.

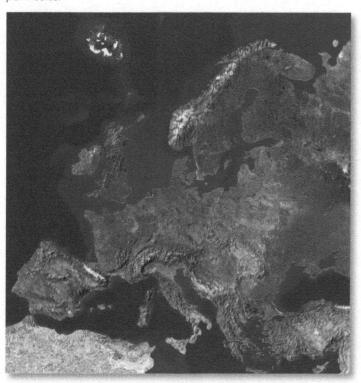

But the geological or physical geographical definitions of Europe are not as relevant as the human geographical ones. Historically, culture has been the most important marker used by Europeans to distinguish themselves from neighbors, and while still important, economic and political characteristics are increasingly used. Is Turkey part of Europe? In this textbook, it is not. Russia? It has its own chapter. Yet Turkey has applied for membership in the European Union, and most Russians view themselves as Europeans. For many of us learning geography in school, the most important borders of the European "continent" passed through Russia (the Ural Mountains), and Turkey (the straits linking the Black Sea with the Aegean Sea, the Bosporus, and the Dardanelles). As visitors to either Turkey or Russia will note, these physical features are more convenient lines on the map than imposing physical barriers to trade or movement, or cultural dividing lines.

This lack of clear borders and identity makes the story of Europe as a world region compelling. We all "know" of Europe, its centrality to world affairs over centuries, its high culture, marvelous cities, and economic prosperity. It is by and large the most developed world region, with the highest literacy rates, longest lifespans, and smallest disparity between rich and poor. But its varied human and physical geography makes the region full of surprises even to the seasoned traveler and student.

Environmental Setting: Physical Geography Enables Development

Europe's physical geography helps to explain the remarkable trajectory of this region as a major population center and global economic and political powerhouse. Navigable waterways and relatively easy access to the seas promoted trade, while a mild climate and productive agricultural land allowed the region to sustain a large population.

Landforms

Europe is characterized by mountainous zones, plains, and river valleys (see chapter opener map). Most of Europe receives ample precipitation for agriculture, so that very little agricultural land in Europe requires irrigation outside of some areas of the far southern Mediterranean climate region. The major river systems of the region historically have been transportation and communication arteries and sources of drinking water or water for industrial uses rather than the source of vital irrigation waters, as in the case of the Nile, Colorado, or Mekong rivers. Europe's most extensive river system is the Danube, which begins in Germany and passes through eight other countries before emptying into the Black Sea in Romania. The Danube drains a basin encompassing 315,000 square miles (816,000 square kilometers), an area slightly larger than Turkey. Europe's best-known rivers flow mostly toward the Atlantic (Rhine, Seine, Loire, Elbe) or Mediterranean/Adriatic Seas (Ebro, Po). Managing river flow as well as traffic on the navigable river systems has always required some degree of cross-border cooperation, and a recent development that illustrates the political and economic integration of Europe through the European Union is the challenge of cross-border river management, as two-thirds of EU land lies within river drainage basins that cross international borders (see *Exploring Environmental Impacts*: Transboundary Water).

EXPLORING
ENVIRONMENTAL
IMPACTS

Transboundary Water

Take a look at the map showing Europe's major river basins (Figure 4-2-1). Where do the river basins match up with the borders of countries? Can you even make out where the Netherlands and Germany lie? The border between Spain and France corresponds fairly well with the watershed separating the Ebro and Catalonia basins from the Adour-Garonne and Rhone in France, but where is Portugal? Is Northern Ireland (a part of the United Kingdom) clearly separate from the Republic of Ireland on this map? Of course there is no particular reason why the borders of states should align with the map of river basins—very few in the world do. But the management of rivers and water quality was until recently solely within the jurisdiction of the separate nation-states in Europe. Surface water is considered a flow resource, meaning it is constantly in motion, and in the case of rivers and streams, water is constantly crossing human-imposed political borders (Figure 4-1-2). When there is heavy rainfall and a river floods, such as occurred along the Danube River during the summer of 2010 following an extended wet period across Central and Eastern Europe, the impacts and efforts to mitigate those impacts

likely touch people in neighboring countries. If a factory in the Czech Republic is releasing pollution into the Elbe River, that pollution will almost inevitably make its way downstream and possibly impact local populations in Germany.

The issue of cross-border management of shared resources, like rivers, is a vexing one for politicians and managers because nature does not acknowledge human constructs such as sovereignty and political borders. Fresh water is a particularly sensitive issue because it is one of the essential needs of all forms of life, and its mismanagement can easily result in serious consequences. Acknowledging this, states in Europe have embarked on a number of initiatives to plan how best to manage shared resources. One of the first transboundary water conservation projects in the world was organized in 1959 by jurisdictions in Germany, Switzerland, and Austria bordering Lake Constance. The Rhine River, once plagued by centuries of industrial and agricultural pollution, is also the site of a novel transboundary cooperative agreement between France, Germany, and Switzerland, which coordinates the cleanup of the river and hydroelectric projects as well.

More recently, the European Union has been the major driver of transboundary environmental cooperation and regulation. The EU has adopted more than 200 directives on environmental issues over the last 40 years with transboundary significance on topics ranging from habitats for endangered species to water quality. The EU's "Water Framework Directive" requires that all member states protect and restore their surface water, and for water bodies that cross international borders, plans must be implemented cooperatively by the affected member states. While there are bilateral and multilateral projects elsewhere in the world on rivers, such as the Mekong River Commission in Southeast Asia, the fact that the entire EU is now subject to transboundary regulation of water quality makes it the most ambitious attempt to manage rivers at the scale of river basins, even when they cross international borders, anywhere in the world.

Sources: C. Johnson, "Toward Post-Sovereign Environmental Governance? Politics, Scale, and the EU Water Framework Directive," *Water Alternatives* 5, no. 1 (2012): 83–97; N. Sommerwerk et al., "Managing the World's Most International River: The Danube River Basin," *Marine and Freshwater Research* 61, no. 7 (2010): 736–748.

▲ FIGURE 4-1-1 **The Great Kazan Gorge in the Danube's Iron Gates area between Serbia and Romania.** The Danube is one of the world's most international rivers.

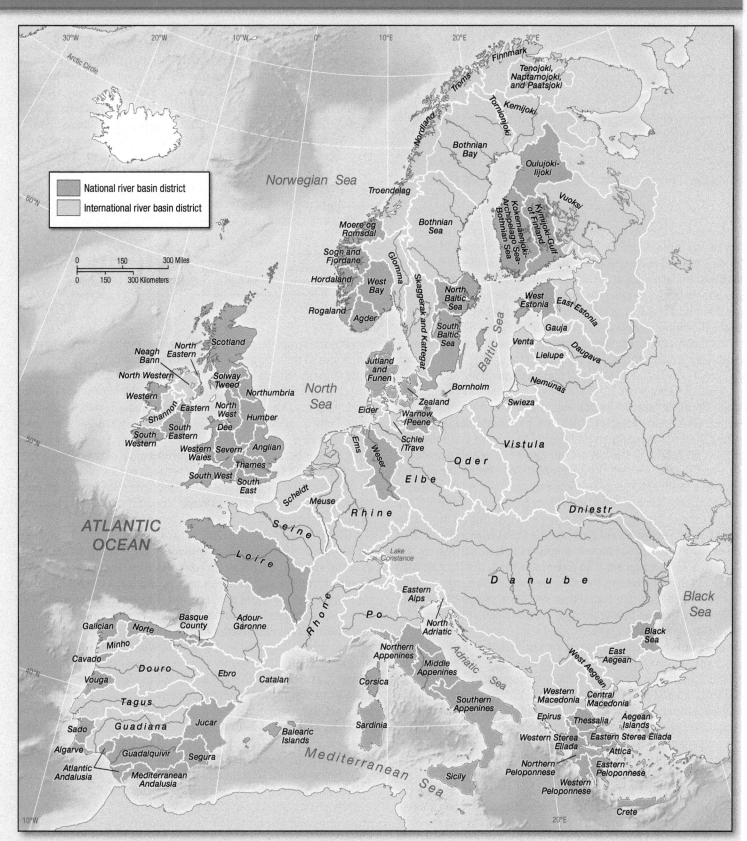

▲ FIGURE 4-2-1 **Major river basins in Europe.** We are accustomed to seeing maps of Europe divided up into nation-states, but this is a different way of visualizing European geography. Can you make out where the political borders of European states would be on this map?

◄ Figure 4-2 **Highlands meet the densely populated midland valley of Scotland.** The flat midland valley of Scotland is what geologists call a *graben*, a structural depression caused by parallel faults. It separates the Caledonian Scottish Highlands from the southern Highlands, and the valley is home to the largest population centers, Glasgow and Edinburgh.

Moderate climates and generally ample precipitation are important factors explaining the more or less even distribution of population in Europe in comparison to other world regions like Central Asia, but the physical terrain of the region is also suitable to large areas of inhabitable land. The tallest mountains are located in the Pyrenees and Alps, and while impressive in their physical appearance and height, these mountainous zones are compact compared to the Rocky Mountains, Andes, or Himalayas. Their east–west orientation also explains why, unlike other major mountain ranges, they do not create significant rainshadow effects as air masses come off the Atlantic.

Spain's Cantabrian Mountains along the country's northern coast, France's Massif Central, and the Jura of Switzerland and France are notable areas of uplift, but punctuate the landscape rather than being major impediments to human settlement or travel. This is largely a result of their age; erosion and deposition have had time to smooth the landscape. The Mediterranean Sea lies atop a major plate boundary between the Eurasian and Hellenic plates and the African Plate, providing (along with Iceland) the region's major tectonically active zone and youngest mountain ranges. Sicily's active Mount Etna and the Greek island of Santorini, site of one of the largest volcanic eruptions in recorded history, are along this subduction zone. In Eastern Europe, the Dinaric Alps define much of the landscape of the western Balkan region, while the Carpathian Mountains wind their way like a backwards "S" from the Polish-Slovakian border through Romania, then merge into the Balkan Mountains of Bulgaria.

Northern and northwest Europe are characterized by areas of very old, highly eroded mountains, now hilly areas with few physical impediments to human activity. This includes much of the southern half of Germany and France, the Scandinavian Peninsula, and the highland areas of northern England and Scotland (Figure 4-2). Some of the most densely populated areas of Europe are located in the valleys of these old, often glacially impacted areas, including Frankfurt, Dresden, and Stuttgart in Germany; Oslo, Norway; and Glasgow, Scotland.

It was not the mountains but rather the plains and lowlands of Europe that formed the focus of economic development over the centuries. The most famous cities are found usually in open areas of flat terrain: London situated in the London Basin; Paris in the Paris Basin, which itself is part of the North European Plain; Budapest on the Great Hungarian Plain; and Venice and Milan in the large

expanse of the Po River valley in northern Italy, one of the few parts of the Italian Peninsula not characterized by mountains or hills. Of these, the largest is the North European Plain, a generally uninterrupted band of gently rolling, formerly glaciated hills stretching from the Pyrenees to the Urals and home to Amsterdam, Hamburg, Berlin, and Warsaw as well as Paris. Historically, the North European Plain is sometimes referred to as invasion highway, since its uninterrupted terrain was easy for armies to traverse and difficult to defend. The Golden Horde of Mongols used this route in the thirteenth century in their incursions into Europe, as did Napoleon in the nineteenth century and Hitler in the twentieth century to invade Russia.

Stop & Think

→ What are some notable features of Europe's physical layout?

Climate

Perhaps the simplest, and most persuasive, explanation for Europe being one of the world's most densely settled regions is its hospitable climate. Paris is located at about the same latitude as Ulaanbaatar, Mongolia, and is about 250 miles (400 kilometers) north, latitudinally, of Montreal, Canada. While Ulaanbaatar, the coldest capital city in the world, experiences a January average high temperature of 3°F (–16°C) and Montreal enjoys a not-so-balmy 21°F (–6°C), Paris basks in the relative warmth of an average high temperature of 44°F (7°C). Similar comparisons can be made for other cities in Europe. As a general rule, Europe is mild for its latitude, and experiences little of the extreme temperature swings one finds in North America and the rest of northern Eurasia.

Given Europe's shape, it is not surprising that proximity to the sea is the main driver of the relative mildness. But it is the nature of the seawater arriving off the western edge of Europe that is key: the **North Atlantic Current** supplies the air masses hitting the west coast of Europe with relatively warm temperatures. This current is an extension of the warm Gulf Stream current that moves warm tropical water up the eastern seaboard of North America before turning out to the mid-Atlantic around Cape Hatteras, North Carolina. At the extreme western edges of Ireland, England, and France, one can find palm trees, normally associated with tropical and subtropical climates, growing at the same latitude as Edmonton, Alberta, Canada! It is the **thermohaline circulation,** or the global "conveyor belt" of ocean currents driven by temperature and salinity differences in seawater, that creates in the North Atlantic a clockwise circulation of seawater that draws warm subtropical and tropical water to Europe's shores. And it is the North Atlantic Current and its impacts on the atmosphere that explain the relatively balmy maritime temperatures (Figure 4-3).

If you were to take a cycling trip from one end of Europe to the other (an endeavor quite common among young Europeans and increasingly popular among visitors to Europe), what climatic conditions would you experience? First, the North Atlantic Current and

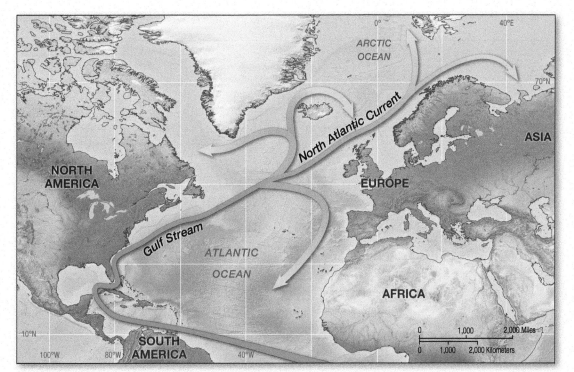

flat terrain of northern France provides some relief for sore legs, and by mid-August the chances of uncomfortably high temperatures further diminish as solar insolation decreases with shorter days. You transition not only to a new language (German) but also about halfway across Germany into the second of the major climate regions in Europe: humid continental (Figure 4-5).

No abrupt line marks changes in climate as you might experience climbing a mountain range (thankfully your route avoids the Alps!). But in the humid continental climate region, the increasing distance from the moderating influences of the Atlantic Ocean result in larger daily and seasonal temperature extremes, and greater variation in precipitation. By the time you reach Passau, along the Danube River and bordering Austria, it is closer to the end of September, a pleasant time of year when the summer heat that can accompany continental climates has passed and before the much colder continental winter arrives. The humid continental climate of Central and Eastern Europe is characterized by competing air masses. Domes of high pressure can lead to summertime highs in cities such as Budapest in the mid-90s Fahrenheit (mid-30s Celsius), interspersed with cooler, wetter periods. Summertime thunderstorms can occur regularly here from the intense heating of the land, but the chances of encountering one of these greatly diminishes by fall. This is the ideal time of year to follow the Danube River from Germany through Austria all the way to Belgrade, Serbia (Figure 4-6).

You must make it to the mediterranean climate region of Greece before the snowy, sometimes brutally cold air masses of early winter move to the northern parts of southeastern Europe. Pulling into Athens in late December, will be a relief in many respects. The mediterranean climate region, which includes not only Greece but also most of Italy, Spain, Portugal, and extreme southeastern France, experiences mild winters, although occasional cold air blasts from the north do penetrate the Italian and even Greek peninsulas; snow is not unheard of in Athens. Summers can be exceedingly hot, with daily maximum temperatures in August regularly above 95°F (35°C) with little or no rain. Most precipitation in the mediterranean climate region falls in the winter months, from November to early March.

This is a selective route through Europe; you bypassed the Alps, with their unique mountainous climate of wet, cool summers and snowy, cold winters, and the Nordic countries of Finland, Sweden, and Norway, with much the same conditions as the Alps. But you will have experienced the most important climate regions that

palm trees in far southwest England notwithstanding, not all of Europe is balmy, nor are the maritime influences uniform throughout the region. In planning this fantasy exercise, it is useful to be familiar with Europe's three large climate regions (Figure 4-4): **marine west coast, humid continental,** and **mediterranean.** You begin your cycling trip in Scotland and continue via London and Reims to Vienna and ultimately Athens. Such a trek allows you to see most of what Europe's climate offers. Marine west coast includes the British Isles and much of France, the low countries of Belgium, the Netherlands, and Luxembourg, as well as western Germany and far northern Spain. The notable feature of this highly maritime-influenced climate is the lack of extreme temperatures, either in summer or winter (due to the moderating influence of the ocean), and an abundance of year-round precipitation, mostly in the form of rain. While London, Frankfurt, and Brussels can experience short outbreaks of heat upward of a daily maximum of 95°F (35°C) in the summer and as low as 5°F (–15°C) in winter, these are rare and usually limited to a few days during the year. Let's start in Scotland in July and bike for six months, at a leisurely pace of around 14 miles (22 kilometers) per day.

The lush green countryside of Scottish summer can mean only one thing: the possibility of rain. Although hitting most of northern Europe in high summer, in the marine west coast climate region it can be rainy and cool even in July, though the summer rains at this latitude rarely come in torrential downpours but rather a gentle, cool drizzle. If you had passed through Liverpool and Manchester in January, the biggest challenge would be the amount of daylight— only around 8 hours, whereas in July, it would be more like 18 hours. Central England has the longest instrumental record of climate data dating to the seventeenth century, and we know here that precipitation during summer months has decreased as a result of climate change (while increasing during winter months), and temperatures have become warmer year-round. After crossing Britain en route to London and the English Channel, at Dover you board a ferry across the Strait of Dover to Calais, France, to continue the ride. The fairly

▲ Figure 4-4 **European climate regions.** The path of your cycle trip passes through the three major climate regions of Europe.

◀ **Figure 4-5 Wine region of Rhineland-Palatinate in southwestern Germany.** In this region the mild marine west coast climate zone transitions to the humid continental climate zone with colder winters.

forced mass migrations to take place, especially from Scandinavia southward. The result was the so-called barbarian invasions of Germanic tribes such as the Huns, Goths, Vandals, and others into territories controlled by the Roman Empire, accelerating its decline. The Celts are perhaps the most famous case of a mobile people; their spread from Central Europe and eventual retreat to the farthest edges of the British Isles, France, and Iberia, is at least in part explained by environmental changes.

As the cases of the Vikings and Romans illustrate, large-scale environmental change is not a recent phenomenon. Farm fields and pastures dominated the European scene in the period leading up to the Industrial Revolution, and these agrarian landscapes were the product of human modifications over many centuries. Forests were felled and domesticated crops planted and harvested. Low-lying marshes were drained across much of northern Europe and in the Po River valley to allow for agriculture. In mountainous areas hillsides were terraced to plant crops and to prevent runoff and erosion. The scale and, above all, speed of that modification reached its height as a result of the **Industrial Revolution,** which began in northern Britain in the late eighteenth century and spread through much of Europe by the late nineteenth century. Massive industrialization of production required raw materials, transportation networks, and housing for the factory workers. The fuel of choice was coal, which was abundant in the areas most closely tied to industrialization such as northwest England, the Ruhr River valley and Saxony in Germany, and Silesia in Poland. Bituminous, or black coal, typically comes from underground mines and so the surface disturbance is relatively minor, but the low-quality lignite or brown coal is typically strip-mined (Figure 4-8). Large open pits, many now recreational lakes, in parts of eastern Germany

define Europe's physical environment. There are few other places in the world in which you could undertake a similar trip yet avoid the environmental perils of other regions, such as monsoons, tornadoes, typhoons, desert heat, blizzards, and more.

Stop & Think

→ How has Europe's climate contributed to its development as a region over time?

Environmental Challenges

It would be difficult to find a region where the human modification of the landscape has been more extensive and dramatic than in Europe. The region's economic success over centuries, even millennia, is often a result of natural resource exploitation and agricultural and manufacturing innovations that profoundly altered the natural environment. Each historical epoch in Europe can be linked to major environmental modifications, and the contemporary landscape of Europe still reflects those changes over time. The Vikings needed wood for their ships and open fields for agriculture and grazing, and they cut down trees wherever they conquered, from the British Isles to Greenland. The Romans did much the same around the Mediterranean; places such as Sicily to this day are largely devoid of forests. Some scholars have attributed the fall of the Roman Empire to self-inflicted economic decline caused by ecological destruction. Five thousand years ago, Scotland's hilly terrain was covered mostly by mixed pine, aspen, birch, and oak forest, but that gave way to the stark, treeless landscapes familiar from films such as *Braveheart* and *Skyfall*. That landscape was largely a result of successive waves of deforestation that continued into the early twentieth century. The legendary forests covering most of the southern two-thirds of Germany, the stuff of Grimms' fairy tales, met a similar fate, and the large tracts of forested land along the *Autobahn* are typically heavily managed, second- or third-growth tree stands.

Not all environmental change is caused by humans. Natural climate cycles in past eras caused both human expansions into previously uninhabited areas as well as retrenchment when the climate became less favorable to established livelihoods. Changes in precipitation and temperature can also be linked to periods of political and social unrest as well as disease outbreaks (Figure 4-7). Archeologists have identified a period from around 300 to 800 C.E., when changing precipitation regimes and below-normal temperatures

▼ **Figure 4-6 Persenbeug castle, Austria.** Cycling along the Danube in Austria is a very popular vacation activity.

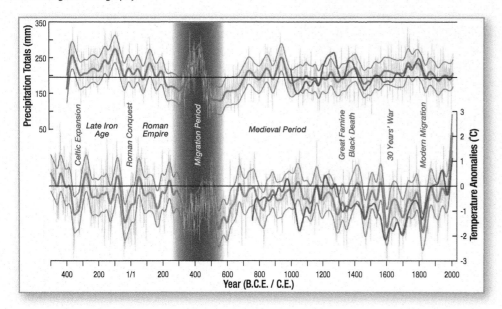

◀ **Figure 4-7 2,500 years of climate change in Europe and its impacts on humans.** As a result of climate change, the last fifty years have been warmer than the long-term average in Europe. Note from this graph the relationships between temperature, precipitation, and major human events.

Source: U. Büntgen, U. et al. "2500 Years of European Climate Variability and Human Susceptibility." Science 331, no. 6017 (2011): 578–582.

are relicts of large-scale coal mining. In the Broads of East Anglia, Britain, the landscape is similarly pockmarked, not as a result of coal mining, but of peat digging. Flying over Belgium and northern France, as well as parts of Eastern Europe, the scarring of the landscape from the mining of ores provides reminders of Europe's industrial past. As Europe has deindustrialized over the last decades, and the region's economic basis has shifted from coal to other fossil fuels such as oil and natural gas, Europeans continue to deal with the lasting environmental impacts of mining.

The burning of coal for manufacturing and power generation (coal remains Europe's largest fuel source for electricity) also created an acid rain problem in many parts of Central Europe. Burning coal emits sulfur dioxide and nitrogen oxides into the atmosphere, which are then reintroduced to the earth's surface when it rains. The high acid content interferes with aquatic organisms' life cycles, changes pH levels in soil causing tree death, and causes decay of buildings, monuments, and statues. Limits on emissions have greatly reduced acid rain since the 1980s, but the cumulative effects of 200 years of acid rain continue to impact ecosystems in parts of Europe (Figure 4-9).

While the Industrial Revolution is typically associated with the mechanization of manufacturing, there was also a revolution in agriculture that dramatically changed Europe's environmental geography. Each successively larger tractor, plow, planter, and harvester demanded larger fields to accommodate their large turning radius. Chemical pesticides designed to protect crops from insects and animals did so at the expense of natural biological cycles, so that farm fields became ecological wastelands (except for the planted crop). Hybridized seed varieties enabled much greater harvests, but required more inputs (especially chemical fertilizers), and these inputs were applied in such quantities that much of them ended up in streams, lakes, and seas, creating unhealthy aquatic environments.

Technological innovations and social change have played out on the environment so that practically nothing you see in Europe looks the way it would have absent the various layers of human activities. But some of the most dramatic and far-reaching strategies to address environmental change have emerged in Europe.

Perhaps the best-known strategies are efforts by coastal cities and countries to deal with sea-level rise and subsidence. Venice, Italy, is famous for its canals and its opulent villas and palaces. The city's historical wealth was a result of the vast merchant empire it controlled from the thirteenth to seventeenth centuries. The entire city, built on a lagoon mostly atop alder piles harvested many centuries ago, is listed as a UNESCO World Heritage Site. While the city's foundation remains largely intact, a combination of seasonal storm patterns, sea-level rise, subsidence, and land-use changes have threatened to submerge Venice. The city is known to flood seasonally, but to date not with catastrophic consequences (Figure 4-10). In 2003, the Italian government approved the construction of a series of underwater, hinged sea barriers that could be raised when there was a threat of inundation. The MOSE Project is scheduled for completion in 2014 at a cost of around US $6 billion.

The Netherlands represents an even more remarkable example of keeping the sea at bay, and could provide the template for other coastal places threatened. The elaborate system of seawalls, dikes, and reclaimed pieces of land called **polder** build on an engineering tradition dating to the Middle Ages. In areas below sea level that were flat as pancakes, the Dutch built mounds called **terpen** for settlements amidst the agricultural fields (Figure 4-11). The iconic windmills were constructed to pump water out of low-lying fields, and the lowland country's wooden clogs were a practical adaptation

▼ **Figure 4-8 A coal power plant in Saxony, Germany.** A strip mine in the foreground supplies lignite for the large power plant in the background.

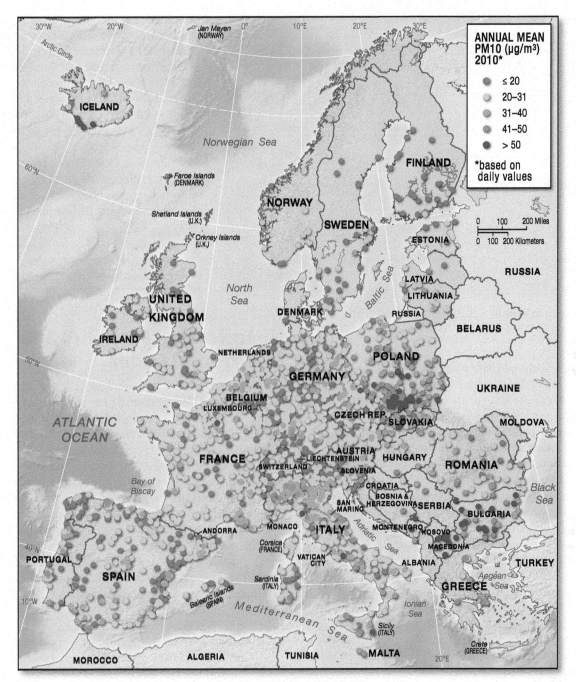

▲ Figure 4-9 Fine particulate matter from burning fossil fuels. Fine particulate matter is a known health hazard. An average annual mean of greater than 40, shown in red and purple on this map, is above European Union air quality standards.

to the perpetually wet environment. There were also political innovations that came into existence as a result of the Dutch dealing with their water. "Water boards," dating to the thirteenth century, were the country's first democratically elected bodies. Catastrophic floods in 1916 and 1953 were the impetus for the world's largest flood control projects, the Zuider Zee and Delta Projects, which protected most of inland Netherlands from storm surges by means of huge sea dikes, while still allowing the major river systems of northwestern Europe, the Rhine and Maas, to empty into the North Sea. The Dutch government has been planning for another large flooding event since the 1950s, and has undertaken extensive studies in

recent years to understand how climate change threatens to undermine these engineering feats, and what if anything can be done to combat not only sinking land but also rising seas (Figure 4-12).

Creating and dealing with environmental change is nothing new to Europeans, but anthropogenic climate change raises new challenges. Scientists estimate that the expected rise in global temperatures would cause significant coastal flooding in the British Isles, Southern Europe, and along the North Sea and Baltic Sea. More people would also be exposed to river flooding. Scientists linked the United Kingdom's wettest year on record in 2012 to climate change. Heat-related deaths would likely increase due to

▲ Figure 4-10 Floods in Venice. Occasional high water is one reason why the Italian government is spending billions of euros to construct the MOSE project.

built offshore. Spain, another country famous for windmills dotting the landscape, derived around 18 percent of its electricity from wind power in 2012, and the northern region of Navarre obtained nearly half of its electricity from wind (see *Geography in Action*: Renewable Landscapes). On a particularly windy night in September 2012, the entire country derived its electricity from wind power. When Germany decided to close 22 nuclear power plants after the 2011 Fukushima accident in Japan, the country's solar power capacity and an unusually sunny February in northern Europe a year later kept the German power system from breaking down.

Stop & Think

> How has Europe's economic development, agriculturally and industrially, impacted its physical environment?

hotter summers. Agricultural yields would likely increase in northern Europe, but decline due to drought in southern Europe. While the mechanisms of climate change are fairly well understood, the precise nature of its impacts are uncertain. One question is whether the melting of polar ice packs could actually cool or divert the North Atlantic Current to such an extent that it would rob Europe of the relatively mild air masses that keep London and Amsterdam winters mild while Edmonton and Ulaanbaatar sit in the deep freeze. European governments are at the forefront globally in promoting a global response to reduce emissions of **greenhouse gases (GHGs)** and preventing land-use changes that are contributing to warming. Greenhouse gases such as carbon dioxide, methane, and nitrous oxide occur naturally in the atmosphere, but are also produced in significant quantities by human activities such as power generation, agriculture, land-use changes, and transportation. The EU has committed to cutting its GHG emissions to 20 percent below 1990 levels by 2020, and has offered to raise its target from 20 to 30 percent if other large countries, such as China and the United States, commit to reductions. The EU's tentative target by 2050 is an 80–90 percent reduction from 1990 levels.

Central to the goal of lowering GHG emissions is burning fewer fossil fuels in electricity generation and transportation sectors. The wind-power generation sector in Denmark employs about 25,000 people and supplies more than 25 percent of the country's electricity use. The small island of Samsø put up 21 wind turbines in the early 2000s to provide power for its 4,000 residents, and it now bills itself as the largest carbon-neutral community on Earth. More and more wind turbines are being

Historical Background: Shaping a Territorial Order

The destructive forces of nationalism that culminated in World War II and the Holocaust tend to be the dominant historical images of Europe in our minds, kept alive in movies, mass media, and high school history classes long after the events themselves. But over the longer term other forces have served to integrate European space. First the Greeks and then more extensively the Romans conquered territories and built infrastructure and cities throughout southern and western Europe. In the wake of the Romans, it was the Roman Catholic Church (and in the East the Byzantine Orthodox Church) that provided a sense of geographic unity in otherwise chaotic circumstances and gave common points of orientation to the varied populations. Later it was secular empires, such as the Austro-Hungarian, the German, and Ottoman, that united large stretches of territory under their jurisdictions.

Europe's Modern Political Map

There is nothing natural or enduring about the political map of Europe. A time-lapse image of the evolution of this map would cause dizziness, and the borders of nation-states continue to change,

▶ Figure 4-11 An elevated terpen from the Middle Ages near Dokkum, Netherlands. The province of Friesland in northern Holland is a good place to see examples of many centuries old terpen, or human-made mounds where settlements were built to avoid seasonal floods.

though at a much slower rate than at almost any point in the past. In 2008, Kosovo became Europe's newest independent state to be recognized by most of the major world powers. There was no independent, sovereign Poland, Finland, Ireland, or Bulgaria in 1900. There was no such entity as Belgium until 1830, Germany or Italy until around 1870. It remains to be seen whether the evolution of the map of sovereign states over the next 200 years will exhibit comparably dramatic transformations as the prior 200 years, or if the "sovereign state" that was invented in Europe will even be a common frame of reference in 200 years. It does appear, however, that the means by which many new states came into existence—war—has been blunted. **Nationalism,** the political cause of intergenerational groups along ethnocultural lines seeking to control their own affairs, is not dead in Europe but has become decidedly outmoded. There are political movements seeking to redefine state territories as being exclusive to one nation, but few with any lasting success. The most persistent territorial disputes involve ethnic groups that do not fit within the dominant framework of national states. Typically they do not seek their own sovereign states, but rather more autonomy to control their own affairs. This process of granting additional powers from central governments to subnational regions is called **devolution.** There are devolutionary movements among the large Hungarian and Roma populations in Romania, as well as Catalan and Basque minority groups in Spain and to a lesser extent, France. In the United Kingdom, the province of Scotland has been granted substantial devolved power by the central government in London. A Scottish parliament was created in 1998 for the first time since 1707, and a referendum for full independence may be held soon.

Few things are as important to people's cultural identities as language and religion, which provide common points of reference within the family, among neighbors, in the community, and beyond. The modern political map of Europe was most profoundly influenced by language and religion over time, and these two issues remain touchstones of community making to this day. Even if only a small percentage of Europeans attend church regularly, particular religious cultural values still help to shape behavior and attitudes. In spite of multilingualism, the vast majority of Europeans are still most

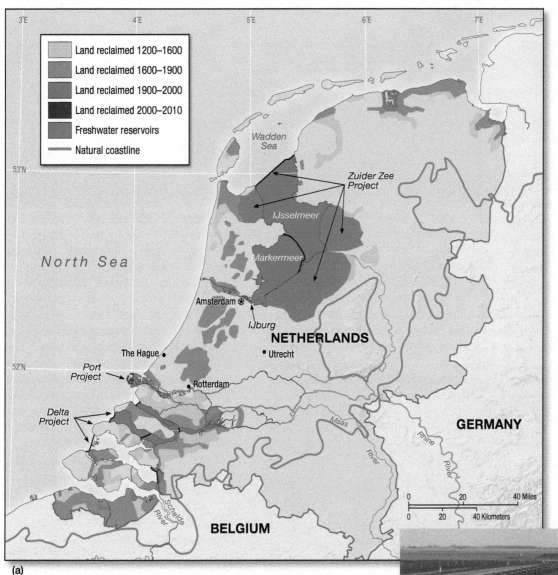

(a)

▲ **Figure 4-12 (a) Land reclamation in the Netherlands.** Where would the coast of the Netherlands be without land reclamation? Data from http://earthobservatory.nasa.gov/IOTD/view.php?;http://en.wikipedia.org/wiki/IJburg. **(b) A sea wall in the northern Netherlands.** Climate change is an issue of particular concern to the Dutch because most of the country lies below sea level. Walls such as these help to keep the sea at bay, but sea-level rise could threaten the massive public works designed to protect the Netherlands from storm surges.

(b)

GEOGRAPHY IN ACTION

Renewable Landscapes

It is said that the great Spanish author Miguel de Cervantes was inspired by the windmills of La Mancha to write his masterpiece *Don Quixote*. Tourists continue to be inspired by the iconic white silos with long wooden arms—*molinos* in Spanish—spreading out over the windswept, dry steppe of La Mancha in central Spain. These old windmills were used to mill the grains, such as wheat and barley, that were grown on the dryland fields.

Today, Spain's landscape is dotted by electricity-generating wind turbines (Figure 4-2-1). In 2011, Spain had the fourth-largest installed capacity of wind power in the world, behind the much larger countries United States, China, and Germany. The consistent winds in many parts of Spain help to explain why producers install wind turbines, but government policies have also been instrumental. Like Germany and other leaders in renewable energy, Spain has a system of incentives that rewards producers for generating electricity from renewable sources. These incentives are called "feed-in tariffs." Spain's economic challenges since 2009 have prompted the government to rethink its generous renewable energy subsidies that attracted substantial international investment in the sector. A new law taking

▲ FIGURE 4-2-2 **Rapeseed crops in northern Germany.** Rapeseed, or canola, blooms brilliantly in the late spring.

effect in 2013 drastically cut subsidies, including 300 million euros that would have gone to the wind sector. With governments around Europe pursuing **austerity measures**, or cutting budgets in an attempt to slow the growth of government-owned

debt, renewable energy around the region faces an uncertain future even as more stringent targets for using renewable instead of fossil fuels take effect.

Windmills and solar panels are found across Europe. The increasing use of biofuels to replace petroleum-based fuels has also had a profound impact on the region's landscapes. Several European countries require that a certain percentage of all transportation fuels sold come from renewable sources, mainly plant proteins (hence the term "biofuels"). Perhaps the most visually dramatic biofuel source is rapeseed (called canola in North America). The seed of the flowering plant produces an oil that can easily be converted into biodiesel. In the landscape, however, it is most noticeable for its brilliant yellow flowers that blanket farm fields in northern Europe during late spring (Figure 4-2-2). Just as some Europeans object to the proliferation of windmills and solar panels, villagers in Somerset, England, have objected to the yellow rapeseed flowers disturbing the traditional landscapes of the region.

Source: http://www.enerscapes.eu/

▲ FIGURE 4-2-1 **A wind farm in Malaga Province, Spain.** Wind turbines dot the Spanish landscape.

comfortable conducting business, telling jokes, and raising their children in the first language they learned. But there are other parts of culture that create shared meanings, and one that is near and dear to many in Europe is food systems.

Religion

Cultural features define and distinguish Europe from neighboring regions. One of these features is the legacy of Christian religious traditions. During that cycle trip from Glasgow to Athens, the most imposing and architecturally noteworthy structures are the churches and cathedrals you encounter in every village, town, and city. You will also have passed through the three dominant Christian regions: **Protestantism, Roman Catholicism,** and **Eastern Orthodoxy.** Although the dominant trend in Europe today is **secularism**—many of those churches will be more or less empty on any given Sunday—the location, size, and opulence of European churches is a good indication of the centrality of religion to everyday life over two millennia. Christianity originated in the Levant (Eastern Mediterranean) but spread rapidly to Europe thanks to the Roman Empire's infrastructure networks and the traveling proselytism of the Apostles and other early Christians. The Romans, once deeply skeptical of Christianity, came to recognize the sway it carried among the empire's downtrodden peasants. Fearing their own demise at the hands of popular insurrections, the Roman elite eventually adopted Christianity as the official religion. The period following the decline of the Roman Empire witnessed power struggles between Christians and non-Christians, but by around 1000 C.E., the Roman Christian Church was once again well established throughout much of Europe, and it was around this time that it split into the Roman Catholic Church in the West (centered in Rome), and the Eastern Orthodox, or Byzantine, church in the East (centered in Constantinople, now Istanbul). In spite of key doctrinal and administrative differences between the two, the basic teachings and cultural features were similar. Much later, in the sixteenth century, another schism occurred when the Protestant movements in Germany, Switzerland, and England resulted in new Christian faiths such as Lutheranism, Anglicanism, and Presbyterianism, which are still prevalent in much of northern Europe and the British Isles (Figure 4-13).

Religion shaped the human geography of Europe in many ways. During the early Middle Ages, Roman Catholic clergy began to use religion as the key feature distinguishing Europe from non-Europe. They were asking many questions, but a key one was how to engender allegiance to the church in Rome among a diverse and geographically expansive area. As intellectuals with limited knowledge of the world beyond a fairly small area, they were also interested in creating a logic for how the world was organized. An enduring artifact of this period is the T+O map showing the world neatly divided into land masses, rivers, and seas, with Europe occupying one quadrant (Figure 4-14). Another is the idea of *christianitas*. To early European Christian elites, Europe was wherever the Christian realm, or christianitas, had spread. God goes wherever Christians went, according to widely held belief, and over centuries the expansion of Catholic and Orthodox influence by means of missionaries and military expeditions sought to establish christianitas as far as possible. Far from simply a historical curiosity, this mentality endures today: Turkey cannot become an EU member, argue some, because its people do not share the same cultural values. Likewise,

no one is seriously proposing that a country such as Morocco be admitted to the EU, even though it is practically a stone's throw away across the Strait of Gibraltar.

Europe is not a continent. Geologically, it is more a subcontinent. Culturally, it is more an idea, rooted as much in Europeans' attempts to define themselves against outsiders as in anything else. While christianitas may have been compelling to European leaders over the centuries, Europe has never been religiously homogeneous. While a monk named Beatus was drafting the first T+O map in Asturias in northern Spain, southern Spain was controlled by the Moors, North Africans who brought Islam with them to Iberia and created a flourishing civilization until they were largely forced out by 1492. Southeastern Europe reflects the influences not only of the Eastern Orthodox Church, but also of Sunni Islam in places such as Bosnia-Herzegovina and Kosovo, where recent armed conflicts have highlighted both the region's diversity and just how elusive peaceful coexistence between different ethnoreligious groups can be.

> ### Stop & Think
> Why is Europe not really considered a continent by geographers?

Most large cities in Europe, including London, Antwerp, Paris, Venice, Frankfurt, Berlin, and Warsaw, had large communities of Jews, and rural areas of what is today eastern Poland (once part of the Russian Empire) were home to a sizeable Jewish population in the area sometimes called the **Jewish Pale.** Life was rarely easy for Europe's Jews, since they were a minority in an emerging territorial logic of cultural nations, but the most tragic part of Jewish history was the Holocaust, during which the German government and its collaborators systematically persecuted Jewish populations and murdered 6 million Jews, most of them from Eastern Europe. In spite of the Holocaust, called Shoah in Hebrew, there are still vibrant Jewish communities in some of the larger cities of mainland Europe, including Berlin and Paris, and especially in London.

One of the dominant trends of recent times related to the cultural geography of religion is secularism. Secularism can mean both the societal trend away from the active practice of religious traditions and the belief in a separation of church and state. In a 2010 Eurobarometer poll less than half of Europeans surveyed trusted religious institutions, and religion was ranked very low among the "most important values," below such things as human rights, peace, democracy, solidarity, and even self-fulfillment. In an earlier poll taken in 2005, some European countries such as Romania, Portugal, and Poland registered large majorities who believe in God, but only 23 percent of Swedes and 16 percent of Estonians said the same. This is reflected anecdotally in visits to the religious structures across the continent, especially in northern Europe. Christmas and Easter in cities like Hamburg, Germany, or Oslo, Norway, are viewed as days of family and rest, but services in those cities' many lovely Protestant churches are likely to be sparsely attended. As in other world regions there is a rural-urban divide, with more rural residents likely to be believers and regular churchgoers than urban dwellers. Standing in the naves of the Durham Cathedral in northern England, Rouen Cathedral in France, or Cologne Cathedral in Germany one marvels at the engineering advances,

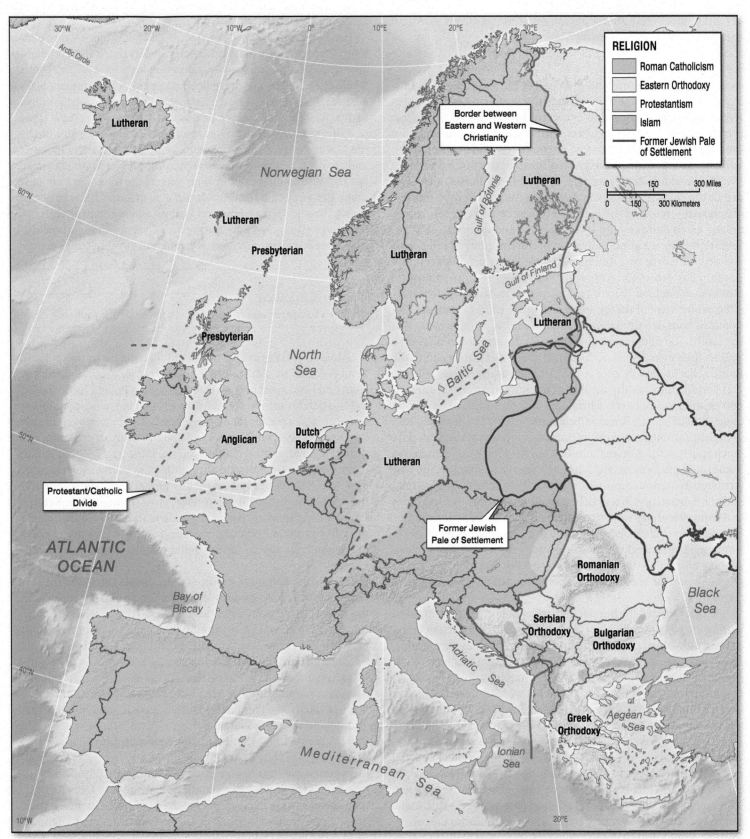

▲ **Figure 4-13 Major religions of Europe.** There are three dominant religious traditions in Europe, but also several important smaller religions. The greatest religious diversity is found in the large cities (largely as a result of immigration) and in southeastern Europe. Data from http://www.jewishvirtuallibrary.org/jsource/History/pale.html.

▲ **Figure 4-14 T&O map from the eighth century.** Early scholars neatly divided the known world into landmasses, rivers, and seas, with Europe occupying one quadrant.

labor, and resources that went into these structures. Seeing them mostly empty on a Sunday morning provides stark evidence of Europe's move away from religion, especially when considered in light of its deeply religious past.

> **Stop & Think**
> ⇨ What is the role of religion in Europe today, and how has religion made its mark on the landscape?

Language

If you are beginning to sense a trend of "threes," language provides yet another example. Your diagonal trek across Europe traverses through the major language regions of Europe: Germanic, Romance, and Slavic languages, three subdivisions of the Indo-European language family (Figure 4-15). There is no perfect alignment of the borders of climate, religion, and language regions. It is too simplistic to say that an area with Eastern Orthodox religion matches up with the Slavic language family (Poland, Czech Republic, and Slovakia, for instance, do not match such a division). But dividing Europe geographically into three is a useful way to conceptualize European space. It is also necessary to acknowledge the climate zones, religions, and languages that do not fit the pattern of threes. As with religion, some languages spoken in Europe do not fit the dominant traditions. These include Basque, spoken in an area straddling the Spanish-French border, and the two languages of the Finno-Ugric language group. Basque, Finnish, and Hungarian are not part of the Indo-European language family. Celtic languages (Welsh, Gaelic), and Latvian, Lithuanian, Albanian, and Greek are

Indo-European in origin but do not fall into the three big groupings. Striking recent developments in the geography of European language are the rise of **multilingualism** and the increasing use of English as a **lingua franca.**

The linguistic geography of contemporary Europe is a snapshot in an evolutionary process of language development. The map of principal languages in Europe shows significant correspondence between the nation-state borders and languages spoken, but there are also many places where this is not at all the case. Why is this? First, the logic of nation-states marries a cultural idea of a nation based language or ethnicity, and a political idea of a sovereign state with well-defined borders. If a state controlled a particular territory, it sought to make sure that the same basic language was spoken within that territory by whatever means at its disposal, such as public schools. Second, rulers sought to ensure that territories speaking a common language were under its control during the process of state formation. In order to understand the language map of Europe, it is essential to consider what existed prior to modern states.

The Roman Empire's armies spread Latin throughout much of southern Europe, often coexisting with languages spoken there before, such as Celtic. Germanic tribes existed throughout much of northern Europe, speaking related tongues. Slavic-speaking tribes were found throughout much of Eastern Europe. Over hundreds of years, as state power was consolidated, languages also became consolidated and standardized. The highly standardized French spoken in France, for example, began as a dialect in the Paris Basin. It would not have been understood by a villager in Marseilles in what is today southern France 400 or 500 years ago. But as the rulers brought territories under their control, they introduced a standard French, their French, to their newly conquered lands. Over time, regional variations in the Latin-based dialects gave way to what is now one of the more standardized languages in Europe, though there are still some variations. In spite of the kings' best efforts, political authority was not able to enforce linguistic conformity everywhere within France's growing territory. Brittany, at the far western edge of France, has a significant population of speakers of Breton, a Celtic language. In 2011 the United Nations Educational, Scientific and Cultural Organization (UNESCO) classified Breton as "severely endangered" because of the precipitous decline in speakers, now numbering only around 250,000. As with other Celtic-speaking populations in Wales and Ireland, there is some effort to increase the number of young people speaking Breton.

One can find comparable linguistic histories in other parts of Europe, though the role of political authority is not always equally strong. Serbo-Croatian, which is spoken in much of the former Yugoslavia, was standardized by a movement of writers and linguists in the nineteenth century, and then adopted officially when a unified kingdom was established in the early twentieth century. As a result of standardization efforts, the various Slavic dialects spoken in the Western Balkans—Serbian, Croatian, Montenegrin, Bosnian—are generally considered one language but with some internal differences. Serbo-Croatian also provides evidence of how language is political. The Serbian state adopted Cyrillic script as the official writing system, while in Bosnia-Herzegovina and Croatia the Latin alphabet is used. Thus Срећан poђeнgaн! and Sretan rođendan! (Happy Birthday!) sound identical and mean the same thing, but obviously look very different on paper.

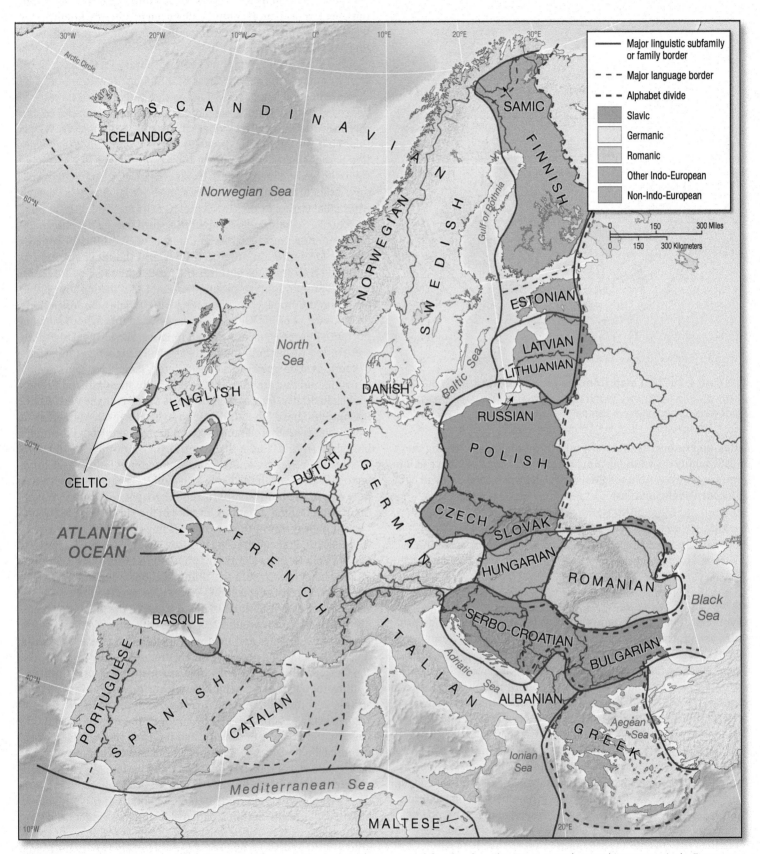

▲ **Figure 4-15 Principal languages of Europe.** There are three major language subfamilies found in Europe, and several important Indo-European and non-Indo-European smaller languages.

◄ **Figure 4-16 Finland has two official languages, Finnish (upper left) and Swedish (upper right), and one unofficial one, English.** This pharmacy in central Helsinki provides a stark reminder of linguistic diversity in Europe. Finland, once partly controlled by Sweden, has a sizable Swedish-speaking minority. More recently, the lingua franca of English has made significant inroads as Finland has become thoroughly integrated in the European and global economies.

English is considered by linguists to be a composite language, combining characteristics of both Germanic languages (especially sentence structure and verbs) and Romance, or Latin, languages (especially nouns). Because the underlying structure of English more closely resembles the Germanic languages, it is classified as such here. Riding from Glasgow to Athens 30 years ago, you would have been hard-pressed to find many English speakers after you crossed the English Channel. This has changed quite dramatically. English is a required school subject in most European countries. Some universities in non-English-speaking countries now offer extensive courses in English, especially in science and engineering. The reason for this is not primarily to cater to English native speakers, but rather to provide common ground for people from around the world. Settling on a lingua franca, or a common language as a medium of communication among people of different language backgrounds, can be both an official act and a natural evolution. European political and economic integration has provided additional incentives for a common language in Europe. How the most widely used lingua franca came to be English relates to economic and geopolitical factors—especially after World War II, English came to be a global language of commerce, and American, Canadian, and British soldiers were stationed in Central Europe for many decades. While frustrating to many nonnative speakers for its intricacies and inconsistencies, English is nevertheless widely perceived to be easy to learn in basic form, at least when compared to some alternatives. Presumably even the Finns themselves would not suggest that Finnish be a lingua franca in Europe (Figure 4-16); who other than the Finns would be up to the task of learning it?

Food and Agriculture

What we eat is determined by a number of factors, but it is food availability and learned habits that are perhaps most crucial. The availability of certain foodstuffs in modern times has much to do with what we can find in the neighborhood grocery store or local restaurants. Prior to large-scale industrial production of food staples, long-distance transport, and refrigeration, climate, soils, and available species were the most important factors determining geographical differences in diet. In northern Europe, preserving summer harvests for a long winter was a question of life or death. Where soils were unproductive but seas were full of fish, the diet reflected this. Like other parts of European cultural geography, the geography of food and diet reflects local and regional traditions as well as extensive borrowing from far afield. In much of Europe, an industrial revolution in agriculture occurred in tandem with the revolution in manufacturing, and the labor that would have otherwise been needed on the farm was used instead in factories. There were also massive environmental changes that accompanied the agricultural revolution in Europe, and recent decades have seen movements countering industrial food production.

Take a traditional Estonian meal, for example (Figure 4-17). The Baltic Sea and a landscape dominated by glacial lakes mean that fish is a dietary staple. Harsh winters historically meant that fishing was not always possible, so smoking or pickling fish—preserving fresh-caught fish for future consumption—was a matter of survival. Vegetables well suited to the region's short summers and marginal soils such as beets and potatoes (originating in South America but brought by traders hundreds of years ago), are grown along with cold-weather grains, especially rye. Fresh or preserved cabbage (sauerkraut) accompanies most traditional meals. Some cooking traditions were borrowed from others, including Germans to which Estonians were linked by a trading network, the Hanseatic League, dating back 800 years. More recently, Estonia was part of the Soviet Union from 1940 until 1991, and Estonians acquired a taste for spicy Georgian and Azerbaijani foods. Yet the typical student in Tartu or Tallinn, the two largest university cities, now has a range of cuisines

▶ **Figure 4-17 A typical summer Estonian meal.** While physical environment does not dictate diets, it shapes cultural traditions. Roadside stands along Lake Peipus, which separates Estonia from Russia, sell smoked perch, which goes well with dark bread, cucumbers and pickles, summer fruits, and of course vodka.

to choose from as a result of globalization. Pizza restaurants and kebab stands outnumber pork and potato joints on most town squares.

The Mediterranean diet is perhaps best known to North Americans. The rugged Iberian terrain can support large animals, such as cows and pigs, but not on the same scale as on the North European Plain. Therefore the most common fat used in food preparation on the Iberian peninsula, and throughout most of southern Europe, is not lard or butter (animal fats), but rather olive oil. Olive trees are well suited to the marginal soils and climate conditions of Spain and Portugal. Seafood is the most common protein source and the Portuguese consume the most fish per capita in Europe. The Portuguese once had a vast sea empire, and brought back a variety of foods and especially spices from abroad including peppercorns, cinnamon, and saffron. Grapes grow well in the mediterranean climate, and thus a culture of wine-making emerged. Port wine and *vinho verde* are familiar around the world, but there is a rich selection of other wines. A typical supper might include a stewed preparation of beef, chicken or pork, wine, garlic, and spices. Other parts of the Mediterranean have their own variations, but with similar themes reflecting the availability of certain agricultural products. Italy's famed spaghetti and tomato sauce is, in fact, a product of borrowing: pasta was brought by Marco Polo from China and tomatoes are native to Mesoamerica and were introduced to Europe by the Spanish.

The wave of agricultural mechanization that swept Europe did not reach all parts of the region with equal force. Until a decade or so ago, it was not uncommon to see horses working in farm fields in rural Poland, which had been part of the Soviet-dominated Eastern Bloc from 1945 to 1989. During that time much of the agricultural production was organized into state-run "collective farms" in an attempt to coordinate production and increase efficiency. The end of the Cold War brought an end to collectivization, but one can still see evidence of collective farms in the landscape across postsocialist Europe (Figure 4-18). Even in the twenty-first century, parts of the former Yugoslavia and Moldova still have largely agrarian economies, and labor-intensive agricultural practices can still be observed (Figure 4-19).

Food and agriculture are deeply emotional topics to Europeans. One only need observe the debates over EU attempts to reduce farm subsidies, or devise food safety rules that offend local sensitivities about a specific type of cheese or meat product, to see how food and agriculture rival—to some surpass—language and religion

▶ Figure 4-19 **Industrial agriculture has not yet reached all corners of Europe.** In rural Moldova, Europe's poorest country, it is not uncommon to see agriculture being practiced as it was a century ago, but European integration is making these scenes less and less common.

as sacred. However, as in many other world regions, globalization, affluence, and contemporary lifestyles have distanced Europeans from traditional diets. Europe is the largest region in terms of revenue for the hamburger chain McDonald's. The typical Bavarian teenager would just as soon eat pizza than a typical meal of *schweinshaxe* (roast pork knuckle) with sauerkraut (Figure 4-20). Perhaps more traumatic for the Bavarian self-understanding, the meal is likely to be washed down with a coke rather than a liter of beer. Per capita beer consumption has been steadily decreasing in Germany, and hundreds of breweries have closed their doors. Tastes have changed, as seen from the selection of restaurants in any city, or from the large greenhouses that provide fresh produce to consumers during the long, dark winters.

At the same time, there is a notable trend of "back-to-the-land" movements, organic farming, and slow food in Europe. The slow food movement was founded in Italy in the 1980s by Carlo Petrini to protest the opening of fast-food chain restaurant McDonald's adjacent to the Spanish Steps in Rome. The goals of the movement, which has now spread to many other parts of the world, are largely point-by-point rejections of the innovations of industrial agriculture: growing heirloom seed varieties instead of hybridized or genetically modified, often tasteless varieties, from industrial seeds; promoting food that is appropriate for the local ecosystem; local food systems rather than long-distance trade of food; and reducing the use of artificial fertilizers and pesticides. Growing cooperatives are now common sights in the outer-ring suburbs of large European cities.

Origins of the Nation-State

The world's modern map of nearly 200 independent states owes its organizing logic to Europe, where the modern state system emerged. The idea that a state has control over a well-defined piece

▲ **Figure 4-20 Bavarian sausages and beer.** A trip to Munich, home of the Oktoberfest, would not be complete without trying a plate of Knackwurst (pork sausage) on pureed potatoes with fried onion garnish, washed down with a Munich lager.

of territory, demarcated by borders, and that outside powers should have no jurisdiction within another state's territory—a principle called sovereignty—is central to international law. The idea that a particular nation, or intergenerational group based on ethnicity (language or religion), should have control over its own affairs and should govern its own territory is also a European invention. This logic of nations and states, typically combined in the all-encompassing term **nation-state,** emerged in Europe over hundreds of years and was exported to the rest of the world largely as a result of imperial expansion and decline. If our modern political map owes its existence—for good and for bad—to Europeans, you might expect that the map of European nation-states is both stable and uncontested. But persistent and ongoing challenges to the supremacy of the nation-state ideal in Europe suggest how messy and imperfect this organizing principle is.

Take a look at modern Italy. For many non-Italians, it is easy to imagine a collective Italian identity. In most of what is colored as Italy on the map, Italian is spoken. We speak of "Italian cuisine," really a highly varied set of typical diets, but with common basic elements such as pasta. A central government in Rome sets rules, collects taxes, and spends those tax revenues on projects largely confined to the territory of Italy. No Spanish citizen living in Barcelona expects the Italian government to provide health care or fix a crumbling street there, just as no one living in Italy expects a tax collector from Japan to show up at the door. The list could go on, but most of us have internalized the nation-state map of Europe as

being based on natural divisions. These divisions along cultural and political lines appear deeply rooted and may seem unchallenged.

But the case of Italy shows how problematic such a simple approach to Europe is, and by extension to the rest of the world. Italy as an independent nation-state within its modern borders did not exist until the mid-nineteenth century. Before that the Italian peninsula consisted of several independent city-states, kingdoms, and semi-independent papal duchies. The Italian that most citizens speak today existed mainly in written form in the nineteenth century, as a literary language. Most people spoke local or regional dialects, and the language spoken in the north would have been nearly incomprehensible in the far south. When the Italian state was created in the 1860s, one of the leaders of the movement wrote "Italy has been made; now it remains to make Italians."

Making Italians proved easier said than done, but economic and political unification proved to be most difficult. To this day, the **Mezzogiorno** (literally, "midday"), the southern half of Italy south of Rome, is criticized by many northern Italians as being economically backward and controlled by organized crime (the Mafia). A movement in the wealthier northern regions has in recent decades been mobilized by resentment over paying the bills for the Mezzogiorno, and the Northern League political party has in the past sought independence for a large chunk of northern Italy, a made-up region they call "Padania," so that they might stop subsidizing the southern part of the country. The Italian example illustrates the cracks in European unity by showing how seemingly rock-solid nation-states were often born out of political expedience, and challenging the notion that shared attributes unite Europe into a cohesive region.

Stop & Think

▶ What are nation-states, and why are they an imperfect reflection of reality?

Population Contours

The greatest economic and social challenge facing Europe is demographic change. Europeans are having fewer and fewer children, and populations are aging and in many cases shrinking. The demographic transition in Europe is comparable to that in Japan. The **total fertility rate (TFR)** in no EU country is greater than 2.1, which is considered the replacement rate for a population. Even when immigration is factored in—the only factor that keeps the overall population increase in the region slightly positive—the EU predicts that the population will increase until around 2025, after which it will decline (Figure 4-21). Countries such as Bulgaria and Romania, formerly Communist states undergoing structural changes and still sending emigrants to other European countries, are forecasted to shrink by 21 percent in the case of Bulgaria and 11 percent in Romania by 2030.

The reasons for these demographic shifts are complex. There is some correlation between economic development and fertility. Ireland had the highest TFR in Europe 50 years ago, but rising prosperity and more women in the workforce have contributed both to delayed childbearing and fewer children. Ireland still has the EU's highest TFR but it is still less than the replacement rate. Urban dwellers typically have fewer children than rural residents, and Europe is one of the world's most highly urbanized regions. Individual decisions have also changed over time. In Communist countries,

▲ Figure 4-21 **Population distribution of Europe.** Take note of the areas of highest and lowest population density.

children were the ticket to larger apartments; under capitalism, children are seen as blessings but also expensive commitments. Self-realization and exploring the opportunities afforded by prosperous societies with universal education also shape decisions, and increasingly there is little shame in either waiting until age 30 or 40 to bear children, or bypassing traditional family formation altogether.

Such demographic realities have implications for the social models as they have developed in Europe. Economic growth is strongly tied to population growth. Universal socialized health care, which is the norm in Europe, depends on sizeable populations of healthy individuals to finance it. Old-age pensions, in which current retirees are financed by current workers, require that the working population be several times larger than the nonworking population. This dynamic is approaching critical levels in many parts of Europe. There is growing realization among governments that measures must be taken to maintain current standards of living. In addition to immigration policies, **pronatalist policies** are being enacted to encourage childbearing. These include generous maternity leave, heavily subsidized daycare and schooling, and outright cash payments to mothers.

Stop & Think

▷ What is the demographic situation in Europe today, and how are Europeans adapting?

The Economic and Social Development of Europe

According to any number of measures, Europe is one of the most highly developed regions in the world. It is home to some of the wealthiest economies, both in absolute and per capita terms. Germany is the world's fourth-largest country by Gross Domestic Product (GDP), and the EU if treated as a country would be the single largest economy. According to the United Nations, the top four countries ranked by GDP per capita in 2011 were Liechtenstein, Monaco, Luxembourg, and Norway, all in Europe. These are each unique, small population countries with special economic circumstances, of course. What is most striking is not necessarily the wealth, but rather the lack of abject poverty in most of Europe. It is quite rare to see a homeless person in a European city, even in the poorest parts of Europe. In some respects, this is the great success story of Europe—the region's economic prosperity is not accompanied by widespread social distress or poverty. While some would argue that Europe's prosperity came at the expense of other parts of the world, it is worth noting that European countries also rank highest in the world in the amount of development aid to poor countries. As a percentage of the national income, the top nine countries in development aid and charitable giving to poor countries in 2011 were European. There are many indicators for human well-being and development, and by most measures Europe consistently has some of the healthiest, best educated, and longest-lived people. Violent crime and gun deaths are rarer here than almost anywhere else. Nevertheless, unemployment in many parts of Europe is chronically high, and many of the industries that fueled the Industrial Revolution over a century ago no longer employ large numbers. This section considers the economic and social development in Europe in tandem, and it makes clear that they are closely linked.

Core and Periphery in Economic and Social Development

Earlier, Italy was used to illustrate how Europe's cultural geography is not a perfect mirror of the national boundaries that developed over time. The economic geography of Italy exhibits different development paths as well. The Mezzogiorno lags behind in terms of economic development when compared to northern Italy. A **core-periphery dynamic** exists in Italy, in which the northern parts of the country around the industrial centers of Turin and Milan and the historically wealthy trading cities of the Po River valley, form the core while the still largely agrarian Mezzogiorno forms the economic periphery. Both money and labor flow between the core and periphery; the Mezzogiorno has been receiving transfers of wealth from the north for over a century, and many Sicilians and Puglians can be found living and working in the large cities in northern Italy. Southern Italians also have historically moved farther afield: most Italian immigrants in the United States, for example, came from the Mezzogiorno, and the original wave of guest workers to Germany during the 1950s also came from southern Italy.

Geographers look at such core-periphery dynamics at multiple spatial scales. There is not just a core-periphery dynamic in Italy, but also within Milan (at the scale of the city) and within the EU, and even globally. The economic core of Europe has shifted over time. During the Middle Ages, the wealthiest places were concentrated in the Po River valley and the Low Countries, with cities such as Venice and Bruges still reflecting the opulence of a much earlier era. The Industrial Revolution shifted the economic center of gravity to a zone stretching from northern England through northwest Europe across the Alps into the industrial north of Italy, a region often referred to by geographers as the **"Blue Banana."** The term was coined to refer to the area of dense population, infrastructure, and industry (shaped like a banana) that transcended national borders and served as an economic motor for the world. Even though much of the heavy industry once found in this belt has long since been closed or relocated, the large labor pool, along with capital investments in infrastructure (highways, railroads), schools, and other parts of the built environment have legacy effects. There are also parts of the economic core that were not geographically located in the Blue Banana but are nevertheless important. Some of the world's wealthiest countries on a per capita basis are Nordic: Sweden, Norway, and Finland. Bavaria and the conurbations of Paris and Madrid are also notable.

Then where is Europe's economic periphery? The Balkan peninsula is home to the poorest countries such as Romania, Moldova, Montenegro, and Bosnia-Herzegovina. The far southern part of the Mediterranean zone never experienced large-scale industrial development and remains relatively poor except for a few tourist enclaves. Although there are powerful forces of inertia in terms of regional development, the core-periphery map of

Europe is not entirely static. Until the 1990s, Ireland would have been considered squarely in the economic periphery, but in the last few decades (at least up until the financial crisis of 2009) the Republic of Ireland was a "tiger economy" experiencing rapid economic growth. This growth coincided with economic liberalization policies pursued in the 1980s, and the country also benefitted from large fiscal transfers from the European Union, of which it has been a member since 1973. Greece, meanwhile, has faced substantial economic bad fortune in recent years. In October 2012 the country registered an official unemployment rate of almost 27 percent, the highest ever recorded in the EU. Nearly 57 percent of young Greeks between the ages of 15 and 24 were without work that month.

While there is no perfect correlation between economic performance and social well-being within a region, there are notable links between the two. Most of Europe enjoys low infant mortality rates, very high literacy and school completion rates, and long life expectancies. Yet in a number of postsocialist countries, including Poland, Moldova, Romania, and Hungary, more than 5 percent of all deaths are attributable to alcohol abuse; in every western European country, it is less than 2 percent. Lithuania, Latvia, and Hungary have among the highest suicide rates in the world.

The integration of Europe has addressed the region's economic and social disparities. Part of the guiding philosophy of the EU was to encourage trade and other types of exchange between the richer and poorer parts of the Union, but there was also an acknowledgment that disadvantaged areas would need a hand up. The EU's **Common Agricultural Policy (CAP)**—representing nearly half of the EU's budget until 2013 when its share decreased substantially—was originally designed to help small farms in countries such as France and Spain. More recently, though, it has provided a large cash stimulus to rural areas in the new member states of Central and Eastern Europe (although the magnitude of subsidies remains tilted heavily toward old member states). Large sums have also been spent on regional development and infrastructure projects in the new member states as part of a far-reaching strategy to reduce the role of uneven economic development in the EU. Infrastructure includes not only road and railroad improvements, but also high-speed Internet connections, viewed as essential to future economic competitiveness. Areas eligible for such funds are generally the postsocialist new member states, parts of Greece, the Mezzogiorno, Wales in the United Kingdom, and the southern half of Iberia. New infrastructure projects are usually accompanied by signs prominently displaying the EU flag and the words "Financed by the European Union" as a constant reminder to local citizens.

Poland: From Periphery to Core?

Until 1989 a Communist country, Poland has been an EU member since 2004 and is now its sixth largest economy. It is the only EU country that did not see a decline in GDP following the 2009 economic crisis. In many respects, Poland is the most remarkable economic success story of Europe's formerly Communist countries, but its current state of development is even more noteworthy when one considers an even longer historical period. As countries elsewhere in Europe were consolidating into nation-states, Poland was partitioned in the late eighteenth century by powerful neighboring empires, Russia, Germany (then Prussia), and Austria-Hungary. Poland effectively did not exist again as a sovereign state until the end of World War I, but by World War II the country's fate was impacted by conflict, this time between Germany and the Soviet Union. Poland's current borders resulted from Germany's defeat in 1945, but the end of the war ushered in a four-decade period of Soviet-style communism.

The introduction of free market capitalism and liberal democracy in the early 1990s created yet another upheaval. Unlike neighboring East Germany, which after its 1990 reunification with West Germany received over US $1 trillion in government investments in infrastructure and economic development, Poland was more or less left to its own devices. In a relatively short time and with the help of large amounts of foreign investment, Poland's manufacturing sector developed into one of the strongest in Europe, while its banking sector became the strongest of any formerly Communist state. There is a core-periphery dynamic to Poland's economic development since 1989. While Warsaw, Gdansk, and the regions bordering Germany are the most prosperous and have the lowest unemployment, areas along the eastern border with Ukraine and Belarus are still largely agricultural and have persistently high unemployment rates of over 20 percent. Unlike heavy industrial regions in parts of Western Europe, Poland's coal and steel sectors continue to be important employers. Katowice, the largest city in the major industrial urban agglomeration in Silesia in southwest Poland, is home to major steelworks and car factories and has one of the lowest unemployment rates of any Polish city.

The most dramatic evidence of the country's transformation is seen in the capital, Warsaw. Its skyline was still dominated in the late 1990s by the Palace of Culture and Science, a Stalinist, Soviet-style skyscraper (Figure 4-22). The city has since experienced a

▼ **Figure 4-22 Palace of Culture and Science, Warsaw, Poland.** The building in the center is referred to irreverently by locals as "Stalin's syringe" and some even less polite names. Most Poles do not lament the new glass and steel high-rises sharing the skyline with the Stalinist building, since it symbolized Soviet domination during the post–World War II era.

building boom. Now the Palace of Culture and Science is surrounded by glass and steel high-rise buildings and cranes, and the daily farmer's market that once existed on the large concrete expanse around the palace is nowhere to be seen.

Industrial Development

For centuries, Europe's colonial empires ensured a high degree of economic development for countries such as Portugal, Spain, Great Britain, France, and the Netherlands. However, it was the early and massive industrialization—partly fed by raw materials from overseas colonies—that gave the region economic advantages that persist to this day. Many of the innovations in manufacturing and transportation that we take for granted came from Europe, as did many of the social and institutional innovations designed to deal with the impacts of industrialization on such a large scale. Steam engines, invented by Scotsman James Watt, first found application on factory floors and then later as a means of locomotion, moving coal along tracks from mine to factory and finished products to ports in northeast England. Railroads replaced horse-drawn wagons and canal boats, and spread quickly to mainland Europe along with the industrial production model. The modern automobile is largely credited to a tinkerer in Mannheim, Germany named Karl Benz, and his wife Bertha, who financed his tinkering habit and promoted the invention on a long-distance journey through southwest Germany in 1888.

The industrial model was based on access to raw materials, energy, markets, and labor. The major commodities were those that could be mass-produced in large factories: textiles and iron (later steel, an alloy of iron and other metals) and whatever could be made out of those things, such as clothing and blankets, ships and rails. Where there were raw materials and energy, but very little labor, people were brought in and resettled with the promise of factory work and a daily wage. The map of European industrialization is very strongly linked to the map of coal, since it was expensive to move the fuel long distances. Sometimes industries located where coal was not readily available but where there were good transport linkages and access to markets, in cities such as Paris, Lyon, Berlin, and Madrid. Unlike farming or small-scale production, industrial factories needed to be in cities, with the labor pool close by.

Thus the Industrial Revolution was also an urban one, and soon new cities sprang up, and historical cities burgeoned in size with the arrival of factories, transport linkages, and masses of humanity. Workers who crowded in cities such as Brussels and Barcelona sought wages in return for their tedious, unglamorous, often dangerous, work on the factory floor. The rise of many influential social movements of the last 200 years can be linked to the Industrial Revolution, as workers sought rights, higher wages, and more control over their fate. It is no coincidence that Karl Marx and Friedrich Engels based their critiques of capitalism on their observations of the extreme social problems they witnessed in Manchester and Brussels, centers of British and Belgian heavy industry, respectively. Although Communist revolutions did not occur where Marx and Engels might have predicted, governments across Europe introduced a number of social reforms during the height of industrialization and urbanization that continue to form the backbone of the social welfare state to this day. Child labor exploitation was a common feature in the early industrial factories, and child labor restrictions were a government response. Old-age pensions, workplace injury insurance, and health insurance all came out of the new social situations created by industrialization.

Rising standards of living for the new "middle class" also created new types of consumption and new places to do it. Whereas a farmer's hands were never idle, factory jobs eventually provided work weeks limited to six or even five days, and regular hours. In exchange for 40–50 hours of labor, the urban factory worker had pockets full of wages. "Leisure time" enabled working men (and wage laborers were mostly men) to engage in activities such as watching others play sports, sitting in the pub quietly or standing rowdily (the lack of moderation in drinking was a constant refrain), or visiting amusement parks with their families. The term "revolution" is apt, and Europe's industrial geography increasingly came to define its cultural, political, economic, and even entertainment landscapes. And it extensively fouled the environment of the region while also giving rise to social and political movements seeking to overcome these ills.

The Industrial Revolution resulted in economies that were ever less based on agriculture and ever more urban, but this model was copied elsewhere. Vast colonial empires that supplied raw materials, additional labor, and markets for finished goods eventually shrank, especially after World War II. Workers seeking higher wages and better working conditions also opened up possibilities for competition from outside Europe. Europe's economy faced decades of deindustrialization and is now largely postindustrial.

As industries moved out of traditional industrial zones, some regions never recovered and have remained as areas where fiscal transfers form the financial basis for the economy. Others have adapted and retooled. Often this is not simply a choice, but rather reflects structural factors that either enable or stand in the way of economic adaptation. The area around Bilbao, in Spain's Basque region, became a major shipbuilding and steel city thanks to iron ore deposits nearby and coal brought by sea from nearby Asturias. The small fishing village was transformed into the largest city on Spain's Atlantic coast, and its industrial growth lasted until the 1970s when a dramatic deindustrialization began. During the 1980s and 1990s, Bilbao's leaders pursued policies to retool the local economy toward service sector industries such as banking and insurance, knowledge-based high tech, and culture-led regeneration. It commissioned a very expensive Guggenheim Museum, designed by the American star architect Frank Gehry, which became the city's most famous landmark and attracted millions of tourists to the city interested as much in seeing the building as the art inside (Figure 4-23). Old industrial sites on Bilbao's waterfront were redeveloped, old buildings either repurposed for the postindustrial economy or torn down. Bilbao is often held up as an example of successful postindustrial economic adaptation strategies, and these strategies have been copied in other parts of Europe with varying levels of success.

Other postindustrial cities have not been as fortunate. In spite of decades of central government investments in shifting

▲ **Figure 4-23 The Guggenheim Museum in Bilbao, Spain.** The building has become a tourist magnet in Bilbao, a city once known mainly as an industrial port.

the economies of northern England's industrial powerhouse cities, Manchester, Liverpool, Hull, and Blackpool have been struggling. Liverpool's first ring of industrial suburbs, neighborhoods like Toxteth, are some of the poorest places in Europe, with very high levels of poverty, poor health, and drug use (Figure 4-24). While the port city has recently capitalized on its legacy in the arts, especially as the home of the Beatles, and its amenities such as soccer teams FC Liverpool and Everton FC, Liverpool's unprecedented expansion over centuries has been accompanied by a recent steady, painful economic decline. Glasgow, Scotland, exhibits similar characteristics. Britain's wealth is more concentrated in the areas around London, including high-tech university cities like Cambridge and Oxford, than at any time in the country's history. Wallonia, the French-speaking southern part of Belgium, and the far north of France around Lille, have also struggled with deindustrialization, with the once mighty coal-and-steel Walloon city of Charleroi now facing high unemployment, outmigration, and derelict factory landscapes. Gelsenkirchen, Germany, in the Ruhr River region, is among the poorest cities in Germany, and in the wake of German reunification in the 1990s and 2000s when hundreds of billions of euros were being spent to rebuild eastern Germany, many in the Ruhr Region expressed their frustration that they were being left out to dry as mines and factories closed. The corner pub (*Eckkneipe* in German), the symbol of leisure time, was the factory worker's refuge, and perhaps the most potent symbol for deindustrialization is the shuttering of thousands of pubs and bars across Europe. The traditional pub has been labeled an "endangered species" in Britain.

Postindustrial Geographies

In spite of decades of deindustrialization, Europe's economy is not in the cellar. So what do people do, and how has postindustrial Europe's economic geography changed? (see *Geography in Action*: The Ruhr River Region's Transformation to a Postindustrial Economy) It is important to point out that not all manufacturing has left Europe. Areas with highly specialized manufacturing sectors that rely on advanced science and engineering have flourished, especially in the western parts of the region. Baden-Württemberg, a province in southwest Germany, is home not only to Porsche,

Daimler (of Mercedes fame), and Bosch, but also hundreds of **small- and medium-sized enterprises (SMEs)** that supply those big companies. None of these industries would thrive without the extremely high level of training provided by state-run universities, technical colleges, and apprenticeship programs. The industrial economy of contemporary Europe is now focused on innovation rather than coal and steel, and the comparative advantages of most European places tend to be their well-trained workforces. Toulouse, a formerly rather sleepy regional administrative and trading city in the far south of France, capitalized on its educational institutions to become a major medical research hub and the center of Europe's aerospace industry as home to Airbus. Its space and aerospace industry clusters have given the area the nickname Aerospace Valley. French-speaking Wallonia was once the major wealth-creating region of Belgium with its heavy industry, but the economic dynamic of the country has been practically inverted with Dutch-speaking Flanders now home to the country's main high-tech clusters. Some European innovation and manufacturing hubs are located in old heavy industry areas (for example Turin, Italy), but these are more likely to be quite some distance from old industrial centers.

The most notable feature of the contemporary European economy is the role of services. These include producer services, tertiary sectors such as banking and finance, insurance, real estate, and business consulting, as well as quaternary sectors such as universities, media, and information technology. In the country that gave the world the Industrial Revolution, the United Kingdom, about 80 percent of the working population is employed in service sector jobs. Similarly high percentages can be found in much of Western Europe, while in the eastern half of the region, countries such as Romania, which entered the EU in 2007, have much lower service sector employment, on the order of 40–45 percent. This reflects a much larger primary sector, mainly agriculture and mining, in the new EU member states. Even Germany, which is best known for its high-tech manufacturing (a secondary economic sector), employs over 70 percent of its workforce in service sectors. The growth in Europe's service sector employment reflects the evolution of the

▼ **Figure 4-24 Toxteth neighborhood of Liverpool.** These condemned row houses are an indication of the structural economic changes in northern England.

economy from primarily agrarian, to industrial powerhouse, to now postindustrial. But it also reflects a highly globalized economy in which Europe's comparative advantages relative to other parts of the world tend to be in service sectors.

Take finance, for example, and the fact that 90,000 bankers work in the City of London, the tiny area at the core of the London metropolitan area. By many measures, London is the largest financial center in the world, where 40 percent of all currency trading and a substantial proportion of the derivatives trading take place. In a country one-fifth the size of the United States and one-twentieth the size of China in terms of population, this suggests that much of the financial heavy lifting in the world is being performed in London. Switzerland and the small state of Luxembourg also have highly specialized and very lucrative banking sectors. Frankfurt am Main's role as the biggest banking center in mainland Europe largely came as a result of geopolitics. After World War II when Germany was divided into West Germany and Communist East Germany, the prewar banking sector, which had been spread through many of Germany's cities, was consolidated in Frankfurt, which happened to be in the West. The most famous German bank, Deutsche Bank, now headquartered in Frankfurt, had been founded in Berlin and maintained its headquarters there until after the war. Another bank that dominated Frankfurt's skyline, Dresdner Bank, as the name suggests was founded in Dresden, but moved to Frankfurt in 1950. It was acquired by competitor Commerzbank in 2009, yet another bank to call Frankfurt home only since just after World War II. Frankfurt is also the home to the European Central Bank, which governs monetary policy for the **Eurozone** (Figure 4-25). Management consultancies, companies that advise other companies on improving profits and efficiency, are also big players in the service sector economy in Europe. The prime real estate in some of the larger cities in Central and Eastern Europe, such as Warsaw, Bucharest, and Belgrade, is often taken by consultancies from abroad who make money by advising private enterprises and governments. But increasingly, high-skilled, English-speaking

▲ **Figure 4-26 Bizkaia Technology Park near Bilbao in the Basque Country of Spain.** Development strategies in Europe often include technopoles.

labor forces in countries such as Hungary and the Czech Republic are making these attractive places for business services to locate to serve their global customer base.

Services also include industries such as tourism and entertainment. Tourism in particular is a major driver in some parts of Europe, and one that leaves its mark on landscapes. Europe is the most visited world region, with over 500 million international arrivals in 2011. Spain, Italy, France, and Greece, in particular, are major tourist draws, with their historical sites, mild winter climates, beaches, and cuisine drawing northern Europeans, Americans, and increasing numbers of Asians for extended stays. Tuscany in northern Italy, Spain's Costa del Sol (Mediterranean coast), and the Greek islands of Crete, Corfu, and Santorini are perennial favorites. Tourism employs local populations in hotels, restaurants, transportation, and a host of other services. In addition to vacationing students and families, amenity-based tourism among retirees is extremely important. Retirees with long productive lives ahead of them, but not compelled to spend long dreary winters in Stockholm, Birmingham, or Hamburg, will spend months in second homes or rental properties in southern Europe. This has given rise to expatriate communities in Spain, Portugal, and Greece where one hears as much English and German on the streets as the local language and where bakeries and butchers cater to expats' longings for what they left behind. Back-to-land vacations, called **agritourism,** have become one of the most lucrative tourism types, with urban dwellers paying tidy sums to harvest grapes or learn cheese-making on farms in Tuscany or in the Dordogne Valley of France.

Another economic trend is the rise of **technopoles,** which are districts with a high concentration of technology-related manufacturing, research, and design. Technopoles often surround large universities or government research institutes, and are perhaps the quintessential manifestation of the rise of a quaternary economy. The M4 corridor near London; Grenoble, France; Milan, Italy; and Munich, Germany are all home to technopoles, and local and regional governments across Europe actively seek to attract businesses and institutions in the technology realm, since technopoles typically bring high-skilled, high-paying jobs with them (Figure 4-26). The area around Dresden in the region of Saxony in eastern Germany has marketed itself as "Silicon Saxony"

▼ **Figure 4-25 Frankfurt am Main skyline, sometimes referred to as "Mainhattan".** Frankfurt is a major banking and service center for mainland Europe, as well as home to one of Europe's busiest international airports. Unlike most large European cities, Frankfurt's central business district houses high-rise skyscrapers. The near 90 percent destruction of the urban core by allied bombing during World War II decimated the urban fabric, and city leaders after the war opted to allow such skyscrapers.

GEOGRAPHY IN ACTION

The Ruhr River Region's Transformation to a Postindustrial Economy

The largest urban area in Germany is not Berlin, Hamburg, or Munich, but rather the Ruhr Region (in German *Ruhrgebiet,* but often called colloquially "Ruhr Pot") (Figure 4-3-1). With around 5 million people living in an area about the size of the Great Salt Lake, the Ruhr Region is also the third-largest urban region in Europe. For much of the twentieth century,

the Ruhr Region was the largest and most productive industrial region in Europe. Its fortunes over the last several decades have reflected the trends of deindustrialization. In recent years there have been signs of economic rebirth, as well as a cultural renaissance, in the region. The postindustrial landscape has been central to these developments.

The Ruhr owes its original boom to a combination of favorable geographical features. First and foremost was the presence of sizable coal seams in the region. There was then the river itself, which emptied into the Rhine River near Duisburg and thus provided an easy route to move goods to market and to bring in enough metal ores to feed the

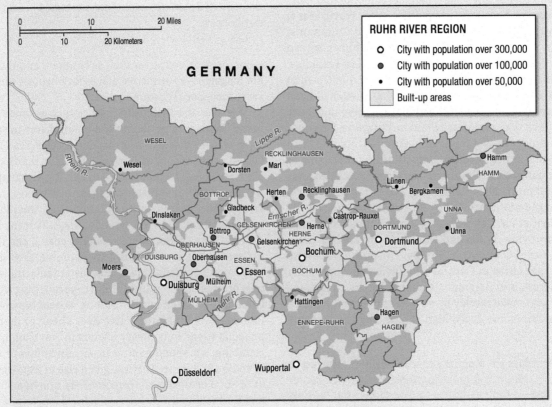

▲ FIGURE 4-3-1 **Past as an industrial hearth, but what does the future hold?** The cities of the Ruhr River Region have struggled with deindustrialization.

(a play on the original "Silicon Valley" in California), in order to capitalize on its research strengths in nanotechnology and microelectronics.

Economic and Social Features of Europe's Cities

More than 70 percent of Europe's population lives in cities, and the region is one of the most highly urbanized in the world. By 2050, the UN expects that well over 80 percent of Europeans will live in cities, although the rate of urbanization is actually slower than in most other regions. This reflects the mature state of Europe's economic development.

Economic Activity

The share of European populations living in cities largely reflects the past 200 years of industrialization, and those parts of Europe that were not heavily industrialized exhibit much lower percentages of people living in cities. The United Kingdom and Belgium have very high percentages of urban dwellers, while Bosnia-Herzegovina and Albania are still largely rural with about half their populations living in cities. Iceland and several microstates, including Malta and Monaco, also have highly urbanized populations. The massive size of European cities also reflects the huge growth in overall population in the region. While the demographic situation in today's Europe is one of slow or even negative population growth, this is a recent phenomenon.

huge factories that sprang up over a relatively short time starting around 1870. As described by the geographer Jean Gottmann, the Ruhr Region was unique in its "vertical integration" of all stages of production in a single confined area (from coal and ore to finished product), as well as its horizontal integration of many types of industries, from textiles to chemicals and iron, steel, and munitions. The small market towns of Essen, Dortmund, Duisburg, Mühlheim, Bochum—the list goes on—swelled into major industrial cities in their own right, specializing in particular industries. Essen was a steel town, and the Krupp family was to Essen what Andrew Carnegie was to Pittsburgh. Wuppertal became a textile town, while Gelsenkirchen was known for its coking plants. The material and symbolic importance to Germany's economy, including its war-making ability, during the first half of the twentieth century is captured by the fact that French and Belgian troops occupied the entire Ruhr Region five years after the end of World War I as a way of extracting reparations from the German government in the form of coal and factory-produced goods. Allied bombers decimated the region's cities and factories during World War II. Although the Ruhr Region was central to West Germany's post–World War II "economic miracle," by the 1960s there were clear signs that a classic case of deindustrialization was afflicting the area: falling employment in the main industrial sectors, higher unemployment, migration of people from the region to other parts of Germany, and a declining image.

While the structural changes were dramatic, there are recent signs that a reorientation of the region's economy to services and even cultural tourism is creating a new vibrancy in the Ruhr Region. Former industrial sites have been converted to art museums and office parks in a process called **industrial reuse,** and many are protected as historical landmarks, preventing them from being torn down. The iconic Dortmunder Union beer brewery in Dortmund is now a high-rise art museum, event and exhibition space, while the Zollverein in Essen, once referred to as the "most beautiful colliery in the world," is now a UNESCO World Heritage Site, home to a museum, a university of the arts, theaters, and restaurants (Figure 4-3-2). A former steel plant complex in Duisburg is now a landscape and amusement park, where tourists can scuba dive in an old natural gas storage facility and rock climb on old factory walls. Today there are 200 museums in the region. The Ruhr Region was selected as a European Culture Capital in 2010. Signs of life in this postindustrial region are encouraging for those who live there, but ongoing challenges include relatively high unemployment and over-extended public financing.

Sources: A. B. Raines, "Wandel durch (Industrie) Kultur [Change Through (Industrial) Culture]: Conservation and Renewal in the Ruhrgebiet," *Planning Perspectives* 26, no. 2 (2011): 183–207.

▲ FIGURE 4-3-2 **Zollverein in Essen.** Once described as the "most beautiful coal mine in the world," the Zollverein now is a site of industrial reuse as an office and entertainment complex.

Through history, the defining feature of cities has been as home to activities other than agriculture. The earliest European cities were Greek, and these were often intricately tied to the sea as trading nodes around the Mediterranean. Merchant and naval fleets defined their character, and what distinguished a city from a village was not only size but also function. Those living in the city typically were employed in specialized trades, the military, as merchants, or administrators. This division of labor between urban and rural persisted and later became even more pronounced with the Industrial Revolution. A nearly uninterrupted trend in the human geography of Europe for two millennia has been rural-to-urban migration. The reasons for moving from country to city have changed over time, but there are common threads. Just as "push" and "pull" factors prompt individuals to migrate from one country to another, there are also such factors at play in the decision to move to cities. The prospect of a factory job might have "pulled" a landless peasant to move to nineteenth century Paris or Bucharest. But changes in land tenure regimes in the countryside that affected the ability to feed a family could have been a "push" for that same person. Being kicked off of land was very common throughout Europe during the eighteenth and nineteenth centuries as land was enclosed and became part of private property markets.

As cities grew in size, almost inevitably the distinction between urban and rural blurred. Nearly all European cities that predate the nineteenth century had defensive fortifications, usually walls that once signaled very clearly the distinction between city and countryside. The few that remain are historical relics, now popular with tourists. Torun, Poland; Carcassonne, France; Tallinn, Estonia; and Lucca, Italy, all have noteworthy remains. Even where remains are no longer standing, it is often obvious where they once stood. Paris is encircled by a ring road, *Boulevard Périphérique,* that was built along the route of the Thiers Wall, the most recent and last wall built to protect Paris from invading armies. A major rail line and city park uses land once occupied by the Hamburg city wall. The fact that these former borders between the city and countryside are often today at the heart of bustling urban areas gives some indication of the massive urbanization that has occurred. Urban and suburban areas can stretch for 6–12 miles (10–20 kilometers) around the largest European cities. In the most densely populated parts of the Blue Banana, cities simply bleed together, such as in Germany's Ruhr Region or the Randstad **megalopolis** of the Netherlands.

Historic Cityscapes

The landscapes of modern European cities are fascinating thanks to the layering of various historical periods. Rome's nickname is the Eternal City, and a long stroll through the city takes one through the centuries, from capital of the Roman Empire to a center of religious power for most of Europe as the home of the Vatican, to more recent indentations on the built environment from the Fascist period of the 1930s and 1940s. In a city such as Rome—or Bucharest, Berlin, London, and Lisbon—it is not just a city's economic role as a merchant trading hub or manufacturing center, but also its political history that leaves a mark on the urban landscape. Religious structures often dominate the central city skyline, as in the case of St. Peter's Cathedral in the Vatican, Lisbon's Estrela Basilica, or Saint Sava, the largest Orthodox cathedral in the world, in Belgrade, Serbia. But especially in former imperial capitals, churches must share the spotlight with administrative or other secular structures. The Louvre, now the world's most famous art museum, was originally a palace, only turned into a museum after the French Revolution when palaces were no longer needed because there were no more kings and queens in France. The castle of São Jorge imposes over Lisbon and is a reminder of the city's Moorish past and the country's imperial wealth.

Cities that were part of the Communist-controlled zone following World War II, such as Warsaw, bear evidence of that period. Often, the unmistakable socialist architecture is found where whatever preceded it had been destroyed by the war. Few cities in Central and Eastern Europe were spared widespread destruction by retreating Germans or advancing Soviet forces. After the war, and facing material shortages and large homeless populations, much was constructed quickly, with little concern for aesthetics. Although the period of Communist rule ended decades ago, the bland, utilitarian apartment buildings are still ubiquitous from Sofia to the Baltic Sea. Complementing imperial and royal palaces in Prague or Budapest were socialist structures, where workers could gather to learn or watch the arts. Bucharest, Romania, was once called the "Paris of the East" and boasted of the world's first electric streetlights by 1900. The country's megalomaniacal leader during much of the Communist era, Nicolae Ceauçescu, changed the urban fabric by ordering neighborhoods bulldozed, new high-rise apartment buildings constructed, and a new home for the government built. That "home," the Palace of the Republic, required over a thousand acres of neighborhoods, including literally dozens of churches and

▲ Figure 4-27 **Palace of the Republic in Bucharest, Romania.** The country's megalomaniacal leader until 1989, Nicolae Ceauçescu, changed the urban fabric of Bucharest by ordering neighborhoods bulldozed to make room for the Palace of the Republic, now called the Palace of the Parliament.

synagogues, to be razed so that it could be built. It is the most expensive administrative building ever constructed, but Ceauçescu never got to move in because it was completed just around the time he and his wife were executed after the government was overthrown in 1989. The building, now officially the Palace of the Parliament, is still in use and stands as a monstrosity on Bucharest's urban landscape (Figure 4-27).

What characterizes European cities today? Age has its virtues, and despite their great size European cities are some of the most compact and easily navigable in the world. Transportation networks consist of long-distance hubs such as airports and train stations, as well as the dense public transportation lines that connect them. London's Heathrow Airport and Charles de Gaulle Airport outside Paris are the two busiest in the region and among the busiest in the world, reflecting Europe's role in the global economy. Every major airport in Western Europe is linked by rail to the center city. Trams, subways, and extensive bus systems can be found in most European cities. While auto traffic is certainly an issue, by virtue of population density and conscious planning, walking, cycling, and riding public transit are widely used—and accepted—means of getting to and from destinations. Car commuters are often faced with multiple layers of expense: some of the highest fuel prices in the world, precious few and costly parking spaces, and in the case of London, Milan, and Riga, Latvia, even congestion charges on cars entering the center city. Major efforts to reduce pollution have yielded positive results in cities once notoriously dirty, but in southern and Eastern Europe, particulate matter from burning fossil fuels frequently reaches unhealthy levels in places such as Athens, Milan, and Belgrade. Cities with unique historical flair are especially popular tourist sites. Prague, Venice, and Bruges exhibit the compactness you might expect from places that flourished during the Middle Ages, but that also means that streets in old city centers during the peak summer tourism months look more like an uninterrupted pavement of picture-snapping tourists than of quaint cobblestones (Figure 4-28).

Urbanization and industrialization helped to create a very large middle class in much of Europe, and the quality of life in many places is the envy of many. Strong labor movements, often arising from the

◀ **Figure 4-28 The Charles Bridge with the Castle of Prague in the background.** Located on the Vltava River in the central Czech Republic, Prague's medieval cityscape makes it a magnet for international tourists. With more than 5 million visitors in 2011, tourism is among Prague's most important industries.

capitalist excesses of the Industrial Revolution, help to ensure that there is a fairly strict division between work and leisure time. Most official workweeks in European countries add up to fewer than 40 hours, and vacation time amounts to more than a month off for most who are employed. This helps to explain not only the importance of tourism, but also why sidewalk cafes are usually filled with patrons at the first sign of warm sun (Figure 4-29). In France and Germany, locals are accustomed to labor stoppages protesting wages, working hours, or benefit cuts; it is simply part of the political economy, and labor unions and social safety nets provided by the state enjoy widespread approval. The recent EU crisis over economic growth and debt has, however, highlighted some of the economic challenges facing this traditional model. Inflexible labor markets in particular make it difficult in countries such as Spain and France to hire and fire workers through cycles in the economy, and at least some of the extremely high youth unemployment rates in southern Europe can be traced to this inflexibility. Whether liberalizing rules would improve the employment situation is a hotly debated question. Beginning in the 1990s, several Nordic countries followed by Germany in 2003 overhauled their tax and welfare systems and liberalized their labor markets. It is likely no coincidence that these countries have experienced economic booms relative to other European countries in the intervening years. The flip side is that economic inequality has risen in Germany since these reforms took effect.

Social Disharmony

Cities in most of Europe reflect the overall economic prosperity of the region, but also tend to magnify social and economic inequalities. First-generation immigrants typically are the most marginalized, and in cities such as Madrid, Paris, and Stockholm, immigrant communities are typically found in the least desirable housing on the urban periphery. Social unrest such as protests and sometimes riots are not unheard of in such communities. Large industrialized areas, meanwhile, struggle with the demographic and economic impacts of deindustrialization. **Brownfields,** or areas abandoned as industry leaves, can be ripe for redevelopment and repurposing in more prosperous cities, but can result in the dislocation of poorer residents. In places such as Manchester, Liverpool, and Gelsenkirchen, whose lifeblood was heavy industry and where adjustment to postindustrial realities was most difficult, large tracts of housing, warehouses, and factories themselves have been virtually abandoned. The question of "what comes next?" is not nearly as easy as

in a city like Munich or London, where property is in short supply.

The more prosperous parts of western and northern Europe have been largely rid of **informal settlements** (settlements of often temporary structures that exist outside of formal legal structures and property markets), but these still exist in southern and southeastern Europe. On the outskirts of Sofia, Bucharest, and many towns and cities in the former Yugoslavia, large informal settlements of Romani (formerly called "gypsies") can be found (Figure 4-30). Conditions in these communities are often quite poor and lack running water and sewage, reliable electricity, and other infrastructure. The human rights of informal settlement dwellers are often violated, since there is little legal recourse when living on land outside formal legal structures. Romani settlements also exist in the suburbs of large Italian and French cities; many have come to Rome or Paris from southeastern Europe. Rome has around 100 camps, and a 2010 study showed that around 24 percent of Romani children living in these informal settlements were malnourished.

That is not to say that everyone within cities such as Berlin, Barcelona, or Paris is part of the formal housing economy, as squatting, or occupying a residence, is still fairly common and the facades of such buildings often show both decay and signs of life and political activism, such as banners with anarchist or leftist slogans. England and Wales made squatting illegal in 2012 but in much of mainland Europe, many squats enjoy formal legal status since housing is considered a basic right and eviction is quite difficult.

Another manifestation of the postindustrial economy in Europe is the number of individuals employed in or actively pursuing arts as a profession. Most large European cities enjoy extremely active artistic communities, both formal and informal, and the arts are heavily subsidized by the state in most of the region. Europe is well known for

▼ **Figure 4-29 Sidewalk café in Antwerp, Belgium.** Mild climates, an absence of pesky insects, and generous amounts of leisure time combine to make scenes such as this one commonplace across Europe.

▲ **Figure 4-30 Romani settlement on the outskirts of Sofia, Bulgaria.** Romani are an ethnic group descending from peoples who arrived in Europe from India about 1,000 years ago. Like some other ethnic minorities in Europe, Romani have faced sustained prejudice and economic isolation, but they have managed to maintain a sense of cultural identity in the face of adversity, especially in southeastern Europe.

its professional orchestras and operas, ballet troupes, and art collections, but this is only part of it. Live music is everywhere, in traditions ranging from punk to jazz. Berlin is rightfully famous for its techno and rave scenes—techno there was originally heavily influenced by the underground techno scenes in Chicago and Detroit, but soon took on a life of its own with new artistic contributions. On the urban landscape, techno and rave were closely tied to postindustrial buildings and to industrial decay more broadly. The most famous techno dance club in Berlin was E-werk, a former power generating station located a stone's throw from the heart of the city, while one of Germany's most famous electronic music groups was Kraftwerk, which is another name for power station. Jazz, blues, and bluegrass, all American imports, enjoy enthusiastic followings as well. Street art, murals, installations of one sort of another are ubiquitous in European cities, and there is often some sort of public subsidy behind projects.

Sports are also an important element of the economy as well as leisure in European cities. The most valuable athletics franchises in the world are Europe's soccer clubs. According to *Forbes* magazine, in 2012 Manchester United and Real Madrid were ranked first and second in the world in terms of their market value, with Barcelona and Arsenal also in the top ten. Local loyalties are strong, and often still reflect ethnocultural geographies. FC Barcelona's motto is *Més que un club* (more than a club)—the "more" expresses the club's association with the broader Catalan independence movement. During the reign of Spain's dictator Francisco Franco, who died in 1975, Catalonians were not allowed to speak or write publicly in their language. The soccer team became an outlet for expressing feelings of identity and nationalism, and that continues long after Franco's death. Glasgow has two main professional teams, and religion and class divide loyalties among them. The Rangers tend to draw followers from conservative Scots of Protestant background, while Celtic fans tend to be of Irish Catholic descent and socialist political leanings. Perhaps the strangest politicized football club is FC Sheriff Tiraspol, an internationally remarkably successful team in the separatist republic of Transnistria, which formally is part of Moldova but has de facto sovereignty. In this throwback society, where the flag and official buildings still have the Soviet hammer and sickle on them, banners are

often hoisted at games in favor of formal recognition of Transnistria's independence. Tiraspol like every city in Europe has at least one large soccer stadium, and their names are determined by corporate sponsorship—Allianz Arena in Munich, Emirates Stadium in London, and Sheriff Stadium in Tiraspol, named after the mafia-style corporation that controls most of Transnistria's formal economy.

Resource Scarcity and Innovation

It is impossible to capture all of the complexities of Europe's economy here. One of the most important factors shaping the development of the region over the past decades has been scarcity of natural resources. While Norway, the Netherlands, and the United Kingdom all still produce significant quantities of oil and natural gas, largely from offshore deposits in the North Sea, for the most part Europe is poor in fossil fuels and raw materials. Even prior to recent times, scarcity has been a perennial factor that led Europeans to explore, expand, and even start wars. Today, more than half of the EU's energy needs are imported, mainly in the form of petroleum coming by ship from the Middle East and Russia, and natural gas, much of it coming from Russia via pipelines. Even coal, for which Europe was once largely self-sufficient, now must be imported from Russia, Colombia, Australia, and the United States.

Obviously, Europeans have not allowed relative scarcity of natural resources to impede their economic development (see *Focus on Energy*: Europe's Drive to Energy Security). Many would argue that in a world of increasingly constrained resources, Europe will be at the forefront of technological developments to reduce consumption. In fields such as transportation, it already is. In April 2012 Dutch consumers paid on average 1.80 euros per liter for gasoline (around US $9 per gallon). The raw crude oil costs roughly the same for European consumers as for consumers anywhere, but the fact that much of it is imported adds societal and governmental costs. Such high prices have numerous effects on consumers, ranging from demanding more fuel-efficient cars to being more likely to drive only when absolutely necessary. Buildings also are typically well insulated and are smaller than in some other world regions, since heating costs are high. Europe is a global leader in renewable energy development as well. Europe's response to scarcity has been energy efficiency, which pays dividends for the European economy as well as for the environment.

Stop & Think

▶ What are some ways in which Europe has adapted to being a resource-scarce region?

Germany: Scarcity and the Mother of Invention

For consumers worldwide, the slogan "Made in Germany" stands for high-quality goods. People are willing to pay a premium for German-made products ranging from automobiles to high-tech machinery, and the export of these goods forms the basis for the German economic success (Figure 4-31). In 2012 Germany had the largest trade surplus of any country in the world, exporting over US $200 billion more in goods than it imported. Sustaining such productivity requires many different raw material and energy sources, and Germany itself is resource poor, so companies must import most of what goes into every Mercedes-Benz. Historically, European manufacturers could depend on colonial possessions for their raw materials, but Germany never had the vast empires of Great Britain and France. When Nazi Germany sought to expand its access to oil fields by advancing its army to the Caspian Sea, it set itself up for the most significant defeat of World War II, at Stalingrad. Germany's postwar economy has depended on reliable supplies of oil, natural gas, metals, and other raw materials, and the country is a leader in strategies to use less of these expensive imports. Buildings use 40 percent of all energy consumed in Germany, and stringent codes ensure that new structures are energy efficient. To address climate change, Germany has committed to reducing the energy demand of buildings by another 80 percent by 2050, primarily through retrofitting existing buildings. Partly to reduce its carbon footprint and partly to reduce dependence on imported energy, Germany has embarked on one of the most ambitious renewable energy programs in the world. In the first half of 2012, 26 percent of the country's electricity was supplied by renewable sources, including solar panels, wind turbines, and biomass. This program is not cheap, and there is still some debate about whether subsidies for renewable energy technologies are worth the expense considering that consumers pay very high rates for electricity.

Recycling has reduced the country's requirements of imported raw materials. Ninety percent of steel from buildings, cars, and tools is recycled, and the recycling rate for copper is the highest in the world at 54 percent. Urban mining, a process by which discarded televisions, mobile phones, and other devices are harvested for the gold, cobalt, gallium, and numerous other precious metals they contain, is a growing phenomenon in Germany. The government also has a Raw Materials Strategy to help ensure uninterrupted supplies of the critical materials that go into German products. While such a strategy enhances the economic health of a country so dependent on exports, there are larger questions about the social, political, and environmental consequences of resource extraction. Many of the raw materials that make Germany a wealthy country come from some of the poorest places on Earth, and often benefits do not accrue to local populations. There are active public debates about such questions as the role of valuable raw materials fueling conflict in places such as the Democratic Republic of the Congo, or to what extent the economic success of a German manufacturer depends on environmental destruction in Papua New Guinea.

▼ **Figure 4-31 Made in Germany.** Audis such as these symbolize high quality to many consumers.

Integration, Disintegration, and Globalization in Europe

Recall your cycle trip from earlier in the chapter. As you travel across Europe you will see signs when crossing a border between, say, Germany and Austria, but you will not be stopped. You only need to show your passport at national borders between Schengen and non-Schengen states. The **Schengen Zone** is the area of the EU that has agreed to completely passport-free travel across international borders. There is no active state of hostility between any European states, nor is any on the horizon. As recently as 1999 this was not the case. To travel by train from Germany to Greece was then only possible by completely bypassing the former Yugoslavia because of the war between Serbia and NATO over Kosovo. The contrasts between Romania and Bulgaria, in 1999 not yet part of the EU, and Austria and Germany were striking. How times have changed. Not only is there no war, but the levels of development in the new member states, and even non-EU members such as Moldova, Serbia, and Montenegro, have achieved heights unimaginable at the turn of this century. This is not to say that southeastern Europe's economic development is at the same level as northwestern Europe's, but the gap has closed considerably in a short amount of time. When one considers twentieth-century history, and the world wars whose origins lay in the nationalist aggression of European states, the status quo seems even more astonishing.

Forces of Integration

Forces of integration bring modern Europeans closer together than at any time but there are also tendencies of separatism—forces of fragmentation. There were many incentives for Europeans to move past the destructive political ideologies of the past. These ideologies were not just about nationalism and territorial aggression, but also reduced life to national identities that usually did not match the more complicated realities on the ground. European integration sought to move Europe past that—and it serves as a lesson for others around the world. The forces that integrate Europe can be divided into three categories: geoeconomic, geopolitical, and cultural.

As the original nation-states, European countries tend to reflect economic and cultural territories of the times when they came into existence. Trade and travel over long distances during the nineteenth century, when Italy, Germany, and several southeastern European states came into existence, was difficult and expensive. France, Spain, and Great Britain, even older as states, had even greater challenges unifying national territories under one economic, political, and cultural flag. By the twentieth century, especially after World War II when the United States and the Soviet Union rather than European nation-states were the superpowers, the relative compactness of Europe's states became an issue of economic competitiveness. Geoeconomically, a trading area with 200 million and eventually 500 million people was better fit to compete globally than a United Kingdom or a Spain with 50 million citizens. The postwar reality also meant that European states' visibility waned on the global stage they had once dominated. Colonial empires dwindled as countries in Southeast Asia, South Asia, and Sub-Saharan Africa declared their independence. Europe was no longer

Europe's Drive to Energy Security

Fueling Europe's economy requires a tremendous amount of energy. Not only manufacturing but transportation, domestic heating, and electricity generation are highly energy-intensive activities. The energy mix in Europe includes a wide range of fossil fuels, nuclear power, hydroelectric power, and, more recently, renewable energy derived from wind, the sun, and biomass. The share of renewables in Europe's energy mix will continue to grow as policy mandates and incentives in the individual European states as well as at the EU level encourage the development of non–fossil fuel resources. However, it is also clear that fossil fuels will continue to play a role in the energy mix in the future. Natural gas is the primary fuel in much of Europe for heating buildings, and a major fuel source for electricity generation (Figure 4-4-1). EU countries import more than half of energy consumed, and nearly 85 percent of the most precious fossil fuel, oil, was imported in 2009. Fossil fuel production has been decreasing in the EU over the last decades. North Sea oil and gas beds are slowly being depleted and what remains is harder to extract. There are still ample reserves of coal in countries such as Poland, but coal use is becoming ever less desirable due to its contribution to air pollution and global warming. As recently as the late 1990s one could still smell coal smoke in much of Central Europe, since it was used as a home heating fuel. Even that has nearly disappeared as countries enter the EU and adopt more stringent air quality measures.

This means the European region will continue to depend on other regions for a large portion of its energy requirements. Excluding Norway, a major supplier of both natural gas and oil to other European countries but not an EU member, Russia and Algeria are the largest suppliers of natural gas to Europe, while Russia and Libya are the largest oil suppliers. But the 2011 Arab Uprising in Libya caused oil supplies to Europe to be interrupted for many months. In 2013, al-Qaeda linked terrorists held hostage several hundred workers at a natural gas facility in eastern Algeria, prompting a major government operation to free them. Following the attack, which resulted in multinational oil and gas companies pulling out their staffs from Algeria, many in Europe wondered about the reliability of energy sources in politically tumultuous regions. Some also critically asked whether European efforts to secure energy supplies were partly to blame for the unrest. Perhaps the greatest level of political concern is reserved for Russia, the single largest supplier of fossil fuels to Europe. Major natural gas pipelines run from Siberia to Central Europe via countries such as Belarus and Ukraine, and a number of times in the last decades the Russian gas monopoly Gazprom has turned off supplies over disputes with intermediary countries. While Europe was not necessarily the target of these politically motivated shut-offs, some Europeans certainly felt the impacts. The incidents left European policy makers asking how to diversify energy supplies to reduce vulnerability to such events.

In recent years, **energy security** has been a political priority for the EU and in nearly all European states. Energy security refers to the accessibility, affordability, efficiency, and environmental sustainability of the fuels in the energy mix. The debate over energy security highlights how central energy is to the economies and modern lifestyles in Europe (and elsewhere); without it, very little of what we do on a daily basis would be possible. Given this reality, governments in Europe have sought to make consumers more efficient energy users, but also have pursued policies to increase the availability of energy by developing new technologies (renewables, for example) and pursuing diverse non-European energy supplies. New natural gas pipelines have been built and others proposed, especially to link the Caspian Sea with Europe via a "Southern Corridor." New, very expensive liquefied natural gas (LNG) terminals are being built in several locations across Europe in order to import gas from distant suppliers such as Qatar and the United States. And there are debates about whether Europe should exploit unconventional sources of fossil fuels such as shale gas and shale oil.

Sustainability is also a component of energy security, and Europeans derive more energy from renewable sources than most world regions. But there are also discussions about the tradeoffs associated with different kinds of non–fossil fuel energies. Nuclear power supplies France and Belgium with a majority of their electricity needs, but Germany decided in 2011 to close its remaining nuclear plants, citing safety concerns in the wake of the Fukushima disaster in Japan. It remains to be seen whether a comprehensive Europe-wide approach to energy security can be developed, but clearly there are still disagreements about how best to ensure that future energy requirements are met.

Sources: B. Söderbergh, K. Jakobsson, and K. Aleklett, "European Energy Security: An Analysis of Future Russian Natural Gas Production and Exports," *Energy Policy* 38, no. 12 (2010): 7827–7843; D. Yergin, *The Quest: Energy, Security and the Remaking of the Modern World* (New York: Penguin Press, 2011).

▲ FIGURE 4-4-1 **Natural gas pipelines in Europe.** Deciphering strict boundaries between Europe and neighboring regions is difficult not only on the basis of physical or cultural barriers, but also as a result of the connectivity provided by natural gas pipelines linking Europe with sources of energy far to the east. Map by Matthew Derrick.

synonymous with global hegemony, and colonial powers could no longer claim exclusive access to markets in their former colonies.

Another result of the new bipolar world was a cultural awakening to the fact that what divided Europeans culturally was less pronounced than what divided Europeans from other parts of the world. The shared Judeo-Christian heritage provided a common frame of reference for those in all quarters of the region. Although the Cold War led most European governments to choose between an alliance with the Soviet Union or with the United States, under the surface there was also widespread ambivalence toward the superpowers as not reflecting core values of Europe. While rugged individualism and social mobility that purportedly defined America found admirers in Europe, there was also little appetite for the lack of a social safety net and a perceived lack of refinement in American culture. The imported Soviet communism was very popular in many parts of Eastern Europe after the traumas of World War II, but soon disgruntlement grew about the oppressive hand of the state stifling any meaningful public debate or dissent. Uprisings in Hungary in 1956 and in Czechoslovakia in 1968 were brutally crushed with the help of Soviet troops, further alienating eastern Europeans from the ideals of the communism as practiced by the U.S.S.R. and its allies.

Since roughly 1950, integration in Europe is most commonly associated with the **European Union (EU),** an association of member states that have voluntarily given up some decision-making rights to serve a greater good of European integration. Now the defining political and economic feature of the contemporary European landscape, the European integration project was born out of the changing power dynamics in the world, and was also urged by outside powers, especially the United States. In 1949, the North Atlantic Treaty Organization (NATO) was founded as a mutual defense treaty among the United States, Canada, and its European allies (at that time mainly northwestern European countries plus Portugal). Less than a decade later, West Germany became a member of NATO, which prompted the Soviet Union and its allies to form its own mutual defense organization, the Warsaw Pact. Following the dissolution of the Warsaw Pact in the early 1990s, many of its members joined NATO. The precursor to the EU was the European Coal and Steel Community, founded in 1951 as a trading regime designed for members depending on each other for heavy industries, namely the coal and steel essential to the two world wars. The **Benelux** countries (Belgium, Netherlands, and Luxembourg), France, Germany, and Italy were the original members. Based on the success of the Coal and Steel Community, cooperation was extended into other realms—mainly economic realms—with the creation of the European Economic Community (EEC) in 1958 and later a European Community, then the European Union.

By the early 1970s the original membership had grown to include the United Kingdom, Ireland, and Denmark. The main administrative center for the EU as well as for NATO was Brussels, which is widely viewed as a compromise location between the two main drivers of the European integration project, France and Germany. It is not entirely coincidental to this decision that Belgium is divided among French speakers in Wallonia and Dutch speakers (a Germanic language) in Flanders, and that Brussels is officially a bilingual city.

Like NATO, the EU's most ambitious enlargement came after the end of the Cold War, when central and eastern European countries were included and 12 formerly Communist states joined. Estonia, Latvia, and Lithuania had been part of the Soviet Union,

while the remaining countries had been close Soviet allies. The EU is comprised of 28 member states; Croatia, the newest member, joined on July 1, 2013. Extending from Ireland in the northwest to Cyprus in the southeast, and from the Strait of Gibraltar to the Finnish-Russian border, the EU also includes a number of **extra-territorial possessions**—legacies of the vast colonial empires of European powers—such as French Guiana in South America, the Portuguese Azores, and the Spanish Canary Islands.

Integration is not meant to suggest that French people suddenly decided that French language and cuisine were not culturally unique and worth preserving, but at least at the political level, there was a growing sense of European identity based on commitment to human rights, shared economic prosperity, and a particular social model. Early postwar leaders in the movement to unite Europe were keenly aware that national identities and loyalties were so thoroughly ingrained in people's self-image that simply declaring a postnational Europe would not work. The map of European states still exists, and there is no active proposal to do away with nation-states. There is, however, a vocal minority in the EU that rejects the idea of integration altogether. This **euroskepticism** is most pronounced in the United Kingdom and Ireland, but also in Denmark, Hungary, and somewhat surprisingly, Latvia. Their arguments mainly revolve around giving up national sovereignty to unelected bureaucrats in Brussels and some of the culturally most emotional issues, such as food and agricultural policy and immigration. The ongoing Eurocrisis that began with the economic crisis of 2008 has undermined people's confidence in EU institutions, with a 2012 poll showing that only 31 percent of EU residents have a positive image of the EU, down from 57 percent in 2007. It is worth noting that trust in national governments has also taken a hit as a result of the crisis.

European integration had lofty goals, and in spite of crises including the most recent one, it is difficult to imagine the political and economic geography of Europe without the EU (Figure 4-32). What are the practical geographies of European integration? First is the single market among the 28 member states, which removes for all practical purposes barriers to trade such as tariffs. Second is the Schengen Zone, in which border checks are removed. Cooperation has been particularly active along international borders where there were always cross-border relations, but with the advent of passport-free travel and no tariffs or taxes, the remaining barriers tended to be social and linguistic. Third is a monetary union—more than half of member states use a common currency, the euro, and share a monetary policy through a central bank located in Frankfurt. More than 300 million people in 17 of the 28 member states use the euro as their currency, including the Greek island of Crete, where the euro's namesake Europa once ruled according to Greek myth. It is noteworthy that an additional 175 million people in the world use currencies pegged to the euro, in circulation only since 2002, and second only to the US dollar as a global **reserve currency.** A common currency makes travel and trade much easier in the Eurozone, but as the Eurocrisis showed, it also makes it difficult for individual states to use monetary policy to stimulate their economies or address inflation.

A successful integration strategy has been to actively encourage cross-border social and political cooperation. The earliest examples of cross-border cooperation come from the war-ravaged areas where Germany, Benelux, and France come together, with the first formalized cross-border region established along the Dutch-German border in 1958. Leaders along these borders saw the potential for

▲ **Figure 4-32 EU Member States and the Eurozone.** Most European countries are part of the EU, but several of the wealthiest (Liechtenstein, Switzerland, San Marino, Norway) and several southeastern European countries are not members.

economic cooperation, cultural exchange, and shared governance in realms of mutual interest, such as transportation. Coordinated European regional policy originated in the 1970s, when leaders of the European Community realized that areas along borders were often structurally disadvantaged. By the 1980s, "Euroregions" were springing up throughout the EU; very few border regions were not part of a Euroregion by the 1990s (Figure 4-33). Formal partnerships among areas with similar economic or cultural characteristics located in different parts of Europe have also been successful. A good example of this is "Four Motors for Europe," in which heavily industrial regions in France (Rhône-Alpes), Italy (Lombardy), Spain (Catalonia), and Germany (Baden-Württemberg) began cooperating on education, scientific research, and environmental issues in the late 1980s; the organization is still active more than 20 years later.

Stop & Think

Why is it challenging to create a "Europe of regions"?

There are many other manifestations of European integration, but two merit brief mention here. A remarkable success story has been an effort to encourage European students to study at universities outside their home country. Erasmus, as the major program is called, allows students to study abroad without paying fees greater than those paid at home. It ensures that nearly any EU institute of higher learning is attended by many students from different European countries, and it has created a level of shared experience and mobility among young people that further integrates European space.

There has also been growth in subnational regionalism. A number of ethnic or linguistic regions have found the EU a good venue to express their desires for increased autonomy. Devolutionary tendencies have always existed, but those regions that for economic or ethnocultural reasons seek autonomy often find sympathy from other EU regions with similar grievances. They have also found success in petitioning Brussels for decisions favorable to their cause. Wales, Scotland, the Basque region, and Catalonia all maintain offices in Brussels to keep direct lines open to the EU, even though the United Kingdom and Spain, respectively, are the official member states.

Geographically, one of the more interesting aspects of European integration has been how the EU and its member states have handled relations with neighboring countries, including application for membership. For many reasons—economic, prestige, political clout—many countries would like to be part of the EU. Turkey, of which part—including most of the country's largest city, Istanbul—is on the European side of the Bosporus strait, is officially considered a "candidate country" for membership in the EU. Yet most observers of EU politics do not anticipate Turkey becoming a member anytime soon for several reasons: cultural (Turkey's population is mostly Muslim); economic (its GDP per capita is less than half the average of Eurozone members, though higher than that of member states Bulgaria and Romania); and, least persuasively, physical location, in that it borders seemingly distant countries such as Syria, Iraq, and Iran. The EU has an official policy toward countries bordering the region, called the **European Neighborhood Policy,** which has largely been concerned with encouraging trade relations, especially of energy; coordinating border controls; and encouraging European-style democratic institutions.

Sweden: EU's Model Country?

Sweden is often held up as a model of the possibilities for combining free market capitalism with social welfare and equity. After being an imperial power in the Nordic region for much of its history, occupying parts of both its neighboring countries, Sweden has not been at war for 200 years and maintained neutrality during both world wars. Most of its 10 million inhabitants live in or around Stockholm, or in the Öresund region in the far south of the country. Sweden has developed global visibility for its multinational success stories, such as Volvo and furniture maker IKEA, but also for its generous welfare benefits, ABBA, and Julian Assange, founder of Wikileaks. It has a large natural resources sector, particularly in timber and paper pulp, but about half of the country's economic output is from its very successful global engineering firms.

The particular brand of social policy, sometimes called the **Nordic Model,** combines heavy taxation and redistribution of wealth with low public debt, good schools, and an emphasis on transparency and individual autonomy. It is based on work-life constitutions put into place in the early twentieth century, which defined the rights and responsibilities of employees and employers in order to reduce conflicts between labor and capital. National-scale human resource management as adopted in the Nordic countries would be hardly imaginable in most parts of the world, but in Sweden it has meant workplace satisfaction, low levels of economic and gender inequality, and profitable enterprises (Figure 4-34). After a debt crisis and economic slowdown in the early 1990s, Sweden cut taxes and reduced government spending, and with a few brief interruptions has enjoyed faster growth than most EU countries, with average annual GDP growth of 2.2 percent between 2002 and 2012.

Sweden became part of the EU in 1995, relatively late and by means of a referendum in which only 52 percent of the voters chose membership. In 2003 another referendum on adopting the euro as the country's official currency went down in defeat, with 56 percent voting no. Thus Sweden (along with the United Kingdom, Denmark, and a handful of other EU countries) has retained its historic currency, the kronor. Support for EU and Euro membership was strongest in the urban areas around Stockholm and Malmo and weakest in rural areas. One of the reasons cited by voters reluctant to approve of membership was that it posed a threat to Sweden's sovereignty and to its historic neutrality, similar to Switzerland's reluctance to join the EU. Although proud nationalists in certain respects, it would be difficult to imagine a conservative politician in the United Kingdom or France make a statement like that of Carl Bildt, Sweden's prime minister in the early 1990s and later foreign minister. In supporting the EU referendum he stated that the nation-state as an independent actor in world affairs was basically dead.

Swedish voters saw the issue of euro adoption differently. Part of the treaty Sweden signed to become a member of the EU also obligated it to join the euro when it was eligible, which it did not and has no intent to do. But this raises the question of whether European integration can function when states can opt out of such major components as a common currency. Does this make Sweden a "free rider"? It does in the eyes of some politicians in other member states, but public chiding as a form of peer pressure generally has hardened opposition to joining the common currency. Recent crises make it even less likely that Sweden and other non-euro countries will adopt the common currency anytime soon.

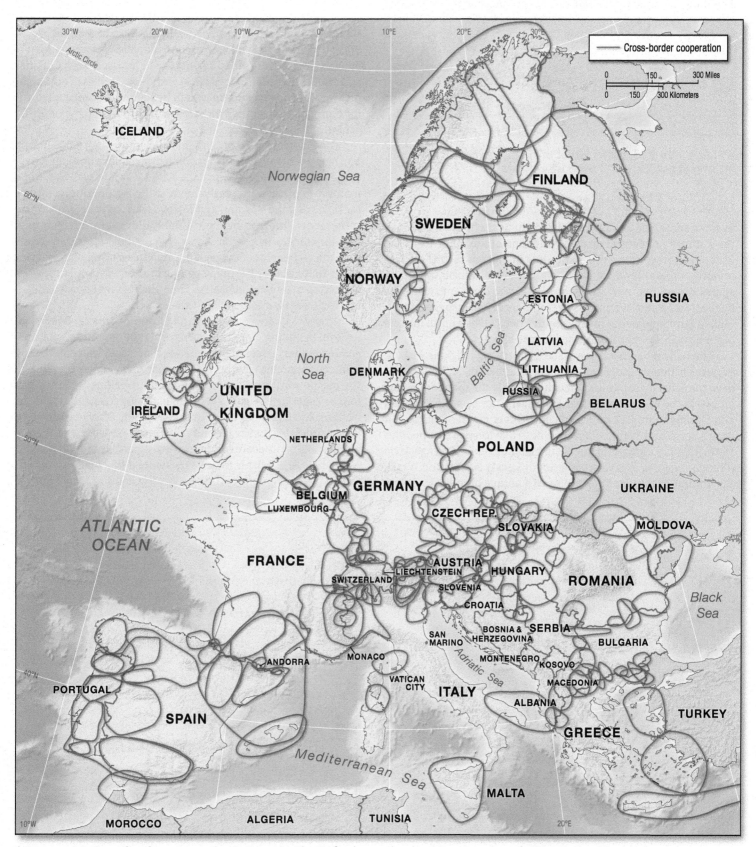

▲ Figure 4-33 **Cross-border cooperation in Europe.** The EU funds cross-border regions in order to facilitate trade and cultural exchange across national borders.

▲ Figure 4-34 **Pedestrian Street in Malmo, Sweden.** The Nordic Model stresses social equity and quality of life.

Internal Disunity

In some sense, the ability of localized political action to challenge the authority of central states is evidence of the maturity of the European integration project. Borders in most of the EU are less important than they once were, but they are also fixed by mutual agreement so that "new" states are unlikely to come into existence by means of war. Nevertheless, it would be naïve to suggest that a unified Europe somehow means unified perspectives on all matters big and small. Former Yugoslavia is a case in point.

The Balkan Peninsula is a place where many religions—Islam, Eastern Orthodox Christianity, Roman Catholicism, and Judaism—have coexisted for centuries. Often they do so in the span of a few city blocks. In the case of the former Yugoslavia, the Austro-Hungarian and Ottoman Empires exercised control over much of this region until after World War I, when the South Slavic-speaking peoples were united in a kingdom that would later be renamed Yugoslavia (literally meaning "the land of the South Slavs"). After World War II, the Communist leader Tito developed a unique brand of socialism and held together the ethnically diverse country through a mix of socialist idealism and brute force. Tito died in 1980 and the socialist state ceased to exist a decade later. In the wake of the collapse, political leaders appealed to old ethnonationalist instincts among the peoples—Serbs were pitted against Croats and Bosnians, Christians against Muslims, Catholics against Orthodox. A crisis of ethnic conflict was brewing in Europe, which by then was preoccupied with furthering the integration project and incorporating the newly free central and eastern European countries.

The massacre at Srebrenica, in which more than 8,000 Bosnian Muslims were slaughtered, has come to symbolize both the ravages of the region's deadliest war since World War II as well as Europe's inability to prevent war at the heart of the region. Intervention by NATO during the Bosnian and Kosovo wars, as well as negotiated settlements enforced by the United Nations, eventually brought the ethnoterritorial conflicts to an end, but not before they exacted their toll on peoples of the region. Largely as a result of the violence of the 1990s and 2000s, many people in southeastern Europe have wanted to distance themselves from the regional term "Balkans" due to its negative association with conflict.

After wars between Serbia and Slovenia, Croatia, Bosnia, and Kosovo, including ethnic cleansing and rape committed mainly by Serbian military forces in Bosnia, several new countries emerged. Recent developments have been somewhat encouraging. Slovenia and Croatia, now EU members, are the most prosperous countries in the former Yugoslavia (Figure 4-35). Croatia's long coastline and historic cities make it a very popular tourist destination. Serbia became an official candidate country for entry into the EU in 2012, and there are bright spots on its economic landscape, especially around Belgrade. Nevertheless, it is likely to be years before it is actually admitted to the EU (see *Geography in Action*: Growing Pains in "New" Europe).

▲ Figure 4-35 **Yugoslavia (1990) and former Yugoslavia today.** Political borders have undergone many changes in southeastern Europe since 1990.

For countries that were Soviet allies prior to 1990, the last decades have been a period of transitions (Figure 4-5-1). First came the challenge of creating new government structures and processes, including parliaments and multiparty democratic elections. Simultaneously, postsocialist capitalism was introduced across Central, Eastern, and southeastern Europe, in places where the dominant ideology for nearly 40 years was premised on the inherent corruption of free market capitalism. The economic and political geographies of postsocialist Europe reflect in many respects remarkable successes against long odds. Countries such as Slovakia, Croatia, Romania, Hungary, and Bulgaria have developed functioning parliamentary systems that resemble those in parts of Europe that were never under Communist rule. While economic output lags behind other European countries, most countries show consistent growth and there is a notable lack of widespread unrest. Far more of the economic output in these countries is in the informal sector—and thus is untaxed and unregulated—than in France or Belgium, but the welfare states, including public pension schemes, largely conform to the European social model found throughout the region.

If the collapse of communism represents the biggest upheaval in the political and economic lives of central, eastern, and southeastern European countries in the last half century, it is perhaps the changes associated with European integration that pose the greatest opportunities and challenges of recent years. In 2004, 10 new countries (eight of them postsocialist) joined the EU, followed in 2007 by Romania and Bulgaria. Along with EU membership came new rules and regulations, the promise but also risks of a new currency, the euro, and the close watchful eye of the EU and its long-standing member states in seeking to ensure that EU expansion would not turn out to be a liability for the rest of the union. Many postsocialist countries are perceived to have high levels of corruption in government and business.

A key aspect of the EU is the Schengen Zone, removing barriers to cross-border movement. Although part of the EU since 2007, Bulgaria and Romania have not been allowed to join the Schengen Zone as of 2013, even though they are certified to comply with the technical requirements of membership. In practice this means that the movement of goods and people from these two countries to other EU countries—part of the basic premise of European integration—is more difficult than it would otherwise be. The reasons cited include concerns about the rule of law and justice systems in the two countries, but political concerns about immigration in countries such as the Netherlands and Germany also played a role. It is frustrating for the governments and people of Bulgaria and Romania, since they made very expensive investments in border security to comply with EU rules only to be told in effect they are not good enough yet to be members of the exclusive Schengen club.

Another growing pain of sorts in the new member states is that of extremist political movements. Hungary and Slovakia have been in the news in recent years for electing governments whose policies stand in direct conflict with EU norms, and in some cases also in conflict with basic human rights. Communist parties are still present in many postsocialist countries, but it is the far-right "law and order" parties, often with anti-immigrant, antiminority tendencies, that are singled out for the most concern. The Slovak National Party (SNP) has won votes by making racist

▲ FIGURE 4-5-1 **The most widespread car in Communist Eastern Europe was the humble East German Trabant, or "Trabi."** These are now collector items, but after the end of the Cold War eastern Europeans quickly disposed of their Trabis in favor of Western makes and models.

comments about the country's Hungarian and Romani minorities, but ended up losing its seats in parliament in 2012. When the far-right Fidesz Party won a large majority in the Hungarian parliament in 2010, it proceeded to rewrite the country's constitution in a way that most outside observers agreed was antidemocratic. EU institutions and other EU member states swiftly condemned the actions. Recent proposed constitutional changes include requiring that homelessness be declared illegal, and banning not only same-sex marriage but any partnership between consenting adults that is not a marriage. Governments change as political parties' fortunes change, but in many new EU member states, what seems to be a constant is the ongoing growing pains in the region.

Sources: European histories (2): Concord and conflict. 2011. In *Eurozine,* http://www.eurozine.com /comp/focalpoints/eurohistories2.html.

Kosovo is Europe's newest sovereign state, though peace is still monitored by an EU police force and its status as a sovereign state was still disputed by Serbia and Russia as of 2013.

There are a number of separatist movements in Europe (Figure 4-36). These movements are unlikely to result in actual independence of nations within the EU, for example the Basque, or

▲ **Figure 4-36 Political separatist movements in Europe.** Wars were fought in the 1990s over secessions, but recent movements have been debated and discussed largely without violence.

Slovakia's Hungarian minority, or even in Belgium where disagreements are ongoing between Dutch-speaking Flanders in the north and French-speaking Wallonia in the south. The reasons for this are twofold: first, European integration provides an outlet for many groups that have felt disadvantaged by their minority status in a national context, but second, the formal structures of the EU make the logistics of forming a new state, or even altering the borders between existing states, extremely complicated. While Scotland may indeed vote to be fully autonomous from the United Kingdom, maintaining its membership in the EU is far from certain, since there is no precedent for this.

The second decade of this millennium has revealed new fractures in European unity. The financial crisis impacted Greece more than any other EU country. The back and forth between Athens and the core of Europe showed that the economic crisis revealed not only economic or policy divisions but also cultural ones. Greece's economy greatly depends on tourism and shipping, two areas dramatically hit by the original recession, but the lingering crisis was largely related to the inability of Greece's government to pay its bills. The sovereign debt crisis was exacerbated by a bloated government payroll and an inability or unwillingness of Greek authorities to collect taxes. A 2012 estimate indicated that up to 5 percent of Greece's GDP could be added to the national budget if taxes were collected effectively and tax evaders brought to court. The debate between Greece and the countries and institutions that would bail it out was largely over what sorts of measures would prevent such a situation from arising again: more austerity, or spending cuts (as Germany and the EU argued) or more stimulus (argued Greece, with the money to come from the wealthier EU states). The rise of ultranationalist movements in Greece also revealed the dark side of the dispute. A xenophobic party called Golden Dawn essentially blamed Greece's woes on immigrants from North Africa, the Middle East, and elsewhere, many of whom arrive in Greece via Turkey in order to reach the Schengen Zone.

Other member states, including Portugal, Spain, and Ireland, all have their own sovereign debt issues that required EU intervention. In a sense this is a success of European integration, in that institutions exist to address such issues, but it is also clear that the notions of shared interests and shared responsibilities have taken hits. German taxpayers have shown a willingness to bail out Greece and Ireland, and Greeks and others have compromised on conditions put on their government by Germany and the EU. Whether such a core-periphery dynamic is sustainable is another question. Many in the historic core of the EU wonder if EU expansion, by inviting a number of structurally challenged countries, came at the expense of progress on other fronts, such as deepening cooperation on security and economic issues.

Serbia: From Pariah State to Prospective EU Member

During the century preceding World War I, a number of power struggles and wars between the Republic of Serbia and the Ottoman Empire—with the heavy involvement of the Austro-Hungarian Empire and Russia—kept the political map constantly in flux. It was only after World War I that the country later called Yugoslavia was carved out of the western Balkans uniting Serbs, Croats, Slovenes, and a few smaller ethnic groups under one government. That marriage was never an easy one. World War II brought with it invasion and occupation at the hands of Germany and Italy, and if the

brutality of war were not enough, interethnic brutality, especially between Serbs and Croats, occurred as well. As many as 500,000 Serbs were killed as part of a targeted campaign of ethnic cleansing by the puppet state of Croatia, while as many as 200,000 Croats were killed during the war, many by Serbs. When the ashes of war had settled, Yugoslavia was once again technically an independent state, but was now a federation of six Socialist Republics.

With the collapse of Yugoslavia in 1991, several republics declared themselves independent of Serbia. Under the pretense of protecting ethnic Serbs living outside Serbia proper, the state under the leadership of Slobodan Milosevic and his generals provided military and logistical support to forces fighting in Croatia and Bosnia. When the first wave of Balkan Wars came to a close with the signing of the Dayton Peace Accords in 1995, Serbia was half the size of Yugoslavia, but maintained control over Montenegro and Kosovo. Kosovo had a majority Muslim Albanian population but a significant Serbian population. Another war ensued, and NATO airstrikes forced Serbia to withdraw its troops from Kosovo in 1999. In 2006 Montenegro voted in a referendum to secede from Serbia, thereby making Serbia a landlocked state; two years later, Kosovo declared its sovereignty, which was subsequently recognized by most European countries and the United States, but not Serbia or Russia. Talks between Serbia and Kosovo to normalize relations were ongoing in 2013, with the key sticking point being the status of the Serbian minority in Kosovo. The Serbian government, however, remained keen to normalize relations in order to speed negotiations with the EU over its membership.

A period of relative calm has settled on this part of Europe, including Serbia. Belgrade, sitting at the confluence of the Danube and Sava rivers, is the largest city in the former Yugoslavia, and a major service and employment center for the country. As a result largely of the Balkan Wars, Serbia has the largest population of refugees and internally displaced persons (IDPs) in Europe, estimated to be over 7 percent of the population, many of whom live in Belgrade. The capital city is known throughout southeastern Europe for its nightlife and cultural offerings, and in an interesting twist to the Balkans' ethnic geography, the city swells with young people from neighboring countries seeking to party in the urbane, open-minded city. There is something reassuring about Croats, Bosnians, and Serbs dancing together on one of the many party barges parked on the shores of the Danube (Figure 4-37).

▼ Figure 4-37 **Party boats on the Danube.** Serbia's capital Belgrade has entered a period of relative peace and prosperity.

Immigration

Immigration is another hot-button issue in much of Europe. Immigrants contribute much to European societies, and it is difficult to imagine London, Stockholm, Stuttgart, Paris, and Madrid without large and vibrant immigrant communities. As controversy over mosques and minarets shows, though, there are limits to the tolerance shown to others. Wealth disparities between Europe and neighboring regions create large incentives for immigration, and as in other world regions where this dynamic exists, one response has been to physically prevent people from entering European territory. The Spanish exclaves of Ceuta and Melilla on Morocco's northern coast were favored places for migrants from across Africa to enter the EU, since the two are the only pieces of EU territory on mainland Africa. Spain's government with EU support constructed 10-meter high fences, with barbed wire and various motion-sensing devices, around Ceuta and Melilla more than a decade ago. Efforts to further reduce immigration intensified after the Arab Spring in 2010. The EU stepped up patrols of the Mediterranean Sea seeking boats carrying migrants, and Greece built a fence along part of its border with Turkey in order to prevent migrants from arriving on EU soil.

Migrants from North Africa, South Asia, and Southwest Asia have brought their religious traditions to Europe, including Islam, Hinduism, and Buddhism. Their arrival has dramatically changed not only the demographic makeup of cities and suburbs but also the built environment. Islam is of particular note. In 1990 there were around 30 million Muslims in Europe, 44 million in 2010, and the number is expected to approach 60 million, or 8 percent of the European population, by 2030. France and Germany have Europe's largest Muslim populations, each greater than 4 million. France's population is about 7.5 percent Muslim, while Belgium, Switzerland, Austria, Germany, and the Netherlands each have more than

▼ Figure 4-38 **The new Central Mosque in Cologne, Germany.** Tall minarets share the skyline with one of Europe's largest Gothic cathedrals in this city on the Rhine.

5 percent. Large cities provide the biggest draw for new immigrants due to economic opportunity, relative tolerance and openness, and existing immigrant populations. Due to expensive rents in city centers, the densest populations of immigrants are found in large apartment buildings (often government housing for socially disadvantaged residents) in peripheral suburbs. Europe's cityscapes, while still dominated by Christian churches, are being changed by the addition of mosques. Cologne, for example, is home to a massive central mosque completed in 2013, the largest in Germany with a 180-feet (55-meter) tall minaret and Ottoman architectural style familiar to its predominantly Turkish worshippers, of which it can accommodate 1,200 (Figure 4-38). The mosque shares the Cologne skyline with one of Europe's largest Catholic cathedrals with its 515-feet (157-meter) high twin towers, and the mosque has not been without controversy with some locals protesting its location, design, color, and size. Switzerland's voters in 2009 passed a referendum banning the construction of new mosque minarets, but the law has not been tested and its constitutionality is not clear.

Migration complicates the nation-state model. Nowhere is this more apparent than in France, with its large population of mostly North African Muslims. France's official secularism emerged during the Enlightenment of the eighteenth century, when writers such as Voltaire and Montesquieu wrote scathingly of religion's role in French society. In the early twentieth century, the French government decreed the official separation of church and state and took a position of neutrality in religious questions. Catholics were allowed to practice their faith and the church permitted to operate largely devoid of government interference, but schools and other public institutions were treated as strictly religion-free zones. The arrival of new populations of practicing Muslims presented a cultural challenge to this French tradition: was the wearing of a headscarf to public school by a young Muslim woman an acceptable expression of personal fashion, or tacit approval of religious expression in the classroom? Even more controversial, should the state be able to ban the wearing of *burqas* (full-body covering, including the face) anywhere in public as it did? France's parliament overwhelmingly approved the ban, with just one no vote in both the upper and lower houses. It remains a touchy subject, especially within the country's large and growing Muslim community.

In spite of increased secularism and greater awareness of the perils of **xenophobia,** fear and loathing of the "other" is still a fact of life in Europe. Most European countries have "nation-first" political movements that gain votes with promises to sideline ethnic minorities, prevent new immigration, and send immigrants back to their country of origin. Religion is the identity marker most used to define immigrants' otherness. These movements include True Finns in Finland, the Swiss People's Party, and the Front National in France. Xenophobic political movements can take root even in places considered open and tolerant. From 2010 to 2012, the governing coalition of the Netherlands came to power only through the support of the Freedom Party, which had 24 seats in the parliament. The Freedom Party promoted closing Islamic schools, restricting new immigration, and banning ritual slaughter of animals as practiced by Jews (*shechita*) and Muslims (*dhabiha*). Few of their more extreme proposals were ever enacted, but the Freedom Party's role

▲ **Figure 4-39 A bakery in Brussels reflects changing neighborhood demographics.** Signs in both French and Arabic suggest how North African immigrants have added to the diversity of Brussels, Belgium's capital and the main administrative center of the European Union.

in government provoked extensive debate in a society better known for its permissiveness and tolerance than its xenophobia. In spite of xenophobic attitudes, a walk through any large European city reveals in subtle and overt ways how immigration adds new layers of culture, language, food, and religion to the region (Figure 4-39).

Since Poland, Slovakia, and Romania entered the EU, formerly sleepy borders with the Ukraine and Belarus are intensely monitored to prevent illicit trade and immigration. Attempts to keep immigrants from entering the EU overland have resulted in comparing Europe to a fortress or "gated community," where entry is restricted to narrow categories of desired migrants such as high-tech workers. But the reality of European demographics also means that without immigration, populations will continue to decline, as will economic growth. Such is the great paradox of modern Europe: Even as awareness of the demographic problems rises, reactions to immigration grow louder in many places (see *Visualizing Development*: Europe's Immigration Patterns).

Europe as a Geopolitical and Geoeconomic Power

Europeans were early to globalize. Venetians established distant trading relations, as did the Hanseatic League. Colonial powers such as Portugal, Spain, Great Britain, Holland, and France got rich from trade, raw materials, and labor exploitation. And while one could argue that Europe's impact on the non-European world was more profound and far-reaching than the impact of the non-European world on Europe, it has by no means been a one-way street. It would be difficult to imagine London or Birmingham today without the ubiquitous curry shops that exist because of the flow of immigrants from Britain's former colonial holdings in South Asia. The

constant interchange of music, film and television, and sports stars between North America and Europe also enriches cultural offerings on both sides of the Atlantic, and those connections would likely not be as strong as they are without the two world wars—started in Europe—that propelled the United States to become involved in European affairs.

Many European countries enjoy large trade surpluses with non-European countries, especially Germany, the region's largest exporter. Countries that had large colonial empires continue to have in many cases ongoing economic and political relationships with former colonial possessions: France with North Africa; Spain and Portugal with Latin America; and even Germany with Namibia in Sub-Saharan Africa. In 2013, for example, France intervened in its former colony Mali to expunge alleged Islamic extremists from the northern part of the country. The region had devolved into near anarchy following a government coup the previous year. In addition to maintaining an active political and economic interest in former colonies, colonial relationships also help to shape migration patterns to Europe. It is no coincidence, for example, that most of France's sizeable immigrant population comes from former colonies in North Africa, and most of Britain's immigrants originate in South Asia. The economic success story of postwar, postcolonial Europe cannot be explained without acknowledging the role that laborers from beyond the region played in fueling the boom.

Europe's economic strength is suggested by the number of non-European states that use the euro or with currency pegged to the euro. Banking and other financial services in London, Zurich, Paris, and Frankfurt serve not only Europe, but the world. The landscapes of Europe's trading relationships are evident across the region. Rotterdam at the mouth of the Rhine River in the Netherlands, and Hamburg at the mouth of the Elbe River in Germany, are the region's two busiest shipping ports and among the busiest in the world. Whereas once the docks were close to the city centers, the requirements of ever-larger container ships have moved the massive loading and unloading facilities closer to the sea. Containerization and the rise of supermax container ships have rendered some formerly major port facilities practically useless for trade in goods. London's Docklands, as the name suggests, was once the center of the world's largest port but now is a redeveloped area of banks and residences because it is too far up the Thames River for modern ships. However, London is hardly peripheral to world trade, both in terms of financial deals in the City of London, as well as at its major airports.

When Europe speaks with one voice in matters of trade, other economies must listen, as illustrated by the EU's disputes with the United States over genetically modified organisms (GMOs) in food. The issue of postcolonial relations and transatlantic trade disputes converged over the seemingly mundane issue of bananas. While Central American bananas are often grown on large plantations owned or formerly owned by the U.S. banana behemoths Chiquita (formerly United Fruit Company) and Dole (Standard Fruit Company), bananas from small farms on Caribbean islands are given preferential treatment in European markets. European countries argued that their preference for Caribbean bananas was rooted in better flavor, more sustainable growing practices, and better social

visualizing DEVELOPMENT ▲   Europe's Immigration Patterns

Immigration is a sticky topic in most of Europe, but there are important regional differences. International migrants comprise nearly 9 percent of Europe's population, but some countries, such as the United Kingdom (11 percent) and Switzerland (23 percent) have a much higher percentage of foreign-born individuals. The main sources of Switzerland's immigrants are other European countries, such as Italy, Germany, and Bosnia Herzegovina, suggesting that such a wealthy country draws people seeking job opportunities even within Europe. As of 2012, just 64 percent of Swiss residents did not have an

"immigration background" in the previous two generations. The United Kingdom has been receiving large numbers of immigrants for the better part of five decades; India is the top source, with Pakistan, South Africa, and Bangladesh also in the top 10. Spain has an even larger percentage of population foreign-born, but as recently as the early 1990s had virtually no immigrants. Some European countries still send more emigrants abroad than they receive immigrants. Albania, several of the former Yugoslavian states, and the Baltic states of Estonia, Latvia, and Lithuania are net providers of migrants.

It is difficult to generalize where and why backlashes against immigrants will occur, or where coexistence occurs with few problems. The social model that has emerged in Europe over centuries entails an active state role in ensuring that basic needs such as housing, medical care, and unemployment or disability allowances are met. However, that model emerged at the scale of the nation-state, meaning that mutual aid and assistance were meant for citizens (often defined along hereditary lines)—that is, social welfare was not a universal good but a national good. While this has changed

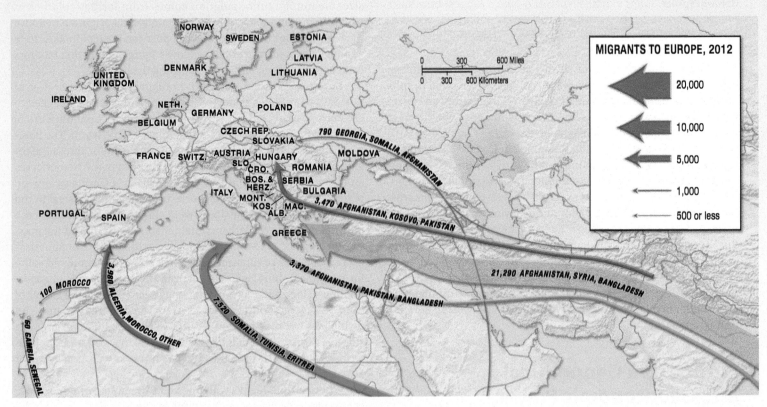

▲ **FIGURE 4-6-1** **Recent Immigration to Europe:** Migration patterns reflect historical relationships, such as colonialism, and economic and political situations in both home and destination countries.
Source: http://www.frontex.europa.eu/trends-and-routes/migratory-routes-map.

conditions for the growers. The United States countered that their two big producers were being unfairly targeted for tariffs on their products. The World Trade Organization (WTO) gave the United States a venue through which to challenge such arrangements by European countries as enshrined in the Lomé Convention, which was signed between the EU (then called the European Communities) and 71 African, Caribbean, and Pacific countries, many of them former European colonies, in 1975. After a series of rulings

and sanctions against the EU for its practices that violated the free trade principles set up by the WTO, the trade war over bananas ended in 2009 when the United States dropped the case after the EU agreed to lower tariffs on Latin American bananas while keeping their import preferences for former colonies. An interesting twist in the story is the recent rise of "fair trade" certification schemes, which have resulted in a renaissance of sorts for the Windward Islands banana industry in the Caribbean.

somewhat since the advent of state welfare systems over a century ago, there is still a roiling debate in many European countries about for whom such programs are intended. Switzerland, with its large immigrant population, has some of the strictest laws in Europe on naturalization (becoming a citizen), which means that some of the benefits of the welfare state are not available to a large segment of the population. However, Switzerland's prosperous economy means that unemployment among immigrants is very low. The European countries with the largest absolute numbers of immigrants are Germany, France, and the United Kingdom. Their large cities are among the most diverse in the world, and in spite of incidences of xenophobia, they are perceived as the more tolerant destinations in Europe.

By now many parts of Europe have become immigrant destinations (Figure 4-6-1). In recent years, however, governments have redoubled efforts to make transit to Europe more difficult for immigrants, especially along the external borders of the EU (Figure 4-6-2). While individual member states are still largely in charge of patrolling borders and controlling citizenship and residency permission, the EU has taken on new responsibilities for border enforcement in the Mediterranean region and along the EU's eastern border with countries such as Turkey, Ukraine, and Belarus. Frontex, an EU agency based in Warsaw, was created in 2004 to coordinate border enforcement; one of its main targets is preventing migrants from crossing the Mediterranean by boat or other means and landing on the shores of Italy, Spain, or Malta (an island state that is one of the newest members of the EU). These efforts have been criticized by human rights advocates because of a number of incidences

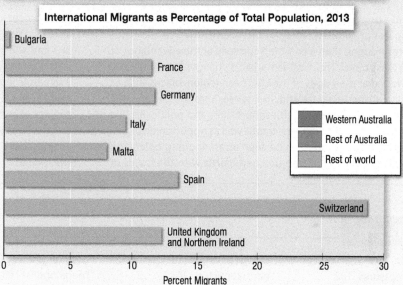

▲ **FIGURE 4-6-2 Foreigner guesthouse in Bulgaria.** Along the external border of the EU, such as here in Bulgaria, governments are constructing facilities to deal with newly arriving migrants and hold them while determining their refugee status. Many are sent back to their home countries. The accompanying chart shows some percentages of international migrants making up the total populations of select European countries.

Source: United Nations.

where boats sank and migrants were killed or injured, and because of the basic obligations of states to protect the dignity of human life regardless of origin or class.

Sources: http://euobserver.com/fortress-eu; R. Zaiotti, *Cultures of Border Control: Schengen and the Evolution of European Frontiers* (Chicago: University of Chicago Press, 2011).

The economic might of Europe is not in question, but many observers claim that Europe, particularly the EU, is an economic giant but a political dwarf. As China's military budgets expand and the United States maintains the most sophisticated and expensive military in the world, European military budgets as a share of GDP are small compared with many other countries. The counterargument is that spending reflects values, and spending on humanitarian aid by European countries abroad is among the highest in the world as a share of GDP. As the global geopolitical order shifts away from a transatlantic alliance and toward the Pacific as the key realm for power politics, how Europe will react remains an open question.

Stop & Think

▶ Why have some labeled Europe, particularly the European Union, an economic giant and a political dwarf in the world?

Summary

▶ Although one of the most highly developed and wealthy regions in the world, Europe is still a work in progress. Recent years have seen one political crisis after another, and call into question the entire project of European integration, which is the most significant development in Europe's recent history. But when looking instead at where Europe is in its long history, the story is much more nuanced—and on balance more positive—than recent media coverage would suggest. Most of Europe has enjoyed peace since World War II. Even in southeastern Europe, a period of relative calm has emerged out of the bloody 1990s, while in Britain, Germany, and Spain, economic distress and how to address it has seemingly supplanted concerns about Islamic terrorism in recent years.

▶ The European Union has raised environmental and socioeconomic standards for a territory now covering most of the European region and a population of some 500 million people. The "invisible" geographies of European integration receive little attention, but are more enduring than a common currency or the advantages of passport-less travel. The most remarkable changes in recent years can be found at the edges of the European region, places such as Romania, Estonia, even Ireland. In a sense, Europe's periphery is converging with the core both economically and politically, in a process that is fraught with tensions as well as opportunities.

▶ Developments in Europe can be seen as an ongoing balancing and counterbalancing of tradition and inertia with the forces of European integration, globalization, environmental change, and technological change. One can witness these tensions all over the map and in all realms of life. Nationalist movements continue to make themselves heard, often reacting both to European integration and immigration. At the same time, Europe is more secular and environmentally aware than at any time in its past. Cities and their residents adapt to economic change and deindustrialization, but there are also plenty of neighborhoods where the structural changes have been so dramatic as to defy easy solutions. Post-socialist Europe has shown remarkable resilience and adaptive capacity over more than two decades since the fall of the Iron Curtain, so that much of that region is unrecognizable to those who knew places such as Warsaw, Sarajevo, or the Black Sea Coast not too many years ago.

▶ Much of Europe has become a destination for immigrants, and these processes have changed the fabric of cities and countryside alike. The population makeup of Europe is changing in both obvious and subtle ways, but it is clear that Europe's current revolution is more demographic than ideological or political. Europe is a dynamic world region seeking to find its place in its neighborhood and in the world. The region has embarked on an ambitious project of political and economic integration. That project has faced a nearly existential crisis in recent years, but the ruptures in unity have largely been handled between politicians and in the media.

Key Terms

agritourism 195

austerity measures 182

Benelux 204

Blue Banana 191

brownfields 199

Common Agricultural Policy (CAP) 192

core-periphery dynamic 191

devolution 181

Eastern Orthodoxy 183

energy security 202

European Neighborhood Policy 206

European Union (EU) 204

euroskepticism 204

Eurozone 195

extraterritorial possessions 204

greenhouse gases (GHGs) 180

humid continental climate 175

industrial reuse 197

Industrial Revolution 177

informal settlements 199

Jewish pale 183

lingua franca 185

marine west coast climate 175

mediterranean climate 175

megalopolis 198

Mezzogiorno 189

multilingualism 185

nationalism 181

nation-state 189

Nordic Model 206

North Atlantic Current 174

polder 178

pronatalist policies 191

Protestantism 183

reserve currency 204

Roman Catholicism 183

Schengen Zone 201

secularism 183

small- and medium-sized enterprises (SMEs) 194

technopoles 195

terpen 178

thermohaline circulation 174

total fertility rate (TFR) 189

xenophobia 212

Understanding Development in Europe

1. What are the main physical and cultural traits that define Europe?
2. How has European integration changed the ways in which environmental challenges are addressed in Europe?
3. What are some ways in which climate and physical geography have influenced diet and agriculture in Europe?
4. What are the three major Christian subdivisions of Europe and where is each located?
5. What are the three major Indo-European linguistic subdivisions of Europe and where is each situated?
6. How does immigration to Europe challenge the idea of the nation-state as it evolved over time in Europe?

7. Where would you most likely find postindustrial landscapes in Europe, and what would they look like?
8. What are the sources of social disharmony in Europe, and what strategies have been used to address them?
9. What is the relationship between immigration to Europe and the demographic issues the region faces?
10. How is European integration impacting the political map of Europe in modern times?

Geographers @ Work

1. Climate change is likely to impact Europe in a number of ways. If you were planning adaptation strategies for Europe, what would you focus on?
2. Much of the tensions in contemporary Europe are between the core and periphery over issues such as austerity, corruption, and redistribution of wealth, but there are also underlying cultural tensions. If you were to identify cultural, political, and economic features of Europe's geography that are likely to cause challenges for policy makers in the future, what would they be?
3. Europe is highly resource dependent, both for energy and other raw materials. Europe also depends on labor from other parts of the world. What are the advantages and disadvantages of relying on other places for such things as energy, raw materials, and sources of labor?

4. Most countries' economies in Europe have transitioned away from heavy industry to postindustrial services. What lessons would a geographer draw from Europe's experience, for urban policy, economic planning, politics, that might have applicability in other parts of the world, particularly for places like China, India, and Brazil that increasingly rely on manufacturing and industry?
5. Newly arriving migrants to Europe face challenges of finding suitable employment and integrating into societies that are not always very welcoming. How might factors such as the geographical location of immigrant communities in cities, histories of nationalism and xenophobia in European countries, and economic changes contribute to these challenges? What are some examples?

MasteringGeography™

Looking for additional review and test prep materials? Visit the Study Area in MasteringGeography™ to enhance your geographic literacy, spatial reasoning skills, and understanding of this chapter's content by accessing a variety of resources, including MapMaster® interactive maps, videos, RSS feeds, flashcards, web links, self-study quizzes, and an eText version of *World Regional Geography*.

Northern Eurasia

Robert Argenbright

Sochi Olympics: Window on a New Russia?

The Russian government hoped that the Sochi Olympics would usher in a new image for the country as a modern, open, and diverse society and put an end to stereotypes of Russia as authoritarian, inefficient, and corrupt. But sometimes stereotypes capture significant aspects of reality.

Post-Soviet Russia suffers from deteriorating infrastructure, and the Olympics provided an opportunity to modernize a crucial region of the country. Furthermore, though Russia has some of the world's richest and most diverse natural resources, its economy is overly dependent on oil and gas exports. Diversification is urgently needed. Expanding the tourist industry—by some accounts the global economy's largest sector—promised to do just that. In particular, the Olympics were meant to stimulate the economy of the North Caucasus, Russia's poorest region and source area for violent Islamist resistance to the Kremlin.

But the Sochi Olympics needed to attract foreign tourists in record-breaking numbers just to break even economically. By the summer of 2013 the estimated cost of holding the Olympics had reached $50 billion, by far the most expensive Olympics ever. And it did not diminish Russia's reputation for corruption given that several of President Putin's personal friends directed major projects which incurred large, unaccounted cost overruns.

The constraints imposed by the region's climate and geography also came with a high price. Sochi's weather can be balmy even in February, so the organizers had to store snow from the previous winter. As to geography, Sochi basically is a narrow, 90-mile long strip between mountains and the Black Sea. An innovative road/train combination was constructed to improve mobility, but it cost almost $9 billion to build, which was about 150 percent of the total cost of the Vancouver Olympics.

Security was another major expense. Terrorism continues to claim hundreds of victims annually in the North Caucasus. Islamist groups have also struck at Moscow directly by bombing airplanes, metro cars, an express train, and a rock festival. In summer 2013, Doku Umarov, "Russia's most wanted man," called on Russia's Muslims to use "maximum force" to disrupt the Games.

The Sochi Olympics did provide a window on Russia, but the view was often unflattering. Environmental groups documented destruction of alpine landscapes and damage to coastal environments. Human Rights Watch revealed widespread illegal exploitation of migrant labor. Also, Russia's law against "gay propaganda" provoked international condemnation. The Sochi Olympics may have changed Russia's image, but they also showed how far the country is from becoming truly modern, open, and tolerant of diversity.

Read & Learn

- ▶ Outline how the opportunities and constraints of Northern Eurasia's resources and environment affect development.

- ▶ Describe how the history of Russia and the Soviet Union has shaped the context of development today.

- ▶ Explain how the "transition" from Communism diverged from expectations and led to the current situation.

- ▶ Account for the significance of "the power vertical" and "crony capitalism" in Northern Eurasian development.

- ▶ Characterize Russia's relationships with its neighbors and the rest of the world.

- ▶ Understand Ukraine's predicament, located between Europe and Russia and divided internally between east and west.

- ▶ Identify the significance of oil and gas, both for the "haves" and the "have-nots."

- ▶ Explore the reasons why development is a spatially uneven process in which some areas excel, while others flounder.

▲ Sprawling 90 miles (145 kilometers) along the Black Sea coast, greater Sochi, host of the 2014 Winter Olympics, offers a relatively mild climate plus convenient access to the mountains north and east of the city that shield it from the Russian winter.

Contents

Figure 5-0 The major physical features and countries of Northern Eurasia.

N orthern Eurasia comprises six former Soviet countries—Russia, Belarus, Ukraine, Georgia, Armenia, and Azerbaijan (Figure 5-1). These countries still have not overcome the Soviet legacy, in contrast to most ex-Communist countries to the west that have become distinctly European in terms of identity and political orientation. Northern Eurasian countries have yet to fully redefine or redevelop themselves; their economies are not completely reformed, their governments at best are vulnerable and flawed democracies, and they lack full-fledged civil societies. But much has changed for the better, as well.

The Soviet Union disappeared with a few strokes of a pen, yet the Communist empire's legacies continue to haunt the region. Russia inherited the U.S.S.R.'s (Union of Soviet Socialist Republic's)

permanent seat on the United Nations Security Council as well as the Soviet nuclear arsenal, thus guaranteeing that Russia would be treated as a great power, if no longer a "superpower" comparable to the United States. Although Russia's economy no longer ranks with global leaders, income from oil and gas exports has provided relative stability and enabled President Vladimir Putin to reassert the Kremlin's authority both at home and abroad. From Russia's standpoint, the other former Soviet countries constitute the "near abroad," a realm where Russia seeks to have unparalleled influence.

However, not all of these countries, especially Ukraine and Georgia, welcome the Russian embrace. Ukraine has indicated the desire to become a fully European country but it cannot ignore its

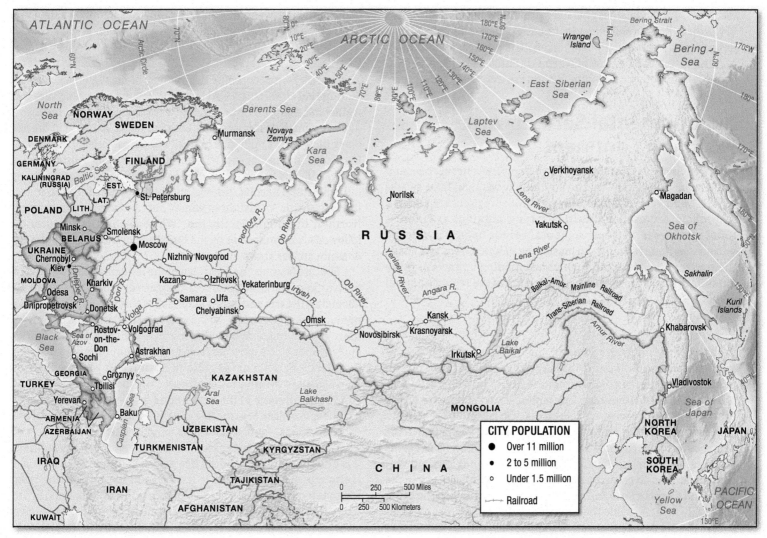

▲ **Figure 5-1 Northern Eurasia.** The countries of Northern Eurasia cover a large area of the world. About a quarter of that area is considered part of Europe; the remaining three-fourths lie in Asia.

massive neighbor to the north, particularly because Russians form a very large minority within the country. Georgia lost a short but shocking war against Russia, but two major territorial disputes remain unresolved. Belarus as a nation is hardly distinct from Russia and its current political-economic system is a throwback to Soviet times. Armenia's economy has rebounded modestly from a major slump in the 1990s, but the struggle with Azerbaijan over Nagorno-Karabakh has only been suspended, not resolved. Azerbaijan, like Russia, faces a problem many countries would be glad to have: how to manage windfall revenues from petroleum exports.

Whatever their individual issues, without exception all of the countries in the Northern Eurasian region face the same pair of challenges. They must further restructure their economies to make them more diverse and robust in order to profit from globalization while simultaneously providing for citizens' needs. They must also firmly establish and enforce the rule of law, for without this, economic development is distorted by corruption and the growth of civil society is stunted by repression. So long as these countries lack

economic stability and healthy civil societies, they will remain vulnerable to political extremism.

Northern Eurasia's Environmental and Historical Contexts

Karl Marx contended that "man" makes his own history, but not under conditions that he has chosen. Today we recognize that "woman" also makes history, but Marx's argument still is worth bearing in mind. The vast space and largely inhospitable climate of Northern Eurasia have offered both opportunities and limits to the development of human societies for thousands of years. Today technology makes possible the exploitation of once-unattainable resources, but the relatively harsh environment and great distances of Northern Eurasia still increase the costs of development and hinder competitiveness in global markets. The physical environment and the historical legacies of Northern Eurasia enable and constrain contemporary development efforts as well.

Environmental Setting: Vast Northern Continent

Northern Eurasia is a realm of great extremes: sweltering heat and brutal cold; excessively dry and abnormally wet; and huge swaths of forest and vast expanses of steppe. Most regions are affected by pronounced oscillations in climatic conditions at different times of the year. The enormous scale of the natural landscape encourages grand schemes to master the natural forces and shape them to human needs. And the corresponding impacts of those alterations affect the viability and function of the region's natural systems.

Landforms

In the west the immense Eurasian landmass (see chapter opener map) presents no obstacles to the eastward movement of air masses or to the intrusion of cold air from the Arctic and Siberia in winter. Even the **Ural Mountains**, conventionally considered to divide Europe and Asia, reach a maximum of 6,250 feet (1,900 meters) in the remote north, but rarely exceed 5,000 feet (1,500 meters) in the settled area of the country. As a result, there are immense regions characterized by relative uniformity of climatic conditions and vegetation patterns. In contrast, the **Caucasus Mountains** between the Black and Caspian Seas and the Transcaucasian lands to the south are highly diverse, while the southern end of the Crimean Peninsula in the Black Sea and western Ukraine's slice of the Carpathian Mountains also stand out as exceptional. Eastern Siberia consists of rugged, eroded plateaus bounded on the east and south by substantial mountain ranges. Narrow strips along the Pacific shore and in coastal river valleys benefit from the moderating influence of the sea, but most of the immense area east of the **Yenisey River** is isolated and inhospitable.

West of the Urals, the main rivers radiate outward from the central area dominated by Moscow. The Volga stands out among them as the longest river in Europe and the one most treasured in Russian culture. Unfortunately, the Volga flows to the landlocked Caspian Sea, which contributed to Russia's historical isolation, although a canal enables access to the Sea of Azov today. East of the Urals, four major river systems rank among the world's greatest. Yet their usefulness is limited by the harsh climate and by their northward orientation, which results in southern, upstream reaches thawing before downstream sections. This "ice-plug" effect is most damaging in the West Siberian lowland; the Ob frequently floods vast areas in early summer. The Yenisey, in contrast, runs a relatively straight course along a great escarpment, making it the most useful for generating electricity. The mighty Lena flows far from densely settled areas and empties into the Arctic, as do the Ob and Yenisey. Only the Amur reaches the Pacific after a northward turn. Upstream, the Amur serves as the border with China; lack of cooperation between Russia and China has hampered development of the area.

Climate

Three main factors combine to give most of the region a vigorously **continental climate**, marked by a long, relatively dry, and very cold winter and a short but surprisingly warm summer. First, the formidable mountain systems to the south in Central Asia and in the east block the Pacific's influence from all but a small area of the Russian Far East. The one ocean where Russia has vast frontage is the Arctic, which is frozen much of the year and generally contributes little moisture. Second, the region's high-latitude location means it receives little insolation in winter but has long days in summer. North of the Arctic Circle, the "White Nights" of summer are followed six months later by the dark days of winter, when the sun does not appear above the horizon.

The last factor is the great size of the landmass. The territory of the Northern Eurasian countries stretches over 6,200 miles (10,000 kilometers) west to east and as much as 1,200 miles (2,000 kilometers) north to south, and the rest of the vast Eurasian landmass also affects the region's climate. When sending an e-mail to Vladivostok from St. Petersburg, one must remember the 11-hour time difference—St. Petersburg is almost as close to Boston as it is to Vladivostok! Between these two great Russian port cities lies a huge territory that enjoys little or none of the moderating influence of the ocean, except near the Black Sea, the Sea of Azov—the shallow northern arm of the Black Sea—and, to a lesser extent, the Caspian Sea.

Winter is the longest season throughout most of the region, and it is brutally cold. Only Antarctica gets colder than East Siberia. It is also rather dry, especially to the east and southeast of the **Volga River**. Whereas Moscow receives 24 inches (600 millimeters) of precipitation on average, the Siberian urban center of Irkutsk receives 15 inches (380 millimeters), while Astrakhan on the Caspian Sea's north shore receives just 8 inches (200 millimeters). The crucial factor is the relationship between precipitation and evapotranspiration. A strong moisture surplus in the north shifts to a pronounced moisture deficit in the south and southeast.

The cold climate adds to the price of development. Consider just the cost of heating for Moscow's 12 million people—nowhere else in the world is such a large city located away from the sea at such high latitude. Now consider Yakutsk, which with 200,000 people is the largest center in East Siberia. In January the average daily low temperature is –42.7°F (–41.5°C). Massive energy consumption is a given. Ordinary steel and rubber tires will shatter at –30°C, so all equipment and machinery must be specially made to endure the Siberian winter.

Permafrost, the permanently frozen earth that underlies part of European Russia and nearly all of Siberia, presents costly engineering challenges. The active layer of permafrost near the surface thaws in summer and is saturated because the frozen ground below blocks drainage (Figure 5-2). Jello would make a better foundation! Even in winter, care must be taken with permafrost to make sure no source of heat causes melting. The ground may simply give way or buckle unpredictably. These risks are compounded in eastern Siberia by frequent earthquakes, which add to engineering costs and may cause massive destruction.

▲ **Figure 5-2** **Mud season in Siberia.** The spring thaw brings with it deep mud, which turns any road without a hard surface into a nearly bottomless bog. This western Siberian village can be cut off for weeks at a time from surface contact with the outside world.

Shipping suffers because all of Russia's and Ukraine's ports are hampered by ice in winter, with the exception of Murmansk on the Kola Peninsula's northern coast. The Gulf Stream keeps the port open, but Murmansk's eccentric location adds to the cost of transport. Shipping in the Arctic Ocean must contend with ice. Atomic-powered icebreakers can extend the shipping season, but these are costly. Global warming is benefiting Russia by making the Arctic more easily navigable. But because fresh water freezes more readily than saltwater, Siberia's rivers will remain a problem.

Natural Regions

The result of the northern location, continentality, and topography is a pattern of exceptionally large bands of essentially uniform vegetation and natural regions: tundra, taiga, mixed forest, deciduous broadleaf forest, forest steppe, steppe, semidesert, and desert (Figure 5-3). Starting in the north, the **tundra** region stretches all across Russia's Arctic shore, and extends southward for hundreds of miles in parts of Siberia. Although summer days are long, the intensity of the insolation is weak—no month averages 50°F (10°C). No trees grow in the tundra due to the short growing season and infertile soil, and also because the active layer of ground is too shallow for tree roots (Figure 5-4). Lichens and moss, and sometimes shrubs, feed herds of reindeer, which in turn have supported indigenous peoples of the north for ages.

Taiga is the Russian word for boreal forest—northern forest dominated by conifers—and Russia contains more of it than any other country. Taiga covers much of northern Russia west of the Urals as well as most of Siberia (Figure 5-5). Where permafrost is not as extensive and more moisture is available, a variety of conifers flourish. In the West Siberian lowland, much of the ground is poorly drained and the forest is interspersed with bogs and meadows. In spring snow begins to melt in the south, and the tributaries of the **Ob River** thaw out. But the ice in the lower Ob acts like a giant plug, sometimes until late in the summer, and this causes extensive flooding. Taiga continues to the east, but its

diversity diminishes to the point that great swaths of forest consist of just one species: the larch, which is uniquely adapted to the bitter winter, short growing season, and constraints imposed by the underlying permafrost. Taiga soils are acidic and not very fertile. Because of the great ice sheets of past glaciations, drainage systems in the taiga region west of the Urals are chaotic. Bogs and marshes abound—and so do ravenous mosquitoes in summer! From about 60°N and increasingly to the south deciduous trees begin to appear, creating a band of mixed forest. This band, like the forest steppe and steppe further to the south, is much broader in the west than the east. Historically, these three natural regions have been best suited for human settlement. At the western extreme of the region, agriculture is possible generally from the St. Petersburg area in the north to the Black Sea coast in the south. In Ukraine, roughly 58 percent of the land is arable, compared to only 7–8 percent in Russia. Increasingly as one moves eastward into the depths of the immense continent, the northern area with a growing season that is too short and the southern area with a growing season too dry put the squeeze on agriculture. Mixed forest provides a much greater array of resources than does taiga, and the leaf litter builds up humus in the soil. But drainage can still be a problem, as it is in much of Belarus. Another type of forest, mostly Asiatic deciduous trees, grows in the Amur River basin and along the coast in the Russian Far East. With commercially valuable species and good accessibility, these forests are extremely vulnerable to excessive logging, given the burgeoning demand for wood products in Japan and China.

To the south of the mixed forest the growing season lengthens but the moisture surplus becomes a deficit. Grasslands grow more extensive, first in the mixed area of the forest steppe, and take over entirely in the **steppe** proper. The more grass, the more fertile the soil. The Tatars who once held sway in the steppe are said to have been able to mount their horses and still be concealed by the tall grass! The soil is so rich with organic matter that it appears as dark as compost—this is **chernozem**, "black earth." Soil fertility and the relatively long growing season, up to seven months in southwestern Ukraine, would make the steppe region perfect for farming, if only precipitation were more plentiful and reliable.

Mountain areas hold a virtually infinite array of microenvironments. Mountains also create conditions for two exceptional natural areas. The southern tip of the Crimean peninsula, shielded from the cold winter air masses, enjoys a mediterranean climate. South of the Caucasus Mountains there is much variety, but all the lower elevations enjoy longer growing seasons and milder winters than the lands to the north. The eastern shore of the Black Sea has a humid subtropical climate, the only area of this type in the region. Moving away from the Black Sea, the climate becomes more arid.

Stop & Think

How might development be affected by Russia's extreme climate?

Environmental Challenges

Some of the world's worst environmental horror stories emerged from the U.S.S.R., including the **Chernobyl** nuclear reactor

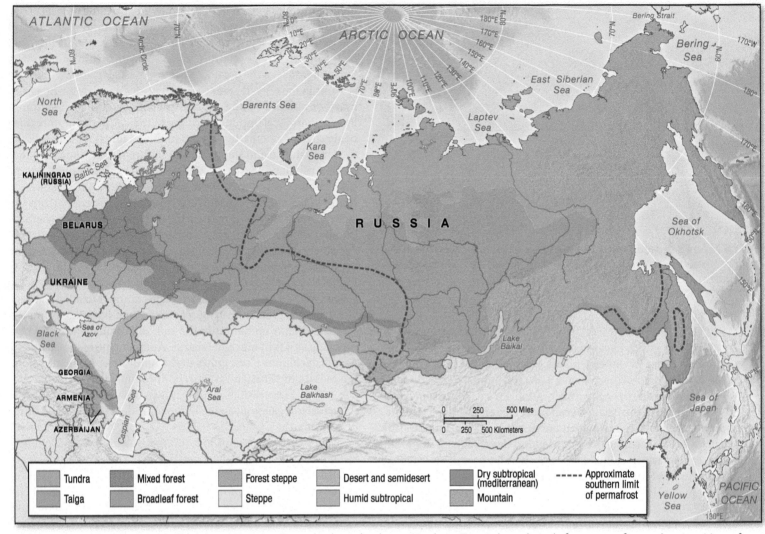

Figure 5-3 Natural regions of Northern Eurasia. For such a huge landmass, Northern Eurasia has relatively few types of natural region. Most of the area is inhospitable to human settlement because of the seasonal extremes of the continental climate, the short growing season, and environmental hazards, such as permafrost.

meltdown and the destruction of the Aral Sea. Countless other environmental problems are known to exist in the region, involving air, water, and ground pollution, erosion, threats to biodiversity, and other issues. All the major rivers in the European part of the area suffer from diverse and numerous sources of pollution. The Black Sea and the Sea of Azov are heavily polluted. Forests have been clear-cut in cold areas where it takes many decades for them to regenerate. The Ural Mountains include a broad array of environmental hotspots, including the site of another nuclear radiation disaster near Chelyabinsk. Oil pipelines from western Siberia leak into fragile tundra and taiga landscapes, harming both wildlife and indigenous people. The Arctic city of Norilsk, with its massive smelters for nickel and other ores, has destroyed a forest the size of Britain with acid precipitation. Nobody in cities drinks tap water—St. Petersburg's water is infamous for giardia, a tenacious intestinal parasite. There are far too many hotspots,

◄ Figure 5-4 The tundra's sublime, but short-lived beauty. In midsummer the sun does not set on this treeless landscape. But the long winters are frigid and dark.

◄ **Figure 5-5 Summer in the taiga.** The mighty Yenisey River dominates the landscape near the city of Krasnoyarsk. "Taiga" is the term for the vast coniferous forests that cover almost half of Russia's surface area.

Belovezhskaya Forest, one of the last undisturbed representatives of Eastern European mixed forest, which is home to lynx, wolves, and European bison. Ukraine stretches to the Carpathians in the west and the Black Sea in the south. But nothing surpasses the Crimean peninsula, particularly its southern area with its mediterranean climate. Georgia also has a relatively mild climate, as well as outstanding vineyards. Azerbaijan offers access to mountains and the Caspian Sea, and Armenia's Lake Sevan is the world's largest alpine lake. These and other natural attractions could become economic assets in the future, if stability can be achieved.

but the political will to deal with these environmental challenges is lacking (see *Exploring Environmental Impacts:* The Unreality Show).

Given the size of the region, vast tracts of generally unpolluted wilderness remain as well. Russia has established 95 national natural reserves and 32 national parks so far. They include Wrangel Island (Vrangelya) with its polar bears and walruses, a central Siberian reserve in the heart of the taiga, the Samara Bend National Park on the Volga River, and the Dagestan Marine Reserve on the Caspian Sea. Wrangel Island and nine other natural areas in Russia have been designated World Heritage Areas by UNESCO. Infrastructure and the types of services needed by most tourists are lacking, but the growing popularity of ecotourism offers hope that the reserves and parks can increasingly pay for their own protection.

One area in particular stands out as a mecca for nature lovers. **Lake Baikal** is one of the greatest natural wonders in the world (Figure 5-6). Situated in a geologic rift zone, Baikal's maximum depth of more than a mile (1.6 kilometers) easily makes it the world's deepest lake. It holds approximately 20 percent of the world's freshwater, comparable to all of the Great Lakes combined. Naturally isolated by mountain ranges from its surroundings in East Siberia, the ecosystem of Baikal is unique. Perhaps the most attractive of Baikal's native species is the nerpa, a freshwater seal. Baikal was the focus of the first environmental protection activity in the U.S.S.R. and, although there are some sources of pollution in the area, it remains relatively undisturbed.

Russia's neighbors also have potential ecotourism assets. Belarus shares with Poland the

Historical Background: Territory and State-Building

Northern Eurasia has been shaped by forces coming both from the "East" and the "West." As Christian outliers in the Caucasus since the fourth century, Georgia and Armenia have always persevered between more powerful neighbors. As for the rest of this vast land, its story begins with the movement of Slavic tribes through forested river valleys out into the steppe. There, Slavic farmers encountered nomadic herding peoples with whom they traded and often warred. By the ninth century, the Slavs had developed a complex society based on towns and trade, as well as agriculture.

From Kievan Rus to Tsarist Russia

Ukraine, Belarus, and Russia share a common state ancestor, Kievan Rus. The **Dnieper River** was the axis of an extensive trade network that linked the Baltic and Black Sea regions. Kievan Rus's close relationship with Byzantium, based in Constantinople (today Istanbul), brought the Cyrillic alphabet and the Orthodox Christian

► **Figure 5-6 Lake Baikal—"Siberia's Gem."** The world's oldest and deepest lake, Baikal holds as much freshwater as the five Great Lakes combined. Its unique ecosystem sustains dozens of endemic species. It was designated a UNESCO World Heritage Site in 1996.

EXPLORING ENVIRONMENTAL IMPACTS

The Unreality Show

The biggest environmental challenge facing Russia may well be the mentality of the ruling elite, especially Vladimir Putin. Since 2000, Putin has displayed a highly politicized perspective on the environment. He seems to view environmentalism, at best, as an impediment to economic development or, at worst, as sabotage. Rather than backing concerted efforts to improve the environment, Putin indulges in widely publicized displays of his personal style of "applied ecology."

Putin's disregard for environmental protection was evident from the beginning. Shortly after taking office as president in 2000, he abolished the Russian State Committee on Environmental Protection and the Russian Forest Service, and transferred their functions to the Ministry of Natural Resources. The latter's function is to facilitate exploitation of natural resources, not to protect them. The following year the Federal Ecological Fund was eliminated. In 2002, the "polluter pays" principle was dropped from environmental law. Although fines were later reinstated, they are too small to matter. In 2011, when the environmental organization Greenpeace publicized 14 oil spills in West Siberia, the government levied a $37,000 fine on Lukoil, one of the country's largest energy companies with annual sales exceeding $100 billion.

Existing environmental legislation lags well behind European standards, but enforcement is an even greater problem. Russia's bureaucracy is notoriously inefficient, and since Putin came to power it has nearly doubled in size. Still more damaging is corruption among government officials. Transparency International's 2012 Corruption Perception Index lists Russia in a tie for 133rd place; this means that the public sectors of more than 130 countries are seen as less corrupt. Environmental regulations generate extra income for many officials, who see themselves as above the law. A 2009 helicopter crash in the mountains of southern Siberia illuminated the regime's hypocrisy in regard to the environment: seven officials from the federal, regional, and local levels were killed in the crash, including the region's top environmental

protection official. The party had been hunting wild argali sheep, an endangered species, from the helicopter. This sort of poaching is punishable by up to two years in prison, but for that to happen the law must be enforced rather than flaunted.

Instead, the law is more often turned against environmental activists. In 2012 two activists were sentenced to three years in prison for graffiti on a fence surrounding the construction site for a mansion belonging to the governor of Kuban region. The activists claimed that the site is in a federally protected nature preserve. The Putin administration is especially antagonistic toward foreign-based environmental groups (Figure 5-1-1) or Russian groups receiving foreign funding. Since 2012 the latter, like all NGOs and individuals receiving support from abroad, must register with the state as "foreign agents."

President Putin occasionally pays lip service to environmental issues. In 2010 he pledged to begin a cleanup of the Arctic region. But in 2012 the government dissolved the group that was most concerned about the Arctic environment, the Russian Association of Indigenous Peoples of the North, Siberia, and Far East. In 2011 Putin gave an impromptu lecture on the environmental risks of the "fracking" process used to exploit shale gas deposits, which threatens Russia's position as the world's leading natural gas exporter. Putin has televised a series of adventures to dramatize environmental issues. Originally, these minidramas were broadcast as if they were "reality TV," but later Putin acknowledged

▲ **FIGURE 5-1-1 Greenpeace activists approach a Gazprom oil platform in the Russian Arctic.** Gazprom, the world's largest natural gas company, is controlled by the Russian government. Thanks to a law passed in 2013, Greenpeace is considered a "foreign agent."

that they had been staged. In 2008 he shot a rare Ussuri tiger with a sedative dart, ostensibly saving a film crew. Other clips in the series involve a polar bear, a snow leopard, a gray whale, and Siberian cranes.

Russian history is replete with examples of deceptive appearances. Catherine II's chief advisor created fake ("Potemkin") villages to lead the empress to believe that Ukraine's development was well advanced, when in fact it had hardly begun. Stalin held "show trials" of former comrades to convince society that "enemies of the people" abounded, when actually his victims were loyal to Communism. Putin today stages "reality shows" to signal the regime's concern for the fate of iconic species, when Russia's environmental problems continue to proliferate. At the same time, Russia's genuine environmentalists face increasing constraints on their activities.

Sources: Environmental Issues and the Green Movement in Russia, *Russian Analytical Digest* 79, 27 May 2010, http://www.css.ethz.ch/publications/pdfs /RAD-79.pdf; Laura A. Henry, *Red to Green: Environmental Activism in Post-Soviet Russia* (Ithaca, NY: Cornell University Press, 2010).

faith. Kievan Rus from the tenth to the thirteenth century was not at all "backward" compared to other European realms. Although the state was a monarchy, towns usually managed their own affairs. The merchants of Novgorod, for example, set up an elected assembly to rule their town, limiting their prince to the role of military leader.

Because of its size, competition among the princes, and a lack of centralizing institutions, Kievan Rus began to fragment from within. The European Crusades shifted trade away from the Dnieper network, causing the Kievan Rus economy to decline. In the second half of the twelfth century the frontier region of Vladimir-Suzdal in the northeast emerged as the main power center. Finally, attacks in the early thirteenth century by the Golden Horde of Ghengis Khan proved unstoppable. In 1240 Kiev itself was sacked. The Golden Horde was led by ethnic Mongols, but most of the troops were Turkic-speaking warriors, whom the Russians knew as "Tatars." Their descendants today form the largest non-Russian nationality in Russia. Kievan Rus's demise was swift and violent; only 80 of 300 towns survived the invasion. History offers few better examples of "backwardness" imposed by defeat in war. Over time Galicia (now western Ukraine) came under the influence of Poland, while the lands to the north (including most of present-day Belarus) came under Lithuanian rule.

In Vladimir-Suzdal, the town of Moscow gradually came to dominate the region. Moscow's princes sought to gather the lands of Kievan Rus together again, but they remained subservient to the Tatars until the 1470s, when the horde split three ways and Russia was able to begin expanding to the east. During this period, the foundations for a society much different from Kievan Rus's were established. The new state's leaders were determined to overcome the political fragmentation that had undermined Kievan Rus. Steadily Moscow's Grand Princes increased their power over lesser princes and other aristocrats. In conquered areas, the Grand Prince installed new elites composed of landlords who swore to serve the state. To support loyal servitors, peasants were bound to the land and required to serve these landowners, and over time this practice of **serfdom** turned into slavery. Russian peasants were losing freedoms at the same time that feudalism was breaking down in Europe. In the West, the emergence of capitalism and the cultural ferment of the Renaissance led to a focus on the individual. But in Russia there was a renewed emphasis on the greater community, as embodied by the state and the Orthodox Church (Figure 5-7). It fell to Russia to uphold the Orthodox faith after the fall of Byzantium in the fifteenth century. As leader of the "Third Rome," Russia's Grand Prince became Tsar, or "Caesar." Thus emerged Russian society's fundamental principle, from the rise of Moscow until the collapse of the U.S.S.R.: the duty of all people, of whatever rank, is to serve the state.

In the mid-sixteenth century, Ivan IV ("the Terrible") took this principle to extremes when he attacked the nobility, seeking to replace them with hand-picked servitors. This effort failed in the end, but Ivan greatly expanded state territory. Russian fur hunters pushed into the East and reached the Pacific a century later. Ivan defeated the Tatars of Kazan and Astrakhan to acquire the Volga region. He also tried to gain a foothold on the Baltic, which would have completed a vast trade network under Moscow's control, but 25 years of war gained nothing. When Ivan died the exhausted state collapsed and chaos reigned in the "Time of Troubles." In 1613 an assembly of nobles elected a new Tsar, Mikhail I, the first of the

▲ **Figure 5-7 Russian Orthodox Cathedral at Zvenigorod.** Kievan Rus adopted Christianity over 1,200 years ago. The Communists tried to eliminate religion after the 1917 revolution, but today the state embraces the Orthodox Church and formally respects Islam, Judaism, and Buddhism as traditional faiths in the country. But newcomers, such as Baptists and Mormons, are not welcome.

Romanov dynasty. The centralized autocracy was restored and territorial expansion resumed. Russia reached the Dnieper and won Kiev in 1654.

The Russian Empire

Russia in 1700 was big (Figure 5-8), but backward in terms of trade, technology, and modern culture. Tsar Peter I ("the Great") transformed the country by introducing new ideas and technology from the West. But he did not reject the fundamental principle of service to the state. To the contrary, he pushed the service principle to an extreme not seen again until the reign of Josef Stalin in the twentieth century. Peter even subordinated the church to the state. Having modernized the army and created the Russian navy, Peter defeated Sweden and gained access to the Baltic. Now "Emperor," he chose a marshy, flood-prone site on the Gulf of Finland for St. Petersburg, Russia's "Window on the West" (Figure 5-9). A dazzling European city was built by serfs, thousands of whom died due to deplorable conditions on the site. Peter I also established a new industrial zone in the Ural Mountains to exploit the plentiful iron ore deposits. He built canals to link St. Petersburg with the Volga and to gain access to the Urals' metals. Transportation remained difficult, but the state did not have to depend on foreign suppliers for cannons or bayonets.

Peter the Great cast a long shadow, but German-born Catherine II ("the Great") in the latter part of the eighteenth century made her own mark. Attracted by the values of the Enlightenment, she nurtured the arts and education in Russia. But she did not allow new ideas and sensibilities derived from the West to radically affect her governance of the empire. The position of the serfs in society continued to deteriorate as the power of the landlords grew. In the West, the American and French national revolutions plus "revolutions" in industry and agriculture were ushering in the modern age, but in Russia the state-society regime remained fundamentally the same. Autocratic and state-centered, Russia relied on expansion, not innovation.

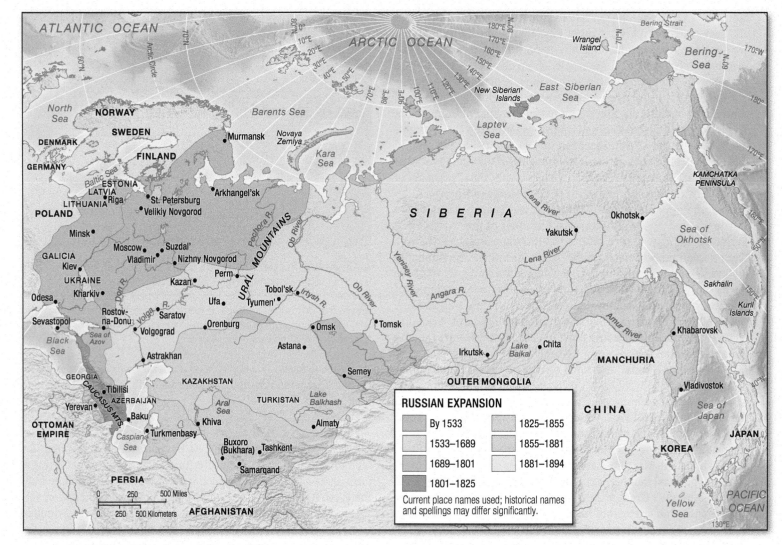

▲ **Figure 5-8 Russia through the ages.** Russia began as "Muscovy" near the upper reaches of the Volga River. Today Russia is the largest country in the world, but the U.S.S.R. was larger, and the late Russian Empire larger still.

Catherine II acquired an area the size of Spain that would later allow Russia to become an agricultural giant and an industrial power. First to fall were the Crimean Tatars, which gave the empire control over the north shore of the Black Sea. With this conquest, Russia acquired the black earth of the Ukrainian steppe, with its longer growing season, as well as maritime access to world markets.

In time, even more valuable resources were found beneath the soil, especially hard coal and iron ore. Catherine had the beautiful port of Odesa built to celebrate the success of her reign and to support the grain trade (Figure 5-10). Russia also played a principal role in the partitioning of Poland, which caused that country to disappear from the map. By annexing eastern Polish areas, Russia acquired the territory that would become Belarus, as well as the rest of Ukraine, except for Galicia, which went to the Austro-Hungarian Empire.

◀ **Figure 5-9 St. Petersburg during "White Nights."** For a few weeks around the summer solstice the sun seems to slip along the horizon late at night, creating an otherworldly glow that enhances the beauty of the former capital's historic buildings and canals.

◀ Figure 5-10 **Odesa on the Black Sea.** Odesa was founded by Catherine the Great to be the Russian Empire's main southern port. Today Odesa is Ukraine's largest port and a major tourist attraction. Here visitors enjoy the view from the top of the staircase made famous in the classic film, "Battleship Potemkin."

Russia acquired Finland and Moldova in the early 1800s, and began to move into the Caucasus. By 1828 Russia controlled Dagestan, Azerbaijan, Georgia, and eastern Armenia, but Chechens and other Caucasian peoples resisted for decades. Georgia was

an ancient country, Christian from the fourth century. Stuck between the Iranians and the Turks, Georgia may have found Russia the lesser evil. The Armenians, also Christians from the fourth century, were in the same predicament. In 1800 Armenia had long been divided into western and eastern parts, held respectively by Turkey and Iran. Azerbaijanis were Turks who settled west of the Caspian in the eleventh century. For a long time they too were ruled by Iran, from which experience they acquired the Shi'ite understanding of Islam. Russia expanded and ruled over an ever-wider range of culturally diverse peoples (Figure 5-11).

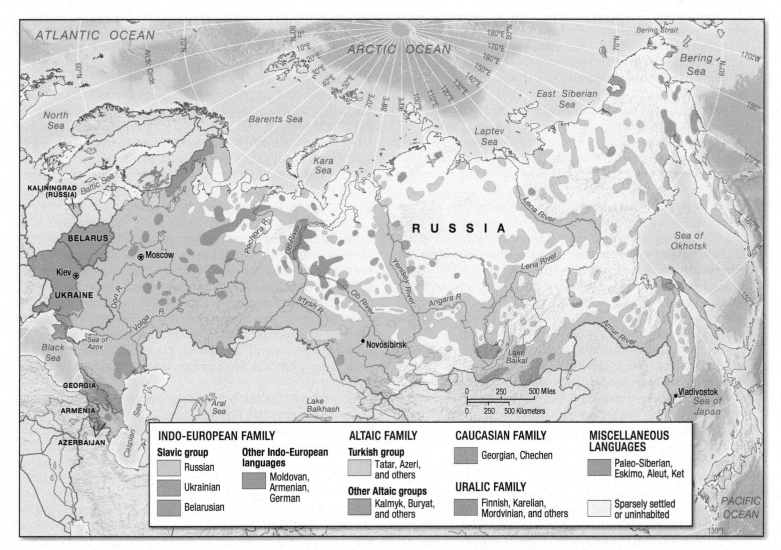

▲ Figure 5-11 **Major language groups of Northern Eurasia.** Two words are used for "Russian": a "russikii" is an ethnic Russian, a "rossiyanin" is a citizen of Russia. This map of languages shows the ethnic diversity in Russia.

In 1800 Russia produced more iron and steel than any other country in the world. But the Industrial Revolution soon left Russia behind. In the Crimean War of 1853 Russia was defeated on its own territory, due largely to the technological edge enjoyed by Britain and France. This shock energized the reform movement. Serfs were formally emancipated in 1861, although most remained bound to their villages by law. But some escaped to the cities to find jobs in industry. The government began to industrialize in order to catch up with the West. The focus was on steel, railroads, and textiles, with grain exports paying for technology imports.

The development program was a mixture of government control and capitalist enterprise, the latter mostly foreign. The Urals languished because modern steel production required coal, which the region lacked. Instead, two linked industrial districts emerged in the south—the coal-rich Donbas, in the Donetsk River basin, and the mid-Dnieper River region with its plentiful iron deposits. Railroads linked the complementary regions and consumed much of their steel output. Azerbaijan's leading city, Baki (Baku), became an oil-boom town. St. Petersburg was the country's largest industrial center and most important port. Moscow grew on the basis of textiles and its centrality in the growing railroad network. Kiev, Kharkiv, and Odesa in Ukraine all flourished thanks to industry and trade.

The rural population more than doubled in 50 years, but agriculture remained backward. Peasants paid 10 times as much tax per unit of land as landlords. Desperate peasants plowed marginal lands, causing erosion and exacerbating poverty. They thought the solution was to obtain more land, but the landlords and the state stood in their way.

For industrial workers, working and living conditions often were very harsh. Life expectancy was lower in urban slums than in primitive villages. But the city offered possibilities for remaking oneself. New urban workers encountered a range of ideologies, some radical, that advocated progress by means of reforming or overthrowing the regime.

Russia may have had a developing country's economy, but it had the ambitions of a superpower. In the 1860s the empire expanded into Central Asia, soon building a railroad to bring the region's cotton to Russia's textile centers. The **Trans-Siberian Railway** was completed to link west and east and to propel Russian expansion. As originally constructed, it cut across Chinese Manchuria, which Russia sought to develop and dominate. Subsequent efforts to project Russian influence into Korea led to a disastrous war with Japan in 1904–1905.

The regime ruled by tradition and force—the people were subjects, not citizens. When the war went badly a massive uprising broke out. Peasants seized landlords' estates and workers began to organize their own governments—"soviets" (councils). The regime overcame the 1905 Revolution with a mixture of violent repression and compromise. Many workers and peasants were executed or imprisoned. But the Tsar did agree to a representative body, the **Duma**, which reformers hoped would lead to a democracy.

Industrial development resumed, with strong government involvement and dependence on foreign capital. The government attempted to replace traditional communal agriculture with detached family farms, hoping to create a conservative class of rural property owners. Perhaps Russia would have successfully modernized, but World War I intervened.

Stop & Think

▶ What continuities can be found in the relationship between state, society, and territory in the five centuries preceding the Bolshevik revolution?

The Soviet Union

Russia's economy broke down during World War I, and breadlines turned into riots. In 1917, the Tsar was forced to step down. The new Provisional Government was democratic, but it continued fighting the war. The people wanted what **Vladimir Lenin**, leader of the **Bolshevik** ("Majoritarian") party, promised to give them: "Bread, Peace, and Land." Lenin charted a radical course, extreme even by Marxist standards. Karl Marx argued that capitalism would have to be completely developed and the world's workers become the majority before the revolution could succeed. Lenin felt that a dedicated vanguard party could free the masses, which would inspire the workers in more advanced countries to rebel and come to the aid of socialist Russia. The 1917 October revolution soon led to chaos. Lenin resolved to build a new type of state, one that would restore and maintain order while at the same time transforming society. In theory, governing was the job of a hierarchy of soviets—hence the name Soviet Union—while the party, soon renamed the **Communist Party**, was to advance the social revolution. In practice, the party took control over the soviets.

The new regime's first great challenge was to survive civil war and foreign intervention. Although there were pitched battles, probably more people were executed in captivity. The worst scourges were hunger, cold, and disease, which killed about 10 million people. In the civil war the state tried to control all economic activity. But in 1921, with the economy in shambles, Lenin announced the New Economic Policy (NEP), which allowed some marketing of agricultural produce and consumer goods. Political opposition was not tolerated, but culturally this was a period of relative freedom, except with respect to religion. Throughout the Soviet period, no one who was not an atheist could rise in society.

Lenin died before he could flesh out the NEP or anoint a successor. In the end, a Russified Georgian, **Josef Stalin**, overcame all rivals. Stalin argued that Russia would be conquered if it did not rapidly develop. He aimed to create "Socialism in one country." Stalin launched collectivization to directly take charge of all farming. Peasants who resisted were killed and many more were exiled to remote labor camps—the Gulag labor camp system. The government exported much of the harvest to pay for imports of advanced machinery. The result was a massive famine in Ukraine, along the middle Volga River, and in other regions. Again, millions died.

The other thrust of Stalin's strategy was rapid industrialization under the banner of the **Five Year Plan**. In reality, Soviet planning meant micromanagement down to the level of specifying how many loaves of bread would be baked. The plan was backed by force from the start (Figure 5-12)—economists and engineers who voiced doubts were shot—but the industrialization drive also generated tremendous enthusiasm, especially among younger workers. They believed they were building a new world. Unprecedented avenues of social mobility opened up for those who could achieve results, and the government provided education, health care, pensions, and other benefits. Under Stalin, everybody served the state, and the

▲ **Figure 5-12 FSB (formerly KGB) Headquarters in Moscow.** The Lubyanka building serves as headquarters of the Federal Security Service (FSB in Russian) just as it housed the KGB, the feared Soviet secret police. Visible in the foreground is a stone that serves as a memorial to the millions of victims of political repression in the U.S.S.R.

state promised to take care of the people. At the same time, millions were identified as enemies of the people and sent to the Gulag.

"Socialism in one country" aimed for self-sufficiency. The regime expanded existing industrial centers, such as the Donbas, the mid-Dnieper region, Moscow, and St. Petersburg (Leningrad) (Figures 5-13 and 5-14). But the Soviets also invested massively in the development of northern and eastern regions to achieve self-sufficiency and to base industry beyond the reach of invaders. One key project found coal for the Urals in the Kuznetsk Basin (Kuzbas), 1,500 miles (2,400 kilometers) to the east. The government built two huge metallurgical centers, sending Siberian coal to the Urals and the latter's iron ore to the Kuzbas by rail. Only a state that took no account of transport costs could contemplate such a project! Labor costs also were held down on difficult projects—such as the Pechora coal mines, the Kolyma gold fields, and the smelting center of Norilsk—thanks to the forced labor of the Gulag.

The Soviet industrialization campaign was a huge success in the heavy-industry sector, but light industries such as textiles and food processing lagged far behind—in the U.S.S.R., there was no consumerism. Heavy industry supported a military buildup in the expectation of war, which came when Nazi Germany invaded in

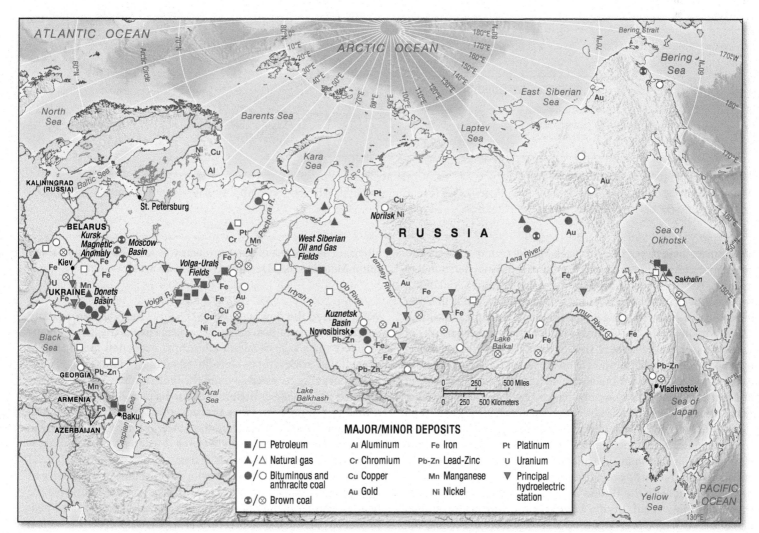

▲ **Figure 5-13 Natural resources.** Russia and Ukraine are rich in minerals. High-quality coal is found in the Donets and Kuznetsk basins. Oil and gas are produced in Azerbaijan and the West Siberian and Volga–Urals fields. The Volga and Yenisey rivers provide hydroelectric power.

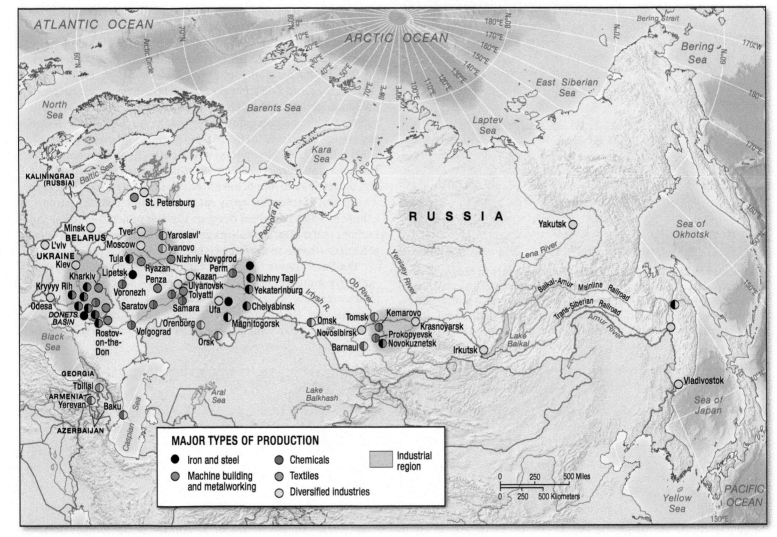

▲ **Figure 5-14 Industrial regions.** Industry in Northern Eurasia is concentrated in six regions. Soviet industry was oriented toward capital goods. In recent years, production of consumer goods has increased.

1941. Soviet losses were astronomical, but the country managed to evacuate 1,200 facilities and over a million industrial personnel to the East. The people rallied in support of the war effort. Millions of women moved into industry, some working double shifts in freezing conditions, even sleeping on the shop floor until the next shift.

Victory in World War II confirmed for Stalin the superiority of his repressive, closed system. The "liberated" countries of Eastern Europe became Soviet puppets. The Soviet military's strength lay not just in sheer numbers, but also in technology. The U.S.S.R. soon joined the nuclear "club" and were the first to put a human into outer space in 1961. The Soviet system excelled at focusing science and industry on the regime's top priorities, but most people endured persistent scarcity of consumer goods and sometimes lacked necessities.

After Stalin's death in 1953, his successors maintained political control without resorting to massive terror campaigns. The Soviet bloc competed with the United States and its allies for influence throughout the world. In this **Cold War**, the Soviets developed maximum self-sufficiency. Huge projects—depicted in a heroic and patriotic light to attract the participation of young Communists, the leaders of the future—transformed remote regions, especially in

Siberia. The greatest of these projects was also the last, the Baikal-Amur Mainline (BAM) railroad (see Figure 5-14).

The Disintegration of the U.S.S.R.

What caused the downfall of the U.S.S.R.? Several factors stand out from a geographical perspective. Scale heads the list—in many respects the scope of Soviet ambitions exceeded management capabilities. Soviet troops were based en masse in Eastern Europe and an even larger force covered the long border with China. The entire length of the Soviet border had to be patrolled because it was a closed country—airplanes that accidentally strayed into Soviet airspace were shot down. And the Soviet military was fighting in Afghanistan for 10 years, while trying to keep up in the military-technology race with the United States.

The **Kremlin**, seat of the Soviet government in Moscow, was trying to maintain control over everything and everybody from the heart of Europe all the way to the Bering Strait. Although some people, mostly Russians of the World War II generation, saw themselves primarily as Soviet citizens, national identities grew increasingly more important, not just in the countries of the Eastern European bloc such as Poland, but within the Soviet Union itself, especially in the Baltic lands, Ukraine, Georgia, and Armenia.

Following the Chernobyl nuclear catastrophe in 1986, outrage over Soviet despoliation of the environment swept the country, often fueling the flames of nationalism. Chernobyl contaminated much of Belarus, and the effects included devastating the health of thousands of children. Ukraine had suffered from pollution dating back well over a century. None of the 15 Soviet Socialist Republics that constituted the U.S.S.R. was exempt from the ecological destruction of Soviet forced-march development. In Russia itself, every day seemed to bring another depressing revelation of ecological distress.

People were tired of having to stand in line for a limited supply of low-quality products that lacked diversity and style. As the populace came into contact with more and more foreigners and foreign goods, some Western products—Levis, Marlboros—became fetish items. The West seemed the land of plenty, while Soviets put up with rationing.

When **Mikhail Gorbachev** became General Secretary of the Communist Party in 1985, he was the youngest leader the country had known in 60 years. He first tried a moderate course of reform based on the historical precedent of the NEP, but soon found that more radical measures were required because the party itself was the main obstacle to progress. The reforms were deployed under the slogan *Glasnost, Perestroika, Demokratiia* (Openness, Restructuring, and Democracy). Chernobyl became a test of glasnost that the regime initially failed when it denied that anything serious had happened. But soon the floodgates of information were forced open and all the current environmental and political issues were exposed along with all the bloody secrets from the past.

Demokratiia made considerable progress. When it became clear that the East European nations would eagerly leave the Soviet bloc if given the chance, Gorbachev let them go without firing a shot. At home, Gorbachev sponsored the First Congress of People's Deputies, an elected representative body that was beyond party control. Glasnost became reality when the Congress's sessions were broadcast nonstop to the entire country. Nobody could remain unaware of human rights issues, national minorities' grievances, economic shortcomings, and environmental problems.

The result was traumatic. The Soviet media previously had not even reported traffic accidents or fires. Now everything bad that had happened over 70 years, plus all the current problems of the immense multiethnic empire, came flooding in all at once. It did not help that Gorbachev, who became the first and only elected president of the U.S.S.R., never managed to work out a coherent plan for perestroika. He allowed cooperatives to operate on a market basis, hoping they would efficiently provide services to the population, but he was unable to protect them either against the bureaucrats or the extortionists who formed the first Russian mafias. Industry and agriculture remained under bureaucratic control.

In 1991 Gorbachev was kidnapped by conspirators who sought to restore the old regime. The coup was poorly organized and proved incapable of overcoming the opposition of the Russian government, led by Boris Yeltsin. Gorbachev was freed, but he no longer mattered. The Soviet regime had lost its legitimacy. While Gorbachev struggled to hold the Soviet Union together, the leaders of Russia, Ukraine, and Belarus met and agreed to declare their countries' independence. No referendum was held to ratify their decision, but neither was there much protest. The immense anticapitalist empire

that changed the course of history was finished because three men had the courage to say so and 280 million people followed their lead.

Putting an end to the political regime was surprisingly simple, but breaking up the tremendous economic system, "U.S.S.R., Inc.," was much more problematic. The Soviet economy to a large extent was a unified, closed system. To separate Belarus, for example, from Russia was to intervene in chains of production for many key products. Most important in this regard were the unified networks for providing natural gas, petroleum, and electricity, as well as the railroad system. Nationalist politicians, especially in Ukraine, overestimated both the market value of their countries' products and the economic benefits of separation from Russia. There was also the expectation of massive aid from the West, which proved elusive.

Other problems carried over into the post-Soviet period. Territorially, the U.S.S.R. was an ethnic patchwork. The Russian Federative SSR was bordered by 14 other SSRs that were based on non-Russian nationalities and formally granted the right of secession by the Soviet constitution. Nobody expected that any would be allowed to secede—the constitution had rarely been observed in the regime's practices—yet when the country did disintegrate, the fissures followed the lines of the Soviet map. But this breakup did not put an end to all ethnic tension. In all 14 of the former SSRs, there were sizable Russian populations that were made into minorities overnight. Where they were viewed as former colonizers, for instance in Estonia and Latvia, their status became that of second-class citizens. In Ukraine and Kazakhstan, the large concentration of Russians in industrial districts and other key areas complicated domestic politics and gave cause to worry that Russia might one day seek to regain these territories.

Stop & Think

> The U.S.S.R. was supposed to be bringing the world a better way of life; it also was a "superpower." Why did it collapse?

Northern Eurasia in Transition

Since 1991 Western observers have spoken of the Northern Eurasian countries as being in transition. Usually it is assumed that after the transition the countries will resemble the United States or Western Europe in having democratic states and market economies. But overuse of the "transition" label obscures some unpalatable realities. None of these countries has an economy that is fully governed by the market. Georgia and Armenia have made the most progress in this direction, while Belarus lags the most. Georgia and Ukraine had "revolutions" that many hoped would pave the way for full-fledged democracy, but Ukraine today is headed in the opposite direction. Azerbaijan and Belarus have authoritarian governments, while Russia has shifted in that direction. Even in Georgia and Armenia, democracy cannot be taken for granted and its survival is not guaranteed. It is important to look at current realities in the Northern Eurasian countries from a geographical perspective (Table 5-1) and consider the gap between what is happening there today and what is needed for them to succeed in this era of globalization.

Population Contours

The most densely settled areas of Northern Eurasia are southwest of the Ural Mountains, where the most favorable conditions for productive agriculture occur in this region (Figure 5-15). Here densities of over 100 people per square kilometer can exist. But the most

Table 5-1   The Countries of Northern Eurasia

		Area		Natural Increase Rate (%)	Population, 1999 (in millions)	Population, 2012 (in millions)	Largest Ethnic Group (% of total)	Second Largest Group (% of total)	Principal Religion
	Capital	Sq. Miles (km) (in thousands)							
Russia	Moscow	6,593 (17,076)		−0.1	146.5	143.2	Russian (80)	Tatar (4)	Eastern Orthodox
Ukraine	Kiev	233 (604)		−0.4	49.9	45.6	Ukrainian (78)	Russian (17)	Eastern Orthodox
Belarus (Belorussia)	Minsk	80 (208)		−0.3	10.2	9.5	Belarusan (81)	Russian (11)	Eastern Orthodox
Azerbaijan	Baki (Baku)	33 (86)		1.3	7.7	9.3	Azeri (91)	Russian (2)	Islam
Georgia	Tbilisi	27 (70)		0.2	5.4	4.5	Georgian (84)	Azeri (6.5)	Georgian Orthodox
Armenia	Yerevan	12 (31)		0.5	3.8	3.3	Armenian (98)	Kurds (1)	Armenian Orthodox

Sources: Population Reference Bureau, *World Population Data Sheet* (Washington, D.C.: Population Reference Bureau, 1999 and 2012); Central Intelligence Agency, *CIA World Fact Book 2008* (New York: Skyhorse Publishing, 2007).

▲ **Figure 5-15 Population distribution of Northern Eurasia.** The most densely settled areas of Northern Eurasia are where arable land attracted agriculturalists. But large settlements in forbidding locations also exist in the north and east, a legacy of the largely self-sufficient Soviet economy.

◀ **Figure 5-16 Billboard promotes campaign against HIV/AIDS.** After the Soviet Union ended, health care in Russia deteriorated, and life expectancy for Russian men and women dropped. Today the decline has been arrested, although problems remain. HIV/AIDS is spreading rapidly, especially among young people.

densely settled areas today are urban centers where up to three quarters of the region's population is found. The national density average in Russia of 8.4 people per square kilometer (22 per square mile) is quite low because of vast spaces with climatically harsh conditions. But the country is also characterized by large settlements in forbidding locations in the north and east, a legacy of the largely self-sufficient Soviet economy.

Russia has been trying to get ethnic Russians abroad to "come home" because the country's population has been shrinking since the breakup of the U.S.S.R. Ukraine and Belarus have also experienced population decline, while in Armenia and Georgia the rates of natural increase have declined to well below the global average (1.2%), but remain positive. Only Azerbaijan, with nearly half its population rural, has a natural increase rate near the world average.

Why are the Slavic countries in the region losing population so rapidly? A single root cause may be difficult to identify. Many young women have left the region, including up to half a million who have been lured or coerced into prostitution in foreign countries. With respect to birthrates, a fundamental factor is that about 10 percent of women are sterile due to complications from abortions, which during the Soviet period were the primary means of birth control. Although abortion rates are declining as more modern means of contraception become available, all three countries rank near the top worldwide in abortions per capita. According to a 2007 UN study, Russia has the dubious distinction of being the world leader with 53.7 abortions per 1,000 women of childbearing age. By European standards, the three countries have very high infant and maternal mortality rates, so the quality of basic health care is a major problem (Figure 5-16). The weakness of the nuclear family and of social institutions leaves many children vulnerable. In Russia there are 2–4 million children (out of a total of 35 million!) living on the streets.

Death rates are the other part of the equation. Environmental pollution in too many places is a threat to health and longevity. In coal-mining towns, for example, excess mortality due to air pollution ranges as high as 19 percent. Lifestyle factors also play a part; in particular, all three countries have high levels of tobacco and alcohol consumption. The rate of infection with HIV/AIDS is growing rapidly, mostly among intravenous drug users but increasingly through

heterosexual contact as well. Russia has about a million people infected with HIV. Violent death is all too common in the region. The number of automobiles is growing rapidly, as is the death count from accidents—currently the fatality rate per automobile is 8–10 times that of West European countries. Russia is infamous for its murder rate, but all three Slavic countries are among the top five for homicides per capita in Europe. This is in part a consequence of alcohol abuse, as are the very high suicide rates.

Russia: The Eurasian Power

In late 1991 there was much excitement in the new Russia. For the first time in their lives people were free to voice their opinions in public, and many did. It appeared that all of the old, repressive institutions would be replaced by new democratic agencies of the people's will. Russian President Boris Yeltsin even launched a fundamental reform of the dreaded KGB, the secret police. Despite some initial successes, euphoria faded and the institutional bases for democracy were slow to materialize. The KGB fought off major reform and regrouped as the Federal Security Service (FSB). Many people who had been part of the old elite found ways to enter the new one—Communists became "democrats" overnight, by donning an imported suit and organizing a photo op at a church.

Russia has undergone a transition, but not exactly the one that was expected in the early 1990s. That chaotic decade gave way to a period of relative stability—some would say "stagnation"—under the authoritarian leadership of Vladimir Putin. Although there has been much talk of innovation and reform in various fields, little has been accomplished. The revenues from oil and gas exports have bolstered the economy and enabled Russia's elite to cling to power. Yet the outbreak of major protests in 2011–2012 and the emergence of civil society in Moscow and other major centers may be harbingers of changes to come.

Post-Soviet Economic Policy

President Yeltsin himself had been a Communist Party boss for decades. He was skilled in the arts of deal-making and political maneuvering, but he knew little about economics. Russian economists and some Western advisors were convinced that the quickest transition to capitalism would be the least painful, but hyperinflation was a grave threat. There were too many rubles in circulation, in savings accounts, and in mattresses. Yeltsin's advisors were afraid to just eliminate state controls and let the

◀ **Figure 5-17 Shopping complex in Moscow.** The retail sector continues to expand in Russia, often without regard for preexisting geographies.

market determine prices and wages because there might be "too much money chasing too few goods," the classic recipe for inflation. Instead, the government launched "shock therapy," the first stage of which was a government-controlled price increase on goods, including basic foodstuffs—as much as 800 percent overnight. People called the policy "shock with no therapy," as they spent their life savings to put food on the table. They felt they had been robbed.

Worse yet, inflation was not averted. Trade barriers were knocked down and the global economy surged in, in the form of consumer goods that were often inferior by European standards but marked up in price in Russia, where there was a huge repressed demand for imports. Few Russian industries were competitive in world markets. Consumption was globalized, but production declined radically.

Demoralization was not the only negative effect of "shock with no therapy." Corruption existed before, but now it was epidemic. Police, judges, medical personnel, teachers, and others who provide vital services to the public all struggled against instant impoverishment as best as they could. Some former professionals, and more than a few college students, helped form the "mafias" that proliferated in the cities.

Had this been the last such reform, perhaps "democrat" would not be a dirty word for most of the public today. But it was not. A currency reform followed that again penalized people who had saved rubles. Even more damaging was the persistent practice of failing to give people their pay. Some employees, and pensioners, went many months on end without receiving their rightful income. What better way to prevent "too many rubles" than to stop paying salaries! But perhaps the most controversial stage of the transition was **privatization of industry**. Under the Soviet system, the country's economy in theory was the property of the people, although in practice it belonged to the state. Privatization reform promised to make all citizens shareholders in corporations, but in reality the elite concentrated control over all valuable assets. A social-Darwinian struggle of survival between the new entrepreneurs led to the emergence of the **oligarchs**, a few individuals who controlled vast economic empires. Whereas under Communist rule nobody could accumulate private wealth, today Russia is second only to the United States in the number of billionaires.

After the initial crash the economy slowly began to recover until the 1998 financial crisis struck. Foreign investment nearly dried up and the government was forced to devalue the ruble. The period was painful for many, but the ruble devaluation made foreign goods more expensive, which helped Russian producers recapture domestic markets. Growth resumed, and now it was more diversified. Foreign investment returned—not just in retail, but also in production for the Russian market, particularly in the vicinity of Moscow and St. Petersburg, which are the largest markets with the best accessibility (Figure 5-17). Beginning his first term in 2000, President **Vladimir Putin** called for reforms to put business on a more stable footing. Several oligarchs, including oil billionaire Mikhail Khodorkovskii, began improving the transparency of their companies' operations and urged development of appropriate institutions to regulate the economy legally.

But soon the pendulum swung in the other direction. The government imprisoned Khodorkovskii, ostensibly for tax evasion. Interpretations vary, but in reality Putin's government did not welcome the oligarchs' involvement in politics. The regime also began to take control of the crucial oil and gas industries, including Khodorkovskii's giant firm. Natural gas is virtually monopolized by the government-controlled **Gazprom**, the largest natural gas company in the world. Until he became president in 2008, Dmitrii Medvedev served simultaneously as Deputy Prime Minister and CEO of Gazprom. Although this combination of political power and economic control would be considered unacceptable in many countries, it is characteristic of Putin's system. It is well known that top officials, especially the **siloviki**, men from "power ministries" such as the FSB, enrich themselves through **rent-seeking**, the use of the resources of the state for the benefit of private interests.

Stop & Think

▷ How did "democracy" lose popularity in Russia?

Post-Soviet Changes in Society

State policy may be stuck in transition, but on the street the explosive growth of consumerism has changed the Russian way of life. Although most people earn paltry salaries by Western standards, housing costs are very low because most people privatized their Soviet-era apartments for a nominal fee. Now people can buy new TVs or laptops, or even cars, such as they could only have dreamt of in the Soviet past. In major cities, shopping malls and megastores are proliferating, while even in the poorest neighborhoods small groceries offer a much greater range of goods than in the past. The message is clear and most urban people, at least, have gotten it— make money and you can buy all kinds of things.

Some people have prospered. Although only a few became billionaire oligarchs, the so-called **New Russians** emerged as the first class of wealthy Russians since 1917. Their conspicuous consumption and uncultured ways make them the target of scorn—few believe that such riches could be accumulated overnight by ethical means. However, another group, a **new middle class**, has also appeared on the scene, at least in the larger urban areas. Unlike the New Russians, new middle class people are professionals who draw salaries in fields valued by the global economy, such as investment counseling, marketing, and real estate. They earn more than the Russian average, but they work long, intense hours, often in fields that did not exist in Soviet times. For years it seemed that new middle class people were satisfied to live their lives in peace and ignore the government's authoritarian tendencies. But in the winter of 2011–2012 tens of thousands took to the streets, especially in Moscow, to protest elections they felt were rigged in favor of the ruling party.

With respect to Russian society today, three types of difference deserve special attention: old/young, male/female, and Russian/non-Russian. There is a significant generation gap in Russia today, a profound difference in perspectives, which has resulted from the old and young growing up in different worlds. Middle-aged people and seniors are uneasy with the continuing erosion of social-security institutions. Many older people react on principle to what they see as a pervasive dog-eat-dog mentality, while young people are more comfortable with a rapidly changing world in which their own individual merits and efforts appear to determine their success (see *Geography in Action*: Young People: Living in a New World).

Communism was a patriarchal system, but the state did provide substantial maternal and child-care benefits. Women were not exploited as sexual objects in the mass media. And quotas were set in legislative and other bodies to ensure significant female representation. The quotas are entirely gone now, and the family-welfare benefits are negligible. Sexual exploitation of women is rampant, and not just in the media. Prostitution is a major industry, both at home and abroad. Many women have been lured by offers of legitimate employment in exotic locations, only to end up trapped and forced to work as sex slaves, with their documents held by their "agents."

On the other hand, life for women under the Soviet system was stifling in many respects, not least because the regime's ideology was antithetical to style—female tractor operators and "heroic mothers" with 8–12 children got their pictures in the news. By and large, Russian women are indifferent to feminism. Traditional domestic values coexist with an interest in expressing and displaying femininity (Figure 5-18). The flood of fashionable clothes and

(a)

▲ **Figure 5-18a To the table, please!** A Russian hostess surveys the dishes she has prepared for the first course of a feast for family and friends. Today such extravagant meals are rarely prepared because of the increasingly hectic pace of life. But women still do virtually all the cooking as gender roles have been slow to change.

(b)

▲ **Figure 5-18b Shops on Tverskaya Ulitsa, Moscow.** Fashionable shops line pedestrian streets in downtown Moscow.

Young People: Living in a New World

A 20-year-old citizen of one of the Northern Eurasian countries has no memory of Communism (Figure 5-2-1). In Russia, Ukraine, Armenia, and Georgia, the young citizen has been exposed to some freedom of the press, freedom of assembly, and contested elections enough for such practices to be considered normal. In these countries most young people growing up have known that this post-Communist world offers much greater scope for individual freedom but also much less security than their parents knew in their youth.

It is tempting to consider the younger and older generations as living in different worlds, but in reality they often share the same apartment (Figure 5-2-2). It is very difficult for young people to afford their own accommodations. Even when people reach middle age, they may have elderly parents living with them because there are virtually no affordable elder-care facilities. Young people in these countries have much less privacy than Americans of the same age. Perhaps this explains the tendency to marry young. Living together without marrying is rarely an option, and neither is "hooking up" casually. Russian men and women marry at the average ages of 24 and 22, respectively. Gender roles generally remain traditional, so young women are expected to have children. Yet few women have more than one or two children because of the conditions they face. Apartments are small and often are shared with grandparents or others. Benefits for mothers and children are not as generous as they were in Soviet times. Marriages frequently are unstable and women tend not to be paid as well as men, but carry the primary responsibility for raising children.

Young people generally are forging their own paths rather than following in the footsteps of the older generation. Increasingly, students are becoming more individualistic, which can be a problem for the educational system, long focused on inculcating group loyalty and a collective spirit. Young people today seem, at least to their elders, to be less interested in the "big issues" and more concerned with looks and style. In contrast to the older generation, young people have been living with consumerism all their lives.

In Russia education is mandatory for nine years. Most students then either go to vocational schools or continue in secondary school for two more years if they intend to go on to higher education. Russia has 570 state universities and about 400 private colleges and institutes—the latter are all fairly new and of highly variable quality. In the public universities an awkward situation has arisen. The state still promises free education to those who pass stringent

▲ FIGURE 5-2-2 **Russian family life.** Family housing in Russia is in short supply, and in many families multiple generations reside under one roof.

entrance exams. Those who do not pass are welcome to attend for a fee. Once they enter they may pay for extra tutoring, second chances on exams, and other benefits. Such payments provide essential revenue for the universities and much-needed extra income to poorly paid instructors. But this system fosters corruption and establishes a class system among students that can only be harmful in the long run.

Students do not "work their way through college," so the transition into the workforce after graduation can be stressful. The careers that students frequently find most attractive often did not exist in the Soviet period. The author once asked a class of graduating seniors at a first-rate library science institute what sort of job they wanted. Unanimously, they answered *menedzher*, "manager." Business, finance, accounting, law, and real estate are particularly attractive for young people today. This contrasts with the late Soviet period, when the U.S.S.R. had more engineers than any other country.

Sources: Hilary Pilkington, et al., eds., *Looking West: Cultural Globalization and Russian Youth Culture* (University Park: Pennsylvania State University Press, 2003); Aleksandar Štulhofer and Theo Sandfort, eds., *Sexuality and Gender in Post-Communist Eastern Europe and Russia* (Binghamton, NY: Haworth Press, 2005).

▲ FIGURE 5-2-1 **The Soviets proclaimed, "Lenin lived, Lenin lives, Lenin will live!"** But these young skaters seem indifferent to the founder of the U.S.S.R. They are too busy enjoying a summer afternoon at the All-Russian Exhibition Center in Moscow.

beauty products has been extremely popular. Although industrial jobs for women have declined drastically, young professional women find diverse and challenging opportunities in new fields such as information technology and real estate.

From the perspective of Western societies, exploitation of Russian women has not ended, but merely changed to different forms. Yet privately few Russian women would confess that they feel inferior to men—to the contrary! Russian men often are depicted as lazy and irresponsible alcoholics, which is an unfair and misleading stereotype. Men ultimately may suffer as much as women from the rigid structure of gender roles. Men often spend long, hard hours trying to provide for their families, which is the main role the prevailing gender ideology allows them. Women, on the other hand, are supposed to be beautiful and nurturing. If they also work very hard, they get little credit for it. In all cases the mother is always considered the primary parent; it is unthinkable for a father to gain custody of children following a divorce. Sometimes in larger cities today, young men may be seen pushing strollers and playing with their children, so perhaps change is under way. But most Russians act out gender roles as they have

been acculturated to do—it seems important to be a "real" man or woman. Under the current, highly stressful conditions these rigid gender roles too often turn male and female partners into antagonists.

The U.S.S.R. became 15 independent countries, of which only Russia has attempted to create a federation. By the government's count there are some 160 nationalities and ethnic groups in Russia. Twenty-one of these nationalities inhabit "autonomous republics," which have certain rights with respect to the preservation of their cultures and economic development of their territories (Figure 5-19). Another 12 nationalities inhabit lesser-order territorial units based on ethnic criteria. Relations between the non-Russian nationalities and Russians vary considerably. Some nationalities, including the Chechens and Ingush, suffered historically and still have bitter grievances today. Others, such as the Udmurts, appear satisfied with their situation. Sometimes economic matters complicate ethnic politics. Tatarstan, for example, has highly profitable oil wells and substantial reserves, and the Tatars compose the largest non-Russian nationality, with about 5.6 million people.

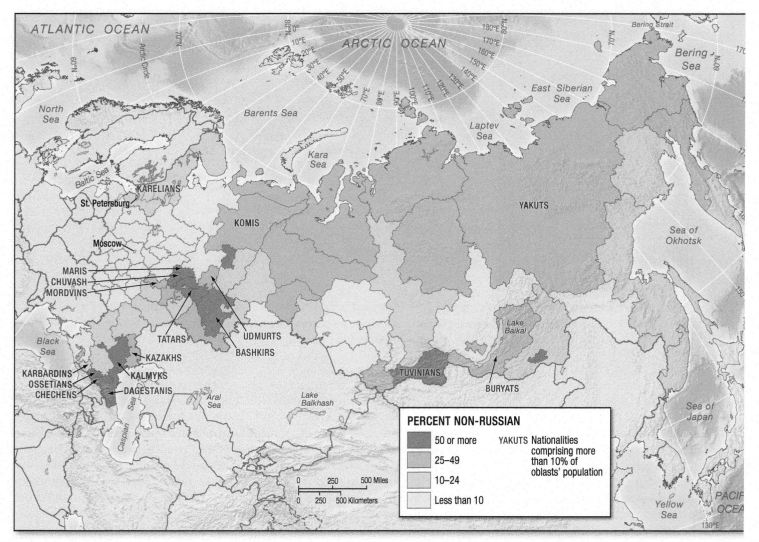

▲ **Figure 5-19 Percentage of non-Russian ethnic groups in Russia.** Russia is home to about 160 different ethnic groups. The largest non-Russian group is the Tatars, who along with twenty other groups have their own "republics" within the Federation.

North Caucasus: Mosaic and Minefield

Just to the east of Sochi and south of Russia's Kuban breadbasket lies the culturally rich but materially poor North Caucasus region, the most restless area in Russia (Figure 5-3-1). Here one finds over 50 different languages spoken; Dagestan alone is home to over two dozen, some spoken by peoples numbering in the low thousands. Through the centuries when left to themselves, these small nations usually managed to find a way of living with each other. But periodically forces from the outside world have combined with local differences to produce tragic outcomes.

In 1994 the Chechen government declared independence from Russia, which led to a Russian invasion. Groznyy, the capital, was virtually destroyed (Figure 5-3-2). But the Russian army was bedeviled by low morale and corruption. In 1996 the two sides agreed to disagree peacefully. The Chechens again proclaimed independence, which Russia declined to acknowledge, but Russian troops withdrew in ignominy. Yet independent Chechnya, landlocked and small, was no paradise. Crime flourished, as did radical Islamist ideology. In 1999 fundamentalist fighters swept into neighboring

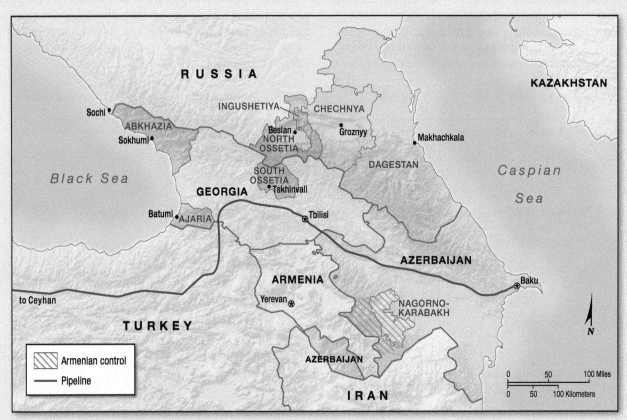

▲ FIGURE 5-3-1 **Conflict zones of the Caucasus region.** Ethnic warfare has resulted in the loss of tens of thousands of lives, the flight of hundreds of thousands of internally displaced refugees, and physical and economic ruin for many inhabitants of the region.

Russian attitudes have hardened against many of the non-Russian nationalities, particularly the Caucasians or others who resemble them in Russian eyes (Figure 5-20). Preexisting prejudices have intensified in the wake of terrorist attacks in Moscow and the North Caucasus (see *Geography in Action*: North Caucasus: Mosaic and Minefield). Denigrated as "blacks," Caucasians are frequently attacked by skinheads and detained by the police in Russian cities. Immigrants from former Soviet Central Asia also face discrimination and racist violence. Most Russians are tolerant of the "foreigners" in their midst, but even the most liberal minded sometimes find the pace of change unsettling. As many as 3 million non-Russians live in Moscow today.

Dagestan. Again Russia intervened, chasing the guerillas back into Chechnya and reigniting the war.

Some of the Chechen leaders spread the struggle beyond Chechnya in the form of terrorism. Two airplanes flying out of Moscow were blown up, killing everybody on board. Also in Moscow, in 2002, Chechen terrorists took over the Dubrovka Theater during a performance, holding patrons hostage until Russian security forces overcame the terrorists with a strong gaseous drug, which also killed about 130 hostages. In 2004, armed rebels seized a school in North Ossetia in the North Caucasus, on the first day of classes when parents traditionally accompany their children. Over 1,100 parents and children were held for three days, when an attempt by the authorities to end the standoff resulted in the deaths of about 335 hostages, nearly 200 of them children. These events made the global news, unlike the brutal treatment of Chechen people by Russian troops and their local supporters. We may never know how many were killed in the second war—probably between 25,000 and 50,000—but nearly one-third of the population fled the tiny country. Today Chechnya is ruled by Razman Kadyrov, son of a rebel commander in the first Chechen war who switched sides in the second and was later assassinated. President Kadyrov is by all accounts ruthless, but he has restored order to a degree not seen since the Soviet period.

But the situation in the North Caucasus remains highly unstable. In several cases, local leaders have been given a free hand, with which they line their pockets. They are also intolerant of dissent and any practice of religion which is not under their control. In these poor lands, leadership of this sort is bound to provoke discontent. There are many small armed groups, some with foreign militants involved. Assassinations and kidnappings of officials and their families are all too frequent, especially in Ingushetia and Dagestan. A Supreme Court judge in Dagestan was assassinated on the street in January 2013. Such circumstances make economic development of the region highly problematic. There are few natural resources, although the majestic mountains and the fascinating cultural diversity could be great assets.

Federal government officials on numerous occasions have announced plans to boost development of the North Caucasus, but actual progress has been slow. The leading sector in the development program is tourism, with four new "ski clusters" planned. A number of problems have emerged, but by far the worst is the combination of the threat of terrorism with the brutal "antiterrorist" operations by the power ministries. Development prospects, whether from tourism, food processing, or other sectors,

▲ FIGURE 5-3-2 **Recovery from war damage in Chechnya.** Some 300,000 Chechens, out of a million, fled their homeland and the capital city of Groznyy to escape violence and destruction. Improved security has led to the return of displaced people and efforts to create new housing to accommodate them.

often rekindle ethnic disputes, which are exacerbated by the lack of formal property rights in many areas.

Stop & Think

What makes the North Caucasus Russia's most restive region?

Sources: The North Caucasus Crisis, *Russian Analytical Digest* 70, 21 December 2009, http://www.css.ethz.ch/publications/pdfs/RAD-70.pdf; John O'Loughlin, Edward C. Holland, and Frank D. W. Witmer, "The Changing Geography of Violence in Russia's North Caucasus, 1999–2011: Regional Trends and Local Dynamics in Dagestan, Ingushetia, and Kabardino-Balkaria," *Eurasian Geography and Economics* 52, no. 5 (Sept.–Oct. 2011): 596–630.

Russia's Economic Geography

The more the Russian economy is subjected to market forces, the more important become typical economic geography factors, such as costs of production and transportation. Hardly anything is produced more cheaply in Russia than anywhere else because of harsh climatic conditions, distance to markets, inadequate transportation and communications infrastructures, obsolescent technology, corruption, and other factors (see Figure 5-13; Table 5-2). Places where such costs can be minimized may do relatively well under market conditions, but other centers of production in Russia may not survive in a competitive economic environment.

◀ **Figure 5-20 "Keep the blood pure!"** Blatant xenophobia and anti-immigrant sentiment are frequently on display in public places in Russian cities, such as this graffiti at a bus stop in Moscow.

Industry

The brightest spots in the economic landscape are the Moscow and St. Petersburg regions (see *Visualizing Development*: Moscow City, New Moscow: Whither Moscow?), along with the oil and gas producing regions of West Siberia and Sakhalin. But the economic structure of even the "winner" regions is changing. Textiles are in decline everywhere, even in their traditional core area of the Moscow region. Machine building and metalworking industries are not as hard hit as textiles, but the transition to competitive conditions has been painful. Because St. Petersburg and Moscow have the largest concentrations of population—close to 20 percent of the Russian population lives in these two metropolitan areas—and the best transportation facilities, they have attracted industries that produce consumer goods for the Russian market, including automobiles and parts, household appliances, and a variety of food products. With the transition to consumerism, urban markets have become crucial in determining what is produced and where production is located (Figure 5-21).

Several years ago, the government resolved to promote the development of information technology (IT) parks. The Soviet Union invested heavily in science and technology; whole towns were established as closed "Science Cities." Today their original consumer, the Soviet military complex, is gone and they lack the resources to diversify research and develop marketable products. Considerable human resources are still in these communities, although many professionals emigrated as part of a brain drain.

Table 5-2 Economic Resources of the Countries of Northern Eurasia

	Per Capita GNI	Major Resources	Principal Industrial Cities	Principal Manufactured Goods	Major Agricultural Products
Russia	19,240	Oil, natural gas, coal, iron ore, hydroelectric power, gold, polymetallic ores, aluminum ore, timber, diamonds, platinum, fertile soils	Moscow, St. Petersburg, Nizhnii Novgorod (Gorki), Yekaterinburg, Novosibirsk, Perm, Chelyabinsk	Steel, machinery, metal working, textiles, chemicals, armaments, paper, food products, transport equipment, electronics, woodworking	Wheat, rye, potatoes, dairy products, beef, swine, corn, barley, oats
Belarus	13,590	Peat, timber	Minsk	Machinery, food, woodworking, electrical goods	Potatoes, dairy products, flax, beef, swine, wheat, rye
Ukraine	6,620	Coal, iron ore, manganese, uranium, fertile soils	Kiev, Kharkiv, Dnipropetrovsk, Donetsk, Odesa	Steel, machinery, metal working, chemicals, food products, agricultural equipment	Wheat, sugar beets, corn, barley, beef, swine, sunflower oil
Armenia	5,660	Copper, molybdenum	Yerevan	Chemicals, cut diamonds, aluminum, food, machinery	Wine, cognac, grains, fruits, vegetables, tobacco
Azerbaijan	9,270	Oil, natural gas	Baki (Baku)	Petrochemicals, food, machinery	Cotton, tea, subtropical fruits
Georgia	4,990	Manganese	Tbilisi	Machinery, food, building materials, steel, chemicals	Tea, subtropical fruits, wine, cognac, tobacco

Source: Population Reference Bureau, *World Population Data Sheet* (Washington, D.C.: Population Reference Bureau, 2012). Data are calculated using the GNI PPP method.

▲ **Figure 5-21 Passenger service on the Ob River in Novosibirsk.** Ice can cause problems on the Ob even in May. Novosibirsk is Siberia's largest city and an important industrial and educational center.

Under Medvedev, construction of "Russia's Silicon Valley" began in Skolkovo, a small settlement to the west of "Old Moscow" that was annexed by the capital in 2012.

Other areas with potential include the Volga region. Although some industries have failed, the automotive sector appears to be recovering (Figure 5-22). The general diversity of the region's industrial base is an asset as well. Rostov-on-the-Don also shows promise. Long called the Gateway to the Caucasus, post-Soviet Rostov is Russia's southern gateway to the sea, and it serves Russia's most fertile agricultural region. Don River traffic is important, and a canal connects Rostov with the Volga and the Caspian Basin. Beyond the Urals centers of oil and gas production are prospering, such as Tyumen in western Siberia.

Two other areas deserve attention due to their potential to succeed in the global economy, although in both cases the geopolitical situation complicates the picture. One is the exclave Kaliningrad on the Baltic Sea, formerly Prussian territory which the Soviets seized in 1945. As a Special Economic Zone, Kaliningrad offers manufacturers several advantages, including a European location but Russian labor costs. The area is known especially for its production of television sets and automobiles. But Kaliningrad's location also has military implications. Russia has threatened to locate

nuclear weapons in Kaliningrad and has placed a strategic radar installation there.

At the other end of the country, the Russian Far East also stands to benefit if Russia becomes more integrated into the global economy. Traditionally Russia's oil and gas exports have been oriented toward Western Europe. But recent policy decisions to diversify market access have resulted in the construction of the Eastern Siberian-Pacific Meridian oil pipeline to the port of Kozmino near Kakhodhia east of Vladivostok (see *Focus on Energy*: Profits and Prospects on page 247). The Russian Far East also has valuable natural resources, including oil, gas, gold, diamonds, fish, and wood products. The port of Vladivostok, with a population of 600,000, is the terminus of the Trans-Siberian railroad and Russia's main link with the Pacific Rim. Long hampered by political turmoil, the Russian Far East today is poised to reap the benefits of globalization. Both Japan and China are eager to tap into the region's oil and gas. Japan especially has a great demand for wood products as well. Freight traffic on the Baikal-Amur Mainline (BAM) railroad, which took decades to build only to nearly collapse when the U.S.S.R. ended, is finally picking up. A project to build a major ice-free container port south of Vladivostok on Troitsa Bay has stalled, but if reanimated would enable Russia, and perhaps northeastern China, to greatly expand participation in Pacific Rim trade. But cooperation with Japan is always hampered by a dispute over the Kuril Islands, which were seized by the U.S.S.R. at the end of World War II. And although formal relations with China are better than they have been for almost 50 years, there is great concern in Russia that China could colonize the Russian Far East.

The big losers in the transition are regions that relied on a narrow base of heavy industry, especially those in remote areas. The coal mines of the Pechora region—above the Arctic Circle west of the Urals—appear to have no prospects, yet the miners struggle on, hoping for a miracle. Many towns in the Urals, once the U.S.S.R.'s industrial heartland, have slumped due to obsolescent technology, world competition, and difficulty accessing global markets.

▶ **Figure 5-22 Headquarters and main production facility of Russia's largest automaker.** Based in Tolyatti in the Volga region, AvtoVAZ produces the Lada, Russia's best-selling domestic car. A new standard hatchback Lada sells for about $10,000.

visualizing DEVELOPMENT ▲   Moscow City, New Moscow: Whither Moscow?

In 1980 Moscow was the model city for Communism and headquarters for the U.S.S.R.'s autarkic economy. There were fewer than 700,000 automobiles. Commercial advertising did not exist—placards on the streets hailed the achievements of the latest Party Congress and called on workers to exceed the targets of the Five Year Plan. Stores closed early, most restaurants were closed to the general public, and the city at night was dark and quiet.

▲ **FIGURE 5-4-1 Moscow, old and new.** Looking from the capital's northwestern district, this picture captures a historical-geographical sequence. In the foreground the housing is from the 1930s. In the mid-ground apartment blocks from the 1980s are visible. In the distance to the far right are typical post-Soviet skyscrapers. And looming over the city in the distance is the massive Moskva Siti complex.

Now Moscow is a vibrant, connected global center, twenty-first-century Russia's window on world culture and markets. In contrast to the drab concrete block buildings of the Soviet era, colorful, diverse residential and office towers dominate the cityscape, along with flashy superstores and shopping centers (Figure 5-4-1). The architect of the capital's transformation was Yurii Luzhkov, Moscow's mayor from 1992 to 2010. Luzhkov set out to help create a new middle class and make Moscow a global city. He tried to make the capital more "convenient" and "civilized" for business people and professionals, which largely meant rebuilding much of the city to favor an automobile-centered lifestyle, which in fact has made Moscow extremely inconvenient.

Luzhkov's favorite project was the Moscow City complex, which is still under construction next to the Moskva River just a few kilometers upstream from the Kremlin. The name "City" evokes London's famous financial center and indicates Moscow's grand ambitions. City's skyscrapers are designed to attract the world's attention; in fact, in 2011 the "Moscow" tower was named one of the "most aesthetic" skyscrapers in the world. The plan provided for moving city government agencies to the complex and attracting foreign firms to a new trade center. Another of Luzhkov's pet project, the Third Transport Ring, provides automobile access to City.

But was it wise to promote more development, and more driving, downtown? Moscow is among the world's worst cities for "commuter pain" and its air pollution, largely due to automobiles, can be lethal. Moscow

Agriculture, Changing in Opposite Directions

Some large-scale crop production continues in Russia (Figure 5-23), but much of the land most suitable for grain, sugar beet, and sunflower production now belongs to Ukraine and Kazakhstan (Figure 5-24). Only the Kuban region, to the north and east of the Sea of Azov, has a comparably rich soil and long growing season. The Volga and Don basins also have rich soil, but the growing season is shorter in the mid- and upper reaches of the rivers. In most of the other areas devoted to diversified agriculture, soil quality is inferior, poor drainage is often a problem north of the Smolensk-Moscow axis, and the growing season is short. Yet because self-sufficiency was vital for the Soviet regime, the government invested heavily in developing agriculture in difficult areas.

The Soviet ideal was large-scale **state farms** that were operated as much as possible on industrial principles. In the early 1930s, the regime set up **collective farms**, also very large scale, which were seen as precursors of state farms. The farmers themselves were supposed to be the owner-operators, but in practice the state maintained strategic control over them. Stalin tried unsuccessfully to abolish farm families' private home gardens. As it turned out, Soviet

today holds over 4 million cars, and another million commute from Moscow Oblast, the province that surrounds the capital. Luzhkov tried to solve congestion by building more highways but because these encouraged more development, congestion only worsened. Luzhkov became so desperate that he was planning to build highways on top of buildings! The end came in 2010: the official explanation was that Luzhkov was fired because he publicly second-guessed then President Medvedev over a highway project. In addition to directing the new mayor, Sergei Sobianin, to attack traffic congestion, in 2011 Medvedev proposed a more radical solution: a massive expansion of the city.

The idea was shocking—to take an area about 1.5 times as large as the capital to be "New Moscow," and to move both the federal and city governments there (Figure 5-4-2). Medvedev also envisaged a new international business center, as if Moscow City had been just a mirage, plus various other new nuclei for scientific and educational complexes, including Skolkovo, a new town close to Old Moscow's western edge, which is meant to become Russia's "Silicon Valley."

How to go about building a new capital next to the old capital? A global competition was held to select New Moscow's designers. But by the time the "winners" were announced in 2012, Putin was president again and much of the momentum behind the New Moscow project had dissipated. The authorities did not confirm that ideas from the design teams' proposals would be used. Highly placed unofficial sources gave word that the government would not move beyond Old Moscow's city limits. It seems

▲ **FIGURE 5-4-2 New Moscow.** In 2012 Moscow increased its size by 150 percent. Originally the plan was to relocate the federal government to New Moscow, but by 2013 President Putin was reconsidering the project.

the inevitably exorbitant costs and quiet resistance by government officials turned the tide against relocation. Putin put off announcing a final decision, but at the same time more "leaks," which are very rare under Putin, indicated that a new government center would be built near the Kremlin.

Moscow has nearly 12 million official residents and it is a magnet for migrants, both legal and illegal. No other place in the country offers a comparable range of opportunities to energetic, ambitious people,

especially the young and well educated. Old Moscow's population density is more like Delhi's than London's. An innovative metropolitan-regional plan is desperately needed. But the choice seems to be between an impractical panacea and a "doubling down" on the old pattern.

Sources: Robert Argenbright, "Moscow on the Rise, from Primate City to Mega-Region," *The Geographical Review* 103, no. 1 (January 2013): 20–36; Robert Argenbright, "New Moscow: An Exploratory Assessment," *Eurasian Geography and Economics* 52, no. 6 (Nov.–Dec. 2011): 857–875.

agriculture depended heavily on **private-plot production**, especially for fruits and vegetables.

In the 1990s, Russian agriculture was privatized. Western advisors expected enterprising families to withdraw from the huge state and collective farms and set up family farms on a commercial basis. More than a quarter million family farms exist in Russia today, but their contribution to overall output is only about 4 percent. Families have been reluctant to leave state and collective farms, which are now independent corporations. Dividing the land raises issues, but a greater problem is distributing other assets, such as farm

machinery. Potential family farmers lack the capital to get started in agribusiness. Credit is weakly developed in Russia generally and especially in the countryside, where family farmers have little collateral and no credit histories. Farmers who stay with the corporation have access to its machinery, fertilizer, and other assets, which they often use on their private plots. In these cases, the corporation may fare badly while the farm families survive thanks to the private plots. However, this does not seem viable in the long run. Over half of the corporate farms run at a loss. Outside interests have gained shares steadily, to the point that the farmers themselves

◄ **Figure 5-23 Wheat harvest in the Kuban region.** The Kuban, blessed with chernozem soil and a relatively long growing season, is Russia's most productive agricultural area.

▼ **Figure 5-24 Agricultural zones.** Only about 8 percent of Russia's land is suitable for crops; and another 5 percent is used for pasture. The Kuban region east of the Sea of Azov is Russia's most productive agricultural area. In Ukraine about 60 percent of the land is arable.

AGRICULTURAL ZONES

Diversified agriculture
Dairying, flax, potatoes, milk and meat livestock, swine production, grains (rye, oats, barley, wheat)

Large-scale grain productions
Wheat, corn, barley, oats, rye, sugar beets, sunflowers, milk and meat livestock, sheep

Urban truck farming
Milk, potatoes, eggs, chickens, vegetables

Humid subtropical specialized agricultural production
Tea, subtropical fruits, vineyards

Cotton, fruits, vineyards, sheep, wheat

ZONES OF LITTLE OR NO AGRICULTURE

Tundra—very little agriculture; crops require protection (hothouses); grazing for reindeer

Taiga—agriculture widely scattered in small areas serving local needs

Drylands—extensive grazing of sheep and cattle; some scattered irrigated agriculture

Mountains

FOCUS ON ENERGY Profits and Prospects

In 2011 exports of natural gas, oil, and oil products accounted for 65 percent of the federation's export revenue (Figure 5-5-1). That represents a growing share of increased export revenue: in 2000 the same products accounted for just over half of a total export income that was only 28 percent of the 2011 total. Russia clearly depends on exports of oil and gas and, so far, that dependence has brought growing prosperity.

Oil exports bring in more revenue than any other commodity—oil exports account for over half the country's export revenues. However, there are problems. The largest known western Siberian oil fields are past their peak, making increasing oil production highly unlikely. Russia would like to complement its exports to Europe by tapping markets in East Asia, and in fact a new pipeline has opened for that purpose. But Russian production in the short term has plateaued, while domestic demand is rising, thanks to the ever-growing number of automobiles. Production may increase but only at great cost, whether it moves offshore in the Pacific or Arctic, or into the vast wilderness of East Siberia. At the same time, producers with vast reserves of unconventional types of petroleum—such as Venezuela with its "heavy" oil and Canada with its tar sands—may find ways to lower their high costs of production sufficiently to out-compete Russia.

The natural gas picture is brighter. Russia's proven reserves amount to about one-fourth of the world total. The holder of the second-largest reserves, Iran, is a minor factor in the world market. Russia is likely to remain the top producer and exporter for some time. Demand is expected to rise, not least because natural gas is by far the cleanest-burning fossil fuel, although the new technology of hydraulic fracturing of shale, "fracking," may make much more natural gas available in other countries, perhaps even doubling global reserves. China in particular stands to gain from developing this new energy source, and it is likely to be less hindered by environmental concerns than other countries with great shale gas potential, especially since its main alternative is to burn more coal.

▲ FIGURE 5-5-1 **Oil and natural gas development in Siberia.** Despite its harsh physical environments, Siberia has become Russia's most important source region for petroleum and natural gas. Almost all of Siberia's production is transported by pipeline to the more densely settled European areas of Russia or beyond to European export markets.

Russia is well supplied with coal, but its best deposits are in the Kuznetsk Basin in southwestern Siberia. With oil and gas, the initial costs of developing pipelines and port facilities are high, but once the infrastructure is in place transport costs are minimal. Most coal is shipped by rail, which also is costly to build, while each shipment of the bulky, heavy commodity is expensive and burdens infrastructure. Distance to markets affects the price of coal more than oil and gas. Russia's domestic use of coal is not increasing, but the country plans to increase exports to China. Yet even China which favors rapid development over environmental protection cannot burn more and more coal indefinitely and ignore its horrendous air pollution problem. Therefore, it is doubtful that Russia's coal industry will expand significantly; it is more likely to contract.

Gas and coal produce most of Russia's electricity, but about one-third comes from nonfossil fuel sources, split evenly between hydroelectric and nuclear. The supply of hydroelectric power dates back to Soviet-era construction of massive dams on the Dnieper, Kama, Volga, Angara, and Yenisey rivers. No significant expansion of hydropower seems likely at present. Nuclear power is back in favor as memories of the Chernobyl disaster have dimmed. Russia is building several new domestic nuclear plants and refurbishing old ones, including some of the same model as the one at Chernobyl that melted down in 1986. Russia is also building reactors abroad, with China and India among its customers. Other alternative sources of energy are insignificant in Russia. There has been some development of geothermal power on Kamchatka. During Medvedev's term as president there was interest in solar and wind power, but President Putin seems unlikely to champion their development.

Stop & Think

▷ Identify the current bright spots and problem areas in the Russian economy.

Sources: Russian Energy Policy, *Russian Analytical Digest* 100, 26 July 2011, http://www.css.ethz.ch /publications/pdfs/RAD-100.pdf; Energy, *Russian Analytical Digest* 113, 15 May 2012, http://www.css .ethz.ch/publications/pdfs/RAD-113.pdf.

typically form a minority of the shareholders. Bankruptcies and liquidations are becoming more frequent—especially where physical conditions are unfavorable or where farms are remote from markets. Current trends are contradictory. New types of commercial operations are emerging. These are not relabeled collectives; their only aim is to make a profit. These new agribusinesses are found where climate, soil, and market accessibility are most favorable. This is normal under market conditions and so may be considered progress, even if it causes hardship for farmers in collectives that fail. At the same time, the area of arable land devoted to private plots at least doubled in the 1990s. More than 13 million people are employed on household plots; some 10 million of them are subsistence farmers (Figure 5-25). About half of the world's countries have populations of 10 million or less, so Russia has a "nation" just surviving through subsistence gardening. This is not a sign of progress for a developed country.

Putin's Political System: The Power Vertical

Vladimir Putin, previously head of the FSB, was named prime minister by Yeltsin in 1999 and subsequently was twice elected president. Barred by Russia's constitution from serving a third consecutive term, Putin selected his former deputy, **Dmitrii Medvedev**, to succeed him as president. Medvedev subsequently won an election that was arranged to produce just that outcome. Soon after the election, Medvedev appointed Putin prime minister. In 2012 Medvedev stepped aside and Putin again was elected president. While Medvedev was president, the term of office was extended from four to six years beginning in 2012. Should Putin win a second reelection in 2018, he could be in power legally for a quarter of a century.

Since Yeltsin's second victory, Russian presidential elections have been managed by the political elite. When Putin came to power he did not run as the candidate of a party; instead a new party, United Russia, formed around him and now dominates the Duma, the legislative branch. Both central and local authorities

▼ **Figure 5-25 Village near Lake Baikal.** The large gardens behind the houses, called "private plots" in the Soviet period, still are often essential for households' well-being in rural areas.

have tended to make it more difficult for opposition groups to demonstrate publicly. In the judiciary there undoubtedly is corruption, but worse is the ingrained habit of serving as an agency of the executive branch rather than upholding the law impartially. "Telephone law," the practice of judges obeying calls from political bosses rather than the law, is deeply entrenched. Selective law enforcement has frequently been employed against citizens as well as foreign firms and NGOs that have run afoul of officials.

During Putin's first presidency the major television stations and most radio stations came under government control. They rarely mention anything unflattering to the government, nor do they allow much airtime to opposition parties. There remain some independent newspapers and magazines, and one independent radio station in Moscow. The Internet is largely unregulated as well, although that may be changing. NGOs with foreign affiliations, and religious faiths that are not held to have sufficiently deep roots in Russia, have come under increasing pressure. In 2012, any groups or individuals receiving foreign funding were required to register as foreign agents.

The massive protests against rigged elections that occurred in the winter of 2011–2012 were mostly confined to Moscow, with significant protests in St. Petersburg and Kaliningrad as well. Since then the "opposition leadership"—a motley group that included a former chess champion, a television celebrity, a blogger, and a former neo-Stalinist turned community activist—has come under severe pressure, including arrests and prosecutions. But repression of the leadership seems to have less impact on the opposition movement than the essential fact that it lacks a coherent program and has very little strength in most of the country. The majority of the Russian population appears to value stability above all else, and Putin appears as the guarantor of order.

This was vividly illustrated by the 2012 controversy surrounding the feminist punk group "Pussy Riot" and their unauthorized "performance" in Moscow's Christ the Savior Cathedral. The affair was a major topic in Moscow and St. Petersburg, on the Internet, and in the Western media. In Moscow, it is widely recognized that the cathedral was rebuilt in the 1990s largely for political reasons and frequently serves as the stage for political ceremonies. As a symbolic space the cathedral is at least as much political as religious. But outside the major cities, most Russians perceived Pussy Riot's violation of that space as sacrilegious, if not in a strictly religious sense, then as an affront to Russian culture and *kulturnost* (civilized behavior).

It is not just ordinary people that value order. When Yeltsin launched Russia on its journey away from Communism he intentionally sought to emulate the West in having a market economy and a democracy. With respect to the latter, he accepted the American view that holding competitive elections overshadowed everything else. Relatively little attention was paid to the less dramatic, but no less essential, task of establishing the institutional basis for the rule of law. Without rule of law property is not secure. Based on this faulty foundation a new system emerged: "**crony capitalism**."[1] Entrepreneurs and their property are not safe without political protection. Officials make money from their offices and may simultaneously be involved

---

[1] Gulnaz Sharafutdinova, *Political Consequences of Crony Capitalism Inside Russia* (Notre Dame, IN: University of Notre Dame Press, 2010).

in business as well—conflict of interest is recognized in theory, but ignored in practice. Networks of "cronies" are highly motivated to make sure their side wins elections. Failure may mean not only loss of office, but also confiscation of property, and perhaps imprisonment. The situation is fundamentally the same at all levels: from the Kremlin down to small towns, clans of "cronies" rule. President Yeltsin, because of his precarious political situation, cultivated relations with regional, especially non-Russian, leaders. Some called this an overdue decentralization, which promised to stop the Kremlin from stifling local initiative and monopolizing the country's resources. But in fact regional leaders built up their own networks for monopolizing political power and maximizing the "rent" taken from local economic activity. With their people in important positions, with the local media under control, and with financial resources available through back channels, these leaders could win elections and build up fiefdoms.

President Putin set out to recentralize control over the country by establishing a layer of supergovernors, hand-picked by Putin himself, to oversee the elected governors. Putin's selections were all *siloviki*, men from the "power" agencies. This move was not popular in the West, but occasioned little protest in Russia. Many people were disgusted with corrupt officials and economic stagnation and believed a firm hand was necessary.

Immediately after the 2004 Beslan massacre in North Ossetia, when hundreds of innocents died after Islamist fanatics seized a school full of children, Putin declared that henceforth he would appoint governors. Political commentators argued that the president cynically used the tragedy as an excuse to grab more power. Again ordinary Russians raised no protest. If Putin—sober and serious—believed he could enhance security by appointing governors, most people were willing to let him try. Putin's supporters call the structure of central control "the power vertical." Yet this "top down" control through Federal Districts and appointed governors seems not to have improved security or reduced corruption. From top to bottom, "crony capitalism" prevails.

Near the end of his term as president, Dmitrii Medvedev reformed the law to resume electing governors, as well as the mayors of Moscow and St. Petersburg. But after Putin reclaimed the presidency he modified the law to allow for the appointment of governors in certain circumstances. However, in September 2013 an election was held to elect Moscow's mayor. Although the election was won by Sergei Sobianin, who was close to Putin, Aleksei Navalny, a businessman and anti-Putin blogger, came in second with a much higher vote tally than had been expected. With this "moral victory," Navalny cemented his position as the most prominent leader of the diverse opposition to Putin's regime. But in the aftermath of the election Navalny's ability to lead the opposition was in doubt because he had been convicted of corruption in a case that was widely seen as politically motivated.

Stop & Think

▷ Is democracy compatible with "crony capitalism?" Explain.

Russia and the "Near Abroad"

Compared to the Yeltsin period, foreign policy under Putin has been much more assertive, especially in regard to the 14 former SSRs, which Russians now call the "Near Abroad." This focus on the "Near Abroad" is not just due to nostalgia for the Soviet empire or concern for Russians in these countries. There are also lingering physical connections, most importantly the oil and gas pipeline networks, as well as well-entrenched trade ties. Russia expects to have special influence for various reasons, including the geographical fact that it shares a border with most of these countries. This can be irritating, or worse, for former Soviet countries that seek to chart their own course.

Russia's position on some issues is what might be expected from any government attending to its national security interests. Most significant is the question of "Near Abroad" countries joining the North Atlantic Treaty Organization (NATO). Six countries that were in the Soviet bloc have joined, along with Estonia, Latvia, and Lithuania, which were part of the U.S.S.R. itself. Ukraine and Georgia also have shown interest in joining the organization. Western leaders try to assure Russia that NATO is no longer an anti-Russian alliance, but Russian leaders well know that NATO was created to militarily oppose the U.S.S.R. They can point to the map and ask, "If NATO is not against us, why does it keep expanding on our doorstep?"

Ukraine is the most important "Near Abroad" country for Russia, and not just because of the size of its population or the wealth of natural resources. Russians typically have not considered Ukrainians as constituting a different nation. Ukrainians have always been seen as "little brothers" or perhaps, "country cousins." The very word "Ukraine" means on the border, highlighting the land's location as part of the Russian empire. The loss of the beautiful Crimea, where Russians make up a strong majority of the population, is especially painful. Russia's Black Sea fleet is still based at Sevastopol. Early in 2008 President Putin bristled at the suggestion that Ukraine might force the fleet's removal. He threatened not only to retake Crimea by force but also occupy eastern Ukraine, saying that Ukraine was not even a real state.

There are conflicts with other "Near Abroad" countries as well. In 2006 Russia declared an embargo on wine from Moldova. This hurt because Russia is the main market—cheap, sweet Moldovan wine does not sell well in Europe. There appeared to be political motivations. Transdniestria, a strip of land on the east bank of the Dniester River, is officially part of Moldova, but functions as an independent republic, thanks to Russian military protection. Moldova seeks reunification, while Transdniestria wants the world to recognize it as an independent country. Its ethnic composition, with Moldovans, Russians, and Ukrainians each making up about 30 percent, complicates the issue. A similar embargo was placed on Georgian wine and cognac.

The incidents with the broadest repercussions occurred when Russia cut off natural gas supplies to Ukraine in 2006 and 2009. These disputes caused great concern in Europe because of its "downstream" location in the pipeline network. Ukraine argued that Russia was attempting to wreck its economy, while Russia complained that Ukraine was receiving too much of a discount for gas, and stealing a large amount. By the winter of 2012–2013 this conflict had cooled thanks in part to the 2010 election of Ukrainian President Viktor Yanukovich, who favors close relations with Russia. More important in the long run, in 2011 the Nord Stream gas pipeline opened. Because the pipeline runs beneath the Baltic—it is the world's longest undersea pipeline—it reduces the leverage of Ukraine and other countries through which Russia's pipelines run to Europe.

Russia's relations with the "Near Abroad" countries of Central Asia have been much less contentious (see Chapter 6). Russia is comfortable with authoritarian regimes that devote themselves to preventing the spread of Islamist fundamentalism, even at the cost of civil liberties and human rights. Russia also wants to participate in the economic development of the region, especially the oil and gas resources. Gazprom has a vital interest in Turkmenistan, where its monopoly control of the pipeline network enabled it to buy gas at one-third the price it charged Europe. But Turkmenistan signed a contract to supply China with gas as well, and a pipeline was first opened in 2009, with expansion and improvements continuing at least until 2013. Kazakhstan and Uzbekistan have also begun using the pipeline and there is talk about connecting Afghanistan to the network as well. Talks also continue concerning possible construction of a pipeline directly from Russia to China.

Northern Eurasia's Western Edges: Between Russia and Europe

The remaining five Northern Eurasian countries have long dealt with being in-between more powerful neighbors. Ukraine and Belarus, like Russia, are descended from Kievan Rus's, but they were long under the influence of Poland and Lithuania. For centuries Georgia and Armenia persevered as Christian outposts between Turkey and Persia. In contrast, Azerbaijan has incorporated both Turkic and Persian elements into its national culture and now is positioned between Europe and Asia.

Belarus: Stuck in Transition

Belarus is about two-thirds as large as neighboring Poland, or about the size of Kansas. With just under 10 million people, there should be plenty of room, but good quality land has always been scarce in Belarus. The last glacial epoch left Belarus with 11,000 lakes and a chaotic drainage pattern. The Soviets' strenuous efforts made almost 30 percent of the land arable, but very little compares to Ukraine's chernozem. Belarus contains few natural resources apart from its forests. Moreover, the Chernobyl nuclear accident contaminated about one-fourth of the country's territory (Figure 5-26).

After World War II the Soviets systematically developed industry in Belarus. But with no resource base, Belarus was bound to be dependent on the U.S.S.R. Energy dependence is extreme. Belarus does produce some machinery, vehicles, and appliances, but these are not in demand in the West, so Russia remains the main market—another form of dependence. Hyperinflation in 2011 led to a drastic devaluation of the Belarusan ruble. The situation remained precarious until Russia sponsored $4 billion in loans, while Gazprom paid $2.5 billion for the country's natural gas pipeline network. In 2010 a customs union came into effect between Belarus, Russia, and Kazakhstan, but so far the main economic changes have affected the latter two partners, not Belarus.

(b)

◄ **Figure 5-26 Contemporary contamination from the 1986 Chernobyl nuclear disaster.** The radioactive contamination of sizable areas of Belarus, Ukraine, and Russia by the Chernobyl nuclear meltdown was one of the worst environmental catastrophes in world history. Nearly three decades later the effects of the disaster on human health and regional land use continue to be felt. (a) Radioactive contamination zones. (b) The contaminated ghost town of Pripyat.

Chernobyl aroused nationalist feelings in Belarus, but this moment was atypical. For decades Belarusans had gradually been assimilating into Russian culture. Today a majority speak Russian as their first language, and almost all Belarusan speakers are also fluent in Russian. Belarus's nationalist first government accomplished little. The economy is still much closer to the Soviet model than to capitalism, and the country's landlocked location is a disadvantage in this era of globalization. Belarus's economy is isolated institutionally and politically as well, largely because of its government.

First elected in 1994, President Aleksandr Lukashenko has consolidated his hold on power. Supporters laud him for eschewing shock therapy and maintaining a social-security safety net to keep ordinary people out of abject poverty. Detractors fault him for curtailing civil liberties. Several outspoken opponents have disappeared or died under mysterious circumstances. Lukashenko is hostile to the West and deals with countries deemed rogue nations by the United States. Non-Russian foreign investment in Belarus is insignificant.

In 1999, Belarus and Russia agreed in principle to a union of the two countries. Russia has neither repudiated the agreement nor moved to implement it. Russians and Belarusans are closely related to each other, so unification seems natural. But Russia has little reason to proceed, because Belarus economically already is its colony and politically has nowhere else to turn.

Ukraine: Between Europe and Russia

Although dwarfed by Russia, Ukraine is almost as big as Texas and larger than any country in Europe, though its population of nearly 46 million is slowly decreasing. Mostly it consists of rolling or flat steppe land with extremely fertile chernozem soil. Except for the

▼ Figure 5-27 **Cropland in central Ukraine.** This patchwork of seasonally inactive fields (purple and brown) shows the intensity of mechanized agriculture near the Dnieper River. Hedges separating the fields and forests along the river appear in yellow lines in this false-color satellite image.

Crimea's southern strip, Ukraine has a continental climate similar to the upper Midwest of the United States. Ukraine has tremendous agricultural potential—nearly 60 percent of the land is arable. A century ago it was the "Breadbasket of Europe." Ukraine is better suited to growing wheat than most of Russia since its climate is sufficiently mild to permit fall planting (Figure 5-27). Also, maize can be grown to maturity, as well as soybeans, along with the traditional crops of sunflowers and sugar beets.

Ukrainian industry was also world-renowned. Ukraine has coal, iron ore, and manganese—the basic ingredients for steel production. The metallurgical centers of the Donbas and mid-Dnieper areas were the mainstays of Tsarist industrialization and leaders in the Soviet period as well. Machine-building and chemical industries were also developed extensively. The giant industrial complexes, sizable hard-coal reserves, and rich agricultural potential led many people to feel that Ukraine would prosper immediately upon attaining its independence.

But post-Soviet, Ukrainian industry and agriculture have declined and stagnated. Ukraine attempted to avoid shock therapy, but the gradualist approach to reforms made little progress. Agriculture has been privatized in theory, but most farms continue to function as in Soviet times, although not as productively because essential inputs are too expensive. Ukrainian industry was undergoing privatization piecemeal via insider deals that outraged the population. The market is not supreme in Ukraine as individuals, enterprises, and institutions have all relied on nonmarket survival tactics. Vegetable gardening and barter are pervasive, as are many activities that would be illegal in law-governed societies. The Ukrainian state was occupied with rent-seeking even more than its Russian counterpart until the public finally rebelled in 2004. The **Orange Revolution** (the opposition party's color) led to a relatively fair presidential election which was won by Viktor Yushchenko, who promised to root out corruption and restructure both state and economy. Yushchenko's victory was seen by many in the West and Russia as a blow against the latter and as an expression of the Ukrainian people's desire to move closer to Europe. But Ukraine cannot literally move away from Russia. Ukraine is indebted to Russia, it benefits from Russian investments, and it depends on Russia for oil and natural gas. Although Ukrainian nationalism thrives in western Ukraine, most Ukrainians are little concerned about separating themselves from Russians. Russians make up the largest minority in Ukraine, with local majorities in the Crimea and some parts of eastern Ukraine. More Ukrainians consider Russian rather than Ukrainian to be their first language, while many in rural areas speak a mixture of the two (Figure 5-28). Intermarriage is very common.

Although many Ukrainians supported efforts by Yushchenko and Prime Minister Yulia Tymoshenko to improve relations with Europe and the rest of the developed world, the two leaders began to struggle with each other and failed to cope with the economic crisis of 2009 that caused a 15 percent shrinkage in the national economy. Consequently, Viktor Yanukovich, who supports closer ties with Russia, was elected president in 2010. The International Monetary Fund provided support in 2010 and the economy began to rebound, but the following year the program was discontinued due to Ukraine's lack of progress with economic reform. Relations with Europe have also cooled down, not least because Yanukovich

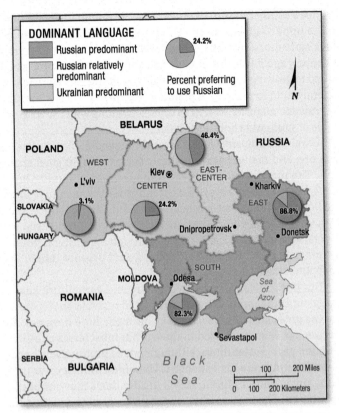

DOMINANT LANGUAGE
- Russian predominant
- Russian relatively predominant
- Ukrainian predominant

Percent preferring to use Russian

▲ **Figure 5-28 Russian influence in Ukraine.** Russians constitute the largest minority in Ukraine, and even most Ukrainians regard Russian as their first language.

prosecuted Tymoshenko on charges related to the 2009 deal that ended the pipeline crisis with Russia. Tymoshenko received a seven-year sentence. She was on trial again for tax evasion in 2013, when the chief prosecutor also accused her of murder. Many view her persecution as the death knell of Ukrainian democracy.

The Ukrainian economy remains sluggish. Ukrainian coal, for example, is vital for metallurgy and potentially important for export, and could be substituted for petroleum and natural gas in electricity production, thus reducing Ukraine's dependence on Russia

(Figure 5-29). But few of the mines turn a profit, while the government doles out billions in subsidies. Over 100 mines have been closed and production is about half the 1991 total. But remaining mines are old; the average depth is 700 meters. Ukraine's mines are not only inefficient but highly dangerous—on average 317 coal miners have been killed annually since 1990. Despite all these sacrifices Ukraine is a net importer of coal, much of it from Russia. Ukraine needs fewer mines, but more productive and safer ones, which rely on advanced technology rather than muscle power.

Similar dilemmas persist in other industrial sectors and in agriculture. Most of the economy is highly inefficient and getting started in business is extremely difficult. After a precipitous economic decline in the 1990s, the following decade witnessed a three-steps-forward-two-steps-back growth pattern that left Ukraine in better shape than in 1991, but, with respect to the global economy, Ukraine's relative share of production has declined. Difficult economic decisions must be made that will be painful for many. At the same time Ukraine must try to overcome disunity and guard its independence, while not antagonizing Russia.

Stop & Think

> Ukraine has a much greater resource base than Belarus, yet the latter enjoys a per capita income level over twice that of Ukraine's. What are the main obstacles to development in Ukraine?

Armenia: Precarious Position

Armenia occupies a territory nearly as large as Maryland, with a population of 3 million. It is mostly a plateau between 3,300 and 6,600 feet (1,000 and 2,000 meters) in elevation with a dry, continental climate. The bulk of Armenia's ancestral homeland—and Mt. Ararat, the Armenian national symbol—lie just to the west in Turkey. In the fourth century the kingdom of Armenia was the first state to adopt Christianity. Armenians speak an Indo-European language and developed their own alphabet in the early fifth century. Because of its location at the crossroads between Europe and Asia, Armenia experienced occasional invasions and nearly constant pressure from powerful neighbors, including Arabs, Turks, and Persians even before Russians came on the scene. But despite the 1915–1916 "ethnic cleansing," which killed about 1 million Armenians and drove the rest out of Turkey, Armenia has endured. Today it seeks recognition as a European nation and membership in the European Union.

In the late 1980s the Armenian majority in the **Nagorno-Karabakh region** began agitating for independence from Azerbaijan (see Figure 5-3-1). The Soviet government failed to resolve the conflict, which

◄ **Figure 5-29 Coal mine in eastern Ukraine.** Coal fueled the emergence of one of the world's greatest industrial clusters in Ukraine and southern Russia in the latter half of the nineteenth century. Today Ukrainian coal-mining is inefficient and dangerous.

contributed to its downfall. With the breakup of the U.S.S.R., the fighting spread to Azerbaijan, where Armenians were attacked, and to Armenia, where the same fate befell Azerbaijanis. Both minorities fled and full-scale war began between the two new states. Armenia seized a corridor between itself and Karabakh, while both countries endured economic decline and political instability. Fighting ceased in 1994, but the fundamental issues have yet to be resolved.

Armenia won independence for Karabakh, although no other country recognizes its sovereignty. Karabakh adds very little to Armenia's relatively poor resource base. For power, the country depends on a Soviet-era nuclear facility that many consider to be unsafe. Although its GNP per capita has improved, much of Armenia relies excessively on investment and remittances from the large Armenian population abroad.

Armenia is a democratic republic, but its political situation is unstable. In 2008 the presidential election sparked demonstrations, which were suppressed violently, with the loss of 10 lives. A state of emergency was declared, which entailed a media blackout. Another stormy presidential election was held in 2013—one candidate was shot just before it was held. Serzh Sargsyan of the Republican Party, the strongest party since 2000, claimed the victory, but his opponents claimed that the election was rigged.

Georgia: Territorial Struggles

Georgia is more than twice as large as Armenia, with a population of 4.5 million. Georgia's history parallels that of its neighbor. Both go back to kingdoms formed over two millennia ago, both adopted Christianity in the fourth century, and developed alphabets in the fifth century. The Georgian language, however, is part of the Caucasian family. Georgia's history also was shaped by the country's crossroads location and its centuries-long effort to survive intact while squeezed between great empires.

Today Georgia's domestic problems are comparable to Armenia's, yet in contrast, Georgia has cordial relations with Azerbaijan but two extreme disagreements with Russia. Georgia's 2003 **Rose Revolution** (demonstrators carried roses to emphasize nonviolence) ousted Eduard Shevardnadze, the former Gorbachev ally whose corrupt regime failed to revive the economy or keep the country intact. Subsequently, President Mikheil Saakashvili has tried to reform the economy while restoring sovereignty over all the territory claimed by the Georgian state. The economy has grown at a respectable rate, and international groups have applauded anticorruption reforms. But many Georgians remain dissatisfied—Saakashvili with great difficulty managed to survive a political crisis in late 2007 that nearly toppled his government. In October 2012 an opposition coalition won a majority in parliament. In October 2013 Saakashvili's second and final term ended with the election of a new president.

Territorially, the Georgian government's biggest success was regaining control over the breakaway region of Ajaria (see Figure 5-3-1). It also managed by the end of 2007 to persuade Russia to remove military bases from Ajaria dating back to the Soviet period. But the Georgian government has not succeeded in regaining control over Abkhazia, which fought for its independence and expelled 250,000 Georgian residents in 1993. Bitter feelings persist on both sides. Russian "peacekeepers" remain in Abkkazia, mainly to deter Georgia from retaking the region by force.

▲ Figure 5-30 **Georgian tank destroyed by Russian forces in South Ossetia.** The short war in 2008 was a disaster for Georgia. South Ossetia and the other breakaway "republic," Abkhazia, remain under Russian protection.

South Ossetia has been at least an equally painful problem for Georgia. It remains independent of Georgia's control, perhaps gravitating toward unification with North Ossetia, which is part of Russia. Ossetians speak a language that is in the Iranic branch of the Indo-European family and are predominantly Russian Orthodox. They have tended to side with Russia in the ever-complex political context of the Caucasus region, much to the displeasure not only of the Georgians, but also such Islamic Caucasian peoples as the Chechens and Ingush. In August 2008 Georgian troops entered South Ossetia and attempted to capture the capital, Tskhinvali. The Russian military immediately crossed the border and engaged the Georgians, who soon fled (Figure 5-30). Russian forces also crossed the border to Georgia and briefly threatened Tbilisi, the Georgian capital. They also invaded Georgia from Abkhazia and occupied the port of Poti temporarily. Georgia expected the West, especially the United States, to come to the rescue, because the United States had been advocating that Georgia (and Ukraine) should be accepted as members of NATO. But there was no chance of the United States or NATO attempting to repel Russia by force in this part of the world. Abkhazia and South Ossetia can remain independent from Georgia as long as they enjoy Russian protection.

Azerbaijan: Caucasian Oil Power

Azerbaijan, which is nearly as large as Armenia and Georgia combined, traces its roots to "Caucasian Albania" of the fourth century B.C., but its crossroads location has affected Azerbaijan no less than its two neighbors. Today's Azerbaijanis are predominantly Turkic Muslims, although a strong Persian influence is indicated by their adoption of Shi'ism. Some 12–13 million Azerbaijanis form Iran's largest minority group, while the population of Azerbaijan itself is but 9 million.

The Karabakh war took a great toll on Azerbaijan; one of the losses was its democratic government, which was replaced by the authoritarian rule of Heydar Aliyev, the former leader of Soviet Azerbaijan. He ruled until 2003 when, like a king, he transferred power to his son Ilham. Ilham Aliyev's position later was bolstered by a questionable election, but he lacks the legitimacy of his long-serving father. There are many opposition groups in Azerbaijan, and some are radical Islamist. The Aliyev regime violently suppresses protests and arrests journalists. Corruption is rife in Azerbaijan; it may be the worst among Northern Eurasian countries. Although official statistics indicate that GDP has been growing rapidly, that does not mean that most Azerbaijanis are prospering today. Continued corruption, repression, and poverty may provoke more unrest. A patriotic war to recover Karabakh could be appealing as a means of rallying, or at least distracting, an aggrieved populace.

The sole reason for economic growth in Azerbaijan is petroleum. Baki, Azerbaijan's capital, has been a major oil-producing center for about 130 years—in 1900 it accounted for half of world production. Offshore drilling technology promises to make Azerbaijan a major producer once again. In May 2005 the Baki–Tbilisi–Ceyhan oil pipeline opened, forging a new connection between the Caspian and the Mediterranean, 1,100 miles (1,770 kilometers) away, a link that avoids Russian territory (see Figure 5-3-1). The health of Azerbaijan's economy depends on the world price of oil, since other sectors suffer from inefficiency and corruption. Oil revenues have enabled Azerbaijan to increase military spending since 2010, while tension with Armenia over Karabakh remains high.

Summary

▶ Geography is an especially important force in economic activity. The formidable physical conditions found in Northern Eurasia and the remoteness of most locations inevitably add to the costs of development. Just how development will proceed, in this age of globalization, remains to be seen. History also matters, especially with respect to political culture. How people understand history affects what they expect from their leaders, how they see the role of the state, and how they judge their institutions. Democracy played little part in the history of Northern Eurasia, so the cultural environment is not particularly hospitable for a transition to democracy.

▶ Northern Eurasian countries have yet to complete the dual transition that has been expected of them—to become democracies with market economies. All six former Soviet countries left the Five Year Plan behind, yet they have not restructured their economies so that the market can function as it does in developed countries. All six countries have held elections, a few of which have been relatively fair. But they have not developed the necessary institutions to make the rule of law supreme. They lack sufficient separation and balance of powers within their governments. They need territorial administrations that honestly serve their diverse populations. They have just begun to create civil societies and democratic political cultures.

▶ Georgia and Armenia still have a long, hard climb to attain prosperity, and both must avoid future violent conflicts if they hope to succeed. The August 2008 hostilities between Georgia and Russia illustrate how abruptly and devastatingly Russian power can be unleashed in defense of its perceived vital interests. Azerbaijan has been dealt a wild card—how the forthcoming oil money will be used is crucial. Ukraine has little room to maneuver; the Yanukovich regime is alienating Europe and the International Monetary Fund, while moving closer to Russia. But that political strategy lacks an economic counterpart that can revitalize development in the country. Belarus is not only landlocked but apparently "time-locked" as well as a holdover from the Soviet era; it is unlikely to change unless Russia provides some leadership in that direction.

▶ Russia's course will be crucial for the whole region. The oil-export windfall poses opportunities and dangers. The government has bolstered its financial position by setting aside billions in a stabilization fund. The appearance of financial stability should help attract investment. But Russia needs systematic reform of crony capitalism from top to bottom. All areas of public life and economic activity would benefit from more transparency and even-handed enforcement of the law. Then the government and people would have the opportunity to decide the best strategy for developing the country's vast resources. The danger is that the oil money may allow more entrenchment of crony capitalism and permit an increasingly authoritarian central government to assert itself more forcefully abroad.

Key Terms

Bolshevik 230	Duma 230	oligarchs 236	taiga 223
Caucasus Mountains 222	Five Year Plan 230	Orange Revolution 251	Trans-Siberian Railway 230
Chernobyl 223	Gazprom 236	permafrost 222	tundra 223
chernozem 223	Josef Stalin 230	private-plot production 245	Ural Mountains 222
Cold War 232	Kremlin 232	privatization of industry 236	Vladimir Lenin 230
collective farms 244	Lake Baikal 225	rent-seeking 236	Vladimir Putin 236
Communist Party 230	Mikhail Gorbachev 233	Rose Revolution 253	Volga River 222
continental climate 222	Nagorno-Karabakh region 252	serfdom 227	Yenisey River 222
crony capitalism 248	new middle class 236	siloviki 236	
Dmitrii Medvedev 248	New Russians 236	state farms 244	
Dnieper River 225	Ob River 223	steppe 223	

Understanding Development in Northern Eurasia

1. What are the obstacles to economic development in the Yakutsk region?
2. Are the current developmental prospects of Belarus promising or flawed?
3. To what extent is Ukraine's agricultural potential realized? Why or why not is that the case?
4. Who are the winners and losers among Russian regions in the economic transition to capitalism?
5. Why is the Nord Stream pipeline significant?

6. Do current trends in agriculture tell us anything about the economic and social transition in Russia?
7. What is "crony capitalism" and how does it factor into the development of democracy?
8. In what ways has the era of Putin (and Medvedev) differed from the Yeltsin period?
9. What is Moscow's role in Russia and how might its position change in the future?
10. Where in Russia are non-Russian groups the most discontented? Why?

Geographers @ Work

1. Imagine traveling from Murmansk to Yalta in the southern Crimea in June. Describe the natural regions through which you would pass.
2. Describe the fundamental principle underlying the relationship of state and society from the rise of Muscovite Russia to the end of the U.S.S.R. Draw contrasts with the Western experience. Identify the remains of this traditional political culture in the current political system.
3. Explain the geographical dimensions of the downfall of the Soviet Union and demonstrate the extent to which these problems are still significant in Northern Eurasia.

4. Think of a spectrum of possibilities with the Soviet command economy on one end and unregulated market economy at the other. Consider Russia's environment and resource base. Work out the political-economic strategy that would be best for Russia, place it appropriately on your continuum, and justify your decision.
5. Thanks to relatively high prices for oil and gas, the Russian government has paid off debts and saved money in a stabilization fund. Assume it is your job to spend this windfall to develop the economy and, given your knowledge of Russia's economic geography, determine where you would invest.

MasteringGeography™

Looking for additional review and test prep materials? Visit the Study Area in MasteringGeography™ to enhance your geographic literacy, spatial reasoning skills, and understanding of this chapter's content by accessing a variety of resources, including MapMaster interactive maps, videos, animations, RSS feeds, flashcards, web links, self-study quizzes, and an eText version of *World Regional Geography*.

▲ Figure 5-31 **Northern Eurasia from space.**

Central Asia and Afghanistan

William C. Rowe

Astana, Kazakhstan: City of the Twenty-First Century?

Traveling over the steppes of Kazakhstan, a visitor might be forgiven for thinking that the steppe grasslands would simply go on forever with no hint of change. However, over the past decade there has arisen the "only city of the twenty-first century" according to Kazakhstan President Nursultan Nazarbayev. After independence from the U.S.S.R., President Nazarbayev unilaterally decided to move the capital from Almaty, a city he cited as too near the Chinese border and in a tectonically active area. He wanted the new capital to be more centrally located in Kazakhstan and renamed the northern city of Aqmola "Astana," or "capital" in the Kazakh language.

President Nazarbayev then began to pour expansive revenues from oil and gas sales into building his grand new capital. No expense was spared and famous architects from Norman Foster to Manfredi Nicoletti created large and often whimsical buildings along the city's main thoroughfare that links the new Presidential Palace (seven times the square footage of the White House) to Foster's Khan Shatyr Entertainment Center, shaped like a collapsing tent. These two signature monuments along with others, like the pyramid-shaped "Palace of Peace and Reconciliation," the new Kazakh museum (based on the Parthenon in Athens), and the Kaz Munai Gaz Complex (built on a plan similar to that of the Hotel Atlantis in the Bahamas) are interspersed between quickly constructed high-rise buildings by Chinese firms that are already starting to show signs of wear and deterioration. Most of the new residents of Astana have embraced this city as the new brash and brassy emblem of their country and hope that the world will come to know Kazakhstan because of cutting-edge architecture and bold city planning, and not as the home of the fictional character, Borat.

▲ The landscape of Astana, Kazakhstan's capital, is changing quickly and dramatically with landmark buildings such as the Aq Orda ("white horde") Presidential Palace at the center of the photo and the twin golden towers of the House of Ministries, which harken back to pre-Soviet Muslim architecture in Central Asia.

Read & Learn

▶ Explain the central role played by Central Asia and Afghanistan in the historic culture and economy of the Eurasian landmass.

▶ Express the importance of river valleys within the context of Central Asia and Afghanistan's population and economy.

▶ Indicate why the Aral Sea has shrunk dramatically and assess what might be done to reverse this process.

▶ Describe how Central Asian urban spaces are and are not changing.

▶ Show how economic development without proper attention to its effect on the environment can provide useful lessons to help improve future development policy.

▶ Identify the reasons why so many rural Tajik men must migrate to Russia to earn a living and discuss the effect that migration is having on village life.

▶ Portray the plight of Afghan farmers and explain why they find it difficult to stop growing poppies for opium.

Contents

Central Asia and Afghanistan

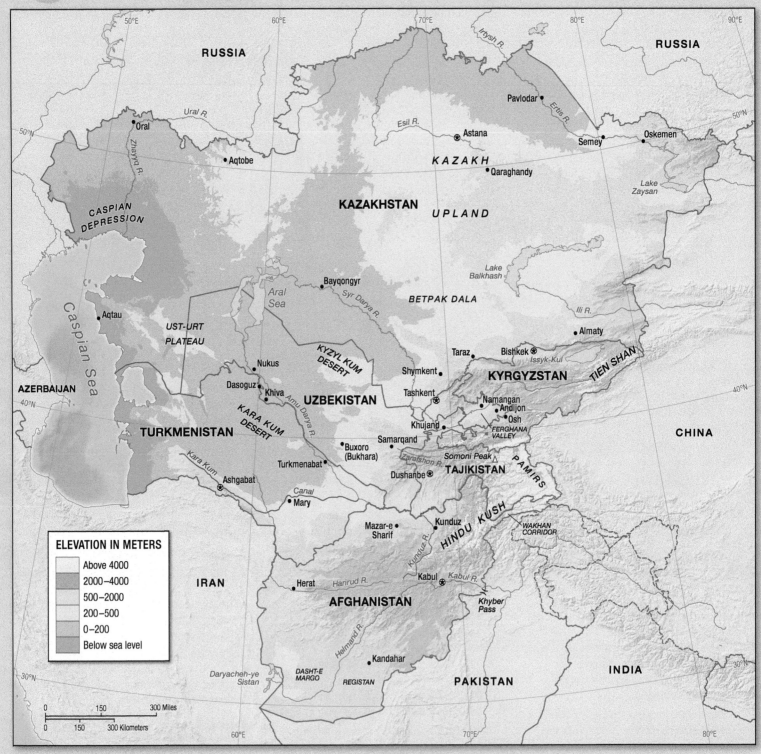

Figure 6-0 The major physical features and countries of Central Asia and Afghanistan.

From 1920 to 1991, the Central Asian countries of Kazakhstan, Kyrgyzstan, Tajikistan, Turkmenistan, and Uzbekistan were part of the Union of Soviet Socialist Republics (U.S.S.R.). The U.S.S.R. was a major international political player in the 1900s and the counterpart to the United States during the Cold War in the second half of the twentieth century. As such, the area was viewed as a part of a superpower and grouped with the "developed" world along with the other Soviet republics, such as Russia and Ukraine. That Central Asia was not as advanced as the Slavic areas of the U.S.S.R. was not widely acknowledged until after the fall of the U.S.S.R. and the independence of the five countries in 1991. Since then, Central Asia has had a more difficult time adjusting to independent status and to the difficult economic changes than any other region of the former U.S.S.R. These countries have shown economic renewal in the twenty-first century, but exhibit both authoritarianism and chaos politically.

Prior to the Soviet period, the region was more closely tied to southern and western Asia because of the Islamic heritage it shares with the Middle East and western South Asia. Afghanistan has been an independent country since the nineteenth century, but had close trade and cultural ties with the region absorbed by the U.S.S.R. The region's cultural history has been dominated by two linguistic groups that include various ethnic groups of Persian and Turkic origins. As a linguistic case in point, all six countries in this region have the same suffix, "-istan," which is old Persian for "place of." Tajikistan, for example, means "place of the Tajiks." These cultural ties link many groups and cross the borders that would eventually separate Afghanistan from the other five countries. Half of the Tajiks live in Afghanistan and the other half in Tajikistan and Uzbekistan. Significant minorities of Uzbeks and Turkmen also live in Afghanistan. For historic, linguistic, and cultural reasons, Afghanistan is included in this chapter (see chapter opener map).

Central Asia and Afghanistan connect four of the world's great "realms": Russia, and through it Europe; the Middle East and North Africa; South Asia; and East Asia. This world region was at the crossroads of some of the most important ancient empires, such as Mesopotamia (in modern Iraq), the Indus River valley (now in Pakistan), the Iranian Plateau, Greece, Rome, and China. This very centrality gives Central Asia both its name and its place in history and geography, and its strategic significance ensures the region's pivotal role in many political and economic issues the world faces today.

Central Asia's and Afghanistan's Environmental and Historical Contexts

The contemporary scene in Central Asia and Afghanistan cannot be appreciated without considering the natural and human contexts of the region. The seeming paucity of natural resources sharply contrasts with a rich diversity of cultural groups, often complexly intermixed and established in place for centuries if not millennia. The rhythmic ebb and flow of nature and society provide a framework that profoundly influences present patterns of economic and political development.

Environmental Setting: A Challenging Locale

The area of Central Asia and Afghanistan crosses latitudes equal to the distance between Austin, Texas and Edmonton, Alberta in Canada, and roughly three-quarters the longitudinal distance of the continental United States (Figure 6-1). Despite this large area, the region is sparsely populated, ranging from a low of 15 people per square mile (6 per square kilometer) in Kazakhstan to 173 people per square mile (67 per square kilometer) in Uzbekistan. This is attributable in part to the natural setting, which encompasses many landform areas but only four natural regions (Figure 6-2). The region is dominated by mountain climates in the southeast and deserts and midlatitude steppes in lower-lying areas to the north and west.

Landforms

With soaring mountain peaks over 19,700 feet (6,000 meters) high and depressions that sink below sea level, Central Asian landscapes are extremely varied. Extensive lowland deserts and semiarid plains contrast with intensive islands of verdant vegetation in deep river valleys and narrow floodplains, creating a difficult, often extreme, setting for human life and livelihood.

Like the Himalayas to the east, the tectonic upthrust caused by the Indian plate sliding under the Eurasian plate created the mountainous region of Central Asia. Due to continuing tectonic activity in the region, Central Asia and Afghanistan are prone to frequent earthquakes. The major mountain chains are the **Hindu Kush** (in Afghanistan), the **Pamirs** (in Tajikistan and Afghanistan), the Fan

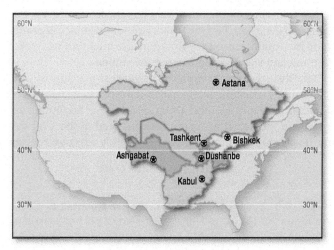

▲ Figure 6-1 **Central Asia and Afghanistan.** A size and locational comparison with the conterminous United States and Canada shows how large the central Asian region actually is.

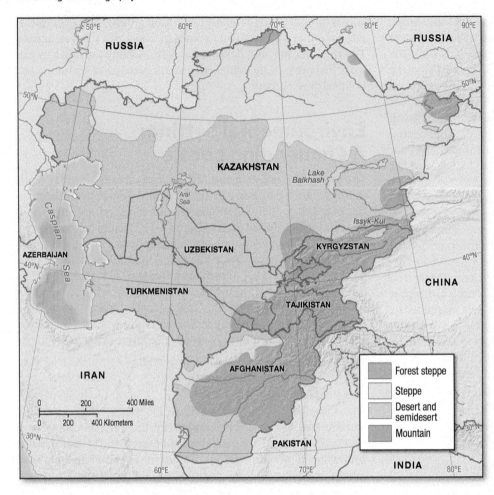

◀ **Figure 6-2 Natural regions of Central Asia and Afghanistan.** Aridity dominates the natural regions of Central Asia and Afghanistan. Only in the mountainous uplands where significant precipitation occurs, and along the streams that drain those highlands, is woody vegetation common. Elsewhere grasslands of varying richness depending on rainfall are interspersed with arid regions with little permanent vegetation. People are found in large numbers only in places where water is abundant, slopes are gentle, and soils are rich.

agriculture, in particular to the populations of southern Kazakhstan, Uzbekistan, and Turkmenistan, or around the Caspian Sea where minerals, especially natural gas and oil, are extracted. Outside the river valleys xerophytic plants—species tolerant of aridity—predominate, although scattered grasslands have historically provided for nomadic herders such as the Turkmen, who ranged particularly across the Kara Kum desert, or the Baluchi of the Registan and Dasht-i Margo deserts in southwestern Afghanistan (Figure 6-4). The **Kyzyl Kum desert** (east of the **Aral Sea** in Uzbekistan and Kazakhstan), the **Kara Kum desert** in Turkmenistan, and the Registan and Dasht-e Margo in western Afghanistan are the most prominent in the region. To the

(in Tajikistan and Uzbekistan), the Tien Shan (in Kyrgyzstan), and the Altai (in southeastern Kazakhstan) (see chapter opener map; Figure 6-3). In Afghanistan, Tajikistan, and Kyrgyzstan particularly, most people live in river valleys, as well as foothills and meadows located in and at the base of these mountain chains. The streams on which they depend are fed by spring and summer runoff from nearby mountains as well as by natural springs that contribute water flow from the mountainous areas into the lower, desert environments of Uzbekistan, Turkmenistan, and southwestern Afghanistan. In the mountain complexes summers are mild, but extreme cold and abundant snowfall characterize the winters.

The desert areas comprise the largest portion of Central Asia and Afghanistan, where precipitation is generally less than 10 inches (250 millimeters) a year. Most major economic activities within the desert areas are concentrated in the river valleys that flow from the mountains and provide extensive irrigated

▶ **Figure 6-3 A young Tajik herder in the Fan Mountains.** Mountains are a distinctive feature of Afghanistan, Kyrgyzstan, and Tajikistan. This herder practices transhumance, a way of life whereby he drives his cattle to mountain pastures in the summertime, where melting snow and ice create lush vegetation, then takes them back to the river valleys in the winter when the snow falls in the mountains.

(a)

(b)

▲ **Figure 6-4 Desert land use.** (a) Aridity and great temperature fluctuations between summer and winter characterize the desert regions of Central Asia and Afghanistan. Extensive grazing of animals provides milk for the daily subsistence of the *Kuchis* (nomads), as well as animals for sale in cities along the seasonal migration route. (b) This nomadic encampment, located 25 miles (40 kilometers) west of Herat near the village of Ghurian in July, spends winters near the Iranian frontier and the late summer and early autumn in the Safid Kuh mountains north of Herat.

west of Central Asia is the **Caspian Sea,** the surface of which lies 92 feet (28 meters) below sea level, and west and north of the Kyzyl Kum and Kara Kum deserts respectively lies the Aral Sea, which is fed by the **Syr Darya** and **Amu Darya rivers.** Other major rivers are the **Zarafshon River** flowing from Tajikistan into Uzbekistan and disappearing into the Kyzyl Kum, and the Harirud, Helmand, Kabul, and Kunduz rivers in Afghanistan.

Climate

Because of its position at the heart of the Eurasian landmass and the massive mountain complexes that cut off the monsoonal airflows from the south (see Chapter 9) most of Central Asia exhibits a continental climate characterized by intensely hot summers, cold winters, and arid conditions (potential evaporation is greater than potential rainfall). Another contributing factor to the climate is that the region is entirely landlocked. Although the Caspian Sea moderates the climate to some extent on the region's far western edge, there are no oceans nearby to mitigate the extremes of the continental climate. Without irrigation, this environment can support only xerophytic vegetation, while northeastern Kazakhstan is largely grassy steppe that is also very dry. The aridity is mitigated by

orographic precipitation in the spring that discharges into alpine valleys (where natural springs contribute to the water supply) before flowing in a generally westerly direction—thus allowing increased habitation and agriculture in this dry region.

The soils and vegetation of the mountain areas are diverse, reflecting the location of these mountains, their local relief, and, most importantly, their altitude. To the north and east of the desert regions lie the steppes of Central Asia. This natural region dominates much of Kazakhstan (Figure 6-5) and is characterized by grasslands that extend both north and west into Russia with only a few trees located in river valleys. Although the soils are generally rich, much like those of the prairie lands of the U.S. Great Plains, a steppe environment differs from a prairie environment in its generally lower and more variable precipitation. This has made for ideal conditions historically for herding cultures such as the Kazakhs and before them the Mongols who invaded the region in the thirteenth century. But agriculture in this region is more precarious and susceptible to drought than in either the Great Plains or the forest steppe in southern Russia, some portions of which overlap the Kazakhstan border. A steppe grassland more dominated by small woody shrubs is found in parts of central and western Afghanistan and forms the basic resource for a substantial pastoral nomadic population both historically and today.

Environmental Challenges

The environmental challenges facing Central Asia and Afghanistan today are largely a result of Soviet planning and overreach that had no place for either sustainability or the welfare of the local inhabitants. In Central Asia, these challenges include the Aral Sea crisis (see *Exploring Environmental Impacts:* The Aral Sea), erosion from the Virgin Lands Program, and the remnants of the Soviet nuclear program. The nuclear program has affected both eastern Kazakhstan and the Fan Mountains of Tajikistan. In eastern Kazakhstan around the Soviet city of Semipalatinsk (now Semey), the Soviets exploded approximately 460 nuclear bombs

▼ **Figure 6-5 Steppe grasslands in Kazakhstan.** The steppe regions of Central Asia are natural grasslands, whose primary historical use has been animal grazing. Efforts to extend dryland farming into these "virgin lands" in Central Asia during the Soviet era resulted in serious land degradation.

The Aral Sea

Mismanagement of the Aral Sea ecosystem represents one of the world's saddest intersections of economic, environmental, cultural, medical, and physical geography. Economic decisions made by Soviet planners in the 1960s created a situation that has devastated an environment, its population, and its climate, which is greatly affected up to 60 miles (100 kilometers) from the former seabed. In the 1950s, there was a huge push throughout the U.S.S.R. to expand agriculture.

In Central Asia the goal was to produce more cotton, so land was shifted from the production of food staples. But this conversion still was not enough and more marginal areas were brought into cultivation. Massive new irrigation canals were dug, including the 1,000-mile- (1,600-kilometer-) long Kara Kum Canal, the longest diversionary channel in the world, which did not feed drainage water back into the river system but took water away from the Aral Sea drainage system toward Turkmenistan.

Although Soviet planners knew that this additional irrigation and diversion would impact the Aral Sea, what happened is one of the worst ecological disasters in modern history. Following construction of the new irrigation works, the Aral Sea started to shrink. The government tried to dredge a canal for fishing ships to get to the Aral Sea from the ports, but soon this was no longer feasible. The port of Muynoq in Uzbekistan is more than 50 miles (80 kilometers) from the sea (Figure 6-1-1). By 2013, the surface level of the southern section of the Aral Sea had fallen by more than 72 feet (22 meters) and its volume by 87.5 percent. From 20 fish species present in 1960, the sea now has no native fish. Salt concentration has intensified due to sea shrinkage, and now the salinity of the Aral Sea is second only to the Dead Sea in Israel/ Palestine and Jordan. More than 60,000 fishing and fish canning jobs have literally evaporated.

The vegetation has changed radically. Native hydrophytic (water-loving) plants are all gone, replaced by halophytic (salt-tolerant) plants. Near the

(a)

(b)

▲ **FIGURE 6-1-1** **The transformation of the Aral Sea.** (a) Looking out from the "coast" of Muynoq across the former seabed, sand and salt stretch to the horizon. (b) This ship is part of the graveyard of ships around the former port of Muynoq in Uzbekistan. Scenes of abandonment like this are common around the former shoreline of the Aral Sea.

in an area of the nearby steppe. Radiation remains a problem and over the years has destroyed the health of both people and livestock in the area. In Tajikistan, uranium mining has caused severe damage to many streams that flow out of the Fan Mountains. The Soviets allowed the runoff (or tailings) of the mines to flow into nearby streams, and from where these tailings enter the water, no life can be found downstream for many miles. This is a point of contention with the Uzbeks who claim it is Tajikistan's duty to clean up these tailings as it affects the health and economic output of Uzbeks.

Aral Sea, plants become stunted and many cannot grow because of the high soil salinity. **Salinization** due to overuse of irrigation water in an arid environment occurs further away from the Aral as well. Salinization is a constant worry in this arid environment; when farmers put too much water on the land and do not provide adequate drainage, some of the water evaporates and salts left behind accumulate, inhibiting the ability of plants to absorb water and nutrients.

Moreover, residual salt from seawater evaporation became airborne, was transported considerable distances from its origin, and settled over thousands of square miles of irrigated land. Along with poor drainage and salinization, this has caused cotton productivity in Uzbekistan to decline drastically and may in fact endanger cultivation of cotton, the crop that caused the disaster in the first place. Airborne particles from the seabed contain many other substances, including fallout from nuclear and biological testing carried out during the Cold War. The Soviets also buried their supplies of anthrax, plague, smallpox, and typhus manufactured as potential biological weapons on an island in the sea. This has caused a catastrophic drop in health among ethnic Karakalpaks who live nearest to the former seashore. Kidney diseases are rampant among the population, strontium (a by-product of nuclear testing) has been found in the thyroids of 23 percent of Karakalpaks tested, and anemia, respiratory diseases, and cancer are up 600 percent since 1980; in port cities a large majority of residents have cancerous or precancerous growths

in their throats, and birth defects have skyrocketed.

Because the Karakalpaks have little power in Uzbekistan, virtually nothing has been done for them except by a handful of international agencies. The Central Asian countries have signed many agreements, but no real progress has been made in returning more water into the sea from the Amu Darya river. As the sea has divided into two parts (Figure 6-1-2), the Kazakhs (who have a considerably stronger economy than the Uzbeks) have built a dam to protect the northern segment. The results of this strategy are promising. The northern portion of the Aral Sea has stabilized and expanded, which decreased salinity and allowed the reintroduction of fish. But the culture, environment, and health of Karakalpaks in Uzbekistan have been too effectively destroyed for any future short-term development to improve the regional environment. Perhaps in the long term, efforts by Kazakhstan could expand and bring hope to the whole region, but for now the world sees the continued destruction of a culture through poor economic choices with no regard for the environment.

Source: Aharon Oren et al., "The Aral Sea and the Dead Sea: Disparate Lakes with Similar Histories," *Lakes and Reservoirs: Research and Management* 15, no. 3 (2010): 223–236.

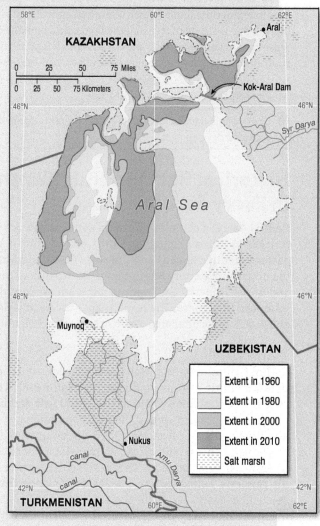

▲ Figure 6-1-2 **Aral Sea shrinkage 1960–2010.** The Aral Sea has been reduced to a small fraction of its former size. *Source:* http://www.mapsofworld.com/tajikistan/shrinking-aral-sea-map.html.

In the 1950s when the Soviets planned the expansion of agriculture that would result in the Aral Sea crisis, they also expanded agriculture on the steppes of Kazakhstan. Millions of acres were plowed to expand wheat production. The steppe area is more prone to drought than the plains area of the United States (the agriculture

the Soviets tried to duplicate), and this resulted in major erosion of the soils in the region and intense dust storms in Kazakhstan, especially around the new capital of Astana.

In Afghanistan, the number one hazard is the detritus of more than three decades of war, especially landmines. The Soviets placed millions of landmines in unmarked areas throughout the country, and the warring factions that have fought for control of Afghanistan have since laid many more. This has had a disastrous effect on the Afghan economy and the Afghan population. Many thousands of farmers and travelers have been killed or maimed in spite of recent efforts to remove the mines; much of the best agricultural land has been unavailable because of fear that the land could contain mines. All of these challenges began with the U.S.S.R., but it is up to the individual countries today to try and mitigate the effects of these Soviet policies.

Stop & Think

→ Does a single-minded focus on the economy risk undermining future economic development?

Historical Background: "Central" in Every Sense

Central Asia is a complex transition zone, heavily influenced historically by forces originating in the west (Southwest Asia and the eastern Mediterranean) and the east (China and Mongolia). By the nineteenth century Russian imperial expansion from the north and British imperial ambitions to the south exposed the region to European influences. The result is an intricate mosaic of peoples and practices accumulated over two and a half millennia.

Ethnic Origins

Within Central Asia and Afghanistan **Turkic ethnic groups** and **Persian ethnic groups** predominate (Figure 6-6). The Persians were the earliest known settlers in Central Asia and included the two most important settled Persian populations, the Bactrians and the Sogdians. Alexander the Great conquered this area in the fourth century and chose as his wife a Bactrian princess. After Alexander's death the Greeks ruled the region under one of his generals, Seleucus, and his descendants. By the end of the Seleucid dynasty's reign, the region reverted to Persian rule until the Arab conquest in the eighth century, which brought Islam to the region. Unlike Iranians to the west, most Central Asians converted to Islam early and consequently are predominantly Sunni Muslims (Figure 6-7). The major exceptions are the Shi'ite **Hazaras** in central Afghanistan and the Ismailis in the Pamir Mountains. Arab political hegemony lasted in Central Asia until A.D. 899, when Persians under the Samanid dynasty ruled much of today's Central Asia and Afghanistan. Persian dominance continued until Turkic tribes conquered Buxoro (Bukhara), the capital, in A.D. 999. The Turkic peoples came from what are now Siberia and Mongolia. While the Persians speak an Indo-European language (the Indo-European language family also includes English, French, Russian, and most other European languages), the various Turkic people speak languages from the Altaic family, which some linguists believe is related to Mongolian and Korean.

Successive waves of Turkic peoples swept through Central Asia. Some like the Turkmen stayed; others like the Seljuks moved on to other places such as the Anatolian Plateau in contemporary Turkey. But they dominated Central Asia from the eleventh to the thirteenth centuries until the Mongol Empire under Genghis Khan conquered them. The Mongols sacked and destroyed the major cities of Buxoro (Bukhara), Samarqand, Khiva (all three in contemporary Uzbekistan), and Balkh (in northern Afghanistan) before moving toward Iran. The worst damage of this campaign was the disruption of the irrigation systems, which destroyed both agriculture and the water supply. Balkh remains a ruin to this day, but Buxoro (Bukhara), Samarqand, and Khiva were all rebuilt over the next centuries and flourished on the **Silk Road.** This caravan route, stretching from China to the Middle East and ultimately on to Europe, was the main procurement corridor for goods from the east, carried mostly on camel caravans through the Central Asian river valleys and oases and on to the markets of the Mediterranean. Although called the Silk Road, silk was not the only product to be traded in this fashion. Paper, for example, was first used outside of China in Samarqand (Figure 6-8) and later made its way to the West. In addition to trade goods, ideas also traveled the Silk Road. Buddhism made its way from South Asia first through Afghanistan, then Central Asia on its way eventually to China and Japan. The giant Buddhas destroyed by the extremist Taliban government in 2000 are a testament to this movement.

Turkic groups continued to move into the region after the Mongol interlude; by the sixteenth century the Uzbeks dominated the river valleys, which were mostly inhabited by eastern Persians known as Tajiks. Later still the Kazakh-Kyrgyz moved into the steppe and northern mountainous regions and a related group, the Karakalpaks, occupied the region east of the Aral Sea. Although Turkic people lived in the area we know as Afghanistan, they did not dominate it in the same way as the rest of Central Asia. Linguistically Persian Pashtuns rose to dominance in Afghanistan except for two short interludes in the twentieth century when Tajiks led the country. From about the time of the arrival of the Uzbeks, the Silk Road began to be eclipsed by the sea route pioneered by the Portuguese, which bypassed Central Asia and the Middle East and allowed Europeans to trade directly with India and China and carry their goods back around Africa to Europe.

Stop & Think

→ How has migration affected the ethnic composition of Central Asia and Afghanistan?

The Great Game and the Soviets

By the nineteenth century, Central Asia had stagnated to the point that it became a pawn in the imperial land grabs made by Russia and Great Britain. This period became known as the Great Game, as Great Britain, fearing Russia was a danger to its Indian colony, engaged in three wars in Afghanistan and extensive exploration in Central Asia and Afghanistan. By the latter half of the nineteenth century a truce was declared; Afghanistan was considered part of the British sphere of influence, and Central Asia part of Russia's. The two governments set the boundaries of modern-day Afghanistan. To separate the powers, a narrow strip of land, called the **Wakhan Corridor,** was created between British India and the Russian-controlled Pamir Mountains. This feature is still on the map today and connects Afghanistan to China.

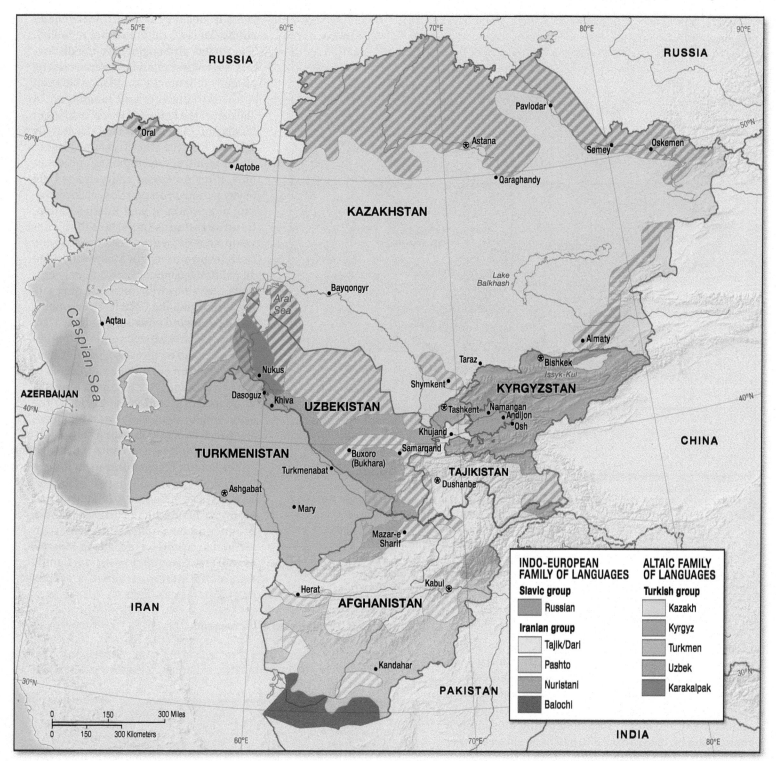

▲ **Figure 6-6** **Major language groups of Central Asia and Afghanistan.** When the borders of the six Central Asian countries were drawn by Great Britain and Russia, and later redrawn by the Soviet Union, no consideration was given to linguistic or ethnic boundaries. Although each country has a titular majority, all have minority populations of ethnic groups found in adjacent states. In the Soviet republics, this result was produced by design so no republic could secede from the Soviet Union without leaving significant numbers of its linguistic community behind in other republics.

Although the British never colonized Afghanistan, the Russians did colonize much of Central Asia in their march south toward India, specifically the areas that would become the republics of Kazakhstan, Kyrgyzstan, and Turkmenistan, along with Samarqand, the **Ferghana Valley,** and the Pamir Mountains. Russia directly administered these areas; the rest of the region, centered around Buxoro (Bukhara) and Khiva, remained independent, but vassals of the Russian Tsar. The 1917 Russian Revolution did not have an

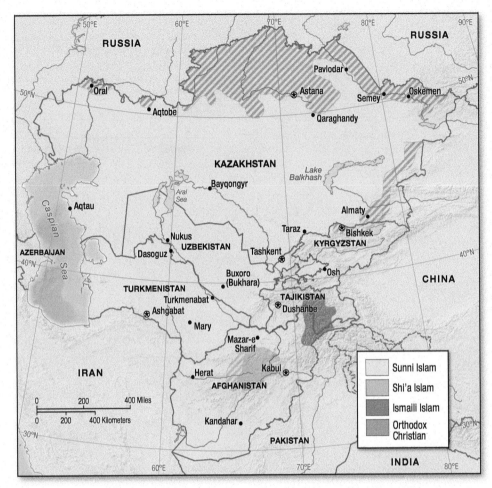

▲ **Figure 6-7 Major religious communities of Central Asia and Afghanistan.** The dominant religion in Central Asia is Islam. Minority Shi'a and Ismaili sects are largely confined to the mountains of central Afghanistan and the Pamir Mountains of eastern Tajikistan and Afghanistan, where historically these minorities have found refuge. Christianity is represented by Russian Orthodox minorities in Kazakhstan and Kyrgyzstan. To a considerable degree, this map is a religious heritage map in the former Soviet republics, since a large part of the population considers itself to be secular, or even atheist, rather than religious in matters of faith.

U.S.S.R. because it would leave many ethnic Uzbeks behind in other republics. The Soviets also augmented the distinctiveness of these republics by accentuating as many differences as possible between the Turkic dialects and languages. In Tajikistan, the language was of a different family and therefore already structurally different from the other Turkic languages.

In terms of development and economy, three processes characterized the Soviet period: centralization, collectivization of agriculture, and Russianization. **Centralization** is important to the development of Central Asia because the U.S.S.R. government in Moscow decided all political, economic, and cultural questions. Decisions were then put into practice throughout the U.S.S.R. The economy was a **command economy,** and the state planned nearly all economic activities. This was realized through a purely quantitative view of equality: as all people of the Soviet Union were theoretically equal, all republics would get statistically the same employment as well as cultural options such as entertainment. Every republic capital, for instance, had an opera and ballet theater (Figure 6-9) even though neither opera nor ballet is part of the cultural heritage of Central Asia. Other issues as diverse as education curricula choices, production decisions, and even architectural decisions were made in Moscow as well (see *Geography in Action:* Central Asia's Distinctive Urban Geography

immediate impact on most of Central Asia. Initially, an attempt was made to create a new country allied with Soviet Russia, one that was independent and Muslim, called Turkistan. However, the area was too important agriculturally to the new U.S.S.R. to permit only indirect control, and by 1925 the U.S.S.R. had annexed all areas of Central Asia except Afghanistan.

One of the first moves by the Soviet government was to carve the region into separate republics, loosely based on ethnicity, which would not allow the inhabitants to unite easily again as Turkistan. Each republic had a titular majority, but the Soviets made sure that small populations of each ethnic group would be in other republics. Hence, there are many Uzbeks in all five republics. Likewise there are many Kazakhs, Turkmen, Kyrgyz, and Tajiks in Uzbekistan. This created a situation whereby Uzbekistan, for example, presumably would not try to secede from the

▼ **Figure 6-8 The Registan of Samarqand.** This image shows the famous madrasah (Quranic school) and mosque complex that make up the Registan in Samarqand. The descendants of Tamerlane built this complex, and it has become the symbol of the renaissance of Central Asia after the destructive Mongol interlude.

◀ Figure 6-9 **Opera and ballet theater in Central Asia.** The Soviet government determined that all republics should have the same access to cultural events. This opera and ballet theater in Tashkent, Uzbekistan, is an architectural example that fuses both European and Central Asian features.

important in light of the equality issue. All citizens of the Soviet Union, to be equal, had to be classified as laborers. Having "farmers" would create another (and potentially "inferior") category of workers. The newly defined laborers were paid in produce from the farm, a periodic monetary stipend, or a combination of the two. In the U.S.S.R., these large tracts of land were known as collective farms or, where specialized monocrops were produced, state farms. Prior to collectivization, the major crop in Central Asia was wheat, followed by other food crops, orchards, vineyards, and—if the farm was large

and Housing). Industrial activities were distributed throughout the various republics. Because Tajikistan had ready supplies of hydroelectric power, a major aluminum smelter was located there. The alumina (the main component of aluminum) came from Ukraine and much of the workforce came from Russia and Belarus. The aluminum was sent to Russia, where it was used as a component in many items. This type of industry, however, was unusual in Central Asia. The central government primarily utilized the region in ways that other European powers such as France and Great Britain used their colonies in Africa and Asia. The primary objective was to extract Central Asia's natural resources (Figure 6-10) such as natural gas and oil (from Turkmenistan and Kazakhstan), gold and uranium (from Kyrgyzstan and Tajikistan), as well as agricultural products, most importantly cotton. The U.S.S.R. also centralized transportation and communications. All highways, railroads, and services linked Russia to each republic, and goods traveled to Russia to be manufactured into products that would be shipped back to the republics.

Collectivization of agriculture is an important hallmark of Communist regimes. This is the process whereby all land, goods, equipment, and crops are appropriated by the government. The numerous medium-sized and small farms were consolidated into large farms that the former owners worked collectively to produce a certain crop or a variety of crops. People were no longer classified as "farmers," but as "laborers." This distinction is

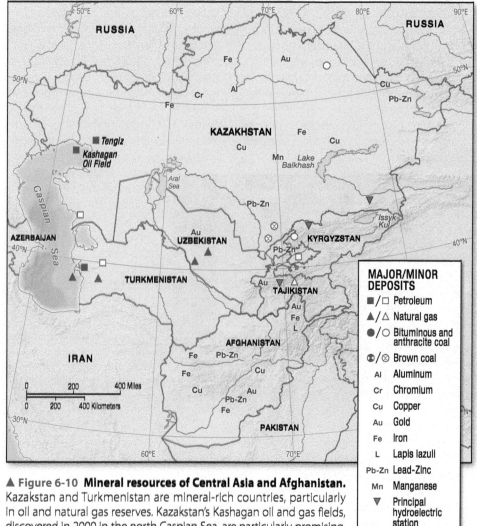

▲ Figure 6-10 **Mineral resources of Central Asia and Afghanistan.** Kazakstan and Turkmenistan are mineral-rich countries, particularly in oil and natural gas reserves. Kazakhstan's Kashagan oil and gas fields, discovered in 2000 in the north Caspian Sea, are particularly promising. High-quality coal in the Kazakh Karaganda deposits, combined with significant iron deposits, as well as a scattering of precious metals throughout Central Asia, provide the basis for an active mining and industrial sector. Two decades of war and civil conflict have inhibited Afghanistan's mineral deposit development.

GEOGRAPHY IN ACTION

Central Asia's Distinctive Urban Geography and Housing

The urban environment in Central Asia is an amalgam of different cultures and spatial ideas that include both indigenous and Soviet constructs. The Soviet Union had definite ideas about urban spaces based on quantifiable egalitarianism as well as centralized planning. This meant that each urban area would theoretically have similar layouts, architecture, and entertainment activities; therefore, each capital had a republic university, an academy of sciences, a Communist Party headquarters, and a writers' union as well as cinemas, theaters, and opera and ballet companies.

Soviet architecture was generally built to impress and at times overwhelm, resulting in urban landscapes where major buildings often occupied an entire block. Tashkent is famous for these structures; much of the downtown was destroyed in a 1966 earthquake and the central city was rebuilt along these open, low-density lines. The Soviets allocated considerable space to green areas and large outdoor gathering spaces. Main roads frequently have large medians with trees, benches, and walking areas and extensive parks were built in and around the cities with cafes, amusement parks, and walking trails (Figure 6-2-1). Larger boulevards end in squares with typically grandiose statuary. In Soviet times, the main square held a large statue of Lenin; today these public sculptures have been replaced by more regionally important figures.

While public buildings embody the Soviet penchant for monumental architecture, private housing is very different and diverse. Soviet-built housing was constructed to maximize function rather than style (Figure 6-2-2). High-rise apartment buildings dominate the urban and suburban landscapes of all major cities in Central Asia. These structures are generally built along the edges of blocks with the interior space (where the buildings' entrances are located) planted with trees to minimize the rather brutal nature of the construction and provide a place of shade away from streets.

In some ways this style mimics the traditional layout of a Central Asian home, which also seeks to block the outside world's gaze from the living area. Traditional homes are found throughout the countryside (except where nomadic herders were settled; these homes were built along Russian standards by the Soviets), but also in older, less central neighborhoods in cities. Unlike American or European residences where the home is open to the gaze of passersby across a stretch of lawn, Central Asian homes are constructed to maximize the amount of private space away from the public gaze. A family will construct their home around a central courtyard planted with fruit and walnut trees and small vegetable gardens. This makes an attractive place in warm weather for family and friends to congregate on a *kot* (an elevated platform with a pile of thin handmade mattresses and pillows). The kot serves as a place to nap and to eat family meals (Figure 6-2-3). Family living quarters and a guesthouse take up two sides of the courtyard; a third side typically houses the kitchen (away from the living quarters for fear of fire). A small barn for a milk cow, chickens, and perhaps a goat, and an outhouse occupy the fourth side. The courtyard entrance is covered by a grape arbor to give visitors shade and a feeling of enclosure. This juxtaposition of western

◄ FIGURE 6-2-1 **Use of green space in urban planning.** This photo shows the park area in the wide median between opposing lanes of traffic on Rudaki (formerly Lenin) Boulevard in central Dushanbe, Tajikistan. Soviet urban planners made extensive use of green spaces and parks in their plans for Soviet republics in an attempt to make them more aesthetically pleasing.

enough—market crops like cotton or tobacco, depending on the terrain and environment (Figure 6-11). After collectivization, the most important crop in the lower river valleys became cotton for textile manufacturing, primarily in Russia (Figure 6-12). Central Asia became a colony, not unlike India or Egypt, where as much land as possible was put into cotton production and the government shipped in food items the Ukraine or northern Kazakhstan could grow more efficiently. In the mountains, the major crop was tobacco in the narrowest and highest of the river valleys (Figure 6-13). The U.S.S.R. also collectivized herds, although many

herders, rather than turning over their animals to the government, either slaughtered them or drove them to Afghanistan. The number of Turkmen and Uzbeks grew in that country and a small population of Kyrgyz became established. Herders who stayed faced a situation similar to that of the farmers: they continued to look after herds as laborers, but now the state owned those herds. The settled Russians who led the collectivization movement felt that nomadic herders were too hard to control and quantify and made them settle and fence in their beasts. This proved disastrous in the early years of collectivization as soon as the first harsh winter began.

◀ FIGURE 6-2-2 **Apartment block in Tashkent.** These typical apartment buildings were built throughout the Soviet Union to meet an increased need for housing, particularly in the larger cities. The structures are almost always uniformly nondescript both inside and out, and reflect the Soviet mentality of equality for all citizens of the country.

religiosity in the region. Expensive high-rise buildings with expansive condos and ornate shopping malls have gone up especially in oil-rich cities in Kazakhstan and Turkmenistan, while all cities across former Soviet Central Asia have seen the addition of mosques and churches. In Tajikistan and Uzbekistan, much of the money for mosque construction comes from Iran and the Arabian Peninsula—two areas that would like to extend their influence into the region by cultural means.

Source: Natalie Kock, "The Monumental and the Miniature: Imaging 'Modernity' in Astana," *Social and Cultural Geography* 11, no. 8 (2010): 769–787.

(Soviet) and traditional architecture shows the region as central to Eurasia with its mix of architectural styles, from Islamic and southern Asian (this housing style is found throughout Afghanistan as well as into the Middle East) to more European.

Since independence little has changed in the way leaders of the post-Soviet republics approach cities. Astana, Kazakhstan's new capital, is a fully planned city—not unlike the cities of the U.S.S.R. President Nazarbayev seems to base his ideas for his capital on St. Petersburg, Washington, D.C., and Brasilia—all planned capitals—combined with Soviet urban planning. Nazarbayev claims that Astana will be "finished" by 2030, but its grandiosity will depend on the price of oil and Nazarbayev's continued presence as head of the government.

Changes can be found, however. While layout and decision making remain similar, new construction expresses both increasing economic activities and a new

▶ FIGURE 6-2-3 **Breakfast in a central courtyard space, Upper Zarafshon River Valley, Tajikistan.** These young men are sitting in a typical outdoor eating area known as a *kot*. In warm weather, rural families prefer to sit outside to take their meals. This feature of traditional domestic architecture is gaining in popularity with urban families. A breakfast normally consists of home-baked bread, green tea, seasonal fruits and nuts, and yoghurt.

Between 1929 and 1933 the number of sheep shrank by 95 percent and the number of cattle by as much as 80 percent in Kazakhstan alone. This caused widespread hardship and demographers of the time speculate that as many as a million Kazakhs died in the ensuing famine.

Stop & Think

▶ Did Soviet economic development significantly change the traditional economy of Central Asia?

The U.S.S.R. also tried to control culture as much as industry. There was a generic ideal of a *Homo sovieticus* envisioned for all citizens. The main instrument in this transformation was **Russification.** Russian culture and language along with European ideas of society and culture were expected to become the norm throughout the U.S.S.R. Teachers taught Russian along with local languages, but increasingly emphasized the Russian language, Russian literature, and Russian culture as the decades wore on. The government also mandated that universities emphasize Russian, and for most students fluency in Russian was required for a college degree. To get a good job,

◀ Figure 6-11 Agricultural zones of Central
Figure 6-11 Agricultural zones of Central Asia. Except in the highlands, where shorter growing seasons and steep slopes limit agricultural expansion, agriculture is dependent on irrigation. Where intensive crop production is possible, large human populations congregate. Soviet efforts to exploit Central Asia's agricultural potential turned the region into a massive cotton farm, drew huge amounts of water away from the Amu Darya river to feed the Kara Kum Canal, and deprived the Aral Sea of the water necessary to sustain its aquatic ecosystems. Despite independence, agricultural diversification and increased food production have proven difficult to attain throughout the region.

and Brahmaputra River valleys, whereas Tajikistan is 93 percent mountains, with most of the Pamir Plateau reaching elevations over 12,000 feet (3,600 meters). Afghanistan and Uzbekistan have by far the largest populations, but these are densely settled in river valleys with only minor populations living in the extensive desert regions.

Demographically, Central Asia was the fastest growing area in the former U.S.S.R. Statistics today still show major growth rates for all five countries, although much lower than Afghanistan's (Table 6-2, Figure 6-15). Compare these numbers to Russia (with a birthrate of 13 and a death rate of 14) or Ukraine (with a birthrate of 11 and a death rate of 15). Infant mortality (an important indicator of a country's level of development) is still high in these countries, most of the population continues to live in rural areas (with the exception of Kazakhstan where roughly one-third of the population is Russian), and per-person purchasing power is lower than in the European republics. When a Soviet woman had 10 children or more, she was presented with a van and a medal and declared a "heroine of the

either in industry or government, one had to be at least proficient in Russian. The idea behind this initiative was that a manager or government official would have to speak to people from all over the Soviet Union and should not be responsible for learning all the languages of the country, rather all people should learn Russian. This created a mostly bilingual culture, in which local languages, literature, culture, and art were all seen as second class to Russian. It also greatly privileged Russians because the linguistic and social abilities intrinsic to their culture were considered the norm and not something they had to learn.

Stop & Think

→ Architecture is an expression of culture. How would you describe the culture of Central Asia through the lens of its architecture?

Population Contours

The population of Central Asia and Afghanistan is not large (Table 6-1) when compared to South Asia or China, but considering that much of the land is arid or mountainous, the region is nonetheless densely settled in its arable parts (Figure 6-14). Tajikistan and Bangladesh, for example, both comprise just over 55,000 square miles (142,000 square kilometers). But Tajikistan has only about 7.1 million people, while Bangladesh has roughly 152.9 million people. Bangladesh, however, lies in the highly fertile plains of the Ganges

▼ **Figure 6-12 Women and children loading cotton.** Scenes like this are prevalent throughout Kyrgyzstan, Tajikistan, Turkmenistan, and Uzbekistan. Both during and after the Soviet period, Central Asian children were taken from school during the cotton harvest and sent to the collective farms to work on the harvest along with the women of the farm.

◀ **Figure 6-13 Upper Zarafshon River valley.** Farmers along the upper reaches of the Zarafshon River were forced by the Soviet government to put all available land in this narrow valley into tobacco, as the climate and physical features were favorable to this crop.

Forging Postindependence Identities

With the fall of the Soviet Union, Central Asia's population was faced with difficult cultural questions. What parts of their culture were intrinsically "Soviet" and did they want to embrace them or renounce them? Where would they look for ethnic or cultural examples to supersede Soviet examples? For each of the five countries, the answers are different and show the growing divide among them. Unlike the Baltic nations of Estonia, Latvia, and Lithuania, Central Asia did not want to break away from the Soviet Union. Independence more or less caught them by surprise and in all but one country (Kyrgyzstan), the ruling Soviet elites stepped directly into power. Since then, these countries have continued to underscore the constructs forced on them by the earlier Soviet government and to preserve the uniqueness of their cultures and ethnicities vis-à-vis each other and neighboring Afghanistan by creating new architecture, art, heroes, and even history.

Uzbekistan and Turkmenistan's approach was to distance themselves as much as possible from Soviet culture. In Kazakhstan, Kyrgyzstan, and Tajikistan, the strategy was to integrate the Soviet past with an often government-controlled sense of ethnic sensibility. They also dealt differently with Russification. Kazakhstan and Kyrgyzstan, which continue to have a large Russian presence, still try to maintain a bilingual society. Turkmenistan and Uzbekistan have de-emphasized Russian, mostly for cultural purposes. Tajikistan declared Tajiki the first language of the country, with Russian an important second language, but due to the hardships caused by their postindependence history, they have de-emphasized Russian. Although Lenin still proudly stands in the main square of Kyrgyzstan's capital, Bishkek, many iconographic Soviet statues have been replaced by historical figures, like Ismail Somoni, the founding emperor of the last Persian (or Tajik) empire in Central Asia, in Tajikistan (Figure 6-16) and Tamerlane in Uzbekistan. Tamerlane, a descendant on his mother's side of Genghis Khan, and his descendants ruled from Samarqand in the fourteenth and fifteenth centuries until the coming of the Uzbeks, so it is an interesting cultural point that the Uzbeks have built some of their identity around this better known historical figure. All countries have also changed the names of streets, cities, provinces, and even mountains from important Communists and hallmarks of Communist ideology to locally important people

Soviet Union." Most of these "heroines" were in Central Asia. The social benefits of large families disappeared with the U.S.S.R., so today it is unusual to see very large families among younger people. Most young families want two or at most four children because now they must pay for education, health care, and other services formerly provided by the U.S.S.R.

> **Stop & Think**
>
> ▷ Why are population indicators an important component of economic development?

Central Asia: Landlocked and Remote

Culturally and ethnically different from the core areas of Russia, Central Asia was regarded by Czarist administrators and Soviet planners as a peripheral, underdeveloped region. Their efforts to exploit Central Asia's resources, modernize the region's peoples, and integrate the area's economy into a larger national social structure shaped the uneven patterns of development that characterize the post-Soviet scene.

Table 6-1 The Countries of Central Asia and Afghanistan				
Country	**Capital**	**Area (sq mi)**	**2012 Population (millions)**	**Largest Ethnic Group (% of total)**
Afghanistan	Kabul	251,772	33.4	Pashtuns (40)
Kazakhstan	Astana	1,049,151	16.8	Kazakhs (54)
Kyrgyzstan	Bishkek	76,641	5.7	Kyrgyz (64)
Tajikistan	Dushanbe	55,251	7.1	Tajiks (71)
Turkmenistan	Ashgabad	188,456	5.2	Turkmen (81)
Uzbekistan	Tashkent	172,741	29.8	Uzbeks (80)

Source: Data from Population Reference Bureau, *World Population Data Sheet* (Washington, DC: Population Reference Bureau, 2012).

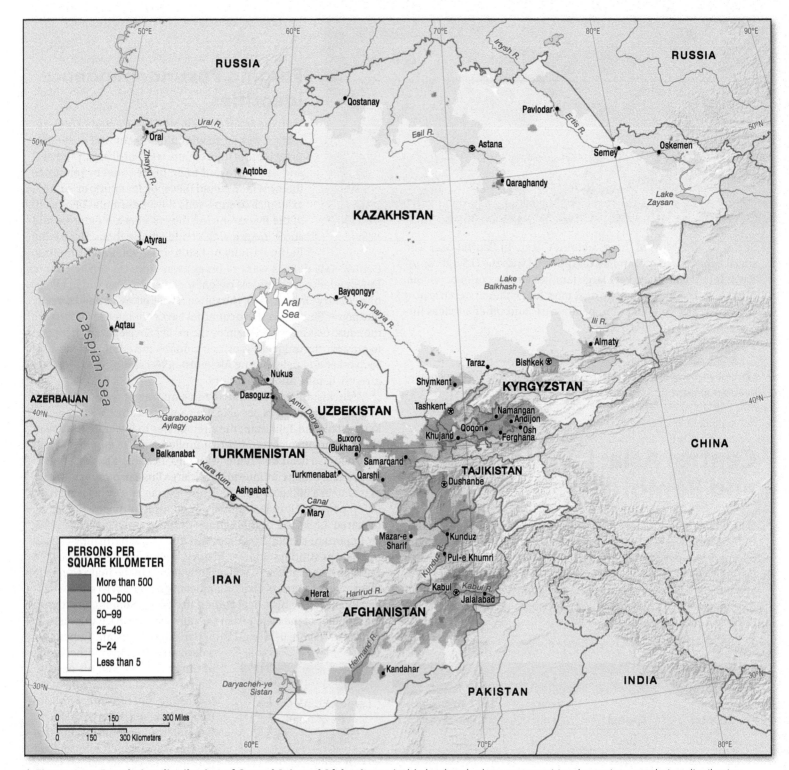

▲ **Figure 6-14** **Population distribution of Central Asia and Afghanistan.** Arable land and urban opportunities determine population distribution in this region.

Table 6-2 Demographic Data for Central Asia and Afghanistan

Country	Birth Rate*	Death Rate*	Infant Mortality Rate*	Percent Urban
Afghanistan	43	16	129	23
Kazakhstan	23	9	17	55
Kyrgyzstan	27	7	25	35
Tajikistan	27	4	53	26
Turkmenistan	22	8	49	47
Uzbekistan	23	5	46	51

Note: * = per 1000

Source: Data from Population Reference Bureau, *World Population Data Sheet* (Washington DC: Population Reference Bureau, 2012).

and historic figures, although most people on the street still refer to them by their Soviet names. The highest point in the former Soviet Union is now in Tajikistan's Pamir Mountains and was called Stalin Peak until the 1960s, when it was changed to Communism Peak until independence; now it is Somoni Peak (24,590 feet [7,495 meters]), named after the creator of the Samanid Dynasty, or as the Tajiks see it, the last Tajik Dynasty before Central Asia began to be ruled by Turkic groups and Russians.

Turning toward Islam

Another major cultural question is that of religion. The Soviet Union took to heart Karl Marx's dictum that religion was the "opiate of the masses" and created a state adamantly secular in nature. But this policy was neither totally nor consistently applied. Antireligious fervor peaked in the 1930s, but was relaxed during World War II. From that point on the combined processes of secularization and Russianization attempted to marginalize Islam. Only at the end of the Soviet period did a full picture of what happened to Islam in Central Asia become clear. Educated and ambitious people had to join the Communist Party wherein one had to be officially an atheist. Although some citizens took this to heart, most did not. What occurred, especially in rural areas, was a wholesale personalizing of

▼ **Figure 6-15 Village family in the Fan Mountains.** In the countryside particularly, large families are common. Note the women's distinctive dress patterns that were maintained throughout the Soviet period as expressions of their culture.

religion. Islam became not a public religion of mosque and clergy, but a familial one where people maintained Islamic traditions within their own homes and empowered "unofficial" clergy (people considered pious but who had no formal religious training) to guide their faith. As meeting at mosques would cause suspicion, men, especially in the countryside, would meet at teahouses, many with "private meeting rooms" in which they could pray and congregate. For women, the spacious guestrooms of a typical Central Asian home provided privacy for their meetings. In the urban sphere where Russianization had a better hold, many people became "cultural Muslims": they considered themselves secularized, but followed Islamic practice for important "life ceremonies"

▼ **Figure 6-16 Statue of Ismail Somoni in Dushanbe, Tajikistan.** The newly independent countries of Central Asia wished to differentiate themselves from each other by showcasing historical people in art forms. In the case of Tajikistan, the government harkened back to the tenth century A.D. when people of Persian descent last ruled Central Asia. They built a colossal monument of the founder of the Somonid Dynasty (A.D. 899–999) to emphasize not only their heritage, but their claim as the original inhabitants of the region.

like weddings, circumcisions, and death rituals. On those occasions clergy would be present to perform appropriate ceremonies in time-honored Islamic ways. But Islam had lost much of its meaning in daily life. This phenomenon was not an isolated development in Central Asia; the same reactions to secular communism can be seen in the Caucasus as well as in Bosnia and Herzegovina, in the former republic of Yugoslavia.

Following independence some Central Asians wanted a renaissance of religion and again include it as a core part of the society. But the secular leadership (especially in Tajikistan and Uzbekistan) dealt harshly with these citizens, causing the more hard-core elements to take up armed resistance. These moves by the government were backed by many citizens to prevent further disruption in their lives. In recent years, religiosity is becoming more common in the lives of Central Asians. As the trauma of the Soviet breakup receded and citizens see that local governments cannot provide the same kinds of social safety nets as the U.S.S.R, more people are turning to Islam as the central guiding force in society. This mirrors to a certain extent the move away from secularism in the Middle East and North Africa where disenchantment with Western, secular ideals have occasioned a strong movement toward Islamism.

Stop & Think

▶ Will greater religious expression reflect Central Asians' rejection of their countries' politics?

Economic Challenges: Landlocked Countries and Resource Extraction

The U.S.S.R. always considered Central Asia its most backward area. Much of this was due to a typical European colonial mentality toward Asia but also toward Islam, and the fact that the area had only very limited industry and relied mostly on agriculture. By the end of the Soviet era, the government felt there was much they could be proud of. Literacy rates had skyrocketed to 98 percent from less than 5 percent; extensive transportation and communication links between the Central Asian republics and Russia were built; many goods and services were now available (even if people had to stand in line for them) along with universal and free health and child care; everyone had a job and was a laborer for the state and the good of the people; and together they formed a superpower. That these statistics masked a very different reality became extensively apparent after independence.

It is difficult to characterize the economic reality that affects Central Asia in its postindependence years. Some authors have used the term deindustrialization, but this is to imply that Central Asia was industrialized. Central Asia was part of an industrialized country, the U.S.S.R. But the main economic activities in Central Asia were resource extraction and agriculture. Demodernization is another term that has been suggested and although it also has problems, specifically when related to well-educated societies, economically it does conjure up the idea that output significantly slowed down as equipment failed, infrastructure fell apart, and each government was faced with an immediate cessation of central governmental funding and subsidies for most economic activities.

Bureaucrats suddenly had to stop concentrating on maximizing subsidies from Moscow and figure out from their meager economies and limited transportation routes how to forge their economic future.

All five former Soviet countries are landlocked (as is Afghanistan) and all transportation routes went to Russia. This meant at the onset of independence that trade had to continue with other former Soviet countries because none of the Central Asian countries had the resources to build railroads or highways to other countries. For Kazakhstan, Kyrgyzstan, and Tajikistan, the extremely high mountains separating them from China and South Asia compounded this connectivity problem. In the case of Turkmenistan and Uzbekistan, the neighbors were war-torn Afghanistan and theocratic Iran, both of which worried those governments. The country that first stepped into this trade void was Turkey, a country that is linguistically and culturally related to the Turkic nations of Kazakhstan, Kyrgyzstan, Turkmenistan, and Uzbekistan and has positioned itself as the model to emulate: a secular Muslim nation that uses Islam as a social guide. Turkish plans, however, were much more expansive and audacious than was their economic capability and many of these ventures collapsed because of poor planning and not enough money. This left some bitterness in Central Asia, particularly in Uzbekistan and Kyrgyzstan, where the investment was badly needed and land and resources had been dedicated to these purposes. Another problem was the attitude of many early entrepreneurs toward Central Asia. They viewed the areas as "developing," a term that, economically, is true. With this perspective usually comes a patronizing attitude that did not sit well with Central Asian bureaucrats and businessmen, many of whom are very well educated and did not like foreigners telling them how to run their affairs.

The Turkic Republics

Postindependence Central Asia displays a wide array of developmental issues, and these countries can no longer be classified as fully economically developed. Kazakhstan has perhaps the greatest potential, both in resources and potential productivity. It has the region's strongest economy and a large agricultural base, in addition to one of the world's largest oil and natural gas deposits in the northern Caspian Sea (see *Focus on Energy:* Energy Potential in Central Asia and Afghanistan). Turkmenistan also has large oil and gas deposits near its share of the Caspian Sea region and although the terrain is arid, agriculture is possible because extensive Soviet-era irrigation projects diverted large amounts of water from the Amu Darya that normally would have gone to the Aral Sea. But Turkmenistan continues to be hampered economically and politically by its authoritarian government. When the previous dictator died, there were hopes that the new president would open the country to more foreign investment, but he has continued the previous dictator's policies.

Privatization of land has been slow to nonexistent in most of Central Asia, with the exception of Kyrgyzstan. The changes that have occurred reflect a balance between historical usage (both Soviet and pre-Soviet) and the contemporary desire for private ownership. Because cotton is the most important crop within the region, dominating the large agricultural sector and constituting the region's most

Energy Potential in Central Asia and Afghanistan

Central Asia has abundant oil and gas reserves, much of which have only recently been tapped. This abundance is concentrated in two countries, Kazakhstan and Turkmenistan. Many multinational firms have gone to Kazakhstan to help exploit these resources, although they have not been developed at a pace hoped for in the West because of continued government corruption and bureaucratic wrangling. Some experts have suggested that by 2020, Kazakhstan will have the potential to produce nearly as many barrels of oil per day as Saudi Arabia, particularly with the addition of the Kashagan field, projected to come on line in 2014, in the northern Caspian Sea.

Because Kazakhstan, like all the Central Asian countries, is landlocked, getting the oil to a deepwater port presents many economic and political problems. Oil produced in Kazakhstan is shipped to foreign buyers through Russian pipelines. This is in line with the Soviet strategy that ensured that all transportation routes and goods had to flow to the "center," that is to Russia, to then be fairly redistributed. Attempts to bypass Russia have caused diplomatic problems with this powerful neighbor, and so far Russia has been able to control most of the oil coming from these promising oil fields. Another problem is that the Kashagan oil is difficult to access because it is located in shallow water where seasonal temperatures can range from −31°F (−35°C) in winter to 104°F (40°C) in summer. The Soviets knew of the vast oil reserves in Kazakhstan, yet preferred to drill the much more easily accessed Siberian oil. After the fall of the Soviet Union, American oil companies began cutting deals with Kazakhstan's President Nursultan Nazarbayev. These deals led to a wide-ranging scandal in 2003, when it was revealed that foreign oil companies had essentially bribed the government for oil concessions. Nazarbayev has not been prosecuted; he has seen to it that, as president, he has immunity from prosecution.

The other country with a sizeable energy potential is Turkmenistan. Although oil reserves are small, natural gas is abundant with 102 trillion cubic feet in gas reserves—ranking Turkmenistan third in the world

behind Russia and Iran. An underwater, trans-Caspian pipeline has been proposed to link Turkmen gas to the South Caucasus Pipeline that transports natural gas from Azerbaijan's Caspian coast to Turkey; however, both Iran and Russia have heavily criticized the construction and blocked attempts to negotiate a potential route. This means that all gas from Turkmenistan must flow to Russia or Iran to be transported out of the region.

Uzbekistan has small reserves of natural gas, but the potential for Kyrgyzstan, Tajikistan, and Afghanistan lies in hydropower. This is reflected in their energy consumption patterns (Figure 6-3-1). Mostly mountainous, the river systems of Central Asia originate in these countries and plans are under way to extend the network of dams and reservoirs to augment electrical

capacity in each country, with a potential to export electricity to neighboring countries (such as Pakistan, China, and India). Uzbekistan has strongly protested this increase because it would disrupt water distribution from all three countries into Uzbekistan and has even gone so far as to block material from reaching Tajikistan and Kyrgyzstan from Russia that could potentially be used in dam construction. Tajikistan and Kyrgyzstan believe they need these dams for developmental purposes, not least because Uzbekistan also controls the natural gas pipelines that supply energy to each country.

Sources: Jakob Granit et al., "Regional Options for Addressing the Water, Energy and Food Nexus in Central Asia and the Aral Sea Basin," *International Journal of Water Resources Development* 28, no. 3 (2012): 419–432; Richard Pomfret, "The Economic Future of Central Asia," *Brown Journal of World Affairs* 19, no. 1 (2012): 59–68.

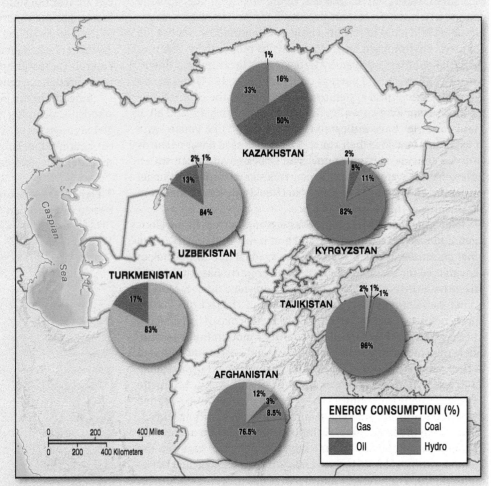

▲ FIGURE 6-3-1 **Energy consumption in Central Asia and Afghanistan.** Hydropower, gas, and oil are the major energy sources in the region.

important export, national and local governments tightly control arable land and especially the best irrigated land. During the Soviet period, families were allowed to have a small private plot adjacent to their homes (what would be considered yards in the United States). On this land, people could grow whatever they wanted either for their own consumption or for trade. Families often had a number of fruit trees, depending on the size of their plot, around which they would plant vegetables and herbs. Over time in the Soviet Union, these **kitchen gardens** became the basis for some of the most productive agriculture in the republics. These gardens constituted as much as 23 percent of land use in some provinces and, because of personal incentive, were some of the most productive land in Central Asia. After the fall of the U.S.S.R., people retained these kitchen gardens and used them initially to augment their families' food supply. Gradually, they have maximized these gardens where possible to grow fruits and vegetables for sale in local markets.

Uzbekistan has the largest population and the largest market-driven agricultural base among the former Soviet republics of Central Asia. The country's diverse resources and people have resulted in less of an economic slide than in the other non–oil-producing countries. The Soviets invested more into industry, relatively speaking, because of the size of the population, although cotton remains the largest sector of the economy. Uzbekistan does not have the large oil deposits of its northern and western neighbors and has only small reserves of natural gas. In the long term, this puts the country in a more precarious position because its industry is rapidly aging and it has to rely on a crop that continuously erodes the country's environment and, ultimately, its productivity. Another potential conflict is driven by Uzbekistan's location as a downstream country. Only 9 percent of the water used in Uzbekistan's extensive agricultural systems originates within the country. The rest flows from Kyrgyzstan, Tajikistan, and Afghanistan, and all of these countries have indicated that they would like to use more of what they consider their water. This has caused a war of words with the considerably more powerful Uzbek government, which insists that the water-sharing agreements signed during the Soviet period—overwhelmingly beneficial to Uzbekistan, Kazakhstan, and Turkmenistan—must remain in place.

Kyrgyzstan has taken a more Western approach to its development and has privatized more land and industry than any of the other countries. While piecemeal and gradual due to the lack of enough land for everyone, the Kyrgyz government has encouraged private enterprise and, after a shaky start, provided rights to land for former members of collective and state farms. The lack of a market infrastructure to absorb agricultural production makes this difficult. At issue also is the nature of the country's mountainous terrain, where transhumance herding is prevalent and agriculture is limited to river valleys and upland meadows. Water is generally available for irrigation in those instances and the terrain is well suited to the dams built along the country's rivers, which produce much-needed hydroelectric power. But Kyrgyzstan's economy has shrunk considerably since independence. As indicated by the coup against the only non-former Soviet leader in the region, Askar Akaev, in the Tulip Revolution of 2005 (a term chosen in part because that flower is indigenous to Central Asia and in part to link conceptually to color and flower revolutions in Ukraine and Georgia), the population has grown weary of economic hardships and the obvious income disparity between government officials and the majority of the population.

Stop & Think

➤ Is there anything the Uzbek government can do to address an environmental problem as great as that of the Aral Sea?

Tajikistan

As difficult as the postindependence years have been in the four Turkic-dominated countries, nowhere were these problems more apparent than in Tajikistan. The poorest republic during the time of the Soviet Union, Tajikistan fell even further behind in the postindependence years as the country immediately sank into a civil war (Figure 6-17), based not on ethnic differences but on regionalism and tribalism. During the Stalin administration, many people from overpopulated river valleys east of the capital of Dushanbe were forcibly moved to another valley where the U.S.S.R. had recently expanded irrigation and the agricultural base. This caused conflict between the immigrants and the people already inhabiting that region, most notably around the cities of Kulob and Kurghon-Teppe. The effects of the civil war were the worst in this agriculturally important area, but the impacts were felt widely as fighting broke out all over the southern part of the country. A peace accord was reached by the victorious Kulobis, their allies, and their opponents in 1997, although the peace was tentative until 2000, when the government cracked down on many of the **warlords** and gained better control. These warlords gained financing for their weapons and soldiers through the **opium** drug trade with Afghanistan.

Since then the economy has grown enormously from a near standstill, recording an 8 percent growth in GDP each year until the global recession of 2008. The country continues to rely on resource extraction, especially cotton production; a growing mining sector that includes gold and uranium; and foreign aid. More recently,

▼ **Figure 6-17 Bus top turned fence.** Throughout Central Asia, there was an extensive Soviet-era transportation network; however, during the civil war in Tajikistan, many buses used in the countryside sat idle and rusted because of insecurity and a lack of spare parts and gasoline. The former overseer of transportation to and from the villages west of the capital of Dushanbe decided that a better use for these rusted bus tops was a goat pen.

this has been greatly augmented by remittances from Tajik workers who have migrated mostly from the agricultural river valleys to Russia for work (see *Visualizing Development:* Tajikistan's Remittance Economy). All goods coming to and from Tajikistan had to go through Uzbekistan. Considering the severely strained relations between the two countries over the civil war, pressures exerted by Islamic rebels operating in the mountains of Tajikistan, and issues pertaining to water distribution, this transportation reality put the entire Tajik economy at the mercy of relations with Uzbekistan. This dependency changed when a highway to China opened in 2004, which has had a positive economic effect on the whole country. The U.S. government also provided aid to build bridges across the Amu Darya that serve as links to Afghanistan and ultimately Pakistan's ports to overcome the constraints of a landlocked location.

Continued Authoritarian Politics: Little Progress toward Democracy

The U.S. administration had forged closer ties with the government of Uzbekistan in particular, as well as with Kyrgyzstan and Tajikistan in the wake of 9/11. With political chaos engulfing Kyrgyzstan, and authoritarianism in Tajikistan and Uzbekistan, the United States continues to see these countries as geostrategically important on the northern border of Afghanistan, but as the war in Afghanistan winds down, how much U.S. attention will be directed at these countries remains to be seen. The initial attention placed some strain on U.S.–Russian relations because Russia still views this area, along with the Baltic and the Caucasus, as the "near abroad" (see Chapter 5) and within its sphere of influence, both politically and economically. The Central Asian administrations, however, see greater diversity in international contacts and relationships beyond those with the former Soviet countries as a positive move. But these developing relationships often are not smooth. Tajikistan still has a Russian army presence in the country and cannot afford to offend Russia or President Putin. Although Kyrgyzstan was initially viewed as the "Switzerland of Central Asia" because of its mountains and stability, riots in 2010 underscored the political fragility of this country.

These followed similar riots that ousted former President Askar Akaev in 2005. In an election marred by widespread irregularities, cronies and family members of Akaev won every seat for which they ran. This sparked a revolution in the volatile southern half of the country that spread throughout Kyrgyzstan and eventually caused the president and his family to flee. Although Kyrgyz citizens initially protested for a less authoritarian government and less fraud, the issue at the root of the people's problems was economic. The government had continually promised better economic conditions but never delivered on these promises. The economy continued to stagnate, and the obvious wealth of the president and his family fueled the feelings of economic desperation among many people. This feeling was supported by the government's inability to effectively reform the economy through privatization. Kyrgyzstan has progressed furthest in Central Asia in privatizing land. Like the other countries of Central Asia, Kyrgyzstan has yet to create an effective "land market" that could provide credit and other services, establish a more viable rural economy, and give farmers more latitude

in their decision making about production, investment, and crop choice. The riots of 2010 highlighted these same issues and exposed ethnic tensions between the Kyrgyz and Uzbeks as well. These riots led to new elections in 2011 which, though marred by irregularities, showed that Kyrgyzstan now has a thriving multiparty political system and that the rewritten 2010 constitution significantly reduced the president's power and created a strong parliament in the hopes of stemming further political violence.

The world community thought a change in government might occur in Turkmenistan when Supurmurat Niyazov, the dictator generally known as Turkmenbashi, or "father of the Turkmen," died in 2006. Niyazov was well known throughout the international community for his Stalin-like personality cult that caused him to make bizarre and egotistical decisions. He renamed the month of January after himself, the month of April after his mother, and the month of September after his book of sacred myth and the modified history (to exclude anything "Soviet") of the Turkmen people. This book, the *Ruhnama*, became the basis for Turkmen education and is a compulsory reading for all Turkmen students. After his death, Turkmenistan elected a new president, Kurbanguly Berdymukhamedov, in elections widely derided as rigged. In his inauguration, the new president categorically stated that Niyazov's policies would continue. Turkmenistan continues to be hampered by a severely authoritarian government, remains adamantly state run, and largely discourages foreign involvement and the application of Western development concepts. It also wishes to remain strictly neutral in all regional conflicts and has not participated in the war in Afghanistan.

Politically, a similar scenario to Kyrgyzstan's seemed likely to arise in Uzbekistan in the Ferghana River valley just across the border from where the Kyrgyz revolution occurred. But Uzbekistan's President Islam Karimov called for a violent and bloody crackdown on the protesters that left numerous dead and caused extreme concern among the United States and its allies over the future of the country. Uzbekistan used the war on terror and the fight against Islamism to justify the crackdown. The United States and the European Union did not look the other way this time and condemned the crackdown, leading Karimov's government to withdraw its support from U.S. efforts against the Taliban insurrection in Afghanistan. This attitude has not been significantly modified in recent years and relations between Uzbekistan and the West remain strained as Karimov continues his grip as an authoritarian dictator.

This crackdown highlights the most important cause for the popularity these authoritarian governments enjoy: their continued fight against what they call the excesses of religious fundamentalism. As a concept, fundamentalism is strictly a Christian one, but the most common local acknowledgment of this phenomenon employs the term *Islamism*. For most people in Central Asia, the term for and the idea of Islamism is one of religious renaissance (Figure 6-18). During the heyday of the U.S.S.R., the government persecuted obvious displays of all religions and particularly Islam. After independence, many people saw this as the chance to reawaken religious feelings in the region. On no level does this imply that they became terrorists or were planning violence or advocating a forcible renewal of the religion. However, it became clear that the former Communists who had taken over the region's governments saw supporters of Islamism as a threat to their power and began to persecute them anew. Some of the Islamists turned to

Visualizing DEVELOPMENT ▲ Tajikistan's Remittance Economy

From 1970 to 1989, the level of rural-to-urban migration in Tajikistan was relatively low (4.64% of the natural increase) compared to other republics. This low mobility was attributed to rural laborers' lack of skills outside of agriculture and the few available jobs open to laborers with poor command of Russian since most managers were ethnically Russian. With these economic restraints came social constraints. Rural Tajiks preferred to stay in the same locale or village, relying on the stability of family and cultural comforts such as language and less government interference with a family's adherence to Islam. Families in the early years of independence moved from isolated villages to Dushanbe to escape the Tajik civil war. Dushanbe's population began to stabilize between Russian immigration and Tajik migration, with the new Tajik migrants settling into apartments and houses of Russians and other minorities who left the newly formed country, moving in with relatives, or building small compounds on the city's periphery.

Urban life brought new perils. Newly arrived migrants had trouble finding food in the city due to their inability to access goods from their old state or collective farms. The collapse of urban industry, due primarily to demodernization following independence, further dampened any desire to migrate to the cities, as did the privatization of homes. Migrants could not afford to abandon their homes in the countryside (which they owned outright) and hope to buy a place in a city, even if a job awaited them.

Although the civil war and demodernization period of the 1990s created a flow of rural-to-urban migrants, the number of migrants looking for work outside Tajikistan dwarfed that trend. With reduced civil war-related violence, Tajik men felt safer about leaving their families for periods of time and, in increasing numbers, searched for work outside the country. For most Tajiks the destination of necessity became Russia, where these migrants have made employment in that country, along with subsequent remittances, a structural part of the Tajik economy (Figure 6-4-1).

According to the last Russian census, 120,136 citizens of Tajik nationality, 78 percent of them male, live in Russia, with 64,165 citizens of Tajikistan registered in the country, of whom 70 percent were 20–39 years of age. It is remarkable that among citizens of Russia with Tajik ethnicity such a large majority are men. These numbers continued to grow exponentially through 2007, the most recent statistical survey of migrants arriving in Russia, when 250,200 Tajiks were in the labor force—an increase of 151,500 over the previous year. This represents legal migration recognized by the Russian Federation; the Asian Development Bank estimated that 400,000–500,000 Tajik men worked either permanently or temporarily in Russia in 2005 and by 2010, the World Bank estimated that 1 million Tajik men worked in Russia.

If these numbers are correct, more than 13 percent of the estimated 7.1 million Tajiks work in Russia. Given that the migrants are overwhelmingly male and that two-thirds of the population is rural, the typical Tajik migrant to Russia is a rural male between the ages of 20 and 39. Lack of skills limits these young men to short-term employment in overwhelmingly labor-intensive, physical work. This has left a dearth of men in the countryside of Tajikistan and also changed the social dynamic. Women now outnumber men working in the markets and nearly half of rural households are headed by women, resulting in increased divorce rates and abandonment to greater household poverty (at least until the remittances start to arrive). The Tajik government has not announced any plans for increased rural job creation or eased its economic hold on agricultural land. Rural workers do not want to lease this land as the government forces them to grow cotton, which must be sold back to the government at reduced rates so that the government can realize a profit. The only privatized sector of this economy is the commodity market around fertilizers, seeds, and pesticides—all necessary

---

violence, especially during the Tajik civil war and after crackdowns in the late 1990s in Uzbekistan. Central Asian governments, particularly after 9/11, used the cloak of the fight against Islamist terrorism to perpetuate their authoritarian regimes.

The area that has garnered the most attention in this fight is the Ferghana Valley (Figure 6-19). The Ferghana Valley has been economically and politically important throughout the region's history. It is an isolated area, cut off from the rest of Central Asia to the north, east, and south by high mountains that are only passable

◀ **Figure 6-18 Muslims in Central Asia.** Islam, which diffused from the Middle East, is the dominant faith of Central Asia. Although most Central Asians never abandoned Islam during the Soviet period, they are witnessing a renaissance of the religion in many places and the practice of Islam is becoming more public. Here women are dressed in traditional silk dresses and headscarves that have long identified the Muslim women of Central Asia.

for growing cotton and sold at market prices. This has placed a majority of households in debt, and the only way to pay it off is to get out of agriculture and migrate to Russia. Rather than address this problem, the Tajik government has worked with Russia's President Putin to increase the number of Tajik migrants who can legally work in Russia. For the near future, migration remains a central and expanding part of rural life for most Tajik men.

Sources: Zvi Lerman and David Sedik, "Agricultural Development and Household Incomes in Central Asia: A Survey of Tajikistan 2003–2008," *Eurasian Geography and Economics* 50, no. 3 (2009): 301–326; Don Van Atta, " 'White Gold' or Fool's Gold? The Political Economy of Cotton in Tajikistan," *Problems of Post-Communism* 56, no. 2 (2009): 17–35.

▶ **FIGURE 6-4-1 Tajikistan's labor migration.** The migration streams from Tajikistan's Raion of Republican Subordination (RRS) region reflect the magnitude and significance of the need of Tajikistan's population for labor opportunities outside of the country and remittances for families to make ends meet. **Sources:** Larissa Jones, Richard Black, and Ronald Skeldon, *Migration and Poverty Reduction in Tajikistan* (Falmer: Development Research Centre on Migration, Globalisation, and Poverty, 2007); Saodat Olimova, *Migration and Development in Tajikistan: Emigration, Return, and Diaspora* (Moscow: International Labor Organization, 2010); Author's fieldwork.

in summer. The valley is also quite small, comprising only about 4 percent of Uzbekistan's land area. Yet it contains 25 percent of the country's population and 35 percent of its arable land. This concentration of population and productive agricultural potential extends across the border into Kyrgyzstan. The Ferghana Valley and adjacent mountains form 40 percent of Kyrgyzstan's land area but contain 51 percent of its population. The Ferghana Valley was an independent entity, under the Uzbek Khanate of Qoqand, which existed from the late sixteenth century until Russia annexed the Khanate in 1876. This was the only independent entity in Central Asia that the Russians annexed outright before the Bolshevik Revolution, in order to secure an important agricultural area for cotton production, and in the next 100 years the Ferghana Valley was turned over almost exclusively to the cultivation of that crop.

Because of the continued importance of cotton to the Soviets, and amid the fear that the area might break away to form an independent

entity again, Stalin's government created the most gerrymandered borders imaginable to break up the valley. The center part of the valley became part of Uzbekistan; the northern, eastern, and southern edges became part of Kyrgyzstan; and the western part, including the only year-around access point to the valley, became part of Tajikistan. Although residents were not happy about this, it was not until the Gorbachev era of perestroika that ethnic and religious tensions came to the fore. These first involved violent conflicts within Uzbekistan between Uzbeks and Meshketian Turks, a group forcibly moved into the region from the Caucasus by Stalin's government. In 1990, a conflict in Osh, Kyrgyzstan over housing and the lack of Uzbek political representation caused the death of 200–500 Uzbeks and Kyrgyz in nearby Üzgen.

Islam did not factor as a force in the region until after independence. Pious Muslims in Ferghana had strongly opposed the accession of Islam Karimov to the presidency of Uzbekistan because he had risen through the ranks of the Communist Party.

▶ **Figure 6-19 The Regional Setting of the Ferghana Valley in Uzbekistan.** The Ferghana Valley is a microcosm of highland life in Central Asia. Divided in complex spatial patterns between Tajiks, Uzbeks, and Kyrgyz, the region's rich soils, abundant water, and hydroelectric power provide livelihood opportunities as well as the motivation for political manipulation and competition.

These people were further aggrieved when the official clergy—those who had been appointed and tolerated by the Soviet government—endorsed Karimov as president. This action laid bare the fact that the "official" clergy did not (and never did) really speak for those who continued to practice Islam. Tension between the government and Islamists continued to simmer throughout the Tajik civil war, when Uzbek Islamists hid in the southern mountains of the Ferghana Valley across the border with Kyrgyzstan and Tajikistan, both of which were too weak to permanently eject them. The situation finally came to a head in the city of Andijon, where an uprising against the government provoked a violent crackdown on protestors and perceived Islamists. Hundreds of people were killed by government forces while protesting Uzbekistan's actions against Islam and attempting to start a Kyrgyz-style revolution. The revolt in the Ferghana Valley did not motivate any other areas to join the Islamist movement; many Uzbek citizens remain secular and oppose a potential Islamic state or any situation that could destabilize the country or region. This situation has caused those who wish to explore their religiosity more publicly to become increasingly frustrated with the government and fearful of its tactics against them.

Afghanistan: The Impact of Modernization

Long sheltered by its interior, landlocked, mountain setting, more peripheral to trade routes than its Central Asian neighbors, and thought to lack abundant resources desired by distant powers, Afghanistan has nonetheless episodically been thrust center stage by its strategic location. From Alexander the Great to the U.S.S.R.'s Leonid Brezhnev, Afghanistan's mountain passes have lured invaders with wider regional strategic interests. The resulting engagement with globalizing forces and distant powers has drawn Afghanistan into often unwilling but increasingly challenging, intense, and violent interaction with the modern world.

Modern Afghanistan

Historically, the Hindu Kush Mountains naturally divided Afghanistan (Figure 6-20). The areas north and west were allied closely with Iran and Central Asia, and the areas to the south and east with India. When the competing Russians and British set the borders of Afghanistan at the end of the "Great Game" in 1895, a very disparate and fiercely independent population was thrown together to attempt to make common cause. They have gotten together to fight off invaders, both British and Russian, but when left alone, they have too frequently fought amongst themselves along ethnic and tribal lines. In the ethnic composition of the country (see Figure 6-6; Table 6-3), the Pashtun are the largest group. This has resulted in the Pashtun community enjoying near-continuous power in running the country, which has caused strife with the other ethnic groups. Only twice has the Tajik minority run the government, once briefly in the early twentieth century and, most disastrously, during the traumatic civil war years before the rise of the Taliban.

Rise of the Taliban

An attempt at economic liberalization occurred during the 1960s and 1970s, but a Communist coup in 1978 turned the economy and government firmly toward the U.S.S.R. When many citizens rebelled, the U.S.S.R. invaded Afghanistan in December 1979 when the Soviet leadership believed the Western countries to be distracted by the Christmas holiday, and occupied the country until 1989 when Soviet troops withdrew in defeat. The Soviets, both before and during the occupation, did set up some industry and expanded transportation and communication. But they also caused extensive damage in rural areas, both economic and environmental, in their attempts to pacify the country. The most egregious act was the distribution of more than a million land mines throughout the countryside, a figure enlarged further by land mines placed by the **mujahideen,** or insurgents, fighting the occupation. This has caused untold deaths and maimed more than a million people, many of them children, who stumble across the mines while doing chores or playing. An international campaign

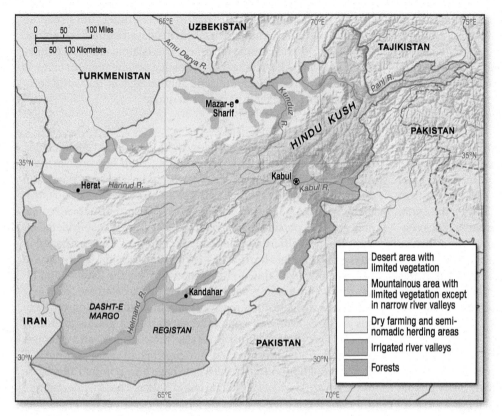

◄ **Figure 6-20 Physical and agricultural regions of Afghanistan.** Land use in Afghanistan is dominated by the country's major river valleys, especially the Kunduz, Harirud, Helmand, and Kabul river valleys, and the central spine of the Hindu Kush Mountains across the central part of the country with interspersed mountain ranges and small river valleys. These river systems constitute high-potential agricultural environments and support the largest portion of the population. The surrounding areas are mostly dominated by nomadic Kuchi herders, largely Pashtun ethnically, except in the southwestern desert areas where Baluchi herders predominate.

to mark the minefields and especially remove the mines from agricultural land is under way—a process that will take decades to complete.

After the Soviet retreat Afghanistan continued under its Communist government until 1992, when the rebels led by Ahmed Shah Masud (a Tajik who became Defense Minister) captured Kabul. The presidency of the country was supposed to revolve between the various ethnic groups, but the Tajik, Burhanuddin Rabbani, refused to relinquish the presidency and the country was plunged into civil war (Figure 6-21). It is difficult to understand and describe the ensuing catastrophe because different groups continuously changed sides whenever they felt they would get a better deal. The Uzbek warlord Rachid Dostum was famous for this as he changed sides on what seemed like a weekly basis. Masud was rumored to have inquired each morning whether he was fighting with or against Dostum on that particular day. The ultimate consequence of this situation is that to this day the different ethnic groups

and tribes are extremely mistrustful of each other, and interethnic dialogue continues to be marked by rancor and suspicion.

The **Taliban movement** began in the madrasahs, or Quranic schools, financed by the Saudi Arabian government and located on the border of Pakistan and Afghanistan. The students were mostly Pashtun refugees whose families were extremely poor and many sent their sons to the schools for no other reason than that they would get at least one meal a day. Saudi Arabia follows a very strict interpretation of Islam known as Wahhabism, named after Muhammad ibn 'Abd al-Wahhab (1703–1787) whose main goal was to get rid of all jurisprudential innovations that occurred later than the third century of Islam (the 900s of the Christian calendar); he was particularly and vehemently opposed to Shi'a Muslims, whom he considered heretics. The Wahhabism of the Saudi donors combined with the local, South Asian Deobandi movement that originally arose in reaction to the impure effects of British colonialism in India. These two movements had profound effects on the students (from Arabic *talib*, plural *taliban*) at the madrasahs that formed the core recruiting grounds for the followers of mullahs, or religious leaders, who led the movement. Taliban financing also mainly came from Saudi Arabia and Pakistan (although the latter vehemently denied it) and they began to militarize around 1994.

Table 6-3 Ethnic Composition of Afghanistan (Estimate)

Ethnicity	Percent
Pashtun	40
Tajik	28
Hazara	15
Uzbek	9
Baluchi	2
Turkmen	2

Note: There are many more ethnic groups in Afghanistan, which is why the total is less than 100 percent.
Source: Composite from numerous sources, average of what analysts believe is the ethnic composition of Afghanistan. Because there has not been a census, there is no hard data available.

▼ **Figure 6-21 Destruction in Kabul, Afghanistan.** Scenes of destruction such as this are very typical throughout Afghanistan.

The Taliban got their break when a warlord in southern Afghanistan held for ransom a convoy of trucks owned by the powerful Pakistani trucking union. The Taliban managed to defeat the warlord and release the convoy so the truckers could continue on to Central Asia. The union, along with the Pakistani government, began to funnel money to the Taliban to keep the roads open for transportation. In this way, the Taliban began their conquest of the south and west of the country and subsequently turned their attention toward Kabul. The various non-Pashtun ethnic groups joined together to defeat this "outside group," a characterization they made because the Taliban were financed by Pakistan, although the movement was mostly made up of displaced local Pashtuns. In 1996, the Taliban captured Kabul and began their assault on the warlords in the north. They were particularly brutal to the Hazara in the mountainous center of the country because they are Shi'a, and many of the atrocities of this period were perpetrated by both sides. The major city in the north, Mazar-e Sharif, controlled by Dostum, fell in 1998. It was assumed that Tajik resistance in the northeast would soon collapse as well, but the Northern Alliance under Masud managed to survive.

The Taliban in Power

Although the north continued to live in a constant state of war, the rest of the country, especially the Pashtun regions, mostly rallied around the Taliban. In early newspaper interviews, many people expressed happiness that a single group had gained power, that all the bickering and infighting had stopped, and that they could get on with their lives. When asked about their views, most people either sympathized with the movement or thought that the extreme positions held by the group were just rhetoric. When the war began to bog down in the north, however, the Taliban showed that their movement was not just rhetoric. They unleashed a steady terror on the population that was particularly aimed at women and minorities and paid for it (at least initially) through the taxing of the opium crop (see *Geography in Action:* Warlords and Opium).

Many restrictions pertaining to women had existed in Afghanistan; many more, especially the mandatory wearing of the burka, had been enacted by the Pashtun prime minister, Gulbaddin Hekmatyar, of the Afghan government during the civil war. Under the Taliban the burka became compulsory. Women were not allowed an education or an occupation with the sole exception of women's health care. Women could not go out unless dressed in the heavy, confining burka and accompanied by a male relative. Men were required to wear beards and closely cut their hair. Any infractions were dealt with swiftly and publicly. All sports were stopped, and the soccer fields converted to public disciplinary centers where people accused of thieving had their hands chopped off, women accused of prostitution were whipped, and all executions took place. Music and television were banned, as were dancing, kite flying, gambling, and any images that included human or animal figures. During this period, for example, the American movie *Titanic* was very popular and many young men copied Leonardo di Caprio's hairstyle. This infraction was seen by the Taliban as doubly egregious as the young men had not only let their hair grow out, but also had seen a banned movie. Their heads were shaved in public and they and their barbers were publicly whipped.

Some of the Taliban's greatest acts of destruction occurred at Bamiyan, where they destroyed the famous standing Buddhas (Figure 6-22), and at the Kabul Archaeological Museum, then considered one of the important repositories of human history in the world. Taliban members destroyed anything that predated Islam, represented another religion, or displayed a human or animal form. Much of the country's historic, pre-Islamic material culture was destroyed. As Afghanistan was such a unique place of cultural fusion between South, Central, and Southwest Asia, the world community lost a remarkable part of its heritage.

Whither Afghanistan?

The events of 9/11 radically changed the situation in Afghanistan. In looking for allies, the U.S. government soon saw the Northern Alliance and (initially) Uzbekistan, with its nearby army bases and airports, as the best potential partners in the area in its war on terror that began in Afghanistan. Prior to 9/11, the Northern Alliance held only about 10 percent of the country; however, with coordinated air assaults and help from American special forces, they managed to retake the strategic northern cities of Kunduz and Mazar-e Sharif before capturing Kabul and eventually driving the Taliban and al-Qaeda into the eastern mountains along the Pakistan border.

In this struggle, the Americans and their allies supported various warlords, who had ruled different parts of Afghanistan during the civil war, in their efforts to retake "their" regions and establish order. This trend has continued and has created various ethnic fiefdoms throughout the country, which have only nominal allegiance to the government of President Hamid Karzai. These warlords were the main reason the civil war continued throughout the 1990s until the Taliban defeated most of them. The international community has questioned putting power back in their hands. But the U.S. government and its allies no longer have the will or the means to break

▼ **Figure 6-22 Great Stone Statues of Buddha in the Bamiyan Valley.** Afghanistan's historic role as a geographical crossroads for south, central, and southwestern Asia was long evident in the great stone statues of Buddha in this overwhelmingly Islamic nation. (a) The statues, carved some 16 to 19 centuries ago out of a sandstone cliff in the Bamiyan Valley 90 miles (145 kilometers) west of present-day Kabul, were considered one of the world's great archaeological treasures and functioned historically as a sacred Buddhist pilgrimage site. (b) In March 2001, despite worldwide pleas that they be spared, they were destroyed by the Taliban government, which then ruled Afghanistan. The left image shows one of the statues before it was destroyed, and the right photo shows the cliff after the statues were destroyed.

(a)

(b)

GEOGRAPHY IN ACTION — Warlords and Opium

Poppies—and the opium they produce—do very well on marginal, low-fertility soils and need minimal space to yield a crop that gives a good payback for the resources expended (Figure 6-5-1), something farmers in better endowed habitats do not have to consider (Figure 6-5-2). As such, the poppy is the ideal cash crop for many parts of Afghanistan. Although a living could be made from trans-humance herding, it is virtually impossible to grow enough crops in the narrow river valleys descending from the Hindu Kush Mountains. Farmers typically plant a variety of crops (including poppies) on tiny plots of land in a checkerboard pattern alongside sources of water. This makes poppy eradication by aerial spraying extremely difficult as this destroys all other nearby crops, leaving the farmer with nothing.

Another factor favoring poppy production is the region's periodic droughts. Vegetables and wheat require a great deal of water to grow; poppies require less. With a good poppy crop farmers can afford to buy food and animals rather than trade their meager wheat and vegetables for what they could not grow or raise. The United Nations quotes the "high sales price of opium" as the predominant reason why 71 percent of farmers surveyed raise opium to provide against poverty. UN data also show that 73 percent of surveyed villages that did not receive agricultural assistance from the Afghan government

▼ FIGURE 6-5-1 **Poppy plots in the upper Kunduz River Valley, Afghanistan.** Poppies are generally grown along creeks and small rivers on very small plots in the Hindu Kush Mountains or on larger plots in flatter areas in the southwest of the country. The limited acreage makes this high-value crop more desirable to mountain and desert dwellers.

cultivated opium in 2012. This supports data from previous years showing that government protection and assistance leads to crop choices away from opium.

During the early Taliban era, the only major export products from Afghanistan were opium and heroin, opium's major refined form. The Taliban did not encourage the trade, but did tax it heavily for much-needed hard currency to continue their war with the Northern Alliance. When by 2000 they had succeeded in taking over 90 percent of the country, the Taliban bowed to pressure and enforced a ban on growing poppies to gain recognition by the international community. Given the amount of production, they were incredibly successful and the amount of opium coming from Taliban-held areas fell by over 90 percent. This did not cause a downward spike in the international heroin market because much of this shortfall was made up by the Northern Alliance. Primary production shifted to the Alliance-controlled northeast. The product was carried overland to Pakistan, refined into heroin in mobile labs, and carried back across northern Afghanistan into Tajikistan. Tajik warlords trafficked the heroin to the Russian Mafia, who sold it on to Europe, further destabilizing Tajikistan and threatening the other Central Asian countries.

After 9/11, farmers throughout Afghanistan began growing poppies again, production levels reached heights never before attained, and multiple trade routes became operational. The Afghan government, with help from the U.K. and U.S. governments, has an ongoing eradication program. But destroying the crop means that the worst affected are the poorest farmers, who many times lose not only their poppy crop but also all crops because of the piecemeal planting used in the mountain valleys. This puts them in debt to warlords, many of whom are allied with the resurgent Taliban and have expanded production especially in the fertile Helmand River valley. Often the only way out of this situation is a form of "debt slavery," whereby farmers must grow poppies to repay the warlords or in some extreme cases, according

▲ FIGURE 6-5-2 **The lower Kunduz River Valley, Afghanistan.** In the extensive lower Kunduz Valley, farmers have considerably more land available and can grow market crops such as rice (pictured), cotton, and melons below their villages.

to the Revolutionary Association of the Women of Afghanistan, turn over a daughter (often as young as 10 years old) for marriage or servitude to the warlord or dealer.

Although many farmers benefit, most grow the crop for sheer survival, and most of the money that stays in Afghanistan goes into the pockets of local warlords who finance and arm their militias, or is collected by the resurgent Taliban. Opium is the fuel that keeps the countryside of Afghanistan unstable and largely out of the reach of the elected government, hampering efforts to aid farmers and wean them off opium production. Farmers are aware of the use of the crop, but plead, not without justification, that they have no alternative. Alternatives suggested (and to a small extent implemented) have been the high-value crops of almonds and saffron. But neither have proven markets yet and farmers fear they would not be able to sell these crops (and thus feed their families) without direct assistance and help from the Afghan government while markets remain available for opium. The United Nations has proven the correlation of assistance with crop change, but until the government can give aid to all farmers, opium will remain Afghanistan's largest economic sector.

Source: United Nations Office on Drugs and Crime, "Afghanistan Opium Survey 2012," April 2012, http://www.unodc.org/documents/crop-monitoring/Afghanistan/ORAS_report_2012.pdf.

▲ **Figure 6-23 Herat, Afghanistan.** Billboards at the Chawk-e Cinema traffic circle in Herat extol the virtues of Coca Cola and the communication benefits of cell phones. The movie theater on the square was torn down during the era of Taliban control, but the name remains in the landscape. The billboards' popular culture products are a sharp contrast to the traditional street dress of the burka-clad women in the foreground.

their power or to at least force them to acknowledge the authority of the elected president.

This leaves Afghanistan in a very precarious position. The Karzai government must depend on the warlords to keep order in the countryside, but cannot control them and can only hope for their continued goodwill in the future. Given their pre-Taliban history, this is extremely unlikely, and the rejuvenated Taliban resurgence of recent years keeps the government and its allies focused on that threat. This inattention to the law-and-order threat posed by the warlords has allowed various criminal gangs to arise, particularly in the western province of Herat. Here children of prominent citizens, particularly doctors, have been kidnapped and held for high ransoms. Between the corrupt warlords and criminal lawlessness in the "safe" provinces, many Afghans fear that their country will

never turn the corner in either security or development. But despite a weak security environment, the influence of the global economy, with values and material culture that sharply contrast with traditional norms, is a widely encountered challenge to traditional values and practices (Figure 6-23) throughout urban Afghanistan.

This influence may take a dramatic turn in the near future as reports indicate untapped mineral wealth in Afghanistan. Now governments and businesses are drawn to Afghanistan for something other than conflict. Vast deposits of copper, cobalt, gold, iron, and lithium (used in making batteries for cell phones and laptops) have been discovered in northern Afghanistan especially and China has already opened bilateral trade with Afghanistan to begin exploration of these minerals. A Pentagon memo mentions that Afghanistan could become the "Saudi Arabia of lithium." Ongoing talks with both the United States and China for development funds could help Afghanistan further maximize water resources once the United States has withdrawn troops. But development experts fear that new resources could simply become a shift in emphasis and that Afghanistan will continue to suffer from the curse of resource extraction just like the other countries of Central Asia. Though this is a legitimate concern, the world would view a shift away from heroin with profound relief and anticipation. This will also affect Central Asia, as these resources may lead to developing more trade routes between adjacent regions and a legitimate economy that would provide both infrastructure and jobs.

Stop & Think

▶ Can Afghanistan develop an economy based on something other than resource extraction?

Summary

▶ Central Asia and Afghanistan are in a very precarious place. In Central Asia, the revolutionary fever has already hit Kyrgyzstan and the leaders of the other countries fear that they could be next. This has caused them to be increasingly authoritarian in their actions and to be suspicious of any overtly religious people. The civil war in Tajikistan has made the population wary of any force that could again destabilize their lives.

▶ Economically, the former Soviet republics of Central Asia were devastated after independence, and while all have made great economic advances since demodernization, only Kazakhstan is moving toward substantive economic development that has created jobs and wealth. Turkmenistan and Uzbekistan have the ability to move in this direction, but their economic and political policies have hampered their economic potential. Tajikistan and Kyrgyzstan are economically in the most precarious positions, and have experienced the most political violence. It has only been through mass migration of their men to Russia that they can alleviate the poverty that afflicts much of the countryside and even many urban residents. Except in Kyrgyzstan, privatization of land continues to proceed at an extremely slow pace, further impoverishing the large rural populations, while increasingly outdated industries make even blue-collar jobs scarce.

▶ Positive trends include the recent opening of many of the economies, especially those bordering China, to international trade and to competition from Russia and other former Soviet countries, and the start of negotiations to build oil pipelines that bypass Russia. The latter, however, could set up an economic situation like that of the Middle East, where countries with large oil reserves (Kazakhstan and Turkmenistan) manage to develop while the rest remain poor and continue to struggle. When looking at the bright new lights and buildings of Astana, Kazakhstan and anticipating the end of U.S. and NATO involvement in Afghanistan, one might think that the region has turned the corner in terms of economic and political stability. But the economic strides in the decade after demodernization were hampered by the global economic recession of 2008. This underlines the region's vulnerability. Although Astana beguiles with its petro dollars and high-rises, Kazakhstan's countryside, like that of all the countries, has stagnated, causing high levels of migration. In Afghanistan, the United States and its allies will have withdrawn most combat troops by 2014, but will efforts employed over the last dozen years to strengthen the elected government of Afghanistan hold once the troops leave? Can there be peace between this government, the various warlords, and the Taliban? A politically secure Afghanistan could trigger an economic renaissance for the region as a whole as recently discovered mineral wealth within the country and the chance for increased trade between the adjacent regions would increase both wealth and jobs.

▶ Historically, Central Asia was at the center of Eurasian trade. The main trade today is in narcotics and cotton, which continues to harm the health and environment of the region. Prospects for a better future lie in the ability of the various governments to address a range of economic, political, and environmental issues that will allow Central Asia to once again become a center for transcontinental trade and permit the opening of their economies to world markets, which may promote investment in the future of these people.

Key Terms

Amu Darya river 261
Aral Sea 260
Caspian Sea 261
centralization 266
collectivization of agriculture 267
command economy 266
Ferghana Valley 265

Hazaras 264
Hindu Kush Mountains 259
Kara Kum desert 260
kitchen gardens 276
Kyzyl Kum desert 260
mujahideen 280
opium 276

Pamir Mountains 259
Persian ethnic groups 264
Russification 269
salinization 263
Silk Road 264
Syr Darya river 261
Taliban movement 281

Turkic ethnic groups 264
Wakhan Corridor 264
warlords 276
Zarafshon River 261

Understanding Development in Central Asia and Afghanistan

1. To what extent is settlement in Central Asia accurately described as "linear" and what resource(s) might account for this characterization?
2. In what ways did the imposition of colonial borders on Afghanistan exacerbate political problems and increase the potential for conflict?
3. How did the Soviets change agriculture in Central Asia and what effect did that have on rural farmers?
4. Why do many experts believe that Kazakhstan has the greatest potential for economic development in Central Asia?
5. How did the economy of Central Asia change during the decade after independence and how did it differ thereafter?

6. What is the geographic significance of the Ferghana Valley?
7. Are there major obstacles to trade and economic development in Central Asia and Afghanistan and, if they exist, can anything be done about them?
8. Is the landlocked situation of Central Asian countries a development challenge?
9. What effect did the reduction of violence from the civil war and the ability to migrate have on the economic prospects of Tajikistan?
10. What is the relationship between the Taliban and the opium crop grown by many Afghan farmers?

Geographers @ Work

1. Assess how the environmental and political conditions in Afghanistan conspire to create a haven for illegal opium production, and discuss how the world community can aid Afghanistan in its struggle for both stability and peace.
2. Freshwater is arguably the most important resource in the world. Show how Central Asia's aridity restricts settlement and economic expansion, and develop planning principles that can reduce the problem.
3. The majority of citizens of Afghanistan, Kyrgyzstan, Tajikistan, and Uzbekistan rely on agriculture as their main occupation. Explain how this employment structure exemplifies a typical colonial construct in the region and suggest a plan that improves prospects

for development of these four countries in the short and long term.
4. Show how problems of land ownership cause hardships for the people of Central Asia. Construct a plan to overcome those political and ecological land-use problems.
5. In some parts of Central Asia, cotton is the most important economic resource; in other parts oil and natural gas play the same role. Contrast the impact of cotton as opposed to oil dependence, and discuss how dependence on and development of one major resource have cultural and political ramifications in addition to economic implications.

MasteringGeography™

Looking for additional review and test prep materials? Visit the Study Area in MasteringGeography™ to enhance your geographic literacy, spatial reasoning skills, and understanding of this chapter's content by accessing a variety of resources, including MapMaster interactive maps, videos, animations, RSS feeds, flashcards, web links, self-study quizzes, and an eText version of *World Regional Geography*.

7

The Middle East and North Africa

Douglas L. Johnson and Viola Haarmann

The Timeless Role of the Suq

Deep in the traditional heart of every Middle Eastern city is the suq (market), or, more correctly, an array of markets, each performing a specialized function. To enter the suq is to encounter a different world of narrow, winding streets roofed to provide protection from sun and heat. In this cooler atmosphere, each short street or segment of a longer street sells the same product: household goods here, then shoes; jewelry followed by rugs giving way to leather goods; cameras and mobile phones yielding place to watches; leading to a small square where everyone sells cooking pots and brass trays! This retail clustering of similar enterprises reminds the Western visitor of banks grouped in a financial district or car dealerships arrayed along an auto strip. Only the scale is different. The same principle of finding a deal, engaging in comparison shopping, never wasting a trip because the only vendor is sold out motivates the suq. Someone always has what you want at a price you can hope to negotiate! Near the main square are the prestige shops: books, jewelry, carpets. At the distant ends of the suq are the humble and grimy trades: metal working, wool dying, leather tanning. Often the people who staff the store live above the shop, while family members and neighbors make or repair the product in a room or courtyard in the rear. Suq merchants are vital members of local communities despite employing too few workers and family members to appear in national statistics. They offer jobs and teach skills to individuals who otherwise might be unemployable. Selling products new and old in a time-honored way, the suq lives on, always making a vibrant, essential contribution to local economies.

Read & Learn

- ▶ Explain the role that aridity and water availability play in the Middle East and North Africa.
- ▶ Identify the different types of water resources accessible in the region.
- ▶ Contrast the region's different types of agriculture and herding activities.
- ▶ Account for the crossroads location of the region and the resulting cultural mosaic.
- ▶ Outline the significance of the Atlas Mountains for the development options of the Maghreb.

- ▶ Explore the motivations behind the uprisings of the "Arab Spring."
- ▶ Compare Egypt's role in North Africa to Saudi Arabia's role in the Middle East.
- ▶ Analyze Turkey's unique potential in the region as a bridge between East and West.
- ▶ Describe how oil wealth or its absence influences the course of development in individual countries.
- ▶ Characterize the persistent conflicts in the region that are impeding cooperation and development.

▲ The suq of the Damascus madina displays the traditional adaptability of all such market's— offering traditional crafts and imported goods to both local and visiting shoppers.

Contents

7

The Middle East and North Africa

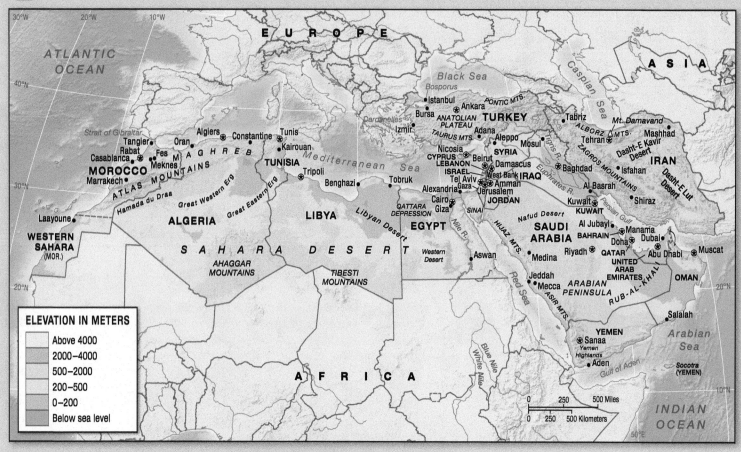

Figure 7-0 **Major physical features and countries of the Middle East and North Africa.**

The Middle East and North Africa (sometimes abbreviated as MENA) region is composed of countries that cluster around the Mediterranean Sea and the Persian Gulf (see chapter opener map). These countries are characterized as much by the unifying bonds of Islam and Arab culture as by deep internal contrasts and severe conflicts. The contrasts are evident in levels of technological development, the extent of dependence on oil for national revenues, cultural differences, religious divisions, the influence of militant Islamic factions, ethnic diversity, and uneven progress toward stable societies.

These contrasting elements create major areas of tension and conflict within the region. The most notable is the struggle between Israel and the Palestinian Arabs, who are supported by most of the region's Arab states. Even when neighboring states have recognized Israel's existence and signed formal peace treaties, substantial elements of their population do not accept their government's official policy. Unresolved, this conflict impacts the domestic politics of Middle Eastern and North African countries as well as regional and international relations and adversely influences economic investment strategies, development opportunities, and human rights practices. Major ethnic divides within many states in the region further magnify these issues.

Interstate rivalries are another area of conflict and sparked the inconclusive eight-year war in the 1980s between Iraq and Iran over resources, territorial claims, and regional leadership. Regional rivalries as much as principles governed which Arab states participated in the 1991 military opposition to Iraq's invasion of Kuwait. Weak involvement by Arab countries in the American- and British-led invasion of Iraq in 2003 and its aftermath of violence reflects their different assessment of the regional political realities and extreme discomfort with the future impact of events in Iraq on their own internal stability.

The geography and future of this complex region are also influenced by the physical and cultural environments of the Middle East and North Africa. Aridity dominates the region and renders much land of limited utility for agriculture. The uneven distribution of oil reserves explains many differences in development, standards of living, and political influence. The Islamic faith remains a powerful force in structuring society and state, despite the presence of nationalist forces and non-Islamic minorities. Finally, this world region has long functioned as a crossroads for Europe, Asia, and Africa, enabling an exchange of goods and ideas throughout history. The colonial influence of the late nineteenth century and the continuing American and European economic and military presence in

Israel and a number of Arab states have contributed to widespread resentment of Western political ideas and foreign policy. But Western technology, educational systems, and business practices have taken root among the region's countries and peoples, and offer a promising foundation upon which future cooperation and positive interaction can develop.

▲ **Figure 7-1 Sand dunes in the Great Western Erg of Algeria.** A small caravan crosses an area of massive, linked, crescent-shaped barchan dunes.

The Middle East and North Africa's Environmental and Historical Contexts

In the MENA region, physical and cultural environments are closely connected. Rhythms in agriculture and animal husbandry follow the rhythms of wetter and drier seasons, years, and decades. Concentrations of people mirror the long-term, local availability of resources. Patterns of housing, styles of clothing, and technological developments reflect age-old adaptations to the hazards and opportunities present in local habitats. Although in many ways diverse, the Middle East and North Africa has become one of the world's most distinctive geographical regions, bound together by circumstances of aridity and a common Islamic cultural heritage.

Environmental Setting: From the Mediterranean to the Desert

The Middle East and North Africa forms a vast region that stretches from the Atlantic Ocean eastward to the Persian Gulf and the lands bordering South Asia (see chapter opener map). Our study of this pivotal region begins with an overview of its varied physical environments.

Landforms

People often think sand dunes are the most common landforms in the deserts of the Middle East and North Africa. There are large sandy areas (*erg*) of mobile sand dunes, such as the Great Western Erg in Algeria or the Nafud Desert in northwest Saudi Arabia (see chapter opener map). But these spectacular environments are the exception rather than the rule. Dominated by crescent-shaped barchan dunes or long, linear *seif* (sword) dunes, sandy areas are largely devoid of vegetation except when groundwater happens to be close to the surface (Figure 7-1). Stony *hammada* (rock) and *serir* (pebble) surfaces are more common. Both sandy and stony areas also lack the soil and water needed to sustain vegetation and human life.

Of much greater importance for human settlement are the mountain ranges and upland plateaus that provide a basis for livelihood and economy. These highlands have lower temperatures, higher rainfall totals, and more reliable agricultural resources than many coastal districts and interior areas. Rainfall totals are often substantial. At Ifrane in the central Middle Atlas Mountains of Morocco, as much as 60 inches (1,500 millimeters) of precipitation can fall in a year. Much of this moisture occurs in the form of snow since the mountains are high, and the wet season occurs in the coolest part of the year. In the area of Midelt, less than 65 miles (100 kilometers) further inland, average annual precipitation is less than 10 inches (250 millimeters). A

similar contrast exists between Beirut, Lebanon on the Mediterranean coast and Damascus, Syria only 50 miles (80 kilometers) to the east. Beirut averages 37 inches (940 millimeters) of rainfall per year, while Damascus is functionally an oasis that receives less than 9 inches (230 millimeters) annually. The intervening Lebanon and Anti-Lebanon mountain ranges absorb much of the moisture produced by the winter storms passing through the Mediterranean, leaving the interior regions dry but creating humid zones along the coast and in the uplands.

These uplands are of tremendous importance to the nearby lowlands. Historically the uplands supported large forests that provided building materials for the cities and fleets of the coastal states. Most of these forests have been removed, and only remnants survive in the higher and more inaccessible places or in protected stands near sacred sites. The rivers that drain the uplands transfer large amounts of water as runoff to the nearby lowlands and provide water for irrigated agriculture, the lifeblood of the settled populations of the lowlands (Figure 7-2). The seasonal contrasts in temperature and

▼ **Figure 7-2 The city of Hama in the Orontes River valley of Syria.** Rising in Mount Lebanon and the Anti-Lebanon mountains, the Orontes River flows through Hama before reaching the Mediterranean Sea. The river's water nourishes irrigated fields and provides domestic water supplies for the settlements along its route.

▲ **Figure 7-3 Land in the Haraz Mountains, Yemen.** Rainfall in the mountains is trapped by precisely contoured terraces on steep slopes to sustain viable agricultural communities with a sophisticated village architecture.

precipitation between the uplands and lowlands have historically constituted the basis for important nomadic movements of livestock. Today much reduced in scope and practiced by fewer people in many countries, this upland–lowland form of migratory animal husbandry is still practiced as in former times when it helped support the great lowland civilizations by supplying meat and other animal products.

The highlands have also supported substantial farming populations throughout recorded history (Figure 7-3). In many respects these uplands are good places to live. In a region where aridity is widespread, uplands offer reasonably secure water supplies. Here minority religious and ethnic groups have often concentrated. The same mountain environments that produced food also provided protection from the central governments of the region, typically located in lowland or

coastal sites. Maronites in Lebanon, Druze in southwest Syria, Berbers in North Africa, and Zaidis in Yemen all took advantage of the defensive features of their mountain homelands. With only limited ability to expand spatially, these groups often invested in labor-intensive agricultural improvements such as terracing, converting steep slopes with thin soils into deep, narrow, and flat pockets of well-watered soil. With soil trapped behind stone walls, these field systems were erosion resistant and stable for centuries. But when better employment opportunities appeared in adjacent lowland areas during the last century, the loss of laborers often led to a collapse of agricultural terraces. As terrace walls crumble due to the lack of maintenance, the soil accumulations of centuries run the risk of washing downslope to the sea.

Climate

Aridity (dryness) is the dominant climate feature of the Middle East and North Africa. Much of the region is so dry that it can support little or no vegetation, and these extensive areas of barren rock, gravel, or sand support very few people. Where water is present, people are found, often in large numbers. These more densely populated settings include upland areas that receive substantial rain and have cooler temperatures, coastal lowlands watered by runoff from the uplands, inland plateaus exposed to seasonal storms, and oases.

Much of the region, particularly the portion bordering the Mediterranean Sea, experiences a dry summer subtropical climate with less than 20 inches (500 millimeters) of annual rainfall (Figure 7-4). Precipitation falls mainly in the late autumn through early spring. During the summer and autumn seasons, typically there is very little rain. Winter rainfall throughout the greater Mediterranean basin is produced by frontal or cyclonic storms that move across the region from west to east. Whenever these storms encounter a mountain barrier, the air is lifted and cooled, often leading to condensation and precipitation. The Atlas Mountains in northern Africa, Al-Jabal al-Akhdar in eastern Libya, Mount Lebanon and the Judean

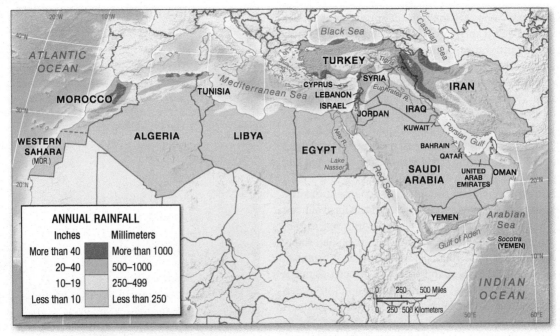

▲ **Figure 7-4 Precipitation patterns in North Africa and the Middle East.** Most of the region receives less than 10 inches (250 millimeters) of precipitation annually. Areas receiving more than 40 inches (1,000 millimeters) annually are centered on the high mountain zones.

hill country to the east of the Mediterranean, and the Zagros Mountains of Iran are all associated with localized **orographic,** or mountain-induced, **precipitation.** Where mountains are high, this precipitation falls as snow and remains on the ground throughout the winter. Some of the mountain ranges close to the sea, such as the Pontic Mountains along the Black Sea coast of Turkey or the Alborz Mountains on the Caspian coast of Iran, generate huge quantities of rain or snowfall. Even the ranges deep in the Sahara, such as the Ahaggar Mountains in southern Algeria, serve as water catchment areas for the surrounding regions and support "islands" of vegetation that are very different in type and productivity from the plants of the surrounding lowlands. These mountain barriers force air to rise, cool, form water droplets, and rain on their windward sides. Their interior (leeward) sides are much drier, because air becomes warmer as it moves downslope, stops releasing moisture, and creates a **rainshadow** effect. Rainshadow drylands, including the High Plateau of Algeria, the central Turkish or Anatolian Plateau, and the Syrian Steppe, often are important cereal and livestock production zones. Where the blocking mountain ranges are very high and extensive, as in central Iran, these interior environments become extensive deserts.

The moisture that nourishes the upland mountains and plateaus also sustains the adjacent lowlands. While some of this moisture falls directly onto the lowlands as the frontal storms cross them, much of the water comes from runoff from the nearby uplands. The streams that flow to the sea all support rich agricultural environments in the coastal lowlands.

Some **rivers** are **exotic,** or foreign, to the area where they supply water. They derive their water supply from outside the region and cross dry areas on their journey to the sea, collecting little if any additional runoff along their downstream courses. The Nile is the best-known example. Rising in the mountains of Ethiopia and East Africa, the Nile flows northward across the Sahara Desert before draining into the eastern Mediterranean. In Egypt the Nile constitutes a long, densely vegetated oasis on a narrow floodplain. A short distance away from its banks, plant growth is impossible (Figure 7-5). The Euphrates and Tigris rivers, rising in the highlands of eastern Turkey, are also exotic streams as they flow through the desert lowlands of Iraq. South of Baghdad, practically no water is added to either stream as they proceed southward to the Persian Gulf.

Similar "islands" of life exist deep within the desert in **oases** (singular: oasis), where erosion has lowered the land surface and a relatively high groundwater table results in water existing close to the surface. In some cases, saline lakes, fed by groundwater from below and diminished by evaporation from above, are found in the lowest central portion of the oasis. Near-surface groundwater is often recharged by rainfall entering regional **aquifers** far outside the desert. Lateral movement of subterranean recharge water is always very slow, and withdrawal of groundwater at rates greater than is possible using traditional, animal-powered technology invariably leads to drawdown and eventual depletion of the aquifer. Deeper aquifer layers are composed of fossil water trapped underground during wetter epochs thousands of years in the past. This fossil water is a finite resource, and its use constitutes a form of mining. Oases historically served as staging points for the caravan trade that linked more productive regions outside the desert across the generally barren arid wastelands.

Living in the desert requires special adaptations by animals, plants, and people. No animal is better suited to the desert than the camel, whose endurance, capacity to travel long distances with

▲ **Figure 7-5 Satellite photo of the Nile River and Delta.** The dark-green, cultivated areas along the Nile River form a linear oasis with human settlement concentrated along the river's banks. The delta, densely occupied by farmland, is the primary agricultural production area of Egypt; the heart-shaped area southwest of the delta is the Faiyum depression, a major farming area nourished by water diverted from the Nile. Similar oasis basins west of the Nile, once dependent entirely on groundwater, now are fed in part by water stored in Lake Nasser behind the Aswan High Dam in the south. The Red Sea separates Africa from the Arabian Peninsula, while its two projecting arms, the Gulf of Suez and the Gulf of Aqaba, define the Sinai Peninsula.

heavy loads, and ability to survive without drinking more than once every five to eight days in the dry season traditionally made it the animal of choice for desert dwellers. To a lesser degree, donkeys and goats also tolerate water shortages and poor fodder conditions. Only in places where local water and grazing conditions improve do cattle, sheep, and horses appear as important livestock. Plants adapt to aridity by evading dry conditions or by adjusting to particular local conditions. Wheat, barley, and rye are domesticated grasses native to this region that avoid drought by completing their life cycle quickly and by producing seeds that can survive for many years until the next rainy period produces good growing conditions. Most of the wild grasses that animals eat have adapted in the same way. Other plants sink deep taproots into the ground where they

▲ Figure 7-6 Climate-adapted housing in El Oued, Algeria. Domed and barrel-vaulted roofs allow more interior roof space into which hot air can rise. Openings allow hot air to escape, setting in motion relatively cooling air movement.

can find water or, like olive and almond trees, withstand drought by limiting their activity to the wet season or to wetter years. Some plants thrive under conditions of extreme environmental adversity. The date palm is able to use saltier water than most plants and thus is a major component of oasis agriculture. Its fruit is a vital source of food for desert dwellers, and its palm fronds and woody trunk are important fencing and building materials.

The peoples of the Middle East and North Africa have also adapted to heat and dry conditions. The traditional flowing robes reduce moisture loss through sweating by insulating the wearer from the intense desert sun and by allowing free airflow to cool the skin. Their light color reflects rather than absorbs sunlight, keeping body temperature lower, and the use of head coverings reduces the risk of sunstroke. Housing also reflects climatic adaptation (Figure 7-6). Vaulted roofs, high ceilings, and the use of tiles promote cooler conditions by allowing warm air to rise or by absorbing less heat. Shutters and sunscreens cover or shade window openings to prevent direct sunlight from penetrating and heating interior spaces. Building houses around a central courtyard provides both shade and private outdoor space. In some places, houses are constructed partially in the ground or extend back into caves to moderate daytime temperatures. Cooling towers promote air movement that produces cooling breezes. Nomads make tents that are easy to shift seasonally from place to place. Often a heat-absorbing black tent is used in winter for warmth, while a light-colored, more reflective tent is typical of summer, when cooling is the chief concern. People also adapt economically to seasonal or perpetual aridity, with nomads moving their herds between wet and dry season pastures to find grass and water for their animals. Many farmers compensate for limited local water supplies by constructing irrigation systems that bring in water from distant sources to grow their crops. Today many farmers use glass or plastic greenhouse-like structures to control temperature and water loss and create a suitable environment to grow vegetables, fruit, and flowers. These high-value specialty products are then sold in the region's major urban markets or exported to Europe as "out-of-season" crops that command high prices to offset transport costs.

Environmental Challenges

People in the Middle East have lived in agricultural villages for 10 millennia, and in cities for half that time. When local populations were small and widely scattered, environmental impacts were limited. But as populations grew, those impacts appeared both locally and in more distant places. Many of the issues faced today are continuations of ancient problems, but are made more serious because the populations depending on the environment are much larger and the power of technology is much greater.

Cities are a focus of today's settlement landscape, but the impact of sheltering and feeding urban populations is felt far beyond city borders. Because forests provide many useful products including food (Aleppo pine yields the pine nuts that feature in Middle Eastern cooking, for example), preserving and extending forest cover is important for habitat protection and local economies. Natural forests have always been the primary source for wood, but dry places seldom have extensive forests. The Middle East and North Africa is no exception. Only 2.4 percent of the region's land surface contains forest, mostly found in the uplands. Maintaining existing forest habitats in the Atlas ranges of the Maghreb, as well as the coastal mountain uplands of the eastern Mediterranean and south coastal mountains of the Black and Caspian Seas bordering Turkey and Iran is important, but difficult to achieve. New trees are planted in existing forests (afforestation) and in historically forested but now largely denuded districts (reforestation), but it is expensive to do successfully. Seedlings require water and human care to establish themselves, need protection from premature harvesting by man and beast to grow, and take a long time to reach maturity. Fire in these forests is a constant danger, and the effort and expense of decades can be wiped out in few days by a sudden blaze, as was the case in Israel's Mt. Carmel fire in early December 2010 (Figure 7-7). The fire occurred during a very dry

▼ Figure 7-7 Mt. Carmel forest fire near Haifa, Israel. The smoke plume trailing westward from the coast of Israel south of Haifa testifies to the intensity of what became the worst forest fire in Israel's history. Drought, high winds, and human carelessness created a volatile fire hazard situation that burned over 7,000 acres (2,800 hectares) and killed 42 people in early December, 2010.

period, involved a forest with a high proportion of pine trees containing very flammable resins, had an apparent human cause, and resulted in serious environmental and economic damage and loss of life. In Lebanon only a dozen stands of the famous Lebanon cedar still survive. These clusters are preserved either because they are hard to reach or because they are located at sacred sites. Efforts to protect and extend forests often encounter local opposition because people who have traditional rights to graze animals in the forest and collect wood are deprived of important elements of their domestic economy. Growing trees for the material and aesthetic benefit of distant populations is not a simple or easy task!

Limited ground cover, whether of trees or grass, means increased potential for erosion. In settled uplands, terraced fields retained soil and water and substituted for reduced forest cover in protecting the land surface from erosion. Substantial rural-to-urban migration means abandoned fields and a loss of erosion protection. How to stabilize and repair terraces and plant crops on them that require limited labor but return maximum profit is unclear. Sediment-laden runoff from uplands deposits silt, decreasing water transport capacity and increasing maintenance costs in lowland irrigation systems. It also reduces the storage capacity of dams and reservoirs, increases flooding in the downstream course of rivers, and clogs coastal harbors. All of these environmental impacts are difficult and expensive to correct. In grasslands, tractor power has permitted the expansion of cereal cultivation into districts once the exclusive preserve of nomadic herders. Often left after harvest with a bare soil surface that is exposed to wind, these fields can contribute significantly to wind erosion and more frequent dust storms. Cropping systems that incorporate fodder grass cover are urgently needed for fields susceptible to wind erosion between growing seasons.

As nomadic herders are pressed into smaller and less productive rangeland spaces, they are forced to use these areas more intensively. Reduced access to river floodplains, now converted to year-long cultivation, also denies herders access to critical dry season pasture areas. The result is **overgrazing** in the remaining rangeland areas, reducing the quality of available fodder. Efforts to improve the productivity of rangelands by introducing wells and surface water collection projects often backfire. These water provision facilities lead to concentrations of animals whose fodder needs in dry years can exceed available local resources. Without control over the number of animals, the resulting vegetation deterioration leads to serious herd losses when droughts reduce naturally occurring fodder.

The drawdown of groundwater reserves is another serious problem. In the Jefara Plain west of Tripoli in Libya, overzealous exploitation of groundwater for irrigation purposes has led to a dangerous decline in the groundwater table and the imposition of severe constraints on agriculture. In many oases, the introduction of diesel-powered pumps lowers the groundwater to depths that can no longer be exploited by traditional lift technologies. Often traditional farmers as well as formerly productive agricultural districts are driven out of production as a result. In coastal districts where urban expansion has heavily increased the extraction of groundwater, the invasion of saltwater into the lowered groundwater table has reduced the quality of water available for drinking and other purposes (see *Exploring Environmental Impacts:* Jeddah's Urban Water Crisis).

Tourism, often promoted in many areas as a boost to local economies, can have undesirable impacts on local resources as well. Particularly threatened are Red Sea coral reefs, which are subject to physical damage from diving and are often swamped with sediments and effluents produced by nearby residential and industrial areas.

Offshore fishing resources of the region's oceans often are indirectly hurt by development efforts. Dams that capture and divert water for irrigation also cut the flow of organic materials that form the basis for marine food chains. This results in the decline of fish stocks. Heavy fishing in the region's few prime marine fishing areas also places finfish stocks under severe pressure. Storage of floodwaters for irrigation reduces the amount of freshwater reaching the sea. This increases the salinity of nearly enclosed water bodies such as the Mediterranean Sea, the Red Sea, and the Persian Gulf, thereby reducing susceptible fish stocks. Efforts to establish fishing in the lakes formed by the region's dams have not compensated for the decline in offshore fishing. But recent signs indicate that fish stocks in the eastern Mediterranean may be recovering. The generous use of fertilizers in Egyptian agriculture might be substituting for the natural nutrients that used to feed the base of the aquatic food chains. This is a counterintuitive development that shows how complex management of the region's many environmental problems can be.

Stop & Think

→ Aridity and limited water resources are a characteristic of the Middle East and North Africa. List environmental problems that occur when land use does not take sufficient account of this reality.

Historical Background: Home of Many Civilizations

Long a center of human settlement and civilization, the MENA region occupies an intermediate location between Europe, Sub-Saharan Africa, and southern and central Asia. In this spatial relationship to neighboring regions, the MENA region has experienced an ebb and flow of peoples, ideas, value systems, and technologies, creating a diverse human environment. The historical origin of the factors shaping the contemporary Middle East and North Africa is the focus of this section.

Economic and Cultural Crossroads

The Middle East and North Africa are a common meeting ground linking Europe, Sub-Saharan Africa, and Asia. The term "Middle East" to designate the eastern portions of this larger region is a historical accident. In the nineteenth century Europeans divided the "orient," the world east of Europe, into near, middle, and far east. In this division, the Near East comprised the area now generally regarded as the Middle East and North Africa. Early in World War II, difficulties sending troops quickly from Britain to defend Egypt from attacks by Italian and German forces resulted in shifting troops and command staff from India to Cairo. Dispatches from this Middle East command retained their traditional byline, as did the articles filed by journalists. Over time identification of the Middle East with the eastern Mediterranean took deep root. Today this is reflected in the widespread use of the term Middle East in popular literature, by academic area studies programs, as well as most aid and international organizations. Because the region is in an intermediate location between three major cultural and continental-scale areas, use of the term "Middle East" retains a certain serendipitous geographic logic.

The region's interactive position reflects its historic connections with the cultures of adjacent land masses. With Europe, the region shares many common scientific and literary traditions; much of the

EXPLORING ENVIRONMENTAL IMPACTS

Jeddah's Urban Water Crisis

Jeddah is the second largest city in Saudi Arabia, and the country's commercial center and major port. Its metropolitan area with a population of more than 5 million people stretches for more than 30 miles (50 kilometers) along the Red Sea coast on the Tihama Plain at the foot of the Asir Mountains (Figure 7-1-1). Jeddah is also the gateway for more than 2 million pilgrims each year to Mecca, Islam's holiest site just 50 miles (85 kilometers) to the east.

Jeddah's coastal location is both an enormous asset that has fueled its growth and a critical liability now that so many people depend on the sustainable use of the water resources in this desert environment. Three major water-related problems plague the city at an intractable scale: a high water table; inefficient water delivery, drainage, and wastewater removal systems; and floods. These water problems are the product of multiple causes.

Jeddah's natural setting on the Tihama coastal plain close to sea level creates a naturally high water table. With limited freshwater recharge from rainfall, groundwater with a high salt content is typical. In a rapidly growing urban environment, one might expect that groundwater levels would decline as domestic and industrial water demands have

grown. But water in Jeddah is supplied by desalinating Red Sea water, and the groundwater level continues to rise with increased water consumption and wastewater discharge! Watering of public spaces at high levels to create a green, oasis environment also contributes to the rising groundwater table. This results in structural damage to buildings and road surfaces as saltwater seeps into building materials.

Urban growth has continued to outpace the ability of urban planners to provide centralized water delivery, drainage, and sewage treatment facilities, which raises the water table further as heavily polluted effluents leak into the ground. Much of the domestic water delivery system, as well as the wastewater removal system, is faulty. More than two-thirds of households are not connected to wastewater pipelines and use below-ground septic tanks to collect their wastewater. As groundwater levels rise, these septic tanks are frequently below groundwater level, making cross-contamination inevitable. Overburdened, perpetually full septic tanks must be pumped frequently. The sewage effluent is removed by tanker trucks and dumped into a large surface

▲ **FIGURE 7-1-2** **Flash flood in Jeddah.** On 29 January 2011 heavy rain east of the city created massive runoff and flooding in the intermittent streams that crossed the urban area on their way to the Red Sea. Inadequate warning of the danger and absence of infrastructure to contain or divert the floodwaters sent torrents flowing through the city streets, tossing cars and other loose objects around like plastic toys.

wastewater lake northeast of Jeddah. Most of the wastewater removed by pipes ends up in the Red Sea, where it spreads far and wide and contaminates the Red Sea's ecological system.

Finally, Jeddah is located in an arid environment with an average annual rainfall of just 2.5 inches (60 millimeters). But on the rare occasions when rain does fall, it tends to be as torrential downpours that cause severe flash floods from rainwater mixed and polluted with overflowing wastewater. Jeddah experienced two such floods in close succession in 2009 and 2011 (Figure 7-1-2). Massive rivers ran through the city, sweeping everything away before them, and causing loss of life and the destruction of thousands of homes and businesses.

The magnitude of these environmental problems is hard to overstate and the scale of the investments required to address them is enormous. A massive effort is under way to construct a sewer network, which ambitiously aims to connect all Jeddah's residents by 2015. This might be overly optimistic!

Sources: K. E. House, *On Saudi Arabia: Its People, Past, Religion, Fault Lines—and Future* (New York: Knopf, 2012); A. al Sharif, "Sewage Lake Threatens Jeddah," *The National*, December 4, 2008, http://www.thenational.ae/news/uae-news/environment/sewage-lake-threatens-jeddah; P. Vincent, "Jeddah's Environmental Problems," *Geographical Review* 93, no. 3 (2003): 394–412.

▼ FIGURE 7-1-1 **The Red Sea port of Jeddah, Saudi Arabia.** The sprawling, rapid growth of Saudi Arabia's second largest city has overwhelmed both the Tihama plain's environment and the urban planner's ability to provide essential water management services.

ancient cultures of Greece and Rome were preserved by Islamic scholars. Overland trade across the Sahara, as well as between the Mediterranean Sea, the Red Sea, and the Persian Gulf connected European economies with those of Africa and Asia. The expansion of Islam from its origin in western Saudi Arabia into adjoining regions also served to link the surrounding areas together in patterns of conflict and accommodation. Islam's expansion eastward into Pakistan, India, Bangladesh, Indonesia, and the Philippines promoted trade contacts between the Middle East and South and Southeast Asia and provided a network of connections maintained by pilgrimage and a sense of religious brotherhood. The contemporary expansion of Islam into Sub-Saharan Africa maintains the connections established by the Saharan caravan trade and Omani mercantile contacts with East Africa many centuries earlier. The legacy of Muslim connections with Europe is found not only in architectural and cultural survivals in Spain and Portugal, but also the Muslim communities in Albania, the former Yugoslavia, and Bulgaria. These latter products of crossroads contact historically have generated serious political and cultural problems within the Balkans, which resurfaced in the 1990s. European colonization in the nineteenth and twentieth centuries had a highly disruptive effect on MENA societies. Conflicts sparked by that contact continue to disturb the region despite the end of the colonial era. The decline of Beirut, Lebanon, as a regional banking and commercial center linking Europe and the Middle East is just one illustration of the difficulties associated with a **crossroads location,** where contrasting cultures meet and often clash.

Ethnicity and Language

Much of the world views the MENA area as a region inhabited primarily by **Arabs** and they are the largest single ethnic group in the region, accounting for more than half of its 475 million inhabitants. Speaking a common Semitic language, and possessing a shared history and culture, the Arabs are a readily identifiable group. But this sense of uniformity is misleading. While Arabic is a shared language, regional dialects, often mutually incomprehensible, exist. Despite the existence of an umbrella political organization in the form of the Arab League, Arab politics are highly variable. Monarchies (Morocco, Saudi Arabia, Oman, Jordan, Kuwait, Bahrain); one-party, secular states (Algeria, Syria, and until recently Tunisia, Egypt, and Iraq); a military dictatorship (Yemen), a new democracy struggling to replace an overthrown dictator (Libya); a religious community–based republic (Lebanon); the influential Islamic advocates of religiously based governance; and the relatively powerless secularist proponents of multiparty democracy all vie for attention and compete for influence. Believing that they share a common culture and history, most Arabs desire a political unity that reflects their perceived oneness. Yet this pan-Arab nationalist goal has proven very hard to achieve. Attempts to form meaningful unions among different combinations of Arab countries have consistently failed.

Many other ethnic groups are also found in the Middle East and North Africa (Figure 7-8). The Arabs occupy the central core of the region, having spread out from their ancestral homeland in

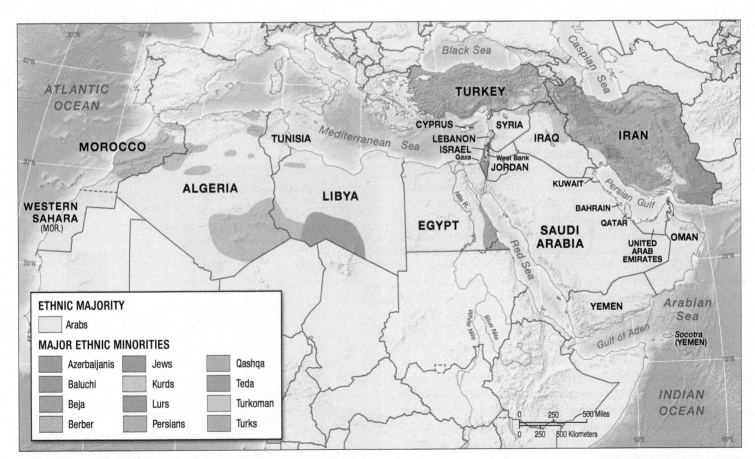

▲ **Figure 7-8 Major ethnic minorities in the Middle East and North Africa.** While Arabs are the largest ethnic group of North Africa and the Middle East, the region contains numerous ethnic minorities. Most of these are found in the higher mountainous areas where physical isolation has protected them from cultural assimilation.

▲ **Figure 7-9 Kurdish bazaar, Dahuk, Iraq.** Located in the mountainous area of northern Iraq, Dahuk is in a region predominantly occupied by Kurds. Together with their ethnic kin in Turkey and Iran, Kurds have long sought autonomy or independence and a national identity.

the Arabian Peninsula. Large minorities of Turks, Persians, Kurds, Berbers, and Nilo-Hamitic groups are found in more peripheral locations.

Turks and **Persians** are the two largest non-Arab groups, dominating the countries of Turkey and Iran, respectively. A substantial majority within their own country, Turks are important minorities elsewhere in the region and in Europe. The presence of Turks in the Balkans, for example, reflects the former extent of the Ottoman Empire, the predecessor state of modern Turkey. Turks and Persians dominate the national political and economic scenes in Turkey and Iran respectively, but both countries contain sizable linguistic and ethnic minorities.

In southeastern Turkey reside large numbers of **Kurds** (about 18%of the Turkish population), with smaller numbers of Arabs (mostly around Adana on the Turkish Mediterranean coast near Syria), Greeks and Jews (in and around Istanbul), Circassians and Georgians (in the northeast), and Armenians (in the southeast) also part of Turkey's ethnic mix. A largely pastoral and agricultural people historically, Kurds are split among several adjoining countries, with concentrations in northeastern Iraq (17% of the Iraqi population) and in the northwestern border zone of Iran (7% of its population), where they occupy mountainous areas that are difficult to reach (Figure 7-9). Smaller numbers of Kurds are found in northern Syria. Speaking an Indo-European language as do Iran's Persian speakers, Kurds have long agitated—sometimes violently, but always unsuccessfully—for autonomy within Iraq and Iran and for independence from Turkey. Iraq's Kurdish political parties seem prepared to work within the structure of a unified Iraqi state, but it is unclear what degree of internal autonomy might be needed to satisfy their longer-term aspirations. Since both substantial autonomy within Iraq and outright independence for Iraqi Kurds are likely to be viewed as a potential long-term threat to the internal unity of both Turkey and Iran, the bounds of what is desirable to Kurds and what might be permitted by their larger and more powerful neighbors remain obscure.

Iran, although dominated by speakers of Farsi and related Persian dialects (65% of the total), has an equally diverse population. Azerbaijani Turks, concentrated in the northwest corner of the country, are the largest minority (16%), while Turkish-speaking Turkoman and Qashqa communities are regionally prominent in the northeast and southwest, respectively. Baluchi are widespread in the arid southeast of Iran, while Lurs and other smaller ethnic groups are scattered through the rugged Zagros Mountains of western and southwestern Iran. Arabs, largely concentrated in the lowlands east of the Tigris River, make up about 3 percent of the Iranian population. The result is a core of cultural homogeneity surrounded by substantial regional diversity.

A smaller but regionally prominent ethnic minority in the Middle East is the Jewish populace. Concentrated in Israel, **Jews** constitute 76 percent of that country's population. Jews of European origin and from North Africa and the Middle East migrated to Israel after it gained independence in 1948, a process of return that continues today from diverse sources. Small Jewish populations are still found in most of the region's other countries. Most non-Israeli Jews live in urban areas, the remnants of much larger and commercially more important populations predating Israeli independence (Figure 7-10). The remainder of Israel's population is composed of largely Muslim Arabs (21%), and an immigrant population whose ethnic and religious origins are not known.

Other small minority communities are scattered through the region. Armenians, once an important element in the population of eastern Turkey, are now only a residual community. Large numbers of Armenians perished in a series of genocidal "ethnic cleansing" attacks in World War I. Survivors are dispersed throughout the region and are prominent in urban areas, particularly in Lebanon

▼ **Figure 7-10 The Jewish quarter in Djerba, Tunisia.** One of the largest populations of Jews living in the Middle East outside of Israel is found on the island of Djerba in southern Tunisia. Here a Jewish family stands in the doorway of their house, which is protected from the evil eye by the painted fish and menorah symbols on the wall.

and Syria. Circassians, Muslim refugees from nineteenth-century Russian imperialism in the Caucasus Mountains, are found in Syria and Jordan. Many of the Gulf States, including Kuwait, Qatar, and the United Arab Emirates, have transient populations of non-Arab foreign workers (Iranians, Pakistanis, Bengalis, Filipinos, and Indians), as well as Palestinians, Egyptians, and Yemenis, living among them.

In North Africa much of the indigenous population has, over many centuries, been assimilated through the adoption of the Arabic language and culture. Still, Berber speakers remain a significant minority, numbering more than 35 percent of Morocco's total population. **Berbers,** concentrated in mountain districts or in the central Sahara, are a less prominent ethnic group in Algeria (approximately 20%), Tunisia (2%), and Libya (2–5%). In south-central Libya, the nomadic Teda, speaking a Hamitic language, dominate the northern slopes and foreland of the Tibesti massif.

Religion and Secularism

Religion is an extremely important part of self-identity for many people. Just as not all Middle Eastern people are Arabs, so too not all are Muslims. Nonetheless, **Islam** is the dominant religion in the region. Revealed to the Prophet Mohammed, who began to have divine visions in A.D. 610, the community of believers practicing Islam (which literally means submission to the will of God, Allah) formally began in 622 on the Arabian Peninsula, when the Hegira (flight) occurred. This event was the forced departure of Mohammed and his followers from the city of Mecca to Medina to the northwest. From his new base in Medina, Mohammed eventually conquered Mecca and expanded the area of Islamic control to include, at the time of his death in 632, most of the western half of the Arabian Peninsula. His successors, the Caliphs, rapidly extended Islamic political control to the east and west until, little more than a century after the Prophet's death, the Umayyad Caliphate stretched from Spain to beyond Afghanistan (Figure 7-11).

The faith established by Mohammed was and is appealingly simple. At its root are the **Five Pillars of Islam**—the creed, prayer, charitable giving, fasting, and pilgrimage—that govern the behavior and belief of the faithful. The fundamental tenet of faith is the *shahada,* the creed, which states that "There is no God but God, and Mohammed is his messenger." **Muslims** believe that Allah (God) is a transcendent, divine being, surrounded by angels and prophets, and Mohammed was the culmination of a tradition of prophetic ministry that began with the Jewish prophets, continued through the teachings of Jesus of Nazareth, and reached its fullest form in the *sura* (books or chapters) revealed to Mohammed and collected in a sacred text, the **Quran.** The second pillar of faith is the offering of regular prayer, which takes place five times a day. *Zakat* (alms) is the third pillar, and expresses the responsibility of more prosperous Muslims to aid those less fortunate. Fasting during the month of Ramadan, the ninth month of the lunar calendar, is the fourth pillar. During this period, practicing Muslims avoid eating, drinking, and smoking between sunrise and sunset to exercise command over bodily cravings and to submit to spiritual control. Finally, at least once, if physically and financially able, Muslims are expected to perform the *hajj*, the pilgrimage to Mecca, where visits to sacred sites and religious devotions predominate. From the Quran is derived the **Sharia,** the rules and laws that govern the society in which Muslims live. To devout Muslims there is no distinction between the religious

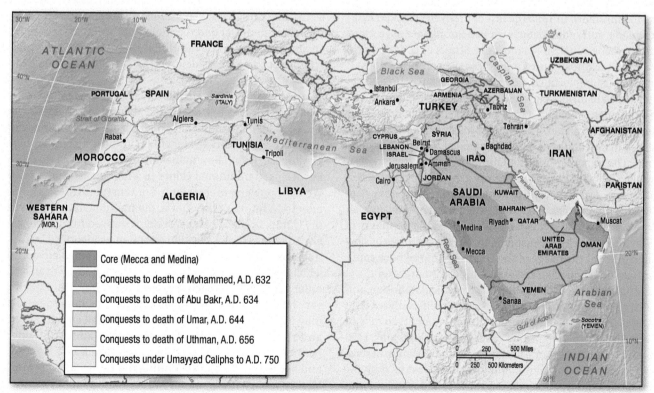

▲ **Figure 7-11 The expansion of Islam from A.D. 622 to 750.** Islam expanded rapidly from its core in the western Arabian Peninsula to dominate most of North Africa and the Middle East by the mid-eighth century.

and the secular, between church and state; both spheres are part of one integrated whole, which should be governed by the Sharia.

It is this belief in the religious state that is at the root of conflicts between secular and fundamentalist forces within the region, just as in earlier times when a dispute over power and authority within the Muslim community led to a division into **Sunni** (the orthodox, who believe that potentially any practicing Muslim can exercise power) and **Shi'ite** (those who believe that only descendants of Mohammed should hold political power). Today Sunnis constitute about 90 percent of all Muslims, although Shi'ites are particularly important in Iran (where 9 out of 10 Muslims are Shi'ites), southern Iraq (more than 50% of the population), southern Lebanon (30% of the population), Yemen (60% of the population), the interior of Oman, and the Gulf coast districts of Saudi Arabia (approximately 6% of the population). Most Shi'ites in Iran are Persian; in the rest of the Middle East they are Arabs. Similarly, most Christian communities, regardless of the rite practiced, are Arabs. The most important of these are the Maronites of Lebanon (30% of the population) and the Copts of Egypt (about 9% of the population).

The resurgence of Islamic values in the Middle East and North Africa today represents a reaction to centuries of increasing contact with and pressure from Western European and North American values and institutions. This pressure has deep historical roots. It began more than a millennium ago when the great westward surge of Islamic expansion crested in the western Mediterranean basin. When Islamic rulers were unable to establish permanent political control north of the Pyrenees in the ninth century A.D., small Christian principalities reemerged in northern Spain, grew stronger, and expanded southward over the next four centuries against decreasingly effective Muslim states. In 1492 the last of these entities, the kingdom of Granada, surrendered. Spanish and Portuguese sovereigns extended the reconquest drive across the Straits of Gibraltar and along the Mediterranean and Atlantic coasts, capturing coastal cities and establishing fortified bases. When Christian attempts to expand inland were turned back and most of the European coastal footholds ultimately were lost, a stalemate developed in the central and western Mediterranean. The region's Muslim states were relatively weak, and functioned as bases for piratical raids on European commerce. The expansion of the Ottoman state in the eastern Mediterranean followed a similar trajectory. Despite rapid expansion in the Balkans, and the extension of nominal authority over much of North Africa in the fourteenth and fifteenth centuries, the Ottoman sultans failed to eliminate Christian rivals both on the periphery of the empire and within it. This ushered in several centuries of slow decline and gradual shrinkage of the Ottoman Empire back south under pressure from emerging and expanding Christian states.

The initial consequences of declining political cohesion and reduced military superiority were to erode the self-confidence of Islamic states and societies and undermine their ability to defend themselves. Rapid industrialization of Western societies after 1700 gave them technological superiority over traditional Middle Eastern rivals. As a result, the Ottoman Empire gradually lost its outlying provinces and tributary states as the caliphate—the secular and religious leadership of the Islamic community—proved incapable of defending the House of Islam against the encroachments of Austria, France, and Russia. Christian populations reemerged to form non-Islamic states (Serbia, Bulgaria, Romania, and Greece in the Balkans); foreign powers asserted themselves as the protectors of Christian minorities within the boundaries of the shrinking Ottoman Empire (Lebanon and Armenia); missionaries, both Protestant and Catholic, attempted to win converts; and French and Italian colonists arrived to settle the best agricultural lands in Algeria and Libya. French penetration began in 1830 with military occupation of coastal Algerian cities and, despite often fierce resistance, gradually extended control into the interior. Italy won a brief war with Turkey in 1912, claimed Libya as a reward, and began sending settlers to occupy prime agricultural land. Both France and Italy considered these North African territories to be integral parts of the mother country, rather than colonies, and had limited respect for indigenous Islamic institutions and values.

Where military power led, economic penetration followed. By the early twentieth century, most of the Middle East and North Africa was divided into political and economic spheres of influence or control. After World War I, the formerly Ottoman eastern Mediterranean (the Levant) was divided between France and England into mandated territories. Many of the emirates in the Persian Gulf entered into protectorate relationships with Britain. Oman was heavily influenced by Britain but remained nominally independent. Southern Yemen was more directly tied to and administered from British India by lease and treaty in the nineteenth century before becoming a formal colony called Aden. Only Yemen, Saudi Arabia, and Iran escaped formal political control. Regardless of formal political status, the modern sectors of the economy, particularly the oil industry, were dominated by Western interests. Most Muslims who sought a modern education found it in schools permeated with Western values, emphasizing individualism; separation of church and state; equality of the sexes; material and technological progress; rationalism; and secular nationalism. Traditional Islamic knowledge and understanding of state and society were set aside, despite protest and resistance.

Traditional voices were never completely silenced. After World War II movements that advocated political independence emerged, and proponents of more traditional Islamic values gained increasing influence. Successful leaders of that time—such as Mohammed V in Morocco, Habib Bourguiba in Tunisia, or Gamal Abdel Nasser in Egypt—used traditional values in a nationalist context to defy the Western powers. Increasingly fundamentalist Islamic forces demanded a return of the Sharia, the traditional legal code based on the Quran, in place of legal systems derived from European models. Adherents of this position can often be identified by their use of traditional dress (Figure 7-12). In public, women cover their head with a scarf, their face with a veil, and their body with an enveloping outer garment that reaches to their feet. Men wear a skull cap and very full, untrimmed beard, although more modern-style clothing may be worn outside the home. Living by an austere personal and religious code, these committed Islamic fundamentalists (sometimes called Salafists because they draw inspiration from and see themselves as defending the values of their pious, early Islamic ancestors, or *salaf* in Arabic) advocate intensified resistance to Israel and increased support for the **Palestine Liberation Organization (PLO)** and **Hamas,** the two most widely recognized political organizations that represent most Arab **Palestinians.** The

▲ Figure 7-12 **Shoppers cross a street outside the Khan al-Khalili suq in Cairo, Egypt.** A variety of clothing styles, from traditional to modern, are displayed by Egyptian pedestrians on their way to or from the market.

more fundamentalist the Islamic group or nation-state, the more likely it is to oppose the prospect of peace and compromise with Israel. To Islamic purists, Israel is an alien entity in the Islamic body politic. Israeli control of Islamic holy places in Jerusalem is a constant source of humiliation and a reminder of past political, technological, and military inferiority.

The impact of the fundamentalist revival has differed from country to country. In Iran in the 1970s, the Shi'ite clergy, led by the Ayatollah Ruhollah Khomeini, spearheaded resistance to the secular policies of the Shah (ruler) and opposed the inequitable distribution of wealth generated by the oil economy. After driving the Shah from power, the Shi'a clerics established an Islamic state in which their values continue to dominate the political and cultural scene, although the presidential elections in 2013 were won by a more moderate candidate acceptable to the conservative-dominated Guardian Council (12 theologians and jurists who rule on the constitutionality of laws and acceptability of candidates). In Egypt, radicals on the fringes of the Muslim Brotherhood assassinated President Anwar Sadat in 1981, in part because he signed a peace treaty with Israel. After a revolution ended the rule of the National Democratic Party and its leader, Hosni Mubarak, the Muslim Brotherhood and allied Salafist parties dominated the post-revolutionary political scene, only to be overthrown in a coup in early July 2013. In Tunisia religious groups were the primary opposition to the country's first president, Habib Bourguiba, and his military successor, Zine El Abidine Ben Ali, and the Ennahda Movement, a moderate Islamist party, emerged as the largest political grouping in parliamentary elections after Tunisia's largely peaceful revolution in 2011. Thus far the movement has avoided pushing an Islamist agenda and establishing Sharia law. In Syria, resistance to the secular Ba'ath Party was maintained by orthodox Sunni movements. Brutally suppressed in a revolt in the city of Hama in 1982 in which upward of 10,000 people were killed, Islamist Sunni groups were not destroyed, reappeared actively beside secular demonstrators in March 2011, and became increasingly dominant as a vicious cycle of violence developed into civil war. And in Algeria the Islamic Salvation Front, the political arm of fundamentalist groups, won the first round of the 1991 elections, which were then cancelled by the National Liberation Front party that controlled the one-party socialist state. A violent revolt by Islamist groups, ultimately won

by government forces, continued for a decade with over 40,000 people dying before an uneasy peace emerged.

No country in the Middle East is without the tension generated by an increasingly intense struggle between secular and religious forces. The outcome of this struggle will determine the nature of both the region's internal politics and its international relationships.

Stop & Think

➡ Explain why the value system of Islam makes it difficult for secular states to assert themselves in the Middle East and North Africa.

Population Contours

The MENA region experienced rapid population growth after 1950 as medical and public health care improved. Infant mortality declined sharply, maternal mortality diminished, and overall life expectancy increased. From staggeringly high levels of infant deaths—20 percent of infants not surviving the first year—Morocco's infant mortality rate fell to 30/1,000 births and Saudi Arabia's to 17/1,000 by 2012. Even Yemen, the poorest country in the region, saw its infant mortality rate fall from nearly one in four to 48/1,000 by 2012. Thus the regional demographic transition began with a spectacular drop from high birth and death rates to one in which birth rates stayed high while death rates declined. Population totals soared as a result.

Where this population lives is equally dramatic. Wherever water is found, people are concentrated. This relationship is reflected in Figure 7-13. The densely populated Nile River oasis meanders northward from Sudan to the Mediterranean coast, and a similar configuration outlines the courses of the Tigris and Euphrates rivers across southeastern Turkey, Syria, and Iraq to the Persian Gulf. Equally dense populations inhabit the southern coastlines and coastal plains of the Mediterranean, Black, and Caspian seas as well as the foothills and valleys of coastal mountain ranges. Much more spatially limited concentrations occur in oases deep in the interior. Considerably sparser population densities appear on semiarid interior plateaus such as central Anatolia and the High Plateau in the interior Atlas Mountains of Algeria. Vast population voids typify expansive areas of sand and rock in the Sahara in North Africa, the Rub al-Khali (Empty Quarter) in Saudi Arabia, and the central desert zones of Iran.

More recently, there are clear indications that the overall population growth rate is in decline even as population size continues to grow. The fertility rate (the average number of children born to a woman during her reproductive years), which hovered between 7 and 8 for most countries in the region in 1950, sank to 2.8 in Saudi Arabia and 2.3 in Morocco by 2013. Even Yemen, which retained a consistent rate around 7.8 throughout the second half of the twentieth century, dipped to 5.2 by 2013. The fertility rate for MENA countries remains above replacement level (2.1 per female) except in Qatar and Tunisia, where it is at replacement, and Lebanon (1.9), Turkey (2.0), and the United Arab Emirates (1.8) where it is below replacement. The apparent trend toward smaller families eventually will result in a decrease in the number of people below 15 years old. But in 2012, only Qatar and the United Arab Emirates had less than 20 percent of their population under 15 years of age. In all other countries the percentage was above 20 percent and was highest in Iraq (43%), the Palestinian Territory (42%), and Yemen

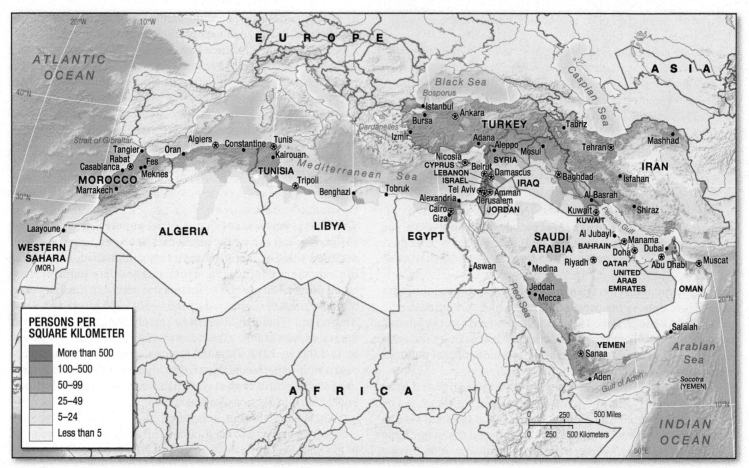

▲ **Figure 7-13 Population distribution in North Africa and the Middle East.** The densest populations are located in those places where adequate water for agriculture and other activities can be found.

(44%). A direct relationship exists between low per capita income and a large population under age 15. As yet the percentage of people over 65 is not high in most MENA countries, reaching 10 percent only in Israel. But the over-65 population can be expected to grow in all countries at the same time the percentage of those under 15 decreases, more people leave rural areas for cities, educational levels rise, and the use of modern methods of birth control increases.

The region's population boom is slowing, but the issue of supporting aging populations that characterize many industrialized countries is not acute. Middle Eastern and North African countries encounter a different set of difficulties associated with their current population structures. Teens and young adults, particularly those with college degrees, often have extreme difficulty finding jobs that reflect their skill sets and aspiration levels. Without regular employment, it is difficult to marry or start a family, even with help from extended family members. Marriage is long delayed or impossible for many individuals trapped in this situation.

Contemporary Livelihood Patterns

A region of great spatial extent and environmental diversity, the Middle East and North Africa support a wide variety of land-use activities (Figure 7-14), which sustain the region's ecological trilogy of farmer, herder, and city dweller. Farming and herding were traditional occupations of most people, but mineral wealth has transformed the region.

Agriculture: Traditional and Modern

Despite the region's aridity, much contemporary agriculture consists of **dry farming** in which farmers rely exclusively on rainfall to produce their crops. Concentrated in the region's better-watered semiarid environments, dry farming typically centers on the cultivation of cereal crops, primarily wheat and barley. Tree crops such as almonds and olives and grapevines are also important, particularly along the Mediterranean coast. Cereal cultivation is an extensive activity; use of tractors and other mechanized farm equipment has often allowed farmers to expand into rangeland areas that were once dominated by nomadic herders. The major zones of dry farming are the coastal plain south of Rabat in Morocco; the High Plateau in Algeria; the Anatolian Plateau in central Turkey (Figure 7-15); the interior steppes of Syria and other portions of the Fertile Crescent, an arc of higher-potential land extending from the eastern Mediterranean to the Persian Gulf; and semiarid zones in interior Iran between the mountain rimlands and the deserts. Cereal cultivation in these areas, however, is very vulnerable to drought.

Run-on farming in which rainwater from a larger area is collected in valley bottoms and on terraced slopes, permits agriculture in drier regions. These run-on systems are best known in the northern Negev in Israel, where their development is associated with the ancient Nabateans. But run-on farming is widely practiced throughout the Middle East and North Africa. Terraced agriculture in mountain districts, such as the Atlas Mountains or the highlands of Yemen or Mount Lebanon, is another important traditional form of dry farming.

Irrigated agriculture is concentrated along major rivers and in oases. Oasis irrigation is always of limited extent because it depends on raising limited groundwater resources to the surface for irrigation, which requires large amounts of human and animal power or fossil fuel energy to accomplish. The shift to gasoline-powered pumps lifts large volumes of water to the surface, permitting cultivation of more land but increasing the risk of drawing too much water from groundwater stores.

Irrigation based on river water is different. Large amounts of water are usually available, at least seasonally. But although temperatures permit year-round cropping in many places, until recently no way existed to store water on a significant scale. Consequently, most irrigation systems relied on retaining the annual flood in small-scale basins maintained with local labor. This was the system typical of the Nile Valley and Delta. Water-lifting devices, such as the *shaduf* (a counterweighted, lever-mounted bucket), the *noria* (waterwheel), and the *tambur* or Archimedes' screw (a wide, hand-turned, threaded screw enclosed in a cylinder that raises water a few feet from one level to another), allowed limited double-cropping in the dry season near the river. Canal construction that diverted Tigris and Euphrates river water permitted more extensive and integrated irrigation in ancient Mesopotamia (contemporary Iraq). These developments never allowed over-season water

storage and were always constrained by low levels of dry season stream flow and the need to practice alternate year fallow to avoid waterlogging and **salinization** (the accumulation of salts in the soil through evaporation).

Only the ***qanat*** system in Iran (*falaj* in the Arabian Peninsula; *foggara* in North Africa) permitted year-round irrigation (Figure 7-16). Qanats tap groundwater, but are fundamentally different from wells, which require lift devices or pumps. Because available groundwater and rich soils are often not located close together, the problem is how to extract and move the water to a site with good soil that farmers can cultivate. A qanat is an underground tunnel that moves water from source to field; where these are widely separated, qanat systems can be dozens of kilometers long. A great deal of labor and skill is invested to construct the tunnel, which starts by digging a vertical shaft until water is encountered. Usually this takes place near the base of an upland at the upper edge of an alluvial fan (deep sedimentary material deposited by surface runoff) where water collects between bedrock and looser, eroded material. From this point the tunnel is extended horizontally at low gradient so the water can move by gravity along the tunnel floor. Vertical shafts are dug at intervals to make removal of the tunnel's excavated material easier. This material is piled in a ring around the shaft opening to block surface water laden with debris from pouring down the shaft after a rainstorm and clogging the tunnel. When passing through loose material in order to maintain gradient, the

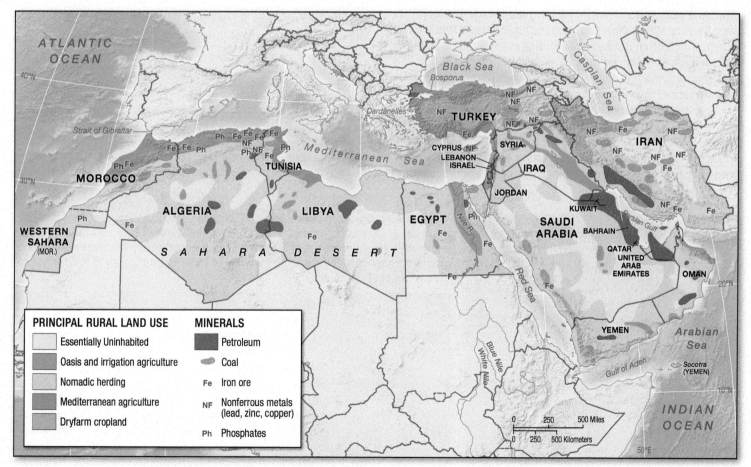

▲ **Figure 7-14 Land use in the Middle East and North Africa.** Drier parts of the Middle East and North Africa are used for herding, but oases based on groundwater and Mediterranean agriculture in higher rainfall areas play the most prominent roles. Oil is the most important mineral resource.

▲ Figure 7-15 **Agriculture in Central Anatolia.** Rolling wheat fields west of Konya are typical of the extensive rain-fed cereal cultivation undertaken in the drier central interior of Turkey.

tunnel is lined with large pottery rings. Gradually the tunnel's floor approaches the surface as each successive shaft is somewhat shallower than its predecessor. Eventually the tunnel emerges at the land surface and groundwater has been brought to ground level by gravity.

Qanats avoided the very expensive and energy-consuming animal and human labor needed to raise water from a well before steam or gasoline pumps were invented. Only occasional maintenance is needed to clear debris from the tunnel and keep the water flowing. When the water reaches the surface, it is moved by a ditch, which often is covered to reduce evaporation loss, to the field where the farmer can use it. Although the volume of flow in a qanat shows some seasonal variation, in most cases a remarkably constant rate of flow is produced. Until the mid-nineteenth century, year-round irrigated agriculture was unknown in most of the Middle East except in areas supplied by a qanat or a traditional lift device.

All that changed with the construction of man-made barriers (barrages), first on the Nile and later on the region's other perennial

streams. In that part of the year when a river's water level is low, it is difficult to remove the water from the stream without expending enormous amounts of energy in the form of animal, and human labor, or fossil fuel power. A **barrage** is a low structure placed across the stream channel that blocks the flow of water in this low-flow period. The goal is to raise the water level to a height that is much closer to the level reached when the stream is in flood. Once ponded behind the barrage, water is diverted into canals where gravity flow carries it to nearby fields. This innovation made it possible to irrigate much larger areas than could be cultivated by the use of lift devices. Gradual expansion of the area cultivated by barrage water enabled year-round cultivation of more and more of the Nile's floodplain and delta. In these areas of continuous cultivation, a complex succession of crops could be grown and the yields of food and fiber per unit of land could be dramatically increased. This development trend culminated in the construction of the High Dam at Aswan in southern Egypt. Since the dam's completion in 1970 most of the Nile's floodwaters, which previously drained into the eastern Mediterranean, are stored in Lake Nasser behind the dam and released as needed throughout the year. Two or three cultivated crops are now possible, and the primary problem has become how to drain the excess irrigation water to avoid soil waterlogging and salinization. The large-scale perennial irrigation system of the Nile is copied throughout the Middle East. Most notable are the large-scale dams on the tributaries of the Tigris constructed by the Iraqis; the major dam built by the Syrians at Tabqah on the Euphrates; the series of dams built or under construction on the Turkish portion of the Euphrates and Tigris rivers and their tributaries; and the Moroccan development of Oued Sebou north of Rabat.

Stop & Think

→ How does the qanat water management system work and what is its significance in drylands?

Animal Husbandry

Although many animals such as cattle, oxen, donkeys, and mules are kept by settled people in the Middle East and North Africa for

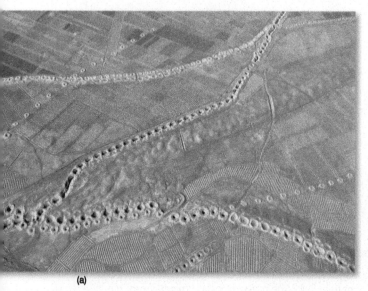

(a)

▼ Figure 7-16 **Qanats in central Iran.** (a) Thirty miles (50 kilometers) southeast of Yazd, a number of qanats and subsidiary branch collection galleries crisscross the alluvial fan and bring water to the farms of Mehriz. (b) The schematic diagram reveals the underground components of the system that enable water to be brought to the surface by gravity flow once the subterranean channel is excavated.

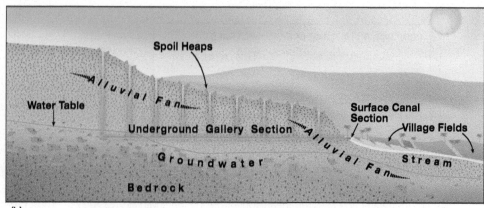

(b)

▲ Figure 7-17 **Nomadic pastoralism in northern Iraq.** Nomadic Kurdish herders traditionally move their animals from northeastern Syria to higher elevations in northeastern Iraq every summer. One such herder watches his sheep while tanker trucks wait to export oil.

food, plowing, and transportation, the bulk of the animal stock traditionally was held by pastoral nomads. **Nomadic herding** of sheep, goats, and camels covers large areas, a system encouraged by aridity and the large amount of land unsuited for agriculture that characterizes the region (see Figure 7-14). Nomadic pastoralists practice rotational grazing on a grand scale. Often moving hundreds of miles in an annual migration cycle, nomads bring their animals to grass and water that is only available on a seasonal basis in each district. Movement is essential to the nomad's existence (Figure 7-17).

Two major types of nomadic herders exist: vertical and horizontal. Vertical nomadism alternates between winter pastures in the lower elevations and summer pastures in the uplands. These herders follow regular routes through the mountain valleys and passes between their major grazing areas, and often plant cereal crops in their winter, lowland grazing areas. Wealthier nomads frequently buy agricultural land along the group's migration route, while less prosperous herders sell their labor to farmers during the harvest season. Many linkages connect farmers and herders. Nomadic herds contribute manure that helps fertilize the farmer's fields in autumn, while the stubble on those fields provides fodder for the herder's animals, primarily sheep that are highly valued in urban markets. These same herds are a major source of meat and animal products for the settled population, and much of the food and equipment used by herders come from farmers and urban merchants.

Horizontal nomadism uses variations in the availability of grass and water and generally operates in relatively flat areas. Such nomads move from wells that provide water for their herds in the dry season to distant regions where rainfall produces surface water and grass. This movement pattern rests the grazing areas around the dry season wells and ensures herd survival except when the most severe droughts occur. These herders tend to keep camels as their major animal, with goats and smaller numbers of sheep maintained for subsistence purposes. Camels, not particularly valued as a source of meat, once were important beasts of burden in the caravan trade. Now largely replaced by trucks, camels are still valued as racing animals.

Never the largest segment of the Middle Eastern and North African population, except in specific areas such as the Red Sea Hills of Egypt or the Empty Quarter of Saudi Arabia, nomadism

declined in importance throughout the region after 1950. Much land once reserved for grazing was converted to agricultural use. The borders of nation-states now cut across the routes of many nomadic groups, making annual migrations more difficult. The reluctance of many countries to see large groups of nomads moving from one place to another has put a lot of pressure on pastoralists to settle. Many changes have taken place in the nomadic way of life, and many herders have become sedentary (Figure 7-18).

Mineral Resources

While a large number of minerals are found in the Middle East and North Africa, they generally occur in small deposits (see Figure 7-14). These deposits were once important to local populations, but they are quite insignificant today on a global scale and are often uneconomic to exploit. Some that are still mined include chromite in Turkey and mercury in Algeria. Moroccan phosphate exports are significant and command a substantial share of the world market. Turkey is the only country in the region with sufficient domestic coal and iron ore reserves to support a broad industrial economy. In other countries where steel mills have been constructed, such as Egypt and Saudi Arabia, the raw materials are imported. This has not discouraged countries from looking for

▼ Figure 7-18 **Bedouin gathering in the Wadi Rum, Jordan.** Bedouin settled in towns and villages periodically return to their customary dryland grazing areas to affirm their traditions and identity. The formerly ubiquitous camel is now largely replaced by pickup trucks and jeeps.

▲ **Figure 7-19 In Salah gas field in the Algerian Sahara.** The plant in this picture removes carbon dioxide equal to the annual emissions of 200,000 cars from the gas produced and returns it to the ground for storage.

new deposits in the Sahara Desert and on the Arabian Peninsula. Saudi Arabia, Egypt, Libya, and Algeria have all found rich iron ore deposits in ancient pre-Cambrian shield formations, but their exploitation is hindered by the extremely arid climatic conditions where they are found and the great distances from the deposits to the coastal locations where they can be utilized. On the whole, the region is not well endowed with most minerals.

The big exception is the widespread occurrence of petroleum and natural gas. Many parts of the Middle East and North Africa have abundant oil and natural gas deposits, but the distribution of this wealth is very uneven (see Figure 7-14). The richest deposits are found in and around the Persian Gulf, and in the Saharan territories of Libya and Algeria (Figure 7-19). These areas export to the rest of the world, earning enormous revenues for the producing countries. A few other countries have found modest supplies of oil that are sufficient for domestic consumption but insufficient for large-scale export; Egypt, Syria, and Yemen are in this category. The remaining countries, including Jordan, Israel, Lebanon, Morocco, Tunisia, and Turkey, are basically "have not" countries, outsiders to the region's major foreign exchange-earning resource. The significance of petroleum is discussed in more detail later in the chapter. So far most countries have shown limited interest in developing what is potentially their most widespread energy resource—solar power (see *Focus on Energy:* Fossil Fuel Consumption and the Search for Renewable Energy).

Urbanization

Nearly 70 percent of the population in the Middle East and North Africa now lives in urban areas, although a number of countries still have very large rural populations: Yemen (71%), Egypt (57%), Syria (46%), and Morocco (42%). The MENA region has a higher level of urbanization than many other regions, including China, South Asia, and Sub-Saharan Africa. Urban areas grew rapidly in the second half of the twentieth century, and some of the region's cities—most notably Cairo, Egypt; Istanbul, Turkey; and Tehran, Iran—now rank among the world's 50 largest urban areas.

This rapid urban growth is the result of high rates of natural increase and rural-to-urban migration. The latter has had several consequences. One is the development of squatter settlements on the outskirts of many urban areas. These technically illegal neighborhoods are usually situated on underused private or state land without the approval of the landowners. Few if any municipal services, such as electricity, water, sewerage, health-care facilities, or schools, are provided, at least initially. Occasionally governments try to remove such settlements, but the general pattern is for these communities to become stable and permanent. From an initial stage of transitional construction with flimsy materials, most houses are transformed into stone or concrete structures. The more durable the structure, the more likely that urban authorities will recognize the reality of occupancy and grant legal tenure to the inhabitants. In the oil states, national governments are able to afford the costs of providing subsidized housing for all who seek it. The non-oil states have fewer resources and often large populations to care for, and are unable to provide sufficient housing, infrastructure, and other services. Many of the rural migrants possess considerable resources and, if given tenure to the land that they occupy, are able to erect their own housing. When recognized by the state, squatter community occupants are often able to build decent neighborhoods.

Many Middle Eastern cities have expanded considerably around their ancient *madina*, or urban core (Figure 7-20). Often this growth has extended into nearby agricultural areas, resulting in the loss of valuable farmland. Other land is lost when it is mined for sand, gravel, and other materials for building construction. As a consequence, urban growth has absorbed some of the best farmland in a region where prime farmland is at a premium.

Since World War II, urbanization has also brought large numbers of rural migrants into the madina as many of the madina's wealthier professionals, merchants, and entrepreneurs have left the old central district for the nearby quarters once occupied by foreign colonial officials and settlers. Organized on a more European pattern, streets in these sections are wider, radiate from generous squares, and are lined with multistory buildings with street-level shops, restaurants, and cafes topped by several floors of apartments, often with balconies (Figure 7-21). In countries that did not experience colonialism directly, such as Saudi Arabia, many people have also left the central city for more spacious urban and suburban districts with larger and more modern housing. The vacated inner-city properties often are filled by large numbers of rural migrants, turning ancient palaces and other structures into multifamily dwellings. Higher population densities have produced extremely crowded living conditions in buildings that frequently are deteriorating (Figure 7-22). Lax inspection practices and poor maintenance by absentee landlords have occasionally resulted in spectacular structural failures. In the older urban areas, streets not designed for large numbers of automobiles and trucks have become increasingly congested, producing both poor safety conditions for drivers and pedestrians and serious air pollution problems. In the much older madina streets, pedestrian traffic (with the occasional pack animal) is the norm, and in some districts the conversion of former palaces or Quranic schools into hotels or condominium-style residences have sparked an interest on the part of young professionals in a return to areas abandoned by their grandparents.

Cairo exemplifies the problems and prospects of cities in the Middle East and North Africa. With an estimated 17.8 million inhabitants in 2012, more than 20 percent of Egypt's population,

Fossil Fuel Consumption and the Search for Renewable Energy

Fossil fuels provide 95 percent of the energy consumed in the Middle East and North Africa (Figure 7-2-1). Abundant oil and natural gas endowments in Algeria, Libya, and the Gulf States make it likely that fossil fuels will continue to dominate the future consumption mix. Coal is an important energy source only in Turkey, based on domestic sources, and Israel, which imports its coal from Australia and South Africa. In recent years natural gas has replaced oil as the region's largest energy source. In economic terms this is a sensible development. Global demand for oil is high and many countries with growing economies, such as China, India, and Brazil, need more oil-based energy to supplement local energy sources. Oil exporters receive a much higher price by selling to international customers than by burning the oil at heavily subsidized prices at home. Saudi Arabia is willing to subsidize unrealistically low domestic oil prices, and forego increasing exports to capture increased profits. Other oil exporters have shifted to natural gas to generate electricity and desalinate seawater while increasing oil export volume.

Only 2.4 percent of the region's energy comes from hydroelectric power, reflecting the region's severe water scarcity. Turkey derives 10 percent of its energy from hydropower, a product of its higher rainfall. Hydropower's share of the energy mix is unlikely to grow significantly because most of the undeveloped potential would come from dams yet to be built in the country's drier southwest. Egypt produces hydropower from one source—the Aswan High Dam—which accounts for just 4 percent of energy consumed. Although many countries are considering electricity generation using nuclear reactors, only Iran has proceeded beyond the thinking stage. Fear that Iran might pursue a nuclear weapons objective disguised as power generation has poisoned relations with its neighbors and the international community. Even if nuclear proliferation issues are settled, the issue of hazardous waste disposal makes nuclear power a problematic contributor to power needs. So far interest in solar- and wind-powered energy systems is limited. But this is changing. The aridity and sunny skies that make agriculture difficult are an asset for solar-generated electricity. The potential is enormous for both large-scale fields of solar collectors and small-scale, disaggregated systems that generate and store solar energy for scattered populations. Light, heat, air conditioning, and telephones could become widely available without connecting to a national electricity grid. Israel is the region's most advanced country in putting photovoltaic (PV) technology to practical use; 85 percent of Israeli homes power their water heaters with PV technology. Where larger, urban populations are concentrated, more elaborate and costly technology provides solar energy on

▲ **FIGURE 7-2-2 Morocco constructs a concentrating solar power (CSP) system.** Morocco's King Mohammed VI examines a model of the country's first CSP during groundbreaking ceremonies in May 2013. By 2015 Morocco hopes to complete the planned 160 megawatt solar/thermal generating plant and greatly improve power supply and reliability in the Ouarzazate area south of the High Atlas Mountains.

a grander scale. Morocco plans to expand its existing PV-based rural electrification program to include concentrating solar power (CSP) systems that can supply large urban populations; between a half-dozen CSP projects and coastal wind farms, Morocco envisages more than 40 percent of its national energy mix being supplied by renewables in 2020 (Figure 7-2-2). Deserts are also windy places because dramatic temperature differences between day and night, uplands and lowlands, sea and land generate air movement that can turn propellers, drive turbines, and produce electricity. Investing the profits of oil sales in renewable energy systems, sponsoring research on innovative new technologies, and educating citizens to maintain and improve existing equipment would help move the region's energy consumption toward a diversified and sustainable footing.

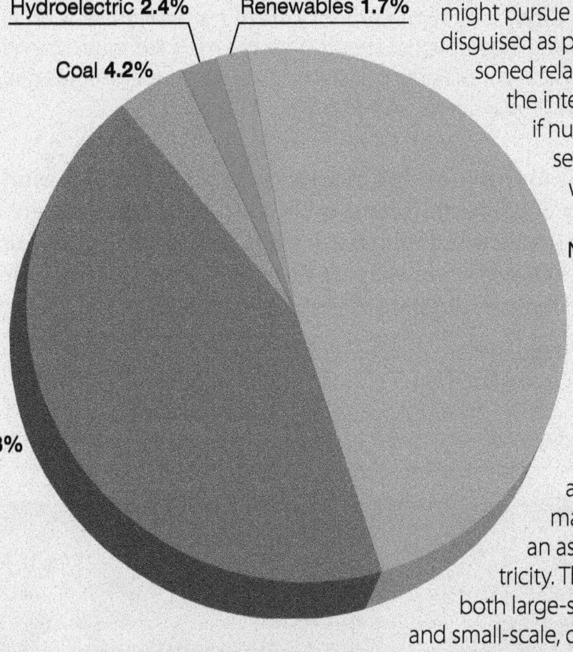

Sources: International Energy Agency (IEA), *World Energy Outlook 2012* (Paris: OECD/IEA, 2012); Michael Deaves, "5 Middle East Solar Markets to Watch," http://www.renewableenergyworld.com/rea/news /article/2013/03/5-middle-east-solar-markets-to watch; "Keeping It to Themselves: Gulf States Not Only Pump Oil; They Burn It, Too," *The Economist,* March 31, 2012, 81–82, http://www.economist.com /node/21551484; "Morocco Launches World's Largest Solar Power Project," *Morocco on the Move,* May 11, 2013, http://moroccoonthemove.wordpress. com/2013/05/11/morocco-launches-worlds-largest-solar-power-project-middle-east-online/

Energy consumption by fuel source (Figure 7-2-1):
- Hydroelectric 2.4%
- Renewables 1.7%
- Coal 4.2%
- Natural Gas 47.9%
- Oil 43.8%

▲ **FIGURE 7-2-1 Energy consumption by fuel source.** Despite abundant sunshine, the Middle East and North Africa overwhelmingly relies on fossil fuels to meet its energy needs. Investment in solar or wind technology is a development strategy more characteristic of those countries lacking access to fossil fuel resources.

Sources: BP Statistical Review of World Energy June 2012 (http://www.bp,com/statistical); IEA (International Energy Agency). World Energy Outlook 2012. Paris: OECD/IEA, 2012.

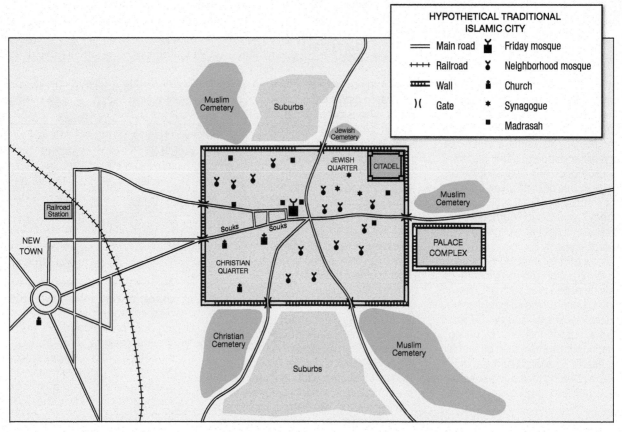

HYPOTHETICAL TRADITIONAL ISLAMIC CITY

===	Main road		Friday mosque
+++	Railroad		Neighborhood mosque
:::	Wall		Church
)(	Gate		Synagogue
			Madrasah

▲ **Figure 7-20 Diagram of a hypothetical traditional Islamic City.** The major components of the traditional Islamic city include the main residential quarters; market (*suq*) located along a complex of streets near the main mosque; Islamic schools (*madrasah*); a citadel; suburbs and cemeteries outside the main gates; and a palace complex where the ruler resides. Attached to the madina, or urban core, is a modern town organized on the European pattern.

Cairo is the largest metropolitan area in the region and the eleventh largest in the world. The city is a prime example of the difficulty of managing a sustainable metropolis in a dryland country with limited resources. Egyptians affectionately call their capital *umm ed-dunya,* "Mother of the World," much as Bostonians refer to their home town as the "Hub of the Universe." But reality is often not so pleasant.

Urban Cairo is very densely settled, and metropolitan expansion is difficult because the agricultural delta region to the north is needed for food production and the expanses of the desert to its east and west are unattractive and expensive to develop. Traffic congestion from the daily peaks of urban commuting are only partly alleviated by the construction of an underground light rail system, and gridlock is a common experience. Air quality is poor, waste disposal is a constant problem, and long-term health issues are endemic. Cairo is among the world's most polluted cities, although there is progress in relocating noxious industrial facilities away from the city and applying tighter emissions regulations.

Affordable housing is one of Cairo's major problems. The gap is widening between those who can afford high-quality residences and those who cannot find adequate housing. Greater Cairo is dotted with informal settlements consisting of illegally built houses, most lacking running water, proper sewage disposal, and electricity. Many of the very poor live illegally in tombs in Cairo's vast City of the Dead. At the same time, many better-quality apartments in the suburbs and satellite towns around Cairo remain empty as entrepreneurial

landlords wait for more affluent renters. A reasonable quality of life for the majority of Cairo's residents, who are steadily increasing in numbers, is hard to attain. Infrastructure and social services are not keeping pace. But given the slightest choice, no Cairene would want to live anywhere else!

▼ **Figure 7-21 Casablanca, Morocco.** Like many cities in the Middle East and North Africa, Casablanca's Mohammed V Boulevard is lined by European-style, art-deco apartment buildings dating from 1920 to 1940. Now a pedestrian street, shoppers can access ground level stores and restaurants without coping with traffic or air pollution.

▲ Figure 7-22 **Dolapdere Quarter of Istanbul.** Rural-to-urban migration to the cities has resulted in extremely high population densities and has produced crowded living conditions.

Stop & Think

Consider the problems of intense urbanization in a dryland country like Egypt.

The Mediterranean Crescent: Maximizing Limited Resources

The Middle East and North Africa together constitute a distinctive region within which we recognize two subregions: the countries of the Mediterranean Crescent along the Mediterranean coast and the Gulf States around the Persian Gulf (see chapter opener map). North Africa and the countries along the eastern end of the Mediterranean Sea share a long history of cultural and economic interaction with Europe, whereas the Arabian Peninsula and the countries along the shores of the Persian Gulf became more prominent in the world with the discovery of oil.

The Mediterranean Crescent countries encompass the North African states of the **Maghreb** (Morocco, Algeria, Tunisia), as well as Libya and Egypt. The subregion also includes the smaller states of the eastern Mediterranean, the **Levant,** as well as Turkey, a unique Middle Eastern country in that it possesses a physical foothold in Europe, has associated membership status in the European Union, and seeks full membership. All of the countries in this broad spatial and political grouping share strong historic ties to Europe. They were once part of the Roman Empire and part of a regionwide culture and political economy. During the Ottoman Empire, most of North Africa was grouped with the Balkan lands of southeastern Europe and the eastern Mediterranean into one political unit. These historical ties are symbolized in the contemporary landscape not only by extensive Roman monuments and ruins but also by similarities in house types, traditional clothing styles, irrigation systems and other agricultural production technologies,

legal arrangements, and patterns of trading and raiding over centuries. Today many of the citizens of these countries, most notably the Maghreb states and Turkey, send migrants to Europe to find employment. They constitute an important source of labor in France, Germany, and other countries. The remittances that these migrants send to families in their homeland are a significant source of income for the recipient countries. Reverse patterns of movement are also prominent. Many European tourists visit Morocco, Tunisia, Egypt, and Turkey, attracted by the warmth and sun of the Mediterranean winter and contributing to local economic prosperity. The Mediterranean Crescent is also linked to southern Europe by a shared climatic and agricultural regime. Dry in the summer and autumn, precipitation is concentrated in winter and spring. Throughout the subregion, mountains close to the coast trap this moisture and make it available for the traditional crops of wheat, olives, and grapes (Figure 7-23). A rich variety of contrasts and similarities link the Mediterranean Crescent countries one to another.

The Maghreb: The Islamic World's Far West

Far to the west of the historic core of the Islamic world in Mecca and Medina is the Maghreb, the "place of sunset," the Arab West, long the most western of Islamic lands where the Atlantic meets northern African shores (Figure 7-24). Three contemporary countries, Tunisia, Algeria, and Morocco, are the major components of this region. Libya, which we include as part of the Maghreb, is something of an exception. Tripolitania, its western section, is most akin to the Maghreb countries, while Cyrenaica, its eastern section, is culturally connected more closely to Egypt.

Throughout the subregion, population distribution mirrors available water. Because rainfall is concentrated in the northern and coastal portions of all Maghreb countries, the bulk of each country's population is located in a limited part of national space (see Figure 7-13). Algeria is typical of this pattern, with 91 percent of its population of 37 million living on only 12 percent of the national territory. Most Moroccans are found on the western side of the crest line of the Middle and High Atlas ranges. Three-quarters of Tunisians live in the northern third of the country and Mediterranean coastal districts. The vast

▼ Figure 7-23 **Mediterranean agriculture in Tunisia.** Olive groves and wheat fields dominate the rolling landscape of north central Tunisia, the bread basket of ancient Carthage and Rome more than 2,000 years ago.

▲ **Figure 7-24 Hassan II mosque in Casablanca, Morocco.** Completed in 1993, the mosque is Morocco's largest and can accommodate 25,000 worshipers in its prayer hall. A location at the edge of the Atlantic shore guarantees a peaceful, pollution-free atmosphere as well as an aesthetically pleasing artistic and spiritual environment based on traditional designs.

majority of Libyans occupy two small pockets of land: the coastal plains and northern slopes of the Jabal Nafusah in Tripolitania and the coastal and mountain districts between Benghazi and Tobruk in Cyrenaica. Agriculture is only possible where water is reliably found, and only the northern parts of all Maghreb countries can produce substantial yields.

Even in these northern agricultural areas, farmers face significant constraints. Much labor in mountainous areas historically has been invested in terraces to keep soil and water in place for plant growth. The most productive areas are river floodplains and deltas, intermontane valleys, and drier plains where more extensive cereal cultivation is possible, such as the coastal plain south of Casablanca, the interior High Plateau of eastern Algeria, and in central Tunisia around Kairouan. Europeans considered these areas of high potential but limited productivity under traditional local agricultural and pastoral techniques. With French and Italian (Libya) colonization in the nineteenth and twentieth centuries, farmers of European origin were drawn to these areas and sought to make the land more productive and to grow products not found in the home territory of France or Italy. But all of these areas were periodically affected by drought, soils were less productive than imagined, and were susceptible to water and wind erosion. Managing them successfully and productively proved a difficult task. Most successful were the vineyards of Algeria, which produced large quantities of table wine for export, and the citrus producers of Morocco, whose oranges found a ready market in France. Cork for bottle stoppers and other uses was a valued output of the Mediterranean oaks that were a major component of low elevation forests in Morocco, Algeria, and Tunisia. Modest agricultural activity occurs in the Saharan portions of the Maghreb, largely date production in oases based on groundwater or near the course of wadis (periodically flowing streams) that drain southward from the High and Saharan Atlas ranges.

Where agriculture is productive, cities have been prominent in the Maghreb for over 2,000 years. Many traditional urban centers were distant from the coast, the hubs of local agricultural zones and political power: Fez, Meknes, and Marrakesh in Morocco; Constantine in Algeria; and Kairouan in Tunisia. Others were coastal, supported by maritime trade: Tunis, heir to ancient Carthage, as well as Algiers, Tangier, and Rabat. Together with cities that grew dramatically in the colonial period, all urban centers have become steadily more prominent since independence in the mid-twentieth century as people have flooded into urban areas from the countryside. Casablanca, the Maghreb's largest city, has close to 4 million inhabitants in its metropolitan area; Algiers is a close second. Metropolitan Tunis and Tripoli both have more than 2 million residents. All of these cities are by far the largest urban agglomerations in their respective countries, dominating each state's industrial, economic, and political life, and are the major gateway for people and goods into and out of their national territories. The influence and importance of these primate urban centers will only grow as they continue to serve as magnets for internal migration.

The contemporary imbalance between "haves" and "have nots" is the most important aspect of the development scene in the Maghreb and the Mediterranean Crescent as a whole (Figure 7-25). Algeria and Libya are "have" nations because their territories contain significant quantities of oil and natural gas. It is the interior desert portion of these countries that is the source of this bounty, a geologic accident that changes a barren and unproductive negative into a potentially dynamic development engine. Algerian and Libyan oil is shipped to the coast where the crude oil is refined and exported. About 40 percent of Algeria's production is sent to the United States, and nearly as much goes to Italy and Spain, whereas almost all Libyan oil is consumed in Europe. Natural gas in Algeria is largely located in association with oil, follows the same general route to the coast, and is liquefied for easier shipment. Libya's natural gas fields are located offshore north of Tripolitania. Both countries primarily export liquefied natural gas (LNG) via underwater pipelines to southern Europe. In Algeria, 98 percent of the country's export income comes from oil and LNG, and Libya's dependence on hydrocarbons is even greater.

The oil and gas income that dominates the budget in Algeria and Libya funds development and social spending. Investment strategies have not always been wise, although efforts to capture downstream oil and gas income by building refinery and derivative production industries have been largely successful. Investments in agriculture and domestic industrial growth have not been particularly successful, and both countries have had difficulties with waste and corruption in resource allocation. Libya has had a particularly serious problem generating its own food supply, reflected by the fact that only one-sixth of its wheat consumption is produced domestically. Although Libya's oil reserves are abundant (among the top 10 globally), Algeria's reserves are projected to run out within the first half of this century at current levels of production. A rise in Algeria's electricity consumption also is capturing an increasing share of LNG for domestic use, imperiling export earnings and threatening the domestic budget. Libya basically stopped exporting oil and gas during the 2011 revolution. But since Libya's oil industry was the only segment of its economy allowed to operate in a modern and professional manner during the 42-year regime of Muammar al-Gadhafi, post-revolution restoration of exports was relatively quick after the chaos and damage of the conflict.

Morocco and Tunisia are the subregion's "have not" countries. Efforts to locate oil and gas resources have not paid off. But phosphates, an essential component for inorganic fertilizer, are an abundant resource. Morocco is the world's leading exporter of phosphates, and Tunisia is the fifth largest. Neither country, however, derives the income from phosphates that Algeria and Libya

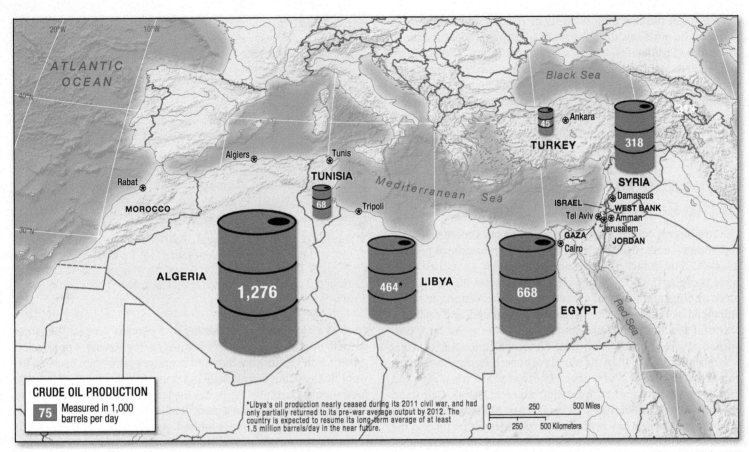

▲ **Figure 7-25** **Oil production in the Mediterranean Crescent countries.** Algeria was the Maghreb's largest oil producer in 2012, but Libya should soon resume its position as the region's largest oil producer. Egypt's production is barely sufficient to meet domestic needs, while Syria was only a modest exporter before its civil war impacted export levels.
Source: Oil & Gas Journal, v. 111. no. 3a (March 11, 2013), p. 31.

realize from oil and gas. In addition to its long-standing contention that the former Spanish Sahara was once a part of Morocco, and should be returned, the presence of large phosphate deposits was a prime rationale for Morocco's decision to occupy the region in 1975. It is no accident that both Morocco (42%) and Tunisia (34%) have considerably larger rural populations than Algeria and Libya. Lacking the oil and gas income of their neighbors, it is impossible to provide comparable levels of social welfare support and employment to lower-income citizens in urban areas. Moroccan GNI PPP per capita is just over half of Algeria's and a quarter of Libya's; Tunisian GNI PPP is less dramatically lower, but only 60 percent of Libyan GNI PPP. Both Morocco and Tunisia have large underemployed and unemployed populations, with Morocco claiming an 8.7 percent unemployed rate at the end of 2012 and Tunisia a 16.7 percent rate. In rural areas in general and among urban youth (14–24 years) in particular, unemployment rates probably exceed 30 percent. The difficulties associated with finding work are significant, endemic, and verge on the intractable. Morocco's efforts to improve infrastructure through high-speed train and highway links between major cities, as well as efforts toward agricultural intensification through land improvement grants and loans, contribute to better economic growth conditions (Figure 7-26). But access to European markets in competition with Spain and Italy, who produce much the same array of agricultural products, is difficult for all Maghreb countries.

Beneath the surface, discontent, disillusionment, and despair simmer, flaring unexpectedly whenever a seemingly minor event galvanizes a widespread response. The December 2010 suicide of Mohammad Bouazizi, a Tunisian street vendor subjected to harassment and seizure of his wares by local authorities, launched protests against unemployment, rising food costs, government corruption, and lack of basic freedoms that quickly drove out Zine El Abidine Ben Ali, an entrenched dictator for 24 years. The Arab Spring movement spread throughout the Maghreb, with different results in each country. In Algeria, the military-supported government faced down demonstrators quite easily, perhaps because a brutal civil war (1991–2000) in which government forces defeated

▼ **Figure 7-26** **Intensive greenhouse cultivation on the Atlantic coast of Morocco.** Farmers near Kenitra on the coastal plain north of Rabat use plastic greenhouses to cultivate vegetables for nearby urban markets.

Islamists left too few people ready to risk a similar trauma. In Morocco the *makhzen,* the governmental monarchy of Mohammed VI, co-opted protesters, issued a new constitution, and accepted and learned to live with the moderate Islamist forces who won election in November 2011. In Libya, protests in Benghazi in support of victims of the Gadhafi regime grew into a national rebellion that took nine months to dislodge the dictator. In no part of the region is the final verdict on the shape of governance, the balance of secular and religious forces, and the proper distribution of resources settled.

Egypt and Turkey: Anchor States of the Eastern Mediterranean

The two largest countries in the Mediterranean Crescent—Egypt and Turkey—have many basic similarities. Both have large populations—82.5 million and 74.9 million, respectively—and their cities serve complex functions. They are administrative, industrial, cultural, trading, and often religious centers. Their economic role is enhanced by their countries' large, well-integrated, and relatively powerful economies and a human resource base with a large proportion of well-educated individuals. A second commonality is that a significant proportion of the populace of each country lives in urban areas, the product of a deep history of urban life. Egypt and Turkey contain several of the region's largest cities—Cairo and Istanbul dominate their national landscapes. Massive immigration from the countryside has swelled urban populations far beyond the ability of urban planners to calculate accurately and of urban systems to absorb satisfactorily. Uncounted underprivileged people have ballooned the population of Cairo's metropolitan area to an estimated 17.8 million inhabitants and Istanbul's to nearly 13.8 million. Other urban centers have attained major size and importance: Ankara, Izmir, and Adana in Turkey; and Alexandria and Giza in Egypt.

A third similarity is a rich agricultural base in both countries, which is essential to feed their large populations and to earn foreign exchange in overseas markets. Egypt's cotton has been and Turkey's citrus fruits remain important export commodities; domestic food-processing industries are a significant development focus. In both countries (but particularly in Egypt, where population pressure is especially high), maintaining and increasing agricultural productivity is a continuous struggle.

Appreciable mineral resources is a significant common feature. Turkey's coal, iron ore, and chrome provide the base for the largest heavy-industry complex in the Middle East. Egypt's oil and phosphates meet internal needs and, together with hydroelectric power from the Aswan Dam, are important bases for economic growth.

Finally, each country is characterized by large semiarid and arid territories of limited agricultural value. Thus, agriculture is concentrated in limited areas and is more intensive than it is in many other countries of Africa and Asia. Egypt—with only the narrow, fertile strip of the Nile Valley and its delta—is the extreme example.

Egypt: A River Oasis Surrounded by Desert

Egypt is a country of contrasts. Cairo and Alexandria (Figure 7-27) are two of the largest cities in Africa, but many peasants in Egypt live under many of the same conditions their ancestors experienced. The Egyptian economy is diversified, with a wide range of basic and processing industries and consumer-oriented enterprises, but the

▲ **Figure 7-27 Alexandria, Egypt.** Founded by Alexander the Great in 321 BC, and named after him, Alexandria was for a millennium Egypt's premier city. Today it is the second largest city in Egypt and Africa behind Cairo, and is a sophisticated intellectual and economic contributor to Egyptian life.

country's population is increasing by 2 percent per year. Constant gains in productivity are necessary just to stay abreast of population growth, and significant effort and resources have been invested toward this goal; but increasingly Egypt must import much of its basic food supply. Currently, Egypt has the largest population of any Arab state, yet slightly less than 3 percent of its land surface is arable.

Egypt's resource base is limited. Agriculture is important, but only a narrow strip of land along the Mediterranean coast and a few pockets of land in the Sinai receive enough rainfall for crop growth. All other cultivation requires irrigation and is concentrated in the valley and delta of the Nile River, where both adequate water and good soils coincide. Outside the Nile Valley, significant cultivation is possible only in a few oases in the Western Desert and the Faiyum basin south of Giza. Alfalfa, cotton, rice, maize, and wheat traditionally are the main crops in Egypt. Long-staple cotton, once the primary export crop and foreign exchange earner, now has significance only for the domestic textile industry. Land use is already intense in Egypt and, without significant changes and investments, major breakthroughs to higher levels of productivity are unlikely.

Egypt has only a few commercially viable mineral resources, but has sufficient oil reserves to meet domestic demand. But domestic oil consumption has grown by 30 percent since 2000, shows no signs of diminishing, and increases pressure on local production, which has begun to stagnate. New discoveries of significant natural gas deposits have changed Egypt's energy scene, and gas use now equals oil consumption. Because most oil and gas is used by the transportation sector, expanding gas consumption at the expense of oil can help reduce the air pollution burden borne by Egypt's urban residents. Egypt plans to use wind and solar sources to better meet future energy needs; this would make good use of unproductive desert zones. But few practical developments have materialized.

Construction of the **Aswan High Dam,** completed in 1970 (although it took six more years for the reservoir to fill), was a dramatic effort to promote agricultural and energy development. Since the dam was completed, no floods with resulting property damage have occurred in Egypt. Water has been available to irrigate the Nile Valley and Delta all year so cropland can produce two or more crops annually, diverted into the Faiyum oasis basin, and channeled into the Western Desert to develop areas in and around the traditional oases of Dakhla and Kharga, the so-called New Valley Project.

Electricity produced at the Aswan High Dam has promoted industrial growth; rural electrification in Upper Egypt has increased significantly as well. A fertilizer industry has emerged to restore fertility to the soils now deprived of nutrients from Nile silt that used to be deposited with the annual floods. The high dam has become an important symbol of Egypt's determination to develop and has had a positive psychological effect that may be as significant as its economic impact. But not all aspects of this gigantic scale development project have been positive: soil salinization and waterlogging have increased; more standing water year-round creates public health problems; aquatic weeds and algae have become much harder to control in the low-sediment, fertilizer-rich canal and river water below the dam; and sediments deposited in Lake Nasser behind the dam will in the long term reduce the lake's water storage capacity. But over time most problems have proved to be more manageable than feared.

In 1970, Egypt was basically self-sufficient in the fundamental components of the Egyptian diet: wheat (bread), corn (maize), sugar, vegetables, and edible oils and milk, while rice was a prestige cereal and important export. Today Egypt imports 40 percent of its food, and 60 percent of its wheat. Egypt imports more wheat than any other country in the world—10 million tons each year! About half comes from Russia, and nearly as much from the United States. Rice production has declined as government policy shrinks acreage allocated to the crop because it is a large water consumer. This water is diverted to other crops, particularly fruit and vegetables, which have replaced rice as the largest food export (Figure 7-28). The water "saved" by importing wheat rather than growing it in Egypt—**virtual water,** in effect using some other place's water to grow the crop—can be allocated to other crops and domestic and industrial uses. The virtual water that comes with the imported wheat constitutes a form of subsidy to Egypt's people.

Population growth in part explains the plight of Egyptian agriculture, since there are many more people to feed in 2013 than there were in 1970: 82.5 million versus 35 million. Continued growth at the present rate would produce a projected population of 102 million in 2025 and 135 million in 2050, but the growth rate is declining as later marriage, greater access to and use of modern birth control,

and smaller families become more common. But in the short term, many more people must be supported on the same amount of arable land with a fixed amount of water. Strenuous efforts to expand the amount of cultivated land by bringing water to arid areas along the Mediterranean coast, near the Suez Canal, and in the Western Desert's New Valley are essential to future well-being. But in many cases these efforts have yet to move beyond the planning stage or have not had the anticipated results. For every new hectare of land brought into production, other forces conspire to remove a hectare from production: waterlogging, salinization, and construction of urban housing on agricultural land as cities expand remain problems that frustrate agricultural development efforts in this arid environment.

Still worse is Egypt's overall water availability. In practice Egypt has maxed out its water credit card. The Nile River supplies almost all of Egypt's water. The allocation of water between countries in the basin is governed by a treaty negotiated in 1929 when Britain controlled or influenced the region. Nile water originates in the highlands of Ethiopia and East Africa, before crossing the Sudan and entering Lake Nasser for distribution to Egypt's fields and towns. Headwater countries increasingly see every drop of water that flows out of national territory as a waste of "their" water that their own country could productively use. Ethiopia would like to develop the hydroelectric and irrigation potential of water resources in the Lake Tana basin and other tributaries to the Blue Nile, and is beginning to move from plans to implementation as financing is arranged. The other upstream countries also have development plans that would involve actually using all of their treaty-allotted share rather than watching the water flow northward. Since Nile water is an absolutely vital resource for Egypt, conflict rather than cooperation is a future possibility.

Cotton, Egypt's "white gold," was the country's major export item and foreign exchange earner since 1860. Its abrupt decline in the twenty-first century is a reflection of the general problems afflicting Egyptian agriculture and industry. The cotton grown in Egypt is particularly valuable because its long staple (fiber) can be woven into an especially high-quality cloth. Between 1970 and 2012 the area planted in cotton declined by 85 percent, reflecting a declining market for top-of-the-line fiber. Less cotton exports mean smaller foreign exchange earnings, and a smaller luxury cloth market produces fewer foreign high-end clothing sales for Egypt's textile industry. Greater market demands for cheaper cloth do not result in more medium or short staple cotton acreage planted in Egypt because production costs are too great to be profitable. Instead shorter staple cotton is imported from Greece and Sudan to meet the textile industry's need to produce cheap denim for jeans. Plans to increase government subsidies to long staple growers encounter many obstacles: textile worker demands for higher pay compromise industry competitiveness; other countries such as Mali subsidize their cotton farmers and flood the international market with cheaper high-quality fiber; demand is for cheaper, not more expensive cloth; and Egypt has a generic debt and deficit crisis in its budget finances. International lenders are unlikely to approve rescue loans if the country's subsidy policy continues.

When the Arab Spring reached Egypt in January 2011, 18 days of massive street demonstrations, civil disobedience, and violence drove the regime of Hosni Mubarak from power. Adoption of a new constitution by popular referendum, election of a president, Muhammad Morsi, and a parliament in which the Muslim Brotherhood and

▼ **Figure 7-28 Intensive fruit and vegetable cultivation in Egypt.** The line between cultivated fields and the desert on the western bank of the Nile is sharp in an agricultural area slowly being invaded by urban expansion.

more extreme Salafist deputies constituted a majority, moved Egypt closer to apparent democracy. But the unwillingness of the Islamist groups to compromise with their more libertarian and moderate opponents failed to resolve many fundamental constitutional and economic issues. Political and economic life continued to be extremely unsettled (see *Geography in Action*: The Arab Spring's Troubled Summer). Industrial growth remained stagnant. Security was poor, which discouraged tourists from visiting and alienated minority Christians and Shi'ites who were often the targets of attacks. Since tourism normally provided 20 percent of Egypt's foreign exchange earnings, continued political unrest hurt many sectors of the economy, particularly the 10 percent of the workforce that depended directly on visitors. Estimates indicate that 15 million Egyptians lacked enough money to buy adequate food. A long tradition of subsidies kept basic food prices low. Removing these subsidies proved difficult for the Islamist regime of President Morsi, given its strong, religiously based social conscience. But Egypt's very heavy debt burden and budget deficit, in part generated by past subsidy policies, made it difficult for the Morsi government to raise foreign capital and jump-start the economy. Debt restructuring and loan guarantees were linked by international lenders to changes in government safety-net policies that subsidized low food prices. The Morsi government's inability at the same time to continue basic food subsidies for the poor and to refinance the Egyptian economy undermined popular support and created a sense of policy drift and administrative incompetence. A stagnant economy, social unrest, extremist violence, and frightening Islamist rhetoric sparked massive anti-Morsi demonstrations. The military saw this as a mandate to remove the Morsi government, press for constitutional revision, and seek new presidential elections. Massive pro-Morsi demonstrations organized by the Muslim Brotherhood resulted in ruthless pushback by the military, and foreshadows a prolonged period of turmoil or worse. On almost every level, the post-Morsi transition to a stable, prosperous, and democratic society promises to be a long and difficult road.

Stop & Think

▸ How does the use of "virtual water" as part of the total mix of available water make sense for a dryland country such as Egypt?

Turkey: Realizing Broad Potential

When World War I ended with Turkey on the losing side, and the Ottoman Empire's long decline ended in complete collapse, a smaller, more homogeneous Turkish state eventually emerged. A revolution led by Kemal Atatürk in the 1920s ruthlessly cut ties to the past in order to achieve modernization. The army played a crucial role in this transformation, serving as an integrating institution, promoting national regeneration, creating a secular state, and driving industrial and agricultural development.

Turkey is one of the few oil "have not" states with a GNI PPP that compares favorably with some of the oil-rich states. At US$15,530 its GNI PPP is almost twice that of Algeria's and falls just under Libya's, although it cannot compare with Gulf State figures. Turkey has a powerful, integrated, and diverse economy compared to other "have not" countries. Sufficient coal and iron resources allowed the country to develop its own heavy industry without having to import basic materials (see Figure 7-14), and it has a number of valuable minerals. Turkey also has a greater percentage of usable, arable land (29.1%,

not including degraded forest areas) than any other country in the region. So there are good reasons why Turkey is in a favorable position.

The Turkish economy is based on three sectors: agriculture, industry, and services. The traditional bedrock of the economy is agriculture, but its relative importance has declined significantly in recent decades. Today agriculture makes up roughly one-third of national employment but only 9 percent of GDP. Nonetheless, a diversity of natural habitats makes possible a substantial array of agricultural land-use options (Figure 7-29). Olives, vines, citrus, and deciduous fruits flourish along the western and southern shores of the country. The eastern Black Sea coast has a uniquely wet and warm climate and produces specialty crops such as tea. Turkey's drier central core is a prime district for cereal crops, mainly wheat and barley. Semiarid and drought-prone districts in the southeast have traditionally been a major livestock production zone.

For political and social welfare reasons, the government invests heavily in infrastructure for agriculture. The Southeast Anatolia or **GAP Project** (Guneydogu Anatolu Projesi in Turkish) in semiarid southeast Turkey is specifically targeted to the poorest part of the country. By building 22 dams, 19 with hydroelectric potential, on the Tigris and Euphrates rivers and their tributaries, the plan is to achieve a win-win situation. When fully completed, a process taking longer than expected, the country hopes to double its agricultural output through the plan's 1.7 million irrigated hectares (4.2 million acres). Already able to meet its domestic food needs, dramatically increased agricultural output will permit far larger exports, particularly of grains and specialty fruits and vegetables, generally within the Middle East. Increased agricultural exports on the scale envisaged would go a long way toward correcting Turkey's balance of payments deficit, which would improve its credentials for full European Union membership. Turkey's southeast is also home to most of its Kurdish population. An early 2013 agreement with Kurdish rebels promises to create a better political and social climate for Kurds, but the need for rural electrification and secure irrigated farming opportunities in the emerging GAP schemes is of great pragmatic importance and makes delivery on the GAP's promise even more important. That these same developments may cause a drop in both the quantity and quality of water supplied to downstream irrigation farmers in Syria and Iraq has not loomed large in Turkish political debate or practical planning.

Modernization of Turkish industry based on local mineral inputs was one of the goals of the Ataturk regeneration program. Concentrating on basic metallurgy and textiles stimulated the development of the country's primary resource base through mining and cotton farming. Never the dominant economic sector, domestic industry today produces iron and steel products that are turned into everything from household appliances and consumer electronics to ships, trains, and motor vehicles. Manufactured commodities provide more than three-quarters of the value of Turkish exports, despite the fact that Turkey remains a global-scale food producer and exporter. But the bulk of Turkey's significant industrial output continues to be domestically consumed, and only rare minerals—such as chromite, meerschaum, and manganese—enter world trade.

Only part of the energy needed to run Turkey's domestic and industrial activities comes from within its own borders. Low-quality lignite coal makes up most of Turkey's coal resource and was sufficient to launch the industrialization process, and to provide the bulk of electric power generation into the twenty-first century. Gas-fired

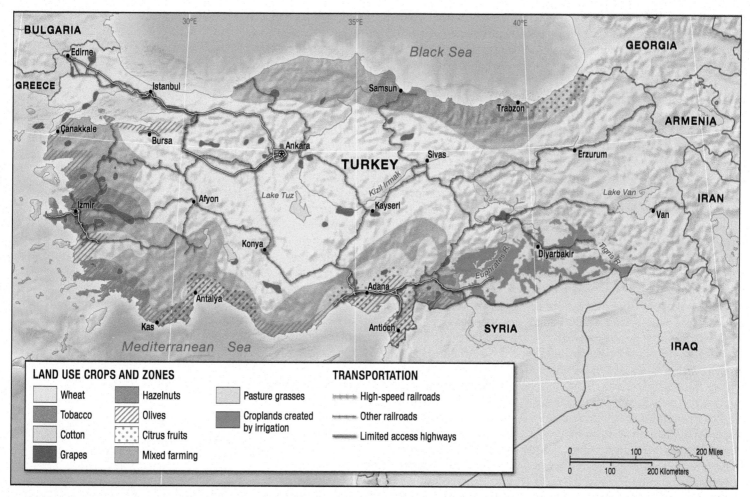

▲ Figure 7-29 **Land use in Turkey.** Traditionally intensive, specialized agriculture was confined to the better-watered coastal districts. But the expansion of irrigated farming in the drier southeastern quarter has encouraged intensification. Emerging high-speed highway and train connections make movement of people and products within the country faster and easier.

power stations relying on imported gas have become the generators of choice since then. But a series of power supply cuts from, and price disagreements with, suppliers in Iran, Azerbaijan, and Russia have made Turkey determined to return to local coal as the main power source despite its low thermal value and pollution threat. Coal imports constitute about half the energy mix. Oil is produced in Turkey, but its annual output is limited and has not grown appreciably since 1980. New oil finds have been announced for the border area near Iraq and Syria, and new natural gas finds in the Black Sea off Turkey's northern coast. Whether these announcements have more substance than similar ones in the past remains to be seen. Using Russian reactors and modest domestic uranium deposits, Turkey hopes to have the first in a series of nuclear power plants on line by 2021, and to accomplish this without encountering the same nuclear weapons concerns that have accompanied Iran's development plans. Turkey is not a heavy energy user, but demand is rising and future economic growth will depend on addressing its energy deficit.

Turkey is completely modern in the sense that the service sector accounts for the bulk of its GNP—64 percent. Tourism is a huge and growing contributor to the service sector's dominance (Figure 7-30). Warm winters, clear skies, attractive coasts and beaches, and magnetic urban settings attract visitors. Turkey lures more tourists than any other country in the Middle East and

▼ Figure 7-30 **Tourism in Turkey.** The small, Mediterranean coastal town of Kas provides a specialized tourist environment featuring scuba diving and accessible underwater Greek ruins as well as more typical café life and shops featuring local handicraft products.

High hopes for change in government, society, and economy were generated by the Arab Spring uprisings in late 2010 and early 2011 (Figure 7-3-1). At best these expectations were only partially realized by late 2013. In Tunisia and Egypt, regime change produced elections and various combinations of new parliaments and presidents. But conservative Islamist parliamentary majorities proved no more competent in promoting democracy and economic growth than their predecessors, and in Egypt in mid-2013 widespread anti-Muslim Brotherhood demonstrations prompted a military takeover and intensified violence (Figure 7-3-2). In Yemen, the incumbent president who ruled for 33 years was replaced by his vice president, but little else changed.

When dictators refuse to surrender power, conflict erupts. Ideological supporters as well as minority ethnic, regional, and religious communities often view the dictatorship as legitimate (or preferable to the opposition) and are willing to fight and die for the regime. Because the central government cannot tolerate heavy casualties among its supporters, it expands tactics that fail to meet Western expectations of fair play. Militias, secret service agencies, and goon squads try to terrorize potential opponents into passivity with kidnappings, random disappearances, sniper attacks, and massacres in small villages and urban neighborhoods. Groups in rebellion carry out

similar revenge tactics against government supporters. An escalating cycle of violence grows. Government military forces isolate hostile areas and destroy them from a distance using bombs and artillery. Casualties multiply and increasing numbers of refugees are the result; misery, fear, and dislocation spread. This hellish process afflicted much of Syria in 2013, with no clear indication of the outcome. Without intervention by NATO air forces, the same process would have taken place in eastern Libya in 2011; only air cover permitted the rebels ultimately to overthrow the Gadhafi regime.

In the troubled summer of the Arab Spring, the road forward will be rocky for all countries. When a change of government is prolonged and violent, the most radicalized and extremist groups gain power, energy, and cohesion from the struggle. They have the best chance to shape the regime that follows. Even when the authoritarian regime collapses quickly, groups persecuted during the dictatorship are the best organized and most successful initial representatives of popular will. Generally these groups are fundamentalist, conservative, and Islamist. Shaped by persecution, they are often intolerant, seek a society that restores the structure and values of early Islam, and see little merit in alternative points of view. Initially, a strongly Islamist ideology conveys legitimacy, although the costs often are social repression, economic stagnation, and growing internal opposition.

Where regime change did not occur, the future is equally problematic. Saudi Arabia, Algeria, and other oil-rich countries gave monetary bonuses to disadvantaged citizens and increased the number of government jobs while using a show of military force to repress demonstrations and maintain the status quo. This strategy may partially address immediate security problems, but does not stimulate economic growth or provide jobs for new job

▲ FIGURE 7-3-2 **Demonstrations turn violent in 2013.** Supporters of ousted President Muhammad Morsi are removed by force from a protest encampment near the Rabaa al-Adawiya mosque in central Cairo on 14 August 2013, and hundreds are killed throughout the country in the violent aftermath.

seekers. Large state-owned industries and big companies owned by the business elite dominate the economic scene. These enterprises exist in a protected domestic market, and are uncompetitive globally. Potential domestic competitors found it difficult to establish new companies before the Arab Spring; the same corrupt, crony-dominated bureaucratic systems remained in place two years later and slowed the pace of development. Youth unemployment rates are far larger than the official figures. Those with higher degrees often lack skills for the jobs available and are uninterested in the manual work on offer. Government jobs are the default, but cannot be expanded indefinitely. Thus widespread frustration is common. In the Maghreb, hundreds of young adults have set themselves on fire in protest against the hopelessness of their situation, and more than half have died in a martyrdom that has failed to spark economic renewal and social justice. This overall discontent will generate major future difficulties unless root causes are addressed and solutions are found.

Sources: Kjetil Selvik and Stig Stenslie, *Stability and Change in the Modern Middle East* (London: I. B. Tauris, 2011); "Arab Spring Cleaning," *The Economist*, February 25, 2012, p. 90; Karin Laub, "Vendor's Suicide Reflects Despair of Mideast Youth." *The Associated Press*, May 11, 2013.

▲ FIGURE 7-3-1 **Demonstration in Egypt during the Arab Spring.** A large, peaceful demonstration estimated to include 2 million people heads toward Tahrir Square as opposition to Hosni Mubarak's government gathers momentum in February 2011.

North Africa. Spectacular surviving cultural artifacts of the ancient Greeks, Romans, Byzantines, and Ottomans provide a plethora of tourist sites. Visitors, primarily from Western Europe (Britain and Germany) and Russia, pour into the country in astonishing numbers—31 million in 2012. Tourism supports 1.7 million jobs directly, almost as many jobs indirectly, and in 2012 made Turkey the sixth most important tourist destination in the world with total revenue estimated at US$25 billion. The success of any tourist industry directly depends on economic prosperity in the tourist source area, and security and safety for tourists in the destination area. As long as Turkey can provide a safe and hospitable environment for visitors, tourism should remain a major contributor to its economic success.

Turkey's population growth rate has begun to slow. The 2.0 total fertility rate is slightly below replacement level and one of the lowest in the region. Despite this, population is still growing at 1.2 percent per year, and the population bubble that occurred from the 1960s to the 1980s will continue to send waves of increased population coursing through the age cohorts for many years. Today over 4 million Turkish citizens reside in Europe, mostly in Germany, where they migrated to provide semiskilled labor and often stayed. About 20 percent of this Turkish population was born in Europe and many of them view themselves as living in both places regardless of the formal niceties of citizenship. Turks still engage in labor migration, but their destinations are more likely to be other Middle Eastern countries or Russia than Western Europe.

The Turkish presence in Europe, both twentieth-century migrants and residual population islands in the Balkans surviving from the Ottoman era, as well as its physical foothold west of the Bosporus, gives a peculiar validity to its interest in EU membership. Turkey's application has not been greeted everywhere with enthusiasm. But as Turkey's economy has grown at a high rate in the last few years (8.9% in 2012), its inflation rate has dropped to 5 percent, and it ranks as the seventeenth-largest economy in the world, its status as an advanced emerging economy makes the country and its market increasingly an asset rather than a liability. Small signs of slow progress toward full membership have emerged, but addressing several issues will be essential to accelerating the process. First and foremost is a long-term solution to the status of Turkey's Kurdish minority. One in every five Turkish citizens is Kurdish, but recognition of their existence as a separate but equal element in Turkish society has proven elusive. The March 2013 deal struck with the major Kurdish opposition party to abandon violence and realize their identity and autonomy goals within the structure of the Turkish state is a promising beginning. If this seemingly intractable issue can be resolved and Kurds can find justice and development in Turkey, many of the cultural and humanitarian issues that have plagued Turkey's EU membership aspirations may more easily be resolved. The second issue is Turkey's foreign policy role in the Middle East and the extent to which it is sufficiently European so that it will not embroil the European Union in Middle Eastern adventures. Resolving the physical division of Cyprus between its Greek and Turkish citizens is important in this regard. Turkey promoted the emergence of the Turkish Cypriot republic and is its sole protector internationally. Reuniting Greek and Turkish Cypriots in a unitary state with autonomous guarantees might serve as a model for Kurds within Turkey and a symbol of shared values between Turkey and

mainstream European political culture. The civil war in Syria poses significant problems for Turkey. Its opposition to the Assad regime is partly a product of humanitarian concerns related to the 330,000 Syrian refugees in Turkey! Turkey also has security concerns along its border with Syria that are both pragmatic (random cross-border attacks on its villages) but also involve Kurdish populations that straddle the border and have at times provided save havens for Turkey's Kurdish rebels.

Great tension exists in Turkish society, culture, and politics between secular and modernist versus religious and traditionalist trends and communities. This tension occasionally bursts into view in policy debates, public demonstrations, and sometimes violent confrontations. In mid-2013, Taksim Square in Istanbul was the scene of confrontations between police and protestors opposing government plans to transform the square from a treed social space into a religious space dominated by a new mosque. A sudden drop in the value of Turkey's currency also demonstrated the country's vulnerability to foreign investment decisions linked to internal stability. Turkey's future trajectory will be determined by how well it engages with regional and internal religious and political issues, harnesses and expands its booming economy, and resolves its ambiguous position as a transitional state between Europe and the Middle East.

Small States of the Eastern Mediterranean: Uneasy Neighbors

The small countries at the eastern end of the Mediterranean Sea—Syria, Lebanon, Israel, Palestine, and Jordan—present a different set of issues than Egypt and Turkey. Discussion of agricultural and mineral resources, basic development factors for many countries, is superseded by questions of human resources, service functions, and the inflow of remittances or foreign aid. Syria is significantly larger than the other states in land area and population. But by spring 2013 its civil war had fragmented the country into a patchwork of smaller fiefdoms controlled by various ethnic and opposition groups united only by their hostility to a truncated and wounded central government.

The eastern end of the Mediterranean is a strategic area that controls important nodes of communications and trade. Routes linking Africa with Asia and the Persian Gulf with Europe and East Asia pass over, through or near the area. Airlines flying to and from East Africa and the Persian Gulf must transit the air space of coastal countries although refueling stops are no longer essential. The Suez Canal remains a vital chokepoint for maritime traffic between the eastern Mediterranean and countries in East Africa and Asia. Jordan provided a vital transport corridor between Aqaba and Iraq throughout the 10 years of American military engagement in Iraq. And if peace and normal relations ever return to the countries of the region and their neighbors, oil pipelines from the Persian Gulf to the Mediterranean's eastern shores may again become a reality.

The focal point of three major world religions, the religious significance of this area is remarkable. It is the source of the Jewish religion, out of which grew the various Christian denominations, as well as possessing significance to Muslims (Figure 7-31). Although Mecca in Saudi Arabia is the central holy place of Islam, Jerusalem is the site of the Prophet Muhammad's ascension into heaven and is a major Muslim pilgrimage center. Medieval maps show Jerusalem

▲ **Figure 7-31 Jerusalem, a focus for three religions.** Jews, Christians, and Muslims find spiritual meaning in Jerusalem as evidenced by the western wall of Solomon's temple, the gold dome of al-Aqsa mosque, and the distant Mount of Olives, all visible in this view.

as the center of the world, with Asia, Africa, and Europe all focused toward it. Present power politics have greatly modified but not totally destroyed that view. From a religious and a geopolitical viewpoint, the area remains extremely important to the world at large.

Syria: Catastrophic Developments

Syria is endowed with important natural resources, primarily agricultural, but also including small-scale mineral deposits and sufficient oil fields to meet domestic requirements and obtain 40 percent of its foreign exchange earnings from oil exports to Europe in 2010. Most Syrian territory lies in the rainshadow of the Lebanon and Anti-Lebanon mountains, but sufficient rainfall here supports dry farming. Aridity increases to the south and east, making agriculture unreliable in two-thirds of the country. The Euphrates River crosses Syrian territory from north to southeast on its course from Turkey to Iraq, providing irrigation and electric power-generating opportunities. A line of ancient cities stretches north-south through the country's traditional core from Aleppo (Halab) near the Euphrates through Hama, Homs, and the oasis of Damascus to the border town of Dara'a. Together with the major port of Latakia, these urban centers form the commercial and political core of Syria.

Since the Umayyad Caliphate established itself in Damascus in 661 A.D., the city has played a major political and intellectual role in the Arab world. Many Syrians expected to exert similar leadership in the pan-Arab nationalist movement after full independence following World War II. The struggle for dominance among and within contending parties ended when a coup within the Arab Socialist Ba'ath Party brought Hafiz al-Assad to power in 1970. Progressive, socialist, and sectarian, the Ba'ath was determined to modernize and transform Syria quickly, and ruthlessly if need be, just as its sister branch was committed to doing in neighboring Iraq.

Initially the Ba'ath provided much needed stability, discipline, and direction with development plans that expanded access to health care and education as well as rationalized many aspects of agriculture and economy. Major investments were made in irrigation, first in the Orontes River valley and later in the Euphrates River valley. The massive **Tabqah dam** was constructed on the Euphrates to irrigate 1.5 million acres (0.6 million hectares) and provide reliable hydroelectric power. Further dams on the Khabur River, a

tributary to the Euphrates in the northeast, were constructed to upgrade cultivation from rain-fed to irrigated. A system of linked cooperatives was established to regularize the flow of livestock from eastern rangelands to cereal cultivation zone fattening facilities to slaughterhouses to urban consumers at affordable prices. Eventually sufficient oil was found in the far northeast, tied in by pipelines to the country's major cities, and linked to refining and export facilities on the Mediterranean coast.

Several problems have marred this potentially hopeful development trajectory. A strongly nationalist regime, controlled by military officers, the government opposed Israel and strongly supported the Palestinian Arabs. This resulted in several losing conflicts with Israel, loss of territory on the Golan Heights southwest of Damascus, and a great deal of expense on armaments. The Ba'ath tolerated no opposition and gradually established a pervasive security apparatus to seize opponents before they could act. This authoritarian approach alienated many people who might have been supportive. Many leading figures in the Ba'ath, including Hafiz al-Assad, were Alawites, adherents of a Shi'ite sect comprising about 11 percent of the Syrian population. Many Sunni Muslims, particularly the more fundamentalist, regard Shi'a in general and Alawites in particular as infidels. The Syrian Muslim Brotherhood included many hardline groups who began in the 1970s to oppose the regime using tactics as extreme as those employed by the Mukhabarat (secret police). The escalating violence peaked in a revolt in early 1982 in Hama, Syria's third largest city and a center of Muslim Brotherhood fundamentalism. The Syrian government called in regular army and elite special forces units, surrounded the central madina (old city), and destroyed it. Thousands of people died, which broke down overt, coordinated opposition.

The Arab Spring inspired renewed opposition in January 2012. Government reaction was to terrorize demonstrators by random violence, usually with imperfect results. Demonstrations spread and encountered resistance, and increasing attacks were made on Ba'ath Party officials, uniformed soldiers (particularly officers), and supporters of the regime. Soldiers deserted rather than fire on crowds, and the opposition, never united, slowly occupied territory. When opponents overran police posts and captured neighborhoods, these areas were closed off and attacked remotely by shells and bombs. Violence escalated into civil war, with massive death tolls, catastrophic levels of internal displacement and transborder refugees, and widespread destruction of critical infrastructure (Figure 7-32). The toll in human suffering and the costs of reconstructing cannot be calculated or even imagined until a clear victor emerges, but Syria's future is bleak.

Lebanon: A Fragmented Society

Lebanon shares the eastern Mediterranean coast with Israel and Syria, is small in area and population, but is culturally complex. Its society is fragmented along religious, but not ethnic, lines. Political parties and armed militias are based on Christian or Muslim communities or sects. Historically many groups came as refugees,

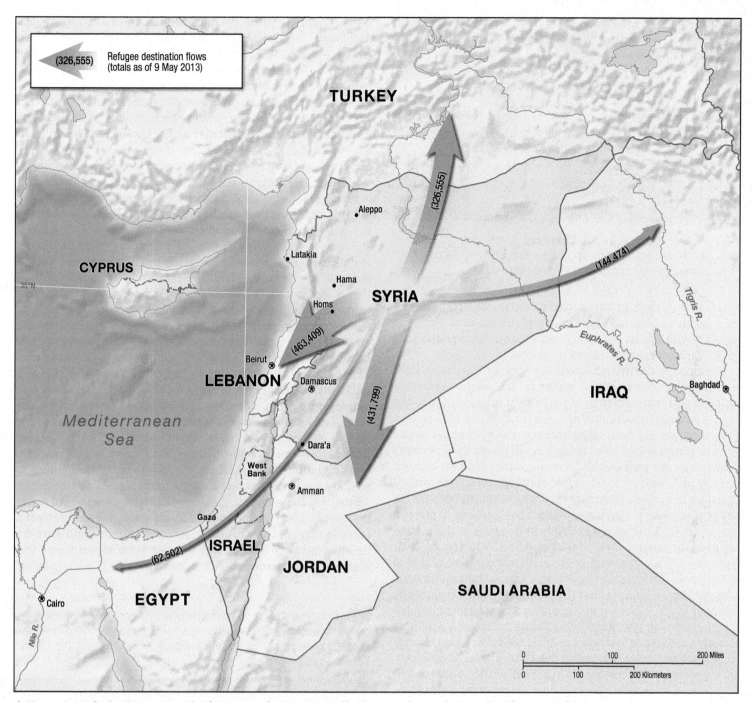

**Refugee destination flows
(totals as of 9 May 2013)**

(326,555)

▲ **Figure 7-32 Syrian international refugees as of 9 May 2013.** The Syrian civil war, which evolved from peaceful demonstrations in 2011 to brutal intercommunal warfare by the middle of 2013, sent waves of refugees spilling into neighboring countries.

Source: United Nations High Commission for Refugees, Syria Regional Refugee Response Portal http://data.unhcr.org/syrianrefugees/regional.php.

attracted to Lebanon's rugged mountain interior, or as traders to profit from Lebanon's strategic economic location. The same protective mountains today pose serious challenges to agricultural development, and Lebanon's regional location confers fewer advantages than it once did.

Mount Lebanon, a ridge that runs north-south through the full length of the country, dominates Lebanon. Spurs from this ridge break up the coastal plain into a series of segments. The Beqaa Valley, a prime wheat farming district, parallels Mount Lebanon and separates it from the Anti-Lebanon range along the Syrian

border. Both ranges present a formidable barrier to movement. Farming in the mountain foothills and lower elevations is small-scale. Today farming only contributes 6 percent to GDP, and remittances from relatives working outside Lebanon are indispensable for many mountain villages: these funds constitute 20 percent of the national economy. Industry is concentrated in very small-scale assembly activities, and most of the workforce is found in the service sector, an indication of the continued vitality of Lebanon's traditional crossroads location and the tenacity of its banking and tourism industries (Figure 7-33). Lebanese entrepreneurship may

▲ **Figure 7-33 Central Beirut, Lebanon.** The fashionable Hamra district in central Beirut appears completely recovered and restored from the trauma and destruction of Lebanon's 15-year civil war.

receive a boost from an unexpected direction. Reports of oil and gas deposits in Lebanon's maritime exploitation zone along the boundary with Cyprus and Syria could wipe out the excessive debt under which Lebanon's economy is laboring.

Historically Christian Lebanese, primarily Maronites, have controlled Lebanon's political system, which assigns leadership positions and representation in parliament in proportion to the size of each religious community as determined in a 1932 census. No new census has been taken because relative changes in size between groups would affect the balance of power. Because many Maronites sought work outside Lebanon, but few Muslims followed the same path, resident Muslims gradually began to equal or exceed Christians in number. Communal and religion-based hostility, the very condition the parliamentary system was intended to prevent, grew as Shi'ite Muslims felt excluded from access to representation and jobs equal to their growing numbers. A 15-year civil war began in 1975; a peacekeeping occupation by Syria from 1976 to 2006 followed. Palestinian refugee support for PLO attacks on Israel prompted Israeli interventions that devastated infrastructure and cost lives in the Lebanese south, and prompted armed opposition by Hezbollah (Party of God), a Shi'ite militia whose military successes resulted in increased political clout. Whether the present renegotiated balance of power that divides jobs and positions equally between Muslims and Christians will provide a better basis for cooperation and development than the previous skewed formula, and whether Lebanon can avoid involvement in Syria's conflict, remains unclear.

Israel: A Small Powerhouse

Israel, established from the British mandate territory of Palestine in 1948, was the culmination of Zionist dreams—a Jewish homeland where no state had existed for 2000 years. Israel's creation was opposed by neighboring Arab states and Palestinian Arabs, whose ancestors had lived nearly as long in Palestine as the ancestors of most Jews had been absent from it. Born successfully amidst conflict, and 50 percent larger than proposed by a UN partition plan, Israel has not been a stranger to strife. Twenty years of antiguerrilla conflict after 1948 culminated in Israel and neighboring Arab states fighting a six-day war in June 1967. Israel attained more defensible

borders from this conflict at the territorial expense of Egypt (Gaza and Sinai), Jordan (**West Bank** [of the Jordan River]), and Syria (Golan Heights). Palestinian Arabs, who lost part of their lands in the 1948 hostilities, found all of their traditional territory in Israeli hands after the 1967 war. The 1973 Yom Kippur War ended in preservation of the territorial status quo, although Israel later negotiated a withdrawal from the Sinai. Subsequent rounds of low-intensity hostilities on its northern border resulted in Israel's 1982 invasion of southern Lebanon, sending 1948 refugees into a new round of forced migration. Two lengthy periods of **intifadah** (uprising) by West Bank Palestinians, a less successful second war in Lebanon against Hezbollah in 2006, and two brief but violent major conflicts with Hamas in Gaza in 2008/2009 and 2012 have continued the string of recurring formal conflicts.

Israel is a relatively rich state, with 92 percent of its people living in cities (Figure 7-34). In a region where per capita GNIs are generally low, Israel's GNI in 2011 was US$28,930, nearly 4.5 times that of Jordan and five times that of Egypt. Israel's inflation rate is low and has declined steadily in recent years. Another striking feature of the country is its high level of education. Between the graduates of its universities and the many people who immigrate to Israel from developed Western countries with advanced academic and technical training, 45 percent of the population has a university or college degree. Israel is one of the few countries with employment issues due to the high education level of its human resources; providing satisfying employment that matches existing skills is not always easy. The quality of Israel's technical and academic achievements is unsurpassed in a number of fields. Israel's work on both water-use efficient hydrology and horticulture as well as in some electronic fields is particularly important (see *Visualizing Development:* Dealing with Water Deficits). An active high-tech industry, particularly in software, has made Israel a leading player internationally. A home-grown, sophisticated defense industry has become a major arms exporter and foreign exchange earner. A high standard of living, technological

▲ **Figure 7-34 The Tel Aviv coastline.** An iconic view of the city's Mediterranean coast displays high-rise buildings concentrated along the coast and the defensive barriers and groins needed to control coastal erosion.

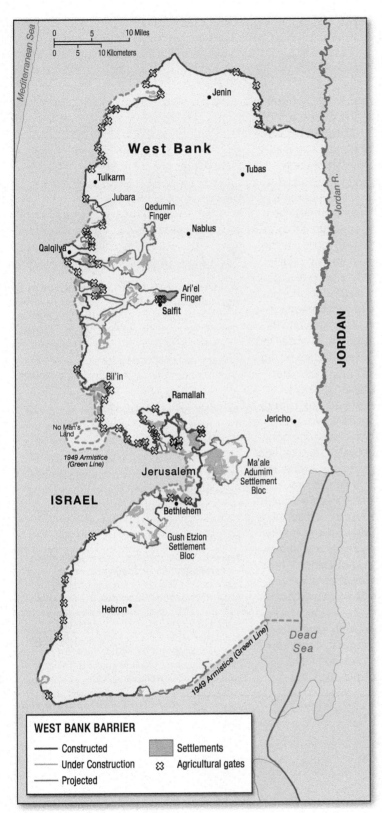

▲ **Figure 7-35 Israel and its West Bank border.** For security reasons, Israel has chosen to separate itself from the West Bank by constructing a barrier. A wall in urbanized areas and a barbed wire fence in more rural settings, the barrier does not adhere rigorously to the internationally recognized boundary.

Source: United Nations. Office for the Coordination of Humanitarian Affairs Occupied Palestinian Territory. "The Humanitarian Impact of the Barrier." July 2012. http://www.ochaopt.org/documents/ocha_opt_barrier_map_july_2012_english.pdf

sophistication, and a well-educated population make Israel an economically and technologically developed country.

Israel is at odds with its neighbors in two major ways: political conflicts are reinforced by a clash of cultural values. Cultural dissonance also exists within Israeli society; Jews who migrated from Middle Eastern countries (**Sephardim**) often find themselves at odds with Jews whose origins are European (**Ashkenazim**). Other discordant notes also exist. Orthodox Jews try to live a life that is consistent with their understanding of Jewish law; they both irritate and are irritated by secular Jews whose commitment to religious observance is weak. Ultra-Orthodox Jews so far are exempt from military service with the Israel Defense Forces (IDF), which has manpower implications for Israeli and commitment issues for the ultra-Orthodox. Security concerns prevent the 20 percent of Israel's citizens who are Arabs from service in the IDF. Israel's Arab population is growing rapidly and may constitute 25 percent of its total population by 2025, a prospect that concerns extreme nationalists who believe this threatens the character of the state.

More significant is the ongoing struggle between Israel and Hamas, which controls the **Gaza Strip,** and the PLO, which dominates the West Bank. Terrorist attacks within Israel and concerns about the security of Jewish settlements on the West Bank have led to the construction of a defensive barrier/wall to separate Israel from the West Bank (Figure 7-35). Suicide bombings have markedly declined, but Palestinian frustration has risen. Achieving a settlement of Israel's conflict with Palestinian Arabs and many of its Arab neighbors is essential to securing a future that is not dominated by, but may never be entirely free from, security issues.

Palestine: Struggling for Statehood

When the state of Israel was established in 1948, no comparable state representing the aspirations of the Palestinian Arabs emerged. The more than six decades that have passed since have not seen a resolution to that problem. Over time it has become harder to discover a way forward that resolves enough fears and grievances to have a chance of success.

Palestinian Arabs have not controlled their own space and destiny in the modern era. Part of the Ottoman Empire until after World War I, Palestine then became a British mandate territory. The end of hostilities in 1949 left the West Bank and Jerusalem occupied by Jordan, the Gaza Strip held by Egypt, and hundreds of Palestinian Arab villages vacant as their occupants fled or were forced out. Palestinian refugee populations were clustered into camps, subsisted on relief handouts administered by the United Nations High Commissioner for Refugees (UNHCR) and the United Nations Relief and Works Agency for Palestinian Refugees in the Near East (UNRWA). Unabsorbed into the structures and economies of their fellow Arabs and unreconciled to the loss of their homes and property, Palestinian Arabs developed a distinct national consciousness. Most Arab governments were unwilling to accept them as full members of their states; only Jordan, and to a lesser extent Lebanon, granted citizenship to substantial numbers of Palestinians. Palestinians often refused to be absorbed because they only wished to regain their ancestral homeland. Most of the current refugee population is composed of descendants of the original refugees, and additional refugees displaced after the 1967 war (Figure 7-36). The

Visualizing DEVELOPMENT ▲

Dealing with Water Deficits

The Middle East and North Africa region is on the threshold of a major water gap. The average annual water resource availability of the region is estimated as 1274 cubic meters per person. Planners believe that social and economic development is difficult to achieve below 1,700 cubic meters per person. By this standard, the region is the most water-stressed in the world (Figure 7-4-1). Egypt, Morocco, and Lebanon are severely water-stressed, and almost all others suffer from water scarcity. The potential for this deficit to worsen over time is very high. Based on current demand, only Algeria and Tunisia will escape a significant gap in their available water supply by 2030 unless concerted efforts are undertaken to deal with the problem.

One way to cope with existing and increasing water deficit is to reduce demand for water. Many wealthy, oil-rich countries use large amounts of water per capita. Comprehensive efforts to educate people to use less water would increase awareness of the problem and of water conservation options. Financial incentives to adopt low water-use household technologies would help. Raising the price of water as the volume of water used increases should promote conservation. Subsidies that encourage a shift from open field irrigation systems with high evaporation rates to water-conserving drip irrigation farming would improve water-use efficiency (Figure 7-4-2), as would moving from open field to closed environment (greenhouse) cultivation, although this requires significant capital outlays for infrastructure. Reusing

water multiple times in industrial processes is worth the cost of installing equipment to cool and clean the water. Above all, because major rivers and aquifers are transnational, cooperation to regulate use, share the resource fairly, and minimize water pollution and degradation is essential.

Much of the water resource does not directly benefit humans, although it may be essential to nature. Capturing more precipitation and surface flow would help people cope with scarcity. Cisterns and barrels to capture water falling on roofs and hard surfaces would provide more local storage. All urban structures should contain underground basement cisterns instead of parking garages. Large surface reservoirs behind big dams lose much of the impounded water through evaporation; smaller underground community storage cisterns are much more efficient alternatives. Treated wastewater from urban treatment plants is a large potential water source. Renovating and reusing this water for agriculture and industry would stretch the water resource significantly.

Desalinating seawater is a viable source of new water. The Middle East contains 53 percent of the world's desalination capacity, in part because many countries have the revenue to fund and fuel their operation. Cost limits desalination to production of drinking water. Major plans to increase desalination capacity exist in many places. Qatar, which has only a two-day emergency water supply, plans to increase its capacity to seven days and generate 36 million additional gallons of water per day by mid-2015 through

▲ FIGURE 7-4-2 **Closed environment agriculture in the Negev.** Greenhouses make it possible to control temperature and the water use efficiency of the plants under cultivation in a very dry habitat. In this way a limited amount of water can produce high-value, out-of-season crops that can earn a profit in distant markets.

desalination. Most desalination plants use fossil fuels for power, and produce sea salt and electricity as byproducts. Substituting solar energy as a power source would help reduce pollution and increase sustainability. Desalinating seawater without producing brine as a waste product that is returned to the sea, would help protect marine environments.

Sources: David Michel, Amit Pandya, Sayed Iqbal Hasnain, Russell Sticklor, and Sreya Panuganti, *Water Challenges and Cooperative Response in the Middle East and North Africa* (Washington, D.C.: Brookings Institution, 2012), http://www.brookings.edu/research /papers/2012/11/water-security-middle-east-lwf; The World Bank, MENA Development Report, *Renewable Energy Desalination: An Emerging Solution to Close the Water Gap in the Middle East and North Africa* (Washington, D.C.: World Bank, 2012).

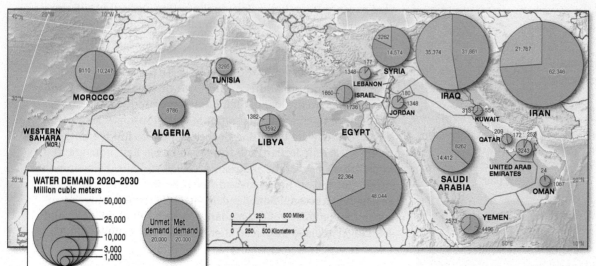

▲ FIGURE 7-4-1 **Future water demand in the MENA region.** For most countries in the Middle East and North Africa, the ability to develop economically will be severely constrained by lack of water in the 2020–2030 decade.

Source: World Bank. *Renewable Energy Desalinization: An Emerging Solution to Close the Water Gap in the Middle East and North Africa* (2012).

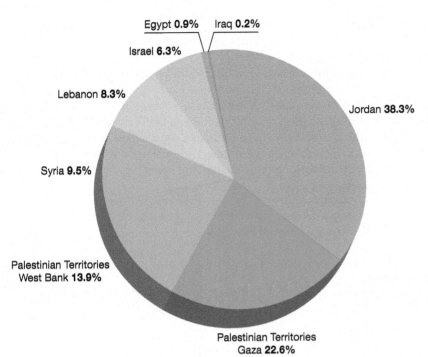

▲ Figure 7-36 Estimated Palestinian Refugee Populations 2012. The common feature linking these numbers is the inability of the descendants of Palestinian refugees dislocated by the 1948 war and subsequent conflicts to return to their ancestral homeland or town. Only Palestinians living in countries geographically close to Israel are included in this calculation.

Sources: UNRWA (United Nations Relief and Works Agency for Palestine Refugees in the Near East), total registered refugees, 1 January 2013, http://www.unrwa.org/userfiles/2013042435340.pdf; UNHCR (United Nations High Commissioner for Refugees), http://www.unhcr.org/pages/49e486426.html; IRIN (Integrated Regional Information Networks, UN Office for the Coordination of Humanitarian Affairs), http://www.irinnews.org/Report/89571/MIDDLE-EAST-Palestinian-refugee-numbers-whereabouts

▼ Figure 7-37 The security barrier between Israel and the West Bank. A section of the wall snakes downslope and separates the Shuafat refugee camp (right) from Pisgat Zeez (left), an area annexed to Jerusalem after the 1967 war and subsequently settled by Israelis.

1982 Israeli invasion of Lebanon further increased the refugee total. Gaza and West Bank Palestinian residents, following a brief period of control by their Arab neighbors, entered a new era of occupation by Israel and management by Israeli Defense Force regulations and checkpoints in 1967.

Since the 1967 war, economic development in Palestinian Arab populated areas has been difficult. At times non-Israeli Arabs have found agricultural and semiskilled work in Israel, but whenever the border has been closed for security reasons access to job sites has been difficult. The Palestinian economic situation has become so dire that collecting statistics on income has become impossible. Gaza is physically severed from Israel by a barrier and cut off from maritime contact by a naval blockade. A nearly completed barrier/wall separates Israel from the West Bank, leaving more than 8 percent of the West Bank on the Israeli side (Figure 7-37). Movement around the West Bank has long been particularly difficult, with regulations adversely impacting the pace and frequency of travel. The wall has complicated the situation, particularly for villagers cut off from access to their fields or jobs. The growth of the Israeli settler movement absorbs land resources and fragments West Bank Arab territory. Israeli settlements require security, which imposes more regulations and restrictions on Palestinian activities. The first period of widespread intifadah (1987–1993) reflected the Palestinian sense that they were stuck in a dead-end situation. For a time after the PLO and Israel signed the Oslo Accords in 1993 it appeared that Palestinians would achieve limited sovereignty over Jericho and other zones, and would begin a gradual process of expanding territorial control and self-determination. But neither this process nor status negotiations have produced meaningful results. The PLO's inability to make progress, the corruption of PLO management of local institutions, and the stagnant state of the West Bank economy has fueled the rise of Hamas, a militant Islamic resistance movement that rejects accommodation with Israel and seeks confrontation regardless of the power differential between Israelis and Palestinians. A second period of intifadah followed (2000–2004). When Hamas seized control in Gaza in 2007, rocket attacks from Gaza on Israel provoked two particularly violent IDF responses in 2008–2009 and 2012. Recognition of Palestine as a nonmember observer state by the UN on November 29, 2012 is more symbolic than practical, since without a negotiated settlement with Israel nothing meaningful can occur.

Funds raised from the global Palestinian diaspora trickle back to support families or organizations in Palestinian territories and refugee camps. But in none of those places is there an economic context where development initiatives can bear fruit. Small acts of kindness and assistance between individuals, villages, or social organizations across the yawning chasm that separates communities do take place, yet their cumulative impact is negligible. An Arab–Israeli resolution would only be the first step in erecting a viable state with a functioning economy. But if such a Palestinian entity should emerge to share the space of the former Palestine mandate with Israel, its appearance would help reduce the regional tensions that impede development.

Jordan: A Precarious Refugee Haven

Jordan is wedged between Syria, Iraq, and Saudi Arabia, east of the Jordan River and the Dead Sea. A modest but rapidly growing population of 6.3 million people struggles to use a limited natural resource base. Aridity is the major constraint, increasing in intensity eastward and southward from the capital, Amman. Limited water resources make agriculture a precarious proposition in most of the country. Only occasionally in the Yarmuk River valley, the largest tributary to the Jordan, and on the eastern bank of the Jordan River is irrigated agriculture feasible. Modest rainfall occurs in the uplands, but is adequate for agriculture only in the northern third of the highland and drought risk is high everywhere. To the east only seasonal grazing is possible. The Dead Sea is a source for potash (potassium salts used in fertilizer and soap) and phosphates are mined; both are important foreign exchange earners. But so much fresh water is extracted from the Jordan River system that the Dead Sea is a dwindling resource, lacking recharge as the Jordan shrinks to a trickle. Absence of meaningful petroleum resources makes Jordan one of the "have not" countries, although modest natural gas discoveries in the eastern desert panhandle provide about 10 percent of the country's electrical energy. Abundant oil shales in central Jordan, undeveloped in part because water is crucial to exploitation, might make a significant contribution in the future.

Jordan is in many respects a refugee haven. The Hashemite dynasty, the sharifan (descendant of the Prophet Mohammad) clan that rules the country, for centuries provided the emirs of Mecca and rulers of Hijaz in Saudi Arabia. An ally of the British in World War I, they were driven out of Mecca by Ibn Saud in the 1920s and were compensated by the British with lordship of the Transjordan mandate. Circassian, Chechen, and Armenian minorities all came in the late nineteenth and early twentieth centuries. After 1948 many Palestinian Arabs arrived, and the 1967 war with Israel caused another wave of refugees. About a third of Jordan's population is Palestinian; all received Jordanian citizenship, although over 300,000 continue to live in UNWRA refugee camps. Recent regional conflicts have pushed yet more refugees into Jordan. Varying estimates place the Iraqi refugee population at between 750,000 and 1,200,000 (Figure 7-38). As many as 500,000 of these may be Assyrian or Chaldean rite Christians. Perhaps 15,000 Lebanese migrated to Jordan after the Second Lebanese War in 2006. And the UNHCR counted nearly 432,000 Syrians seeking refuge from that country's

civil war in May 2013. Whether or not these recent refugees stay permanently, invariably they place strains on economy and social service institutions.

Jordan's limited resources place severe constraints on economic development. International aid supports a debt-ridden economy with subsidized prices for basic necessities. Without economic linkage to the potentially richer and better developed West Bank, an economically viable Jordanian state is hard to envisage, whereas settlement of the Palestinian issue will present Jordan with new opportunities for growth and development.

Stop & Think

> Consider the refugee burden that regional conflict places on the small countries of the eastern Mediterranean.

The Gulf States: Living on Oil

No region of the world is more intimately associated with oil than the Middle East. That image is reasonable because more than one-fourth of the world's known petroleum reserves are found there. But the "black gold" is unevenly distributed within the region; some nations, by the accidents of geology and discovery, have much greater oil reserves than others (Figure 7-39; see also Figure 7-14). So abundant are proven reserves and the present or potential output of such oil producers as Iran, Iraq, Kuwait, and Saudi Arabia that they deserve independent treatment as a subregion within the Middle East. The oil industry dominates the economies of these states, pouring more money into their national treasuries than the countries can spend quickly on productive internal enterprises. With prosperity so closely tied to one resource, these countries must grapple with a complex array of development prospects and problems. And in Iraq, the past and potentially future oil-fueled prosperity of the country is enormously complicated by the turmoil and chaos that has characterized Iraq since the fall of Saddam Hussein's Ba'ath regime.

The overwhelming impact of oil on the countries that surround the Persian Gulf has dramatically altered their economies and shaken their social systems. Nations such as Iran and Iraq, which once had viable agrarian economies and a rapidly modernizing economic structure, were destabilized by the social changes that the expanding oil economy unleashed. Even countries that produce limited amounts of oil (such as Yemen, where oil production exceeds local demand but where reserves are small) or that lack the resource entirely (such as several of the sheikhdoms of the United Arab Emirates) have become enmeshed in the oil economy of the region. Often their citizens migrate to neighboring countries in search of employment, complicating development initiatives in both the labor-exporting and the labor-absorbing societies.

Below we explore similarities and differences within the region and review how petroleum-generated wealth has been put to use. We then look at the impacts and changes that have followed from

▼ **Figure 7-38 Iraqi refugees in Jordan.** Christian Iraqis, in uncertain but very substantial numbers, are major victims of violence in Iraq. Often living in mixed religious neighborhoods in Iraq, many have fled to Jordan to escape death threats.

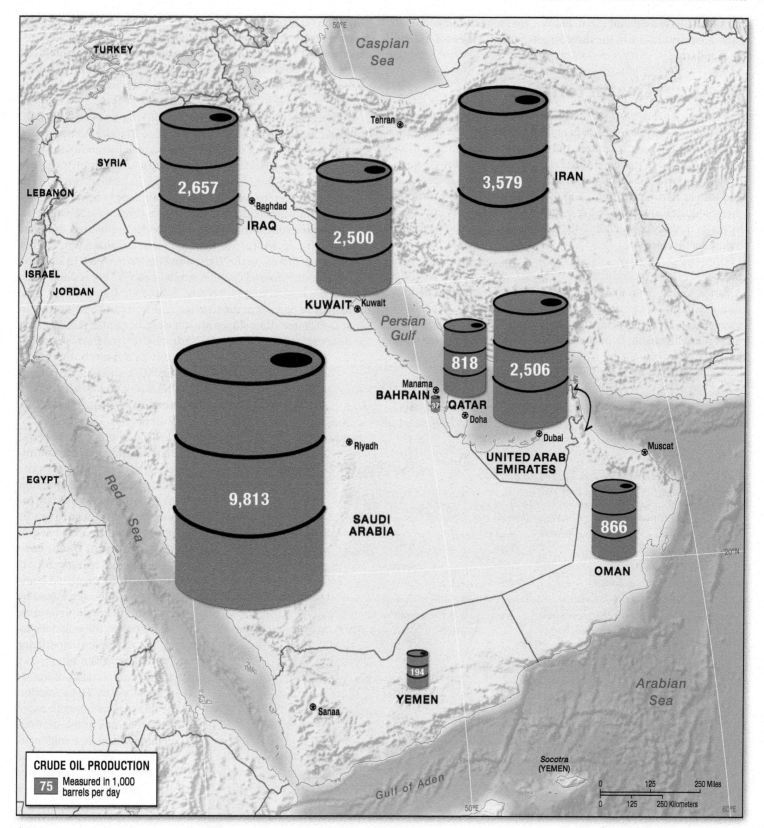

▲ Figure 7-39 **Oil production in the Persian Gulf countries in 2012.** Saudi Arabia is the regional leader in oil production and export, while Iranian oil exports have declined as a result of international sanctions. Iraq, which has very large oil and natural gas resources, is beginning to recover from the damage inflicted by war and internal conflict and become a major petroleum exporter.

Source: *Oil & Gas Journal,* v. 111, no. 3a (March 11, 2013), p. 31.

petroleum-based development. Finally, we examine the nature of contemporary change in the three largest Gulf States: Saudi Arabia, Iraq, and Iran.

Regional Characteristics

The Gulf States (Figure 7-40) share many, often limiting, characteristics that are part of a similar geographic setting and shared history. Within that generic context, the smaller political entities of the region have had to pay careful attention to the oscillating currents of political power emanating from the region's larger countries, Iran and Iraq in particular. Protecting their oil resources, overcoming their liabilities, and developing their societies in an area impacted by major military conflicts and the Arab Spring are the intertwined themes of this section.

Limited Resource Base

A basic characteristic shared by the Gulf States is a limited natural resource base. Only petroleum and natural gas, often in deposits of staggering size, are of great significance to the area (Figure 7-41); other minerals are not found in sufficient quantities to make mining profitable under present conditions. Only Iran possesses enough mineral reserves to support a modern metallurgical industry. Because much of the Gulf region, particularly its more mountainous districts, has not been explored intensively by modern methods, assessments of the mineral resource potential of the region may change.

Agricultural resources in most Gulf States are also limited. Much of the food consumed in the Gulf States today is imported, a situation that increases dependence on and vulnerability to global economic conditions. Most of these countries suffer from extremely limited water supplies concentrated either in highland areas, such as those in Yemen or western Saudi Arabia, or in lowland oases, where groundwater is close to the surface (Figure 7-42). Only Iraq and Iran have substantial areas suitable for cultivation. Iraq has both rain-fed agriculture in the north and irrigation potential in its arid south, based on the Tigris and Euphrates rivers. In Iran, the western mountains and Caspian seacoast receive appreciable rainfall, but the central core is very dry. Yet even in Iraq and Iran, little more than 10 percent of the total land surface can be cultivated. In other Gulf States, most slopes are too steep or rainfall is too limited for nonirrigated agriculture. Irrigation by surface flow is possible only in a limited number of isolated sites, usually near the base of mountains, where seasonal runoff provides water.

The traditional qanat system of tapping deep groundwater and bringing it to the surface by gravity flow that originated in Iran is an ingenious response to water shortage and is still used in this subregion. Qanat construction and maintenance is the work of specialized communities, often organized by clan or village associations, who possess the engineering skills and practical experience to engage in the often risky subterranean tasks of building and repair. The attraction of this skilled labor to more lucrative jobs in the modern economy, and their migration to locations where those jobs are found, often undermines the long-term productive

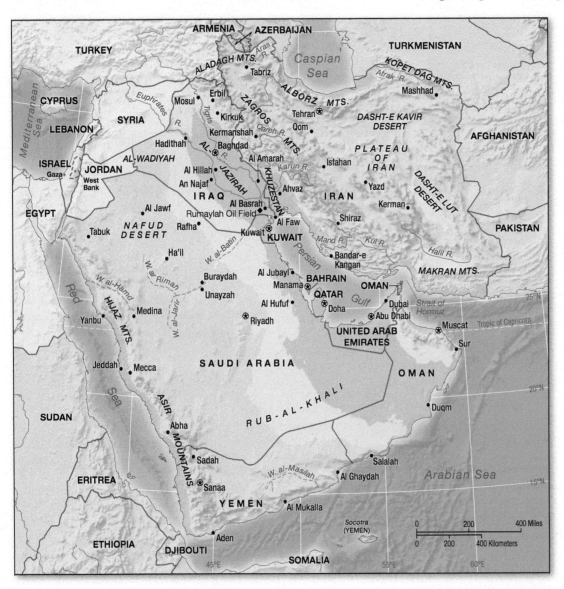

◀ **Figure 7-40 The countries of the Persian Gulf.** Iran, Iraq, and Saudi Arabia dominate the Persian Gulf region.

▲ **Figure 7-41 Oil tankers at Kharg Island, Iran.** Gas and oil dominate the economies of the countries bordering the Persian Gulf. These offshore oil-loading facilities are the major export outlet for Iranian oil.

agricultural base of their home communities. Population growth, agricultural intensification through irrigation, and urbanization all place great pressure on local water supplies. Overexploitation of groundwater, waste disposal, the economic limitations of desalting ocean water, and salinization of irrigated soils are all problems related to the region's limited water supplies.

Low Population Density

While most Gulf States are not densely populated and have small total populations, Iran, Iraq, and Saudi Arabia all have large populations that dominate the regional scene, although large parts of each country's territory is almost devoid of people. Even in countries with low overall population densities, the effective density is much higher because the population is concentrated in small, highly productive zones. Thus, the metropolitan area of Kuwait City contains 2.4 million of the country's 2.9 million people (Figure 7-43). The contrast between small nodes of dense population with high levels

▼ **Figure 7-42 Date harvest in Iran.** In the oasis city of Bam, Kerman province, sorted dates rejected for export are hauled away for use as animal fodder. Qanats provide the water for the date palms, and women and children sort and pack the dates while the men concentrate on the heavy field work.

▲ **Figure 7-43 Kuwait City.** Rapid urbanization and modernization characterizes the cities of the Gulf countries. Expenditure of massive oil revenues has created modern urban centers where often in the smaller states a majority of the country's population lives.

of economic activity and the vast expanses of essentially uninhabited space characterizes the landscapes of the Gulf States.

How to bring development to these low population spaces and maximize the benefits of oil revenue before the wells run dry is the critical question. Oman has launched a major development initiative in its poorest, least populated area, Wustah governorate. A small fishing village, Duqm, midway between Muscat and Salalah, is being transformed into a modern port. The second largest drydock in the Middle East is already established, housing for an eventual 100,000 population has begun, a refinery and petrochemical complex are intended as the focus of a special economic zone, an airport and coastal hotels are in the works, and a railroad and pipeline are envisaged as links to the interior. Why here? In part Duqm's geostrategic location is the answer. Close to sea lanes crossing the northern Indian Ocean, a major port could capture some of the business now concentrated in Aden, Yemen. But more important is the site's potential role as an alternative to tanker passage through the Strait of Hormuz. Any sign that a dispute with Iran might close the Strait to oil shipments causes oil prices and insurance rates to skyrocket. Linked to Gulf oil fields by pipeline, oil exports could bypass a risky passage through the Gulf, Duqm would boom as a transshipment center, and jobs would be brought to a region where few modern options presently exist.

Prominence of Religion

Equally distinctive is the role played by religion. While Islam plays a fundamentally important part of people's lives in all of the Middle East, in the Gulf countries it assumes a particularly pervasive role. In what is now Saudi Arabia, Mohammed first revealed himself as a prophet and launched the preaching mission that, by persuasion and conquest, came to control the region both spiritually and politically. Here are the major centers of pilgrimage for Muslims: Mecca and Medina (Figure 7-44). Never directly controlled by European imperialism, the inhabitants of the Arabian Peninsula and of Iran, surrounded by mountain barriers and barren deserts, remained aloof from substantial contact with non-Muslim cultures and their influence.

▲ **Figure 7-44 The Great Mosque in Mecca, Saudi Arabia.** Pilgrims from all over the Muslim world gather both inside and outside Mecca's major mosque during the 2012 hajj (pilgrimage).

Oil revenues have greatly transformed local societies and economies. No state in the subregion has been able to isolate itself from oil wealth and the stress it places on indigenous social and cultural systems. As a result, controversy has arisen over the role of traditional values in governing contemporary life. In Iran, where secular trends progressed rapidly under the monarchy, opposition led by Shi'ite Muslim clerics ultimately drove the Shah from power in 1979 and substituted a much more traditional set of norms, values, and customs in behavior and dress. This fundamentalist Islamic revival stressed the importance of traditional values and behavior in opposing Western-style modernization. The more rapid and extreme the pressure for change, the more violent is the reaction from traditional centers of authority and belief. The dominance of the more traditional supporters of President Mahmoud Ahmadinejad during his two terms in office demonstrate the resilience of traditional patterns and the toughness of their advocates. Control of election institutions by traditionalists makes it difficult for reformist candidates to stand for election, let alone win, and ensured a conservative parliamentary majority in the March 2008 elections and Ahmadinejad's reelection in 2009. A two-term limit on holding the presidency insured that the 2013 election would produce a new leader. But few expected that a relatively centrist candidate such as Hassan Rouhani, a more moderate cleric in a largely conservative field of candidates, would emerge as the winner. Iran's serious economic difficulties (in part due to sanctions that prevent oil exports, imposed by the international community in an effort to change Iran's nuclear acquisition policy) and widespread support for a relaxation in social restrictions led slightly more than 50 percent of voters to support Rouhani. With considerable international experience as Iran's negotiator on nuclear issues, Hassan Rouhani's election offers hope that progress on domestic reform and a peaceful resolution of worries about the purpose of Iran's nuclear development program can occur during his term in office.

In Saudi Arabia, secular law has never replaced the Sharia law of Islam. The Ulama, the religious teachers and scholars, have maintained influence over daily life. While traditional Muslim women veil in the countries of the Mediterranean Crescent, it is not required for all women; in most of the Gulf States, it is. The separate social spheres of men and women are strictly maintained, females participate much less actively in public life, educational and employment opportunities for women are more limited, and males dominate household life and decision making to a greater extent than in the Mediterranean countries (see *Geography in Action:* Women in the Middle East and North Africa: Breaking Down Opportunity Constraints). In the countries of the Gulf, strict adherence to accepted norms, sometimes enforced by morals police, is the common pattern of social life to which foreigners as well as locals are expected to conform.

Weak Urban Traditions

Until oil development created new employment opportunities and sparked massive population movement to a few urban centers, most people in the subregion were engaged in traditional livelihood activities: oasis agriculture, rain-fed farming in the more mountainous areas, and pastoral nomadism. Thus, urban traditions are generally weak in the oil-rich states, and until recently the educational and technological sophistication of the bulk of the population was limited. That characteristic stands in sharp contrast to the urban sophistication and historical importance of the main urban centers of the large states in the region—such as Tehran, Esfahan, Shiraz, Mashhad, and Qom in Iran and Baghdad, Basrah, and Mosul in Iraq. In the smaller, more traditional states of the Arabian Peninsula, the traditional skills of most people are unsuited to employment in many aspects of the oil industry.

In addition to agriculture and animal husbandry, traditional skills often relate specifically to trade. Many of the smaller towns in the Arabian Peninsula grew along trade routes that linked the eastern Mediterranean overland via camel caravan to the Persian Gulf, the Indian Ocean, and South Asia. The Prophet Muhammed was engaged in this trade. Muscat was the base for Omani mercantile sea routes to East Africa and South and Southeast Asia, in competition with the Portuguese for control of the Indian Ocean, before accepting the presence of a Portuguese fort at Muscat from 1507 to 1650.

The shortage of many of the skills needed for modern economic life has serious consequences. Gulf States recruit skilled personnel from industrialized countries—as well as from India, Pakistan, and Palestinian areas—to fill specialized job niches. Most domestic household workers come from the Philippines, Ethiopia, India, and Sri Lanka, while Pakistanis have replaced Egyptians and Yemenis as the preferred labor for agriculture, construction, and transportation. Foreign worker domination of the labor market is particularly overwhelming in some of the smaller Gulf countries. Native UAE citizens comprise only 18 percent of Abu Dhabi's population of 2.4 million. In Qatar, foreign workers constitute 94 percent of the active labor force. This is an unhealthy economic and social situation, and the size of the foreign labor population compared to Qatari citizens can constitute a potential security problem. Iran—by virtue of its size, diverse resource base, and large population—has been more immune to these pressures. Dissatisfaction with this labor force

Women in the Middle East and North Africa: Breaking Down Opportunity Constraints

The level of women's personal freedoms, access to educational and economic opportunities, and participation in social and political processes, vary widely from country to country in North Africa and the Middle East. They range from the secular parliamentary democracy of Turkey, where about 15 percent of the members of parliament are women (Figure 7-5-1), to the theocratic monarchy of Saudi Arabia, where women hope to be able to vote and run in municipal elections for the first time in 2015. While there is progress throughout the region, it is slow and full of small triumphs and frustrating setbacks in a struggle against deeply rooted social restrictions on women's self-determination at all levels of society. This constitutes a colossal waste of human potential, both socially and economically. The most extreme interpretations of Islamic and cultural traditions insist on women's invisibility to uphold moral rectitude and family honor, be it behind anonymous robes and veils or behind impenetrable household walls.

The Arab Spring was a defining moment for women as they took to the streets together with men. As new constitutions are in progress in countries like Tunisia and Egypt, women are pushing for the expansion and protection of their rights. But long-established social norms cannot be overcome in one giant leap. Conservative forces are pushing back throughout the region and it will take a widespread change in attitude toward the role of women from the bottom up, starting in the family, as well as from the top down where women are being elected in greater numbers to new parliaments and participating in writing new constitutions.

Education remains key to empowering and launching women on a path of greater self-determination and diverse contributions to their respective societies. While the overall female literacy rate is

▲ FIGURE 7-5-1   **The Turkish Parliament in Ankara.** Women play an increasing role in Turkish politics, appearing in important party and committee leadership positions.

still much too low, particularly in rural areas, great strides have been made toward universal schooling at least at the six-year primary level in most countries of the MENA region. The literacy rate for females ages 15–24 across the region is around 85 percent as educating young girls is becoming the norm, particularly in urban areas. As younger women enter their prime adult years, they will have a positive impact on advancing women's education more generally. Much still needs to be done to promote and support secondary and continuing education. Many women have to leave school too early because of family constraints, whether financial or customary—even more enlightened fathers and husbands often do not see the point of further education in the face of marriage, child rearing, and all manner of family responsibilities. And even well-educated women have a much higher unemployment rate than men. Women make up little more than 25 percent of the region's formal labor force, as they are either forbidden or discouraged from engaging in income-generating

work outside the home in favor of the traditional domestic roles that control the interactions between men and women and keep women respectably segregated.

Yet education and employment options are important to improving families' well-being as well. As women gain access to income-earning opportunities, they help their families attain a higher, more middle-class quality of life. This is enhanced by broader knowledge and understanding of available options, which allow women to promote family planning, improve family health and nutrition, and set a role model for the next generation of females (and males) in closing the gender gap.

Sources: S. El Feki, *Sex and the Citadel: Intimate Life in a Changing World* (New York: Pantheon Books, 2013); World Bank, *Middle East and North Africa: Women in the Workforce* (Washington, D.C.: World Bank 2010), http://go.worldbank.org/S8N5AO53H0; F. Roudi-Fahimi and V. M. Moghadam, *Empowering Women, Developing Society: Female Education in the Middle East and North Africa* (Washington, D.C.: Population Reference Bureau, 2003).

▲ **Figure 7-45 University library in Al Ain, United Arab Emirates.** A professor engages in an informal discussion with students. Gulf States are investing in state-of-the-art expertise in science, medicine, business administration, and many other fields.

situation has led most oil-producing states to engage in vigorous educational and job-training activities in an effort to replace foreign experts with local personnel (Figure 7-45). But high-paying, responsible jobs in the modern economy are not unlimited, and males often are unwilling to fill positions they regard as menial.

International Relationships

The oil resources of the Gulf States are of overwhelming importance for the global economy. More than two-thirds of the known oil reserves outside North America, the countries of the former Soviet Union, and Eastern Europe are found around the Persian Gulf. Saudi Arabia's oil reserves are second only to Venezuela, and represent about one-fifth of the world's reserves. Saudi Arabia is also the world's second largest oil producer, after Russia.

Beginning with the discovery of oil in Iran in 1908, oil discovery and development spread southward through the Persian Gulf regions of the Arabian Peninsula, and knowledge of the extent and extractable quantity of oil and gas in existing deposits has grown rapidly. In the early years of oil development in the Gulf, Western oil companies controlled exploration, extraction technology, transportation, refining, management expertise, and market access. Gulf governments received only a fraction of the revenue generated from exploitation of their oil and gas, and lacked the capital to develop the resources themselves. Gradually that situation changed as countries developed the technical and administrative skills of their own citizens. Many countries nationalized at least a controlling interest in the firms that were developing their resources. In the world market, long-term demand for oil outstripped the supply, so nationalization and dramatic increases in oil prices took place in tandem.

During the 1980s, it appeared that the Gulf States' dominance in petroleum production might diminish. Demand decreased due to energy conservation in some industrialized countries and new petroleum deposits were discovered in other parts of the world. War between Iran and Iraq damaged oil production, refining, and

transportation facilities in both countries. At the same time, other Gulf States reduced their output to maintain higher prices and conserve oil reserves for possible development of petrochemical industries. The first Gulf War (1991) severely damaged Kuwait's oil facilities, and it took time to bring its production back to the global market. By the time Kuwait had doubled its pre–Gulf War I production, in 2004, sanctions crippled Iraq's production. Even after the Ba'ath regime was overthrown, Iraq's aging oil infrastructure was repeatedly damaged by rebel attacks and output stagnated. Rising demand in China, India, and other developing countries with rapidly expanding economies helped to push oil prices steadily upward. By early 2008, oil prices were temporarily in excess of $100 per barrel. Concerns among UN member states that Iran's push to build a nuclear energy capability would also include nuclear weapons acquisition has resulted in sanctions that reduced export of Iranian oil. Uncertainty about the security of existing oil production and delivery systems in the Gulf contributed to volatile global oil prices. Concern was heightened because so much of the world's production capacity is concentrated in one relatively small and potentially vulnerable region. Given improvements in recovery technology, coupled with the traditional tendency to understate or otherwise obscure reserve capacities, the continued prominence of the Gulf States in the world petroleum economy seems likely for the foreseeable future.

Development and the Petroleum Economy

Large petroleum outputs and soaring oil prices have moved such former backwaters as Kuwait, Abu Dhabi, and Oman into international prominence. Formerly poor, oil-producing states face the task of coping with an embarrassment of riches. Although abundance has been the watchword of the recent past, the long-term future is by no means assured. Some states, such as Oman and Yemen, have sufficient revenues for the present but must use their money wisely to prepare for the day when oil revenues might not be available. Diversification of oil-based economies, development of industrial processes that do not depend on oil, and rejuvenation of often sluggish agricultural sectors are top priorities.

The Gulf States are dealing with the eventual decline in oil revenues in two ways. The first approach invests petroleum income internally, in development projects that generate long-term income and employment—such as expanding irrigated agriculture, building a petrochemical and plastics industry, creating tax-free zones that encourage foreign investment; in designing higher-education institutions with regional appeal; and in social services that improve citizens' material well-being in the short term. The second approach invests capital that cannot be absorbed internally into overseas enterprises.

Internal Investment of Petroleum Income

The Gulf countries have spent large sums of petrodollars to improve infrastructures in the Gulf States. Ports have been built, such as Doha in Qatar, where none existed before. Investments in airports, highways, sewer systems, water systems, and pipelines have drastically changed the appearance of many oil states. Housing projects

▲ **Figure 7-46   The Mutthra district in Muscat, Oman.** New Housing and public buildings, funded by oil revenue, have almost totally replaced traditional housing stock in Muscat. The potential for floods is high when runoff from surrounding mountain slopes encounters densely packed housing on the valley floor.

are also common. Oman has rebuilt its capital, Muscat, destroying much traditional architecture in order to construct modern housing units for its growing urban population (Figure 7-46). Most citizens regard such changes as an inevitable and progressive aspect of modernization.

Substantial sums are also invested in agricultural improvements. In many oil states, desalinization plants now operate, providing water primarily for drinking but also for greenhouse irrigation of vegetables to help feed burgeoning urban populations (Figure 7-47). Agricultural operations, while often heavily

▼ **Figure 7-47   Desalinization plant in Dubai, United Arab Emirates.** In an effort to cope with limited surface and groundwater resources, most Gulf states operate desalinization plants to provide water for drinking purposes.

subsidized, are one way in which abundant financial resources can help to reduce food imports. Investing in sustainable activities to produce marketable goods without harming the environment is difficult. Achieving an operational scale that is commercial rather than experimental is particularly problematic. Unfortunately, few of these agricultural development schemes are practical in economic terms, and most require skills that many local residents lack. The technology employed frequently brings with it serious land-management problems. Many irrigation schemes have caused serious soil salinization due to mismanaged application and drainage of irrigation water. Most have required financial subsidies that cannot be maintained indefinitely.

A number of Gulf countries are making major efforts to create twenty-first century tourist and business environments that will have global economic impact. Dubai is one of the most extravagant actors on the scene, transforming a desert coastline into an urban landscape replete with eight-lane highways and thrusting skyscrapers. Most spectacular is the creation of offshore islands in the form of palms that have drastically increased the Dubai coastline through land reclamation at an unprecedented scale with the help of Dutch expertise (Figure 7-48). "The World," a 300-island archipelago in the shape of a world map, is also being created. Millions of tourists each year already fly in to shop in Dubai's malls, sleep in opulent hotels, and amuse themselves in exotic settings such as the indoor, acclimatized ski slope in the Mall of the Emirates or an underwater hotel and water world where one can swim (safely isolated) among sharks. That coastal ecosystems are destroyed as sea bottoms are dredged, islands are created, and seawater becomes opaque is simply considered the cost of projects that are expected to generate massive levels of return to the owners of development rights. From the local investment perspective, building houses on the sand is a triumph of ingenuity rather than a prescription for potential disaster in a global warming future. This investment, however, may not prove sound as tentative reports surface of sinking of these land reclamation projects by several millimeters a year.

▼ **Figure 7-48   Land reclamation offshore in Dubai, United Arab Emirates.** Artificial offshore islands create development opportunities on a massive scale: the Palms. These creations are part of a larger effort to make Dubai a commercial and entrepreneurial center for tourism and finance for the larger region.

Little effort is directed toward development that builds on areas of local expertise. Especially neglected is the pastoral sector, where considerable traditional experience still exists and where past use of low-productivity rangeland was once well managed. Today most Gulf States import their meat from industrialized countries, such as Australia, even though the population prefers to consume local sheep, now largely unattainable at a reasonable price.

In Oman, investing oil revenues to develop agriculture and maritime fishing and to diversify industry is especially crucial because existing oil reserves of 5.5 billion barrels are relatively modest (24th in the world). At current rates of production, there are concerns that Oman's oil and gas reserves will run out sooner rather than later. Although some new oil and gas fields have been developed in southern Oman (Dhofar), and an active exploration program has been pursued, the prospects for major new discoveries are uncertain. Development of natural gas resources has generated an important new domestic energy source, but has been overwhelmed by rapidly increasing demand—to the point where periodic electricity blackouts have become common. Because agriculture is limited by low and unreliable rainfall, great expansion of Omani farm holdings is not possible. Consequently, Oman aims eventually to achieve food self-sufficiency both by improving the efficiency of existing agricultural systems and by developing underground water resources in a series of small-scale projects. Investment in the agricultural sector is necessary in the oil-producing states to ensure economic viability in the future, but is hard to achieve when lower-cost food supplies can be imported from abroad.

Industrial development is also taking place in the Gulf States. First priority is frequently given to the building of cement factories to support the construction industry. Also common are petrochemical plants that produce fertilizers, chemicals, and plastics, as well as plants that liquefy gas for export. There are many consumer-oriented enterprises, including soft-drink bottling plants, flour mills, fish-processing plants, and textile firms. Craft traditions often give focus to small-scale industrial development that can survive if there is a tourist and export trade to which it can cater.

The Gulf States are also attempting to overcome their shortages of skilled workers and professionals. Almost every country has invested in modern universities (see *Geography in Action: Science and Higher Education in the Middle East and North Africa*) as well as the rapid development of primary and secondary school systems. To staff those systems, numerous foreign teachers have been hired, often recruited from more educationally advanced Middle Eastern countries such as Egypt. Large numbers of students also study abroad under governmental contract. When they return, they are channeled into decision-making positions at all levels of the government. Newly acquired expertise makes it increasingly possible to replace foreign personnel in the oil-producing states.

Petrodollar Investment Overseas

Unable to absorb all of their oil revenues in internal investment projects, oil-producing states seek investment opportunities overseas. These transfers of funds represent the actions of both wealthy private investors and the region's governments. Increasingly, oil money is used to purchase property, banks, farmland, or industrial enterprises in industrialized countries. The governments of oil-producing countries have become active investors in the industrial and service infrastructure of oil-consuming countries.

Investments by sovereign wealth funds are not easy to track, since neither private parties nor public agencies seek to publicize the specifics of their activities. Real estate purchases in London and Paris are an investment focus, but less visible sites also attract investment. In 2012, several investments made headlines. The Abu Dhabi Investment Authority bought an office park in Ghent, Belgium for €110 million. Abu Dhabi's government-owned Mubadala Development Company sank US$2 billion into Brazil's Centennial Asset Equity Fund. Qatar National Bank paid a "depressed" price of US$1.9 billion to take over the Egyptian branch of Nationale Societé General Bank, France's number two lender. It did this precisely because Egyptian political unrest and uncertainty made the acquisition a bargain purchase. The goal of all such investments is to make money on the investment and multiply current petroleum income for economic support in the future.

Changing Power Relationships

Oil wealth has altered the world's balance of power in several ways. Most of the oil states have used their wealth to purchase arms to modernize their military forces. Possessing sophisticated weaponry made it possible for Iraq to attack Iran in 1980 and prompted other Arab states to purchase arms to counterbalance Iraq's power, as well as that of Israel. The result has been a Gulf arms race, in which the industrialized nations sell weapons to friendly countries. The fact that those weapons can also be used on brother Arabs was demonstrated by Iraq's invasion of Kuwait in 1990.

The region's dominance in the **Organization of Petroleum Exporting Countries (OPEC)** also gives it substantial global political power. Proposals to raise prices or threats to embargo oil shipments to supporters of Israel have sent tremors throughout the industrial world. The oil-producing states are clearly aware of their position in the global economy and are not afraid to use their influence and pursue their own political and economic strategic objectives. The psychological satisfaction derived from such a superior political position explains much of the contemporary behavior of these states. Thus Saudi Arabia supports Sunni rebels in the Syrian civil war, because it expects a much more acceptable regime to emerge than the Ba'athist government that has controlled Syria since 1963. Iran exerts its influence in similarly indirect ways, by providing support to Shi'ite groups such as Hezbollah in Lebanon, the Assad Ba'ath regime in Syria, several of the Shi'ite militias in Iraq, and Hamas (a Sunni group) in Gaza. Qatar intervened directly in the Libyan civil war, sending special forces to provide basic tactical training to the rebels and air force units to buttress the UN no-fly zone, perhaps with the hope of gaining an inside track on future business contracts and political influence. The United Arab Emirates hope to transform into a major entrepôt for business and tourism. And Abu Dhabi, Doha, and Dubai are all pushing airport expansions that will, if realized, make the Gulf and its airlines a hub of the global air transport system.

The emergence of the **Gulf Cooperation Council**—including Saudi Arabia, Kuwait, Bahrain, Qatar, the United Arab Emirates,

Science and Higher Education in the Middle East and North Africa

The pursuit of scholarship and learning has powerful roots in the Middle East and North Africa. The richness of Egyptian civilization, symbolized by such magnificent institutions as the ancient Library of Alexandria, was a wonder even in Greco-Roman times, and the splendor of its scholarly institutions increased further after the Arab conquest in A.D. 640. Arab scholars were largely responsible for preserving the writings of the classical world and transmitting them to the Christian West. When the Dark Ages enveloped Western Europe, scholarship and invention characterized the universities of the Middle East. A major center of scholarship was established in "The House of Wisdom" in Baghdad a thousand years ago. With the Quran's admonition of "the scholar's ink is more sacred than the blood of martyrs" in mind, Arab-Islamic scholars have contributed to all fields of science for many centuries, but most institutions declined and many disappeared with the collapse of the Abbasid empire in the thirteenth century.

Today the MENA region is seeing a revival and resurgence of academic institution building and scholarship at unprecedented levels. Today's Bibliotheca Alexandrina, near the site of its venerable predecessor and inaugurated in 2003, is a state-of-the-art library and cultural center on the shore of the Mediterranean. Renowned institutions such as the American University in Cairo (established in 1919) and the American University of Beirut (established in 1866) have been joined by numerous new developments. Some of these are American-model universities, such as Al-Akhawayn University in Ifrane, Morocco. Others pursue major institutional partnerships, as in the case of

▲ FIGURE 7-6-1 **The Weill Cornell Medical College at Education City in Doha, Qatar.** One French, one British, one Qatari, and six American universities are clustered together in a group campus on the outskirts of Doha. The goal is to create a center of educational excellence that will contribute to local development and attract students to the region from far afield.

Education City in Doha, Qatar, a complex spanning several square miles that hosts branch campuses of six American and two European universities (Figure 7-6-1).

This is particularly important around the Persian Gulf where countries want to prepare for the future as their oil-based economies need to transition into knowledge-based economies. Students from all over North Africa and the Middle East as well as South Asia pursue quality college and university educations to prepare for the professional needs in their home countries, which are competing in an increasingly globalized world. For now these opportunities are limited to an elite segment of Middle Eastern society that can afford this privileged education. But as higher education at all levels improves and the middle class expands in the region,

there will be ever-increasing demand for high-quality degrees and professional skills. In the process, interaction between cultures—involving students, teachers, and increasingly popular study abroad programs—is inevitable. Both forces of promoting preservation of regional traditions and encouraging participation in global opportunities are strong in the Middle East and North Africa, and these expanding institutions promote and support multicultural encounters that advance mutual understanding.

Sources: "Islam and Science: The Road to Renewal," *The Economist,* 26 January 2013, 54, 55–56; T. Lewin, "Global Classrooms, Made in the U.S.A.," *New York Times,* A1; and February 11, 2008, A1; M. E. Falagas, E. A. Zarkadoulia, and G. Samonis, "Arab Science in the Golden Age (750–1258 C.E.) and Today," *The FASEB Journal* 20 (2006): 1581–1586.

---

and Oman—in 1981 as a regional organization was a local response to the need for a unified approach to common trade and security concerns. But the council's relative lack of independent military power means that it must rely on support from other countries.

Changing Social Conditions

Developmental growth fueled by oil wealth has brought massive changes to the societies of the oil states. One important change has been rapid urbanization. Attracted by new job opportunities, social welfare programs, and an atmosphere of excitement and diversity,

▲ **Figure 7-49 The northern suburbs of Tehran, Iran.** The Alborz Mountains provide a spectacular backdrop for Tehran's northern suburbs. Rapid urbanization fueled by oil wealth has produced many social changes as well as a spectacular growth in Tehran's size.

rural inhabitants have flocked to urban centers. Doha, the capital of Qatar, an insignificant village in 1960, recorded 796,000 inhabitants in its 2010 census. Similar growth has taken place in the capitals of most of the region's states and sheikhdoms: the population of Baghdad, Iraq, now exceeds 7.6 million; the metropolitan area of Tehran, Iran, had grown to at least 12.2 million people by the 2011 census (Figure 7-49); and Riyadh, Saudi Arabia, has passed the 5.2 million mark.

Other expanding cities also are a focus of regional migration in the large Gulf States—Mosul and Basra in Iraq; Esfahan, Tabriz, and Mashhad in Iran; and Mecca and Jeddah in Saudi Arabia. Much of the growth of those cities has been accompanied by the erection of shantytowns and other temporary housing on the outskirts of the urban areas. The sanitary and health problems that have resulted from excessive crowding are gradually being relieved as oil money is invested in improved housing and infrastructure.

As centers of economic growth and change, cities are also the scenes of cultural conflict. Many rural migrants have difficulty adjusting to urban life in the Gulf States, just as their counterparts do in other parts of the world. Traditional Islamic and customary values are receiving their strongest challenge in urban areas. Bombarded by a variety of exotic stimuli—ranging from the material possessions of the industrialized West to the clothes, movies, and behavior patterns of the foreign employees of oil and construction firms—the citizens of the oil-producing states have been forced to confront a new social setting.

Values conflicts can produce serious social strains. Traditional leaders often react forcefully to apparent violations of cultural and social norms. Traditional values are upheld by religious leaders who object to the pace and direction of social change. The resurgence of fundamentalist Islamic approaches to social organization represents a profound challenge to the secular, modernist, Western-influenced segments of society and to the aspirations and struggle of women for equal treatment. Work habits and social priorities

also differ. New urban arrivals find it difficult to accept Western notions of allocating time and resources in the increasingly globalized business environment; they often cannot reconcile the need to apply financial gains to social expectations, such as brideprice or their kinship obligations, with the competing drive to apply resources to savings or job advancement. Poor mutual understanding of cultural values results in tension between native workers and foreign employees.

Great distinctions in income and status have also appeared as a result of the oil boom. Although some of the newfound wealth has filtered down to lower social levels, much of the income gained is eroded by an inflationary spiral triggered by oil development and accelerated imports. Individuals close to the import trade, local representatives of foreign firms, individuals with contacts in high government positions, and professionals have benefited the most from petroleum growth. The unskilled are left behind. Some states—Kuwait, for example—provide free social services to improve living conditions, but others have been less farsighted. Often the gap between the ruling elite and the majority of the population is wide, a condition particularly evident in states with low population densities and massive oil revenues, but also found in the region's more populous states (Figure 7-50).

Changing Political Allegiances

Social unrest casts shadows over the political future of many oil states. Opposition to the central governments in many countries comes from a variety of sources, including ethnic and religious minorities, traditional leaders alarmed at changes that are taking place, rural people who feel neglected by distant governments, and modernist forces that wish to accelerate the pace of change and radically restructure society. And the large numbers of foreign workers within many oil states' jurisdictions feel no particular loyalty to the existing

▼ **Figure 7-50 Camels and cars on the desert-urban edge.** Camels take precedence over cars at this camel crossing at Nad al-Sheba in Dubai. Here the "ship of the desert" remains valued for its racing speed and young boys hold sway as jockeys.

governments. The result is a highly unstable set of conditions that could explode into conflict at any time.

The Kurds are an example of an ethnic minority in Turkey, Iraq, Iran, and Syria that has struggled for greater autonomy, either within the structure of existing states or through complete independence. In Iraq, the Kurdish population of the north was engaged for years in armed rebellion against the Arab-dominated central administration. Although a desire for greater regional autonomy was the apparent reason, greater revenue from the northern oil fields in or near Kurdish territory was an underlying motivation.

Revolutionary Iran has many sympathizers in the Shi'ite religious minorities in the Gulf States. Iran's growth as a regional power has implications for regional stability. Equally significant were the military and technological capabilities developed by Iraq during the 1980s and 1990s, which forced its Arab neighbors to ally themselves with outside political and military powers to confront Iraqi aggression. Caught between a Shi'ite dominated Iraq, a religiously zealous Iran, and the economic interests of the Western powers, the traditionalist societies of the Gulf States find themselves in a difficult situation. The U.S.-led 2003 intervention in Iraq had an unintended consequence: fundamentalist Sunni fighters from Saudi Arabia, Yemen, and other Gulf countries joined many Sunni Iraqis in opposing a continued U.S. military presence. Similar fighters have appeared in growing numbers and with increased effectiveness among the opponents of the Ba'ath regime in Damascus. However the Syrian civil war ends, the return of these Islamist volunteers to their home countries in the Arabian Peninsula is an unsettling security issue for Gulf governments.

Equally significant are governmental changes that might result from potential military coups. Young officers in many of the oil states often possess nationalistic and pan-Arab ideals. Impatient with the pace of economic development, the pronounced social and economic inequalities in many oil states, and the corruption of a boom economy, they could use their positions to seize power. Concern over excessive social permissiveness and the resentment toward foreign workers also could spark unrest in the lower levels of the social and military hierarchy, and fuel opposition from traditional religious authorities and their supporters.

Stop & Think

▶ What development strategies might oil-wealthy countries in the Gulf pursue to maximize benefits for all their citizens?

Petroleum Powerhouses: Isolation and Globalization

The Gulf States exhibit many contrasts. While the smaller political entities of the Gulf share many of the dilemmas of their larger neighbors, their issues are relatively tractable because of the smaller scale at which these difficulties are encountered. For the large states of the Gulf—Saudi Arabia, Iraq, and Iran—the policy dimensions of large national territories, uneven distribution of population and resources, rapid economic development, and environmental change pose profound possibilities and problems. In each country, patterns of isolation contrast with pressures for linkage to distant parts of the globe. The outcomes of these pressures are distinct local responses to change, and in each case, considerable isolation of local society from the global community.

Saudi Arabia: A Desert Ruled by Islam and Oil

Saudi Arabia occupies the bulk of the Arabian Peninsula, separated from the African continent by the flooded Red Sea rift valley. Saudi Arabia resembles a giant block that is raised on one side and sunken on the other. The block's western side rises abruptly from the coastal lowland, with some peaks in the Asir and Hijaz mountains approaching 10,000 feet (3,000 meters), heights that result in the country's coolest temperatures and greatest rainfall. From the highlands, the block gradually decreases in elevation toward the center of the peninsula. This central region is very dry, but contains numerous seasonal streams and depressions where groundwater is close to the surface and supports oases. In the Nafud Desert in the north and the Rub al-Khali (Empty Quarter) in the south, surface deposits of shifting sand dunes dominate the landscape. Along the Gulf coast, the block becomes a broad, featureless coastal plain, dotted with small oases and underlain by large oil and natural gas deposits. Saudi Arabia has less arable land than Egypt without the compensating benefit of a large river, so historically trade rather than agriculture has been a pathway to prosperity. Mecca and Medina became economic centers because of locally favorable agricultural settings and strategic positions at the junction of major caravan routes. Elsewhere nomadic pastoralism was the most common livelihood activity, providing animal products to the peninsula's trade centers and camels for its caravans.

Discovery of oil along the Gulf coast in the late 1930s transformed the Saudi economy. Although other minerals, including zinc, iron, copper, and silver, are known to exist, they are not found in large deposits. Discovery of gold, bauxite, and phosphates in sufficient quantity to justify exploitation and export is a recent development. Saudi Arabia's oil reserves are the second largest in the world and exploitation of these resources has pulled the country out of its relative pre-1945 isolation. Great efforts have been made by the Saudi government to invest oil profits in productive development projects, but the results have not been uniformly successful.

One consequence of the changes initiated by oil revenues was rapid population growth. In 1970, population grew at a rate of 4.6 percent per year, and was still growing at nearly 4.0 percent in 2000, although growth declined to 2.8 percent by 2012. This rapid growth resulted in today's large, youthful population and the government's concern about its ability to feed its population. Afraid that imported food could potentially be cut off, Saudi Arabia embarked on a policy of subsidizing domestic agriculture. The goal was independence in basic food supplies. By 2003 Saudi Arabia was the world's sixth largest exporter of wheat! But this rapid ascent to major player in the global grain market invariably proved unsustainable. The irrigated circles of wheat that flourished in the desert depended on government subsidies for fuel and equipment as well as fossil groundwater. Aquifers that sustained the system rapidly shrank. With every grain of wheat shipped overseas a considerable amount of virtual water went along. For a very dry country to ruin a vital, essentially nonrenewable, water resource for a temporary, unsustainable financial gain made little sense. By 2009 the effort was abandoned. Saudi Arabia committed to becoming a food importing country again, but with a twist. It plans to seek agricultural development opportunities outside its borders, potentially in Sudan, Ethiopia, the Philippines, and Vietnam (Saudis prefer long-grain rice to all other cereals), where it can invest capital and expertise in equipment, land, and technology in an effort to promote yield improvements that will help all parties. At home, a concentration on water-efficient agriculture, producing vegetables and other crops with limited water, is now the goal.

Efforts to diversify the economy have also encountered problems. Some elements of public infrastructure development have been reasonably successful. Thermal electricity generating plants, now fueled mainly by natural gas rather than oil, have brought reliable electricity to most parts of the country. Saltwater desalinization plants generally have kept pace with the needs of an increasingly urbanized population. Sewage treatment plants have not matched urban population growth. Most industrial investment has gone into petrochemical plants and oil refineries, all tied to the continued availability of oil and natural gas as their raw materials. Bauxite and phosphate mining is increasing and a north-south railroad line to carry the raw materials to the Gulf coast is in the last stages of construction. Beyond these developments, few successes exist; none provide employment for many people.

Saudi Arabia's other distinctive attribute beside oil is religion. Islam began in the Hijaz where Mohammed received his spiritual mission. From small beginnings Islam rapidly spread to neighboring regions. Mecca and Medina, crucial to Mohammed's early prophetic period, are the destinations for millions of pilgrims every year—3 million in 2012—to perform the hajj, an essential obligation for believers (Figure 7-51). Protecting the holy sites visited by these pilgrims is a fundamental responsibility of Saudi Arabia's ruling family and government. But religion is much more deeply embedded in Saudi society than facilitating the annual pilgrimage. Saudi Arabia's ruling dynasty came to power through an agreement between Abdul Aziz bin Saud and an extremely fundamentalist Islamic sect, the followers of ibn Abd al-Wahabb. This crucial decision has given Saudi Arabia's Wahabbi-dominated Islamic scholars control over many aspects of contemporary Saudi life. To an extent unprecedented in other Islamic schools of thought, Wahabbism claims direct control over public and private morals, condemns as heretics and backsliders anyone who does not share their views, promotes *jihad* aggressively, and rigidly interprets Islamic legal rules.

The stranglehold this places on opportunities for economic and social development in Saudi Arabia is significant. Morals police intervene to prevent men and women from participating in mixed gatherings in public. Schools are segregated by gender, and, despite a surplus of qualified female teachers, women cannot teach boys. Most women are unable to travel outside the home without permission of their husband or a male family member, and often lead a restricted existence. From a Wahabbi perspective, the sole purpose of schools is to teach knowledge of and respect for the Koran and religious purity. The initial Saud-Wahabbi compact resulted in Wahabbi dominance of the education system and teachers, administrators, and the curriculum must conform to Wahabbi doctrine. Educational reforms that would prepare Saudi youth with the skills,

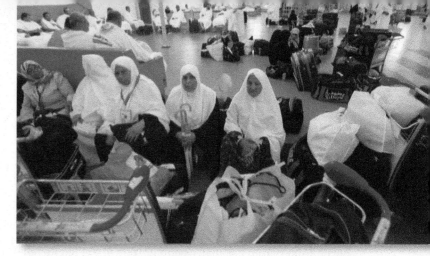

▲ Figure 7-51 **Pilgrims at King Abdulaziz International Airport, Jeddah.** Located a short distance outside of Jeddah, the King Abdulaziz Airport is the major entry for pilgrims to Mecca, which is only 40 miles (64 kilometers) away.

knowledge, critical thinking, and problem-solving capabilities needed for jobs in a diversified modern economy are lacking. Men often are not qualified for the jobs they seek, and women, who often have adequate qualifications, are not allowed to work because they might come into contact with men.

Unemployment is widespread among young adults (20–24 years old); 10 percent is the unofficial national statistic. But targeted surveys suggest that 45 percent of female and 30 percent of male young adults lack work. Most males shun manual labor and aspire to government jobs, which promise a desk and potentially an office; most women cannot find jobs outside the home, and, when family wealth and status are substantial, are even barred from many domestic tasks. This creates a huge number of unfilled job openings for imported labor. One out of every three Saudi residents is a foreign national, and two-thirds of all jobs are filled by expatriates. This is a huge drain on the Saudi economy because foreign workers try to send the maximum monthly sum possible back to their homelands. Estimates suggest that up to $20 billion a year may be remitted. Were funds of this magnitude available for purchases from Saudi businesses, more economic activity would take place, and more young Saudis might find job opportunities in the private sector.

Saudi government is theoretically an absolute monarchy capable of making and applying difficult decisions. In practice decisions are constrained by Wahabbi theology, social custom, conservative efforts to limit innovation and alternative value systems, bureaucratic inertia, patronage networks, and consultative councils. This environment slows social development, makes it difficult to develop the human capabilities of Saudi citizens, wastes resources, and fails to address endemic problems such as poverty. Widespread discontent exists but bubbles beneath the surface; oil revenue is handed out as bonuses whenever the government fears unrest might burst into the open. Income from oil provides breathing space to phase in social and economic change over sufficient time to minimize a violent outburst, but only if the aging al-Saud family leaders invest in planned, long-term change.

Stop & Think

➤ Is it possible for Saudi Arabia to achieve sustainable economic development without comparable development of its human resources?

Iraq: Struggling to Rebuild

Iraq is physically divided into three parts. The eastern third northeast of the Tigris River is an area of foothills and mountains, where peaks reach 10,000 feet (3,000 meters) along the Iranian border. There up to 40 inches (1,000 millimeters) of precipitation fall, largely as snow. Farming of grains and deciduous fruits is possible, and most of the inhabitants are Sunni Kurds. In the central third, between the Tigris and Euphrates rivers, is located **al-Jazirah,** "the Island." Much less precipitation occurs in al-Jazirah than in the mountains, and rainfall decreases from north to south. The landscape becomes very flat 60 miles (100 kilometers) north of Baghdad and elevation decreases gradually thereafter to the Persian Gulf. Both rivers begin in central and eastern Turkey, flow southward across Iraq, and drain into the sea through a common channel, the **Shatt al-Arab.** In southern al-Jazirah, low stream gradient, poor drainage, and frequent floods produce marshlands. South of Baghdad all agriculture depends on irrigation. Most of Iraq's population lives in al-Jazirah, and the area around Baghdad has been the political core of the region for millennia. Both Sunni and Shi'a Arabs live in al-Jazirah, with Shi'ites dominant in the south. West of the Euphrates lies al-Wadiyah, a stony desert cut by wadis that drain toward the Euphrates. Few people live in al-Wadiyah, but winter grazing exists based on scattered rain-fed pastures and surface pools.

The productive alluvial soil of al-Jazirah is a major natural resource, but irrigation is essential. Unlike the narrow Nile Valley, al-Jazirah is wide. The rivers' annual flood can only reach a small part of the area. To use Tigris and Euphrates water effectively, floods must be controlled and the water made available all year. Ancient engineers found ways to divert water from the rivers during the low-flow part of the year. They also created canal systems that took advantage of elevation differences between the riverbeds of both rivers and shifted water by gravity flow from the Euphrates to the Tigris north of Baghdad and from the Tigris back toward the Euphrates south of Baghdad. This created the world's first significant interbasin water transfer system. Modern dams in the mountain headwaters of Iraq's Tigris tributaries capture much of the annual flood. This water is released throughout the year to produce electricity and support irrigation. Flood water that cannot be controlled by these dams is diverted into natural depressions in or near al-Jazirah and used later in the crop growing season.

Ancient farmers avoided salinization by cultivating a field every other year. This allowed irrigation water to flush salts below the crop root zone and prevented groundwater from rising and bringing salts back. Today population growth makes it necessary to cultivate every field each year. To do this, the land must be drained by ditches that reach below the root zone, collect used irrigation water, and remove this salty water from agricultural areas. Because saline waste water cannot be reused to irrigate fields downstream, two completely different management systems must be maintained: water delivery canals and waste water drainage canals. Keeping these two systems functioning, preventing canals from becoming blocked with silt, avoiding waterlogged soils, ensuring that clean water is available for crop growth, and using only the right amounts of fertilizer, herbicide, and pesticide are complicated, interconnected tasks.

In the 1990s the Ba'ath government of Saddam Hussein was confronted with serious problems that resulted from its defeat in the first Gulf War in which a U.N.-sanctioned intervention reversed Iraq's seizure of Kuwait. The Ba'ath regime's severe postwar repression of internal opponents paralleled a provocative defense of sovereignty—insistence on the regime's right to control its own space and refusal to allow international arms inspections—that produced frequent small military and diplomatic confrontations with the United States as the prime enforcer of peace terms. Ultimately, in the wake of the 2001 World Trade Center attack, these disputes resulted in claims (later found to be erroneous) that Iraq's government supported international terrorism and possessed hidden weapons of mass destruction. A military intervention led by the United States and Britain followed in 2003. The rapid overthrow of the Ba'ath regime, the execution of Saddam as a war criminal, and a prolonged struggle between Iraqi ethnic, religious, and political factions as well as with the coalition army followed. The final outcome of this struggle remains unresolved, although outside military forces no longer are part of the equation.

Southern Iraq's swamps, once the home of half a million **Marsh Arabs,** demonstrate the interplay between land management and political issues. The 1992–2003 sanctions imposed on Iraq damaged its economy and impacted ordinary citizens, but also provided justification for government efforts to increase efficiency and productivity in irrigated agriculture. The Ba'ath government tried to expand irrigated land by diverting water that fed the swamps into new irrigation projects and by directly draining many swampy areas. The more water was diverted to irrigation, the less water was available to sustain the swamps. Most Marsh Arabs were Shi'ites and had rebelled in 1991; to the Ba'ath, destroying their homeland seemed justified revenge. When a second military intervention removed the Ba'ath regime in 2003, the swamps were largely destroyed and many Marsh Arabs were refugees in Iran.

Efforts then began to restore the marshes. The goal was to provide the basis for a return of the Marsh Arabs and to restore important habitat for wild animals and migratory birds. These efforts have made only modest progress and are not likely to succeed. Much Tigris and Euphrates water is diverted for hydroelectricity generation and irrigation in southeast Turkey and therefore does not reach the marshes. Many more dams and irrigation projects are planned in the next decade. Iran has equally big plans for the Tigris tributaries that it controls. So substantially less water will be available for irrigation in Iraq, and wetland restoration is likely to become a low priority. Water quality also is declining. Both rivers already show increasing salt levels over several decades due to reuse of drainage water downstream. Contamination with agricultural chemicals is also certain to increase. Iraq's great soil resources are diminished by a poor location at the end of the water distribution system because no agreement exists that divides the water resource fairly between user countries. This problem will only grow worse over time (Figure 7-52).

Oil is Iraq's other abundant resource. Iraq's reserves are substantial, trailing only Saudi Arabia, Venezuela, and Canada globally. Petroleum development is concentrated between Mosul and Kirkuk in the north; south of Basrah near Kuwait; and in scattered fields between Basrah and Baghdad near Iran. Many of Iraq's oil fields pose security problems; located close to national borders, they are exposed if hostilities arise. Northern fields are near or in regions where Kurds want expanded autonomy or complete independence.

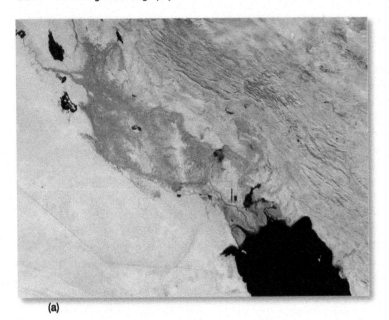

(a)

(b)

(c)

◄ **Figure 7-52 Marshes of southeastern Iraq.** (a) The false-color blue areas in the satellite image show snow in the Zagros Mountains, while the true-color green areas between the Tigris and Euphrates rivers (Al-Jazirah) reflect irrigated districts. The Persian Gulf is visible in the lower right, while similar-colored areas on land often reflect man-made floodwater impoundments. (b) Many of the marshes were drained by Saddam Hussein's government as a consequence of Iraqi political and economic policies. One of these drainage canals appears as a vertical black line north of the Gulf. The massive shrinkage of the marshes eliminated support for agriculture and fishing, left both boats and people high and dry, and destroyed vital habitat for local wildlife and migratory birds. (c) For millennia the wetlands of the lower Tigris and Euphrates floodplains supported a people known as the Marsh Arabs. Efforts to restore the marshes since the collapse of the Ba'ath regime have met with modest success, such as this area in al-Hamar marsh south of al-Basrah. But competing demands for irrigation water throughout the river basins make only partial recovery likely.

Central government control is weak in Kurdish districts and the Kurds increasingly have begun to manage the oil resources of their area. Recently Genel, an Anglo-Turkish oil company, announced the discovery of a new field in Iraqi Kurdistan, based on exploration contracts with the Kurdish regional government. Kurdish oil is exported by tanker trucks to southeastern Turkey, but direct pipeline connections are anticipated. Turkey unexpectedly agreed in 2013 to settle its differences with Kurdish political groups who have been in rebellion against the Turkish government. A flood of direct Turkish business contacts with Iraqi Kurdistan, bypassing the government in Baghdad, has been under way since early 2012. An end to internal Turkish-Kurdish conflict may pay significant dividends for Turkey by providing access to petroleum in Kurdistan not available in its own territory. It may also promote the emergence of an independent Kurdish state in northern Iraq.

This prospect does not appeal to the Iraq government. Oil revenues support 95 percent of Iraq's budget, so any oil controlled by

an autonomous or independent Kurdish state would have serious implications. Iraq faces major problems with its oil infrastructure (pumps, wells, pipelines, refineries, port facilities). Much of the basic equipment was heavily damaged due to neglect, military conflict, guerilla attacks, sanctions, limited investment, and increasingly obsolete technology. Since 2010, gradual improvements have occurred. Iraq is now the second largest OPEC exporter after Saudi Arabia. Some of the earnings from greater oil exports have been reinvested in infrastructure improvements. Major upgrades have occurred in electricity generation using oil, and electricity blackouts have become less frequent in 2013. But energy wastage remains serious. Gas, frequently a byproduct of oil extraction, is still flared (burned) off at the well rather than captured to help pressurize oil deposits to maximize yield or converted to liquid natural gas. Use of gas in electricity generation is almost nonexistent, but would conserve oil and reduce urban pollution. And violent civil conflict between Shi'a and Sunni, which has increasingly driven people out of mixed into religiously homogenous urban neighborhoods, as well as turned hundreds of thousands of Iraqi Christians into international refugees, continues to undermine domestic peace and security.

Iran: In Search of an Identity

Iran's physical environment resembles a donut—mountainous areas surrounding a sparsely populated to uninhabited central area. The central core is a low plateau approximately 3,300 feet (1,000 meters) above sea level in which little life exists. Cut off by the Zagros Mountain ranges from moisture-bearing cyclonic storms, the central plateau is a barren region; summer temperatures often exceed 113°F (45°C). The northern half of the desert plateau, the Dasht-e Kavir, contains extensive salt pans and salt crusts, while the southern Dasht-e Lut is equally barren but covered with mobile sand dunes. The Zagros Mountains reach heights over 10,000 feet (3,000 meters) and receive winter precipitation over 40 inches (1,000 millimeters). Streams from the Zagros drain westward to the Tigris River and provide water to fertile valleys and basins. The Alborz Mountains south of the Caspian Sea are formidable barriers; Damavand Mountain (18,602 feet/5,670 meters) northeast of Tehran is the highest mountain in the Middle East. The north-facing slopes of the Alborz receive as much as 80 inches (2,000 millimeters) of precipitation annually. This contrasts with the semiarid conditions around Tehran less than 30 miles (50 kilometers) to the south. Lower mountain ranges rim the plateau's eastern side, while the southern Makran Mountains complete enclosure of the plateau. The intensely arid south, where annual precipitation totals rarely exceed 4 inches (100 millimeters), is the country's driest region.

Oil, Iran's most important resource, is abundant in Khuzestan, the eastern extension of the Tigris River's alluvial floodplain, and in the basins between Zagros Mountain ridges paralleling the Persian Gulf. Iran's oil reserves rank among the world's top five. Iran is OPEC's third largest producer, but export totals have declined since 2006 due to international sanctions. Iran is also has huge natural gas reserves, the world's second largest, although except for exports to Russia little gas enters international trade. The largest natural gas fields are found along the Persian Gulf near Bandar-e Kangan, but significant deposits also exist near the Caspian Sea and in the extreme northeast near Mashhad. Minerals such as coal, iron, copper, zinc, and bauxite are exploited for domestic use; only chrome is exported.

With the exception of a few oases on the central plateau and near the Makran Mountains, most of Iran's population is located in association with the mountain rim. Rich soils occur in many mountain basins. Much of upland basin agriculture is based on wheat and barley, or sugarcane, vegetables, and tree crops irrigated with groundwater brought to the surface by qanats or pumps (Figure 7-53). Gasoline-powered pumps have drawn down the water table below levels that many qanats tap, and may both destroy traditional technology and eventually prove unsustainable. Dam construction in the uplands retains water in cooler zones for irrigation use. These dams have also altered the agricultural system by permitting the introduction of new crops, such as sugar beets, and by diminishing the impact of drought. Two regional specialty crops are particularly important: rice, cultivated on the Caspian Sea coastal plain using paddy irrigation techniques supported by the abundant water of the Alborz Mountains, and pistachios grown on the dry coastal plain of Khuzestan and near Kerman. Animal husbandry is also important, particularly in the western mountains. Pastoral nomads still migrate from the lowlands of Khuzestan and the Persian Gulf to summer pastures high in the Zagros. But government pressures that promote sedentarization and agricultural expansion along migration routes have reduced grazing land essential for successful pastoral migrations.

The population of these upland basins and valleys has a rich urban tradition, and each substantial basin is dominated by a major city. Most cities have ancient foundations, and all have grown rapidly since 1950. Tehran has grown to be one of the world's largest cities as rural residents and middle-class professionals from the provinces have flocked to the capital. Tehran is more than three times the size of Mashhad, the next largest urban place. Isfahan, Tabriz, Shiraz, and Karaj are the only other cities with an official population over 1 million. Despite the limited number of large cities, close to 70 percent of Iranians live in urban centers, and this figure is predicted to reach 80 percent by 2030.

Economically the cities are dominated by the **bazaar** (Figure 7-54). This is more than a cluster of streets with covered roofs where stores are located; it is also an informal network of family, craft, and professional associations that manage and control major industrial and commercial activities of the economy, such as textiles, carpets, metalworking, and food processing. Support of the bazaar is exceedingly important for the success and stability of the political elite (government bureaucrats, military, financiers and industrialists, large landowners, and professionals). Another crucial group in society is the Islamic clergy (**mullahs/mujtahids**), who traditionally are responsible for interpreting religious law, providing basic education, serving as notaries for legal documents, and administering religious endowments. Support from the religious hierarchy, and particularly from the most prestigious religious scholars, the mujtahids, some of whom hold the honorific title of Ayatollah, is crucial to any government because these religious figures are regarded as the primary protectors of both spiritual values and secular justice.

The 1979 revolution that overthrew Iran's monarchy was inspired by traditionalists for whom the mullahs were the primary voice. Rapid but poorly distributed economic growth after 1950, authoritarian central government, repressive secret police, abuses of power, rampant corruption, waste, and mismanagement sparked protests from progressive leaders, intellectuals, students, and the bazaar. These groups found common ground with the mullahs in their opposition to rapid changes in economy, government, and society. Alarmed by the spread of values, ideas, and lifestyles that they regarded as alien and not Islamic, many antimonarchists sought a return to more traditional norms. They achieved their objective when the monarchy collapsed and was replaced by a republic based on Islamic principles.

This republic had a sense of its spiritual purity, believed Western governments and its Islamic neighbors were morally deficient, and found little to admire in the world around them. Internally, strong controls were placed on religious, social, and political behavior; limited tolerance was displayed toward alternative beliefs. The Baha'i faith, founded in Iran in the nineteenth

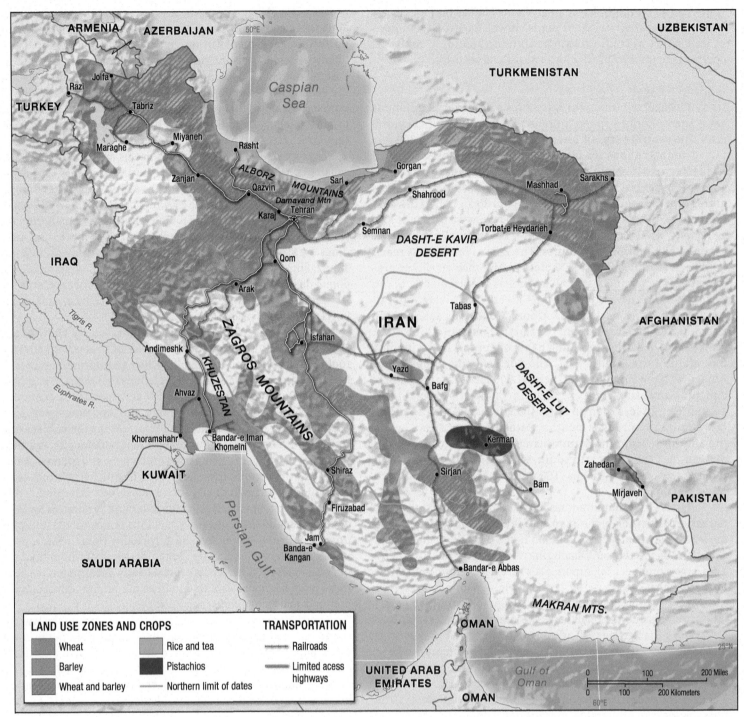

▲ **Figure 7-53 Land use in Iran.** Agricultural land use in Iran follows basic variables such as temperature and rainfall: water-demanding crops such as tea and rice along the Caspian Sea; dryfarmed cereal crops in the cooler uplands and irrigable lowlands; dates in the warmest central and southern parts of the country; pistachios and other dryland tree crops in the warmer lower foothills. High-speed rail and road links form an emerging network linking major cities as well as ports to interior locations.

century, was viewed as a heretical sect and its practitioners were persecuted intensely. Armenian Christians, long a prominent component of Tehran's business community, gradually dwindled under restrictive pressures. Critics of the new order, from newspaper editors to reformist politicians, found it difficult to gain traction for their views.

Diplomatically the Iranian government has opposed the influence of Western countries and values in the region. Iran has done this through indirect support of what it considers to be legitimate resistance movements, such as Hamas in Gaza/Palestine and Hezbollah in southern Lebanon, but which many Western governments regard as terrorist organizations. These

▲ **Figure 7-54 Iranian bazaar.** The bazaar is both a commercial setting, such as this collection of retail shops in Tehran, as well as a powerful political and financial force.

groups find favor in Iranian eyes because they are locked in conflict with Israel, regarded as a proxy for the United States. Iran supports most regime changes produced by the Arab Spring because they are seen as a wave of Islamist resistance to Western imperialism. But in the struggle taking place in Syria, Iran supports the Ba'ath government, in part because much of the rebel opposition is provided by Wahabbi fundamentalists with support from Qatar and Saudi Arabia. Hezbollah's decision to send some of its fighters to support Syria's government raises the prospect that Syria's civil war might spread into increasingly polarized and engaged communities in neighboring countries. More direct confrontation between

Iran and the Syrian rebellion's Western and Middle Eastern supporters could result as Iran becomes increasingly concerned about the survival of its major ally among Arab governments.

More serious is the struggle over Iran's nuclear program. Iran seeks control of nuclear technology and fuel in order to base its electricity generation on nuclear power. Many countries believe Iran also wants to build nuclear weapons as a prelude to an attack on Israel. The anti-Israeli rhetoric of Iran's former President Ahmadinejad lent credibility to this viewpoint. Sanctions imposed to influence Iran's nuclear policies intensified in 2006. These sanctions have not achieved their goal of changing Iran's perceived objective, although nuclear power not weapons has always been Iran's stated purpose in acquiring nuclear capability. But sanctions have impacted Iran's economy. Iran's currency has lost half its value, inflation is rampant, oil exports have slowed dramatically, and the price of food for ordinary Iranians has skyrocketed. Assassination of Iranian nuclear scientists and cyber attacks on computers and equipment enriching nuclear fuel have slowed the program. A "red line," which some countries believe Iran should not be allowed to cross, indicating nuclear weapons capabilities is possession of enough medium-grade nuclear material to move quickly to build one bomb. Recently Iran began to process this material into energy-grade materials to keep its "prebomb" stock below this critical level. Whether this will be read as a signal of genuine compromise or as a short-term expedient remains unclear. Election of a less rigidly conservative president, Hassan Rouhani, in 2013 suggests that a more flexible position might develop in Iran's approach to nuclear issues. But until this conflict is resolved, it will be very difficult for Iran to develop its economy and improve the lives of ordinary citizens.

Summary

- ▶ The Middle East and North Africa world region appears in popular image to be a sandy desert wasteland inhabited by one people, the Arabs, who are uniformly Muslim in religion. In reality, the region is environmentally diverse with considerable productive potential, and in these more productive zones, great population densities are found.

- ▶ Yet aridity dominates almost all parts of the region and places severe constraints on all economic activity. Finding ways to overcome this water constraint and manage water resources efficiently will strongly influence the success of development.

- ▶ The cultural diversity of the region's human population is as great as its environmental variation. More than one-third of the region's population belongs to ethnic, linguistic, or religious minorities that are neither Arab nor Muslim.

- ▶ Mediterranean Crescent countries have attained significant levels of economic development. Except for Algeria and Libya, that development has occurred without the aid of massive oil revenues. Egypt, Israel, Turkey, Morocco, Lebanon, and Jordan have made progress by utilizing their individual and unique agricultural, industrial, and environmental/cultural resources.

- ▶ Despite economic progress, the stability and future of the Mediterranean Crescent countries are jeopardized by severe internal tensions. Authoritarian governments overthrown in

Tunisia, Libya, and Egypt and assaulted in Syria have not been replaced by governments much better able to solve deep-seated problems of social justice and unequal access to jobs, education, and economic stability. Increasingly sharp debate between Islamic revival and resistance movements and supporters of greater tolerance and diversity has difficulty discovering common ground on many social, political, and economic issues. And the Israeli-Palestinian dilemma contains such divergent claims and aspirations that pursuit of a compromise solution may be illusory.

- ▶ The petroleum economy has changed and continues to alter many features of the political, economic, and social fabric of the Gulf States. Gleaming skyscrapers have replaced Bedouin tents without engaging much of the population in productive enterprise; imported labor barely outnumbers the indigenous unemployed.

- ▶ The challenge facing the Gulf States is how to harness petroleum wealth to promote the enduring development of their economies and societies. Traditionalists are not reconciled to the changes often associated with globalization and rapid modernization and find support in rigid religious regulations that stifle women and youth. Cultural attitudes that denigrate manual work and discourage innovation produce social and economic patterns that undermine enduring development.

Key Terms

aquifers 291
Arabs 295
aridity 290
Ashkenazim 319
Aswan High Dam 310
barrage 302
bazaar 337
Berbers 297
crossroads location 295
dry farming 300
exotic rivers 291
Five Pillars of Islam 297
GAP Project 312

Gaza Strip 319
Gulf Cooperation Council 330
Hamas 298
intifadah 318
irrigated agriculture 301
Islam 297
al-Jazirah 335
Jews 296
Kurds 296
Levant 307
madina 304
Maghreb 307
Marsh Arabs 335

mullahs/*mujtahids* 337
Muslims 297
nomadic herding 303
oasis/oases 291
Organization of Petroleum Exporting Countries (OPEC) 330
orographic precipitation 291
overgrazing 293
Palestine Liberation Organization (PLO) 298
Palestinians 298
Persians 296
qanat 301

Quran 297
rainshadow 291
run-on farming 300
salinization 301
Sephardim 319
Sharia 297
Shatt al-Arab 335
Shi'ites 298
Sunnis 298
Tabqah Dam 316
Turks 296
virtual water 311
West Bank 318

Understanding Development in the Middle East and North Africa

1. What are the major sources of water in North Africa and the Middle East and how are they distributed across the region?

2. What is the difference between dry, run-on, and irrigated farming and where is each technique typically found?

3. What are the advantages of pastoral nomadism and what are the major problems of sedentarization?

4. How can salinization be avoided in irrigated agriculture?

5. What are the values embodied in the Five Pillars of Islam?

6. How have the countries of the Mediterranean Crescent increased the livelihood options of their inhabitants without the benefit of oil wealth?

7. What are the benefits and costs of major dam projects in the region?

8. What problems arise when a major river system such as the Tigris-Euphrates has to be shared by several countries?

9. What development options has oil wealth made possible in the Gulf States?

10. What are the most pressing problems of present-day North African and Middle Eastern cities?

Geographers @ Work

1. Explain why extending irrigation into desert areas is a risky development strategy and outline the long-term limitations to success.

2. Evaluate what water resources are available in the region and what strategies are needed to avoid overexploitation.

3. Explore how oil revenues are benefiting the Gulf countries and what investment options will help secure their future as the world moves beyond the dominance of fossil fuels.

4. Characterize the events of the "Arab Spring" and the obstacles that make the road to true democratization across the region so difficult.

5. Compare the varying social strictures in different parts of the region and how this supports or impedes development and gender equality.

MasteringGeography™

Looking for additional review and test prep materials? Visit the Study Area in MasteringGeography™ to enhance your geographic literacy, spatial reasoning skills, and understanding of this chapter's content by accessing a variety of resources, including MapMaster™ interactive maps, videos, RSS feeds, flashcards, web links, self-study quizzes, and an eText version of *World Regional Geography*.

▲ Figure 7-55 **The Middle East and North Africa from space.**

8

Africa South of the Sahara

Samuel Aryeetey Attoh

The Paradox of Sub-Saharan Africa's Mineral Wealth

Sub-Saharan Africa is endowed with vast mineral reserves. Even countries perceived to be "poor" have strategic reserves that draw global mining companies. South Sudan's oil reserves, Niger's uranium deposits, and Mali's gold mines have attracted attention from foreign investors. In an ideal world, such vast reserves would yield substantial revenues for investments in rural clinics and schools, improving sanitation services and infrastructure, and developing businesses across all sectors of Africa's economies. But such is Africa's paradox: a wealth of mineral reserves juxtaposed against a landscape of deprivation and poverty. African governments generate very little revenue from mining because of a lack of meaningful and beneficial contracts with foreign companies. The average royalty rate assessed on the sale of minerals produced is 3–4 percent. Some countries, like the Democratic Republic of Congo, offered a 20-year exemption from royalties and corporate income taxes. Liberia had to cancel 65 percent of mining contracts that were poorly negotiated between 2003 and 2006. There is also a shadow economy. In Ghana, the term *galamsey*, "gather them and sell," refers to unregulated and illegal mining activities conducted by unskilled workers who use their hands and simple tools to dig pits and tunnels in search of gold. This problem is compounded by Chinese "gold diggers" involved in clandestine operations that exploit villagers in their quest to market gold through the global supply chain. This risky activity has resulted in environmental destruction and numerous accidents and deaths, prompting arrests and increased regulation from the Ghanaian government.

Resource nationalism is on the rise in Africa. Ghana assesses a 5 percent royalty on mining output and is considering raising corporate taxes on mining companies from 25 percent to 35 percent as well as imposing a windfall tax of 10 percent on super profits. South Africa is contemplating a 50-percent windfall tax on super profits and a 50-percent capital-gains tax on the sale of prospecting rights. South Africa has also announced that it will assess levies on exported raw materials, but none on finished products, to encourage the production of finished metals to promote job growth. Africa can capitalize on comparative cost advantages in labor to develop value-added manufacturing to boost investments, strengthen linkages in other economic sectors, expand exports, and generate more revenue to support development.

Read & Learn

▶ Gain knowledge of the ancient, medieval, and colonial history of Africa South of the Sahara.

▶ Understand the human causes of environmental degradation in sub-Saharan Africa.

▶ Develop appreciation for the environmental, sociocultural, and economic diversity of Africa's regions.

▶ Identify the multidimensional aspects of human development and the underlying factors accounting for patterns of income inequality and poverty.

▶ Discover the development strategies that address poverty and human development issues in Africa South of the Sahara.

▶ Acquire an understanding of the unique human and physical geographies that characterize the countries of West, Central, East, and Southern Africa.

▶ Know how intraregional trading groups in West and Southern Africa have evolved and how they have contributed to regional development.

Contents

Africa South of the Sahara

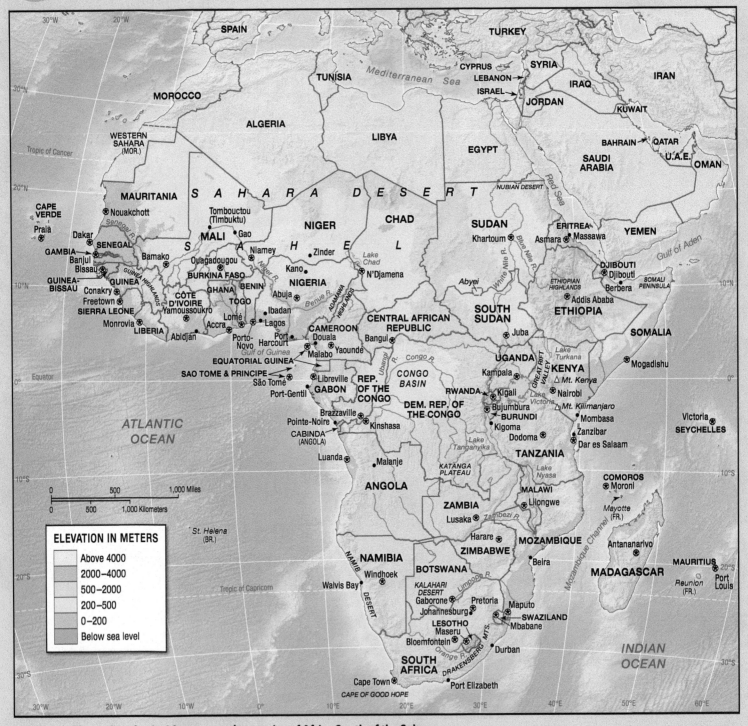

Figure 8-0 **The major physical features and countries of Africa South of the Sahara.**

For most people, Africa evokes a variety of images: rain forests, deserts, drought, famine, wars for independence, and great stores of natural resources. More recently, images of civil wars in southern Sudan and Mali come to mind. Africa South of the Sahara is a vast and varied region, home to more than three-quarters of a billion people.

Most black African peoples have achieved self-governance. Even the Republic of South Africa has instituted a multiracial government. Yet despite Africa's long cultural history, most of the countries in this world region have been independent states for little more than half a century or less. The boundaries established by the colonial powers during the late nineteenth and early twentieth centuries have complicated the effectiveness of those states.

The tremendous variety that characterizes the African continent can be difficult to convey. Some African countries are among the poorest in the world, with limited agricultural or industrial resources. Others are making significant progress and have the resource potential for even greater growth. The sections that follow set forth the geographical basis for the development of this diverse and increasingly important part of the world. They provide insight into the challenges and progress of black Africa and illustrate the varied approaches that are being pursued in the quest for national development.

Sub-Saharan Africa's Environmental and Historical Contexts

Africa is the world's second largest continent, and Africa South of the Sahara covers 9.8 million square miles (25.3 million square kilometers), over 80 percent of the continental land mass. Sub-Saharan Africa includes 42 mainland and 8 island countries. The mainland countries can be divided into 4 subregions: West, East, Central, and Southern Africa (Figure 8-1). The sub-Saharan region has a rich and diverse physical and human resource base and a resilient cultural heritage. It is also characterized by such contrasts as extreme wealth and abject poverty, gender inequality, pristine forests interspersed among degraded and desert terrain, rich mineral deposits buried under desolate landscapes, and European values and institutions superimposed on traditional cultural settlements. Not so long ago, the tendency was to characterize the subcontinent in terms of the "*d's*": destitute, doomed, dark, disaster-prone, drought-stricken, disjointed, disconnected, and debt-ridden. However, perceptions are changing, and people are becoming more aware of the "*p's*": the potential, prospects, and possibilities inherent in sub-Saharan Africa's evolving socioeconomic geography.

Environmental Setting: From Rain Forests to Deserts

The region's diverse physical environments are created by the dynamic interaction of several natural processes. Physical landscapes are complex and the climatic, vegetational, and biogeographical elements contain both assets and liabilities. Assets include majestic mountains, many minerals embedded in pre-Cambrian rocks, rich volcanic soils in East Africa, scenic and economically valuable lakes, the biodiversity and commercial value of rain forests, and great hydroelectric power potential (see *Focus on Energy:* Renewable Energy Resources in Sub-Saharan Africa, page 349). Among the liabilities are straight coastlines that limit opportunities for natural harbors and narrow continental shelves that restrict potential offshore oil exploration and fish habitats. Leached soils in many rain-forest areas inhibit agricultural development. The physical environment is threatened by careless human activity in ecologically sensitive areas. The magnitude of this threat depends on the ability of communities and governments to develop strategies to manage and conserve the environment.

Landforms

Geologically, much of tropical Africa consists of a great plateau that tilts downward from east to west. This plateau is fractured and scoured by several major river systems, leaving large gorges and undulating surfaces. Highland East Africa averages about 4,000–5,000 feet (1,200–1,500 meters) above sea level, while the lower plateau of the western and central regions averages 1,000–1,500 feet (300–450 meters) in elevation. East Africa has several prominent mountain landscapes, such as the extensive East African Plateau, which features the two highest points in Africa: Mount Kilimanjaro (19,340 feet; 5,895 meters) and Mount Kenya (17,057 feet; 5,199 meters) (Figure 8-2). Further north is the Ethiopian Massif, which has its highest point at Ras Dashen (15,157 feet; 4,620 meters). East Africa also features some extensive plains, such as the Serengeti Plains of Tanzania. West and Central Africa are not entirely low-lying regions. Mount Cameroon (13,435 feet; 4,095 meters), the Jos Plateau (5,840 feet; 1,780 meters) in Nigeria, and the Fouta Djallon Highlands of Guinea are examples of major uplands that rise above the surrounding plateau.

In Southern Africa, the plateau is framed by a narrow coastal plain. The plateau reaches its highest point in the eastern sections where the Drakensberg Mountains (over 11,000 feet; 3,350 meters) are located. The plateau slopes downward toward the interior savanna and steppe plains and the arid regions of the Kalahari and Namib deserts in the west (Figure 8-3). The southwestern sections of Africa are rimmed by the Cape Fold Mountains, which rise to about 6,500 feet (1,980 meters), and the Karoo rock series, which contains coal deposits.

Sub-Saharan Africa's plateau consists primarily of ancient crystalline rocks that have been created by immense heat, pressure, and chemical changes, creating a wealth of minerals in the process. The old, geologically stable core or shield areas of Africa are rich in chromium and asbestos, and areas of West Africa and the Gabon-Congo region have rich reserves of gold, diamonds, and manganese. Oil and gas deposits are associated with younger sedimentary rocks that occur along linear zones of the Atlantic front stretching from the Niger River delta to the Democratic Republic of Congo.

Another unique aspect of the region's physiography is the **East African Rift Valley** which begins in the north with the Red Sea and extends through Ethiopia to the Lake Victoria region, where it divides into eastern and western segments and continues southward through Lake Malawi (Nyasa) and Mozambique (Figure 8-4).

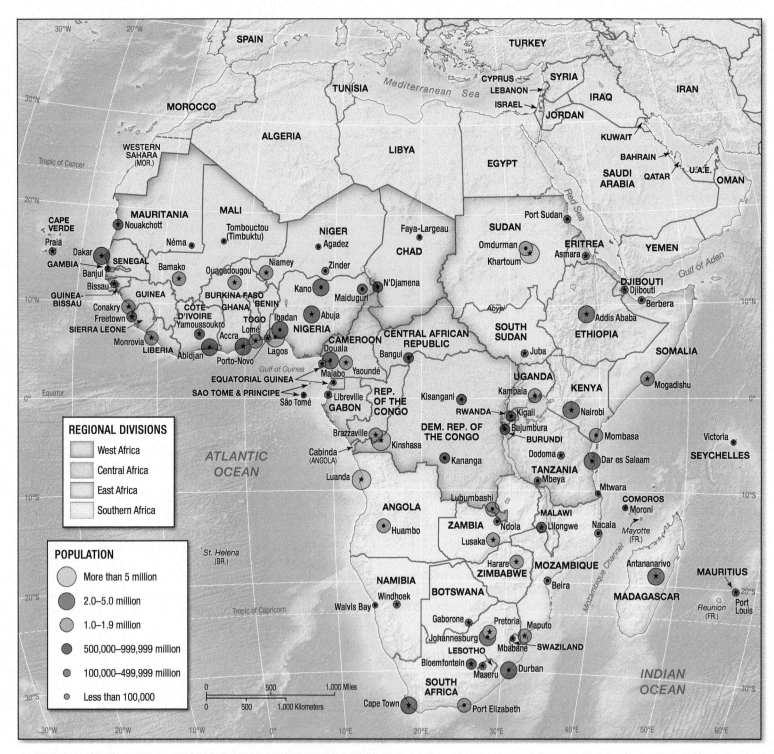

▲ **Figure 8-1 Countries and major cities of Africa south of the Sahara.** Sub-Saharan Africa consists of four regional divisions: West, Central, East, and Southern Africa. Most of the region's major cities are located close to coastal areas and to regions well-endowed with natural resources.

▼ **Figure 8-2 Major landforms in Africa.** (a) Most of tropical Africa consists of a great elevated plateau that is broken by large inland basins. The volcanic East African Rift Valley is also a prominent feature. (b) Eastern Africa's Rift Valley system, spectacularly revealed in this satellite image, is highlighted by long, narrow lakes and extends northward through the Red Sea.

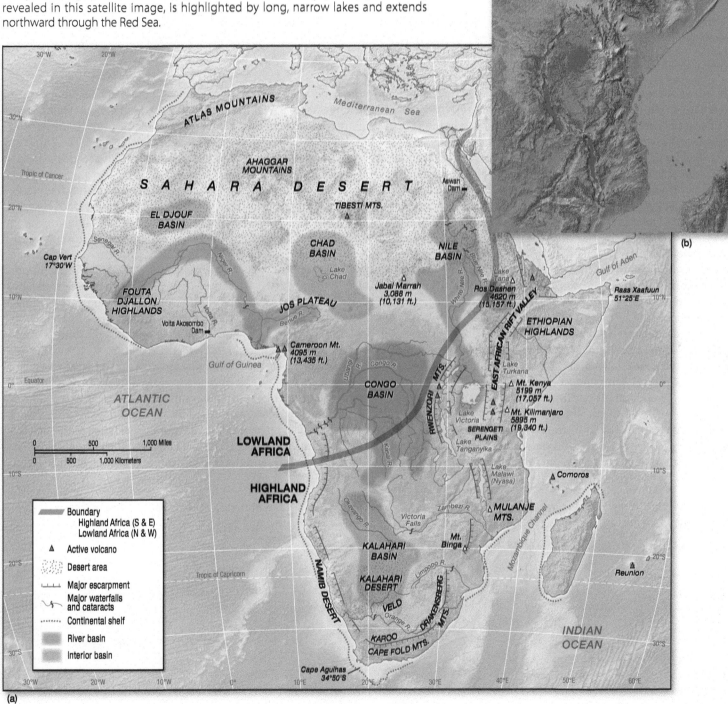

(b)

(a)

(a)   (b)

▲ **Figure 8-3 The Kalahari and Namib in Southern Africa.** Southern Africa has large desert regions with unique sets of microenvironmental characteristics. (a) The animals grazing are springbok (*Anticorcas marsupialis*), a gazelle noted for its sudden and random jumps into the air as well as its adaptability to dryland conditions. (b) The Namib is a cool coastal desert fronting the Benguela Current of the southern Atlantic Ocean. Gemsbox oryx graze among drought-tolerant camel thorn acacia.

The rift valley was created by faulting, as tensional forces associated with continental drift began to pull the eastern sections of Africa away from the rest of the continent, leaving in the area of separation a great valley that subsided. Important features associated with the rift valley are the **Great Lakes of East Africa** (with the exception of Lake Victoria). Especially unique are the crater lakes (Figure 8-5) and the elongated lakes that occupy deep trenches in the rift valleys, such as Lake Malawi (Nyasa), Lake Tanganyika, and Lake Turkana. Lake Victoria, the world's second largest lake in terms of area, is nestled between the two arms of the rift valley. The rift belt, along with the offshore islands of Réunion and the Comoros in the Indian Ocean, as well as the Canaries in the Atlantic Ocean, constitute the major volcanic regions of Africa. There are several explosive craters around the Uganda–Democratic Republic of Congo border.

The interior plateau is drained by major river drainage basins—the Nile, Congo (Zaire), and Niger. Other important rivers in West Africa are the Senegal, Volta, and Benue (a tributary to the Niger), while the Zambezi, Limpopo, and Orange rivers are particularly significant in Southern Africa. Historically, the Zambezi River separated white-ruled Southern Africa from independent black Africa. Today it separates Zambia (formerly Northern Rhodesia) from Zimbabwe (formerly Southern Rhodesia) and divides highland northern Mozambique from lowland southern Mozambique. Along the Zambia–Zimbabwe river boundary, the Zambezi plunges over the spectacular Victoria Falls into a series of gorges (Figure 8-6). Further downstream, two major hydroelectric projects are located at Lake Kariba and at Cahora Bassa in Mozambique. The Zambezi River has a total annual flow double that of all of South Africa's rivers combined.

African rivers have limited navigability because they frequently are interrupted by falls and rapids, and their water levels vary greatly from season to season. The Congo is navigable only up to 85 miles (137 kilometers) inland before its course is broken by a series of rapids just west of Kinshasa. But African rivers have much potential for hydroelectric power generation, owing to their swift flow and often steep falls. The Congo River carries the second largest volume of water in the world (only the Amazon carries more) and has enormous potential for hydroelectric power. African rivers also are home to many aquatic species, and inland fishing is a major activity.

Major drainage basins associated with the region's largest rivers were formed in part by tectonic forces that downwarped parts of the plateau. They became repositories of sediments eroded from plateau surfaces and mountain massifs and deposited by rivers that converged on the basins. These drainage basins are associated with mountain ranges such as the Fouta Djallon Highlands, where the Niger begins, and the East African Highlands, which separate

▼ **Figure 8-4 East African Rift Valley.** Active and dormant volcanoes are associated with the East African Rift Valley system.

▼ **Figure 8-5 Lake Mutanda in Uganda.** This crater lake in Uganda was formed when the volcano exploded and collapsed.

FOCUS ON ENERGY

Renewable Energy Resources in Sub-Saharan Africa

When the United Nations declared 2012 as the International Year of Sustainable Energy for All, many plans emerged to harness sub-Saharan Africa's enormous renewable energy resource: hydro, geothermal, solar, and wind power. Africa's current energy consumption is primarily focused on biomass (wood, charcoal, agricultural residues, and animal waste) and oil, and to a much lesser extent on electricity, natural gas, and coal (Figure 8-1-1). The International Energy Agency's World Energy Outlook (WEO) indicates that 31 percent of sub-Saharan Africans are without access to electricity and over 650 million people rely heavily on traditional use of biomass for cooking (Figure 8-1-2). The IEA's Energy Development Index monitors a country's progress in transitioning toward modern energy access in terms of clean cooking facilities, electricity, and commercial energy. The index ranks most sub-Saharan African countries in the bottom half of developing countries. This has prompted the UN Advisory Group on Energy and Climate Change to call for the goal of

▲ FIGURE 8-1-2 **Merchant selling wood.** Woodfuel consumption accounts for as much as 90 percent of energy consumption in some rural parts of Africa.

universal access to modern energy services by 2030.

A region-wide renewable energy resources strategy that promotes viable, low-cost technologies must minimize negative environmental impacts, improve living standards, and provide employment opportunities for impoverished households. But implementing such a strategy has its challenges. African governments do not have the investment funds and technical capacity to effectively and efficiently adopt renewable energy technologies. This goal is usually overshadowed by other government priorities: jobs and food that focus on industry and agriculture. This dilemma explains why over 90 percent of Africa's hydroelectric power remains untapped, and why geothermal energy is underexplored and underresearched. Given this handicap, small-scale investments in solar photovoltaic (PV) power, mini-electric grid systems, and biogas and advanced cookstoves will require a coordinated effort at all levels, including public/private and bilateral/multilateral partnerships.

African and European Union leaders are committed to improving

the reliability, efficiency, and quality of electrical power supply systems through unified regional electricity markets or regional power pools such as the West, East, Central, and Southern African Power Pools. Other projects in progress include a 250 MW Bujagali hydropower facility in Uganda, partially operational in 2012, and the Olkaria geothermal plant in Kenya. Ethiopia plans to become one of Africa's major green power exporters and has plans to invest in six wind-power projects, as well as geothermal, ethanol, and solar PV cell facilities. Efforts are under way in several African countries to push development of efficient biomass energy technologies. These innovations would reduce indoor pollution, environmental degradation, and the social burdens on women and children who must gather firewood for cooking fuel. Since 2010, South Africa has made it a priority to create more green jobs, install more solar home systems in rural areas, begin more local renewable energy projects, and enhance research and development in support of a green economy.

Sources: International Energy Agency, *World Energy Outlook 2010* (Paris: Organisation for Economic Co-operation and Development, 2010), http://public.eblib.com/EBLPublic/PublicView.do?ptiID=615983; "About the Bujagali Hydropower Project," *Agha Kahn Development Network*, http://www.akdn.org/photos_show.asp?Sid=199.

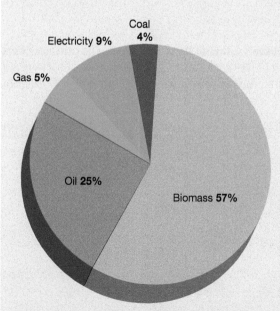

Coal 4%
Electricity 9%
Gas 5%
Oil 25%
Biomass 57%

▲ FIGURE 8-1-1 **Fuel consumption in Africa.** Africa is unusual in that most of its energy demand is still met by biomass consumption.
Source: International Energy Agency, 2010.

▲ **Figure 8-6 Victoria Falls (of Zambia–Zimbabwe).** Victoria Falls lies on the Zambia–Zimbabwe border near Livingstone. It averages more than 5,500 feet (1,700 meters) in width and more than 300 feet (90 meters) in depth.

the Congo (Zaire) Basin from the coastal plains on Tanzania's east coast. The middle portion of the Nile Basin is positioned between the Ethiopian Highlands in the east and the Marra and Ennedi ranges in the west. The Nile's headwaters are located in the highlands of East Africa. The White Nile drains Lake Victoria in the Rift Zone, whereas the Blue Nile originates in Lake Tana in northeast Ethiopia. The Chad Basin is centered on shallow Lake Chad, which receives most of its water supply from rivers that originate in the wetter regions of Nigeria, Cameroon, and the Central African Republic to the south. Historically, Lake Chad has provided significant water resources to farmers, herders, and fisherfolk living near the lake and in surrounding countries. But droughts, high evapotranspiration rates, and increased human demands for water, including diversion of water for irrigation from the Chari river system that supplies the lake, have greatly diminished its depth and surface area.

Southern Africa's Kalahari Basin features two major physiographic landscapes: the Okavango Delta and the Makgadikgadi (Makarikari) salt pans. The Kalahari Basin lacks surface water in most of its southern sections. Its northern portion receives perennial stream flow mainly from the Okavango River, which rises from the highlands of Angola and drains into the dry expanses of Botswana, forming a vast inland delta that covers about 400,000 acres (162,000 hectares) (Figure 8-7). This region is a haven for one of Africa's most diverse wildlife areas and is developing its ecotourism potential.

Among the liabilities inherent in Africa's physiography are its coastlines and continental shelves. Since the continent is mainly a plateau, its coasts tend to be straight and smooth with very few

indentations, unlike the Scandinavian peninsula of northern Europe or the northwestern coast of North America, which have deep river valleys or fjords. Africa's coastlines are also exposed to erosion by offshore currents. In West Africa, sandbars frequently front the coasts of Nigeria, Ghana, and Senegal. This was especially problematic during colonial times when ships had to anchor some distance from the coast, and surf boats had to be employed to unload and transport cargo to the coast.

Africa has few natural harbors. Most are artificial harbors that have been constructed at considerable expense. Among the most significant of these are the ports of Dakar (Senegal) (Figure 8-8), Abidjan (Côte d'Ivoire), Tema (Ghana), Durban (South Africa), and Mogadishu (Somalia). Natural harbors in West Africa include Freetown (Sierra Leone) and Banjul (Gambia). The southwestern shores of the Atlantic Ocean include major ports at Lobito and Luanda (Angola) and Libreville (Gabon). Important railway terminal ports in Mombasa (Kenya), Maputo and Beira (Mozambique), Dar es Salaam (Tanzania), and Cape Town (South Africa) were all developed at sheltered natural harbors.

The region's limited continental shelf extends only for a short distance from the coastline before dropping abruptly to the ocean depths. Countries with extensive continental shelves (like the United States, Peru, Chile, and Argentina) have taken advantage of the inherent opportunities such as offshore oil drilling, mineral exploration, and fishing. Another disadvantage of deep shorelines is the poor development of beach areas and other shoreline resources that are the basis for recreation and tourist development.

▼ Figure 8-7 **Okavango Inland Delta, Botswana.** The core area of the Okavango Inland Delta covers about 400,000 acres (162,000 hectares). It is a haven for a variety of mammals, fish, birds, and reptiles and a boost to Botswana's ecotourist industry.

▲ **Figure 8-8 Dakar, Senegal on Soumbedioune Bay.** Dakar (2.5 million people) is Senegal's political and economic heartland and is a major regional port.

Climate

To the casual observer, sub-Saharan Africa appears uniformly hot and humid since much of the region falls between the Tropics of Cancer (23.5° N) and Capricorn (23.5° S). But several factors account for variations in temperature and rainfall regimes across this vast continent. The cool Benguela ocean current in southwest

Africa and the warm Mozambique ocean current in eastern Africa modify the climates of their coastal environments. The snowcapped, high-altitude environments in the Ethiopian highlands and around Mount Kilimanjaro contrast with much warmer temperatures in the lower elevations of Tanzania's middle belt and the forest zones of central Africa. Cities and villages that border Africa's large inland lakes and extensive coastlines benefit from the modifying maritime effects, in contrast to the hotter and drier continental locations of the African Sahel. Regions located near the lower latitudes of the equator are much warmer and wetter than the higher latitude regions of Southern Africa.

Thus altitudinal, latitudinal, maritime, and continental effects help explain climatic diversity in Africa. But two pressure systems—the **Intertropical Convergence Zone (ITCZ)** and the Subtropical High Pressure System (STHP)—have dominant effects on wind patterns and rainfall regimes. The ITCZ is a low-pressure system created by high temperatures along the equator. Winds converge toward the ITCZ and then rise and cool, creating high levels of instability in the atmosphere that result in rainfall (Figure 8-9). Mean monthly temperatures in rain-forest zones remain above 64.4°F (18°C) all year, and substantial precipitation occurs in every month, with most stations averaging between 60 and 80 inches (1,500–2,000 millimeters) annually. Yearly precipitation has reached 160 inches (4,000 millimeters) in Douala (Cameroon) and 140 inches (3,500 millimeters) in Freetown (Sierra Leone). Abundant rainfall in tropical rain-forest regions is associated with high temperatures, low atmospheric pressures, and often luxuriant plant growth.

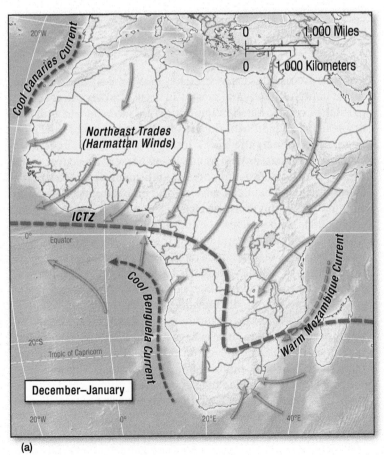

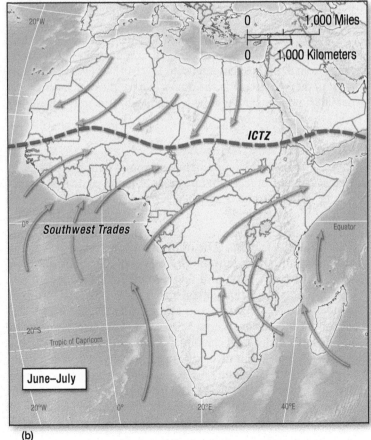

(a)   (b)

▲ **Figure 8-9 Seasonal shifts of the Intertropical Convergence Zone (ITCZ).** The ITCZ is a major determinant of sub-Saharan Africa's wind patterns and rainfall regimes.

The air that rises from the equator diverges northward and southward toward the poles, descending between 25° and 40° north and south respectively and generating dry and calm conditions. These STHP cells account for semiarid and desert conditions in the Sahara, the Sahel, and the Kalahari. Coastal deserts in Mauretania and Namibia benefit from cool ocean currents over which the prevailing westerly winds blow before reaching the land and moderating temperature and humidity conditions.

The location of the ITCZ at any time of the year is dictated by the seasonal movement of the sun. During the summer season in the Northern Hemisphere, maximum heating occurs around latitude 23° N as the ITCZ gradually shifts northward toward the Tropic of Cancer. This shift brings along rain-bearing southwesterly winds from the Atlantic Ocean and small but intense rain cells that move across the landscape as the ITCZ moves northward. Average annual rainfall totals drop to about 40–60 inches (1,000–1,500 millimeters) in the heavily wooded grasslands adjacent to the rain forest and to around 20–40 inches (500–1,000 millimeters) in the grassland regions of the drier margins. During the summer season in the Southern Hemisphere, the ITCZ shifts southward toward the Tropic of Capricorn. This movement draws with it dry northeasterly winds (Harmattan, in West African local dialect), which create dusty and hazy conditions across much of West Africa.

At the outer margin of ITCZ influence is a zone of low rainfall (the Sahel on the southern edge of the Sahara). This averages from 10 to 20 inches (250–500 millimeters) annually, but varies greatly in amount from year to year and from place to place. In N'Djamena (Chad), which is located in one of the moister sections of the Sahel, the average monthly rainfall exceeds 2 inches (50 millimeters) only from June to September. The average maximum daily temperatures range from 87°F (31°C) in August to 107°F (42°C) in April. Because most of the rain occurs in the hottest part of the year, its effectiveness in supporting plant growth is reduced. Droughts lasting several years are common.

Environmental Challenges

Four major natural environments or biomes define sub-Saharan Africa: humid tropical rain forests, subhumid tropical savannas, semiarid steppes, and deserts. Each region is characterized by its own set of ecological advantages and vulnerabilities.

Tropical Rain Forests. The tropical rain-forest biome is found in the equatorial portions of the Democratic Republic of Congo (DRC), much of Gabon and the Republic of Congo, south and central Cameroon, the coastal strips of West Africa, and portions of Kenya, Tanzania, and Madagascar (Figure 8-10). The rain-forest biome covers about 7 percent of the total land area of sub-Saharan Africa, and constitutes 20 percent of the world's rain forests. It is biologically diverse, with more than 8,000 plant species, 80 percent of the continent's primate animal species, and 60 percent of its songbirds (Figure 8-11). The forest is dominated by broadleaved evergreen trees that rise to heights of 165 feet (50 meters) and sometimes 300 feet (90 meters). A middle canopy layer, made up of trees from 80 to 115 feet (25–35 meters), forms a continuous cover as tree crowns interlock; this deprives lower layers of direct sunlight. Trees up to 50 feet (15 meters) high form the lowest layer. In addition to their role in preserving Earth's biodiversity, rain forests contain many high-value timber species, numerous wildlife species of economic value, medicinal resources (the vast majority known only to local traditional societies), and a setting for a growing ecotourism industry.

The region's tropical rain forests face many threats. Massive land-use conversion and modification following deforestation (see *Exploring Environmental Impacts:* Tropical Deforestation and Loss of Resources) seriously impacts global species diversity and undercuts the basis for ecotourism. Most rain-forest soils are intensely weathered and difficult to manage. Their surface layers are characterized by iron and aluminum oxides, which gives them a reddish color. Many of their soil nutrients have been washed down into lower horizons through a process known as soil leaching. The seeming fertility suggested by luxuriant forest growth is actually a myth. Rain-forest nutrients are primarily found in vegetation and not in the soil. Farmers who enter deforested areas burn the remaining vegetation and timber-cutting wastes to gain a brief nutrient burst for their crops. But soon soil fertility falls and crop yields decline. Chemical fertilizers, managerial skills, capital, and Green Revolution technological inputs invested into these relatively poor soils can support a variety of plantation crops. But small farmers can seldom afford these inputs, and often the land is reclaimed by a grass-dominated shrubland.

> **Stop & Think**
>
> Discuss the role that human factors play in the degradation of forests and soils in Africa.

Tropical Savannas. Large stretches of West, East, and Southern Africa contain substantial grassland areas. During the year, two periods of concentrated precipitation are interspersed with two dry seasons aligned with the seasonal movements of the ITCZ. Vegetation reflects the area's lower rainfall and ranges from perennial tall grasses with about 10–49 percent tree cover to areas of open grassland interspersed with small trees. Most of the plants, including baobabs, acacias, and mopane species, are well adapted to fire and moisture stress. The region provides expansive habitats for both livestock and wildlife development. Coping with both dry seasons and more prolonged periods of drought requires special adaptations. Traditional pastoralists, such as the cattle-herding Fulani of West Africa and the Masai of East Africa, adapt by moving seasonally between districts that have adequate grass and water. Nomadic movements have the additional benefit of moving animals and herders away from tsetse flies in the wetter woodlands and river valleys during the rainy season. But as farmers increasingly move into savanna rangeland during wet years, herders become limited to the drier savanna. This puts serious pressure on available rangeland as herd animals and wildlife concentrate in less space, resulting in overgrazing and leading to erosion and degradation.

This tropical savanna environment also is home to some of Africa's most famous national parks and game reserves. Tanzania has more than 95,000 square miles (246,050 square kilometers) of

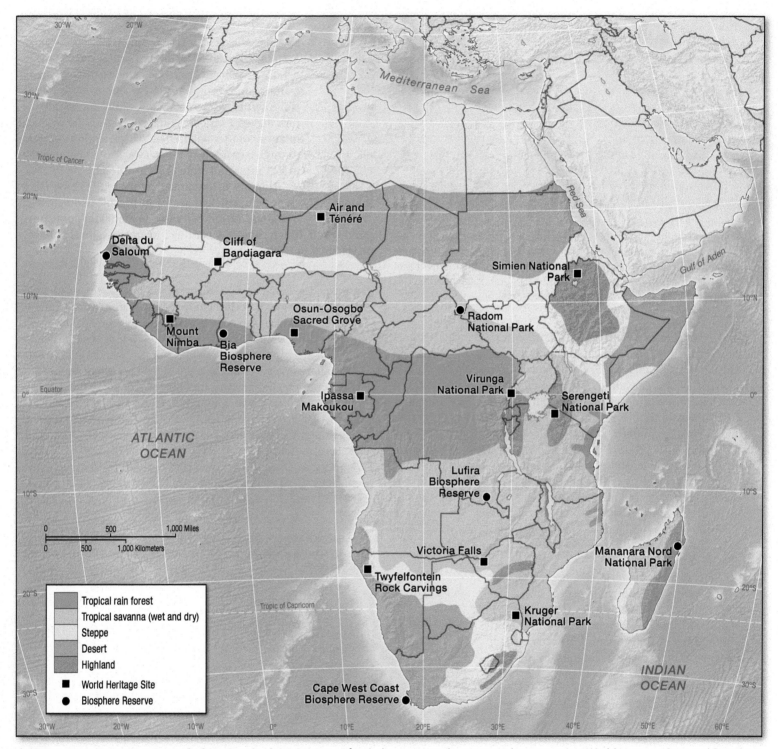

▲ **Figure 8-10 Natural regions of Africa South of the Sahara.** Africa's diverse natural regions are home to many World Heritage Sites and Biosphere Reserves.

▲ **Figure 8-11 African rain forest.** The tropical rain forest consists of four distinct vertical layers of vegetation and is characterized by an extraordinarily high number of plant and animal species. Despite its lush appearance, the tropical rain forest is a fragile natural environment that has been receding rapidly in Africa and other equatorial regions. This scene is from the Bwindi National Park in Uganda.

land devoted to parks and reserves. The famous Serengeti National Park (Figure 8-12) has the largest concentration of plains game in Africa, including millions of wildebeest, zebras, and gazelles. It is a UNESCO (United Nations Educational, Scientific, and Cultural Organization) **World Heritage Site** with cultural and natural properties of outstanding and universal value. **Ecotourism**—travel to natural areas to understand the physical and cultural qualities of the region while producing economic opportunities for the local inhabitants and conserving the natural environment—is becoming an important industry for the people who live in or near these national parks and game reserves. But illegal killing of wild animals for meat, folk medical benefits, and ivory, as well as the elimination of lions and other animals viewed as predators on herds or pests, put many species under great pressure both inside and especially outside national parks.

Soils in the subhumid savanna region have a light color, limited organic matter, and a high clay content. They are more fertile than the soils of the rain forest, are widely cultivated, and support a range of crops, including cocoa, rubber, bananas, maize, cassava, sorghum, and millet. Management of these soils to prevent erosion, particularly by wind, can be a serious problem in both traditional and commercial crop production areas.

Steppes. Poleward of the subhumid savannas are the semiarid steppes. These grasslands are climatic transition zones between the savannas and true deserts, receive limited rain, are prone to prolonged droughts, and characterized by sparse, shrubby vegetation (Figure 8-13).

Low native trees and shrubs have developed adaptations to long periods of aridity, such as thick leathery leaves, needlelike leaves, and long roots. Important native food plants include wild yams and grain sorghum, which are well adapted to local conditions. The steppe margins of the Kalahari in Southern Africa and the Sahel zone on the southern side of the Sahara are often threatened by desertification and soil degradation. This occurs because people gather wherever water resources appear to be secure. With people come cattle, farming activities, and vehicles that disturb the land surface. Without limits on the type and intensity of use, adapted native vegetation degrades through overuse, soil blows away, and patches begin to look more and more like a true desert.

The soils of the African steppe tend to be shallow, stony, and deficient in surface organic matter. Low in nutrients, they have limited potential for agricultural production, which is usually confined to areas where small-scale irrigation is possible.

Deserts. The Sahara is the largest of the desert regions of Africa (Figure 8-14). Other deserts include the Ogaden of eastern Ethiopia and Somalia, the Kalahari of Southern Africa, and the Atlantic coastal zones of Namibia and Angola. In the southern fringes of the Sahara, drought is a near-permanent condition. But substantial groundwater is present in some areas. Some of this subsurface water is recharged by underground aquifer layers that move water northward from the tropics. Other groundwater supplies are derived from fossil water created 10,000 years ago when the Sahara was a wetter place. All underground water is a precious resource, which can be, and all too often is, easily overused.

As in the Sahel, drought is a major problem in Southern Africa, particularly in its western and central sections. At least 70 percent of Namibia and Botswana experience chronic aridity that negatively

▼ **Figure 8-12 The Serengeti Plains in Tanzania.** The Serengeti Plains in Tanzania comprise the largest concentration of grazing mammals in Africa and have been designated as a World Heritage Site.

EXPLORING ENVIRONMENTAL IMPACTS

Tropical Deforestation and Loss of Resources

The United Nations Food and Agriculture Organization's Global Forest Resource Assessment Report indicates that Africa suffered a net loss of 3.4 million hectares of forests between 2000 and 2010, second only to Latin America which registered a loss of 4.0 million. Coastal West Africa has experienced the most drastic loss of moist lowland tropical forests. Côte d'Ivoire, Ghana, and Nigeria formerly contained the largest moist tropical forests in the region but less than 33 percent of the original forest vegetation remains (Figure 8-2-1). Most of the existing forests are in isolated protected reservations. Limited logging is still permitted in some of the most productive reservations. The Central African nations of Gabon, Congo, and the DRC have the lowest rates of deforestation in Africa, with more than 65 percent of their original forests still intact. This region has some of the lowest densities of rural populations, and any form of forest exploitation is very localized.

The causes of **deforestation** are mainly human induced, although there are also important physical factors such as natural fires, floods, volcanic eruptions, pests, and disease vectors. Primary human-related causes are agriculture, logging, and fuelwood consumption. In the food crop sector, the bush fallow (slash-and-burn) system of cultivation creates immense pressures on standing vegetation. Growing demands from increasing populations cause fallow periods to be reduced, limiting opportunities for soil replenishment and successful forest transition to a secondary stage. Where fallow periods are reduced, poorer woodlands and savanna grasses tend to colonize abandoned lands and inhibit forest regrowth. Logging causes much damage to the forest floor, creating gaps in previously well-structured rain forests and leading to invasions of weedy species. While logging may not lead directly to the loss of contiguous tracts of forest space, it undermines forest preservation efforts and sets the stage for other direct onslaughts such as road construction and agricultural land use. The harvesting of wood for fuel also leads to negative environmental impacts. In Burkina Faso and Chad, wood accounts for more than 90 percent of total national energy consumption. Any energy policies in place in the region have generally excluded marginal rural and urban populations. Increased demand for fuelwood by the urban poor places additional burdens on the rural environment. As the costs of electricity and other energy alternatives continue to be prohibitive for marginal urban and rural populations, the demand for fuelwood will likely continue to increase and threaten the stability of forest preserves.

Forests provide a wide range of goods and services that can be substituted for manufactured goods in rural subsistence economies. Fibers, wood poles, wild honey, medicinal herbs, and bark are just a few of the benefits or by-products of the standing forest. Tropical forests also provide optimal environments for shade-loving tree crops, such as cocoa, rubber, and coffee. From an ecological and socioeconomic standpoint, the forests are central to Africa's survival. Because the value of forests is immeasurable, they need to be managed efficiently and wisely to support Africa's drive toward sustainable development.

Sources: Michelle Kovacevic, "Demand for Chocolate Drives Deforestation in West Africa," *CIFOR Forest News*, 2011, http://blog.cifor.org/2372/deforestation-west-africa/; A. I. R. Cabral, M. J. Vasconcelos, D. Oom, and R. Sardinha, "Spatial Dynamics and Quantification of Deforestation in the Central-plateau Woodlands of Angola (1990–2009)," *Applied Geography* 31, no. 3 (2011): 1185–1193; G. von Maltitz, "Potential Impacts of Biofuels on Deforestation in Southern Africa," *Journal of Sustainable Forestry* 31, no. 1–2 (2012): 80–97; E. T. A. Mitchard, et al., "Measuring Biomass Changes Due to Woody Encroachment and Deforestation/Degradation in a Forest-savanna Boundary Region of Central Africa Using Multi-temporal L-Band Radar Backscatter," *Remote Sensing of Environment* 115 (September 2011): 172–183.

◀ FIGURE 8-2-1 **Clearing part of a forest reserve in southeast Nigeria.** Nigeria has lost over 55 percent of its primary forest due to intense logging and other human activities.

▲ **Figure 8-13 African Sahel in Mali.** The Sahel is characterized by a semiarid climate and is primarily for animal grazing. Recent overgrazing has resulted in considerable vegetative loss and, in some areas, in desertification and other forms of environmental degradation.

impacts cereal production and occasionally uproots people from villages facing water shortages. Southern African governments have instituted emergency food assistance and food-for-work programs in an attempt to alleviate these crises, but logistical difficulties related to transportation and aid distribution have hampered relief efforts.

The lack of soil moisture within deserts restricts vegetation to a few favored sites where water occurs near the surface, such as in intermittent river channels and oases. In such zones, perennial plants may survive. Outside these moist zones, most of the plants and animals of the desert exhibit permanent physiological adaptations to drought. Soils in dry zones are low in organic matter and often have hard layers of salt built up near the surface. The date palm is a commonly cultivated crop, and sorghum and millet are grown on a limited basis. Short-lived plants often sprout, flower, and die within a few weeks following the sporadic rains, serving as a resource for those nomadic herding groups who are able to reach them quickly.

Historical Background: Deep Traditions and Contemporary Struggles

Africa experienced a long and rich history prior to the era of formal colonialism, which began in the late nineteenth century, after the Berlin Conference of 1884 ratified the European powers' division of the continent. Many aspects of Africa's native cultures were lost or destroyed during the colonial era, and numerous misconceptions about Africa arose in Western societies. It became habitual for colonialists to deny almost any degree of social or political achievement to Africans themselves. Some scholars at the time even questioned whether native technologies were really indigenous to Africa or whether they had been imported by external agents. Other Westerners even went so far as to deny the status of "civilization" to the early African urban centers because their peoples lacked writing or organized social and political structures.

Geographers and historians are reconstructing Africa's past through the study of folklore, poetry, archeological sites, agricultural cropping systems, art, and architecture. We now recognize that ancient African civilizations were characterized by rules of social behavior, codes of law, and organized economies. Towns and cities with diverse sociopolitical organizations existed for several millennia. Cities such as Napata, Meroe, Axum, Djenne, Timbuktu, Gao, and Great Zimbabwe (Figure 8-15) were major

centers of cultural and commercial exchange, religion, and learning. These cities had clearly defined divisions of labor, class structures, communication networks, and spheres of influence. The diffusion of technological innovations, including iron technology, stonemasonry, and other crafts from the early eastern African civilizations of Kush and Axum was widespread to both the west and the south.

Ancient Civilizations

The earliest known civilizations of Africa emerged in the central part of the Nile River Valley (Figure 8-16). One of the most notable was the black kingdom of Kush and its capital Meroe, which flourished from about 2000 B.C. to the fourth century A.D. Meroe was largely influenced by the Nubian Kingdom. Archeological evidence from Meroe reveals a civilization that thrived on stone and iron technology. Elaborate stone walls, palace buildings, swimming baths, temples, and shrines indicate a culture with an organized social, religious, and political order. It also had an organized agricultural economy based on pastoralism and cultivation complemented by advancements in irrigation agriculture. Meroe and other Kushite cities reached a level of technological sophistication unmatched in precolonial Africa.

Other prominent cities in this region of East Africa were Axum, a metropolis of the ancient kingdom of Ethiopia, and the Red Sea port of Adulis. Classical writers and ancient Axumite coins provide evidence of a powerful Axumite kingdom that lasted from the first to tenth centuries, when Axum's influence extended across the Red Sea to southern Arabia. Historical evidence from Axum suggests a culture with remarkable engineering and architectural skills applied to quarrying, stone carving, terracing, building construction, and irrigation. Archeological artifacts also indicate a high level of achievement in metallurgical technology and manufacturing. Axum's trade network extended to the Roman provinces of the eastern Mediterranean, southern Arabia, and India, and centered on the export of ivory, gold, emeralds, and slaves in exchange for iron, precious metals, clothing, wine, vegetable oils, and spices.

Prominent civilizations that emerged between 700 and 1600 A.D. in the West African savanna were Kumbi (the political center) and Saleh (the commercial center) in Ghana, Timbuktu and Djenne in Mali, and Gao in Songhai. These cities thrived as major centers of the trans-Saharan trade. Gold mining, iron technologies, pottery making, and the production of textiles indicate a high level of technological development in the region. Kumbi Saleh was the capital of the first powerful state—Ghana—which flourished from about 700 to 1100 A.D. It was a major commercial center known for gold,

▼ **Figure 8-14 The Sahara Desert.** The Sahara accounts for about 10 percent of the African continent. Temperatures can reach as high as 125°F (52°C).

▲ **Figure 8-15 Ruins of Great Zimbabwe.** Great Zimbabwe was once the center of an advanced, precolonial civilization in Southern Africa. The surviving ruins are a national monument and a World Heritage Site.

gemstones, copper, and iron. An elaborate economic system was developed along with a system of taxation.

Timbuktu, Djenne, and Gao emerged as great centers of learning and trade in the Mali and Songhai empires of the West African savanna, flourishing as intermediate trade centers between the forest zone to the south, and the desert regions of North Africa and Egypt. A salt trade route developed between Teghaza in northern Mali and Timbuktu, and a gold route between Djenne and the forest regions of the south.

Cities in the West African forest region had well-developed kingdoms and urban civilizations, including Benin, Yorubaland, and Hausaland in Nigeria, and the Ashanti kingdom of Ghana. These kingdoms had multicentered urban networks and organized social and political systems, and they functioned as political, commercial, and spiritual centers. Ife, the cradle of the Yoruba civilization, was the spiritual capital of the Oyo empire, as were Sokoto for Hausaland and Kumasi for the Ashanti. The architectural design of Yoruban cities like Ife consisted of a series of concentric city walls with the royal palace at the center. The walls provided prestige, intimidated potential intruders, and enhanced a ruler's ability to command his subjects. Inner walls protected the privacy of rulers, and outer walls provided refuge for the masses. Other examples of urban symbolism were passageways and alleys that intersected market plazas and the creation of intimate urban spaces to encourage social interaction and cohesiveness. Cities of the forest region had diverse technologies in iron and metalwork, stone building, coarse mud architecture, gold mining, and glass making.

In the central African equatorial region, urban civilizations arose in what became the modern states of the Congo, DRC (Zaire), Angola, Zambia, Rwanda, and Burundi. These areas were the least urbanized in precolonial Africa because they lacked the large centralized empires and intricate urban networks that characterized the savanna and forest regions of West Africa. A number of significant cities did evolve, including Musumba, the capital of the Lunda empire; Mbanza-Congo, the capital of the Kongo empire; and Kibuga, the capital of the Buganda kingdom. Archeological evidence from this region suggests high levels of craftsmanship and professional artisanship in iron, copper, ivory, pottery, metalwork, and mining.

In coastal East Africa, some of the well-known historic centers were Mogadishu in Somalia; Malindi, Gedi, and Mombasa in Kenya; and Zanzibar and Kilwa in Tanzania (Figure 8-17). Gedi had elaborate walls, palaces, mosques, and well-designed homes. Kilwa was a well-constructed city with a palace and commercial center, domed structures, open-sided pavilions, and vaulted roofs. A Swahili culture that was urban, mercantile, literate, and Islamic emerged in this region. Technological developments included coin minting, copper work, building craftsmanship, boatbuilding, and the spinning and weaving of cotton. External trade was very active; ivory, gold, copper, frankincense, ebony, and iron were among the goods traded for Chinese porcelain and glazed wares from the Persian Gulf.

Much has been written about the stone-built ruins of Great Zimbabwe in Southern Africa (see Figure 8-15). Monoliths, altars, and a stone tower were discovered at this site and scholars suggest that the city originated as a vital spiritual focal point. Thirty-two-foot high elliptical walls dating back to at least the fourteenth century have been discovered, along with the "Great Enclosure," which contains about 182,000 cubic feet (5,100 cubic meters) of stonework. The technological achievements of Great Zimbabwe go beyond building and stone construction to include mining and metallurgy, manufacturing of pottery, wood carving, and cotton spinning.

High levels of social and technological achievement existed among precolonial African civilizations. The advent of European colonialism saw the destruction of much of Africa's rich material heritage. But there is renewed effort to discover the richness and elegance of past African civilizations.

Most cities in Africa today are colonial creations. The majority of African capitals have a coastal orientation because they were created as "headlinks" designed to funnel the resources of the interior to overseas destinations. Most precolonial cities are now virtually extinct or, as in the case of Timbuktu or Gao, are just relics of their former selves. A few ancient cities have survived the ravages of colonialism and still thrive today as historic centers. These include Kumasi in Ghana, Ibadan and Ife in Nigeria, Mogadishu in Somalia, Mbanza-Congo in Angola, and Mombasa in Kenya. Historians place the founding of Mombasa as early as 500 B.C. It functioned primarily as a center for trade in agricultural products, gold, and ivory. Mombasa is now known for its main tourist attraction, Fort Jesus, a World Heritage Site built by the Portuguese in the 1590s to house slaves. Today, Mombasa is Kenya's second-largest city with more than 1 million inhabitants and serves as a major port city for Kenya, Uganda, and Rwanda. Key industries include tourism, oil refining, metal working, sugar processing, handicrafts, and cement works. Mbanza-Congo's roots can be traced back to the fourteenth century, when it served as the spiritual and political center of the Kongo Kingdom. Today, it serves as the regional capital of Zaire province in northwest Angola, as well as a center for oil production and the distribution of agricultural products.

European Colonialism

Colonialism imposed foreign values on indigenous institutions. It was largely paternalistic, exploitive, and inhumane. The era of European colonialism can be divided into four main periods: the age of initial contact, the period of enslavement, the period of land exploration, and the era of formal colonialism. The first three periods

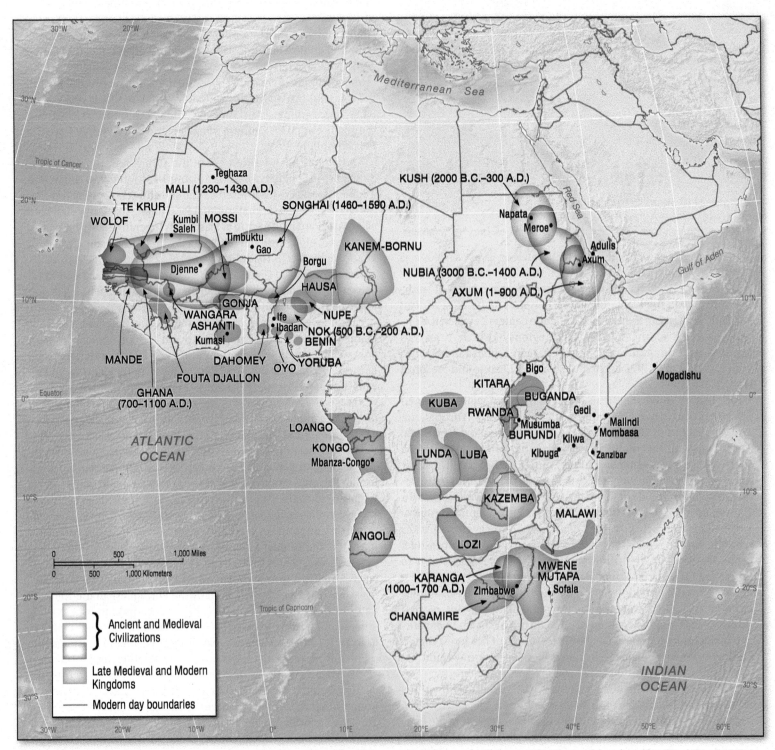

▲ **Figure 8-16 Ancient, medieval, and late medieval sub-Saharan African civilizations.** Sub-Saharan Africa has a rich cultural heritage. Some civilizations date to the fifth century B.C. The zones that supported the greatest number of civilizations and kingdoms were West Africa and south-central Africa.

▲ **Figure 8-17 Stone Town, Zanzibar, Tanzania.** Zanzibar has functioned as an urban and commercial center of eastern coastal Africa since precolonial times. Stone Town is the old trading center on the island where Omani merchants once dominated commerce and the production and export of cloves.

focused on trade in commodities and slaves, while the formal colonial period emphasized the extraction of mineral and agricultural resources to meet the industrial needs of Europe. One of the most serious consequences of European colonialism was the limitations imposed on African industrial development.

Early Periods: 1400–1880.

The first meaningful European contact with Africa South of the Sahara came in the fifteenth century through Portugal's search for a sea route to India. The Portuguese confined their trading activities to the coast, leaving inland trade to African merchants. European products such as copper and brass were exchanged for African gold, and trading posts and forts were established along the coast of West Africa (Figure 8-18). A second objective was the establishment of world empires. After Portugal's initial contact in 1420, the Spanish, English, French, and Dutch followed.

Trade in goods and commodities continued in the early to mid-1400s until the discovery of the Americas in 1492 and the establishment of New World tobacco, sugarcane, and cotton plantations. The Native American population was subsequently decimated by European diseases, and those who survived often fled their Spanish and Portuguese overlords. Lacking sufficient laborers, the period of enslavement began as New World Europeans turned to African slaves to work the huge estates. Estimates of the number of Africans transported across the Atlantic Ocean range from 8 to 12 million people. The impact of slavery on African societies included the disruption of cultural institutions; the reduction of industries, crafts, and other forms of manufacturing; and the increased incidence of tribal wars—prisoners were often sold as slaves. The major sources of African slaves were the Senegambia coastal region between Guinea-Bissau and Liberia, and Congo and Angola. Other key source areas included the Dahomey and Yoruba kingdoms, the Gold Coast (Ghana), the Niger delta, and Mozambique.

The shipment of African slaves was abolished in 1808, owing largely to the humanitarian efforts of philanthropists and religious leaders such as Englishmen Granville Sharpe and William Wilberforce, as well as Western-educated and freed slaves such as Olaudah Equiano of Nigeria and Ottabah Cugoano of Ghana. With the end of slavery, European economic activities in Africa reverted to trading in commodities such as gold, coffee, cocoa, and palm oil. Missionaries also renewed their efforts to convert Africans to Christianity and assist in their social and economic development.

By 1840, scientific and geographic curiosity about Africa had grown to such an extent that European explorers were intensifying efforts to access the region's hinterlands and interior resources. The period of exploration between 1840 and 1890 is characterized by the explorations of Henry Stanley (along the Congo River), David Livingstone (along the Zambezi River), and Mungo Park (along the Niger River). Their accounts of Africa's potential resources led to a "scramble for territory" and set the stage for the 1884 Berlin Conference that partitioned Africa among the European powers, and ushered in the formal era of colonialism. This last phase was precipitated by the shift from mercantile to industrial capitalism in Europe and the Industrial Revolution.

Formal Colonialism: 1880 to the Mid-1950s.

The **Berlin Conference** marked the beginning of the formal era of colonialism in Africa. It was a significant turning point in Africa's history because it shaped the social, economic, and political futures of the continent's countries. The European powers met in Berlin—without African consent or participation—to establish procedures for the allocation of African territory among themselves. Territories were assigned on the basis of the principle of effective occupation, resulting in the imposition of arbitrary political boundaries on the diverse cultural landscape of African societies. European colonizers felt free to impose their values, policies, and institutions on a fragmented and disunited continent. Figure 8-19 shows the distribution of European colonies in sub-Saharan Africa in the mid-1950s, just prior to the wave of independence movements. Liberia and Ethiopia were the only states never to be colonized.

British colonial policy was based on **indirect rule**. Lord Lugard, the governor general of Nigeria, was instrumental in formulating this policy with the explicit purpose of incorporating the local power structure into the British administrative system. African chiefs and kings were to function as intermediaries,

▼ **Figure 8-18 Elmina Castle, Ghana.** Coastal West Africa has as many as fifty forts, which were used by Europeans as trading posts and for housing slaves.

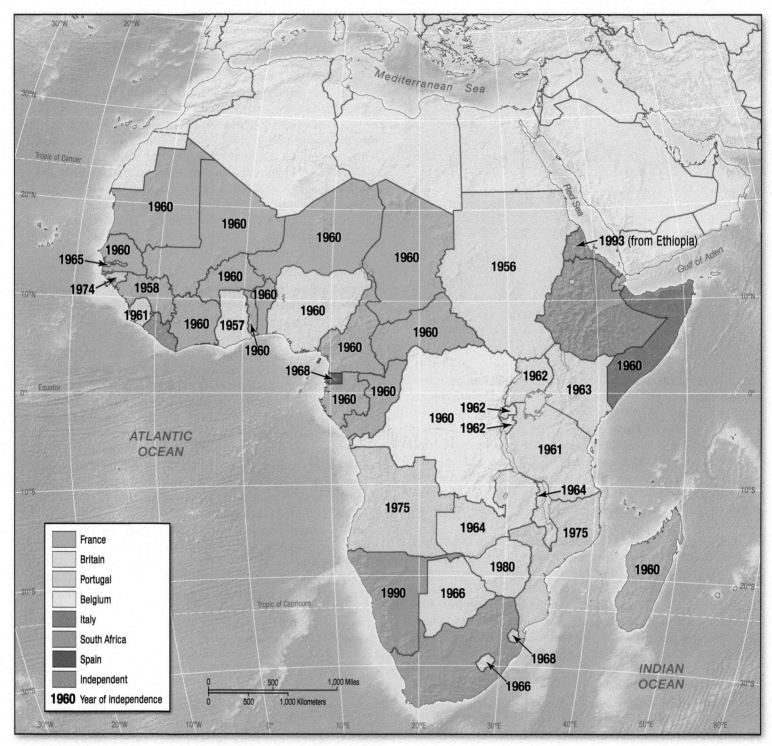

▲ **Figure 8-19 Colonial map of sub-Saharan Africa, 1956.** The French dominated West Africa in the late colonial era, and the British ruled East Africa. Belgium controlled the Congo Basin, and Portugal governed portions of the Atlantic and Indian Ocean coasts of Southern Africa. Liberia and Ethiopia remained independent.

acting as links between their people and the colonial authorities. The chiefs were responsible for enforcing local ordinances, collecting taxes, and carrying out the day-to-day affairs of the colonial authority.

The French enacted a highly centralized administrative structure based on a policy of **acculturation** that sought to encourage Africans to adopt and assimilate French culture, language, and customs. Two French "overseas" federations were created: French West Africa, with Dakar (Senegal) as its headquarters, and French Equatorial Africa, centered around Congo-Brazzaville. Directives originated from Paris and were funneled through the administrative centers. Policies initiated in the overseas federations had to be approved by the French national assembly in Paris. The Portuguese policy of assimilation was similar to France's acculturation policy. Angola, Mozambique, Guinea-Bissau, and São Tomé and Príncipe were regarded as overseas provinces of Portugal. Portuguese policy focused on developing a social hierarchy, which was dehumanizing to Africans trapped at the bottom of the social ladder. Africans who aspired to Portuguese citizenship were granted *assimilado* status. All other people, the so-called *indigenas*, were relegated to working agricultural estates that produced cotton and other cash crops for Portugal. The Belgians controlled DRC, Rwanda, and Burundi. Belgian policy was blatantly exploitive and paternalistic, extracting mineral and agricultural wealth from the colonies while investing almost nothing in the social development and education of the native peoples. At independence in 1960, the entire DRC had only 16 university graduates.

The formal colonial era waned in the early 1950s. World War II had taken its toll on European economies, African soldiers who had fought side by side with European "Allies" had returned with ideas of freedom, the United Nations promoted the principle of self-determination, and African scholars who studied at foreign institutions returned to organize political platforms and parties to challenge colonial rule. These developments, combined with the dehumanizing and exploitive nature of colonialism, encouraged the decolonization process, although the scars of colonialism remain.

The Impact of Colonialism

Even if we assume that colonialism had some positive effects, the negative impacts far outweighed the positive ones. The political boundaries that Europeans imposed sometimes forced hostile ethnic groups to share the same territory and, at other times, divided the same tribe between two or more countries. Civil wars and nationalistic military campaigns have repeatedly severely destabilized parts of the region,

resulting in a constantly recurring refugee crisis. African countries are still searching for appropriate solutions to defuse sources of hostility.

Another legacy of colonialism is the number of small-sized and landlocked states that present-day Africa inherited. The potential for agricultural and industrial development is limited in **microstates** like Gambia, Lesotho, Swaziland, and Djibouti, which have small domestic markets and labor pools. Africa also has close to 40 percent of the world's **landlocked states,** whose trade with the outside world must pass through neighboring, sometimes hostile, countries. Because colonial development efforts centered on the coastal zones, these landlocked countries also tend to lack adequate rail and road networks, limiting the spatial interaction between their rural and urban sectors.

The lack of an extensive transportation network also helps explain the limited trade among African countries. Sub-Saharan Africa's transportation systems were developed to link interiors to coasts, not to create links between countries. Intra-African trade accounts for less than 15 percent of total African trade. Colonialism may have ended, but **neocolonialism** is prevalent. Most African countries continue to provide much of their raw material exports to the European industrialized countries (Table 8-1). Although politically independent, many African nations continue to depend

Table 8-1	Principal Primary Exports and Trading Partners of Selected Sub-Saharan African Countries		
Country	**Principal Primary Exports**	**Percent of Total Exports**	**Principal Trading Partners**
Benin	Cotton, cashews	58	France, China
Burundi	Coffee	76	Belgium, Germany
Chad	Petroleum	91	France, Portugal
Congo Republic	Petroleum	87	United States, France
Gabon	Petroleum and petroleum products	70	France, United States
Niger	Uranium	70	France
Nigeria	Petroleum	86	United Kingdom, United States
Rwanda	Coffee	52	Germany, Belgium
Sudan	Crude petroleum	89	China, Japan, Saudi Arabia
Tanzania	Machines and transport equipment	36	United Kingdom, India, Spain
Togo	Phosphates	53	France, Ghana, Burkina Faso
Uganda	Coffee	31	Belgium, France, United Kingdom
Zambia	Copper	64	United Kingdom, South Africa

Source: Based on *Africa Development Indicators 2011* (World Bank: Washington, DC, 2011).

economically on their former colonizers for trade, technology, and other goods and services, which discourages economic diversification. Another challenge to economic development is the fact that most African economies are still monocultural—that is, they rely on one or two primary products for export revenue. A sectoral imbalance also persists, with weak internal linkages between agriculture and industry, both of which focus on exporting products to overseas markets rather than providing for local needs. Creating linkages between key economic sectors within a country is a prerequisite for developing a self-sufficient and self-sustaining economy.

Perhaps the most significant impact of colonialism is that African countries must cope with a dual or, in some cases, triple heritage. Governments are confronted daily with conflicts and contradictions between traditional African and imposed European value systems. The inherent contradictions pervade the social, cultural, economic, and political lives of Africans. Countries wrestle with the question of whether or not to adopt a common non-Western language, school systems debate replacing their European curriculum with one that is Afrocentric, and politicians search for the "ideal" constitution that incorporates aspects of traditional authority into largely Western-based political systems. In the northern regions of West, Central, and East Africa, a third element, Islam, further complicates the political and cultural environments. In Sudan, Nigeria, and Somalia, in particular, Islamic fundamentalism has emerged as a potent force. The legacies of colonialism are still felt in the daily lives of Africans.

> **Stop & Think**
>
> ▷ Evaluate the economic and political impacts of colonialism in sub-Saharan Africa.

Contemporary Sub-Saharan Africa

Sub-Saharan Africa includes linguistically and religiously diverse populations bound by shared family customs and traditions. These shared value systems are often juxtaposed by sharp contrasts in living conditions and economic lifestyles between rural and urban households.

Language Diversity

One of the most intriguing aspects of sub-Saharan Africa's cultural geography is the more than 1,000 languages that exist in the region. Most of these languages do not have a written form or tradition, and approximately 40 are spoken by 1 million or more people. Linguistic scholars have identified four major linguistic families: Niger-Kordofanian, Nilo-Saharan, Khoisan, and Afro-Asiatic (Semitic-Hamitic).

Niger-Kordofanian, the largest linguistic family in this world region, is divided into two related but distinct branches, the Kordofanian and the Niger-Congo. The Kordofanian branch is centered in a small area in the Nuba hills of Sudan and consists of about 20 languages. The Niger-Congo branch is spoken by more than 150 million people and stretches across half of sub-Saharan Africa, extending from West Africa to the equatorial and southern regions. It includes the west Atlantic languages of Wolof and Fula in Senegambia; the Guinean languages of Kru (Liberia), Akan (Ghana), Yoruba (western Nigeria), and Igbo (eastern Nigeria); the central and eastern Sudanese languages of Azande (Central African Republic) and Banda; and the most widely spoken subfamily, the Bantu (Figure 8-20). Linguistic studies trace the origins of Bantu to the southeast margins of the Congo (Zaire) rain forest, although some evidence suggests linkages to the West African forest and savanna regions. Bantu languages are spoken from the equatorial zone to South Africa. Those with 10 million or more speakers include Lingala (DRC), Tswana-Sotho (Botswana and South Africa), Zulu, and Swahili (Kenya, Tanzania), which has also been subjected to Arabic influences. Other languages spoken by more than 1 million include Bemba (Zambia), Luba (southern DRC), Shona (Zimbabwe), and Buganda (Uganda).

The Nilo-Saharan linguistic family stretches in a west-to-east direction, from the Songhai language in southwest Niger to the Nilotic languages of Nuer and Dinka in southern Sudan and Luo and Masai in southwestern Kenya. It also includes the Saharan languages of Kanuri (which can be traced to the Kanem and Bornu kingdoms of Lake Chad), Kanembu, and Teda.

The Khoisan linguistic family is confined to the Kalahari Desert region in Namibia, Botswana, Zimbabwe, and South Africa. It dates back several millennia and once dominated the entire territory of southern and eastern Africa from Somalia to the Cape of Good Hope. But over the centuries, the Khoisans have been ravaged by diseases, Bantu invasions, and the expropriation of land by European settlers. Today the Khoisan language is associated with the Nama, Hottentots, and Bushmen (San) communities occupying areas unwanted by other peoples.

The Afro-Asiatic (Semitic-Hamitic) linguistic family, widespread in North Africa, is found in sub-Saharan Africa in Mauritania and the East African Horn, where Semitic languages such as Amharic and Tigre are spoken. This language family also includes Cushitic languages such as Somali and Mbugu (Tanzania), and Chadic and Berber languages.

Besides the four major linguistic families, two other language groups of non-African origin are spoken in the region: the Malay-Polynesian family was introduced to Madagascar about 2,000 years ago, while Afrikaans, a derivative of Dutch, has Indo-European origins dating back to 1602 when the Boers arrived in South Africa.

In many parts of the continent, people whose native languages are mutually incomprehensible use a common language—a **lingua franca**—for communication. Swahili, which developed along the East African coast from a fusion of Arabic with local Bantu, is now spoken widely in East Africa. It is also becoming increasingly popular in neighboring Central Africa. In West Africa, Hausa, which is spoken by more than 50 million people, is the most important lingua franca, particularly in Nigeria and Niger. Other tongues that have the potential to develop into major regional languages include Amharic in the Horn of Africa, Mandinke in West Africa, Lingala in Central Africa, Sotho-Tswana in the central parts of Southern Africa, and Zulu in southeast Africa. The colonial languages of English, French, and Portuguese are still widely spoken and remain a lingua franca in many parts of Africa.

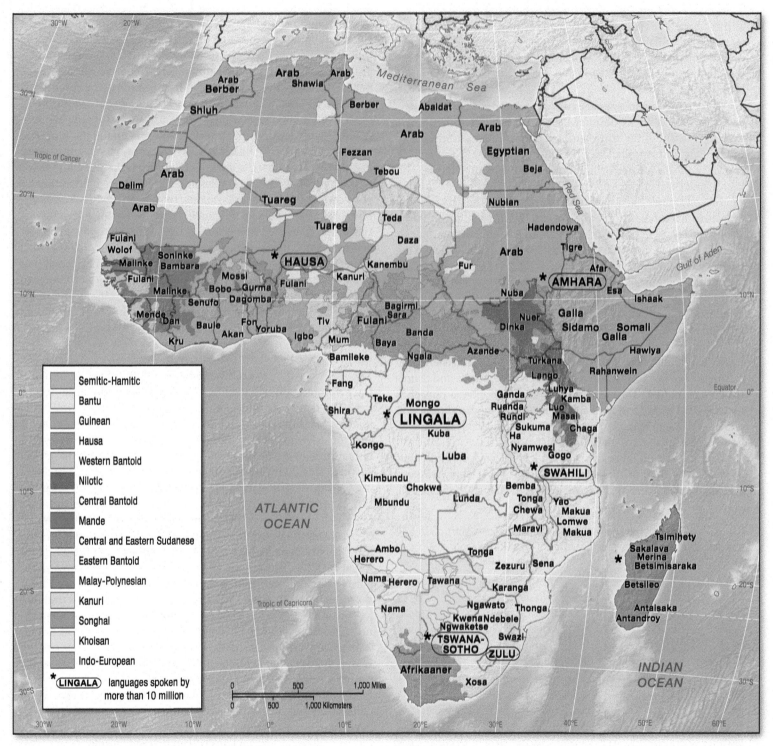

▲ Figure 8-20 Distribution of languages in Africa. Although more than 1,000 languages are spoken in sub-Saharan Africa, these can be grouped into linguistic families that have distinct geographic patterns.

Religion

The legacy of colonialism and the diffusion of Islam from the Arabian Peninsula and North Africa have profoundly influenced the religious landscape of Africa South of the Sahara (Figure 8-21). Islam is concentrated in the northern Sahelian regions, the Horn, and the coastal corridors of Kenya, Tanzania, and Mozambique (Figure 8-22). Significant Muslim minorities also exist in the central regions of several West African states, resulting in recent insurrections in northern Nigeria and northern Mali from extremist Islamic groups.

Christianity is widespread in the central and southern sections of Africa, with a strong Roman Catholic presence in the former Portuguese-ruled territories of Angola and Mozambique and former Belgian-controlled Rwanda, Burundi, and DRC; an Anglican presence in the former British holdings of Ghana, Nigeria, and Kenya; and Presbyterians common in Malawi. The Coptic Christian Church in Ethiopia dates back to the fourth century and has resisted pressures from Islam since the seventh century. Judaism had a strong following in the Gondar region of Ethiopia, where blacks, who came to be called **Falashas** (black Jews), were converted to the faith by Semitic Jews who emigrated to Ethiopia between the first and seventh centuries. Almost all Falashas migrated to Israel in a series of airlifts between 1984 and 1991. Other religions, such as Buddhism and Hinduism, are practiced by small numbers of people in southern and eastern Africa.

Religious traditions in Africa often blend, producing mixed beliefs and rituals. Many Christian churches are Africanized. African carvings have been substituted for church decorations and drums for church bells in Catholic masses, and Pentecostal churches feature a wide range of African music, including drumming and dancing. Worship in these churches is vibrant and lively, with much dancing, reminiscent of traditional celebrations.

African traditional religions often have been wrongfully perceived as animistic or atheistic. Most Africans are deeply religious and acknowledge the existence of a supreme being. In the Congo, Nzambi is the all-powerful creator of the sky, earth, and man. God is perceived as having two sides: Nzambi Watanda (God above), who is good, and Nzambi Wamutsede (God below), who is wicked. This belief reflects life's realities, with its alternating periods of good and bad fortune. In Sierra Leone, God is referred to as Meketa (the everlasting one) and Yataa (the omnipresent one). Traditional African religion ascribes to a hierarchical order, with God at the top, followed by ancestral spirits and divinities, human beings, animals, plants, and inanimate objects. Departed ancestors are believed to serve as intermediaries between the living community and God. Deities and ancestral spirits are honored in sacrifices and special ceremonies. Traditional religionists in the Congo believe in two types of ancestral spirits: the Binyumba inhabit the kingdom of the dead, while the Bakuyu, who have not yet been admitted to the abode of deceased spirits, wander around and are appeased through offerings. Small temples and spirit shrines to honor nature gods dwelling in rivers, mountains, hills, and lakes are commonplace. Yoruba traditional religion in Nigeria has a four-tiered structure of spiritual or quasi-spiritual beings. At the top of the hierarchy is the supreme being, Olodumare, and his subordinate ministers, called *orisha*, constitute the second tier. The third tier consists of deified ancestors, called *shango*, followed by spirits associated with natural phenomena such as rivers, lakes, mountains, and trees. Divination and fortune telling are popular activities in traditional African religion.

Family and Kinship Relations

The African family is usually an extended unit. Each member of the extended family has obligations and ties to the other members. The family includes both the living and the dead. Unlike the bilateral **kinship relations** characteristic of Western societies, most African societies are of unilineal descent, tracing lines of descent either through the father's side (patrilineal system) or through the mother's side (matrilineal). About three-fourths of African societies are patrilineal. These include inhabitants of the pastoral savannas of West and East Africa, such as the Fulani, the Nuer in southern Sudan, the Masai (west Kenya and Tanzania), the Kikuyu (central Kenya), and the Buganda of Uganda. Matrilineal societies include the Akan of Ghana and the Lamba and Bemba of Zambia. In matrilineal societies, links to the father's family are secondary with regard to the inheritance of property. Resources are inherited from the mother's brother, who is the matrilineal authority. This tradition has resulted in wives and children being left with little or no property after the death of a husband/father. Countries such as Kenya, Zambia, and Ghana have considered reforming inheritance rules to ensure that wives and children have access to land and other family property upon the father's death, although these "civil" laws often conflict with customary law. In patrilineal societies, widows and children are not always guaranteed property rights either, especially in polygamous marriages.

In traditional African societies, marriage is a union, not of two individuals, but of two extended families. Marriage is perceived as a civil contract between two families. This contract calls for the transfer of goods or money, or both, from the bridegroom's family to the bride's family in the form of bride wealth. The payment of

▼ **Figure 8-21 Mosque and market in Mali.** Islamic religion, trade, and culture diffused into the savanna states of West Africa around 900, and today is a major force in the drier parts of West, Central, and East Africa.

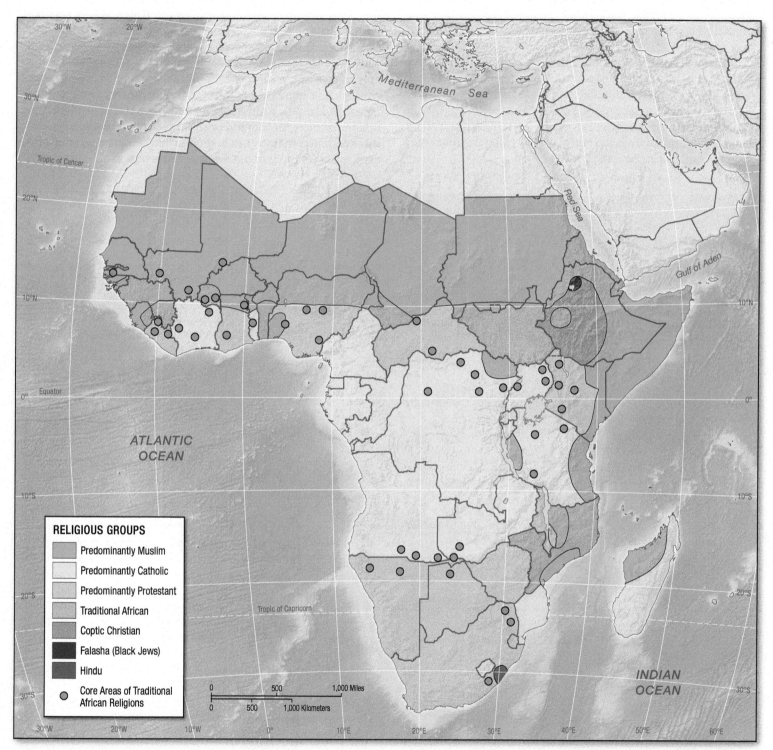

▲ Figure 8-22 Distribution of major religions in sub-Saharan Africa. Islam has diffused southward into the northern regions of black Africa. Roman Catholicism predominates in central Africa, and Protestantism in Southern Africa.

RELIGIOUS GROUPS

- Predominantly Muslim
- Predominantly Catholic
- Predominantly Protestant
- Traditional African
- Coptic Christian
- Falasha (Black Jews)
- Hindu
- Core Areas of Traditional African Religions

bride wealth compensates the bride's family for the loss of her labor since she then has to devote all her time and allegiance to her husband's family. Although divorce is infrequent, it can be granted for a number of reasons including adultery, infertility, impotence, and an inharmonious relationship with a mother-in-law, but only after counseling and efforts at family intervention and conflict resolution have failed.

In spite of language and religious differences, Africans share a number of common customs and traditions with respect to their family and kinship relations. The extended family, respect for the elderly, socialization between the elderly and the young, and the significance of ancestors are all attributes that most Africans share. Another important trait is the role played by cultural symbols as a means of expression (Figure 8-23).

Village Life and Traditional Agriculture

The majority of sub-Saharan Africans, 63 percent, live in rural areas and rely on agriculture as their principal economic activity. Their villages constitute the basic units of social and community organization. Most villages in the forest and grassland regions are nucleated settlements, and such compact arrangements encourage communal interaction. Most dispersed or scattered settlements are associated with decentralized social and political systems. The smallest unit of residence within villages is the village compound or homestead, consisting of small houses occupied by members of an extended family (Figure 8-24). Although their design and arrangement may differ from one ethnic group to another, family compounds share some common characteristics such as one entranceway with a reception house in close proximity, a courtyard to enhance social

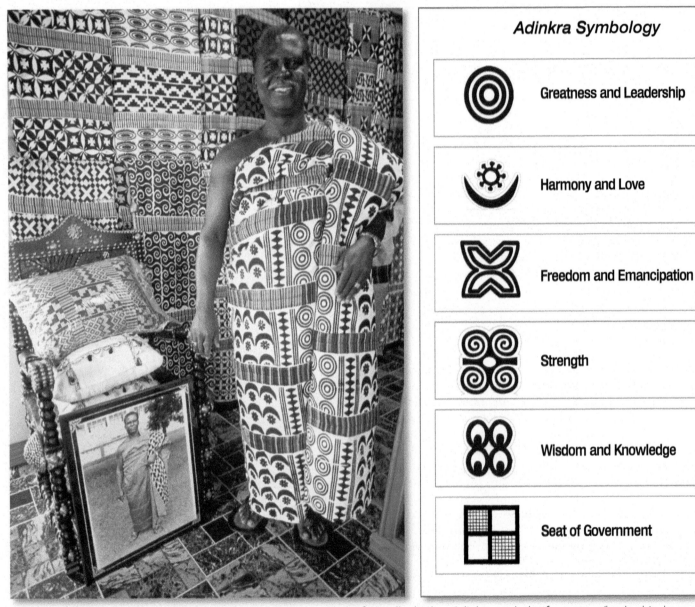

Adinkra Symbology

Greatness and Leadership

Harmony and Love

Freedom and Emancipation

Strength

Wisdom and Knowledge

Seat of Government

▲ **Figure 8-23 Cultural symbology in Ghana.** This Ghanaian subchief proudly displays Adinkra symbols of greatness/leadership, harmony, governance, experience, and wisdom. The colors of the Kente cloth also have symbolic relevance. Gold/yellow signifies royalty and wealth; blue indicates harmony and love; green denotes nurturing of the land, growth, and spiritual revival; and red illustrates a heightened political and spiritual awareness.

▲ **Figure 8-24 Masai village in Kenya.** Extended family compounds in compact villages are the basic units of residence within many sub-Saharan rural communities. The arrangement facilitates social and economic interaction and strengthens traditional kinship ties. Animals leave the family corrals at the beginning of the day to graze in the surrounding pastureland.

interaction, and a number of multifunctional rooms. The Somolo of southern Burkina Faso live in circular, multistory houses enclosed by walls of puddled mud. The center of the house has a small courtyard, and the thatched roof is supported by posts. There are as many as 20 rooms, which include one room for each wife, children's rooms, kitchens, storerooms, and granary and grinding areas. The Yoruba compounds of western Nigeria are typically square or rectangular in shape with houses built around one large courtyard. The main courtyard may be linked to a series of subsidiary courtyards that allow air and light into surrounding rooms and that enable rainwater to be collected easily into pots and tanks.

Rural residents rely on agriculture for their livelihood and sustenance. In a context of rural dualism, a larger traditional, subsistence farming sector coexists with a commercial, export-oriented economy. Commercial agriculture is dominated by the export of cash crops such as cocoa, coffee, tea, sisal, oil palm, and peanuts, and the production of raw materials for the urban-industrial sector. Traditional agriculture encompasses a variety of cultivation, pastoralist, and fishing activities. Traditional agriculture is labor intensive and utilizes simple tools such as hoes, machetes, and dibble sticks.

Among the more common farming practices in the African forest zones is **shifting cultivation**, where farmers move every few years in search of new land after the fertility of the soils of their existing plots become exhausted. Once identified, the new plot is cleared and burned. The nutrients contained in the ashes temporarily increase the fertility of the soil. A few trees are often left standing to shelter crops from excessive sunlight and to decrease erosion. After the nutrients from the ashes are exhausted, the land is abandoned and soon reclaimed by the forest. A variant of shifting cultivation is the rotational bush fallow system, where the cultivated area rotates around a fixed area. Fallow periods are shorter than those of shifting cultivators. Both farming practices have the potential to threaten the ecological stability of forested areas if the growing population exceeds the carrying capacity of the land. Little long-term damage is likely, however, if population levels remain relatively low. Other rural farming practices include the intensive growing of fruits and vegetables within the confines

of compounds and combining the raising of livestock with the cultivation of crops (mixed farming). The major staple foods that are grown include tropical tubers such as yams, cassava, cocoyams, and sweet potatoes; grain crops such as sorghum (guinea corn), millet, and maize; and fruit crops such as bananas and plantains (cooking bananas).

Pastoralism is practiced in the grassland and semiarid regions of Africa. The Fulani nomads of the West African Sahel, the Masai of Kenya, and the nomadic Tswanas of Botswana are among the best-known pastoralists (Figure 8-25). Pastoralists face many challenges, including the environmental constraints of drought, land degradation, and loss of grazing land; the subsequent economic loss of herds and wealth; competition from other land uses for limited grazing land; and the attempts of some governments to pursue policies that limit grazing land and promote the settlement of nomads.

Fish production is concentrated in riverine, coastal, and lake areas. Marine fishing is still undeveloped and involves only few relatively large fishing vessels. Inland fishing resources are the backbone of many rural communities located near coasts, rivers, lakes, and lagoons. Important inland fishing areas include the

▼ **Figure 8-25 Fulani nomadic pastoralists.** By migrating from one area to another, sub-Saharan pastoral nomads are able to utilize the seasonal resources of adjoining natural regions. The cattle of this Fulani man, wearing a traditional hat, are drinking from a seasonal water hole in Mali.

Niger delta, Lake Chad, and some coastal areas of Eastern and Southern Africa.

Although subsistence agriculture is often perceived as being static and inactive, it is a dynamic and enterprising economic sector with well-organized marketing systems. The traditional African market is a focal point for the exchange of goods, services, and ideas among subsistence farmers and pastoralists (Figure 8-26). It also serves as a forum for political activities and for social functions such as marriages. One common form of traditional market is the periodic market, which is held about every four to eight days to serve the needs of dispersed populations and mobile traders.

The traditional farming sector faces environmental, economic, and institutional challenges. Prolonged droughts and soil degradation have restricted the amount of arable land in many regions. Scarce arable land is often monopolized by commercial agriculture. African governments continue to focus their research and policy efforts on producing one or two primary products for export. Traditional farmers are frequently denied access to credit, extension services, and the industrial technologies of the Green Revolution. Diffusion of Green Revolution technologies is hindered by the limited development of high-yielding strains of the staple tuber crops (such as cassava, yams, and cocoyams) that constitute the daily diet of African farmers. The Green Revolution in Africa has functioned as a classic case of hierarchical diffusion, in which new technologies and innovations "leapfrog" from one large-scale farm to the next, before eventually descending to the level of smallholder farmers. This is a reflection of the dual structure of rural Africa, in which wealthier and more influential farmers benefit from foreign innovations much more than traditional farmers. In 2006, however, the Bill and Melinda Gates Foundation joined forces with the Rockefeller Foundation to create the Alliance for a Green Revolution in Africa (AGRA) with the goal of developing 100 new crop varieties in 5 years, and tripling farm yields in 20 years. AGRA now has many government and private sector partners working to help small farmers in a number of African countries.

Africa's system of land tenure, or access to land, also presents some challenges to agricultural productivity. Throughout most of traditional sub-Saharan Africa, land is held communally rather than individually: land belongs to the living, the dead (ancestors), and the yet-to-be-born (future generations). An example of a traditional land tenure system is family land, which is simply passed on through a lineage with rights to the land held jointly by a number of heirs. Usually, no monetary transactions can take place with such land, but destitute families are sometimes forced to sell family holdings. Problems associated with family land include the increased fragmentation of land resulting from rapid population growth and the subdivision of family land among multiple heirs, and the inequities associated with matrilineal inheritance. Communal land belongs to the lineage, village, or community; under ideal circumstances every member of the community has equal right to as much land as needed. In almost all cases, the head of the village or clan is in charge of the land and its disposition. Stool land is vested in the stool, or skin, which is the symbol of kingship among the Yoruba in Nigeria, the Mossi in Burkina Faso, the Baganda in Uganda, and the Ga and Ashanti in Ghana. The traditional leader or chief has a sacred duty to hold the land in trust for the people. Subjects of the stool can access land for farming and shelter requirements, but in return they must provide customary services and pay homage to the stool.

Urbanization

Rapid population growth has triggered the movement of people from neglected rural areas to urban centers. About 346 million people, 38 percent of sub-Saharan Africa's total population, now live in urban areas. Sub-Saharan Africa is among the least urbanized world regions, but its urbanization growth rate is among the highest. Cities are growing at 4.4 percent per year, compared to 3.3 percent for Asian cities and 2.5 percent for South American cities. Some of the fastest growing cities are Abuja, Nigeria (8.3%); Ouagadougou, Burkina Faso (7.3%); Luanda, Angola (6.0%); and Lilongwe, Malawi (5.4%). Rural-urban migration accounts for more than 50 percent of urban growth in Gambia, Sierra Leone, Liberia, and Côte d'Ivoire.

Factors that push migrants out of rural areas include benign neglect by governments, lack of economic opportunities, prolonged droughts, the desire to escape from the social constraints of the extended family, short-term cash needs for bride wealth, marital instability, birth order, and inheritance laws. Economic opportunity, although limited, is a strong pull factor for cities. Instead of being prime movers of innovation, change, and development for their hinterlands or surrounding areas, most sub-Saharan cities develop a parasitic relationship with the surrounding rural hinterlands, consuming most of the resources and benefits generated in the countryside while giving little in return.

Typically, African cities are primate cities that receive a disproportionate share of the economic, cultural, and political resources of their nations. Many cities have failed to provide a sufficient number of formal sector job opportunities, and most migrants end up working in the informal sector, which operates outside the mainstream of government activity and benefits. This sector includes a variety

▼ **Figure 8-26 Traditional village market in Mozambique.** Traditional markets offer considerable insight into rural life in sub-Saharan Africa. In addition to supplying foodstuffs, they are often the source of medicinal substances, fuel, and clothing, and provide valued social interaction.

of jobs, ranging from artisans to basket weavers, goldsmiths, and garment makers. These talented individuals operate without official recognition; they receive few benefits, but also seldom pay license fees or taxes. The informal sector contributes to the formation of human capital by providing access to training and apprenticeships at a cost much more affordable than formal training institutions. It also encourages the recycling of local resources, such as tires from automobiles (which are used to make footwear), and it caters to the customized needs of residents who cannot afford to purchase items in bulk.

Rapid urbanization has left many cities with inadequate water supply and sewage disposal systems, limited solid-waste disposal mechanisms, and inefficient public transportation. The basic infrastructure of roads, highways, and other public services is lacking in most cities as well. The United Nations Center for Human Settlements (UNCHS) has developed the **City Prosperity Index (CPI)** to measure productivity in cities in terms of economic growth and job generation; infrastructure development (water, sanitation, and information and communication technology); quality of life reflected in the use of public spaces to promote community cohesion and civic identity; equity and social inclusion to promote gender equality and civil rights; and environmental sustainability to ensure efficient management of natural assets and energy resources. African cities, particularly Monrovia, Liberia, and Conakry, Guinea, rank among the lowest in urban prosperity, while cities like Nairobi, Kenya, and Capetown, South Africa, perform relatively better. The CPI is accompanied by a policy matrix, the Wheel of Urban Prosperity, which provides policy prescriptions based on legal, institutional, and planning regulations.

Contrary to the popular perception of African cities exhibiting an inverse concentric pattern, with the rich residing in the center of downtown and the poor living on the periphery, more recently upper- and middle-income districts are proliferating in suburban and exurban regions in cities such as Accra, Ghana (Figure 8-27). This growth has occurred in an uncontrolled and uncoordinated manner, with little effort to implement growth management strategies designed to coordinate the rate, character, quality, timing, and location of residential development. These shortcomings are further compounded by complex land tenure systems, lack of comprehensive land surveys to inventory land uses and parcels, and failure of governments to collaborate with community-based organizations, including informal networks, to find viable solutions to urban problems.

Many African governments work closely with the World Bank and the UNCHS to provide urban sites and services, upgrade squatter settlements, and encourage self-help initiatives. In these programs, governments purchase and assemble plots of land, install the necessary infrastructure or services (roads, water and sewer lines, electricity), and sell the plots to low-income households at low interest rates. The purchaser of the plot is responsible for designing and building a home to conform to his or her needs. Building, design, and finance costs are kept low to ensure easy recovery of costs. The program, if carefully implemented, could be a successful partnership between international agencies (who provide low-interest loans and technical advice), national governments (who assemble and develop the sites), universities (where engineering and architectural faculty and students offer technical advice), and residents who have an opportunity to participate in the decision-making process. The program empowers local residents and adds a sense of self-esteem and purpose to their initiatives and efforts.

Slum and squatter upgrading schemes have also taken on new meaning in African cities. Early attempts to eradicate these settlements were rendered futile. Displaced households ended up relocating to other sites, where they duplicated their previous lifestyles. Most governments now accept the existence of squatter settlements and are trying to provide the necessary facilities to upgrade them by relaxing rigid building codes and sponsoring

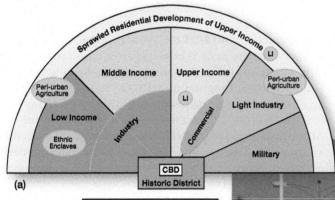

▲ **Figure 8-27 Accra, Ghana.** (a) Land use in Accra follows a sectoral pattern of development. Suburban areas are experiencing uncontrolled residential sprawl. (b) Accra's Independence Square features the Black Star monument arch with the inscription "Freedom and Justice."

more research into alternative building materials that are both biotic and fire resistant. The UNCHS has developed a "best practices" database, which draws on more than 4,000 initiatives from 140 countries to illustrate innovative and creative approaches to urban problem-solving in developing countries. Suggestions range from strengthening local urban governance, decentralized decision making, increased transparency and accountability, to disaster prevention.

Development Trends

Development is a multidimensional phenomenon involving a broad set of economic, social, environmental, institutional, and political factors. In sub-Saharan Africa, this includes providing adequate educational, health, and nutritional benefits to enhance human capabilities, creating opportunities for women and other traditionally neglected groups, upgrading physical infrastructure to facilitate the exchange of goods and services, conserving vital nonrenewable resources and sustaining stable environments, and advocating for human rights and creating avenues for the free expression of ideas. These opportunities and services are especially needed by women, children, the urban poor, and rural peasant farmers. Successful development requires a holistic approach, integrating economic and human dimensions, assessing impacts of development technologies on various segments of society, and analyzing the spatial patterns of uneven development between the more modernized corelands and more traditional peripheral regions.

Economic Dimensions of Development

The nations of Africa South of the Sahara generated a combined Gross National Income of approximately $1.96 trillion in 2011. This is less than Brazil ($2.25 trillion), and only 13 percent of the U.S. total ($15.2 trillion). The average per capita GNI PPP for sub-Saharan Africa was $2,238. Immense variations among countries ranged from a high of $25,620 in oil-rich Equatorial Guinea and $14,550 in diamond-wealthy Botswana to a low of $340 in conflict-ridden DRC. Other countries at the high end of the economic scale include South Africa, Gabon, and Namibia. These countries are well endowed with oil and strategic mineral resources, although the uneven distribution of wealth within each country is problematic. Countries at the low end of the scale include Burundi, the Central African Republic, and the war-torn societies of Liberia and Sierra Leone.

While these economic indicators may seem gloomy, there are encouraging signs. The region is recovering from low levels of investment and productivity, as well as high debt burdens in the 1980s and 1990s, to record high rates of economic growth in recent years. Business climates are improving in such countries as Burkina Faso, Ghana, Mali, and Rwanda; the middle class is growing throughout the region; trade and investment relationships are increasing with Brazil, China, and India; the number of democratic governments is growing; and expanding economies recently have recorded economic growth rates in excess of 6 percent per year. Non-oil-producing countries such as Ethiopia and Burkina Faso have recorded growth rates as high as 9 percent due to investments

in livestock and cotton, respectively. These emerging trends need to be sustained by investments in physical (roads, rail, power plants), social (health and education), and technological (fiber optics, information and communication technologies) infrastructure. More institutional reforms in banking, taxation, and governance (including reducing corruption) are needed to deliver critical services to productive sectors of African economies and boost entrepreneurial activity.

Human Dimensions of Development

As a group, the nations of sub-Saharan Africa rank lowest in the world in measures of human development. About 70 percent of the lowest-ranked countries in the Human Development Index (HDI) are found in Africa South of the Sahara. These low rankings reflect the relatively low regional life expectancy of 52 years and fairly low levels of educational attainment. There are some encouraging signs, however. The region's HDI has improved from 0.365 in 1980 to 0.463 in 2011. Recently HDI improvement rates have been better than in the Arab states, Central Asia, and Latin America and the Caribbean. Other encouraging indicators include an increase in the number of children completing primary school from 51 percent in 1991 to 69 percent in 2008, and improved access to safe water and sanitation. Improvements in meeting the basic human needs of sub-Saharan Africa's peoples must be achieved through better nutrition, improved medical care, more equitable income distribution, increased levels of education and employment, and expanded opportunities for women (see *Geography in Action:* Achieving Gender Parity in Sub-Saharan Africa).

The Impact of HIV/AIDS on Development. An ongoing threat to the development of human resources in sub-Saharan Africa is HIV/AIDS (Acquired Immune Deficiency Syndrome). The Joint United Nations Program on HIV/AIDS (UNAIDS) reports that an estimated 34 million people worldwide were living with HIV in 2010. Of these, 22.9 million or 68 percent were in sub-Saharan Africa. In the same year, UNAIDS estimated that 1.2 million African adults and children died of AIDS and 14.8 million children were orphaned. Five percent of all adults aged 15–49 in the region were living with HIV/AIDS.

Because of HIV/AIDS, the average life expectancy in sub-Saharan Africa has fallen from 62 to 52 years. Southern Africa has the world's highest prevalence of adult HIV/AIDS, with rates exceeding 20 percent in Swaziland, Botswana, and Lesotho (Figure 8-28). In the Islamic countries of Mauritania, Mali, and Niger in West Africa, adult HIV/AIDS prevalence is less than 1 percent. There are also significant intraregional variations in HIV/AIDS prevalence. In South Africa, HIV prevalence among prenatal clinic attendees is as high as 35 percent in KwaZulu-Natal and as low as 18.5 percent in the Western Cape (Figure 8-29). There are also extreme gender differences. In Lesotho, 14.2 percent of young women aged 15–24 have HIV, compared to 5.4 percent of young men. Some scholars question the validity of AIDS estimates in Africa because of the potential for a high proportion of underreporting, and others believe that the reports are exaggerated. The latter argue that AIDS estimates by the World Health Organization are not necessarily

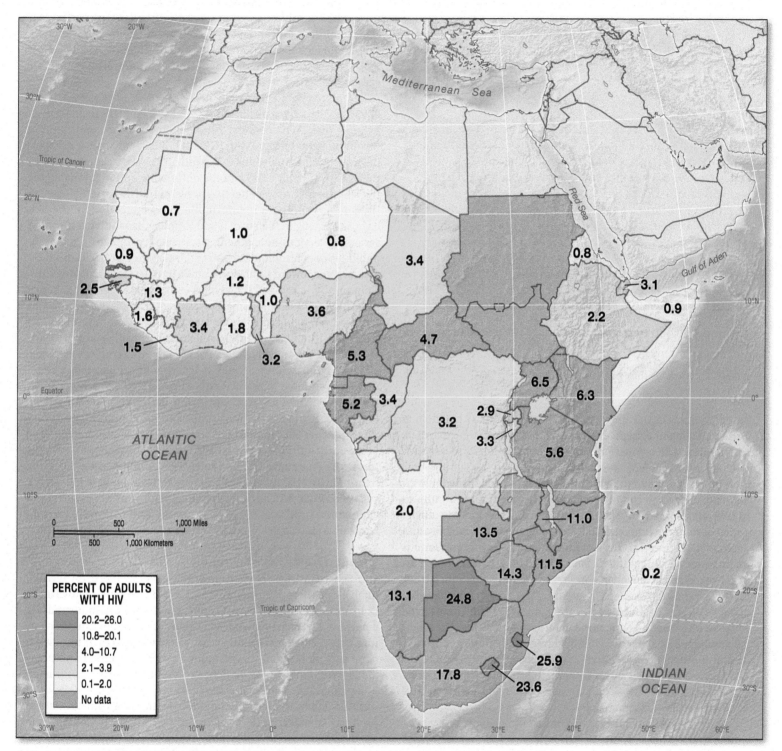

▲ **Figure 8-28 Percent of adults with HIV in Africa south of the Sahara.** The nations of Southern Africa have the world's highest rates of adult HIV. Life expectancies have fallen dramatically and human and economic development have been greatly impacted.
Source: UNAIDS Global Report 2010, http://www.unaids.org.

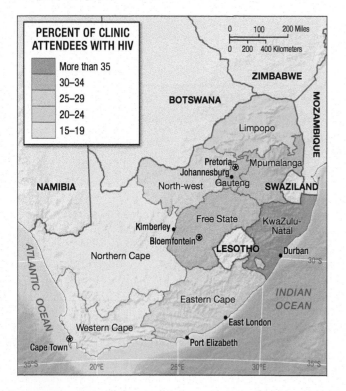

▲ **Figure 8-29 Intraregional variations in HIV prevalence, South Africa.** Adult HIV prevalence in South Africa varies greatly by region, but has increased significantly in all provinces.
Source: South African Department of Health Study, 2010.

based on lab tests but on a list of clinical symptoms that include persistent coughing, high fever, weight loss, and chronic diarrhea, all symptoms that overlap with those of other serious diseases such as tuberculosis, cholera, and malaria.

While accurate numbers of sub-Saharan Africans living with HIV may be hard to compile, it is clear that the disease continues to have widespread impacts. Contributing factors include official denial in some government circles, inadequate health facilities and personnel, illiteracy, the lack of preventive programs, and serious misconceptions about the disease. High-risk groups in Africa include sexually active workers, migrants, military personnel, truck drivers, and drug users who share needles. Large cities have become peak areas of infection, and wars and civil turmoil have forced refugees into areas that are prone to infectious diseases. The medical system, already overwhelmed by numerous tropical diseases, cannot cope with the magnitude of the continuing AIDS crisis. Lack of adequate blood screening equipment and unhygienic medical practices, including the lack of sterilized equipment, puts recipients of health services at further risk.

The AIDS epidemic is having a devastating impact on the economic and social lives of sub-Saharan Africans. Because AIDS affects adults in their productive years, the disease has a direct impact on the development of human capital. A World Bank study in Tanzania estimated the cost of replacing teachers dying from AIDS at $40 million through 2010. In the same country, rural households affected by AIDS deaths spend approximately $60 (the equivalent of annual rural income per capita) on treatment and funerals. A study of South Africa estimated the direct cost of AIDS in the year

2000 was between $1.2 billion and $2.9 billion. Economic growth is compromised as household savings and resources are diverted away from productive investment. From a social and psychological standpoint, AIDS has a devastating impact on children who lose their parents. Millions of orphaned children face limited educational opportunities, and many are left to fend for themselves or are cared for by other children or by the elderly.

There is some encouraging news. Slight improvements in access to antiretroviral therapy resulted in a 20 percent decrease in AIDS-related deaths between 2004 and 2009. Between 2001 and 2009, new infections declined by 22 percent and HIV incidence fell by more than 25 percent in 22 sub-Saharan African countries. But the problems associated with HIV/AIDS still loom large as substantial investments are needed to expand preventive treatments. So far, there is no cure for AIDS and the only viable option for managing the disease is through effective prevention. Necessary measures include providing more information and education on the disease, improving the safety of blood supplies, expanding testing and screening services, strengthening surveillance and institutional capabilities to control the disease, and integrating prevention strategies targeted toward youth and women into development programs. Community-based and social marketing approaches can enhance information dissemination efforts. Community-based approaches focus on removing financial, bureaucratic, and communications barriers in localities and rural districts and on visiting homes to provide counseling and education. Social marketing employs commercial marketing techniques such as incorporating information about AIDS prevention into popular radio or TV shows. On a larger scale, the African Union has proposed a Pharmaceutical Manufacturing Plan to coordinate the production of local medicines, while PharmAccess Foundation has partnered with Dutch insurance companies to create a health insurance fund that subsidizes insurance premiums and copayments for poor households seeking treatment for HIV, tuberculosis, and malaria. There also has been a call for a single African Medicines Regulatory Agency to safeguard against the proliferation of poor quality and counterfeit drugs, boost pharmacological research, and expand access to treatment.

Uneven Patterns of Development

Sub-Saharan African economies are characterized at every level by dualistic patterns of uneven development. At the international level, there are extreme gaps between Africa and the more-industrialized countries, and at the national level, there are core-periphery disparities between the more modern urban centers and their traditional rural peripheries. It has been argued that these gaps are perpetuated by a dominance-dependence relationship, which results in an unequal exchange of resources between more- and less-developed nations, such as the exchange of raw materials for industrial products. Thus, Africa South of the Sahara ends up in a subordinate position and becomes vulnerable to the rules and regulations of a global economic system controlled by the technologically advanced countries. In this relationship, there is an implicit harmony of interest between privileged elites in the industrialized countries and their less-industrialized suppliers. The ruling elites in Africa that profit from this arrangement can be seen as facilitating the unequal exchange of resources by investing their own money in foreign goods and services, enabling rich countries and

** CHAPTER 8 Africa South of the Sahara 373

Achieving Gender Parity in Sub-Saharan Africa

The 2012 World Development Report on Gender Equality and Development focuses on three key dimensions of gender equality: the accumulation of endowments (education, health, and physical assets), the use of those endowments to take advantage of economic opportunities and generate incomes, and the application of those endowments to take action and exercise control—the ability to make effective choices in the household and in society at large. This is consistent with the 2010 Millennium Development Goal (MDG) Summit's call for action to ensure that women have equal access to education, health care, and economic opportunities, and can be involved in development policy decision-making.

The need for improved gender equality in sub-Saharan Africa is demonstrated by the relatively poor performance of these countries on the United Nations Development Program's (UNDP) Gender Inequality Index and the Gender Empowerment Measure. Most of the progress made on gender parity has been with primary and secondary education, but with few gains in tertiary or post-secondary education. In 2009, the majority of sub-Saharan African countries had a Gender Parity Index (GPI) of more than 0.90 in primary and secondary education, putting them on track to achieve parity by 2015. Malawi (100 girls in school for every 100 boys), Rwanda (100.3), Namibia (103), and Lesotho (107) have already achieved parity. In tertiary education, GPIs over 0.90 have been achieved only by Botswana, Cape Verde, Lesotho, Mauritius, and South Africa. Not much progress has been made toward women's economic empowerment as measured by the share of women employed in the nonagricultural sector. The highest shares are in Ethiopia (47%), South Africa (44%), and Namibia (42%) (Figure 8-3-1). Further gender gaps remain in terms of access to assets (land rights and credit) and in wage earnings. Limited progress has occurred with women's representation in national parliaments. The highest performing countries in 2011 were Rwanda (51%), South Africa (43%), Mozambique (39%), Uganda (37%), and Burundi (36%) (Figure 8-3-2). In two-thirds of sub-Saharan African countries less than 20 percent of the parliamentarians are women.

The Convention on the Elimination of all Forms of Discrimination against Women (CEDAW) and the MDGs show that women's empowerment and gender equality are both global and African priorities. Various programs aimed at empowering African women have been instituted. Microfinancing and cell

▲ FIGURE 8-3-2 **Gender parity in Rwanda.** At 52 percent Rwanda has the highest percentage of women parliamentarians in the world.

phone–based mobile banking programs now provide women with access to credit, business start-ups, and money transfer services through such organizations as World Women's Banking (WWB), Women's Opportunity Network (WON), Access Bank in Gambia and Rwanda, and Sero Lease and Finance in Tanzania. Ethiopia, which previously did not have any land rights for women, now issues joint land titles for wives and husbands through its land certification program. Women in Burundi, South Sudan, and Uganda have increasingly engaged in peace and reconstruction efforts following conflict. The Adolescent Girls Initiative (AGI), a public-private partnership, has designed a set of programs in Liberia, Rwanda, and South Sudan to assist adolescent girls in their transition from school to productive employment through job and vocational training, mentoring, and basic business skills training.

Sources: World Bank, *World Development Report (2012): Gender Equality and Development* (Washington, DC: International Bank for Reconstruction and Development/World Bank, 2011); United Nations Development Program, *Assessing Progress in Africa Towards the Millennium Development Goals: MDG Report 2010* (New York: United Nations Development Program, 2011).

▲ FIGURE 8-3-1 **Gender reforms in Ethiopia.** Ethiopia is engaging in a number of policy reforms to improve gender equality. This is a call center in Addis Ababa with a largely female workforce.

multinational corporations to appropriate much of the wealth generated by African economies. Reports of scandal, corruption, and embezzlement in African governments are not uncommon. Episodically, we hear about African heads of states and government officials funneling hard-earned money generated from rural areas into foreign banks. The global north–south wealth gap is a function of both one-sided foreign economic relationships and poor judgment and bad internal management on the part of African governments and urban elites.

National core-periphery disparities are intensified by the uneven pace of economic activity. Figure 8-30 shows the distribution of major cores of economic activity, including manufacturing, agricultural export/processing, and mining centers. Key manufacturing enterprises such as oil refining, cement production, machinery, and transport equipment often locate in major cities to take advantage of cost savings derived from close proximity to skilled labor, larger markets, utilities, and better transportation infrastructure. Major islands of modern development are clustered along the petroleum belt of West Africa; the "ring of diamonds" that encompasses southwest Namibia, northeast South Africa, Botswana, southwest DRC, and northeast Angola; the Luanda-Malanje corridor of Angola; the high-tech and mineral-rich region of Witwatersrand near Johannesburg in South Africa; the copper belt of DRC and Zambia; as well as the rich agricultural region bordering Lake Victoria.

This dual economic structure is also characteristic of both urban and rural areas. In urban areas, the formal/informal sector dichotomy works to the disadvantage of small-scale enterprises that operate outside the mainstream of government regulation and benefits. In rural areas, relatively wealthy commercial/cash-crop farmers of cocoa, coffee, cotton, rubber, and tea plantations coexist with subsistence farmers who produce just enough food for themselves and their families. These patterns of dualism create a three-way interactive process between the commercial sector in rural areas, the formal institutions in urban areas, and the industrialized world economy. The commercial farmers benefit from urban-based government incentives. Formal urban institutions benefit from the foreign exchange derived from the overseas sale of rural cash crops. The foreign exchange earned is then internalized in urban areas. This three-way interactive process reinforces the disparities that exist between core and peripheral regions at all scales, and polarizes people who work in the urban informal and rural subsistence sectors. The constituents from these disadvantaged communities have limited opportunities to realize their human potential and to participate in the global exchange economy.

Development Strategies

Trying to find solutions to reduce the patterns of uneven development is a difficult task. More recent development strategies in sub-Saharan Africa have focused on poverty reduction and the expansion of trade and investments networks. **Microfinancing** has emerged as a key strategy with potential to mitigate the social and economic challenges that confront more than 350 million sub-Saharan Africans who live in poverty. There are about 530 microfinance institutions (MFIs) in Africa South of the Sahara that provide basic financial services, including savings, loans, microinsurance, and money transfer services (mobile money). Some formal MFIs are large commercial banks (Capitec in South Africa), state banks

(Equity Bank in Kenya), and credit and savings institutions (ASCI in Ethiopia) with a reach of over 500,000 people. Other formal MFIs include community-based organizations (rural financial institutions in Tanzania) and self-help groups (Harambee in Kenya). There are also smaller, informal MFIs that serve as rotating savings and credit associations (ROSCA), such as *Ekub* in Ethiopia, *Esusu* in Nigeria, and *Tontines* in Cameroon. These are based on the premise of group lending, where people join together and guarantee each other's loans to minimize risk.

The Foundation for International Community Assistance (FINCA) pioneered the concept of **village banking** as a way of providing very poor families with small loans to start microenterprises as well as microinsurance packages to help families defray costs associated with illness, natural disasters, and funerals (Figure 8-31). FINCA has extended its products to over 56,000 people in Uganda to include loans for home improvement, solar energy systems, and mobile phones. Several MFIs specifically target women, including WWB and the Women of Africa Fund for Micro Enterprise (see *Visualizing Development:* Sub-Saharan Africa's Cell Phone Revolution).

Stop & Think

How has the expansion of mobile phones impacted the delivery of basic services to sub-Saharan Africans?

Microfinancing has economic advantages such as employment creation, poverty alleviation, asset building, providing entrepreneurial opportunities, and capitalizing on local ingenuity. Sociopsychological benefits result from the promotion of self-worth, empowerment, and human dignity. But there are also challenges. The Microfinance Information Exchange in Washington, DC, estimates that only 9.6 million of Africa's poor have access to MFIs. In countries like Somalia, Central African Republic, Eritrea, and Liberia, an average of 2,000 people have borrowed from MFIs. Infrastructure must be upgraded and the institutional capacity to deliver financial services to the rural poor must improve. To overcome this barrier, Safaricom in Kenya launched M-PESA, a mobile banking service that provides money transfer services to support community development projects.

In 2001, the United Nations launched a critical set of poverty-reduction strategies, the **Millennium Development Goals (MDGs)**, to respond to eight global development challenges through programs aimed at eradicating extreme poverty and hunger, achieving universal primary education, promoting gender equality and empowering women, reducing child mortality, improving maternal health, combating HIV/AIDS and endemic diseases, ensuring environmental sustainability, and building a global partnership for development. Progress has been slow and uneven toward achieving the MDGs in Africa, but there are some encouraging signs. A 2011 report prepared by the Economic Commission on Africa, the African Union, and the Africa Development Bank, *Assessing Progress in Africa toward the Millennium Development Goals*, reveals that progress has been made on primary school enrollments, gender parity, women's empowerment, and access to improved sanitation and safe water. Countries making significant progress toward many of the goals include Burkina Faso, Ghana, Ethiopia, Rwanda, Uganda, Tanzania, and Malawi. The African Union's **New Partnership for**

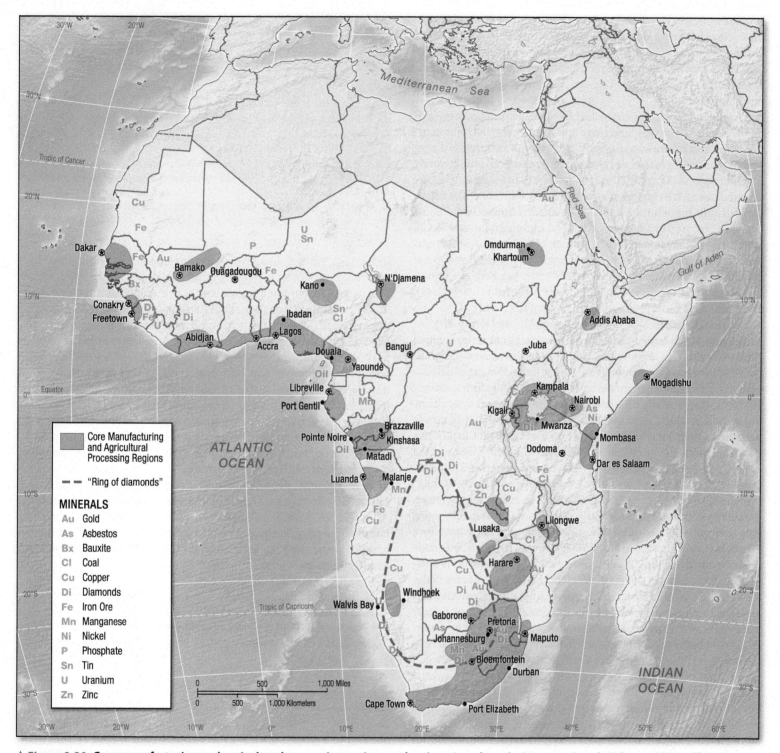

▲ **Figure 8-30** **Core manufacturing and agricultural processing regions and major mineral-producing areas in sub-Saharan Africa.** There is a strong correlation between population distribution and the location of manufacturing activities. Manufacturing tends to be concentrated in the most densely settled and urbanized regions.

Africa's Development (NEPAD) program was designed in 2001 to complement the activities of the MDGs in alleviating poverty, accelerating regional integration, and promoting good governance. Both the World Bank and the International Monetary Fund (IMF) have launched Poverty Reduction and Growth Facility Programs to break the cycle of poverty. Countries that participate in the program have developed poverty reduction strategies that focus on universal, informal, and technical-vocational training; primary healthcare improvements; agricultural and rural development efforts; and good governance.

Sub-Saharan African countries have broadened their development strategies by expanding trade and investment networks in order to expand business opportunities, promote sectoral (industry and agriculture) growth, develop human potential, and diversify their economies. Since 2000, the United States has strengthened its bilateral ties with a number of African countries through the **Africa Growth and Opportunity Act (AGOA)**. Exports from AGOA-eligible countries have grown by more than $65 billion and over 300,000 jobs have been created in Africa. A range of African businesses and manufacturers now have access to U.S. markets, including apparel producers in Botswana (Caratex), textile firms in Kenya (Chandu EPZ), seafood companies in Tanzania (Pecheries Frigorifiques), and a bird seed supplier in Ethiopia (Prospre International).

Trade agreements have extended beyond U.S. and European partners. South-South trade (trade among developing countries south of North America and Europe) and cooperation has expanded significantly as the emerging economies of China, India, and Brazil have invested increasingly in Africa's infrastructure, manufacturing, energy resources (particularly oil), telecommunications, and agriculture. These partnerships have been consolidated through diplomatic initiatives and multilateral forums such as the Forum on China-Africa Cooperation (FOCAC), the India-Africa Forum Summit (IAFS), and the India, Brazil, South Africa Initiative (IBSA). Sub-Saharan African countries find the policies of "non-interference" and "no strings attached" associated with such agreements appealing. In 2009, China overtook the United States as Africa's second-largest trading partner, accounting for 14 percent of Africa's total trade. Africa's trade with China grew from $20 billion in 2001 to $120 billion in 2009. Between 2004 and 2009, trade between Africa and India quadrupled from US$9.9 billion to $39.5 billion, but more recently China's investments in Africa have come under scrutiny because of its relationships with corrupt African governments and its questionable working conditions. India has invested heavily in Zambia's copper mines, Liberia's iron ore, South Sudan's hydrocarbon sector, and East Africa's telecommunications sector. Its oil companies are also engaged in Angola, Cote d'Ivoire, Equatorial Guinea, and Senegal. Brazil's trade with Africa exceeds $20 billion. Its large metal and mining corporation, Vale, has invested $1.7 billion in Mozambique's coal mining industry, and its multinational energy corporation Petrobras is engaged in deepwater drilling for oil in Angola.

Non-Governmental Organizations (NGOs) provide a wide variety of aid packages to sub-Saharan countries, particularly in the social, health, and education sectors. NGOs operate neither as government nor as for-profit organizations. They are a diverse group of largely voluntary institutions that work with local organizations to provide technical advice and economic, social, and humanitarian assistance. They can be professional associations, religious institutions, research institutions, private foundations, or international and

▲ Figure 8-31 **Village banking in East Africa.** Microcredit and microfinancing opportunities are helping groups in Africa to open small businesses and credit accounts.

indigenous funding and development agencies. The World Bank is collaborating with an increasing number of grassroots NGOs to address rural development, population, health, and infrastructural issues pertaining to the human dimensions of its structural adjustment programs. Kenya, Uganda, and Zimbabwe have a large number of NGOs. Most of the "indigenous" NGOs in Africa are community-based grassroots and service-based organizations. They usually fill a void where governments have been largely ineffective and out of touch with local needs. Some of their specific developmental objectives include tackling poverty by providing financial credit and technical advice to the poor, empowering marginal groups, challenging gender discrimination, and delivering emergency relief. The Zambia Center for Social Development, an umbrella NGO, coordinates over 100 organizations that focus on building community capacity, broadening networks, and advocating stronger policy frameworks. In Uganda, two organizations remain active in assisting the victims of AIDS. The AIDS Support Organization (TASO) provides community-based counseling, social support, and medical services to persons with AIDS and their families, while the Uganda Women's Efforts to Save Orphans cares for many of the million-plus orphaned children.

Stop & Think

▷ What are the challenges and opportunities facing sub-Sahara African countries that have expanded their trade networks to include China, India, and Brazil?

Population Contours

Sub-Saharan Africa's estimated population of 902 million people continues to grow at the rapid rate of 2.6 percent per year. Its population will increase to over 1.2 billion by 2025 and about 2.0 billion by 2050, if current growth rates are maintained. By then, the region may have more than four times as many people as the United States and Canada, and three times as many people as Latin America. Sub-Saharan Africa is second in regional population only to Asia, which will likely continue as the world's most populous region. In spite of these rapidly growing populations, 20 sub-Saharan African

Visualizing DEVELOPMENT ▲

Sub-Saharan Africa's Cell Phone Revolution

A technological revolution is sweeping sub-Saharan Africa, driven by an exponential growth in cell phone use. The number of mobile cellular subscriptions has increased dramatically from 17 million in 2001 to 463 million in 2011 (Figure 8-4-1). Africa is the fastest-growing and second-largest cell phone market in the world, drawing increasing attention from telecommunication giants such as Britain's Vodafone, South Africa's MTN Group, China's Zhongzing Telecom (ZTE), and Kenya's Safaricom. Countries with the highest rates of cell phone usage include Botswana (143 subscriptions per 100 people), South Africa (127), Gabon (117), and Namibia (105). Individuals often have multiple subscriptions and poorer households tend to share the cost of purchasing a unit. This emerging revolution has the potential to transform Africa's social and economic landscape with far-reaching effects that extend into the remotest villages.

Cell phone usage has several major economic benefits: increased access to and use of information; improved coordination among businesses which leads to greater efficiency; creation of new jobs in the information and related sectors; reduced exposure to risks and vulnerabilities from natural disasters through improved communications among social networks; and improved delivery of financial, agricultural, health, and educational information and products and the promotion of mobile-based development, or m-development. New mobile phone platforms provide access to populations previously without local banks (Figure 8-4-2).

These new, mobile phone–based customers are now able to transfer funds, pay bills, open bank accounts, and access basic insurance products. Safaricom's M-PESA (*pesa* means money in Swahili) in Kenya has 15 million active users engaging in 3 million transactions a day and transfers of nearly $700 million a month. Mobile phones are used to advance immunization programs for children located in Mozambique's remote villages, to track malaria treatments in Tanzania's rural clinics, and to access and update patient records in South Africa's remote areas. Mobile tutoring and learning increasingly have an impact on Africa's children, who now have access to digital books through the efforts of nonprofit organizations such as Worldreader. The professional development and training of Africa's teachers and educators can be enhanced through immediate access to educational materials.

In countries where national currencies are weak and often grossly inflated, mobile phones can create alternative forms of mobile money. In situations where stronger banknotes, such as the U.S. dollar or the euro, have an important local economic function, coins for these currencies are almost nonexistent. Mobile phone airtime is increasingly used to fill this gap. The airtime "funds" that a phone owner accumulates instead of small change have real value because they can be transferred to other phones, exchanged for cash or goods in stores, or used to pay small debts. New start-up companies have emerged to promote and facilitate these transfers and extend phone technology into economic activities formerly dependent on face-to-face contact.

Mobile phone technology is not a panacea for Africa's problems. For example, cell phones can provide farmers with instant information about food prices from distant locations and

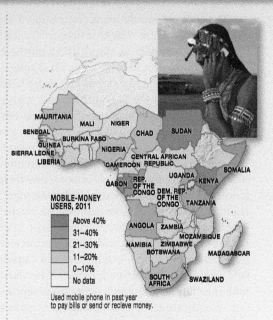

▲ **FIGURE 8-4-2 Mobile phone use in rural Africa.** Mobile phones allow herders and farmers in remote locations to access information about livestock and farm products. The map portrays the geographical distribution of Africans who have used their cell phones to conduct bank transactions or pay bills.
Source: http://isthisafrica.com/mobile-money-banking-in-africa

help them make optimal decisions about sales and productivity, but they still need an upgraded infrastructure to store, transport, and market their products. Mobile financial services can only be effectively disseminated to people without access to banks if the proper regulatory and legal frameworks are in place to strengthen cybersecurity and consumer protection laws exist to safeguard against identity theft, fraud, hacking, and money laundering. Instant information on preventive health care can be implemented with some degree of success only if adequate facilities and follow-up services are available. But the potential to improve development in all of these areas is enormous.

Sources: Jenny C. Aker and Issac M. Mbiti, "Mobile Phones and Economic Development in Africa," *Journal of Economic Perspectives* 24, no. 3 (2010): 207–232; Jenny C. Aker, "Information from Markets Near and Far: The Impact of Mobile Phones on Grain Markets in Niger," *American Economic Journal: Applied Economics* 2 (July 2010): 46–59; "Airtime is Money," *The Economist*, January 19, 2013, 72, 74, www.economist.com/news/finance-and-economics/21569744-use-pre-paid-mobile-phone-minutes-currency-airtime-money.

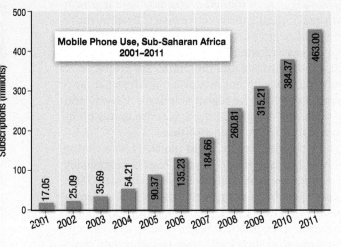

▲ **FIGURE 8-4-1 Mobile phone use in sub-Saharan Africa.** For millions of Africans mobile phone use has become a significant tool to improve their everyday lives.
Source: World Bank Group, *Africa Statistical Yearbook.* Washington, 2012.

countries have populations of less than 5 million, and half have populations under 10 million. Only nine countries have populations of more than 25 million, specifically, in rank order, Nigeria, Ethiopia, DRC, South Africa, Tanzania, Kenya, Uganda, Sudan, and Ghana.

The region's population is concentrated in two major zones of dense settlement: the West African coastal belt stretching from Dakar (Senegal) to Libreville (Gabon); and a north–south belt stretching from the Ethiopian highlands down through Lake Victoria, the copper belt of DRC and Zambia, and to the Witwatersrand region of South Africa (Figure 8-32). Three broad, sparsely populated zones include, from north to south, the Sahel region extending from Dakar in the west to Mogadishu in the east, the west-central forest regions of DRC and Gabon, and the arid/semiarid region of southwest Africa. These spatial distributions coincide with environmental (vegetation, soil, climate, topography), developmental (levels of urbanization, industrialization, and agricultural development), and sociopolitical characteristics (oppressive regimes, ethnic disputes, resettlement schemes). The West African coastal strip contains most of West Africa's urban, economic, and political centers. Other economic centers, such as the copper belt of DRC and Zambia, the diamond and gold mining centers of South Africa's Witwatersrand region around Johannesburg, and the rich agricultural lands of the Lake Victoria borderlands, attract large population clusters as well.

Patterns of uneven distribution are also evident on a microscale. In Tanzania, the sparsely settled central region, which has a high concentration of tsetse flies, is surrounded by dense population clusters around the fertile slopes of Mount Kilimanjaro in the north, the shores of Lake Victoria in the northwest, the shores of Lake Malawi in the southwest, and the economic center of Dar es-Salaam on the east coast. In DRC, the population density in the great forest region is about half the national average of 76 people per square mile (29 per square kilometer). Chad's southern region is the country's agricultural heartland—the center for cotton and groundnut (peanut) cultivation, with twice the annual rainfall of the northern Sahel.

Sub-Saharan Africa continues to have the highest fertility and mortality rates in the world, along with the highest proportion of young dependents. Families have an average of five children, although there is considerable variation by region, socioeconomic status, and place of residence (rural versus urban). While the majority of countries continue to have high fertility rates, most countries in Southern Africa have below-average rates. In South Africa, Botswana, and Zimbabwe, comprehensive family planning programs, coupled with improvements in female literacy, have slowed birth rates to 2.4, 2.8, and 4.1, respectively.

Cultural factors are the strongest forces driving high fertility rates. Most females in traditional societies marry at an early age. High rates of remarriage and polygamy (having more than one wife at a time) compensate for any potential effects that divorce or widowhood might have on fertility rates. Furthermore, belief systems, customs, and traditions have a significant impact. The predominantly patrilineal societies place a high premium on lineage and spiritual survival. The family lineage is seen as an extension of the past and a link to the future, so family planning is strongly resisted. Fertility is equated with virtue and spiritual approval. Barren women are treated with disdain and ostracized from society. Also, in a male-dominated society, decisions about reproduction and family size are usually made by the husband, which may explain the low levels of contraception practiced. Children are regarded as economic assets—a source of wealth and prestige and a labor reservoir for household chores. They are also required to offer tribute to their parents. This flow of wealth to the elderly is a socially sanctioned and religiously expected tribute. Since a high premium is placed on children, African women aspire to elevate their status by complying with their husband's request to have more children. Fertility is further enhanced through child fosterage, in which children are sent from their natural parents to be raised and cared for by their grandparents or foster parents. About one-third of children in West Africa are fostered. Benefits of fosterage include stronger family ties, companionship for widows, better educational opportunities for the children, and assistance with domestic chores. The practice also lessens the economic burden placed on parents who would otherwise struggle to provide adequately for all their children. Another fertility-enhancing factor is the ethnic rivalry that exists among many sub-Saharan societies. In countries such as Nigeria, Kenya, Ethiopia, and Uganda, where ethnic tensions run high, there is intense competition for economic and political resources. Communities generally view reductions in fertility as the equivalent of committing ethnic suicide, since larger populations are seen as generating greater resources and power.

The high fertility rates explain why the population in Africa South of the Sahara is so young. More than 40 percent of the population is under 15 years of age. The population pyramids in Figure 8-33 compare the proportions of males and females in various age groups in Africa and the United States. The graph shows that Africa's population pyramid is markedly broader at the base—that is, younger—than that of the United States. The region's workforce faces the economic burden of supporting a large proportion of youth. Large government expenditures are required for health care, education, and job training programs to accommodate future employment. The impact of young and rapidly growing populations will carry over into the future, bringing increased demands for housing, employment, and job benefits. It also will likely translate into continued overall population growth as today's youth reach their reproductive years. Even in those countries where fertility rates are beginning to decline, populations will likely increase substantially before leveling off. A few countries, such as South Africa and Botswana, are projected to reach the population replacement rate of 2.1 children per family by 2030. These countries have higher standards of living and higher literacy rates, and are promoting aggressive family planning campaigns to control population growth.

There are some positive trends. Child mortality is declining twice as fast as it was a decade ago and fertility rates have gradually declined in most countries in the last few decades. Demographers and economists are keeping a close eye on the so-called **demographic dividend,** which was critical to the recent expansion of East Asian economies. The combination of declining fertility, economic growth, and investments in health care and education could result in larger working-age African populations and fewer dependents, and yield economic returns or a demographic dividend.

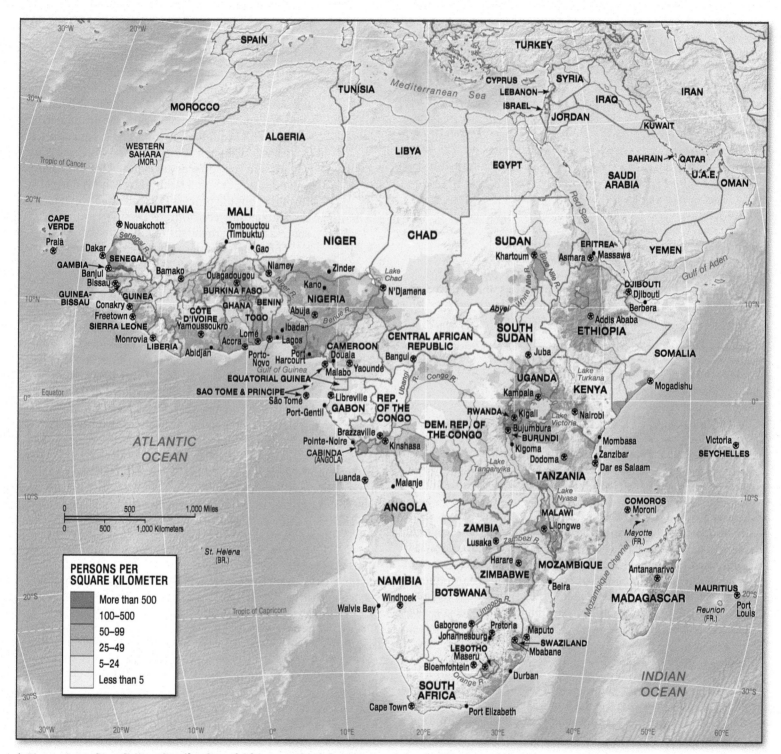

▲ **Figure 8-32 Population distribution of Africa south of the Sahara.** Saharan Africa's population is most concentrated in the Guinea coast of West Africa, the highlands of Ethiopia, the Great Lakes of East Africa, and the southeast coast of Southern Africa. Dry zones are sparsely settled.

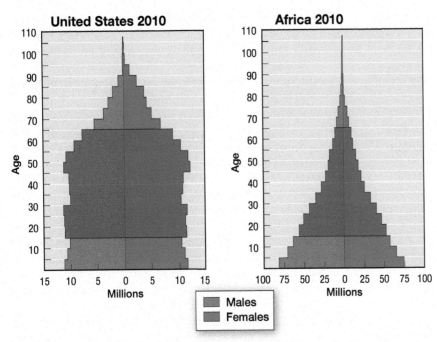

▲ **Figure 8-33 Age–sex population pyramids of Africa and the United States, 2010.** The population of Africa is much more youthful than that of the United States. More than 40 percent of Africans are below the age of 15.
Source: United Nations, Department of Economic and Social Affairs, Population Division, *World Population Prospects: The 2010 Revision* (New York, 2011), http://esa.un.org/wpp/population-pyramids/population-pyramids.htm.

West, Central, and East Africa: Diverse Development Paths

The regions of West, Central, and East Africa cover about three-quarters of sub-Saharan Africa's land area and account for more than 80 percent of its population (see Figure 8-1, page 346). Compared to Southern Africa, these three regions have a more tropical environment because of their proximity to the equator, lower levels of European settlement, and less industrial capacity. The countries in these subregions exhibit immense geographical, cultural, and economic diversity.

West Africa

West Africa's physical geography is dominated by the Niger, Senegal, and Volta rivers and the Fouta Djallon and Jos highlands. Coastal West Africa has a humid equatorial climate, while the more inland countries experience semiarid and arid conditions. Agricultural potential is considerable at the wetter end of this gradient, but the interior landlocked Sahelian states are water constrained and vulnerable to drought. Ethnic conflicts and a legacy of colonial boundaries and policies have often complicated and disrupted development. Regional cooperation in transportation, intraregional trade, scientific research, natural disaster mitigation, and security issues is notable, if often difficult to realize fully.

The Sahel States: Coping with Drought

The **Sahel** region is a transition zone between the savanna and desert environments of sub-Saharan Africa. It centers on latitude 15° N and is 125–250 miles (200–400 kilometers) wide, stretching from Mauritania and Senegal in the west to central and southern Sudan, Ethiopia, and parts of Somalia in the east. The ecology of the region is adapted to an extended dry season lasting up to 9 months and annual rainfall averaging between 10 and 20 inches (250 and 500 millimeters). Both plant growth and species diversity are limited by the semiarid conditions. In most areas, agriculture is not feasible without irrigation. In the dry season, grass withers and soil surfaces are exposed to compaction and erosion. Some soil removal by wind is normal at this time, but the process is accelerated when farmers and herders concentrate animals around water sources and disturb the surrounding areas.

The region has periodically experienced devastating droughts (Figure 8-34), causing significant loss of life and reductions in cattle stocks. Because most forms of vegetation and animal life in the Sahel are capable of adapting to periods of prolonged drought, it is the lack of human adaptation that creates the greatest hazard—that of desertification. **Desertification**, the degradation of nondesert areas to more desertlike conditions, is linked to human mismanagement and drought. Lack of rainfall leads to diminished surface streamflow, loss of water in surface impoundments such as freshwater lakes and wetlands, and reduced recharge of aquifers. This affects underground inflow to springs and streams. Agriculture is affected when the soil lacks the necessary moisture for plant intake or for effective plant growth. Human activities that cause soil erosion, surface compaction, and surface runoff can limit water infiltration into the soil and produce drought-like conditions even when rainfall is adequate. This implies that sound soil and water management practices, which often are lacking in many parts of the Sahel states, could reduce the impact of drier years.

▼ **Figure 8-34 Sahel landscape.** The Sahel ("shore of the Sahara") is a highly variable environment that poses many challenges to the people who live there. Recurring droughts, poor soils, often sparse and degraded vegetation, epidemics, and famines are common occurrences. Here a woman separates millet seed from chaff before grinding flour.

Desertification and soil degradation in the Sahel are made worse by many factors. Population pressure on land leads to overcultivation of marginal lands and overgrazing. Most of the scarce arable land in areas with more favorable moisture conditions is monopolized by cash crops and other commercial agricultural activities. Subsistence farmers are usually relegated to marginal areas, which are intensively cultivated without adequate crop rotation. These agricultural practices are usually unable to produce enough surpluses to support farmers during drought periods. Pastoralists are in an equally perilous position because expansion of dry farming and irrigation agriculture into critical dry season pastures denies herders access to fodder for their animals. Massive herd die-offs and severe overgrazing near water sources are a common result when drought periods increase stress on the agricultural environment.

Inappropriate technologies, such as drawing too much water from wells with power pumps, and inadequate governmental responses to drought may also contribute to desertification. At times, governments disassociate themselves from any event or occurrence that has the potential to tarnish their image. Generally, governments have adopted a reactive rather than a proactive approach, too often allowing the problem to reach crisis proportions before acting.

The U.S. Agency for International Development (USAID) has developed a Famine Early Warning Systems Network (FEWS-NET), which uses satellite and remote sensing technologies to assess the extent to which people become vulnerable to food insecurity and famine, and provides alerts before an agricultural or economic crisis peaks. Figure 8-35 shows food insecurity vulnerability for the West African Sahel between July and September 2012. South-central Mauritania, eastern Mali, northern Burkina Faso, and southwest Niger had crisis levels of food insecurity. Vulnerable households experienced short-term instability that resulted in losses of assets and/or significant food shortages. Adjacent areas experienced very stressful but not crisis conditions. The Sahel Working Group estimated that more than 10 million people suffered drought-induced food shortages.

Countering desertification and food insecurity requires a planning and management approach with long-term objectives. This includes an emphasis on reforestation and revegetation, soil conservation and soil fertility improvement, water conservation, and organizational responsiveness. A farmer-managed natural-regeneration (FMNR) system in Niger developed a low-cost approach to "regreen" the Sahel by terracing slopes to stop soil erosion and by replanting trees and shrubs that yield food, fodder, and fuel. The *Acacia albida* tree, which is widely distributed throughout Senegal, has been recommended as an appropriate species

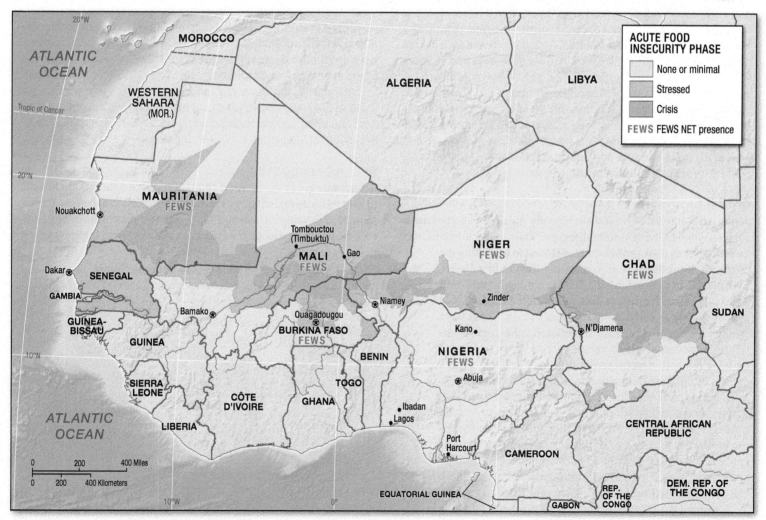

▲ **Figure 8-35 Food insecurity in the Sahel region, 2012.** The countries of the West African Sahel are vulnerable to food insecurity and famine. Food insecurity is a function of both physical and human conditions.
Source: FEWS NET. www.Fews.net.

to spearhead reforestation efforts because of its reverse foliation properties. It is leafless during the agricultural season, so does not compete with cultivated plants for light and moisture. During the dry season, it produces leaves that provide needed shade and shelter against wind erosion. The leaves it sheds add nitrogen and organic matter to the soil. Fuelwood, construction and fencing material, and medicinal substances are also derived from *Acacia albida*.

Vegetative cover and windbreaks are also effective soil conservation methods. Farmers in Burkina Faso utilize local ingenuity through the *Zai* technique to dig a grid of planting pits that are filled with manure. Water conservation methods such as water retention dams, water catchments, and salt barriers complement soil conservation efforts. In Gambia, low ridges of earth are used to retain freshwater and to keep saltwater out of cultivated lands. On the slopes of mountains and plateaus, such as in central Mali, terracing is practiced to capture runoff, retain soil moisture, and redirect fertile soil. These water and soil conservation techniques support the production of staple foods such as sorghum, millet, and rice, as well as food and fodder trees.

These efforts toward environmental rehabilitation cannot be successful without proper organizational and policy initiatives. A number of international NGOs have supplemented grassroots efforts with technical and financial support. In Niger, a group of national and international technicians developed effective site-specific techniques to assist local village cooperatives with water catchments, windbreaks, and reseeding. In Chad, Action for Greening Sahel has been active in constructing communal water sources and supporting rice and soybean production. The provisions of Agenda 21 of the United Nations Conference on Environment and Development have specifically targeted solutions for desertification and drought. Policies designed to create early warning systems, to improve job, income, and educational prospects, and to mobilize collaborative efforts at the local, national, and international levels will likely reduce the impact of desertification in future years.

Côte d'Ivoire: African Capitalism Gone Awry

Although elements of capitalism are very much alive in countries such as Nigeria, Ghana, South Africa, and Kenya, Côte d'Ivoire (Ivory Coast), prior to 1990, probably came closest to exhibiting the classic characteristics of a capitalist state. Buoyed by a diverse agricultural base dominated by cocoa, coffee, timber, and palm oil, Côte d'Ivoire was able to develop an indigenous capitalist class as early as the 1920s and 1930s. These were wealthy plantation capitalists who utilized the entrepreneurial attributes they acquired from French colonists to develop an effective, well-organized political machine that gained control of the state during the decolonization process. Following independence in 1960, the government instituted a form of state capitalism dominated and co-opted by state bureaucracies. Further efforts were made to stimulate the private sector, such as creating a stock market, providing incentives to encourage public-sector employees to become private entrepreneurs, and instituting counseling and financial programs to stimulate the development of local capital.

Despite Côte d'Ivoire's relative success in encouraging local entrepreneurship and a vibrant private sector, numerous constraints hampered the emergence of African capitalists (Figure 8-36). The largest holders of industrial capital continued to be the state or multinational corporations. African entrepreneurs tended to be small-scale, with few moving up to the next level, and they were usually threatened by large-scale, more efficient, enterprises. Indigenous capitalists were not always supportive of structural change, with many frequently refusing to invest in rural areas. Côte d'Ivoire's history of economic prosperity, entrepreneurialism, and political stability came to a halt with the death of benevolent dictator Felix Houphouet Boigny in 1993. In the following decade, the country became mired in a coup d'état and civil war that had xenophobic undertones and divided the country between the rebel-controlled Muslim north and the government-controlled south.

Following the signing of the Ouagadougou Peace Accords in 2007, a reunited Côte d'Ivoire reengaged with the international community and reactivated programs that focused on debt restructuring, infrastructure rehabilitation, governance, and institutional development. The country financed these initiatives with revenues from coffee and cocoa exports, as well as oil, gold, and iron ore. The tourist industry was also upgraded to take advantage of the rich savanna fauna and an elaborate national park system. Another civil war broke out in 2011 when then President Gbagbo refused to step down after election results had declared Alassane Ouattara the winner. Debilitating economic, political, and diplomatic sanctions forced Gbagbo to yield his office and face the International Criminal Court in The Hague on charges of crimes against humanity.

Côte d'Ivoire's economy has yet to recover from decades of military and civil unrest. Coffee, cocoa, oil, and gold production levels have dropped. Private sector investment that was once a hallmark of the economy has declined steeply. The north-south divide has increased, with poverty levels reaching new highs in the north. As it did in 2007, the country is trying to reengage with regional partners and open up new trade networks with Asian countries such as South Korea, Singapore, and the Philippines.

Ghana: Economic and Political Resurgence

Ghana was the first colonized country in Africa South of the Sahara to gain its independence in 1957. Ghanaian intellectuals were instrumental in developing the necessary political platforms in and outside Ghana to challenge the colonial authority. With independence

▼ Figure 8-36 **Abidjan, the economic center of Côte d'Ivoire.** Abidjan is a major economic, cultural, and political center in French-speaking West Africa.

came much optimism. Ghana, "the black star of Africa," was poised to make its move as a leader in the Pan-African movement. Expectations were very high; Ghana (formerly known as the Gold Coast) had significant gold, manganese, bauxite, and timber reserves (Figure 8-37). It was also a leading world producer of cocoa. It had a well-educated human resource base, and its population was supported well by its 92,000-square-mile (238,000-square-kilometer) area. Ghana was also endowed with one of the world's largest artificial lakes, Lake Volta, which was dammed at Akosombo to develop a multipurpose river project that included the harnessing of hydroelectric power to smelt alumina and to develop inland fishing.

Ghana's drive to development began well. In the early 1960s, its economy was growing at a rate of more than 5 percent a year; it had one of the highest per capita incomes of the region; and it had adequate cash reserves to invest in its physical, social, and technological infrastructure. But from the mid-1960s to the mid-1980s, Ghana went through a period of economic recession, declining investments, five military coups, and a debilitating brain drain.

In 1983 the government, under the auspices of the World Bank and IMF, launched an Economic Recovery and Structural Adjustment Program with a projected $4.2 billion investment that was intended to get the country out of its economic rut. Like other countries that comply with World Bank and IMF standards and rules, Ghana began the process by devaluing its currency to lower

▼ **Figure 8-37 Gold bodyware in Ghana.** The gold ceremonial regalia of this Ashanti chief bears witness to the important role that gold has played in the history and development of Ghana—formerly known as the Gold Coast.

the external price of its exports (with the expectation of increased foreign demand for its goods), trimming overstaffed state bureaucracies and padded payrolls, improving institutional management, encouraging privatization, eliminating price controls and subsidies, and removing import quotas and high tariffs that protected uncompetitive firms. The Ghanaian economy has since responded with a healthy annual growth rate of 7 percent and achieved an all-time high of 20 percent in June 2011. The gold mining industry has revived, and foreign and domestic investments have increased. Now likened in some respects to Brazil, Ghana is frequently cited as the most successful model of modern-day economic reform in Africa.

Ghana's recovery is not complete, since current growth rates reflect a recovery from 20 years of depression. Inequalities still exist between its north and south, the social and technological infrastructure needs upgrading, and a number of negative social consequences resulting from structural adjustments have emerged. Currency devaluations and strict credit restrictions have severely harmed small-scale farmers and traders as well as women. Many victims of cutbacks in the public sector remain unemployed, since no effort was made to retrain them for new jobs.

To counter the adverse effects of structural adjustments, Ghana has instituted PAMSCAD—the Program of Action to Mitigate the Social Costs of Adjustment. The program is designed to supplement a basic-needs strategy to improve primary health care, reduce childhood diseases, supplement child nutrition, and compensate for adjustment-related job losses. In 2006, Ghana signed the Millennium Challenge Compact, receiving $540 million to boost the production of high-value cash and food staple crops, promote private-sector agribusiness development, and enhance its export base in the horticultural sector. The country now has expanding economic sectors in agriculture, industry, mining, and telecommunications boosted by oil discoveries and expanding partnerships with Mongolia, Ukraine, India, and Brazil. It is one of a few countries expected to meet the Millennium Development targets set for 2015. Barring any political misfortune, Ghana is poised to establish itself as a leader in social, economic, and technological reform in sub-Saharan Africa.

Nigeria: The Geopolitics of Oil and Ethnic Strife

Nigeria was once perceived as sub-Saharan Africa's economic giant. As Africa's most populous state, with a population of 170 million, Nigeria was expected to lead the region toward economic prosperity. Once hailed by the West as a "cradle of democracy" and "the golden voice of Africa," Nigeria is now mired in political turmoil, ethnoregionalism, and economic mismanagement.

Boosted by the development of its petroleum industry in the late 1960s, Nigeria recorded better than average economic growth rates up to 1980. Nigeria is now categorized as a low-income country—its per capita GNI PPP barely exceeds $2,000—with poor human-development capabilities. About 54 percent of its population lives in conditions of multidimensional poverty, deprived of health, education, and basic living standards.

A number of interrelated factors have stalled Nigeria's economic progress. Among them are the geopolitics of oil, deep-seated ethnic divisions, and the excesses of military rule, greed, corruption, and mismanagement. Oil still dominates the economy, accounting for 20 percent of the GDP, 75 percent of government revenues, and 86 percent of total export earnings (Figure 8-38). This dependency on

▲ **Figure 8-38 Nigerian oil and gas terminal, Niger River delta.** Nigeria continues to rely heavily on its abundant oil production for government revenue.

oil makes Nigeria vulnerable to the volatile world petroleum market. Nigeria also relies heavily on foreign investors such as Royal Dutch Shell, whose holdings account for a third of the 2 million barrels of crude oil produced daily. The Nigerian National Oil Corporation was deregulated in 1989 and now engages in joint ventures with Shell Development Company of Nigeria. Together the two companies control more than 60 percent of land concessions in Nigeria, with other corporations such as Elf Aquitaine of France, Gulf, Mobil, and Texaco accounting for the remainder. In addition to oil, Nigeria has the largest reserves of natural gas in Africa, as well as significant reserves of lignite coal, tin, and iron ore.

Nigeria's climatic diversity makes possible a broad range of food and cash crops (Figure 8-39). The northern savanna zone supports cash crops such as peanuts and cotton as well as food crops such as guinea corn, millet, and cassava. In the southern forest and wooded zones, principal cash crops include cocoa, rubber, oil palm, and coffee, while food crops include sorghum, rice, maize, and yams and other tropical tubers. Yet, Nigeria's agricultural output has not kept pace with its annual population increase of 2.6 percent, and the country is an importer of cereal grains. Oil could have been utilized as a unifying force in support of improved agricultural performance, economic diversification, and cultural integration. Instead, it has been a divisive influence, as various ethnic groups and political factions compete for material wealth and political power.

Nigeria's population is extremely diverse, consisting of more than 250 ethnic groups and more than 500 languages. The Hausa-Fulani in the north, the Yoruba in the southwest, and the Igbo in the southeast collectively constitute 65 percent of the population. At

independence in 1960, Nigeria consisted of three broad regions: the northern region, with 79 percent of the land area and 54 percent of the population; and the eastern and western regions, with 22 percent and 24 percent of the population, respectively. In 1963 a midwestern region representing the Edo and Ijaw ethnic groups was added. Initially, the North was reluctant to join the South, but compromises were reached when the North received some legislative concessions. In 1964, amidst controversy, the first elections since independence were held, and the Nigerian National Alliance, a coalition of Hausa and Yoruba parties, won. This victory signaled the North's ascendancy within Nigerian politics. Today, the North still seeks to maintain its relative hegemony in Nigerian politics while the South seeks to neutralize it. In the midst of this long-term North–South struggle for dominance are several regional groups seeking greater local autonomy.

To accommodate regional demands for autonomy and to neutralize northern domination, the Nigerian federation has undergone considerable restructuring since 1963. It was "balkanized" into 12 states in 1967, 19 in 1976, 21 in 1987, and 30 in 1991. **Balkanization** (a term referring to the Eastern European experience) involves dividing a country or region into smaller and often hostile units. This fragmentation has intensified regional divisions and ethnic dissension.

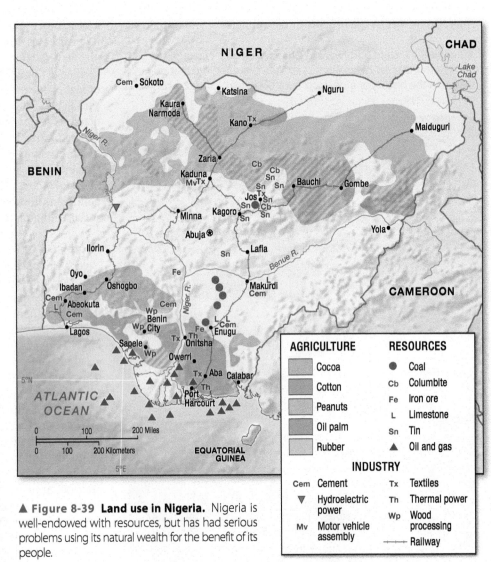

▲ **Figure 8-39 Land use in Nigeria.** Nigeria is well-endowed with resources, but has had serious problems using its natural wealth for the benefit of its people.

The dissension has been punctuated by military intervention, threats of secession, and a devastating civil war between 1966 and 1970 that cost millions of lives. The Nigerian military dominated the political scene for more than 25 years following independence and completely lost its perspective, seeking political opportunity and material wealth. Military academies became training grounds for future political leaders, and army officers aspired to political rather than military appointments. Civilians have been unable to unite to form a cohesive political platform to challenge military rule.

Nigeria's most embarrassing moment internationally occurred in 1993 when the presidential elections were nullified—after Nigeria had spent the previous seven years developing a transition program to restore democracy. Military rule was reimposed, and political parties were banned. The military government, in an attempt to redeem itself, called for a national constitutional conference to determine Nigeria's political future and discuss issues of power sharing, the distribution of oil revenues, political party reform, and decentralization of government. Nigeria's present underdeveloped state is almost entirely a result of wasted human resources and corruption. The country's abundant mineral and agricultural wealth has served only to heighten, not diminish, ethnic divisions.

In May 1999, Nigeria took a major step toward democratic reform with the inauguration of President Obasanjo and his civilian government. President Goodluck Jonathan assumed the presidency in May 2010 and went on to win a disputed election in April 2011. The government has the difficult tasks of rebuilding a devastated infrastructure and weakened economy, and unifying a divided population. As the winds of change sweep across Nigeria, a number of threats persist. Impoverished residents of the oil-rich Niger River delta continue to press for a greater share of the region's wealth, and radical Islamic fundamentalists in the North pursue their political agenda.

Recently, Islamic Sharia law, based on the Quran, was instituted in 12 states in northern Nigeria, resulting in such drastic penalties as stoning for adultery, amputation for theft, and flogging for alcohol consumption. Women tend to be more victimized than men. Sharia law is diffusing to southern states and threatens to trigger yet another constitutional crisis in Nigeria.

Stop & Think

▶ Why has Nigeria, with its wealth of natural resources, not become an economic giant in sub-Saharan Africa?

Regional Economic Integration

The **Economic Community of West African States (ECOWAS)** was created in 1975 and is the largest of Africa's regional organizations. Only Chad and Mauritania are not part of ECOWAS. The organization's objectives include establishing a common customs tariff, a common trade policy, the free movement of capital and people, and coordinating agricultural, communications, energy, and infrastructural policy. Members have also signed a pact of nonaggression and mutual defense.

Calls for African unity have recurred since the early 1900s. The first Pan-African conference was convened in London in 1900 to protest colonial rule in Africa and the themes of Pan-Africanism and regional cooperation have been given increasing attention.

The 1980 Lagos Plan of Action reaffirmed Africa's goals of establishing a common market and enhancing collective self-reliance and self-sustained development. Intraregional linkages need strengthening to enhance economies of scale in production, implement advanced production technologies, and promote greater functional specialization.

ECOWAS has made some progress in the areas of transportation and telecommunication development, the free movement of persons, and regional security. Much of the transcoastal West African highway network from Lagos in Nigeria to Nouakchott on the coast of Mauritania and the trans-Sahelian highway from Dakar on the coast of Senegal to N'Djamena in Chad have been completed. Efforts to enhance the region's telecommunications with multimedia and broadband technologies are being realized. ECOWAS travel certificates have been introduced to facilitate intraregional travel, and an ECOWAS Peace Monitoring Group (ECOMOG) is in place to mitigate conflicts and preserve peace and stability in the region. Intraregional exports among member countries increased significantly from 2.7 billion in 2000 to 7.3 billion in 2009 and there are plans to adopt a single currency, the Eco, by 2020. But the challenge of achieving greater economic integration is hampered by uneven levels of development between countries and the unequal distribution of benefits and costs. Smaller member states of ECOWAS are overwhelmed by the predominant influence of larger states like Nigeria and Côte d'Ivoire. Other conflicts have arisen over religious, linguistic, and ethnic differences, and dependence on foreign markets continues. Chad, disadvantaged by its landlocked location, still sends most of its exports to France, as does Sierra Leone to Britain. Other problems include a lack of regional complementarity because many countries produce the same goods as other members, and regulatory and procedural barriers.

Prospects for a fully integrated Pan-African Economic Community are greater once these challenges are overcome. Further collaborative efforts in the areas of transportation and communication, research and technology, environmental management, resource conservation, and food security have growth potential. The African Economic Research Consortium is one example of a successful regional capacity-building venture. It provides a forum for African researchers to discuss and evaluate research on a variety of economic topics ranging from debt management to taxation policy and structural adjustment.

Central Africa

Much of Central Africa lies in the equatorial region where tropical humid conditions prevail. The states in the region account for 17 percent of sub-Saharan Africa's land area and only 11 percent of its population (see Figure 8-1, page 346). Gabon, the Congo Republic, and Equatorial Guinea have capitalized on the wealth of mineral and forest resources. Cameroon, the Central African Republic (CAR), and the Democratic Republic of Congo (DRC) have been plagued by economic mismanagement and/or political instability. Overall, the pattern of development has been very uneven. Rapid population growth throughout the region, together with political instability and human welfare problems compromise future prospects for improvement.

Central African Republic: Dilemma of a Landlocked State

Like most landlocked countries in Africa, the Central African Republic (CAR) has significant development challenges. The northern region lies in the watershed of Lake Chad and its landscape is characterized by semiarid savanna grasses, bushes, and acacia trees. The south supports denser vegetation as it merges into true rain forest near the Congo Basin. The CAR is doubly dependent—on both its immediate neighbors, and international markets for the import and export of crucial goods. The principal route that external trade must take is a 1,110-mile (1,800-kilometer) journey south along the Ubangi River to Brazzaville on the Congo (Zaire) River, and then by rail to Pointe Noire on the Atlantic coast. In addition to the high freight and line-haul costs associated with shipment between different modes of transport, exports must also absorb very high insurance costs associated with safeguarding goods against theft, vandalism, and perishability at coastal ports.

Economic development in the republic is hampered by a poorly developed social and physical infrastructure, economic mismanagement, and political instability. Life expectancy in the CAR is only 48 years and the literacy rate is among the lowest in Africa and the world. The country has no railroad, and less than 2 percent of its estimated 13,700-mile (22,000-kilometer) road network is paved. Its telecommunications system is substandard, consisting mostly of low-powered radios. Agriculture accounts for about 55 percent of the republic's economic output and 60 percent of the workforce, and is dominated by coffee (its major export crop), cotton, and, to a lesser extent, tobacco. It has valuable species of hardwood that are underutilized, as well as untapped reserves of petroleum and uranium. Diamonds now account for about 50 percent of export earnings, despite an increase in smuggling activities.

Aside from its landlocked location, the meager economic performance and limited development of the republic can also be attributed to the 15-year despotic regime of President Jean-Bedel Bokassa. He interfered with the republic's attempts to preserve civilian rule after independence in 1960 and forcefully took over the reins of government in 1966, heralding one of the cruelest, most corrupt, and inhumane regimes in postindependence Africa. He developed delusions of grandeur as he declared himself first president-for-life in 1972, then marshal in 1974, and finally emperor (with Napoleonic overtones) in 1977. It is estimated that he squandered one-quarter of his country's annual income on a lavish coronation ceremony. Bokassa's excesses were tolerated by France, which continued to bail him out with loans and economic assistance because of its heavy involvement in the country's diamond and gold mining industries.

Today this fragile and postconflict state still struggles to build an economy that, while having much potential, remains greatly constrained by the country's landlocked location, low levels of human capacity, and limited ability to sustain a stable political environment.

Gabon: Forest and Mineral Wealth amidst Inequalities

Gabon remains one of the most prosperous countries in Africa South of the Sahara because of its extensive forest and mineral resource base. About 75 percent of its 103,347 square miles (267,669 square kilometers) of land is covered with forest and woodland. Its forests contain several species of tropical hardwoods and softwoods; most notably, it is the world's largest producer of Okoume, a softwood used to make plywood. As in most African countries, its timber exports consist mainly of raw logs (Figure 8-40). In recent years, there has been an increase in the export of sawn wood, veneers, and plywood though the scale of activity remains small. Another problem that Gabon faces is the depletion of its coastal forests. To curtail deforestation, the government has instituted a log licensing program, which permits selective logging in three distinct logging regions.

Gabon's main source of wealth is its diverse mineral base of petroleum, manganese, high-grade iron ore, uranium, and gold. In 2010 the petroleum industry accounted for 48 percent of GDP and 80 percent of export revenues. Most of the oil deposits are concentrated in sedimentary rock beds along the coast around Port Gentil—"the petroleum city." Private foreign investment predominates; the French multinational Elf Aquitaine owns 57 percent of Elf Gabon in partnership with the Gabonese government (25%). Other foreign petroleum companies operating in Gabon include Occidental, Amoco, Arco, and Shell.

Although Gabon is one of sub-Saharan Africa's richest countries, with a per capita GNI PPP of $13,060, not everyone shares in its prosperity. Extreme rural–urban disparities are at the root of a continuing exodus from rural areas. Libreville, the capital, and Port Gentil together account for almost half of the country's 1.6 million people. Underpopulation in rural areas inhibits the development of the agricultural sector that employs 37 percent of the labor force but accounts for only 8 percent of economic output. Gabon must rely on migrants from neighboring countries to work on cocoa, coffee, and oil palm plantations. **Agroforestry**, the management of trees and shrubs integrated with agricultural systems for multiple products and benefits, is a viable development strategy that can reduce economic pressures for deforestation while slowing the rural exodus by providing opportunities for farming communities to meet crop, fuelwood, and other related needs. Gabon does, however,

▼ **Figure 8-40 Tropical logging in Gabon.** Logging is a leading economic activity in Gabon, but deforestation threatens its long-term sustainability.

rank second among sub-Saharan African countries on the HDI (0.674) with a life expectancy of 62 years, and the average length of schooling tripled from 2.3 years in 1980 to 7.5 years in 2011.

The Democratic Republic of Congo (formerly Zaire): Development Potential Wasted

Sub-Saharan Africa's largest country has yet to recover from years of brutal Belgian colonial rule, followed by autocratic rule under its longtime dictator, Mobutu Sese Seko, who was overthrown in 1997 by rebel forces led by Laurent Kabila who was succeeded by his son, Joseph Kabila. The country has reverted to its original name, the DRC, which was adopted after independence and used from 1964 to 1971. The DRC faces serious challenges as it tries to unite several social and political factions, rebuild a fledgling economy, and assert itself as a stabilizing force in Central Africa.

Along with Nigeria, the DRC has been described as a nonfunctioning giant in sub-Saharan Africa. The country is endowed with abundant natural resources but has not lived up to its development potential. The vast Congo Basin, drained by Africa's largest river, has a hydroelectric power potential of 100,000 megawatts, the largest on the continent. The basin is also a repository for extensive timber resources and a major supplier of plantation crops, including coffee, oil palm, rubber, and bananas (Figure 8-41). Other significant crops are cotton, peanuts, cassava, plantains (cooking bananas), maize, and rice. The DRC has abundant mineral resources and is a

▼ **Figure 8-41 Oil palm fruits.** Oil palm cultivation has expanded rapidly within the Congo Basin. The crop is grown for the vegetable oil that is obtained from the kernels of the nuts.

world-class producer of copper, cobalt, coltan (columbite-tantalite, a mineral used in transistor manufacturing), industrial diamonds, and zinc. Other significant reserves include gold, manganese, uranium, and petroleum. The mining industry accounts for about 25 percent of the DRC's economic output and about 75 percent of total export revenues.

Despite a large and diverse resource base, the DRC's 69 million people remain one of the poorest populations in Africa, with a per capita GNI PPP of only $320. The nation is plagued by low educational attainment, high female illiteracy, and high rates of child malnutrition. The majority of its population lacks access to health services, basic sanitation, and safe water. *Foreign Policy* magazine's Failed States Index places the country in the critically failed category. The situation has not improved much in the wake of an international boycott of President Kabila's second inauguration in December 2011. Perhaps the single most important factor to explain the DRC's social and economic malaise is the more than three-decades-long despotic and self-indulgent leadership the country endured under Mobutu. His regime was the epitome of corruption, greed, and incompetence. Mobutu did little to improve the country's social and physical infrastructure, but chose to enhance his personal wealth by building castles, chateaus, and villas in Spain, France, Belgium, and Switzerland. His net worth was estimated at $3–$7 billion. Initially little international pressure was exerted on Mobutu because of the importance of the DRC's resources. The country's cobalt reserves are vital for the manufacture of aircraft engines, and coltan is vital for computers, mobile phones, and video games. In the present-day DRC, however, access to these mineral resources and their productive use for the good of the country remains problematic. The struggle between ethnic militias and local warlords to control access to valuable minerals in the northeast of the DRC is symptomatic of the country's larger problems. It reflects how weak central government control has failed to provide an adequate environment for meaningful development.

The country has never really severed the connection with its former colonial authority. The DRC has expanded its trade networks to include China, Brazil, and members of the Southern African Development Community (SADC), but a client-patron, neocolonial relationship persists as Congolese economic and corporate institutions continue to be firmly linked to the Belgian economy.

The social and economic conditions in the DRC have worsened under the leadership of Joseph Kabila as tensions prevail between his government and rebels in the eastern sections of the country, where Congolese Hutus and Tutsis have long coexisted with the Hunde and Nyanga ethnic groups, who consider themselves to be natives of the DRC. These conflicts not only have destabilized the region, but also have drawn the international community's attention to the high incidence of rapes that occur in remote towns and villages, prompting the country to be labeled the "rape capital of the world." International efforts to encourage dialogue between the Kabila government and opposition groups, promote the disarming of militias, and achieve power-sharing have largely failed, as skepticism about the legitimacy of the Joseph Kabila government and its willingness to share power with opposition groups remains. A destabilized DRC does not bode well for the future development prospects of the East African Great Lakes region and beyond.

East Africa

East Africa contains 26 percent of sub-Saharan Africa's land area and 37 percent of the population (see Figure 8-1, page 346). Physical environments are quite diverse, with semiarid and arid conditions dominating Somalia, northern Kenya, and Sudan and temperate conditions in the fertile Lake Victoria agricultural borderlands of southeast Uganda, northwest Tanzania, and southwest Kenya. Tensions based on ethnic rivalries afflict every nation in the region, and have resulted in civil wars and secession efforts (see *Geography in Action: South Sudan: Birth of a New but Fragile State*). Socialist and communal development strategies have been followed and largely found wanting, but neoliberal approaches have not succeeded either. Pockets of success in ecotourism and agriculture are encouraging but distressingly limited.

Somalia: Turmoil in a Failed Clanocracy

Somalia is strategically located astride the Horn of Africa, facing a key international sea route for oil shipments originating in the Persian Gulf. The port of Berbera in northern Somalia is also a major base for the U.S. military's rapid deployment force, which responds to emergencies in the Middle East. Somalia has long feuded with its neighboring countries. Its national flag has five stars, representing its claims to Somali-populated areas in eastern Ethiopia, eastern Kenya, and Djibouti and the consolidation of northern and southern Somalia.

Somalia is one of the few countries in sub-Saharan Africa that comes close to the Western concept of a nation-state in which the national or ethnic boundary of a group of people with common customs, language, and history coincides with the state boundary. Ethnically, the Somalis belong to the Hamitic group. They share a common language, although with differences in dialect; predominantly follow a pastoral way of life; have a shared poetic literature that is a vital part of public and private life; support a common political culture; and share the religious heritage of Sunni Islam. The Somalis believe that they are drawn together by kinship and genealogical ties, descending from a common founding father, Samaale. This ethnic unity is offset by the Somalis' division by kinship into clans. There are four major pastoral clans (Dir, Daarood, Isaaq, Hawiye) and two agricultural ones (Digil, Rahanwayn). The clans are patrilineal descent groups that define the political and legal status of people and assign them specific social claims and obligations. A person owes allegiance first to his or her immediate family, followed by the immediate lineage, the clan of lineage, and then a clan-family that embraces several clans.

This **clanocracy** acts as both a unifying and divisive force. Mohammed Fara Aidid and Ali Mahdi, both from the Hawiye clan, collaborated to overthrow the government of former president Siyaad Barre (from the Marehan clan, a subclan of the Daarood) in 1991. The coup was a reaction to Barre's brand of scientific socialism, which clashed with Islamic and clan-based traditions. Scientific socialism attempted to abolish kinship influences and clan-based activities in Somali life. All forms of tribal or clan association, rights, and privileges—including those over land, pasture, and water—were abolished and claimed by the state. Private funerals, family marriage ceremonies, and payment of blood money were banned. Administrative boundaries were reorganized and renamed to exclude reference to clan names. There were proposals to replace the family as the basic unit of organization with settlements and collective farms. "Orientation centers" were set up throughout the country to foster the formation of a new community and individual identity based on loyalty to the ruling party, state, and president. Scientific socialism threatened the underlying foundations of Somali culture and was abandoned after the 1991 collapse of the Barre government. Following a successful coup, Aidid and Mahdi became rivals, as their allegiances shifted to their subclans.

Between 1991 and September of 2012, Somalia was essentially without a president or formal government. A Transitional Federal Government (TFG) was instituted in 2004, but its influence was largely confined to the central portion of the country including the capital, Mogadishu (Figure 8-42). The remaining territory was controlled by

▼ **Figure 8-42 The fragile state of Somalia.** Somalia's Transitional Federal Government has limited influence. Its control is confined to the central regions surrounding Mogadishu. The remaining regions are controlled by clans, secessionist groups, and radical Islamic movements.

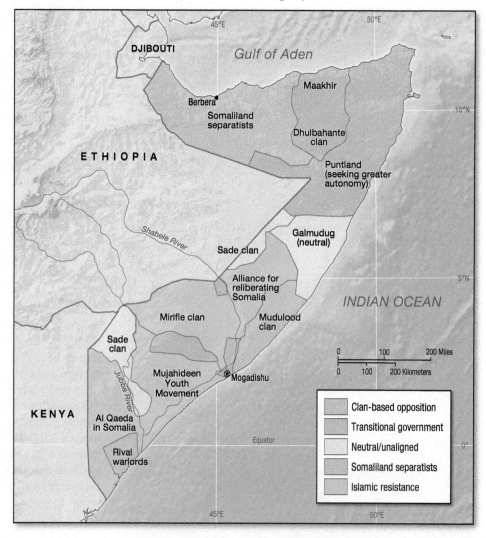

GEOGRAPHY IN ACTION

South Sudan: Birth of a New but Fragile State

South Sudan became Africa's newest independent country on July 9, 2011 (Figure 8-5-1), after battling decades of institutionalized ethnocentrism, racism, and colonialism that began during Anglo-Egyptian rule and continued under a northern-dominated government after independence in 1956. South Sudan is ethnically, religiously, and geographically distinct from its northern neighbor. In contrast to the predominantly Arab-Muslim north, South Sudan's 9.4 million people practice Christianity and traditional African religions. The south is populated by 200 ethnic groups, including the Dinka (the largest), Nuer, Azande, Shilluk, and Bari. The forested regions of South Sudan are packed with valuable timber species and contrast sharply with semiarid and desert north Sudan. The country is also home to one of the largest wetlands in the world, the Sudd, which is a haven for migratory bird species and wildlife. This large wetland is often compared with the Okavango delta in southern Africa.

South Sudan's independence came amidst much hope for a bright future. But immense challenges must be overcome: sharing oil revenues, resolving border disputes, and easing lingering ethnic tensions. With independence, landlocked South Sudan controlled three-quarters of the previously united country's oil reserves. But oil export requires transport through oil pipelines that cross its northern neighbor to distribution facilities on the Red Sea. A 2005 agreement split oil revenues 50–50 between south and north, but expired after independence and was not replaced. Tensions escalated as South Sudan accused the north of embezzling oil revenues and charging exorbitant transit fees. In retaliation, despite total dependence on oil revenues to fund its budget, the South Sudan government stopped oil exports. The shutdown crippled the south's economy, pressured the north's economy, sparked street demonstrations against the Khartoum government, and fueled simmering border conflicts between the two states. In October 2012 the leaders of the two states met in Ethiopia and signed a partial agreement settling some disputed issues. South Sudan agreed to restart oil shipments via northern pipelines at drastically lowered rates for processing and transport ($9.48 instead of $36 per barrel). In return, the south agreed to pay a $3 billion compensation fee to the north to offset lost revenue from the oil fields now located in South Sudan.

Other boundary issues remain unsettled. A referendum to determine whether the contested border region of Abyei will be in the south or north has been delayed. Abyei contains vital oil fields and fertile grazing land and is the site of a dispute between the subsistence farming Ngok Dinka, who are committed to joining the south, and the nomadic Arab Misseriya who have allegiances to the north but graze their herds in Abyei for several months of the year. Aside from ethnic tensions, arguments between the new countries are invariably centered around the sharing of mineral, agricultural, and water resources, particularly Nile water, in territories straddling the as yet undemarcated border between the two new states.

South Sudan is a young, yet fragile state. It has enormous growth potential. But the newly independent country can only achieve political and economic sovereignty with cooperation from the north and support from the international community. It can follow the example of Angola, Mozambique, and Rwanda, overcome immense obstacles, and launch a process of nation building, democracy, and economic prosperity. Or it can collapse into isolation, conflict, and chaos.

Sources: "Unhappy Birthday," *The Economist*, July 14, 2012, 40; "Better than Nothing," *The Economist*, October 6, 2012, 60–61; Mireille Affa'a-mindzie, "Negotiating Peace in the Sudans: The Addis Ababa Agreement," *Global Observatory*, October 9, 2012, www.theglobalobservatory.org/analysis/365-negotiating-peace-in-the-sudans-the-addis-ababa-cooperation-agreement.html.

▼ FIGURE 8-5-1 **Juba, South Sudan.** The capital of South Sudan is located along the banks of the White Nile and is emerging as a regional financial and transportation hub.

secessionist and radical Islamic groups. Northern Somaliland (formerly British Somalia), including the regionally dominant Isaaq clan, declared its independence from Somalia in 1991. Officially called the Puntland State of Somalia, its leaders, mainly from the Darood Majeerteen subclan, declared the northeast region an autonomous state in 1998 without seceding from Somalia. Between Somaliland and Puntland is Maakhir, another self-declared autonomous state inhabited by the Warsangali and other Darood clans. South of Puntland is Galmudug, where the Saleeban subclan of the Darood resides. The region controlled by the TFG was contested by radical Islamic groups such as Al-Qaeda, Al-Shabbab, and Al-Ittihad al-Islamiya. The latter, along with the Islamic Courts Union, emerged with intentions of imposing Sharia law in Somalia. In 2007 this fundamentalist alliance pushed the TFG out of Mogadishu, which sparked military intervention by Ethiopian troops who withdrew in 2009. This period of political and economic instability was further exacerbated by sporadic incidents of piracy along the Gulf of Aden which threatened international shipping, forcing NATO to launch antipiracy operations.

In 2011, the Somali government, African peacekeeping, and Kenyan forces wrestled control over Mogadishu away from Al-Shabbab. In September 2012, Hassan Sheikh Mohamud was installed as president along with a new parliament, which has since been legitimized and recognized by the United States. It is too early to tell whether the new Somalia government will endure. Islamic extremism remains a threat and underlying clan differences remain a strong divisive force. Clan warfare is typically provoked by disputes over water and pasture rights or political control. Although a person belongs to a network of clans, primary sociopolitical loyalty is to the *diya*-paying group, a close-knit group of kinsmen united by a contract (*heer*) that specifies the terms of blood compensation for murder, feuds, and any illegal acts. This community group usually constitutes the basis for social and political action. This tradition is important because it means that the clan system has an established framework for resolving conflicts. Clan elders play a significant role in mediating disputes over local affairs, including blood money, rights to watering holes, and access to pasture. That is the principal reason why Somalis reacted so adversely to U.S. military efforts to capture clan leaders during the U.N. famine relief effort of 1992–1993, which was prompted by food shortages in the wake of the political turmoil of 1991–1992. Until a strong and unified government emerges, untainted by the interests of a specific clan, conflict and economic stagnation will continue to afflict Somalia.

Stop & Think

What are the contributing factors that explain why Somalia is plagued by political instability and pervasive poverty despite its homogenous culture?

Tanzania and Ethiopia: Working on African Socialism

African socialism was designed as the antithesis to European domination. It was viewed as a mechanism for achieving balanced, self-reliant development based on social justice and mutual cooperation. Advocates considered socialism to be consistent with traditional African values, which emphasized the moral virtues of economic equality, reciprocity, and mutual aid. In Tanzania, President Julius Nyerere spent the 1960s and 1970s building a society based on socialist principles, articulating the justification for African socialism in the 1967 Arusha Declaration. That year marked the implementation of a socialist policy geared toward eliminating poverty and income inequalities and improving the quality of rural lives. Major commercial, industrial, and financial institutions were nationalized, and public officials were restricted in their ownership of property.

In 1968, the *Ujamaa* (familyhood or brotherhood) policy was instituted to promote self-help, self-esteem, and local initiative along family, communal, and cooperative lines. The Ujamaa village, modeled along the same lines as the Soviet collective or the Chinese commune, was a consolidated, self-sufficient settlement that provided educational, health, and social services to meet the basic needs of its resident population. The villages were intended to become the basis for social organization and cooperative activity in Tanzania, pooling labor, land, and capital resources for agricultural and other ventures. Proceeds from the cooperative efforts financed a variety of development projects. To encourage rural participation in the decision-making process, Tanzania embarked on a policy of decentralization in the early 1970s, as regional and local district offices were dispersed into rural villages.

Tanzania's socialist strategy had mixed results. Despite some success in achieving social equity, the country continues to be one of the poorest in sub-Saharan Africa. Tanzania has since undergone a structural transformation in its productive sectors boosted by increased private sector investments in mining and tourism. It has become the fourth-largest producer of gold in Africa. Tourism generates about 25 percent of export earnings and the country's network of national parks and game reserves, including Kilimanjaro, Serengeti, and the Ngorongoro reserve, are major world destinations (Figure 8-43).

Ethiopia adopted a socialist policy in 1975, when a military council replaced Emperor Haile Selassie and immediately nationalized all banks, insurance companies, and several industrial and commercial enterprises. A new land reform program was instituted to bring all rural and urban land under government control. This program had

▼ **Figure 8-43 Ngorongoro Conservation Area, Tanzania.** Ngorongoro Crater is a UNESCO World Heritage Site and popular safari destination, where blue wildebeests and zebras as well as many other animals can be observed in their natural habitat.

devastating effects on communal and traditional forms of land ownership in villages, as rural farmers were forced into collectives. Years of socialist policies have weakened the country's financial sector and regulatory capacity. The government is implementing market-driven policy prescriptions set by the World Bank and African Development Bank, which aim to boost production in livestock, maize, coffee, and other agricultural and horticultural activities. Chinese investments in Ethiopia have increased dramatically in the last decade, particularly in road construction, telecommunications, and power generation, as well as textile, garment, shoe, and leather manufacturing.

Tanzania and Ethiopia are typical of what happens to African countries that toe the "socialist" line. Tanzania achieved some measure of social equality in health care and education. But nationalizing industries slowed down productivity owing to mismanagement, production inefficiency, and the lack of capital and technological investments. A major problem was that the policy of socialism was imposed from the top by a bureaucracy that was out of touch with the needs of rural people and inexperienced in handling the complexities of rural development. Self-reliance at the state level implies total autonomy and independence in decision-making and the allocation of resources. No single sub-Saharan country is in a position to break away and develop a totally autonomous stance in our globally interdependent world. Countries such as Tanzania and Ethiopia have been firmly entrenched in the global exchange economy since colonial times, and are learning to redefine self-reliance from the perspective of interdependence and mutual cooperation.

Kenya: Transition of a Settler Colony

Much of the economy of present-day Kenya has been molded by British intrusion and colonization. The British were attracted to Kenya because of its temperate highland environments and the rich volcanic soils in the southwest highland region, which is one of the most productive agricultural zones in Africa. From a strategic standpoint, Britain was able to prevent German East Africa from extending its influence from Uganda to the fertile highland regions and the coastal areas beyond. Kenya, like Zimbabwe, was transformed into a settler colony. An administrative and economic structure was superimposed to marginalize the indigenous population and preserve permanent European settlements. Monetary taxes were imposed on local populations to finance a settler-oriented infrastructure and force Africans to offer cheap labor on settler farms. Africans were displaced from their lands and relegated to crowded and inhospitable reserves, where social and economic conditions were deplorable. Having acquired the best land, the British proceeded to establish a plantation economy based on tea and arabica coffee.

At independence in 1963, Kenya inherited an agricultural system based on large-scale commercial estates owned primarily by Europeans and Asians. This occurred despite earlier nationalistic efforts to reclaim land from European settlers. Jomo Kenyatta organized Kenya's largest ethnic group, the Kikuyu, to demand access to white-owned land. This culminated in the Mau Mau rebellion of 1952, followed by a devastating and costly three-year war. After the war, the colonial authorities granted Africans access to land and to political authority. Land reforms, land consolidation, and resettlement schemes provided Africans with opportunities to improve farm management practices and advance from subsistence to commercial farming. The Kikuyu and

Luo played active roles in preparing the country for independence. In 1963, Jomo Kenyatta became the first prime minister.

For a long time Kenya remained a fairly stable but authoritarian country. Kenyatta's government lasted from 1963 to 1978 and was dominated by Kikuyus, many of whom emerged as an elite class and became wealthy. The Kikuyus are concentrated in the Central Province north of the capital of Nairobi and reaped the benefits of land ownership in the coffee- and tea-producing areas of that region. Arap Moi, who succeeded Kenyatta, held a steady course in spite of attempted coups, threats from Muslims, and African-Arab rivalries. The Kenya African National Union's (KANU) dominance since independence in 1963 came to a peaceful end when Mwai Kibaki's National Alliance Rainbow Coalition (NARC) secured a majority vote in 2002. Violence broke out after the 2007 elections when Kibaki claimed victory after a controversial vote count. Former U.N. Secretary General Kofi Annan stepped in to broker a power-sharing arrangement, which includes an ethnic relations bill and a bill establishing a Truth and Reconciliation Commission to address the violence and long-term injustices. Kenya has since developed a new constitution that preserves the presidential system, establishes additional checks and balances on executive power, and institutes a process of political decentralization.

Agriculture is still the leading economic activity, providing 75 percent of employment and 55 percent of export earnings. High-grade tea, coffee, and horticulture (fruits, flowers, and vegetables) dominate exports. The country is the world's second largest exporter of tea (behind Sri Lanka). Other cash crops include sisal, cotton, sugarcane, and maize. Foreigners still own and manage several large-scale commercial farms, and there is some smaller-scale indigenous production. Kenya also has a relatively well-developed industrial sector in food processing, tobacco, textiles, oil refining, plastics, pharmaceuticals, electrical cables, rubber, ceramics, industrial gases, and car assembly. Industry is supported by a strong entrepreneurial community, which remains largely European and Asian. The tourist industry continues to thrive and is a major source of foreign exchange.

Nairobi, the capital, dominates the urban landscape, with an estimated population of 3.2 million (Figure 8-44). The city accounts

▼ **Figure 8-44 Nairobi, Kenya.** Nairobi's high-rise city center serves as the commercial and communications hub of East Africa. Nairobi is home to more than 3 million people, most of whom live in sprawling squatter settlements.

for almost 40 percent of the country's urban population. To curb rural-to-urban migration and offset the dominance of Nairobi, the government is relying on a regional decentralization strategy to develop and expand rural trade and smaller urban production centers. These are expected to stimulate market and employment expansion in areas of unrealized agricultural and livestock production potential.

Rwanda and Burundi: Coping with the Legacy of Ethnic Genocide

Genocide in Rwanda and Burundi destabilized the east and central African region, killed more than half a million people, and displaced countless refugees. The Tutsi and Hutu, who share the same language and traditions and have lived in the same territory for more than 500 years, were involved in a tragic genocidal war. The Tutsi, who constitute about 10 percent of Rwanda's population of 10.8 million and about 14 percent of Burundi's population of 10.6 million, are believed to be descendants of Nilotic migrants from Ethiopia. The majority Hutu are descendants of Bantu migrants from West Africa. A third group, the Twa hunter-gatherers, are the original inhabitants and constitute only 1 percent of the population in the region.

Tensions between the Hutu and Tutsi grew during the colonial era, when racial stereotyping and social categorization was common. Prior to colonial contact the Hutu and Tutsi were governed by a kingship institution based on Tutsi lineage that presided over a centralized court. Most of the king's representatives were Tutsi, responsible for agricultural production, grazing land usage, taxation, and defending the king's court. This governance structure set the stage for social distinctions between the cattle-herding class of Tutsi aristocrats and the servant class of Hutu farmers. From the fourteenth century to the end of the nineteenth century, the Tutsi-dominated monarchy developed into a cohesive social and political system. This monarchy was reinforced under German rule from 1896 to 1916. The Germans, under a system of indirect rule, preserved the monarchy and collaborated with it to carry out policies that were mutually beneficial. The Tutsi monarchy, in turn, utilized the German presence to consolidate its position and subjugate Hutu rebels in the northern region.

After World War I and the subsequent collapse of German East Africa, Ruanda-Urundi was entrusted to Belgium under a mandate from the League of Nations. The Belgians also supported the privileged Tutsi position and created a racial hierarchy that was degrading to the Hutu and Twa. The Tutsi were designated as the superior race, and Belgian colonialists provided them with the best jobs and education. This intensified hatred between Hutu and Tutsi; it also led to an awakening of Hutu consciousness in the 1950s. By 1959, the oppressed Hutu majority, with the support of some Belgian missionaries, organized a political revolution that led to the massacre and exile of hundreds of thousands of Tutsi. Following independence in 1962, the already small territory was further balkanized into two new states: Rwanda and Burundi. The Tutsi monarchy collapsed and the Hutu monopolized power in Rwanda. In the early 1990s, a new round of genocide directed at the Tutsi intensified an existing civil conflict, resulted in the victory of the Tutsi-dominated Rwandese Patriotic Front (RPF), and sparked a flood of Hutu refugees into neighboring countries.

Rwanda and Burundi are recovering from a past mired by ethnic genocide. An International Criminal Tribunal for Rwanda (ICTR) and a state court system were established to prosecute genocide suspects.

▲ **Figure 8-45 A gas methane extraction platform in Lake Kivu, Rwanda.** Lake Kivu contains approximately 65 billion cubic meters of dissolved methane gas. This gas extraction facility is part of the Rwandan government's effort to explore long-term, sustainable energy solutions.

Failures in the courts and long delays prompted Rwandans to explore indigenous peacemaking methods such as the Gacaca court system, which emphasizes reconciliation over retributive justice. Modeled after South Africa's Truth and Reconciliation Commission, and drawing on traditional African values that stress contrition, forgiveness, and compensation to wronged individuals, this restorative form of justice was instituted in 2002 to encourage community participation, promote unity, restore social harmony, and expedite trials. Much of the punishment was in the form of community service such as rehabilitating homes or restoring schools and hospitals. Human rights organizations such as Amnesty International and African Rights are concerned that suspects might not receive a fair trial, but the system seems to have gained popularity among most Rwandans.

Rwanda is recovering remarkably from its history of ethnic genocide and is quickly becoming a regional hub for information and communication technologies with key investments in its fiber optics program, wireless Internet, and mobile TV sectors. Crop intensification programs have boosted farm production in maize, wheat, cassava, and coffee, and the manufacturing sector is showing signs of recovery through food processing. The country has strengthened bilateral ties with India to encourage investments in renewable energy projects aimed at electrifying rural schools with solar energy and constructing hydropower plants (Figure 8-45). Burundi's agricultural sector is growing at a healthy rate with increased export earnings coming from coffee and tea. It has expanded its economic partners to include Saudi Arabia, China, India, and Thailand who have focused on upgrading Burundi's infrastructure and telecommunications.

Southern Africa: Development in Transition

Southern Africa (see Figure 8-1, page 346) lies in a subtropical environment and has a wealth of mineral resources. The region is engaged in an ongoing transformation of its social, political, and economic structures. Angola and Mozambique are recovering from decades of civil strife and are seeking a peaceful transition to democratic and economic reform. Namibia, after years of South African rule, is still coping with the challenge of social and political reconciliation. Botswana has transformed its economy from one of the poorest in the world to one of the most prosperous in Africa and is now being examined as a model of political stability. Zimbabwe gained its independence much later than most African countries and is wrestling with issues of land reform, political unrest, and an economic recession. South Africa is potentially a major player in the economic and political transformation of not only Southern Africa but also the rest of Africa South of the Sahara.

Angola and Mozambique: Emerging from Conflict

Both Angola and Mozambique endured a long history of colonial exploitation and civil strife, rendering their economies stagnant and, at times, nearly inoperable. When most French and British African colonies gained their independence in the early 1960s, Portugal resisted granting freedom to its "overseas provinces," maintaining that they were to remain an integral part of Portugal forever. It took years of brutal warfare, inhumane policies, forced labor, blatant exploitation, and the collapse of a fascist government in the "motherland" to set the Portuguese African colonies free in 1975. During the colonial era, Angola and Mozambique were divided into districts. Land was seized from Africans, and men were forced into labor camps to grow coffee in the fertile northern provinces of Angola and cotton in northern Mozambique. Forced contractual labor separated men from their families for as long as a year. Only a few Africans were afforded a decent education; the overwhelming majority had to fend for themselves or seek assistance from missionaries. At independence, both colonies had among the lowest literacy rates in Africa.

Deplorable socioeconomic conditions prompted the formation of liberation movements in these two **insurgent states**. An insurgent state is one marked by revolt, insurrection, and guerrilla activity inspired by national liberation movements that oppose the established authority. In Angola, three movements were formed along regional, ethnic, and ideological lines. Urban elites teamed up with workers in Luanda to form the MPLA (Popular Movement for the Liberation of Angola) with support from the Mbundu tribe; refugees from coffee plantations in the north formed the FNLA (National Front for the Liberation of Angola); and rural agriculturalists from the southeast, with support from the Ovimbundu of the central highlands, formed UNITA (the National Union for the Total Independence of Angola). By independence in 1975, the FNLA threat had faded, leaving the MPLA and UNITA to vie for political power. At the time, Angola could best be described as a **shatterbelt**—a region characterized by intense political discord and subjected to external pressures. The socialist MPLA was backed by Soviet and Cuban troops, while UNITA received support from South Africa and the United States, who were concerned about the spread of communism. South Africa justified its unlawful occupation of Namibia on the grounds that Cuban forces were present in Angola. Most of the revenues derived from Angolan oil sales to the United States were used to purchase military weapons from the Soviets to fight UNITA.

Although the MPLA was able to gain political control at independence, it spent the next 16 years trying to deal with the threat from UNITA and South Africa. Angola spent more than half of its export earnings on defense and twice as much on the military as on health care and education combined. The MPLA government and UNITA agreed to a cease-fire in 1991 and set the stage for multiparty elections, which the MPLA won by a wide margin.

Angola's prolonged civil war ended formally with the signing of the Luena Accords in 2002. With the UNITA threat diffused, the MPLA government began to concentrate efforts on peacebuilding and economic reform initiatives. Angola has emerged as one of sub-Saharan Africa's richest countries, at $5,460 ranking in the top 10 in GNI per capita, second in oil production, and third (seventh in the world) in diamond production. It is also well-endowed in phosphates, iron ore, copper, manganese, and uranium. These mineral resources are supplemented by a diverse agricultural sector based on coffee (the main cash crop), cotton, sisal, and oil palm. An extensive network of national parks and game reserves supports a fairly vibrant tourist industry. Angola must still contend with high levels of income inequality, low human development, and lingering problems associated with land mines from civil wars. But the government can mobilize revenues from its massive oil and diamond reserves, in conjunction with humanitarian aid, to invest in public works programs, build human capacity, and develop institutional capacity to deliver social services and remove land mines.

Mozambique's experience in many ways parallels Angola's. The Socialist FRELIMO (Front for the Liberation of Mozambique) challenged Portuguese colonial authority and assumed political power after independence in 1975. It was no coincidence that both the MPLA and FRELIMO started off along socialist lines. Socialism was largely a reaction against the mercantilist policies of the Portuguese, and both movements had organized masses of people to fight for basic human rights. FRELIMO's key opposition was RENAMO (Mozambican National Resistance), a movement supported earlier by Rhodesia (now Zimbabwe) and later by South Africa. After a series of peace initiatives, a General Peace Agreement was signed in 1992, formally ending 16 years of intense warfare and initiating a cease-fire. Under terms of the agreement, more than $390 million of humanitarian and electoral assistance were pledged. A United Nations Verification Mission was established to ensure implementation of the pact, and in October 1995, the country's first-ever free elections were held. FRELIMO won the presidential vote and the larger share of seats in the National Assembly.

After years of conflict, Mozambique has concentrated its efforts on restoring a deteriorated infrastructure, increasing investments in education and health, encouraging privatization, and implementing broad structural reforms in banking, finance, and commerce. The country maintains one of the highest economic growth rates in sub-Saharan Africa (8% in 2010) which is supported by increased coal production and exports. Tax incentives promote mining of bauxite, graphite, gemstones, gold, and marble. An expanding network of Chinese, Indian, and Brazilian investors helps underwrite resource development projects and infrastructure improvements. Mozambique is one of the world's largest producers of cashew nuts; cotton and fish are also major exports.

Namibia: Dilemmas of a Young Independent State

Namibia, formerly known as South West Africa (SWA), finally gained its independence in March 1990 after 75 years of South African domination. South Africa left a legacy of racial and socioeconomic segregation, divided ethnic loyalties, a limited manufacturing base, high debt burdens, high unemployment, and a highly extractive and poorly integrated economy. The country continues to cope with the challenges of economic reform and social and political reconciliation.

Namibia is a sizable country, with an area of 318,291 square miles (824,374 square kilometers), but a small, multiracial society of only 2.4 million people. Most of the population is concentrated in the central plateau region, which rises east of the Namib Desert coastal plain. The northern plateau is primarily home to the largest ethnic group, the Ovambo (50% of Namibia's population), and the Kavangos and Caprivians. Most of the Europeans, who account for about 6 percent of the population, live in the central and southern highlands along with other ethnic groups such as the Hereros, Namas (Hottentots), San (Bushmen), and coloreds (people of mixed racial heritage). The European population consists mainly of Afrikaners, British, and Germans. The German influence began as early as 1884 when SWA became a colony. After Germany's defeat in World War I, Namibia became a League of Nations trust territory, which was administered by South Africa from 1921 to 1945.

South Africa never made a sincere commitment to safeguarding the well-being and interests of South West Africans as prescribed by the League of Nations and later by the United Nations. South Africa was more interested in imposing its system of apartheid, consolidating its power base, controlling strategic resources, and annexing SWA territory. Namibia has large uranium and diamond reserves and a deep-water port at Walvis Bay. It also functioned in the postcolonial period as a psychological buffer between Cuban-influenced Angola and South Africa. In 1946 South Africa attempted to incorporate SWA as its fifth province, a request that was denied by the United Nations.

To counter South African aggression, the Ovambo founded a nationalist movement in the late 1950s. The movement, initially called Ovamboland People's Congress, was renamed the South West Africa People's Organization (SWAPO) in 1960 to broaden its political base. In 1968 the United Nations Security Council declared South Africa's presence in SWA illegal, and in 1974 it recognized SWAPO as the authentic representative of the Namibian people. With the end of the Cold War and the withdrawal of Cuban troops from Angola, South Africa eased its grip and agreed to end its administration of Namibia. Namibia held elections in 1989 and SWAPO won the majority of votes.

Contemporary Namibia faces many challenges. The constitution ensures basic freedoms and human rights, but the government derives most of its support from the Ovambo. The government must broaden its political base to include other ethnic minorities in the interest of political stability. Namibia's GNI PPP per capita of $6,420 is one of the highest in sub-Saharan Africa, but extreme racial inequalities exist. Namibia has a strong mining sector anchored by diamonds, uranium, zinc, and copper. It is one of the world's leading producers of gem-quality diamonds. More than 90 percent of Namibia's diamonds are produced by NAMDEB, a 50–50 joint venture between the government of Namibia and DeBeers, the mining

▲ **Figure 8-46 Walvis Bay, Namibia.** Walvis Bay is one of the leading ports of Southern Africa, a potential gateway to the global economy for much of interior Southern Africa.

and distribution cartel that has dominated the industry for more than a century. In addition to the mining sector, which accounts for more than 80 percent of export earnings, Namibia potentially has the richest fishing zone in Africa South of the Sahara, owing to the nutrient-rich waters of the cool Benguela Current. In 1994, Namibia ended the last vestige of South African domination—Walvis Bay—by incorporating the area into its territory. Originally a British exclave in German SWA, and then a separate South African-administered area, this prized possession represents a strategic gateway linking Southern and Central Africa with Latin America. The port is Namibia's only deepwater harbor and has been declared a free trade zone to attract foreign investment and stimulate development in the region (Figure 8-46).

Namibia's natural resource endowments and relatively stable political environment bode well for its economic future. Growing investments from India in the country's health and educational sectors and China in the social service and construction sectors, as well as Russia in uranium and natural gas, have helped spur economic growth. But the country needs to overcome social and economic problems emanating from high unemployment rates and severe income and spatial inequalities. Development policies that promote education and training and expand trade are on the right track toward promoting economic growth and social mobility.

Botswana: Model of Economic and Political Stability

Botswana serves as a model of economic and political stability in Africa. At $13,700 the country has the highest GNI PPP per capita in Southern Africa, one of the highest economic growth rates, one of the best business climates, and is the least corrupt country based on Transparency International's corruption perception index. It is now categorized by the World Bank as an "upper middle income" country, a far cry from its status as one of the poorest countries in

▲ **Figure 8-47** **Diamond mine in Botswana.** When it gained independence from Britain in 1966, Botswana was one of the world's poorest countries. A year later, industrial diamonds were discovered. Diamonds earn more than 60 percent of export revenues, and Botswana has one of Africa's fastest-growing economies.

diversify its economic base by extending its infrastructure, upgrading its manufacturing base, developing its vast coal reserves, facilitating private sector development, and promoting ecotourism in the Okavango delta basin area.

Stop & Think

➤ Describe the special problems faced by landlocked states and the strategies they have employed to cope with their location.

Zimbabwe: Land Reform and Economic Transformation

Zimbabwe derives its name from the Shona words *dzimba dza mabwe*, meaning "houses of stone." It is linked to the famous stone ruins of Great Zimbabwe, which date back to the thirteenth century (see Figure 8-15). Artifacts discovered from this great civilization reveal a well-organized social and political structure and an elaborate trade network based on gold, ivory, beads, ceramics, and copper objects.

Modern Zimbabwe, formerly known as Southern Rhodesia, gained its independence in 1980. Southern Rhodesia had opted for internal self-government rather than succumb to Afrikaner domination under a union with South Africa. The British, who administered the settler colony, had excluded Africans from trade and commerce and maintained a cheap and compliant labor force. The ethnic makeup of this country of 12.6 million people is quite diverse. The largest African ethnic group is the Shona, followed by the Ndebele, Tonga, Venda, and Sotho. At independence, extreme inequalities existed in land ownership and other means of production; Europeans controlled more than 75 percent of the most productive land, while most African subsistence farmers were packed into densely settled marginal lands known as communal lands (formerly referred to as Tribal Trust Lands). Since the early 1980s, the Zimbabwe government has tried in several phases to resettle African families through a number of programs, including the Land Reform and Resettlement Program (LRRP). One of the objectives of these programs was to acquire land from large-scale commercial farmers for distribution to landless people, overcrowded families, agricultural college graduates, and poor people with some farming skills. Other goals included diverting populations away from highly concentrated communal areas, enhancing agricultural productivity, reducing rural poverty, and sustaining economic development initiatives. Lands identified for acquisition included underutilized, derelict, and foreign-owned farms, and properties in close proximity to communal areas. Farms exempted from the program included those belonging to churches or missions, agro-industrial properties involved in meat and dairy production, properties within Export Processing Zones, and plantation farms engaged in large-scale production.

The land reform program had its share of problems and fell consistently below its land acquisition and resettlement targets. A report by the UNDP revealed that by 1997 the government had purchased only 42 percent of its intended target of 20.5 million acres (8.3 million hectares) of land and settled 44 percent of the 162,000 communal families it had expected to resettle. Many of the resettled

the world at independence in 1966. This impressive economic performance was spurred early on by the beef cattle industry and later by developments in the diamond and copper-nickel industries that account collectively for 85 percent of the country's export earnings (Figure 8-47). Botswana's record on human development and gender equality has been equally impressive. Its HDI value of 0.633 is among the highest in sub-Saharan Africa. Its Gender Inequality Index of 0.507 (among the best in Africa) is indicative of advancements that women have made with respect to health, education, income, and legal rights. It further demonstrates the government's commitment to universal education and the extension of social services and health clinics to rural areas.

Botswana is a multiparty democracy that has held free and fair elections since its independence in 1966. The political situation has been relatively stable and avenues exist for the expression of political ideas. Legislative power resides in a 40-member parliament whose members are elected to five-year terms. A 15-member House of Chiefs provides advice on ethnic and constitutional matters. The latter body is especially relevant in view of the dominance of the Tswana (75% of the population) over other ethnic groups such as the Shona, San, and Ndebele. "Freedom squares" are available as forums for the public to discuss and debate political issues. These public spaces reflect a long African tribal tradition of open discussion of public issues in group meetings. Civil liberties inherent in the political system and the inclusion of traditional authorities in the decision-making process offer lessons for more autocratic governments in Africa.

Botswana is not without its problems. Spatial and income inequalities remain, with most of its 1.9 million people concentrated in the southeastern part of the country. Drought and water shortages, accompanied by food insecurity, are common in the western and southern sections. But to the government's credit, emergency relief efforts are timely and well coordinated. The limited population (and domestic market) is a constraint, especially with much skilled manpower migrating to South Africa. This limitation has negatively affected Botswana's manufacturing sector, which accounts for only 4 percent of the Gross National Income. The government continues to

▲ **Figure 8-48 Industrial plant in Zimbabwe.** Zimbabwe's once thriving manufacturing sector continues to be hampered by political and economic uncertainty.

families are located in marginal areas not suitable for cultivation, lack basic infrastructure, and are distant from schools and farms. In 1992 the government renewed its efforts to redistribute land acquired from commercial farmers by passing the Land Acquisition Act. By 1999 much of the prime farmland remained in the hands of commercial farmers or had been acquired by senior government officials and elites, with little or no land trickling down to the intended beneficiaries.

In response, the government introduced the Fast Track Program (FTP) in 2000. This was designed to accelerate the land reform program. The FTP further polarized both the Commercial Farmers Union, which insisted on protecting its members' property rights, and the veterans of the liberation movement, who advocated access to even more land. The most controversial aspect of the FTP was the government's attempt to revise the Land Acquisition Act and extend its power to forcibly acquire land. At independence, the government had agreed to adhere to the Lancaster House Agreement, an agreement negotiated with the British government that protected white Zimbabweans from forced land acquisitions for a period of 10 years. After 1990, the government moved aggressively to accelerate the land acquisition process by forcibly acquiring land without just compensation. This resulted in many court litigations and created a rash of illegal land occupations and violent confrontations. The issue of land reform and resettlement was a focal point of intense debate in the hotly contested 2008 election, which ended with incumbent president Robert Mugabe winning amidst much controversy. The land crisis negatively impacted tobacco production, a major source of government revenue, as well as cotton, sugar, livestock, and maize production.

While the majority of African countries support democratic reform, Zimbabwe's 2013 election serves as a grim reminder of the country's history of an uncompromising, nonconformist political dictatorship. Mugabe and his Zimbabwe African National Union-Patriotic Front (ZANU-PF) zealously protected their despotic autocracy by rigging the election, intimidating opposition supporters, and disrupting efforts by opposition leader Morgan Tsvangirai and his Movement for Democratic Change (MDC) to engage in a free and fair electoral process. The election results have been certified by the Zimbabwe courts and recognized with some trepidation by the African Union and by the Southern African Development Community. The United States has expressed concerns about the fraudulent nature of the elections, acknowledging irregularities in the provision and composition of voter rolls.

While Zimbabwe remains fragile in the face of high unemployment and inflation, there have been encouraging signs in the economy. Tobacco production is increasing and sugar, tea, and coffee are contributing to an agricultural recovery. While the usually resilient manufacturing sector suffered from economic mismanagement and low investments, there has been some resurgence in mining activity associated with gold, platinum, and diamonds (Figure 8-48). Much uncertainty remains about Zimbabwe's future in the face of high levels of corruption and questionable government regulations. Economic success is undermined by periodic droughts, a polarized society, and a government that is currently disengaged from the global community.

South Africa: Postapartheid Challenges

In 1994, South Africa emerged from the vestiges of its ill-conceived apartheid philosophy. Efforts to overcome the legacies of racial and economic segregation remain quite daunting. But South Africa continues to press on with reforms aimed at overcoming spatial and economic inequalities, adjudicating land disputes, and empowering black men and women.

Geostrategic Location

South Africa is a pivot area of the Southern Hemisphere relative to three major zones of peace: the Indian Ocean, which was declared a "zone of peace" by the United Nations in 1971; the Antarctic region, which under the provisions of a 1959 treaty is restricted to research and scientific activities; and Latin America, which was declared a Nuclear Weapon Free Zone under the terms of the Treaty of Tlatelolco, signed in Mexico City in 1967. South Africa also is located on a major alternate route for international shipments of petroleum and minerals. When the Suez Canal was closed briefly after the Mideast War of 1967, the ocean route around South Africa became the principal route for oil shipments from the Persian Gulf to European and American markets. Although the canal has long since reopened to smaller tankers, South African ports remain significant support stops for supertankers as well as destinations for petroleum shipments to Southern Africa.

South Africa's mineral wealth and economic importance make it a major power, not only in Southern Africa, but also widely in Africa (Table 8-2). South Africa makes up only 6 percent of sub-Saharan Africa's total population and 5 percent of its surface area but accounts for 30 percent of the region's gross national income, 40 percent of industrial production, 80 percent of crude steel production, and 58 percent of installed electricity capacity. South Africa also has the

Table 8-2 Mining in South Africa

Major Mineral Reserves		
	World	
Mineral	**Rank**	**%**
Platinum Group Metals (PGMs)	1	88
Manganese	1	80
Chromium	1	72
Gold	1	30
Alumino-Silicates	1	38
Others: Coal, Diamonds, Vermiculite, Nickel, Antimony, Zirconium		

Major Mineral Production		
	World	
Mineral	**Rank**	**%**
Platinum Group Metals (PGMs)	1	58
Vermiculite	1	38
Chromium	1	36
Vanadium	1	34
Titanium minerals	2	22
Others: Diamonds, Gold, Coal, Zirconium		

Sources: National Environmental Research Council, World Mineral Production, 2012; British Geological Survey, Nottingham, UK; Republic of South Africa, *South Africa Yearbook: Mining and Minerals*, 2011.

densest road, rail, and air networks in sub-Saharan Africa. Its rail and harbor network is the only reliable trade link with the outside world for the landlocked countries of Botswana, Lesotho, Swaziland, Zimbabwe, Zambia, and Malawi, as well as for much of the DRC. Each year, several million migrant workers are employed in South Africa's mining, industrial, and agricultural sectors.

During most of the Cold War era, South Africa used its economic, strategic, and political leverage to sustain its policy of racial segregation, called **apartheid**. At the time, South Africa was seen as a "bastion" against the spread of communism. With the end of the Cold War, South Africa's political leverage faded, setting the stage for social, economic, and political reform in the country. With one-man, one-vote black rule firmly established in South Africa, and with relative stability and political reform in neighboring countries gaining ground, South Africa now has the opportunity to play a positive and enabling role in the social and economic development of sub-Saharan Africa.

Historical Background

Of South Africa's 51.1 million inhabitants, 79 percent are African, 9 percent European, 9 percent "colored" or of mixed race, and 3 percent Indian or Asian. The African majority is composed of numerous ethnic groups; the largest are the Zulu, Xhosa, Tswana, Bapedi, and Basotho. The European population is about 60 percent Afrikaner and 40 percent British. The Afrikaners are the oldest European community in South Africa. They are descended from Dutch farming people (Boers) who settled in Cape Town after 1652,

when a fort and vegetable garden were established for the Dutch East India Company by Jan van Riebeeck. The location also served as a way station for ships engaged in trade with Asia. The Dutch farmers proceeded to suppress and alienate the indigenous Khoisan pastoralists in order to acquire tracts of land. Slaves were imported from the East Indies (Indonesia and Malaysia) to work on plantations. In 1806 the British took over the Cape Colony and ended slavery. Some Afrikaners stayed while others went on the Great Trek to the high veld (grassland) region of the northeast interior. The Afrikaners encountered resistance from the Ndebele and Zulu during their trek. After initial losses, they eventually defeated the Zulu in the Battle of Blood River in 1838. The Afrikaners regarded the Great Trek as a predestined journey to "the promised land."

Gold and diamond discoveries in the high veld attracted the British to the region. The Orange Free State and Transvaal territories were annexed, and the British suppressed the Dutch Boers in the Anglo-Boer War from 1899 to 1902. In 1910 the Union of South Africa was established, legitimizing dual English and Afrikaner domination. In 1912, African elites countered by forming the African Native National Congress (ANNC). In the following year, Africans were relegated to reserves located in the least desirable areas, amidst protests from the ANNC. The 1913 Native Land Act restricted Africans to only 10 percent of the country's land area and denied them the right of land ownership outside the reserves. This act set the stage for South Africa's official apartheid policy, which was later implemented under the Afrikaner government.

Legitimization of Apartheid: 1948–1989

In 1948 the Nationalist Party prevailed over the South African Unified Party—a union of British and Afrikaners—signaling the formal beginning of apartheid in South Africa. Apartheid is an Afrikaans word meaning "apartness" or "separateness." The Afrikaner government policy was a simple strategy of divide, conquer, and subdue. The architects of apartheid were the Broederbond (union of brothers), a secret, extremist nationalist society whose main agenda was to preserve the identity of Afrikaners and promote nationalist indoctrination through churches, schools, colleges, and the press. The Afrikaner government created a hierarchy of rigid laws designed to keep races effectively apart. Apartheid was legislated at three levels: personal, urban, and national.

On the personal level (petty apartheid), legislative acts were instituted to discourage races from using the same facilities and to prevent mixed marriages and social interaction. In urban areas, buffer zones such as railroads, waterways, highways, and cemeteries effectively separated destitute black townships (Figure 8-49) from affluent white suburbs.

Superimposed on these was the policy of grand apartheid. Its ultimate purpose was to create independent black homelands on the basis of ethnicity (Figure 8-50). These "homelands" or "bantustans" never stood a chance of gaining even the semblance of statehood, as they lacked the necessary economic and human resources required to become economically and politically self-sufficient, and were never recognized by the international community. The strategy was designed to diffuse any potential threats from the African majority by keeping it divided, fragmented, and disunited. In many respects, the tensions that exist today between Zulu and Xhosa can be attributed to this strategy of divide and

▲ Figure 8-49 **Soweto Township, Johannesburg, South Africa.** The black township of Soweto, located in suburban Johannesburg, is a vivid reminder of racial segregation and income inequality associated with apartheid South Africa.

conquer. About 75 percent of Africans were relegated to small portions of inhospitable lands (13–14% of all land) without any significant mineral reserves and without any potential for cultivation. The boundaries of these fragmented and landlocked territories were drawn in such a way that they conveniently avoided the major core regions in South Africa such as the Witwatersrand gold mining district near Johannesburg, or the coastal ports of Durban, Port Elizabeth, and Cape Town. Severe backwash effects were created as able-bodied males frequently left the marginal homelands to work in these core economic centers (Figure 8-51), leaving the homelands with meager revenues to construct an appropriate infrastructure. The minority nationalist government was able to maintain its dominance by stripping the opposition of political power; restricting African access to education, jobs, and the military; imposing harsh penalties for violations such as the failure to carry a passbook; strictly enforcing the apartheid laws; and using intimidation.

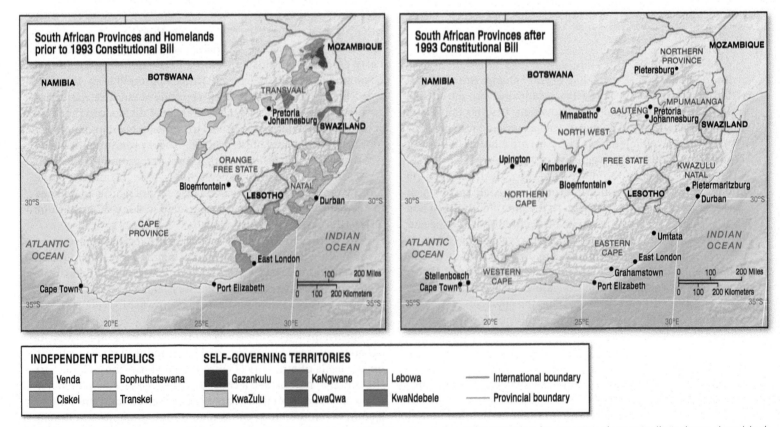

▲ Figure 8-50 **Restructuring of provinces in South Africa.** Apartheid promoted racial separation by creating theoretically independent black homelands based on tribal and ethnic affiliation. After the collapse of the apartheid regime, a new administrative structure of nine provinces expanded political representation and allowed for greater regional autonomy.

▲ Figure 8-51 **Johannesburg, on the South African Veldt (plateau).** Johannesburg, known as the "city of gold," is the economic and industrial hub of South Africa.

The Transition Period: 1989–1993

By 1989, external and internal events prompted the South African government to engage in economic and political reform. The Cold War was ending, and Cuban forces had withdrawn from Angola, assuring Namibian independence in 1990. The frontline states of Angola and Mozambique were beginning to seek peace agreements, and the international community stepped up efforts to disinvest and disengage from South Africa's economy. Internally, the South African economy was hurt by sanctions and years of massive expenditures on internal and regional security. Africans had organized more cohesive political strategies to press for change and reform. In 1990, President Frederik W. De Klerk released longtime black militant Nelson Mandela from prison and lifted the ban on political organizations, including the **African National Congress (ANC,** successor to the ANNC), the Pan African Congress, and the South African Communist Party. By mid-1991 most of the legislative instruments of apartheid had been dismantled. In late 1991, the Convention for a Democratic South Africa (CODESA) brought several political factions together to discuss procedures for a new constitution. An interim constitution was drafted and approved in late 1993, to ensure civil, social, and political liberties for all South African citizens. The constitution embraces a Bill of Rights promising equality regardless of race, gender, sexual orientation, physical disability, and age. This Bill of Rights further guarantees free speech, movement, and residence; freedom of religion, politics, and conscience; freedom to use any of 11 languages; a fair trial with no torture or forced labor; economic, social, and cultural rights; and the right to work and to education.

Postapartheid South Africa

The transition period was characterized by intense negotiations and efforts by several political factions to steer the country toward peace and stability. With an interim constitution in hand and a Bill of Rights drafted, South Africa was ready for its first-ever free elections in April 1994. Nelson Mandela and his ANC were handed a mandate to lead South Africa by winning more than 12 million votes (62.6% of the total). The provisions for forming a government of national unity entitled any party winning more than 20 percent of the vote to a deputy presidential position and any party winning 5 percent or more of the vote to a national cabinet position. As a result, National Party leader De Klerk and Thabo Mbeki of the ANC became deputy presidents, while Mangosuthu Buthelezi, leader of the Inkatha Freedom Party, was appointed minister of internal affairs. In 1993 the number of provinces was increased from four to nine to allow for expanded political representation and to accommodate regional autonomy (see Figure 8-50). Under the new system, 10 senators were elected from each province. The boundaries were realigned in such a way that each new province had at least one major metropolitan area to encourage functional relations with their hinterlands. Other criteria considered were the development potential of each region, the availability of an adequate infrastructure and resource base, and the distribution of people and physical landscapes.

A new constitution followed in 1996 based on the principles of a liberal democracy, providing for majority rule in a multiparty political system, the protection of civil liberties, an independent judiciary, and a free press. President Mandela maintained social and economic stability in South Africa and successfully guided the country to its second democratic elections in June 1999. He also secured the continued dominance of the ANC in South African politics with the election of Thabo Mbeki to the presidency in 2004, and Jacob Zuma in 2009. Numerous challenges face the South African government, which seeks to broaden participation in social, economic, and democratic reform.

To address the issues of nation building and integration, the government launched an ambitious **Reconstruction and Development Program (RDP)** in 1994. The program was particularly aimed at meeting basic human needs such as water, electricity, health, education, transportation, and telecommunications to all people; and promoting the democratization of South African society. Some accomplishments of the program include extending electricity to about 1.7 million rural and urban households; providing more than 5 million people with access to health-care facilities; increasing access for more than 1 million rural households to safe drinking water; and granting affordable housing to 6 million households (about half of the intended target). Critics of the program argue that the government did not meet its intended targets, and provided low-quality services and facilities to its intended beneficiaries.

The Government has since launched an equally ambitious Accelerated and Shared Growth Initiative for South Africa (Asgi-SA) program, which is aligned with the MDGs and is aimed at reducing unemployment to 15 percent and halving poverty by 2014. Asgi-SA has been painstakingly slow to implement and is widely criticized for its lack of accountability and transparency. South Africa's unemployment rate remains at 25 percent, and 23 percent of its population still live in poverty. Complementing the Asgi-SA are the policies of **Black Economic Empowerment (BEE)** and the Spatial Development Initiative (SDI), both designed to address social and spatial imbalances. Black Economic Empowerment is a moral and economic imperative directed at achieving human resource development and employment equity for blacks through ownership and control of enterprises and assets, access to senior management positions, preferential procurement from black companies, and corporate social investment. But so far little has been achieved with BEE. Blacks occupy just 12 percent of senior positions in private business compared to 75 percent of whites. The situation is much worse for black chief executive officers (4%) and chief financial officers (2%); BEE has created a small black elite, but does not have the intended broad-based impact.

The SDI is a geographically focused industrial policy intended to restructure the space economy of South Africa through targeted, strategic investments in areas endowed with significant high potential. The 11 SDIs and 4 Industrial Development Zones (IDZs) chosen exhibit potential in the industrial, agricultural, technology, and tourism sectors (Figure 8-52). The SDI strategy focuses on developing a first-rate infrastructure, implementing marketing and investment strategies, building human and resource capacity, promoting public-private partnerships, stimulating growth in surrounding regions, and strengthening links to global markets. Implementing such programs is a major challenge in view of the extraordinary financial and international resources required. The intended outcomes are hard to realize. South Africa still has one of the widest gaps of income inequality in the world (see *Geography in Action: Seeking Social Justice in Contemporary South Africa*). Millions live in poverty, including millions of black women who continue to work in the informal sector. In November 2010, the government unveiled a New Growth Plan with a goal of creating 5.5 million jobs by eliminating bureaucratic red tape and prompting skill development in key "job-drivers" focused on a green economy, agroprocessing, manufacturing, and tourism.

Another major challenge confronting the South African government is the land reform required to address the injustices of forced removals and years of denying Africans access to suitable land. The three major components of the government's land reform program are land restitution, land redistribution, and land tenure reform. For land restitution, the greatest concern is linked to the magnitude of forced removals and the administrative and financial implications. In 1994 the **Restitution of Land Rights Act** was enacted to permit the government to investigate land claims and restore ownership to those who unjustly lost land. Restitution is justified if the claimant was dispossessed of land after June 1913. A Land Claims Court was established to adjudicate disputes. Aside

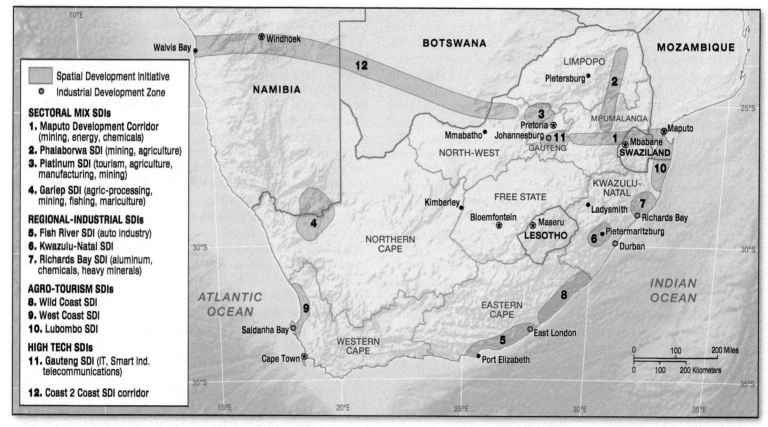

▲ Figure 8-52 **South Africa's Spatial Development Initiative (SDI) areas.** SDIs were designed to boost development in surrounding regions and strengthen links to global markets.

Seeking Social Justice in Contemporary South Africa

The collapse of formal apartheid in the early 1990s and the efforts to achieve national reconciliation and restitution for past injustices encouraged hope for a better future for all South Africans. No one expected change to be instant or fast, but everyone envisaged a future in which basic necessities of life—clean water, electricity, decent housing, work, access to education, and an income adequate for fundamental needs—would become available to all.

This promise has been difficult to achieve. The ANC-dominated government has developed plans designed to move society and economy toward greater and more complete social justice. But for many reasons these plans have been only partially realized. Overall economic growth has been slow, investment in mining and industry is sluggish, wages have languished, unemployment rates are high, and discontent is widespread. Some of this tension burst out among mine workers in 2012 in a series of strikes, some violent, for better wages. Few strikes have resulted in substantial pay raises and benefits, so further miner unrest is likely and symptomatic of a deeper problem. A growing gulf exists between those left behind in the postapartheid period and those benefiting from the new system and access to power sources within it, particularly the new political and business elite linked to the ANC and a small but growing black middle class.

While labor relations in many of South Africa's mines may continue to deteriorate and produce conditions that are difficult for management to control, this situation is not an isolated occurrence. Deeply rooted problems also affect the agricultural sector. One notable example is the plight of agricultural workers in the wine-producing districts of the Western Cape (Figure 8-6-1). Agricultural workers are among the lowest paid workers in South Africa, but the contrast between the amount of labor needed to produce a bottle of wine and the value of the final product makes the low wages unsettling. Feeling pressure to cut costs, estate owners also are increasingly mechanizing their operations. The contrast between white estate owners and black and brown laborers, many the descendants of former slaves, magnifies the problem. The physical beauty of the natural landscape, centuries-old estate buildings, and quaint towns attract tourists and Cape Town commuters. Their presence projects the image of an affluent upper class shadowed by poverty and unemployment. Strike threats to increase wages and improve working conditions could, if carried out, trigger a prolonged period of unrest and serious economic problems for both labor and management.

But there are positive developments. Some wineries do manage their labor relations well, incorporating workers into decision making and shareholding, and gaining ethical trade certificates along the way that help with sales abroad. There is also a growing South African middle class, and as younger people benefit from better educational and skill-acquiring opportunities this middle class will expand. If general economic conditions improve, this growth is likely to be faster. Development opportunities for women are difficult to find, but selling cosmetics for Avon is one example of a small-scale business opportunity that has opened up for some women. For the successful, it can mean income levels nearly equivalent to those of an average black male, which translates into female empowerment. Whether relatively small-scale, local initiatives, in conjunction with larger-scale governmental policy incentives and infrastructure projects, can improve social and economic conditions for the majority of the population will determine the future course of South Africa's democratic institutions and their ability to deliver on their promise.

Sources: "It's not Just the Mines," *The Economist*, September 8, 2012, 45–46; "In the Pits," *The Economist*, August 25, 2012, 38; "Wrong Vintage," *The Economist*, January 26, 2013, 45; "Cosmetic Difference," *The Economist*, August 18, 2012, 58.

▲ FIGURE 8-6-1 **Wine estate in South Africa.** This vineyard in the Western Cape Province is one of several in South Africa that produce over 250 million gallons of wine per year, placing the country consistently among the top 10 wine producers in the world.

from land restoration, other forms of restitution include providing alternative land, payment of compensation, and priority access to government housing and land development programs. With respect to land redistribution, the main concerns are how to respond to people's needs in a fair, equitable, and affordable manner; how to address urgent cases of landlessness and homelessness; and how to provide credit, grants, and services for land acquisition and settlement. Although the government has already spent over $4 billion in land purchases, it is still far behind its target of buying 30 percent of land from commercial farmers to give back to blacks. The majority of redistributed lands are operating below productive capacity due to corruption, ineffective bureaucracies, and limited resources.

The government has also initiated a number of reforms in rural and urban areas. In the rural areas, a two-tiered rural local government structure consisting of district and local councils was established to promote local economic development initiatives and ensure environmental and social sustainability. The government has stepped up training and capacity-building efforts to maximize opportunities for development. The Development Facilitation Act has been enacted to address reform in urban and regional planning. The act seeks to expedite land development projects, promote efficient and integrated land development, and manage the rate, quality, and character of urban growth. Another strategy of urban development is the Masakahane Campaign ("let's build each other"), designed to enhance community development initiatives, service delivery, and financial needs. This campaign also mobilizes state, private, and community resources to deliver basic housing and infrastructural services; improves the capacity of local, metropolitan, and district councils to deliver and administer services more effectively; and supports community initiatives to pay for services and promote economic development.

South Africa's government is working hard to mobilize existing resources to build a strong and sustainable economy that is beneficial to all citizens. However, questions surrounding the viability of the multiracial republic still remain. Although the number of provinces was increased from four to nine to decentralize government and diffuse possible regional tensions, there are still threats of secession from different groups. Extremist white nationalists are seeking a "volkstaadt," while Zulu hard-liners refuse to come to terms with a government of national unity. There are also rumblings from Tswanas who at one time wanted no part of a federation. Although these are difficult and complex challenges, the government continues to push its agenda for change and unity through peace initiatives and economic reforms. As South Africa confronts these issues internally, it also has to deal with the challenge of defining its role in Southern and sub-Saharan Africa.

Stop & Think

▶ What are the prospects for South Africa becoming a catalyst for socioeconomic development in other regions of sub-Saharan Africa?

Regional Cooperation in Southern Africa

As South Africa addresses its internal problems, it is eager to improve its regional and international image. Opportunities exist for South Africa to enhance trade, development, and cooperative relations with regional and continent-wide organizations. South Africa is also expected to become a major player in the African Union (AU), the African Development Bank (ADB), and the United Nations Economic Commission for Africa (ECA). On a regional scale, South Africa already has trade and monetary relations with the Southern African Customs Union (SACU) and the Common Monetary Area (CMA) and is actively engaged with the SADC, and the Common Market for Eastern and Southern Africa (COMESA).

The **Southern African Development Community (SADC)** was founded in 1980 to reduce southern African dependency on South Africa for rail, air, and port links, imports of manufactured goods, and electrical power. Unlike SACU and CMA, the emphasis was on economic cooperation instead of integration. Areas of cooperation include transportation and communications, food security, industrial development, energy conservation, mining development, and environmental management. The initial priority was transportation, with an emphasis on making Mozambique an alternative outlet for the subregion's landlocked states. SADC invested in improving road and rail links to Mozambique and upgrading the port facility at Beira. Each country is responsible for one sector. Angola is responsible for energy while Botswana coordinates agricultural research, livestock production, and animal disease control. Malawi coordinates inland fisheries, forestry, and wildlife; Namibia takes care of marine fisheries and resources; South Africa manages finance and investment; Tanzania handles industry and trade; and Zambia coordinates mining, employment, and labor. South Africa, Botswana, Zambia, and Zimbabwe dominate intra-SADC trade, accounting for more than 75 percent of exports and more than 65 percent of imports. Intra-SADC trade has intensified with merchandise exports increasing from US$4.5 billion in 2000 to $16 billion in 2009. Major exports to world markets include mineral and fuel oils, precious metals, and iron and steel. A joint agreement has been signed with Brazil to promote cooperation in information and communications technology, energy, and food security. China has established business interests in finance, banking, telecommunications, and agribusiness.

COMESA, originally founded in 1982, was transformed in 1992 into the Common Market for Eastern and Southern Africa. COMESA facilitates trade by granting preferential tariffs on selected goods, especially intermediate goods and those that promote local economic development efforts. Members seek to enhance cooperation in agriculture, industry, telecommunications, and monetary affairs. A major problem for COMESA is the varying levels of economic development among its member countries and the disparities in wealth and resources between the largest and smallest ones. Interaction and cooperation are hampered by weak transportation and communication links, fledgling economies, refugee problems, a lack of regional complementarity (duplication in production), administrative mismanagement, lack of information, and regulatory and procedural barriers.

Duplication and overlap between SADC and COMESA create problems. There is the potential for a merger that could eventually have pan-African implications. But considerable differences in levels of industrial and economic development, government policy, and political stability severely limit the prospects of such a merger.

For South Africa, the choice is whether to opt for SADC or COMESA. SADC seems a logical choice, since it is economically more viable and confined to Southern Africa. COMESA offers South Africa an opportunity to extend its influence beyond Southern Africa. Whatever the outcome, South Africa's integration into the Southern Africa region can only enhance regional development, which could have positive ripple effects in all of sub-Saharan Africa.

Summary

▶ Sub-Saharan Africa has an enduring legacy marked by ancient civilizations that were major centers of commerce, culture, learning, and technological innovation. This period of economic and political prosperity was interrupted by more than 500 years of colonial intrusions that disrupted cultural institutions, created extractive economies, and stifled meaningful social and political organizations. While contemporary sub-Saharan Africa may seem far removed from the vestiges of colonialism, it still faces important challenges in development policy and planning. Although the socioeconomic problems at times seem insurmountable, opportunities exist to tap into Africa's vast reservoir of human and natural resources and create policies that are not solely growth-inducing, but also pay more attention to human equity and development.

▶ An immediate challenge confronting sub-Saharan African countries is the severity of environmental and forest degradation caused by such human-related factors as slash-and-burn agriculture, logging, and fuelwood consumption. Sub-Saharan Africa must develop comprehensive sustainable energy policies that harness its enormous renewable energy resources such as hydro, geothermal, solar, and wind power.

▶ The regions of West, Central, and East Africa, like Southern Africa, exhibit considerable diversity in their environmental and ethnic characteristics, a diversity compounded by the unequal levels of development among the constituent countries. In terms of economic and human development, the Sahel region includes some of the poorest and least-developed countries in the world, countries that have long been subjected to periods of extreme drought and food insecurity. Countries such as Ghana and Gabon are making economic progress, although internal disparities among income groups and between rural and urban areas remain problematic. In the area of sociopolitical development, countries such as the DRC and Zimbabwe continue to be mired

in uncompromising dictatorships and ethnic turmoil. These countries must come to terms with an international order that is increasingly intolerant of human rights violations, socioeconomic injustice, and ethnic genocide. Countries like Ghana and Benin are riding the wave of multiparty politics. As more countries join the trend toward greater democracy and institute needed political changes, the prospect that sub-Saharan Africa will meaningfully participate in the economic and technological achievements of the global economy rises.

▶ The Southern Africa subregion is undergoing radical transformations in its social, political, and economic institutions. Angola and Mozambique are experiencing some semblance of peace and stability after years of civil and ethnic strife rooted in Portuguese exploitation. Namibia is moving away from years of domination by South Africa, and Zimbabwe is still seeking solutions to the sensitive issues of land and political reform. In the midst of these transformations, Botswana, buoyed by its diamond industry, has emerged as one of the fastest-growing economies in the world. South Africa remains the key player in the economic and political transformation of Southern Africa.

▶ Regional organizations such as ECOWAS, SADC, and COMESA already provide a mechanism for countries in the subregion to coordinate their socioeconomic policies and to pursue avenues for greater regional cooperation and stability. Sub-Saharan Africa has expanded its trade networks beyond these intraregional organizations to include China, India, and Brazil to enhance investments in telecommunications, manufacturing, and energy resources. International efforts to revitalize sub-Saharan Africa, such as the UN-sponsored MDGs, are welcome but must be well intended, address the extreme inequalities that exist between the core and the periphery, and embody strategies that provide the poor with opportunities to achieve their full human potential.

Key Terms

acculturation 361

Africa Growth and Opportunity Act (AGOA) 376

African National Congress (ANC) 399

agroforestry 386

apartheid 397

balkanization 384

Berlin Conference 359

Black Economic Empowerment (BEE) 400

City Prosperity Index (CPI) 369

clanocracy 388

colonialism 357

deforestation 355

demographic dividend 378

desertification 380

East African Rift Valley 345

Economic Community of West African States (ECOWAS) 385

ecotourism 354

Falashas 364

Great Lakes of East Africa 348

indirect rule 359

insurgent states 393

Intertropical Convergence Zone (ITCZ) 351

kinship relations 364

landlocked states 361

lingua franca 362

microfinancing 374

microstates 361

Millennium Development Goals (MDGs) 374

neocolonialism 361

New Partnership for Africa's Development (NEPAD) 376

Non-Governmental Organizations (NGOs) 376

pastoralism 367

Reconstruction and Development Program (RDP) 399

Restitution of Land Rights Act 400

Sahel 380

shatterbelt 393

shifting cultivation 367

Southern African Development Community (SADC) 402

Ujamaa 390

village banking 374

World Heritage Site 354

Understanding Development in Africa South of the Sahara

1. What are the attributes of the principal natural regions of sub-Saharan Africa?

2. Why are both the savanna and the tropical rain forest areas in sub-Saharan Africa potentially attractive to ecotourists?

3. What cultural and economic factors are most responsible for the continuing high fertility rates of sub-Saharan Africa?

4. Why was Cote d'Ivoire's economic success in the first two decades after independence not continued successfully thereafter?

5. If wildlife contributes significantly to economic development, why is the survival of so many species at risk in sub-Saharan Africa?

6. How successful have regional organizations such as ECOWAS and SADC been in achieving economic integration, mutual cooperation, and economic progress objectives?

7. Who are the Tutsi and the Hutu and how would you explain their historical, social, and economic relationship in the countries of Rwanda and Burundi?

8. What are some of the problems inherent in redistribution of land in Zimbabwe?

9. How does the history of instability and poverty in Angola and Mozambique stem from both Portuguese colonial practices and governmental policies enacted since independence?

10. In what ways has Botswana become a model for socioeconomic development for other countries of sub-Saharan Africa?

Geographers @ Work

1. Evaluate the impact of colonialism on the social, economic, and political underdevelopment of the countries of sub-Saharan Africa, and describe how geographers might help overcome the legacies of colonialism in the postcolonial era.

2. Describe the geographical expertise that might be drawn upon to address the problem of AIDS in the countries of sub-Saharan Africa?

3. Give examples of how the geographic concepts of spatial distributions and human-environmental interrelationships, as well as traditional sub-Saharan African values might contribute to appropriate rural development strategies for the peoples of sub-Saharan Africa.

4. Identify the regions that are most affected by desertification and drought in sub-Saharan Africa and demonstrate how geographers can contribute to improved responses by national governments and international agencies to these recurring environmental challenges.

5. Explain how the colonial boundaries imposed on sub-Saharan African states and the grouping together of distinct tribal and ethnic groups that have long distrusted one another affected the political and economic success of countries after they achieved independence.

MasteringGeography™

Looking for additional review and test prep materials? Visit the Study Area in MasteringGeography™ to enhance your geographic literacy, spatial reasoning skills, and understanding of this chapter's content by accessing a variety of resources, including MapMaster™ interactive maps, videos, RSS feeds, flashcards, web links, self-study quizzes, and an eText version of *World Regional Geography*.

▲ Figure 8-53   **Africa from Space.**

South Asia

Christopher A. Airriess

Energy and the Development Challenge

In July 2012 India experienced the largest power blackout in human history, affecting some 600 million people in 20 northern states that comprise the traditional economic core of the country. Trains ground to a halt, traffic signals went haywire, and hospitals lost essential power for a day or more. While the national power company restored 90 percent of the power grid within half a day in some areas, this event was a catastrophic culmination of years of daily power cuts and outages. India, the world's second-fastest growing economy and the sixth greatest consumer of electricity, cannot produce enough power. The blackout was the result of one or more states drawing excess electricity from the national power grid, leading to the collapse of the entire system.

India's blackout tells us much about larger development issues directly or indirectly associated with globalization. In addition to energy needs for industrialization, growing affluence of Indians benefiting from the country's engagement with the global economy means increased use of energy-intensive household appliances, especially air conditioners, which have become the new status symbol. But the main challenge is state-owned power firms that sell electricity at low, subsidized prices. Without assurances of acceptable profits by power providers, there is no incentive to invest in needed infrastructure to secure a dependable transmission network. The process of globalization is generally accompanied by the neoliberal philosophy of privatizing state-owned assets to create adequate financial incentives for private capital to thrive. The moral quandary in India, and other poor countries, is what happens to the hundreds of millions of poor people who cannot afford higher electricity costs resulting from privatization. It is difficult to envision India keeping pace with the other BRIC countries (Brazil, Russia, and China), defined as newly advanced economies, as long as a basic economic good such as electricity is in short supply.

▲ Girls reading by candlelight in Assam state during the massive 2012 power outages in India.

Read & Learn

▶ Describe the climate characteristics that explain the influence of wet and dry monsoons.

▶ Identify the region's primary environmental and energy challenges.

▶ Explain how historical movements of peoples and cultures resulted in South Asia's current cultural conflicts.

▶ Link British colonial rule to the current economic contours of South Asia.

▶ Identify social and economic forces that reinforce gender inequality in India.

▶ Describe how the process of globalization affects the lives of India's poor rural population.

▶ Compare India's urban-industrial regions and link their development to globalization and the government's economic policies since 1990.

▶ Contrast the political and economic constraints to greater levels of economic development in Pakistan, Nepal, Bangladesh, and Sri Lanka.

Contents

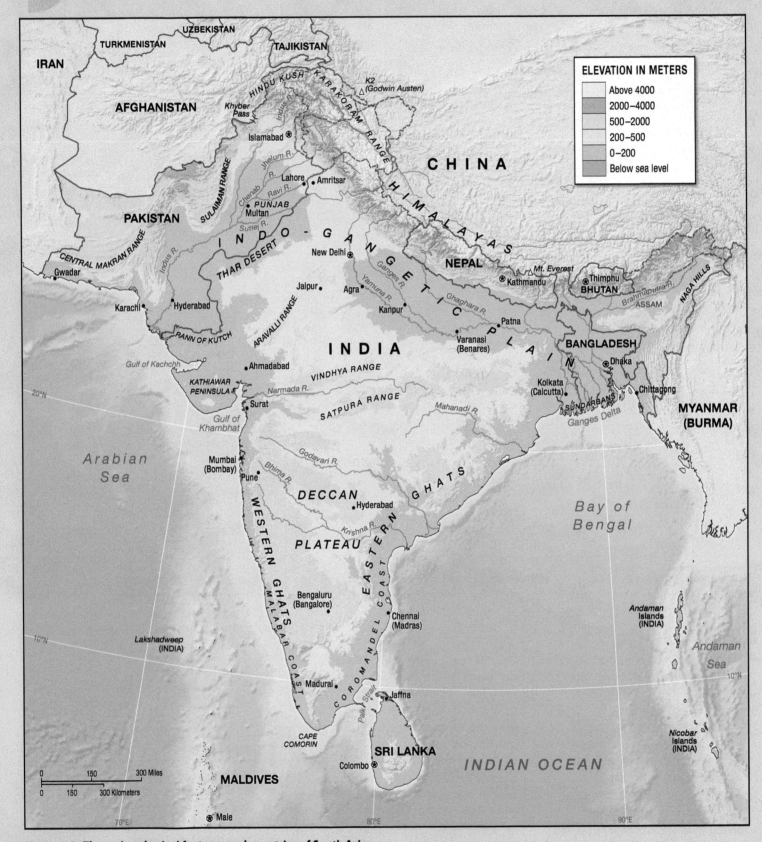

ELEVATION IN METERS

Above 4000
2000–4000
500–2000
200–500
0–200
Below sea level

TURKMENISTAN

UZBEKISTAN

TAJIKISTAN

IRAN

AFGHANISTAN

HINDU KUSH

KARAKORAM RANGE

K2
△ (Godwin Austen)

CHINA

Khyber
Pass

Islamabad ✹

Indus R.

Jhelum R.

Chenab R.

Lahore • Amritsar

PUNJAB
Multan

Ravi R.

HIMALAYAS

PAKISTAN

SULAIMAN RANGE

Sutlej R.

INDO-GANGETIC

THAR DESERT

New Delhi ✹

Ganges R.

NEPAL

Mt. Everest △

✹ Kathmandu

✹ Thimphu
BHUTAN

CENTRAL MAKRAN RANGE

Indus R.

Jaipur •

Agra •

Yamuna R.

Ghaghara R.

Brahmaputra R.

NAGA HILLS

ASSAM

Gwadar •

Karachi •

Hyderabad •

ARAVALLI RANGE

Kanpur •

PLAIN

Patna •

BANGLADESH

Varanasi
(Benares)

✹ Dhaka

RANN OF KUTCH

INDIA

Kolkata
(Calcutta)

Chittagong

Gulf of Kachchh

Ahmadabad •

VINDHYA RANGE

SUNDARBANS

20°N

KATHIAWAR
PENINSULA

Narmada R.

SATPURA RANGE

Mahanadi R.

Ganges Delta

MYANMAR
(BURMA)

Surat •

Gulf of
Khambhat

*Arabian
Sea*

Mumbai
(Bombay) •

Pune •

Bhima R.

Godavari R.

DECCAN

• Hyderabad

GHATS

*Bay of
Bengal*

WESTERN GHATS

PLATEAU

Krishna R.

EASTERN

*Andaman
Islands
(INDIA)*

10°N

*Lakshadweep
(INDIA)*

Bengaluru
(Bangalore) •

COROMANDEL COAST

• Chennai
(Madras)

*Andaman
Sea*

MALABAR COAST

Madurai •

Jaffna •

Palk Strait

10°N

CAPE
COMORIN

SRI LANKA

INDIAN OCEAN

Nicobar
Islands
(INDIA)

0   150   300 Miles

0   150   300 Kilometers

MALDIVES

Colombo ✹

✹ Male

70°E

80°E

90°E

Figure 9-0 **The major physical features and countries of South Asia.**

South Asia is one of three subregions of Asia, and is often called the "Indian subcontinent" because of the territorial dominance of its largest country, India (see chapter opener map). On India's periphery are the smaller countries of Pakistan, Nepal, Bhutan, and Bangladesh; the island country of Sri Lanka, and the even smaller archipelagic country of the Maldives. South Asia is the second poorest world region after sub-Saharan Africa. Although India has made substantial strides toward achieving greater levels of economic growth, South Asia as a whole shows little of the dynamism characteristic of its Asia–Pacific Rim neighbors. Several factors contribute to this state of relative poverty. South Asia (1.64 billion) has passed East Asia (1.58 billion) as the most populous world region. High population concentrations do not of themselves cause poverty or weak economic development, for people are both producers and consumers of resources. But when human productivity is low, high population densities frequently are associated with widespread poverty.

Because South Asia is primarily rural and the majority of the people are subsistence farmers, poverty is greatest in the countryside, despite the dramatic increases in agricultural output achieved through the Green Revolution (see Chapter 1). Another characteristic distinguishing South Asia from the rest of Asia is the sometimes divisive roles of ethnicity, religion, and politics in the economic development process. Cultural diversity is not necessarily a barrier to forging national unity, but when differences are exploited by competing political groups whose interests do not address the well-being of the larger community, development is slowed. Lastly, the South Asian tradition of state control of industry does not provide the required competitive environment and economic growth to absorb the huge pool of surplus labor, though India has made significant strides by privatizing industry and engaging the capitalist forces of globalization.

South Asia's Environmental and Historical Contexts

South Asia includes a wide range of both physical and human environments because of its substantial latitudinal and longitudinal reach (see chapter opener map). South Asia is also characterized by one of the world's oldest civilizations when Hinduism was first established, and through subsequent invasions Islam was introduced. Following a succession of politically and economically powerful empires, South Asia was transformed by British colonial rule. Post–World War II independence brought the breakup of South Asia into five different states primarily based on the distribution of Hindu and Muslim faiths, plus two others that were not colonial possessions.

Environmental Setting: Mountains and Monsoons

South Asia is structured by mountain ranges and major plateaus. More important for South Asia's people are the fertile lowland areas of the great river valleys and deltas. Of paramount importance is the seasonal monsoon rainfall system, which provides the water that supports rural South Asians' livelihoods.

Landforms

The world's most imposing boundary between world regions is South Asia's northern mountain rim, which encompasses the Karakoram in northern Pakistan and India and the even loftier Himalayas that separate lowland India from the Tibetan Plateau. This formed in ancient geologic times with the subduction of the northeast-migrating Indian plate under the Asian plate. The edge of the Asian plate became upthrusted and folded into an imposing mountain wall. Because of this active tectonic plate environment, the northern mountain rim has been subject to numerous earthquakes. The most recent occurred in 2005 in Kashmir where 80,000 people died and over 100,000 were left homeless (Figure 9-1).

The second large-scale physical region encompasses the alluvial plains of the Indus, Ganges, and Brahmaputra rivers, called the **Indo-Gangetic Plain**. The Indus and Ganges rivers are separated by the uplifted Thar Desert. About 300 miles (480 kilometers) wide, the Ganges is the largest of the three river plains. The nutrient-rich Indo-Gangetic alluvial soils support about half of South Asia's people. Much of the southern half of India consists of an elevated surface called the Peninsular Plateau, which is underlain by a core of ancient igneous and metamorphic rocks. Its most extensive portion, the **Deccan Plateau**, averages between 2,000 and 3,000 feet

▼ **Figure 9-1 Upper Indus River Valley in Kashmir, India, near Leh.** There exists a stark contrast between the Indian lowlands and the Himalayas. Perceived as religiously sacred, no other mountain chain in the world possesses as many peaks exceeding 20,000 feet (6,000 meters) as the Himalayas.

(600 to 900 meters) above sea level and possesses the largest share of India's minerals. The edges of the Deccan Plateau abut the Western and Eastern Ghats, meaning "steps." The Western Ghats are a steep mountain range with average elevations of 5,000 feet (1,500 meters), below which is a narrow, lushly vegetated coastal plain (Figure 9-2).

Climate

Monsoon comes from the Arab word *mausin,* meaning season. More specifically, monsoon refers to the prevailing winds that occur during certain times of the year, bringing pronounced wet and dry seasons. Human activity has long been influenced by these dramatic seasonal wind shifts, which are caused by the differential heating of ocean and land surfaces during the summer and winter, in combination with the position of the jet stream and the Inter Tropical Convergence Zone (ITCZ). These shifts in wind direction are a gigantic version of the more localized land and sea breezes that many coastal residents experience daily. During the northern hemisphere summer, a huge low-pressure cell develops over southwestern Asia in response to the gradual heating of the landmass (Figure 9-3). As the air over the landmass is heated, it

▲ **Figure 9-2 Western Ghats of India.** Southern India is dominated by the elevated Deccan Plateau. The Western Ghats are the mountainous edge, or escarpment, of the drier highland plateau.

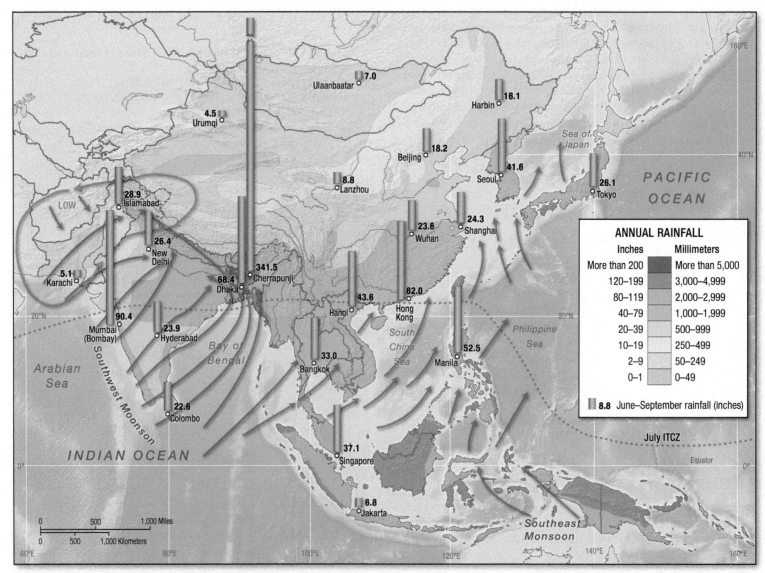

▲ **Figure 9-3 Annual rainfall and dominant atmospheric wind patterns over Asia during the summer.** Asia's summer monsoon rains are fed by southerly wind flows. The bars reflect average summer rainfall totals in selected locations.

rises, drawing in warm, moist air from the Indian and Pacific oceans to the south and east. April and May mark the beginning of the rainy season or wet monsoon. Sustained and heavy rainfall then occurs when the ITCZ, a belt of low pressure, migrates northward during the northern hemisphere summer months. During this four-month period substantial amounts of precipitation fill rivers and saturate dry soils. During the northern hemisphere winter the Asian landmass becomes much colder than the adjacent oceans, forming a strong high-pressure cell over east-central Siberia (Figure 9-4). A dry northerly or easterly wind then spreads over much of southeastern continental Asia, and greatly reduced rainfall characterizes the dry monsoon. This dramatic seasonal shift of wind and precipitation produces stark landscape changes in agricultural areas (Figure 9-5).

The monsoon expresses itself differently in each of the three subregions of Asia. In South Asia, the summer, or southwestern, monsoon is divided into two branches, one originating in the Arabian Sea and the other in the Bay of Bengal to the east. As winds from the Arabian Sea strike India's southwestern coast, they release between 60 to 100 inches (1,500 to 2,500 mm) of rain from May through September. The mountain wall of the Western Ghats

captures much of this moisture for the narrow coastal plain and thus reduces the rainfall in the near interior Deccan Plateau, which remains relatively dry and unsuited for intensive agriculture without irrigation. By June, winds from the Bay of Bengal sweep across the Bengal lowland toward the Himalayan Mountains, where tremendous tropical downpours drench the southern slopes as the warm, moist air masses rise and cool. Mountain-induced **orographic precipitation** (see Chapter 2) is greatest in the Cherrapunji region of the Assam Plateau in far eastern India, where total annual precipitation averages 420 inches (10,700 mm). The Himalayan Mountains then direct the wet air masses westward into the Ganges River Valley and eastward into the Brahmaputra River Valley. For many South Asian locations, more than 75 percent of annual precipitation occurs during those four wet season months. The arrival of the rains is welcomed—both for the life-giving moisture and for the associated cooling, which relieve the 90°F–100°F (32°–38°C) temperatures typical of much of interior India at the end of the dry season. Late in the monsoon season, **cyclones** (as Indian Ocean hurricanes are called) frequently devastate the lowland coastal areas of the Bay of Bengal. Heavy rain and occasionally storm surges are destructive

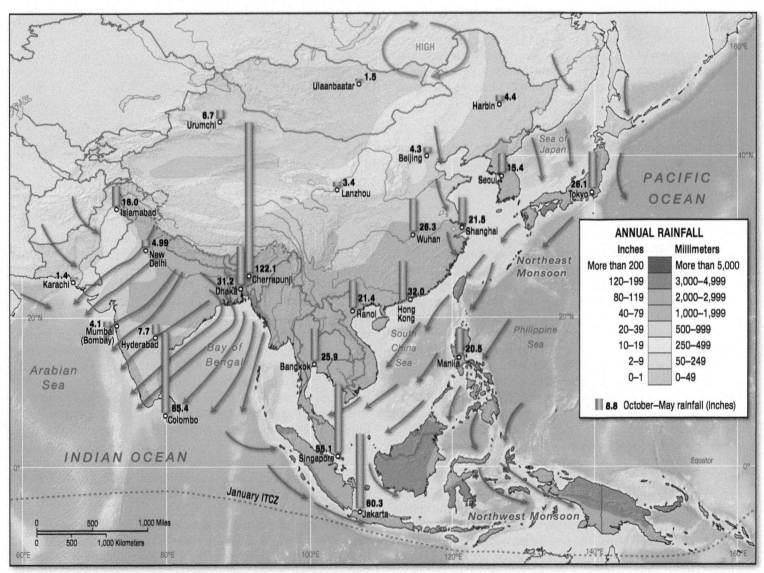

▲ **Figure 9-4 Annual rainfall and dominant atmospheric wind patterns over Asia during the winter.** The dry winter season is associated with northerly continental wind flows. The bars reflect average winter rainfall totals in selected locations.

(a)

(b)

▲ **Figure 9-5 Monsoon contrasts.** (a) Peasant women planting rice in Odisha state on the northeastern coast of India during the wet summer growing season. (b) In an attempt to combat an unusually hot and dry period before the onset of monsoon rains, Odisha villagers try to improve their drinking water supply by deepening a village well.

forces associated with the storms lashing the densely settled portions of southern Bangladesh. The dry and cooler winter monsoon from October through March brings less than 10 inches (250 mm) of rain to all but the extreme tip of India and Sri Lanka, with temperatures averaging 60°F (15.5°C) in northern India.

Stop & Think
Explain the monsoon process and why the Ganges River Valley is wetter than the Indus River Valley.

Environmental Challenges

South Asia is characterized by growing environmental problems. Air pollution is the most serious atmospheric environmental problem, particularly across the northern part of the region. Both in urban areas and broad swaths of densely settled rural areas, the skies are obscured by a pea soup-like haze (Figure 9-6). Pollution levels are especially high during the dry monsoon season when the lack of rain to clean the air is coupled with descending air associated with a dominant high pressure cell. Pollution sources include the usual culprits of dirty coal-fired plants and leaded gasoline. Often ignored are microscopic, suspended particles that originate not only from fossil fuel use but also from burning biomass—wood and cow dung used by the poor for both urban and rural cooking stoves. These fine particles enter the lungs and the bloodstream to cause much higher rates of heart disease, asthma, and lung disease. In some urban areas the levels of suspended particulates are 5 to 10 times higher than the World Health Organization deems acceptable. Women and children who spend the majority of time in the indoor spaces of home are particularly vulnerable. In Bangladesh, some 90 percent of the population use some form of biofuel for cooking, and particulate matter, along with malnutrition and unsafe drinking water, are leading causes of premature death.

The solution to indoor particulate pollution, a problem China has aggressively tackled, is stoves with a much improved ventilation system.

Sustainable water is the most serious challenge to economic and social development. Growing water scarcity is directly linked to rapid population growth. With approximately 21 percent of the world's population and the second highest rate of natural population increase of any world region, South Asia possesses only 5 percent of the world's renewable water resources. Nepal and Bhutan are the only South Asian countries with above world average per capita water availability. Annual water availability declined 70 percent between 1950 and 2005, leading some experts to identify South Asia as the world's most water-stressed region. Although rapid population growth is an important factor in explaining water scarcity, poor infrastructure, government water policies, outdated water extraction technologies, the interstate politics of water, and wet monsoonal patterns and global climate change also affect sustainability.

Most of South Asia's water goes toward agricultural production. Central and regional governments have not maintained irrigation ditches and pipes, resulting in the leakage of millions of gallons of water each day. Governments also have been slow in introducing drip irrigation technologies, forcing farmers to continue the water-wasting flood irrigation method. In urban areas, decaying pipes and lack of water treatment facilities also leads to the loss of precious water. Indian cities lose up to 40 percent of their freshwater resources as a result.

Due to polluted surface water resources and the lack of government-mandated conservation measures, water sustainability is severely threatened by extraction of groundwater from aquifers, the last source of regional water security. Many experts view groundwater use as one of the major environmental issues facing the region. Groundwater extraction in India accounts for approximately 24 percent of the total global groundwater extraction. Aquifers are accessed by sinking tube wells or pipes into the ground until the aquifer is reached, then the water is pumped to the surface and into storage reservoirs (Figure 9-7). About 70 percent of Bangladesh's and 50 percent of India's agricultural irrigation water comes from aquifers. The popularity of tube wells is in part explained by government rural electricity subsidies to operate pumps. Groundwater extraction is critical in urban areas as well. In Dhaka, the capital of Bangladesh, 87 percent of residents use groundwater; Lahore, Pakistan's second most populous city, is almost totally dependent on aquifer-derived water. Questions of groundwater sustainability are paramount because extraction has been so rapid that water tables have been dramatically lowered over the past few decades. In Lahore, extraction has been so rapid that some districts of the city have witnessed a water table drop of 65 feet (20 meters). In northwestern India, one of the country's most productive agricultural regions, the water table dropped approximately 1 foot between 2002 and 2008, leading some experts to predict that groundwater resources will be entirely exhausted in a few decades.

Sustainability of scarce water resources has become important at the larger transnational scale in the form of rivers crossing international boundaries. Two of the three great rivers of South Asia, the Indus and the Brahmaputra, as well as their important tributaries, have much of their upriver basins in India. Therefore India could control the volume of downriver water flow in both Pakistan and Bangladesh, which derive 75 percent and 91 percent of their respective river water resources from beyond their borders. There have been periodic water disputes between the three South Asian countries, but long-term water treaties have mediated more serious potential conflicts, even during wars and the absence of any meaningful economic integration and cooperation among the three countries. For example, the Indus Water Treaty, which has been in effect for more than 50 years, guarantees Pakistan 80 percent of river volume flow and India is prohibited from using river water from tributaries that feed the Indus for storage purposes.

Although various transborder river management agreements have prevented serious military confrontations, this may not be true in the future. India's thirst for power generation has initiated plans for a number of upstream dam projects to generate hydroelectric power. These projects are "run-the-river" in nature, that is, no water storage is required to generate power, but Pakistan remains mistrustful of India's ultimate intentions. Future global climate change and monsoon-generated rains also introduce potential conflict related to water scarcity and security. The water volumes of major rivers, for example, depend on glacial melt from the Hindu Kush, Karakoram, and Himalaya mountains. But warmer temperatures due to global climate change have already caused significant glacial melting and will reduce water supplies in future decades. Climate change research also indicates that wet monsoon periods will become shorter in length with more intense precipitation events and result in drier river basin environments. This is critical because 73 percent of India's annual precipitation totals and 55 percent of nonirrigated agriculture occur during the rainy season. Better management and sustainability of transborder water resources would dramatically improve if the respective governments cooperated by sharing technologies and resource data, which would decrease always-present suspicions that one country is stealing water from another.

▲ Figure 9-6 **Satellite image of the Ganges and Brahmaputra river valleys.** The area is often blanketed in a pea soup-like haze from a wide variety of pollution sources. Poor air quality is most pronounced during the dry monsoon season.

▲ Figure 9-7 **Tube wells in India.** A tube well brings aquifer-sourced water to the surface in eastern India. While children play, adults apply the water to rice padi fields at the beginning of the growing season.

Historical Background: Empires and Identity

South Asia has been subjected to numerous external influences from the west, many of which penetrated through Afghanistan's famed Khyber Pass. These successive waves of cultural infusion came from invasions rather than peaceful means, and shaped the major linguistic and religious contours of present-day South Asia.

The Precolonial Heritage

South Asia's earliest civilization was centered on the Indus River, from which the word "India" derives. Dating to approximately 2350 B.C., the Indus Valley empire, commonly referred to as the **Harappan culture**, was centered on the three cities of Harappa, Ganweriwala, and Mohenjo Daro. Each city was well planned, dominated by a citadel, surrounded by massive walls, and laid out using a grid street pattern. Farmers working the rich alluvial soils cultivated wheat, barley, legumes, dates, and cotton and supplemented their diets with meat from domesticated sheep and goats. Substantial long-distance trade with lands further west came through various Arabian Sea trading ports. Harappan culture began to decline about 1900 B.C. possibly triggered by waves of Indo-Aryan invaders from the west or from human- or naturally induced environmental changes in this delicately balanced desert environment.

Beginning about 2000 B.C. a new culture group, the Aryans, invaded from the west to produce a mixed Indo-Aryan civilization, which at first was primarily nomadic but adopted sedentary agriculture over time. The Indo-Aryan culture subsequently spread eastward into the then forested regions of the Ganges River Valley, where large urban settlements were established. Indo-Aryans also introduced **Hinduism**, the world's oldest major religion, which now dominates India (Figure 9-8). Integral to Hinduism is the **caste system**, wherein individuals are expected to remain throughout their lives within one of four major socioeconomic groupings, called **jati**. In Hindu belief, just as each individual possesses a duty to the larger society, so too do entire *jati* and the hundreds of sub-*jati*. Only when individuals fulfill their own duty within their *jati* will an idealized Hindu world function harmoniously (Figure 9-9). The caste system institutionalized both social status and economic roles within the larger society, and only through cyclical rebirth, or reincarnation, is mobility to a higher caste believed possible. To achieve this upward spiritual mobility, one's soul requires the accumulation of good karma, or good deeds, over many generations. At the apex of this highly stratified socioeconomic system is the religious Brahman caste. One group, known as the untouchables or *dalit* (meaning broken, depressed, or downtrodden), is relegated to performing economic activities deemed dirty or polluting and stands altogether outside the caste system.

Indo-Aryan culture did not completely transform southern peninsular India. Hinduism and the caste system were adopted, but the languages of the native Dravidian peoples were not displaced (Figure 9-10). In part based on language, southern India even today exhibits distinctive cultural, political, and economic attributes when compared to the north. In the many centuries following the establishment of Indo-Aryan civilization, a succession of empires ruled over parts of South Asia. In the third century B.C., the **Mauryan Empire** became the first political state to control most of South Asia.

South Asia is also the source region of **Buddhism**, another major world religion. Emerging in what today are the southern lowlands of Nepal, Buddhism is anchored in the teachings of Siddhartha Gautama or Buddha (meaning the "awakened one") in the fifth century B.C. The core of Buddha's teachings includes the four noble truths. These principles aim to achieve individual enlightenment through the elimination of the basic human conditions of suffering, anxiety, and dissatisfaction to better understand the basic nature of the world. Although successful in spreading throughout India and becoming an important faith, especially during the Mauryan period, the popularity of Buddhism rapidly declined by the twelfth century A.D. But a host of Buddhist beliefs, such as *ahimsa* (the principle of doing no harm to any living being), were adopted into Hinduism. Before its decline in India, Buddhism spatially diffused to East and Southeast Asia beginning in the second century B.C. In South Asia, Buddhism remains an important faith in some Himalayan Indian states as well as Bhutan, Nepal, and Sri Lanka.

The Spread of Islam

As with the Aryans earlier, Islam (see Chapter 7) entered South Asia through the mountain passes of the Hindu Kush. Virtually all the population of the Indus River Valley converted to Islam which then spread into the Gangetic Plain, where some of the Hindu population converted. Islam was embraced more widely further east in the lower Ganges River Plain that today makes up much of Bangladesh.

The most powerful of all the Islamic empires was the **Mughal Empire** of the sixteenth and seventeenth centuries (Figure 9-11). After conquering the Gangetic Plain, the Mughal Empire shifted its seat of power to Lahore in the upper Indus Basin (now Pakistan), which subsequently became one of the most celebrated Islamic cities in the world. The Mughals then conquered all but the far southern tip of India.

Islam was especially attractive to untouchables and Hindus of lower caste because it partially freed them from the rigid social order of Hinduism. Among the rural middle class of the strategic Punjab region in the northwest, rejection of the caste system produced the blended Hindu and Islamic religion of **Sikhism.** While conversion to Islam brought marginal improvement in their lives, converts continued to be viewed by the higher castes as a convenient pool of exploitable labor. A less severe caste system remained in place, even among later Christian converts. Unlike Buddhism, Sikhism, and Jainism (a small nonviolent religion with ancient Indus valley roots), the diffusion of Islam brought about sharp conflict with established Hindu society because it was associated with a nonnative conquering culture. This in part suggests why Islam, despite the political and economic power of the Mughal Empire, failed to convert most South Asians. Approximately 13 percent of Indians today profess the Islamic faith.

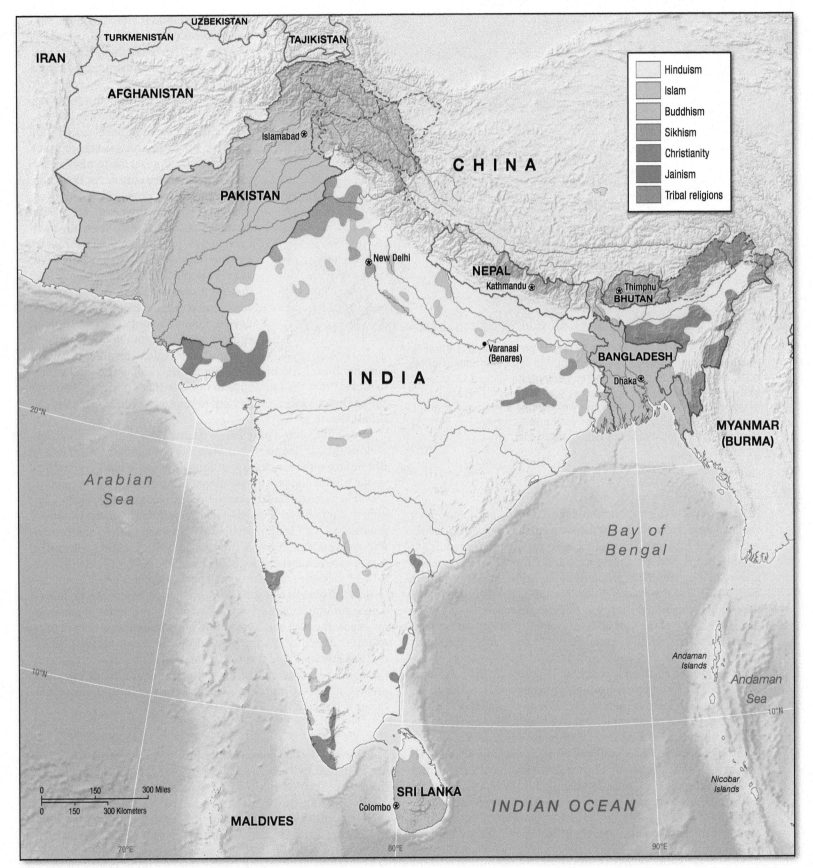

▲ **Figure 9-8 Major religions of South Asia.** Hinduism dominates in most of India, while Islam is the principal religion of Pakistan and Bangladesh. Buddhism prevails in most of Sri Lanka and in portions of the northern Himalayan regions. Sikhism, Christianity, and Jainism form scattered enclaves.

▲ Figure 9-9 **Ritual bathing in Varanasi on the Ganges River in northern India.** As one of seven Hindu holy cities in India, pilgrims come each year to bathe in the sacred waters of the Ganges. Originating in the Himalayas, the river's waters symbolically represent the purity of heaven and thus wash away the worldly sins of religious pilgrims. Varanasi is also a spiritual center for Buddhists because Buddha formulated his principles in this vicinity around 500 B.C.

The Colonial Transformation

In the competitive environment characterizing early European trading forays into Asia during the sixteenth and seventeenth centuries, India provided an ideal intermediate trading location between Europe and Southeast Asia and China. The Portuguese arrived in the early 1500s, followed by the British and Dutch in the early 1600s, and the French in the late 1600s. By the beginning of the eighteenth century, European trading stations dotted the coast of South Asia. Over the next 250 years the British eventually gained political and economic dominance over a large portion of South Asia, leaving some regions as independent princely states, many of whom formed alliances with Britain (Figure 9-12). Both Portugal and France continued to control small possessions along the coast, but surrendered them to India in the 1950s and early 1960s.

The early economic exploitation of South Asia was left to the quasi-private **British East India Company**. Expressive of an economic philosophy of **mercantilism**, whereby the goals of trade and tribute were more important than the direct control of territory and people, the British East India Company gradually replaced Mughal rule. A monopoly on trade granted by the British crown enabled the company to control indirectly about two-thirds of South Asia. After annexing much of the east coast by 1760, the East India Company was able to gain control of the prized lower Ganges Plain (Bengal) by 1765. Then by slowly annexing Indian states up the west coast, eventually the company came to control the Bombay region in 1818. This process signaled the end of the company's role as just a

trading concern as it morphed into a surrogate colonial government.

The British East India Company replaced native administration by giving land deeds to local higher-caste *zamindar* (landlords), who had functioned as tax collectors during the Mughal period. Through this new landlord–tax collector class, the company was able to exact high taxes from peasants to pay for the cost of colonial administration. Taxes had to be paid in cash rather than in kind, so peasants were forced to cultivate a wide variety of cash crops, such as cotton, peanuts, indigo, jute, and opium, in order to pay their taxes. Unable to pay both high land rents and taxes, many peasants lost their land, creating a huge underclass that contributed to the region's future underdevelopment. The dramatic reduction of subsistence cultivation in the company-administered regions resulted in recurring famines, particularly during the first half of the nineteenth century. From the late 1830s to the late 1850s, only 1 percent of the company's huge profits went to improving social and physical infrastructure.

Beginning in the nineteenth century, the company also took actions that decimated South Asia's textile industry. Although it did not have the machinery of Europe's textile industry, Indian textile production far surpassed that of Britain in terms of quality, which in part explains the company's attraction to the region. Unable to compete with inexpensive British machine-made cotton goods, and with Indian textile exports banned from the British domestic market, the once thriving domestic handloom industry disappeared almost overnight. By the mid-1800s, when South Asia formally became a British colonial possession, this newly "deindustrialized" region had begun taking on the attributes of an underdeveloped country. Cash crops were grown in the interior hinterlands for export through colonial trading ports. Cotton was the primary export commodity of Karachi and Bombay (now Mumbai), while rice and jute were shipped through Madras (Chennai) and Calcutta (Kolkata), respectively. These trading centers gradually assumed greater economic power, leaving interior and indigenous settlements to stagnate.

British Rule

The move to formal colonial rule was, in part, a response to the Sepoy Rebellion in 1857. The Sepoys were Indian troops used by the British. Their rebellion was an expression of the intense resentment toward the British that had emerged in all quarters of Indian society. British commanders and their loyal Indian troops swiftly crushed the Sepoys and, realizing that India might be lost, created a colonial possession in 1858. This marked a transformation from the mercantile colonialism of the British East India Company to the territorial colonialism of the British Crown, resulting in a deeper transformation of South Asian society.

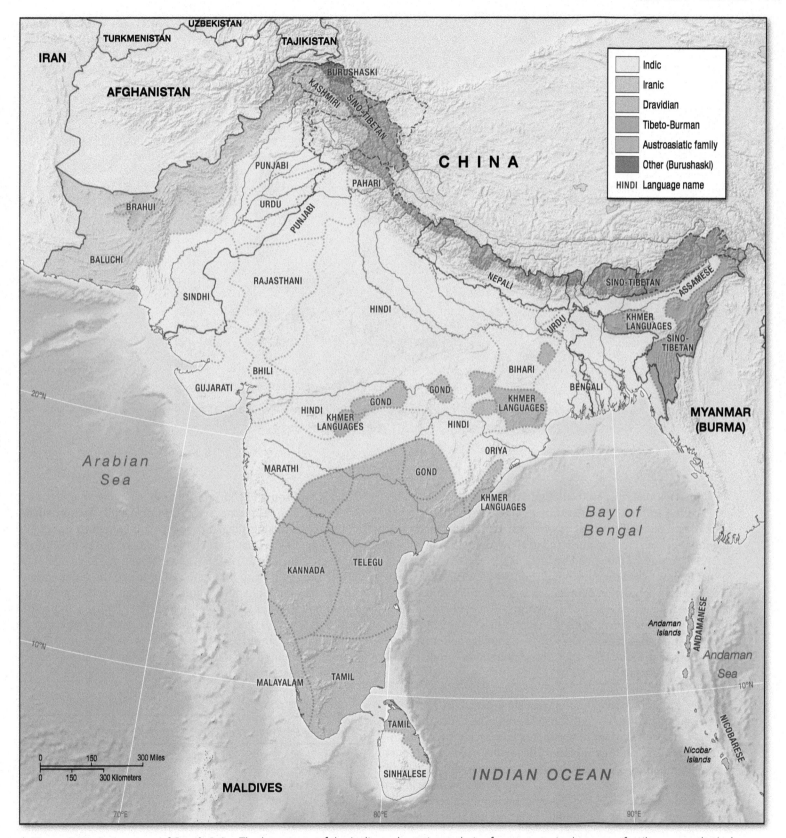

▲ **Figure 9-10 Languages of South Asia.** The languages of the Indian subcontinent derive from two major language family groups: the Indo-European languages, which are dominant in the central and northern regions, and the Dravidian languages, dominant in the south. The diversity of languages and cultures on the subcontinent contributes to the difficulty in forging unified national identities in India and Pakistan.

◄ **Figure 9-11** **Taj Mahal, Agra, India.** The Taj Mahal, meaning "crown palace" in Persian, was completed in 1647 as a royal mausoleum for the wife of the king. Although an Islamic structure in a Hindu majority state, the Taj Mahal is India's most popular visitor attraction.

Under colonial rule, cultivation of cash crops expanded to include coffee and tea. The peasantry was taxed even more to pay for road and railroad construction needed to transport increasing quantities of cash crops to ports and to allow for rapid deployment of British troops for perceived security needs. Increased taxation further impoverished the rural population. By the 1940s, 40 percent of the farming population in India was landless, approximately triple the percentage at the turn of the century. Famines became more frequent and widespread, with the second half of the nineteenth century witnessing a famine or severe food shortage about every 2.5 years. Despite the food shortages, wheat was still exported to England from the Punjab.

Although the textile industry did recover, particularly around Bombay, industrial development was slow-paced; by 1940, less than 1 percent of India's population was employed in the industrial sector, which was limited to producing goods that did not compete with British factories at home. The relative absence of industry meant that the population remained predominantly rural and that much of the limited urban-industrial growth took place in the few coastal cities that had become true metropolitan centers. The administrative capital of colonial India was not relocated from Calcutta to New Delhi until 1911.

The British did not cultivate cultural and political homogeneity but engaged in a systematic policy of divide and rule so that an indigenous unified front would not emerge to threaten British rule. The list of cultural and political entities in colonial India was almost endless, and some 113 "princely states" never came under direct British rule. The caste system remained rigid, and **communalism**—an uncompromising allegiance to a particular ethnic or religious group—persisted. Despite such divisions, the desire for independence grew steadily in the 1920s and emerged as a force in the early 1930s. The leader and inspirational figure of the independence movement was Mohandas Gandhi. He organized opposition to British colonial policies, advocated truth and justice in personal and civic activities, and practiced nonviolence in contesting imperial control. Gandhi's political goals were realized in 1947 when colonial rule ended and a number of independent states emerged from the former British India.

Stop & Think

▷ Describe the economic impact of British rule on South Asia.

Independence and Nation-State Building

At the largest scale, territory was divided into predominantly Muslim and Hindu regions. Owing to the historic clustering of Muslims at the western and eastern margins of South Asia, Islamic Pakistan was created in 1947 in the form of West Pakistan and East Pakistan, separated by more than 900 miles (1,500 kilometers) of Indian (Hindu) territory. The union of West and East Pakistan was tenuous from the beginning; not only did the respective populations

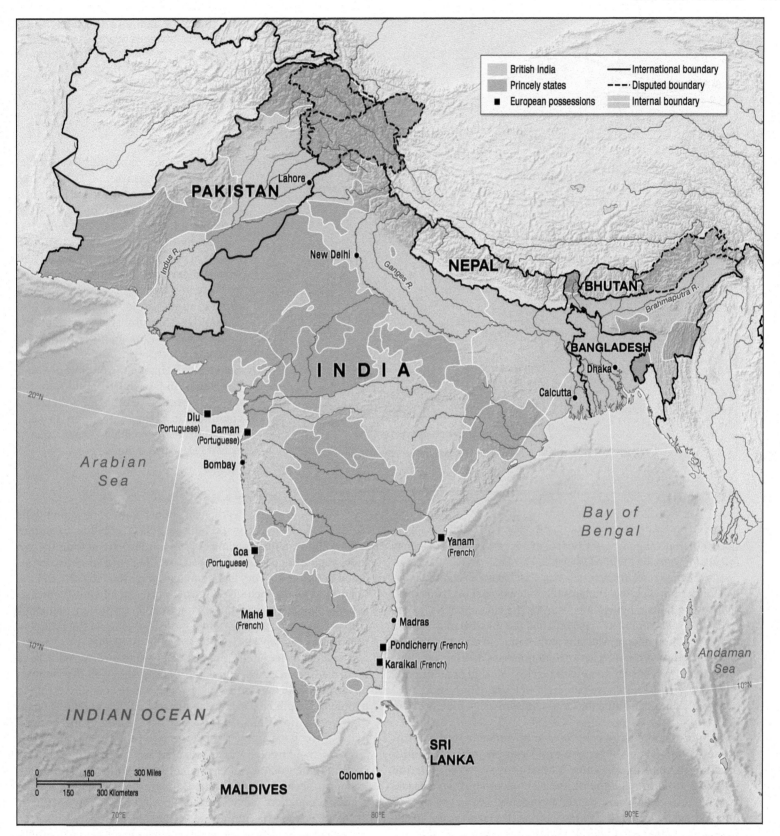

▲ **Figure 9-12** **The British Empire in South Asia in 1947.** British occupation of South Asia focused initially on coastal trade stations, with later penetration into the interior regions.

◀ **Figure 9-13 Jammu and Kashmir.** The rugged mountainous region of Jammu and Kashmir is subject to conflicting territorial claims from India, Pakistan, and China.

U.S. government pressured Pakistan to rein in Kashmiri insurgents. This became even more pronounced after a Kashmir-inspired terrorist attack on the Indian parliament later that year. The devastating 2005 earthquake in Kashmir hastened the peace process between the Indian and Pakistani governments. How to divide the territory in a way that is acceptable to all three groups—Indians, Pakistanis, and Kashmiris—is the problem. Public opinion polls reveal that the majority of Muslim Kashmiris want an independent state rather than being politically part of either Pakistan or India, while Hindu Kashmiris desire to remain within India. Peaceful conditions in Kashmir are critical to a stable political relationship between nuclear armed Pakistan and India.

Population Contours

South Asia has become the most populous world region. India (1.25 billion), Pakistan (180 million), and Bangladesh (152 million) are among the ten most populous countries with relatively high growth rates. Population densities are very high in many agricultural regions and the region as a whole includes some of the largest and most crowded urban areas in the world. Dense settlement generally coincides with well-watered regions, whether through natural precipitation or irrigation (Figure 9-14). Much like other populous and territorially large countries, India's population distribution is varied. A large multistate region of high population density comprises the country's historical core in the Ganges River Valley, where one-third of India's people live. Both coasts are also densely populated. In Pakistan, high population densities are found primarily in the eastern half of the country in the agriculture-rich Indus River Valley. The largest urban population clusters are Karachi (21 million) in the far south, the dual city of Islamabad-Rawalpindi (4.5 million) in the north, and Lahore (10 million) in Punjab. Bangladesh has the highest population densities of any South Asian country; virtually the entire country has more than 500 people per square kilometer. In Nepal the majority of the population lives in the central foothills, and in Sri Lanka, the highest population densities are found in the southwestern quarter of the country where the capital, Colombo, is located.

speak different languages, the people of West Pakistan were largely oriented toward the Middle East, while East Pakistan was more culturally oriented toward Southeast Asia. East Pakistan achieved independence from West Pakistan in 1971, following a short (two-week) civil war, and was renamed Bangladesh. The creation of a separate India and Pakistan in 1947 also entailed the massive migration of some 14 million people; Muslims and Hindus, in about equal proportions, crossed the new national boundaries, with the greatest movement from border areas with no clear Muslim or Hindu majority. Many Muslims chose to remain in India; today, India's 162 million Muslims constitute the largest Muslim national minority in the world.

Hindu-Muslim strife continues to be endemic in the northernmost Indian state of **Jammu and Kashmir**, where a Hindu-dominated government rules a Muslim majority population (Figure 9-13). Although the United Nations established a cease-fire line in 1949, Pakistani-backed insurgents have waged a recurrent war against the Indian army, which has taken some 40,000 lives. It is unclear whether Pakistan desires to annex Kashmir or whether its motive is to gain exclusive control of the headwaters of the Indus River, which is vital to Pakistan's irrigation-dependent agriculture. After the September 11, 2001 attacks in New York, the

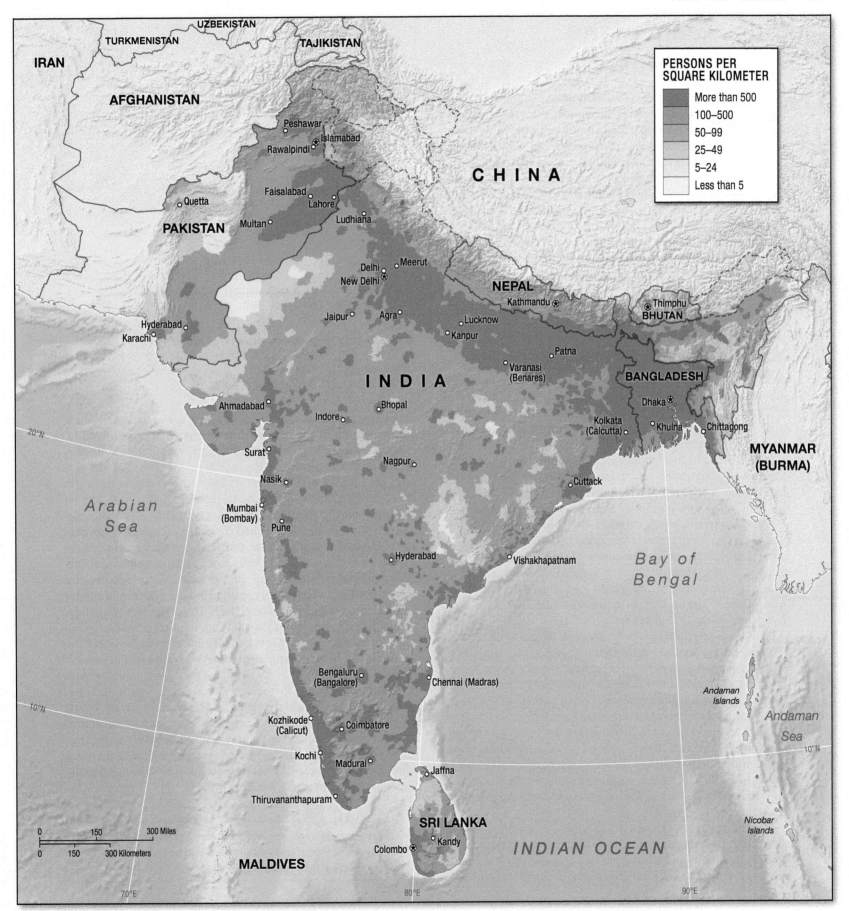

▲ **Figure 9-14** **Population distribution of South Asia.** Population distribution is characterized by high densities in the river valleys and in coastal areas.

India: Giant of the Subcontinent

Population growth is a central challenge to India's economic development. While Green Revolution technologies as well as a reduction in growth rates have dramatically improved India's ability to feed its people, far too many of the country's rural poor remain hungry. After decades of government-controlled industry, India began liberalizing its economy in the early 1990s in an effort to engage the forces of globalization. Some of the country's industrial regions have been transformed because of substantial foreign direct investment coupled with private domestic investment that has thrust them onto the global economic stage. Others remain focused on heavy industries in an environment of continued government regulation. India contains some of the most populous cities in Asia and the world, and although globalization is transforming these urban landscapes to express newfound wealth, far too many inhabitants remain part of a growing urban underclass.

Managing Population Growth

Inhabited by more than 1.25 billion people (17.7% of the world's population), India is the world's second most populous country after China. In India, as well as the rest of South Asia, development progress will require slowing population growth while increasing economic productivity.

Since 1921, India's population has more than quadrupled (Figure 9-15). Population growth was relatively low until the 1920s, then steadily rose to rates above 2 percent in the 1960s due to falling death rates. This continued until the 1980s when both birth rates and death rates began to decline. India's fertility rate remained around six children per woman until the mid-1960s but fell to less than three by 2011. India's annual rate of population increase has dropped to 1.5 percent, which is still triple that of China. This decline in the growth rate can be seen in the changing age structure of Indian society. The proportion of the population under age 15 fell from 41 percent in 1971 to 33 percent in 2011.

This transition from high to moderate fertility in India since the 1970s has several causes. As in other developing countries, better educated women have resulted in reduced fertility levels. Female literacy rates remain lower than male rates, but government programs have dramatically improved women's access to education. In 1971, only 22 percent of women were literate; by 2011, that proportion had almost tripled to 65 percent, or less than 10 percentage points behind the male literacy rate. Increased educational attainment tends to delay marriage, reducing the number of reproductive years. In 1971, the average age of marriage for females was 17 but increased to 22 years by 2011, despite the continued practice of child marriages.

Declining fertility is also a direct result of the central government family planning programs. Despite logistical constraints, low levels of financing, and a history of unpopular mandatory sterilization programs, various initiatives have cut birth rates. In 1970,

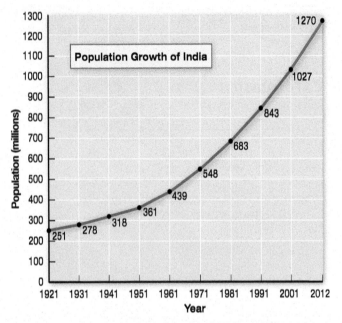

▲ Figure 9-15 **Population growth of India: 1921–2012.** India's population has more than quadrupled in less than a century.
Source: Census of India, various years.

only 13 percent of women of childbearing age used some form of contraception; this increased to more than 50 percent by 2011. However, the most popular contraceptive method is voluntary female sterilization, which does not dramatically reduce fertility rates because couples will quickly have several children early in the marriage before the planned sterilization procedure. Another factor influencing population growth rates is unequal gender relationships. In a culture where men control key economic resources such as land, dowries, income, and savings, women want sons so that some of their offspring will have access to wealth. Fertility rates among these women are likely to be relatively high because they and their husbands will continue to have children until they have at least two surviving sons. The relationship between gender inequality and population growth is not uniform across India but varies by community and location (see *Visualizing Development: Community, Gender, and Regional Variations in Population Growth*)

The recent decline in fertility rates indicates that India is approaching stage three of the demographic transformation model (see Chapter 1). This does not mean that India's population growth will soon end, because even if every couple has only two children, the population would still not level off for another 60 to 70 years. Assuming that replacement levels are reached by 2020, India's population is projected to reach 1.7 billion by 2050.

Stop & Think

➤ How is gender bias connected to population growth, patriarchy, caste, and religion to produce different development outcomes between southern India and northwestern India?

visualizing
DEVELOPMENT ▲

Community, Gender, and Regional Variations in Population Growth

Population growth rates within India vary considerably by ethnic, religious, and caste groups. Although culture certainly influences fertility rates, economic status better explains these differences—Hindus, for example, have fewer children than the lowest castes and Muslims but regional population patterns can also be explained by economic factors. The states with the most rapid growth during the 2001–2011 period were a handful of poorer tribal northeast states. Above-average growth rates also characterize the highly populated and poorer northern "Hindi belt" states of Bihar, Uttar Pradesh, Madhya Pradesh, and Rajasthan. This is in sharp contrast to the richer southern states of Andhra Pradesh and Kerala, where growth rates are significantly below the national average. As an indicator of relative levels of poverty, urban fertility rates in 2010 stood at 1.9 per woman, while in rural regions, 2.8 was the norm.

These statistics express regional variation in population growth but mask the regional **gender bias** in India's population structure (Figure 9-1-1). Gender bias exists when one or the other sex represents

an abnormally larger percentage of the population. In 2010, there were only 940 Indian females per 1,000 males. Whether for reasons of infanticide, abortion, or nutritional and medical neglect, India has, like China and a number of other developing countries, a deficit in females. In India the problem is extreme. A 2011 United Nations report claims that India exhibits the highest female child mortality rates of any country in the world, and that girls between the ages of one and five are 75 percent more likely to die when compared to boys. At the regional level, the degree of bias favoring males is substantial. Northwest and northern India exhibit stronger gender bias than do southern states. In 2010, the northwestern states of Punjab, Haryana, and Uttar Pradesh were together characterized by a ratio of 892 females per 1,000 males, whereas the far southern states of Karnataka, Kerala, and Tamil Nadu averaged a 1,015 females per 1,000 males. While male in- and out-migration rates impact the

female ratio of a given state, female gender bias is strongly correlated with high rates of female literacy and labor force participation. In southern states, women tend to have greater social and economic freedoms, such as owning land and engaging in a wide variety of empowering economic activities. In the northwest, a patrilocal social structure exists whereby a bride moves to her husband's parents' village, is secluded within the household, and denied access to land and political participation.

Sources: *Census of India*, 2011; Ravinder Kaur, "Across-Region Marriages: Poverty, Female Migration and the Sex Ratio," *Economic and Political Weekly* 39, no. 25 (2004): 19–25; Mahendra Premi, "Religion in India: A Demographic Perspective," *Economic and Political Weekly* 39, no. 39 (2004): 4297–4302.

▼ **Figure 9-1-1 Gender bias by Indian political units, 2011.** Much of India is experiencing a severe deficit in females. This map shows the number of females per 1,000 males.

Source: Census of India, 2011.

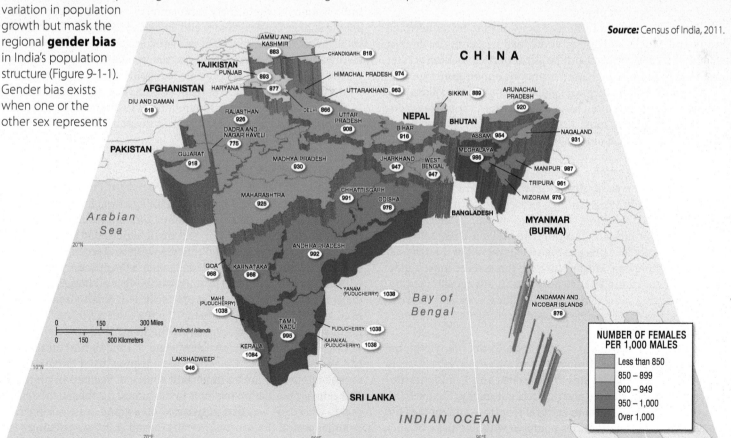

NUMBER OF FEMALES PER 1,000 MALES	
	Less than 850
	850 – 899
	900 – 949
	950 – 1,000
	Over 1,000

Accommodating Diversity

India's cultural diversity requires the government to accommodate the aspirations of ethnic and religious groups, as well as socioeconomic classes, in this most populous democracy in the world. Because cultural and economic differences are often perceived as being religiously based, India was created as a secular state containing constitutional guarantees that the government will not seek to promote nor interfere with any particular religion. While secularism has been instrumental in reducing communal antagonisms, it has neither eliminated discrimination toward the lower castes nor stopped the increasing incidence of Hindu–Muslim confrontations. Religious identity has become a preeminent political issue, partly due to fear on the part of Brahmans and other upper castes that reform would eliminate their political and economic power.

To improve economic opportunities for the lowest of the four major castes, the government has designated them as **"scheduled castes"** (of which the *dalits* or untouchables are a part), "scheduled tribes," and "other backward castes" eligible for special consideration in university admissions and government-sector employment as clerks, railroad laborers, postal workers, and sometimes middle-level administrators. This "affirmative action" philosophy has been reasonably successful considering the deep-seated nature of casteism. Scheduled castes and tribes made up 24.4 percent of India's population (approximately 295 million people) in 2010, but the vast majority remain desperately poor. Members of this unskilled labor force work as scavengers, agricultural laborers, miners, leather workers, sweepers, and sanitary workers (see *Geography in Action: Contesting Casteism*, on page 429).

Communal unrest in India also continues to fester between Hindus and Muslims. Hindus and Muslims generally coexist peacefully in rural villages, but political parties seeking greater power tend to promise increased economic opportunities to Muslims, the poorest of the major religious communities. Political appeasement of Muslims often leads to sometimes violent reactions by some Hindu groups, most often where Muslims have achieved some degree of political or economic advancement. In this sense, the destruction of a Muslim mosque by Hindus becomes a political and economic, rather than a religious, act. The increased political power of the conservative, Hindu-based **Bharatiya Janata Party (BJP)** since the early 1990s has only aggravated communal unrest because of its claim that India is a Hindu-only country and all other religious communities are thus not true Indians. Discriminatory practices generally take place at the state level rather than the national scale, where non-Muslims occupy some of the most important government administrative positions in this secular country. Societal fracturing along religious and political lines only deprives India of energies that could be focused on constructive economic development.

The central government accommodates the many regional cultures and economies through the administrative framework of **federalism**, which provides the 28 states and 7 union territories a measure of economic and political autonomy (Figure 9-16). Because language is central to cultural identity—and most of the state and territorial boundaries coincide with linguistic boundaries—the government recognizes 22 different official languages (including English) that account for some 90 percent of Indian speakers. Indo-European **Hindi**, the native tongue for 43 percent of Indians, is the principal official language. The colonial language of English continues to function as an informal official language. Most middle- and upper-class adult Indians are multilingual, speaking their regional language plus some Hindi or English.

The strongest challenge to national unity originated in Punjab, where Sikh separatists, seeking an independent "Khalistan" in the 1960s, forced New Delhi to redraw boundaries so that Punjab became a Sikh majority state. New Delhi also granted statehood to the handful of tribal regions of the northeast after separatist movements emerged during the 1960s and 1970s. In 2000, regional dissent prompted the government to create three new states. Jharkhand was created from the southern half of Bihar, Chhattisgarh from the eastern third of Madhya Pradesh, and Uttarakhand from northwestern Uttar Pradesh (see Figure 9-16). Jharkhand is populated primarily by forest-based tribal peoples desiring greater economic self-determination. Southern Bihar was mineral rich, but its mines and processing plants were controlled by interests in northern Bihar, part of the Hindi-speaking heartland. The Jharkhand people, many of whom worked as poorly paid mine laborers, saw little economic benefit from being part of a larger Bihar state. The Jharkhand movement is just one example of tribal resistance to larger hegemonic forces at work in India and other developing nations, where the economies and environments of traditional societies are being threatened or destroyed.

Agricultural Development in India

India is an overwhelmingly rural country, with 52 percent of its people engaged in agriculture and 69 percent living in villages and small towns. Contrasts with China's agricultural landscape are noteworthy. In India the percentage of agricultural workers is far higher, and 46 percent of India's land area is classified as productive cropland, compared to only 10 percent of China's. With approximately 1,300 and 3,200 people per square mile of cultivated land in India and China, respectively, India's agriculture is far less intensive overall than China's (Figure 9-17).

Agricultural Regions

The spatial distribution of agricultural systems and crop types in India is very much influenced by the availability of water (Figure 9-18). More land is dedicated to rice production than to any other crop. In drier regions, wells, canals, and small impounded ponds called "tanks" supplement rainfall. From the western Deccan to the edge of the Thar Desert, drought-tolerant millet is the primary grain crop, with pockets of cotton and peanuts assuming local importance. India is the world's largest millet producer. This small-sized grain is used as flour for making flatbreads. In the drier north-central and northwest, wheat and chickpeas are the staple food crops, with the politically strategic northwestern state of Punjab functioning as India's "breadbasket." Chickpeas are a legume used in curries or to make flour and are an important source of vegetarian protein. The primary plantation region is in the humid northeast, the same region where tea was first introduced on a grand scale during the colonial period (Figure 9-19). Maize-based shifting agriculture is also practiced by the tribal hill peoples. As in China, most of India's

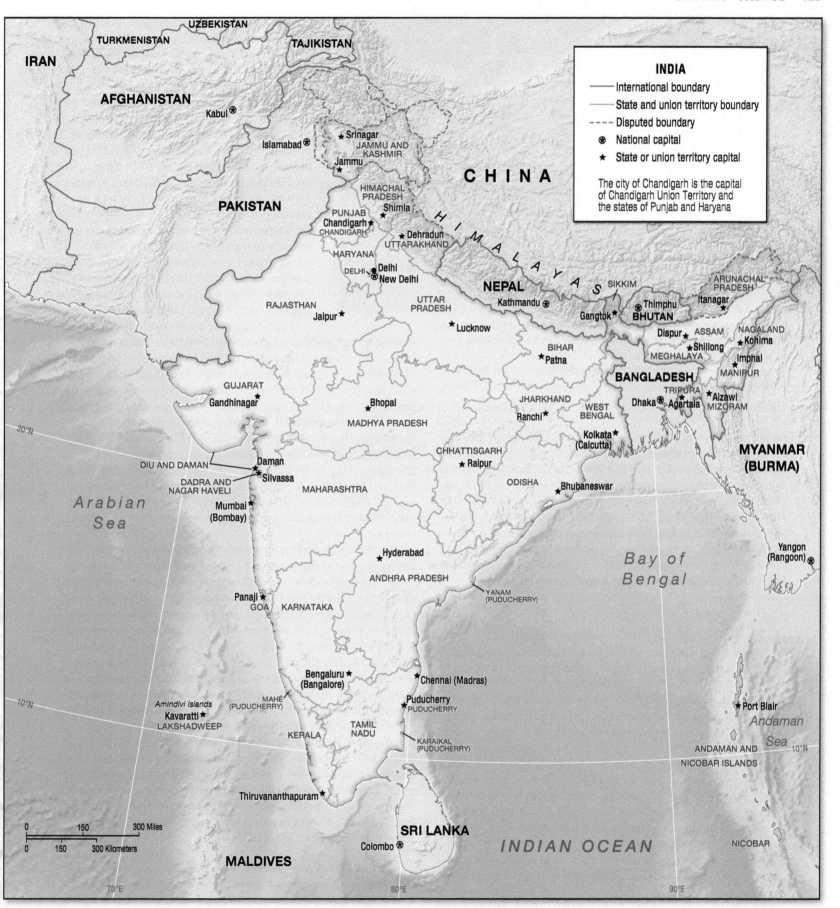

▲ **Figure 9-16 Political units of India.** India, one of the world's largest countries, is divided into 28 states and seven union territories.

▲ **Figure 9-17 Extensive agricultural resource use on the Deccan Plateau.** Drier than the coastal mountain ranges and plains that surround it, the Deccan Plateau is used less intensively. Goats and cattle graze on fallow land and postharvest residues on the fields that surround village centers.

farmers supplement their primary crops with other crops, whether for subsistence or commercial purposes. But unlike land-scarce China, villagers also rely heavily on livestock. The most important is cattle, which are rarely consumed for their meat because of the traditional Hindu taboo against consuming beef. Dairy products in the form of milk, yogurt, and "ghee" (clarified butter), however, are important sources of animal protein. Cattle dung is gathered for cooking fuel in many regions where fuelwood is scarce (Figure 9-20). Cattle are used for plowing and for short-distance transport. Cattle hides sustain India's thriving leather and tanning industries.

Agricultural Productivity and Change

Many experts believe that modernizing Indian agriculture is the key to achieving sustained economic development. Compared to China and other more industrialized countries, the productivity of Indian agriculture is relatively low, particularly for rice (Figure 9-21). The dramatic yield differences between China and India are partly due to China's greater proportion of land under irrigation, greater use of fertilizer, and more widespread practice of double-cropping. The poor relative position of agriculture in the economy is stark; 52 percent of India's labor force is in agriculture, but this sector contributes only 17 percent to national GDP. Despite relatively low yields per unit area, high-yielding hybrid-grain wheat varieties associated with the Green Revolution have dramatically boosted Indian agricultural productivity since the 1970s. Between 1973 and 2010 the production of cereal grains increased from 119 million to 234 million metric tons. Food production has kept pace with population growth and India now exports selected agricultural commodities, mainly milled rice, various edible oils, and chickpeas and lentils, primarily bound for the European Union, Southeast Asia, and

the United States. Despite these surpluses, too many of India's poor are hungry. A 2011 Ministry of Health and Family Welfare report found that 40 percent of children and 36 percent of women were undernourished. According to the 2011 Global Hunger Index, India ranked lower than its neighbors Pakistan, Nepal, and Sri Lanka as well as some sub-Saharan countries. This situation is not so much a reflection of the availability of food, but rather the quality and frequency of food intake. Studies have also shown that undernourished rural areas often have high agricultural productivity.

The sources of continued rural poverty are many. Although the government has dedicated much energy and money to improving village education, electrification, health access, and other social services, meaningful land reform has hardly begun. Inheritance laws mean that most farm plots are small, discouraging mechanization and cooperative use of irrigation. The average size of a household plot is only 6.5 acres (2.6 hectares), but 34 percent of households cultivate plots measuring only 0.5 acre (0.2 hectare). Cash cropping and the Green Revolution have only widened the gap between the rural rich and poor. The hybrid seeds, chemical fertilizers and insecticides, and mechanized irrigation that might improve yields are either too expensive or fail to provide the harvest security and self-reliance so prized by peasant farmers. Unable to adequately support their families, many peasants lease their lands to Green Revolution farmers, only to lose their self-employed status. Fully half of India's peasants are landless and work as agricultural laborers or as sharecroppers on the margins of the agricultural economy.

The rural poor have not benefited from Green Revolution technology or government economic policies implemented in the early 1980s, designed to force the agricultural sector to be more market oriented to globalize India's economy and comply with World Trade Organization rules. These structural adjustment policies particularly hurt rural women, many of whom live below the poverty line (Figure 9-22). Under a neoliberal market-based economic system, government-subsidized food and social service programs have been cut, forcing women to spend more time on earning income to make up for the increased cost of food and social services once provided by the government. In addition, as agroprocessing industries have become more mechanized in response to greater international competition, female agricultural laborers, who depend on these critical off-season employment opportunities, are losing their jobs. Increasing rural debt among small farmers is another concern. The social marginalization of poor farmers, debt associated with switching to genetically modified seeds, and the risks and costs of shifting from food to cash crops have resulted in more than 200,000 suicides among farmers since the late 1990s.

Stop & Think

▶ Explain why the processes of globalization have not helped poor farmers in India.

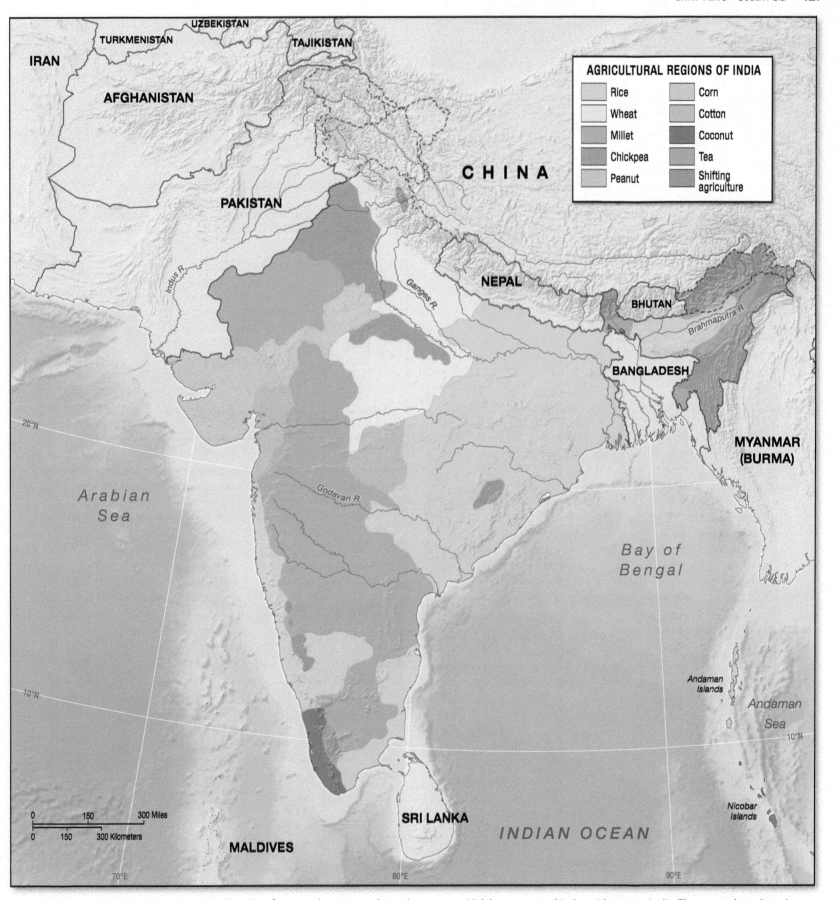

AGRICULTURAL REGIONS OF INDIA

Rice	Corn
Wheat	Cotton
Millet	Coconut
Chickpea	Tea
Peanut	Shifting agriculture

▲ **Figure 9-18 Agricultural regions of India.** Rice farming dominates along the western Malabar coast and in humid eastern India. The more drought-tolerant wheat and millet prevail in the more arid central and western zones.

▲ **Figure 9-19 Harvesting tea leaves, state of West Bengal.** Tea is the primary plantation crop of the humid Himalayan foothill regions of northeastern India. Here female laborers climb a steep slope to pick tea on an estate in Darjeeling.

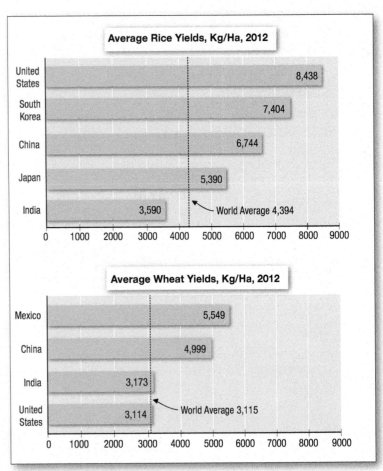

Average Rice Yields, Kg/Ha, 2012

United States	8,438
South Korea	7,404
China	6,744
Japan	5,390
India	3,590 → World Average 4,394

0 1000 2000 3000 4000 5000 6000 7000 8000 9000

Average Wheat Yields, Kg/Ha, 2012

Mexico	5,549
China	4,999
India	3,173
United States	3,114 → World Average 3,115

0 1000 2000 3000 4000 5000 6000 7000 8000 9000

▲ **Figure 9-21 India's grain yields compared to selected countries, 2012.** Despite the expenditure of many hours of work per land unit, rice yields in India are significantly below the world average. South Korea and Japan have rice yields substantially higher than the world average.
Source: Food and Agriculture Organization, http://faostat.fao.org/default.aspx?PageID=567#ancor

▲ **Figure 9-20 Dried cattle dung.** In much of India, cattle dung is collected and dried to be used as cooking fuel.

▲ **Figure 9-22 Female agricultural laborers in West Bengal.** Women laborers harvest potatoes for shipment to local wholesale markets.

Contesting Casteism

The caste system has been questioned for centuries, although seriously contesting the unfair nature of casteism first began under British rule by both Christian missionaries and concerned upper-caste Indians. The British system of divide and rule actually often encouraged greater rigidity of the caste system. The British passed the 1935 Government of India Act that included reserving seats in state governments for "scheduled castes," and several laws and amendments to the Indian constitution since independence provide protections and greater economic opportunities for these oppressed groups. In 1950, the legal designation of an individual as an "untouchable" was outlawed. A Commissioner for Scheduled Castes and Tribes was appointed in the late 1970s to advise the government on caste issues, followed in 1990 by the establishment of the more powerful National Commission for Scheduled Castes and Tribes.

Despite laws against discrimination by caste, the institution remains pervasive, particularly in rural areas where scheduled castes experience residential and educational segregation, lack access to public water wells, and are barred from entering certain Hindu temples. Dalit Christians commonly have their own churches or sit in segregated sections of the church. The government policy of reserving 25 percent of university places for scheduled castes and tribes has met with public protests by upper castes claiming reverse discrimination. Some conservative political groups maintain the caste system is central to Hinduism and any attempt to dismantle it is anti-Hindu. In response, spokespersons for the backward communities claim that their struggle is against higher-caste groups such as the Brahmans, who exploit the poor, and has nothing to do with Hinduism. The Vedas, the oldest holy texts of Hinduism, only mention the caste system in a passing manner.

Many anticaste activists have emerged to represent a variety of interests. Foremost are political parties that actively seek a greater voting constituency by promising more attention to the poorly enforced positive discrimination laws. Another source is domestic and international human rights groups pushing for better working conditions for child and bonded laborers and agricultural workers, and legal assistance for those abused by the police and their upper-caste "masters." The third source of anticaste activism is of a self-help nature, whereby local lower-caste organizations unite with similar groups to call for better working conditions (Figure 9-2-1). Public protests and work stoppages as a form of civil disobedience are increasingly common ways for low-caste workers to express their frustration. Double disadvantaged by caste and gender, some women self-help groups have established small revolving loan schemes in which members may borrow money to establish small-scale, labor-intensive businesses.

Sources: Sonalde Desai and Amaresh Dubey, "Caste in 21st Century India: Competing Narratives," *Economic and Political Weekly* 46, no. 11 (2012): 40–49; Ghanshyam Shah, Harsh Mander, Sukhadeo Thorat, Satish Deshpande, and Amita Baviskar, *Untouchability in Rural India* (New Delhi: Sage, 2006); Frank de Zwart, "The Logic of Affirmative Action: Caste, Class and Quotas in India," *Acta Sociologica* 43, no. 3 (2000): 235–249.

◀ FIGURE 9-2-1 **Lower caste demonstration in India.** Lower caste protestors of the All India Confederation of Scheduled Caste and Scheduled Tribe organizations in New Delhi demand that affirmative action policies of the public sector should extend to the private sector as well.

India's Industrial Economy

India's postindependence industrial economy grew slowly because industrial production was initially state-planned and controlled. Since the early 1990s the industrial sector has grown dramatically as a result of the government promoting both domestic and foreign private investment, more private-sector control, and more market-oriented policies. Progress has been slow: in 2011, industry accounted for only 29 percent of GDP (compared to 47 percent in China) and manufacturing accounted for only 17 percent of the total workforce. India's economy has not relied on exporting manufactured consumer goods to the same extent as China; its industries are anchored in information technology and business services that are skills-based and capital intensive but not substantially labor-intensive. Information technology industries make up about 7 percent of India's GDP, but only 2 percent of the country's workforce. This situation poses problems for absorbing unskilled rural-to-urban migrant labor, as experienced in China. But India is one of the fastest-growing economies in the world, experiencing a healthy 7.3 percent average annual GDP growth rate from 2000–2010, and was the world's fourth largest economy in 2010. India's GNI PPP (Gross National Income in Purchasing Power Parity) increased dramatically from $1,649 in 1995 to $3,400 by 2010.

The Industrial Resource Base

From independence to the acceleration of globalization in the early 1990s, India had a sufficient resource base that could provide raw materials and energy for industrialization (Figure 9-23). With few exceptions, India's energy resources are found in the southern two-thirds of the country, and mostly located in the Deccan Plateau. In 2010, India accounted for approximately 10 percent of the world coal reserves and in 2009 about 42 percent of its energy production came from lower-quality bituminous coal. India is also a major producer of iron ore, accounting for 6 percent of world production and 5 percent of world reserves. India is not well endowed with oil and natural gas, despite oil fields in Assam, Gujarat, and Punjab, plus newly developed offshore fields around Mumbai. But oil and natural gas comprise 31 percent of India's energy consumption. To supplement commercial energy as well as to promote irrigation and flood control, the government has constructed numerous hydroelectric facilities, which are sustained by the wet summer monsoons. In 2011, hydroelectric power provided 14 percent of the country's power. Another 24 percent come from biomass such as wood and cow dung, commonly used in India's villages. India's future energy demands far outstrip domestic energy resources and the country must aggressively secure foreign sources of energy (see *Focus on Energy:* India's Energy Challenge).

From Centrally Planned to Market-Based Industrial Economy

Because Great Britain left India only a rudimentary industrial infrastructure, postindependence governments attempted to develop a number of key industries to attain greater self-sufficiency. Democratic India adopted the Soviet model of industrial development anchored by five-year plans and invested heavily in many state-operated firms processing cotton, jute, sugar, and other agricultural raw materials. The government also developed heavy industries, including iron and steel, chemicals, shipbuilding, and automobile manufacturing. While a handful of large private-sector corporations and small-scale manufacturing and commercial enterprises thrived, the state, rather than the market, determined what and how much was produced. The industrial sector was effectively isolated from the more competitive forces of the global economy.

By the late 1990s, there were approximately 300 central government-owned corporations and thousands of corporations owned by state and municipal governments. Many state-owned companies were efficient, profitable, and oriented to the domestic market, but most relied on outdated technology and a huge pool of surplus labor, slowing India's development at a time when the economic growth of its Pacific Rim neighbors was skyrocketing. Most exports remained tied to lower-value textiles and leather, as well as to gems.

In response to its declining global economic standing, the government instituted sweeping reforms in 1991 to increase the involvement of private domestic and foreign firms in the economy. Much of this new investment took the form of joint ventures between domestic and foreign corporations. These reforms increased GDP growth, but unlike China and many other Pacific Rim economies, India did not attract much foreign investment in consumer goods-based export manufacturing. This is attributable both to India's continued reliance on import-substitution industries protected by high tariffs and also to the small-scale nature of India's low-tech, labor-intensive industries such as apparel and toy manufacture. The government felt that large-scale, higher-technology firms would reduce employment opportunities, and unlike China, blocked entry to iconic U.S. multibrand retail firms such as Walmart due to fear over the loss of employment in the millions of small, family-run general stores. In a global economic environment where multinational firms mass-produce identical products, small-scale Indian firms simply cannot compete. In 2012, however, the government did pass laws allowing these multibrand retail corporations to enter the domestic market. Further deregulation of industry might attract larger-scale investment, and some Indian states have been very aggressive in deregulating their economies to attract both private domestic and foreign investment.

Evolving Urban-Industrial Regions

The geographical distribution of India's industrial base continues to reflect colonial influences, but a more diversified industrial sector is emerging as a result of economic reforms in the 1990s. Some industrial regions have grown dramatically based on private domestic investment, public investment from both the federal and state governments, and foreign direct investment (FDI). India has become the third largest recipient of FDI in the world after the United States and China (Figure 9-24). The economic sector that received the largest FDI from 1990–2009 were services (31%). The greater Mumbai region attracts the most investment, followed by New Delhi, the state of Gujarat north of Mumbai, and the southern states of Karnataka and Tamil Nadu. Although dwarfed by inflowing FDI, outflowing FDI by Indian corporations has dramatically grown as well. Between 2000 and 2010, FDI outflow increased from US$514 million to US$14.5 billion. The largest Indian multinational corporation is the Tata Group, a company with diverse overseas investments including automobiles, chemicals, and telecommunications.

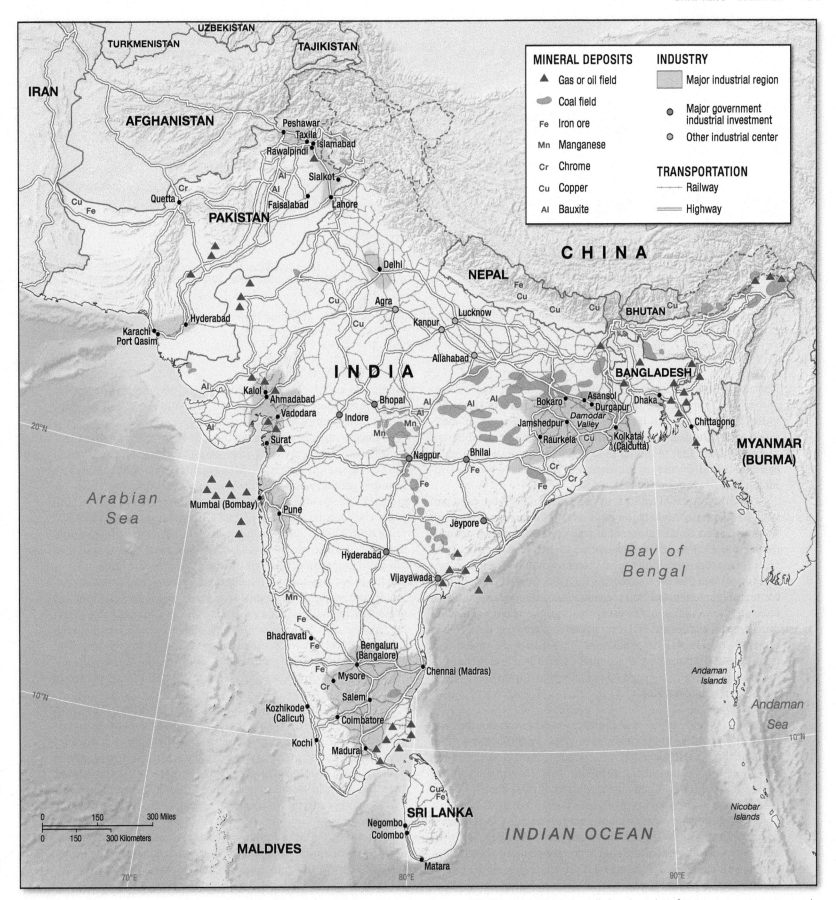

▲ **Figure 9-23 Primary mineral and industrial regions of South Asia.** Although poor, the subcontinent has a well-developed surface transportation network, inherited in part from British colonial investments. The major industrial regions are centered on the principal cities.

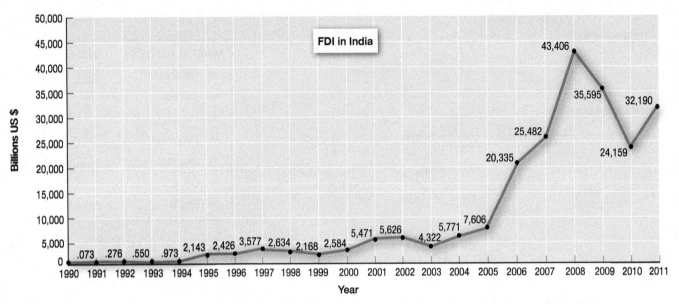

▲ Figure 9-24 **Foreign Direct Investment (FDI) in India, 1990–2011.** As a result of the economic reforms of 1991, FDI has dramatically increased.

Source: World Bank, data.worldbank.org/indicator/BX.KLT.DINV.CD.WD

Kolkata/Damodar Valley Since independence, Kolkata has anchored the premier national industrial region of the **Damodar Valley**, where coal and iron ore promoted the growth of state-owned metallurgical, machine, fertilizer, and chemical industries. The Damodar Valley is similar in its products and national significance to the Ruhr Valley of Germany and is one of the world's largest industrial regions.

Once the capital of colonial British India and still the capital of West Bengal state, Kolkata is a poor city, in part because the state was controlled by a democratically elected Communist government from 1977 to 2011 and directed investment to agriculture rather than industry. While rural income equality improved, private investment fled the state during the 1980s because of tight state control of manufacturing and recurrent labor stoppages. The 1990s ushered in a more friendly investment atmosphere, but Kolkata has not attracted significant FDI. From 2000 to 2009 West Bengal only received 1.2 percent of total national FDI. Much of the investment from 1995–2010 came from the federal government and private domestic capital, which focused on the energy and transport sectors, rather than manufacturing.

Mumbai As the capital of Maharashtra state, Mumbai has become the showcase of the new and progressive India and is often dubbed the "Maximum City" after a 2004 novel describing the city's meteoric economic and social fortunes. During the colonial period this peninsula-sited city attracted a diverse array of South Asian ethnic entrepreneurs; even today, over half of its residents are non-Maharashtra natives. Mumbai has become India's most cosmopolitan city, and symbolizes the new era of national economic growth linked to the global economy. Modern skyscrapers, hotels, and apartment buildings dot the cityscape, evidence of a highly trained workforce of professionals and entrepreneurs flush with new investment capital. The city is India's busiest international seaport as well as the country's primary air-transport gateway.

Mumbai is also the headquarters of major domestic corporations, the national stock exchange, and the center of the Hindi-language film industry known as "Bollywood."

Maharashtra state has taken the lead in injecting greater competition into the remaining state-owned industries as well as in attracting foreign investors. Maharashtra state and the greater Mumbai urban region has received more FDI than any other region of India; greater Mumbai, a huge urban region of 20.5 million people, accounted for much of the 34.1 percent share of India's FDI value from 2000 to 2009. Despite a severe decline in its huge textile industry, Mumbai still boasts a healthy industrial base in automobiles and petrochemicals. But the driving economic force today is service industries such as telecommunications, real estate, banking, insurance, and other business services. It is employment in these sectors that provides the basis of India's vaunted and growing "middle class" connected to the global economy. New satellite suburban development, particularly Navi Mumbai or New Mumbai, a huge planned new city across the river from the older peninsular Mumbai, is expressive of this emerging middle-class landscape (Figure 9-25).

Despite Mumbai's prosperity, a significant slice of the city's residents are poor rural-to-urban migrants, as well as those left behind by Mumbai's engagement with the global economy, which has resulted in the closing of state-owned industry. About half are slum dwellers, or squatters living in self-constructed shanties. Even in Navi Mumbai, more than a third of residents live in shanty settlements interspersed between upscale housing developments.

As with other large Indian cities experiencing dramatic economic growth, new industries have located distant from the urban core. Pune, a city of 4 million inhabitants located almost 100 miles to the southeast of Mumbai, has attracted a large number of multinational and domestic manufacturing plants. Pune's FDI is primarily concentrated in automobiles and auto parts.

India's Energy Challenge

Compared to the other BRIC countries China, Russia, and Brazil, India's domestic energy resources are severely inadequate for a country that aspires to being a major global economic player. Between 2007 and 2011 India's average per capita electricity consumption was a very low 571 kWh compared to China (2,631 kWh), United Kingdom (5,692 kWh), and the United States (12,914 kWh). But demand is dramatically increasing and India now must import the majority of its energy. Imported energy costs rose from US$20 billion to US$100 billion in the first decade of the twenty-first century. By 2030, India will need to import 80 percent of its energy, including coal. India has 20 operating nuclear reactors with seven more planned, but even though these plants help to diversify the country's domestic energy sources, nuclear power will not become an important energy source despite promotional claims from the government.

China must also import energy to sustain its rapid economic growth, but India cannot compete with China in the global market for fossil fuels. China is more successful at signing oil exploration and production agreements with sub-Saharan African governments because it can provide government-backed financial incentives. India is also beset by a number of domestic energy production and distribution problems. Much of its coal reserves are located in isolated regions and rail transport cannot efficiently handle long-distance bulk transport. Its largest coal company, once a state corporation but privatized in 2010, cannot keep up with demand and is endemically corrupt. Many private power-generating firms choose to import more expensive coal. The country's oil production and distribution infrastructure is outdated, and the industry's knowledge-based personnel are aging. India thus must import 70 percent of its power needs, making it the fifth-largest oil importer in the world. Government policy also inhibits the improved provision of energy. While most countries have a single coordinating ministry, India has seven government bodies involved in

▲ FIGURE 9-3-1 **Small-scale energy generation in rural India.** Using cattle manure, villagers produce methane-based biogas as a localized alternative energy source to generate power for lighting, cooking, and pumping water.

energy. This is compounded by conflicting federal and state government energy policies. Lastly, while federal and state governments talk about privatizing state-owned companies, only 23 percent of energy-generating capacity was in private hands in 2011.

India's promising economic future will be severely constrained unless it solves these basic institutional and infrastructural energy problems. Approximately one-third of Indians lack electricity because they are too scattered and isolated to be reached by conventional transmission lines. Even those connected to the power grid regularly experience frequent blackouts and power shortages, like the country's massive power blackout in 2012. Bringing power to scattered rural populations requires the use of point-specific energy sources such as wind, solar, and biomass generators. Local, small-scale methane gas production, for example, is a substitute for electricity because it uses animal manure to generate fuel for lamps and cookers (Figure 9-3-1). Methane gas production leaves a mineral- and nutrient-rich residue, which can be spread on fields as an organic fertilizer, thus meeting both

basic agricultural and energy objectives. But these technological efforts to provide energy require government subsidies and the political will to bring basic development benefits to all citizens rather than just those populations favored by income or geographic location.

As an emerging economy that will only consume ever greater amounts of energy in the near future, what does all of this mean for India's role in global climate change? Unfortunately, fossil fuels will anchor the country's economic growth trajectory. India's CO_2 emissions doubled between 1990 and 2010 and are projected to increase another 250 percent in the next two decades. At this rate, India will account for about 10 percent of the total global CO_2 emissions increase between 2010 and 2030. While India in 2010 was the world's third largest CO_2 emitter, it must be remembered that its current per capita emission level of 1.5 percent is low when compared to the United States (16.9%) and even China (6.8%).

Sources: "The Future Is Black," *The Economist,* January 21, 2012, http://www.economist.com/node/21543138; "India," U.S. Energy Information, http://www.205.254.135.7/countries/cab.cfm?fips=IN.

▲ Figure 9-25 **A section of the skyline of Navi Mumbai (New Bombay).** India's professional and middle classes dominate the emerging residential landscape of Navi Mumbai, a massive new satellite settlement across the river from peninsular Mumbai.

Bengaluru The third and most dynamic urban-industrial region is centered on Bengaluru (Bangalore), the capital of Karnataka state in south-central India. The city-region has been dubbed the "Silicon Valley of India" because of the growing presence of domestic and foreign electronics and software production in this more temperate and higher-elevation garden city whose population has grown from 4.1 million in 1991 to 8.5 million in 2011. Traditionally anchored in state-owned defense and electronics industries as well as science and engineering universities, Bengaluru has attracted the likes of Texas Instruments, IBM, Google, and Intel, as well as domestic computer software firms, in addition to the homegrown and world-class information technology (IT) firms of Infosys and Wipro (Figure 9-26). India has become one of the global centers of the IT industry and business process outsourcing, and the growth of this sector has outpaced all other economic sectors. In 2012, revenue reached US$100 billion and exports accounted for 69 percent of this total, with 87 percent of exports directed to North America and Europe. The industry is critical to India's economic growth as this sector in 2012 directly employed 2.8 million people, indirectly employed 8.9 million people, and accounted for 7.5 percent of national GDP and 25 percent of export value. Although Karnataka state ranks third in FDI (6%), and the IT industry is reasonably well distributed across a number of Indian cities, Bengaluru has attracted the largest share of FDI in IT of any city for two decades. The stark contrast between Bengaluru's IT industries and high-tech parks and its poorer neighborhoods, however, provides a perfect illustration of the concept of "dual economies," so common in the developing world (see *Geography in Action:* "IT for All" and the Digital Divide).

Stop & Think

▶ Examine how the processes of globalization have produced different development outcomes in India's three primary urban-industrial regions.

Other Regions As is the case with China's capital, Beijing, India's national capital of New Delhi is not a national industrial center. It is the center of government power and home to some of the best world-class universities in the country, in addition to a broad spectrum of industries. Chennai in the state of Tamil Nadu also includes a broad spectrum of industries, especially textiles and light manufacturing. Both cities are particularly strong in automobile manufacturing, a rapidly growing industry in India. Although still in an infancy stage because the government only deregulated the auto industry in the mid-1990s, India now attracts most of the global automobile brands in the form of assembly and parts plants. Of the five cities attracting most of the automobile FDI between 2000 and 2009, New Delhi (33.5%) and Chennai (11.0%) captured 44.5 percent of the total. In an ironic twist vis-à-vis its former colonial master, the Indian conglomerate Tata, India's second largest automobile producer, purchased Land Rover and Jaguar from Ford in 2008. In the same year, Tata also unveiled the "Nano" car with a fuel-efficient, two-cylinder engine in hopes of attracting the young and the growing middle class to become car owners as well as tapping into a potential export market. Quickly building a loyal clientele for

▼ Figure 9-26 **High-tech facility in Bengaluru.** One of the most dynamic urban-industrial regions in India, Bengaluru is among the global centers of information technology.

GEOGRAPHY IN ACTION
"IT for All" and the Digital Divide

India's meteoric rise as a global center of IT research and services in Bengaluru, Mumbai, and Hyderabad has prompted the government and media to promote an "informational economy" as a solution to the country's endemic development problems through improving economic growth, productivity, and competitiveness. It has become clear that managing knowledge and information is critical to prosperity in the global economy, but it is equally certain that this choice of promoting an informational economy is based on the unfounded belief that technology leads directly to human progress. The question to be asked in the Indian context is whether the government's promotion of an "IT for All" (primarily via the Internet) program as a development tool to empower even the poor is an illusion or is working. Whether at the local or national scale of analysis, the geographical spread of IT in its various forms has led to a deepening of uneven development.

The success of the program at the more local scale depends on the adoption of Internet technologies in rural areas where most poor people live. The first problem is educational attainment. The Internet provides greater economic opportunities for those with college and high-school educations, but the technology is of little use to those with primary schooling or less. Although the quality of a college education in India is quite good, elementary school education, particularly in rural areas, is poor, partly due to the low quality of teacher training (Figure 9-4-1). With half of students dropping out before completing the eighth grade, not that many Indians have met the minimum education threshold required for the perceived positive benefits of the technology.

▲ FIGURE 9-4-1 **Village school in Rajasthan, India.** Students receive poor quality instruction because of the lack of infrastructure and dedicated teacher training.

A second barrier to implementing an "IT for All" program is inadequate teledensity (the number of telephones per person), particularly in rural areas. It is estimated that only 8 percent of the population can afford a minimum annual expenditure on telephone and Internet connections. While the small farmer might be able to afford Internet access, the farm-labor class and all those below the poverty line would remain out of touch.

At the larger geographical scale, the digital divide is expressed between the rich and poor, between urban and rural areas, and between the IT enclaves and the poor in the surrounding hinterland. As a model of development, "IT for the entire country" has only reproduced greater economic and social inequalities at all three scales. Meaningful social and economic change to the lives of people in an impoverished agricultural economy must provide a radically new program for access to a higher-quality universal education, one of the most basic of all development goals, and indeed rights.

Sources: Anthony P. D'Costa, "Geography, Uneven Development and Distributive Justice: The Political Economy of IT Growth in India," *Cambridge Journal of Regions, Economy and Society* 4 (2011): 237–251; Joyojeet Pal, "The Developmental Promise of Information and Communications Technology in India," *Contemporary South Asia* 12, no. 1 (2002): 103–119.

the Nano has proven difficult despite its modest price tag. Many Mumbai citizens apparently believe that a "cheap" car lacks status and fails to project the positive personal image they desire.

A more market-driven economy that is subject to foreign competition will almost certainly increase the rate of industrial growth and promote long-term technological modernization. Despite low overall literacy rates, India has a large core of well-educated university graduates. Significant numbers have migrated overseas in a process that is a serious loss of human capital, although many of these overseas Indians continue to be engaged with their home country and are an economic benefit (see *Geography in Action:* Globalization and India's Diaspora). India's industrial transition will also likely mean that millions of workers at state-owned enterprises must find work within the private sector. While affirmative-action policies for backward castes are enforceable in government employment and government-owned industries, this is not the case in the private sector. India's industrial geography may also change substantially as a result of government deregulation. The more dynamic western half of the country, especially the triangle formed by connecting New Delhi, Chennai, and Mumbai (see Figure 9-23), appears poised to benefit the most from new private-sector investment, while the older Damodar Valley heavy-industry region may gradually decline and assume a "Rust Belt" economic position.

Table 9–1 Urban India, 1901–2011			
Year	Total Population (millions)	Urban Population (millions)	Percentage Urban of Total Population
1901	238.3	25.7	10.9
1911	252.0	26.6	10.6
1921	251.3	28.6	11.4
1931	278.9	33.8	12.1
1941	318.6	44.3	13.9
1951	361.0	62.6	17.3
1961	439.2	78.8	18.0
1971	548.1	108.8	19.9
1981	683.3	163.2	23.7
1991	843.3	214.7	25.6
2001	1,027.0	304.0	28.0
2011	1,210.1	377.1	31.8

Source: Data from Government of India.

Urban India

India has always been characterized by stark contrasts between its urban and rural worlds. In the colonial period, this contrast became even sharper as the European-controlled centers of trade and administration became islands of modernization. Since independence, as the prosperity gap between cities and rural regions widened, India experienced increasing rural-to-urban migration. The stream of urban migrants has exceeded the ability of the urban economy to provide adequate employment, housing, and services.

Urban Growth

Levels of urbanization in India, like those of other developing countries, remained relatively low into the middle of the twentieth century. As recently as 1941, only 13.9 percent of the population lived in cities. Since then, the proportion of the population residing in urban places more than doubled to 31.8 percent in 2011 (Table 9-1). The most urbanized states are in the west and the south: Gujarat, Maharashtra, Karnataka, Kerala, and Tamil Nadu. Although these figures are substantially lower than those of most developing countries, they nonetheless represent massive increases in real terms, given the country's enormous population base. It is significant that the Indian government requires a settlement to have 5,000 or more residents to be classified as urban, that is a census "town." This eliminates thousands of communities that, in many countries, would be called small urban places. These particular urbanization statistics, therefore, do not provide a clear picture of the extent of urban population growth.

For most of India's modern period, urban growth rates were based on natural increase and were relatively slow. Since 2001, urban growth is explained by dramatic increases in rural-to-urban migration. Much of this growth has taken place not in the largest urban places, but in medium and small-sized urban places where dispossessed farmers have migrated as a result of the deepening agrarian crisis. The number of cities with 1 million or more inhabitants increased from 35 in 2001 to 53 in 2011 and the number of "towns" increased from 5,161 to 7,935. However, the growth rate of India's largest cities substantially slowed or declined during this same 10-year period. Compared to the urbanization in the West, India's experience is one of "urbanization without industrialization." The three largest metropolitan regions are Mumbai (21.8 million), Delhi (Dilli) (24.2 million), and Kolkata (16.9 million), 3 of the 10 most populated cities in the world. Unlike Tokyo, Shanghai, Hong Kong, and Singapore, however, India's three megacities are not classified as "world cities" in the economic sense because of insufficient finance, transport, and telecommunications linkages with the global economy.

The Urban Poor

Because of the endemic poverty and limited cultural and educational resources of the countryside, first- and second-generation rural-to-urban migrants perceive the city as a place of greater opportunity. Members of the lower castes are particularly attracted to cities because the greater anonymity of urban life frees them from the socioeconomic constraints of village culture. But a severe shortage of wage-based jobs forces most migrants into unskilled work: transport and construction work, food vending, domestic services, and small-scale cottage industries are common sources of income (Figure 9-27). While inhabitants are poor, they are hardworking, proud of their dwellings, and contribute to the urban economy; underemployment rather than unemployment is the norm.

GEOGRAPHY IN ACTION Globalization and India's Diaspora

India has long been an important source of emigrants worldwide, with 30 million Indian citizens and those of Indian ancestry living overseas. During the colonial period, Indian workers provided the backbone of plantation and railroad labor in many British colonial possessions. Two types of Indian emigrants predominate: unskilled and semiskilled laborers working primarily in the Persian Gulf states, and Indian professionals who migrate to primarily English-speaking countries, especially the United States, Great Britain, Canada, and Australia (Figure 9-5-1). Whether historic or modern day, Indians living abroad make up the **Indian diaspora.** The term "diaspora" is traditionally used to describe the dispersal or permanent migration of people from their home regions or countries. Many diaspora Indians today are able to maintain close contact with relatives, friends, and business contacts in India through the time- and space-collapsing technologies of globalization. These include the Internet, e-mail, low-cost telephone connections, more frequent and less expensive air-passenger service, and satellite TV. In an era of globalized interconnections and linkages, the elite segment of nonresident Indians (NRI) plays a significant role in the economic, social, and political development of India.

The economic impact of the Indian diaspora is substantial because NRI wealth flows back home. NRIs in 1991 sent $2.1 billion back to India, but that sum dramatically increased to $55 billion by 2010. India ranked first in the world in remittance volume. Remittances made up 3.1 percent of GDP and exceeded the value of total FDI in 2010. Pakistan and Bangladesh rely even more on remittances. The Indian government promotes NRI investments in stock shares of Indian companies. Numerous successful software firms in Bengaluru, many of which are owned by members of the Brahman caste, are financially linked to NRI software engineers in America's Silicon Valley. Developers of exclusive residential complexes in India explicitly advertise to potential NRI investors. Economic change

▲ FIGURE 9-5-1 **South Asian immigrant commercial district in Chicago.** Many Indian and Pakistani immigrants in the United States are educated professionals, but an equal number are small-business owners as illustrated in a large commercial district on the near north side of Chicago that serves the larger South Asian diasporic population of the city.

at a smaller scale is also common, as half of remittances are used for family needs. Villages in Punjab, Gujarat, and Kerala states, where emigration rates are high, have sections of improved housing as a result of NRI remittances.

The social exchange between the diaspora population and India is profound as well. NRIs remain immersed in Indian popular culture via DVDs and satellite television programming. Indian popular media cater to the NRI population. Bollywood, India's film industry, which is the largest in the world, derives a substantial portion of its annual revenue from its diaspora customers. Bollywood movies increasingly feature the Westernized, affluent lifestyles of NRIs to moviegoers in India through name-brand designer fashions and youthful fun rather than the traditional singing and dancing common to most Indian

films. Many NRI-influenced films are also responsible for promoting "Hinglish" (a combination of Hindi and English words in the same language) among educated young Indians.

Long-distance migration, diasporas, and space-collapsing technologies are all characteristics of globalization. No longer are we able to conceptualize impenetrable national boundaries, closed national economies, or cultures that are exclusively tied to a particular place. For the Indian as well as other ethnic-based diasporas, technology allows people to simultaneously inhabit multiple "identities" and "homes."

Sources: Sunil Batia, *American Karma: Race, Culture, and Identity in the Indian Diaspora* (New York: New York University Press, 2007); N. Jayaram, *The Indian Diaspora: Dynamics of Migration* (Thousand Oaks, CA: Sage, 2004); Pamela Shurmer-Smith, *India Globalization and Change* (New York: Oxford University Press, 2000).

▲ Figure 9-27 **Indian street vendor.** Millions of Indian laborers in cities subsist on low-paying and unskilled jobs.

Without money for durable housing, tens of millions of India's urban poor live in substandard dwellings that vary from single-room rentals to makeshift shelters built of assorted discarded materials. Because many migrants live on land that does not belong to them on the outskirts of the city, their squatter settlements resemble a "village in the city" (Figure 9-28). Squatter settlements are common at railroad right-of-ways, riverbanks, coastal margins,

▲ Figure 9-28 **Shanty community of Dharavi.** Located in central Mumbai, over one million people live in one of Asia's largest slums.

land prone to flooding, and even open spaces adjacent to airport runways. Public services such as electricity, sewerage, and clean water are uncommon. More than 60 percent of the urban population lacks municipal sewerage and water systems, leading to high levels of intestinal diseases, particularly among children, because water extracted from near-surface sources with hand pumps is often contaminated. At the bottom of the poverty chain is a large pool of sidewalk or pavement dwellers who do not even have a roof over their heads. Mumbai is home to 1.2 million pavement dwellers. India's recent economic growth has produced ever wider disparities between the urban rich and poor.

Attempts to "clear" shantytowns have not been very successful for a host of reasons. Squatters comprise a significant voting bloc, and in a democratic country such as India, politicians are eager to garner their votes with plans to improve squatter settlements. There also are many shantytown advocacy groups who lobby the government on behalf of residents. With land prices dramatically increasing with the globalization of the urban economy, developers are eager to clear shantytowns to construct shopping malls and housing for the emerging middle class and the rich, and governments want to construct new highways and parking lots. Both groups commonly promise squatters new living accommodations elsewhere in the city if they decide to relocate. But for good reason, squatters distrust those attempting to remake landscapes for the globalized affluent.

India's Neighbors: Diverse Development Challenges

Economic development challenges for India's neighbors revolve around different problems associated with rural income inequalities and political instability. In rural Pakistan, landlords control most resources, and the flood of refugees from neighboring, war-torn Afghanistan coupled with the rise of extremist Islam has created highly unstable political conditions. In Nepal, extreme rural poverty fueled a 10-year-long Communist insurgency that ended with the overthrow of the monarchy and the recent election of a democratic government. As an icon of developing world poverty, Bangladesh has traditionally relied on huge quantities of foreign aid. Despite political instability, Bangladesh more than most poor countries has attempted to address basic rural development needs, particularly those targeting females. In Sri Lanka, where great strides have been made in terms of reaching basic development goals, the legacy of a 25-year civil war between the government and Tamil rebels remains.

Pakistan

With 180 million people and a 2010 per capita GNI PPP of $2,790, Pakistan ranks as one of the world's most populous and poorest countries. Industry accounts for only 25.8 percent of GDP value added, and over half of the country's export income depends on textile and clothing production. Pakistan has extensive energy resources including natural gas, but lacks the domestic investment

▶ **Figure 9-29 The Indus River–centered state of Pakistan.** As illustrated by the location of major urban places, the most productive economic regions are those paralleling the Indus River valley.

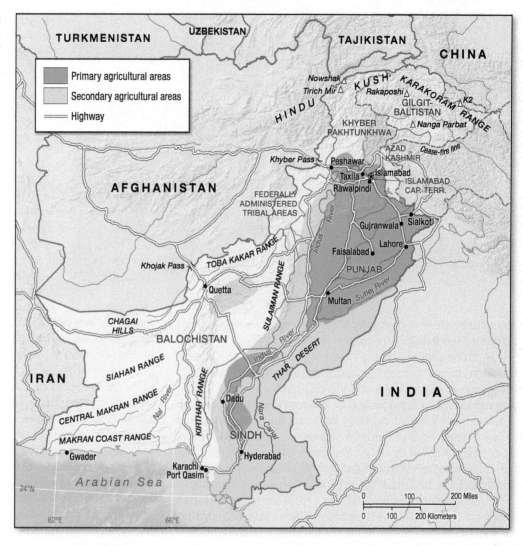

capital needed to develop these resources. In 2011, Pakistan had to import 80 percent of its oil needs, which represented 30 percent of its total energy supply. The government plans to privatize energy exploration and production in the hopes of becoming more energy self-sufficient. A dry climate and frequent cloud-free days make the development of solar energy promising. However, FDI comprised less than 2 percent of GDP during the 2000–2008 period, which suggests that Pakistan is not an attractive country for multinational corporations.

Agriculture centers on wheat, rice, and cotton cultivated by irrigation in an otherwise dry climate within the confines of the Indus River Basin. This district is the largest contiguous irrigated block of land in the world, but it has not fared well in increasing productivity (Figure 9-29). The average annual growth rate of agricultural production has declined steadily and Pakistan has been forced to import a significant amount of foodstuffs for several decades to make up the difference. The main reason for the lack of progress in agricultural development and the resulting food insecurity is the power of the conservative feudal landlord class. A mere 2 percent of Pakistan's richest landlords control a quarter of the country's land, and millions of serf-like peasant farmers depend on landlords for survival. Agricultural productivity is less of a concern for the landlords than the social prestige and political power that land ownership conveys. Maintaining soil fertility and preventing the accumulation of yield-reducing salts in the soil has also proved a serious challenge.

Another obstacle to economic growth is the substantial ethnic and regional tensions resulting from the country's transitional location between the Islamic Middle East and Hindu India. Cultural diversity does not preclude the achievement of national unity, but in Pakistan efforts to forge a national identity have been largely symbolic. Chief among these efforts was moving the national capital from the port city of Karachi to the interior city of Islamabad after the 1947 partition (Figure 9-30). Pakistani nationhood also has been strengthened by its outward religious homogeneity. Some 96 percent of Pakistanis are Muslims and Pakistan is the second most populous Muslim country in the world after Indonesia. Islam is the state religion—the country's official name is the Islamic Republic of Pakistan. Although not a true theocracy, Pakistan has incorporated elements of Islamic law, or Sharia, into its legal system. The early postindependence government used Islam to counter ethnic and linguistic differences and promote national unity. The three dominant ethnic groups are the Punjabis (42%) in the northeast, the Pashtuns (17%) in the northwest, and the Sindhis (14%) in the south; each group speaks a different language.

Pakistan is a federal state, but the provinces often resent central government power largely because provincial governments lack autonomy and the central government in Islamabad controls economic resources. Punjabis produce much of the wheat for the country and dominate the country's heavy-handed military and civilian bureaucracy. Sindh province is a major agricultural and industrial region and includes the country's largest urban area and leading financial center, Karachi, which is also Pakistan's main seaport. Despite the economic and political power these assets provide, Sindh also feels economically shortchanged by the central government. Bloody riots in Karachi slum neighborhoods broke out in 2010 between Pashtun and Mujahir (Urdu speakers who migrated from India in 1947) migrants. More recently, sectarian violence has

◀ **Figure 9-30 Islamabad, Pakistan.** Students and other visitors view the planned national capital of Islamabad from an overlook in Daman-e-Koh Park.

broken out between the majority Sunni and minority Shia as their religious leaders developed extremist views.

It is impossible to explain Pakistan's social, economic, and political problems without discussing the long-term civil conflict in neighboring Afghanistan, and the unintended role the United States has played in these problems. Pakistan's northwest frontier has traditionally been inhabited by Pashtuns, an ethnic group that also occupies portions of southern Afghanistan. Islamabad has historically been wary of Pashtuns, and this wariness only increased when millions of refugees flooded into Pakistan's northwest and Balochistan after the former Soviet Union's invasion of Afghanistan in 1979. Before the invasion, Pakistan's relations with the United States were poor because of Pakistan's dismal human rights record, dictatorial governments, and its desire to become a nuclear power. After the 1979 invasion, this relationship took a positive turn as the United States wished to cultivate friendly relations with Pakistan in order to halt the spread of communism. As a result, the U.S. government lifted its economic embargo against the country, and Pakistan became the recipient of advanced military hardware and economic aid. Pakistan became the world's fourth largest recipient of U.S. military aid and the second largest recipient of U.S. economic aid during the 1980s. In an effort to combat communism, both the United States and Pakistani governments promoted Islam. This program resulted in the increased establishment of Islamic schools, or **madrasahs**, many of which corrupted their original purpose of providing a decent education to promote hatred toward the United States and to create distrust between the various sects of Islam. Because much of the foreign economic aid was spent on consumer goods rather than investments in education, health, and rural development, the growing poverty in the country only fueled this orientation to an extremist form of Islam.

Following the September 11, 2001 terror attacks on New York City and a new rush of Afghan refugees into Pakistan as a result of the U.S. military invasion of Afghanistan to overthrow Taliban rule, this uneasy U.S.-Pakistan relationship entered a new stage. To prop up the regime of General Pervez Musharraf, the United States provided even more military, political, and economic support to this frontline state in the "war on terror." Musharraf, however, could not significantly reduce the role of the country's Khyber Pakhtunkhwa (Northwest Frontier Province) as a home base for terrorists crossing the Afghan border. He assumed greater power by increasing the role of the military in this democratic country, suspending the constitution, imposing a state of emergency, curtailing press freedom, and reducing the role of the judicial branch of government. His unpopularity only increased when Benazir Bhutto, an opposition politician and former Prime Minister, was assassinated in December 2007, purportedly by individuals connected to Musharraf. The negative view of the United States on the part of the average Pakistani was fueled by the U.S. government's continued support of a corrupt and dictatorial government as a means to further its goal to stamp out terrorism in one of the world's most geostrategic regions. The U.S. liquidation of Osama bin Laden, the al-Qaeda leader, on Pakistan soil without the advance knowledge or permission of the Pakistan government further alienated an already hostile popular mood. An extremist form of Islam has become deeply entrenched in Pakistan and further radicalization threatens public security in India, as evidenced by the high-profile 2008 Mumbai bombings.

By most measures, Pakistan has become a dysfunctional country with little evidence of a national civil society. Few foreign or domestic investors have sufficient confidence in the nation's political and social environment to make any substantial long-term financial commitments. The continued economic stagnation has created a pool of more than 2 million Pakistani migrant workers, who have sought employment from Japan to Saudi Arabia.

Stop & Think

▷ In what ways have both domestic ethnic tensions and the conflict in Afghanistan led to unstable political conditions in Pakistan?

Nepal

Nepal, much like neighboring Bhutan to the east, is a landlocked country that occupies a physical and cultural transition zone between Tibet to the north and India to the south. This Himalayan state is divided into three east-west–oriented landform regions, each characterized by distinctive culture groups. The southern region, Terai, is part of the subtropical Gangetic Plain and is populated primarily by Hindi-speaking Hindus. Paralleling the lowland is the central foothill region, where some two-thirds of Nepal's 31

► **Figure 9-31 Buddhist monastery in Nepal.** The Dingboche Monastery is both a spiritual retreat and a destination for visits from a growing number of tourists.

million people reside in the more temperate mountain valleys and adjoining slopes. The most populated portion of the central foothills region is the Kathmandu Valley where Kathmandu is located, the nation's capital and largest city (1 million), as well as the second and third largest cities. Although ethnically diverse, the peoples of the central foothills practice a Hindu-Buddhist syncretic belief system and speak the national language of Nepali, which is a fusion of Hindi and several Tibeto-Burmese languages. The cultural mix in the central foothill region produces colorfully textured scenes of religious structures that represent some of the most recognizable tourist landscapes in the world (Figure 9-31). To the north lies the sparsely populated and spectacular high-mountain Himalayan zone, inhabited by peoples of Tibetan descent.

Nepal has experienced worldwide success in marketing its natural and cultural heritage. Adventure tourism has become a key component of the country's development plans and is a leading generator of foreign exchange. The tourist industry creates substantial domestic employment opportunities. The long-term consequences of adventure tourism in Nepal still remain to be assessed. Environmental degradation from trekking activities includes soil erosion, trailside litter, and deforestation associated with the fuel needs of tourists and the construction of lodges. Loss of the social and cultural integrity of the most heavily affected local populations is also an issue of concern.

As prominent as tourism has become, its significance pales in comparison to the host of problems facing Nepal. With a per capita GNI PPP of only $1,210, Nepal is the poorest South Asian country and one of the poorest in the world. Almost 66 percent of the population is engaged in agriculture, but even this most basic economic activity is threatened by severe soil erosion associated with deforestation. With virtually no industry to gainfully employ the population, which is growing by 1.8 percent annually, land-hungry peasant farmers are cutting the remaining forests at alarming rates. Under such conditions, the female gender burden increases dramatically on two fronts: women are forced to spend more of their daily lives walking farther to obtain water and firewood, and they must work longer because eroded upland soils have forced many males to migrate to the plains in search of work. About 25 percent of Nepalese live below the poverty line, 39 percent of children under the age of 5 are malnourished, and literacy rates are extremely low. Industry is so poorly developed that the value of imports is more than double the value of exports. Nepal's landlocked location results in increased transport costs associated with external trade. With a high dependence on foreign aid as well as on remittances from Nepalese working abroad that account for as much as 24 percent of GDP, Nepal's development challenges appear overwhelming.

Political instability has been the norm for decades and is a byproduct of the economic problems. In 1990, an absolute monarchy was replaced by a constitutional monarchy, but this more democratic system of government did not reduce calls for the complete abolition of the monarchy. In the countryside, an agrarian-based Maoist Communist movement emerged by 1996 and found support among subsistence farmers living in near feudal conditions. In response, the king suspended parliamentary rule in 2005, but later that year an alliance of democratic political parties, plus the main Communist Party, reached an understanding to promote peace. By 2006, the king granted power back to the parliament, and by 2008 the monarchy was abolished. After the 2008 elections, the Communist Party of Nepal controlled 37 percent of the parliamentary seats and the chairman of the party was elected Prime Minister. The 2011 elections resulted in a Communist prime minister who signed an agreement with other political parties to conclude the peace process. While revitalizing the valuable tourism business is critical to economic growth, it is imperative that the government address a variety of issues relating to the rural economy, especially in former conflict regions. These include equitable access to environmental resources in a sustainable way, particularly for women and members of the lower castes, greater local governance, and improved road and irrigation infrastructure. Perhaps Nepal's strategic location between its giant neighbors, China and India, will allow the government to attract economic investment and development assistance that might provide increased political stability. Although not ecologically wise, investment from Nepal's neighbors in hydropower dams would provide the Nepalese government with some of the revenues needed to improve the lives of its population.

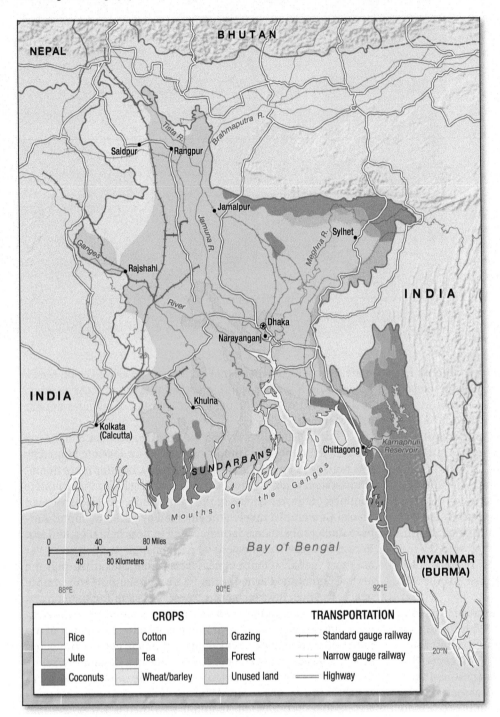

◀ **Figure 9-32 Land utilization in Bangladesh.** Bangladesh is focused around the main drainage channels of the Ganges and Brahmaputra rivers. Almost the entire country is a large deltaic plain, much of which produces rice for the country's large and growing population.

Forty-five percent of the population live below the poverty line, 41 percent of children under age five are malnourished, and only 25 percent of the population is urbanized. Some development indicators, however, do point to sustained progress in meeting people's basic needs. Unlike other South Asian countries, especially Pakistan, female access to education is equal to that of males, total fertility rates have declined dramatically from 3.3 in 2000 to 2.3 in 2012, and child immunization rates are high. Bangladesh is also home to the **Grameen Bank**, a microlending institution that targets poor women and has over 2,000 branches in the country. For this model of rural development, which has been adopted in many developing countries, its founder, Muhammad Yunus, received the Nobel Peace Prize in 2006. The Grameen model has been used to spatially expand communications in Bangladesh's villages. Small loans are provided to women to purchase a cell phone. The phone owner then charges other women a small usage fee. Women use the proceeds to pay back the original loan and pay for their children's education or start a small business. This village "Grameen Phone" model has been successfully adopted in other developing countries where the poor cannot afford their own phone.

Occupying large portions of the Ganges and Brahmaputra river deltas, Bangladesh contains some of the most fertile soils in the world, but exploiting this rich agricultural environment has its risks. Because much of Bangladesh lies barely above sea level, annual cyclones (hurricanes) and their associated tidal surges occasionally cause great loss of crops, animals, and human life. In 1991, an unusually strong cyclone resulted in more than 100,000 deaths (Figure 9-33). Cyclone death tolls are high because the country's 153 million people occupy a territorially small state, making it one of the most densely settled countries in the world (1,062 inhabitants per square kilometer). The poorly developed transportation and communications network does not allow for large-scale evacuation ahead of a storm's landfall. The ecologically

Bangladesh

Few countries in the world have symbolized the developing world's poverty as much as Bangladesh. The country emerged from its 1971 civil war with West Pakistan as a desperately poor nation with few natural resources (Figure 9-32). The war devastated the country's roads, bridges, ports, and electrical power plants. After more than four decades as an independent country, Bangladesh continues to rely heavily on foreign aid, and its per capita GNI PPP of $1,810 remains one of the lowest in the world outside of sub-Saharan Africa.

sensitive Sundarban swamp forest is especially vulnerable to human-induced environmental change (see *Exploring Environmental Impacts:* Threats to the Sundarban).

The economy of Bangladesh rests almost exclusively on agriculture, which absorbs 63 percent of the country's labor force. The primary crops are rice and jute. The industrial economy of Bangladesh is poorly developed, and contributes only 28 percent of GDP value. The relative absence of industries is partly explained by the lack of mineral resources, but more importantly because the privatization of state-owned industries has been slow. Like India, Bangladesh has become an important player in the global garment industry. Bangladesh has become the world's fourth largest exporter of inexpensive ready-made clothing such as t-shirts and jeans, produced primarily in privately owned mills and employing about 4 million women from rural areas. Garment exports account for almost 80 percent of total export value and 45 percent of total industrial employment.

Unlike Pakistan, where the potential for ethnic strife makes foreign investors wary, Bangladesh is more culturally homogeneous. Some 87 percent of Bangladeshis are Muslim, with the rest primarily Hindu. Both religious communities share a common ethnic heritage as well as the Bengali language. Political instability, however, has been a cause for concern on the part of foreign investors. Bangladesh has a functioning constitutional democracy with multiple political parties, but constant charges of election fraud and economic corruption obstruct national political consensus. Political paralysis makes it difficult to move toward greater levels of development.

Sri Lanka

Gaining independence from British colonial rule in 1948 and changing its name from Ceylon to Sri Lanka in 1972, this teardrop-shaped and mountainous country, separated from southern India by the 21-mile-wide (34-kilometer-wide) Palk Strait, is roughly the size of West Virginia (Figure 9-34). Economically, Sri Lanka stands apart from its South Asian neighbors. It is considered a lower-middle income developing country with a per capita GNI PPP of $5,010, substantially higher than that of India. Government programs have significantly bettered the quality of life of the Sri Lankan population: average life expectancy of 75 years matches that of some developed countries and about 94 percent of Sri Lankans are literate. Sri Lanka's Human Development indices are the highest in South Asia, higher than all but three much richer Southeast Asian countries, and slightly higher than the world median, the result of a basic-needs approach to development adopted by a moderate socialist government.

The cultural contours of Sri Lanka are also significantly different from those of its South Asian neighbors. Some 74 percent of Sri Lanka's 21 million people are Sinhalese-speaking Buddhists, with much of the balance being **Tamil**-speaking Hindus of Dravidian descent. This group is divided into "Northeastern Tamils" (13%)

▲ Figure 9-33 **Flooding in the Ganges River Delta.** Farming opportunities in the Ganges delta are matched by the equally great liabilities associated with flooding.

and "Upcountry Tamils" (6%). Migrating from northern India during the first millennium B.C., the Sinhalese established an advanced agricultural state in the drier northern half of the island. Successive Hindu migrations and the extension of southern Indian political influence by A.D. 1000 forced the Sinhalese to migrate into the forested central and southern mountains. They established successive kingdoms until the arrival of the British in the early 1800s. The British imported hundreds of thousands of Tamils to work on rubber and tea plantations established in the central and southwestern areas of the island, increasing the Tamil share of the island's population.

Owing in part to their feelings of racial superiority and having experienced discrimination at the hands of the British, the newly independent Sinhalese-dominated government began to discriminate systematically against the Northeastern Tamil minority whose incomes were slightly higher than incomes in the Sinhalese population. The policy of discrimination included reserving a greater percentage of university places as well as government jobs for Sinhalese to better reflect their proportion of the total population. Northeastern Tamil incomes that had been 20 percent higher than the national average in the 1960s had declined to below average by 1980. Provoked by a Sinhalese-language only policy, government-sponsored Sinhalese settlement in Tamil areas, development projects focused on Sinhalese areas, and the continued poverty of Upcountry Tamil plantation workers, younger Tamil males in the early 1980s formed a guerrilla army that demanded a separate Tamil state in the northern and eastern regions of the island. The quarter-century long civil war that resulted between the "Tamil Tigers of Eelam" and Sri Lankan government forces took 68,000–80,000 lives, mostly those of innocent civilians, and displaced 200,000 inhabitants. After

▶ **Figure 9-34 Sri Lanka, a large tropical island.** Sri Lanka's substantial agricultural economy is based on both estate (cash) crops and subsistence farming. Much of the island's interior remains underdeveloped, supporting either shifting cultivation or forest vegetation.

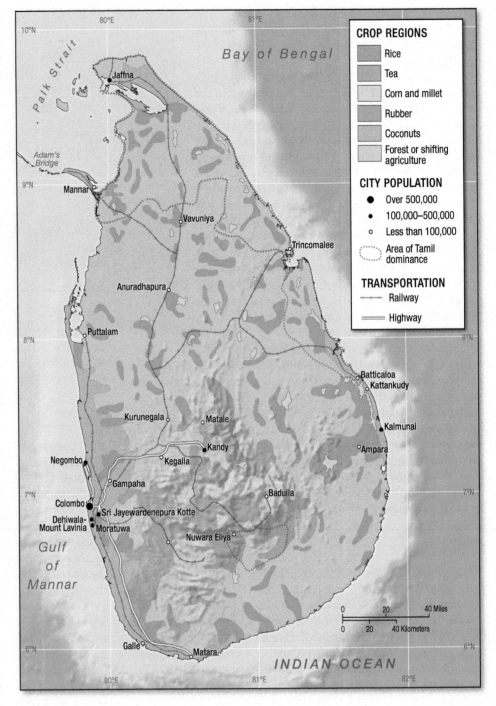

numerous ceasefires since 2002, military activity ended in 2009 and almost all those displaced by the war have been resettled.

The tragedy of ethnic conflict in Sri Lanka was a serious decline in national economic progress, particularly in rural areas where half the population lives. Rural poverty remains high at 25 percent and half of the poor farming households are small-scale farmers. Sri Lanka's export economy continues to rely on a handful of products. Agriculture accounts for 13 percent of GDP and 20 percent of export value, primarily in the form of high-quality tea. Industry accounts for 29 percent of GDP and 71 percent of export value; about half of this comes from textile and garment production. Modern industry is limited to the capital, Colombo. Although the government has been slow in privatizing state-owned industries, it perceives the economic potential of Sri Lanka to be similar to that of the Pacific Rim's economic dynamos. The government has created 10 export-processing zones that have attracted a handful of foreign investors. The resurrection of the once-thriving tourist industry, which was based on the spectacularly beautiful highlands and beaches as well as eight UNESCO World Heritage sites, would be a welcome source of renewed revenue if infrastructure is updated and international tourist interest rekindled.

Stop & Think

▶ Compare the positive basic-needs development accomplishments in Bangladesh and Sri Lanka.

EXPLORING ENVIRONMENTAL IMPACTS

Threats to the Sundarban

The Sundarban is a 220-mile (354-km) wide region comprising much of the combined delta formed by the Ganges, Brahmaputra, and Meghna rivers in southern Bangladesh and a portion of West Bengal state in India (see Figures 9-6 and 9-32). The mostly saline Bay of Bengal part of the delta is home to the single largest mangrove forest in the world, and other brackish and freshwater tree species comprise the tree cover further inland. In a country with only 16 percent forest cover, Sundarban accounts for half of Bangladesh's reserved forest lands and 45 percent of its timber and fuelwood output. This labyrinth of tidal flats and creeks is also home to rare and threatened animals including large cats, dolphins, crocodiles, otters, marine turtles, plus 248 species of birds. The poster animal of the Sundarban as a global biodiversity hot spot is the Bengal tiger; only about 350 remain in this wet environment (Figure 9-6-1). The Sundarban has been populated by humans for centuries and today is home to 12 million people who intensively farm or exploit its rich and diverse fauna and flora through padi rice farming and shrimp farming, wood collection for household purposes, honey and wax collection, and fishing (Figure 9-6-2).

The long-term ecological sustainability of the Sundarban is uncertain because of human-induced environmental changes. The most serious problem is the reduced flow of freshwater and sediment to keep the delta at least stable in size. Increased water extraction by India from the water-hungry upstream portions of the Ganges and Brahmaputra has reduced stream flow. Especially critical was the construction by India in 1975 of the Farakka Barrage across the Ganges River, just before it enters Bangladesh, to divert water during the dry season to the Hoogly River to reduce sedimentation in Kolkata, eastern India's primary international port. Reduced freshwater flow and sedimentation in the delta result in increased saltwater intrusion and degradation of vegetation that anchors these low-lying tidal lands. Global climate change will also cause problems as rising sea levels flood the delta, pushing brackish

▲ FIGURE 9-6-2 **Cast net fishing in the Sundarban.** Fishing is one of the many ecologically sustainable strategies in the struggle to provide households food and shelter.

water further inland and bringing more intense and destructive cyclones further onshore. Sea level rise in the Sundarban is significantly higher than the global average and even a small rise will inundate at least 40 percent of the Sundarban in the next 50 years. As a result, the remaining habitat will become even more overexploited by humans, accelerating environmental degradation. Accommodating the millions of displaced persons over the next few generations will become a major challenge to the Bangladesh and India governments. It is difficult to imagine a positive outcome for this deltaic ecosystem. Although larger-scale forces are difficult to control, important first steps include more local participation in conservation projects, targeted government financing for better resource management, and enforcing protections for this natural resource.

Sources: *Living with Changing Climate: Impact, Vulnerability and Adaptation Challenges in Indian Sundarbans* (New Delhi: Centre for Science and Environment, 2012), http://www.cseindia.org/userfiles/Living%20 with%20changing%20climate%20report%20low%20 res.pdf; Brij Gopal and Malavika Chauhan, "Biodiversity and Its Conservation in the Sundarban Mangrove Ecosystem," *Aquatic Sciences* 68, no. 3 (2006): 338–354; *Sundarbans: Future Imperfect. India: World Wildlife Federation*, 2010, http://www.wwfindia.org/about_wwf/ critical_regions/sundarbans/

▲ FIGURE 9-6-1 **Bengal Tiger in the Sundarban swamp forest.** Adapted to this watery environment, this magnificent large cat is now a threatened species.

Summary

- South Asia is one of the most culturally diverse regions in terms of religion, ethnicity, and language, all of which are a byproduct of centuries of migration into the region. The colonial experience did little to dampen communal strife and the persistent threat of communal violence within countries during the modern period; a broader grassroots development philosophy would help to address this problem. Ethnicity and religion are not the ultimate source of communal tension; unequal access to human advancement is the primary problem. Each ethnic or religious community contains its own economic elite, middle class, and poor underclass. Lasting development will occur only through a significant reduction of class inequalities within individual communities rather than by treating any particular ethnic or religious group as a completely homogeneous entity.

- Relatively high population growth rates hamper economic and social development, but are not the cause of poverty. Greater government attention to reduce the injustices associated with gender inequality in deeply patriarchal cultures would advance economic opportunities for females, which would result in lower fertility rates. Only in Bangladesh and Sri Lanka have gender issues anchored government development policies. Female empowerment is especially low in rural areas where the majority of poor South Asians live. The rural economy is faced by a number of serious problems. Issues of land tenancy, the lack of access to loans, unequal adoption of Green Revolution technologies, and poor educational opportunities are fundamental barriers to the economic advancement of poor farmers. Long-term government inaction to solve the problems of the rural poor has in part led to the rise of religious extremists in Pakistan and communists in Nepal.

- The absence of economic opportunities in agriculture has led to substantial rural to urban migration and this movement in part explains the growth of some of the largest urban regions in the world. Urban economies have been unable to provide adequate employment opportunities for many as evidenced by the proliferation of urban slums without basic public services.

- In India, the unequal spread of the benefits of globalization only exacerbates these income inequalities as capital-intensive information industries rather than more traditional and labor-intensive, export-based consumer product industries have captured the overwhelming attention of state and national governments. Government policies promoting the greater privatization of national economies and engagement with globalization must also consider the many poor who have lost traditional sources of income on which they so desperately depend.

- Equally critical to development growth is the future state of South Asia's environment. In part explained by the rapid growth of population, South Asia's dwindling water resources pose serious problems to agricultural and industrial development. Improving national water distribution infrastructure to prevent massive resource wastage would be a viable alternative to the potential for interstate military conflicts over this precious resource.

Key Terms

Bharatiya Janata Party (BJP) 424
British East India Company 416
Buddhism 414
caste system 414
communalism 418
cyclones 411
Damodar Valley 432

Deccan Plateau 409
federalism 424
gender bias 423
Grameen Bank 442
Harappan culture 414
Hindi 424
Hinduism 414

Indian diaspora 437
Indo-Gangetic Plain 409
Jammu and Kashmir 420
jati 414
madrasahs 440
Mauryan Empire 414
mercantilism 416

monsoon 410
Mughal Empire 414
orographic precipitation 411
scheduled castes 424
Sikhism 414
Tamil 443

Understanding Development in South Asia

1. When do the dry and rainy monsoon seasons occur in South Asia and what specific landform features cause orographic precipitation?
2. Why is water scarcity South Asia's primary environmental concern?
3. What are India's major energy challenges?
4. What is the source of conflict in Kashmir?
5. How has the Indian government addressed the issue of caste?
6. What are the shortcomings of information-based economic development policies in India?
7. What is the role of the state in India's engagement with the global economy?
8. How have non-resident Indians (NRIs) impacted India's economy?
9. What are the primary economic, political, and religious conflicts that hamper economic development in Pakistan?
10. In what ways have gender inequality issues been addressed in Bangladesh and Sri Lanka?

Geographers @ Work

1. Connect British East India Company activities and direct British colonial rule to South Asia's economic development.

2. Explain how India's government has accommodated diversity in terms of religion, caste, and region or political territory.

3. Evaluate the reasons for India's persistent low agricultural productivity and rural poverty.

4. Compare the communal and regional differences in India's population growth with the population patterns of India's neighbors.

5. Describe the economies of India's three primary urban-industrial regions and show how each region has engaged the global economy.

MasteringGeography™

Looking for additional review and test prep materials? Visit the Study Area in MasteringGeography™ to enhance your geographic literacy, spatial reasoning skills, and understanding of this chapter's content by accessing a variety of resources, including MapMaster® interactive maps, videos, animations, RSS feeds, flashcards, web links, self-study quizzes, and an eText version of *World Regional Geography*.

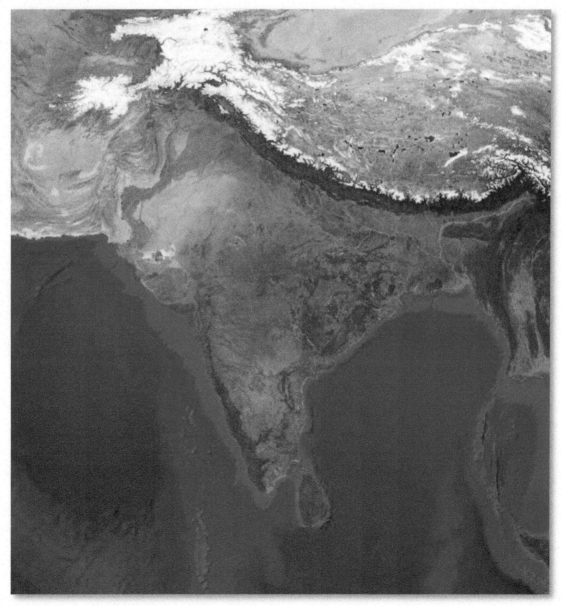

▲ Figure 9-35 **South Asia from space**

10 East Asia

Christopher A. Airriess

Apple, China, and Globalization

Production of components for and assembly of iPads and iPhones, in addition to many other brand-name consumer electronic products such as Kindle, Wii, and PlayStation, take place in 13 Foxconn facilities across nine provinces in China. Foreign electronics companies contract out production to Taiwanese-owned Foxconn, the world's largest electronic components manufacturer, with 1.2 million workers in China. Its largest production facility is located in the far southern city of Shenzhen, adjacent to Hong Kong, where 250,000 employees work at a manufacturing campus that includes dormitories, supermarkets, canteens, gyms, and healthcare facilities. This campus made global news in late 2009 when worker suicides spiked. In response, independent inspectors prompted Foxconn to make changes, and the company agreed to increase wages, reduce overtime, and establish shorter work weeks. Six-day and 60-hour work weeks were commonplace. Problems have continued to emerge at other Foxconn facilities. In May 2011, Foxconn admitted that it used "student interns" from a local trade school to work on assembly lines. In October 2012, several thousand workers walked off the job because of work strain associated with quickly getting the new iPhone5 to market.

These events illustrate the hidden underbelly of globalization and undercut the "cool" image of the Apple brand. Working conditions in China's industries are a byproduct of the geographical flexibility of globalized production chains. Innovative technology design takes place in rich countries and product manufacture takes place in countries with abundant low-cost labor. Solving labor and human rights abuses in China is not easy. Consumer-brand corporations in rich countries operate in China because of the most basic capitalist logic: they can make greater profits because they can be removed from concerns about the lives of workers and their working conditions. Foreign governments could pressure their retailers and manufacturers to enforce labor rights, but business interests insist that this would restrict free trade. Perhaps the ultimate solution is for the Chinese government to aggressively enforce labor laws at the same time that consumers pressure brand-name corporations to do the same with their suppliers.

Read & Learn

▶ Compare and contrast the physical environments of countries in the region.

▶ Understand the sources and causes of China's environmental degradation.

▶ Know how China's agricultural economy has been transformed since the late 1970s reforms.

▶ Recognize how the late 1970s reforms have developed China's industrial economy.

▶ Describe the different economic characteristics of China's three primary and globalized urban-economic regions.

▶ Explore how China's rural and urban population contours have changed as a result of the late 1970s reforms.

▶ Explain the success of both South Korea's and Taiwan's economic development strategies.

▶ Identify the reasons for Japan's past economic successes and for its current economic decline.

▶ Outline the reasons for and impacts of China and Japan's population decline.

▲ The Foxconn plant in Shenzhen is populated primarily by a female workforce, while male supervisors are charged with keeping the assembly process running efficiently.

Contents

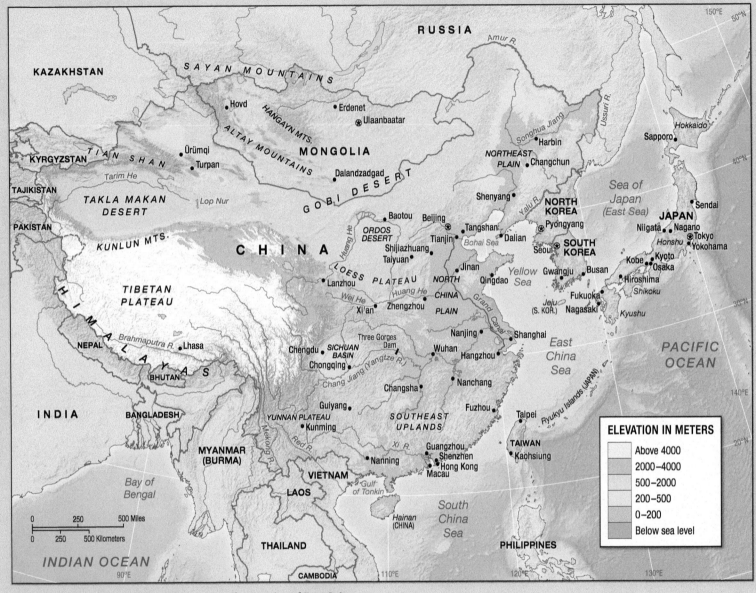

Figure 10-0 **The major physical features and countries of East Asia.**

East Asia is the second of the three subregions of Asia, anchored by the very large state of China (People's Republic of China) (Figure 10-1). North of China is medium-sized and landlocked Mongolia, to the northeast are the small-sized peninsular states of North Korea and South Korea, and to the southeast is the island state of Taiwan (Republic of China), to which China lays claim politically. Also included in the region is the archipelagic country of Japan. For much of ancient history, East Asia's cultural, political, and economic contours have been greatly influenced by China, whose civilization stretches back some 4,000 years. Compared to the rest of the non-European world, East Asia was spatially marginal to the process of European-based colonialism. This is not to suggest that East Asian societies have always been static or unchanging.

Since World War II, East Asian nations have experienced social and economic change at an unprecedented pace. After more than 30 years of adhering to Cold War socialist economic principles that were inward-looking and self-reliant, the Chinese government changed its socialist economy and engaged in the process of globalization. This decision has not only transformed China's own economy but the world economy as well. Japan assumed developed country status by globalizing its economy relatively soon after the devastating effects of World War II. While the pace of Japan's economic growth has faltered since the early 1990s, it remains one of the world's most important centers of economic power. During the Cold War, South Korea and Taiwan became allied with the United States, while China and North Korea aligned themselves

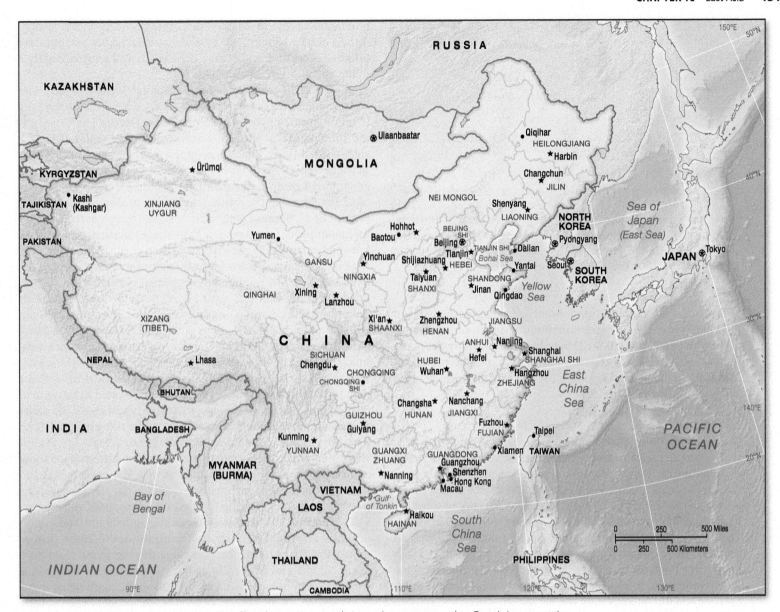

▲ **Figure 10-1 Political units of East Asia.** China's vast size stands in stark contrast to other East Asian countries.

with the Soviet Union. South Korea and Taiwan have subsequently experienced rapid economic growth by following an export-based industrial development philosophy that links their national economies to the global economy. China, Taiwan, Japan, and South Korea together comprise the world's most powerful and dynamic economic region.

East Asia's Environmental and Historical Contexts

A wide range of habitats, from cold, dry grasslands to warm, humid subtropical forests, are found in East Asia. Distinctive societies have developed in these places over thousands of years, using the resources available and transforming their local environments. The

lives and livelihoods of the people and societies of East Asia's nation states illustrate stark contrasts between cultural continuity and rapid economic and social changes.

Environmental Setting: Vastness and Isolation

East Asia includes dramatic and diverse environments because of its great longitudinal and latitudinal extremes. Situated on the east side of a large continent and with similar latitudinal dimensions, comparisons with the lower 48 United States are warranted (Figure 10-2). China and the United States are the third and fourth largest countries in the world, with China's area slightly exceeding that of the United States. If only the lower 48 states of the United States are considered, East Asia and the United States share a similar longitudinal (east–west)

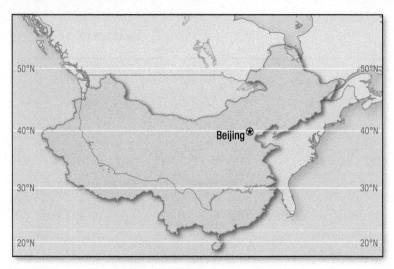

▲ Figure 10-2 **Size and locational comparison of East Asia and the conterminous United States.** China occupies an area of about the same size and latitude as the United States.

dimension. The latitudinal (north–south) dimension exhibits slight differences in that a small portion of southeastern China extends as far south as Cuba and its northeasternmost territory lies at the same latitude as Canada's Quebec Province. In many respects, East Asia shares landform and climate attributes with the lower 48 of the United States as well.

Landforms

China and the United States' east-west similarity extends to eastern river basins and uplands contrasted with drier, but elevated uplands and highlands in the west (see chapter opener map). Perhaps the most human-impacted environmental region in China is the Loess Plateau, an elevated tableland 4,000–5,000 feet (1,200–1,500 meters) above sea level situated between the Ordos Desert and the North China Plain. Loess is a term used to describe a fine, yellow, dust-like soil deposited thousands of years ago by winds originating in the Mongolian grasslands to the north of China. Loess is very fertile, but at the same time highly susceptible to erosion. Infrequent but strong summer thunderstorms in this otherwise dry climate region have caused gullying on the slopes of hills and mountains. For centuries this has resulted in soil carried away to streams flowing into the Huang He, appropriately translated as the "Yellow River" because of the yellow hue of its loess sediments. Farmers have for centuries terraced the hillsides of the Loess Plateau in an effort to create level areas on which to cultivate crops as well as keep the soil in place (Figure 10-3).

Downstream from the Loess Plateau is the North China Plain, an extensive riverine surface built up from the silt deposited by the Huang He on an ancient shallow continental shelf. Through the centuries, the river has deposited loess-derived silt, gradually elevating the riverbed and the water the river carries above the surrounding plain. This necessitated the construction and continual raising of levees or dikes to hold the river within its channel. Periodic levee breaks and river-channel changes have brought frequent devastating floods to this densely settled region for thousands of years. It is no wonder that the river's nickname is the "river of sorrow."

The Sichuan Basin, located in south-central China is a large interior basin. Surrounded by high mountains and plateaus, the Sichuan Basin is densely inhabited by an agricultural population situated primarily on the Chengdu Plain. It was the northwestern

mountainous edge of the basin that experienced a horrific 7.9 magnitude earthquake in May 2008; more than 80,000 people were killed, 4,000 children became orphaned, and approximately 5 million people were left homeless.

Between the eastern edge of the Sichuan Basin and the flatter riverine plains of south-central China is a section of the Chang Jiang (Yangtze), or "Long River," called the **Three Gorges**. Here, river water is forced to flow through a narrow, 150-mile (240-kilometer), steep-walled valley, which is no greater than 350 feet (107 meters) in width (Figure 10-4). Within this hydrologic context the Chinese government completed construction of the Three Gorges Dam in 2012, the largest in the world in terms of installed power capacity (Figure 10-5). Because of the size, cost, and the potential environmental impact of the dam, few large-scale infrastructure projects have engendered such great debate. Government arguments for the dam include much-improved river navigation by large ships. The hydroelectric power generated is expected to help reduce dependency on high-sulfur coal, which produces intolerable levels of air pollution and acid rain. The government also claims that the dam will dramatically reduce serious downstream flooding. The dam's detractors, both domestic and foreign, rightly claim that with technological advances in energy generation, mega-hydroelectric projects are now seen as uncompetitive dinosaurs. Coal-fired plants with pollution controls, for example, are estimated to be twice as cost-effective. Critics have been extremely vocal about the dam's tragic environmental and human impact. The rising reservoir level has submerged habitat for many endangered animal species, including migratory cranes and river dolphins, and it destroyed about a thousand culturally important sites as well. The direct human impact was relocating approximately 1.5 million people because rising waters required submerging 13 cities, 140 towns, 1,352 villages, and 115,000 acres (47,000 hectares) of valuable agricultural land.

After leaving the constricted confines of the Three Gorges region, the Chang Jiang meanders sluggishly across the flat surface of the Middle and Lower Chang Jiang Plain. The middle plain is surrounded by numerous low mountains and hills, dotted by many shallow lakes. These lakes act as flood reservoirs for the Chang Jiang

▼ Figure 10-3 **China's Loess Plateau.** Loess is a very fertile soil, but prone to erosion. To cultivate on steep slopes, terracing is essential if erosion is to be controlled. Where terracing is absent, heavily dissected slopes often are the result.

▲ Figure 10-6 **Lower Chang Jiang Plain.** Agriculture in this area features a mixture of rice paddies, fishponds, and field or tree crops. In the foreground rice paddies are separated by narrow footpaths. The more rectangular shape of fishponds is observable in the middle ground. Raised dikes separate fishponds and are the site of vegetable, grain, and fruit cultivation. The haze reflects pollution generated by nearby urban areas.

▲ Figure 10-4 **The Three Gorges region of south-central China.** Water levels of the Chang Jiang were significantly raised with the completion of the Three Gorges Dam.

during the high-water summer monsoon season. The lower plain of the Chang Jiang sits less than 10 feet (3 meters) above sea level, and this wetland environment is characterized by a patchwork of rice paddies and fishponds, laced by a dense network of streams and canals (Figure 10-6).

One of the most physically stunning of China's eastern regions is the Yunnan Plateau, which occupies the environmental transition zone between the cold Tibetan Plateau and eastern monsoon China. With elevations ranging from 5,000 to 9,000 feet (1,500–2,700 meters), this dissected upland is laced by mountainous spurs of the Tibetan Plateau. Between the mountain ridges are deep river gorges and small upland valleys dotted with agricultural settlements. Stretching along China's southeast coast are the Southeast Uplands, which average 3,000–4,000 feet (900–1,200 meters). With rugged hills and low mountains reaching down to the narrow

coastal plain, the region includes most of China's deepwater harbors. The last important environmental region of eastern China is the Northeast Plain, an extensive rolling hill surface that is an important grain farming region. Although threatened by deforestation, the surrounding uplands are China's most important source of timber.

Two major environmental zones occupy the vast area of western China, both of which are expressions of the Indian plate colliding with the Asian plate. The first and largest is the **Tibetan Plateau**, an area that encompasses approximately 25 percent of China's territory. It is the largest, most elevated plateau in the world, and is sometimes referred to as the "rooftop of the world" or the "third pole" of the world because of its cold and inhospitable environment (Figure 10-7). Averaging 13,200 feet (4,000 meters) in elevation, most of the population is found on the wetter and warmer southern edge. North of the Tibetan Plateau is the Tarim Basin, much of which is occupied by the Talka Makan Desert. It is a region of internal drainage where rivers flowing from the lofty Tian Shan and Kunlun Mountains disappear into the desert sands, sometimes forming salt lakes such as at Lop Nor or Turpan, the latter lying at 500 feet (150 meters) below sea level. Pockets of agricultural settlement are found along a belt of river-fed oases located at the mountain–desert edge (Figure 10-8).

Mongolia's landforms can be divided into two generalized regions. Much of the southern half of the country is a broad flat or

◀ Figure 10-5 **The Three Gorges Dam, opened in 2012.** The world's biggest hydropower project, located in Hubei Province, provides flood protection and improved navigation deep into interior China.

▲ Figure 10-7 **Tibetan Plateau.** The Tibetan plateau is a remote and sparsely settled region of southwestern China. Summer nomadic movement of herders with their yaks to high elevation pastures is a traditional way to access seasonal grazing; animal herding is one of the ways in which many of Tibet's people exploit their limited resource base.

Like South Asia, East Asia has a monsoon climate, but is modified by its east coast location and its more northerly and continental position (see Figures 9-3 and 9-4, pages 410 and 411). The wet summer monsoon envelops southernmost China by April and lasts approximately six months, but it does not reach northeastern China until June. Rainfall amounts are greatest in southern China where the annual average exceeds 50 inches (1,250 millimeters). As a result, the lowlands of eastern China south of the Chang Jiang River experience warm and sultry summers similar to those of the southeastern United States. Interior regions receive substantially less precipitation because of the shorter summer monsoon and their greater distance from the sea. A comparison of summer monsoon precipitation totals for Hong Kong, Wuhan, and Beijing illustrates this phenomenon (see Figure 9-3). Similarly, western China and Mongolia receive very little precipitation during their short wet season. Like much of coastal East Asia, China is also subjected to hurricane-like **typhoons.** Originating in the Pacific Ocean and lashing the Philippines, typhoons most frequently make landfall between Shanghai and Hong Kong with high winds and heavy rain (Figure 10-10). Compared to Atlantic hurricanes, Pacific typhoons are more numerous and powerful because the Pacific Ocean is larger and warmer.

By October, the summer monsoon retreats in advance of the northerly winds associated with the dry winter monsoon. Compared to the wet summer monsoon season, the dryness of the winter monsoon is stark, particularly in north central China. Beijing receives only 20 percent of its annual precipitation during the nine-month dry season (see Figure 9-4). During the winter monsoon north-central China experiences cold, dry winds with clear and bright days, broken by occasional light snowfalls. Because of the Qin Ling mountain barrier, the middle Chang Jiang River Valley rarely experiences hard and long freezes. The average January

undulating elevated surface; the most southern stretches comprise the Gobi Desert with the southeast being short or tall grassland. In the far northwest are the tall Altay Mountains and the lower and more eroded Hangayn and Tian Shan Mountains in north-central Mongolia. North and South Korea occupy a peninsula with rugged, but not extremely elevated north-south mountains occupying the eastern two-thirds of both countries. Only along the western side of the peninsula are there unbroken narrow plains and lowlands.

The east to west trending mountain chains and basins characterizing the East Asian mainland are a result of active tectonic plate collisions along the western edge of the Pacific, known as the **Pacific Ring of Fire.** Japan, a long north to south archipelago, was created by the meeting of four tectonic plates: the Pacific and Philippine plates, which are subducted or thrust under the Eurasian and North American plates, create crustal folds that express themselves as mountains. A young and dynamic geological environment, these mountains and hills comprise 80 percent of Japan's land surface. Mountains are rugged, with steep slopes, but they are not very high by world standards. Most peaks are below 6,000 feet (1,800 meters), but 10 peaks higher than 9,000 feet (2,700 meters) dot the Japanese Alps in central Honshu. Mount Fuji, perhaps the world's most famous volcanic cone, reaches up to 12,388 feet (3,776 meters). The balance of Japan's land surface is composed of flat surfaces found either as terraces at the downslope edge of mountains or along relatively narrow coastal plains (Figure 10-9) Most of the largest pockets of flat surfaces are found along the Pacific side of Honshu Island. Taiwan too is formed at tectonic plate margins with a mountainous east coast and a broad north to south plain along the western third of the island.

▼ Figure 10-8 **Oasis settlements in Xinjiang.** Farmers capture water from mountain-derived subsurface streams to make the soil bloom at the mountain–valley edge. Many settlements date to the Han period, when they were critical travel links along the great Silk Road.

▲ Figure 10-9 **Narrow coastal plain in Japan.** Because usable land is so scarce, Japan's coastal plain is intensively utilized, both for agricultural and industrial purposes.

temperature in Wuhan is only about 40°F (4.5°C). In contrast, the winter monsoon brings primarily dry and frigid winds across the northern and western grassland and desert basins of this region.

Because Japan assumes an elongated north-to-south orientation, has a mountainous landscape, and is surrounded by water, its climate is varied and complex. Although the range of summer temperatures is not substantial, Hokkaido experiences cool summers, and from northern Honshu to southern Kyushu, summers become increasingly warmer. Winter temperatures vary more; the average January temperature in northern Hokkaido is 28°F (−2°C), in central Honshu 39°F (4°C), and in southern Kyushu 47°F (8°C).

The wet summer monsoon is characterized by Pacific air masses from the southwest flowing across the archipelago. Summer is wetter than winter, and southern Japan is wetter than the north, but there is no distinct dry season, as Tokyo's precipitation totals attest (see Figures 9-3 and 9-4). The summer monsoon rarely penetrates as far north as Hokkaido. Much like the warm Gulf Stream off the east coast of the United States, the Kuroshio Current heightens humidity levels in southern and eastern Japan. Japan averages three typhoons per year, far fewer than China. With river length between the mountains and coast so short and river gradients so steep, the potential for flooding on the densely populated coastal plain is ever present. The winter monsoon in Japan is peculiar because of the Sea of Japan and Japan's mountain backbone. Cold and dry winds that sweep off the Asian mainland and then across the Sea of Japan pick up moisture and precipitate large amounts of snow along the west coast and western-facing slopes of the mountains. On the more populous eastern side of the mountains there is little snowfall because of drier and milder conditions. Niigata on the west coast receives an average annual snowfall of 78 inches (2,000 millimeters), but Tokyo's total is just 5.5 inches (140 millimeters).

Stop & Think

Why is western China so much drier than eastern China?

▲ Figure 10-10 Typhoon Rananim in August 2004 moves to the northwest to make landfall along China's densely populated central coast. The northwest Pacific experiences three times the number of such storms when compared to the northwest Atlantic.

Environmental Challenges

A large population coupled with rapid economic growth raises the question of whether China's environment has suffered as a result. Overall the government's drive to modernize has progressed at the expense of the environment, whether during the earlier Communist period or the present reform era. Most industries during the prereform era were dominated by high energy and resource consumption, where quantity rather than quality and efficiency mattered. Since the reform period, environmental degradation has only worsened because many more and different types of polluters now exist. Pollution has become even more serious as the average household becomes more affluent and consumes more. Although environmental protection officially became a government priority in the early 1980s, legislation and enforcement have been lax. Enforcement of environmental protection has been weak because local officials are judged by the central government on their ability to promote economic growth, and enforcement of environmental laws is perceived as antibusiness. Many current and future environmental problems face the government, but only two of the most serious are discussed here.

Perhaps the most basic environmental challenge facing China is the provision of clean freshwater. With only 7 percent of global water resources and approximately 20 percent of the world's population, clean water is in short supply. To aggravate the problem, 80 percent of annual precipitation occurs in the south and 70 percent falls during the summer months. In addition, per capita natural water resources are low by developing country standards with only 28 percent of the global average. Between 1980 and 2005, per capita

455

water availability declined by 25 percent. Because of high demand levels coupled with wastage, two-thirds of the country's cities experienced water shortages in 2010. The 2008 Beijing Olympics consumed 300 million cubic meters of water from surrounding provinces without financial compensation. While the majority of China's water resources are in the form of lakes and rivers, the dramatic rise in subsurface freshwater extraction in coastal cities has resulted in substantial saltwater intrusion during the dry season.

Water wastage is substantial because the low water prices paid by users provide few financial incentives to repair distribution systems or reduce consumption. Factories in China recycle less than 30 percent of water as compared to up to 80 percent in developed countries. Mismanagement also causes severe shortages in some regions. Because so much water is diverted for irrigation, hydropower generation, and industry use in the upstream stretches of the Huang He, water shortages are frequent in the North China Plain. The lower courses of the river near the delta have in some places been running dry since 1985 and flow has been insufficient to transport silt to nourish coastal estuarine environments. These water problems dramatically impact agricultural production in provinces north of the Chang Jiang. The river supports approximately 64 percent of China's cultivated land but supplies only 17 percent of exploitable national water resources. The planned construction of a north–south water transfer project to transport water from the Chang Jiang to the dry north further complicates the issue.

Water pollution is also a serious problem as examples from 2010 show. One-quarter of China's surface water is too polluted even for industrial use and less than half of total water supply is drinkable. The World Health Organization (WHO) estimates that almost 100,000 people die annually from water pollution-related illnesses and that 75 percent of diseases are linked to substandard water quality. Approximately 320 million rural inhabitants do not have access to safe drinking water and approximately 190 million use drinking water containing excessive levels of hazardous substances. Agricultural activities and industrial and municipal wastewater contribute equally to the problem. Chemical pesticides, fertilizers, and livestock waste in surface runoff are the primary forms of rural-based water pollution. Animal wastes have increased dramatically because of the rising demand for meat among a more affluent population. Untreated residential and industrial wastewater enters 75 percent of rivers flowing through urban areas, making their water unfit for both human consumption and fishing. Eighty percent of China's sewage and other wastes enter rivers and lakes untreated. Many township and village enterprises that have become critical to China's economic growth drive are the sources of much of this pollution. The government has set five-year targets to improve water quality, but how long it will take to substantially improve the country's water resources is debatable.

The other major environmental problem is air pollution in older urban areas and in those regions experiencing dramatic postreform growth. China surpassed the United States as the largest producer of greenhouse gases in 2006, but two important observations qualify this fact. China's per capita CO_2 emissions are relatively low by global

▲ **Figure 10-11   The signature CCTV building in Beijing shrouded in air pollution in 2013.** China's major cities often experience dangerous levels of air pollution, particularly during winter months.

standards. The offshore movement of manufacturing from the rich world to China in part explains high CO_2 emissions. Nevertheless, the skies of China's major cities, especially in the north and during the dry season, are some of the most polluted in the world. Beijing residents are literally choking in a city that experienced only 25 "healthy air" days in 2006 (Figure 10-11). One important source of pollution is high-sulfur coal; Beijing's residents and factories are highly dependent on coal as a heating and energy source. An estimated 70 percent of the coal used fails to meet the government's stated environmental standards. Transport is another major source of air pollution. The Beijing city government has mothballed thousands of older polluting buses and taxis, only to see old vehicles replaced by a dramatic increase in privately owned motorcycles and automobiles. Respiratory problems are widespread because one-third of China's population are exposed to harmful levels of pollution. Government officials estimate that air pollution levels will dramatically increase by 2020 if the consumption of dirty energy and automobile use is not curtailed. A barrier to reducing auto emissions is the importance of the auto industry to the national economy and growing consumer affluence. Although there were only seven automobiles per thousand people in 2005, that number is projected to quadruple by 2020. Problems of global warming aside, the various negative impacts of air pollution cost China 3.8 percent of its GDP in 2007 and in 2010 coal emissions contributed to some 400,000 premature deaths.

Japan's status as a postindustrial economy does not protect it from many serious environmental problems. A clear consequence of Japan's early development path was intensified environmental pollution. Until the late 1960s, Japan had done less than any other major industrial country to protect its natural environment from the effects of uncontrolled industrial and urban development that favored corporate profit over public health. Air and water pollution became acute. Beloved Mt. Fuji could rarely be seen from Tokyo, which was shrouded in a brownish smog most of the time. Fishing was halted in Tokyo Bay. Poisonous chemicals, such as organic mercury and cadmium, entered the food chain in certain localities, which produced horrible birth defects and suffering.

Spurred by growing public protest in the early 1970s, the Japanese government finally recognized the seriousness of the problem and created the National Environment Agency (similar to the Environmental Protection Agency established in the United States at about the same time). Billions of yen have since been invested in environmental cleanup, education, and protection and the results are beginning to show. With its immense wealth and human resources, the government has managed to turn the tide and Japan is unquestionably much cleaner today than it was 40 years ago. Nevertheless, hundreds of toxic hot spots remain around the country in addition to many illegal waste dumps that dot the countryside as urban places rid themselves of waste. Although environmental groups have been excluded from the overly bureaucratic government decision-making process, thousands of grassroots environmental organizations have emerged to contest both government and corporate power by framing the nature–society debate around local issues.

Stop & Think

> Based on levels of economic growth, how and why are China's air pollution problems different from India's?

Historical Background: Ancient Roots and Global Reach

Long histories of separate, isolated development, mixed with occasional bursts of economic, political, or cultural contact, characterize the current nation states of East Asia. The result is a distinctive web of regional diversity, combined with deeply rooted cultural and political systems. After a brief overview of these historical processes and population dynamics, the details of distinctive patterns are outlined in the country presentations.

East Asia's Historical Profile

Individual countries of East Asia possess their own historical experiences, but common historical attributes do exist. China has heavily influenced the religious, philosophical, and linguistic traditions of all East Asian countries except Mongolia. Unlike the regions of South Asia and Southeast Asia, Western colonial rule was rarely imposed in East Asia. The long-term process of westernization of indigenous cultural, social, and political values before the modern era was far less pervasive compared to other non-western regions.

First emerging in north central China, successive dynasties conquered the more humid far south before annexing the drier lands to the far west. With this geographical spread of Chinese civilization came the application of uniform systems of law, political administration, and written language that evolved into a unified Chinese cultural space. Almost 2,000 years of imperial rule came to an abrupt end with the arrival of Western merchants and militaries in the mid-nineteenth century and the surrender of sovereignty to Western economic interests in enclaves along China's coast. The breakdown of a unified China caused internal political fragmentation and the eventual rise of competing political visions of a modern China. Communists and Nationalists engaged in a protracted civil war that ended in a Communist victory after the defeat of the Japanese invasion in World War II.

Located along the northeast coast of the Asian continent, Japan was able to remain distant from the larger changes brought about by westernization. Unified by the emperor and a hierarchically structured feudal society, Japan's military rulers engaged in a self-imposed isolationist policy until the end of the nineteenth century when the arrival of Westerners challenged the political status quo. Unlike China, Japan's rulers embraced those aspects of Western culture and technology that served national interests and became East Asia's first industrialized country. Japan followed the development pattern of the West by participating in the international economy and by becoming a regional colonial power until its defeat in World War II. Throughout much of their histories, both Korea and Mongolia remained in the shadow of more powerful neighboring political powers. For much of its early history, the Korean Peninsula occupied the edges of imperial China's influence, but later became economically integrated into Japan's colonial empire. Mongolia was even more marginal to China's imperial ambitions, and became a client state of the Soviet Union during the early twentieth century.

Population Contours

With a population of 1.58 billion and 22 percent of the world population, the countries of East Asia comprise the second most populated world region. China (1.35 billion) and Japan (128 million) are the first and tenth most populated countries in the world. South Korea (48.9 million), North Korea (24.6 million), Taiwan (23.3 million), and Mongolia (2.9 million) round out the balance of East Asian countries. In addition, some regions of individual countries possess high population densities (Figure 10-12). Even with large expanses of open spaces in China and Mongolia, the population density of East Asia is about three times the world average. The region also has some of the most populated urban areas in the world.

With about 85 percent of East Asia's population, China is the obvious population giant of the region. Population densities are highest in the eastern third of the country; some 94 percent of China's population resides in the humid eastern region, which constitutes only 43 percent of China's land area. Within this eastern region, about 40 percent of the population occupies some 10 percent of the land area, coinciding with the alluvial valleys of the lower Huang He and Chang Jiang and the coast. Population densities in some of these more crowded rural regions may easily reach 250 persons per square mile (100 persons per square kilometer). Population densities progressively decrease with increasing distance from the coast and some provinces are far more populated than others. Much of the drier and elevated western half of China is sparsely populated. Likewise, drier Mongolia to the north of China is sparsely populated. Environmentally dominated by desert and grasslands, Mongolia's average density is three persons per square mile, which makes it one of the lowest population density countries in the world. Population concentrations in North and South Korea are found in the western foothills and lowlands where both national capitals are located. In Japan, approximately 80 percent of the national population is found on the largest island, Honshu, and the majority of this population is strung along the coastal plain and foothills of the Pacific or the southeast coast where the largest urban regions are located. Dominated by an eastern mountainous backbone, the vast majority of Taiwan's population is concentrated along the western alluvial plains dotted with large urban regions.

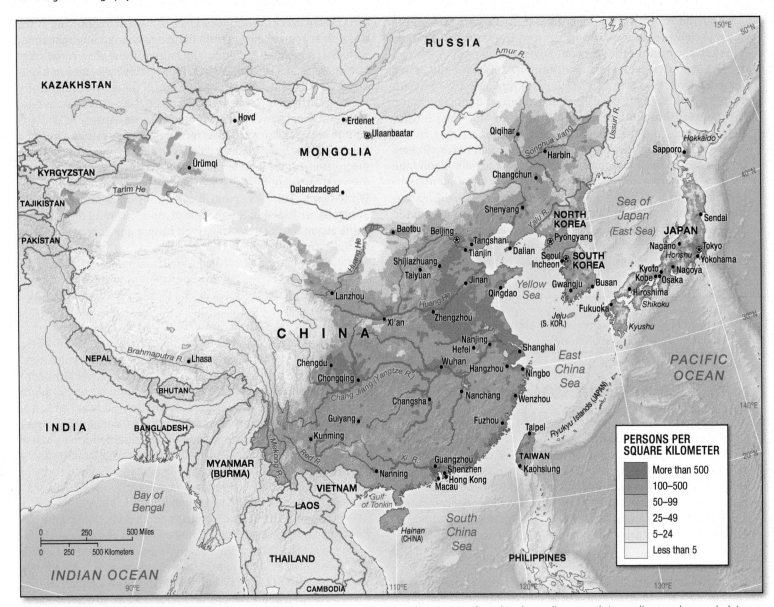

▲ Figure 10-12 **Population distribution of East Asia.** The greatest population densities are found in the well-watered river valleys and coastal plains. Eastern and western China exhibit one of the sharpest population density transitions of any country in the world.

China: Emerging Global Powerhouse

Perhaps no country attracts more global media attention than China. This is for good reason! China's fast growing economy has directly impacted most other national economies around the world. Think about all the consumer products, manufactured domestically just a generation ago, that are now imported to the United States from China—furniture, sports shoes, clothing, toys, and a wide array of electronic goods ranging from DVD players to laptop computers. As a result, China is perceived in some Western circles as a serious economic threat. Global media attention is also focused on China because it seems to lack the political changes that should hypothetically accompany this economic transformation. It is important to remember that China has not adopted a form of capitalism common in the West. Beijing has accepted capitalism as the primary engine of economic growth, but within the context of strong government control. The implications of China's increasingly globalized economy include growing internal regional economic disparities, the transformation of agriculture and food habits, and massive population movements.

Spatial Evolution of Early Chinese Culture

Chinese cultural, economic, and political institutions have been greatly modified over time, but their long-term persistence explains why Chinese civilization is often referred to as the world's oldest surviving culture. This distinct culture was perceived by the Chinese themselves as being indigenously derived, without substantial external influences. The Chinese, hemmed in by mountains to the north, west, and south and the Pacific Ocean to the east, came to think of themselves as the bearers of a superior civilization that possessed a deep sense of cultural unity. The Chinese rulers believed they lived in a special place, the **Middle Kingdom** or *Zhongguo*.

China's culture hearth emerged on the great bend of the Huang He that includes the Wei River Valley and adjacent loess lands. From 1766 B.C. to A.D. 1912, 15 major dynasties ruled China, each important in its own right. During some dynasties, very important cultural, political, and economic behaviors emerged, and it is these dynasties that are given attention here (Figure 10-13).

The first authentic Chinese dynasty was the Shang (1766–1122 B.C.), centered where the Huang He enters the North China Plain. The Shang contributions to Chinese cultural identity included the development of metal working and the distinctive character-based written language. Most importantly, the Shang period marked the transformation of Chinese society from one based on

egalitarian agricultural communities to one oriented toward socially and occupationally stratified urban centers. Government officials had the power to forcibly enlist peasants for unpaid labor during times of war or when government construction projects required large inputs of labor.

The militaristic **Zhou dynasty** (1027–256 B.C.) originated farther upstream on the Huang He and established their capital at a site close to modern-day Xi'an. The Zhou extended the imperial domains southward beyond the Chang Jiang, farther west up the Huang He and Wei River valleys, as well as to the northeast. One of the most important cultural traditions that emerged during the end of the Zhou period was the ethical philosophy of **Confucianism**, based on the writings and teachings of Confucius. Confucianism emphasized that humans were moral by nature and that immoral

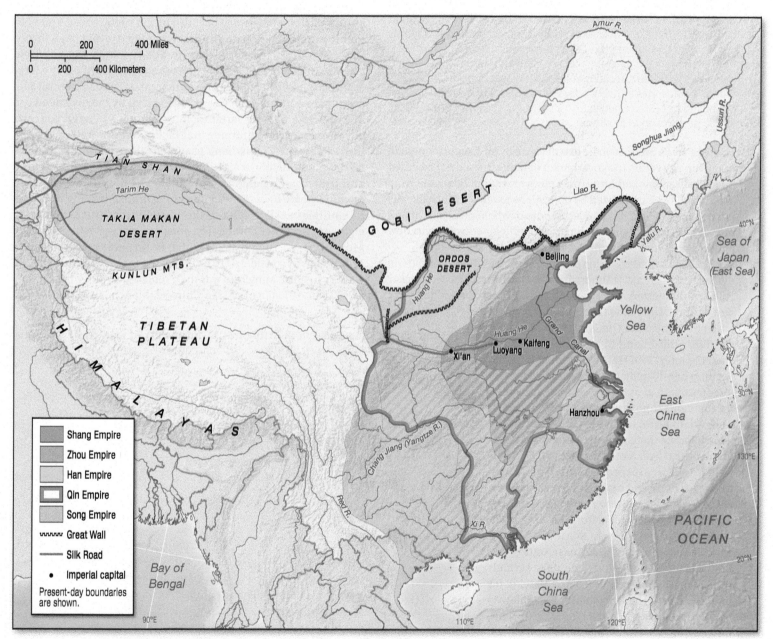

▲ Figure 10-13 **Spatial evolution of Chinese political territory.** Gradually digesting more territory in the south over many centuries, Chinese political control and accompanying Han culture spread out over thousands of years from its core at the great bend of the Huang He River.

▲ **Figure 10-14 Great Wall of China.** Completed during the third century B.C., this 3,000-mile-long (4,800-kilometer-long) structure was constructed to separate the pastoral nomadic life of the northwest from the sedentary agricultural life of the south. Essentially meaning "long wall" in Chinese, sections of the wall were designated as UNESCO World Heritage sites in 1987. Today many sections of the Great Wall are crumbling because of neglect. This particular section remains intact, and some are open to tourist excursions.

behavior was a product of the loss of virtue by all. Confucius argued that authority within the government or family should not be based on formalized law and brute force punishment but rather on living a virtuous life characterized by obligations to others. Much like the philosophies of Socrates and Jesus, both of whom preached "man's moral duty to man," the teachings of Confucius constituted a new ethical system that has endured in East Asia to the present.

The next dynasty to bring in a new imperial "order" was the Qin dynasty (221–207 B.C.). Although it was short-lived, the Qin dynasty ruled in the period during which China became a unified state and culture. The name China probably originated from the word Qin. The totalitarian Qin emperor Shih-huang-di forged a unified and spatially integrated state by dividing imperial territory into 40 military regions, each administered by a staff of officials who were appointed by the central government based on merit rather than birth. He has become universally known for the thousands of terracotta warriors found buried with him to protect him in the afterlife. Another aspect of Qin rule that defined Chinese culture and space was the completion of the Great Wall, which served to spatially separate China's political core from the drier northwestern grasslands of Inner Asia (Figure 10-14), which posed a barrier to Chinese settlement. Lower rainfall prevented sedentary agriculture, which was a defining characteristic for being Chinese. The Qin dynasty was followed by the **Han dynasty** (206 B.C.–A.D. 220), a period so important that the Chinese today still refer to themselves as "people of the Han." As organizers of the first large-scale empire in East

Asia, the militarily powerful Han conquered a wide corridor of far western lands in Inner Asia from pastoral nomadic kingdoms and brought these new territories within the Chinese orbit. The development of the great **Silk Road** indirectly connected the Chinese and Roman empires through traders who traveled Inner Asia as merchant intermediaries. The Han then turned their energies to the wetter southern frontier, where opportunities to secure an abundant source of food were greater. This south-central region became key to the economic and political power of the northern-centered government. Although constructed during the Sui dynasty that followed the Han, the 1,400-mile-long (2,250-kilometer-long) **Grand Canal**, between modern-day Hangzhou (just south of Shanghai) and the heart of the North China Plain, facilitated the much increased trading of commodities among the regions of a now more integrated empire (Figure 10-15).

The **Song, or Sung, dynasty** (A.D. 960–1279) was a distinctive period of Chinese history because of the many social and economic development patterns that emerged. Having been forced south by Mongol invasions, the new Song capital was Hangzhou, a city of huge physical proportions that easily outsized the largest urban settlements in Europe at that time. The commercial success of the Song dynasty was characterized by the expanded use of early-ripening rice varieties, improvements in irrigation, and government-printed money that facilitated interregional trade throughout the Chang Jiang basin. For the first time in China's history, long-distance maritime trade developed, bringing Arab and Indian merchants and products to the coastal cities and encouraging Chinese merchants to establish trading networks in Southeast Asia. China in the thirteenth century resembled eighteenth-century Europe. The Song dynasty was replaced by the Mongol-based Yuan dynasty which in turn was replaced by the Ming dynasty, considered to be one of the most stable dynasties in China's history. Ming rule was then replaced by the Qing dynasty (A.D. 1644–1911), a non–Han Chinese, Manchu-led government

▼ **Figure 10-15 Grand Canal in the city of Wuxi, northwest of Shanghai.** This canal connected many regions of the empire, improving transport. Today it continues to provide the means for low-cost movement of bulk goods.

from the forested region of present-day northeastern China and the last great Chinese dynasty.

East Meets West

The arrival of Western traders during the early eighteenth century signaled the beginning of the collapse of the world's oldest culture. Fundamentally challenged by Western technologies, sciences, and economic systems, China's traditional view of the world with the Middle Kingdom as its center was shattered. The meeting of the East and the West was not an equal, mutually interactive environment in which China could reject things foreign or, as in the past, transform them to suit spiritual and intellectual traditions.

The East India Company, under the authority of the British government, was in competition with other European colonial powers for the potentially lucrative China trade. With great hesitation, the Qing dynasty government restricted foreign trade contacts to Guangzhou (Canton) in 1702. The British bought tea, silks, porcelains, and medicines, but had to pay in silver because, with the exception of cotton from colonial India, the Chinese government thought English goods were inferior. Insistence on this requirement seriously drained British silver reserves and created a

deficit-payment problem. The British solution was to sell Indian **opium** (a dangerously addictive drug from which heroin is derived) to the Chinese in exchange for silver bullion. The opium trade became extremely profitable, and by the end of the nineteenth century, an estimated 40 million Chinese (roughly 10% of the population) were addicted. China experienced a dramatic outflow of silver and attempted to stamp out opium use by confiscating opium chests from British ships docked in Guangzhou, resulting in the First Opium War (1839–1842) in which the British soundly defeated Qing government forces.

Treaty Ports

China's defeat in the First Opium War resulted in further spatial penetration by Western interests. In 1842 China was forced to sign the humiliating Treaty of Nanking, which opened up five coastal ports to Western powers. Although many more ports were opened by 1920, only about 15 ports experienced substantial economic activity (Figure 10-16). These enclaves essentially became foreign-owned territories where foreign, not Chinese, laws governed human behavior. Known as **extraterritoriality**, or the imposition of foreign laws to the exclusion of local laws, many Chinese essentially

▲ **Figure 10-16 Treaty ports of coastal and riverine China.** With the European occupation of major port cities, China was forcibly opened to Western influences.

▲ **Figure 10-17 Shanghai Bund.** The Bund is the former European commercial district of Shanghai, associated with the treaty port era. Facing the Huangpu River, it is now surrounded by modern skyscrapers but continues to function as a popular recreation and tourist district.

became foreigners in the European-controlled portions of treaty ports. Treaty ports became the primary centers of foreign trade and large-scale industry. Shanghai was the most important of the treaty ports and one of the busiest ports in the world during the early twentieth century (Figure 10-17).

Westerners viewed China as a market for manufactured goods at a time when the mass-production innovations of the Industrial Revolution were already beginning to saturate markets in Europe. Treaty ports also allowed Western banks and shipping companies to expand their business by dominating both the import and export trades. Western-owned factories could manufacture low-cost goods at treaty ports because of inexpensive domestic labor and then market those goods in China or export them to other colonial territories. The impact of the treaty ports on China's economic development was mixed. Although a class of treaty-port Chinese came to own a substantial share of factories and founded Western-style banks and shipping companies, the vast majority of the Chinese residing in the enclaves were poor and exploited factory workers. But most importantly, treaty ports did little to promote the economic development of regions beyond the shadow of these enclaves.

Nationalism and a New China

In response to the humiliation of losing territorial and economic sovereignty, various nationalistic movements soon developed after the 1911 collapse of the Qing government. One of these movements was led by **Sun Yat-sen**, a medical doctor referred to as the "father of the republic." Sun revived his earlier Nationalist Party, Guomindang, in 1919 and established its headquarters in Guangzhou. Its platform, the "Three Principles of the People," centered on nationalism, democracy, and livelihood. Sun died in 1925 and was replaced by General Chiang Kai-shek, who took the party in a different direction. Enlisting the financial help of treaty-port Chinese capitalists and conservative rural landlords, as well as the United States government, those closest to Chiang became corrupt while amassing great personal fortunes.

Founded in Shanghai in 1921, the **Chinese Communist Party (CCP)** was also a nationalistic movement. **Mao Ze-dong** emerged as its leader in 1935 and received both financial and moral support

from the Soviet Union. Although the two parties were once united to replace warlord rule, the Nationalists drove the urban-based CCP into the countryside in the late 1920s. The CCP began to mobilize mass support from the landless rural peasantry who constituted 80 percent of China's population. Sensing the CCP's success among the peasantry, Chiang's Nationalist forces embarked on military campaigns against the Communists, but were halted in 1937 by the invasion of China by the Japanese military. The civil war resumed after the end of World War II, with the Communists eventually emerging victorious. By the end of 1949, Chiang and his followers were forced to flee to the island of Taiwan, leaving the Communists in total control of the mainland.

Transformation under Communism

The new **Communist Party of China (CPC)** chose socialism as its primary development philosophy and soon came to monopolize political power, reserving the right to make all decisions pertaining to economic and social policy. In agriculture and industry, the government controlled all capital investment through state-owned industrial and financial institutions and restricted the influence of foreign capital and external economic forces in the socialist pursuit of economic self-reliance. State control of social development was equally pervasive and political dissent was ruthlessly suppressed. In the late 1970s and early 1980s, the government began to realize the limitations of these guiding economic principles in building a modern state and embarked on a road of economic reform that was as revolutionary as the events of 1949. With the death of Mao in 1976 and the rise to power of the more pragmatic leader **Deng Xiaoping**, China began to emulate the capitalist economies of many of its East Asian neighbors. Testing the waters of capitalism entailed decentralizing government economic power and promoting greater interdependence in the global economy.

Transformation of Agricultural Production

With about 60 percent of the labor force directly or indirectly engaged in agricultural activity, China remained a predominantly rural country until the 1990s. The achievements of Chinese governments, both past and present, must be measured largely on the basis of agricultural productivity, because about one-fifth of the world's population must be maintained on only 7 percent of the world's arable land—only half of which is considered of good quality.

Agricultural Regions and Dominant Crops

In contrast to agriculture in Europe or North America, China's agricultural energies are, for the most part, still focused on growing food crops rather than plants for industrial or animal feedstock purposes. China may be divided into three broad agricultural regions based on both farming systems and dominant food crops (Figure 10-18). The first and most obvious regionalization of agriculture is between the wetter eastern and drier western halves of the country. Pastoral nomadism has long dominated in western China but during the past century, Han Chinese have settled oases and river valley bottoms to cultivate wheat, barley, and maize

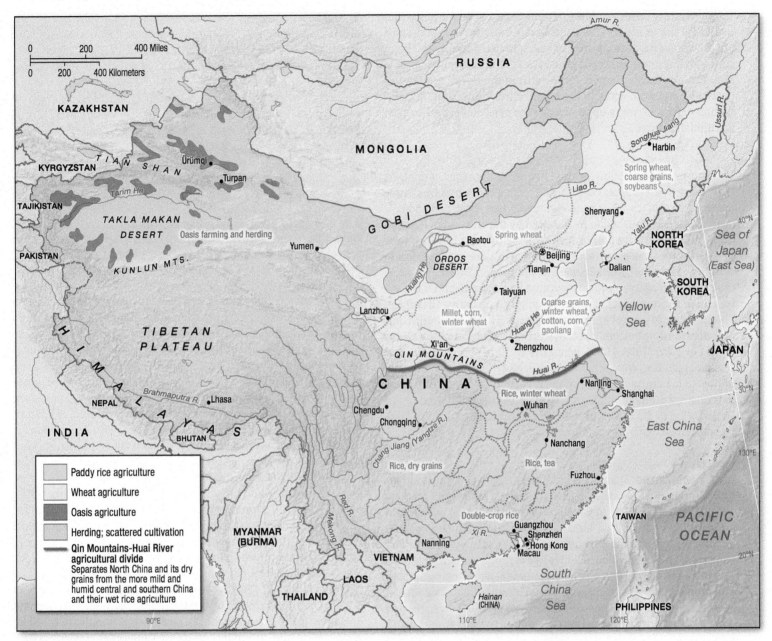

▲ **Figure 10-18 Agricultural regions of China.** China can be divided into three broad agricultural regions: rice in the southeast, wheat in the northeast, and oasis agriculture in the west.

in addition to fruit and nut crops. Even within the eastern region of China substantial differences exist between the south, which is moist with mild and relatively short winters, and the north (except the northeast), which suffers from precipitation shortages and great seasonal variations between long, harsh winters and short, mild summers. In the far south and southeast, paddy rice is the dominant grain crop. Because of an almost year-round growing season, many farmers practice double-cropping of rice with tea as a dominant nonfood crop. The middle and lower eastern Chang Jiang basin represents the agricultural transition zone between north and south where the common mix of crops just north of the river is the double-cropping of summer paddy rice and winter wheat.

The dominant crop of the North China Plain is winter wheat and, because of shorter summers, double-cropping with other crops is difficult. When double-cropping takes place, it is usually with maize or with *gaoliang,* a variety of millet. The northeast is a productive and highly mechanized agricultural region, which was settled largely during the twentieth century by pioneering Han Chinese. In the river plains, spring wheat is grown along with maize and potatoes, while the elevated slopes support wheat and millet farming.

Superimposed on these agricultural regions, which are based largely on the types of grain produced, is the widespread process of agricultural intensification, whereby a broad range of supplementary crops are cultivated for commercial purposes and to add variety to the diet. Vegetables and various types of melons are grown everywhere, as are soybeans, to make traditional bean curd—an important source of protein in a traditionally meat-scarce diet. Fruit orchards also are important; in the north, apples and pears are common, while in the south, citrus dominates. Supplementary crops increase opportunities for peasant farmers to earn additional income.

Agriculture during the Prereform Years

In the years directly following the 1949 revolution, the Communist Party launched a program to radically reconstruct agricultural production. During the 1950s land ownership was abolished and collectives were established to be followed by the creation of People's Communes. The state determined which crops and in which proportions were to be grown and distributed to state agencies. While supplies of subsistence grains were adequate, and few went hungry, the diet of the average peasant became less diverse, and there existed little regional specialization based on comparative advantage. The inability of the government to effectively control production and distribution became evident during the period of the **Great Leap Forward** (1958–1961), when the spirit of communalization was greatest. Bad weather, poorly conceived incentives for communes to maximize output, and the diversion of human resources to industrial production based on inflated agricultural production statistics led to the deaths of some 14–26 million people from famine-related illness and starvation.

Post-Mao Agricultural Reforms

In an attempt to boost both financial incentives and enthusiasm for agricultural production, Beijing introduced the **household responsibility system** in 1978, in which peasant households are allotted a piece of "responsibility" land and obliged to produce a specific amount of grain to be sold to the state at a regulated price. Once that contract is fulfilled, the household is free to produce cash crops to be sold privately, at local markets, or to the state at above-market prices. The production of major crops increased dramatically (Table 10-1). Most increases occurred, despite decreased grain acreage, through greater use of fertilizers and Green Revolution hybrid varieties. Land dedicated to wheat and rice declined in area by 14 percent and 11 percent, respectively, between 1981 and 2010.

▲ **Figure 10-19 High-value urban agriculture.** Located at the urban edge of Shanghai, farmers cultivate high-value green vegetables and are able to market their produce to urban consumers because of superior road transport.

The responsibility system has boosted agricultural production, creating greater general prosperity in many rural areas and a parallel reduction in the rural-urban income gap. The growth rate of cereal-grain production has declined, while the production of vegetables and fruits increased dramatically (Table 10-1 and Figure 10-19). Unlike cereal grains, fruits and vegetables yield greater profits for peasant households. Land dedicated to vegetables and fruits, much of it converted from grain fields, increased 393 percent and 83 percent, respectively, between 1981 and 2010. In addition, the eating habits of the growing number of more affluent urban consumers demanded greater variety in the diet, prompting both suburban and distant producers with good transport access to cultivate higher-quality vegetables and fruits. Especially linked to rising urban affluence is increased livestock production. Pork accounts for about half of the meat consumed. Although pigs are still part of peasant agriculture, large-scale factory farm operations are becoming more common. Per capita consumption of meat is only half that of the United States, but the per capita consumption of meat in China increased from 12 pounds in 1980 to 62 pounds in 2010. Meat consumption rates in rural areas are far below urban areas. Because urban incomes are higher in the coastal provinces, greater opportunities to specialize and increase rural incomes characterize China's eastern provinces.

In the world's most populous country, with a declining percentage of the labor force employed in agriculture and with limited arable land, the question of food security is always a concern for the Chinese government. Two factors account for the absence of future food security problems: wheat and rice consumption is declining because of changing food habits, and the government provides subsidies to farmers to ensure that sufficient amounts of these two grains are available to consumers. Imports of corn and soybeans have dramatically increased, but these commodities are used primarily for the growing commercial livestock industry. Given the low population growth rate, there is no need to substantially increase production.

Table 10-1	Output of Selected Chinese Agricultural Products—1991, 2000, and 2010		
Agricultural Product	**Metric Tons (millions)**		
	1991	**2000**	**2010**
All cereals	398.5	441.5	590.7
Rice	185.7	189.8	197.2
Wheat	95.9	99.6	115.1
Soybeans	9.7	15.0	15.0
Vegetables	130.7	356.0	473.0
Fruits	241.0	645.0	1,221.8

Source: faostat.fao.org/site/567/DesktopDefault.aspx?PageID=567#ancor

Stop & Think

➔ How did China's late 1970s reforms transform the country's agricultural economy?

Industry and Regional Economic Growth

As with agriculture, the primary objective of China's industrial policy has been to promote regional self-sufficiency, and the government was relatively successful in meeting that goal until the transformative policy reforms of the late 1970s. Since then, the influence of the global economy and the parallel rise of both domestic and foreign investment in manufacturing have made China a much richer nation. At the same time, China's engagement with globalization has benefited some regions far more than others.

Mineral Resource Endowment and Distribution

China possesses almost all the mineral resources it needs to reach its industrial goals. To fuel rapid economic growth, China has relied heavily on coal. Since the mid-1990s China has ranked as the world's largest coal producer, accounting for 14 percent of the world's proven reserves, and relies on coal for 70 percent of its energy consumption and 79 percent of electricity production. While coal is produced in most of China's provinces, the greatest concentrations are located in northern and northwestern provinces (Figure 10-20). China also was ranked fifth in world oil production in 2010, with much of its production located in the northeast, especially Heilongjiang Province. With its rapidly growing economy, China also has become the second largest consumer of oil in the world behind the United States. Following decades of aggressive dam construction, China ranks first in the world in hydroelectric power production, accounting for about 16 percent of the world total. China's rich endowment of coal resources in addition to hydropower favors interior regions. But because most of the benefits accrue to the high-energy demand coastal provinces, this demand/supply imbalance creates a domestic-scale core-periphery relationship. Despite its well-endowed resource base, China's energy needs, related to its growing industrial

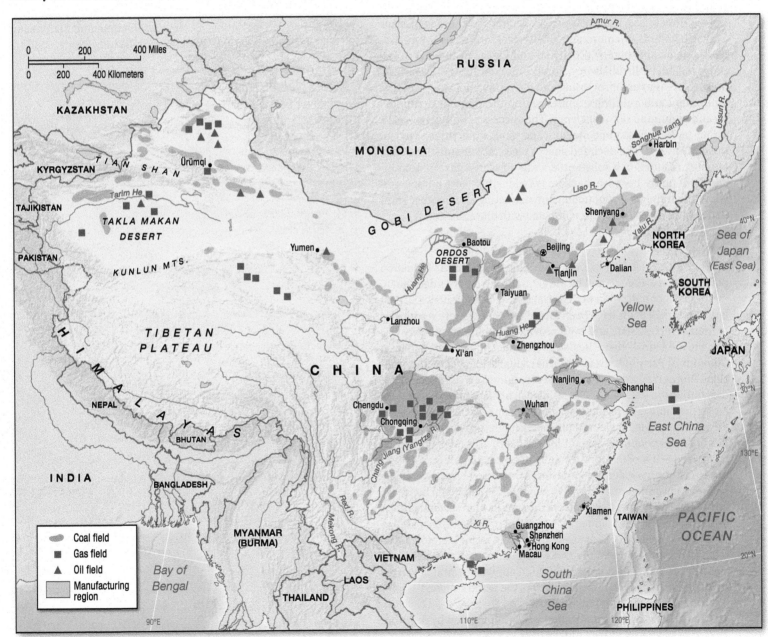

▲ Figure 10-20 **Major mineral resources of China.** China is richly endowed with mineral resources, particularly in interior provinces.

base as well as the rise of middle-class energy consumption, presents serious problems for the country's future energy security (see *Focus on Energy:* China's Energy Challenges).

Stop & Think

▷ Describe the ways China has met its energy needs.

Modernization of Industry

Following the centrally planned Soviet pattern characterized by state-owned enterprises (SOEs), much of China's early industrial development was in the heavy industries, such as iron and steel, chemicals, electricity generation, and textiles. New industrial development, based on large individual plants, was primarily concentrated in inland areas where mineral resources were more abundant (see Figure 10-20). Because much of this industry was military related, an interior location was important because sites distant from the coast were perceived as less vulnerable to external attack during the threatening Cold War period. Apart from these megascale industrial plants, medium and small-scale industrial enterprises were also established by provincial and city governments to improve regional self-sufficiency. Rural commune authorities focused on such areas as agroprocessing and the production of agricultural equipment and building supplies to improve peasant productivity and commune self-sufficiency. The success of both large-scale, urban-based industry and medium and small-scale, rural-based industry is mixed. The declared goal of regional self-sufficiency and equality resulted in a substantial waste of scarce government investment funds, because managers of industries were bureaucrats with little training in production and marketing and because similar facilities in the same area competed with one another.

Industrial Reform and Rapid Growth

The late 1970s ushered in a dramatic change in China's industrial policy, which was anchored by the central government's decision to increase decentralization of financial and decision functions. Most large and small SOEs have been forced to become more responsive to market forces. State and even provincial planning authority was given to individual enterprises, so government bureaucrats and local Communist Party officials function as capitalist managers. This transition from a rigid, centrally planned economy to a free-market economy during the 1980s and 1990s resulted in millions of factory workers losing their jobs. The region most impacted by SOE closures has been the northeast—Heilongjiang, Jilin, and Liaoning provinces—because this region's economy was traditionally based on a large number of megasized SOEs, with a long tradition of central planning. SOEs still remain strong in what the central government perceives as strategic sectors: transportation, energy, telecommunications, petrochemicals, and banking. The transformation of the state-owned sector has been dramatic; in 1995, it accounted for 78 percent of GDP but decreased to 40–50 percent by 2010. Conversely, the private sector has grown dramatically. Some world-class private companies such as Galanz and Haier produce "white goods," household appliances, and now have a global market. Another firm is Lenovo, the world's second-largest producer of personal computers in 2011.

The process of decentralization was accompanied by the rapid growth of **town and village enterprises (TVEs)**, which are collectives owned by towns and villages but often with private capital

investment that produce a wide variety of products for both domestic consumption and export. Their contribution to the new economy of China is important. By 2003, TVEs employed 135 million people and these enterprises also significantly contributed to absorbing surplus rural labor and transforming rural incomes. In the past two decades most TVEs have become stand-alone private firms producing a wide variety of lower and medium-quality exports based on spatial clusters of production that focus on a specific product. In Zhejiang Province just south of Shanghai, small-scale manufacturing around the city of Wenzhou produces 70 percent of the world's lighters and Zhuji produces 35 percent of the world's socks.

Foreign Investment in Industry

In its rush to create a more industrialized country, China's government reversed its hostile attitudes toward foreign investment by adopting an open-door policy that welcomes foreign capital in certain areas. **Foreign Direct Investment (FDI)** began in 1979 when the government established four **Special Economic Zones (SEZs)**, centered on Zhuhai, Shenzhen, Shantou, and Xiamen, and in 1988 designated a fifth SEZ on Hainan Island (Figure 10-21). In

▲ **Figure 10-21 Special Economic Zones, open cities, and open economic regions.** The Chinese government designated many coastal areas for accelerated industrial and economic development during the postreform era.

FOCUS ON ENERGY

China's Energy Challenges

China measures its success by its progress in economic growth, and energy is a critical factor in this larger development process. Between 1980 and 2010, China's total energy consumption quintupled; U.S. energy consumption during the same period did not even double. China must construct 12 new power plants every year just to keep up with energy demand, but in the United States only one or two power plants are built each year. Despite its abundant energy resources, since 1996 China consumes more energy than it produces. China is the world's largest coal producer, but in 2011 became the world's largest coal importer. China consumes more coal than the United States, Japan, and the European Union combined.

An important reason for such high energy use is that China's industrial economy has made a structural shift from lower-energy consumer product industries associated with light manufacturing to far more energy-intensive industries that include glass, steel, cement, and aluminum. These industries account for about one-third of total energy consumption. This high consumption is due in part to the technologies used in these energy-intensive industries, which are 20 percent less efficient than the international average. China uses four times the energy to produce one dollar in GNP value than the United States and 12 times that of Japan. This means that China's "energy intensity," or the amount of energy needed to produce a unit of GNP is very high, although the energy to GNP dollar ratio has greatly improved over the past two decades. A related cause for high energy consumption in industry is government-subsidized energy production. When global prices for energy rise, consumers are barely affected. This policy is intended to reduce the potential for social unrest.

China's nonfossil fuel energy production has great potential, but only accounted for 13 percent of total energy production in 2010. China is the world's largest manufacturer of wind turbines and possesses about one-quarter of the world's installed capacity. But problems of geographical mismatch exist between the windy, distant producing regions of the west and the consuming regions of the east (Figure 10-1-1). Private developers rushed to develop wind farms, but grid connectivity with state-owned utility companies has been uncoordinated. The government is now pushing for localized, smaller-scale wind power development where access to the local grid is available. A similar story exists for solar energy. China is a world leader in solar panel manufacturing, but energy production is concentrated in the drier west and technology is relatively unsophisticated. Nuclear energy is also in its infancy stage with 14 existing plants along the eastern seaboard and 25 plants in the construction or planning stage. With insufficient uranium supplies, China is expected to account for 20 percent of global uranium demand in 2020.

One well-developed source of renewable energy is hydropower. In an effort to reduce pollution from coal, China's installed hydroelectric capacity dramatically increased from 75 gigawatts (GW) in 2000 to 200 GW in 2010. The plan is to increase capacity to 330 GW by 2020. Much of this new construction will take place in southwestern China along the Chang Jiang and the Chinese portions of the Salween and Mekong Rivers. This flurry of hydropower dam construction is full of potential problems. In addition to ecological damage and population displacement, future power generation is questionable because of unreliable rainfall patterns, potentially produced by global warming. In 2011, substantially reduced rainfall caused a 25 percent reduction in hydroelectricity generation. Between 2000 and 2009, China experienced a 13 percent reduction in water resources as a result of decreased rainfall and snowmelt.

Unable to satisfy its voracious demand for energy from domestic sources, China has aggressively developed international sources of energy supply, especially in oil and natural gas. Relying on Persian Gulf countries for a quarter of its oil imports, China's state-owned energy firms have developed trade ties with

▲ FIGURE 10-1-1 **Wind turbines generate energy in far-western Xinjiang province.** China is the world's largest producer of wind energy, but long-distance power transmission to eastern consuming regions as well as connection to more local utility grids remains a problem.

oil-rich sub-Saharan African countries, as well as with countries that control rich deposits of militarily important mineral resources such as cobalt, chromium, and titanium. This resource foray is so substantial that China replaced the United States as sub-Saharan Africa's largest trading partner in 2009. China's "resource diplomacy" often involves development loans to sub-Saharan governments to construct infrastructure such as roads, railroads, and telecommunications. Unlike Western loans, China's loans do not involve interference with domestic government affairs or reforms. Western governments perceive China's resource diplomacy as undermining efforts to promote good governance. This may be partially true, but Western mineral resource corporations are also guilty of promoting corrupt government practices in mineral-rich developing countries.

Sources: Christopher Alessi and Stephanie Hanson, *Expanding China-Africa Oil Ties* (New York: Council on Foreign Relations, 2010), http://www.cfr.org/china/expanding-china-africa-oil-ties/p9557; The Economist Intelligence Unit, *A Greener Shade of Grey* (2012), http://pages.eiu.com/rs/eiu2/images/EIU_ChinaRenewableEnergy.pdf

1984 the government also established 14 **open coastal cities**, which have operated much like SEZs but with lower levels of government funding for site improvement. Three **open economic regions** were established a year later—two around SEZs and one encompassing the large hinterland of Shanghai. In the 1990s, an open economic region centered on the Bohai Sea (an arm of the Yellow Sea) in the north was promoted as an additional region for FDI.

The location of SEZs is not surprising because they were viewed as social and economic laboratories, geographically restricted to the country's margins. This experiment with SEZs, open coastal cities, and open economic regions has been very successful. Since 1992 China has ranked first among developing countries in FDI, and by 2010 China trailed only the United States in world FDI ranking. While China continues to produce lower-value consumer products such as toys and clothing that characterized its FDI investments during the 1980s, the value of consumer exports as a result of FDI has dramatically increased. China has become the global leader in electronic-product exports such as DVD players, digital cameras, personal computers, mobile phones, and color televisions. High-tech exports as a percentage of total exports increased from 11 percent to 29.1 percent between 1998 and 2008. Over 90 percent of global laptop and tablet assembly takes place in China using components supplied by Taiwanese vendors. As China's economy matures, FDI in the service sector also has grown dramatically, particularly in retail, real estate, finance, and tourism. Walmart has 370 Supercenters, Sam's Clubs, and Neighborhood Markets in 140 cities in 21 provinces (Figure 10-22). Since 2008, FDI in the service sector has surpassed the manufacturing sector.

Before 2008 Hong Kong and Taiwan contributed the largest share of FDI. In 2010, the United States consumed 19.2 percent of China's exports, followed by Hong Kong (for reexport), Japan, South Korea, and Germany. Between 1990 and 2010, China's share of global exports more than quadrupled from 1.8 percent to almost 10 percent, overtaking Germany as the world's greatest exporter. The regional share of FDI in China has been spatially uneven, with eastern provinces (mostly coastal) accounting for more than

80 percent. This imbalance in part explains the spatial disparity in per capita provincial GDP (see *Visualizing Development:* Spatial Inequalities in Development).

Stop & Think

▷ Explain the differences between China's and India's experiences with industrial growth as a result of FDI.

Dynamic Urban-Economic Regions

Three large-scale and dynamic urban regions have dominated China's postreform economy. In the Asian context, geographers refer to them as **Extended Metropolitan Regions**. Each is anchored by a multitude of large-, medium-, and small-sized urban places separated by rural spaces that are often an important source of labor, and connected by a well-developed transport and communication network that promotes functional interaction. These "city-regions" of the Pearl River Delta, Yangtze River Delta, and the Bo Hai Rim have become territorial platforms for FDI and thus most engaged with the process of globalization. Together the three regions in 2010 accounted for 65 percent of national GDP and 70 percent of FDI. Each has different experiences with globalization.

Pearl River Delta. The earliest urban-economic region to receive foreign investment was the Pearl River Delta (PRD) region anchored by Hong Kong and Guangzhou, the capital of Guangdong province, and eight other cities, some of which are connected by high-speed train service. As a result of FDI inflow beginning in 1979, the PRD became the world's 16th largest economy and 10th leading exporter by 2006. In 2010, the PRD accounted for 19.2 percent of China's total FDI and 27 percent of China's export value. With a 2010 population of 60 million in the 10 major urban centers, the growth of the PRD was highly dependent upon the transborder migration of Hong Kong industries to take advantage of inexpensive land and labor (see *Geography in Action:* Hong Kong's Past, Present, and Future). A common ethnic-Cantonese culture and pre-1949 kinship ties, account for the more than 50,000 factories supported by Hong Kong-sourced investment. This cross-border investment is a small-scale core-periphery or "front shop/back factory" relationship because Hong Kong retains the higher-value design, financial, and marketing functions while Guangdong is relegated to providing lower-value infrastructure, land, and labor. The first region in China to experiment with reforms, the PRD was also first to receive lower-value FDI in garment, textile, toy, and electric appliance manufacturing. With little planning and few environmental controls, the "gritty" PRD earned the "sweatshop of the world" label. In the 1990s, the regional economy experienced a "second industrial revolution" anchored in capital-intensive electronics manufacturing for export. Based on newly established economic and technology zones where parts are sourced locally, Hong Kong and Taiwan-linked investment accounted for 65 percent of the PRD's exports in 2006. In 2010 the PRD, without Hong Kong and Macau, generated 9.4 percent of national GDP, 10.3 percent of national industrial output, 19.2 percent of national FDI, and 27.4 percent of China's exports. The city of Shenzhen just across the Hong Kong border provides an ideal example of this global-economic induced transformation at the local scale. The most successful of the SEZs, Shenzhen was a medium-sized fishing center of 30,000

▲ **Figure 10-22 A Wal-Mart Supercenter in Beijing.** With some 370 Wal-Mart facilities in China, this global retailer has become a common urban landscape feature throughout the country.

The prereform era was characterized by a relatively uniform interprovincial distribution of per capita GDP. Greater geographical concentration of FDI, private domestic investment, and the economic impact of privatized TVEs in some regions and not others have since produced pronounced interprovincial economic inequalities (Figure 10-2-1). Coastal economic growth has outstripped that of the central, western, and northeastern provinces. This is not surprising because of the east's early advantage in establishing SEZs, open coastal cities, and open economic regions. Aided by superior infrastructure and human capital, eastern China's economic growth was rapid. From 1990 to 2010, eastern provinces increased their share of per capita GDP from 49.5 percent to 53 percent while the remaining central, western, and northeastern regions registered small declines in their shares of per capita GDP. The per capita GDP

of Guangdong, for example, is almost triple that of the interior province of Gansu. The large autonomous territory of Nei Mongol has a relatively high per capita GDP for an interior province because of the dramatic increase in coal mining as well as development of other mineral resources in select locations. As a whole, Nei Mongol remains quite poor.

Since China launched its economic reform policies, noncoastal provinces have been economically handicapped because their economies were anchored in primary sector agriculture and mining. Industrial production in interior provinces was also more dependent on SOEs. Interior province per capita GDPs have increased, but not as fast as the GDPs of coastal provinces. This interior province growth has been due to foreign and domestic companies increasingly seeking cheaper labor. Coupled with growing consumer demand and improved long-distance transport

infrastructure, interior provinces now are more attractive to investment. Much of this interior per capita GDP growth takes place in and around urban centers. China's leaders continue to hope that a spatial "trickle-down" effect will occur and eventually enable these interior provinces to benefit economically from the process of globalization. To promote economic development in western provinces, the government launched its "Go West" program in 2000, which entailed substantial investments in basic transport infrastructure. The response has been mixed and has been limited to manufacturing investment in only a handful of urban regions such as Chongqing, Chengdu, and Xian. Much of this investment goes to the manufacture of intermediate products that are then transported to coastal provinces for final assembly and export.

Source: Gregory Veeck, Clifton W. Pannell, Christopher J. Smith and Youkin Huang, *China's Geography: Globalization and the Dynamics of Political, Economic, and Social Change* (Lanham, MD: Rowman and Littlefield, 2011).

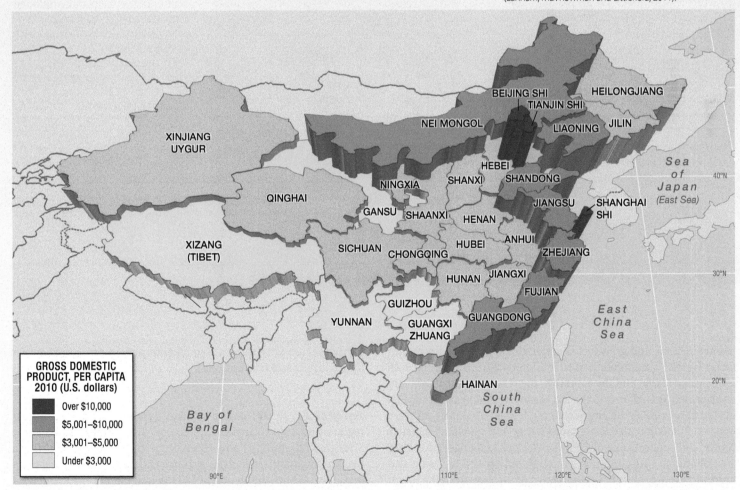

▲ **FIGURE 10-2-1 Provincial per capita gross domestic product.** Despite economic prosperity at the national level, China's interior provinces lag seriously behind the coastal provinces.
Source: 2011 China Statistical Yearbook.

Hong Kong Island was first ceded to Britain in 1842 as part of the spoils of the First Opium War, and the adjoining mainland areas of Kowloon and the New Territories were added by 1898 (Figures 10-3-1 and 10-3-2). In 1997, the British Crown Colony of Hong Kong, often referred to as a "borrowed place on borrowed time," was returned to China. The nearby Portuguese colony of Macau was returned in 1999. Hong Kong's economic functions have changed over time. Before 1949 Hong Kong served as an important entrepôt, or transshipment point, for much of south China's external trade with the rest of the world. After the Communists gained power, Hong Kong lost most of its commercial links to mainland China, and the colony was forced to shift its economic energies to the manufacturing and export of cheap "Made in Hong Kong" consumer goods such as textiles, clothing, footwear, and electronics. It soon became a global manufacturing center fortuitously based on the post-1949 migration of mainland Chinese capitalists and refugees who provided an endless source of cheap labor.

With much of its manufacturing base relocated to mainland China in the 1980s, Hong Kong's economy has become well-grounded in the higher-value service sector—banking, insurance, real estate, and shipping—which in 2011 represented 93 percent of GDP. Tourism, especially from the mainland, is the second most important income generator.

With a 2010 per capita GNI PPP of $47,480, Hong Kong's economy remains strong. It ranks first in the global Index of Economic Freedom, and continues to be a place where fortunes are made. But

◄ FIGURE 10-3-1 **The Special Administrative Region of Hong Kong.** Comprised of Hong Kong Island, Kowloon, and the New Territories reaching up to the border of the mainland where the city Shenzhen is located, much of Hong Kong is mountainous with dense settlement in valleys or on reclaimed land. Located in the west is the international airport on Lantau Island built on land reclaimed from the sea.

people in 1980, and grew into a major global platform for manufacturing with a population of 10 million by 2010. Beginning with lower-value textile and toy assembly, the restructuring of industry transformed Shenzhen into the PRD's most technologically sophisticated city. In 2008, the city accounted for half of China's software exports. Shenzhen also is a major finance and transport logistics center, and supports one of China's two stock exchanges (the other is in Shanghai). Its container ports ranked second in China and fourth in the world in 2010. Indicative of the growth of the larger

service sector, Shenzhen is also an emerging international center for trade fairs and conventions.

Yangtze River Delta The second dynamic urban-economic region is the lower Yangtze River Delta (YRD) region anchored by Shanghai (23 million) and 15 medium and small-sized cities in neighboring southern Jiangsu province and northern Zhejiang province. With a 2010 population of 100 million, the YRD accounted for

several serious problems detract from this sense of prosperity. One is a growing economic disparity between rich and poor because blue-collar workers lost their jobs when factories relocated to the PRD and have not found replacement jobs or employment at equal wage rates. This is exacerbated by the government's reduction in needed social welfare services as part of a growing neoliberal economic philosophy. As a result, Hong Kong is characterized by the highest income inequality among developed economies. Another problem is poor environmental quality, particularly the high level of air pollution. There were only 41 healthy air days in 2010, and the air is three times as polluted as the skies of New York or Paris. The two primary domestic sources of air pollution are dirty coal-fired power plants and vehicle emissions. Approximately half of Hong Kong's air pollution originates from factories and power plants in the PRD that blow emissions south into Hong Kong during the winter monsoon. Aside from the unhealthy living conditions for permanent residents, this obstacle to attracting and retaining foreign multinational companies and their employees is threatening future economic growth.

One present and future issue on the minds of many Hong Kongers is the state of political freedom. Hong Kong was designated a Special Administrative Region (SAR) by Beijing after the 1997 handover and was promised that it would be

▲ **FIGURE 10-3-2 Victoria Harbor and the skyline of Hong Kong.** Hong Kong is one of the world's leading financial and shipping centers. Its return to China in 1997 did not negatively impact the city's strong global economic position. Hong Kong's central district, viewed from The Peak on Hong Kong Island, appears in the foreground with Kowloon visible across the harbor.

allowed to retain its economic and social system, with Beijing controlling the functions of foreign affairs and defense for the next 50 years. Essentially, the arrangement is a one-country, two-systems relationship. It should be noted that as a colonial possession, Hong Kongers enjoyed only limited representative government, but in the few years before the handover, the colonial government did increase the level of representative democracy. Hong Kong's leader, called a chief executive, is appointed by Beijing and his or her power over the legislative and judicial branches of government has increased through time. Press freedom, the ability of citizens to engage in public demonstrations, and the formation of political parties still exist,

but such democratic institutions are not guaranteed in the future. One example of Hong Kong's eroding political autonomy occurred in 2012 when Beijing attempted to force Hong Kong's education system to adopt a school curriculum that cultivated political loyalty to Beijing. After months of public protest against this educational policy, which many claimed was pure political propaganda and brainwashing, it was shelved by the Hong Kong government.

Sources: Kent Ewing, "Patriots and Protests in Hong Kong," *Asia Times,* August 7, 2012, http://www.atimes .com/atimes/China/NH07Ad01.html; Michael Ingham, *Hong Kong: A Cultural History* (Oxford: Oxford University Press, 2007); Steve Tsang, *A Modern History of Hong Kong* (New York: I.B. Tauris, 2004).

18 percent of national GDP, 21 percent of the country's industrial output, 43 percent of national FDI, and 37 percent of total exports. During the pre-Communist period, Shanghai was the jewel among all treaty ports. It was the eighth busiest port in the world in the 1930s and it earned the name "Paris of the East" because of its cosmopolitan culture. The symbolic center of the city was the Bund, where a sweep of Victorian and Art Deco commercial buildings fronted the Huangpu River. Although interior cities garnered the majority of the national government's industrial investment during the Communist period,

Shanghai remained the country's leading industrial center. While still retaining a traditionally strong manufacturing base, especially in automobile manufacturing, Shanghai's economic fortunes were transformed in 1990 with the establishment of the Pudong development area (Figure 10-23) across the Huangpu River from the older core of Shanghai. With preferential tax breaks, FDI in high-tech manufacturing, and financial services, Shanghai has grown rapidly in nine development zones. Pudong, the face of the new Shanghai, is the most spectacular of these growth foci. As a symbol of this new prosperity

▲ Figure 10-23 **The Pudong Area of Shanghai.** With the older core of Shanghai in the foreground, the "new" Shanghai is located across the Huangpu River in Pudong where high-tech manufacturing and the growing financial service sectors are concentrated. Pudong symbolizes the globalized landscapes of urban China.

electronic products sector that includes telecom products, computer manufacturing, integrated circuits, auto electronics, and cell phones. Critical to the rise of the high-value electronics industry, particularly software and integrated circuit design, is the BRR's educated workforce or what might be called human capital. Beijing is home to three world-class universities and research institutes. The cities of Tianjin, Qingdao, and Dalian possess either technological development zones, science parks, or high-tech development zones to attract investment in the highly competitive global electronics and information industries. Beijing's growing prominence in the BRR as well as in the rest of China is also directly linked to the 2008 Beijing Summer Olympics.

The PRD, YRD, and BRR urban-economic regions are characterized by an efficient transport infrastructure of limited-access expressways and high-speed trains, but interior China is not. The national government has embarked on an aggressive program of national highway and high-speed rail construction to link interior regions with the coast so that economic opportunities associated with globalization spread inland and promote national economic integration (Figure 10-25).

Beijing commenced its national trunk highway system in 1990 and by 1995 some 2,000 kilometers (1,200 miles) had been constructed. The next 16 years witnessed frantic highway construction as the total length of this trunk highway system reached 74,000 kilometers (46,000

and globalization, Pudong's skyline is punctuated by the futuristic 95-story Shanghai World Financial Center completed in 2008.

Although Shanghai's functional relationship with the YRD is similar to Hong Kong's gateway relationship with the PRD, important differences exist. When Shanghai began attracting high-tech investment, many state-owned industries relocated to YRD cities because of cheaper land and labor. Shanghai's competitive relationship with YRD cities was one of "leader versus followers" because it was the source of technology transfer and investment. Since the late 1990s, a handful of YRD cities such as Suzhou and Wuxi began establishing very successful industrial parks and technology zones to attract high-tech foreign investment. Unlike the PRD, where the light manufacturing of lower-value consumer goods dominates, a substantial amount of FDI in the YRD region focuses on computers, mechanical and electrical products, and chemical products. Especially important are Taiwanese firms that have relocated to the YRD to manufacture semiconductors, personal computers, digital cameras, and LCD monitors for global brands such as Apple, Dell, and Canon. Although Shanghai's future economy rests on financial and other types of services, its relationship with its YRD cities in terms of attracting high-value FDI is now as a competitor rather than a "leader versus followers."

Bohai Rim. The Bohai Rim Region (BRR) comprises the third and latest dynamic urban-economic region to emerge along China's east coast. With 86 million people and more territory than either the PRD or YRD, the arc-shaped region around the Bohai Sea comprises the provinces of Hebei, Shandong, and Liaoning, plus the municipalities of Beijing (22 million) (Figure 10-24) and its port, Tianjin (8.6 million), as well as 10 additional cities. The BRR's economy has always been anchored by state-owned heavy industry, but because it is the last of the three regions to globalize and was aggressively promoted by the national government, its experiences with globalization are somewhat different when compared to the PRD and YRD. Attracting only 7.6 percent of China's FDI share in 1990, the BRR garnered 26.2 percent of FDI by 2009. Much of this investment is in the

▲ Figure 10-24 **Beijing skyline.** The urban landscape of Beijing has both a traditional and contemporary flavor. In the foreground are the rooftops of the ancient Forbidden City and beyond is the egg-shaped National Theater built for the 2008 Olympic Games.

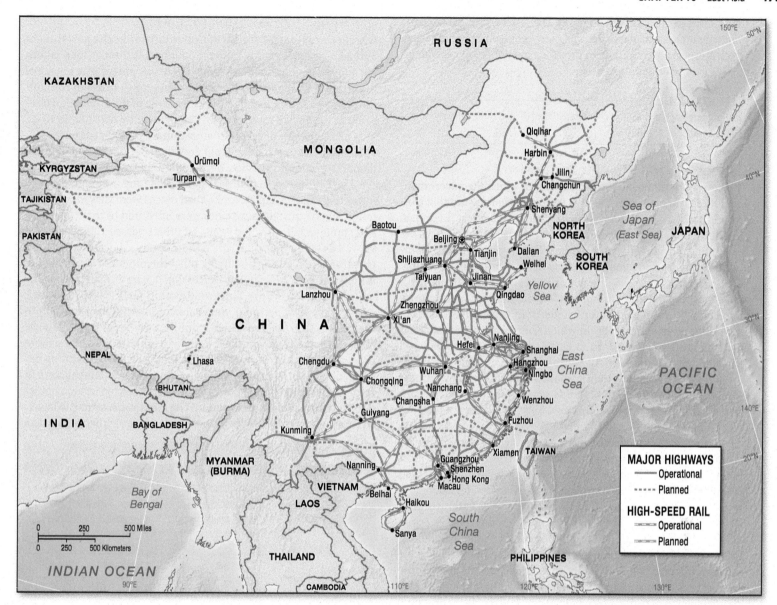

▲ **Figure 10-25 National highway and high-speed train routes in China.** The government embarked on the massive construction of a modern transport surface system in the 1990s. Especially important was the construction of east–west highways to assist economic growth in interior provinces.

miles) by 2011, almost equal to the 76,000 kilometers (47,000 miles) of interstates in the United States. By 2020, the length of China's trunk highway network will reach 85,000 kilometers (53,000 miles). Plans call for all provincial capitals and cities of 200,000 or more people to be connected. These tolled "expressways" are limited-access roads; China has more tolled highways than any other country in the world. Especially important to national economic integration are those planned for China's far west. China is also investing in high-speed rail transport; tracks will more than double to almost 10,000 miles (16,000 kilometers) from 2012 to 2020 (Figure 10-26). Trains with speeds up to 220 miles (350 kilometers) per hour will run on dedicated tracks while regular high-speed trains traveling up to 155 miles (250 kilometers) an hour will share tracks with regional, commuter, and freight trains. This ambitious high-speed rail program has short-term liabilities and long-term benefits. The pace of investment has meant substantial government debt. The benefits include cutting travel time in half between cities less than 600 miles (1,000 kilometers) apart because of long

▲ **Figure 10-26 A high-speed train departs for Beijing from the provincial capital of Shijiazhuang, Hebei.** This is one of the 35 stops on the Guangzhou to Beijing route.

transit times associated with air travel. The result is greater economic productivity as urban places become more geographically connected, which is especially important for providing interior provinces with increased economic opportunities.

> **Stop & Think**
>
> ▶ Compare and contrast the characteristics of China's three primary urban-economic regions.

Gender Impacts of Economic Reforms

The current period of economic reform, in which the government has adopted capitalism but retains many facets of government control and planning, also incorporates policies that harness traditional Confucian philosophy to gender the development process. This hybrid philosophy directly impacts the different employment opportunities of males and females engaged in rural-to-urban migration. Rural migrants comprise the majority of labor for urban growth and export-led manufacturing, and females account for one-third of the total. The large outflow of males from villages has created a "feminization of agriculture" in which women are responsible for 60 percent of agricultural work. Farming females are considered "virtuous" by the government because by remaining in the kitchens and fields they allow their husbands to pursue money-earning opportunities in urban areas. In this sense, traditional social construction of gender roles is reinforced through Confucian ideals of women's position within the family.

These traditional female roles are sustained by women who do migrate to urban places in search of wage income work. The most common work for females is in manufacturing that requires detailed assembly work—as garment workers or seamstresses, for example—or as domestic servants, restaurant servers, and hotel workers. Males are heavily represented in more physically demanding factory work, and as construction laborers, porters, and manual laborers. This gendered division of the migrant urban labor force reflects the situation whereby migrants do not choose jobs, but the jobs choose the migrants. Age characteristics and time spent by migrant laborers in urban places also differ by gender. Female workers are primarily younger, between 15 and 24 years old, and single. Urban employers operating under the capitalist model of production efficiency prefer these "maiden workers" because these women are perceived to be more docile and better at detail work. Migrant males are able to perform demanding physical work well into their 40s, and family responsibilities are less disruptive for husbands than for wives. Work in urban places is far more an "episode" in women's lives, as marriage generally means returning to a rural place of residence. Poorly paid migrant work is an extension of women's family roles because these migrants have simply replaced the constraints that they left at home for a different form of subordination in the city.

Urbanization and Migration

In the world's most populous country, issues of urbanization and migration become central to the processes of economic planning and social development. The agricultural and industrial reforms of the modern period have dramatically increased rural-to-urban migration and accelerated social change never witnessed in the world over such a short period of time.

Controlled Urbanization under Mao

During the opening decades of Communist rule, the government limited peasant migration from the countryside through food rationing and restrictions on urban industrial development. The impact of these policies in reducing the unplanned growth of urban communities was so substantial that the percentage of the population classified as urban had reached only 12.8 percent by 1978 (Table 10-2). This extremely low level of urbanization was significant because it enabled China to avoid the rapid growth of peripheral squatter settlements so common in most developing country urban places.

The Chinese variety of Marxist philosophy indirectly favored restricting rapid urban growth. Urban places were viewed with contempt and distrust because cities were inhabited by more educated and commercially oriented capitalist urban classes. Cities were also viewed as being "parasitic" or "nonproductive" because the urban elites were perceived to be living off of the products of rural labor. A bold and disruptive policy expressing this antiurban philosophy materialized during the **Cultural Revolution** (1966–1969), which was marked by a period of deurbanization. Millions were sent to the countryside to learn "correct political thinking" from the rural masses. In reality, government policies were anchored in the more practical concerns over the ability to feed a rapidly expanding national population if rural-to-urban migration was left unchecked. In response, national population growth rates exceeded urban population growth rates during this period. The administrative tool to legally control the movement of people between rural and urban areas was a

Table 10-2	Urban Growth in China, 1949-2012	
Year	**Total Urban Population (millions)**	**Percentage of Total Population Classified as Urban**
1949	49.00	9.1
1953	64.64	11.0
1957	82.18	12.7
1960	109.55	16.5
1964	98.85	14.0
1969	100.65	12.5
1978	122.78	12.8
1982	152.91	15.1
1986	200.90	18.9
1994	345.39	29.0
1998	372.75	30.0
2002	486.67	38.0
2007	579.92	44.0
2010	672.95	50.0
2012	688.5	51.0

Sources: Terry Cannon and Alan Jenkins, *The Geography of Contemporary China* (New York: Routledge, 1990), 210; World Bank, *World Development Report 1998/1999* (New York: Oxford University Press, 1998/1999); Population Reference Bureau, *World Population Data Sheet 2002, 2007, 2012* (Washington, D.C.: Population Reference Bureau, 2002, 2007, and 2012).

▲ Figure 10-27 **Grandmother and grandson in rural Fujian Province.** Among urban families child gender preference is far less important when compared to more traditional rural families.

place-of-residence, or *hukou* **registration system**, wherein rural and urban dwellers were issued registration booklets that restricted their access to food, employment, housing, and health services to their home area. Without being registered as urban inhabitants, rural-to-urban migrants found it very difficult to live permanently in the city.

Economic Reform and Rapid Urban Growth

Part of the late 1970s economic reforms included relaxing the hukou registration system. As a result, the surplus rural labor created by agricultural reforms was free to move from villages to small, medium, or large urban places, swelling the urban proportion of the country's population. Between 1978 and 2012, the proportion of China's population living in urban areas more than tripled, from 12.8 percent to 51 percent. China's urban population of 672 million is more than twice the total population of the United States. Much of the increase in China's urban population is a result of rural-to-urban migration, in particular the country's burgeoning "floating" population (see *Geography in Action:* China's Urban Floating Population). Longer-distance interprovincial migration has favored the more economically dynamic coastal provinces.

China's Population Profile

The magnitude of China's huge 2012 population of 1.35 billion people can be appreciated through the simple observation that the number of people added to China's population since 1975 is slightly greater than the entire population of the United States. Although growth rates have substantially decreased over the past 30 years, the population is projected to increase to approximately 1.45 billion by 2030 before beginning to decline. The spatial distribution of China's population is uneven. The Han are China's majority culture group, accounting for 92 percent of the population and primarily found in the eastern two-thirds of the country. The remaining 8 percent of China's population are comprised of ethnic or cultural minorities primarily located in interior regions. The government officially recognizes 55 different minority groups, of which 18 have populations greater than

1 million people. With the exception of Xinjiang and Tibet (Xizang), every province or autonomous region has a Han Chinese majority.

Population Growth

China's large population reaches back many centuries (Figure 10-28). China was the most populated country in the world in the mid-1700s, when the population was already approaching 200 million, up from only 20 million during the mid-1600s. This massive population increase is attributed to a century of relatively peaceful conditions, improved intensive farming technologies, and the introduction of New World crops such as maize and sweet potato. By the time the Communists assumed power in 1949, the population had ballooned to approximately 548 million people, and during the next 30 years the number of people nearly doubled again, to 1 billion.

Population growth increased dramatically during the first two decades of Communist rule. In general, Mao Ze-dong held "pro-natalist" views based on the Marxist notion that more workers produce more economic goods. This relationship between production and population was perhaps best captured in Mao's popular saying that "every stomach comes with two hands attached." Rapid population growth was also the outcome of a successful healthcare delivery system. It was not until the early 1970s, when the government came to view people as consumers rather than producers, that a systematic national birth-control program was implemented.

Policies to Reduce Population Growth

The policies first introduced to reduce population growth were extremely successful. Between 1970 and 1978, the total fertility rate was more than halved, from 5.8 to 2.7, and the natural rate of increase declined from 2.6 percent to 1.2 percent. The family planning programs, which took on a revolutionary flavor, convinced people to marry later, extend the interval between births, and have fewer children. People were expected to adhere to birth quotas, and educational programs specifically designed to modify traditional attitudes toward children became widespread. Two specific programs were very important: distribution of contraceptives, and improvement of educational opportunities for women as an avenue to reduce fertility rates.

While these population-control policies were very effective, the recommended two-children policy throughout much of the 1970s was replaced by a mandatory one-child policy in 1979. Attempting to reach the one-child goal has required a complex system of economic rewards for those who conform and penalties for couples who have two or more offspring. The economic rewards remain in place through the child's teenage years and include healthcare subsidies, free primary and secondary school, better housing, and higher retiree benefits. Parents who have two or more children are heavily fined by aggressive local family planning committees and, in many cases, the mother is forced to have an abortion. Since 1980, the government has collected over $300 billion in fines. There are exceptions to the one-child policy, predominantly among the country's non-Han minority groups, who are exempted. Under special conditions—physical disability of the first child, special needs to preserve the family line, and sometimes in cases where the only child is a girl—the one-child restriction may also be waived. The number of exceptions to the policy increased in 2001 when the central government allowed provinces to enact their own policies to meet local circumstances. Some experts claim that the one-child policy applies to only 40 percent of the population.

GEOGRAPHY IN ACTION

China's Urban Floating Population

Millions of rural inhabitants have found their way to cities in hopes of improving their lives and those of families who remain behind in rural areas. In 2010, approximately 211 million rural migrants, or one-third of China's population with rural residence permits, spent much of the year working in urban places. By 2030 an additional 300 million, or half of the rural population, is expected to migrate to cities. The average age of migrant workers is 27.3 years, which reflects both the desire on the part of younger rural people for a new life as well as the demand by employers for a more flexible and malleable workforce (Figure 10-4-1). Working far longer hours than permanent urban residents, migrants often work 55 hours per week and the majority work at least six days a week. Depending on the industry, Chinese factory workers earn approximately $300–$450 per month. Living conditions for migrants are very basic and precarious. Approximately 60 percent of migrant workers are provided housing by their employer due to insufficient public housing; accommodations are most often small rooms with multiple occupants, in factory or construction site dormitories. The rest who earn enough to rent housing often rent rooms from farmers located at the urban fringe. The vast majority of migrant workers experience low morale because of low wages, overwork, and very little leisure time.

The proportion of migrant workers in a single city's population is astounding. In the multicounty urban region comprising the export-oriented city of Dongguan in the PRD, 4.3 million or 73 percent of the 5.9 million total urban inhabitants are temporary migrants. In very large cities such as Shanghai, the 9 million migrants in 2010 accounted for 39 percent of the total urban population of 23 million. While some temporary migrants can obtain a "blue" hukou permit and thus access to a limited number of government services, the vast majority remain as temporary migrants with little or no education and welfare rights. Because local public schools receive no funding from the central government, migrant families must pay for public school. As a result, only 6 percent of migrant children have ever attended school. In addition, only 5 percent of migrants have any form of medical and unemployment insurance. Although migrant workers are looked down upon by urban residents, authorities generally have not enforced residency laws because cheap labor is critical to urban economic growth. A small number of human rights groups, both domestic and international, have called for better treatment of migrants. The government is sensitive to negative international newspaper accounts of the underpaid and ill-treated 30,000 migrant workers who were involved in constructing venues for the 2008 Beijing Summer Olympic games. While China's floating population is supposed to return home to the countryside, it is inconceivable that most will. China's government must seriously rethink the

▲ FIGURE 10-4-1 **Male construction workers in Beijing.** The emergence of new and gleaming urban landscapes in Chinese cities has only been possible with cheap labor provided by rural to urban migrants.

hukou system as an institutional mechanism to control rural–urban migration so all citizens have the right to obtain decent work and living conditions. Otherwise, the government must contend with an increasingly large urban underclass.

Sources: Yu Chen, "Rural Migrants in Urban China: Characteristics and Challenges to Public Policy," *Local Economy: The Journal of the Local Economy Policy Unit* 26, no. 5 (2011): 325–336; Wenshu Gao and Russell Smyth, "What Keeps China's Migrant Workers Going? Expectations and Happiness Among China's Floating Population," Monash University Department of Economic Discussion Paper, 2010, http://www.buseco.monash.edu.au/eco/research/papers/2010/1410chinagaosmyth.pdf; Jianfa Shen and Yefang Huang, "The Working and Living Space of the 'Floating Population in China," *Asia Pacific Viewpoint* 44, no. 1 (2003): 51–62.

Although the one-child policy has substantially reduced population growth, it has not been completely effective because of less-strict enforcement in rural areas where two children are allowed, particularly when the firstborn is a girl; families still desire more than one child, due to cultural tradition (Figure 10-27). This slight upward trend in annual natural increase may significantly threaten long-term population-reduction goals.

Many rightly believe that reduced fertility associated with the one-child policy has resulted in improved living conditions for the majority of Chinese citizens, but they also acknowledge negative moral, social, and economic consequences. The Chinese, like some other Asian cultures, have always favored male offspring. In China, female infanticide (the systematic and intentional killing of infants) was a problem until the Communist revolution when the practice all but disappeared because of the government's ideology of male-female equality. After the one-child policy was enacted, the deficit of females skyrocketed through not only infanticide but also sex-selective abortions using prenatal sex detection technologies such as ultrasounds. Some sources suggest that since the one-child policy went into effect, some 200 million females are missing from

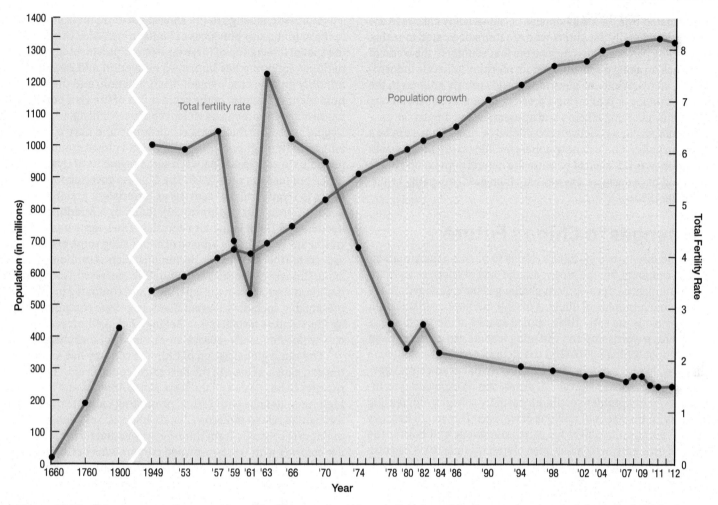

▲ **Figure 10-28** **Chinese population growth and fertility trends.** Although China's total fertility rate has dropped dramatically in the last several decades, population growth has continued in part because of two-child families in rural areas and older segments of the population living longer.

the population. The consequence is that China's sex ratio stands at 118 males for every 100 females, compared to the world average of 101 to 100. Despite the government's many campaigns to convince families that a single-daughter family is virtuous, female abortions and infanticide are most common in families having only daughters or more daughters than sons.

The direct consequences of single-child families as well as a shortage of females are many. The existence of fewer women has led to a "marriage squeeze" in which there are too few females for men wanting to marry. While this situation can sometimes lead to an increase in female status because more women can now exercise greater choice in mate selection, poor men are ultimately disadvantaged. Coupled with the impact of globalization and capitalism, the shortage of females has fueled the illegal trafficking of hundreds of thousands of girls and women to support a growing sex work industry. Another economic and social consequence is future high dependency ratios. Those 60 years of age or older made up 10.3 percent of China's population in 2000, but rose to 13.3 percent in 2010. In a one-child society, the number of elderly dependents increases dramatically relative to the number of children, and the ability of sons and daughters to fulfill their traditional social safety-net obligations to their parents is threatened (Figure 10-29). This problem

▲ **Figure 10-29** **Elderly women playing cards in a Guangzhou park.** Because of the one-child policy, the proportion of elderly in the population has steadily risen.

is exacerbated by a deficit of females because in any culture in the world, it is generally daughters who are more dedicated to taking care of aging parents. China may be the first country in the world to experience an aging population before reaching full-scale industrialization status. While the government still strongly adheres to the one-child policy, it is allowing a growing chorus of debate among some family planning officials and academics, and among the general population, to relax what is considered by many Chinese to be a draconian policy. Some demographers in China claim that the one-child policy was not needed because the natural population growth rate would have declined anyway with increased prosperity beginning in the 1980s.

Challenges to China's Future

Few would disapprove of China's choice to pursue a more market-oriented economy, but the break with the past has been so substantial that numerous development challenges have emerged. While it is true that hundreds of millions of Chinese have been lifted out of poverty since the early 1980s, and the ranks of the middle class are growing, a worrisome and widening income gap exists between rich and poor. China's 2010 Gini coefficient, a measure of income inequality, is approaching levels so severe that even central government officials claim that social unrest can be expected. The income inequality problem could be partially solved by increasing the wages for hundreds of millions of workers. This would improve workers' purchasing power for consumer goods and bolster the domestic economy, which has been slumping since 2008 because U.S. and European consumers have been purchasing fewer export goods from China as a result of the prolonged global recession.

A second cause for concern is the absence of any movement toward political reform that includes a more representative government, which might allow individuals to speak out against perceived injustices. Mass incidents such as labor strikes, riots, and public protests more than doubled between 2006 and 2010. Some of these protests are in reaction to land grabs on the part of local government officials for commercial development. These corrupt practices are the indirect result of reforms that transfer economic decision-making power away from the central government and gives greater authority to local government officials who are now responsible for economic growth. Although Beijing needs local officials to implement and manage central government policies, these officials have opportunities for personal financial gain when promoting economic growth. In effect, the local-scale Communist Party that once joined with workers against the capitalist class is now allied with the growing capitalist class against workers. To limit growing popular resentment, Beijing restricts the use of Internet and social media sites not approved by the government. Facebook and Twitter have been blocked since 2009 and Google moved its data center to Hong Kong in 2010 after surveillance interruptions in China. To better control information that could encourage social unrest, Beijing has promoted the blogging site *Weibo*, which blocks particular keywords linked to events that would cast the government in a bad light. A country wishing to join the ranks of global economic superpowers cannot limit its people to only government-approved information.

Another challenge to the central government is ethnic unrest in the autonomous provinces of Xinjiang and Tibet (Xizang). While the spatial integration of these far western provinces into the larger national economy has improved education and health care for minority groups, employment discrimination, and the loss of cultural identity because of the rapid influx of the Han population, it has resulted in sometimes violent conflict. Xinjiang, where Muslim Uygurs constitute the largest ethnic group and have close cultural affinity with neighboring Central Asian culture groups, has seen periodic violent clashes between some segments of Uygur and Han urban populations since 2008. The Chinese government too readily refers to Uygur political activists as terrorists, a practice that marginalizes problems (Figure 10-30). Ruled by a Buddhist theocracy headed by a Dalai Lama for centuries, Tibet came under Chinese rule in the 1950s. In the process of eliminating some of the negative aspects of Tibet's theocratic institutions, such as serfdom and forced labor, Chinese authorities also brutally suppressed Tibet's religious traditions. Uprisings in the late 1980s, and continued protests to the present day, including Tibetan Buddhist priests committing suicide by fire, stand as testaments to Beijing's failed ethnic minority policies, which frequently reflect Han-centered cultural chauvinism.

The rapid globalization of China's economy has and will continue to cause, at least in the near term, concern among its trading partners in both Asia and the West. North America and Europe have high trade deficits with China, prompting calls for trade reforms through increased tariffs on Chinese imports. Yet a substantial slice of imports originate from third-party manufacturing facilities under contract with Western-owned corporations or Chinese-owned plants with production contracts with Western retailers such as Walmart. Anger is better directed at foreign corporations rather than China itself. The West, and particularly the United States, is also concerned about China's growing military and specifically naval strength, and the potential threat this poses to the larger geopolitical stability in the western Pacific Rim region. Such Western concerns should be tempered because the Chinese government believes

▲ Figure 10-30 **A Muslim Uygur Family shopping at a market in the capital of Urumqui, Xinjiang.** Such outdoor markets offer a wide range of traditional and modern consumer goods.

that military confrontation would jeopardize its current rapid economic growth trajectory and engagement with the global economy. This does not mean that China is afraid to throw its weight around on control of the South China Sea. Recent diplomatic and military confrontations have been less compromising, particularly over two island groups also claimed by its neighbors. Displaying new naval hardware, China has claimed the Diaoyu Islands just northeast of Taiwan in the East China Sea, an island group also claimed by Japan (which calls them the Senkaku Islands), as well as a handful of island groups in the South China Sea quite distant from China's shores that are also claimed by various Southeast Asian countries.

China's Neighbors

China's immediate neighbors include Mongolia to the north, and North Korea, South Korea, and Taiwan to the east. After decades of economic malaise, Mongolia has a chance for greater levels of economic growth, if revenues from its mineral resource exports are used for basic needs development. Although North Korea remains a reclusive Communist dictatorship with an economy that

is practically lifeless, South Korea and Taiwan, anchored by export-oriented manufacturing sectors, possess two of the most robust economies in the world. Like Japan and emerging Pacific Rim economies, both South Korea and Taiwan adhere to a development philosophy of **state-led industrialization** in which the central government directs the structure and orientation of the national economy through various economic policies. Pacific Rim economies are classified as capitalist, but the role of the state is more pervasive than in the free-market capitalism of the West. Two explanations for the emergence of this "managerial capitalism" exist. One suggests that because private capital was very scarce at the onset of industrialization, national governments largely raised and coordinated the allocation of investment capital. The other maintains that strong state control was deemed necessary to survive in an intensely competitive global economy.

Mongolia

Mongolia, physically isolated and landlocked, is about three times the size of California, but is only populated by about 3 million people (Figure 10-31). Much of southern Mongolia is desert or steppe (treeless grassland) and it is in the better watered but still arid north

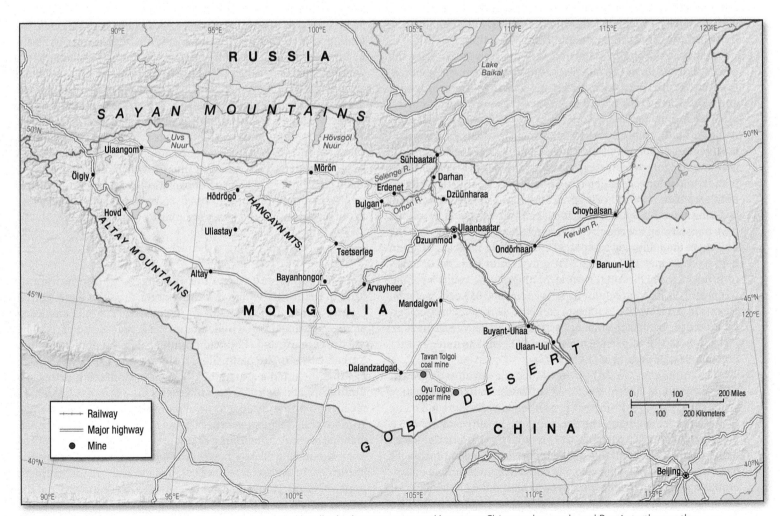

▲ Figure 10-31 **Mongolia.** Mongolia is an isolated and landlocked country situated between China to the south and Russia to the north.

and west where the densest population concentrations are found. Historically, Mongolia is perhaps best known as the home of a succession of great pastoral nomadic empires that conquered much of the central Asian grassland belt during the thirteenth and fourteenth centuries. The most heralded was the Mongol Empire ruled by Genghis Khan and later by his grandson Kublai Khan, whose land-based empire was the largest in human history. It stretched from the edge of Central Europe to the Korean Peninsula. Over the next 500 years various weaker Mongol leaders engaged in frontier military conflicts with Chinese dynasties. Economically sustained by nomadic herding, Mongolia was influenced by Tibetan Buddhism in the sixteenth century. A stronger, independent China in 1911 annexed southern Mongolia (the modern Chinese province of Nei Mongol), but Mongolia's resistance to the extension of Chinese control helps explain the close alignment of the country with the former Soviet Union during much of the twentieth century. The new Mongolian People's Republic in 1924 adopted a Soviet-style government, replete with the collectivization of livestock and the repression of Tibetan Buddhism. Following Moscow's lead, Mongolia adopted a modern democratic and parliamentary system of government in 1993, although many of those in power are former Communist Party notables.

▲ Figure 10-32 **A yurt squatter settlement on the outskirts of the national capital of Ulaanbaatar.** The round-shaped yurts are evidence of former pastoral herders engaged in rural to urban migration in response to poor economic opportunities in the countryside.

Mongolia's economy stagnated until the early 2000s when the country opened up its vast and recently discovered mineral resources to development. These resources include copper, gold, uranium, coal, and other strategically important minerals. The first major mineral project currently under development is the Oyu Tolgoi copper mine in the far south, a site discovered in 2001 and considered to be one of the three largest copper reserves in the world. Also being developed is the Tavan Tolgoi coal complex, one of the largest single coal deposits in the world. Mongolia is perceived as one of the world's last mining frontiers and its importance is based on a huge demand for resources by China as well as South Korea and Japan. Mineral resource exports have transformed the country's economy. Mongolia's per capita GNI PPP more than doubled from $1,997 in 2000 to $4,360 in 2011, and between 2003 and 2011 annual GDP growth rates averaged 6.95 percent, one of the highest in the world. Mining now accounts for about 30 percent of GDP and mining exports comprised 85 percent of total exports in 2011. Without sufficient domestic capital and technological expertise, much of the mining sector involves heavy FDI. In 2011, FDI accounted for 39 percent of national GDP and the mining sector accounted for 68 percent of this total.

The mining mania that powers Mongolia's economic growth comes with some current and future liabilities. Even though mining wealth has only recently increased government revenues, too many Mongolians remain poor. The decline of the poverty rate from 35.2 percent in 2008 to 29.8 percent in 2011 is proof that the "trickle-down" to the needy is slow and that greater effort is needed to provide for basic needs such as health care and education. An example of government neglect is the meteoric growth of squatter settlements on the edge of the national capital of Ulaanbaatar (Figure 10-32). The dispersed nomadic herding households that comprise 42 percent of the population are not benefiting from the mining boom. These nomadic people deserve help because drought and harsh winter conditions over the last decade have decimated livestock herds. Water-intensive mining activities have worsened the ecological sustainability of the fragile grassland ecosystem for nomads, their herds, and wild fauna alike because of polluted surface water and significant drawdowns of regional water tables. Much like other resource-rich developing countries, government corruption has become a serious problem. Mongolia must avoid becoming a resource periphery for the richer nations importing its valuable mineral resources and develop a more broad-based economic development strategy. Mongolia is critical to resource-hungry China; this powerful neighbor to the south consumes 85 percent of Mongolia's mineral exports, which has prompted recent "resource nationalism," or greater government control of resources, in the country's mining FDI legislation.

The Two Koreas

Between Japan and China lies the rugged and mountainous Korean Peninsula which is a little larger than the state of Utah (Figure 10-33). Although native kingdoms such as Silla and Koryo emerged during the first millennium, the Japanese and particularly the Chinese exerted significant influence on the development of Korean culture. The peninsula received a Chinese cultural package that included a system of writing, art, Confucian principles, Buddhism, and political norms. Japan's influence has been more recent and primarily economic in nature. From 1905 to the end of World War II, the peninsula became an integral part of the Japanese colonial empire, functioning as a source of minerals for resource-poor Japan, as well as a market for Japanese consumer product industries. In serving Japanese colonial interests, roads, railroads, ports, factories, shipyards, and oil refineries were constructed that dramatically altered the economic landscape of the peninsula. But the often cruel treatment of Koreans by Japanese colonial masters continues to be a nagging issue in present-day political relations.

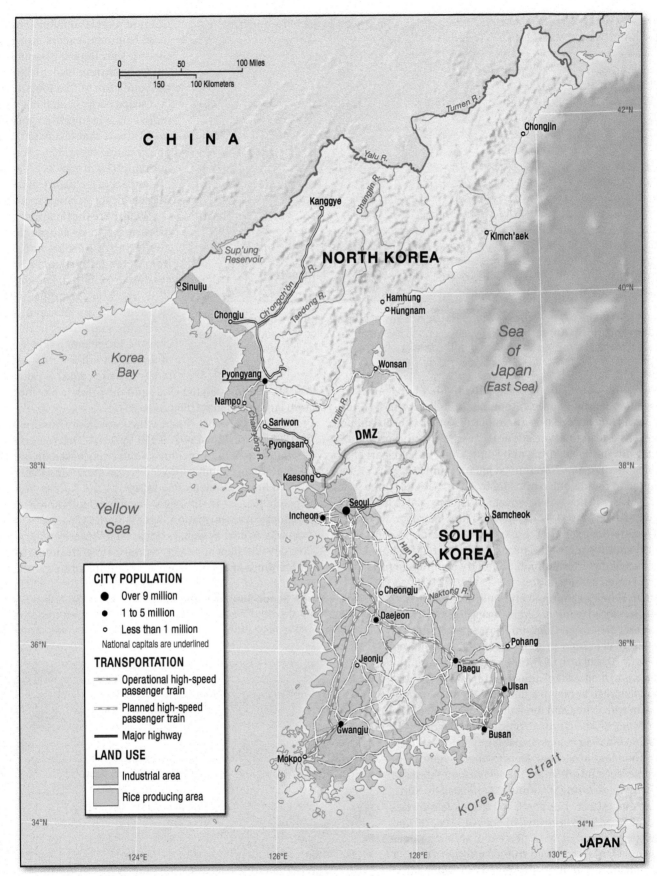

▲ **Figure 10-33 The two Koreas.** The Korean peninsula is divided into the countries of North Korea and South Korea, with widely diverging agricultural and industrial development.

▲ **Figure 10-34 The Grand People's Study House or Central Library, and Kim Il Sung Square in Pyongyang, North Korea.** Flanked by other central government buildings, the square provides a venue for political speeches and military display parades often seen on T.V.'s worldwide.

Chinese population. Ruled by the dictatorial "supreme leaders" Kim ll Sung and his son Kim Jong-il between 1948 and 2011, the current leader is Kim Jong-un, the young son of Kim Jong-il. The urban landscape of the capital of Pyongyang is replete with many government-inspired monuments that testify to the absolute power of this supreme leader-based dictatorship (Figure 10-34). There is little evidence that the nature of despotic rule will significantly change soon.

South Korea has followed an outward-oriented path of economic development. South Korea's transformation from a war-ravaged and impoverished country in 1953 to the proud host of the 1988 Summer Olympics was expressive of one of the most radical national economic makeovers in the twentieth century. Its gleaming skyscraper capital of Seoul (23 million people live in the larger urban region) is now one of the world's most populated cities and the primate city for the world's 15th largest economy (Figure 10-35).

Today the peninsula is politically divided between North Korea (Democratic People's Republic of Korea, with 24.6 million people) and South Korea (Republic of Korea, with 48.9 million people). As a result of World War II, postwar North Korea gravitated toward the Soviet orbit, and South Korea became part of the United States–led anti-Communist alliance. In 1950, North Korea attempted to reunify the peninsula by force, but after three years of war, a ceasefire line was drawn at the 38th parallel to function as a demilitarized zone (DMZ) as well as a new international boundary. Despite the events of the Korean War and the continuing political subdivision of the peninsula, many older Koreans on both sides of the international boundary continue to feel a deep sense of shared cultural unity.

North Korea's economic and political systems resemble the closed world of China during the Mao years. A bleak and dictatorial Communist regime controls every aspect of economic production and social relations in the North. Based on a relatively good mineral resource base, state-owned heavy industries remain tied to the domestic market, although a limited amount of trade is carried on with China and the states of the former Soviet Union. Much of North Korea's productive energies have been spent building military capability, particularly missile and nuclear programs. Depressingly symbolic of these misplaced priorities are the severe food shortages and, sometimes, outright famine experienced by North Koreans since 1994. The death toll from a famine between 1994 and 1998 is estimated between 300,000 and 800,000 people, and up to 62 percent of children under the age of seven experienced stunted growth as a result of malnutrition. The food shortage is so serious that some 30,000–100,000 North Koreans, primarily from the northeast, have illegally crossed the border into China to seek refuge among the substantial ethnic Korean

The GNI PPP per capita, which was in the hundreds of dollars in 1963, skyrocketed to $29,110 by 2010. This remarkable transformation to a high-income country was based on both external and internal economic forces. Foreign grants and loans—from the United States immediately following the Korean War and from Japan beginning in the 1960s—enabled South Korea to finance its economic transformation. Access to the U.S. and Japanese markets was crucial as well. Internal factors were equally important. In a development approach reminiscent of the mid-1800s strategy of Meiji-era Japan, the government transformed the agricultural

▼ **Figure 10-35 Seoul, South Korea.** South Korea's prosperity is expressed in the contemporary skyline of Seoul. This primate city has grown to occupy much of the limited coastal plain between the Yellow Sea and the peninsula's mountainous backbone.

sector by dismantling large feudal estates and returning the land to the farmers. The financially compensated landlords were recruited to become the country's business leaders. Mechanization of agriculture also reduced the need for farm workers, enabling many peasants to migrate to urban areas and secure employment in the rapidly expanding industrial sector.

South Korea's initial industrial development was based on an import-substitution strategy that protected industries producing for the domestic market from foreign competition by imposing high import duties. As with Japan, industrial development eventually became focused on export-oriented manufacturing. Labor-intensive industries such as textiles were developed first, but by the 1990s the focus had changed to more capital-intensive industries such as shipbuilding, petrochemicals, heavy machinery, electronics, and automobiles. The export of automobiles to the United States has become very aggressive, as evidenced by the increasing number of Hyundai and Kia cars now on American roadways. Both corporations operate assembly plants in the United States. The electronics and appliance corporations Samsung and LG have also made deep inroads into the American consumer market and have threatened the global dominance of Japanese consumer electronics firms. The government's role as general manager of the national economy was instrumental as it funneled investments into industries deemed most likely to help the country achieve its development goals. The state also controls the spatial distribution of industry through detailed regional planning, and many formerly neglected regions have benefited from this approach. The primate city of Seoul monopolizes much of the country's export-led manufacturing, but urban areas in the far southern provinces are centers of prosperous export-oriented production as well. One of the five busiest container ports in the world is in the southern city of Busan. To further the goal of spatial equity, the government in 2002 announced plans to develop a new national capital in Sejong City, 75 miles south of Seoul. After much criticism of the plan, specifically based on efficiency concerns, a compromise was worked out to transfer 36 agencies to the new capital by 2015, but keep the president's office and the important defense and foreign ministries in Seoul.

Government control over business activity has also engendered substantial corruption through bribes to obtain contracts. In 1996, two of South Korea's recent presidents were found guilty of accepting hundreds of millions of dollars in corporate bribes. Intimate company-government ties have promoted the economic fortunes of a few select family-owned megacorporations called **chaebol**. In 2010, the country's four largest chaebols accounted for half of national GDP. In many respects chaebols are much like the Japanese keiretsu.

The first quarter of the twenty-first century will likely prove to be a critical period for the Korean Peninsula. In the post–Cold War world order, North Korea has become increasingly isolated and is considered by some governments to be a "rogue" country because of its secret development of nuclear weapons. In 2002, the North Korean government expelled International Atomic Energy Agency inspectors and withdrew from the Treaty on the Non-Proliferation of Nuclear Weapons. Desperately wanting increased trade opportunities and various forms of aid, the government disabled its sole nuclear reactor in late 2007, but foreign governments remain wary of North Korea's nuclear intensions. Continued testing of rockets whose primary role appears to be military rather than peaceful space exploration increases alarm and apprehension.

While still a totalitarian state, the North Korean government has in the past few years instituted a limited package of economic reforms that include increased wage levels, reduced rationing of basic commodities, less government control over industrial production, and the opening of foreign trade zones. Reunification of North and South seems highly unlikely in the prevailing political atmosphere. But the South Korean government in 2000 reduced tensions between the two countries by instituting the **Sunshine Policy** to promote reconciliation and peace on the peninsula. The policy has led to the reconnection of rail lines between the two countries, the reuniting of some cross-border families, and closer economic cooperation. One expression of closer cooperation is the 2003 establishment of the Kaesong Industrial Complex just across the DMZ (the demilitarized zone between North and South Korea) where South Korean companies using North Korean labor in 2011 produced $370 million of manufactured goods destined for South Korean consumers. However, due to North Korea's numerous provocations and internal South Korean politics, the Sunshine Policy was terminated in 2008. North Korea's continued development of nuclear arms, despite a United Nations ban on such activity, only creates greater distrust and fear on the part of South Korea and other major global powers, including China. The resolution of tensions will not materialize any time soon and in fact internal tensions between the civilian government led by the young and inexperienced Kim Jong-un and the all-powerful military has only created greater political and economic uncertainty. Generals do not want to surrender the central role of the military in national ideology, nor their control over significant sectors of the national economy.

Stop & Think

→ Identify the primary political and economic differences that make South and North Korea very different countries.

Taiwan

The island country of Taiwan (Republic of China) is another East Asian industrial economy (Figure 10-36). Lightly populated when it became an administrative unit of China's Fujian Province in 1683, Taiwan's 23.3 million people now live with a population density of 646 people per square kilometer. A separate province in 1888, Taiwan was conquered by the Japanese in 1895 and remained a part of their empire until 1945. Much like the Koreas, the island's function within the empire was primarily to supply raw materials and consume Japanese finished products. When Nationalist forces fled from the Chinese mainland to the island on the heels of the Communist victory in 1949, Chiang Kai-shek and his followers claimed to be the government of China in exile. Taiwan was recognized by the United Nations as the true representative of China until 1971, when its seat was given to the People's Republic of China. A severe blow to Taiwan's political identity came in 1978 when the United States, its staunchest ally, recognized the People's Republic of China as the legitimate government of China and transferred its embassy to Beijing. China continues to view Taiwan as a renegade province that will be "liberated" sometime in the future.

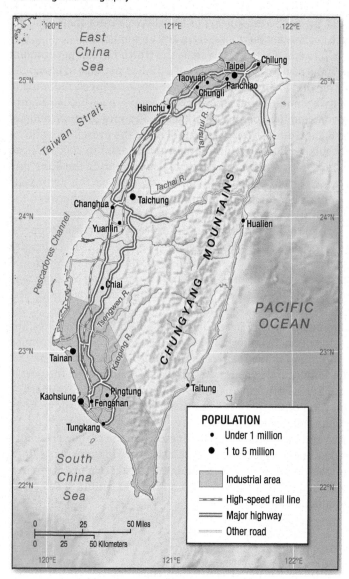

▲ Figure 10-36 **Island of Taiwan.** Taiwan was occupied by Chinese Nationalist forces following the Communist conquest of the mainland. The nation has developed one of the world's most rapidly expanding economies.

Taiwan's economic growth after World War II was spurred by U.S. investment aid and open markets in the United States for Taiwanese exports. Coupled with the foreign aid was a large pool of Nationalist entrepreneurs and a skilled labor supply. Much like South Korea, the development of domestic human resources through government guidance has been key to sustained growth. Because of land reform, Taiwan supports an efficient and productive agricultural base, although an increasing amount of food must be imported. Unlike South Korea, whose export-led economy is primarily based on heavy industry controlled by megacorporations, Taiwan's economy is more diversified, with a strong presence of competitive small and medium-sized firms. In addition to the basic heavy industries of shipbuilding, iron and steel, textiles, and chemicals, the country exports precision instruments, telecommunications equipment, electronic parts, and higher-value computer-related products. There is concern about the offshore flight of Taiwanese capital, particularly to China, which has supplanted the United States as Taiwan's largest trading partner. In 2010, China accounted for 60 percent of Taiwan's total overseas investments and 41 percent of the island's exports. Much of the export volume to the mainland was in intermediate goods or parts for electronics assembly in China. As a result, over 1 million Taiwanese live, and 70,000 firms operate, in China. This export-driven economy, ranked 17th in the world in 2009, has produced the world's fourth largest foreign currency reserves and a 2010 per capita GNI PPP of $37,500.

Much like Japan, Taiwan relies heavily on importing most of the raw materials used to produce finished exports. The vast majority of its industries are located along the broad alluvial lowlands of the west coast, where raw materials are unloaded. The adjacent industrial corridor is anchored in the north by the capital, Taipei and its outport of Chilung and in the far south by the port city of Kaohsiung. In addition to auto expressways, the corridor is also anchored by a 211-mile (340-kilometer) high-speed passenger train between Taipei and Kaohsiung that opened in 2007. This west coast region, which faces the Chinese mainland, is the key to Taiwan's economic future. The integration of Taiwan's economy with the mainland will continue to expand as long as political tensions between Communist China and democratic Taiwan are reduced. Although these political stumbling blocks are formidable, the two governments have become increasingly amicable as the Taipei government has deemphasized its independent country status.

Japan: Tradition and Modernity

In the Western imagination, Japan is a country of contrasts with many contradictions between history and modernity. Much of the cultivated land is small rice paddy fields meticulously tended by household farmers, a sharp contrast to armadas of factory fishing vessels scouring the world's oceans. Japanese culture seems to greatly value nature by virtue of forest-covered hillsides, neatly manicured Buddhist temple gardens dotted with bonsai trees, ikebana (the Japanese art of flower arrangement), and many nature festivals, including the nationwide Cherry Blossom festival. This vision of nature, so bound up in national identity, clashes with the reality of heavily polluted coastal waters during the 1960s and 1970s and the continued whaling activity of fishing fleets and dolphin harvests.

Culturally, kimono-clad women, tea ceremonies, ancient Shinto temples, and centuries-old castles contrast with the punk culture of public spaces, fast-food outlets, anime, modern skyscrapers, and neon-emblazoned entertainment districts, all in close geographical proximity to each other in Japan's largest cities. Socially, Japan's institutions seem to value hierarchy and order in education, business relations, and the factory floor, but in many ways Japanese society is quite permissive. The seeming economic contrasts are stark as well. Following the devastation of World War II, Japan rapidly developed economically into the world's second most powerful country with automobile and electronics corporations pumping out high-quality products that are the envy of the world. In contrast, Japan's economy since the early 1990s has experienced a serious, long-term recession, calling into question the ability of the "Japan Inc." model to meet the challenges of globalization. Although Japan's 2011 per capita GNI PPP of $34,740 remains high, its economy slipped to the third largest in the world, supplanted by China in 2010.

Much of what we perceive as being indigenous or domestic in nature has actually emerged as a result of Japan's government, whether historical or modern, responding to external or international forces. The external has greatly influenced the domestic in that Japan has always made accommodations to the changing international order, and this has modified the domestic scene. While Japanese government leaders have almost always been politically conservative, they have for the most part been pragmatic or realistic, nonpolitical, and opportunistic in the country's interaction with the wider world because of national self-interest. Engaging the world, and the corresponding changes that occur at home, are products of national survival.

Japan's Challenging Physical Environment

Expressed in art and literature, as well as traditionally anchoring ideas of nationhood, nature in its broadest sense has been central to Japanese culture. In what other country in the world does a volcanic peak such as the perfectly symmetrical Mount Fuji function as an icon or symbol of national identity? Japan is also a challenging environment because of earthquakes and the typhoons that accompany the summer wet monsoon each year. With a dense population located on a restricted amount of coastal flat space, such storms create much anxiety among the population. Despite these ever-present environmental hazards, or perhaps because these occur with such regularity and are beyond human control, the Japanese hold these forces of nature in great respect and awe.

Location and Insularity

Located in the northeast corner of Asia, Japan's land area is a little smaller than that of California (Figure 10-37). As an island country located offshore from a large continental mainland, Japan is often compared to Great Britain. But Japan's insularity in the sense of isolation has been far greater in that it was never invaded until World War II. The main body of Japan is composed of the four large islands of Hokkaido, Honshu, Shikoku, and Kyushu. Stretching for approximately 1,150 miles (1,850 kilometers) between 45° and 31° N, Japan's longitudinal reach is similar to the eastern United States from upstate New York to southern Mississippi. Including the long but narrow

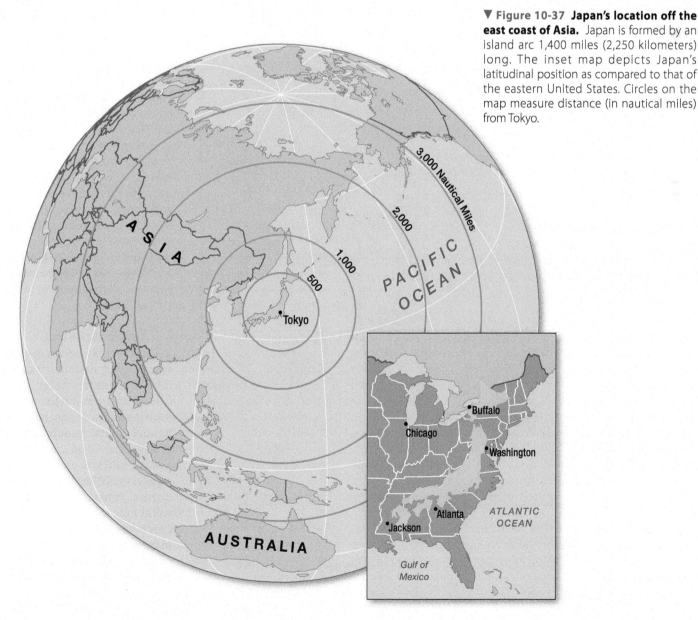

▼ **Figure 10-37 Japan's location off the east coast of Asia.** Japan is formed by an island arc 1,400 miles (2,250 kilometers) long. The inset map depicts Japan's latitudinal position as compared to that of the eastern United States. Circles on the map measure distance (in nautical miles) from Tokyo.

Ryukyu island chain that stretches from southern Kyushu to Taiwan increases Japan's north-to-south distance by an additional 700 miles (1,100 kilometers) to 24° N. In addition to the Bonin Islands 500 miles (800 kilometers) south of central Honshu, Japan claims four of the southern Kurile Islands off northeast Hokkaido. These islands are administratively part of Russia but remain contested because a treaty with Russia following World War II was never signed.

Natural Hazards

The natural hazards associated with this geologically active landscape are many, but earthquakes are the most life threatening because Japan occupies a small but volatile segment of the Pacific Ring of Fire (Figure 10-38). In addition to numerous active volcanos, there are 500–1,000 sensible earthquakes each year; only a handful results in loss of life or property. Earthquake epicenters occur along the offshore tectonic subduction zones and also along active fault zones that run up the centers of Kyushu, Honshu, and Hokkaido islands. When earthquakes occur offshore, the potential for a tsunami, a Japanese word meaning "harbor wave," is possible. A destructive tsunami occurred in March 2011 off the east central coast of Honshu caused by the Fukushima earthquake (Figure 10-39). The earthquake and subsequent tsunami killed 16,000 people and displaced 340,000 people. Two other destructive earthquakes also remain in the Japanese collective memory. The 1923 Great Kanto earthquake killed between 100,000 and 140,000 people in Tokyo and Yokohama and left more than half of the housing stock in ruins. The 1995 Hanshin earthquake was centered on the city of Kobe and resulted in 6,000 deaths, 300,000 homeless, and the destruction of 55,000 structures. The majority of the dead were elderly who occupied older wood-framed houses with heavy tiled roofs in the older part of the city. Infrastructure damage ran to $200 billion (Figure 10-40). In addition to damage to collapsed elevated highways and severed gas lines, the port of Kobe was inoperable until 1997. The container port was Japan's busiest and the sixth busiest in the world before the earthquake and has yet to regain its transport prominence. Most active fault lines are located in central Honshu where the largest urban-industrial centers are found, so the potential for an even more destructive earthquake in the future is high. Because earthquake damage can be substantial, the Japanese government has developed some of the most sophisticated earthquake prediction programs in the world. Building-code standards and disaster response and recovery plans far exceed those in the United States.

Mineral Resources

Japan is severely lacking in mineral resources and must import nearly everything it needs for energy production (except hydropower) and industrial development. Japan has the most limited natural resource endowment of any of the world's major economic powers, and was only 16 percent energy self-sufficient in 2011. As a consequence, Japan must shop around the world to obtain the critical raw materials it needs. The most important resources imported are crude oil, coal, iron ore, nonferrous metals, and wood and wood pulp.

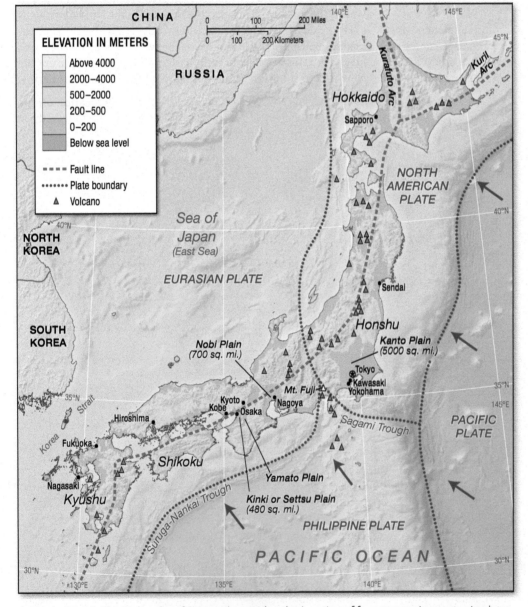

▲ **Figure 10-38 Physiography of Japan.** Located at the junction of four converging tectonic plates, the mountainous Japanese archipelago experiences frequent earthquakes and is dotted by volcanos, which makes for a hazardous living environment.

Source: Adapted from P. P. Karan, *Japan in the 21st Century: Environment, Economy, and Society* (Lexington: University Press of Kentucky, 2005), 11, 13.

▲ **Figure 10-39** **2011 post-tsunami aerial image of the town of Minamisanriku.** Even structures located on designated "higher ground" in the city were washed away by the more than 50-foot (15-meter) high tsunami wave. Only those on the surrounding hills were spared.

Japan's energy resources illustrate especially well the country's dependence and vulnerability. In 2010, the country was the world's largest importer of natural gas, second largest coal importer, and the third largest importer of oil. The Persian Gulf states provide more than 80 percent of Japan's oil. Until the 1950s, the primary energy source was coal. The government has since tried to steer Japan toward greater reliance on petroleum, which peaked at 73 percent of energy needs in 1975. The oil crises of the 1970s made Japan increasingly aware of the political and economic insecurity of world oil supplies. Petroleum as a source of energy declined from 57.1 percent to 42.0 percent between 1990 and 2010. Coal, natural gas, and nuclear power have filled the energy supply gap. Between 1990 and 2010, these three energy sources together increased from 36.3 percent to 53 percent. In 2010, before the Fukushima or Tohoku earthquake, nuclear power accounted for 30 percent of electricity generation (see *Exploring Environmental Impacts:* The Fukushima Disaster and Nuclear Power in Japan). To promote greater energy security through the diversification of sources, the government provides favorable loans to a handful of private Japanese oil and exploration firms to establish overseas partnerships in oil and natural gas fields around the world. In 2011, these Japanese firms accounted for 19 percent of oil imports, but the government's goal is to increase this share to 40 percent by 2030.

▲ **Figure 10-40** **Earthquakes in Japan.** This previously elevated expressway in Kobe collapsed during the 1995 Hanshin earthquake.

Japan's Cultural and Historical Past

Unlike China, Southeast Asia, and South Asia, Japan has no grand monuments testifying to the ancient greatness of its past because, in part, Japan lacked the large river valleys that sustained land-based empires characterized by large-scale and extensive agricultural production, urban places, and centralized and bureaucratic governments. While its insular and isolated location did promote cultural homogeneity, much of the cultural development of Japan is based on multiple external influences.

The Beginning of Japanese Culture

The earliest known culture is the stone-age **Jomon**, a period that involved the migration of peoples from the forests of the northeastern Asian mainland beginning 12,000 years ago. The indigenous Jomon culture was replaced some 2,300 years ago by the **Yayoi** culture, migrants from a region just north of present-day China who arrived in Japan via the Korean Peninsula. From their Kyushu base, it was the Yayoi who introduced sedentary agriculture (paddy rice in particular), bronze and iron technology for making farming tools and weapons, the ancient form of spoken Japanese, and a religion that eventually developed into **Shintoism**, which is anchored in a number of gods or *kami* (Figure 10-41). A great transformation of Japanese culture and politics commenced around 1,700 years ago with the rise of the **Yamato period**. At the eastern end of the Inland Sea near Osaka on the Yamato Plain, the first Japanese state emerged, centered on a military aristocracy and anchored by successive divine kings. From China arrived many innovations that

EXPLORING ENVIRONMENTAL IMPACTS

The Fukushima Disaster and Nuclear Power in Japan

In the late afternoon of March 11, 2011, Japan experienced the world's worst nuclear disaster since the 1986 Chernobyl accident in Ukraine. An earthquake registering 9.0 on the Richter scale occurred 43 miles (69 kilometers) off the coast of northeastern Honshu. This was the most powerful earthquake experienced in Japan and the fifth most powerful in the world since 1900. Tsunami waves generated by the earthquake flooded three reactors of the Fukushima Daiichi Nuclear Power Plant complex operated by the Tokyo Electric Power Company (TEPCO). The tsunami caused a cooling system failure that resulted in a nuclear meltdown and the buildup and subsequent explosion of hydrogen gas in all three reactors (Figure 10-5-1). Some 450,000 inhabitants were displaced and a 12-mile (19 kilometer) exclusion zone was established around the nuclear power plant. Soil tests have revealed contaminated soils up to 15 miles (24 kilometers) from the plant. Referred to as Japan's 3/11, the loss of life, internal refugees, separated families, contaminated agricultural produce, soil pollution, and the still-unknown human health consequences of radiation fallout created a sense of doom throughout much of the country in the following months. Estimates put the total cost of the disaster at between $195 and $305 billion. Government plans for rebuilding the region will further bloat the government's already huge debt; it may take decades to complete the cleanup.

Because the majority of Japan's 50 nuclear reactors were subsequently shut down for safety reasons, the Fukushima disaster also calls into question the role of nuclear power in the country's energy future. Thirty percent of Japan's electricity generation comes from nuclear power; any reduction will challenge the country's future

energy planning. Most scenarios have the country reducing its nuclear power dependency to 15 percent. The Japanese people are very concerned about nuclear power in part because the industry has lied to the public and falsified safety reports submitted to the government. The public does not trust the government to properly regulate the industry because of the traditional cozy relationship between the government and corporations. When the government held hearings on post-Fukushima energy policy, power company executives primarily were invited to testify.

Nuclear power plant safety is of great concern, and fears are exacerbated by the insecure nature of employment in Japanese nuclear plants. Some 88 percent of workers at Japan's nuclear plants are low-paid, unskilled, and temporary employees with little

safety training who do what is referred to as "precarious" work. In Japan they are called "nuclear gypsies"; they work for subcontractors and perform the dirtiest work, often associated with radiation exposure. These workers are often drawn from marginalized populations such as Korean immigrants or other social minority groups. These nuclear gypsies also comprise the backbone of the workforce hired for cleanup work at Fukushima. This dangerous and insecure work is associated with twenty-first century global capitalism whereby a core of well-paid, permanent employees are supported by a periphery of laborers performing risky work in an effort to maximize profits.

Source: Andrew Herrod, "What Does the 2011 Japanese Tsunami Tell Us about the Nature of the Global Economy?," *Social and Cultural Geography* 12, no. 8 (2011): 829–837.

> Water treatment facility has been rendered functionless: intakes are damaged, debris is likely clogging intakes, fuel oil tanks which fuel emergency pumping systems appears to have been damaged.

> Badly damaged reactor containment buildings at Unit #3.

> Reactor building at Unit #1.

14 March 2011: DigitalGlobe's WorldView-2 satellite image of Fukushima nuclear power station. The roof and wall panels around reactor unit #1 have exploded and the containment structure around Unit #3 has been badly damaged. Moreover, at this point the water intake facility has been damaged, along with the fuel oil tanks that probably power the diesel generators that pump the water through the cooling apparatus during power outage. This confirms that the facility's coolant system is completely without function at the time of the image acquisition.

▲ **FIGURE 10-5-1**

would transform Japan. Buddhism fused with native Shintoism, Confucian philosophical and legal concepts were adopted to form the basis of a more centralized imperial state, and Chinese writing was borrowed by the elite.

From the 700s to the 1100s, known as the **Nara** and **Heian periods**, Chinese influences began to mature; Chinese architecture, music, silk-making, painting, and calligraphy became popular among the elite. In 794, at the beginning of the Heian period, the

▲ **Figure 10-41 Shinto shrine in Kyoto, Japan.** The torii gate to the Fushimi Inari shrine marks the boundary between sacred and secular space.

the Tokugawa shogunate (*shogun* meaning "general of the army") directly ruled much of central Honshu and was allied with castle-based regional *daimyo* or feudal lords and their many samurai from southern Kyushu to southern Hokkaido. The **Tokugawa period** witnessed elevated economic growth based on increased rice production, the rise of merchant associations and banking facilities, and commodity trade. Urbanization and economic interaction between settlements increased. A host of urban places with many different functions emerged: castle towns, post or stage towns, market towns, port towns, and religious centers. Japan's first large-scale urban region that set the stage for the country's future urban-industrial core emerged focused on the cities of Edo, Nagoya, Kyoto, and Osaka along Honshu's south-central Pacific coast.

Especially important to the enhanced interaction of urban places was the southernmost of five highways, the **Tokaido Road** or "East Sea Road" between Edo and Kyoto and later Osaka. The Tokaido Road was a government-sanctioned highway with 53 post stations, providing lodging and horse stables for travelers who were primarily members of the elite or government officials (Figure 10-43). While all four cities were castle towns with multiple functions, three had distinctive identities, much of which has survived to the present day. As the center of government administrative control and aristocratic culture, Edo was the largest of the cities. By the late 1700s, its population was over 1 million and the largest in the world. Kyoto and Osaka each had populations of 300,000–500,000. Kyoto was the seat of the emperor and the intellectual and arts center of the Tokugawa period. As a critically important port town located at the eastern end of the Inland Sea, Osaka was the premier commercial and business center of Honshu.

Japan Opens Its Doors

After centuries of indirect contact, Japan experienced direct foreign contact through the arrival of European traders and those Japanese trade vessels venturing to Asian waters. In 1543, the Portuguese were the first Europeans to arrive, followed by the Spanish, Dutch, and English over the next half century. The visitors were impressed by Japanese technological and cultural achievements, while the Japanese were attracted to European trade goods that included guns, tobacco, and Chinese luxuries. Much of the contact was limited primarily to Kyushu. Conversions to Christianity, primarily Roman Catholicism, were rapid; there were 200,000 converts on Kyushu by the end of the 1500s. By the early 1600s, the new Tokugawa government became suspicious of Westerners and their religion and began a systematic policy of discrimination and sometimes persecutions. By 1640, the Spanish and Portuguese were expelled and the Dutch, English, and Chinese were confined to a small island in the port of Nagasaki on the western coast of Kyushu. The government closed Japan over fears that foreign influences would undermine peasant and samurai loyalties.

Japan's self-imposed isolation came to an end in 1853 when American Commodore Perry sailed into Tokyo Bay with a four-vessel fleet. Perry sought trade opportunities as well as western Pacific supply ports for merchant and whaling vessels. The Tokugawa government initially turned down the request for trade privileges, but conceded upon threat of a naval bombardment. Over the next few years, the Tokugawa government adopted Western military technology to modernize their armed forces to confront the West. The

capital moved from Nara to nearby Kyoto and remained there until the mid-1800s. Modeled after the Chinese Tang Dynasty capital of Xian, Kyoto was centered on the imperial palace, which was surrounded by residences of the aristocracy and in turn by shopkeepers and commoners (Figure 10-42). The Heian period is also known as the classical period characterized by the development of an independent Japanese system of writing characters that made indigenous literature possible. The Yamato king was transformed into an emperor with a "mandate of heaven" and a direct line of descent from the Shinto sun goddess Amaterasu. The 1100s to mid-1800s were a distinctive period in Japan's historical geography. Governments were controlled by *bakufu* or "behind the scenes" rulers who were the military leaders of their respective family-based territories, with the emperor in Kyoto limited to a reduced government role. In 1603, the Tokugawa family from the Kanto Plain came to power, reduced the emperor to a sacred, figurehead role, and brought peace and efficient bureaucratic rule for the next 260 years. Based in Edo (modern-day Tokyo), which was distant from the imperial capital of Kyoto but more geographically central,

▼ **Figure 10-42 The Golden Pavilion (Kinkaku) in Kyoto.** This structure is one of the most famous and treasured relics in Japan, illustrative of the influence of Chinese culture on Japan in previous centuries.

▲ **Figure 10-43** **Scene from the Tokaido Road that connected Edo with Osaka.** Located in the historically important city of Ōtsu on the shore of Lake Biwa, this station was the western terminus of the Tokaido Road.

anti-Western reaction from some powerful daimyo as well as the emperor was strong because this engagement with the West meant a loss of Japan's sovereignty as well as an increase in Tokugawa power. Aside from the extraterritoriality rights given to foreigners, the Tokugawa government was seen as monopolizing military technology and other technological innovations adopted from the West. So daimyo resistance was in part based on the emerging spatial inequality in access to economic development opportunities. With the rising power of the merchant class, growing impoverishment of the approximately 2 million samurai, and increased taxation, the demise of the Tokugawa period was imminent.

> **Stop & Think**
>
> What historical factors laid the foundation for Japan's economic-industrial core?

Modernization and the Japan Model

The Tokugawa shogunate was replaced by a group of powerful daimyo, who as an oligarchy restored the emperor to nominal power in what is called the **Meiji Restoration** in 1868. An excellent example of government or political elites transforming the national economy and society to conform to changing international developments, Japan achieved developed world status over the next 60 years. The Meiji government borrowed heavily from the West, adopting what the government perceived each country did best. A constitutional government with a legislature was established, daimyo lands were transformed into prefectures each with a governor chosen from the daimyo, and the imperial capital was transferred from Kyoto to Tokyo. The samurai class was disbanded and many, because of their high levels of education, were funneled into the government bureaucracy or became teachers or military officers. Farmers were released from feudal obligations, but because of land privatization,

the majority became tenant farmers. The government also invested heavily in a modern communications system that included roads, railroads, ports, and the telegraph (Figure 10-44). A free market economic system was embraced, necessitating the establishment of uniform tax laws and a banking system. A modern corporate environment emerged in key industries, some of which were founded by former daimyo.

Japan's transformation from a small feudal East Asian state into the economic giant of today was very much based on what is referred to as the **Japan Model**, a unique adaptation of Western methods to indigenous Japanese culture and values. This model included government guidance, not control; competent bureaucracy; proper sequencing of the development process; focus on comparative advantage and regional specialization; wise investment of surplus capital; development of infrastructure; emphasis on education and upgrading of the labor force; population planning; and a long-range perspective.

This model was a powerful force in the twentieth century as the economies of South Korea, Taiwan, Hong Kong, and Singapore, followed later by China and others, adopted many of these practices or policies, albeit with adaptations to fit each country's unique conditions. Today it might be more appropriate to label this approach the **East Asian Model**. Under rapidly changing circumstances Japan no longer finds its system working well and is trying to reinvent itself once again. To understand the model, it is important to note that Japan is now in the third of a series of transformations that began with the Meiji era.

Japan's First Transformation: Rise to Power

Meiji Japan's economic transformation was very much anchored in the **zaibatsu**, large industrial and financial cliques that provided an effective means of marshaling private capital for investment. The zaibatsu worked through vertical and horizontal integration of the economy. A single zaibatsu might have control of an entire operation, from obtaining raw materials to retailing the final product. The nearest equivalent in the West is a giant conglomerate or multinational corporation. By the 1920s, the zaibatsu—particularly the big three of Mitsui, Mitsubishi, and Sumitomo—controlled a large part of the nation's economic power.

▼ **Figure 10-44** **A 1905 trolley car in Tokyo, Japan.** The Meiji government in late nineteenth century Japan invested heavily in modern innovations.

Japan passed from a traditional society stage to the preconditions needed for economic takeoff at the beginning of the Meiji period. The modernization of industry was very much geared toward the military, because it was military power that would enable Japan to achieve equality with the Western powers and avoid the disaster unfolding in China under colonial intrusion. At the same time, as the need to import raw materials grew, export industries were encouraged, particularly silk-making and textiles. By the end of the nineteenth century, Japan still had a small industrial base, but manufacturing production increased enormously between 1900 and the late 1930s, particularly for inexpensive light industrial and consumer goods. Foreign markets, while important, were secondary to the growing domestic market.

Military victories over China in 1895 and Russia in 1905 encouraged Japan to pursue a course of territorial expansion, motivated by the need for secure sources of raw materials; Taiwan in 1895 and Korea in 1911 became the first victims of a growing, but short-lived, Japanese Empire. By the end of World War I, Japan was a fully recognized—and increasingly feared—world power. Although late to the colonial game, Japan's colonial designs on Asia were in part an adjustment to the prevailing international political system. European countries competed with each other for overseas possessions and, in response, Japan reordered its domestic and military-based industries to do the same. From World War I to the late 1930s, Japan went through its drive-to-maturity stage of development. The transfer of workers from agriculture to industry, hastened by the depression of the 1930s, kept industrial wages down. Textile and food-processing industries gradually gave way to heavy industry, especially as Japan further militarized in the 1930s to eventually conquer parts of China and Southeast Asia. Japan's extreme vulnerability—its lack of domestic raw materials—contributed greatly to its defeat by the United States. Japan's attack on Pearl Harbor demonstrated not only its profound lack of understanding of the United States, but also how insular and inward-looking the nation still was.

Japan's Second Transformation: The Quest to Be Number One

When Japan surrendered in August 1945, the nation was prostrate. Destruction from the war had been catastrophic, especially in urban/industrial areas. The bombing of Hiroshima and Nagasaki had turned Japan into a test case for the era of modern warfare. Stripped of its empire, Japan shrank back to just the main archipelago (the Ryukyu Islands were not returned until the 1970s). The future of the Japanese people seemed bleak. Yet within a decade, Japan was free of foreign occupation and well on the road to recovery because of two important factors. Japan was able to marshall its greatest resource—a well-educated, technically proficient people. Its able administrators and entrepreneurs were eager to seize the reins and rebuild the nation just as fast as U.S. authorities would allow. The role of the U.S. government was the second factor critical to Japan's postwar recovery. During the Korean War (1950–1953) Japan was used as a procurement site for war material; billions were invested into Japan's reindustrialization to fight that war. To help Japanese exports, the United States gave Japan open access to the huge American consumer market and sold American technology at bargain prices to Japanese companies. One of the most famous cases was semiconductors, which spawned a revolution

in consumer electronics and propelled Japan to world dominance in that industry. From the 1960s to the 1980s the United States absorbed 25–30 percent of Japan's total annual exports. Japan was also spared most of the cost of self-defense by being placed under the U.S. nuclear umbrella, although Japan does pay a major share of the cost of maintaining U.S. forces and bases in Japan that still exist today. Japan's U.S.-designed constitution forbids it from redeveloping an offensive military capability, permitting only "Self-Defense Forces." Japan thus could direct virtually all its energies into peaceful economic growth.

The Japan Model Reinvented

Many of the features of the Japan Model reemerged after World War II and new features were added that reflected the changed conditions. Close cooperation between government and business grew and this was particularly important in financing the modernization process. Banking credit was backed by the government, which made heavy capital investment possible. High rates of household savings gave the banks more money to loan to industry. Unlike American businesses, Japanese firms were able to concentrate on long-range strategies and research and could be less concerned about short-term profits. The breakup of the zaibatsu during the occupation was unsuccessful; they returned in a new form, known as the **keiretsu**, after the Americans left. These keiretsu played a critical role in Japan's dramatic postwar growth. Examples of modern keiretsu include Mitsubishi, Sumitomo, and Mitsui which at their core are usually financial institutions, but have alliances or shares in a wide variety of other industries. Under the Mitsubishi corporate umbrella, for example, is the auto company Mitsubishi, the camera company Nikon, the brewery Kirin, and the shipping company NYK Line, in addition to a vast array of companies producing plastic, steel, paper, and petroleum products (Figure 10-45). The government's strategy was to make Japan number one in the world in economic strength and production, and it was the seemingly broad-based support of the Japanese people for this ultimate goal that gave rise to stereotyping the country as a kind of all-encompassing national corporation dubbed "Japan, Inc." Economic growth was impressive, averaging at least 10 percent per year during the 1950s through the 1970s.

Japan's government successfully guided an economy anchored in capitalism, but suited to the Japanese context. Through such key government organs as the Ministry of International Trade and Industry (MITI), since renamed the **Ministry of Economy, Trade, and Industry (METI)**, growth industries were targeted and then supported with generous assistance to maximize the country's comparative advantage. This twinning of government policy and private capital to achieve high levels of economic growth is often referred to as **bureaucratic capitalism**. Textiles and food processing, which had dominated until the war years, shrank rapidly in importance after 1950 because of growing regional competition. Government stimulus shifted to higher-value iron and steel, petrochemicals, machinery, automobiles and other transportation equipment, precision tools, and consumer electronics. Automobiles and electronics are probably the products most familiar to consumers around the world. Heavy industry's share of total industrial production passed the 50 percent mark around 1965.

To protect its domestic industries in the 1950s and 1960s, Japan raised tariffs on imports and foreign investment was also restricted.

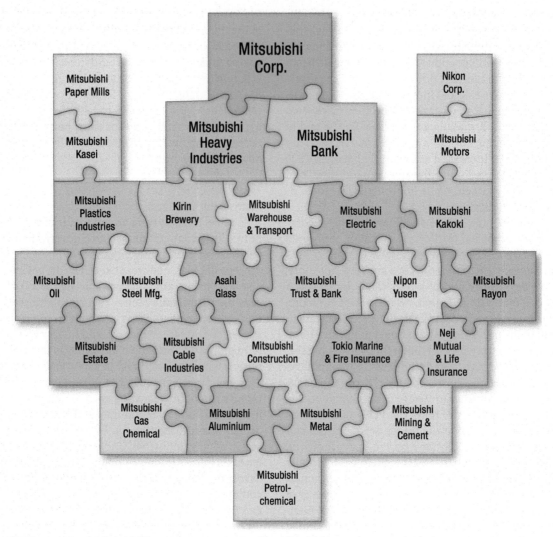

▲ Figure 10-45 **Mitsubishi corporate structure.** The Mitsubishi conglomerate includes many companies with diverse product lines.

The rationale was that Japan's economy was too weak to withstand uncontrolled imports and unlike in other Asian countries, foreign investment would have meant losing economic autonomy. By the mid-1970s, the barriers began to fall, and since then Japan has become much more open economically. Yet many foreign companies attempting to break into the domestic market continue to have difficulties penetrating the complex marketing system and overcoming their ignorance of the Japanese and their culture.

Another feature of Japan's economic system has long been its tiered structure, consisting of a pyramid with a relatively small number of modern, giant companies at the top; a greater number of medium-sized firms in the middle; and thousands of tiny workshops and family establishments at the bottom. That structure had fully emerged by the 1930s. Large firms tend to dominate such industries as transportation equipment, electrical machinery, steel, precision machinery, and chemicals, where economies of scale are needed. The smaller firms are concentrated in consumer goods areas such as leather products, textiles and apparel, and foodstuffs. The larger companies contract out substantial parts of their production to smaller firms. They also tend to have the famous lifetime employment system developed in the postwar years as a way of placating

labor unions. This paternalistic relationship between employer and employees has been much envied and studied by other countries, but is now proving increasingly difficult for many companies to sustain.

Stop & Think

> What is the Japan Model, and how did it help the country to become one of the richest in the world?

Overseas Investment

One of the characteristics of the Japanese economy beginning in the 1970s is the substantial relocation of industries abroad. By the late 1980s, Japan was the world's largest single source of FDI. The three major world regions consuming the greatest share of Japan's global FDI were North America (48.2%), Europe (21.0%), and Southeast Asia (12.2%). Between 2000 and 2010, the recipient regions of Japan's FDI shifted as North America declined to 25.6 percent, Europe remained about the same with 26.2 percent, and Asia, because of China, increased to 26.7 percent. There are several domestic and international factors behind this dramatic geographic shift in production. The earliest stage of FDI in the 1960s was in natural resources. As a resource-poor country, substantial FDI was made in resource-extracting industries to secure supplies of critically important raw materials for domestic manufacturing; this FDI focus was especially strong in Southeast Asia. Eventually these domestic resource-intensive industries such as aluminum-based manufacturing simply moved to Southeast Asia. In the 1970s, labor-intensive manufacturing also relocated to low-cost labor Southeast Asia to assemble final products for the regional market or to produce parts for export back to Japan for assembly.

These regional resource and labor complementarities based on Japanese FDI produced a regional trade triangle. The 1980s witnessed a rise of trade friction based on Japan's huge trade surpluses that prompted Japanese industries to move offshore to circumvent costly import tariffs. Japanese FDI in Southeast Asia and later China increasingly serves as "export platforms" to North America and European markets. But Japanese companies continue to be an important source of FDI in Europe and North America, particularly in auto parts and assembly. In 2010, 85 percent of Honda automobiles in Europe were produced in Europe. In 1980, all Japanese-brand automobiles in the United States were imported from Japan, but by 2011, the number of automobiles assembled in the United States reached 2,422,151. By 2012, Japanese auto companies had 13 vehicle plants in 10 states, 7 engine plants in 5 states, 9 parts plants in 7 states, and 11 research and development centers in 9 states. Including automobile distributorships, Japanese auto companies in the United States employed some 407,451 workers in 2012 (Figure 10-46).

Urban Industrial Regions

With the rapid growth of the post–World War II industrial economy, levels of urbanization increased dramatically. As late as 1960, Japan's urban population accounted for less than half the total population, but increased to 86 percent by 2011. Thus, Japan's transformation to a predominantly urbanized nation has taken place relatively recently. By contrast, the United States passed the 50 percent urbanization rate before the 1920s. In 2010, Japan had 12 cities with a population of 1 million or more, 17 with 500,000–1,000,000 inhabitants, and 43 with 300,000–500,000. A striking characteristic of Japan's urbanization is the extreme spatial concentration of the urban population.

The largest conglomeration of urban-industrial activity is found on the island of Honshu, especially concentrated in three

▼ Figure 10-46 **Japanese auto assembly plant in the United States.** This Subaru manufacturing plant is located in Lafayette, Indiana.

huge urban regions that have gradually coalesced into what is called the **Tokaido Megalopolis**, named after the Tokugawa-era road (Figure 10-47). The urban regions making up the Tokaido Megalopolis are Tokyo–Yokohama, Nagoya, and Osaka–Kobe–Kyoto, also known as the Keihin, Chukyo, and Hanshin industrial regions, respectively (the fourth largest industrial region is associated with the city of Fukuoka on Kyushu and is outside the Tokaido Megalopolis). About 64 million people, or half of Japan's total population, live in these three urban regions. The population growth rate in the central cities of this megalopolis has slowed almost to a standstill, as the fastest growth is occurring in unplanned suburbs and satellite cities of the major metropolitan centers, a pattern analogous to that of the United States.

As in most developed countries, industrialization provided the major stimulus for urbanization. In Japan's case, an important additional factor was the desire of Japan's business and government leaders to concentrate industry, especially heavy industry, in a few areas, most of them near the coast. Concentration takes advantage of economies of scale, and location near the seashore made it cheaper to handle large quantities of imported raw materials, such as iron ore, coal, and oil. Much of the postwar development of industry occurred on reclaimed land along the shoreline. All the major cities, but especially the three largest urban regions, now have largely artificial, expanded coastlines built up over decades.

Keihin. The Keihin region (37 million) centered on Tokyo (8.9 million) is the most populated urban region in the world, and is overwhelmingly dominant at a number of geographical scales. The region contains about 29 percent of Japan's population. Tokyo is the imperial capital, the seat of the Japanese government, the center of media and advertising, and also the country's dominant financial and corporate center (Figure 10-48). In a spatially synergistic fashion, the centralization of political decision making in Tokyo attracts corporations seeking influence. The Keihin region is home to the greatest number of first-class universities in Japan, in addition to Tsukuba Science City, the first and largest of the country's many planned research nodes or "technopoles." The more traditional urban-industrial landscape stretches south through Yokohama, Japan's third largest city (3.6 million) and the primary historic and present shipping port for the larger region. At the regional scale, Tokyo is the principal Asian Pacific Rim economic hub and is embedded in the larger networks of the world economy. Along with New York and London, Tokyo is one of three command centers of global finance, commerce, and production (Figure 10-49).

Hanshin. The Hanshin or Kansai region (17 million) centered on Osaka (2.6 million), Kobe, and Kyoto is the country's second largest urban-industrial region. Unlike the Keihin region, which supports a diverse economic base, Hanshin's regional economy rests on more traditional industry and commerce. Heavy industry dominates, particularly in the chemicals, shipbuilding, and steel sectors, although these industries have become less important as Japan deindustrializes, in part because of offshore production. The aging industrial landscape that stretches from Osaka westward to Kobe testifies to its "rustbelt" fortunes. While the region was the commercial and entrepreneurial center of the country until the Meiji period, many of its homegrown keiretsu have transferred their headquarters to

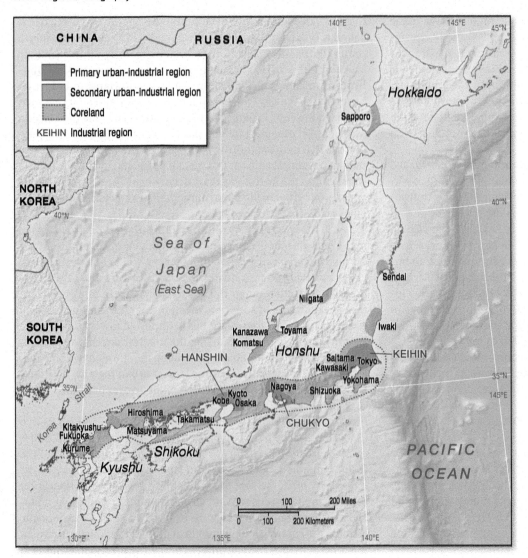

◀ **Figure 10-47 Japan's Tokaido Megalopolis.** The majority of Japan's urban-industrial development is concentrated along the southern coast of Honshu, particularly in the three regions of Tokyo–Yokohama, Nagoya, and Osaka–Kobe–Kyoto.

Consequences of the Japan Model

By the 1970s, Japan began to realize the price it was paying for its development strategy. National attention had been focused on economic growth at the expense of social welfare and the environment. The costs of Japan's so-called economic miracle have been high.

Urban Challenges

As population became ever-more concentrated in the core, a number of urban ills have surfaced or worsened. To accommodate huge urban populations, the Japanese government developed some of the world's best public transportation systems. These complex systems, marvels of modern engineering and design, function with amazing efficiency and punctuality, carrying millions of intra- and inter-urban commuters daily. The crush of people during rush hour is incredible. But since the 1960s Japan also has moved aggressively toward becoming an automobile society. Between 1970 and 1990 passenger traffic doubled and the number of vehicles tripled. Despite massive freeway and viaduct construction in recent decades, the country has witnessed a decline in car ownership, with only 73 percent of households owning a car in 2010 compared to 79 percent during the 1999–2007 period. But private vehicle ownership still leads to problems of traffic congestion, parking shortages, air pollution, noise, strip development, and urban sprawl. Japanese consumers are increasingly turning to more fuel-efficient subcompact cars as a result, and younger Japanese are not purchasing cars as they hold less cachet than in previous generations.

Aside from traffic problems, most foreigners are also struck by the seemingly unplanned sprawl of Japan's cities, especially Tokyo. Cities are characterized by dense siting of buildings, especially in the older sections, lack of architectural unity, and the relative lack of green spaces and recreation areas. On the positive side, Japan's cities are renowned for their low crime rates and general civility of life. Housing remains one of the most critical urban problems. Escalating land costs through the postwar decades led to housing becoming less and less affordable for middle-class families, particularly

Tokyo. Its second-region status has spawned resentment among Osaka's business leaders because of the power of Tokyo's government bureaucracy. In comparisons of Tokyo and Osaka, Tokyo is viewed as sophisticated, Osaka as flashy; Tokyoites are regarded as aloof, Osakans as blunt and sociable.

Chukyo. The Chukyo region (10 million), centered on the city of Nagoya (2.2 million and fourth largest), is Japan's third most important urban-industrial region. Located between Tokyo and Osaka, Nagoya is an important heavy industrial city much like Osaka, but with a noticeable difference: it is the home to Toyota Motor Corporation with its Toyota City and adjacent parts plants. Toyota City was a small silk textile town that changed its name in 1959 when Toyota arrived. The "autopolis" of Toyota City has 420,000 inhabitants and is very much a company town. Early on, Toyota and many of its auto-parts suppliers who were members of the larger Toyota keiretsu engaged in an innovative production system to create a new form of urban-industrial district emulated in other parts of the world. Based on "just-in-time" production methods, assembly plants act as hubs or anchors around which parts suppliers and related services industries provide inputs at the exact time when they are needed by the hub assembly plant.

▲ **Figure 10-48 The Tokyo urban region.** Few industrial countries are characterized by such a great concentration of economic and political power in a single urban region. Much like other large Japanese cities, reclaimed land has allowed urban space to be expanded.

in Tokyo. Typical housing space has always been much smaller in Japan than in the United States, averaging just 92 square meters (990 square feet), compared to 185 square meters (1,991 square feet) in the United States (Figure 10-50). The cost differential is far greater. The cost of a home in Tokyo is roughly 13 times the mean annual income, compared with 10 times in Osaka, but only 3 times in New York City. In terms of the price per square foot, Tokyo is one-third more expensive than New York. This promotes suburban sprawl and lengthy commuting times as people seek lower-cost housing on the metropolitan edge.

▼ **Figure 10-49 Shinjuku district in Tokyo.** Throngs of cars and pedestrians crowd Tokyo's Shinjuku entertainment district.

Regional Imbalances

Every country in the world has regional development imbalances, but especially sharp contrasts have emerged in Japan. The population and modern economy has become increasingly concentrated on the eastern, Pacific side of the island nation at the expense of the western side, which borders the Sea of Japan. Hokkaido, northern Honshu, the part of Honshu facing the Sea of Japan, southern Shikoku, and southern Kyushu all lost population as massive movement toward the core took place. This regional imbalance has bedeviled the government for decades, and attempts to alleviate the problem have had limited success.

Since the 1960s, Japan has developed five National Comprehensive Development Plans and other measures to promote industrialization in a more regionally balanced manner. Tunnels and bridges link all four main islands and are tied in with the Shinkansen high-speed railway system, aimed at facilitating dispersion of people and industry and better integrating the country (Figure 10-51). Various industrial zones have been built outside the core region with government assistance for the same purpose, and there has been some dispersion of industrial production. Aside from relocating to these industrial zones, companies have gradually dispersed to rural areas and small towns as well. This process of domestic industrial dispersion coupled with companies moving production overseas has resulted in a decrease in the share of total manufacturing output for Japan's primary urban-industrial regions.

Rural Challenges

For political and social reasons, farming has been slow to change in Japan and has been the most-protected economic sector in the country. For decades the government has lavishly subsidized farmers with price supports and stiff tariffs on imported food in exchange for loyal political support of the postwar conservative governments led by the Liberal Democratic Party (LDP). Farming and food self-sufficiency are viewed by agricultural organizations and the government as ideologically sacred; household agriculture

▼ **Figure 10-50 Japanese family members eating a meal in their small apartment.** The average Japanese home is only about half the average size of a home in the United States.

▲ **Figure 10-51 High-speed Japanese railway system.** Japan's Shinkansen railway system is among the most efficient in the world and does much to facilitate the movement of goods and people between the major urban centers.

is the essence of Japanese culture. While subsidies have declined slightly since 1995, government financial support during 2009–2011 still accounted for 50 percent of agricultural income. The reality is that based on comparative advantage, Japanese agriculture is one of the most inefficient in the industrialized world. The average farm size is still a mere 1.8 hectares (4 acres), and in many prefectures in the core region the average size is well under 1 hectare (less than 3 acres). No one can make a full-time living from such small plots of land (Figure 10-52). In spite of lavish government subsidies, most Japanese farmers still cannot make a living from farming only and must work part-time in industry or services. Only 17 percent of farming households in 2011 engaged in farming on a full-time basis. Coupled with very high import tariffs on some basic staples such as rice, wheat, sugar, beef, and butter, food prices for Japanese consumers in 2011 were 56 percent above the world average and are double those in the United States.

▲ **Figure 10-52 Combining of rice in Japan.** Miniature mechanization is common in the form of small tillers, tractors, and threshers. In spite of mechanization and high government subsidies, agriculture remains unprofitable as a full-time occupation.

Although many Japanese politicians resist liberalizing or reducing the role of government in agricultural production, both domestic and international factors will slowly bring about change. With a decline in farming voters through time, politicians feel less obliged to cater to the farm vote. The high cost of food has led politicians to seek the support of the far more numerous food consumers, who have become increasingly dissatisfied with the government's privileged treatment of agriculture during hard economic times. External pressure for agricultural policy reforms have also come in the form of free trade agreements with other countries. If Japan wants its high-value exports to enter other countries with relatively low tariffs, Japan must also impose low tariffs on imported food. Otherwise, Japan's export competitiveness is compromised for the sake of a highly protected domestic agricultural sector.

Can Japanese agriculture survive the increased pressure of a more liberalized global economy? Opening the agricultural economy to foreign competition does not necessarily mean the total devastation of the domestic farm economy because Japan currently imports approximately 60 percent of its food needs. The way forward could resemble what took place in South Korea decades ago. This includes eliminating subsidies for inefficient part-time farmers, which would allow full-time farmers to become more efficient by increasing farm size. This in turn would attract younger farmers, which is critical because the average Japanese farmer is 65 years old. Increased farm size might encourage the growth of niche crops and marketing in addition to forcing producers to be more flexible and competitive so that the remaining farmers can compete with the influx of imported agricultural products. There is no doubt that agriculture will comprise an ever-declining share of national GDP. Rather than have agriculture wither on the vine under obsolete policies, it is better to allow remaining and future farmers of Japan to engage in what they do best under conditions of comparative advantage.

A Shrinking Population and Immigration

Another indirect consequence of the Japan Model is dramatically changed social and demographic patterns. Other industrialized countries have experienced similar changes, but the degree of change is even more magnified in Japan. The country has entered stage four of the demographic transformation model, with a population of 127.6 million in 2012, up only slightly from 123.6 in 1990. The total fertility rate declined from 1.6 to 1.4 during this same period. Some projections suggest that the population could sink as low as 95.5 million by 2050. This swift demographic transition is characterized by a dramatic decrease of the percentage of the country's youthful age group and a parallel increase in the over-65 age group (Figure 10-53), both of which have serious economic repercussions in terms of an adequate labor force and elderly care.

There are only two ways to reverse this trend. The first is to boost the birthrate. The government in 1994 initiated the "Angel Plan" to assist young married couples in childrearing by increasing spending on nursery schools and other child-care–related programs in the hope that young women pursuing careers could be enticed to become working mothers (Figure 10-54). These government programs did yield a slight increase in fertility rates and prompted the government to launch a broader program in 2009. Some of these goals include encouraging workers to use 100 percent of their paid

Japan's Changing Population Structure

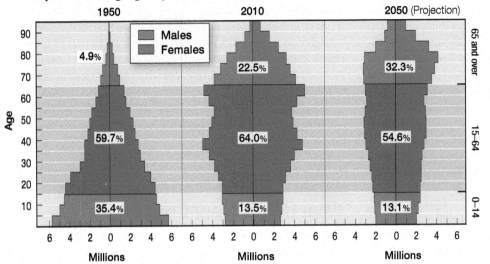

◄ Figure 10-53 **Japan's changing population structure.** Japan's population has aged greatly since 1950 and is expected to continue to do so in the decades ahead.

young professional women do not desire to assume the traditional burdens of marriage that include all housework responsibilities and providing care for her husband's aging parents. The statistics illustrate this social transformation; in 2010, 18 percent of women ages 35–39 had never married compared to 13 and 15 percent in the United Kingdom and the United States, respectively. For many adult female singles, living alone and buying a small cat or dog is the next best thing to supporting a family. There are more pampered pets than children under the age of 15 in Japan (Figure 10-55).

Despite government measures, the entrenched values of employed work as a measure of personal success are difficult to change.

Another looming demographic problem is the aging population. In part because of a good healthcare system, Japan's life expectancy of 83 for both males and females is the highest in the world. In 2010, 22.5 percent of the population was older than 65, tied with Monaco for the highest proportion in the world. By 2050 one-third

annual leave, reducing the percentage of workers who work overtime, and increasing the number of hours fathers spend each day with children. Raising the fertility level will be difficult because the average age of marriage in Japan is 29 and 30 for females and males, respectively. This has led to the phenomenon of "parasitic singles," a less-than-flattering term to describe adult children who continue to live with their parents rent free in part because of the lack of employment opportunities in a long-term recession economy. Many

▼ Figure 10-54 **Japanese day care center.** Government-funded day care centers attempt to help Japanese families balance family life and careers.

▲ Figure 10-55 **A young Japanese woman with her pet dog on a park bench.** Pampered pets function as surrogate offspring among young professional women not desiring a traditional marriage.

of Japan's population is projected to be in this elderly group. The small-city environment outside of the Tokaido Megalopolis provides evidence of this aging population. As younger members of the community move to larger cities in search of employment and educational opportunities, these smaller urban places become geriatric places. Commercial establishments are dominated by healthcare services, few youngsters are seen in public spaces, and neighborhoods are dark by 11:00 P.M. The government is very worried about an aging population because of the huge fiscal burden of providing more elderly social and health services, especially given the reduced tax revenue as the proportion of the working population shrinks. The problem is further complicated as signs emerge that the younger generation is questioning the wisdom of their elders and does not want to be part of the Japan Model. This is primarily expressed as not wanting lifetime employment with a single firm and preferring part-time work or job hopping.

The second solution to the serious repercussions associated with a shrinking labor force is to liberalize the country's immigration policies. The United Nations estimates Japan would have to admit at least 600,000 immigrants annually to stabilize the population and maintain the workforce at current levels. The adoption of a more open immigration policy to meet its labor needs would have profound implications for Japan's much-vaunted "homogeneous" society (see *Geography in Action:* The Outsiders: Historical Minorities in Japan). Even the existing level of legal and illegal residents has been causing anxiety among many Japanese. This is not because the average person is inherently distrustful of foreigners but because people remain influenced by the centuries-long, government-generated myth that cultural homogeneity (a guise for cultural superiority) is a positive attribute of the country. Many Japanese are also negatively affected by sensationalized and overblown crime-related stories in the media in which immigrants are featured.

In 2010, there were more than 2.1 million legal foreign residents living in Japan, representing 1.6 percent of the total population, the lowest of all industrialized countries. This total is led by Chinese (32.1%), Koreans (26.5%), Brazilians (10.8%), and Filipinos (9.8%). Korean and Chinese migration dates mostly from the prewar colonial era as a source of cheap labor. The roughly 317,000 Brazilians are mostly *Nikkeijin,* or foreigners of Japanese descent, whose parents and grandparents left Japan during the early 1900s and in substantial numbers just after World War II because of the war-ravaged economy. Much like immigration to other rich East and Southeast Asian countries, Filipinos started to arrive in the 1970s to seek out wage-earning opportunities absent in their much poorer home country and now number more than 210,181. Males engage in low-wage and unstable employment in the service economy, while females have arrived as mail-order brides for farmers; they are viewed as more "compliant" than Japanese women. They also work in the "entertainment" industry as sex workers often controlled by the *yakuza,* the Japanese mafia.

Many foreigners have come as "temporary" migrant workers. The number of illegal residents is between 300,000 and 400,000 and increasingly consists of Chinese from the People's Republic of China, as well as workers from the Philippines, Bangladesh, Nepal, Pakistan, Vietnam, and Iran who come on tourist or other visas and then overstay their visas. Most do dirty and dangerous jobs that most middle-class Japanese would refuse and live in constant fear of immigration authorities. While the government has created very strict laws on both legal and illegal immigration, the realization

that the economy increasingly depends on immigrants undercuts a climate of strict enforcement. Immigrants, both legal and illegal, do have strong advocates among the more than 200 organizations founded by Japanese citizens that work on behalf of immigrant rights. The economic stagnation of recent years has slowed but not stopped the numbers of foreigners who want to work in Japan.

Stop & Think

➤ What are the negative consequences of Japan's modernization model?

Japan's Third Transformation: Charting a New Course

The first decade of the twenty-first century saw a complex set of internal and external factors nudging Japan toward new paths of national development, a task made more urgent by continuing economic stagnation and the cost of rebuilding following 2011's devastating earthquake and tsunami. Different approaches to Japan's economic malaise have been offered, but nothing has worked so far. During the 1980–1990 period, average annual GDP growth was, for a mature industrialized country, relatively healthy at 3.9 percent. In the so-called lost decade of 1991–2000, average annual GDP growth plummeted to 1.38 percent. While rebounding to 1.4 percent between 2000 and 2008, it still was substantially lower than the sluggish 2.4 percent average annual GDP growth rate in the United States during the same period.

Deindustrialization

It is easy to identify the process of **deindustrialization**, the decreasing importance of domestic industry in the larger national economy, as an important factor in Japan's slowed economic growth rate. The hollowing out of Japanese industry began in the 1970s as the costs of production in Japan escalated. The only way to hold down production costs, besides adopting more advanced technologies, is to seek lower-cost labor countries. By 2000 almost 15 percent of Japanese manufacturing had shifted to other countries, primarily East and Southeast Asia, North America, and Europe. Some in Japan argue that the country should accept the inevitable and learn to live with the status quo and reduced personal consumption as production shifts steadily overseas. Critics of Japan's industrial policy argue that **sunset industries**, those losing their international competitiveness, should be allowed to decline and die. The most important of these are iron and steel, electronics, and motor vehicles. And many wonder if the current lifestyle, with its extreme emphasis on high mass consumption and all the negative environmental consequences associated with it, is the best Japan can do.

Structural Deficiencies

The prevailing neoliberal, free market mainstream opinion points to deeper problems than simply deindustrialization. Some scholars claim that the economic philosophy of state-led industrialization that allowed Japan to beat the West during the post–World War II period is no longer suited to the modern global economy. The industrial culture of lifetime employment, good labor-management relations, acceptance of low return on investment, and the subsidizing of inefficient domestic industries is in a sense "anticapitalist,"

The Outsiders: Historical Minorities in Japan

Japan's human resources and culture are not as homogeneous as popular perception would suggest. Ethnic and social minorities make up about 4 percent of the population, or roughly 5 million people. The ethnic minorities consist of Koreans, Chinese, Okinawans, Ainu, and foreign residents. The social minorities, however, are composed of *burakumin*, persons with disabilities, and children of interracial ancestry.

Historically, Japan's cultural homogeneity stems from practices that limit opportunities for people outside the cultural and social mainstream. The Japanese traditionally regard themselves as a unique people, sometimes referring to themselves as the Yamato people, in reference to the Yamato Plain around Kyoto where the Japanese culture developed in centuries past and from which the ancestry of the imperial family is derived. There is still a strong current in Japanese society to preserve the purity of the Yamato majority; anyone else is an outsider and can never hope to be fully accepted into the mainstream. The 1947 constitution expressly prohibits discrimination based on race, creed, sex, social status, or family origin. But that U.S.-imposed provision has not fundamentally altered centuries-old attitudes and practices.

Chinese make up the largest minority group with about one-fifth being descendants of residents who were in Japan before World War II. The remainder are more recent temporary residents, guest workers, or migrant laborers. The Chinese are especially visible in "Chinatowns" in various cities. The Chinese fare somewhat better in Japanese society than the Koreans, although they are still targets of social discrimination. Koreans represent the second largest minority, first coming to Japan during the colonial occupation of Korea (1910–1945) when thousands were forcibly brought or enticed to move to Japan as low-cost laborers. By the end of World War II, some 2.5 million Koreans were living in Japan. Those who chose to remain after the war were deprived of citizenship when the Japanese government declared them aliens in the 1952 peace treaty with the United States. Birth in Japan does not guarantee citizenship, and the government makes it very difficult for Koreans to obtain citizenship although most have Japanese names, speak fluent Japanese, and have attempted to integrate into Japanese society. The Koreans remain mired at the lower end of the economic ladder, victims of social and economic discrimination, and tend to live in ghettos in the larger cities.

The Ainu were among Japan's earliest inhabitants (Figure 10-6-1). Racially different and almost exclusively a hunting and fishing people, they also were treated as aliens by the Yamato Japanese. Only 25,000 pureblood are left, mostly in a few locations in Hokkaido. The Ainu have been gradually assimilated into Japanese culture since the early 1800s. In the early 1900s, "native schools" were established in Hokkaido in the hope of making Ainu children more Japanese by destroying their cultural identity. Like Native Americans struggling to maintain some of their identity, there has been an upsurge of cultural pride over the past several decades.

The Okinawans, on the Ryukyu Islands south of Kyushu, were not politically incorporated into Japan until early in the seventeenth century, even though they are of basically the same stock as the majority Japanese. Isolated from the main islands and speaking a variant form of Japanese, Okinawans have been treated as second-class citizens ever since their incorporation. Like the Ainu, younger Okinawans have recently reasserted their cultural identity in part because of the social and cultural disruption associated with the post–World War II presence of large U.S. military bases.

The *burakumin* are the largest and most abused social minority. They are physically indistinguishable from other Japanese as they have the same racial and cultural origins. Somewhat like India's untouchables, Japan's burakumin have been discriminated against for centuries because of their past association with the slaughtering of animals and similar occupations. Japan's major religions, Buddhism and Shintoism, regard those activities as polluting and defiling. Thus, a subclass of Japanese was forever branded as unfit for

▲ FIGURE 10-6-1 **Elderly Ainu males in traditional dress at the Marimo Festival at Lake Akan, Hokkaido.** The festival's function is to celebrate nature, and Ainu culture is just one aspect of the larger festival's events to promote tourism.

association with the "pure" majority. That discrimination was formalized and legalized during the Tokugawa period and is still entrenched in Japanese society. Most *burakumin*, who number some 2–3 million, live in ghettos scattered throughout the country. Denied access to better-paying jobs, housing, and other benefits, they eke out a living at the bottom of the socioeconomic ladder. Many burakumin try to hide their origin and quietly integrate into the mainstream of society, but they are usually found out when background checks are made for marriage or employment.

Sources: Tony Fielding, "The Occupational and Geographical Locations of Transnational Immigrant Minorities in Japan," in *Global Movements in the Asia Pacific*, Pookong Kee and Hidetaka Yoshimatsu, eds. (Singapore: World Scientific Publishing, 2010); Michael Weiner, *Japan's Minorities: The Illusion of Homogeneity*, 2nd edition (New York: Routledge, 2009); *Japan Statistical Yearbook*, 2013.

particularly in the context of the hypercompetitive global economy. These experts observe that Japan has not responded well to the external forces of the global economy where neoliberalism is gaining momentum. While this analysis may be partially true, we must also consider Japan's distinctive form of capitalism and its particular structural deficiencies to help identify the country's economic problems.

The first deep-seated structural problem is the traditional cozy relationships between business, government bureaucracy, and banks. These relationships led Japan's economy to stunning success in the past, but may have outlived their usefulness because of corruption, cronyism, and bad decisions behind a wall of inadequate public information. Successive weak prime ministers (eight between 2005 and 2013) were unable to change the system, even those who came into office as reformists. Another problem is the government's increasing reliance on massive deficit spending on public works projects to try to jump-start the economy in what some refer to as the "construction state." As a result, the country is now saddled with many underutilized roads, railroads, bridges, and land reclamation projects.

Consequently, consumer confidence is at a historic low. Although Japan is experiencing its most serious economic difficulties since World War II, the general public still is relatively well off and does not sense the full dimensions of the problem. The outside visitor to Japan gets little impression of any crisis, especially if he or she does not know the language. As one Japanese leader put it, "This has almost become a Japanese disease: the slow strangulation of an affluent and egalitarian society." Because the government has high savings, it can continue to accumulate debt and postpone structural reforms. Critics argue that nothing short of a drastic shake-up of government, and the emergence of a true reformist leader, can reverse this dangerous trend.

Summary

▶ While the countries of East Asia are characterized by export-led economies that are in some form government-administered, each has experienced very different development trajectories. Who would have predicted in the early 1980s that China would grow into the world's second largest economy? Nor would experts have predicted that Japan's economy would slip from the second to the third largest in the world as a result of an extended 25-year recession. Experts would not have imagined in the 1980s that the economies of both South Korea and Taiwan would become even more robust through the aggressive promotion of consumer-based exports. Despite Japan's flagging economy, East Asia has come to occupy a central position in the global economy.

▶ China assumed the rank of a global economic power as a result of its late 1970s reform policies that reduced the commanding role of the state in economic growth. China accepted the processes of globalization by liberalizing both agricultural and industrial production. In agriculture, many farmers now produce higher-value crops and livestock to cater to wealthier urbanized domestic consumers. China's global economic prominence is based on attracting FDI coupled with promoting domestic capital to become the world's most successful export-led consumer manufacturing economy. The meteoric rise of manufacturing centered on the country's three dynamic urban regions would not be possible without the agricultural reforms that freed millions of rural dwellers to seek work in coastal province factories.

▶ Although the globalization of China's economy has brought many material benefits, lingering problems call into question the durability of this economic success. For more than 30 years globalization has primarily benefited coastal regions and left hundreds of millions of poor rural people in interior provinces with little economic opportunity. The situation for the poor has become worse because the economic safety net once provided by the government was reduced. China's rush to become richer has serious environmental impacts as well. Water shortages and air pollution have resulted in some of the world's least healthy cities. In the context of global climate change, current rates of resource consumption are not sustainable. Although rich in energy resources, China must import greater amounts of these resources. This requires Beijing to develop closer political ties with many governments around the world. China's most serious challenge is to match its unprecedented economic growth with political change. Still a one-party state, the Communist government must allow greater political participation and decision making by its people, despite the political instability that might cause. Development is not just about economics but is also based on inclusionary politics, and to date the average Chinese citizen does not have a voice in the development process.

▶ China's neighbors have experienced different economic growth trajectories. Once an isolated and poor country, Mongolia's newfound and abundant mineral resources offer opportunities for sustained economic growth if revenues are used to improve the basic needs of the population. Allied with the capitalist West during the Cold War, both South Korea and Taiwan have experienced decades of interaction with the global economy. This participation was accompanied by rapid rates of economic growth anchored in democratic institutions. South Korea's economic growth has depended on a sometimes unhealthy relationship between the government and large corporations, while Taiwan's economy has focused on small and medium-sized firms functionally linked to larger firms producing for global electronics brands. Both countries have lingering political problems that impact the development process.

▶ Japan was the first non-Western country to industrialize based on its own distinctive model of modernization and development, which evolved out of three major transformations the country experienced after the 1868 Meiji Restoration. The first transformation was the initial period of industrialization, followed by militarization that led to empire, world-power status, and defeat in 1945. The second involved rebuilding after the war and developing export-based global economic opportunities with the assistance of the United States to become the second most powerful economy in the world. The third transformation is marked by an inadequate response to a highly competitive global economy. As a mature industrialized country, Japan has experienced deindustrialization, a stabilized and aging population, changing social and economic values in the younger generation, and structural problems associated with the Japan Model. The current transformation is made more urgent by the economic stagnation that has beset Japan since the early 1990s, a lasting solution to which has yet to be found.

Key Terms

bakufu 489
bureaucratic capitalism 491
chaebol 483
Chinese Communist Party (CCP) 462
Communist Party of China (CPC) 462
Confucianism 459
Cultural Revolution 474
deindustrialization 498
Deng Xiaoping 462
East Asian Model 490
Extended Metropolitan Regions 468
extraterritoriality 461

Foreign Direct Investment (FDI) 466
Grand Canal 460
Great Leap Forward 464
Han dynasty 460
Heian period 488
household responsibility system 464
hukou registration system 475
Japan Model 490
Jomon period 487
keiretsu 491
Mao Ze-dong 462
Meiji Restoration 490
Middle Kingdom 459

Ministry of Economy, Trade, and Industry (METI) 491
Nara period 488
open coastal cities 468
open economic regions 468
opium 461
Pacific Ring of Fire 454
Shintoism 487
Silk Road 460
Song dynasty 460
Special Economic Zones (SEZs) 466
state-led industrialization 479
sunset industries 498
Sunshine Policy 483

Sun Yat-sen 462
Three Gorges 452
Tibetan Plateau 453
Tokaido Megalopolis 493
Tokaido Road 489
Tokugawa period 489
town and village enterprises (TVEs) 466
typhoon 454
Yamato period 487
Yayoi period 487
zaibatsu 490
Zhou dynasty 459

Understanding Development in East Asia

1. What role do landforms, climate, and population density play in explaining development differences between the eastern and western halves of China?
2. How did colonialism impact trade and development in China?
3. What impacts did China's 1978 reforms have on agricultural, industrial, and urban change?
4. What are the past, present, and future social and economic consequences of China's one-child policy?
5. What are the present and future problems associated with Mongolia's mining economy?
6. How have the divergent development philosophies of North and South Korea impacted each country's economic and political success?
7. What has been Taiwan's political and economic relationship with China and how is this interaction pattern likely to change in the future?
8. How did Japan reorder its political and economic system to meet new international circumstances during the Meiji Restoration?
9. What factors account for the economic structures of the Keihin and Hanshin urban regions of Japan?
10. Why has the Japan Model begun to produce negative consequences and how can poor economic results be overcome?

Geographers @ Work

1. Identify China's major environmental problems and explain their causes.
2. Analyze the geographic factors that have influenced the economic growth of coastal China, and suggest ways that the interior regions of China, which have lagged behind coastal regions in their development, might catch up with east coast regions.
3. Describe China's pre- and postreform agricultural economy. Identify how reforms have impacted rural to urban migration, and suggest how problems associated with migration might be overcome.
4. Each of China's three urban-economic regions is structurally different. Describe each region's economic structure, identify how each has engaged the global economy, and speculate on how each region might evolve in the future.
5. Describe how Japan has compensated for its limited physical resource base both before and after World War II to achieve a high level of economic development.

MasteringGeography™

Looking for additional review and test prep materials? Visit the Study Area in MasteringGeography™ to enhance your geographic literacy, spatial reasoning skills, and understanding of this chapter's content by accessing a variety of resources, including MapMaster® interactive maps, videos, RSS feeds, flashcards, web links, self-study quizzes, and an eText version of *World Regional Geography*.

11
Southeast Asia

Christopher A. Airriess

Floods Disrupt Global Production Chains

Southeast Asia is an important player in global production and trade, and Thailand functions as a key production link in the global supply chain for electronics and automobile components. Production came to a grinding halt in late 2011 when devastating floods in Thailand closed many of these facilities, primarily located in the Bangkok urban region and surrounding provinces. With historic rains in upstream northern Thailand, plus an extended wet monsoon season lasting until October in the south, it was the wettest year in 6 decades. With 13 million people affected, 800 dead, and extensive damage to both agriculture and industry, the flood was one of the costliest disasters in the world. More than 10,000 factories around Bangkok were forced to close and 650,000 workers laid off until factories reopened several months later. Insurance estimates put the losses at between $15 and $20 billion.

The economic ripples around the world as a result of Thailand's flood damage were widespread and diverse, especially for Japanese corporations that rely on Thailand as a production base. Toyota slowed production at many of its global assembly plants because Thailand was a key, if not the only source for some 100 auto parts. Honda also postponed the launch of a new model in addition to halting automobile production in its Malaysia assembly plants. Both Sony and Nikon produce all their digital cameras in Thailand and also had to postpone the launches of their new models. California-based Western Digital, which produces one-third of the world's computer hard drives, saw a 40 percent or $6.5 billion decline in its Thai exports. A handful of multinational corporations have constructed flood abatement infrastructure around their local facilities, but the Thai government must address much larger-scale flood prevention infrastructure if the country is to remain an important production link in the geographically flexible, globalized supply chain of electronic and auto parts. Like the Fukushima earthquake and tsunami in Japan, other natural disasters equally disrupt the seamless logistics chains that anchor the global economy.

Read & Learn

- Give reasons for the degradation of Southeast Asia's forests and coastal environments.

- Outline the energy resource endowments and challenges confronting the region.

- Describe the economic role of Southeast Asia in the larger process of colonialism as well as the function of ethnic Chinese in this economic system.

- Explain the economic growth characteristics of modernizer and reformer countries.

- Account for the diversity of development in Southeast Asia in relation to demographic factors.

- Understand how Singapore's economy is spatially embedded in the regional and global economy.

- Specify the various ways race or ethnicity has driven the development process in Malaysia.

- Identify the core-periphery relations between Java and outer island Indonesia as well as recent government measures to relieve these geographic tensions.

- Describe the national and local problems associated with the geographic concentration of economic growth in Thailand.

- Contrast experiences of Vietnam and Myanmar in their respective involvement with the global economy.

▲ The persistence of wet monsoon rains in 2011 flooded most industrial parks in the greater Bangkok urban region, such as the Rojuna Industrial district in Ayutthaya.

Contents

503

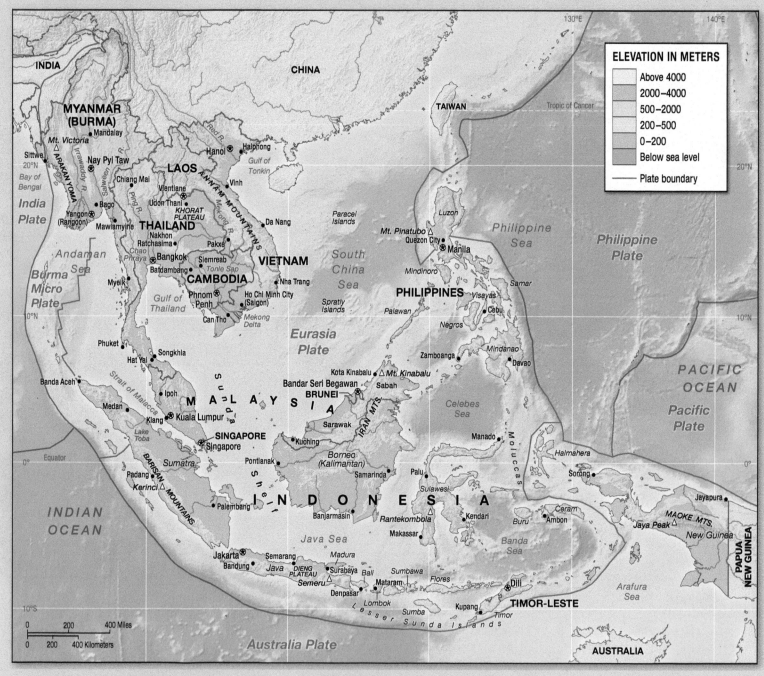

Figure 11-0 **The major physical features and countries of Southeast Asia.**

S outheast Asia is a region of large and small peninsulas and islands, surrounded by East Asia, South Asia, Australia, and the Pacific and Indian oceans (see chapter opener map). While the label Southeast Asia might give the impression that this area of the world is "what is left over" from East Asia and South Asia, or perhaps a watered-down version of these two culture worlds, it is a distinct region that stands apart from its larger neighbors.

The thousands of mountains, basins, and islands that make up this 11-country region have produced a politically, culturally, and economically complex mosaic that defies easy generalization. The physical fragmentation of Southeast Asia may partly account for its control by so many distant Western powers during the colonial period. The multitude of colonial powers further increased the region's cultural diversity. In the modern period, each state has adopted different political and economic systems that include hyperglobalized Singapore, the oil-rich monarchy of Brunei, the socialist states of Vietnam and Laos, which are now making a transition to capitalism, and the once isolationist military dictatorship of Myanmar (Burma), which is now opening the country to a wider globalized world.

To better understand the great physical and human diversity that is Southeast Asia, it is best to divide the region into two constituent subregions in order to make generalizations at a scale larger than individual countries. The first subregion is Mainland Southeast Asia, a group of countries that are physically part of the Asian continent and include Vietnam, Cambodia, Laos, Thailand, and Myanmar. The second is Insular Southeast Asia, comprising the countries of Malaysia, Singapore, Indonesia, Brunei, Timor-Leste, and the Philippines. While the peninsular half of Malaysia is physically part of the Asian continent, its economy and culture warrants its inclusion in Insular Southeast Asia.

Southeast Asia's Environmental and Historical Contexts

Southeast Asia's landforms are primarily comprised of mountains, river valleys, and coasts. In the continental portion of the region, the population has traditionally been concentrated in river valleys separated by mountain chains. In the island half of the region, population clusters are found along coastal plains. The location of precolonial civilizations generally reflects these environmental opportunities, and the physical fragmentation allowed a diverse group of colonial powers to divide up the region.

Environmental Setting: Mountains, River Valleys, and Islands

Southeast Asia stretches more than 3,000 miles (4,800 kilometers) from Myanmar in the west to Papua, the Indonesian half of the island of New Guinea, in the east. Including the oceans and seas that complete the region's territory, its dimensions are approximately equal to those of South Asia. Situated between the Indian and Pacific oceans and located almost entirely within the tropics, Southeast Asia is characterized by tropical forests and monsoon climates.

Landforms

In Mainland Southeast Asia, landforms are composed of alternating bands of mountain ranges and river valleys (see chapter opener map). These north–south aligned mountain ranges are actually lower elevation spurs of the Himalayas. While a handful of

▲ Figure 11-1 **Mountain ranges of Mainland Southeast Asia.** Fog drapes a river valley between two north to south trending mountain ranges in northern Thailand.

mountain peaks reach up to 9,000–10,000 feet (2,700–3,000 meters), most average 3,000–5,000 feet (900–1,500 meters), and are geologically stable (Figure 11-1). From west to east between these mountain ranges run several very big rivers, including the Irrawaddy, Salween, Mekong, and Red. These rivers provide irrigation water for rice-based agriculture, fish for millions, and before the modern period functioned as transport corridors that reached deep into the forested continental interior (Figure 11-2). The geographic center of Insular Southeast Asia is characterized by mountainous cores with narrow coastal plains; in eastern Borneo and Sumatra these coastal plains are fronted by wide, waterlogged forests. While laced by numerous rivers, none of the islands contain rivers as long as those in Mainland Southeast Asia. The geographic center of this island subregion is anchored by the geologically stable and shallow platform called the Sunda Shelf, while its edges are marked by sweeping volcanic arcs pushed up at the edges of the Indian and Pacific plates. In the west, the arc includes the Indonesian islands of Sumatra, Java, Bali, Lombok, and Sumbawa. The eastern margin, which

▼ Figure 11-2 **Mekong River delta, Vietnam.** This floating market is a common scene in the lower reaches of rivers in Mainland Southeast Asia.

505

constitutes one small segment of the Pacific Ring of Fire, follows a line extending from the Philippines to New Guinea. While no single volcano or mountain in this part of Southeast Asia is extremely lofty, many peaks reach 10,000 feet (3,000 meters) or higher (Figure 11-3). At the edge of both volcanic arcs are deep oceanic trenches marking tectonic plate boundaries. These plate boundaries have produced some of the most destructive earthquakes and volcanic eruptions in the world.

Two examples from Indonesia, one a volcanic eruption and the other an earthquake, each producing a devastating tsunami, illustrate this hazardous geologic environment. In 1883 the volcanic cone of Krakatoa between the islands of Java and Sumatra explosively erupted, a sound that was heard thousands of miles away and sent global climate-altering ash 50 miles (80 kilometers) into the atmosphere. The eruption and following tsunami killed 36,000 people from the surrounding coastal lowlands. The most destructive and deadly tsunami in the world since Krakatoa was the December 2004 tsunami off the northwest coast of Sumatra. After a magnitude 9 "megathrust" earthquake, a 750-mile (1,200-kilometer) section of the Indian Plate was thrust up about 65 feet (20 meters) over the Burma Plate. The energy released from this plate slippage was equivalent to 23,000 Hiroshima-sized atomic bombs. Within hours of the earthquake, tsunami waves ranging in height from 33 feet (10 meters) in Sumatra to 13 feet (4 meters) elsewhere made landfall. Confirmed deaths were approximately 228,000 people in the 13 countries impacted, with Indonesia (130,736), Sri Lanka (35,322), and India (12,405) the most affected. In Banda Aceh at the northern tip of Sumatra, about 15 percent of the city's population of 400,000 perished.

Climate

The monsoon climate expresses itself differently in Southeast Asia when compared to South and East Asia (see Figures 9-3 and 9-4, pages 410–411). In Mainland Southeast Asia, the pattern of a wet season from June through September and a dry season from October through May resembles that of its two larger regional neighbors, but because of its tropical location, temperatures do not cool during the dry monsoon. While there are distinct wet and dry seasons, the difference is not as stark because some rain does fall during the dry season as Bangkok illustrates.

Precipitation patterns in Insular Southeast Asia are more complex. During the wet southwest monsoon, much of this subregion receives copious rainfall. During the dry northeast monsoon, however, dry winds off the Asian landmass pick up moisture from the South China Sea and precipitate moisture over some locations of the region as well. Jakarta and Singapore provide good examples of this phenomenon. Much of Insular Southeast Asia's equatorial position makes for a more even annual distribution of rainfall too, as Singapore illustrates. Only the eastern and western edges of Southeast Asia are affected by damaging tropical cyclones. The Philippines and sometimes northern Vietnam are lashed by Pacific Ocean typhoons as these intense low-pressure systems make their way to East Asia. One such storm was super Typhoon Haiyan in November 2013 that is purported to have been one of the strongest tropical cyclones to make landfall anywhere in the world. With sustained winds as high as 195 mph (315 kmph), the storm leveled towns and villages across several island provinces and left thousands dead (Figure 11-4). Bay of Bengal cyclones occasionally make landfall in Myanmar, as evidenced by the devastating Cyclone Nargis in 2008.

Environmental Challenges

Population growth, rapid urbanization, and export-driven economies during the modern period threaten forests and coastal habitats. Forests traditionally played a prominent role in the livelihoods of the Southeast Asian peoples. Much of the region was originally covered by tropical forests, owing to the year-round warmth and humidity. During Southeast Asia's premodern era, forests provided building materials, food, and medicine for local use or long-distance trade. The people of Southeast Asia also have been oriented toward

▲ Figure 11-4 **Typhoon Haiyan damage in the city of Tacloban, Philippines.** Residents of a seafront neighborhood search through the rubble of their homes. Typhoon Haiyan has devastated this city of a quarter million inhabitants on Leyte Island.

the sea for food and transportation. This is especially true for Insular Southeast Asia where monsoon winds and calm, shallow waters make coastal boat travel relatively safe and efficient. While Southeast Asia is confronted with various serious environmental problems, combatting forest loss and coastal environment degradation are paramount.

Southeast Asia has the greatest relative rate of deforestation compared to any other tropical region. Deforestation began in earnest in the 1970s and by 2010 only 49 percent of the forests remained. Many of these forests are located in highlands; most coastal plains have been stripped of trees. Countries with the least amount of forest cover in 2010 were the Philippines (26%), Thailand (37%), and Vietnam (42%), and those with the greatest forest loss between 1990 and 2010 were Cambodia (24.5%), Indonesia (22%), and Myanmar (21%). Half of the region's remaining forests are in Indonesia; although this megasized country still has 52 percent of its territory in forest, it has experienced the greatest amount of deforestation in terms of land area. In 2010, only 10 percent of forested areas in Southeast Asia were classified as "protected," virtually guaranteeing that deforestation will continue. In Myanmar and Cambodia, only 4 and 5 percent of forests, respectively, are classified as "protected."

The many reasons for deforestation differ among countries and have changed through time. An important long-term reason is commercial logging. Southeast Asia has been the dominant global source of timber and pulp for more than 40 years. While most countries have banned the export of raw logs, illegal logging remains rampant, particularly in Cambodia and Myanmar. Enforcement is a problem as local and regional government officers are bribed to look the other way. Realizing that greater incomes can be earned by processing logs into pulp, plywood, and furniture, some governments have promoted domestic wood-based industries. Vietnam is a major exporter of furniture, with much of the wood being smuggled across the border from Laos. Another major cause of deforestation is the growth of large oil palm plantations (Figure 11-5). In 2011, Malaysia and Indonesia produced 86 percent of the world's palm oil, used in a wide variety of food products ranging from chocolate to cookies as well as in soaps and cosmetic products. Palm oil is also in great demand in Europe as a cheap and efficient "green energy" biofuel to reduce fossil fuel emissions. The problem is that burning trees release vast amounts of carbon dioxide into the atmosphere, and the forests destroyed can no longer act as carbon

▼ Figure 11-5 **Oil palm plantation in Sabah, Malaysia.** The clearing of lands in Southeast Asia for the planting of oil palm plantations is a major source of regional deforestation.

▼ Figure 11-6 **Orangutan in Sumatra, Indonesia.** With its forest habitat being rapidly depleted, the orangutan is a threatened primate species.

dioxide sinks (see *Exploring Environmental Impacts: Deforestation and Regional Air Pollution*, page 410). Indonesia has become the third greatest carbon dioxide emitter in the world. Oil palm plantations have spread so rapidly in Borneo that the island's forest cover declined from 73.7 percent to 50.4 percent between 1985 and 2005, and is estimated to drop to less than one-third by 2020.

While some logged forests have regrown into secondary forests or have been planted with fast-growing trees for pulp, the species biodiversity of both flora and fauna is much reduced. This is a critical problem because Southeast Asia supports 42 percent of the world's biodiversity and, compared to other tropical regions, has the highest proportions of threatened vascular plants, reptiles, birds, and mammals. According to the International Union for Conservation of Nature, Malaysia and Indonesia in 2010 ranked third and fourth in the world for the highest number of threatened species. In addition to the Asian elephant and Sumatran tiger, the highest-profile animal victim of deforestation is the orangutan (meaning "forest person"), an orange-haired, tree-dwelling primate found in Sumatra and Borneo (Figure 11-6). The wild orangutan population today is approximately 14,000–25,000, but experts predict that if the present rate of deforestation continues, the orangutan will become extinct by 2025.

With 90 percent of Southeast Asians living within 60 miles (100 kilometers) of the coast, it is understandable why coastal environments are ecologically threatened. Among various degraded coastal environments, the loss of mangrove forests and coral reefs is the most severe. Southeast Asia contains 35 percent of the world's mangroves, the largest stock of mangrove forests of any world region. With the exception of Laos, long stretches of mangrove line every coast that is protected from high-wave action, particularly where rivers reach the sea in the form of deltas and estuaries. Mangroves provide a critical natural nursery environment for both freshwater and deep sea fish species. Humans have exploited mangrove wood for dyes and charcoal for centuries, and the pace of this use has increased in the postcolonial era. Over 80 percent of mangrove ecosystems have been lost. In Indonesia, home to 20 percent of the world's mangrove forests, approximately 50 percent has been

◀ **Figure 11-7 Shrimp farm in southwestern Thailand.** Shrimp farming is environmentally destructive because these ponds replace ecologically productive mangrove forests.

was originally used to stun fish for the global trade in exotic tropical fish, but this method is now used on lobsters and large reef fish for live exports to restaurants throughout East and Southeast Asia. These chemicals are lethal to the coral. Tourists on diving vacations, particularly in Thailand and the Philippines, also do physical damage to coral reefs by breaking off pieces as souvenirs.

lost since 1980. A major factor is the conversion of mangrove to farmed shrimp ponds, where shrimps are intensively raised using commercial feed (Figure 11-7). This industry has grown so quickly that farmed shrimp often exceed wild catches in some countries. Seen as a source of greater income, much of the farmed shrimp is exported, primarily to the European Union, the United States, and China, making Southeast Asia the global epicenter of the industry. During 2006–2011, Southeast Asia—particularly Thailand, Indonesia, and Vietnam—accounted for 56 percent of total U.S. shrimp imports. Mangrove conversion to shrimp ponds also causes saltwater intrusion, coastal land loss, and declining coastal water quality, as well as the loss of traditional coastal livelihoods. This is unfortunate because the sustainable use of mangrove forests yields greater value than the profits earned from exporting shrimp. In response, some governments have become sensitive to mangrove destruction, and replanting efforts in a handful of countries has slowed the increase in mangrove forest loss.

Southeast Asia contains 30 percent of the world's coral reefs, but these have not fared well since the 1970s. All of the Philippines and the eastern half of Indonesia are part of the "Coral Triangle," a marine area that supports 600 species of reef-building corals, 2000 species of reef fish, six of seven sea turtle species, and is vital to the spawning stage for the global tuna industry. In the larger Southeast Asia region, 80 percent of reefs are at high or medium risk as a direct result of human activities. Only 4.5 percent of Philippine reefs are considered in excellent condition. There are many sources of coral reef degradation. Large-scale agriculture, logging, and mining dramatically increase the sedimentation rates of rivers, resulting in poor coastal conditions for reef growth. Urban-sourced pollution runoff and offshore oil spills have also exacted a toll. Extracting reef resources is another serious problem. Artisanal resource extraction has always existed, but increased demand for reef corals and fish put greater pressure on reef sustainability. For example, coral is used in road foundations and as a source of lime for mortar. The export market for home décor and tourists are also important. Reef fish are killed by "blast fishing," a common practice throughout Southeast Asia. Although efficient over the short term, repeated explosive blasts eventually disintegrate coral reefs. Sodium cyanide

Stop & Think

▶ Identify the various human-induced reasons for Southeast Asia's dramatic rates of deforestation.

Historical Background: Empires and Colonial Possessions

Mainland Southeast Asia's great river valley empires were primarily centered on agriculture with rice the primary grain. In contrast, the maritime trade-centered empires of Insular Southeast Asia regionally exchanged forest products. European colonial powers replaced these native political and economic systems and integrated the region into the global flow of natural resources from plantations and mines to fuel industrialization in the West. The legacy of these colonial powers influenced the diverse nature of economic development in modern Southeast Asia.

Pre-European Empires

The emergence of economic, political, and cultural cores in Southeast Asia began with the introduction of Hindu and Buddhist systems of belief and political organization from India. The lone exception was Annam in northern Vietnam, which was influenced by Han China. In Mainland Southeast Asia there were a number of riverine empires (Figure 11-8). The most powerful was centered on the lower valley and delta regions of the Mekong. First occupied by the Indianized kingdom of Funan during the early centuries A.D., the focus of power moved inland to Tonle Sap, the largest natural lake in Southeast Asia, where the powerful commercial Khmer Empire ruled between the ninth and fifteenth centuries. Harnessing the waters of the Tonle Sap and using miles of canals, extensive areas of irrigated rice paddy could be cultivated even during the dry monsoon season. The monumental ruins at Angkor Thom and **Angkor Wat** testify to the agricultural productivity and trade capabilities of this most powerful of all pre-European Southeast Asian empires (Figure 11-9). At its greatest spatial extent, Angkor ruled over a wide area, including portions of modern-day Thailand, Laos, and Vietnam.

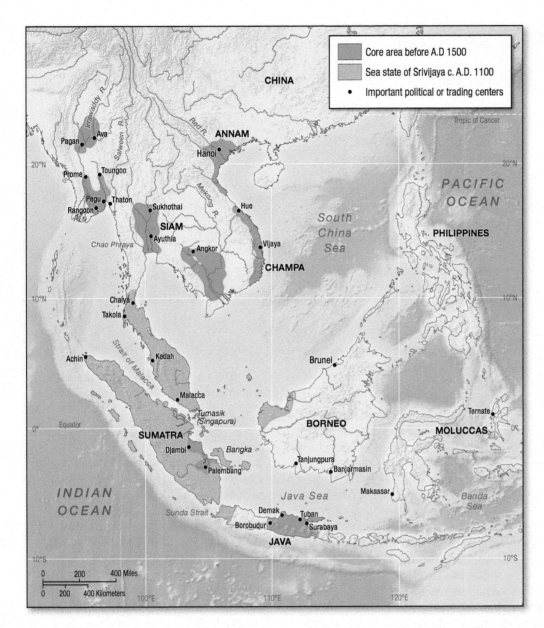

◀ **Figure 11-8 Precolonial states of Southeast Asia.** Southeast Asia supported a number of advanced civilizations prior to European conquest.

latter characterized by village life, oral traditions, folk religions, and shifting agriculture coupled with forest-gathering activities. Even today, there are substantial cultural distinctions between the more populous and powerful lowland peoples and the upland minority groups, such as the Karen of Myanmar and Thailand and the Hmong of Vietnam and Laos.

In Insular Southeast Asia many small kingdoms dotted coastal regions, but only three large-scale pre-European cores are recognized here (see Figure 11-8). The first was the Indianized and Buddhist kingdom of Srivijaya, which flourished from the seventh to the fourteenth centuries, with its capital of Palembang in the freshwater swamp region of southeastern Sumatra in present-day Indonesia. Srivijaya was a thalassocracy—a sea-based state. Like many other smaller maritime kingdoms of the region, its power was derived from the long-distance maritime trade that flowed into and out of its capital of Palembang. Little reliance was placed on agriculture. The second pre-European insular core area was a group of successive Indianized states that first emerged during the early centuries A.D. on the island of Java. The earliest of these Hindu-Buddhist states occupied small fertile valleys of central and eastern Java where, much like

With the exception of Champa, which resembled an Insular Southeast Asia maritime state, all of these early political, economic, and cultural cores were centered on riverine environments, particularly in their lower stretches where alluvial lands could be transformed from forest and freshwater swamp environments to productive rice-growing areas. The power of the state rested on taxing rural production to finance defense, trade, and irrigation. Mainland Southeast Asian cores emerged only after adopting Indian models of state organization based on kinship, the court, secular and religious bureaucracies, and a royal and sacred capital. These lowland-dominant cultures contrasted sharply with upland-subordinate cultures, the

▶ **Figure 11-9 Angkor Wat ruins of Cambodia.** At Angkor Wat, a series of temple cities stands as testimony to advanced urban development in precolonial Southeast Asia. Damaged from years of civil war during the 1970s and 1980s, the United Nations and a handful of other countries provide funds to restore these 900-year-old cultural heritage sites.

EXPLORING ENVIRONMENTAL IMPACTS

Deforestation and Regional Air Pollution

Deforestation not only leads to a dramatic decline in tree cover and species diversity, but because forest loss is often the result of burning, widespread air pollution is the unfortunate consequence. Natural forest fires have always been common in Southeast Asia as a result of prolonged drought caused by a weak wet monsoon season. These fire events were short-lived and did not cause extensive damage. Since 1970 forest fires have increased in frequency, intensity, and geographical extent in part because of the strengthening of ENSO (El Niño-Southern Oscillation),which brings warmer water surface temperatures and higher air surface temperatures and pressure to the western Pacific. These conditions result in prolonged drought even in wet tropical forests. While ENSO strengthening, possibly linked to global climate change, leads to more fires, it is the human-induced changes to tropical forests that have led to the more frequent, intense forest fires. Changes creating fire-prone forests include disturbed logging areas that leave the remaining trees stressed, and commercial oil palm and timber plantations that create standing forest edges susceptible to fire because of biomass accumulation and windier conditions. Two forest types account for 30 percent of forest loss in Insular Southeast Asia between 2000 and 2010: lowland evergreen forest and peat swamp forest. The latter type, found principally in lowland Kalimantan (Borneo) and Sumatra, is highly susceptible to ENSO-related drought because, as groundwater levels drop, the highly organic and combustible peat easily catches fire and continues to smolder until rains return and elevate the water table.

The most destructive fires in recent decades occurred in 1982–1983, 1987, 1997–1998, and 2006, and lasted anywhere from 2 to 6 months (Figure 11-1-1). Each fire created a widespread smoke haze that blanketed large parts of southern Kalimantan, eastern Sumatra, as well as Malaysia and Singapore. The worst fires created a smoke haze that reached the Bay of Bengal and the Philippines. In the most impacted regions, the haze would last for days and sometimes weeks, which not only compromised people's health but also reduced school attendance, closed airports, dramatically decreased tourist flows, and caused many auto accidents. During the worst smoke haze periods, photosynthesis of some hybrid rice plants was obstructed so that harvests in Malaysia declined. Each of these fire events was the direct result of burning forests for commercial oil palm and timber plantations. During the 1980s and 1990s, the Indonesian government's response to these fires was halfhearted in part because it lacked the technical expertise to extinguish fires. The government simply blamed the fires on ENSO or the agricultural activities of shifting cultivators. In 2002, most Southeast Asian countries signed the ASEAN Agreement on Transboundary Haze Pollution to regionally address cross-border pollution resulting from forest fires. While the Agreement does provide for cooperation in fire monitoring,

▲ FIGURE 11-1-1 **Smoke and haze from fires in southeastern Borneo, Indonesia.** These fires (marked by red dots) spread a thick haze over Singapore and Malaysia to the west and are only extinguished with the arrival of wet monsoons.

prevention, mitigation, and technical assistance, it does not explicitly require any form of multinational enforcement because of the importance of national sovereignty. Indonesia only ratified the Agreement in 2012. While there have not been any severe ENSO periods since 2006, moderate-scale forest fires continue to bring week-long smoke haze to Singapore and Malaysia.

Sources: Jukka Miettinen, Chenghua Shi, and Soo Chin Liew, "Deforestation Rates in Insular Southeast Asia Between 2000 and 2010," *Global Change Biology* 17, no. 7 (2011): 2261–2270; David Seth Jones, "ASEAN and Transboundary Haze Pollution in Southeast Asia," *Asia Europe Journal* l4, no. 3 (2006): 431–446; S. Robert Aiken, "Runaway Fires, Smoke-Haze Pollution, and Unnatural Disasters in Indonesia," *The Geographical Review* 94, no. 1 (2004): 55–79.

the mainland riverine states, they gained political control by taxing rural economic activity. The most famous of these empires was the Majapahit whose rule stretched from 1300 to 1500. It was both a maritime and agrarian empire that at its zenith encompassed multiple vassal states throughout much of Insular Southeast Asia.

During the fifteenth and sixteenth centuries, the arrival of Arab, Indian, and Chinese traders dramatically transformed the development of Insular Southeast Asia. Indian merchants introduced Islam, which was adopted by local royalty, who became leaders of "sultanates." Islam subsequently spread rapidly throughout the local population and eventually reached Mindanao in the southern

Philippines. The main exceptions to adoption of Islam were the more isolated interior peoples, who remained animists, and Bali, which remained Hindu. The small coastal and river-mouth trading sultanates procured forest products from their respective riverine hinterlands and shipped them to Malacca on the southwestern coast of the Malay Peninsula (see Figure 11-8). Luxury trade items included nutmeg, cloves, mace, pepper, and gold, in addition to aromatic woods such as camphor, sandalwood, and cassia. Like Srivijaya's Palembang, the entrepôt of Malacca was a cosmopolitan thalassocracy visited by traders from throughout much of Asia. For a short period of time, it was the busiest port in the world.

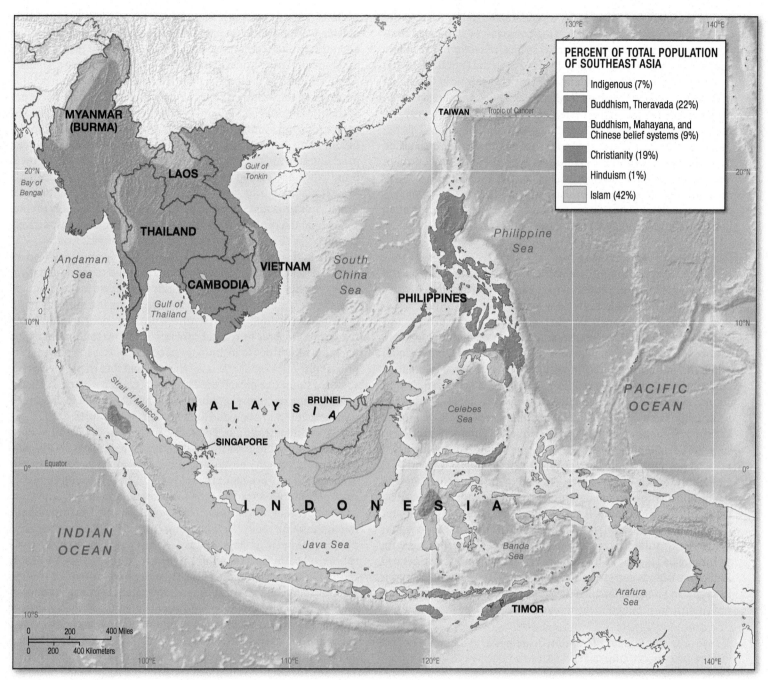

▲ **Figure 11-10 Religions of Southeast Asia.** Mainland Southeast Asia is primarily Buddhist whereas Islam dominates in Insular Southeast Asia. Exceptions are Roman Catholicism in the Philippines and the island of Timor, Hinduism in Bali, and pockets of Christianity and indigenous religions in interior regions.
Source: Adapted from Thomas R. Leinbach and Richard Ulack, *Southeast Asia: Diversity and Development* (Upper Saddle River, NJ: Prentice Hall, 2000).

As a result of these substantial external economic and cultural influences, Southeast Asia became a mosaic of religious beliefs (Figure 11-10). Today, Mainland Southeast Asia is mostly Buddhist with the Theravada branch dominant in Myanmar, Thailand, Cambodia, Laos, and southern Vietnam, and the Chinese-influenced Mahayana branch dominant in northern Vietnam. Islam prevails in most Insular Southeast Asian countries, including the far southern edges of Thailand and the Philippines. There are some notable exceptions to Islam's dominance in this subregion; Spanish and Portuguese colonialism introduced Roman Catholicism to much of the Philippines and Timor-Leste, respectively. Islam never spread to the Indonesian island of Bali where an

earlier form of Hinduism remained intact. There exist pockets of Christianity, mostly in interior regions populated by ethnic minorities, which have become Christian as a result of primarily Protestant missionary activity during the colonial period. Indigenous religions survive in isolated mountainous regions of Kalimantan and especially Mainland Southeast Asia, where these ethnic minority cultures also were influenced by colonial-era Protestant missionaries.

Colonialism and Development

Although colonialism radically changed the traditional economic and social contours of Southeast Asia, its influence varied. The impact of the West can be divided into two time periods. The first period, from about 1500 to 1800, was characterized by **mercantile colonialism**. Private Western trading companies established trading forts and engaged in commerce with local elites for native luxury goods. This form of colonialism involved both the Dutch East India Company centered on Java and the British East India Company with trading forts in Burma, the Malay Peninsula, and Singapore. In the Philippines and Timor-Leste, the Spanish and Portuguese, respectively, engaged in direct colonial rule by the late 1500s. This period was followed by an era of **industrial colonialism**, which lasted from 1800 to 1945 and brought direct Western political control of territory as well as Western private or corporate economic involvement. By the late 1800s, the British and Dutch had brought much of their territories under direct rule, as did the French in French Indochina (modern-day Vietnam, Laos, and Cambodia).

Core-Periphery Exchange

Although short in duration, industrial colonialism had a deep and widespread impact on Southeast Asia. Europe and the United States depended on tropical lands for raw materials to fuel industrialization as well as to establish new markets overseas. The array of raw materials obtained from Southeast Asia was vast and included oil, rubber, tin, hemp, sugar, palm oil, tea, and tobacco. On the global scale, the main impact of industrial colonialism was to create a core-periphery exchange relationship between the Western core countries and their peripheral Southeast Asian colonies. Within the colonial possessions themselves, regional cores where economic activity was focused and regional peripheries emerged; this led to substantial differences in local levels of economic development.

The principal focus of the colonial economy was commercial agriculture, which supplied raw materials to the controlling Western nation and financed colonial administration. One form of commercial agriculture was the large-scale corporate plantation. Rubber and oil palm were the primary plantation crops of the west coast of British Peninsular Malaya and the northeastern coast of Sumatra in Indonesia (Figure 11-11). Sugarcane dominated in the central Philippine Islands and Java, and rubber was the chief plantation crop in southern French Indochina. Because plantations were associated with industrial production technologies, they evolved into economic

▼ **Figure 11-11 Rubber plantation in Malaysia.** Rubber has been a leading commercial crop of Southeast Asia since its introduction from South America's Amazon Basin during the colonial era.

enclaves, with stronger and more extensive economic linkages to the outside Western world than to their surrounding hinterlands. A **dual economic system** emerged, consisting of a modern plantation sector operating in the midst of traditional peasant cropping systems.

In addition to promoting plantation agriculture, some colonial powers frequently forced or urged peasant farmers to dedicate part of their traditional paddy land to cultivating crops for export. This **cash cropping system** was especially common in the Philippine Islands and Java during the 1800s. This greatly lowered grain production, which, combined with continued population growth, eventually necessitated importing vast quantities of rice, much of it from Thailand, Burma, and southern Vietnam. Using local labor, the British in Burma and the French in Indochina purposefully drained substantial tracts of the Irrawaddy and Mekong deltas, respectively, for the purpose of creating "rice bowls" for Southeast Asia.

Colonial activity was focused on the core region or selected enclaves of the colonial periphery, which tended to marginalize minority peoples who did not significantly participate in the economic activities of the core. One reason for neglecting the minority populations was the high cost of connecting their mountainous home regions to the river or coastal core. A second reason for not exposing these minorities to technological advances was the paternalistic philosophy of colonial officials, who decided that these "backward" societies should be preserved. This policy created a historical legacy of marginalizing cultural minorities that has continued to the present day, particularly in Mainland Southeast Asian countries.

The concentration of economic activity in the colonial cores also contributed to ever-increasing differences between the more Westernized urban centers and the indigenous rural hinterlands. Modernization proceeded at a rapid pace in the larger cities, while the rural peripheries attracted little investment in industry, education, or health care. This domestic core-periphery relationship remains much the same today. In the former British possession of Malaya, the small island of Singapore functioned as the primate city for a vast rubber, oil palm, and tin mining periphery on the Malay Peninsula. In the Philippines, Manila was both a regional primate city and a gateway to the rest of world, where raw materials from the southern archipelago were processed and exported to the United States. In Burma, the British established Rangoon in the Irrawaddy delta as their colonial capital, reviving the far south precolonial core that had been replaced by the upstream core centered on Mandalay. In Vietnam, the French reinforced the two core areas of the north and south that dominated the country prior to colonial conquest.

Chinese Immigrant Middlemen

Intimately linked to the formation of colonial cores was the presence of non-Western foreigners—in particular, Indians and **Overseas Chinese**. While Chinese settlers became a significant presence in many Southeast Asian port cities as early as the fourteenth century, their populations dramatically increased when trading, agricultural, and mining activities expanded during the early European colonial era. The civil wars in China were an added inducement for many Chinese to migrate and seek opportunities in *Nanyang,* or "southern seas," a term the Chinese use to describe coastal Southeast Asia. During the late colonial period the flow of Chinese reached its highest levels. By the end of World War II, approximately 10 million Chinese resided in Southeast Asia, with the greatest concentrations in Thailand and Malaya. Most arrived as desperately poor plantation contract laborers. Over time many opened small retail shops and trading concerns, while others engaged in interisland shipping or worked as clerks in Chinese- and European-owned companies. Overseas Chinese became indispensable to the success of Western-based industrial colonialism and hence referred to by the term "middlemen" because they often controlled the retail trade of imported Western consumer products. South Asian Indians in the British colonial possessions of sub-Saharan Africa provided a similar middlemen function. Eventually, some became captains of industrial, banking, insurance, shipping, and corporate agroprocessing empires that competed with European firms. Excluding Dutch Indonesia and Burma, Overseas Chinese accounted for some 38 percent of Southeast Asian capital investment during the late 1930s (see *Geography in Action:* Overseas Chinese in Southeast Asia).

Stop & Think

▶ Identify the various geographical expressions of economic and social change introduced by colonialism.

Decolonization

The colonial empires constructed by the West rapidly crumbled when Japanese forces conquered most of Southeast Asia during World War II. The Japanese coveted the same natural resources that had fueled the factories of Europe and the United States. Despite Japanese propaganda that promised a Southeast Asian economic development model based on Asian values and needs, Japan proved to be an even more paternalistic (if not dictatorial) colonial master than the Europeans and Americans. When the war ended, each Southeast Asian territory was reoccupied by its former European colonial power, hoping to resume the former colonial relationship. However, the Western superiority myth had been shattered by the Japanese and, inspired by the global call for decolonization, many nationalistic movements calling for independence soon surfaced. The struggle for independence was relatively peaceful except in Indonesia and Vietnam, where military conflict ensued. The Philippines was the first to gain independence in 1946, followed by the Mainland Southeast Asian possessions between 1947 and 1954. Indonesia achieved independence in 1949, but it was not until 1963 that Malaya became a sovereign state. Singapore split from the Malayan Federation, which also included Sarawak and Sabah on the island of Borneo, in 1965. Because of its fear of annexation by much larger Malaysia, the tiny oil-rich British protectorate of Brunei did not become independent until 1983.

Modern Economic Growth and Stagnation

A new economic orientation that better suited the needs of these newly independent countries was in order. Two of the most urgent needs were to diversify economic production and to reduce dependence on exports of raw materials to the West. Delivering modernization to rural inhabitants of the economic and cultural peripheries was another daunting task. Yet another challenge was nation-building, the creation of a single national identity out of a heritage of multicultural societies.

When referring to economic growth, geographer Jonathan Rigg identifies Southeast Asian countries as either modernizers or

reformers. Those classified as modernizers together make up the original members of the **Association of Southeast Asian Nations (ASEAN)**, which was established in 1967 to promote regional political stability. These states were the first to adopt economic policies that promoted manufacturing as an engine of economic growth. Because much of this growth in export-led manufacturing was tied to FDI (Foreign Direct Investment), these countries were the first in the region to globalize their economies and were among the first in the world to participate in what today we know as globalization. These modernizer countries include Singapore, Malaysia, Thailand, Indonesia, and the Philippines.

For these modernizer economies, the principal sources of FDI in the 1960s and 1970s were corporations from Western Europe and the United States. Since the 1980s, firms from Japan, South Korea, Taiwan, and Hong Kong account for most of the FDI. These East Asian firms were attracted by the region's diverse natural resource base, relatively abundant and inexpensive labor, and government tax breaks. Another extremely important, but often overlooked, development factor in most countries has been the high level of political stability. Only with these factors in place have the modernizer countries been able to attract foreign-owned manufacturing and become integrated into the global production system.

Countries classified as reformers are those governments during the 1970s and 1980s who were tied to economic policies anchored in strong socialist principles and were averse to FDI: Myanmar, Laos, Cambodia, and Vietnam. War and recovery also had a negative impact on economic growth. Only in the 1990s did some governments begin to open their national economies to globalization, and as an indicator of their openness, joined ASEAN. While Brunei is a relatively rich country, its wealth is derived almost exclusively from oil and natural gas rather than manufacturing exports. The newly independent but desperately poor country of Timor-Leste does not fit into this modernizer and reformer economic growth model either. With the exception of Thailand, all modernizer countries are located in Insular Southeast Asia, while all countries classified as reformer countries are in Mainland Southeast Asia. Only recently have reformer countries attracted FDI in manufacturing. This has been primarily from East Asian firms and Singaporean capital. Economic openness at the regional scale was enhanced by the establishment of the **ASEAN Free Trade Area (AFTA)** in 1992 to reduce tariffs on trade between member countries. AFTA was established in part to compete against the ascendancy of China's highly competitive export economy in a more globalized economic world.

Country differences in per capita GNI PPP (Gross National Income in Purchasing Power Parity) greatly depend on levels of FDI from 2000 to 2010. This in turn impacts Human Development Index rankings; this is a measure based on life expectancy, educational attainment, and standard of living (Table 11-1). Singapore is the richest country in the region and ranks first in these development indicators. The island state garnered 30 percent of the total ASEAN member FDI, but this figure is inflated because a substantial share of regional FDI is funneled through banks in this global financial center. Brunei is an anomaly because its economy is almost totally dependent on oil exports; although its per capita GNI PPP as well as its Human Development Index score are high, much of its FDI has been spent in the form of oil exploration and processing facilities. Thailand and Malaysia are the second and third largest FDI recipients, and as upper middle-income countries, their Human Development Index scores are relatively high. While classified as modernizer

Overseas Chinese in Southeast Asia

Overseas Chinese—ethnic Chinese permanently residing outside the People's Republic of China and Taiwan—today number 50 million, with three-quarters of this total living in Southeast Asia. Like the early migrations of Chinese to the United States, most of the **Chinese diaspora** originated from the two southern Chinese provinces of Guangdong and Fujian. Many "Chinatowns" around the world might better be called "Guangzhoutowns," named after the largest city in the Pearl River Delta. Overseas Chinese continue to play a prominent economic role in many Southeast Asian countries (Figure 11-2-1). Their economic impact is strongest in Singapore, Malaysia, and Thailand, where ethnic Chinese constitute 74.1, 24.3, and 12.6 percent of the respective populations. Even in countries with smaller Chinese populations, their numbers belie their economic strength. In Indonesia, ethnic Chinese in 2010 comprised only 3.7 percent of the population but held significant economic power at the village, regional, and national scales.

The economic success of the ethnic Chinese can be traced to internal and external factors. The widely held view that Chinese are "good at business" relative to indigenous Southeast Asians is a common explanation. Confucian traditions that promote the family as the basic social unit have led the Chinese to use their extensive family connections to engage in a wide range of business activities. The previous experience of some ethnic Chinese in the highly commercial environments of South China is a second internal factor. External factors have been equally important. Colonial governments never accorded the Chinese the same social status as Europeans and never allowed them to fill government positions. The Chinese focused their energies on business, where fewer barriers were laid in their path. Because both colonial and modern governments discriminated against the immigrants when it came to schooling, Chinese families often sent their children to superior universities abroad. The dominance of Chinese doctors, accountants, engineers, and lawyers in many Southeast Asian countries is, in part, a by-product of local educational discrimination.

The nature and extent of the discrimination against ethnic Chinese minorities in Southeast Asia vary from country to country. In the Philippines, discrimination has been insignificant because most Chinese have long intermarried with the local people, converted to Catholicism, and adopted indigenous names. In Thailand, a shared Buddhist religion, similar physical appearance, and adoption of Thai names have reduced the potential for systematic discrimination. Ethnic Chinese have been subject to discrimination in some Southeast Asian countries, however. In the decade following the 1975 unification of Vietnam, most of the slightly more than 1 million ethnic Chinese who were highly concentrated in the south were forced to leave the country because of their capitalist tendencies. In Malaysia and Indonesia, the vast majority of lower- and middle-class ethnic Chinese were until recently legally discriminated against in a number of ways, although the financially and politically connected Chinese elite suffer little from official prejudice. The civil strife experienced in Indonesia in response to the 1997 Asian financial crisis was especially hard on the middle class. Thousands of Chinese businesses were burned or looted as scapegoats for the economic meltdown. Despite latent anti-Chinese feelings, many Southeast Asian countries continue to depend on this minority group as a source of capital and entrepreneurial talent, particularly as their national economies become further intertwined with global markets.

Sources: Hong Liu, ed., *The Chinese Overseas* (New York: Routledge, 2006); Laurence Ma and Carolyn Cartier, eds., *The Chinese Diaspora: Space, Place, Mobility, and Identity* (Lanham: Rowman Littlefield, 2003).

◀ FIGURE 11-2-1 **Chinese shophouses in Penang, Malaysia.** Over the past centuries, millions of Chinese have migrated to Southeast Asia. In many countries, they possess substantial economic power.

economies, Indonesia and the Philippines are still developing countries in the broadest sense of the term, as indicated by all three statistical indicators. As transitional socialist countries, Vietnam, Laos, and Cambodia remain poor, although Vietnam has made great strides in attracting FDI. Its FDI inflows are more than four times those of the remaining reformer countries combined. Although still low, Vietnam's Human Development Index score is higher than that of Laos, Cambodia, and Myanmar because of the greater wealth generated by FDI. Timor-Leste has the highest per capita GNI PPP, but only because of large quantities of foreign assistance for this poor country after gaining independence from Indonesia. Myanmar is characterized by the lowest socioeconomic indicators for this group of reformer countries.

Stop & Think

▶ Why are levels of per capita GNI PPP, FDI, and Human Development Index ranking so different between modernizer and reformer countries?

Table 11-1   Socioeconomic Indicators of Southeast Asia

Country	Per capita GNI PPP ($) 2011	Average Annual FDI Inflows $ (billions) 2000–2011	Human Development Index 2011
Modernizers			
Singapore	59,790	107.728	.86
Brunei	50,180	3.117	.83
Malaysia	15,190	58.348	.76
Thailand	8,390	76.648	.68
Indonesia	4,530	52.831	.67
Philippines	4,160	18.567	.64
Reformers			
Timor-Leste	3,600	na	.49
Vietnam	3,260	47.392	.59
Laos	2,600	1.602	.52
Cambodia	2,260	4.932	.52
Myanmar	1,950	5.590	.48

Sources: *World Population Data Sheet 2012* (Washington, D.C.: Population Reference Bureau, 2012); The World Bank, http://www.data.worldbank.org.

Urban and Rural Transformations

Compared to other major world regions, Southeast Asia has traditionally had relatively low levels of urbanization. In 2011, the region's total population was only 43 percent urbanized; only South Asia and sub-Saharan Africa have lower levels of urbanization. Urbanization levels are much higher in the richer countries of Insular Southeast Asia, with Indonesia, the Philippines, and Malaysia averaging 56 percent and the Mainland Southeast Asia countries of Thailand, Vietnam, Laos, and Myanmar averaging a much lower 28 percent. However, Southeast Asia includes the three megacity (greater than 10 million) regions of Jakarta, Manila, and Bangkok. These are **Extended Metropolitan Regions (EMRs)** that include a core (major city), inner (periurban), and outer zones (*desakota* and densely populated rural) in the larger urban area (Figure 11-12). Evocative of the EMR landscape, the Indonesian term *desakota* is a compound word combining *desa* (village) and *kota* (city). Inclusion of this outer zone as part of the EMR, which often extends 50 miles (80 kilometers) from the urban core, is justified because while land use is primarily agricultural, factories dot the landscape, a significant number of inhabitants work in nonagricultural occupations, and good transport networks allow for high mobility. The 2010 population of the Jakarta EMR was 28.3 million, with 18.2 million in the core and inner zones. The populations of the Manila and Bangkok EMRs were 22 and 14 million, respectively. Partially explaining the growth of EMRs is the overconcentration of FDI in assembly manufacturing as these urban regions become "production platforms" that are integrated into the global economy.

Whether to classify the EMR outer zone as urban raises the larger question of whether it is rural or urban spaces that are being transformed. The rural spaces of Southeast Asia beyond the EMRs are witnessing a dramatic transformation as both globalization and urbanization proceed at a rapid pace. Rural districts are experiencing a "deagrarianization" because young rural people prefer urban occupations as opposed to the drudgery of farmwork. Nonfarm labor accounts for 30–50 percent of rural household income. In Indonesia and Thailand many farmers have diversified their income-generating activities by taking seasonal work in cities or through nonagricultural pursuits such as piecemeal handicrafts. In older and poorer

▼ **Figure 11-12   A hypothetical Extended Metropolitan Region (EMR) in Southeast Asia.** Driven by foreign direct investment (FDI), the geographical mix of urban and rural landscapes of the EMR is a product of economic globalization.
Source: Adapted from Norton Ginsburg, Bruce Koppel, and T. G. McGee, *The Extended Metropolis: Settlement Transition in Asia* (Honolulu: University of Hawaii Press, 1991).

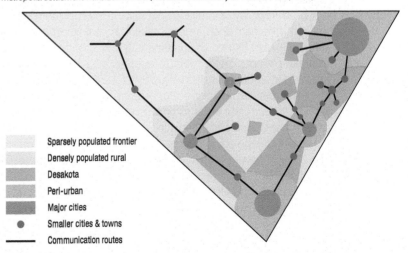

- Sparsely populated frontier
- Densely populated rural
- Desakota
- Peri-urban
- Major cities
- • Smaller cities & towns
- —— Communication routes

households, younger adult members often work in urban factories, on construction sites, or in public transport, while older adult members work the land in a survival strategy that incorporates both the paddy field and an urban place. In richer village families, the income earned by family members in the city is remitted and used to generate even more wealth in the countryside through the purchase of more modern agricultural technologies. While national censuses count individuals based on their permanent place of residence, rural household income-generating activities cross rural-urban boundaries.

Many seasonal migrants to the city are underpaid, exploited, and live in cramped conditions, and rural spaces are increasingly characterized by a "graying" population coupled with a landscape of abandoned agricultural fields. But these conditions are expressions of the larger modernization process associated with globalization. In this process, urban development depends on change and innovation in rural space. Innovations in agriculture depend on remittances from urban wage-earning relatives. It is difficult to imagine rural development or urban development as mutually exclusive processes. Much like villages within the outer zone of EMRs, more distant villages across Southeast Asia are spatially isolated in the physical sense, but less so in terms of their economic and social isolation.

Energy Resources

Compared to other world regions, Southeast Asia's energy resources are relatively poor in quantity as well as unevenly distributed. The region only has 1 percent of the world's oil reserves, 4 percent of natural gas reserves, and less than 1 percent of coal reserves. Indonesia is the resource giant of Southeast Asia; its resource share includes

FOCUS ON ENERGY

Southeast Asia's Energy Resources: Cooperation and Challenges

Faced with the challenge of developing more secure as well as diverse energy sources, Southeast Asian countries have the potential for regional cooperation through ASEAN. One goal proposed in the 1980s has already been met. Taking advantage of the region's abundant liquefied natural gas (LNG) reserves, and striving to reduce the high cost of developing LNG terminals, eight cross-border gas pipelines stretching 3,000 miles (4,800 kilometers) now deliver natural gas between Myanmar, Thailand, Malaysia, Singapore, and Indonesia. Plans call for continued expansion of cross-border gas pipeline infrastructure. Although still in the planning stages, regional energy cooperation could be enhanced by transborder partnerships in an ASEAN power grid. This structure would increase flexible distribution of electricity where needed. Four power grid interconnections already exist between Malaysia and Singapore, Thailand and Malaysia, Thailand and Cambodia, and Vietnam and Cambodia. ASEAN energy ministers signed an agreement in 2009 that provides greater energy security by urging member countries to supply other members experiencing shortages of crude petroleum and other oil products.

Another example of energy cooperation and challenge could materialize in the form of hydropower. In 2010, hydropower accounted for only 1.7 percent of primary energy consumption, but could increase dramatically with dam construction along the upper reaches of the Irrawaddy River in Myanmar and the Mekong River in Laos. Both countries stand to earn large revenues by selling power to energy-hungry China and Thailand. The environmental impacts, however, are huge. The Mekong supports the world's most productive freshwater fishery. Dam construction would radically alter river ecology and threaten the 1,200 species of fish in the river as well as the food security of millions of people living along the Mekong's banks. While regional and local anti-dam activists have been influential, the quest for more energy and the government revenues generated from such projects might be too strong to halt dam construction.

At a broader scale, Japanese, South Korean, and Chinese firms, often in partnership with each other or with Southeast Asian governments, have invested in LNG gas terminals to promote their own energy security needs. Especially important to East Asian states is the secure maritime transport of energy from the Persian Gulf through Southeast Asian maritime space. It is through Southeast Asian waters that 80 percent of the oil imported by East Asia flows, especially through the Strait of Malacca between Sumatra and peninsular Malaysia. Another 60 percent of natural gas and coal bound for Japan and Taiwan traverse this most strategic international waterway.

Regional and extraregional cooperation to manage energy resources is needed when it comes to possible conflicts over potentially large deposits of oil, gas, and mineral resources in the South China Sea. Marine territorial claims by multiple ASEAN members as well as China and Taiwan, which include the small, but habitable Paracel and Spratly Islands, as well as reefs and atolls such as the Scarborough Shoal, have been contested in recent years. China claims much of the South China Sea as its own "security" space, but these claims are questionable because they infringe on the maritime rights of Southeast Asian countries within their 200 nautical mile, exclusive economic zones as defined by the United Nations Convention on the Law of the Sea (UNCLOS) (Figure 11-3-1). While China would rather negotiate claims with individual states, most Southeast Asian governments prefer multilateral agreements, because these treaties increase their collective power in the face of a militarily much more powerful China. Military confrontations, although minor in nature, have been most common between China and Vietnam, but included China and the Philippines in early 2012. China's interest is not only to access energy resources but also to control sea lanes critical to China's resource-hungry economy.

Sources: "Roiling the Waters," *The Economist*, July 7, 2012, http://www.economist.com/node/21558262; Françoise Nicholas, "ASEAN Energy Cooperation: An Increasingly Daunting Challenge," 2009, http://www.ifri.org/?page=contribution-detail&id=5481&id_provenance=87&provenance_context_id=64.

42 percent of the region's natural gas, 24 percent of the oil, and 72 percent of the coal reserves. Malaysia is a distant second, with 32 percent of the region's natural gas and 35 percent of oil. In terms of primary energy consumption, the regional energy mix in 2010 was 40.8 percent from oil, 17.6 percent from natural gas, 14.1 percent from coal, 4.4 percent from geothermal, and 1.7 percent from hydropower. Because the region's relatively poor energy resource endowment is coupled with high GDP growth rate, primary energy consumption will more than double between 2010 and 2030, and even Indonesia and Malaysia will become net energy importers by 2030. Much of this energy consumption increase will be in the form of coal; by 2030 coal will contribute 23.7 percent of primary energy consumption when compared to oil and natural gas which will remain close to their 2010 percentage shares. Because coal is the most carbon-intensive fossil fuel, the resulting increase in carbon dioxide emissions from coal will worsen air quality, particularly in urban areas; indeed, carbon dioxide will triple by 2030.

Other nonfossil fuel energy sources hold promise, but remain in an early development stage. The Philippines is the world's second largest producer of geothermal power, but problems with connecting geothermal sources to the grid remain. Southeast Asia is a late entrant in nuclear power generation with five countries considering building facilities. But following the 2011 Fukushima accident in Japan, only Vietnam is seriously moving forward with plans to construct two reactors. Because Southeast Asia's energy demands will almost double between 2010 and 2020, governments must face some serious challenges in the short term (see *Focus on Energy: Southeast Asia's Energy Resources: Cooperation and Challenges*).

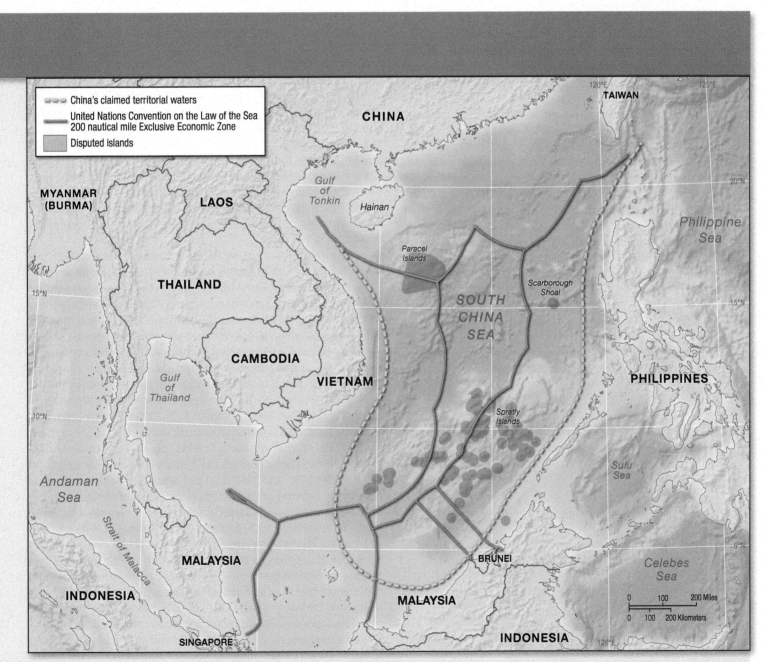

▲ FIGURE 11-3-1 **Territorial claims in the South China Sea.** With large mineral resource potential, China, Vietnam, and the Philippines are the primary contestants in a political dispute that the Association of Southeast Asian Nations (ASEAN) has not been able to solve.

Tourism

As a world region deeply engaged with economic globalization, Southeast Asia is also a major tourist destination. Thailand, Singapore, and Malaysia have emphasized tourism as a dedicated development strategy because tourism activities provide employment opportunities and promote upgrading transport infrastructure to cater to international tourists. In 2011, the travel and tourism industry contributed 10.9 percent or $237.4 billion to regional GDP and directly or indirectly accounted for 8.7 percent of total employment. The importance of the travel and tourism industry to the regional economy is substantial. Total tourism receipts in 2011 were $881.9 billion, equal to approximately 89 percent of the total value of FDI in that same year. While international tourists to Southeast Asia were from more developed regions and countries during the 1980s and 1990s, source regions today are dominated by the growing middle and upper classes in Southeast and East Asia (Figure 11-13). During 2007–2011, tourists from ASEAN and East Asian countries accounted for 63 percent of international arrivals. Shorter travel time partly explains the rise of international tourist arrivals from 29.9 million travelers in 1995 to 81.9 million travelers in 2011, a 270 percent increase.

Countries attracting the greatest regional share of international tourists from 2000 to 2011 were Malaysia (34.0%), Thailand (24.1%), and Singapore (13.7%). Malaysia's high share is in part explained by daily or weekend trips by Singaporeans for shopping and entertainment activities. Singapore's high regional share is also explained in part by international travelers staying a few days in this hypercosmopolitan city state before or after visiting another Southeast Asian country.

It is the diverse mix of touristic activities that makes Southeast Asia an attractive traveler destination. Sun and surf tourism is especially important in Thailand (Phuket), the Philippines (Cebu),

and along the coast of Vietnam. Cultural heritage tourism, with destinations including both precolonial and colonial landscapes, is attractive in a number of countries. The precolonial temple complexes of Angkor Wat in Cambodia and Borobudur in Central Java attract more discriminating and educationally minded tourists. The many Hindu temples of Bali and the plethora of Buddhist temples in the greater Bangkok urban region are frequent stops on the mass tourism circuit. Colonial landscapes in the form of government buildings, churches, grand hotels, and ethnic Chinese shophouse districts are also popular, especially in Hanoi, Ho Chi Minh City, Singapore, and Penang and Malacca in Malaysia. Singapore and Malaysia have aggressively promoted cultural heritage districts for tourist consumption. While not yielding large revenues, adventure or ecotourism activities, particularly among younger tourists from Europe and Australia, are especially common in the forested mountains of Mainland Southeast Asia. Here village-based ethnic minorities have become part of the touristic experience as well.

Population Contours

With some 608 million people (8.6% of the world's population), Southeast Asia, unlike its neighbors of South and East Asia, is not a demographic giant. But Indonesia, with 241 million people, is the world's fourth most populous country and other countries in the region are highly populated by world standards. These include the Philippines (96.2 million), Vietnam (89.0 million), Thailand (69.9 million), and Myanmar (54.6 million), which are ranked 12th, 14th, 20th, and 24th in the world, respectively. Southeast Asia is also characterized by some of the least populated countries in the world; Brunei and Timor-Leste have 400,000 and 1.1 million people, respectively.

Population distribution in most countries is spatially uneven (Figure 11-14). In Mainland Southeast Asia, people are concentrated primarily along major river valleys where growth has followed the historic settlement patterns. River deltas such as those of the Irrawady in Myanmar, Chao Phraya in Thailand, and the Mekong and Red in Vietnam are densely populated because of high agricultural productivity as well as the growth of urbanization and industry associated with globalization. Only Vietnam is characterized by dense populations along much of its coastline. In Insular Southeast Asia, people are primarily concentrated along narrow coastal plains. This density pattern is due to economic inertia associated with colonial-period plantations or with more modern regions of resource extraction. The exception to this coastal orientation of population density is the island of Java and the neighboring islands of Bali and Madura. The population concentrations in Java occur because of the historical presence of ancient empires and dense rural populations cultivating rich volcanic soils. Java's population density of 2,070 persons per square mile is double that of Japan's. Although Java comprises only 18.7 percent of Indonesia's land area, it supports 58 percent of the national population. There is a stark contrast between the population densities of Java and Kalimantan (Indonesian Borneo) and the two Indonesian provinces occupying the western half of the island of New Guinea. Population characteristics and development levels are connected. Examining dynamic variables such as rate of natural increase and total fertility rate help clarify this linkage (see *Visualizing Development: Demographic Diversity and Development*).

▼ **Figure 11-13 Source regions and countries of international tourists in Southeast Asia, 2007–2011.** As evidence of economic integration, the majority of tourists originate from Association of Southeast Asian Nations (ASEAN) and East Asian countries.

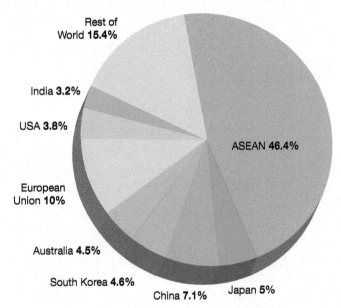

Rest of World 15.4%

India 3.2%

USA 3.8%

ASEAN 46.4%

European Union 10%

Australia 4.5%

South Korea 4.6%

China 7.1%

Japan 5%

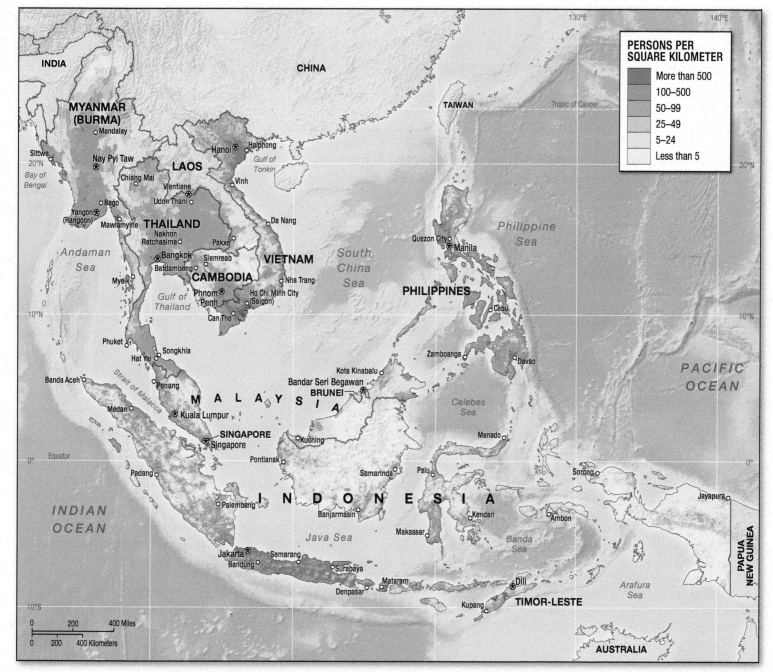

▲ Figure 11-14 **Population distribution of Southeast Asia.** Present-day population distributions reflect both premodern environmental opportunities and modern-day economic opportunities linked to economic growth and globalization.

Insular Southeast Asia

Insular Southeast Asian economies were some of the first in the developing world to engage the economic forces of globalization. Rich Singapore has a thoroughly globalized economy, exports high-value electronics goods, and is a regional center of banking, transport, and service industries. Still a major global producer of plantation products, Malaysia has become a middle-income industrialized country with an economy based on a wide variety of electronics

exports. The economies of Indonesia and the Philippines remain more tied to resource exports and are poorer developing countries confronting problems of economically integrating their respective far-flung archipelagos.

Singapore

Singapore is distinctive for many reasons. With 5.3 million inhabitants, it is often referred to as a **city-state** because it measures only 272 square miles (704 square kilometers), making it the smallest

Visualizing DEVELOPMENT ▲

Demographic Diversity and Development

Much like economic indicators such as FDI, service sector employment, or level of urbanization, demographic indicators tell us about levels of development and in many cases are products of larger government development policies. The most basic population indicators are rates of natural increase and total fertility (Figure 11-4-1). Singapore provides an interesting example. Now one of the richest countries in the world with a healthy service- and information-based economy, its wealth is illustrated by low rates of natural increase and total fertility. Singapore's population is declining and the

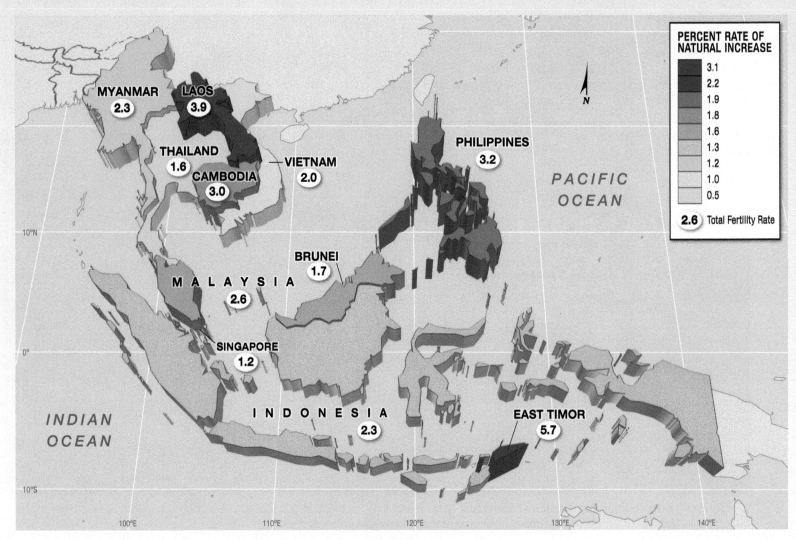

▲ **FIGURE 11-4-1 Rates of natural increase and total fertility in Southeast Asian countries, 2011.** The different contexts of economic growth and development produce highly varying demographic responses.

Source: Population Reference Bureau, 2011 World Population Data Sheet, *http://www.prb.org/Publications/Datasheets/2011/world-population-data-sheet.aspx.*

and most urbanized of all Southeast Asian countries (Figure 11-15). Singapore is also the only developed nation in Southeast Asia where ethnic Chinese constitute the majority. Because of economic and ethnic differences with Malay-majority Malaysia, this ethnic Chinese enclave became a separate independent country in 1965.

Much like Hong Kong, Singapore functioned as an entrepôt for regional maritime trade, becoming the "crown jewel" of Britain's Southeast Asian empire. Blessed with a strategic location at the eastern entrance to the Strait of Malacca and with superior port facilities, Singapore became the leading colonial-period transshipment and processing center for the Malay Peninsula and for a substantial portion of Dutch Indonesia. Today Singapore continues to function as a leading regional maritime trade center, competing with Hong Kong as the busiest container ports in the world. The island generates enough manufacturing exports to fill half of those containers. The remaining containers are transshipped to

government must internationally recruit both low-and high-skilled labor to fill this shortfall so the economy can continue to grow. In 2010, 35 percent of Singapore's population was foreign born, up from 16 percent in 1990. The majority of this nonresident workforce (those that hold passes to reside in Singapore for a limited time period) are in low-skill, low-wage jobs, but one-quarter occupy high-skill and high-wage jobs in Singapore's high-tech economy. Malaysia however is an anomaly to the generalization that richer countries exhibit relatively lower rates of natural increase and total fertility rates. Compared to Thailand and Indonesia, countries with substantially lower per capita GNI PPP, Malaysia's rates of natural increase and total fertility are that of a much poorer country. This is partly due to the Malay-dominated government in the 1980s putting in place pronatalist policies for native Malays. The goal was to increase the Malay population relative to ethnic Chinese and Indian groups. Malay families are not forced to assume increased financial liabilities associated with having additional children because of preferential treatment in university education and government employment.

Indonesia and the Philippines are two developing countries with quite different demographic experiences. Indonesia is the poster country for government-led, voluntary population reduction programs, but the Philippines still is characterized by persistently high total fertility rates. The Philippines is a Roman Catholic country, but high population growth rates are better explained by endemic poverty in that poor people often have many children, regardless of religion. Vietnam has become richer in the past 2 decades, and this country's low

rate of natural increase and total fertility is a reflection of increased prosperity, but also a coercive two-child policy enacted in the 1990s but abandoned in 2003, and the wide availability of abortions. In Laos and Cambodia, persistently high rates of natural increase and total fertility are due to poverty and the lack of significant family planning programs in countries with a dispersed rural population (Figure 11-4-2). Timor-Leste is characterized by the highest total fertility rate in Southeast Asia and one of the highest in the world. Endemic poverty, a patriarchal society, the cultural desire to replace those who died in the

war of independence from Indonesia, and surprisingly few differences between rich and poor as well as urban and rural fertility rates are all factors. While level of prosperity is a primary determinant of population growth, country-specific contexts make for diverse demographic profiles in Southeast Asia.

Sources: Gavin Jones, "Women, Marriage and Family in Southeast Asia," in *Gender Trends in Southeast Asia*, Theresa Devasahayam, ed. (Singapore: Institute of Southeast Asian Studies, 2009); Gavin Jones and Richard Leete, "Asia's Family Planning Programs as Low Fertility is Attained," *Studies in Family Planning* 33, no. 1 (2002): 114–126.

▲ FIGURE 11-4-2 **Children in rural Cambodia.** In the absence of government family-planning programs, and coupled with endemic poverty, fertility rates are high.

Singapore from neighboring countries using small feeder vessels; the containers are then loaded onto enormous oceanic container vessels (Figure 11-16). In addition to its role as a regional and global shipping hub, Singapore's government promotes it as a "global maritime center," offering financially rewarding **producer services** such as banking, insurance, communications, and consulting for maritime transport-related activities. Singapore's global maritime center function is further complemented by the city-state's role as

the world's third-largest oil refiner, where imported oil undergoes additional value-added processing before reexport to neighboring countries.

Its maritime transport-related businesses have helped Singapore achieve a GNI PPP per capita of $59,790, the third highest in the world and the highest in Asia. Singapore's wealth is also grounded in a diversified economic base. Ranked near the top of the Index of Economic Freedom and the world's fourth-largest

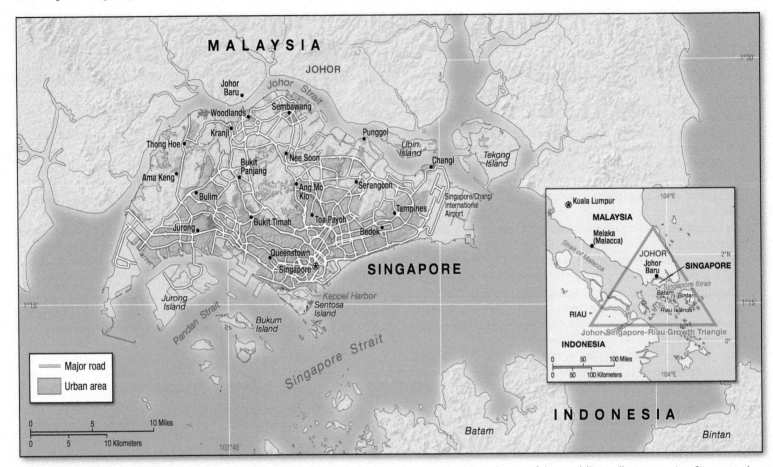

▲ Figure 11-15 **Singapore Island and adjacent Malaysian and Indonesian territory.** Although one of the world's smallest countries, Singapore has one of the highest levels of economic prosperity.

foreign exchange market, Singapore is the regional headquarters for multinational corporations that generate substantial numbers of high-paying service jobs. The government has attracted many foreign high-technology firms to its state-planned industrial parks.

During the 1970s and 1980s, Singapore attracted firms that assembled lower-technology consumer products, such as stereo

▼ Figure 11-16 **Singapore harbor.** Singapore's many container terminals are constructed on reclaimed land and are considered the best operated in the world. Collecting container cargo from as far away as the Persian Gulf, value-added transport services generate substantial income.

equipment, televisions, and video recorders. Because land and labor costs were much higher in Singapore than in neighboring countries, the city-state promoted higher-technology and capital-intensive industries domestically, and enticed lower-technology industries to relocate either long distance—to Thailand, Vietnam, or southern China—or to neighboring Malaysia and Indonesia. This example of a "borderless world" associated with globalization led to the 1989 establishment of a **growth triangle** centered on Singapore, the nearby southern Malaysian state of Johor, and the Indonesian islands of Batam and Bintan, situated just south of Singapore (see Figure 11-15). This economic relationship enables both foreign and domestic investors to base the labor-intensive portions of production in Johor and Batam while Singapore retains higher-value services such as management, financing, and transport logistics. By the mid-2000s, Bintan and Batam supported 500 manufacturing facilities, many of them lower-value textiles and footwear, as well as provided recreation space at beach resorts and golf courses for Singaporeans and international tourists.

Foremost among sectors replacing low-tech assembly was the global electronics industry, which in 2011 accounted for 6.3 percent of GDP. Singapore supplied 40 percent of the world's hard disk media as well as functioned as a global center for integrated circuit design, wafer fabrication, and assembly and testing activities. Also important are interactive and digital design industries associated with video games, animation, and online media. Higher-value manufacturing of pharmaceuticals, biotechnology, and medical devices now assumes great growth potential. Singapore's continual pursuit of higher-value economic activities is supported by its world-class education system and highly trained workforce, as well as aggressive government promotion of an information-technology culture. The role of information technology and higher-value economic activities such as producer services is critical to globalizing the city-state's economy and providing a competitive edge in the region.

The pervasive role of government in economic and social planning has been central to Singapore's success. The state controls many profitable domestic industries and actively seeks out foreign investment. Social planning policies have assured social stability in an ethnically diverse population, adequate housing for all, a relatively crime-free environment, high personal savings rates, and an immaculately clean urban environment. Despite being perceived by some as an overly regulated state, Singapore's government continues to attract FDI because it is virtually free of corruption. Although on paper Singapore is organized as a constitutional democracy, these accomplishments are the product of a "soft-authoritarian" government dominated by a single political party that has handily won every election. Because the government views its primary responsibility to be the promotion of economic growth, political freedoms in the Western sense become less important in legitimizing political power.

Stop & Think

→ How does the Singapore-based Growth Triangle resemble the economic relationship between Hong Kong and the Pearl River Delta in China?

Malaysia

Malaysia is a spatially fragmented country of 29 million, consisting of the more densely populated peninsular portion of West Malaysia and sparsely populated East Malaysia on the northern coast of the island of Borneo (Figure 11-17). Since the 1970s Malaysia has evolved from a traditional developing country with an economy dependent on natural resource exports such as tin ore, rubber, and other plantation crops, to an upper middle-income, industrialized economy whose leading exports are now manufactured products. Guiding this economic transformation has been the goal—laid out in the 1970–1990 master plan, the **New Economic Policy (NEP)**—of increasing the economic contributions of the majority ethnic Malays, called *bumiputras*, at the expense of both Chinese and Western economic interests. State-owned industrial enterprises were established to provide nonagricultural employment for Malays and, by the mid-1980s, when manufacturing replaced agriculture and mining as the most important economic sector, the Malay share of the manufacturing workforce had increased to 50 percent.

A second means adopted to raise the incomes of the predominantly Malay rural workforce was to increase government funding of roads and other infrastructural improvements needed to

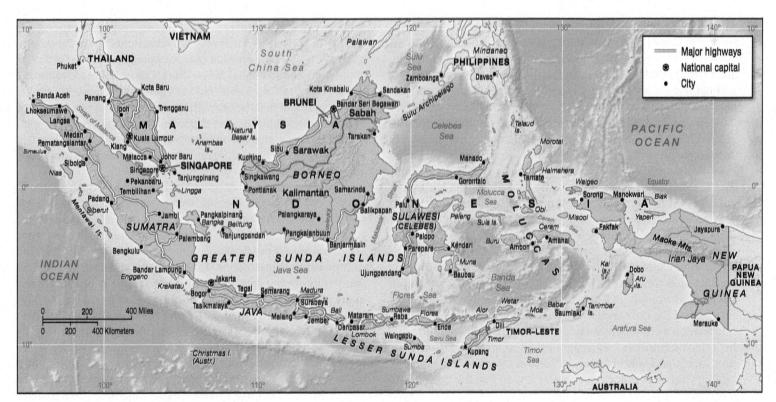

▲ Figure 11-17 **Malaysia and Indonesia.** Malaysia and Indonesia are insular nations that share a number of cultural traditions. Indonesia relies extensively on maritime shipping for economic development..

expand the commercial cultivation of rubber, oil palm, and coffee by smallholder farmers in frontier areas. Anchored by small urban places supporting agroprocessing industries, these schemes have often been touted as the most successful rural land development programs in the non-Western world. In part because of these agricultural schemes, rural poverty rates dropped from 21 percent in 1989 to 8.5 percent in 2009. With the exception of urban Singapore, Malaysia has the lowest rural poverty rate in Southeast Asia. An upper middle-income country, Malaysia's 2011 per capita GNI PPP of $15,190 has also been achieved through FDI in export-based manufacturing. Between 1980 and 2010, manufactured goods as a percentage of export value increased from 19 percent to 76 percent. Originally tied to lower-technology products, Malaysia has increasingly attracted higher-technology, electrical and electronic products, which accounted for 42 percent of total export value in 2010. A substantial portion of higher-value electronics production such as silicon wafer fabrication, semiconductor devices, and information technology products is located on the ethnic Chinese-dominated island of Pinang, which has been dubbed "Silicon Island." Although most of the original workers—employed by such companies as Intel, National Semiconductor, Hewlett-Packard, Sony, and Panasonic—were young rural Malay females, who are referred to as "children of the NEP," increasing numbers of young female Indonesian migrant workers now populate these assembly plants as a part of the larger regional division of labor. Although still part of the west coast economic core, Pinang's electronics industry, coupled with the southern (Singapore-centered) growth triangle in Johor state, contributes to a healthier geographic deconcentration of industrial development away from the already saturated national capital and emerging Kuala Lumpur EMR (Figure 11-18). Despite the growth of the higher-value technology manufacturing sector, recent demand for resources, particularly by the booming economies of East Asia, has increased Malaysia's commodity-based exports from 13 percent to 22 percent of total export value during the 2000–2010 period. Facilities to add value to plantation products, oil, and natural gas are strongly supported by both domestic and foreign investments.

In an attempt to become less economically reliant on revenues earned from foreign-owned assembly plants whose long-term competitiveness is challenged by lower labor costs in neighboring countries, and motivated by a desire to propel the country into the "postindustrial" age, in the mid-1990s the Malaysian government launched a development program cleverly called "Vision 20/20," which set the stage for the country to develop a globally connected and regionally dominant information technology and knowledge-based economy. The first spark for such an economic transformation was the establishment of the **Multimedia Super Corridor (MSC)** in 1996, stretching for 31 miles (50 kilometers) south from Kuala Lumpur. Anchored by "Cyberjaya," Intelligent City, the corridor is envisioned as a "technopole"—an information technology research cluster whose inhabitants reside in "telecommunities." Although linked to increased globalization, MSC's goal was to foster indigenous creativity and expertise among "technopreneurs," and it was hoped that the presence of large multinational firms will help develop a Malaysian-based information-technology culture.

▲ **Figure 11-18 Petronas Towers in Kuala Lumpur.** As a landscape symbol of national success, the Petronas Towers stand in the center of an increasingly globalized city.

While the gestation period for such a large-scale development project is long, some initial observations about Cyberjaya can be made. Generation of information and communication technology (ICT) knowledge has been adequate, but many of the knowledge producers are government-linked institutions and corporations and not the private sector as originally envisioned. Even the government admits that the creation of an acceptable threshold of foreign companies and talent has fallen short. The envisioned residential community has not materialized as 60 percent of the working population commute to Cyberjaya. The political symbol of a globalized ethnic-Malay society has yet to produce significant national-scale multiplier effects and thus may only exacerbate the spatial inequalities between the greater Kuala Lumpur EMR and the more rural and poorer regions of the country.

Despite economic successes, Malaysia still has much to accomplish. Although originally established to increase the Malay share of wealth, state-owned enterprises are a huge financial burden. The government has been wary of rapidly reducing the state-owned sector because it provides a source of employment for the politically powerful ethnic Malays. To address these perceived problems, many state-owned enterprises have been partially privatized, and many policies favoring ethnic Malays were addressed in the 1991 New Development Policy. To further these goals, the 2010 New Economic Model promotes a decreased role of government and an increased role of the private sector in economic growth based on higher-value manufacturing and services. Part of this 2010 model is the 1Malaysia program that promotes greater ethnic harmony. This is intended to forge a more meaningful and unified national identity whereby Malaysians, regardless of their ethnicity, should think of themselves as members of a single family (Figure 11-19). In addition, the long-term goal of spatial equity in development has not been met. Of Malaysia's 14 political units, only a handful have incomes above the national average, and most of these are located in the peninsular west coast core region. As part of the economic periphery still tied to natural resource exploitation, three of the four peninsular east coast states, as well as Sarawak and Sabah in East Malaysia, remain relatively poor. In 2009, Sabah's poverty rate stood at 19.2 percent compared to the national average of 3.8 percent. The colonial inheritance of uneven spatial development persists to the present day.

Indonesia

Measuring some 741,100 square miles (1,920,000 square kilometers), Indonesia is by far the largest country in Southeast Asia (see Figure 11-17). With 241 million inhabitants, Indonesia is also Southeast Asia's most populous country and the fourth most populous in the world. Although five islands support the vast majority of Indonesia's population, the problem of physical fragmentation is pervasive and presents a variety of development challenges. The overarching goal of the newly independent government in 1949 was to forge national unity from the diverse array of ethnic groups that inhabited the far-flung island realm. The achievement of unity was particularly difficult because of the political, economic, and cultural domination of Java, the most populous island. In pursuit of the goal of national unity, the regional maritime trade language of Malay was successfully promoted as the national language.

Although Indonesia is also the most populous Islamic country in the world, religious freedom has generally been respected. While the Indonesian population is 88 percent Sunni Muslim, Hindus dominate on Bali, substantial pockets of Protestants exist

▼ Figure 11-19 **A 1Malaysia billboard in Penang.** Proclaiming that "together we achieve it," the 1Malaysia program promotes a single national identity in a multiethnic society that has traditionally favored the majority Malay population.

on Sumatra and Sulawesi, Roman Catholics predominate in West Timor, animist groups are found in Kalimantan and Papua, and Chinese Buddhists congregate in urban areas. All are officially recognized by the government. Some extremist Islamic groups have emerged in Indonesia; since September 11, 2001, Islamic extremists bombed two tourist nightclubs in Bali in 2002, the Australian embassy in Jakarta in 2004, and two Western hotels in Jakarta in 2009. This extremist movement has produced little of the grassroots political radicalism found in some countries of the Middle East, for several reasons. Islam was primarily introduced from India and was modified by a preexisting Hindu and Buddhist religious culture. Women hold a prominent role in village and household economic exchange, and Indonesia does not exhibit the strong patriarchal or male dominant culture common in the Middle East. Colonial legal systems were adopted after independence, reducing the role of Islamic (sharia) law. Governments are characterized by democratic institutions, allowing for moderate dissent, making a politically extreme form of Islam less likely to emerge in a widespread fashion. Indonesia is ethnically diverse and the government does not want to alienate the non-Muslim minority, especially the economically important Chinese, by allowing greater Islamization of national culture.

Much of Indonesia's early development planning focused on two population-related issues: growth and distribution. With a population approaching 100 million by the late 1960s, the government implemented a noncoercive, volunteer family-planning program promoting the two-child family. The program has been recognized around the world as quite successful, considering the relative poverty of the country. As a result, average annual population growth rates declined from 2.9 percent in 1970 to 1.3 percent in 2011. Reflecting other government initiatives, the proportion of the population living below the poverty line gradually decreased to 13 percent in 2010 with the majority of this total being rural inhabitants. Because the crowded island of Java supports 58 percent of the country's people, the government has sponsored transmigration programs designed to resettle landless Javanese peasants on the outer islands, where they can engage in subsistence and cash-crop agriculture. Resettling approximately 4 million people, particularly on Sumatra, the transmigration programs have not substantially altered population distribution patterns and have not slowed rural-to-urban migration—four of Indonesia's five most populous cities are located on Java. With an estimated population of more than 28 million, Jakarta has become a prototypical EMR and primate city, accounting for 12 percent of the national population and 25 percent of national GDP in the mid-2000s (Figure 11-20).

During the 1960s and 1970s, Indonesia's economy was inward-looking and dependent on a narrow manufacturing base supported by exports of petroleum and a variety of plantation crops. Industrial growth relied on state-owned corporations in response to the perceived domination of ethnic Chinese capital in the national economy. Depressed global petroleum prices during the early 1980s forced the Indonesian government to liberalize the economy, particularly with reference to FDI in export manufacturing. In response, Indonesia was able to attract substantial FDI in lower-value textile, garment, footwear, and electronics parts

▲ Figure 11-20 **The skyline of Jakarta, Indonesia's primate city.** High-rise buildings line transport corridors that radiate into the much larger extended metropolitan area.

manufacturing from South Korean and Taiwanese firms. Much of this investment was spatially concentrated in Java, a process that only magnified Jakarta's primacy. Although the government promoted small and medium-sized industry among the native population or *pribumi*, large-scale private domestic industry favored just a select few ethnic Chinese and pribumi corporations. This favoritism led to the institution of crony capitalism, in which the rich and politically well-connected derive huge financial windfalls. This incestuous economic relationship between high-ranking government officials and businessmen, plus the financial crisis of 1997, led to the downfall of President Suharto. Since the early 2000s, political and economic reforms have begun to transform Indonesia at both the national and regional scales. More inclusive and transparent multiparty national elections, once dominated by the single-party military government, coupled with the greater decentralization of economic decision making to the outer island provincial governments has ushered in a wave of optimism on the part of both Indonesians and foreign investors. For much of the country's modern history, the central government in Java made the decisions concerning outer-island economic development, and this included retaining much of the royalties earned from foreign investment in mining and plantations. This unequal economic relationship, combined with outer island resource extraction by foreign multinationals, is a classic example of the core benefiting at the expense of a less-developed periphery. Coupled with the pervasive influence of Javanese culture, people on the outer islands often felt colonized by the national government on Java, which they perceived as a form of **internal colonialism** (see *Geography in Action:* Indonesia's Changing Core-Periphery Relationships). This political transformation has coincided with the dramatic increase in demand for mineral resources by booming East Asian economies over the past decade, resulting in healthy GDP growth rates and a per capita GNI PPP of $4,530. Indonesia's outer islands, especially Kalimantan and Sumatra, attracted 47 percent of agricultural FDI and 44 percent of mining FDI among all ASEAN countries during the 2005–2010 period. The country has become the world's largest coal exporter and is a major exporter of gold, copper, and other strategic minerals. Taking full

advantage of the East Asian demand for natural resources to promote a more sustained economic growth trajectory requires adding value to these resources domestically before exporting semifinished or finished products. The government has gradually started to impose export taxes on raw materials. Importing countries view these taxes as a form of "resource nationalism," but promoting further value-added processing at home is critical to Indonesia's economic growth.

Although dwarfed by the higher-value production FDI going to Singapore, Malaysia, and Thailand, foreign investment in manufacturing comprised 38 percent of total FDI during 2005–2010. Multinational corporations are attracted to Indonesia because of inexpensive labor for export product assembly as well as a very large and youthful domestic consumer market. The Jakarta EMR receives most of the FDI in manufacturing because of better telecommunications and transport infrastructure relative to other locations in the country. The primacy of the Jakarta EMR is reflected in the spatial inequalities of per capita GDP. The Jakarta EMR's per capita GDP is almost four times that of Indonesia as a whole, five times that of Java, and six times that of Eastern Indonesia, the poorest region of the country.

Indonesia's physical size and fragmentation, its cultural-ethnic diversity, and its historic core-periphery political and economic cleavages present obvious development problems. But greater provincial participation in the development process, a reduced role for the military in politics, and a more transparent government without the endemic corruption that has accompanied political power in Indonesia has provided the average citizen a better foundation for economic and social justice. Sustained economic growth can only be accomplished by developing more infrastructure and improving the workforce's technological skills to attract higher-value FDI in manufacturing. Only then might the country's vast natural resources be harnessed to capture greater value added in manufacturing and increase Indonesia's regional and global economic prominence.

The Philippines

The Philippines, with a per capita GNI PPP of $4,160, includes 7,000 islands and encompasses an area of 115,831 square miles (300,000 square kilometers). The economy and culture of the Philippines is distinctive among Southeast Asia's countries because of its early colonization by Spain (Figure 11-21). After an initial settlement in Cebu, one of the central Philippine Visayan Islands, the Spanish plundered Manila and transferred their base of operations to this small Malay-Muslim settlement on the island of Luzon in 1574. Much as they did in Latin America, the Spanish granted large tracts of land to colonial officers and the Catholic Church. Most Filipinos eventually converted to Catholicism, with the exception of Muslims living on or near the southernmost large island of Mindanao, where the colonial government never gained strong administrative control. With a powerful and exploitive regional

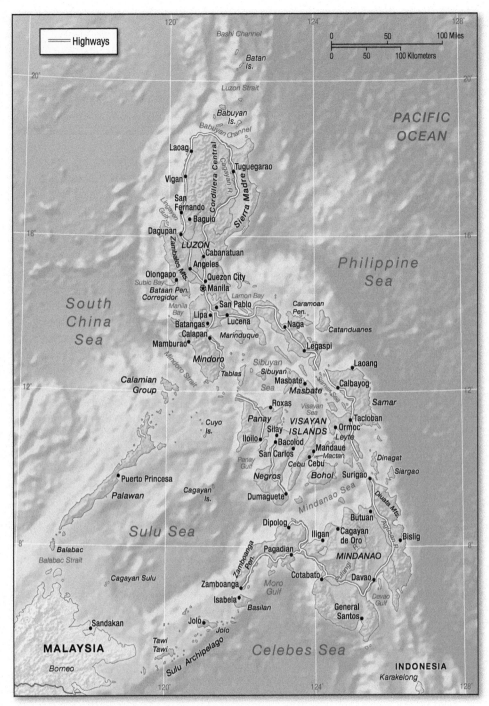

▲ **Figure 11-21 The Philippines.** The country of the Philippines includes more than 7,000 islands, many of them uninhabited. The Philippine Islands have experienced the greatest degree of Western cultural influences compared to other Southeast Asian countries.

corruption—remained. The overwhelming presence of American culture (English, in addition to Tagalog, is the lingua franca) contributed to a failure to develop a distinct national identity, an essential ingredient for long-term economic development. Owing to both Spanish and American influences, the archipelago has long been regarded as an Asian outpost of Western thought and practice, and shares many social and economic commonalities with Latin America.

Following independence in 1946 Philippine economic growth remained tied to agriculture, including subsistence cropping, cash cropping, and industrial agroprocessing. A relatively high 37 percent of the population continues to live in rural areas. Land reform to benefit the vast numbers of poor peasants has been a part of the agenda of every recent government, and was quite successful during the 1990s, although much remains to be accomplished because most political decision makers are large landholders themselves. A manufacturing base did evolve during the early postindependence period based on agricultural processing and textiles, and these two sectors provided most exports. The country's economic growth was also negatively affected during the 1970s and 1980s, when martial law was imposed to strengthen President Ferdinand Marcos's grip on political power and to contain the country's growing Communist insurgency. The latter was reinforced by widespread social unrest arising from deepening economic disparities. The result was the flight of domestic capital overseas, a decline in foreign investment because of political instability, and a large-scale outflow of workers, both educated and undereducated, seeking better economic opportunities overseas (see *Geography in Action:* Transnational Migrants and the Regional Division of Labor, page 531). The national economy began to rebound with the fall of Marcos and the election of Corazón Aquino as president in 1986. The return of funds that had fled in the previous decade and migrant worker remittances injected substantial amounts of investment capital. Foreign investment also returned, particularly at newly established export processing zones where electronic products and garments are important exports. The Philippines has become an important global recipient of "business process outsourcing." The country receives FDI in the form of payroll, billing, and call centers, in part because of the English-language fluency of its educated middle class. The former U.S. naval facility on Subic Bay near Olongapo in central Luzon has attracted numerous offshore manufacturing facilities as

landlord base and Chinese merchants who often intermarried with the Spanish and converted to Catholicism, a highly rigid socioeconomic class structure developed.

The Philippines was ceded to the United States in 1898 as part of the spoils of the Spanish–American War, and the United States proceeded to modernize the almost feudal Philippine economy. Although the transportation network was upgraded, agroprocessing industries were established in the provinces, popular education was instituted, and health and social services were improved, the fundamental sources of income disparities—social class inequity and

GEOGRAPHY IN ACTION

Indonesia's Changing Core-Periphery Relationships

Resentment of the Java-based government by non-Javanese peoples of Indonesia's periphery is related to the desire for greater economic and cultural autonomy. Occupying the eastern and western ends of the archipelago, respectively, the provinces of Papua and Aceh illustrate outer island resentment. Papua, one of two provinces located in the western half of the island of New Guinea, is sparsely populated by ethnic Melanesians. When Indonesia gained independence in 1949, the Dutch retained both provinces until 1962 when the region was administered by the United Nations. In 1969, Papuans participated in a questionable election managed by the Indonesian military and voted to join Indonesia. Immediately after the election, the Free Papua Movement (OPM) emerged calling for independence; low-level warfare between the OPM and Indonesia's military has resulted in human rights abuses against civilians that continue today. One high-profile conflict is between native Papuans and Freeport-McMoRan, a U.S.-based multinational firm operating the Grasburg mine, one of the world's largest gold and copper mining operations (Figure 11-5-1). Papuans have criticized, sometimes violently, the extensive environmental damage caused by mine tailings disposal, extremely low wages of mine workers, and the absence of economic benefits to the province. The Indonesian army has been used to silence these environmental and equity concerns, and the army or its surrogates in the form of "special police" have killed or imprisoned OPM members.

At the other end of the archipelago in the northern Sumatran province of Aceh, much greater levels of violence have occurred since independence. As a powerful maritime sultanate in the 1500s, the Dutch only conquered Aceh in the late 1800s. Practicing a very conservative form of Islam and resisting integration into a larger Indonesian nation in 1949, a civil war ensued until Jakarta granted Aceh "special autonomy" status in 1959 that allowed greater local control over religious law and education. Despite greater autonomy, the Free Aceh Movement, established in 1976, waged a war of independence. A source

of conflict during this early period was the massive Exxon-Mobil natural gas complex, the world's most lucrative natural gas production facility. Sporadic attacks against the facility took place because of the perception that Aceh gained little economic benefit. These assaults led to stationing Indonesian army troops at the facility, and the conflict between the Free Aceh Movement and the Indonesian military dramatically widened, resulting in the deaths of more than 10,000 civilians. With a seemingly never-ending conflict and the massive dislocation caused by the 2004 tsunami in which Aceh was the destruction epicenter, a peace agreement was signed in 2005 providing Aceh local self-government status as well as the right to control 70 percent of oil and natural gas reserves.

Timor-Leste (formerly East Timor) is another example of core-periphery tensions, but these tensions are not directly related to conflicts over natural resources. In 1999 the tiny province of East Timor, which had been a Portuguese colony until 1975 when it was ruthlessly annexed by the Indonesian army, gained independence. The desire for independence is understandable considering that 200,000 East Timorese (who are Catholic and culturally more similar to Melanesian Papuans) lost their lives during Indonesia's brutal occupation of the island. Economic reasons were equally important, because non-Timorese constituted only 5 percent of the population but controlled

75 percent of the economy. Timor-Leste's economic recovery will be difficult. The Indonesian military destroyed 70 percent of the island's economic infrastructure during the 1999 war and less than 5 percent of the country's natural habitat remains. About half of Timor-Leste's 1.2 million people live in poverty and its only important export is coffee. A 2005 agreement, however, between the Australian and Timor-Leste governments to develop rich oil and gas fields in the Timor Sea has produced significant revenues. Although a government-controlled Petroleum Fund to promote wise use of oil and natural gas revenues was established, the absence of processing facilities in Timor-Leste means that all value-added processing takes place in Australia, creating a new form of core-periphery relationship. As a result, the oil and natural gas industry produces little in the way of employment opportunities for a country with depressingly high unemployment rates. The potential for an oil dependent Timor-Leste is high as non-oil exports between 2005 and 2010 accounted for 10 percent or less of government revenues. This absence of a more diversified national economy only exacerbates high unemployment and the potential for serious social unrest.

Sources: Andrea K. Molnar, *Timor Leste: Politics, History and Culture* (New York: Routledge, 2010); Jacques Bertrand, *Nationalism and Ethnic Conflict in Indonesia* (New York: Cambridge University Press, 2004).

▼ FIGURE 11-5-1 **Papua province gold mine.** This mine, situated high in the Maoke Mountains of western New Guinea, western New Guinea, has been the object of prolonged and serious conflict between the Indonesian government and a coalition of international environmental and indigenous tribal interests.

well, as have zones on the southern and northern edges of Manila and Mactan Island near Cebu. Among the modernizer countries of Southeast Asia (except oil-rich Brunei), the Philippines still attracted the lowest amount of FDI during the 2000–2010 period.

Much recent economic growth has been centered on the Manila EMR and central Luzon. Industrial investment in the outer-province cities of Cebu in the central Visayas, and Cagayan de Oro and Davao in northern and southern Mindanao modestly counter Manila's economic dominance. While the Philippines has distinct regional economies the Manila EMR in 2009 accounted for 32 percent of national GDP, and 8 of the 10 richest cities in the country were located in the Manila EMR. Average per capita income of the Manila EMR is 2.4 times greater than the national average and 5.2 times greater than the poorest region. At a larger scale, only three of the 14 administrative regions have per capita incomes above the national average, and these three are located on Luzon. Rural migrants flood into metropolitan Manila seeking a better life. With limited economic opportunities, many live in squatter settlements or other forms of unimproved housing with little access to clean water and utilities. A third of the population in metro Manila occupies substandard squatter settlements, and the EMR has one of the largest squatter problems of any national capital in Southeast Asia (Figure 11-22).

The Philippines must tackle a number of interconnected problems that are very difficult to remedy. High unemployment and underemployment rates are due in part to high population growth rates. Not enough jobs are being created for those entering the workforce, which explains why 29 percent of Filipinos, about 28 million people, live below the poverty line. Closing the income gap between the rich few and the sea of poor is critical to the country's future economic prospects. At the same time, this most democratic of Southeast Asian countries experiences endemic political corruption and instability, which drives potential foreign investments elsewhere. After Spanish and American domination, followed by decades of rule by domestic elites, the people of the Philippines deserve better from their political leaders.

Mainland Southeast Asia

With the exception of Thailand, Mainland Southeast Asian countries have only recently liberalized their economies to engage the forces of globalization. Initially dependent on the export of agricultural and food products, Thailand has become a newly industrialized country with foreign investment in a wide variety of industries, particularly automobile assembly and parts for the larger Southeast Asian market. After a decades-long military conflict, socialist Vietnam has become a major target of foreign investment after introducing market-based economic policies in the 1980s that have resulted in one of the highest economic growth rates in the world. Cambodia's economic rebound after a short-lived, Communist-inspired, and horrifically murderous regime after the collateral damage from the Vietnam War has been very slow, as it remains heavily reliant on foreign aid. Resource-rich Myanmar remains one of the last isolationist military governments in the world, but has recently begun to open both its economic and political systems to outside contact and influence.

Thailand

Thailand occupies the geographic heart of Mainland Southeast Asia and is distinctive among its neighbors because it was never a Western colonial possession and is the only "modernizer" country (Figure 11-23). Changing its name from Siam to Thailand in 1939, the country has been ruled by successive mixed civilian and military governments under a system of constitutional monarchy in which the king has little political power but is much revered and carries substantial moral influence. The country has experienced many military coups, but most have been relatively peaceful. Until the late 1970s, Thailand's economy was primarily based on a healthy agricultural export sector that included large quantities of rice, tropical fruits, seafood, and canned foods, exports that remain important today. Throughout much of the 1960s and 1970s, Thailand also gained economic benefit through servicing the various Western military needs associated with the Vietnam War, as well as from foreign aid provided with the expectation that the country would serve as a bulwark against the perceived spread of Communism.

Beginning in the early 1980s, the government established economic policies promoting export-based manufacturing activity with both foreign and domestic investors. Since then manufacturing has increased rapidly. Aside from the usual factors such as economic incentives for foreign firms, streamlined government bureaucracy, political stability, and inexpensive labor, Thailand offered a friendly business culture to East Asian investors who shared the Buddhist religion. The country was also friendly to ethnic Chinese, a factor important to Hong Kong and Taiwanese investors. Equally important, particularly for the future, is Thailand's central location within mainland Southeast Asia that allows investors access to the resource-rich markets of neighboring countries.

Although primary commodity exports remain important, manufacturing has become the engine of economic growth. As a result, Thailand's GNI PPP per capita in 2011 was $8,390, making it an upper middle-income country. Thailand's 2011 exports accounted for 61 percent of GDP and manufacturing exports accounted for over 76 percent of this total. Unlike Malaysia, Singapore, and the Philippines, where FDI in manufacturing exports is anchored in electronics, Thailand has become the center of Southeast Asia's

▼ **Figure 11-22 Squatter settlement in Manila.** Squalor and sophistication reside in close proximity, yet remain worlds apart, as Manila's downtown towers over the congestion of nearby impoverished neighborhoods.

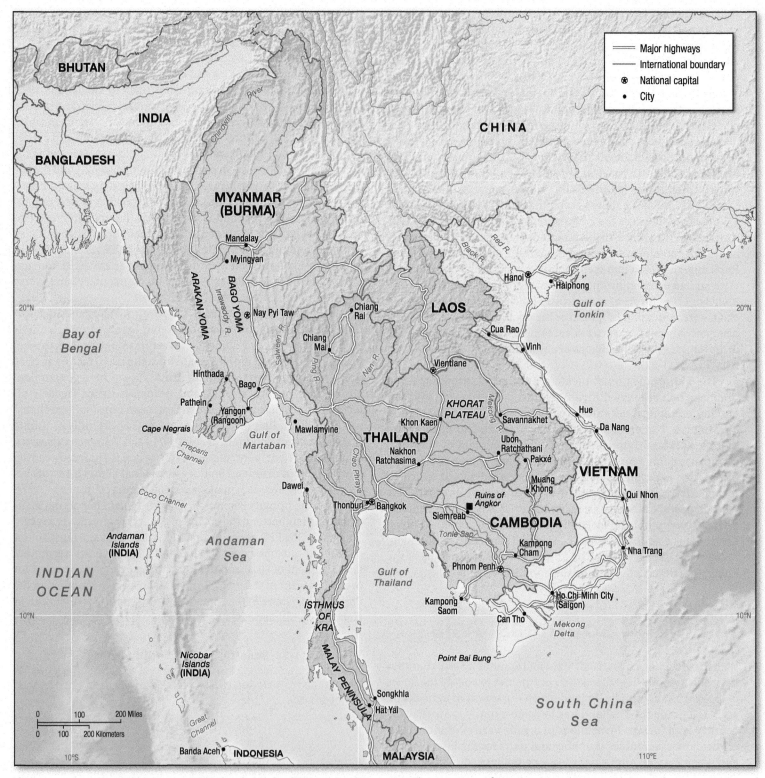

▲ **Figure 11-23 Myanmar (Burma), Thailand, Cambodia, Laos, and Vietnam.** The countries of Mainland Southeast Asia physically consist largely of north–south trending mountain ranges and broad intervening lowland river valleys, which support the majority of the population.

automobile industry; it is the fourteenth largest global producer and the industry accounted for 12 percent of GDP in 2011. While most ASEAN countries produce auto parts or assemble vehicles, Thailand is home to more than 15 foreign corporations assembling passenger cars as well as 1,800 parts suppliers. Thailand is also the world's largest exporter of pickup trucks. Japanese manufacturers arrived in the 1980s, and General Motors, Ford, BMW, and Porsche have established assembly operations. While auto consumption in Thailand has dramatically increased, 43 percent of vehicles produced in 2012 were exports, primarily to markets in Asia and specifically Southeast Asia, where AFTA allows for low tariffs on imports of auto parts or vehicles between member states. Also important

GEOGRAPHY IN ACTION

Transnational Migrants and the Regional Division of Labor

Globalization and more open national borders and economies have promoted far greater transnational labor mobility in what can be described as a regional division of labor. The push to seek employment overseas is driven by both internal and external forces. Source countries are characterized by high unemployment and receiving countries by higher economic growth and labor shortages in jobs that are dangerous, dirty, or demeaning (known as "3D jobs"). These push and pull factors provide the basic conceptual framework to better understand migrant mobility, but in reality the forces at work are far more complex.

With some 14 million people working outside their home Southeast Asian countries, the volume of transnational labor flows is astounding. The greatest labor flow is from the Philippines where 8.2 million migrants, close to 10 percent of the population or 20 percent of the labor force, work overseas. The second greatest volume is from Indonesia with 2.7 million transnational migrants. Both countries contribute a large proportion of female domestic maids to richer destinations such as Singapore, Hong Kong, Taiwan, Japan, and Persian Gulf states where they earn $400–$500 a month (Figure 11-6-1). In 2012, about 300,000 foreign maids worked in Hong Kong and 200,000 in Singapore, mostly from the Philippines and Indonesia. The high-value-added service industry has grown dramatically in Hong Kong and Singapore, and together with fertility rates below replacement level, has created a rising demand for domestic help in two-income professional households. Although not mature postindustrial countries, the middle class in the oil-rich Persian Gulf states can afford to hire foreign maids as well. In the United Arab Emirates alone there are 285,000 foreign maids primarily from the Philippines and Indonesia. Employed not only as domestic servants, but factory workers, health professionals, entertainers (often as illegally trafficked sex workers), and agricultural workers, female migrants make up

at least half of the transnational migrant worker population in what is referred to as the "feminization of migration."

Male transnational migrants also engage in work that is perceived to be gender suited, such as plantation and construction labor. Two million Indonesian migrants work in Malaysia; of these 310,000 (16%) work in oil palm plantations and make between $8 and $12 per day. In the Malaysian state of Sabah where growth of oil palm plantations has been the most rapid, 90 percent of plantation labor is Indonesian. Similarly, 220,000 (11%) Indonesian migrants work in Malaysia's construction industry and make between $8 and $20 per day. This benefits the receiving country: a Hong Kong construction company can pay a Thai worker one-quarter of the salary usually paid to native workers.

Sometimes migrant labor is a factor in global production seeking out cheaper labor. In Thailand, lower-value assembly jobs are increasingly filled by transnational migrants, as Thai migrants do the same jobs but with better pay in richer regional countries, creating a musical-chair form of transnational labor movement. In 2009, approximately 1.3 million "documented" migrants from neighboring Myanmar, Laos, and Cambodia called Thailand their temporary home. Some 82 percent of these were from neighboring Myanmar, a country characterized by serious economic mismanagement and government discrimination against ethnic minorities. As part of a program to decentralize industry away from Bangkok, Thai companies subcontracting for foreign firms have established 200 textile, garment, ceramic, and food-processing factories in Tak Province in northwestern Thailand, which primarily employ Burmese migrants who work below the established minimum wage.

Countries sending citizens abroad benefit greatly. Governments often see labor migration as a vehicle to diffuse social

and political tensions. To better regulate migrant flows, a "migration industry" has emerged, involving recruitment agencies, labor brokers, lawyers, banks, and credit agencies who profit from these labor flows. Because overseas workers send home substantial remittances, governments view "labor exports" as an economic development strategy. Filipino overseas workers remitted 21.4 billion dollars in 2012, equal to 9 percent of national GDP. The Philippines is the world's fourth-largest remittance receiving country after India, China, and Mexico. Remittances contribute substantially to improving villages from which the migrant laborers originate. On the negative side, migrant labor hurts families because the spouse left behind must try to do the work of two and children grow up without both parents. The physical and mental abuse of some migrant laborers, especially young female domestic workers, is also not uncommon in the patriarchal household culture of some receiving countries such as Persian Gulf states. The emergence of a migrant-labor culture as a byproduct of globalization is an issue that most sending and receiving governments would prefer to ignore, because it puts them in the uneasy position of treating human beings as commodities.

Sources: P. Martin, *Migration in the Asia Pacific Region: Trends, Factors, Impacts.* Human Development Research Paper, no. 32 (New York: UNDP, 2010); K. Hewison and K. Young, eds., *Transnational Migration and Work in Asia* (New York: Routledge, 2006).

▼ FIGURE 11-6-1 **Indonesian domestic workers in Hong Kong.** Thousands socially gather during their Sunday off in one of Hong Kong's open public spaces. Filipina domestic workers congregate in other parts of the city.

to Thailand's economy is tourism. With one of the best-developed tourism industries in the world, visitors are attracted to the sun and surf locations in the country's south, cultural heritage sites based on Buddhist temples in the larger Bangkok region, and ecoadventure destinations in the mountainous north.

While the Thai government has attempted to spatially disperse industry to many provinces, the economic growth associated with export-oriented manufacturing has been almost totally restricted to the capital city region and a handful of surrounding provinces. Especially important is the eastern seaboard region where most automobile assembly and chemical plants as well as Thailand's international port are located. It is this industrial growth in adjacent provinces that explains why, as in other Southeast Asian EMRs, population growth has been greatest outside the built-up urban core of the Bangkok Metropolitan Region (BMR) (Figure 11-24). The wealth disparities between Thailand's economic core and the rest of the country are sharp. The 2011 per capita GDP of the BMR was twice that of the country as a whole and seven times that of the northeast region, the poorest part of the country. The BMR bias in government spending is equally great. Although the BMR accounts for 17 percent of Thailand's population and 26 percent of GNP, the region received 72 percent of government expenditures in 2011. The northeast region received only 6 percent of government expenditures, but accounts for 34 percent of the population. The disparities in development between Bangkok and its adjacent industrializing provinces and the rest of the country raise the question whether Thailand is a newly industrializing country or whether this economic core is simply a newly industrializing urban region.

As a result of economic overconcentration, Bangkok is beset by a number of environmental problems, one of the most important being saltwater intrusion from the Gulf of Siam into the city's groundwater table, a process facilitated by the overextraction of freshwater. With the water table dropping, flooding problems are increasing as the city experiences subsidence averaging 2–4 inches (50–100 millimeters) a year. Subsidence problems came to a head in 2011. To prevent rainy season flooding, flood walls have been gradually built around Bangkok's core. With historic rains in upstream northern Thailand and an extended wet monsoon season lasting until October, the flood walls surrounding Bangkok played a part in the backup of southerly, downslope water flow and subsequent flooding of the city's northern suburb squatter settlements, and much of central Thailand.

Stop & Think

▶ Explain how geographical inequalities in economic growth in Malaysia and Thailand differ.

Vietnam

Unlike in other Mainland Southeast Asian countries, Chinese cultural, social, and political influences have been significant in Vietnam. This link was made more than 2,000 years ago when the Viet people who once occupied the southernmost Chinese province of Guangdong were pushed southward into the Red River lowlands of northern Vietnam (see Figure 11-8). After slowly spreading southward as an independent political entity between the 1300s

▲ Figure 11-24 **Bangkok, Thailand.** Islands of traditional Buddhist architecture such as this temple coexist with the new modernist-style buildings of central Bangkok.

and 1700s, Vietnam was conquered by France in the mid-to-late nineteenth century. A coalition of nationalist groups emerged in the 1940s and, under the leadership of Communist and nationalist Ho Chi Minh, succeeded in sequentially driving out Japanese and French colonial forces. The 1954 Geneva Agreement, which formally separated Vietnam into northern and southern states, reflected historic regional divisions. Over the next 2 decades, northern forces waged a war of unification against the south and its U.S.-led allies. The establishment of an independent, unified Vietnam occurred in 1975, following the hasty withdrawal of the last U.S. troops after a physically and emotionally devastating war. Known by the North Vietnamese as the "American War," it resulted in 58,000 American war deaths but 2 million Vietnamese casualties. There was no reconciliation with the South, a region that, from the northern Communist perspective, had been fully "corrupted" by long-term exposure to Western capitalist influences. Similar to the experience in early Maoist China, farms, factories, and businesses were confiscated by the state, reeducation was required of those tainted with Western ideas, and ethnic minority highlanders were forced to settle and engage in the more "civilized" Vietnamese form of intensive agriculture. More than 1 million people were forced from southern cities into new economic development zones, where the government believed they would be more productive. The 2 million or so refugees that fled Vietnam over the following 15 years were a testament to the harsh conditions of conformity imposed on Vietnam's people.

By the early 1980s Vietnam's hard-line Marxist policies had fashioned a country that, within just a few years, had become one of the world's poorest. Agricultural stagnation, lack of investment capital, a virtual absence of consumer goods, and declining support from the Soviet Union all contributed to the economic malaise. Recognizing its decline, and aware of Communism's growing economic challenge in the Soviet Union, Vietnam's leaders began to loosen the strings of the centrally planned economy and adopted a program of **doi moi** (economic renovation). This policy shift entailed dismantling agricultural communes, opening the country to foreign investment, reforming the financial system, and reducing subsidies for state-owned

▲ **Figure 11-25 Ho Chi Minh City (Saigon), Vietnam.** A more globalized economy has transformed the skyline of Vietnam's dominant commercial center.

For economic growth to continue, critical issues must be addressed. While the number of inefficient state-owned enterprises has been halved, these still account for 40 percent of national economic output and consume far too much government investment. The country also must attract higher-value FDI, because just like the Pearl River Delta in China, low-wage, low-skilled manufacturing is relocating to less-expensive countries such as Cambodia or Bangladesh. Confusing foreign investment laws, corrupt officials, and continued suspicion of foreign intentions on the part of the government prevent an even more rapid economic transformation. Government surveillance of the Internet, lack of political freedom, and periodic discrimination against religious institutions, both Buddhist and Christian, can only create social unrest among young Vietnamese. The potential for political, economic, and social tension is great because 60 percent of Vietnam's 89 million were born after the war, and this younger generation lacks the emotional ties to the revolution symbolized by their aging government leaders.

companies. The economy as a whole has responded well to the doi moi program. With a 2011 GNI PPP per capita of $3,260, Vietnam is the richest of Southeast Asia's reformer economies and is now classified as a lower middle-income country. Although starting at a low base, since the 1990s Vietnam has experienced one of the world's highest rates of GDP growth as well as the highest export growth rate. Exports as a share of GDP rose from 24.9 percent in 1994 to 77.5 percent in 2011. Industry in 2011 made up 41 percent of GDP while agriculture decreased to 20 percent. The impact of economic growth on reducing poverty was substantial as a 58 percent poverty rate in 1993 declined precipitously to 18 percent by 2011. Poverty reduction was not geographically even, as the southeast experienced greater declines than the central highlands and northern uplands where most ethnic minorities live. In the traditional rural economy where rice anchors production, increased household property rights supported greater levels of production such that Vietnam shifted from being a rice importer in the mid-1980s to the world's second largest exporter of rice in 2010.

The engine of Vietnam's economic turnaround has been FDI, which accounted for 40 percent of industrial production in 2009 and 7.5 percent of GDP in 2010. Between 2005 and 2010, Vietnam attracted 16 percent of the total manufacturing FDI and 76 percent of all FDI in ASEAN. The greatest share of FDI is invested at the opposite ends of the country, the national capital of Hanoi in the north and Ho Chi Minh City (Saigon), the historic commercial capital, in the south (Figure 11-25). East Asian and ASEAN members, particularly Taiwan, Japan, Singapore, Malaysia, and South Korea, dominate FDI. Aside from financing hotel construction, tourism, and oil development, FDI has specifically targeted garments and footwear for export, and more recently electronics. Vietnam's export market is primarily Asia, but with the lifting of the U.S. trade embargo in 1995, the United States has become Vietnam's most important export destination. Vietnam's prospects for increased exports were realized by its membership in the World Trade Organization in 2007. The country is also an important tourist destination, particularly for sun and surf and ecoadventure recreation.

Cambodia

Nestled between Thailand and Vietnam (see Figure 11-23), Cambodia's past has been even more troubled than Vietnam's. Called Kampuchea by its own people, Cambodia was once a proud empire centered on the magnificent twelfth-century Hindu temple complex at Angkor Wat and sustained by productive rice agriculture nurtured by the Tonle Sap and Mekong River. Since the fall of the Khmer Empire in the fourteenth century, this culturally homogeneous land has been periodically occupied by neighboring Vietnam and Thailand. The French ruled Cambodia from 1863 until World War II, but, following independence, political events once again drew Cambodia into a wider regional conflict.

The war in Vietnam spilled over into eastern Cambodia as the United States attempted to prevent North Vietnamese troops from using this region to launch operations into South Vietnam. In the early 1970s, the monarchy of Prince Sihanouk was overthrown by military leaders who, in turn, lost control of the country to the Communist **Khmer Rouge** in 1975. The Khmer Rouge isolated the country from the outside world and launched an unspeakable reign of terror by emptying cities of their inhabitants, murdering most of the educated elite, and suspending formal education in a process of "reconstruction" that the world has now come to know as the "killing fields." The result was widespread starvation and the loss of some 2 million lives. Once North Vietnam conquered the South, it invaded Cambodia, ousted the Khmer Rouge, and installed its own Communist government in 1979. The various factions entered into a coalition government in 1982, and by 1988 Vietnamese troops had withdrawn. In 1992 the United Nations sent in peacekeeping troops, and this was followed by parliamentary elections in 1993. The Khmer Rouge continued to engage in limited insurgency activities in the far western regions of the country until the late 1990s.The legacy of war and genocide has hampered Cambodia's ability to reach substantially greater levels of economic growth and development. Throughout the 1990s, Cambodia was the recipient of the highest per capita foreign aid in the world and was flooded by an army of nongovernmental organizations that assisted in basic development projects, particularly related to health and education. These organizations remain active and important today (Figure 11-26). With a GNI PPP per capita of $2,260, too many of the country's 15 million people remain poor, but poverty rates declined from 47 percent in 1994 to 28 percent in 2011. National development goals must focus on rural areas as 90 percent of Cambodia's poor are rural inhabitants. Improving the lives of rural people during the 1990s was severely hampered because of the millions of land mines and unexploded bombs used by Vietnamese, American, and Khmer Rouge military forces. Cambodia is one of three countries in the world that account for 85 percent of unexploded land mines. Although far less productive than Vietnam, rice production gradually increased 54 percent between 1999 and 2009 by adopting high-yielding rice varieties, increasing double cropping, and expanding irrigation. Emulating both Thailand and Vietnam, Cambodia now exports rice. While economic growth has been relatively high since the late 1990s, that growth rests on a very narrow base. In 2010, FDI accounted for 7 percent of GDP and much of this investment was in garment manufacturing, a sector that in 2008 employed almost 70 percent of the manufacturing workforce and accounted for 70 percent of manufacturing export value. Approximately 90 percent of garment and textile factories are foreign owned and three-quarters of the factory employees are female. Tourists visiting Angkor Wat and other historic cultural sites also are an important source of foreign earnings.

Cambodia has made great economic strides since the 1990s, but serious challenges remain. Promoting the growth of higher-value manufacturing will need a more educated workforce. Only 20 percent of adults have the equivalent of a high school degree and government expenditure on education is one of the lowest in the world. Government corruption is endemic and seems to have increased; Cambodia's government was ranked the twentieth most corrupt in the world in the 2011 Corruption Perceptions Index and was the second most corrupt country in Southeast Asia behind Myanmar. Corruption ranges from granting land concessions to companies and dispossessing inhabitants of their land and livelihood to small bribes paid to teachers or hospital staff for services. Cambodia also is a major Southeast Asian source and transit country for female sex trafficking between Thailand and Vietnam. Reasons are complex, but include high poverty and unemployment rates, low educational attainment, economic disparities between rural and urban areas causing rural to urban migration, and increased levels of international tourism. With less than aggressive government action, female HIV rates are the highest in Southeast Asia.

Myanmar (Burma)

Surrounded by mountains and centered on the large Irrawaddy River basin, Myanmar's ecological framework is similar to Thailand's (see Figure 11-23). While its eastern neighbor has taken advantage of its resource endowments, Myanmar's development policies have for decades resulted in economic decline. Myanmar is the one reformer country most resistant to the openness associated with globalization. Led by an authoritarian and isolationist regime that embraced an indigenous brand of socialism, Myanmar has seen its economy steadily crumble despite a well-endowed resource base of timber, rice, gems, gold, tin, and petroleum. In 1962 the single-party military dictatorship of Ne Win was installed in

the midst of ethnic rebellion on its national frontiers and a growing Communist insurgency. After a new constitution was enacted in 1974, the government's isolationist policies moderated to allow increased foreign contact. The result was a gradual rise in prosperity, although by the mid-1980s economic growth had stagnated.

The hardening of dictatorial military rule began in the late 1980s in response to an unsuccessful popular uprising against the government, followed by house arrest of Aung San Suu Kyi, the daughter of a national hero, who challenged the government to democratize the political system and received the 1991 Nobel Peace Prize. The 1990 democratic elections were nullified by the government. The government in 1989 also changed the name of the country from the British-imposed Burma to *Myanmar,* the name of an Irrawaddy River ethnic-Burmese kingdom destroyed by the British in the 1800s. In 2005 government functions were also moved to a new capital, Nay Pyi Taw, 97 miles (156 kilometers) north of Yangon (Rangoon). This loss of government functions and economic isolation from the world economy has led to a severe deterioration in Yangon's former economic and political role (Figure 11-27). The relocation of the capital to the interior can be interpreted in two very different ways: the government's view is that some developing countries create new interior capitals because a more central location allows for better administration; foreign observers claim that the new interior capital is expressive of a "bunker" or retreat mentality on the part of the isolationist regime, paranoid about the potential democratizing forces of globalization. In late 2007, Buddhist monks, joined by Burmese civilians, led another popular uprising that was also quickly suppressed. Most expressive of the military's cruel and isolationist rule is its immoral response to Typhoon Nargis in 2008, in which it refused foreign relief workers and aid; at least 138,000 people perished.

Perhaps to reduce Aung San Suu Kyi's influence and to gain a better standing in the international community, a series of political reforms were enacted beginning with a new constitution in 2008,

▼ Figure 11-26 **Road sign outside of Seam Reap, Cambodia.** This international nongovernmental organization provides basic services to poor children. Both the governments of Cambodia and South Korea provided funding for this children's center based on Buddhist teachings.

the 2010 release of Aung San Suu Kyi from house arrest, the 2010 elections that ushered in civilian rule, and greater press and individual freedoms. In the 2012 special elections, Aung San Suu Kyi was elected to the parliament. While these political reforms took the world by surprise, and appear to possess momentum, the 2008 constitution safeguards the military hold on power, and does not alter its substantial control of economic power through military-held monopolies. These reforms do not seem to have addressed the decades-long military conflict with ethnic minorities along the Thai, Laos, and Chinese borders, and the more recent serious human rights abuses against the ethnic Muslim Rohingya peoples on the western border with Bangladesh. The lawlessness along Myanmar's mountainous borders has been especially strong where Myanmar, Thailand, and Laos meet at the Mekong River in a region called the Golden Triangle. Once the second largest source of illegal opium production, the region now is a center for illegal drug trafficking, scenic tourist resorts, sex tourism, and casinos. From an economic perspective alone, ethnic minority conflicts must be resolved as much of the abundant natural resources expected to attract substantial FDI are located on Myanmar's territorial edges.

▲ Figure 11-27 **Streetscape of Yangon, Myanmar.** Because of past isolation from the global economy, low-profile British colonial-era buildings dot the urban landscape.

Stop & Think

Describe the different ways in which the national economies of Vietnam, Cambodia, and Myanmar have interacted with the global economy.

Summary

▶ The many countries that comprise Southeast Asia and the experience of being colonized by so many Western powers make the region very different from South Asia and East Asia. Southeast Asia became a critical source of raw materials to fuel Western industrialization; the region's demographic composition was forever changed by enlisting Chinese immigrants to assist in the colonial process.

▶ In the immediate post-World War II period some countries chose to open their economies to globalization and as modernizers experienced rapid rates of economic growth with varying degrees of government management. Others decided to remain economically isolated based on socialist economic principles. By the 1990s, most of these centrally planned economies adopted some form of capitalism and began engaging in globalization and reform. Differing levels of economic growth are expressed in a wide range of development variables, particularly income but also in some basic demographic measures that contribute to the region's social diversity.

▶ Whether modernizer or reformer, the region's physical environment has suffered as a result of rapid economic growth. The greatest environmental degradation occurs in the form of deforestation; large swaths of tropical hardwoods are disappearing, replaced with plantation crops and commercial forests. Coastal environments have been seriously harmed as commercial shrimp farming replaces mangrove forests, and coral reefs are destroyed from river sedimentation, urban-sourced water pollution, and

coral mining. It is fortunate that the industries driving rapid economic growth are not energy intensive because Southeast Asia's mineral resource endowment, compared to other global regions, is relatively poor. But the region as a whole will require substantial energy imports to fuel future economic growth.

▶ In Insular Southeast Asia, Singapore is the only rich and developed country. Taking advantage of its central location in the region, the city-state's government has implemented policies to successfully embed its economic growth in high-value manufacturing and even higher-value information technology services. Malaysia has gradually developed an increasingly higher-value, export-led economy although with strong government policies favoring the indigenous Malay population. Indonesia has made substantial strides in a number of development measures and in reducing tensions associated with unbalanced core-periphery relationships. The Philippines is the poorest modernizer country and continues to be plagued by income inequalities, high population growth rates, and endemic corruption.

▶ In Mainland Southeast Asia, only Thailand has fully embraced globalization and as a result is an upper middle-income country. Thailand provides an ideal example of a national economy that has gradually added increasingly greater value to its export-oriented industry. Vietnam's economy has also engaged the opportunities afforded by globalization to export lower-value products, but future growth depends on further reducing still-strong government economic control. Rising from the ashes of Khmer Rouge rule in the 1990s and assisted by large quantities of foreign aid, Cambodia has attracted only minimal foreign investment and is saddled by endemic government corruption and extensive rural poverty.

Ruled by a dictatorial and isolationist military government since the 1960s, Myanmar has only recently experimented with democratized rule and allowed foreign investment.

Key Terms

Angkor Wat 508
ASEAN Free Trade Area (AFTA) 513
Association of Southeast Asian Nations (ASEAN) 513
cash cropping system 512

Chinese diaspora 514
city-state 519
doi moi 532
dual economic system 512
Extended Metropolitan Regions (EMRs) 515

growth triangle 522
industrial colonialism 512
internal colonialism 526
Khmer Rouge 533
mercantile colonialism 512

Multimedia Super Corridor (MSC) 524
New Economic Policy (NEP) 523
Overseas Chinese 512
producer services 521

Understanding Development in Southeast Asia

1. How are Mainland and Insular Southeast Asia's landform and population distributions connected?
2. What are the major environmental challenges facing Southeast Asian nations?
3. In what ways were the economic and political characteristics of Southeast Asian states similar or different during the precolonial era?
4. How do enclaves and dual economies fit into the core-periphery model of colonial economies?
5. What are the economic and spatial characteristics of Extended Metropolitan Regions?

6. How is the structure of foreign investment in Singapore changing?
7. How and in what ways did Malaysia's government increase the role of indigenous Malays in the national economy?
8. What problems have prevented the Philippines from achieving higher levels of economic growth?
9. How does Thailand's economic development experience compare with that of Myanmar (Burma)?
10. Why are there historical and modern differences in the political and economic orientation of the northern and southern halves of Vietnam?

Geographers @ Work

1. Describe the role of ethnic Chinese in Southeast Asia's economic development.
2. Compare the different political and economic contexts that explain the economic development trajectories of the modernizer and reformer countries.
3. Describe the transformation experienced by Southeast Asian agricultural economies due to increased urbanization and globalization.

4. Present a geographical explanation for the importance of Singapore in the larger regional economy of Insular Southeast Asia.
5. Chart the changing economic and political tensions between Java and Indonesia's outer islands since the country achieved independence.

MasteringGeography™

Looking for additional review and test prep materials? Visit the Study Area in MasteringGeography™ to enhance your geographic literacy, spatial reasoning skills, and understanding of this chapter's content by accessing a variety of resources, including MapMaster® interactive maps, videos, RSS feeds, flashcards, web links, self-study quizzes, and an eText version of *World Regional Geography*.

▲ Figure 11-28 **Southeast Asia from Space.**

Australia, New Zealand, and the Pacific Islands

12

Simon Batterbury

The "Wild White Man"

In 1802, the British convicted an illiterate former soldier, William Buckley (1780–1856), of petty theft and transported him to a newly established penal settlement in Port Phillip Bay in southeast Australia. The British, only established in Australia for 14 years, were seeking labor for their vast colony. Buckley and two other convicts escaped from the penal camp, which was abandoned soon after because of lack of water and food. Buckley alone survived, and was later accepted into a band of Aboriginal hunter-gatherers, the Watourong. To the Watourong, he was the reincarnation of their chief. For 32 years he lived as they did—learning their language, marrying and fathering children, travelling and hunting. He became a respected elder and met no white men until 1835, when he saw a ship off the south coast and overheard some Watourong men plotting to murder these new arrivals. Torn in his loyalties, Buckley surrendered to the British. There was confusion because he had forgotten how to speak English and resembled a local clansmen. He was known thereafter as the "wild white man." Buckley worked as an interpreter as the British colonized the region and laid the foundation for Australia's second largest city, Melbourne. Unhappy in both the European and Aboriginal cultures, and fearful of how rapid settlement would impact the indigenous population, Buckley eventually settled in Tasmania where he lived for 19 more years.

Buckley's story fascinates Australians. The phrase "Buckley's chance" today means "no chance in the world," since the British gave him up for dead and believed no convict could survive in the Australian bush for so long. Australia was regarded as harsh, alien, and unforgiving by earlier settlers, but for the Aboriginal people, the Australian bush was their home. Buckley warned the Watourong that colonists would come, and seize their resources and land. But he could not prepare them for the full onslaught to come. Buckley has become an important "in-between" figure amidst a clash of cultures and the impacts of economic development in Australasia and Oceania. Not least, he showed a resourcefulness that later "Aussie Battlers"—hard-up settlers making a living and overcoming adversity—value to this day.

▲ The Yarra River splits Melbourne's central business district as it winds its way to Port Phillip Bay on whose shores William Buckley met the first group of English settlers in the area in 1835.

Read & Learn

- ► Outline the environmental advantages and disadvantages of the Australian continent.
- ► Explain the settlement history of Australia and New Zealand.
- ► Compare the situation of Aborigines and Māoris in the contemporary societies of Australia and New Zealand.
- ► Identify why Australia and New Zealand are among the most well-off and stable countries in the world.

- ► Characterize the challenges facing the Pacific Islands.
- ► Show how remoteness influences the region's economic opportunities and choices.
- ► Explain the importance of Australia's growing relationship with its Asian neighbors.

Contents

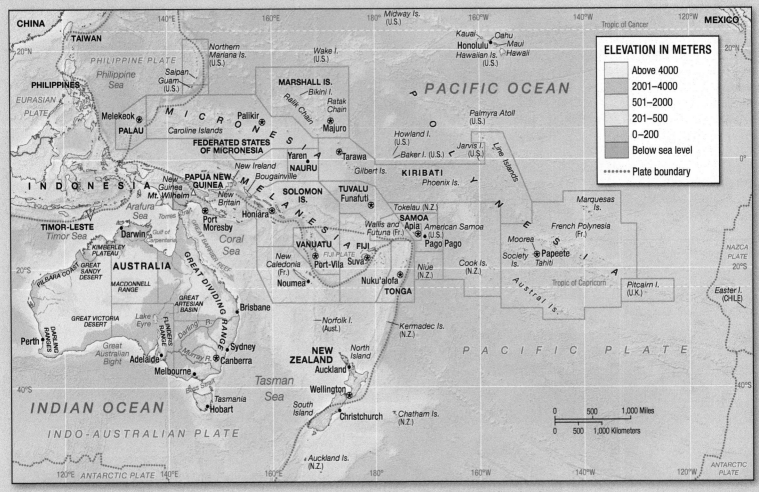

Figure 12-0 The major physical features and countries of Australia, New Zealand, and the Pacific Islands.

Australia and New Zealand are both blessed with abundant space and natural resources, and physical landscapes of great variety. Today both have high levels of material development associated with the transplantation of Western society and economy to territories that, by accidents of geography and history, had been largely untouched by—although known to—the peoples of Asia and the Americas. Although the two nations share common historical development and economic circumstances, they differ greatly in size and have very different landscape features and biotic communities. Both had established indigenous populations, in Australia dating back a startling 50,000 years, but the greatest change in their histories was the establishment in the late eighteenth century of British settler colonies. Today, they have evolved into modern and multicultural nations although the majority of residents have European heritage. Before World War II, both countries operated as "suppliers" of goods for Britain and its empire. That function is no longer dominant as trade has diversified and

internationalized; ties with the British have gradually weakened, and there is little enthusiasm for reversing this trend. Australia and New Zealand still depend on trade, but with a broader range of industrialized nations. There are important links to the United States, Japan, China, and, to a lesser extent, other Asian countries. Australia has a major mining economy linked to Asian and particularly Chinese manufacturing. New Zealand retains a greater reliance on livestock, agriculture, and forestry than Australia. Australia's wealth is more diversified, with rich deposits of metals, coal, and natural gas; it also produces agricultural and some manufactured goods. Both countries have a highly skilled labor force.

The Pacific Islands share some resource endowments and landscapes with their larger and more affluent neighbors, yet they differ greatly in their cultural and economic attributes. The Pacific Islands also are characterized by great natural beauty, scattered across a vast ocean world. A second common feature is a shared legacy of colonialism, although most of the islands were exploited for their resources and only a few saw extensive European, American, or Japanese settlement. Resource exploitation still resulted in a loss or dilution of many components of indigenous culture, and dissatisfaction or open revolt against colonial or more recent independent rule have marked the region's history. Almost every indicator of social and economic development, especially in terms of income, health, and education, casts the Pacific Island states and territories, with few exceptions, as "less developed" than its neighbors in all directions. Pacific Island peoples have responded to their isolation by developing livelihoods attuned to their geography, notably through fishing, tourism, and some primary agricultural production and mining. But they cannot outcompete larger and more technologically advanced nations of the world, and many islanders migrate to earn money or live elsewhere, including in New Zealand and Australia.

Environmental and Historical Contexts of Australia, New Zealand, and the Pacific Islands

Although Australia is a continent, very slightly smaller than the continental United States, the other landmasses of the Pacific Basin are a collection of islands (see chapter opener map), diverse in size, physical geography, population, culture, levels of economic development, and political structure. Combined, the landmass of the Pacific Basin contains less than 0.5 percent of the world's population (roughly 36 million people). Australia and New Zealand are settler nations that were colonized and governed by Britain. Australia has a sizeable economy despite a small population, a thriving mining industry, and several important city regions. New Zealand is more remote and less rich in lucrative resources, but sustains a developed economy from its livestock sector and tourism. With a few exceptions the smaller Pacific islands of Melanesia, Micronesia, and Polynesia are peripheral to the global economy because of their geographical position, small population, and resource endowments. Most have experienced far less intensive colonial settlement and have much lower levels of economic development.

Environmental Setting: Vast Spaces of Land and Sea

One very large and generally arid landmass, Australia, dominates the region and provides the setting for a large assemblage of unique flora and fauna. Islands make up the rest of this vast region. Great distances often separate these spots of habitable land one from the other; the vagaries of wind, current, and human contact often determine what combinations of plants and animals occupy each location. Diversity of habitat and human adaptation in a context of considerable isolation characterize the environmental setting of much of the region.

Landforms

The Australian continent as a whole was once joined to Antarctica, as the easternmost part of Gondwana, which broke up in the distant geologic past. Australia moved northward and later separated from present-day Tasmania, New Caledonia, and Vanuatu. The smallest of the continents, Australia is as remarkably flat as it is dry. The western half of the continental mass is composed of a low elevation, eroded tableland seldom exceeding 1,000 feet (300 meters) above sea level. A few more prominent ranges, the Darling Range inland from Perth for instance, are really escarpments that frame the edges of the tableland (Figure 12-1). Large desert areas occupy most of the western interior. The eastern half of the continent is composed of two main features: a large lowland basin extending from Adelaide in the south to the Gulf of Carpentaria in the north, and a long mountain range inland from and running parallel to the east coast. The Great Artesian Basin is a sandstone layer, recharged by moisture from the northern Great Dividing Range. It underlies over 20 percent of Australia's interior land surface and is the only reliable source of fresh water in the northern interior lowlands. The southern interior lowlands receive surface flow from the higher southern Great Dividing Range through the Darling and Murray river systems. The irrigated Riverina district west of Sydney is a particularly important agricultural region in this area. Short streams flowing eastward from the crests of the Great Dividing Range provide water for a fertile but narrow coastal plain before emptying into the Coral and Tasman Seas.

The continent's native flora and fauna are truly unique—especially marsupials (animals with a pouch to carry their young) like kangaroos, koalas, wombats, numbats, Tasmanian Devils, and possums (Figure 12-2). On other continents, marsupials have largely been displaced by placental mammals. It is thought that many of Australia's mega-fauna, much larger marsupials like the hippo-sized Diprotodon, died out 40,000–50,000 years ago in the early days of human occupancy. Yet despite its uniqueness the continent was never completely isolated. The dingo, a wild dog, arrived about 5,000 years ago from Asia, and camels, sheep, cows, and all manner of unwelcome predators like cats and foxes came with

▼ **Figure 12–1 Darling Range in Western Australia.** The view south from Mt. Vincent, southeast of Perth, reveals the rolling contours of the mountains created along the Darling Fault. Thin soils and dense, bushy woodland characterize the landscape.

(a)

(c)

(b)

◀ Figure 12–2 **Marsupial wildlife in Australia.** (a) Gray kangaroos, quite successfully adapted to living near humans, give a distinctively Australian flavor to the landscape of New South Wales. (b) Koalas are tree dwelling and eat only certain eucalypt leaves. (c) The Tasmanian Devil is a carnivorous animal, currently endangered by a transmissible facial tumor disease.

Climate

Many Australians think of their home as a "sunburnt country" (a phrase from Dorothea MacKellar's 1908 poem, "My Country"). This is largely true, even if there are places where one can be wet, cold, and even snowbound. Australia's climate is wettest and most humid in the north, dry and desert-like in the vast center, and becomes mediterranean in the south and southwest. Only 11 percent of Australia gets more than 40 inches (1,000 millimeters) of rain a year; two-thirds of the country receives less than 20 inches (500 millimeters) (Figure 12–3).

This generally dry climate is largely due to Australia's low latitude, but also is influenced by the El Niño Southern Oscillation (ENSO) in the Pacific that can create drying of the southern part of the continent and wetter conditions in the north. El Niño (Spanish: the little boy) is caused by warming of the ocean in the eastern Pacific which brings heavy rainfall and thunderstorms, but reduces fish stocks due to the presence of nutrient-deficient warm water. At the same time, it deprives the western Pacific of rain. For Australia, drier than normal conditions are also generally observed in Queensland, inland Victoria, inland New South Wales, and eastern Tasmania in the southern hemisphere winter (June to August). Southeast Australia experienced a prolonged drought from 2003–2010, one of the worst in recent centuries, and ENSO may have been in part responsible (Figure 12–4). Saturday, February 7, 2009, is known as "Black Saturday," when 173 people died in Victoria as 300 separate bushfires raced across a parched landscape with strong winds and temperatures of 115°F (46°C). Five towns were almost completely destroyed along with 2,000 homes. The fires were a reminder that the Australian environment is unpredictable, and may become more so under climate change scenarios.

By contrast La Niña (Spanish: the little girl) occurs when cold water intensifies, bringing dry weather and more fish. During La Niña events, easterly trade winds move warmed water and air toward the western Pacific. The collected moisture is dumped in the form of typhoons and thunderstorms. Strong La Niña events affected Australia in 1988–1989, 1995, 1999–2000, and 2009–2011. Heavy rainfall and flooding in the northern East coast occurred as recently as 2013. As a result of global anthropogenic climate change, hotter summers and unreliable, possibly lower, rainfall across the major settled and farmed regions is the prediction for the future. This risk is a current preoccupation of the country's decision makers and its commercial interests.

New Zealand has a humid temperate climate, commonly known as marine west coast, with mild summers and winters (see Figure 12–3). As a result New Zealand escapes the aridity that is such a prominent part of Australia's continental-scale landscape. But in the highlands of South Island, weather conditions are severe enough for glaciers to form, and winter sports bring thousands of tourists to its ski areas. The country's north-south elongation means that average temperatures in the north are at least 10°F (4.5°C) warmer than in the south.

European settlers over 200 years ago, all of these endangering the native fauna. Approximately 80 percent of the 20,000 plant varieties are **endemic species** (not found elsewhere) and most common are drought-tolerant eucalypts (gum), banksia, and acacia (wattle) tree species. Australia shares important flora and fauna with New Guinea to its north, because until 14,000 years ago they were connected by a land bridge.

Like the islands that make up Japan, New Zealand was formed from the Ring of Fire, a section of the belt of unstable crust rimming the Pacific. The islands are the crest of a giant fold in the Earth's crust that rises sharply from the ocean floor. Almost three-quarters of South Island is mountainous, dominated by the Southern Alps, which rise to elevations above 12,000 feet (3,600 meters). North Island is less rugged, but it has peaks that exceed 5,000 feet (1,500 meters) and live volcanoes. The indigenous Maori people referred to the North as **Aotearoa** (The Land of the Long White Cloud), a name now extended to the whole country.

Unlike Australia, New Zealand has few land mammals, except bats and flightless birds, some species of which have been rendered extinct by waves of human settlement. For example, the 12-foot (3.6-meter) high moa bird was overhunted by the Maori people long before the arrival of Europeans. The national symbol of New Zealand is the flightless Kiwi bird.

The physical geography of the Pacific Islands has made conventional Western-style economic development difficult. Many high islands are of volcanic origin with extremely rugged interior cores, but only a few have mineral wealth. Others are relatively flat atolls that have formed on the tops of coral reefs, are low-lying, and are exposed to storms and **sea level rise**. The vast majority of the islands, except mainland Papua New Guinea, are relatively small and limited in the numbers of people they can support. Individual islands within archipelagoes are often situated great distances apart, making communication difficult even for extraordinarily capable traditional seamanship. But the wide dispersion of habitable specks of land also gave each island extensive legal fishing grounds.

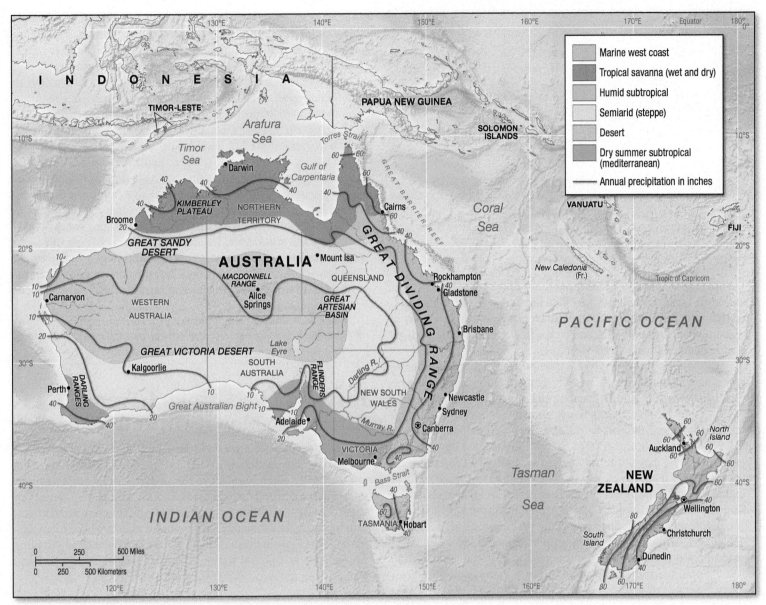

▲ Figure 12–3 **Climate regions and precipitation in Australia and New Zealand.** Much of Australia is little used because water is scarce; a large area of Australia receives less than 20 inches (500 millimeters) of precipitation annually. New Zealand has a much wetter climate.

Most of the islands of both the north and south Pacific have a tropical marine climate, with warm temperatures throughout the year and limited daily and seasonal variation. Rainfall is greater on the high islands that pose more of an obstacle to storms, although this precipitation often has a monsoonal rhythm. Tahiti's wet season, for example, occurs between November and April with January the wettest month at Papeete, the capital, when 13.2 inches (340 millimeters) fall on average. In contrast, the same station records on average only 1.9 inches (48 millimeters) in August. Most of these Pacific Islands can expect to encounter a moderate-intensity typhoon once a year and at least one major typhoon once in a decade. Major typhoons are a matter of particular concern for the inhabitants of low-lying islands. Despite the buffering effect of coral reefs, these islands risk severe destruction to coconut trees and limited agricultural soils if a typhoon storm surge overtops the atoll. These same low islands will be particularly at risk if global climate

change leads to even modest increases in sea level and changes in sea water temperature that impact corals and fish populations. Not even a tropical paradise is immune to natural hazards.

Environmental Challenges

Numerous environmental challenges face Australia, New Zealand, and the Pacific Islands. Not surprisingly, many of these challenges relate to climate change. The general expectation for the region is that global warming overall will produce drier and harsher conditions than those experienced over the last 100 years. Droughts will last longer, if not be more numerous, and will impact both continental-scale Australia and the tiniest inhabited atoll. Drier conditions would place severe constraints on available water for irrigated agriculture, particularly in Australia's Riverina "breadbasket" district in New South Wales, potentially affecting both domestic consumption and exports. Drier conditions and higher

▲ **Figure 12–4 Drought in the Riverina region of southern New South Wales, during an El Niño event, 2007.** When drought causes a potential crop to fail, sheep are often grazed on the scanty plants in an effort to salvage something from the disaster.

temperatures increase the likelihood of wildfires. Higher temperatures and strong winds can generate firestorms that turn drought-parched vegetation into a lethal inferno. Fire is a natural part of Australia's ecosystems, but can be expected to become increasingly damaging to humans and more frequent in the future.

Climate change is driven by greenhouse gas emissions across the globe. Australia's contribution to climate change is not only a result of its own domestic carbon footprint, but is also a product of the coal it exports to feed the engine of growth and industrialization outside its borders. This leads to conflicts over environmental policy on several levels. Environmental groups within Australia have become increasingly vocal about the need to set limits on the expansion of mining and export of coal, a policy that has serious implications for economic prosperity and has gained only limited traction among the general public. Steadily expanding carbon exports place Australia squarely in opposition to the best interests of its regional neighbors. Threatened by rising temperatures and sea levels, many island nations are vocal about the need to curb the excesses that feed global warming. Similar conflicts in values and objectives occur in managing land impacted by the mining industry, whose activities require large amounts of water during active operations and, if not forced to pay for long-term, meaningful remediation, can leave behind a blighted and battered landscape. These issues are a particular source of concern in Western Australia, and are certain to emerge in the future in New Caledonia and Papua New Guinea, where mining of resources clashes with the environmental values and land ethics that are a prominent part of indigenous culture.

Atolls with limited land surfaces will find replenishment of groundwater reserves by natural processes increasingly difficult, thus making continued occupancy problematic. For low islands, global warming is closely linked to higher sea levels and greater hazards of storm damage and erosion. Higher global temperatures contribute to warmer seas as well. The coral reefs on which low islands depend for storm protection are everywhere threatened by water temperature changes that promote species that attack corals or reduce their ability to process sunlight into food (see *Exploring Environmental Impacts*: The Great Barrier Reef). Rising sea levels threaten low islands whose highest point may be only a few meters above present sea levels. But higher sea levels also threaten coastal environments on large islands and continents. Higher sea levels

increase direct wave action against the coast, promoting erosion. The sediments produced are easily disturbed by storm action and ship propellers, which raises water turbidity and is harmful to many marine species. Increased sediment movement and deposition can adversely affect coral reefs, which are sensitive to reduced exposure to sunlight. Siltation is also a serious management problem for port and harbor installations.

In countries with substantial tropical and temperate forests, such as Papua New Guinea (PNG) and New Zealand, deforestation is a significant issue. The temptation to realize substantial profits immediately, rather than manage for sustainable yields over the long term, is considerable. With a rainforest that is the third largest in the world, PNG is also seeing this resource decrease by 1.5 percent each year. Much of the logging exceeds legal limits, and the erosion and habitat change that accompanies clearance has serious consequences. Loss of biodiversity is an important, if somewhat abstract consequence, but the deterioration of an essential resource is devastating for impoverished local populations.

Stop & Think

> Distinguish the different ways in which Australia and the Pacific island nations are particularly vulnerable to climate change.

Historical Background: From Ancient to Modern in 200 Years

Europeans were late arrivals to Australia, New Zealand, and the Pacific Islands, particularly from the standpoint of colonization, control, and settlement. Not so the region's indigenous peoples, who preceded European explorers, administrators, and settlers by thousands of years. Their prodigious feats of exploration and migration by outrigger canoe populated remote islands whose often limited resources were masterfully exploited and developed. Melding the insights, values, and technologies of the region's major cultural groups remains an incomplete process.

Settlement and the Colonial Era

Until 1788, Australia was inhabited by Australian **Aborigines**, the continent's indigenous people, and the Torres Strait Islanders off the northern tip of Queensland. Their numbers are unknown but may have been half a million people. These populations were members of some 300 distinct "nations" and there were multiple languages spoken. Aboriginal people have complex origins, but it is certain that they have been in Australia for over 50,000 years, much of that time as semi-nomadic hunters and gatherers (and very occasionally, settled peoples). This is considerably longer than the human occupancy of North America. Aboriginal beliefs anchor these people to their ancestral land, their "country." Without cultivars that could be farmed or animals that could be domesticated, Aboriginal people led a mobile existence in close harmony with nature, although it is believed their impacts on the land were nonetheless extensive, particularly through the use of fire for bush clearance and to chase out animals for hunting.

The long delay in the European discovery and settlement of Australia was caused by many factors, including the vastness of the Pacific and Indian Oceans, the direction of the prevailing winds and currents, and the lack of any sign, from the sea or from brief

The Great Barrier Reef

The Great Barrier Reef off the northeast coast of Queensland is one of the world's great natural wonders. It stretches for 1,616 miles (2,600 kilometers) over an area of approximately 133,000 square miles (344,400 square kilometers), making it the largest coral reef ecosystem in the world. The reef is so large that it can be seen from space. The **Great Barrier Reef Marine Park**, comprising about a third of the total reef, is the area in which the Great Barrier Reef Marine Park Authority limits fishing, pollutants, and damaging tourism. A "Great Barrier Reef World Heritage Area" was declared in 1981. Ecologically sustainable tourism is the aim, and a 2004 zoning plan brought into effect for the entire area has been widely acclaimed as a new global benchmark for the conservation of marine ecosystems.

The reef system consists of more than 3,000 reefs ranging in size from 2.5 acres (1 hectare) to over 24,700 acres (10,000 hectares), interspersed with tropical islands. There are over 400 species of corals living within the reef and over 1,500 species of fish. The reef is also home to 30 species of whales, dolphins, and porpoises as well as dugongs and turtles. Large numbers of birds visit the reef annually, and up to 1.7 million breed on the islands each year. All this has made the Great Barrier Reef one of Australia's top tourist destinations with an average of 1.9 million visitor days a year, second only to Sydney. Visitors contributed A$59 million a day to the Queensland economy in 2012.

Many in the scientific community worry about the reef's future. The most significant threat comes from climate change. Rising sea temperatures have resulted in mass coral bleaching and ocean acidification (due to absorption of excess CO_2), which destroy the coral itself. A bad bleaching episode in summer 2006 was a warning for the future. Under temperature stress, algae harbored by coral tissue die, food production ceases, and coral bones turn white (Figure 12-1-1). Displaced fish stocks have sought new habitats elsewhere, and this has resulted in more chick mortality among the seabirds that rely on the fish.

More than just bleaching impact corals. Over 50 percent of the coastal wetlands of Australia's northeast coast have been destroyed since colonization. Coastal mangroves act as a natural filter, catching pesticides and sediment from farming that otherwise would reach the ocean. During tropical storms, pollutants and sediments do wash out as far as the reef, aided by the lack of coastal protection. Algae have also increased in recent years, creating more competition for sunlight and oxygen, lowering water quality, and providing perfect conditions for the eggs of the Crown of Thorns starfish (*Acanthaster planci*), which have increased in numbers. One starfish can eat up to 65 square feet (6 square meters) of coral reef in a year. One of the only effective starfish predators, the triton mollusk, has had its numbers diminished by overfishing. Recent surveys indicate that half of the Barrier Reef coral population has died since 1986.

One of the most controversial issues surrounding the Marine Park is the presence of oil shale beds that lie around and underneath it, an expensive source of oil that is high in CO_2 emissions once refined and burned. Mining and drilling licenses existed in and around the Park in the years before all exploitation was halted within its boundaries, and were championed by the politically conservative Queensland government (1968–1987). As the global price of oil rises, some feel the reef is not safe from economically viable shale extraction, and there are serious efforts to restart it. In the last few years, the rapid expansion of the coal and natural gas industries inland in the Bowen Basin has led to port facilities being upgraded to export minerals. Habitats in the area are experiencing significant impacts from the sediment and silt disturbances caused by construction of coal and gas processing and exporting facilities around Curtis Island and the port city of Gladstone, and the environmental effects have attracted international attention and concern.

Managing tourism sustainably and protecting the health of the fragile reef system are great challenges as well. Effective policy and planning is essential for the reef's survival, and yet the threats include global sea temperatures as well as distant inland mining and farming methods over which local regulations have no control. Only time will tell whether the proximate threat of oil shale extraction can be avoided.

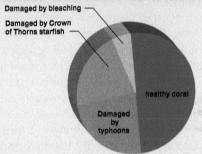

Damaged by bleaching

Damaged by Crown
of Thorns starfish

healthy coral

Damaged
by
typhoons

▲ FIGURE 12-1-1   **Coral health at the Great Barrier Reef, Australia.** Dead whitened corals and partly damaged corals impacted by increased ocean temperatures have increased significantly on the Great Barrier Reef. These were revealed during low tide at Fitzroy Island in the Marine Park in Queensland in April 2011.

Source: http://www.aims.gov.au/latest-news/-/asset_publisher/MIU7/content/2-october-2012-the-great-barrier-reef-has-lost-half-of-its-coral-in-the-last-27-years

Sources: Queensland Government, *Tourism and Queensland's Economy*, March 13, 2003, *http://www.business.qld.gov.au/industry/tourism/tourism-in-queensland/queenslands-tourism-industry/tourism-and-queenslands-economy/*; Shaz, "The Ongoing Threats to the Great Barrier Reef due to Human Activities," January 4, 2011, *http://www.talkingscience.net/2011/01/the-ongoing-threats-to-the-great-barrier-reef-due-to-human-activity/*; Nicky Phillips, "Great Reef Catastrophe," *Sydney Morning Herald*, October 2, 2012, *http://www.smh.com.au/environment/conservation/great-reef-catastrophe-20121002-26vzq.html*

landings on the shore, that the continent possessed worthwhile resources. There were several landings by explorers and shipwrecked sailors, and some failed attempts at settlement around its shores. It was in 1770 that Britain's Captain James Cook became the first European to survey the east coast of Australia, the part of the continent that appeared most suitable for settlement. But the first British ships did not disembark at Sydney Cove until 1788. After losing the American War of Independence, Britain was no longer able to relocate petty criminals and other convicts across the Atlantic. Australia seemed a useful, and suitably remote, alternative and convicts swelled the population from the arrival of the First Fleet until 1868 when the last convict boats arrived in Western Australia.

Exploration and settlement by adventurers, emancipists (convicts who had served out their sentences), and others continued into the nineteenth century. Immigration was encouraged by Britain through land grants to settlers that paid no heed to prior Aboriginal occupancy. For the British authorities, Australia was *terra nullius* (empty land) that they could legally occupy. A great stimulus to development and immigration was the gold rush of the 1850s, which brought large numbers of prospectors and settlers particularly to New South Wales and Victoria, and later to other states. Victoria's capital, Melbourne, had squalid beginnings as a pioneer town, but developed rapidly with opulent Victorian housing and grand architecture financed by ranching, sales of high quality wool, and the gold rush (Figure 12–5). In 1901, the six Australian colonies— Queensland, New South Wales, Victoria, South Australia, Western Australia, and Tasmania—were federated into the Commonwealth of Australia, and the new city of Canberra was first planned in 1908 as the national capital and constructed in subsequent years. The Parliament moved from Melbourne to Canberra in 1927.

New Zealand was one of the last countries in the Pacific to be settled, first by eastern Polynesians who came by ocean-going canoes (*waka*) in separate voyages between A.D. 1250 and 1300. Their language and spiritual beliefs evolved further in isolation from other groups after their mid-thirteenth century arrival in New Zealand. These first inhabitants developed a form of tribal organization based on their origins in Polynesia, and they altered the native plants and wildlife with widespread hunting and introduction of plant species. From the 1500s, with easy food sources like the moa bird exhausted, warfare ensued among tribal groups and settlements were fortified.

Nearly three-and-a-half centuries passed before a Dutch explorer, Abel Janszoon Tasman, became the first European known to have sighted the islands. But it was not until Captain James Cook arrived in 1769 that serious exploration began. European settlement of New Zealand, which occurred at a similar time to Australia, involved the resettlement of some Australian residents eastward as part of the migration flow. Settlement was confined largely to the fringe lowlands around the periphery of North Island and along the drier east and south coasts of South Island. This pattern persists to the present day. The next 80 years saw traders and a range of settlers coexisting, sometimes uneasily, with the **Maori**, New Zealand's indigenous population. Several very early settlers became "wild white men" (Pākehā Maori) and joined the Maori, even fighting the British during the more brutal phases of colonization. The Treaty of Waitangi was signed in 1840 between the British and the Maori and was the foundation of the nation. Technically, it gave Maori equal rights with British citizens in exchange for British sovereignty. The treaty has been used to guarantee rights that were absent in Australia to indigenous inhabitants, although the treaty's precise wording and meaning has been contested. Full integration into New Zealand society and economy remains a problem for both Maori and New Zealand.

All of the Pacific Islands, high and low, were eventually seized and controlled by Western powers in the last half of the nineteenth and early decades of the twentieth century. Spain lost its island possessions to the United States as an outcome of the Spanish-American War. Germany lost its possessions in Papua New Guinea and the Marshall Islands to Australia and Japan as a consequence of defeat in World War I. European and American interest in the region combined geostrategic issues of control over trade routes, an outlet for unwanted domestic populations, and access to resources unique to a tropical habitat. The discovery of gold in New South Wales and Victoria in 1851 sparked an Australian gold rush comparable to that in California in 1848 and changed the thinking of countries with imperial ambitions. Suddenly the sources of wealth and industrial power could potentially be found beneath the soil of at least the larger tropical isles. If globalization had been slow to reach the Pacific until the nineteenth century, the region was irrevocably connected thereafter.

The influence of European colonization is now waning. Most island groups have attained political independence even if economic dependency remains. The economic development of the region is becoming increasingly linked to the powerful countries that ring the

▼ **Figure 12–5 Melbourne's Victorian heritage remains visible.** A former bank at 333 Collins Street is part of a row of opulent structures that preserve a pedestrian-scale streetscape, now linked to modern skyscrapers and offices.

Pacific and trade with it—notably China, Japan, and the United States. Because of its small population and particular resource endowments, the region lacks the strategic importance of other world regions. Despite this, Australian, Papua New Guinean, and New Caledonian mineral exports are vital to the world's manufacturing industries and for energy production. The region's cities and their hinterlands are growing and are increasingly multicultural, and there is two-way flow of migrants with other developed regions that is mutually enriching.

Population Contours

Historical and environmental factors explain much about population distribution in the region. Australia is a classic example of population distribution following basic resources—in this case, water. Aboriginal populations lived where water, game, and plant food were available. Almost all of the continent's central core was devoid of meaningful resources and avoided in terms of significant settlement. Colonial settlement largely replaced indigenous populations,

leaving them with the least desirable habitats from a European perspective. Thus the bulk of the Australian population remains concentrated east of the crest line of the Great Dividing Range in Queensland, New South Wales, and central Victoria, and in isolated, moderately dense nodes around Adelaide and Perth (Figure 12-6). Inland from the Pilbara Coast, in the Kimberly Plateau, in Arnhem Land behind Darwin, and in the flatlands of the interior **Outback** of New South Wales and Victoria where water is scarce, population is limited and widely dispersed.

In Papua New Guinea the reverse distribution occurs: coastal lowlands, particularly on the heavily forested south coast, are very low-density environments. Only the coastal cities of Port Moresby and Lea are exceptions. Instead, the bulk of population is concentrated in the central highlands where cooler temperatures and more favorable agricultural opportunities are found. Most of New Zealand's relatively small population is concentrated on the higher potential North Island, and in a few large urban areas on South Island.

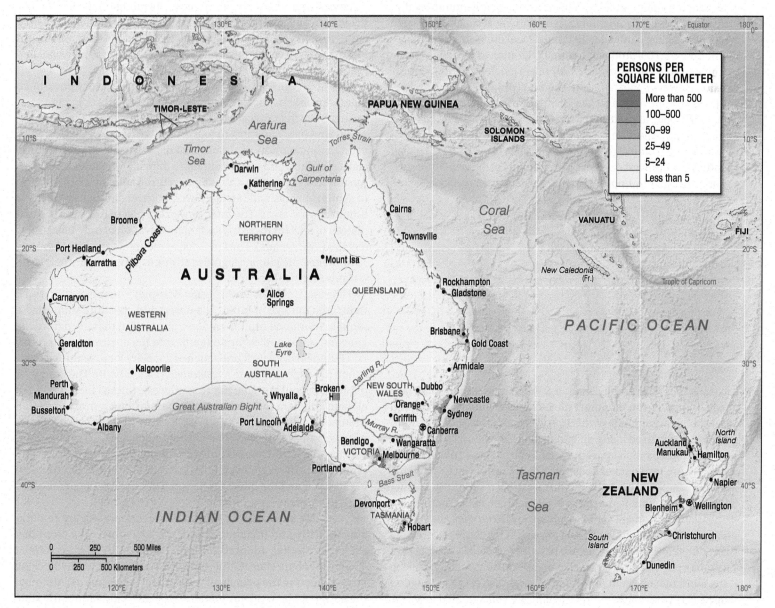

▲ Figure 12–6 **Population distribution of Australia and New Zealand.** The population of Australia is concentrated in a narrow strip along the east, southeastern, and southwestern coasts, but the drier interior is comparatively empty. A mountainous interior, particularly in the South Island, similarly confines New Zealand's population to coastal locations.

Two fundamentally different population dynamics operate in the region. In Australia and New Zealand life expectancy is high, use of modern birth control is widespread, the birth rate is low, and children under 15 make up 20 percent or less of the population. On all these measures the population dynamics of the region's two largest economies are closer to Europe or North America than they are to the regional norm. At the other extreme are countries such as Papua New Guinea, Nauru, Kiribati, and the Solomon Islands with high birth rates, declining death rates, high infant mortality rates, elevated total fertility rates, and young, rapidly growing populations. For these countries a demographic transition to a stable state has yet to occur. Other countries, such as Guam, Fiji, French Polynesia, New Caledonia, and Palau share some of the characteristics of post–demographic transition countries, such as a low birth rate or high life expectancy, but not others. These countries appear to be deep into the demographic transition, and should attain population stability at roughly present levels by 2025. Whether living in emerging post–demographic transition states or in countries that have yet to begin a demographic transition, most citizens of Pacific Island states must contend with limited local opportunities for economic growth. This produces internal migration to the few large urban areas (often the national capital) where greater prospects appear to exist. However, most of these seemingly higher potential cities contain large, poverty-stricken populations struggling, often outside public view, to make ends meet. Migrants also engage in chain migration patterns along pathways provided by political connections to former colonial powers. The presence of over 4,000 Marshall Islanders in northwest Arkansas is an example of the viability of such connections and the opportunities that such options provide.

> **Stop & Think**
>
> ▶ How is the colonization process that occurred in Australia and New Zealand fundamentally different from the one that took place in the United States and Canada?

Australia: A Vast and Arid Continent

Australia is similar to the United States in some respects: the language is English, contemporary life includes low-density suburban sprawl, high automobile ownership, easy access to manufactured goods and processed foods, shopping malls, the information superhighway, and a passion for team sports. With almost 3 million square miles (7.8 million square kilometers), including the large southerly island of Tasmania, Australia extends for 2,400 miles (3,860 kilometers) from Cape York in the north at 11°S latitude to the southern tip of the island of Tasmania at 44°S latitude; it also extends about 2,500 miles (4,000 kilometers) from east to west. Yet, a closer look reveals many differences between the two nations. Although similar in size, Australia's population is far smaller at just under 23 million, making it the 52nd most populous country. Although average population density is very low, the country's people are concentrated in a relatively small portion of the continent (see Figure 12-6). Isolation in the rural interior has led to the survival of Aborigine communities with very distinctive lifeways and belief systems.

The fragility and uncertainties of the Australian climate means the boundaries of climatic regions are difficult to represent. The most important region for humans today is the humid highlands of the Great Dividing Range. This upland feature extends in a belt 400–600 miles wide (640–960 kilometers) up and down the east coast, a remnant of an eroded mountain range marking the separation from New Zealand 80 million years ago. Coastal plains run along the base of the highlands and this is where most of the nation's population, major cities (including Sydney, Melbourne, and Brisbane), and the modern economy are concentrated (Figure 12-7). About 80 percent of the population lives within 20 miles (30 kilometers) of the coast. The climate of Queensland is more humid traveling northward, which meant it was initially believed unsuitable for white settlers and developed later.

Two other regions are important but lie outside Australia's historic core. In the southwestern corner of Australia and in a band along the eastern portion of the southern coast, the climate is mediterranean or dry summer subtropical, encouraging settlement particularly in and around Perth and Adelaide, but the total population is still low. Tasmania, a large southerly island, has temperate rain forests, that are constantly under threat for logging, and a cool and moist climate similar to northern Europe.

The drier northern and central parts of the country pose much greater problems for development. Along the northern fringe of Australia are the tropical savannas, where the climate—intense humidity and heat and three to four months of heavy rain followed by eight to nine months of almost total dryness—has made commercial agriculture and settlement more difficult. The huge interior of Australia has low relief and is desert surrounded by a broad fringe of semiarid grassland (steppe), which is transitional to the more humid coastal areas. The interior western half of Australia is a vast plateau of ancient rocks with a general elevation of only 1,000–1,600 feet (300–500 meters). The few isolated mountain ranges are too low to influence the climate significantly or to supply many perennial streams for irrigation.

The state of Western Australia occupies a massive area, some four-fifths the size of India. The desert and coastline supported a mobile Aboriginal hunter-gatherer population that lived from the land and water bodies for millennia, but the environment is harsh and European settlers found much of it too dry for the rain-fed cultivation to which they were accustomed (Figure 12-8). The European population concentrated in the southwest, with its mediterranean climate and flora rather than the wetter northwest frontier around Broome and the Fitzroy River. Most parts of the state form a low plateau with an average elevation of about 1,300 feet (400 meters) with very low relief, and there is little year-round surface water here. The temperature averages over 99°F (37°C) in the central interior during summer and the distances between the small settlements of the interior are great.

The arid and savanna soils of Western Australia are remarkably infertile and unsuitable for large-scale intensive cropping. Ranching, better adapted to the terrain, occurs across almost 40 percent of the state. Cattle and sheep stations have low stocking rates, reflecting the paucity of grasses and water points. Although the soils may be weak for modern intensive cultivation, the state of Western Australia is immensely rich in mineral resources (see *Visualizing Development*: The Rush for Mineral Resources in Western Australia).

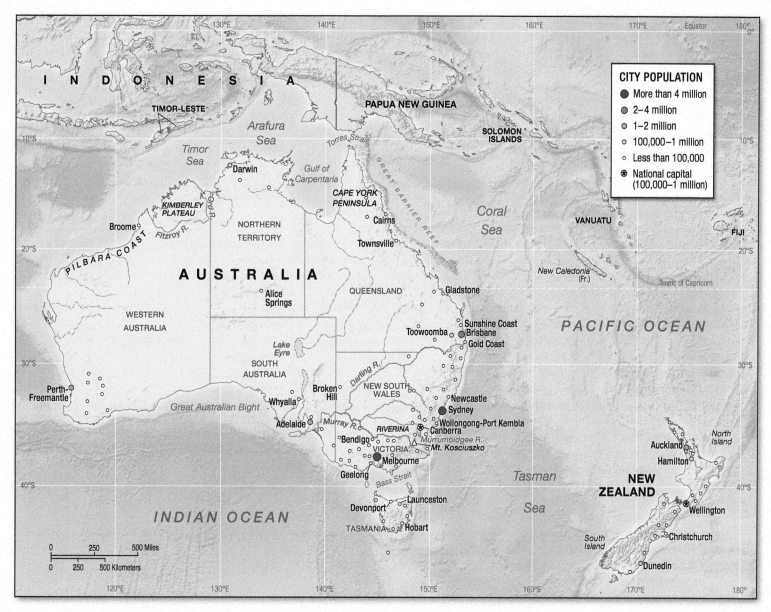

▲ Figure 12–7 **Political and urban structure of Australia and New Zealand.** Australia and New Zealand are distant outposts of European culture, with the major cities concentrated along the coastlines.

The Ins and Outs of Migration

One of the most important developments of the twentieth century was the implementation of the **White Australia policy**, officially termed the Restricted Immigration Policy, which began in 1901 and did not effectively end until the 1970s. Successive Australian governments were concerned that a relatively small white population controlling such a large land area so close to densely populated regions of Asia put the country somehow at risk from invasion

▶ Figure 12–8 **Arid landscapes of western Australia.** Much of interior and western Australia experiences a desert climate. These regions are used primarily for extensive animal grazing. This image shows sheep at the Toorale station in the Outback of New South Wales 800 miles (1,200 kilometers) west of Sydney during a drought period.

The Rush for Mineral Resources in Western Australia

Western Australia is endowed with large deposits of iron, alumina, natural gas, nickel, and gold—wealth in the desert. The state produces more than 20 percent of the world's alumina, and about 17 percent of its iron ore (Figure 12-2-1). The central coast area known as the Pilbara is the nation's engine of iron production. Iron ore mines run 24 hours a day, and in the Pilbara alone there are 16 mines, with more awaiting approval by state and federal environmental agencies. The Argyle diamond mine in the northwest is the world's largest producer of diamonds and the only known source of pink diamonds. Western Australia also extracts up to 75 percent of Australia's gold. The northwest continental shelf is home to huge reserves of natural gas. At James Price Point, Woodside Petroleum is planning the largest liquified natural gas (LNG) processing plant in the world 25 miles (40 kilometers) north of Broome. However, in 2013, this project was put on hold after many disputes between Woodside, its partners, and governmental agencies over economic and environmental issues.

Asian markets have expanded and demand, particularly from China, for Australian minerals has created a boom economy, much of it reliant on Western Australia. China sources 40 percent of its minerals from Australia, and predictions show this demand

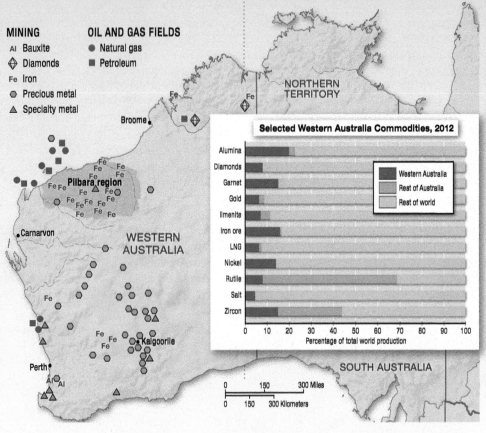

▲ **FIGURE 12-2-1** **Western Australia's contribution to global commodity production for selected minerals.**

Source: Western Australian Minerals and Petroleum Statistics Digest. Rutile is titanium dioxide, *http://www.dmp.wa.gov.au/1521.aspx#1593*

---

nonwhite immigrants was only outlawed by the Racial Discrimination Act in 1975.

Using a rhetoric of "populate or perish," fresh from the unrealized threat of wartime invasion by Japan, other European and North American settlers were accepted from 1947, but they still had to be white. The influx of immigrants from Italy, Greece, Macedonia, and the Balkans was particularly marked. With the eventual elimination of the white-only policy two decades later, more immigration from Asia became possible. More than 4 million people—about one-fifth of the present total population—have moved to Australia from various destinations. There are still substantial Italian and Greek populations—for example Melbourne is home to one of the largest Greek-speaking populations in the world. The new residents have helped to create an increasingly distinct Australian character—generally harmonious, urban-centered, multicultural, and with less social stratification than in Britain—now including over 100 nationalities in Sydney and Melbourne alone.

Most Australians approve this trend, although racist sentiments and occasional violence between ethnic groups have occurred. Australian immigration policy is by no means egalitarian; for several decades it has favored people with education, skills, money, and a potential for adapting to life in Australia (privileging English-language fluency), although one-third of the total (190,000 people in 2012–2013) are admitted in order to reunite families. Acquisition of citizenship for legal immigrants is still much easier than it is in the United States, although Australia's refugee and asylum seeker policy is harsher and has been heavily criticized by human rights activists and the United Nations. Refugees, particularly Tamils from Sri Lanka, Hazaras from Afghanistan and Pakistan, Iraqis and people from other war-torn nations have attempted to travel to Australia by boat from Indonesia, a country where they are unable to gain residency permits and struggle to build a livelihood. From 2001 to 2007 asylum seekers were detained in offshore detention centers on Christmas Island (an Australian island in the Indian Ocean south

continuing despite the global recession. Some industry experts have called it a "once-in-a-lifetime-market." Coupled with increasing demand from Australia's traditional trading partners such as Japan and South Korea, the resource sector is relatively buoyant. Western Australia's economy grew by 4.4 percent in 2010–2011. The mining "boom" of the 2000s started to decline a little by 2012, but demand is still so great that companies have trouble attracting enough workers. One in five workers in Western Australia is employed in the mining or associated industries, and there are spin-offs for other sectors of the economy, notably in construction and communications.

Perth—one of the world's most isolated large cities—is an important mining and oil administration center. Unemployment was just over 4 percent in 2013. Scores of skilled laborers from the eastern states and overseas have relocated to Western Australia to take advantage of jobs and high wages. Up to 2,000 people a week were migrating to the state, mostly to Perth, in the late 2000s. City growth has created traffic congestion, and infrastructure is struggling to handle the influx. Urban water shortages are an ever-present threat in this very dry state; a controversial desalinization plant is in operation and a second one proposed.

This modern day "gold rush" is characterized by many social, environmental, and economic problems. In particular, the mining boom strains the industry's relations with the area's Aboriginal inhabitants. Aboriginal culture is deeply connected to the land and water that mining destroys. Weeli Wolli spring is a place of spiritual importance for the Martu Idja Banyjima near Newman in the Pilbara. The spring, and the seasonal stream that it feeds, is central to local ecology. But a nearby mine not only extracts for its operations large amounts of groundwater, it also may triple the amount of water, now contaminated, it discharges into the Weeli Wolli's outflow. Environmentalists fear the rapid mining expansion will have unforeseen and permanent impacts on already scarce water resources. Rio Tinto, the largest iron ore producer in the Pilbara, has tripled production in the region since 2002 to 290 million tons (metric) a year. This has significantly increased its water usage. Although Rio Tinto is committed to an A$200 million package of remediation efforts over 40 years, concern exists that a future economic downturn might result in the mining giant and other companies walking away from its obligations and leaving the Western Australia government and the Banyjima to cope as best they can.

Although the Pilbara region has boomed, many locals feel that corporate profits are passing them by. The majority of resource sector workers in the region operate on "fly-in, fly-out" contracts and support families in Perth or elsewhere in the country. Short-term workers do not invest locally, although there are growing numbers of Aborigines in the workforce. Many women are involved, notably driving large mining trucks. Local businesses are short-staffed, unable to compete with the salaries offered by the mines. The Pilbara does have a small tourist industry, but mine workers take up hotel beds. The housing sector cannot keep up with demand and prices are very high.

The challenge for the Western Australia government is to continue its strong mining sector, based on Asian demand for raw materials, while managing and minimizing numerous negative environmental and social impacts. This will require close liaison with Aboriginal communities and local governments, and above all diversification of the economy as the current mineral boom wanes.

Sources: Alex Cullen, research assistant. Anthony Halley, "Social Costs of Western Australia's Mining Boom," April 23, 2013, *http://www.mining.com/social-costs-of-western-australias-mining-boom-27467/*; Joe Lopez, "As Mining Boom Fades, Western Australian Government Imposes Austerity Measures," June 7, 2013, *http://www.wsws.org/en/articles/2013/06/07/waps-j07.html/*; Vicky Validakis, "Rio Tinto Looks to Create 2000 Jobs in the Pilbara," February 4, 2013, *http://www.miningaustralia.com.au/news/rio-tinto-invests-$3b-in-pilbara-iron-ore-expansio/*

of Java), Nauru (an independent nation in the Pacific), and Manus Island in Papua New Guinea, as a deterrent to attempting risky sea crossings. A smaller percentage of arrivals came by plane and were also detained. "Offshore processing" of asylum seekers was widely condemned, but began again in 2012 on the same three islands as the numbers of boat people arriving on unsafe vessels rose significantly. Then in 2013, boat arrivals were sent directly to Manus Island with no chance of resettlement in Australia at all. By mid-2013, there were over 8,500 people in immigration detention. The percentage of asylum seekers successful with their claims varies, but can be as low as 30 percent.

Legal immigration peaked at nearly 150,000 a year in the late 1980s, but has since settled at a more modest level of about 120,000 after 2005. More than a third of each year's intake now comes from Asian countries, principally China, Vietnam, and India (Figure 12-9). A quarter of Australia's present population was born abroad, and another quarter is made up of first-generation children of migrants born in Australia. Migration levels are still relatively strong because wage rates and job opportunities still exist, despite the lingering impacts of the global economic recession. Migrants from economically stricken countries like Ireland, Greece, and Spain increased in 2012 and 2013.

Countering this, emigration averages under 90,000 people a year, including many native New Zealanders with whose homeland Australia has mutual residency rights. There are 1 million expatriate Australians of which 200,000 live in the United States. They include Hollywood actors, academics, and business people. The signing of the **Australia–United States Free Trade Agreement (AUSFTA)** allows work permits to be issued between both countries and provides for special visa status. Australians have also flourished in the service sectors of Singapore, Hong Kong, and Dubai when those economies are strong, and further afield in all manner of jobs in Europe and North America. The relative isolation of the country propels many young Australians to "go walkabout" at some time in

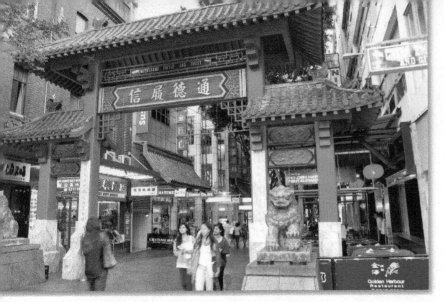

▲ **Figure 12–9 Chinatown in Sydney, Australia.** As the Asian immigrant community increases in Australia, Chinese-dominated commercial districts emerge in the largest cities and establish the entry symbols of gates and lions that provide spiritual protection and cultural identity.

their lives as travelers, students, and settlers overseas. Many "astronaut" households, in which the wage earner generates income from an Asian (often the Chinese) economy and travels there frequently on business, are based in Australia.

Australia's Minorities

Australia's Asian minority totals approximately 10 percent, with two-thirds living in Sydney or Melbourne. It includes many Vietnamese, some of whom arrived as refugees, as well as people from China, Malaysia, Singapore, and the Philippines. The country is more tolerant of Asian immigration than it has been in the past. Money talks: more than 60 percent of the country's exports now go to Asia; only 11 percent are shipped to Europe. Some Australians still argue for reduced immigration, but the tide is shifting toward supporters of the multiculturalism that it has permitted. Nonetheless, there is still debate over what level of population is sustainable, given the continent's environmental challenges. Dreamers in the nineteenth century once speculated on an eventual population of 100 million. Pessimists, seeing the current challenges presented by severe water shortages and uncertain climatic trends, have estimated the optimum population as small as 8 million. Increased water, agricultural, and energy efficiency is certainly required, now and in the future.

Aboriginal people total around 460,000 today or about 2 percent of Australia's population. They are dispersed across the continent, occupying their ancestral land ("country") or settled elsewhere. The "Dreamtime," when the creator-ancestors travelled the land and left their marks, holds particular importance to many of them. Dreamtime stories are vital to culture, and paintings of the Dreamtime by Aboriginal artists are more than representations: they actually embody and are part of these stories. The discovery and promotion of Australian art has brought fame and wealth to prominent artists (Figure 12–10).

The indigenous hunter-gatherer livelihood was largely erased by European settlement, and with it went some of the cultural practices and spiritual attachment to "country." Europeans, dazzled by their new world, sometimes treated Aboriginal people violently and irrationally, and there was mutual incomprehension; several massacres took place from the 1830s onward. Christian missionaries and the

school system brought wholly unfamiliar concepts to the indigenous population, imported from distant lands. The "**Stolen Generation**" refers to at least 100,000 Aboriginal children that were forcibly taken from their mothers to be brought up by whites, a practice that only ended in 1969 (some Native Americans in the United States also suffered this fate). The rationale for this policy differed over time and among different administrations, but it was essentially racial—to submerge the black population into the white majority—and only occasionally was it justified on the grounds of "protecting" these children from alleged poverty or harm. The 2002 movie, *Rabbit-Proof Fence*, provides a graphic illustration.

The process of integration, and reconciliation with the insults of colonial history, has been slow. There are parallels with the United States' civil rights and Native American movements. Aboriginal people were not counted in the census until the late 1960s, and the renowned artist Albert Namatjira and his wife Rubina were the first to obtain Australian citizenship in 1957. Voting rights differed by state, but were in practice inferior to those enjoyed by white Australians.

After more than two centuries, many indigenous Australians, despite the presence of many professionals and public figures among their number who are integrated into modern life, find contemporary Australia is still an unhappy place. Modern, multicultural Australia sometimes sits awkwardly with those anchored to their social relationships and to cultural beliefs stretching back for millennia. Its language, English, may not be their habitual tongue. Local customary laws and the Australian legal system, and forms of social organization, differ markedly. Alcohol did not exist prior to European arrival and the indigenous population, whether in the cities or in rural areas, has low tolerance to its ill effects. Remote Aboriginal settlements rarely have adequate modern services and most housing is in public ownership. Since the demise of hunter-gathering lifestyles, money is required for food and essentials, but

▼ **Figure 12–10 Aboriginal art in Australia.** Contemporary Aboriginal art has become highly desirable and valuable. Here artist William Sandy, a Pitjantjatiarra tribesman from the Northern Territory, is working on a canvas.

there are few jobs for the residents of the remote regions, where Federal welfare payments provide the majority of cash income. In 2007, the Federal government linked these payments to good behavior including alcohol bans and school attendance; these "intervention" policies, applicable in the Northern Territory, have divided indigenous leaders and observers. Life expectancy for Aboriginal people is up to 12 years less than the Australian average, and they are 3 times as likely to be "unemployed" (in Western terms), and 14 times as likely to be in prison.

For many indigenous leaders, resolution of the inferior position of Aboriginal Australians must begin with sovereignty as a response to their historic dispossession by the British. Unlike Native Americans in the United States, Aboriginal people do not have true sovereignty over their ancestral lands, unless they purchase them. This option has disappeared in metropolitan areas. Melbourne, for example, sits on Wurundjeri land and its traditional owners were displaced early in its development. Alienation from "country" has been a sore point, resulting in legal challenges. These cases, pursued through the courts by people whose ancestry ties them to places they have lost, have resulted in successful "Native Title" claims on customary grounds, although "Native Title" is not the same as full "ownership." Aboriginal Land Councils now manage large tracts of remote Australia, and are at least able to negotiate financial settlements with those wishing to use their land, such as mining companies who pay royalties. There is a national "Sorry Day" to Aboriginal people each 26 May, but it was not until February 2008 that, amid a torrent of collective emotion, the Australian government, then led by Kevin Rudd, formally apologized to the "Stolen Generation" of Aboriginal children and their families for past injustices. In 2013, an "Act of Recognition" seeking constitutional recognition of the "unique and special place" of Aboriginal and Torres Strait islander peoples was presented to Parliament by then prime minister Julia Gillard.

An Urbanized Society

For a country with so much land and so few people, Australia's high degree of urbanization is unusual. Some 89 percent of Australia's population lives in urban areas. Around 40 percent of the people live in the two sprawling metropolitan areas of Sydney and Melbourne, with populations of approximately 4.7 and 4.2 million respectively (Figure 12–11). One reason for this is that Australia's agricultural production is extensive rather than intensive, and employs few people. But there was an explicit attempt at the time of Federation to build an urban, rather than rural, society. In addition, especially since World War II, Australia has encouraged industrialization to secure greater economic stability through a more diversified and self-sufficient economy.

All five of Australia's largest cities—Sydney, Melbourne, Brisbane (2.2 million), Perth-Freemantle (1.9 million), and Adelaide (1.3 million)—are seaports, and each is the capital of one of the five mainland states of the Commonwealth. These cities, particularly Melbourne, regularly appear in top positions in global "most livable cities" rankings, given their relatively low crime levels, vibrant art and culture, relatively attractive job prospects, and high-quality housing stock. As in the United States, the trend is toward suburbanization, with urban sprawl extending the boundaries of metropolitan areas. Drought, however, has forced a significant drive

▲ **Figure 12–11 Perth, Western Australia.** A city of 1.9 million, Perth is considered one of the world's most livable cities. This image shows the Swan Bell Tower, completed in 2000. The copper and glass campanile contains 18 bells, 12 from St. Martin's-in-the-Field, a gift of the City of London on the occasion of the country's bicentenary celebration in 1988, and 6 cast from metal mined in Western Australia..

to conserve urban water and to reduce consumption. Melbourne, Sydney, Adelaide, and Perth have had very significant water shortages, with restrictions on unnecessary water use and a real risk of occasional bushfires reaching the outer suburbs. Planning controls have restricted, but not halted, suburban expansion. Rising demand for housing has also led to recent inner-city gentrification and revitalization, notably in Melbourne (Figure 12–12) and Sydney.

Cities were the nodes around which surface and marine transport infrastructure and systems were built. Before Federation in 1901, each state had built its own rail system linking the hinterland to the chief port and international markets; the first rail tracks were built to Port Melbourne in 1854. But different railroad gauges (track widths) were used and there are still three different gauges in operation across the country. After many years of planning, trains could travel all the way from Adelaide to Darwin in 2001, reducing the isolation of the Northern Territory capital and the towns along the rails. Australia's most famous passenger service is the Indian Pacific, operating twice a week between Sydney and Perth and taking 65 hours (Figure 12–13). Container ships are still vital for imports, exports, and internal trade. Air travel, which has significantly fallen in cost since the 1980s, links the major cities, numerous smaller towns, and international destinations. Australia thus reflects a strong degree of urban primacy. This urban dominance grows as rural depopulation continues due to an uncertain agricultural future and the long-term decline of the agricultural sector.

▲ **Figure 12–12 Gentrification in Northcote, Melbourne.** Neighborhoods near the city center, once in decline due to the shift of middleclass residents to the suburbs, have regained popularity based on proximity to urban amenities, attractive older housing that can be renovated, and an innovative population.

The Australian Economy

Australia's high living standards can be attributed to its small population and its reasonably well-developed and diversified export economy, which depends on the extraction of minerals, agriculture, and education and services. At 11.6 million people, the total Australian workforce is smaller than that of the New York

▼ **Figure 12–13 Australia's east–west transcontinental railroad.** The Great Southern Railway runs a transcontinental train, the India Pacific, over 2,700 miles (4,300 kilometers) from Sydney to Perth. Shown here near a siding stop at Rawlinna in Western Australia, the train is an important connecting link to the larger world for small, isolated communities that it passes once each week in each direction.

Metropolitan Area. In 2012, GNI PPP was $42,400, compared to $49,800 in the United States (which has a lower cost of living and lower average salaries), and $36,700 in the United Kingdom, placing Australia in the world's top 20 nations based on this indicator. From 2002 until 2008, the national accounts were in surplus. The Australian response to the global financial crisis that followed was to create jobs by investing in public works, using the surplus to pay for upgrades to public school buildings around the country. Due to the financial crisis, the country is now running a debt of over A$10 billion and growth in its important mining revenues has cooled since 2011. Nonetheless, Australia is one of the best-performing economies in the world and economic growth, averaging 3.4 percent a year, has been almost continuous for the last 20 years. The unemployment rate, at 5.2 percent, is low (highly favorable compared to that of the United States, which was 7.6% in mid-2013).

The services sector is the largest part of the Australian economy, accounting for around three-quarters of GDP and four out of five jobs. This shift has taken place in parallel with urbanization and growing economic maturity. By 2012, services provided 69 percent of GDP, industry (including mining) contributed 26.6 percent, and agriculture generated just 4 percent. The service sector includes banks, financial companies, and insurance, all on a smaller scale than found in North America, and a significant higher education sector that is also attractive, although expensive at current exchange rates, to international students. The Australian National University and the Universities of Melbourne and Sydney are consistently ranked in the world's top 50 institutions.

Agriculture and Trade

For such a large landmass, Australia has a remarkably small amount of land that can be farmed reliably and without irrigation. It is an old continent. A lack of tectonic activity, volcanoes, and substantial mountain ranges means its landscape is not rapidly refreshed by uplift and downcutting, so the soils are ancient and characteristically red, often rich in iron and with other fertility constraints. European settlement, concentrated in the more moderate conditions of the east coast, the southeast, and the southwest, initially had little need of the arid interior.

Settlers did eventually move inland, carving their farms and ranches (**stations**) out of Aboriginal ancestral territory. In total about 40 percent of Australia—to the north, east, and west of the arid interior—now has ranching as its major economic use. Sheep, including the Merino with its fine wool, drew early settlers to Australia, and meat and wool exports were early sources of wealth. Livestock were first concentrated in areas with more reliable water, but beef cattle in the drier parts rely on boreholes and ephemeral surface water when available. The extensive dryland ranches are now, as in the United States, operated with modern technology including helicopters and advanced communication systems.

About 25 percent of the land area receives sufficient rainfall to support agriculture, but rough terrain, salinity problems, and poor soils reduce this further. About 5 percent of Australia's agricultural land is used for food crops, with a further 5 percent devoted to profitable dairying herds grazing on improved pastures and grasses. Wheat and barley are dominant crops, alongside sugarcane, fruits, and animal products, primarily from cattle and sheep (Figure 12–14), and these generate a significant export trade. Sheep rearing became a mainstay of the economy in the nineteenth century, when it provided wool for Britain's textile industry. By 1850, Australia was the world's largest exporter of wool, and it still produces a major share of the world's production, despite now making up a small percentage of total exports. Sheep and cattle stations are usually quite large, often encompassing thousands of acres. Anna Creek cattle station in South Australia is 12,000 square miles (31,000 square kilometers), an area about the size of Maryland and Delaware combined. Development of refrigerated shipping after 1880 enabled Australia to supply European markets with both meat and dairy products. In recent years, much of the increased demand for wool, beef, mutton, lamb, and dairy products has come from the countries of East and Southeast Asia, as well as the Middle East, where Australian meat and dairy products are common in supermarkets and food stores.

Australia's wheat production has also benefited from modern technology. The introduction of mechanization in the twentieth century permitted wheat to be extensively cultivated and about 50 percent of Australia's total cropland, 27 million acres (11 million hectares), is now devoted to wheat (Figure 12–15). Like Canada and the United States, Australia has become one of the great breadbaskets of the world. Although wheat generates 17 percent of the value of all Australia's agricultural exports, and is Australia's largest and most important agricultural export commodity, its share of Australia's total exports by value has declined as mineral exports have risen. Australia's Outback yields millions of tons of grain each year (22 million tons in 2012) at an average of more than 1.8 tons/hectare. Yields can, and do, vary significantly from year to year depending on the severity of drought conditions and the availability of water at critical points in the wheat crop's growth cycle.

Australia produces many other crops and is self-sufficient in foodstuffs. Sugarcane, one of the more important crops, is grown along the northeastern coastal fringe. Annually, Australia produces 3–4 million tons of sugar, most of which is exported to Japan and other Asian markets. It is the world's fourth largest sugar exporter, behind Brazil, the European Union, and Thailand. Other important crops include a wide variety of temperate and tropical fruits for both domestic consumption and export markets.

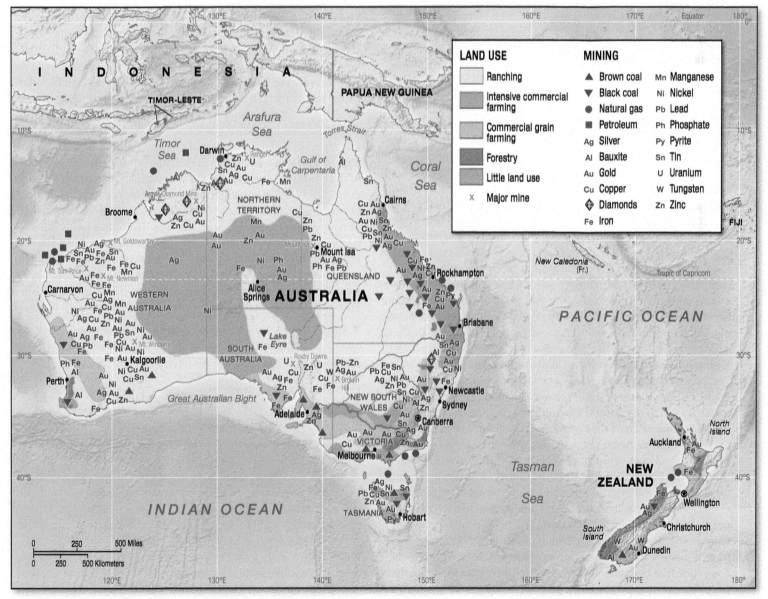

▲ Figure 12–14 **Rural land use and mineral resources of Australia and New Zealand.** Once only occupied by their indigenous peoples, Australia and New Zealand now have extensive forms of agriculture, with ranching dominating much of the area of each country. Other extensive activities include mechanized commercial grain farming in Australia and forestry in New Zealand.

▲ Figure 12–15 Commercial grain farming in southwest Australia. With 50 percent of its cropland devoted to wheat, Australia has become one of the world's largest breadbaskets.

Despite periodic droughts, the agricultural sector fared well until the 1980s owing to increased use of chemical fertilizers, improved water supplies, and success in reducing the population of invasive rabbits, an animal that was brought by early settlers, escaped captivity, and bred extensively in the wild, plaguing cropland across much of the farmed area. But since then the outlook for farming, and for farmers, has been less positive. The potential for further increasing crop yields is limited, and drought conditions in the 2000s led to farm failures and depopulation of the smaller country towns. Drought remains a constant threat, although it has diminished again since 2008 and decent yields returned. In some regions, including southwestern Australia, there has always been a major issue with the buildup of natural salts in the topsoil and even in drinking water.

Crop production is highest in the states of New South Wales, Victoria, and Queensland, and the latter produces the majority of tropical fruits. Australian vineyards (Figure 12–16) have been one notable success, and the country now exports wine worldwide, comparing favorably with longer-established European producers and winning numerous awards. Given the risk of drought, irrigation water derived from rivers and dams is used by farmers where possible, and the continent's major river catchment, the Murray-Darling, supplies irrigation water for five states along its southwesterly course. With echoes of the plight of the Colorado River in the United States and Mexico, water scarcity leads to numerous disputes. Debates over how to allocate water equitably to rural farmers, to urban dwellers, and to the natural environment have been acrimonious. In 2012, after many years of negotiation, the Murray-Darling Basin Plan was signed into law. This gives the Federal (Commonwealth) government control of the whole Murray-Darling system through the Murray-Darling Basin Authority in an effort to monitor allocations and to reduce inefficiencies. The health of the river is extremely poor, especially in riverine wetlands depleted due to upstream dams, and there are low flows and high salinity in the lower reaches of the Murray from where it enters South Australia. With its new powers, the government can release more "environmental" flows down the river to nourish the natural environment rather than only support irrigation farmers. Starting in 2012, 2,750 gigaliters of water (a gigaliter is a billion liters) will be purchased from irrigators' water entitlements over a seven-year period to restore the health of the river system. Communities dependent on farm irrigation are strongly opposed.

The tariffs levied by the European Union have limited the amounts of Australian meat, butter, grain, fruit, and sugar that can be sent to Britain and other European markets since 1972. In addition, the European Union and the United States send their own agricultural surpluses to other markets in which Australia is competing. Nonetheless, illustrating the uncertainties that always prevail in the farming sector, global food price rises of 2007–2008 and 2011 increased demand for Australian meat in Asia, which provides new incentives to maximize yields despite the environmental constraints. The wet tropics of Australia's "Top End," its far north, may in the future see far greater production, given worldwide interest in food crops and biofuels.

Stop & Think

➤ Explain why ranching and meat production play such a major role in the Australian agricultural economy.

Mining, Manufacturing, and Globalization

The growth potential is much greater for Australia's mining than it is for its agriculture. Australia is truly a powerhouse of the world's mining sector, and BHP Billiton, founded in Broken Hill in New South Wales, is the world's largest mining company. Australia has enormous reserves of most of the key minerals needed in today's global economy (see Figure 12–14). It leads in production of bauxite and alumina, diamonds, lead, and uranium (with 40% of the world's uranium deposits); is second in production of gold, nickel, and zinc; is third in iron ore and manganese; and is fourth in coal, copper, and silver. Japan was the leading buyer in the latter decades of the twentieth century and still takes most of Australia's coal exports. But China is the largest mineral export market overall, with a skyrocketing demand for raw materials. Although production of oil and natural gas has improved and new Timor Sea fields (offshore from Darwin) have recently come into production, Australia does import some petroleum. In 2012, the government passed the Minerals Resource Rent Tax levied on 30 percent of the "super profits" from the large companies mining iron ore and coal in Australia (see *Focus on Energy*: Carbon Pricing and Coal in Australia). Although

▼ Figure 12–16 Viticulture in Australia. Coldstream Hills Winery, located just outside the Melbourne growth boundary in the Yarra Valley, produces high-quality wines with a large market at its doorstep. The mediterranean climate of southeastern Australia also encourages the cultivation of vines, and Australia has become an increasingly important producer and exporter of wine.

FOCUS ON ENERGY

Carbon Pricing and Coal in Australia

Among developed countries, Australia is the highest emitter of carbon dioxide on a per capita basis, largely due to carbon-based energy production (77% of total emissions) and transport (15%). It has, however, operated a carbon pricing mechanism (not a tax) since 2012, overseen by the Clean Energy Regulator. This has been the source of extensive political clashes and debate, pitting free marketeers against regulationists and the political left and greens against the right. The aim is actually relatively modest: to cut Australia's greenhouse gas emissions by 5 percent below 2000 levels by 2020; future targets are more ambitious. Carbon was initially priced at A$23 per (metric) ton but this will rise over time. Businesses and public bodies emitting over 25,000 (metric) tons of CO_2 equivalent annually must purchase emissions permits. Power stations, mines, and heavy industry must record their emissions, but agriculture and forestry are excluded at this time. Farmers can join the Carbon Farming Initiative, sequestering carbon by tree planting, and then sell credits to emitters.

Carbon pricing is relatively new, but has already led to CO_2 reduction, particularly through more methane capture from landfills. Carbon pricing will not survive the 2013 national election in its present form; there is a concern that key industries, like energy-intensive aluminum plants, may close or relocate overseas, and price hikes on flights and house construction may sway voters against it. The Australian Labor Party preferred transitioning to an Emissions Trading Scheme by 2015; the Liberal Party, whose coalition grouping won the 2013 election, plans to eliminate carbon taxation and reduce restrictions on mining operations and emissions. Coal produces over 40 percent of Australia's greenhouse gas emissions, but the coal industry receives significant financial compensation under the current pricing scheme, because it is deemed to be a significant sector that was established before concerns over global warming and because it is the major fuel used for Australian energy generation (over 80%). High-quality bituminous (black) coal is mined in Queensland and New South Wales with a small amount

from Western Australia, and lower-quality lignite (brown coal) in Victoria and South Australia. In South Australia, sub-bituminous coal, intermediate between black and brown coal, is the primary product. But a new technology for removing moisture from brown coal and consolidating the material into higher-energy briquettes offers the prospect of developing the region's very large, presently unexploited, lignite deposits.

Australia's largest mines are run by Xstrata, BHP Billiton Mitsubishi Alliance (BMA), and Rio Tinto. Coal is transported from underground mines and vast open pits to power stations and shipping terminals. Production is around 290 million tons per year, dwarfed only by iron ore, although this varies with world economic demand and weather conditions; recent flooding in Queensland halted production temporarily. Most of the exports go to Japan, China, South Korea, and India.

Australia's recent contributions to carbon emissions have resulted in widespread protests, led by environmental groups and Beyond Zero Emissions (BZE) who argue that inside a decade Australia could produce enough domestic energy without using hydrocarbons. Their plan involves renewable energy: photovoltaics, concentrated solar energy systems (CSP), solar thermal plants, wind energy, and biomass fuel, combined with greater energy efficiency.

As with all mining projects, there are local impacts of heavy coal exploitation. The Bowen Basin of Central Queensland is an example (Figure 12-3-1). It covers an area of 29,000 square miles (75,000 square kilometers) with significant black coal deposits layered in sediments. Most of the 48 mines are open-pit with almost 50 more under consideration, including the massive Carmichael Coal and Rail Project that would export coal to India. The region's small towns generally report positive

▲ FIGURE 12-3-1 **Coal mining in the Bowen Basin, Central Queensland.** Queensland's Bowen Basin is the major source of coking coal in Australia, and one of the largest deposits of bituminous coal in the world. The coal is shipped to the coast by rail and from there by ship to other parts of Australia and elsewhere in the world.

impacts on business activity, but suffer from coal dust following blasting and occasional water quality and supply problems. There are many holding ponds containing toxic excess water, particularly with the serious flooding in recent years. Temporary workers contribute little to community-building, and housing costs have risen.

Of great concern to environmental groups and some farmers is coal seam gas extraction—hydraulic stimulation and hydraulic fracturing, or **fracking**, a technology that releases gas from coal seams but involves chemicals and can also lead to groundwater pollution and depletion, and dangerous emissions (see *Focus on Energy* in Chapter 2). In the Bowen Basin 3,200 wells have already been drilled and, supported by the Queensland government, the port of Gladstone is being significantly expanded by constructing three large plants to process natural gas into a liquified form (LNG) before shipment to Asian markets. This location is part of the World Heritage Area of the Great Barrier Reef, and UNESCO has intervened to force more environmental assessments before the project goes ahead.

Sources: Colin Latimer, "Bowen Basin to See New Mining Projects," May 9, 2011, *http://www.miningaustralia .com.au/news/bowen-basin-to-see-new-mining-projects*; Australian Government, "Clean Energy Regulator," *http:// www.cleanenergyregulator.gov.au/Pages/default.aspx*; Beyond Zero Emissions, *http://bze.org.au/*

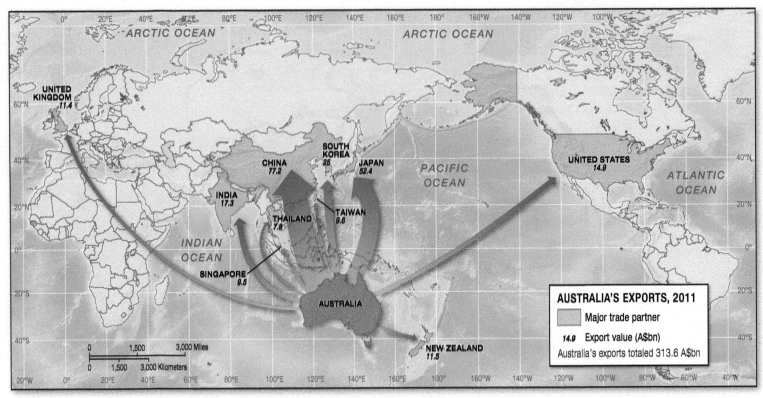

▲ **Figure 12–17a Australia's exports and major trade partners, 2011.** Australia still maintains significant trade activity with Great Britain (U.K.), but its primary trade partners are now Pacific Rim countries, particularly China and Japan.

Source: http://www.dfat.gov.au/publications/trade/trade-at-a-glance-2012.html. Based on ABS trade data.

generally welcomed since the money is destined to support tax cuts for small businesses and infrastructure, faced with an economic downturn and political instability the large mining companies are actively lobbying to overturn this legislation or to reduce its financial burden. The Liberal Party is very receptive to these industry goals and plans to cut the mining tax and abandon carbon pricing.

Manufacturing, by contrast, is not a major strength of Australia's economy. Industrial capacity is primarily geared to **import substitution industries** that produce a small range of consumer goods for the domestic market, as well as to the partial processing of mineral and agricultural products, and some heavy industry, such as iron and steel. Australia produces its own cars, traditionally Holdens (GM) and Fords. But sales of imported vehicles are over three times as high as sales of domestically manufactured vehicles. The Australian automobile industry also has been slow to rise to the challenge of lowering carbon emissions through fuel efficiency and alternative fuels. Sale of a Toyota hybrid vehicle assembled in Melbourne began in 2010, and there are other imported hybrids including the GM Volt and Toyota Prius. Across the manufacturing sector, some tariff barriers exist to protect domestic producers from foreign imports. The 2005 Australia–United States Free Trade Agreement (AUSFTA) has eliminated some tariffs between these countries, but the agreement has not benefited both nations equally. Australian exports to the United States are exceeded by U.S. imports to Australia.

What remains of the manufacturing industry is a source of pride, even though "Australian Made" is glimpsed far less often than "Made in China." Manufacturing is concentrated in the state capitals, where industry has access to available markets, fuel, business

and government contracts, and both overseas and internal transportation systems. The leading industrial state is New South Wales, centered in Sydney; Victoria is second, focused on Melbourne. Other industrial activity occurs in a few large provincial centers on or near the coast, including Wollongong and Newcastle. In 2008, Melbourne's Port Phillip Bay was dredged to provide a deeper channel for larger container ships to access the port facilities.

The overall pattern of imports and exports are shown in Figures 12-17a and 12-17b. Australia's shift toward Asia and the Pacific Rim countries, and the declining role of the United Kingdom, is revealed in the trade data. China (24.6% of total exports) is now ahead of Japan (16.7% of total exports). There are significant exports to South Korea (8%), India (5.5%), the United States (4.8%), and New Zealand (3.7%). In exchange, Australia's markets have become flooded with cheap Chinese manufactured products, of varying quality. China now leads imports (14.9% of Australia's total), compared to 14.3 percent from the United States and 6.8 percent from Japan.

Stop & Think

 Outline the main reasons why manufacturing is much less profitable in Australia than other economic activities.

Tourism

In the 1980s, tourism began to assume new importance as a fourth leg in the Australian economy after services, mining, and agriculture, generating 5.2 percent of GDP and providing about 8 percent of total employment. Although hampered by the long distances that many international visitors must travel, Australia now attracts

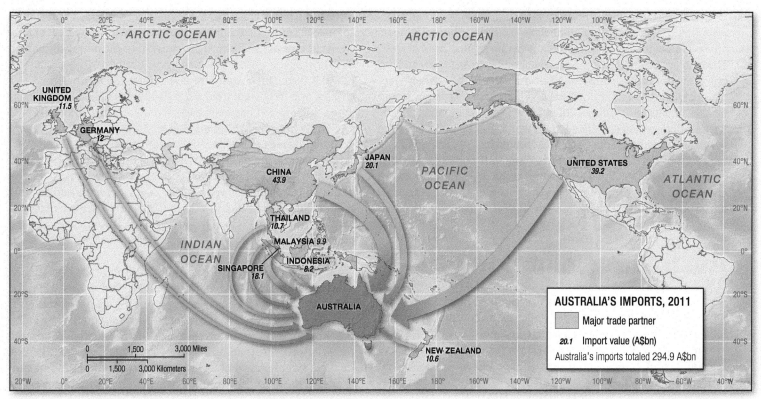

▲ **Figure 12–17b** **Australia's imports and major trade partners, 2011.** Although nearly a quarter of imports by value come from the E.U., Australia now imports primarily from its major export partners in the Pacific Rim countries.

Source: Australian Government, Bureau of Statistics, http://www.dfat.gov.au/publications/trade/trade-at-a-glance-2012.html. Based on ABS trade data.

almost 6 million foreign visitors a year, with the greatest increases coming from Asia (now 36% of annual visitors). New Zealand is the largest individual source of international visitors (20%), followed by the United Kingdom, the United States, and China (10% each). Other important visitor sources include Japan, South Korea, Singapore, Hong Kong, and India. The traditional draws are Sydney, the Great Barrier Reef, and various destinations in the Outback including Uluru (Ayers Rock). The **Gold Coast** in Queensland has emerged as a tourist playground with echoes of southern Florida, and it receives substantial numbers of domestic as well as international tourists. Cairns in northern Queensland has experienced strong growth, partly as a result of adding an international airport, as the jumping-off point for visitors seeking tropical recreation on the Reef. After decades of destructive development, some environmental battles are now being won in Queensland in order to preserve the biodiversity that tourists come to see. Darwin, capital of the Northern Territory is also flourishing and its airport, now a stopping point for budget flights to Asia, is open 24 hours a day. The prospects for tourism have dimmed slightly with the global financial crisis and a fall since 2008 in the value of North American and European currencies relative to the Australian dollar, as well as high global fuel prices. This has increased costs for many visitors.

Australia's Future

Australia is at a critical juncture in its national development. It has enjoyed remarkable economic growth for most of the twenty-first century, maintaining a high standard of living and social benefits for many residents. At the same time, a more long-standing desire to maintain a relatively egalitarian society, through adequate wages, protective tariffs, high tax rates, unemployment benefits, and free health care (all very different from the United States) has been eroding. One reason has been the domination of Australian politics by the Liberal Party (U.S. Republicans equivalent), in power for almost three times as long as the center-left Labor Party (equivalent to the U.S. Democrats) since Federation in 1901. The Labor Party was in power from 2007-2013, but was defeated by a Liberal coalition led by Tony Abbott, which has moved quickly to introduce pro-business policies. Australia's favored trading status with China has brought new, perhaps temporary, economic optimism. Asia clearly is both the challenge and the future of Australia.

From a conventional "economic development" perspective, Australia hosts profitable economic activities and several of its key sectors are in good shape. It must still contend with the problems of a small population relative to land area and resources, versus its Asian neighbors and larger trading partners. Travel by jet and ship are carbon-intensive and likely to get more rather than less expensive. Environmental challenges will be a significant problem for Australia in a warmer world, although it can easily produce power from renewable energy. It was in 2007, with the new Labor government of Kevin Rudd, that Australia finally signed the Kyoto Protocol and in 2012 the country introduced its carbon pricing mechanism.

If we take a less conventional economic perspective, Australia's relative isolation, stunning natural beauty, high wages, low population, postindustrial economy, and a certain cultural individualism and intolerance of social hierarchies are positive features, making it one of the most desirable places in the world to live.

An important, albeit symbolic, step in the minds of many citizens would be for Australia to declare itself a republic, ending its recognition of the British Crown as the nation's titular head of state. The Queen's Birthday is still a national holiday in Australia, and her image appears on its coins. Republicans point to the economic realities of Australia's trade, tourism, and investment, now dominated by Asia, and its multicultural cities. They argue further that becoming a republic would also signal to the world that Australia has come of age as a member of the G20 and accepted its new, emerging role in the Asian-Pacific realm. A referendum on becoming a republic was held in late 1999 but was defeated at the polls. Proponents are expected to work hard to develop a consensus on this issue and call for another vote, although no progress has been made so far.

Australia's Neighbors

Australia dominates the region in size, population, and economic power. But its neighbors—from the edge of the enormous sweep of the Pacific Ocean to the tiniest isolated island surrounded by vast expanses of water—make up in diversity for their relative lack of scale and clout. New Zealand shares much of Australia's colonial experience, but little of its contemporary economic importance. Taken together with the other countries and territories in the region, Australia's neighbors offer markedly different perspectives on development paths into the twenty-first century.

New Zealand: A Pastoral Economy

More than 1,000 miles (1,600 kilometers) southeast of Australia, New Zealand consists of two main islands—North Island, with a smaller area but three-fourths of the population, and the South Island—as well as a number of much smaller islands, including the Chathams 500 miles (800 kilometers) to the east (see chapter opener map). The country is located entirely in the temperate zone, from about 34°S to 47°S latitude.

Settlement and Early Development
The **Maori** people were New Zealand's first inhabitants. New Zealand was one of the last countries in the Pacific to be settled, by eastern Polynesians who came by ocean-going canoes (waka) in separate voyages between A.D. 1250 and 1300. Their language and spiritual beliefs then evolved further in New Zealand, in isolation from other groups. They altered the native plants and wildlife with widespread hunting and introduction of plant species. From the 1500s, with easy food sources like the moa bird exhausted, warfare ensued between tribal groups and settlements were fortified.

Europeans arrived in some numbers from the early 1800s when most of the country was under the jurisdiction of New South Wales in Australia. They soon introduced their own familiar plants, animals, and economic structures. Because the climate is ideal for growing grasses and raising livestock, New Zealand, like Australia, specialized in livestock rearing from the earliest days of British

settlement, but at great cost to indigenous flora and fauna. This was "pioneer" country: 30 million acres (12 million hectares) of forest were cut and burned from 1840 onward to make way for pasture, but little of the timber was used productively. Of 580,000 acres (200,000 hectares) of *kauri* forest (a conifer, the North Island's most striking and important tree), only 12,000 acres (5,000 hectares) remain. Soil erosion has been extensive. The land wars of the 1860s and 1870s between whites and Maori, in violation of the Treaty of Waitangi signed in 1840, saw the steady expansion of the European settler **pastoral economy** over the rural landscape, producing livestock and livestock products (Figure 12–18). New Zealand has one of the world's highest proportions of livestock (cattle and sheep) to human population—a ratio of 14 to 1. Even the 3 percent of the land area that is cropped is devoted in large part to animal feeds. Horticultural crops and more sustainable timber harvesting are also of some importance.

Refrigerated shipping began in the 1890s, permitting massive food exports to Britain. Pastoral industries still dominate exports and, because of the country's small population, make New Zealand a world leader in per capita trade. It is among the world's top two or three exporters of mutton, lamb, butter, cheese, preserved milk, wool, and beef. In exchange, New Zealand imports most of its manufactured goods and some other foodstuffs. It has a substantial viticulture; in Marlborough, the northeastern corner of South Island, New Zealand produces globally acclaimed Sauvignon Blanc wines.

The Need for Industry and Diversification
With such heavy dependence on trade and a relatively narrow economic base, New Zealand is far more vulnerable to the vagaries of world economic conditions than Australia and it is less affluent according to conventional economic measures. New Zealand's per capita income is only about three-fourths that of Australia's. Up until the 1970s its primary production gave it a high standard of living. But Britain's membership in the European Community (now the European Union)

▼ Figure 12–18 **New Zealand sheep in a pasture paddock.** Australia and New Zealand have both been major exporters of meat, dairy products, and wool. New Zealand currently retains the more pastoral economy.

cost New Zealand its privileged importer status, increased competition with European suppliers of agricultural produce, reduced sales to Britain by 75 percent, and set in motion an economic downturn.

Attempts at diversification, primarily through industrialization, have met with limited success. Although New Zealand has coal, gold, natural gas, some iron ore, and a few other minerals, production is dwarfed by Australia. The local market is small and dispersed, which restrains large-scale production and efficient marketing. The cost of skilled labor is high, and competition from overseas producers, such as Japan and the United States, can be severe. Most of the current manufacturing industries in New Zealand are high-cost producers that survive because they are protected by tariffs. Overall, manufacturing contributes 24 percent of the national income in New Zealand and employs about the same proportion of the labor force.

New Zealand's economy has benefited from a free-trade agreement known as "Closer Economic Relations," which was signed with Australia in 1983. That agreement opened Australia's larger domestic market to New Zealand's products and gave a much-needed stimulus to New Zealand's industry, making Australia the country's largest trading partner in both exports and imports.

After a major recession in the 1970s and 1980s, which saw living standards fall significantly, a major debate began over New Zealand's economy. This resulted in the introduction of free-market policies in 1984, sweeping away economic protectionism. The new economic regime has endured changes of government, but it has only been successful in certain sectors. The global financial crisis led to a fall in GNP per capita of 1.5 percent between 2007 and 2012. Several new trade treaties are now in place. In 2008, the government signed the New Zealand China Free Trade Agreement, the first China has made with a Western country, mainly resulting in reduction in tariffs on imports to both countries, and some temporary visas for Chinese working in New Zealand.

New Zealand, like Australia, has experienced a significant "brain drain," losing many **Kiwis** (a nickname for inhabitants of New Zealand), especially skilled migrants, overseas for varying periods. There is a solid tourist industry that capitalizes on skiing, outdoor activities, and the attractive mountain scenery

▼ **Figure 12-19 Lake Waikaremoana.** This famous recreation area is located in Te Urewera National Park in the North Island of New Zealand, 50 miles (80 kilometers) west of Gilborne. The lake is partly encompassed by trails and overnight huts, and the premier "track," which takes three days to complete, is one of New Zealand's best-known "Great Walks."

(Figure 12-19). The mountains of South Island are shown magnificently in the movie trilogy, *Lord of the Rings* (2001–2003), and *The Hobbit* (2012).

New Zealand's People

In spite of the predominantly agricultural economy, most of New Zealand's 4.4 million Kiwis live in cities, similar to the situation in Australia. However, the cities of New Zealand are generally much smaller; the largest is Auckland on North Island, with 1.4 million people. Other major cities are Wellington, the capital, and Hamilton, both also on North Island, and Christchurch (see *Geography in Action*: The Christchurch Earthquakes, 2010–2011) and Dunedin on South Island.

Approximately 68 percent of New Zealanders say they have European origins—mostly British, but including people from the Netherlands, Yugoslavia, Germany, and other countries who arrived in significant numbers once immigration restrictions were eased after World War II. The indigenous Maori population is the largest minority, at about 16 percent. The remaining minority population is made up of other Pacific Island peoples (mostly from Niue, Samoa, Tokelau, and Tonga), who migrated to New Zealand after 1960, and 350,000 Asian residents. The Maori went through a long period of population decline under European settlement after the Treaty of Waitangi until about the 1970s, but are expected to account for 20 percent of the population by 2050. Economically and socially the Maori are disproportionately present on the lower rungs of the socioeconomic ladder (see the movie *Once Were Warriors*, 1994). More than a quarter of working age Maori are on welfare, and their per capita income is well below the national average. In recent years, Maori have been exercising their political muscle in an effort to stand up for what they see as their land rights and other privileges. Te Reo Maori is being revived as an indigenous language. Because of their larger share of total population and their special status under the Treaty of Waitangi, Maori have greater political power than Aboriginal groups in Australia. Nonetheless, Maori social and employment problems remain a key national issue.

Historically, New Zealand has been distinguished by a progressive politics—Maori held seats in parliament from 1867, and women had a right to vote. Since the late twentieth century New Zealand has taken a principled stance, supported by its citizens, against nuclear power and the dangerous nuclear testing that has occurred elsewhere in the Pacific, notably by France and the United States. In 1985, it refused nuclear-powered or nuclear-armed ships access to its ports, and became a nuclear-free zone in 1987. This caused the United States to suspend its alliance with New Zealand under the ANZUS military treaty between the two countries and with Australia, since the United States refused to reveal if nuclear materials were present on its vessels. This diplomatic stalemate continued until 2010 when Hilary Clinton signed the Wellington Declaration, and in 2012 American warships could again enter New Zealand waters. In 1985 the Greenpeace ship Rainbow Warrior, campaigning against French nuclear weapons tests in the Pacific, was sunk in Auckland harbor by the French intelligence service. One person was killed and the bombers were arrested and imprisoned, creating a major diplomatic incident and the resignation of the French Minister of Defense.

GEOGRAPHY IN ACTION

The Christchurch Earthquakes, 2010–2011

A massive earthquake, 7.1 on the Richter scale, hit the city of Christchurch on September 4, 2010, initiating from a fault on New Zealand's South Island where the Canterbury Plains meet the Southern Alps. The region is subject to tectonic activity since it is located close to a system of faults in the earth's crust where the Australian Plate meets the Pacific Plate passing under the island. New Zealanders are aware of earthquake risks, and modern buildings are reinforced against earthquake shocks. There was damage to property, mainly older buildings, and infrastructure, although there were no fatalities from the quake itself. The precise location of the earthquake is uncertain.

On February 22, 2011 a second quake occurred, with its epicenter much closer to the city and only 3 miles (5 kilometers) underground. This second quake was more serious, of the type likely to occur only once in a thousand years, and involved horizontal movement along a fault line right under the built-up area. Prime Minister John Key stated that it "may well be New Zealand's darkest day." Despite being lower on the Richter scale (6.3), it is thought to be one of the most serious earthquakes affecting a major urban area. Millions of tons of soil and silt under the eastern suburbs of the city were liquefied, and 185 people were killed. Further movements and aftershocks occurred in subsequent months.

Many buildings in Christchurch were already compliant with earthquake codes. This helped to avoid greater loss of life and property. Nonetheless, sewage, water supply, and power were disrupted and the two universities were closed for several weeks. The city's ChristChurch Cathedral was severely damaged and much of the central business district was destroyed (Figure 12-4-1). Some 10,000 houses collapsed and about 100,000 more were damaged. The main hospital continued to run with emergency generators. International assistance was forthcoming, including the U.S. Geological Service who were able to provide high-resolution satellite

▲ FIGURE 12-4-1 **Christchurch, earthquake, February 2011.** Damage from the earthquake and aftershocks had a devastating impact on the infrastructure of Christchurch. ChristChurch Cathedral was not immune and suffered severely; massive and expensive repairs will be required to make both city and church operational again.

imagery, and Australian police and search and rescue experts.

The earthquake swarm, New Zealand's biggest natural disasters in the country's recorded history, will cost upward of 30 billion dollars to rebuild entire neighborhoods and the downtown to higher standards. Rebuilding Christchurch is under way. The renewed central business district will include more open space, and a height limit of seven stories on all buildings. New Zealand's tsunami and earthquake readiness is high, but there has been some reluctance to rebuild the city fully given the catastrophe. There have been disputes about the rebuilding program, and outmigration has occurred.

The quake tells us that preparedness and disaster planning saves lives. But disaster planning is expensive as well as effective. Power, sewerage, and fresh water were supplied and systems restored in a matter of days. There are similarities with the San

Francisco earthquake of 1989, which also occurred in an earthquake-ready city. The 1989 earthquake had the same magnitude and also led to extensive damage, but killed 63 people. Compare this with the 2010 earthquake in Haiti, again similar in magnitude, but with over 220,000 deaths. Although earthquakes are natural events, the preparedness level of local populations are social and economic in nature and determined by levels of readiness and investment. Some countries have fewer resources than others. Impacts, and decisions regarding the future of affected areas, depend on organizational capacity and political will.

Sources: Eileen McSaveney, "Historic Earthquakes: The 2011 Christchurch Earthquake," *Te Ara: The Encyclopedia of New Zealand,* June 10, 2013, p. 13, *http://www.teara.govt.nz/en/historic-earthquakes/page-13*; Dominque Schwartz, "Christchurch Rising from the Rubble after Deadly Earthquake," *ABC News,* June 28, 2013, *http://www.3news.co.nz/Chch-earthquake-video-goes-viral/tabid/423/articleID/303789/Default.aspx*

New Zealand faces difficult choices if it hopes to maintain long-term prosperity in the twenty-first century. It does have the advantage of relative self-sufficiency in foodstuffs, and a temperate climate less affected by potentially adverse climate change than most other parts of the world, despite two centuries of extensive local landscape modification.

Stop & Think

➤ How does New Zealand's remote location impact its ability to function economically?

The Pacific Islands

The vast realm of the Pacific Islands, or **Oceania** as it is frequently called, evokes images of swaying palm trees, beaches, and coral reefs. North Americans are most familiar with the state of Hawaii and its islands, because they are a popular tourist destination and a movie and television setting. French citizens are well aware of French Polynesia and Tahiti for similar reasons (Figure 12–20). The more complex geographies of Oceania escape general attention, although they are better known to Japanese, Australians, and New Zealanders. Because the total population of this huge area, which extends over several million square miles of Pacific Ocean, is only about 11.9 million (using the broadest interpretation of the region's boundaries), its economic importance for the modern global economy is slight. Yet the Pacific Islands exemplify humanity's response to the challenges of isolation and distance, and adaptation to a scarcity of land and other physical resources in the light of sea level rise and other major environmental challenges. They show how small and scattered cultures with limited conventional economic opportunities face social and political problems derived in part from a unique geographical setting but also from a unique colonial history.

Regional Groupings

The Pacific Islands, some 30,000 in all, consist of several levels of regional groupings. At the first level, there are the three realms of Melanesia, Micronesia, and Polynesia (Figure 12–21 and Table 12-1). While geographically necessary, we should recognize that these are Western names of convenience that do not fully reflect regional complexities of race, history, and physical environment.

Melanesia includes New Guinea (the largest Pacific island, which is divided into the sovereign nation of Papua New Guinea and the Indonesian provinces of Papua and West Papua), New Caledonia, Vanuatu, Fiji, and the Solomon Islands. The name derives from melanin, the pigment in human skin, and refers to the dark-skinned, Papuan-speaking peoples who predominate in this area. Papua New Guinea, with nearly 7.3 million people, has by far the largest population in the Pacific Island realm and was settled the earliest. It gives Melanesia the largest total population of the three subregions at about 9.3 million.

Micronesia, with more than half a million people and a very high population density of 409 persons per square mile (158 persons per square kilometer), means small islands. Most of these lie north of the equator. They include the Marianas, Guam, Wake Island, Palau, the Marshall Islands, Kiribati, Nauru, and the Federated States of Micronesia.

Polynesia, whose name means "many islands," stretches across a huge triangular area from Midway and Hawaii in the north to New Zealand at the southern extreme, and eastward as far as Easter Island and Pitcairn (the isolated outcrop of *Mutiny on the Bounty* fame, with only 48 residents). Polynesia's total population is slightly over 2 million of which 60 percent live in Hawaii. Culturally, New Zealand's Maori heritage also places the islands as part of Polynesia, well before its more recent sweeping transformation by British colonial settlement.

The second level of groupings in the Pacific Islands consists of the various political arrangements and governmental structures that characterize the region. Several islands lack full "statehood" or "sovereignty"; this is a region in which the colonial legacy has not been fully overturned. Western Samoa was the first entity to be granted independence (from New Zealand) in 1962. Altogether, there are 23 political entities: 11 independent nations; 4 self-governing entities in free association with former colonial rulers; 3 dependencies or collectivities still linked to the United Kingdom, the United States, and New Zealand; and one of the 50 U.S. states (See Table 12-1). None of these structures has been a guarantee of success in terms of modern economic development.

There is a third subregional pattern that consists of extreme variations in levels of development and population distribution within each of the

▼ Figure 12–20 **Tahiti, French Polynesia.** (a) Many of the Pacific Islands possess great scenic beauty, but on the high islands the rugged, picturesque central core often limits development prospects. (b) Fortunately, whether high island or low, as this beach-front resort scene on Moorea adjacent to Tahiti indicates, all islands derive much of their revenue from tourism.

(a)

(b)

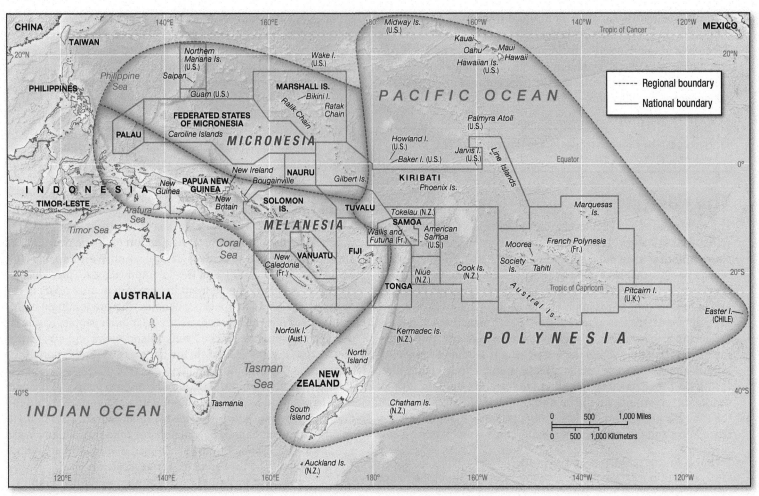

▲ **Figure 12–21 Pacific Island realm.** The Pacific Islands can be divided into three subregions: Micronesia, Melanesia, and Polynesia.

Sources: http://www.peacesat.hawaii.edu/40resources/Maps/index.htm; http://www.geographicguide.com/oceania-map.htm

political entities noted above. This regional imbalance, common to larger nations of the world, may be found both on single islands, such as Papua New Guinea, or Viti Levu in Fiji, or within island chains such as the Marshall Islands and Hawaii.

Oceanic Challenges

Traditional island societies were typically hierarchical and community based, dependent on fishing and subsistence agriculture, which centered on the cultivation of coconuts, taro, breadfruit, and other fruits and tubers as well as fishing and hunting. Agricultural skills sustained relatively high population densities with innovative multicropping. Early European explorers mistakenly interpreted local lifestyles to be an idyllic blend of a tropical climate, beautiful environments, and populations and societies seemingly free of major conflict. Tahiti in French Polynesia was explored in the 1760s and thought to be a tropical "Garden of Eden," populated by "noble savages" who inspired European thinkers disenchanted with their own continent and its ills. But traditional society was not free of warfare, epidemics, and social stratification. In addition, since the sixteenth century European traders and explorers had exploited commercial opportunities, initially for copra (dried coconut meat; crushed to produce coconut oil). Seals, sea cucumbers (*bêche-de-mer* or *trepang*), and sandalwood (*Santalum,* a species of aromatic tree), were traded with China. Early British settlers to Australia were supplied with Pacific island pork in exchange for British-made guns,

from 1793. The whale population of the Pacific was decimated by 1850 by North American and European whalers.

Once the myth of "Tropical Edens" was largely disproved through further sustained contact between Europeans and islanders, Christian missionaries started to arrive in the Pacific in large numbers. Even in the interior of Papua New Guinea (one of the world's most inaccessible regions) customary lifeways have been hybridized with a monetary economy, commercial rather than subsistence production, church-going, and other aspects of a colonial presence. By the late nineteenth and early twentieth centuries, foreign culture and technology—much of it very welcome, like metal goods and cloth—had severely disrupted traditional societal and economic patterns, and some island populations had been decimated through the inadvertent introduction of exotic diseases.

From the mid-nineteenth century the islands were governed through artificial colonial administrative arrangements that often disregarded historic cultural and resource utilization patterns. France, Germany, and Britain were dominant powers in the region. As the subsistence economy was reduced or destroyed by colonizers, the islanders became and have remained dependent on imported foodstuffs and manufactured goods. The missionaries, colonial administrations, commercial interests, and twentieth-century wartime invasions by Japan all helped to destroy or further challenge social structures, belief systems, and traditional foodways. Plantation agriculture concentrated on sugarcane, pineapples,

Table 12–1	**The Pacific Islands**	
Name	**Political Structure**	**2012 Population**
Melanesia		
Papua New Guinea	Independent	7,294,924
Solomon Islands	Independent	577,008
Vanuatu	Independent	252,276
New Caledonia	French Territory	262,658
Fiji	Independent	882,373
	Subtotal	9,269,239
Micronesia		
Palau	Independent	20,794
Fed. States–Micronesia	Independent	110,840
Guam	U.S. Territory	180,000
Northern Mariana Islands	Free Assoc.–U.S.	58,383
Marshall Islands	Independent	68,480
Nauru	Independent	10,377
Kiribati (incl. Line Is.)	Independent	105,283
	Subtotal	553,247
Polynesia		
Hawaiian Islands (Incl. Midway Is.)	U.S. State	1,400,000
Tuvalu	Independent	11,365
Wallis & Futuna	French Territory	13,288
Tokelau Islands	N.Z. Territory	1,411
Samoa	Independent	186,104
American Samoa	U.S. Territory	69,544
Tonga	Independent	106,146
Niue	Free Assoc.–NZ	1,269
Cook Islands	Free Assoc.–NZ	19,569
French Polynesia	French Territory	281,131
Pitcairn	U.K. Territory	48
	Subtotal	2,089,875
	Grand Total	11,901,959

Source: Country Meters, *http://countrymeters.info/en/*; index mundi, *http://www.indexmundi.com/*; U.S. Bureau of the Census. 2010 Census; World Population Data Sheet, 2012.

coffee, tea, copra, and cacao; and large-scale mining of gold, copper, nickel, manganese, petroleum, and natural gas was initiated in Papua New Guinea, New Caledonia, and a few other locations. Over time, the native societies were further transformed by the colonial rulers' systematic importation of alien laborers, including Indians (under indentured labor schemes to cultivate sugarcane)

to Fiji; Chinese to French Polynesia; Koreans to Guam and American Samoa; Filipinos to Guam, Palau, and the Northern Marianas; and Japanese to various parts of Micronesia. The native Hawaiian population has been reduced to a minority by American and Asian settlement. Queensland in Australia ran counter to this trend, aggressively recruiting or capturing Pacific islanders for plantation labor in the late nineteenth century.

Pacific atolls are the front line of the international climate change crisis. The populations of some low-lying islands are particularly impacted by any change in the relationship between sea level and land surface, given rising tides, lack of fresh water, and shrinking landmasses (see *Geography in Action*: Living at Sea Level Surrounded by Water). The people of the Carteret Islands, part of Papua New Guinea, are perhaps the worst affected by sea level change worsened by tectonic subsidence. Since 2007, the Carteret islanders have engaged in a protracted process of abandoning their traditional homes and migrating to Bougainville with government assistance. Deep emotional attachment to the land of their ancestors as well as the gradual pace of worsening conditions contribute to a prolonged and hesitant transitional process. This may be the first, but certainly will not be the last, of such environmental refugee movements in the low islands and atolls of Oceania.

For many islanders, the environmental threat is merely one of several. They also have to contend with low levels of income and health, social inequalities, and fragile governance. Political instability, separatist movements, political/military coups, and other troubles have afflicted several of the island groups at various times, including Bougainville in Papua New Guinea, the Solomon Islands, Fiji, and New Caledonia. Many rural areas and outer islands are experiencing depopulation, thus exacerbating spatial imbalances. It is common for agricultural- and fishing-based livelihoods to give way to extensive migration to urban centers, both domestic and foreign. For some observers these migration streams are sensible adaptations to a lack of opportunities and to increasing environmental problems. But they form part of a **MIRAB** syndrome ([**MIgration, Remittances, Aid, and Bureaucracy] economies**) that is not necessarily a healthy path to Pacific development. Employment opportunities for the island population include very limited manufacturing jobs. The small garment industry in Fiji and the Toyota seatbelt plant in Western Samoa are very much outliers in the general pattern. Many educated employees work in the service sector, including the tourist industry.

Tourism has a patchy record both as a generator of jobs for local residents and as a contributor to long-term economic development. The francophone territories have extremely high prices since most goods are imported from Europe and prices are pegged to the euro, while the vagaries of the Japanese economy affect the Pacific's prime source of tourists. Quality of education also varies widely, but literacy levels and years of schooling are generally low, except where a Western-style school system is funded and maintained, as in Hawaii and Guam. The Pacific has two major universities and several smaller ones. The University of the South Pacific (USP) is headquartered in Suva, Fiji (Figure 12–22), and the University of Hawaii is a United States public university on Oahu. The University of the South Pacific is owned by the governments of 12 Pacific countries and offers programs suitable for U.S. exchange students.

Pacific Islanders today are reappraising their options, and continue to exercise their adaptive skills and creativity. Adaptation to

GEOGRAPHY IN ACTION

Living at Sea Level Surrounded by Water

The Marshall Islands, uniformly small and poor, were first populated 4,000 years ago by Micronesians. After numerous foreign visits, the islands fell under a German trading company in the late 1800s. The Japanese occupied the Marshalls in World War II, until a U.S. invasion ousted them in 1944. The Marshall Islands are a case study of the cultural dilution and economic dependence that so often have followed foreign involvement in the Pacific Island region.

The islands are a tiny string of coral atolls with a total population of 68,000. Most of the residents are crowded into two towns: Majuro on the island of the same name, where landfill has combined reefs and islets to create an airport (Figure 12-5-1), and Ebeye on Ebeye island, part of the Kwajalein atoll. The Kwajalein atoll system itself is distinctive, since the islands and reefs that compose the atoll surround the world's largest lagoon. Ebeye is home to 13,000 inhabitants on 80 acres (32 hectares), and is one of the world's densest settlements. Altogether, the Marshall Islands include just over 1,200 islands and islets, 30 atolls, and over 900 reefs. Yet only 70 square miles (181 square kilometers) of actual dry land are contained in this profusion of island pinpoints! In contrast, the marine Exclusive Economic Zone (EEZ), which extends 200 nautical miles outward from the coast, allows the Republic of the Marshall Islands to control over 730,000 square miles (1.9 million square kilometers) of Pacific Ocean territory, valuable for fishing.

Two particular issues flow from this situation of limited habitable territory and small population, relative isolation, and limited resource base. The first is the problematic power position the Marshall Islands occupies relative to the United States. Granted a United Nations trusteeship over the Marshall Islands in 1947, the United States chose to use some of the islands as sites for nuclear weapons testing. These tests took place on the Bikini and Enewetak atolls in the 1940s and 1950s, where the United States exploded 66 bombs. The atolls were seen as small and marginal economically; their small populations were essentially powerless and

easily persuaded to evacuate to other islands. The result of the test explosions was deep cratering of the impacted atoll lagoons, severe damage to reefs and marine organisms, and radiation pollution of habitat and fish that persists to the present. Partial compensation has been paid to the evacuees, but their full claims have yet to be dealt with and their loss of valued ancestral space will never be reclaimable due to radiation contamination. The United States still maintains a strategic U. S. Army missile test range at Kwajalein Atoll.

Although the Marshall Islands were granted independence in 1986, they are too small to easily fulfill all functions of statehood and operate in free association with the United States. Propped up by continued U.S. subsidies and military expenditures, the free association agreement allows movement of Marshall Island citizens to Guam, Hawaii, and the mainland United States. The traditional culture and lifestyle of the islanders has been severely impacted and changed, and the islands exhibit the environmental and socioeconomic challenges common to Oceania. These are compounded by lingering health problems related to nuclear weapons testing.

The second issue is the precarious setting of island areas only 7 feet (2 meters) on average above sea level. This makes the Marshall Islands the country most dangerously threatened by climate change in the world. Typhoons have always been a threat to low islands in the Pacific, but the intensity of these storms will likely rise as global warming increases; moreover, as glaciers and polar ice packs melt, sea levels will rise. Coral reef growth is unlikely to keep pace with higher sea levels, and warmer seas will increase coral bleaching and reduce the corals'

▲ FIGURE 12-5-1 **Majuro Arnata Babua International Airport, Marshall Islands.** The tiny coral atolls that make up the Marshall Islands contain few natural resources, despite their lush, densely vegetated tropical landscapes. The runway of Majuro Amata Kabua International Airport on the coral island of Majuro takes up almost the entire width of one segment of the atoll.

protective role. Without corals to provide sand and rock to counter coastal erosion and to protect islands from being overtopped by storm surges, the impact of storms will increase. Warmer air temperatures and increasing aridity will reduce rainfall and lead to increasingly severe droughts, particularly in the northern Marshall Islands. Drought and reduced rain will increase crop failure and put greater pressure on freshwater resources, leading to overuse of limited freshwater reserves and the intrusion of saltwater into aquifers. Costly desalination of seawater is at best a partial solution, and unlikely to solve the basic problem—the increasingly precarious nature of life only a few feet above sea level.

Sources: D. D. Turgeon et al., *The State of Coral Reef Ecosystems of the United States and Pacific Freely Associated States: 2002* (Silver Spring, MD: National Oceanic and Atmospheric and Pacific Freely Associated States, 2002), *http://www.rmiembassyus .org/Environ/status_coralreef.pdf*; Rosemary Rayfuse, "Life after Land," *New York Times,* July18, 2011, *http://www.nytimes.com/2011/07/19/opinion/ 19rayfuse.html?ref=marshallislands*; AnaMarie D'Aubert and Patrick D. Nunn, *Furious Winds and Parched Islands: Tropical Cyclones (1558–1970) and Droughts (1722– 1987) in the Pacific* (Melbourne, VIC: Xlibris, 2012).

▲ **Figure 12–22 University of the South Pacific.** The main campus in Suva, Fiji, is a center of Pacific higher education.

geography and history includes new economic ventures, although imports consistently outweigh exports. Fishing licenses across the Pacific have been sold to fleets from Japan, Taiwan, South Korea, and the United States, exploiting the 200 nautical mile marine **Exclusive Economic Zone (EEZ)** of each nation or territory. French Polynesia exports high-value black pearls. Tonga and Kiribati have sold passports to Hong Kong Chinese. Kiribati has sold satellite launch and tracking services. A number of the islands are marketing themselves as loosely regulated global "financial service centers" with favorable taxation regimes. The low-lying atoll countries are vigorous activists in the international debate over the global warming and sea level rise that affect them badly. Most of the Pacific Islands, Australia, and New Zealand formed a supranational organization in 1971, the Pacific Islands Forum, to promote the region's collective interests. The Secretariat of the Pacific Community (SPC) in Nouméa and Fiji conducts development projects and programs, supported by governments and international donors. Numerous international aid organizations are active in the Pacific, including USAID.

In reality, it is unlikely that many of the island entities will achieve accelerated economic growth in the near term, and economic "development" has a checkered history in the region. Long-term improvements to livelihoods and better means to address poverty will depend on strengthening education, increasing regional cooperation, and creating specialty niches in the global economy. The availability of temporary work elsewhere is vital to financial sustainability of the island states, and there are guest worker schemes in both Australia and New Zealand. These help to meet demand for agricultural labor in both countries. For the mineral-rich Pacific economies like Papua New Guinea and New Caledonia, the challenge is to negotiate strong agreements with mining companies that adequately compensate local people for environmental and social impacts.

Stop & Think

▶ Consider the unique difficulties associated with development in the far-flung Pacific Island territory.

Papua New Guinea

Papua New Guinea (PNG) consists of the eastern half of the island of New Guinea, the large islands of Bougainville and New Britain, and many smaller ones. The polyglot collection of clans and cultural groups, speaking more than 700 languages and numbering over 7 million people, are the most heterogeneous on earth in a country a little larger than California. The population is predominantly rural, and many people still practice subsistence farming and have infrequent contact with outsiders. Highland farming systems appear to be some of the oldest in the world, rivaling those of Egypt in age. The population includes indigenous Papuans whose ancestry dates back to the time when PNG was connected to Australia by a land bridge, and more recent arrivals from the Austronesian archipelago. Papua New Guinea was split between Britain and Germany along the east-west mountain crest line along the middle of the island in the late 1800s, but passed to the Australian government in the early twentieth century, gaining full independence in 1975. Colonial control was hardly present in the interior, which still has poor roads and highways and is now reached by air and on foot. The country was the site of bitter fighting between Allied and Japanese forces during World War II.

The country's mountains are extremely rugged, with densely forested peaks reaching more than 14,000 feet (4,200 meters). There is extraordinary biodiversity, not all of it as yet catalogued. Port Moresby, with over 300,000 people, is located on the coast not far from northern Australia. It is at the bottom of the 2012 "most livable cities" index, with high levels of theft, murder, and violence. The main reason for this depressing condition is the lack of jobs for rural-urban migrants who drift to the city, resulting in crime and gang activity.

Papua New Guinea is the largest and the best endowed of the developing Pacific Island states. But its rich resources have sometimes been more of a "curse," failing to generate significant poverty alleviation and development even after political independence. Rich timber and mineral resources (especially gold, copper, and petroleum, which account for nearly 75% of export earnings) have generated considerable conflict. Their easy exploitation is impossible because there are complex land tenure rules and many customary landowning groups, making negotiation over rights and compensation payments extremely difficult. One of the largest copper mines in the world was at Panguna on the island of Bougainville (Figure 12–23). The island was wracked for 10 years by a bitter civil war, led by secessionists who protested the dominance of Port Moresby and New Guinea over their island, and the mine was closed in 1989. As a result, PNG's export earnings from copper plummeted. Bougainville now has partial autonomy, and there are plans to reopen the mine. In contrast, the OK Tedi gold and copper mine in PNG's Central Highlands, reaching the end of its life, provides over 25 percent of the nation's export earnings. Poisonous mine tailings and massive soil erosion have proven highly damaging to local environments, fish, and river systems and provoked an outcry and successful lawsuits by local people in the 1980s. Reparations given to locals affected by this and other mines have set off internal conflicts, and it is believed the effects of OK Tedi's downstream pollution will endure for 300 years.

Other important resources include tropical hardwoods, since the world's third largest tropical forest is found between PNG and

▲ **Figure 12–23 Panguna copper mine, Bougainville.** Although one of the world's largest, the Panguna copper mine closed in 1989 when it was caught up in a bitter civil war that wracked Bougainville Island for over a decade. Equipment has gathered nothing but rust since the mine closed. Now that Bougainville has an autonomous government, negotiations are taking place about reopening the mine.

West Papua. Logging has been extensive and involved many deals, some of them shady, with international companies, politicians, and local leaders. The country is under intense pressure to preserve its remaining forests, now seen globally as vital for locking up carbon and preserving biodiversity. Despite its economic possibilities, PNG had an estimated GDP per capita of only US$2,700, although the majority of its citizens adhere to traditional subsistence livelihoods with only limited cash and occasional migrant remittances. Services, markets, and infrastructure are still very patchy.

Australia is in the uncomfortable position of assisting PNG as its main aid donor and former colonial power, and it invests money and personnel each year to support good governance, better health, security, and economic development. Papua New Guinea exemplifies the difficult challenges of balancing cultural diversity, custom, and tradition, along with the more recent arrival of resource extraction and Western concepts of property and governance.

Hawaii

Hawaii, the 50th U.S. state (1959), consists of a string of mountainous volcanic islands rising to 13,796 feet (4,205 meters) on the northern perimeter of Polynesia. This was the farthest north that Polynesian sailors reached in their epic canoe voyages across the Pacific, and the islands were populated rather late, from A.D. 300 to 500, possibly from the Marquesas Island chain. Powerful chieftaincies emerged, but were subjugated under a single ruler by the time of Western discovery and settlement. The monarchy was illegally deposed in 1893 by a coalition of settlers and businessmen, and links to the United States, the major trading partner, were strengthened. Today the population of 1.4 million enjoys American technology and income levels, although there are marked social disparities across the island chain. In some ways, Hawaii is the most transformed of the Pacific Island groups and it has struggled to maintain its Polynesian culture. The native population, first estimated in 1778 by Captain Cook to more than 400,000, declined to about 54,000 in

1876 due to introduced disease and reduced access to resources. From that low point the population regenerated to almost 240,000 individuals who claimed native ancestry in the twenty-first century. But native Polynesians today are a distinct minority in their ancestral homeland, displaced by successive waves of Portuguese, Chinese, Japanese, Filipino, North American, and other immigrants. Asians now account for 39 percent of the population, whites for 23 percent, Hawaiians and other Pacific Islanders constitute a mere 10 percent. Hawaii has lost much of its unique flora and fauna, including at least 20 flightless birds, to 6,000 introduced species and the plundering of lucrative resources like sandalwood trees, which Hawaiians themselves cut and traded in the early 1800s. Coastal regions have been given over to commercial agriculture, tourism, dense settlement, and several U.S. military facilities.

Some 900,500 people are concentrated in Honolulu and its suburbs on the island of Oahu, the Pacific's second largest city after Auckland, New Zealand. The city is highly dependent on trade with the U.S. mainland and parts of Asia (especially Japan) for food and virtually all the industrial goods needed to sustain its economy. Spending by about 7.3 million tourists, somewhat recovered after a drop in 2011 due to a decrease in Japanese tourism following the Fukushima tsunami, has cushioned a decline in sugarcane and pineapple plantations, and a fall in military spending associated with Pearl Harbor, the headquarters of the U.S. Pacific Fleet. Although Hawaii's multicultural Asian-Pacific-Anglo population appears to have created a harmonious society, there is conflict between socioeconomic classes and over political representation. Native Hawaiians are struggling to preserve their ancestral rights within a highly globalized society and economy.

Even with its relative prosperity derived largely from U.S. statehood, Hawaii faces the same obstacles that confront the other Pacific Islands in the quest for economic success. Hawaii did not reap the benefits of the booming U.S. economy of the 1990s and fell behind in various socioeconomic indicators; a brief revival gave way to recession along with the rest of the United States in the late 2000s. Efforts to revitalize the economy have focused on improving the tourist environment and diversification, difficult to attain with an ongoing global economic downturn. Well over half of the tourist business is concentrated in the Waikiki Beach area near Honolulu (Figure 12–24), although decentralization efforts have increased tourism on other islands, particularly Hawaii, Kauai, and Maui. Plans to develop a high-technology industrial sector have been disappointing, due to Hawaii's physical isolation and small population. Even maintaining tourism is a challenge, as other more affordable and attractive sites lure the dominant market of Japanese and mainland U.S. tourists.

Guam

The largest island in Micronesia, Guam covers 250 square miles (650 square kilometers). It too has been almost entirely transformed by foreign occupation and control. The earliest human settlers were the Chamorro people, from Southeast Asia, arriving about 3,500 years ago. Their population rose to 30,000–50,000 by the sixteenth century and they developed intensive agriculture including paddy rice, resulting in massive changes to the island's original biota. Guam was the first Pacific island that the Spanish explorer Magellan landed on in 1521. It was annexed as part of the Spanish East Indies, and Spanish missionaries arrived in 1668. They were far from

▲ **Figure 12–24 Waikiki Beach, Honolulu, Hawaii.** Once a resting place for Hawaiian royalty, Waikiki Beach is now Honolulu's popular resort district on the south coast of Oahu island. The iconic Diamond Head dominates the beach where tourists gather and surfers roam

▲ **Figure 12–25 Gun Beach, Tumon Bay, Guam.** As Guam attempts to diversify its economy, marine sanctuaries exist in close proximity to tourism and hotel complexes.

welcomed by Chamorro warriors, and warfare resulted. Epidemics and violence reduced the population to less than 5,000 people by 1710. Spanish imports, including corn, cattle, and dogs, combined with deforestation to alter the native ecology and fauna yet further. The United States established a naval base on Guam in 1898 after winning the Spanish-American War and securing the island. The population increased to 23,000 by 1941. New industries included copra exports.

In World War II, Japan's military held the island between 1941 and 1944 before the U.S. military retook it in a bloody battle. This period resulted in severe damage to the ecology and to infrastructure. The brown tree snake was introduced, probably by accident, possibly as a stowaway in the wheel wells of airplanes arriving from Southeast Asia. Since landing, this nocturnal, elusive predator has decimated the population of native birds and reptiles. The brown tree snake has also reduced the population of rats, an earlier invader that used to keep at bay harmful insects that damage the island's coconut trees. An unforeseen consequence of fewer rats has been an explosion in the spider population, 40 times larger on Guam than on other islands. Efforts to protect what remains of Guam's sea turtle population, which lays its eggs on Guam beaches, has led to the establishment of marine wildlife preserves. This has the additional benefit of protecting coral reefs, a major attraction for visiting tourists (Figure 12–25).

Guam did not become independent like other island nations but remains an "unincorporated territory" of the United States; its citizens have American passports. The United States has increased its military presence, claiming the island is an important link to Asia and the Middle East. This surge could add a further 40,000–60,000 military personnel and their dependents, many transferred from bases on Okinawa, Japan, to the island's population over the rest of the decade. If this were to occur as planned, the island's population, which reached an all time official high of 155,000 in the 2010 census, was variously estimated to be above 180,000 in 2012, would easily soar to well over 200,000, and resulting construction projects could threaten coral reefs that are a tourist attraction. Chamorros make up 37 percent of this 2010 total, many with Spanish or Filipino ancestry and surnames, while Filipinos constitute about 25 percent of the population total. A Compact of Free Association with the Marshalls and Micronesia allows their citizens to reside in the United States, including in Guam. Tourism, mainly catering to Japanese, attracted by this "America in Asia," has taken off to create a coastal resort economy similar to parts of Hawaii (including the largest Walmart in the world), but the military presence is the backbone of the economy.

New Caledonia

New Caledonia's challenges are environmental and political. It comprises the large 7,700 square mile (20,000 square kilometer) island of Grande Terre, three large outlying islands, and several smaller ones. Grande Terre is mountainous and has extraordinary biodiversity and many endemic species. It is a piece of Gondwanaland that broke away from Australia 60–80 million years ago. The first humans were Melanesian clans who arrived over 3,000 years ago; later known as the Kanak people, they farmed the land and fished in Grande Terre's coastal lagoons. After Europeans plundered the island's sandalwood trees and Christian missionaries arrived, New Caledonia was officially colonized by France in 1853 and served as a penal colony. The liberated prisoners were encouraged to stay, as in Australia, and allocated land for livestock and for cultivation. But this displaced many native owners whose title to land was customary and not legally recognized. Free settlers later arrived, and some Asian and Polynesian migrants were attracted by mining wealth. Kanak populations plummeted due to disease, and the surviving native people were eventually confined to "reserves," with their movements and use of their languages restricted for decades. Meanwhile the city of Nouméa, the "Paris of the Pacific," grew rapidly. Nouméa now contains some 60 percent of the country's 262,000 citizens, and hosts a center dedicated to Kanak culture (Figure 12–26).

▼ **Figure 12–26 Jean-Marie Tjibaou Cultural Center, Nouméa, New Caledonia.** Inaugurated in 1998 and designed by Italian architect Renzo Piano to resemble traditional Caledonian village huts, this cultural center celebrates the Kanak culture.

(a)   (b)

Figure 12–27 New Caledonia's endemic vegetation and mining heritage. (a) There are 13 endemic species of the tall and distinctive *Araucaria* conifer tree. Many plants in New Caledonia can tolerate high levels of nickel in the soil. (b) Minerals are usually found at high elevations or in laterite soils. These abandoned mineworkings were built by Japanese in 1938, and brought ores from a plateau to the coast, where they were exported by boat. The Japanese attack on Pearl Harbor in 1941 ended all Japanese activity on the islands.

The Kanak have retained their identity and were not assimilated into francophone culture. An organized Kanak pro-independence movement emerged in the 1970s and there were violent uprisings against the white population and France. In 1988, Jean-Marie Tjibaou, a widely respected leader, signed the Matignon Accords with the loyalist mining magnate Jacques Lafleur. The peace deal included further land restitution, rural development activities, and a referendum on independence. Tjibaou was assassinated in 1989, and the referendum failed—France still has a major stake in the future of the territory, and Caledonians hold French citizenship. The Nouméa Accords led to gradual devolution of some administrative powers, and the Kanak now have political and demographic majorities in two of three provinces. Another referendum on independence is scheduled between 2014 and 2018.

This island contains great mineral wealth, including about 25 percent of the world's nickel reserves, which represents over 90 percent of its export revenues. Mining impacts have been severe, and have placed the island's exceptional ecology and indigenous people under great stress (Figure 12–27). The UNESCO-listed reef and lagoon that circles Grande Terre (the second longest in the world after the Great Barrier) is now a threatened habitat. Despite this, the Northern Province has developed its own mining and nickel processing operation, hoping to escape the MIRAB syndrome. The Koniambo project is probably the world's largest mine and processing plant with majority ownership by an indigenous corporation. The aim is to create jobs, infrastructure, and economic opportunities away from Nouméa and spread the country's wealth more widely.

Summary

▶ Australia, New Zealand, and the Pacific Islands are quite unlike any other world regions in the nature of their diversity. In Australia there is far too much land with too little water; in the Pacific Islands there are vast amounts of water, but in most cases far too little land. And in New Zealand, where water exists in frozen, liquid, and thermal-heated states, paradise is threatened by tectonic hazards.

▶ The region occupies a remote but strategically significant part of the planet. The prime agent of change has been the dispersal of Europeans as explorers, then traders, and in some places settlers. Their arrival was in all respects revolutionary and transformative, for the natural environment as much as for indigenous populations that had hitherto managed to exist without significant environmental degradation.

▶ Although far from the Western world in location, Australia and New Zealand are now Western in culture and have remained in an economic orientation toward utilizing agricultural resources—and minerals, in the case of Australia—largely for export to the developed world. Australian, Papua New Guinean, and New Caledonian mineral exports are vital to the world's manufacturing industries and for energy production.

▶ Past trade relationships with the United Kingdom were strong historically. Both Australia and New Zealand, however, are in the process of reorienting their economic relationships, largely toward the Pacific Rim, and importantly with Japan, China, and the United States. That reorientation also includes increased attention to diversification of economic activity, a goal that will be more easily attained by Australia than New Zealand because of its greater resource endowment. Economic changes, along with the significant change in immigration policy during the 1970s, signal that Australia and New Zealand recognize the realities of their location in the Asian world. The prioritization of economic growth in Australia and its long refusal to countenance a slowdown in carbon emissions, however, have made it far from popular, especially for Pacific islands that face much more severe outcomes from the global warming to which the rich nations like Australia contribute significantly.

▶ Papua New Guinea and New Caledonia are endowed with a rich natural resource bases. The other Pacific islands, however, are challenged in their economic development by their scarcity of development possibilities and resources. Many of the island states of Oceania are also characterized by cultural disunity and social inequality. Future economic growth of Oceania will hinge on the achievement of greater regional cooperation, filling specialty niches in the global economy, and the fuller development of the region's human resources, including the indigenous populations whose societies have been seriously disrupted by Western conquest and settlement.

Key Terms

Aborigines 544
Aotearoa 542
Australia–United States Free Trade Agreement (AUSFTA) 551
carbon pricing 557
endemic species 542
Exclusive Economic Zone (EEZ) 567

fracking 557
Gold Coast 559
Great Barrier Reef Marine Park 545
import substitution industries 558
Kiwis 561
Maori 546

Melanesia 563
Micronesia 563
MIRAB (Migration, Remittances, Aid, and Bureaucracy) economies 565
Oceania 563
Outback 547
pastoral economy 560

Polynesia 563
sea level rise 542
stations 554
Stolen Generation 552
White Australia policy 549

Understanding Development in Australia, New Zealand, and the Pacific Islands

1. How has the physical geography of Australia influenced its pattern of settlement?
2. How can one explain the small population of both Australia and New Zealand?
3. What was the White Australia policy and what has happened to it?
4. Why is Australia so highly urbanized, given its extremely low population density?
5. Where is the core region of Australia and why is it located where it is?

6. Why has it been difficult for Australia and New Zealand to develop large manufacturing sectors in their economies?
7. What are the positive and the negative aspects of the mining boom in the region?
8. In what ways have Australia's overseas linkages been changing and why?
9. What are the major environmental challenges facing the small island nations of the Pacific?
10. Why have the environments and cultures of Guam and Hawaii been so altered since the colonial period?

Geographers @ Work

1. Advise the Australian government on how to develop the northern part of the country by preparing a plan that builds on the region's assets and overcomes or avoids its constraints.
2. Given Australia's resources and strengths as an exporter of primary products, outline the reasons why the country might wish to become a major exporter of manufactured products as well.
3. Describe options that Australia might employ to overcome its water shortages.

4. Determine what policies might improve the conditions for Australia's Aboriginal people and New Zealand's Maori. Suggest the types of agreements that should be put in place before new mines are permitted in areas inhabited by indigenous people.
5. Recommend the adaptations you would suggest to the government of a small Pacific island nation, if you were providing advice on how to deal with sea level rise.

MasteringGeography™

Looking for additional review and test prep materials? Visit the Study Area in MasteringGeography™ to enhance your geographic literacy, spatial reasoning skills, and understanding of this chapter's content by accessing a variety of resources, including MapMaster® interactive maps, videos, RSS feeds, flashcards, web links, self-study quizzes, and an eText version of *World Regional Geography*.

Glossary

Aborigines: Descendants of the inhabitants occupying Australia at the time of European settlement.

acculturation: The cultural modification of a group because of contact with other cultures. This is the initial stage of the assimilation process in which a group learns enough about local customs to operate successfully in the larger community, and may go no further.

acid rain: Precipitation that is sufficiently toxic to diminish soil fertility, damage standing crops and forests, poison lakes, kill fish, eat away medieval stone structures, and ruin paint finishes on automobiles. Acid precipitation occurs as various substances, including sulfur and carbon, are released into the air as a consequence of burning huge amounts of fossil fuels such as coal and petroleum products. These substances mix with water vapor in the atmosphere and fall to Earth as acidic precipitation.

Africa Growth and Opportunity Act (AGOA): Signed into law by President Clinton in 2000, this act promotes trade between African states and the United States by giving countries greater access to American markets, credit, and technology.

African American migration: The movement of blacks in the United States from rural to urban areas.

African National Congress (ANC): A black political group in South Africa, committed to majority rule and abolition of apartheid; victorious in the 1994 national elections, the first in which blacks could participate.

African Union (AU): A major cooperative organization that consists of 54 African states and began functioning in 2002.

Agricultural Revolution: A period characterized by the domestication of plants and animals and the development of farming.

agricultural surplus: With domestication of plants and animals, agricultural productivity soared to the point that some farmers produced more than they consumed, creating surpluses.

agritourism: A form of tourism in which city dwellers pay for a back-to-the-land vacation that involves helping farmers with harvesting or food processing.

agroforestry: The sustainable harvesting of food and other forest products.

al-Idrisi, Muhammad: (A.D. 1099-1166) Cartographer, geographer, and traveller famed for creating a map inscribed on a silver ball for King Roger II of Sicily.

al-Jazirah: The flat alluvial plain between the Tigris and Euphrates rivers in southern Iraq.

alluvium: Material, often very fertile, that has been transported and deposited by water.

altitudinal life zones: A sequence of climate zones in mountain areas that vary with elevation as temperature and precipitation change.

Amu Darya river: A river in Uzbekistan that flows into the Aral Sea.

Angkor Wat: Area of Cambodia where monumental ruins stand as testimony to the agricultural productivity and trade capabilities of this once-powerful empire.

Aotearoa: The Land of the Long White Cloud was the name given by the indigenous Maori people to the North Island of New Zealand.

apartheid: The former policy of the South African government that maintained strict white-nonwhite segregation.

Appalachian Highlands: The Appalachian Mountains region of the eastern United States, constituted in part by the Blue Ridge/Great Smokies and Ridge and Valley physiographic provinces.

aquifers: Underground, water-bearing rock strata.

Arabs: A Semitic people of the Middle East and North Africa who share a common language, cultural history, and religion (Islam).

Aral Sea: A sea in the southern drylands of Kazakhstan and Uzbekistan that is rapidly shrinking due to massive water diversion from its two river sources for desert irrigation projects.

area studies tradition: A geographic perspective that emphasizes the study of specific regions and an understanding of the varied aspects of those regions.

aridity: Dryness; the dominant climatic characteristic of the world's desert and semidesert regions. Generally found in interior continental locations or in coastal areas bordering cold ocean currents.

Aristotle: (384-322 B.C.) First Greek geographer to divide the world into three broad climatic zones. Author of *Meteorologica* in which he discussed the physical characteristics of the earth.

ASEAN Free Trade Area (AFTA): Established in 1992 to reduce tariffs on trade between member countries of the Association of Southeast Asian Nations (ASEAN).

Ashkenazim: Jews originating from northern, western, and eastern Europe.

Association of Southeast Asian Nations (ASEAN): A political-economic organization formed in 1967 to promote cooperation among and trade between member nations. Originally composed of Brunei, Indonesia, Malaysia, Philippines, Singapore, and Thailand, the group was joined by Viet Nam, Laos, Myanmar, and Cambodia by 1999.

Aswan High Dam: A large dam project that was begun on the Nile River at Aswan in the 1960s. The project's goals were to make year-round cropping possible on virtually all Nile River valley land, provide water with which to increase cultivated acreage in Saharan oases (the New Valley Project), generate hydroelectric power for industry, and protect the Nile Valley from destructive floods.

austerity measures: Budget cutting, job reduction, or other cost-reducing measures by governments in an effort to slow the growth of government-held debt.

Australia-United States Free Trade Agreement (AUSFTA): Signed in 2005, the agreement has attempted to stimulate increased trade by eliminating some of the tariff barriers between the two countries.

average annual precipitation: The average total precipitation during the year, expressed in inches or millimeters. Precipitation figures are presented in rainfall equivalents but include all forms of precipitation.

Aymara: An important indigenous ethnic and linguistic group of Peru and Bolivia.

Aztec: One of the four high civilizations of pre-Columbian Latin America, centered on the area around present-day Mexico City.

bakufu: Military leaders, who exercised power "behind the scenes" and controlled family-based feudal territories, dominated Japan from 1100-1850 until the Meiji Restoration reestablished the Emperor's authority.

balkanization: The breakup or fragmentation of a large political unit into several smaller units, such as that which occurred in the Ottoman and Austro-Hungarian empires during the nineteenth and twentieth centuries.

banana republic: Phrase coined to describe countries with a single export crop whose economy and political system is controlled by foreign companies and a domestic elite.

barrage: A low structure placed across a stream to block the flow of water and divert the impounded water for irrigation purposes.

bazaar (or suq): A portion of a Middle Eastern city, such as a street or cluster of streets with covered roofs where shopkeepers and merchants are located, but also serving as an informal network of family, craft, and professional associations that manage and control the major traditional industrial and commercial activities of the economy.

Bedouin: Arabs who live by nomadic herding in the deserts of North Africa and the Middle East.

behavioral perspective: An approach that emphasizes human action and the role that behavior plays in society.

Benelux: The countries of Belgium, Netherlands, and Luxembourg.

Berbers: A pre-Arabic culture group of Morocco and Algeria, many of whom now speak Arabic and are Muslims.

Berlin Conference: The 1884 conference of European powers that marked the beginning of the formal era of colonialism in Africa by dividing Africa among these powers.

Bharatiya Janata Party (BJP): A conservative Hindu-based political party in India which often has been reluctant to compromise with Muslim and Sikh interests. This has slowed efforts by some to seek common approaches to development challenges.

birthrate: The number of births occurring in a given year per 1,000 people.

Black Economic Empowerment (BEE): A policy of the South African government that stresses developing enhanced employment opportunities and better business ownership conditions for blacks.

Blue Banana: A banana-shaped zone stretching from England across northwest Europe and into northern Italy with a large concentration of industry, infrastructure, and people.

Bolivian gas wars: A struggle between Bolivia's highland and lowland regions for control of the natural gas resources that provide income for Bolivia's social welfare programs.

Bolshevik: A Russian word meaning "majority," now taken to mean a Communist, or adherent of communism.

Bowditch, Nathaniel (1773–1838) A geographer, mathematician, and navigator who published the first accurate navigation tables for mariners.

BRIC (Brazil, Russia, India, China) countries: A group of countries that are regarded as emerging economies with increasing global economic power.

British East India Company: English company established in 1600 to trade with India. In pursuit of its commercial interests, the company eventually exercised political and military control over large parts of India until it was replaced by direct British rule in 1858.

brownfields: Areas where formerly prosperous industrial zones have experienced industrial decline and job loss with accompanying property abandonment and reduced tax base.

Buddhism: A religion that emerged in India from the teachings of Siddharha Gautama or Buddha ("awakened one") in the fifth century B.C., spread widely in eastern and central Asia, and remains an important faith in Bhutan, Nepal, and Sri Lanka.

bureaucratic capitalism: In China, supportive government policy and private capital are combined to achieve high levels of economic growth.

business process outsourcing (BPO): The practice in which large companies employ specialized companies, often located in distant countries, to provide services such as call centers in order to reduce labor costs.

calcification: A process occurring in dry regions where limited precipitation results in less leaching of soluble materials and thus the accumulation of calcium carbonates in the soil.

Canadian Shield: A relatively smooth glaciated land surface region that nearly encircles the Hudson Bay and extends southward to the Great Lakes area of the United States. Consists of thin soils, stony surfaces, and areas with no soil at all.

carbon pricing: A policy that charges emitters of the waste products that cause climate change for the cost of their pollution, either by a monetary tax on the volume of emissions or the revocation of permits to operate.

cash cropping system: The result of Western colonial powers forcing peasant farmers to dedicate a portion of their farmland to the cultivation of export crops.

Caspian Sea: A large, internally draining body of salty water in southern Russia, into which the Volga River empties, that lies approximately 90 feet (27 meters) below sea level.

caste system: A rigid system of social stratification based on occupation, with a person's position passed on by inheritance; derived from the Hindu culture.

Caucasus Mountains: Two ranges (Greater Caucasus and Lesser Caucasus) that stretch from the Black Sea to the Caspian Sea.

Celts: Descendants of central European tribes of peoples who arrived in the British Isles and the Brittany area of France around 2000 B.C. The Celts developed languages distinctive from others in western Europe; the numbers of native speakers of Celtic languages (Breton, Scots Gaelic, Irish Erse, and Welsh) are in decline today.

centralization: Important to the development of Central Asia during the Soviet era. The government in the U.S.S.R's capital, Moscow, made all political, economic, and cultural decisions and put them into practice throughout the U.S.S.R.

chaebol: A select few megacorporations in South Korea.

chain migration: The practice of initial migrants to a country to sponsor a stream of family members to their new location.

Chernobyl: Site of a nuclear plant disaster in 1986 in the former Soviet Union that has left a sizable portion of Ukraine and Belarus contaminated.

chernozem: A Russian term referring to the fertile black soils of the steppe or semiarid zone of Russia, Ukraine, and Kazakhstan; similar soils in other parts of the world.

chinampas: Mexican term for highly productive raised agricultural beds in former wetland areas.

Chinese Communist Party (CCP): Also known as the **Communist Party of China (CPC)**, this nationalistic and socialist movement was founded in Shanghai in 1921 and came to power after World War II when it defeated its arch rival, the Kuomintang (KMT) Party, in the Chinese Civil War.

Chinese diaspora: The migration of large numbers of Chinese from China to other countries.

chorological framework: The fundamental geographic perspective that emphasizes the importance of place.

chronological framework.: The fundamental historic perspective that emphasizes the importance of time.

City Prosperity Index (CPI): A composite index that measures the productivity, infrastructure, quality of life, equity, and environmental sustainability of urban areas.

city-state: A sovereign country consisting of a dominant urban unit and surrounding tributary areas.

clanocracy: The dominant place of clan families in defining the political and legal status of people in Africa.

climate: The average temperature, precipitation, and wind conditions expressed for an extended period of years; prevailing conditions over time.

climate change: Any shift in temperature, precipitation, and other climatic characteristics that alter prevailing conditions and move the climate system toward a different state.

Cold War: The period of hostility, just short of open warfare, between the Soviet Union and the United States and its allies, essentially from 1945 to the mid-1960s, although its end is often dated as 1989 with the fall of the Berlin Wall.

collective farms: A form of government-organized and supervised large-scale agricultural organization in the former Soviet Union and Eastern Europe; a collective leased land from the government, and workers received a share of net returns to the organization.

collectivization of agriculture: The process of forming collective or communal farms, especially in communist countries; nationalization of private landholdings.

colonialism: The system by which some powers controlled foreign possessions, usually for economic exploitation; most prevalent from the sixteenth through the mid-twentieth centuries; essential to understanding contemporary development in many developing regions.

command economy: Centrally controlled and planned livelihood system; the best example is the communist form of economic organization.

Common Agricultural Policy (CAP):   European Union agricultural policy with the aim to protect small farms, set uniform agricultural standards throughout the EU, and encourage beneficial exchange between the richer and poorer parts of the Union.

communalism: An uncompromising allegiance to a particular ethnic or religious group.

Communist Party: In the former Soviet Union, the Communist Party's original purpose was to advance the social revolution; in practice, the party also took control over the soviets—the governing councils.

comparative advantage: The idea that a given area gains by specializing in one or more products for which it has particular relative advantages; leads to trade to obtain other needed commodities.

Confucianism: The philosophy based on the writings and teachings of Confucius, which emphasized the importance of living a virtuous life characterized by obligations to others.

conquistadores: The Spanish conquerors of America, particularly sixteenth-century Mexico and Peru.

continental climate: A climate defined by the extreme heating and cooling characteristics of land rather than water, with hot summers and cold winters.

continentality: A measure of distance from the oceans. Generally speaking, the more inland, or continental a location, the drier its climate and the greater its temperature range are likely to be.

core-periphery dynamic: A situation where some regions lag behind in economic development and thus form a periphery to a more powerful and prosperous core.

Corn Belt: An extensive region in the central part of the United States with rich soils where corn and hogs once dominated farm output, but where soy beans and other crops now are also common.

creative destruction: An active transformation in which people alter their habitat to produce a new, human-dominated system that meets human needs. Something must be destroyed in order to create the conditions that promote benefits for humankind.

Creoles: Collective name for persons of mixed European (usually French or Spanish) and black heritage who often speak a French or Spanish dialect of the same name and are prominent in Guyana, Suriname, Haiti, and French Guiana.

crony capitalism: A system of government in which networks of friends and allies maintain themselves in power by using their official position to extract "rent" from businesses and property owners in return for protection.

crossroads location: A place where contrasting cultures meet and often clash.

Cultural Revolution: The upheaval in China during the 1960s when old cultural patterns were condemned and new Maoist patterns were strongly enforced.

culture complex: A group of culture traits that are activated together; for example, how clothing is made and distributed to consumers.

culture hearth: The source area for particular traits and complexes.

culture realm: An area within which the population possesses similar traits and complexes; for example, the Chinese realm or Western society.

culture trait: A single element or characteristic of a group's culture—for example, dress style.

cyclones: Indian Ocean hurricanes.

Dairy Belt: An extended region from Nova Scotia to Minnesota where moist, cool conditions and nearby urban markets favor specialization in dairy production.

Damodar Valley: The principal heavy-industrial region of India, located west of Calcutta, with Jamshedpur serving as the region's focus.

death rate: The number of deaths occurring in a given year per 1,000 people in a given area.

Deccan Plateau: An elevated (2,000-3,000 feet/600-900 meters above sea level) surface in southern India, rich in minerals, but limited in its agricultural potential due to its relative dryness.

deforestation: Loss of trees in an area, sometimes caused by physical factors such as fires, floods, pests, etc., but more often human-induced (caused by agriculture, logging, and fuelwood consumption).

deindustrialization: Refers to the severe decline of primary and secondary mass-production industries in industrialized countries after 1950.

demographic dividend: The benefit that results when reduced child mortality, declining fertility, greater economic growth, and improved health care and education produce a larger working-age population, fewer dependents, and greater economic returns.

demographic transition: A theory of the relationship between birthrates and death rates, as well as urbanization and industrialization, based on Western European experience.

Deng Xiaoping: Dominated the Chinese government from 1976 until his death in 1997, during which time China began to emulate the capitalist economies of its East Asian neighbors.

desertification: The process by which desert conditions are expanded; occurs in response to naturally changing environments and the destruction of soils and vegetation brought on by human overuse; takes place on the margins of desert regions.

deserts: Areas of extreme dryness where human settlement and economic development is severely constrained by lack of water.

development: A process of change that leads to improved well-being in people's lives, takes into account the needs of future generations, and is compatible with local cultural and environmental contexts.

devolution: The process by which a country's national government yields some of its powers to local authorities.

diffusion process: The process whereby cultural groups and their goods and innovations move across and settle the landscape (for example settlers moving into new territories).

dikes: Long earthen embankments that are used to shield land from the sea. Dikes have been used extensively in the Netherlands as an effort to reclaim land from the sea for agricultural purposes.

Dnieper River: The axis of an extensive trade network that linked the Baltic and Black Sea regions starting in the tenth century.

doi moi: "Economic renovation"; a program adopted by Vietnam's leaders in the 1980s to attract foreign investment.

dry farming: Agriculture wherein farmers rely exclusively on rainfall to produce their crops.

dual economic system: A modern plantation or other commercial agricultural entity operating in the midst of traditional cropping systems.

Duma: Russia's principle legislative assembly; this body was established first in 1906 by the Tsar to create a more representative base for imperial government, disappeared after the Bolsheviks seized power in 1917, and reemerged after 1993 when democratic forms of governance replaced the Soviet Union.

earth science tradition: A geographic perspective in which emphasis is on understanding the natural environment and the processes shaping that environment.

East African Rift Valley: African valley that begins in the north with the Red Sea, extending 6,000 miles (9,700 kilometers) through Ethiopia to the Lake Victoria region, where it divides into eastern and western segments and continues southward.

East Asian Model: Adaptation of Western methods to indigenous culture and values.

Eastern Orthodoxy: Christian branch created from the split of the Holy Roman Empire in A.D. 1054. In this schism, the western church became Roman Catholicism, and the Greek church became Eastern Orthodoxy. Initially centered in Greek-speaking areas and focused on the city of Constantinople (modern Istanbul), the Eastern Orthodox church has now expanded into a collection of 14 self-governing churches with the Russian Orthodox Church being the largest. It is the dominant Christian branch in southern and eastern Europe and represents more than 10 percent of the world's Christians.

Economic Community of West African States (ECOWAS): Created in 1975, it is one of Africa's largest regional organizations. Among its objectives are the establishment of a common customs tariff, a common trade policy, and the free movement of capital and people.

ecosystems: The assemblage of interdependent plants and animals in particular environments.

ecotourism: The travel to natural areas to understand the physical and cultural qualities of the region, while producing economic opportunities and conserving the natural environment.

El Niño: The warm, irregularly occurring, equatorial surface current whose southward movement along the western coast of South America blocks the upwelling of cold, nutrient-rich water and disrupts typical regional weather patterns. Its name in Spanish (the little boy or the child) derives from its folk association in late December with Christmas.

encomiendas: Grants of authority and responsibility from the Spanish crown to Europeans in Latin America; included control over large parcels of land and became a mechanism by which Europeans and their descendants gained and maintained control over land and indigenours villages.

endemic species: Plants or animals that are found only in one area and nowhere else.

energy security: A state achieved when an energy mix is created whose component fuels are accessible, affordable, efficient, and environmentally sustainable.

environmental stewardship: The principle of managing environmental resources with a view to their long-term sustainability.

Eratosthenes: (ca. 276–ca. 196 B.C.) Greek geographer from Alexandria, Egypt who accurately measured the earth's circumference and recognized the need for maps to show relationships between one region and another.

ethnic minorities: Groups that differ ethnically, linguistically, or religiously from a country's majority, and may seek independence or autonomy on that basis.

European Neighborhood Policy: A policy initiative in which the EU encourages trade relations and other common activities with neighbors on its immediate frontiers.

European Union (EU): An association of member states that have voluntarily given up some decision-making rights to serve a greater good of European integration.

euroscepticism: A view that rejects the idea and process of European integration.

Eurozone: Those countries in the European Union who use the euro as their currency.

evapotranspiration rate: The combined loss of water from direct evaporation and transpiration by plants.

Exclusive Economic Zone (EEZ): An internationally recognized right of control exerted by nation-states over marine resources for 200 nautical miles from their coastline.

exotic rivers: Those foreign to the area to which they supply water; they gain moisture from outside the region and cross dry areas on their journey to the sea, collecting little (if any) runoff along their lower channels.

export processing zones (EPZ): Enclaves in countries where materials are imported, processed, and reexported as finished products.

ex-situ conservation: A strategy for conserving the biodiversity of Earth's food crops by collecting as many varieties as possible of wheat, maize, rice, potatoes, beans, cassava, and other crops and storing seeds and cuttings in seed banks.

Extended Metropolitan Region (EMR): A space economy of regionally based urbanization in which the proliferation of manufacturing facilities in rural locations along urban-centered transportation corridors has caused the social and economic traits of rural spaces and their residents to become urban in character. For example, megacities such as Jakarta, Manila, and Bangkok, whose populations exceed 10 million inhabitants and which include villages and paddy fields from which industries draw labor, are considered extended metropolitan regions.

extraterritorial possessions: European territories that are legacies of the colonial era, such as the Spanish Canary Islands.

extraterritoriality: The imposition of foreign laws to the exclusion of local laws, allowing foreign citizens to operate as if they were on their native soil and making foreigners of indigenous people who live in such jurisdictions, such as the former treaty ports in China.

Falashas: Black Jews of Ethiopia, who were converted to Judaism by Semitic Jews between the first and seventh centuries A.D.

federal state: A state in which considerable powers are vested in the individual provinces, and the central government is relatively weaker than in a unitary state.

federalism: The administrative framework that provides states or territories a measure of economic and political autonomy.

feminist perspective: A perspective that emphasizes the role that women play in economic, political, and social spheres.

Ferghana Valley: A rich agricultural and industrial valley divided between Uzbekistan, Tajikistan, and Kyrgyzstan and drained by the Syr Darya river, which provides the water for cotton cultivation.

Five Pillars of Islam: The tenets that govern the behavior and belief of faithful Muslims: the creed, prayer, charitable giving, fasting, and pilgrimage (specifically to Mecca).

Five Year Plan: A strategy in the former Soviet Union for achieving rapid industrialization through centralized management and the forced achievement of production goals.

Foreign Direct Investment (FDI): The encouragement of investments by foreign capital, often in specially designated areas, in an effort to spur economic growth and development.

fossil fuels: Organic energy sources formed in past geologic times, such as coal, petroleum, and natural gas. The occurrence of fossil fuels is finite and not renewable within a time frame meaningful to humankind.

fracking: Releasing natural gas from rock formations by fracturing the rocks with a combination of hydraulic pressure and chemicals.

free trade zones (FTZ): Enclaves in countries where materials are imported, processed, and reexported as finished products.

French Canada Core: The area in eastern Canada, primarily along the St. Lawrence River, where settlers of French origin established a long-term presence.

friction of distance: The time and effort that is required to overcome physical movement across the landscape. Typically this friction is reduced with the advent of transportation innovations (for example, in 1800, traveling from New York to St. Louis could take one month, whereas today it takes only a few hours).

frost-free period: The period during each year when frost is not expected to occur, based on average conditions.

GAP Project: A massive development project that involves constructing two dozen dams on the Tigris and Euphrates rivers to provide hydroelectric power and water to irrigate large tracts of dry land in southeastern Turkey.

Gaza Strip: A small territory on the southeast coast of the Mediterranean Sea; inhabited by Palestinians and administered since 2007 by Hamas who seized control from the Palestinian Authority.

Gazprom: Government-controlled Russian natural gas company.

gender bias: Population disparity that occurs when one or the other sex makes up an abnormally large percentage of a population. For example, India possesses a gender bias in favor of males.

geographic grid: The system of mathematically determined latitude and longitude lines used to determine the location of every place on Earth's surface.

geographic information science (GIScience): A broadly defined basic research field that combines geography, cartography, geodesy, and information science using geographic information systems as an analytical tool.

geographic information systems (GIS): A computerized tool that allows the gathering, constructing, and manipulating of layers of spatial data for the analysis of a wide range of geographical problems.

geography: A field of knowledge concerned with the study of how and why things are distributed over the Earth; includes four basic approaches: (1) the study of distributions, (2) the study of the relationships between people and their environment, (3) the study of regions, and (4) the study of the physical Earth.

global warming: Increasing atmospheric temperatures, particularly due to the release of greenhouse gases that are produced by the burning of fossil fuels. When these gases build up, they retain heat within the atmosphere.

globalization: The growing integration and interdependence of world communities through a vast network of trade and communication links.

Gold Coast: The coast of Queensland in Australia, which is attracting great numbers of international tourists.

golden horseshoe: An industrial district in Canada extending from Toronto to Hamilton, on the western end of Lake Ontario.

Gorbachev, Mikhail: The only elected President of the U.S.S.R., Gorbachev initiated reforms that ultimately resulted in the dissolution of the Soviet Union.

Grameen Bank: The brainchild of Mohammad Younis, this microfinance organization began in Bangladesh in 1976 as a microlender of money to poor villagers, a majority of whom were women. Younis won a Nobel Peace Prize in 2006 for this significant and successful grassroots development initiative.

Grand Canal: A 1,400-mile (2,250 kilometer)-long canal between modern-day Hangzhou and the heart of the North China Plain; it facilitated the trading of commodities among the regions of the Chinese empire and continues to move bulk goods today.

Great Barrier Reef Marine Park: One third of the total Great Barrier Reef system off Australia's eastern coast is controlled by this authority, whose goal is to limit pollution and control tourism so one of nature's most impressive habitats can be conserved.

Great Barrier Reef: A 1,250-mile (2,000 kilometer)-long reef off the coast of Australia; it is one of the greatest natural wonders of the world.

Great Lakes: Large inland lakes shared by the United States and Canada and drained by the St. Lawrence River/Seaway. The regions surrounding the lakes are characterized by high levels of agricultural and industrial output.

Great Lakes of East Africa: Part of the East African Rift Valley, containing crater lakes and the elongated lakes of Africa's deep trenches.

Great Leap Forward: China's 1958-1961 attempt to socialize agriculture and increase production in both farming and industry.

Great Plains: Western portion of the Interior Plains of the United States. Consist of sediments brought down from the Rocky Mountains; vary in elevation from 2,000 feet (600 meters) near the Mississippi River to more than 5,000 feet (1,500 meters) at the base of the Rockies.

Green Revolution: The use of new, high-yielding hybrid plants—mainly rice, corn, and wheat—to increase food supplies; includes the development of the infrastructure necessary for greater production and better distribution to the consumer.

greenhouse gases (GHGs): Gases, such as carbon dioxide, methane, and others, that trap heat and cause a greenhouse-like effect as they increase in magnitude in the atmosphere.

Gross National Income in Purchasing Power Parity (GNI PPP): A widely used statistical measure of economic output which factors in differences in cost of living from one country to another, so that the resultant figures are more comparable.

growth triangle: An area of industrial growth in Southeast Asia centered on Singapore, the Malaysian state of Johor, and the Indonesian island of Batam.

guest workers: Workers from other countries who seek employment in the urban industrial centers of foreign nations.

Gulf Cooperation Council: A regional organization that deals with common problems in a unified manner; consists of Bahrain, Kuwait, Oman, Qatar, Saudi Arabia, and the United Arab Emirates—countries with large oil reserves, small populations, and a location along the Persian Gulf.

Gulf of Mexico: An extension of the Caribbean Sea and Atlantic Ocean, bordering the southeastern United States, Mexico, and some of the Caribbean Islands.

Gulf-Atlantic Coastal Plain: Land surface region, composed of sedimentary materials, that extends in the United States from New Jersey to Texas, and then southward into Mexico.

hacienda: A Spanish term used in Latin America for a large rural landholding, usually devoted to animal grazing; which had a high degree of internal self-sufficiency and operated under a semi-feudal system dominated by criollo owners who resided primarily in urban centers.

Hamas: A Palestinian political party founded in 1987, Hamas is an Islamic resistance movement opposed to the state of Israel. In addition to carrying out military activities and suicide bombings, Hamas is active in providing a network of social services and programs to Palestinians.

Han dynasty: (206 B.C.-A.D. 220) A militarily powerful dynasty that organized the first large-scale empire in East Asia.

Harappan culture: The Indus River Valley empire, dating back to ca. 2350 B.C.; it centered on the three cities of Harappa, Gan-weriwala, and Mohenjo-Daro. It began to decline ca. 1900 B.C.

Hazaras: In a predominantly Sunni Muslim country, the Hazara are a Shi'a Muslim, Persian-speaking ethnic community living in central Afghanistan.

Heian period: From A.D. 794-1185, at a time when Chinese cultural influences were strong, Japanese imperial court art, literature, and poetry flourished and attained their classical expression.

Herodotus (ca. 484-425 B.C.): One of the earliest Greek geographers to map and name the continents of Europe, Asia, and Africa. Known as the father of geography and the father of history. Author of *Historia* (mid-fifth century B.C.).

Hindi: One of the national languages of India; one of India's several languages of Indo-European origin.

Hindu Kush Mountains: Major mountain chain in Afghanistan.

Hinduism: A formalized set of religious beliefs with social and political ramifications; the dominant religion of Indian society.

holistic discipline: An area of study, such as geography or history, that synthesizes and integrates knowledge from many fields.

home town associations (HTA): Associations formed by migrants, often from the same town or district, to provide assistance in support of community projects in their home country.

household responsibility system: A Chinese policy based on a production contract in which a peasant household is obliged to produce a specific amount of grain or cotton to be sold to the state at a regulated price.

***hukou* registration system:** In China, household residency permits were required by law until the late 1970s. Without such permits, it was difficult for rural residents to move to cities because their access to food, jobs, housing, and health care was limited to their home district.

Human Development Index (HDI): Most widely used indicator of human development levels of countries and regions of the world; derived from life expectancy at birth, educational attainment, and income.

human geography: The study of various aspects of human life that create the distinctive landscapes and regions in the world.

von Humboldt, Alexander: (1769–1859) German geographer, author of *Kosmos*. He combined information from his studies of Greek, archaeology, and various sciences with extensive field work in Latin America to form an integrated geographic composite view of the world.

humid continental climate: Humid climate type that possesses warm-to-cool summers and bitterly cold winters.

humid subtropical climate: Areas of warm, moist climate, seasonally exposed to intense cyclonic storms (hurricanes; cyclones), typically located on the southeastern sides of continents.

al-Idrisi, Muhammad (1099-1166): Famous Arab geographer, traveller, and map maker.

import substitution industries: Manufacturing companies that benefit from policies that favor local production of goods rather than their importation.

import-substitution industrialization (ISI): Strategies intended to protect infant industries from foreign competition, thereby stimulating local demand for local products and expanding local manufacturing employment opportunities.

Inca: One of the four high civilizations of pre-Colombian Latin America, centered on the city of Cuzco, Peru, in the Andes and extending from southern Colombia to central Chile.

income disparities: Significant differences in income between specified groups of people or regions.

Indian diaspora: The migration of large numbers of South Asian Indians to other, largely English-speaking countries.

indirect rule: The policy on which British colonial rule was based; its purpose was to incorporate the local power structure into the British administrative structure.

Indo-Gangetic Plain: The combined lowland alluvial plains of the Indus, Ganges, and Brahmaputra rivers. The fertile soils of this well-watered region support half of South Asia's human population.

industrial colonialism: Western political control of colonial territory and economic interests.

industrial reuse: The conversion of former industrial sites for other purposes, such as converting factory buildings into art museums or office parks.

Industrial Revolution: The period of rapid technological change and innovation that began in England in the mid-eighteenth century and subsequently spread worldwide; accompanied by the development of inexpensive, massive amounts of inanimate energy through the use of fossil fuels.

infant mortality rate: The number of children per thousand live births who die before reaching the age of one year.

informal economy: Jobs not covered by national labor and employment laws, such as street vending, small business operation, and domestic work.

informal settlements: Groups of structures, often of a temporary nature, that exist outside of formal legal rules and property markets.

Information Revolution: In contrast to the prehistoric cave dweller and his wall, our ability to produce, store, access, and apply information is massive and nearly instantaneous—truly revolutionary. It is how the information is produced, stored, and accessed that creates a revolution; in the Information Age it comes down to a single transforming technology: the microprocessor.

initial advantage: Early factors that propel the development and expansion of particular activities.

in-situ conservation: A strategy for conserving the biodiversity of Earth's food crops through smallholder tropical polyculture. By this method, a farmer may cultivate several varieties of 20 to 30 crops, thus preserving 40 to 120 varieties of food crops in a single plot.

insurgent states: Ones marked by revolt, insurrection, and guerrilla activity from national liberation movements that oppose the established authority.

Interior Lowlands: A vast sedimentary accumulation zone between the Appalachian Plateau and the Great Plains, essentially comprising the drainage basins of the Mississippi, Missouri, and Ohio rivers in the United States, and extending northward west of the Canadian Shield as far as the Arctic.

Interior Plateaus: A large part of the western United States situated between the Rocky Mountains on the east and the Sierra Nevada and Cascade mountains on the west, including the Columbia and Colorado plateaus.

internal colonialism: A condition found in many less-developed countries where local elites, often living in urban cores, exploit the masses, many of whom live in outlying peripheral rural regions.

internally displaced people: People who are forced to flee their homes, but who stay in their own country, creating a domestic refugee population.

Intertropical Convergence Zone (ITCZ): A low-pressure system created by high temperatures along the equator, which shifts its position following the seasonal movement of the sun north or south of the equator.

intifadah: Uprising by Palestinian Arabs against Israeli control of the West Bank.

irrigated agriculture: The use of water derived from either surface streams or groundwater sources to grow crops; often found in arid areas where rainfall is insufficient to produce a reliable crop. Center-pivot irrigation tapping groundwater in the North American Great Plains or the use of Nile River water stored behind Egypt's Aswan Dam are examples.

Islam: The dominant religion of the Middle East and North Africa. This universalizing belief system was established by the Prophet Mohammed after he began having divine visions beginning in A.D. 610. Islam (which literally means "submission to the will of Allah, [or God]") has five tenets, or pillars, that must be upheld by its adherents: there is only one God and Mohammed is his Prophet (the creed); the offering of regular prayer; the responsibility of the more prosperous to aid the less fortunate; a fast during the ninth month of the lunar calendar; and the undertaking of a pilgrimage to Mecca.

Isthmus of Panama: The narrow strip of land that connects Central and South America.

Jammu and Kashmir: The northernmost Indian state with mixed Hindu and Muslim populations.

Japan Model: The Japanese approach to development, characterized by efficient government and bureaucracy, a sound currency and banking system, growth of education, and effective harnessing of the abilities of the Japanese people.

jati: The four major caste groups to which all Hindus belong, with Brahmins at the top of the socioeconomic hierarchy and untouchables at the bottom.

Jewish Pale: Former Jewish settlement areas in eastern Poland.

Jews: Members of an ethnic group who claim as their belief system Judaism, a monotheistic religion formed by one of several Semitic peoples who resided in Southwest Asia more than 3,000 years ago. Subjected to repeated episodes of persecution throughout their history, the majority of the Jewish population became scattered throughout the world but eventually established their homeland of Israel in 1948. Today Jews make up a small but regionally prominent ethnic minority in the Middle East by constituting more than two-thirds of Israel's population.

Jomon period: A stone-age cultural period from which Japanese culture began to emerge around 12,000 years ago.

Kara Kum desert: A prominent desert east of the Aral Sea in Uzbekistan and Kazakhstan.

keiretsu: Large industrial/financial cliques that formed in Japan after the American occupation ended following World War II.

Khmer Rouge: The Communist Party of Cambodia, overthrown by the Vietnamese but of continuing importance for many years as an insurgent group and a political force.

kinship relations: The ties and relations of the extended family unit, often used in association with traditional societies.

kitchen gardens: During the Soviet period, families were allowed to have a small amount of land adjacent to their homes for growing whatever they wanted for their own consumption or for trade.

Kiwis: Nickname for the inhabitants of New Zealand.

Kremlin: Seat of the Russian government in Moscow.

Kurds: A pastoral and agricultural people residing largely in Turkey, Iraq, and Iran, with lesser numbers in Syria, Armenia, and Azerbaijan.

Kyzyl Kum desert: A prominent desert in Turkmenistan.

Lake Baikal: The world's deepest lake with a maximum depth of more than a mile (1.6 kilometers), one of the world's greatest natural wonders and a Russian destination for nature lovers.

land concentration: A situation in which a small number of people control most of the productive land.

land degradation: A product of human actions that lower a region's biological productivity or diminish its usefulness to humans.

landlocked states: Countries without a marine coast; their trade with the outside world must pass through neighboring, sometimes hostile, countries. Almost 40 percent of the world's landlocked states are in Africa.

landscape modification: The constant change that landscapes undergo in response to natural and human-induced processes.

laterization: A process of soil formation in the tropics; the leaching of soluble minerals from soils because of abundant rainfall, thereby leaving residual oxides of iron and aluminum.

Lenin, Vladimir: Marxist, leader of the Bolsheviks who started the bloody civil war in 1917 that resulted in the establishment of the Union of Soviet Socialist Republics (U.S.S.R) in 1922.

Levant: The eastern Mediterranean region from western Greece to western Egypt.

lingua franca: A common auxiliary language used by peoples of different languages and dialects; commonly used for trading and political purposes.

loess: Deposits of wind-transported, fine-grained material; usually easily tilled and quite fertile.

lost decade: The decade of the 1980s when Latin America suffered from a serious economic crisis.

Lower Canada: The French-speaking region of Canada found on the lower reaches of the St. Lawrence River and the Gulf of St. Lawrence (present-day Quebec), as distinguished from Upper Canada.

Loyalists: Persons who remained loyal to Great Britain during and after the Revolutionary War. After the revolution, many of these loyalists no longer felt welcome in the lower 13 colonies and so fled to British North America (modern Canada).

madina: The ancient urban core of Middle Eastern cities.

madrasahs: Traditional Islamic schools, often attached to a mosque, where students study the Quran, learn basic reading and writing skills, and are introduced to the spiritual, philosophical, and legal scholarship of the Islamic community.

Maghreb: A region of northwestern Africa that includes portions of Morocco, Algeria, and Tunisia.

Malthusian theory: A theory advanced by Thomas Malthus that human populations tend to increase more rapidly than the food supply.

manifest destiny: The conviction of many nineteenth-century Americans that God willed the United States to extend from the Atlantic to the Pacific oceans; a similar belief claims a special U.S. mission to promote democracy throughout the world.

man-land tradition: A geographic perspective that emphasizes the relationship between people and the physical environment used to support their livelihoods.

Mao Zedong: Emerged as leader of the Chinese Communist Party (CCP) in 1935; died in 1976.

Maori: The pre-European Polynesian inhabitants of New Zealand, who have been more successfully integrated into modern society than the Aborigines of Australia.

maquiladora industry: From the Spanish verb maquilar, meaning "to mill or to process"; used to denote foreign-owned (largely U.S.) manufacturing firms located in Latin America, particularly northern Mexico, to realize the advantage of low-cost labor.

marine west coast climate: Climatic region of northern Europe, northwestern North America, southern Chile, and portions of Australia and New Zealand; characterized by year-round cool and wet atmospheric conditions.

Marsh Arabs: Cultural group residing in Iraq's southern marsh region. In order to maintain their culture, these people combined agriculture, cattle rearing, and fishing. Drainage efforts in the 1990s undermined the Marsh Arabs' livelihood and reversing the process has been largely unsuccessful.

Mauryan Empire: A political state; the first of a series of empires that ruled over parts of South Asia.

Maya: One of the four high civilizations of pre-Colombian Latin America, situated in southern Mexico and northern Central America.

Media Luna: The shape ("half-moon") formed by departments in the eastern lowlands of Bolivia, whose inhabitants share few historical or cultural connections to the indigenous communities of the highlands.

mediterranean climate: A dry summer subtropical climate common to the Mediterranean Basin, southern California, central Chile, and portions of Australia.

Medvedev, Dmitrii: Prime Minister of Russia since 2012. Previously served as the third President of Russia.

megalopolis: Originally the continuous urban zone between Boston and Washington, D.C.; now used to describe any region where urban areas have coalesced to form a single massive urban zone.

Meiji Restoration: Marked the end of Tokugawa rule in 1868 and the beginning of the period when Japanese society and its economy were transformed from feudal to modern; Meiji means "enlightened rule," which in this case meant adopting selected Western traits, particularly education and technology.

Melanesia: Section of the South Pacific consisting of islands stretching along the northern perimeter of Australia, from New Guinea eastward to Fiji. The name derives from melanin, the pigment in human skin, and refers to the dark-skinned, Papuan-speaking peoples who predominate in this area.

mercantile colonialism: The period before A.D. 1800 when European powers exerted control indirectly by supporting private trading companies who established commercial and political relationships with local elites to obtain the goods they wanted.

mercantilism: The philosophy by which most colonizing nations controlled the economic activities of their colonies; held that the colony existed for the benefit of the mother country.

mestizos: Persons in Latin America of mixed European and indigenous ancestry.

Mezzogiorno: "Land of the midday sun," an area comprising the Italian peninsula south of Rome, plus Sicily and Sardinia.

microfinancing: A development strategy that targets economically marginal people with good business ideas to receive small loans, insurance, and money transfers so they can launch and operate a small business.

Micronesia: The groups of islands in the Pacific just north of Melanesia, from Guam and the Marianas in the west to Kiribati in the east. The region's name refers to the small islands that predominate here.

microstates: Very small independent countries whose potential for agricultural and industrial development is limited.

Middle America: Mexico, Central America, and the Caribbean.

Middle Atlantic Core: An early settlement area centered on Pennsylvania and New York that was settled by substantial numbers of German, Scots-Irish, Dutch, and Swedish in addition to the dominant English.

Middle Kingdom: A reference to China, reflecting the traditional Chinese view of China as the center of the known universe.

milkshed: The area serviced by a given milk-producing region.

Millennium Development Goals (MDGs): In 2001 the United Nations launched a program that targeted for eradication the most intractable development challenges, including extreme hunger and poverty, gender inequality, child mortality, and HIV/AIDS.

Ministry of Economy, Trade, and Industry (METI): The Japanese government agency responsible for guiding and directing the development of the Japanese economy.

MIRAB (Migration, Remittances, Aid, and Bureaucracy) economies: A potentially unhealthy development situation in which limited local development opportunities encourage migration, migrants send back remittances that support their families but do not lead to productive investments, most local jobs are found in government bureaucracies, and inability to generate investments, jobs, and economic growth produce a cycle of dependence on foreign aid to support basic services.

mixed farming: The raising of crops for both human and animal feed. Mixed farms often have animals, such as cattle or pigs, on the farm itself. Dairy farms can be viewed as a specialized form of mixed farming.

monsoon: A seasonal reversal in surface wind direction; associated primarily with South, East, and Southeast Asia.

Moors: People of mixed Arab and Berber stock of northwest Africa who invaded and inhabited Iberia from the eighth through the fifteenth centuries, thereby diffusing racial and cultural traits to Spain and Portugal.

Mughal Empire: The most powerful of all the Islamic empires of the sixteenth and seventeenth centuries in South Asia.

mujahideen: Opposition movements in Afghanistan and elsewhere in the Middle East and North Africa, inspired by Islamic fundamentalism.

mullahs/mujtahids: Muslim clergy who have traditionally been responsible for interpreting religious law, providing basic education, serving as notaries for legal documents, and administering religious endowments.

multilingualism: The ability of a person to communicate in more than one language.

Multimedia Super Corridor (MSC): The name given to a region designated by the Malaysian government for development as an information technology research center

Muslims: People who surrender to the will of God (Allah) as revealed by the prophet Mohammed; followers of Islam.

Nagorno-Karabakh region: Part of Azerbaijan with an Armenian majority, whose desire for independence resulted in war between Azerbaijan and Armenia. Fighting ceased in 1994, but the fundamental issues are as yet unresolved.

Nara period: From A.D. 710-794, a time when Chinese cultural influences became stronger and the Japanese adopted Chinese written characters.

nationalism: A territorial expression of loyalty to a state. Nationalism can also be reflected in an ethnic group's desire for political independence in the form of a separatist movement.

nation-state: A political entity that has control over its territory and bases its legitimacy on representing a particular group of people who share common characteristics (ethnicity, language, religion, and so on)

neocolonialism: Retention of the trade relationships and patterns of pre-World War II colonialism. Often cited to explain continuing uneven distribution of wealth.

neoliberal policies and **neoliberalism:** An economic and social approach based on a reduced role for government in many areas of public life and an unrestrained free market economy.

neo-Malthusians: Referring to contemporary modifications of the notions of Thomas Malthus regarding population growth and production capacity.

New Economic Policy (NEP): Malaysia's master plan, the goal of which was to increase the economic contributions of the majority ethnic Malays, at the expense of both Chinese and Western economic interests.

New England Core: The most northern of the American coastal culture cores developed in southern New England where numerous bays and harbors became the focus of maritime trade and fishing.

new middle class: This group has appeared recently in Russia, at least in the larger urban areas. Unlike the New Russians, new middle-class people are professionals who draw salaries in fields valued by the global economy.

New Partnership for Africa's Development (NEPAD): African Union (AU) program designed in 2001 to complement the Millenium Development Goals (MDGs) in alleviating poverty, accelerating regional integration, and promoting good governance.

New Russians: Have emerged as the first class of wealthy Russians since 1917. Their conspicuous consumption and uncultured ways make them the target of scorn.

New South: The more urbanized and industrialized South, which emerged in the United States in the twentieth century.

nomadic herding: Involves herding of sheep, goats, and camels in dryland areas in an annual migration cycle often covering hundreds of miles, to bring animals to grass and water only available on a seasonal basis, particularly in the Middle East and North Africa.

Non-Governmental Organizations (NGOs): A diverse group of largely voluntary organizations that work at the grassroots level to provide technical advice and economic, social, and humanitarian assistance.

Nordic Model: A social policy that combines heavy taxation with broad and generous public services.

North American Free Trade Agreement (NAFTA): Went into effect in 1994. Canada, Mexico, and the United States are now members of a single trade union intended to lead eventually to completely free trade and increased interaction between the member countries.

North Atlantic Current: This warm Gulf Stream extension reaches the coast of northwest Europe and modifies the air masses that dominate European weather patterns.

oasis (pl. oases): An island of life in the desert, where surface erosion and relatively high groundwater levels result in water being found close to the surface.

Ob River: West Siberian lowland river. When its tributaries thaw in the spring, ice in the lower Ob acts like a giant plug, sometimes until late summer, causing extensive flooding.

Oceania: The vast realm of Pacific Islands extending over several million square miles, including the islands of Melanesia, Micronesia, and Polynesia.

oligarchs: Russian privatization of industry has resulted in a struggle for survival between the new "entrepreneurs" and emergence of the oligarchs, a few individuals who control vast economic empires.

open coastal cities: Much like Special Economic Zones (SEZs), but with lower levels of government funding for site improvement.

open economic regions: Larger economic regions around Special Economic Zones (SEZs) and open coastal cities.

opium: A dangerously addictive drug from which heroin is derived.

Orange Revolution: Ukraine's 2004 Orange Revolution, a public rebellion against government practices that took its name from the opposition party's color, and led to the election of Viktor Yushchenko, who promised to root out corruption and restructure both state and economy.

Organization of Petroleum Exporting Countries (OPEC): A group of oil-producing countries that controls 85 percent of all the petroleum entering international trade; a valorization scheme to regulate oil production and prices.

orographic effect: Mountains create a barrier to a moving air mass that must be overcome, which in the process modifies the temperature and moisture characteristics of the air mass. The result is excess precipitation on the windward side and drought on the leeward side of the mountain.

orographic precipitation: Mountain-induced rains such as those of the Cascade Mountains in the American Northwest.

Outback: The interior and isolated backlands of Australia, particularly those areas beyond intense settlement.

overgrazing: Loss of vegetative ground cover resulting from stocking more grazing animals than the land can sustain.

Overseas Chinese: Persons of Chinese ancestry now living in other countries. Many maintain family and economic ties with relatives and associates still living in China.

Pacific Coastlands: A system of mountains and valleys that extends along the western edge of North America.

Pacific Rim countries: Includes those countries rimming the Pacific Ocean in both East Asia and the western Americas; sometimes refers more exclusively to the East Asian rimland countries.

Pacific Ring of Fire: A zone that encircles the Pacific Ocean, which includes Japan, the Philippines, and parts of Indonesia, as well as the Andes Mountains of South America and many of the coastal ranges of Canada, Alaska, and eastern Siberia.

Palestine Liberation Organization (PLO): An umbrella organization created by the Arab states in 1964 to control and coordinate Palestinian opposition to Israel; originally subservient to the wishes of the Arab states, now dominated by the independent, nationalist ideology of El-Fatah, the largest and most moderate of the Palestinian resistance groups.

Palestinians: The Arabs who claim Palestine as their rightful homeland; some 5 million Palestinian Arabs residing in various Middle Eastern states, and another 6 million living in Israel, Gaza, and the West Bank.

Pamir Mountains: Major mountain range in Tajikistan and Afghanistan.

pastoral economy: An economic system dependent on the raising of livestock—sheep, cattle, or dairy animals.

pastoralism: System in which a group depends solely on pastures for their livestock herds; the best-known African pastoralists are the Fulani nomads of northern Nigeria, the Masai of Kenya, and the nomadic Tswanas of Botswana.

payment for environmental services (PES): The practice of charging people who benefit from environmental services such as clean water for the cost of such services.

peninsulares: An alternate term for Iberians, used in colonial Latin America.

Pentacostalism: Protestant groups that focus on the literal interpretation of the Bible.

permafrost: Permanently frozen ground common in high latitudes.

Persian ethnic groups: A group of people who speak an Indo-European language and live in Iran and Central Asia, where they are often called Tajiks.

Persians: Natives of Iran; one of two of the largest non-Arab groups in the Middle East (the other is the Turks).

physical geography: That component of geography that focuses on the natural aspects of the earth, such as climate, landforms, soils, or vegetation.

physiologic density: Population density expressed as the number of people per unit of arable land.

plate tectonics: The process whereby portions of Earth's surface are folded, fractured, and uplifted through the movement of Earth's lithospheric plates. Many of the mountain ranges in Europe (for example, the Alps, Pyrenees, and Carpathians) were formed as a result of the slow collision of the Eurasian and African plates.

podzolization: A process in humid regions whereby soluble materials are leached from upper soil layers, leaving residual soils that are frequently infertile and acidic.

polar climate: Characterized by freezing conditions much of the year, this climate experiences long, dark winters and short summers.

polder: A tract of land reclaimed from the sea and protected by dikes.

Polynesia: A large grouping of islands in the Pacific stretching in a huge triangular area from Midway and Hawaii in the north to New Zealand in the south, and eastward as far as Pitcairn Island. The name means "many islands." Polynesia is the largest in size of the three Pacific Island subregions.

population density: The number of people per unit of area.

population distribution: The placement or arrangement of people within a region.

population growth rate: For a country or region, this rate involves calculation of the birthrate and death rate, modified by subtracting people who emigrate from the area and adding people who immigrate into the area.

postindustrial service industries: Service industries involve neither extraction of resources nor manufacturing but instead include a broad range of activities such as government, transportation, banking, retailing, and tourism. Postindustrial denotes service industry dominance, often coincident with deindustrialization.

postmodern perspective: A theoretical perspective that attempts a radical critique of modernist approaches to culture, economy, and society.

poverty: Material deprivation that affects biologic and social well-being; the lack of income or its equivalent necessary to provide an adequate level of living.

Prairies: In Canada, the great expanse of land between the Great Lakes and the Rocky Mountains, the northern extension of the Great Plains and the Interior Lowlands.

precautionary principle: This rule mandates that whenever significant change is likely to occur as a result of a proposed development or whenever the long-term implications of a proposed change are obscure, implementation of the anticipated project must proceed slowly, with proper examination of likely impacts and available remedial actions.

primary level of economic activity: Focuses on extractive activities, such as agriculture, mining, forestry, and fishing.

primate city: A city where a disproportionate share of a country's population and economic power is concentrated.

Prince Henry the Navigator: (A.D. 1394-1460) Portuguese ruler; sponsored explorers seeking new routes to Asia and the Americas.

pristine myth: The misconception that the New World was a lightly populated, unmodified wilderness when it was first encountered by Christopher Columbus.

private-plot production: In the Soviet Union, Stalin tried unsuccessfully to abolish farm families' home gardens, their private plots, but Soviet agriculture depended heavily on private-plot production, especially for fruits and vegetables.

privatization of industry: The process of transferring partially or completely state-owned industries and farms to private control.

producer services: Businesses such as banking, insurance, communications, and consulting.

pronatalist policies: Policies that encourage larger families by offering support through such practices as generous maternity leave and housing and educational subsidies.

Protestantism: Christian branch formed from the second great schism, the Protestant Reformation during the 1500s, which split the western church into Roman Catholicism and Protestantism. This branch constitutes approximately 24 percent of the world's Christians. A fundamentalist form of Protestantism includes those Protestant religions in Latin America that are growing rapidly at the expense of nominal Catholicism; emphasizes a literal interpretation of the Bible as the basis for Christian life.

Ptolemy: (ca. A.D. 90-168) Greek astronomer and early mapmaker from Alexandria, Egypt. Designed a map of the world (ca. A.D. 150) that used a coordinate system to show the location of 8,000 known places.

Pushtuns: An ethnic group of the northwest corner of India that also occupies portions of Afghanistan.

Putin, Vladimir: Second president of Russia from 2000 to 2008, Prime Minister during the presidency of Dmitrii Medvedev, and fourth President of Russia since 2012.

qanat: An underground tunnel, sometimes several miles long, tapping a water source; natural water flow accomplished by gravity; common in several Middle Eastern countries, where they are used to tap groundwater found in alluvial fans.

quilombos: Communities in the Brazilian hinterland formed by runaway slaves.

Quran: The sacred text of Islam.

rainshadow: An effect produced by descending, drying air masses, causing the leeward or interior sides of mountains to be much drier than the windward sides.

Reconstruction and Development Program (RDP):   The program launched by Nelson Mandela's government of South Africa, based on six major principles of improvement.

regional disparity: Distinctive differences in well-being among the inhabitants of the several regions of a given country; most likely to be seen in economic, social, and biologic conditions.

regional geography: That component of geography that focuses on particular regions and the geographic aspects of the economic, social, and political systems of those regions.

remittances: Money or goods sent back to their home country by migrants.

renewable energy: Power derived from sources, such as water, wind, and solar radiation, that can be infinitely used, as opposed to fossil fuels, which are mined as finite resources.

rent-seeking: The use of the resources of the state for the benefit of private interests. In Russia and some other Eurasian countries, public officials "privatize" their government functions, by seeking "rent" in exchange for allowing development to proceed.

reserve currency: A currency, such as the U.S. dollar or the euro, held by governments or banks as part of their foreign exchange funds and used to conduct international transactions.

Restitution of Land Rights Act: A policy instituted in South Africa in 1994 to permit the government to investigate land claims and restore ownership to those who unjustly lost land.

Ritter, Karl: (1779–1859) Author of *Die Erdkunde*; held the first chair of geography in Germany, at the University of Berlin in 1820.

Rocky Mountains: A great series of mountain ranges in the western United States and Canada.

Roman Catholicism: Christian branch created from the split of the Holy Roman Empire in A.D. 1054. In this schism, the western church became Roman Catholicism, and the Greek church became Eastern Orthodoxy. Roman Catholicism is focused on Rome and the Vatican, with the Pope being the highest authority in the Church. It is the dominant Christian branch in southwestern Europe and constitutes approximately 50 percent of the world's Christians.

Rose Revolution: Georgia's 2003 Rose Revolution, so-named because demonstrators carried roses to emphasize nonviolence, ousted Eduard Shevardnadze, the former Gorbachev ally whose corrupt regime failed to revive the economy or keep the country intact.

run-on farming: A technique that collects water from a larger area and concentrates it in valley bottoms and on terraced slopes; it permits agriculture in regions that are otherwise too dry for crop cultivation.

Russification: A policy of cultural and economic integration practiced in the former U.S.S.R. that required all other Slavic and non-Slavic groups to learn the Russian language.

sacrifice zones: To create sustainable livelihoods in the present, people often create sacrifice zones—sacrificing the future use of potentially sustainable resources or the present productivity of distant areas.

Sahel: The semiarid grassland along the southern margin of the Sahara Desert in Western, Central, and Eastern Africa.

salinization: The accumulation of salts in the upper part of the soil, often rendering the land agriculturally useless; commonly occurs in moisture-deficient areas where irrigated agriculture is practiced.

scheduled castes: The name given to the Hindu and Sikh untouchables by the Indian government.

Schengen Zone: The area of the EU that has agreed to completely passport-free travel across the borders of member countries.

sea level rise: Increases in ocean level caused by global warming that put low-lying coastal areas and atolls at risk of flooding, storm surges, and outright submersion.

secondary level of economic activity: Includes activities that transform raw materials into usable goods.

secularism: Indifference to religion and/or a belief in the separation of church and state.

separatism: The phenomenon in which a group of people seeks to pursue independence from their nominal country. Separatism is usually ethnic (cultural) in nature, and in most instances ethnicity must be linked to some major grievance such as persecution, attempted ethnocide, forced assimilation, domination, lack of autonomy, or denial of access to the country's power structure for separatism to be sparked.

Sephardim: Jews originating from Southern Europe and the Middle East.

sequential occupancy: A concept of historical geography that considers the geography of an area during successive periods of time.

Serbs: Ethnic group of Orthodox Christians, the majority of whom reside in Serbia, which is the largest (in land area and population) of the republics that made up the former Yugoslavia. Serbs speak a Serbo-Croatian Slavic language but use a Cyrillic alphabet.

serfdom: For centuries in Russia, peasants were bound to the land and required to serve the landowners. Over time this condition of serfdom turned into slavery.

settlement frontiers: New territories that are occupied and culturally imprinted by settlers, usually by the process of diffusion.

Sharia: An Arabic legal code based on the Quran.

Shatt al-Arab: Common channel located in southern Iraq through which the Tigris and Euphrates Rivers flow into the Persian Gulf.

shatterbelt: A politically unstable region where differing cultural elements come into contact and conflict.

Shi'ites: One of two main branches in Islam, predominant in Iran and in parts of Iraq and Yemen.

shifting cultivation: A farming system of land rotation, based on periodic change of cultivated area; allows soils with declining productivity to recover; an effective adaptation to tropical environments when population density remains low.

Shintoism: Religion in Japan based on the worship of natural and ancestral spirits.

Sikhism: An Indian religion that includes elements of the Hindu and Islamic faiths.

Silk Road: Caravan route stretching from China to the Middle East and ultimately on to Europe, the main historical corridor for goods from the east.

siloviki: Russian President Putin's appointees.

small- and medium-sized enterprises (SMEs): Smaller companies that supply materials and parts to larger companies.

soil erosion: Soil loss associated with the loss of protective vegetative cover. Once the topsoil is exposed to the forces of running water and wind, it can be carried away in very short periods of time.

Song dynasty: (A.D. 960–1279) A period of Chinese history during which advancements occurred that made the Chinese agricultural economy one of the most sophisticated in the world.

Southern African Development Community (SADC): A group founded in 1980 to reduce dependency on South Africa for rail, air, and port links; imports of manufactured goods; and the supply of electrical power.

Southern Core: Developing first on the coastal plain southward from Virginia, the region developed a plantation economy featuring export crops (indigo; tobacco; cotton).

spatial integration: The process whereby the settlement frontier is eliminated through the creation of trade areas and the establishment of ties with the core areas and the surrounding communities. According to historian Frederick Jackson Turner, European core areas on the Atlantic seaboard provided important early sources of settlers responsible for the spatial integration process in the United States and Canada.

spatial perspective: The geographic approach that places location at the center of research, analysis, and explanation.

spatial relationships: The close association of human and natural phenomena in place and their mutual interdependence and interaction.

spatial tradition: A geographic perspective that emphasizes how things are organized in space, especially spatial distributions, associations, and interactions.

Special Economic Zones (SEZs): Areas in China that function as modern-day treaty ports, receiving a substantial amount of foreign investment.

Specialty Crop and Livestock region: An area from southern New England to Texas in which a variety of livestock and crop areas are based on local environmental characteristics and market opportunities.

Stalin, Joseph: Overcame all rivals after Lenin died in 1924 to take over rule of the Soviet Union. He initiated collectivization of farms, the "Gulag" labor camp system, and the Five Year Plan for industrialization; millions died in the Soviet Union during his rule due to famine, war, and repressive policies.

stages of economic growth: A theory developed by Walt Rostow in which five stages of economic organization are recognized: traditional society, preconditions for takeoff, takeoff, drive to maturity, and high mass consumption.

state farms: Large agricultural systems controlled and managed by the government in the old Soviet Union; workers were paid wages.

state-led industrialization: A development philosophy found in Japan, South Korea, Taiwan, and other Asian countries in which the state either partially owns corporations that operate under commercial principles or directs the structure and orientation of the national economy.

stations: The very large sheep or cattle ranches associated with Australia.

steppe: Grasslands in the midlatitudes and the subtropics.

steppe climate: Intermediate between drier and wetter conditions, these semiarid climate zones are typically characterized by grasslands and high precipitation variability.

Stolen Generation: Aboriginal children that were forcibly taken from their mothers to be brought up by whites, a practice that only ended in Australia in 1969.

Strabo: (64 B.C.- ca. A.D. 23) Geographic scholar, author of *Geographia*, whose fascination with the cultural and natural differences that existed in his world was a precursor to the modern regional studies approach.

structural adjustment programs (SAPs):   Programs initiated to adjust malfunctioning economies, assist in debt restructuring, and promote greater economic efficiency and growth; widely implemented in the Sub-Saharan and Latin American countries.

subarctic climate: Long, cold winters and short summers with limited precipitation characterize this high-latitude climate.

Sun Yat-sen: (1866-1925) Chinese revolutionary and political leader known as the "father of the republic." His goals focused on the "Three Principles of the People": nationalism, democracy, and livelihood.

Sunnis: The largest branch of Islam; predominant throughout most Middle Eastern countries, with the exception of Iran.

sunset industries: Industries in Japan that the government considers no longer competitive and therefore appropriate to be phased out.

Sunshine Policy: A policy of reconciliation between South Korea and North Korea, introduced in 2000.

sustainability: The ability of a resource use system to function well for a long time by taking into account the long-term environmental consequences of change.

suq: see bazaar.

syncretism: Belief systems that result from the combination of introduced and local beliefs.

Syr Darya river: A river in Kazakhstan that flows into the Aral Sea.

systematic geography: An approach to geographic study in which the emphasis is on specified subjects; for example, economic geography, urban geography, climatology, water resources, or population geography.

Tabqah Dam: A large dam on the Euphrates River in Syria, intended to significantly increase irrigated agriculture and hydroelectric power generation.

taiga: The large coniferous forest extending across northern Russia.

Taliban movement: A fundamentalist Islamic student-based movement that originated in the Afghan refugee camps in Pakistan.

Tamil: Language spoken by Sri Lankans of Dravidian descent, who are known as Tamils, and who are also a minority in Sri Lanka. The Tamils feel they have been systematically discriminated against by the majority group, the Sinhalese Buddhists.

technopoles: Areas that concentrate high-tech manufacturing, computer-based information services, and research-and-development activities.

terpen: Human-made mounds designed to protect their inhabitants from seasonal floods.

terracing: Technique in which steep hillsides are cut in a stair-step fashion and shored up by retaining walls for agricultural purposes.

tertiary level of economic activity: Includes activities that focus on the provision of goods and services.

thermal inversion: A condition that develops when warm, stable air overlies cool air, trapping the cool air and any pollutants that are released beneath it.

thermohaline circulation: The global system of ocean currents, driven by temperature and salinity differences, that brings warmer seawater from the tropics to Europe and moderates Europe's climate.

Three Gorges: A region of China where the Chang Jiang ("Long River") flows through a narrow, 150-mile (240-kilometer) long, steep-walled valley no wider than 350 feet (107 meters).

Tibetan Plateau: The largest environmental zone of western China, as well as the largest and most elevated plateau in the world, it occupies 25 percent of China's territory. It is a cold, inhospitable environment referred to as the "rooftop of the world" or the "third pole."

Tokaido Megalopolis: A large, multinuclear urbanized region in Japan extending from Tokyo to Osaka.

Tokaido Road: Historical Japanese road between Edo, Kyoto, and Osaka. The Tokaido Road was a government postal highway with 57 stations with porter stations, horse stables, and lodging.

Tokugawa period: The period (1615-1868) during which the focus of power in Japan shifted to the Kanto Plain area and many of the modern Japanese characteristics were firmly fixed in the culture.

total fertility rate: The average number of children born to a woman during her lifetime.

town and village enterprises (TVEs):   Collectives owned by towns and villages, but often including private capital investment, that produce products for both domestic consumption and export.

trade blocs: Trade agreements between groups of countries.

transnational capitalist class: Globe-trotting executives and professionals, who view their work as part of a globally competitive process, share upscale lifestyles, and see themselves as citizens of the world as well as citizens of their own countries.

transnational corporation (TNC):   A corporation with offices, production facilities, and other activities in multiple countries. It is geographically mobile and can take advantage of lower labor costs in one country or a more lenient regulatory environment in another country to minimize production costs and/or maximize revenues.

transnational migration: Migrant use of modern transport and communication technologies to stay in touch with their home community, remit money and goods, and return periodically for visits.

Trans-Siberian Railway: System that crosses Siberia, beginning at the Ural Mountains and ending at the Pacific Ocean terminus of Vladivostok.

Treaty of Tordesillas: A treaty negotiated between Spain and Portugal in 1494 that divided the New World between those two countries at roughly the 50th meridian, giving Portugal the rights to areas to the east and Spain to areas to the west; this treaty followed a papal bull from the previous year, which had declared the New World as belonging to Spain, and Africa and India to Portugal.

tsunamis: Seismic sea waves triggered by energy released from deep earthquakes and also from massive landslides and volcanic eruptions.

tundra: Vegetation zone found on the margins of the Arctic, particularly in northern Russia and northern Canada. No trees grow because of the short growing season, infertile soil, and shallow layer of thawed ground; beneath lies permanently frozen earth or permafrost.

Turkic ethnic groups: A wide variety of ethnic communities (Azerbaijanis, Kazakhs, Uyghurs, Yakuts) living in southwest, central, and northern Asia and speaking related Turkic languages.

Turkic ethnic groups: Natives of Turkey; one of two of the largest non-Arab groups in the Middle East (the other is the Persians).

typhoons: The equivalent of hurricanes or tropical cyclones; occur in the Pacific, especially in the area of the China seas.

Ujamaa: A Swahili word meaning "familyhood," expressing a feeling of community and cooperative activity; a term used by the Tanzanian government to indicate a commitment to rapid economic development according to principles of socialism and communal solidarity.

Upper Canada: The English-speaking region of Canada found on the upper reaches of the St. Lawrence River and in the Great Lakes region (present-day Ontario), as distinguished from Lower Canada.

Ural Mountains: A chain of ancient, greatly eroded, low mountains in Russia that mark the traditional boundary between Asia and Europe.

urban heat islands: Phenomenon in which cities tend to be characterized by significantly higher temperatures than those of the surrounding areas. This temperature imbalance is largely due to the air pollution generated by cities as well as the large amounts of heat that are absorbed by the pavement and masonry of the urban infrastructure.

vegetation succession: The replacement of one plant community by others over time.

village banking: A type of local-scale microfinancing that helps families defray the costs associated with unforeseen emergencies, such as illness, natural disasters, and funerals, as well as improvements like home improvement, solar energy systems, and mobile phones.

virtual water: Water that is not physically present, but is represented in a commodity because water was needed to grow or produce it; for example, the water invested in a wet place to grow rice that is then exported to a dry place that lacks sufficient water to grow its own rice.

Volga River: With its tributaries, this river forms Russia's major water route, allowing for the development of industrial activities.

voluntourism: The combination of tourism and volunteer work for a good cause.

Wakhan Corridor: A narrow strip of land created between the Russian-controlled Pamir Mountains and British India still connects Afghanistan to China today.

warlords: They use opium profits to finance and arm the militias that keep the countryside of Afghanistan unstable and out of the reach of the elected government.

West Bank: The territory occupied by Israel since 1967 that lies immediately west of the Jordan River and the Dead Sea; claimed by Palestinians.

Wheat Belts: West of the Corn Belt, in the Great Plains, lie the wheat belts (winter wheat and spring wheat), where farmers have converted prairie to cropland.

White Australia Policy: The policy formerly used by Australia in an attempt to exclude nonwhites from migrating permanently to Australia and to encourage whites, especially the British, to settle in Australia; officially termed the restricted immigration policy.

world cities: Centers of global finance, corporate decision making, and creativity that have worldwide reach, such as New York City, Hong Kong, and London.

World Heritage Site: An area with cultural and natural properties deemed to be of outstanding universal value.

xenophobia: Fear of people of foreign origin.

Yamato period: The rise of the Yamato period some 1,700 years ago led to the emergence of the first Japanese state, centered on a military aristocracy and anchored by successive kings.

Yayoi period: Some 2,300 years ago migrants from a region just north of China arrived in Japan via the Korean Peninsula and, among other innovations, introduced sedentary agriculture and paddy rice.

Yenisey River: Marks the boundary between the West Siberian lowland and the plateau of Central Siberia.

zaibatsu: A large Japanese financial enterprise, similar to a conglomerate in the United States but generally more integrated horizontally and vertically.

Zapatista movement: An indigenous movement in Mexico in defense of the rights of indigenous peoples to land, jobs, and political freedom.

Zarafshon River: Flows from Tajikistan into Uzbekistan and disappears into the Kyzyl Kum desert.

Zhou dynasty: (1027-256 B.C.) A militaristic dynasty that followed the Shang dynasty; they established their capital at a site close to present-day Xian.

Credits

Introduction Figure I-1 Mary Evans Picture Li/Mary Evans Picture Library Ltd/Age Fotostock; Figure I-2 Bridgeman-Giraudon/Art Resource, NY; Figure I-3 (left) Robert Huberman/RGB Ventures LLC dba SuperStock/Alamy; Figure I-3 (right) luna/Fotolia; Figure I-4 Megastocker/Alamy; Figure I-5 luoman/Getty Images; Figure I-6 Philipus/Alamy; Figure I-8 Jean-Christophe Bott/EPA/Landov; Figure I-12 Rebecca Hale/National Geographic/AP Images; Figure I-13 U.S. Army Photo/Alamy.

Chapter 1 Chapter Opener Tina Manley/Alamy; Figure 1-1 Florian Kopp/imagebroker/Alamy; Figure 1-2 Tan Jin/Xinhua/Photoshot/Newscom; Figure 1-3 Robert Essel NYC/Corbis/Glow Images; Figure 1-4 Robert Harding World Imagery/Alamy; Figure 1-5 Xiong Jinchao/ZUMAPRESS/Newscom; Figure 1-8 Sean Gallup/Getty Images; Figure 1-12 Zhang Jun/XINHUA/AP Images; Figure 1-13 KHAM/Reuters/Landov; Figure 1-14 Qilai Shen/In Pictures/Corbis; Figure 1-15 Jeff Greenberg/Alamy; Figure 1-16 Mmphotos/Photodisc/Getty Images; Figure 1-2-1 Yang Jianhua/ZUMA Press/Newscom; Figure 1-17 Val Handumon/EPA/Newscom; Figure 1-18 David Noton/David Noton Photography/Alamy; Figure 1-19a DEA/V. Giannella/De Agostini/Getty Images; Figure 1-19b Mark Dyball/Alamy; Figure 1-21 Galyna Andrushko/Shutterstock; Figure 1-24 David Robertson/Alamy; Figure 1-25a Corbis; Figure 1-25b Christian Petersen/Getty Images Sport/Getty Images; Figure 1-26a ftfoxfoto/Fotolia; Figure 1-26b James L. Amos/Getty Images; Figure 1-27 U.S. Air Force Photo/Alamy; Figure 1-28 Mo Fini/Alamy; Figure 1-30 Aurlia Frey/AKG-Images/Newscom; Figure 1-31 Diptendu Dutta/AFP/Getty Images; Figure 1-32 Jose Fuste Raga/Corbis; Figure 1-33 Jeremy Richards; Figure 1-34 Adrenalinapura/Fotolia; Figure 1-35a John Van Decker/Alamy; Figure 1-35b U.S. Army Photo/Alamy; Figure 1-36 Libor Sojka/AP Images; Figure 1-37 Keith Erskine/Alamy; Figure 1-38 Jeff Schmaltz/NASA Images; Figure 1-43 Prakash Singh/AFP/Getty Images; Figure 1-5-1 Bob Sacha/Corbis; Figure 1-49 Guy Harrop/Alamy; Figure 1-50 ImageBroker/AGE Fotostock.

Chapter 2 Chapter Opener Rolf Hicker Photography/Alamy; Figure 2-2 CharlineXia Ontario Canada Collection/Alamy; Figure 2-3 Andre Jenny Stock Connection Worldwide/Newscom; Figure 2-4 Chris R. Sharp/Science Source; Figure 2-5 Stephanie Graeler/Alamy; Figure 2-6 Michele Falzone/Alamy; Figure 2-7 James P. Blair/National Geographic/Getty Images; Figure 2-8 Annie Griffiths Belt/Corbis; Figure 2-9 Lanis Rossi/Alamy; Figure 2-11 Neil Setchfield/Alamy; Figure 2-12 Jonathan Hayward/The Canadian Press/AP Images; Figure 2-2-1 U.S. Geological Survey/U.S. Department of the Interior; Figure 2-2-2 Skip Bolen/EPA/Newscom; Figure 2-3-1 Andre Jenny/Alamy; Figure 2-15 Russ Heinl/Alamy; Figure 2-18 Brian Jannsen/Alamy; Figure 2-19 Brian Jannsen/Alamy; Figure 2-22 Stania Kasula/Alamy; Figure 2-31 Elenathewise/Fotolia; Figure 2-32a Edwin Remsberg / Alamy; Figure 2-32b USDA Photo/Alamy; Figure 2-33 Brian Brown/Alamy; Figure 2-35 Thomas R. Fletcher/Alamy; Figure 2-37 Howard Sandler/Shutterstock; Figure 2-39a Billy Weeks/Reuters; Figure 2-39b WorldFoto/Alamy; Figure 2-40 Mark Elias/Bloomberg/Getty Images; Figure 2-41 Ted S. Warren/AP Images; Figure 2-42 Thomas Kitchin & Victoria Hurst/Alamy; Figure 2-44 Timothy Fadek/Corbis; Figure 2-45 Valerie Macon/Getty Images Entertainment/Getty Images;

Figure 2-49 Ron Chapple Stock/Alamy; Figure 2-50 Cal Vornberger/Alamy; Figure 2-6-1 Clark Brennan/Alamy.

Chapter 3 Chapter Opener el Pavitt/AWL Images/Getty Images; Figure 3-1 Brad Jokisch; Figure 3-2 Cmon/Fotolia; Figure 3-4 Brad Jokisch; Figure 3-6 Brad Jokisch; Figure 3-1-1 Enrique Molina/Age Fotostock/Alamy; Figure 3-8 Brad Jokisch; Figure 3-9 Brad Jokisch; Figure 3-10 Amy Nichole Harris/Shutterstock; Figure 3-12 Jorge Royan/Alamy; Figure 3-13 Brad Jokisch; Figure 3-17 Brad Jokisch; Figure 3-2-1 Victor Ruiz Caballero/Reuters; Figure 3-18 Brad Jokisch; Figure 3-20 Brad Jokisch; Figure 3-21 Milla Kontkanen/Alamy; Figure 3-3-1 Jaime Saldarriaga/Reuters/Landov; Figure 3-22 Brad Jokisch; Figure 3-23 Brad Jokisch; Figure 3-24 Scott Dalton/WpN/UPPA/Photoshot; Figure 3-25 Jason Szenes/EPA/Newscom; Figure 3-29 Brad Jokisch; Figure 3-30 Brad Jokisch; Figure 3-31 Van der Meer Marica/Arterra Picture Library/Alamy; Figure 3-32 Ramon Espinosa/AP Images; Figure 3-33 Enrique de la Osa/EPA/Newscom; Figure 3-37 Patricio Realpe/LatinContent/Getty Images; Figure 3-38 Brad Jokisch; Figure 3-39 Petrut Calinescu/Alamy; Figure 3-40 Enrique Molina/Age Fotostock/Alamy; Figure 3-41 Lightroom Photos/Alamy; Figure 3-42 Y. Levy/Alamy; Figure 3-43 Jbor/Shutterstock; Figure 3-44 Yadid Levy/Robert Harding World Imagery/Alamy; Figure 3-5-1 Diego Lezama Orezzoli/Corbis; Figure 3-45 Michael S. Lewis/Glow Images; Figure 3-46 Fuste Raga/Corbis/Glow Images; Figure 3-6-1 Roussel BernardD/Alamy; Figure 3-6-2 Valeria Pacheco/AFP/Getty Images/Newscom.

Chapter 4 Chapter Opener Radius Images/Corbis; Figure 4-1-1 Aldo Pavan/AGE Fotostock; Figure 4-2 Corey Johnson; Figure 4-5 Corey Johnson; Figure 4-6 Martin Siepmann/Glow Images; Figure 4-8 Arndt Sven-Erik/Arterra Picture Library/Alamy; Figure 4-10 Guillem Lopez/Alamy; Figure 4-11 Corey Johnson; Figure 4-12b Corey Johnson; Figure 4-2-1 Ken Welsh/Alamy; Figure 4-2-2 LOOK Die Bildagentur der Fotografen GmbH/Alamy; Figure 4-16 Corey Johnson; Figure 4-17 Corey Johnson; Figure 4-18 Corey Johnson; Figure 4-19 Corey Johnson; Figure 4-20 Corey Johnson; Figure 4-22 Corey Johnson; Figure 4-23 David Keith Jones/Alamy; Figure 4-24 Cernan Elias/Alamy; Figure 4-25 Corey Johnson; Figure 4-26 Javier Larrea/age fotostock/Alamy; Figure 4-3-2 imagebroker/Alamy; Figure 4-27 Corey Johnson; Figure 4-28 Corey Johnson; Figure 4-29 Corey Johnson; Figure 4-30 Corey Johnson; Figure 4-31 Stephan Goerlich/imagebroker/Alamy; Figure 4-34 Peter Forsberg/Alamy; Figure 4-5-1 Corey Johnson; Figure 4-37 Wojtek Buss/Glow Images; Figure 4-38 Henning Kaiser/EPA/Newscom; Figure 4-39 Corey Johnson; Figure 4-6-2 Corey Johnson.

Chapter 5 Chapter Opener Rolf Richardson/Alamy; Figure 5-2 Ferdinand Hollweck/Imagebroker/Alamy; Figure 5-4 Keren Su/Corbis; Figure 5-5 Robert Argenbright; Figure 5-6 Danita Delimont/Gallo Images/Getty Images; Figure 5-1-1 Denis Sinyakov/AP Images; Figure 5-7 Robert Argenbright; Figure 5-9 Nadia Isakova/Alamy; Figure 5-10 E. D. Torial/Alamy; Figure 5-12 Robert Argenbright; Figure 5-15 Robert Argenbright; Figure 5-16 Robert Argenbright; Figure 5-17 Robert Argenbright; Figure 5-18a Robert Argenbright; Figure 5-18b: Alex Segre / Alamy; Figure 5-2-1 Robert Argenbright; Figure 5-2-2 Kazbek Basayev/AFP/Getty Images; Figure 5-3-2 Uzakov Sergei/ITAR-TASS/Corbis;

Index

World – Political

160°W 140°W 120°W 100°W 80°W 60°

ARCTIC OCEAN

80°N

ALASKA
(U.S.)

GREE
(DE

60°N

CANADA

UNITED STATES
AND CANADA

Aleutian Islands
(U.S.)

St. Pierre
and Miquelon
(FR.)

40°N

UNITED STATES

ATLANTI
OCEAN

PACIFIC OCEAN

Bermuda
(U.K.)

Midway Islands
(U.S.)

MEXICO

Caribbean Sea

20°N

Hawaii (U.S.)

See inset below

GUYANA
SURINAME
FRENCH GUIANA
(FR.)

AUSTRALIA,
NEW ZEALAND
AND THE
PACIFIC ISLANDS

VENEZUELA

COLOMBIA

KIRIBATI   Equator

0°

Galápagos Islands
(ECUADOR)

ECUADOR

LATIN AMERICA
AND THE CARIBBEA

Tokelau
(N.Z.)

PERU

Cook Islands
(N.Z.)

Wallis and Futuna (FR.)

BRAZIL

SAMOA

French Polynesia
(FR.)

American Samoa
(U.S.)

TONGA

BOLIVIA

20°S

Niue
(N.Z.)

PARAGUAY

CHILE

Pitcairn Islands
(U.K.)

Easter I.
(CHILE)

PACIFIC OCEAN

URUGUAY

ARGENTINA

40°S

Falkland Islands
(U.K.)

South Ge

60°S

The Caribbean

90°W   80°W   70°W   60°W

FLORIDA
(U.S.)

0   250   500 Miles

0   250   500 Kilometers

Gulf of Mexico

THE
BAHAMAS

ATLANTIC OCEAN

20°N

Turks & Caicos
(U.K.)

CUBA

Virgin Is.
(U.K.)

Anguilla (U.K.)
St. Barthélemy (FR.)

Cayman Islands
(U.K.)

Puerto Rico
(U.S.)

ANTIGUA AND BARBUDA

MEXICO

JAMAICA

HAITI

DOMINICAN
REPUBLIC

Montserrat (U.K.)
Guadeloupe (FR.)

BELIZE

Virgin Is. (U.S.)
St. Martin (FR. & NETH.)

DOMINICA

GUATEMALA

Caribbean Sea

ST. KITTS AND NEVIS

Martinique (FR.)

ST. LUCIA

HONDURAS

EL SALVADOR

ST. VINCENT AND
THE GRENADINES

BARBADOS

Aruba (NETH.)

Netherlands
Antilles
(NETH.)

GRENADA

NICARAGUA

TRINIDAD
AND
TOBAGO

10°N

COSTA
RICA

PACIFIC
OCEAN

PANAMA

VENEZUELA

GUYANA

COLOMBIA

90°W   80°W

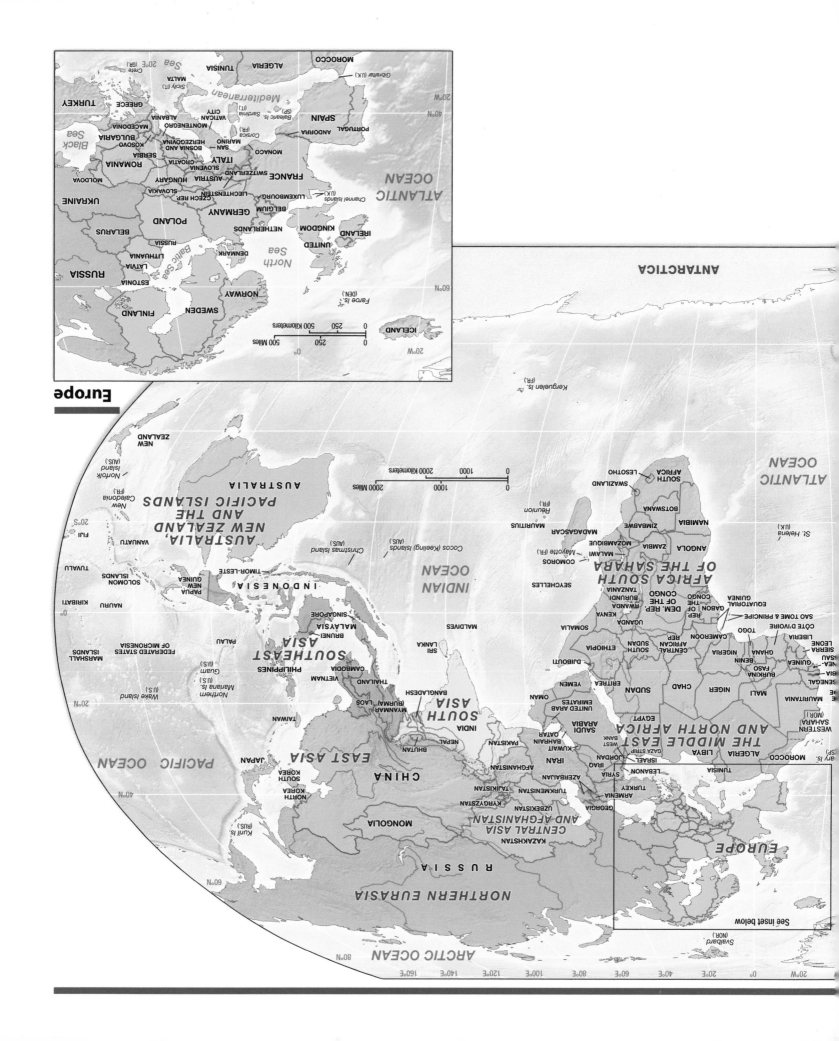

Europe

Europe

ATLANTIC OCEAN

Mediterranean Sea

Black Sea

North Sea

Baltic Sea

MOROCCO
ALGERIA
TUNISIA
MALTA
Sicily (IT.)
Crete (GR.)
20°E
Sardinia (IT.)
Corsica (FR.)
Balearic Is. (SP.)
GREECE
TURKEY
SPAIN
ANDORRA
PORTUGAL
MONACO
SAN MARINO
ITALY
VATICAN CITY
ALBANIA
MACEDONIA
MONTENEGRO
KOSOVO
BULGARIA
SERBIA
BOSNIA AND HERZEGOVINA
CROATIA
SLOVENIA
ROMANIA
MOLDOVA
HUNGARY
AUSTRIA
SWITZERLAND
LIECHTENSTEIN
SLOVAKIA
FRANCE
CZECH REP.
LUXEMBOURG
UKRAINE
BELGIUM
GERMANY
POLAND
BELARUS
NETHERLANDS
Channel Islands (U.K.)
UNITED KINGDOM
IRELAND
RUSSIA
LITHUANIA
LATVIA
ESTONIA
RUSSIA
DENMARK
NORWAY
SWEDEN
FINLAND
40°N
20°W
60°N
Faroe Is. (DEN.)
ICELAND
0°
20°W
0   250   500 Miles
0   250   500 Kilometers

ARCTIC OCEAN
NORTHERN EURASIA
RUSSIA
80°N
60°N
20°W   0°   20°E   40°E   60°E   80°E   100°E   120°E   140°E   160°E

EUROPE
See inset below
Svalbard (NOR.)

THE MIDDLE EAST AND NORTH AFRICA
MOROCCO
ALGERIA
LIBYA
EGYPT
TUNISIA
ary Is. (SP.)
WESTERN SAHARA (MOR.)
LEBANON
SYRIA
ISRAEL
JORDAN
GAZA STRIP
WEST BANK
IRAQ
IRAN
TURKEY
ARMENIA
AZERBAIJAN
GEORGIA
KUWAIT
BAHRAIN
QATAR
UNITED ARAB EMIRATES
SAUDI ARABIA
OMAN
YEMEN

CENTRAL ASIA AND AFGHANISTAN
KAZAKHSTAN
UZBEKISTAN
KYRGYZSTAN
TURKMENISTAN
TAJIKISTAN
AFGHANISTAN
MONGOLIA

EAST ASIA
CHINA
NORTH KOREA
SOUTH KOREA
JAPAN
TAIWAN

SOUTH ASIA
PAKISTAN
INDIA
NEPAL
BHUTAN
BANGLADESH
MYANMAR (BURMA)
SRI LANKA
MALDIVES

SOUTHEAST ASIA
LAOS
THAILAND
VIETNAM
CAMBODIA
PHILIPPINES
MALAYSIA
BRUNEI
SINGAPORE
INDONESIA
TIMOR-LESTE

PACIFIC OCEAN
Kuril Is. (RUS.)
Wake Island (U.S.)
Northern Mariana Is. (U.S.)
Guam (U.S.)
PALAU
FEDERATED STATES OF MICRONESIA
MARSHALL ISLANDS
NAURU
KIRIBATI
TUVALU
SOLOMON ISLANDS
VANUATU
FIJI

AUSTRALIA, NEW ZEALAND AND THE PACIFIC ISLANDS
PAPUA NEW GUINEA
AUSTRALIA
NEW ZEALAND
Norfolk Island (AUS.)
New Caledonia (FR.)
Christmas Island (AUS.)
Cocos (Keeling) Islands (AUS.)

INDIAN OCEAN
Kerguelen Is. (FR.)
Réunion (FR.)
MAURITIUS
MADAGASCAR
Mayotte (FR.)
COMOROS
SEYCHELLES

AFRICA SOUTH OF THE SAHARA
MAURITANIA
SENEGAL
MALI
NIGER
CHAD
SUDAN
ERITREA
DJIBOUTI
SOMALIA
ETHIOPIA
SOUTH SUDAN
CENTRAL AFRICAN REP.
NIGERIA
BENIN
BURKINA FASO
GUINEA
GUINEA-BISSAU
SIERRA LEONE
LIBERIA
CÔTE D'IVOIRE
GHANA
TOGO
CAMEROON
EQUATORIAL GUINEA
SAO TOME & PRINCIPE
GABON
REP. OF THE CONGO
DEM. REP. OF THE CONGO
UGANDA
KENYA
RWANDA
BURUNDI
TANZANIA
ANGOLA
ZAMBIA
MALAWI
MOZAMBIQUE
ZIMBABWE
NAMIBIA
BOTSWANA
SOUTH AFRICA
SWAZILAND
LESOTHO

ATLANTIC OCEAN
St. Helena (U.K.)

20°S
0°
20°N
40°N
60°N

0   1000   2000 Miles
0   1000   2000 Kilometers

ANTARCTICA